Algebra

Properties of Real Numbers

Commutative: $a + b = b + a; \quad ab = ba$

Associative: $a + (b + c) = (a + b) + c;$
$a(bc) = (ab)c$

Additive Identity: $a + 0 = 0 + a = a$

Additive Inverse: $-a + a = a + (-a) = 0$

Multiplicative Identity: $a \cdot 1 = 1 \cdot a = a$

Multiplicative Inverse: $a \cdot \dfrac{1}{a} = 1, \; a \neq 0$

Distributive: $a(b + c) = ab + ac$

Exponents and Radicals

$a^m \cdot a^n = a^{m+n}$ $\qquad \dfrac{a^m}{a^n} = a^{m-n}$

$(a^m)^n = a^{mn}$ $\qquad (ab)^m = a^m b^m$

$\left(\dfrac{a}{b}\right)^m = \dfrac{a^m}{b^m}$ $\qquad a^{-n} = \dfrac{1}{a^n}$

If n is even, $\sqrt[n]{a^n} = |a|$.

If n is odd, $\sqrt[n]{a^n} = a$.

$\sqrt[n]{a} \cdot \sqrt[n]{b} = \sqrt[n]{ab}, \; a,b \geq 0$

$\sqrt[n]{\dfrac{a}{b}} = \dfrac{\sqrt[n]{a}}{\sqrt[n]{b}}$

$\sqrt[n]{a^m} = (\sqrt[n]{a}\,)^m = a^{m/n}$

Special-Product Formulas

$(a + b)(a - b) = a^2 - b^2$

$(a + b)^2 = a^2 + 2ab + b^2$

$(a - b)^2 = a^2 - 2ab + b^2$

$(a + b)^3 = a^3 + 3a^2b + 3ab^2 + b^3$

$(a - b)^3 = a^3 - 3a^2b + 3ab^2 - b^3$

$(a + b)^n = \sum_{k=0}^{n} \binom{n}{k} a^{n-k} b^k$, where

$$\binom{n}{k} = \frac{n!}{k!\,(n-k)!} = \frac{n(n-1)(n-2)\cdots[n-(k-1)]}{k!}$$

Factoring Formulas

$a^2 - b^2 = (a + b)(a - b)$

$a^2 + 2ab + b^2 = (a + b)^2$

$a^2 - 2ab + b^2 = (a - b)^2$

$a^3 + b^3 = (a + b)(a^2 - ab + b^2)$

$a^3 - b^3 = (a - b)(a^2 + ab + b^2)$

Interval Notation

$(a, b) = \{x | a < x < b\}$ $\qquad [a, b] = \{x | a \leq x \leq b\}$

$(a, b] = \{x | a < x \leq b\}$ $\qquad [a, b) = \{x | a \leq x < b\}$

$(-\infty, a) = \{x | x < a\}$ $\qquad (a, \infty) = \{x | x > a\}$

$(-\infty, a] = \{x | x \leq a\}$ $\qquad [a, \infty) = \{x | x \geq a\}$

Absolute Value

$|a| \geq 0$ $\qquad |-a| = |a|$

$|ab| = |a| \cdot |b|$ $\qquad \left|\dfrac{a}{b}\right| = \dfrac{|a|}{|b|}, \; b \neq 0$

For $a > 0$,

$|x| = a \to x = -a \quad \text{or} \quad x = a,$

$|x| < a \to -a < x < a,$

$|x| > a \to x < -a \quad \text{or} \quad x > a.$

Equation-Solving Principles

$a = b \to a + c = b + c$

$a = b \to ac = bc$

$a = b \to a^n = b^n$

$ab = 0 \leftrightarrow a = 0 \quad \text{or} \quad b = 0$

$x^2 = k \to x = \sqrt{k} \quad \text{or} \quad x = -\sqrt{k}$

Inequality-Solving Principles

$a < b \to a + c < b + c$

$a < b$ and $c > 0 \to ac < bc$

$a < b$ and $c < 0 \to ac > bc$

(*Algebra continued*)

Algebra and Trigonometry

GRAPHS AND MODELS

Algebra and Trigonometry

GRAPHS AND MODELS

Marvin L. Bittinger
Indiana University–Purdue University at Indianapolis

Judith A. Beecher
Indiana University–Purdue University at Indianapolis

David Ellenbogen
St. Michael's College

Judith A. Penna

An imprint of Addison Wesley Longman, Inc.

Reading, Massachusetts • Menlo Park, California • New York • Harlow, England
Don Mills, Ontario • Sydney • Mexico City • Madrid • Amsterdam

Sponsoring Editor	Jason Jordan
Project Manager	Kari Heen
Managing Editor	Karen Guardino
Text Design	Geri Davis/The Davis Group, Inc.
Art, Design, Editorial, and Production Services	Martha Morong/Quadrata, Inc. Geri Davis/The Davis Group, Inc.
Marketing Manager	Michelle Babinec
Illustrators	Scientific Illustrators, J. A. K. Graphics, Parrot Graphics, and Gail Hayes
Compositor	The Beacon Group
Cover Designer	Karen Rappaport
Cover Photograph	Takeshi Odawara/PHOTONICA

Photo Credits

134, PF Sports Images **140 (left),** SuperStock **309,** SuperStock **314,** The Topps Company, Inc. **366,** Darryl Jones **371,** courtesy of Goss Graphics Systems **702,** © Mary Dyer

2, Graph on number of births: From THE NEW YORK TIMES, 2/12/95, p. A1. Copyright © 1995 by The New York Times Company. Reprinted with permission.

Library of Congress Cataloging-in-Publication Data

Algebra & trigonometry: graphs & models/Marvin L. Bittinger . . . [et al.].
p. cm.
Includes index.
ISBN 0-201-84888-0
ISBN 0-201-69442-5 (Precalculus)
1. Algebra. 2. Algebra—Graphic methods. 3. Trigonometry.
4. Trigonometry—Graphic methods. I. Bittinger, Marvin L.
QA154.2.A3124 1996
512′ .13—dc20 96-21889
CIP

1 2 3 4 5 6 7 8 9 10—DOW—99989796

Contents

Introduction to Graphs and Graphers 1

R
Basic Concepts of Algebra

R.1 The Real-Number System 20
R.2 Integer Exponents, Scientific Notation, and Order of Operations 25
R.3 Addition, Subtraction, and Multiplication of Polynomials 32
R.4 Factoring 36
R.5 Rational Expressions 40
R.6 Radical Notation and Rational Exponents 49
R.7 Solving Equations 58
R.8 Solving Inequalities 66
R.9 Modeling and Applications 72
SUMMARY AND REVIEW 79

1
Graphs, Functions, and Models

1.1 Functions, Graphs, and Graphers 84
1.2 Functions and Applications 98
1.3 Linear Functions and Applications 111
1.4 Data Analysis, Curve Fitting, and Linear Regression 127
1.5 Distance, Midpoints, and Circles 135
1.6 Symmetry 140
1.7 Transformations of Functions 147
1.8 The Algebra of Functions 158
SUMMARY AND REVIEW 167

2 Polynomial and Rational Functions

2.1 Introduction to Polynomial Functions and Complex Numbers 172
2.2 Quadratic Equations and Functions 180
2.3 Polynomial Models and Curve Fitting 196
2.4 Polynomial Division; The Remainder and Factor Theorems 208
2.5 Theorems about Zeros of Polynomial Functions 215
2.6 Rational Functions 224
2.7 Polynomial and Rational Inequalities 238
SUMMARY AND REVIEW 245

3 Exponential and Logarithmic Functions

3.1 Inverse Functions 250
3.2 Exponential Functions and Graphs 261
3.3 Logarithmic Functions and Graphs 274
3.4 Properties of Logarithmic Functions 287
3.5 Solving Exponential and Logarithmic Equations 296
3.6 Applications and Models: Growth and Decay 303
SUMMARY AND REVIEW 318

4 The Trigonometric Functions

4.1 Trigonometric Functions of Acute Angles 322
4.2 Applications of Right Triangles 333
4.3 Trigonometric Functions of Any Angle 343
4.4 Radians, Arc Length, and Angular Speed 358
4.5 Circular Functions: Graphs and Properties 372
4.6 Graphs of Transformed Sine and Cosine Functions 389
SUMMARY AND REVIEW 404

5 Trigonometric Identities, Inverse Functions, and Equations

5.1 Identities: Pythagorean and Sum and Difference 410
5.2 Identities: Cofunction, Double-Angle, and Half-Angle 423
5.3 Proving Trigonometric Identities 431
5.4 Inverses of the Trigonometric Functions 438
5.5 Solving Trigonometric Equations 448
SUMMARY AND REVIEW 459

6 Applications of Trigonometry

6.1 The Law of Sines 464
6.2 The Law of Cosines 476
6.3 Complex Numbers: Trigonometric Form 484
6.4 Polar Coordinates and Graphs 496
6.5 Vectors 505
SUMMARY AND REVIEW 518

7 Systems and Matrices

7.1 Systems of Equations in Two Variables 522
7.2 Systems of Equations in Three or More Variables 532
7.3 Matrices and Systems of Equations 540
7.4 Matrix Operations 548
7.5 Inverses of Matrices 560
7.6 Systems of Inequalities and Linear Programming 566
7.7 Partial Fractions 577
SUMMARY AND REVIEW 585

8 Conic Sections

8.1 The Parabola 590
8.2 The Circle and the Ellipse 598
8.3 The Hyperbola 608
8.4 Nonlinear Systems of Equations 617
SUMMARY AND REVIEW 625

9 Sequences, Series, and Combinatorics

9.1 Sequences and Series 630
9.2 Arithmetic Sequences and Series 641
9.3 Geometric Sequences and Series 651
9.4 Mathematical Induction 662
9.5 Combinatorics: Permutations 669
9.6 Combinatorics: Combinations 679
9.7 The Binomial Theorem 686
9.8 Probability 695
SUMMARY AND REVIEW 705

Appendixes 709

A Determinants and Cramer's Rule 709

B Parametric Equations 716

Answers A-1

Index of Applications I-1

Index I-5

Preface

Algebra and Trigonometry: Graphs and Models covers college-level algebra and trigonometry and is appropriate for a one- or two-term course in precalculus mathematics. The approach of this text is more interactive than most algebra and trigonometry texts. Our goal is to enhance the learning process through the use of technology and to provide as much support and help for students as possible in their study of algebra and trigonometry. A course in intermediate algebra is a prerequisite for the text, although Chapter R provides sufficient review to unify the diverse backgrounds of most students.

Content Features

- **Integrated Technology** The technology of the graphing calculator is completely integrated throughout the text to provide a visual means of increasing understanding. In this text, we use the term "grapher" to refer to all graphing calculator technology. The use of the grapher is woven throughout the exposition, the exercise sets, and the testing program without sacrificing algebraic and trigonometric skills. We use the grapher technology to enhance, not to replace, the students' mathematical skills and to alleviate the tedium associated with certain procedures. It is assumed that each student is required to have a grapher (or at least access to one) while enrolled in this course.

- **Learning to Use the Technology** To minimize the need for valuable class time to teach students how to use a grapher, we have provided several features that shorten the learning curve while increasing the students' knowledge of the fundamentals of a grapher. The first of these is the section entitled "Introduction to Graphs and Graphers" found at the beginning of the text. It introduces students to the basic functions of the grapher. The others are the *Graphing Calculator Manual* (see p. xiv) and the video series entitled *Graphing Calculator Instructional Videos* (see p. xiv). In addition, a set of programs has been included in the *Graphing Calculator Manual*. All of these features have been specifically written and produced for this text.

- **Interactive Discoveries** The grapher provides an exciting teaching opportunity in which a student can discover and further investigate mathematical concepts. This unique Interactive Discovery feature is used to introduce new topics and provides a vehicle for students to "see" a concept quickly. This feature reinforces the idea that grapher technology is an integral part of the course as well as an important learning tool. It invites the student to develop analytic and reasoning skills while taking an active role in the learning process. (See pp. 114, 226, and 392.)

- **Function Emphasis** The use of technology with its immediate visualization of a concept encourages the early presentation of functions. Graphing and functions are both introduced in the first section of Chapter 1. The study of the family of functions (linear, quadratic, higher-degree polynomial, rational, exponential, logarithmic, and trigonometric) has been enhanced and streamlined with the inclusion of the grapher. Applications with graphs are incorporated throughout to amplify and add relevance to the study of functions. (See pp. 84, 205, and 380.)

- **Variety in Approaches to Solutions** Skill in solving mathematical problems is expanded when a student is exposed to a variety of approaches to finding a solution. We have carefully incorporated three solution approaches throughout the text: algebraic, graphical, and numerical. Chapter openers illustrate an application with a concurrent grapher presentation of both a table and a graph (see pp. 171 and 521). The TABLE feature on a grapher provides a numerical display or check of the solution (see pp. 198 and 329).

  To highlight both the algebraic- and graphical-solution approaches in solving equations, we have used a two-column solution format in numerous examples (see pp. 184, 297, and 454). In the algebraic/graphical side-by-side features, both methods are presented together; each method provides a complete solution. This feature emphasizes that there is more than one way to obtain a result and illustrates the comparative efficiency and accuracy of the two methods.

- **Real-Data Applications** Throughout the writing process, we conducted an energetic search for real-data applications. The result of that effort is a variety of examples and exercises that connect the mathematical content with the real world. Source lines appear with most real-data applications and charts and graphs are frequently included. Many applications are drawn from the fields of health, business and economics, life and physical sciences, social science, and areas of general interest such as sports and daily life. We encourage students to "see" and interpret the mathematics that appears around them every day. (See pp. 134, 304, 371, and I-1.)

- **Regression** Using regression or curve fitting to model data is introduced in Chapter 1 with linear functions. This visual theme is continued with quadratic, cubic, quartic, higher-degree polynomial, exponential, logarithmic, logistic, and trigonometric functions. Although the theoretic aspects of curve fitting cannot be developed in this course, the power of the grapher is very apparent in this area as the technique is applied to real data. Students can quickly make the "what is this used for?" connection between real data and the extrapolated results of the curve fitting, thus giving them a better conceptual understanding of the material. (See pp. 129 and 312.)

- **Verifying Identities** Identities can be partially verified with a grapher using both the GRAPH and TABLE features. This use of the grapher is first seen in the Introduction to Graphs and Graphers and is continued in later chapters in discovery and verification of possible identities (see pp. 11, 287, and 413). This content feature allows a visual answer to such frequent questions as "Why isn't $(x + 2)^2$ equal to $x^2 + 4$?" This approach also provides a unique lead-in to the development of the properties of exponents and logarithms.

- **Optional Review Chapter** Chapter R provides an optional review of intermediate algebra. We purposely placed the Introduction to Graphs and Graphers before this chapter to allow the grapher to be used in the review. The incorporation of technology gives these topics a fresh perspective and sets the tone for the rest of the course (see pp. 41 and 60). Chapter R can also be used as a convenient source of information for a student who needs a quick review of a particular topic.

Pedagogical Features

- **Use of Color** The text uses full color in an extremely functional way, as seen in the design elements and artwork on nearly every page. The choice of color has been carried out in a methodical and precise manner so that its use carries a consistent meaning, which enhances the readability of the text for both student and instructor. (See pp. 93 and 397.)

- **Art Package** The text contains over 1500 art pieces including a new form of art called photorealism. Photorealism superimposes mathematics on a photograph and encourages students to "see mathematics" in familiar settings (see pp. 115 and 340). The exceptional situational art and statistical graphs throughout the text highlight the abundance of real-world applications while helping students visualize the mathematics (see pp. 205, 335, and 447). The design and use of color with the grapher windows exemplifies the impact that technology has in today's mathematical curriculum (see pp. 13, 159, and 448).

- **Annotated Examples** Over 770 examples fully prepare the student for the exercise sets. Learning is carefully guided with numerous color-coded art pieces and step-by-step annotations, with substitutions and annotations highlighted in red (see pp. 182 and 305). The basis for problem solving is a five-step process established early in the text to aid the student in strategically approaching and solving applications. (See pp. 72 and 527.)

- **Variety of Exercises** There are over 5500 exercises in this text. The exercise sets are enhanced not only by the inclusion of real-data applications with source lines, detailed art pieces, and technology windows that include both tables and graphs, but also by the following features.

  Technology Exercises Since the grapher is totally integrated in this text, exercise sets include both grapher and nongrapher exercises. In some cases, detailed instruction lines indicate the approach the student is expected to use. In others, the student is left to choose the approach that seems best, thereby encouraging critical thinking. (See pp. 194 and 302.)

  Skill Maintenance The exercises in this section have been specifically selected to review concepts previously taught in the text that are foundations for the material presented in the following section. They are chosen to prepare the student for the new concept(s) that will be covered next. (See p. 97.)

  Synthesis Exercises These exercises, which appear at the end of each exercise set, encourage critical thinking by requiring students to synthesize concepts from several sections or to take a concept a step further than in the regular exercises. (See p. 195.)

  Thinking and Writing Exercises for thinking and writing, at the beginning of the synthesis exercises, are denoted with a maze icon ◈. They encourage students to both consider and write about key mathematical ideas in the chapter. Many of these exercises are open-ended, making them particularly suitable for use in class discussions or as collaborative activities. (See p. 260.)

- **Chapter Openers** Each chapter opens with an application illustrated with both technology windows and situational art. The openers also include a table of contents listing section titles. (See pp. 83 and 249.)

- **Section Objectives** Content objectives are listed at the beginning of each section. These, together with subheadings throughout the section, provide a useful outline for both instructors and students. (See pp. 111 and 522.)

- **Highlighted Information** Important definitions, properties, and rules are displayed in screened boxes. Summaries and procedures are listed in color-outlined boxes. Both of these design features present and organize the material for efficient learning and review. (See pp. 116 and 123.)

- **Summary and Review** The Summary and Review at the end of each chapter provides an extensive set of review exercises along with a list of important properties and formulas covered in that chapter. This feature provides an excellent preparation for chapter tests and the final examination. Answers to all review exercises appear in the text along with section references that direct students to material to reexamine if they have difficulty with a particular exercise. (See pp. 318 and 404.)

Supplements for the Instructor

Instructor's Solutions Manual

The Instructor's Solutions Manual by Judith A. Penna contains worked-out solutions to all exercises in the exercise sets, including the thinking and writing exercises. It also includes a sample test with answers for each chapter and answers to the exercises in the appendixes. The sample tests are also included in the Student's Solutions Manual.

Printed Test Bank/Instructor's Manual

Prepared by Donna DeSpain, the Printed Test Bank/Instructor's Manual contains the following:

- 4 free-response test forms for each chapter, following the format and with the same level of difficulty as the tests in the Student's Solutions Manual.
- 2 multiple-choice test forms for each chapter.
- 6 alternate forms of the final examination, 4 with free-response questions and 2 with multiple-choice questions.
- Index to the Graphing Calculator Instructional Videos.

Testgen EQ

Testgen EQ is a computerized test generator that allows instructors to select test questions manually or randomly from selected topics or to use a ready-made test for each chapter. The test questions are algorithm-driven so that regenerated number values maintain problem types and provide a large number of test items in both multiple-choice and open-ended formats for one or more test forms. Test items can be viewed on screen, and the built-in question editor lets instructors modify existing questions or add new ones that include pictures, graphs, accurate math symbols, and variable text and numbers.

Additional features in the new Windows and Macintosh Test Generators allow the instructor to customize both the look and content of testbanks and tests. Test questions are easily transferred from the testbank to a test and can be sorted, searched, and displayed in various ways. Testgen EQ is free to adopters.

Course Management and Testing System

InterAct Math Plus for Windows and Macintosh (available from Addison Wesley Longman) combines course management and on-line testing with the features of the basic tutorial software (see "Supplements for the Student") to create an invaluable teaching resource. Consult your local Addison Wesley Longman sales consultant for details.

Supplements for the Student

Graphing Calculator Manual

The Graphing Calculator Manual by Judith A. Penna, with the assistance of John Garlow and Mike Rosenborg, contains keystroke level instruction for the Texas Instruments TI-82, TI-83, TI-85, and Hewlett Packard HP 38G graphing calculators. Modules for the Casio 9850, the Sharp 9200 and 9300, and the HP48G are available on request. Contact your local Addison Wesley Longman sales consultant for details.

Bundled free with every copy of the text, the Graphing Calculator Manual uses actual examples and exercises from *Algebra and Trigonometry: Graphs and Models* to help teach students to use their graphing calculator. The order of topics in the Graphing Calculator Manual mirrors that of the texts, providing a just-in-time mode of instruction.

Student's Solutions Manual

The Student's Solutions Manual by Judith A. Penna contains completely worked-out solutions with step-by-step annotations for all the odd-numbered exercises in the exercise sets in the text, with the exception of the thinking and writing exercises. It also includes a self-test with answers for each chapter and a final examination.

The Student's Solutions Manual can be purchased by your students from Addison Wesley Longman.

Graphing Calculator Instructional Videos

Designed and produced specifically for *Algebra and Trigonometry: Graphs and Models*, the Graphing Calculator Instructional Videos take students through procedures on the graphing calculator using content from the text. These videos include most topics covered in the Graphing Calculator Manual, as well as several additional topics.

Every video section uses actual text examples or odd-numbered exercises from the text to help motivate students while they learn to use the graphing calculator. In the videos, an instructor shows students keystroke level procedures that they will need to succeed with the grapher as they proceed through the course. The videos are correlated to the sections of the text.

A complete set of Graphing Calculator Instructional Videos is free to qualifying adopters.

InterAct Math Tutorial Software

InterAct Math Tutorial Software has been developed and designed by professional software engineers working closely with a team of experienced math educators.

InterAct Math Tutorial Software includes exercises that are linked with every objective in the textbook and require the same computational and problem-solving skills as their companion exercises in the text. Each exercise has an example and an interactive guided solution that are designed to involve students in the solution process and to help them identify precisely where they are having trouble. In addition, the software recognizes common student errors and provides students with appropriate customized feedback.

With its sophisticated answer recognition capabilities, InterAct Math Tutorial Software recognizes appropriate forms of the same answer for any kind of input. It also tracks student activity and scores for each section, which can then be printed out.

Available for both Windows and Macintosh computers, the software is free to qualifying adopters.

Acknowledgments

We wish to express our genuine appreciation to a number of people who contributed in special ways to the development of this textbook. Jason Jordan and Greg Tobin, our editors at Addison Wesley Longman, shared our vision and provided encouragement and motivation. In addition, the production and marketing departments of Addison Wesley Longman brought to the project their unsurpassed commitment to excellence. The unwavering support from the Higher Education Group has been a continuing source of strength for this author team. For this we are most grateful. Mike Rosenborg, Barbara Johnson, Patty Slipher, Irene Doo, and Larry Bittinger provided many constructive comments as well as accuracy checks to the manuscript.

Finally, Professor Bittinger would like to thank his MA 153 students at IUPUI for their productive response to parts of the manuscript that were class-tested in the spring of 1996 using the graphing calculator. This teaching approach resulted in the most satisfying class he has taught at IUPUI in 28 years. Further information regarding this class can be obtained from Professor Bittinger at his e-mail address, exponent@aol.com, or through his home page (see Web Connection that follows).

We would also like to thank the following reviewers for their invaluable contribution to the development of this text:

Sandra Beken, *Horry-Georgetown Technical College*

Diane Benjamin, *University of Wisconsin—Platteville*

Robert Bohac, *Northwest College*

Diane W. Burleson, *Central Piedmont Community College*

John W. Coburn, *St. Louis Community College at Florissant Valley*

Elaine N. Daniels, *Salve Regina University*

Donna DeSpain, *Benedictine University*

Eunice F. Everett, *Seminole Community College*

Rob Farinelli, *Community College of Allegheny County South Campus*

Betty P. Givan, *Eastern Kentucky University*

Allen R. Hesse, *Rochester Community College*

Heidi A. Howard, *Florida Community College at Jacksonville*
Steve Kahn, *Anne Arundel Community College*
Timothy A. Loughlin, *New York Institute of Technology*
Larry Luck, *Anoka-Ramsey Community College*
Joseph D. Mahoney, *Paducah Community College*
Peggy I. Miller, *University of Nebraska at Kearney*
John A. Oppelt, *Bellarmine College*
Tom Schaffter, *Fort Lewis College*
Eric Schulz, *Walla Walla Community College*
Judith D. Smalling, *St. Petersburg Junior College*
Craig M. Steenberg, *Lewis-Clark State College*
Kathryn C. Wetzel, *Amarillo College*
Kemble Yates, *Southern Oregon State College*

Web Connection

Students and instructors can obtain more information about books written by Professor Bittinger and his co-authors by contacting *Marv's Math Corner* on the Internet at the following address:

http://www.math.iupui.edu/~mbitting/

Included on this Web site is information about books and their supplements, as well as study tips and sample practice final examinations. Students and instructors are welcome to e-mail questions, comments, and constructive criticism.

M.L.B.
J.A.B.
D.J.E.
J.A.P.

- **Applied Chapter Openers:** Each chapter begins with a relevant application highlighting how concepts presented in the chapter can be put to use in the real world. These applications are accompanied by numerical tables, equations, and grapher windows to show students the many different ways in which problem situations can be examined.

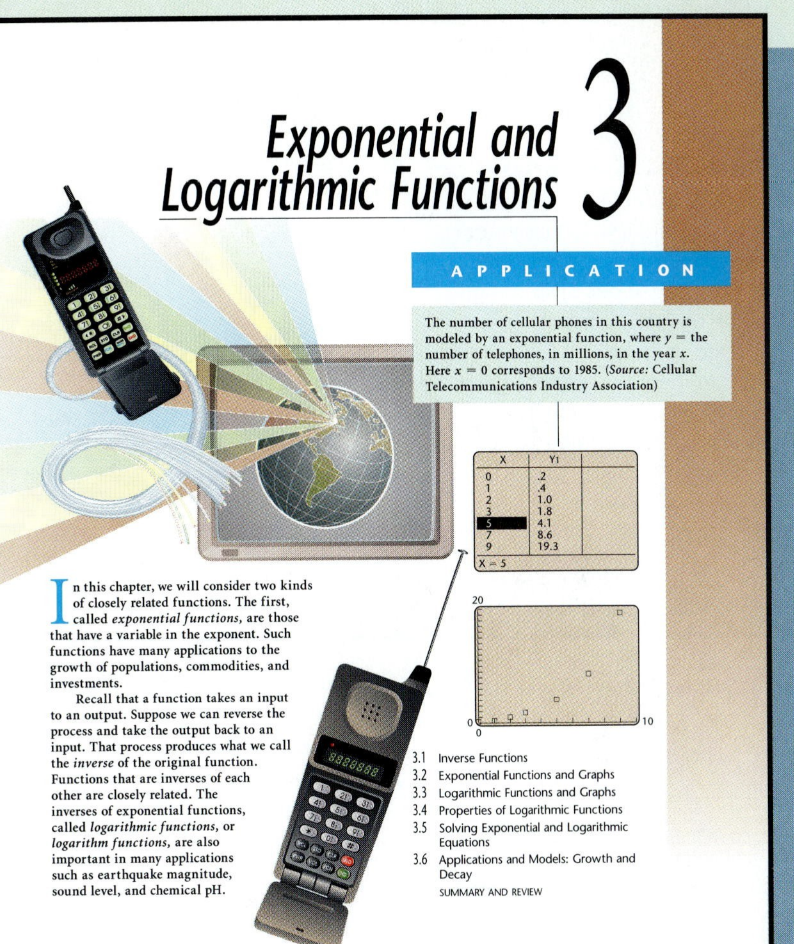

Exponential and Logarithmic Functions 3

APPLICATION

The number of cellular phones in this country is modeled by an exponential function, where y = the number of telephones, in millions, in the year x. Here $x = 0$ corresponds to 1985. (*Source:* Cellular Telecommunications Industry Association)

X	Y1
0	.2
1	.4
2	1.0
3	1.8
5	4.1
7	8.6
9	19.3

X = 5

In this chapter, we will consider two kinds of closely related functions. The first, called *exponential functions*, are those that have a variable in the exponent. Such functions have many applications to the growth of populations, commodities, and investments.

Recall that a function takes an input to an output. Suppose we can reverse the process and take the output back to an input. That process produces what we call the *inverse* of the original function. Functions that are inverses of each other are closely related. The inverses of exponential functions, called *logarithmic functions*, or *logarithm functions*, are also important in many applications such as earthquake magnitude, sound level, and chemical pH.

3.1 Inverse Functions
3.2 Exponential Functions and Graphs
3.3 Logarithmic Functions and Graphs
3.4 Properties of Logarithmic Functions
3.5 Solving Exponential and Logarithmic Equations
3.6 Applications and Models: Growth and Decay
SUMMARY AND REVIEW

Equation S

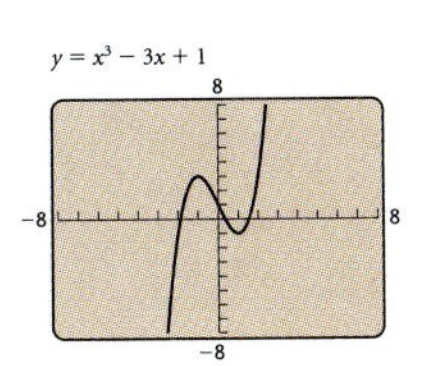

Example 10 tions to thre

SOLUTION T coordinates the x-axis. T x-intercept h which $y = 0$ such x-value

To acqui fast way to d zoom-in feat small section amine the gr zoom in on the gr sired accuracy is a

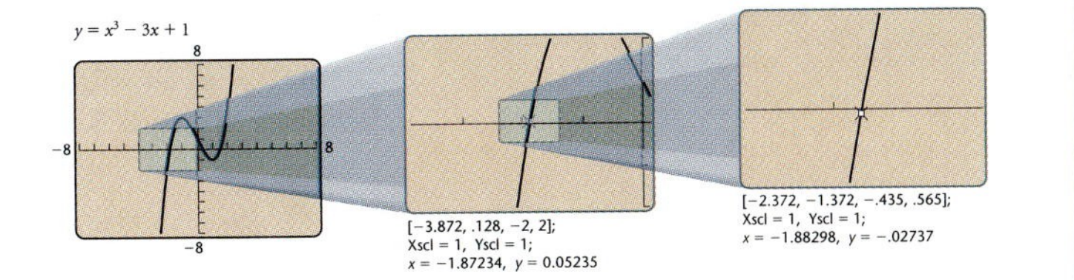

To find the solution to the nearest thousandth, we trace and zoom until the cursor's x-value just to the left of the intercept and the cursor's x-value just to the right of the intercept are the same, when rounded to the nearest thousandth. By using the TRACE and ZOOM features three more times, we find the solution to be about −1.879. In a similar manner, we find that the other solutions of the equation $x^3 - 3x + 1 = 0$ are about 0.347 and 1.532. If available, we might also use the SOLVE or ROOT features to approximate solutions.

There are many ways in which the ZOOM feature can be used. For example, we can zoom in for more precision, as in Example 10, or zoom out on a graph—say, to reveal more of its curvature—adjusting the factors of the zoom in any way we choose. In Example 10, we used zoom factors of 4. We may also be able to use a ZOOM-BOX feature to zoom in on a boxed region of our choosing. All such details can be found by consulting the manual for your particular grapher or the Graphing Calculator Manual that accompanies this book.

- **Grapher Integration:** The author team assumes that the student will use a grapher throughout the course and during homework or group sessions. Numerous grapher windows appear throughout the text, some in a unique ZOOM pattern.

- **Interactive Discoveries:** Throughout the exposition, students are directed to investigate new concepts before they are formally developed. This design invites students to be actively involved with the material in order to identify a mathematical pattern or form an intuitive understanding of a new topic.

114 CHAPTER 1 • GRAPHS, FUNCTIONS, AND MODELS

Interactive Discovery

With a square viewing window (see the Introduction to Graphs and Graphers), graph the following equations:

$$y_1 = x, \quad y_2 = 2x, \quad y_3 = 5x, \quad \text{and} \quad y_4 = 10x.$$

What do you think the graph of $y = 128x$ will look like?

Clear the screen and graph the following equations:

$$y_1 = x, \quad y_2 = \tfrac{3}{4}x, \quad y_3 = 0.48x, \quad \text{and} \quad y_4 = \tfrac{3}{25}x.$$

What do you think the graph of $y = 0.000029x$ will look like?

Again clear the screen and graph each set of equations:

$$y_1 = -x, \quad y_2 = -2x, \quad y_3 = -4x, \quad \text{and} \quad y_4 = -10x$$

and $$y_1 = -x, \quad y_2 = -\tfrac{2}{3}x, \quad y_3 = -0.35x, \quad \text{and} \quad y_4 = -\tfrac{1}{10}x.$$

From your observations, what do you think the graphs of $y = -200x$ and $y = -0.000017x$ will look like?

If a line slants up from left to right, the change in x and the change in y have the same sign, so the line has a positive slope. The larger the slope is, the steeper the line. If a line slants down from left to right, the change in x and the change in y are of opposite signs, so the line has a negative slope. The larger the absolute value of the slope, the steeper the line.

y = 3x, y = 2x, y = x, (1, 3), (1, 2), (1, 1)

y = −3x, y = −2x, y = −x, (1, −1), (1, −2), (1, −3)

y = 0, (1, 0)

, or $y = 0$, as shown in the graph on the right s both the x-axis and a horizontal line.

ical Lines

l, the change in y for any two points is 0. ne has slope 0.

the change in x for any two points is 0. Thus *ned* because we cannot divide by 0.

pe and an undefined slope are two very different

184 CHAPTER 2 • POLYNOMIAL AND RATIONAL FUNCTIONS

The Quadratic Formula

The solutions of $ax^2 + bx + c = 0$, $a \neq 0$, are given by

$$x = \frac{-b \pm \sqrt{b^2 - 4ac}}{2a}.$$

Example 3 Solve $3x^2 + 2x = 7$. Find exact solutions and approximate solutions rounded to the nearest thousandth.

We show both algebraic and graphical solutions. Note that only the algebraic approach yields the exact solutions.

ALGEBRAIC SOLUTION

After finding standard form, we are unable to factor, so we identify a, b, and c in order to use the quadratic formula:

$$3x^2 + 2x - 7 = 0;$$
$$a = 3, \quad b = 2, \quad c = -7.$$

We then use the quadratic formula:

$$x = \frac{-b \pm \sqrt{b^2 - 4ac}}{2a}$$
$$= \frac{-2 \pm \sqrt{2^2 - 4(3)(-7)}}{2(3)} \quad \textbf{Substituting}$$
$$= \frac{-2 \pm \sqrt{4 + 84}}{6} = \frac{-2 \pm \sqrt{88}}{6}$$
$$= \frac{-2 \pm \sqrt{4 \cdot 22}}{6} = \frac{-2 \pm 2\sqrt{22}}{6}$$
$$= \frac{2}{2} \cdot \frac{-1 \pm \sqrt{22}}{3} = \frac{-1 \pm \sqrt{22}}{3}.$$

The exact solutions are

$$\frac{-1 - \sqrt{22}}{3} \quad \text{and} \quad \frac{-1 + \sqrt{22}}{3}.$$

Using the scientific keys on a grapher, we approximate the solutions to be -1.897 and 1.230.

GRAPHICAL SOLUTION

To solve $3x^2 + 2x = 7$, or $3x^2 + 2x - 7 = 0$, we first graph the function $f(x) = 3x^2 + 2x - 7$. Then we look for points where the graph crosses the x-axis. It appears that there are two possible zeros, one near -2 and one near 1. We can use TRACE and ZOOM to approximate these zeros, or we can use a SOLVE or POLY feature.

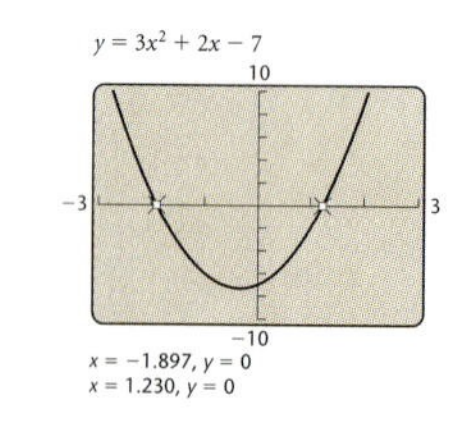

We get the approximate zeros -1.896805 and 1.2301386, or -1.897 and 1.230, rounded to three decimal places. The zeros of the function are the solutions of the equation $3x^2 + 2x = 7$.

- **Side-by-Side Algebraic and Graphical Solutions:** Many examples in the text are presented in a two-column format that shows simultaneous algebraic and graphical solution methods. This balanced approach allows students to compare the efficiency and appropriateness of each method.

- **Art:** Generous amounts of color-coded technical and situational art appear throughout the text to enhance understanding of an example or exercise, to interest students, and to aid in the visualization of concepts.

SECTION 1.3 • LINEAR FUNCTIONS AND APPLICATIONS 115

Applications of Slope

Slope has many real-world applications. For example, numbers like 2%, 4%, and 7% are often used to represent the **grade** of a road. Such a number is meant to tell how steep a road is on a hill or mountain. For example, a 4% grade means that the road rises 4 ft for every horizontal distance of 100 ft if a vehicle is going up; and −4% means that the road is dropping 4 ft for every 100 ft, if the vehicle is going down.

The concept of grade is also used in cardiology when a person runs on a treadmill. A physician may change the slope, or grade, of a treadmill to measure its effect on heart rate. Another example occurs in hydrology. When a river flows, the strength or force of the river depends on how far the river falls vertically compared to how far it flows horizontally.

Example 2 *Ramps for the Handicapped.* Construction laws regarding ... for the handicapped state that every vertical rise of 1 ft ... rizontal run of 12 ft. What is the grade, or slope, of such ...

... de, or slope, is given by

... 0.083 ≈ 8.3%.

154 CHAPTER 1 • GRAPHS, FUNCTIONS, AND MODELS

SOLUTION

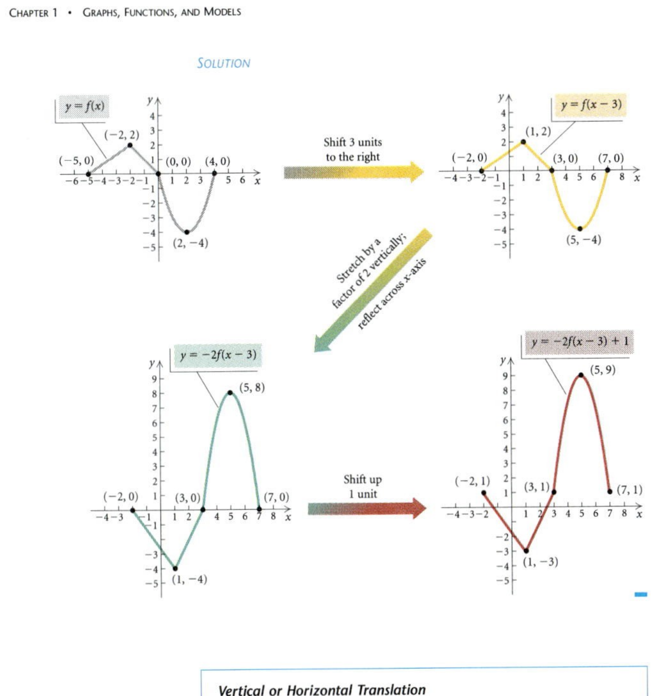

Vertical or Horizontal Translation

For $b > 0$,

the graph of $y_2 = f(x) + b$ is the graph of $y_1 = f(x)$ shifted *up b* units;

the graph of $y_2 = f(x) - b$ is the graph of $y_1 = f(x)$ shifted *down b* units.

For $d > 0$,

the graph of $y_2 = f(x - d)$ is the graph of $y_1 = f(x)$ shifted *right d* units;

the graph of $y_2 = f(x + d)$ is the graph of $y_1 = f(x)$ shifted *left d* units.

- To assist students with understanding, **graphs with multiple curves** use different colors for each curve. Movement on the graph may be indicated by a gradual shift in color on an accompanying shift arrow.

- **Data Analysis and Modeling:** The author team highlights and reinforces the theme of data analysis throughout the text.

1.4 Data Analysis, Curve Fitting, and Linear Regression

- *Analyze a set of data to determine whether it can be modeled by a linear function.*
- *Fit a regression line to a set of data; then use the linear model to make predictions.*

Mathematical Models

When a real-world problem can be described in mathematical language, we have a **mathematical model**. For example, the natural numbers constitute a mathematical model for situations in which counting is essential. Situations in which algebra can be brought to bear often require the use of functions.

Mathematical models are abstracted from real-world situations. Procedures within the mathematical model then give results that allow one to predict what will happen in that real-world situation. If the predictions are inaccurate or the results of experimentation do not conform to the model, the model needs to be changed or discarded.

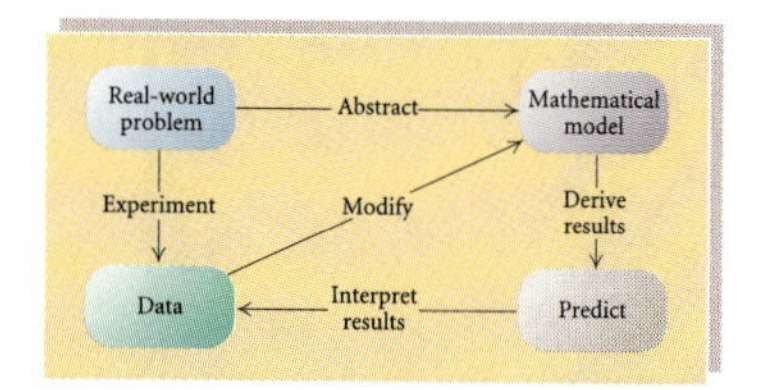

Mathematical modeling can be an ongoing process. For example, finding a mathematical model that will enable an accurate prediction of population growth is not a simple problem. Any population model that one might devise would need to be reshaped as further information is acquired.

Curve Fitting

We will develop and use many kinds of mathematical models in this text. In this chapter, we have considered many functions that can be used as models. Let's look at four of them.

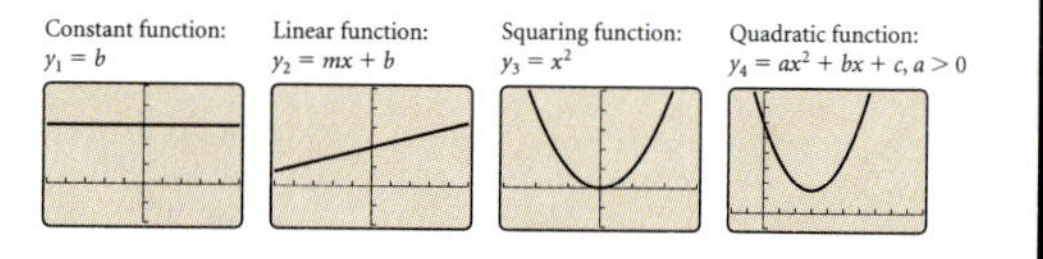

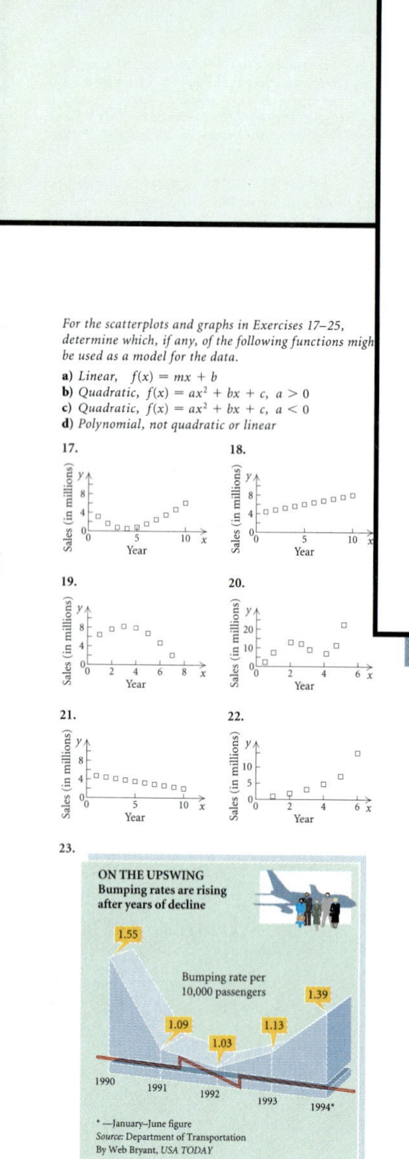

For the scatterplots and graphs in Exercises 17–25, determine which, if any, of the following functions migh[t] be used as a model for the data.

a) *Linear,* $f(x) = mx + b$
b) *Quadratic,* $f(x) = ax^2 + bx + c,\ a > 0$
c) *Quadratic,* $f(x) = ax^2 + bx + c,\ a < 0$
d) *Polynomial, not quadratic or linear*

25.

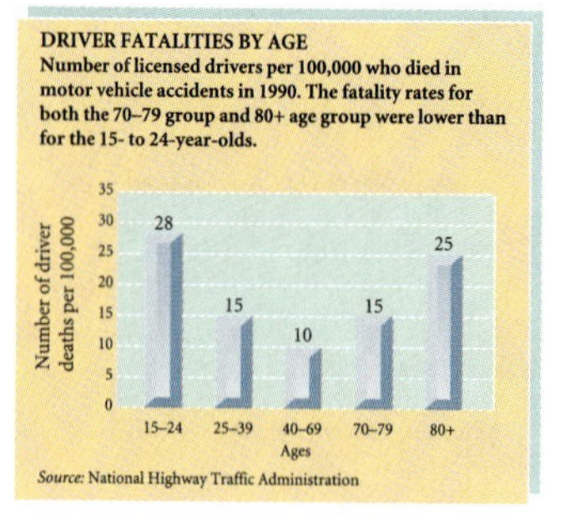

DRIVER FATALITIES BY AGE
Number of licensed drivers per 100,000 who died in motor vehicle accidents in 1990. The fatality rates for both the 70–79 group and 80+ age group were lower than for the 15- to 24-year-olds.
Number of driver deaths per 100,000
28
15
10
15
25
15–24
25–39
40–69
70–79
80+
Ages
Source: National Highway Traffic Administration

- By fitting curves to data and discussing **mathematical models**, students develop an understanding of mathematical patterns as they are seen in the real world and gain insight into using models to make predictions.

- **Direction lines** in the exercise sets instruct students to use both algebraic and graphical methods to solve problems. Sometimes the student is asked to solve with a grapher and check algebraically. In other problems, the student is directed to solve algebraically and check the work with a grapher. Many exercises do not specify a solution method and allow the student to determine the method, thus encouraging critical thinking and analytical skills.

302 CHAPTER 3 • EXPONENTIAL AND LOGARITHMIC FUNCTIONS

3.5 Exercise Set

Solve each exponential equation algebraically. Then check on a grapher.

1. $3^x = 81$
2. $2^x = 32$
3. $2^{2x} = 8$
4. $3^{7x} = 27$
5. $2^x = 33$
6. $2^x = 40$
7. $5^{4x-7} = 125$
8. $4^{3x-5} = 16$
9. $27 = 3^{5x} \cdot 9^{x^2}$
10. $3^{x^2+4x} = \frac{1}{27}$
11. $84^x = 70$
12. $28^x = 10^{-3x}$
13. $e^t = 1000$
14. $e^{-t} = 0.04$
15. $e^{-0.03t} = 0.08$
16. $1000e^{0.09t} = 5000$
17. $3^x = 2^{x-1}$
18. $5^{x+2} = 4^{1-x}$
19. $(3.9)^x = 48$
20. $250 - (1.87)^x = 0$
21. $e^x + e^{-x} = 5$
22. $e^x - 6e^{-x} = 1$
23. $\frac{e^x + e^{-x}}{e^x - e^{-x}} = 3$
24. $\frac{5^x - 5^{-x}}{5^x + 5^{-x}} = 8$

Solve each logarithmic equation algebraically. Then check on a grapher.

25. $\log_5 x = 4$
26. $\log_2 x = -3$
27. $\log x = -4$
28. $\log x = 1$
29. $\ln x = 1$
30. $\ln x = -2$
31. $\log_2 (10 + 3x) = 5$
32. $\log_5 (8 - 7x) = 3$
33. $\log x + \log (x - 9) = 1$
34. $\log_2 (x + 1) + \log_2 (x - 1) = 3$
35. $\log_8 (x + 1) - \log_8 x = 2$
36. $\log x - \log (x + 3) = -1$
37. $\log_4 (x + 3) + \log_4 (x - 3) = 2$
38. $\ln (x + 1) - \ln x = \ln 4$
39. $\log (2x + 1) - \log (x - 2) = 1$
40. $\log_5 (x + 4) + \log_5 (x - 4) = 2$

Use only a grapher. Find approximate solutions of each equation or approximate the point(s) of intersection of a pair of equations.

41. $e^{7.2x} = 14.009$
42. $0.082e^{0.05x} = 0.034$
43. $xe^{3x} - 1 = 3$
44. $5e^{5x} + 10 = 3x + 40$
45. $4 \ln (x + 3.4) = 2.5$
46. $\ln x^2 = -x^2$
47. $\log_8 x + \log_8 (x + 2) = 2$
48. $\log_3 x + 7 = 4 - \log_5 x$
49. $\log_5 (x + 7) - \log_5 (2x - 3) = 1$
50. $y = \ln 3x,\ y = 3x - 8$
51. $2.3x + 3.8y = 12.4,\ y = 1.1 \ln (x - 2.05)$
52. $y = 2.3 \ln (x + 10.7),\ y = 10e^{-0.07x^2}$
53. $y = 2.3 \ln (x + 10.7),\ y = 10e^{-0.007x^2}$

Skill Maintenance

54. Solve $K = \frac{1}{2}mv^2$ for v.

Solve.

55. $x^4 + 5x^2 = 36$
56. $t^{2/3} - 10 = 3t^{1/3}$
57. *Total Sales of Goodyear.* The following table shows factual data regarding total sales of The Goodyear Tire and Rubber Company.

YEAR, x	TOTAL SALES, y (IN MILLIONS)
1. 1991	$10,906.8
2. 1992	11,784.9
3. 1993	11,643.4
4. 1994	12,288.2

Source: The Goodyear Tire and Rubber Company Annual Report.

a) Use linear regression on a grapher to fit an equation $y = mx + b$, where $x = 1$ corresponds to 1991, to the data points. Predict total sales in 1999.

b) Use quadratic regression on a grapher to fit a quadratic equation $y = ax^2 + bx + c$ to the data points. Predict total sales in 1999.

Synthesis

58. ◆ In Example 3, we took the natural logarithm on both sides. What would have happened had we taken the common logarithm? Explain which seems best to you and why.
59. ◆ Explain how Exercises 29 and 30 could be solved using the graph of $f(x) = \ln x$.

74 CHAPTER R • BASIC CONCEPTS OF ALGEBRA

3. **Carry out.**

ALGEBRAIC SOLUTION

We solve the equation:

$$x + 2x = 30$$
$$3x = 30$$
$$x = 10.$$

GRAPHICAL SOLU

Graph $y_1 = x$

point of inters

4. **Check.** When the Australian 10 + 20, or 30 days

5. **State.** After one ye 10 vacation days and Australian employees get an average of 20 vacation days per year.

In some applications we need to use a formula that describes the relationship between variables. When a situation involves distance, speed, and time, for example, we need to recall the **distance formula**:

$d = rt$, where d = distance, r = rate (or speed), and t = time.

Example 2 *Speed.* A 1996 BMW M3 leaves a town on the Autobahn traveling at its top speed of 237 km/h. Fifteen minutes later, a 1995 Aston Martin DB7 leaves the same town and follows the same route at its top speed of 266 km/h. How long will it take the Aston Martin to overtake the BMW? (*Sources: Car and Driver,* August 1995 and February 1995)

SOLUTION

1. **Familiarize.** We make a drawing showing both the known and the unknown information. We let t = the time, in hours, that the BMW

BMW 237 km/h

t hr

d km

Aston Martin overtakes BMW here.

Aston Martin 266 km/h

t − 0.25 hr

d km

- **Examples/Exercises:** The examples and exercises reflect the text's focus. Many use situational or grapher art and many also incorporate recent, source-based data to illustrate applications and concepts. An Index of Applications is included at the end of the text.

- **Synthesis/Skill Maintenance exercises:** Skill Maintenance exercises prepare students for the next section by reviewing and reinforcing skills that are used in the following section. Synthesis exercises require additional thought and encourage students to take the material one step further by combining concepts. Exercises marked with the maze icon are designed to foster critical thinking and writing skills; they may also be completed by small groups.

the formula

$$C(x) = \frac{100 + 5x}{x},$$

where x = the number of people in the group and $C(x)$ is in dollars. Determine $C^{-1}(x)$ and explain what it represents.

69. *Reaction Distance.* You are driving a car when a deer suddenly darts across the road in front of you. Your brain registers the emergency and sends a signal to your foot to hit the brake. The car travels a distance D, in feet, during this time, where D is a function of the speed r, in miles per hour, that the car is traveling when you see the deer. That reaction distance D is a linear function given by

$$D(r) = \frac{11r + 5}{10}.$$

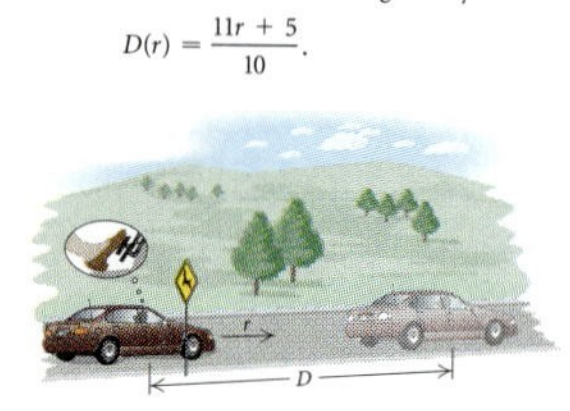

a) Find $D(0)$, $D(10)$, $D(20)$, $D(50)$, and $D(65)$.
b) Graph $D(r)$.
c) Find $D^{-1}(r)$ and explain what it represents.
d) Graph the inverse.

Skill Maintenance

Graph each of the following.

71. $y = x^3 - x$ **72.** $x = y^3 - y$

73. $f(x) = \sqrt[3]{x}$ **74.** $f(x) = \dfrac{8}{x^2 - 4}$

Synthesis

75. ◆ Suppose that you have graphed a function using a grapher and you see that it is one-to-one. How could you then use the TRACE feature to make a hand-drawn graph of the inverse?

76. ◆ The following formulas for the conversion between Fahrenheit and Celsius temperatures have been considered several times in this text:

$$C = \tfrac{5}{9}(F - 32)$$

and

$$F = \tfrac{9}{5}C + 32.$$

Discuss these formulas from the standpoint of inverses.

Using only a grapher, determine whether the functions are inverses of each other.

77. $f(x) = \sqrt[3]{\dfrac{x - 3.2}{1.4}}$, $g(x) = 1.4x^3 + 3.2$

78. $f(x) = \dfrac{2x - 5}{4x + 7}$, $g(x) = \dfrac{7x - 4}{5x + 2}$

79. $f(x) = \dfrac{2}{3}$, $g(x) = \dfrac{3}{2}$

80. $f(x) = x^4$, $x \geq 0$; $g(x) = \sqrt[4]{x}$

. Find three examples of functions that are their own inverses, that is, $f = f^{-1}$.

. Consider the function f given by

$$f(x) = \begin{cases} x^3 + 2, & x \leq -1, \\ x^2, & -1 < x < 1, \\ x + 1, & x \geq 1. \end{cases}$$

Does f have an inverse that is a function? Why or why not?

CHAPTER 3 Summary and Review

Important Properties and Formulas

One-to-One Function:	$f(a) = f(b) \longrightarrow a = b$
Exponential Function:	$f(x) = a^x$
The Number e = 2.7182818284...	
Logarithmic Function:	$f(x) = \log_a x$
A Logarithm is an Exponent:	$\log_a x = y \longleftrightarrow x = a^y$
The Change-of-Base Formula:	$\log_b M = \dfrac{\log_a M}{\log_a b}$
The Product Rule:	$\log_a MN = \log_a M + \log_a N$
The Power Rule:	$\log_a M^p = p \log_a M$
The Quotient Rule:	$\log_a \dfrac{M}{N} = \log_a M - \log_a N$
Other Properties:	$\log_a a = 1$, $\log_a 1 = 0$, $\log_a a^x = x$, $a^{\log_a x} = x$
Base–Exponent Property:	$a^x = a^y \longleftrightarrow x = y$
Exponential Growth Model:	$P(t) = P_0e^{kt}$
Exponential Decay Model:	$P(t) = P_0e^{-kt}$
Interest Compounded Continuously:	$P(t) = P_0e^{kt}$
Limited Growth:	$P(t) = \dfrac{a}{1 + be^{-kt}}$

REVIEW EXERCISES

1. Find the inverse of the relation
$\{(1.3, -2.7), (8, -3), (-5, 3), (6, -3), (7, -5)\}$.

2. Find an equation of the inverse relation.
a) $y = 3x^2 + 2x - 1$
b) $0.8x^3 - 5.4y^2 = 3x$

Given each function:

a) *Determine whether it is one-to-one, using a grapher if desired.*
b) *If it is one-to-one, find a formula for the inverse.*

3. $f(x) = \sqrt{x - 6}$ **4.** $f(x) = x^3 - 8$

5. $f(x) = 3x^2 + 2x - 1$ **6.** $f(x) = e^x$

7. Find $f(f^{-1}(657))$: $f(x) = \dfrac{4x^5 - 16x^{37}}{119x}$, $x > 0$.

In Exercises 8–13, match the equation with one of figures (a)–(f), which follow. If needed, use a grapher.

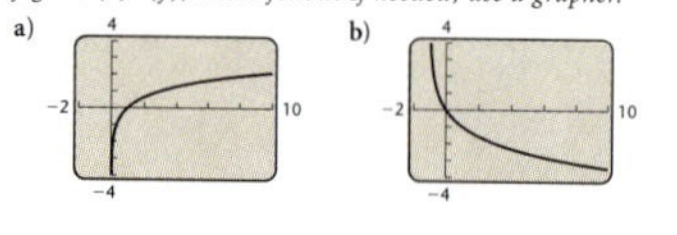

- **End-of-Chapter material:** End-of-Chapter material includes a summary and review of properties and formulas along with a complete set of Review Exercises. Review Exercises also include Synthesis exercises, as well as critical thinking and writing exercises. The answers to the Review Exercises, which appear at the back of the text, have text section references to further aid students.

Algebra and Trigonometry

GRAPHS AND MODELS

Introduction to Graphs and Graphers

- *Review hand-drawn graphs of equations.*
- *Use a grapher to create graphs with various viewing windows, to create x, y tables for equations, and to determine whether an equation is an identity.*
- *Use a grapher to find coordinates of points on a graph, to solve equations, and to find the points of intersection of two graphs.*

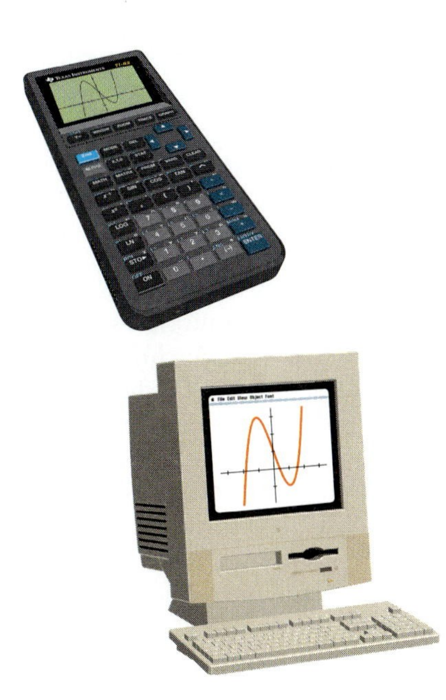

Graphing calculators and computers equipped with graphing software can alleviate much of the toil of graph making. We will henceforth refer to all such graphing utilities simply as **graphers**. As the equations we encounter become more complicated, it becomes increasingly difficult to produce accurate hand-drawn graphs. It can take considerable time to calculate just a few ordered pairs that are solutions. In addition, many ordered pairs are often required to produce an accurate graph. The use of a grapher can not only shorten this process but also perform many other mathematical procedures efficiently. Also, many equations arise that are difficult to analyze without machine assistance. Throughout this text, we will use graphers to enhance the learning process. The following is our philosophy regarding the use of graphers.

The Use of the Grapher

The grapher creates a visual presentation that *increases* understanding and *saves* time. It is used to enhance the mathematics, not to replace it!

Most of our discussion of the grapher is presented in a relatively generic form. Expanded discussion specific to certain graphers is included in the Graphing Calculator Manual that accompanies this book. To determine specific procedures, you should consult this manual, your user's manual, or your instructor.

Graphs

Hand-drawn graphs are reviewed first. With this groundwork established, we then consider the use of graphers. This introduction to graphs and graphers provides the foundation for the remainder of the text.

Graphs provide a natural way to link algebra and geometry. It is not uncommon to open a newspaper or magazine and encounter graphs. Shown below are examples of bar, circle, and line graphs.

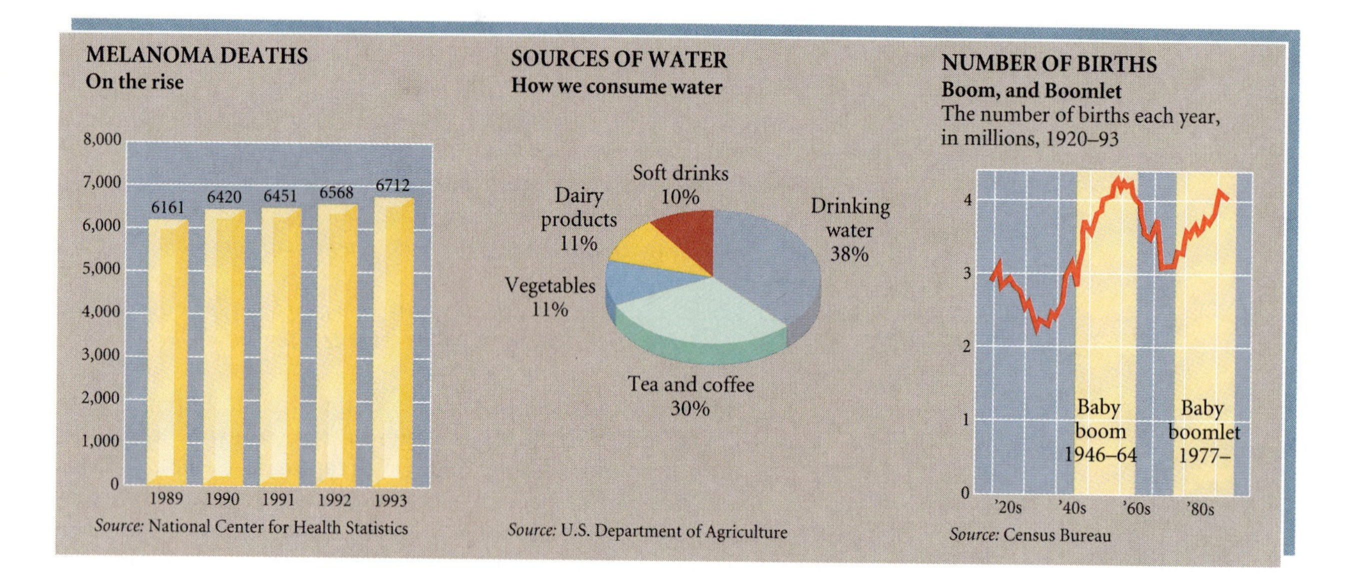

Many real-world situations can be modeled using equations in which two variables appear. We use a plane to graph a pair of numbers. To locate points on a plane, we use two perpendicular number lines, called **axes**, which intersect at 0. We call this point the **origin**. The horizontal axis is called the **x-axis**, and the vertical axis is called the **y-axis**. (Other variables can be used.) The axes divide the plane into four regions, called **quadrants**, denoted by Roman numerals, numbered counterclockwise from the upper right. Arrows show the positive direction of each axis.

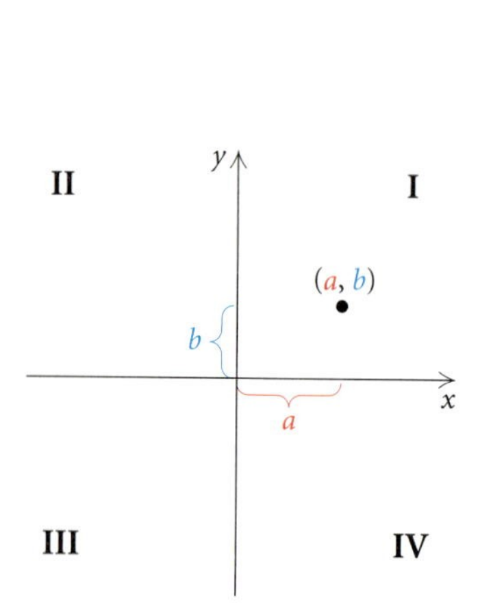

Each point (a, b) in the plane is called an **ordered pair**. The first number, a, indicates the point's horizontal location with respect to the y-axis, and the second number, b, indicates the point's vertical location with respect to the x-axis. We call a the **first coordinate**, **x-coordinate**, or **abscissa**. We call b the **second coordinate**, **y-coordinate**, or **ordinate**. Such a representation is called the **Cartesian coordinate system** in honor of the great French mathematician and philosopher René Descartes (1596–1650).

Example 1 Graph and label the points in the set

$\{(-3, 5), (4, 3), (3, 4), (-4, -2), (3, -4), (0, 4), (-3, 0), (0, 0)\}$.

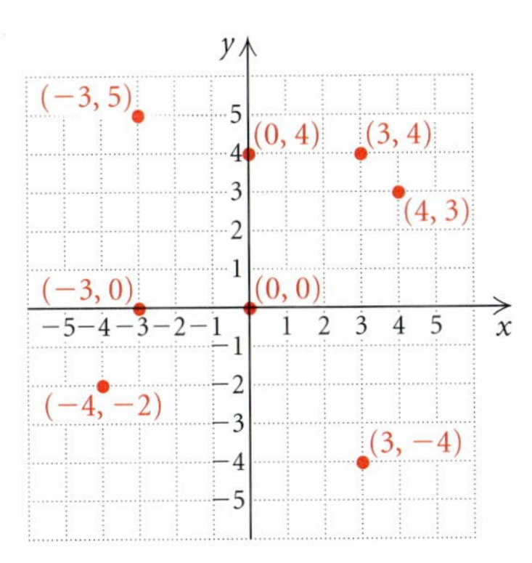

SOLUTION To graph or **plot** $(-3, 5)$, we note that the x-coordinate tells us to move from the origin 3 units to the left of the y-axis. Then we move 5 units up from the x-axis. To graph the other points, we proceed in a similar manner. (See the graph at left.) Note that the origin, $(0, 0)$, is located at the intersection of the axes and that the point $(4, 3)$ is different from the point $(3, 4)$.

A graph of a set of points, as in Example 1, is called a **scatterplot**.

Solutions of Equations

Equations in two variables, like $2x + 3y = 18$, have solutions that are ordered pairs such that when the first coordinate is substituted for x and the second coordinate is substituted for y, the result is a true equation.

Example 2 Determine whether each ordered pair is a solution of $2x + 3y = 18$.

a) $(-5, 7)$ **b)** $(3, 4)$

SOLUTION We check as follows.

a)

$$\begin{array}{r|l} 2x + 3y & = 18 \\ \hline 2(-5) + 3(7) & ?\ 18 \\ -10 + 21 & \\ 11 & 18 \end{array}$$

We substitute -5 for x and 7 for y (alphabetical order).

FALSE

The equation $11 = 18$ is false, so $(-5, 7)$ is not a solution.

b)

$$\begin{array}{r|l} 2x + 3y & = 18 \\ \hline 2(3) + 3(4) & ?\ 18 \\ 6 + 12 & \\ 18 & 18 \end{array}$$

We substitute 3 for x and 4 for y.

TRUE

The equation $18 = 18$ is true, so $(3, 4)$ is a solution.

Graphs of Equations

The equation considered in Example 2 actually has an infinite number of solutions. Since we cannot list all of the solutions, we will make a drawing, called a **graph**, that represents them.

To Graph an Equation

To graph an equation is to make a drawing that represents the solutions of that equation.

Suggestions for Hand-Drawn Graphs

1. Use graph paper.
2. Draw axes and label them with the variables.
3. Use arrows on the axes to indicate positive directions.
4. Scale the axes, that is, mark numbers on the axes.
5. Calculate solutions and list the ordered pairs in a table.
6. Plot solutions, look for patterns, and complete the graph. Label the graph with the equation being graphed.

At left are some suggestions for making hand-drawn graphs.

Example 3 Graph: $2x + 3y = 18$.

SOLUTION To find ordered pairs that are solutions of this equation, we can replace either x or y with any number and then solve for the other variable. For instance, if x is replaced with 0, then

$$\begin{aligned} 2 \cdot 0 + 3y &= 18 \\ 3y &= 18 \\ y &= 6. && \textbf{Dividing by 3} \end{aligned}$$

Thus $(0, 6)$ is a solution. If x is replaced with 5, then

$$\begin{aligned} 2 \cdot 5 + 3y &= 18 \\ 10 + 3y &= 18 \\ 3y &= 8 && \textbf{Subtracting 10} \\ y &= \tfrac{8}{3}. && \textbf{Dividing by 3} \end{aligned}$$

Thus $\left(5, \frac{8}{3}\right)$ is a solution. If y is replaced with 0, then

$$\begin{aligned} 2x + 3 \cdot 0 &= 18 \\ 2x &= 18 \\ x &= 9. && \textbf{Dividing by 2} \end{aligned}$$

Thus $(9, 0)$ is a solution.

We continue choosing values for one variable and finding the corresponding values of the other. We list the solutions in a table, and then plot the points. Note that the points appear to lie on a straight line.

x	y	(x, y)
0	6	$(0, 6)$
5	$\frac{8}{3}$	$\left(5, \frac{8}{3}\right)$
9	0	$(9, 0)$
-1	$\frac{20}{3}$	$\left(-1, \frac{20}{3}\right)$

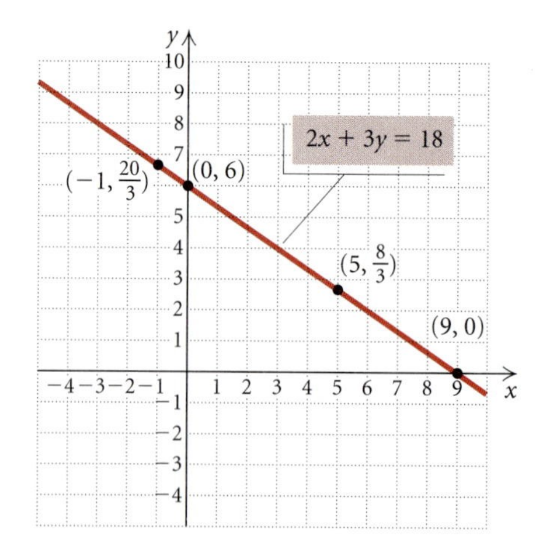

In fact, were we to graph additional solutions of $2x + 3y = 18$, they would be on the same straight line. Thus, to complete the graph, we use a straightedge to draw a line as shown in the figure. That line represents all solutions of the equation.

Graphs of lines are examined in detail in Section 1.3.

Example 4 Graph: $y = x^2 - 5$.

SOLUTION Since y is expressed in terms of x, the easiest way to find solutions of this equation is to replace x with various values and then calculate the corresponding values for y. In cases like this, we say that x is the **independent variable** (because we *choose* its value) and y is the **dependent variable** (because its value is then calculated).

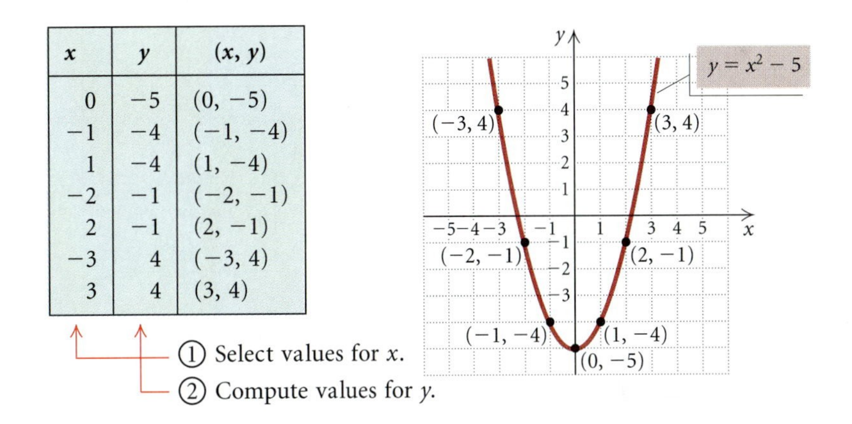

x	y	(x, y)
0	−5	(0, −5)
−1	−4	(−1, −4)
1	−4	(1, −4)
−2	−1	(−2, −1)
2	−1	(2, −1)
−3	4	(−3, 4)
3	4	(3, 4)

Next, we plot these points and connect them. We note that as the absolute value of x increases, $x^2 - 5$ also increases. Thus the graph is a curve that rises on either side of the y-axis.

Graphs of curves similar to the one in Example 4 are examined in more detail in Section 2.2.

Example 5 Graph: $x = y^2 + 1$.

SOLUTION Since x is expressed in terms of y, we select values for y and then find the corresponding values for x. In this case, y is the independent variable and x the dependent variable.

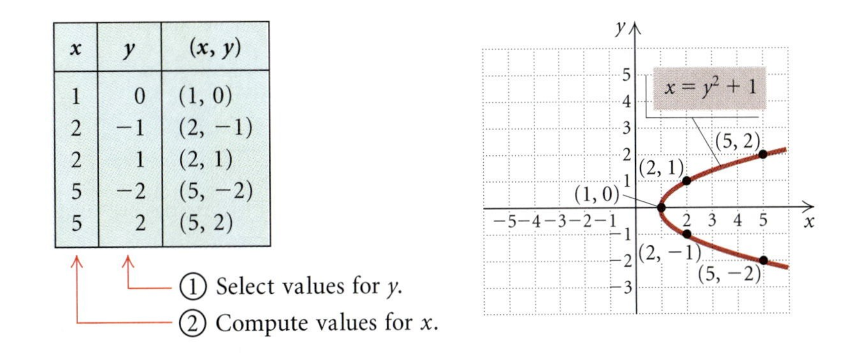

x	y	(x, y)
1	0	(1, 0)
2	−1	(2, −1)
2	1	(2, 1)
5	−2	(5, −2)
5	2	(5, 2)

When plotting, we must remember that x is the first coordinate and y is the second coordinate. Note in the figure that the graph moves farther from the x-axis as it extends farther to the right.

Example 6 Graph: $xy = 12$.

SOLUTION To find numbers that satisfy the equation, it helps to solve for y (that is, $y = 12/x$) or solve for x (that is, $x = 12/y$). Typically, we solve for y if it is convenient. Then we make substitutions.

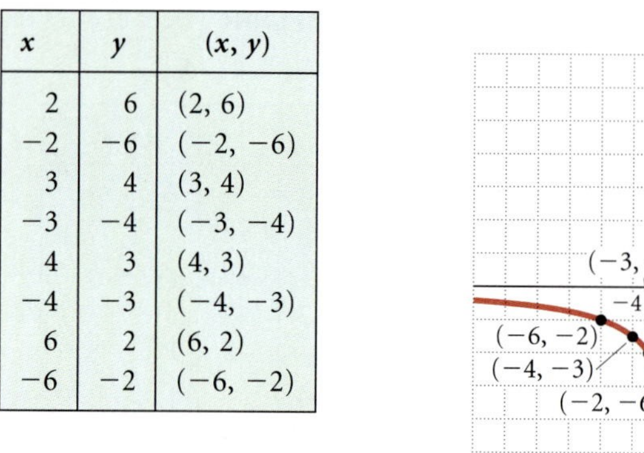

x	y	(x, y)
2	6	(2, 6)
−2	−6	(−2, −6)
3	4	(3, 4)
−3	−4	(−3, −4)
4	3	(4, 3)
−4	−3	(−4, −3)
6	2	(6, 2)
−6	−2	(−6, −2)

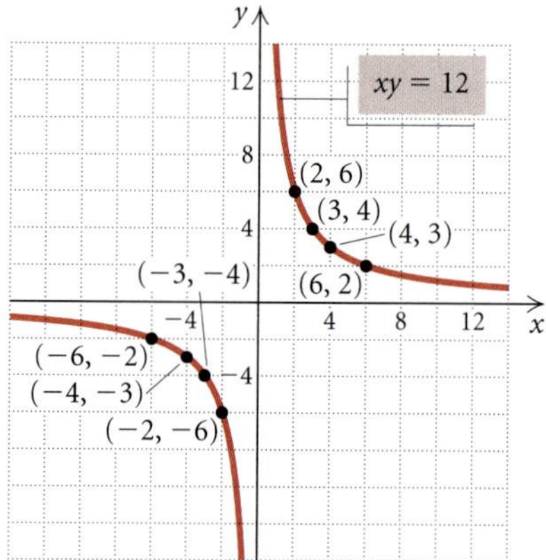

We plot these points and connect them. As the absolute value of x becomes small, the absolute value of y becomes large, and vice versa. Since neither x nor y can be 0, the graph does not cross either axis. Thus the graph consists of two branches.

Example 7 Graph: $y = |x|$.

SOLUTION We find numbers that satisfy the equation. For example, when $x = -3$, $y = |-3| = 3$. When $x = 2$, $y = |2| = 2$, and when $x = 0$, $y = |0| = 0$.

x	y	(x, y)
0	0	(0, 0)
−1	1	(−1, 1)
1	1	(1, 1)
−2	2	(−2, 2)
2	2	(2, 2)
−3	3	(−3, 3)
3	3	(3, 3)

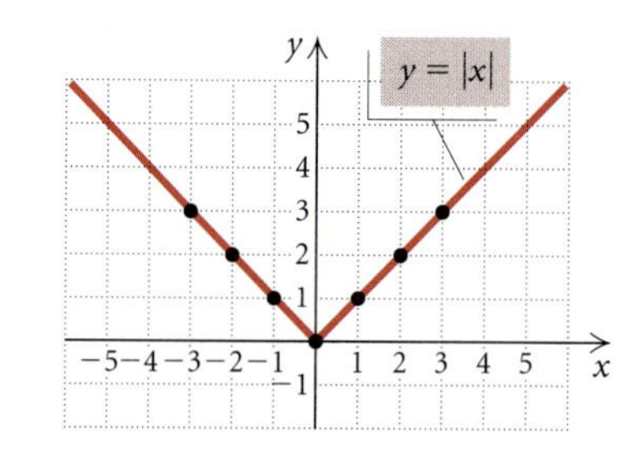

We plot these points and connect them. As the points fall farther from the origin, to the left or right, the absolute value of x increases. Thus the graph is a V-shaped curve that rises to the left and right of the y-axis.

Graphers and Viewing Windows

We now consider the use of graphers. One feature common to all graphers is the **viewing window**. This refers to the rectangular portion of the xy-plane that appears on the screen of a grapher. The notation we

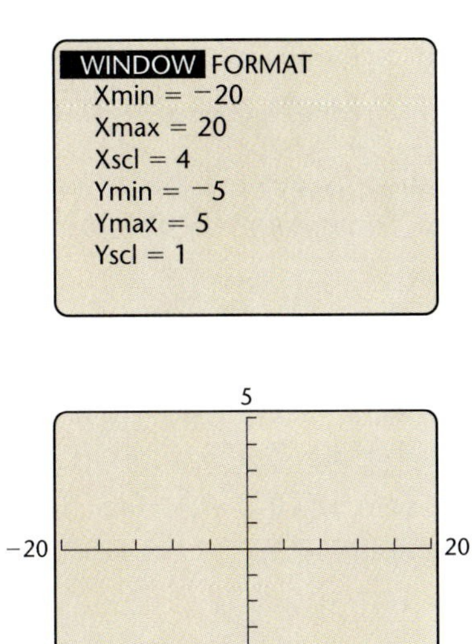

will use to describe viewing windows consists of four numbers, [L, R, B, T] in brackets, which represent the *L*eft and *R*ight endpoints of the x-axis and the *B*ottom and *T*op endpoints of the y-axis. A WINDOW feature might be used to set these dimensions, but the method varies for each grapher. The screen at top left is a window setting of $[-20, 20, -5, 5]$ with axis scaling denoted as Xscl $= 4$ and Yscl $= 1$, which means that there are 4 units between tick marks on the x-axis and 1 unit between tick marks on the y-axis. The screen at bottom left shows the results from this setting.

Axis scaling must be chosen with care, because tick marks become blurred and indistinguishable when too many appear. On some graphers, a setting of $[-10, 10, -10, 10]$, Xscl $= 1$, Yscl $= 1$ is considered the **standard window**. There is usually a procedure to obtain a standard setting quickly. Consult your manual.

The primary use of graphers is to graph equations. As an example, let's graph the equation $y = x^3 - 3x + 1$. The equation might be entered using the notation y = x^3 − 3x + 1. Some software use BASIC notation, in which case the equation would be entered as y = x^3 − 3*x + 1. We obtain the following graph.

$y = x^3 - 3x + 1$

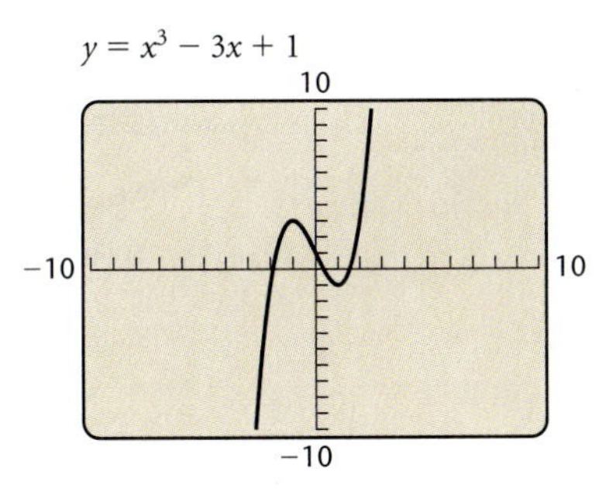

You may need to change the viewing window in order to clearly reveal the curvature of a graph. For example, each of the following is a graph of $y = 3x^5 - 20x^3$. The graph on the right best represents the curve.

$y = 3x^5 - 20x^3$

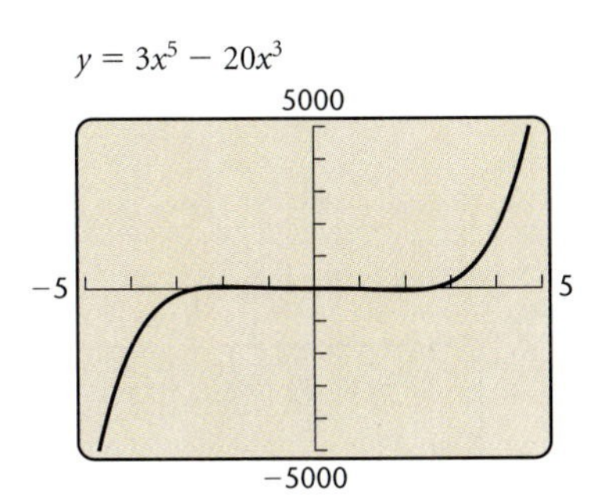

$y = 3x^5 - 20x^3$

500, −4, 4, −500

$y = 3x^5 - 20x^3$

80, −4, 4, −80

To graph an equation like $2x + 3y = 18$, most graphers require that the equation be solved for y, that is, "$y = \ldots$". When we solve $2x + 3y = 18$ for y, we obtain $y = (18 - 2x)/3$, or $y = 6 - \frac{2}{3}x$. As another example, we can write an equation like $x = y^2 + 1$ as

$y = \pm\sqrt{x - 1}$ and graph the individual equations $y_1 = \sqrt{x - 1}$ and $y_2 = -\sqrt{x - 1}$. The combination of the two graphs yields the graph of $x = y^2 + 1$.

Just as the form in which an equation is entered varies among graphers, so too do other types of notation. Some examples of commonly used grapher notation follow.

EQUATION	POSSIBLE GRAPHER NOTATION
$y = 4x^3 - 7x^2 + 5x - 10$	y = 4x^3 − 7x² + 5x − 10 or 4*x^3 − 7*x^2 + 5*x − 10
$y = \sqrt{x - 1}$	y = (x − 1)^(1/2) or √‾(x − 1)
$y = \|x - 5\|$	y = abs (x − 5)
$y = \sqrt[3]{x}$	y = x^(1/3) or ∛‾x
$y = \dfrac{7.8x^2 - 1}{x^5 + 3}$	y = (7.8x² − 1)/(x^5 + 3) or (7.8*x^2 − 1)/(x^5 + 3)

Example 8 Graph each of the following equations choosing a viewing window that best reveals the curvature of the graph. Answers may vary.

a) $2x + 3y = 18$ b) $y = x^2 - 5$ c) $x = y^2 + 1$

d) $xy = 12$ e) $y = |x|$ f) $y = x^4 - 2x^2 - 3$

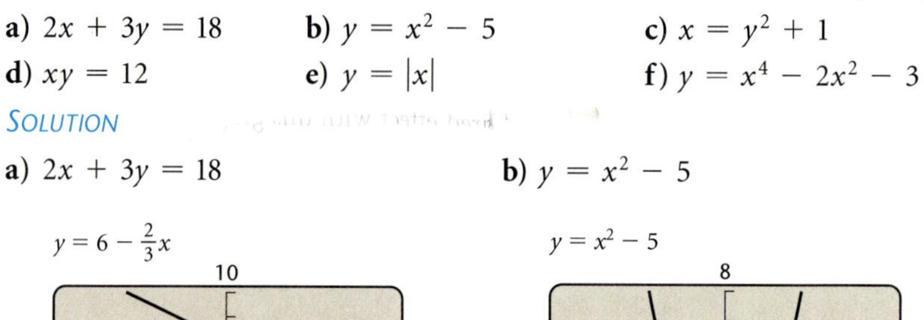

SOLUTION

a) $2x + 3y = 18$

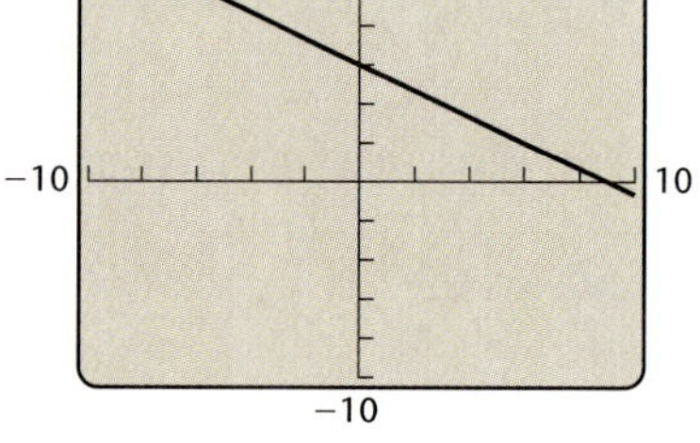

b) $y = x^2 - 5$

c) $x = y^2 + 1$

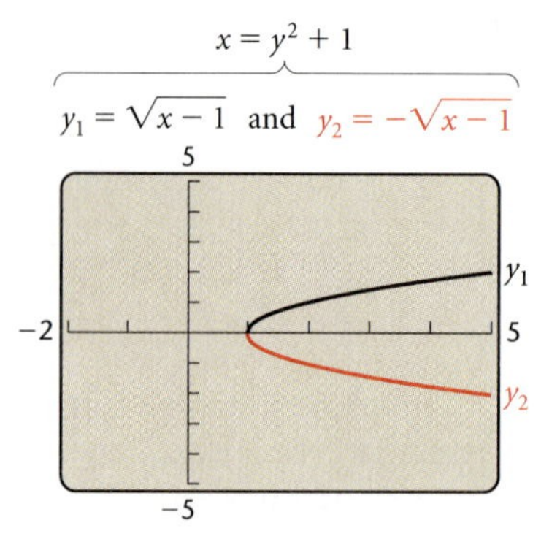

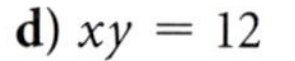

d) $xy = 12$

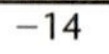

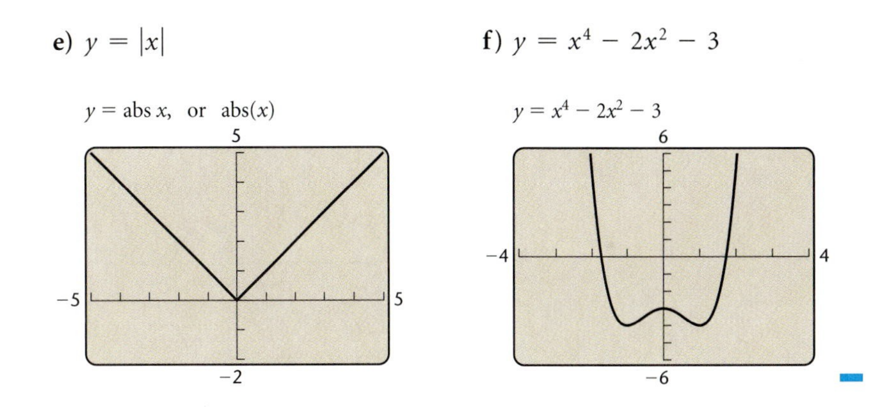

Squaring a Viewing Window

Consider the $[-10, 10, -10, 10]$ viewing window shown first at left. Note that the distance between units is not visually the same on both axes. In this case, the length of the interval shown on the y-axis is about $\frac{2}{3}$ of the length of the interval shown on the x-axis. If we change to the window shown second at left, with dimensions $[-6, 6, -4, 4]$, we get a graph for which the units are visually about the same on both axes. We choose these dimensions so that the length of the y-axis is $\frac{2}{3}$ the length of the x-axis. The window has been *squared*. Any dimensions in this ratio will produce the desired effect with this grapher.

On another grapher, the ratio might be $\frac{8}{15}$, and the window $[-7.5, 7.5, -4, 4]$ would show the units visually the same on both axes.* Creating such a window is called **squaring**. On some graphers, there is a ZSQUARE feature for automatic window squaring. Consult your manual about how to square a window.

Each of the following is a graph of the line $y = 2x - 3$, but the viewing windows are different.

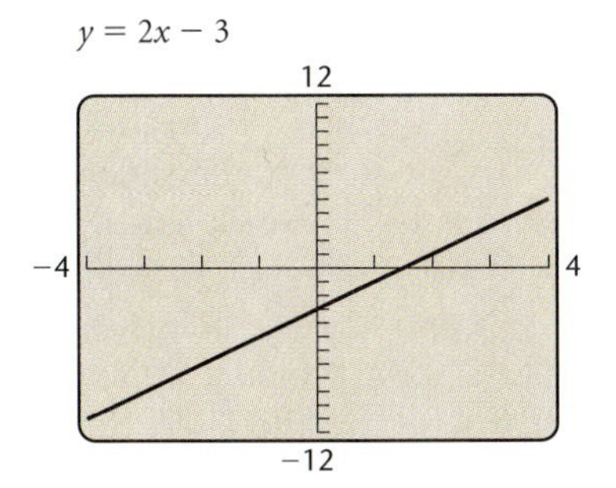

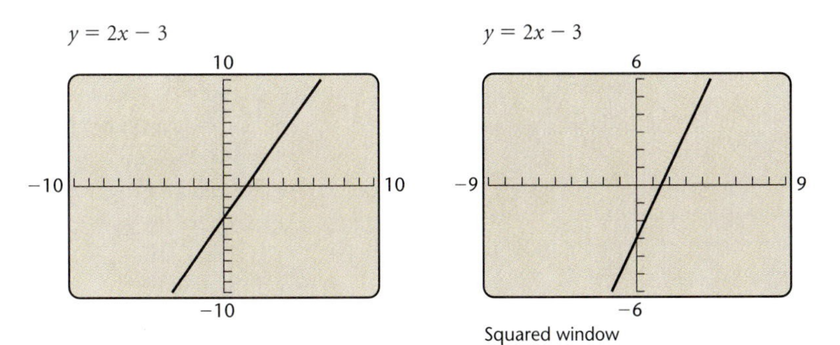

Squared window

When the x-units and the y-units are visually the same, we get an accurate representation of the *slope* of the line. We will study slope in Section 1.3.

*The exact ratio is 63/95 on a TI-82 or a TI-83, 63/127 on a TI-85, and 103/239 on a TI-92.

A squared window also eliminates distortion of the graph. Compare the graph of the circle $x^2 + y^2 = 4$ shown here in both a nonsquared and a squared window.

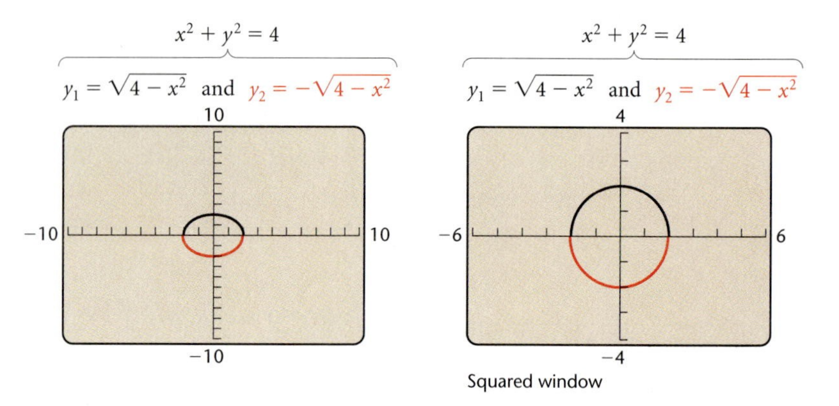

Squared window

Circles will be studied in Section 1.5.

The Table Feature

TblMin = 2, ΔTbl = .25

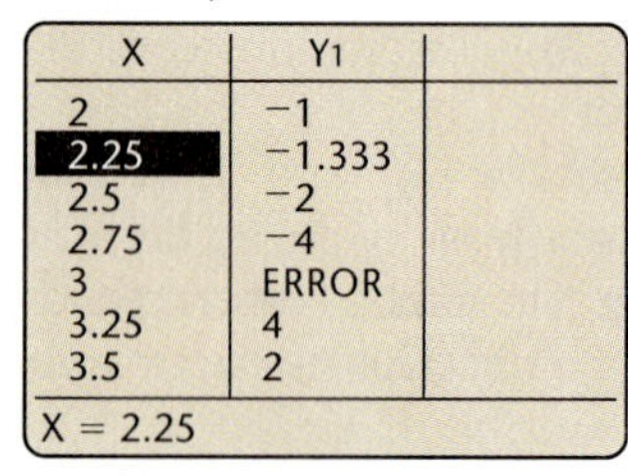

X	Y1
2	−1
2.25	−1.333
2.5	−2
2.75	−4
3	ERROR
3.25	4
3.5	2

X = 2.25

Many graphers can display a table of values similar to the tables we used for hand-drawn graphs. For an equation entered on the "$y =$" screen, we can use the table set-up screen to choose a minimum x-value along with a step-value or increment. The grapher then produces a table of x- and y-values for the given equation. For example, if we enter the equation $y = 1/(x - 3)$ along with a minimum x-value of 2 and an increment of 0.25, the table at left is produced. We see that $y = -1$ when $x = 2$, $y = -1.3333$ when $x = 2.25$, and so on. The ERROR entry indicates that 3 is not an acceptable value for x.

We can use the arrow keys to scroll through more x, y-values for this equation. Some graphers can display both a graph and a table on a split screen.

The TABLE feature can also be used to evaluate an expression in one variable for particular values of the variable. Again the expression is entered on the "$y =$" screen and the table is set so that the independent variable is in ASK mode. This allows us to enter x-values one at a time and see the corresponding y-values for the given expression. Suppose we want to evaluate $2x^5 - 6x^3 + 7$ for $x = -8$ and for $x = 5$. First we enter $y = 2x^5 - 6x^3 + 7$; then, with the table set in ASK mode, we enter −8. The corresponding y-value −62,457 is displayed in the table. Similarly, we enter 5 and read the y-value 5507. We can continue to enter x-values as desired.

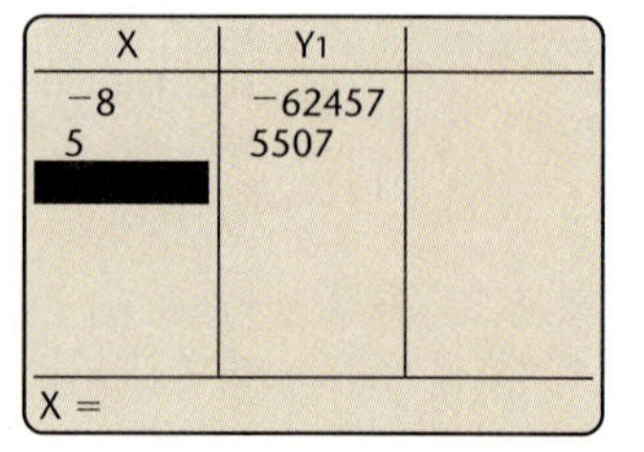

X	Y1
−8	−62457
5	5507

X =

Program

The Graphing Calculator Manual that accompanies this text contains programs that can be entered in the TI-82, TI-83, TI-85, or HP38G graphing calculators. For a complete list of these program references, see the Program heading in the index. The first program is designed for the TI-85.

TABLE: This program provides the TI-85 with a TABLE feature.

Identities

Consider the equation

$$(x + 1)^2 = x^2 + 2x + 1.$$

We know from the laws of algebra that this equation is true for every possible substitution of x by a real number. Let's look at this equation graphically. We consider two equations of the form "$y = \ldots$", one using the expression on the left side of the equation and one using the expression on the right:

$$y_1 = (x + 1)^2 \quad \text{and} \quad y_2 = x^2 + 2x + 1.$$

If we graph these two equations using the same viewing window, $[-5, 5, -1, 6]$, we see that the graphs seem to coincide. You might try other viewing windows to see that they do not differ. The TABLE feature will also confirm that y_1 and y_2 appear to have the same value for a given value of x.

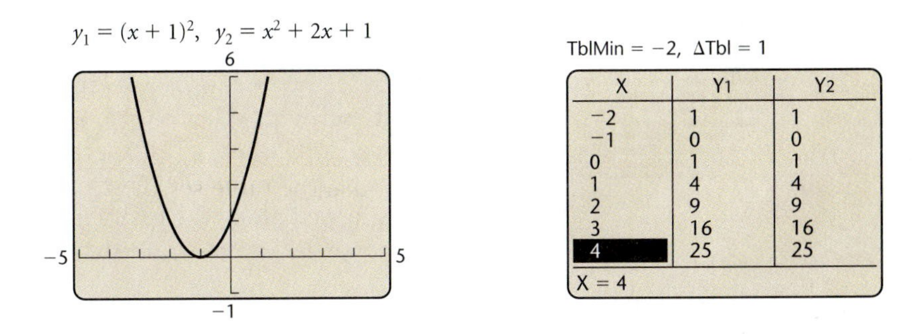

$y_1 = (x + 1)^2,\ y_2 = x^2 + 2x + 1$

TblMin = −2, ΔTbl = 1

X	Y1	Y2
−2	1	1
−1	0	0
0	1	1
1	4	4
2	9	9
3	16	16
4	25	25

X = 4

The equation $(x + 1)^2 = x^2 + 2x + 1$ is called an **identity**. As illustrated above, it is true for every possible substitution. The graphs are the same for the values of x for which both expressions are defined. One way to get a partial check of a potential identity, or to make a conjecture that an equation is an identity, is to graph each side and see if the graphs coincide. Although this procedure can be fruitful, one must be aware of its shortcomings. First, we rarely can draw all or even most of a graph. Thus there is always the possibility that somewhere outside the window, the graphs differ. The second limitation is that the size and scale of the window may deceive us about whether graphs really do coincide. An alternate partial check can also be obtained using the TABLE feature or other function evaluator. Many values of a pair of expressions can be compared very quickly.

Example 9 Determine graphically whether each of the following seems to be an identity.

a) $(x^2)^3 = x^6$ **b)** $\sqrt{x + 4} = \sqrt{x} + 2$

SOLUTION

a) We leave it to the student to graph $y_1 = (x^2)^3$ and $y_2 = x^6$. The graphs appear to coincide no matter what the viewing window. Thus, $(x^2)^3 = x^6$ seems to be an identity.

b) We graph $y_1 = \sqrt{x + 4}$ and $y_2 = \sqrt{x} + 2$ using the viewing window $[-6, 8, -1, 6]$. We see that the graphs differ, so the equation is *not* an identity. Comparing y-values using the TABLE feature also illustrates that this equation is *not* an identity.

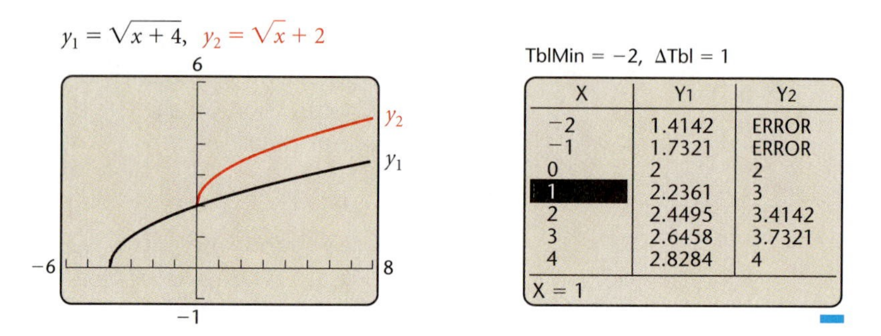

X	Y1	Y2
−2	1.4142	ERROR
−1	1.7321	ERROR
0	2	2
1	2.2361	3
2	2.4495	3.4142
3	2.6458	3.7321
4	2.8284	4

X = 1

The Trace Feature

Once a graph has been created, we can investigate some of its points by using the TRACE feature that most graphers offer. Consider the graph of $y = x^3 - 3x + 1$ in the viewing window $[-10, 10, -10, 10]$ with Xscl = 2 and Yscl = 2. When the TRACE key is pressed, a cursor (often blinking) appears somewhere on the graph, while the x- and y-coordinates are shown elsewhere on the screen. These coordinates will change as the cursor is moved along the graph using the arrow keys (◁, ▷). Some points on the graph are shown in the figures below.

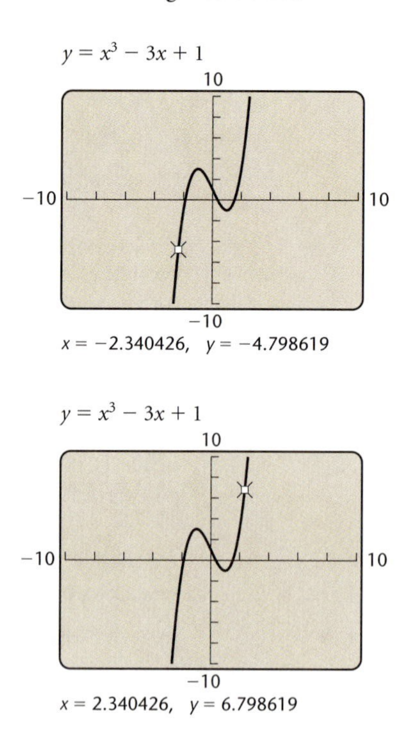

Equation Solving

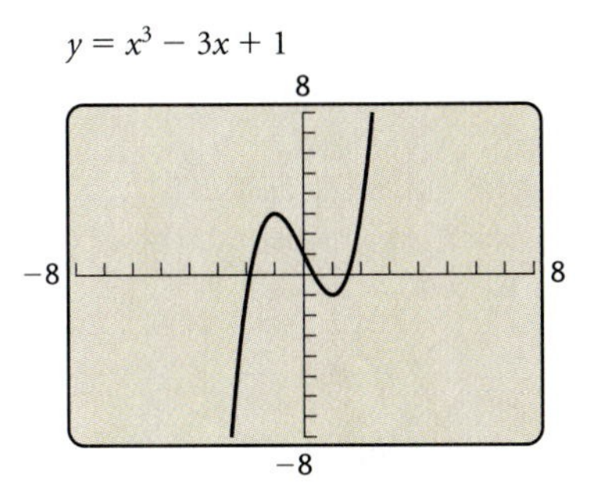

Example 10 Solve $x^3 - 3x + 1 = 0$ graphically. Approximate solutions to three decimal places.

SOLUTION The solutions of the equation $x^3 - 3x + 1 = 0$ are the first coordinates of the points where the graph of $y = x^3 - 3x + 1$ crosses the x-axis. These points are called the **x-intercepts** of the graph. An x-intercept has the general form $(a, 0)$ and occurs at an x-value a for which $y = 0$. It appears from the graph shown at left that there are three such x-values, one near -2, one near 0, and one near 1.5.

To acquire more precision, we can change the viewing window. A fast way to do this on many graphers is to use a feature called ZOOM. The zoom-in feature allows a portion of the graph to be magnified; that is, a small section of the graph is enlarged in the viewing window. Let's examine the graph near $x = -2$. We trace to the y-value closest to 0 and zoom in on the graph at that point. We continue to zoom in until the desired accuracy is achieved.

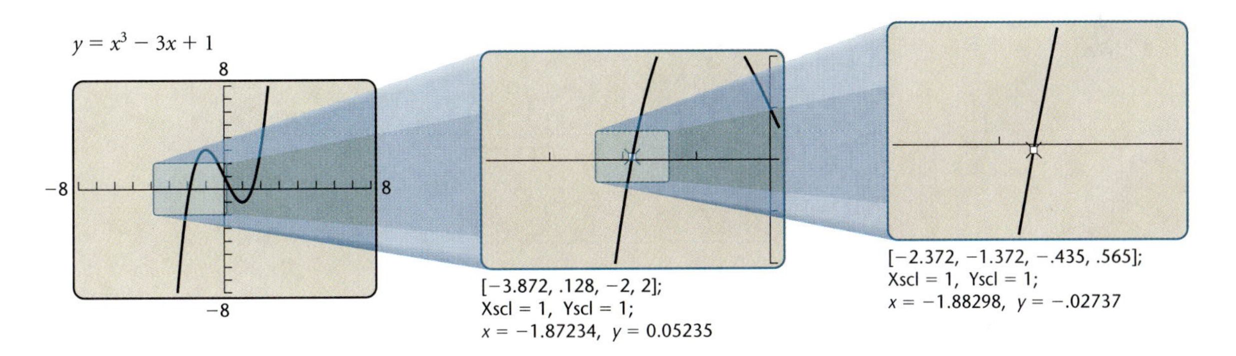

To find the solution to the nearest thousandth, we trace and zoom until the cursor's x-value just to the left of the intercept and the cursor's x-value just to the right of the intercept are the same, when rounded to the nearest thousandth. By using the TRACE and ZOOM features three more times, we find the solution to be about -1.879. In a similar manner, we find that the other solutions of the equation $x^3 - 3x + 1 = 0$ are about 0.347 and 1.532. If available, we might also use the SOLVE or ROOT features to approximate solutions.

There are many ways in which the ZOOM feature can be used. For example, we can zoom in for more precision, as in Example 10, or zoom out on a graph—say, to reveal more of its curvature—adjusting the factors of the zoom in any way we choose. In Example 10, we used zoom factors of 4. We may also be able to use a ZOOM-BOX feature to zoom in on a boxed region of our choosing. All such details can be found by consulting the manual for your particular grapher or the Graphing Calculator Manual that accompanies this book.

Finding Points of Intersection

There are many situations in which we want to determine the point(s) of intersection of two graphs.

Example 11 Find the points of intersection of the graphs of the equations $y_1 = 3x^5 - 20x^3$ and $y_2 = 34.7 - 1.28x^2$. Approximate the coordinates to three decimal places.

SOLUTION

Method 1. We use a grapher, trying to create graphs that clearly show the curvature. Then we look for the coordinates of the points at which the graphs cross each other. It appears that there are three points of intersection. We can use TRACE and ZOOM to approximate their coordinates, or we can use the INTERSECT feature, if available. Experiment to see which you prefer.

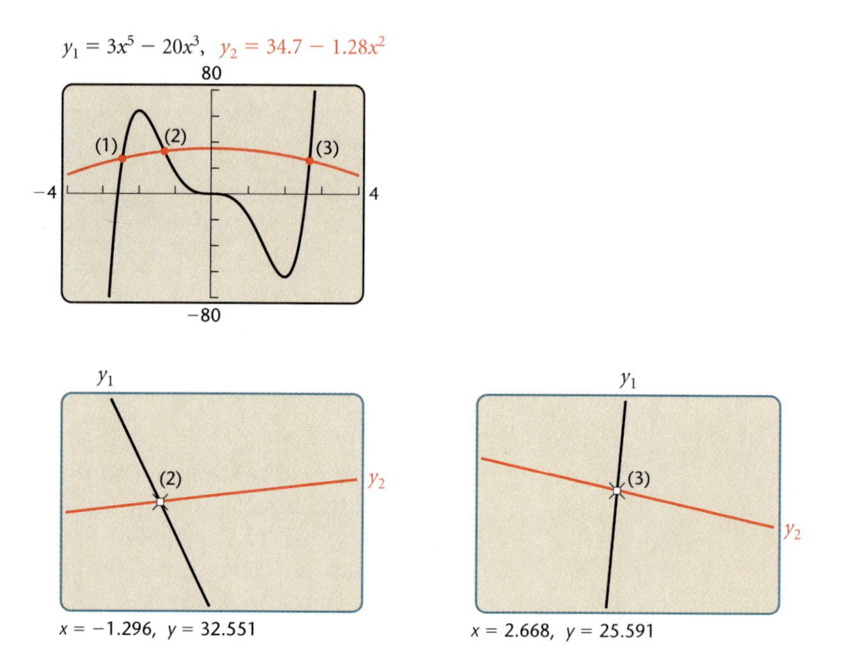

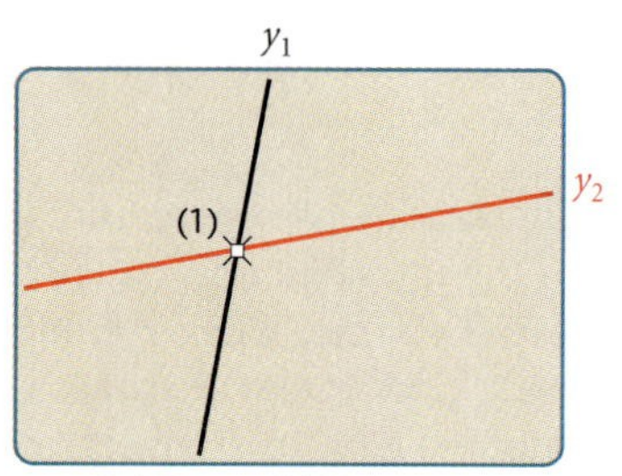

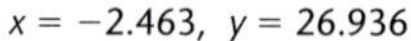

The values found can be checked by substitution. The points of intersection are about $(-2.463, 26.936)$, $(-1.296, 32.551)$, and $(2.668, 25.591)$.

Method 2. Another method is to use the SOLVE feature. We find the x-values that are solutions of the equation $y_1 = y_2$. To do so, we solve

$$y_1 - y_2 = 0, \quad \text{or} \quad 3x^5 - 20x^3 + 1.28x^2 - 34.7 = 0.$$

On some graphers there is actually a POLY or ROOT feature, which allows you to enter this polynomial and gives all the solutions of this equation directly. We round each x-value to three decimal places. Then, to compute the second coordinate, we substitute the x-value into either y_1 or y_2.

Example 12 Solve: $\frac{2}{3}x - 7 = 5$.

SOLUTION

Method 1. One way to solve this equation is to ask ourselves, "For what x-values will the expression $\frac{2}{3}x - 7$ equal 5?" This can be visualized by asking, "Where will the graph of $y = \frac{2}{3}x - 7$ cross the horizontal line $y = 5$?" Thus we graph the equations $y_1 = \frac{2}{3}x - 7$ and $y_2 = 5$ and look for the *first coordinate* of their point of intersection. The INTERSECT feature provides us with an efficient means of finding this point. TRACE and ZOOM could also be used.

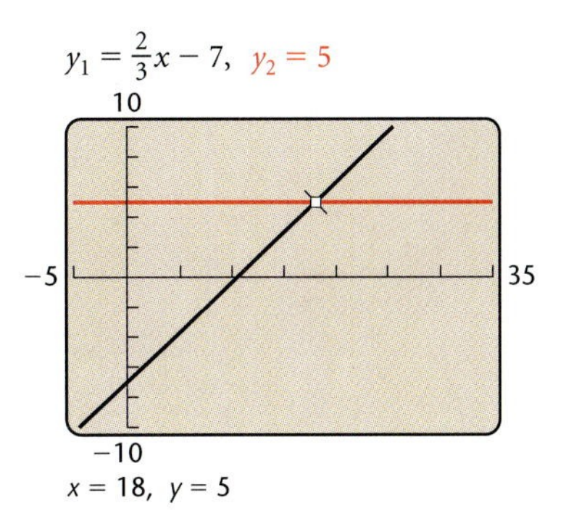

X	Y1	Y2
17.7	4.8	5
17.8	4.8667	5
17.9	4.9333	5
18	5	5
18.1	5.0667	5
18.2	5.1333	5
18.3	5.2	5
X = 18		

We see that the point of intersection is (18, 5). Thus the solution is 18.

We can also confirm the solution numerically, using the table shown at left. The step-value in the table is 0.1. We choose the minimum x-value to be 17.7 and observe the y-values as the x-values increase to 18. The solution can also be checked using the table set in ASK mode.

Method 2. Another way to solve this equation is to solve

$$y_1 - y_2 = 0, \quad \text{or} \quad \tfrac{2}{3}x - 7 - 5 = 0, \quad \text{or} \quad \tfrac{2}{3}x - 12 = 0.$$

We can do this in a manner similar to Example 10. We graph the equation $y = \frac{2}{3}x - 12$ and see that it intersects the x-axis at (18, 0). Thus the solution is 18. We could also use the SOLVE or ROOT features.

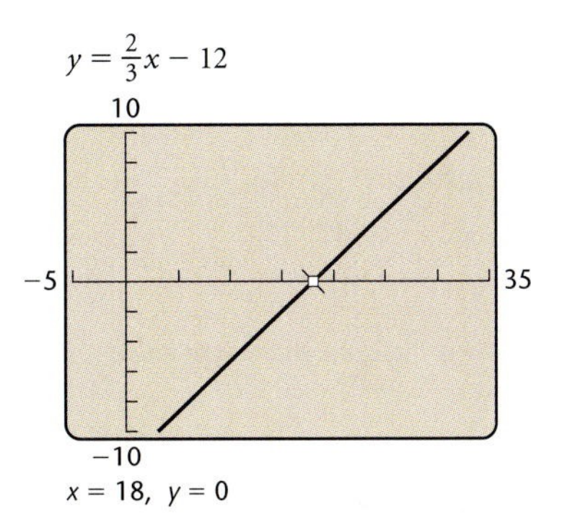

A Special Case

Graphers, for all their advantages, also have their disadvantages, one of which can occur with certain kinds of rational equations. Consider the equation $y = 8/(x^2 - 4)$. The numbers -2 and 2 cannot be substituted into the equation because they result in division by 0. Thus no point with either of these numbers as a first coordinate can be part of the graph. Now look at the two graphs of $y = 8/(x^2 - 4)$ shown below.

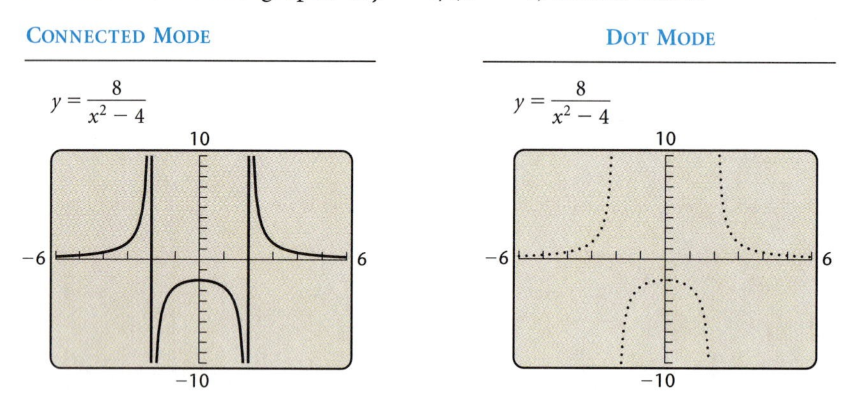

In CONNECTED mode, a grapher connects plotted points with line segments. In DOT mode, it simply plots unconnected points. In the graph on the left, the grapher has connected the points plotted on either side of the x-values -2 and 2 with vertical lines. These lines, however, cannot be part of the graph because substitutions of -2 and 2 are not allowed in this equation. Thus this is not a correct graph of this equation. The graph on the right is more accurate. When graphing equations in which a variable appears in a denominator, use DOT mode if you have a choice.

This kind of difficulty should not deter you from using a grapher. As we will see throughout the text, we will often need to apply our math knowledge to make the best use of a grapher.

Exercise Set

Graph and label each set of points. Make a hand-drawn graph.

1. $\{(4, 0), (-3, -5), (-1, 4), (0, 2), (2, -2)\}$

2. $\{(1, 4), (-4, -2), (-5, 0), (2, -4), (4, 0)\}$

3. $\{(-5, 1), (5, 1), (2, 3), (2, -1), (0, 1)\}$

4. $\{(4, 0), (4, -3), (-5, 2), (-5, 0), (-1, -5)\}$

Determine whether the given ordered pairs are solutions of the given equation.

5. $(1, -1), (0, 3)$; $y = 2x - 3$

6. $(2, 5), (-2, -5)$; $y = 3x - 1$

7. $\left(-\frac{1}{2}, -\frac{4}{5}\right), \left(0, \frac{3}{5}\right)$; $2a + 5b = 3$

8. $\left(0, \frac{3}{2}\right), \left(\frac{2}{3}, 1\right)$; $3m + 4n = 6$

9. $(-0.75, 2.75), (2, -1)$; $x^2 - y^2 = 3$

10. $(2, -4), (4, -5)$; $5x + 2y^2 = 70$

Make a hand-drawn graph of each equation. Use the indicated x-values along with any others you are curious about.

11. $y = 3 - 2x$; $x = -3, -2, -1, 0, 1, 2, 3$

12. $x + y = 4$; $x = -3, -2, -1, 0, 1, 2, 3$

13. $3x - 4y = 12$; $x = -3, -2, -1, 0, 1, 2, 3$

14. $y = x^2$; $x = -3, -2, -1, 0, 1, 2, 3$

15. $y = -x^2$; $x = -3, -2, -1, 0, 1, 2, 3$

16. $y = \sqrt{x}$; $x = 0, 1, 4, 9$

17. $y = |x| - 2$; $x = -3, -2, -1, 0, 1, 2, 3$

18. $y = x^3$; $x = -2, -1, 0, 1, 2$

19. $y = \sqrt{x} - 4$; $x = 0, 1, 4, 9$

20. $y = |x| + 2$; $x = -3, -2, -1, 0, 1, 2, 3$

21. $y = -\dfrac{2}{x}$; $x = -4, -2, -1, -\dfrac{1}{2}, \dfrac{1}{2}, 1, 2, 4$

22. $y = 4 - x^2$; $x = -3, -2, -1, 0, 1, 2, 3$

In Exercises 23–30, use a grapher to match the equations with graphs (a)–(h).

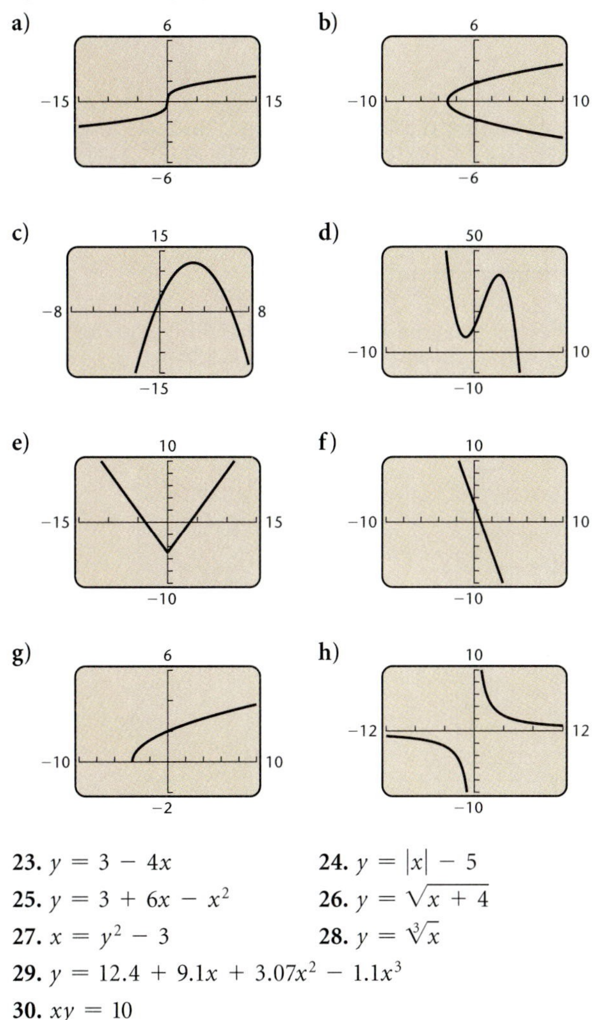

23. $y = 3 - 4x$

24. $y = |x| - 5$

25. $y = 3 + 6x - x^2$

26. $y = \sqrt{x + 4}$

27. $x = y^2 - 3$

28. $y = \sqrt[3]{x}$

29. $y = 12.4 + 9.1x + 3.07x^2 - 1.1x^3$

30. $xy = 10$

Graph each of the following equations, using the standard viewing window, $[-10, 10, -10, 10]$. *Then try other windows until you find one that best represents the graph. Answers may vary.*

31. $y = \frac{1}{5}x + 2$

32. $4x - 5y = 20$

33. $x = y^2 + 2$

34. $y = |x - 1|$

35. $xy = -18$

36. $y = (x + 2)^2$

37. $y = x^2 - 4x + 3$

38. $x = 4 - y^2$

39. $y = \sqrt{x} - 2$

40. $y = x^3 + 2x^2 - 4x - 13$

41. $y^2 = 4 + x$

42. $y = \sqrt[3]{x + 4}$

In Exercises 43–46, an equation is given together with four choices of viewing windows and scaling conditions. Choose the best option. If you reject a situation, explain why you did so.

43. $y = 2x^3 - x^4$

- **a)** $[-1, 1, -1, 1]$, Xscl = 1, Yscl = 1
- **b)** $[-2, 3, -4, 3]$, Xscl = 1, Yscl = 1
- **c)** $[-2, 3, -4, 3]$, Xscl = 0.01, Yscl = 0.1
- **d)** $[-20, 30, -40, 30]$, Xscl = 1, Yscl = 1

44. $y = x^4 - 8x^3 + 18x^2$

- **a)** $[-10, 10, -10, 10]$, Xscl = 1, Yscl = 1
- **b)** $[-1, 1, -10, 10]$, Xscl = 1, Yscl = 1
- **c)** $[-3, 6, -10, 60]$, Xscl = 1, Yscl = 10
- **d)** $[-10, 60, -3, 6]$, Xscl = 1, Yscl = 10

45. $y = \dfrac{8x}{x^2 + 1}$

- **a)** $[-1, 1, -10, 10]$, Xscl = 1, Yscl = 1
- **b)** $[-5, 5, -7, 7]$, Xscl = 1, Yscl = 1
- **c)** $[-3, 6, -10, 60]$, Xscl = 1, Yscl = 10
- **d)** $[-100, 100, -100, 100]$, Xscl = 10, Yscl = 10

46. $y = 4x - x^3$

- **a)** $[-0.1, 10, -10, 10]$, Xscl = 1, Yscl = 1
- **b)** $[-10, 0.1, -10, 10]$, Xscl = 1, Yscl = 1
- **c)** $[-5, 5, -0.1, 1]$, Xscl = 1, Yscl = 1
- **d)** $[-3, 3, -10, 10]$, Xscl = 1, Yscl = 1

47. Using a grapher, complete this table for the equations

$$y_1 = \sqrt{10 - x^2} \quad \text{and} \quad y_2 = \frac{16}{x + 2.8}.$$

Then extend the table from -2.6 to 3.3.

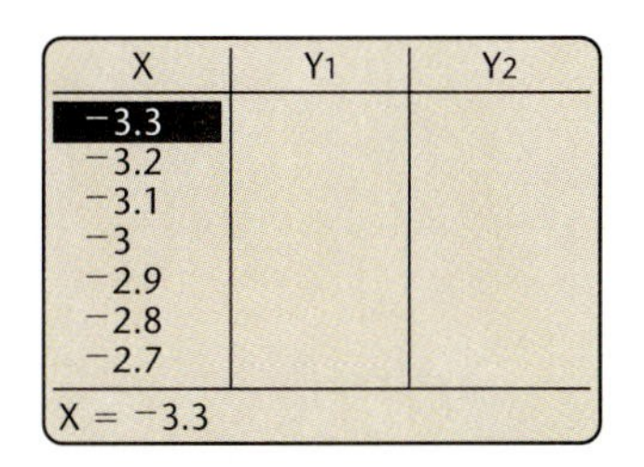

X	Y1	Y2
−3.3		
−3.2		
−3.1		
−3		
−2.9		
−2.8		
−2.7		

X = −3.3

Determine with a grapher whether each of the following seems to be an identity.

48. $7x = x \cdot 7$

49. $(x - 5)^2 = x^2 - 10x + 25$

50. $(x + 4)^2 = x^2 + 16$

51. $x^3 + x^4 = x^7$

52. $(x - 9)^2 = (x + 3)(x - 3)$

53. $x^3 - 1 = (x - 1)(x^2 + x + 1)$

54. $(x + 2)^3 = x^3 + 8$

55. $\sqrt{x^2 - 16} = x - 4$

56. $(x^2)^3 = x^6$

57. $\dfrac{x^5}{x^2} = x^3$

58. $\sqrt{1 + x} = 1 + \dfrac{x}{2} - \dfrac{x^2}{8} + \dfrac{x^3}{16}$

Solve with a grapher.

59. $\frac{3}{4}x + 2 = -4$

60. $-5 = -\frac{3}{2}x - 3$

61. $3x + 4 = -\frac{2}{5}x + 1$

62. $15 - 2x = \dfrac{2x + 5}{3}$

63. $1.4x + 0.7 = 0.9x - 2.2$

64. $-2.6(x - 8.4) + 1.92 = 23x - 0.93(8x + 11.3)$

65. $x - 7.4 = 2.8\sqrt{x + 1.1}$

66. $\sqrt{x - 3.2} + \sqrt{x + 5.03} = 4.91$

67. $x^3 - 6x^2 = -9x - 1$

68. $2.1x^4 - 4.3x^2 = 5$

69. $1.09x^2 - 0.8x^4 = -7.6$

70. $144x + 140 = x^3 - 3x^2$

71. $x^4 + 4x^3 + 300 = 36x^2 + 160x$

72. $x^3 = 2x^2 + 2$

73. $|x + 1| + |x - 2| = 5$

74. $|x + 1| + |x - 2| = 0$

Find the points of intersection of the graphs of each pair of equations using a grapher.

75. $y_1 = x^3 + 3x^2 - 9x - 13,\ y_2 = 0$

76. $y_1 = \sqrt{7 - x^2},\ y_2 = 2.5$

77. $y = x^3 - 3x^2,\ 4x - 7y = 20$

78. $y = x^4 - 2x^3,\ y = -0.7x + 3$

79. $y = \dfrac{8x}{x^2 + 1},\ y = 0.9x$

80. $y = x\sqrt{8 - x^2},\ 93x - 100y = 1$

81. *A Hole in a Graph.* Consider the graph of

$$y = \frac{x^2 - 4}{x - 2}.$$

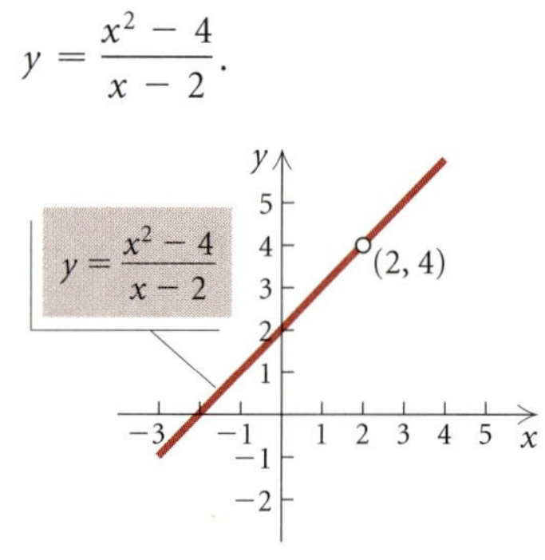

We cannot substitute $x = 2$ into the equation. This creates a "hole" in the graph. The point $(2, 4)$ is not part of the graph. On some graphers it may be very difficult to find the hole. Graph this function and try to find the hole. Experiment with some of the grapher's features.

In Exercises 82 and 83, make a hand-drawn graph. Look for a pattern to find the missing data. Try to determine an equation that fits the data.

82.

x	y
1	4
2	7
3	10
4	13
5	
6	
	31
	34

83.

x	y
0	1
1	2
2	5
3	10
4	17
5	
	82
	122

Basic Concepts of Algebra

R

APPLICATION

The solution of the equation $x + 2x = 30$ gives the average number of vacation days a U.S. worker receives each year.

X	Y1	Y2
7	21	30
8	24	30
9	27	30
10	30	30
11	33	30
12	36	30
13	39	30
X = 10		

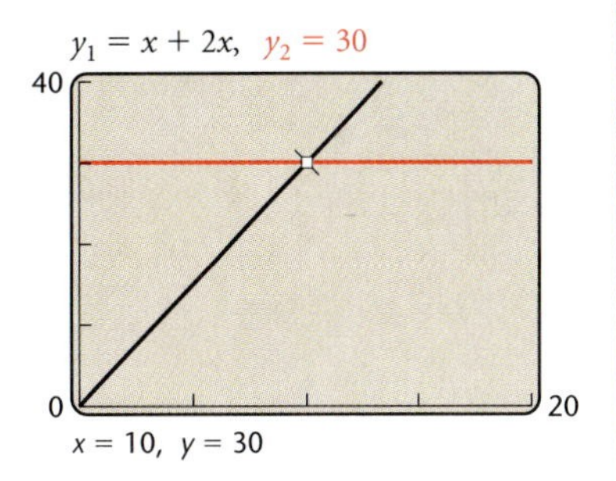

The algebraic concepts reviewed in this chapter provide the foundation for your further study in this text. They include the properties of the real-number system, manipulations of algebraic expressions, the solution of equations and inequalities, and the solution of applied problems, like the one shown concerning vacation time. Grapher technology is used to enhance the understanding and visualization of these topics.

R.1 The Real-Number System
R.2 Integer Exponents, Scientific Notation, and Order of Operations
R.3 Addition, Subtraction, and Multiplication of Polynomials
R.4 Factoring
R.5 Rational Expressions
R.6 Radical Notation and Rational Exponents
R.7 Solving Equations
R.8 Solving Inequalities
R.9 Modeling and Applications
SUMMARY AND REVIEW

R.1 The Real-Number System

- *Identify various kinds of real numbers.*
- *Find the absolute value of an algebraic expression.*
- *Identify the properties of real numbers.*

Real Numbers

In applications of algebraic concepts, we use real numbers to represent quantities such as distance, time, speed, area, profit, loss, and temperature. Some frequently used sets of real numbers are listed below.

Sets of Real Numbers

Natural numbers. The numbers used for counting: $\{1, 2, 3, \ldots\}$.

Whole numbers. The natural numbers and 0: $\{0, 1, 2, 3, \ldots\}$.

Integers. The whole numbers and their opposites: $\{\ldots, -3, -2, -1, 0, 1, 2, 3, \ldots\}$.

Rational numbers. The numbers that can be expressed in the form p/q, where p and q are integers and $q \neq 0$. Some examples are $\frac{1}{4}$, $-\frac{12}{5}$, 0.125 $\left(0.125 = \frac{1}{8}\right)$, 106 $\left(106 = \frac{106}{1}\right)$, and $-0.\overline{3}$ $\left(-0.\overline{3} = -0.333\ldots = -\frac{1}{3}\right)$.

Irrational numbers. The real numbers that are not rational. Some examples are $-\sqrt{5}$, π, $\sqrt[3]{19}$, and $-0.2020020002\ldots$.

Decimal notation for rational numbers either *terminates* (ends) or *repeats*. Each of the following is a rational number.

a) 0 — **$0 = \dfrac{0}{a}$ for any nonzero integer a**

b) -7 — **$-7 = \dfrac{-7}{1}$, or $\dfrac{7}{-1}$, or $-\dfrac{7}{1}$**

c) $\dfrac{1}{4} = 0.25$ — **Terminating decimal**

d) $-\dfrac{5}{11} = -0.\overline{45}$ — **Repeating decimal**

Decimal notation for irrational numbers neither terminates nor repeats. Each of the following is an irrational number.

a) $\pi = 3.1415926535\ldots$ — **There is no repeating block of digits.**

$\left(\frac{22}{7}\right.$ and 3.14 are rational *approximations* of the irrational number $\left.\pi.\right)$

b) $\sqrt{2} = 1.414213562\ldots$ — **There is no repeating block of digits.**

c) $-6.12122122212222\ldots$ — **Although there is a pattern, there is no repeating block of digits.**

The set of all rational numbers combined with the set of all irrational numbers gives us the set of **real numbers.** The relationships among the various sets of real numbers are shown below.

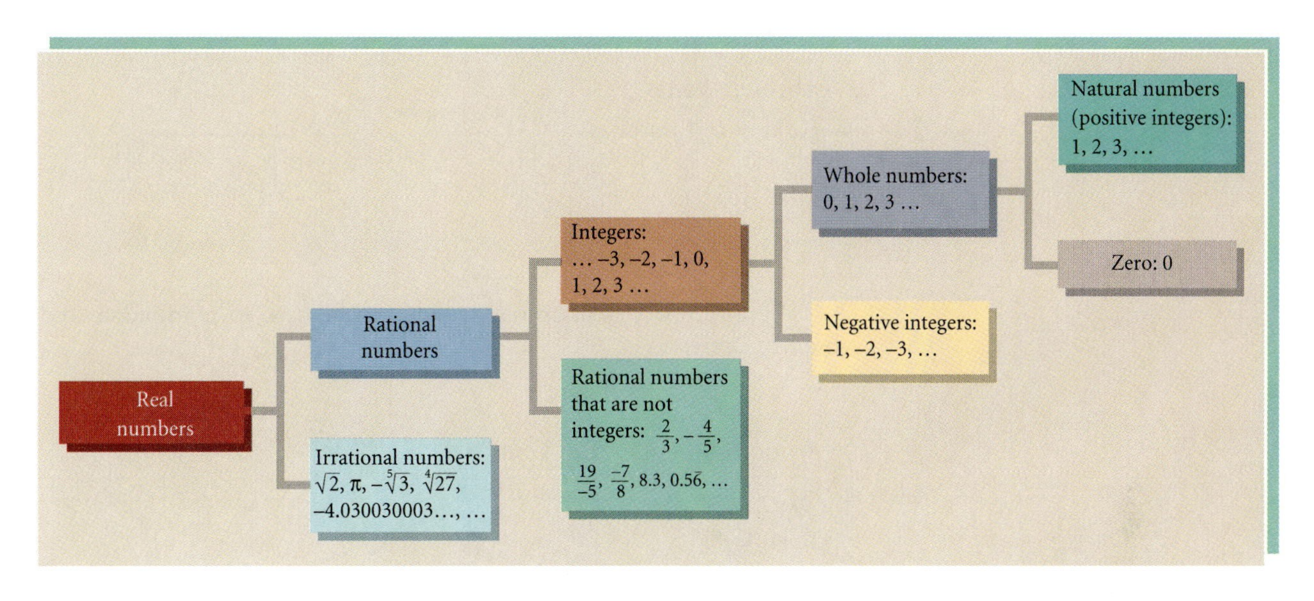

The symbol $\in$ is used to indicate that a member, or *element,* belongs to a set. Thus if we let $\mathbb{Q}$ represent the set of rational numbers, we can see from the diagram that $0.5\overline{6} \in \mathbb{Q}$. We can also write $\sqrt{2} \notin \mathbb{Q}$ to indicate that $\sqrt{2}$ is *not* an element of the set of rational numbers.

When *all* the elements of one set are elements of a second set, we say that the first set is a **subset** of the second set. The symbol $\subseteq$ is used to denote this. For instance, if we let $\mathbb{R}$ represent the set of real numbers, we can see from the diagram that $\mathbb{Q} \subseteq \mathbb{R}$ (read "$\mathbb{Q}$ is a subset of $\mathbb{R}$").

The real numbers are *modeled* using a **number line,** as shown below.

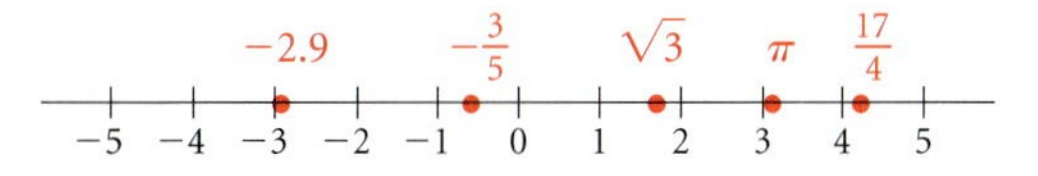

The order of the real numbers can be determined from the number line. If a number a is to the left of a number b, then a **is less than** b ($a < b$). Similarly, a **is greater than** b ($a > b$), if a is to the right of b on the number line. For example, we see from the number line above that $-2.9 < -\frac{3}{5}$, because -2.9 is to the left of $-\frac{3}{5}$. Also, $\frac{17}{4} > \sqrt{3}$, because $\frac{17}{4}$ is to the right of $\sqrt{3}$.

The statement $a \leq b$, read "a is less than or equal to b," is true if either $a < b$ is true or $a = b$ is true.

Absolute Value

The number line can also be used to provide a geometric interpretation of *absolute value*. The absolute value of a number a, denoted $|a|$, is its distance from 0 on the number line. For example, $|-5| = 5$, because the distance of -5 from 0 is 5. Similarly, $\left|\frac{3}{4}\right| = \frac{3}{4}$, because the distance of $\frac{3}{4}$ from 0 is $\frac{3}{4}$.

Absolute Value

For any real number a,

$$|a| = \begin{cases} a, & \text{if } a \geq 0 \\ -a, & \text{if } a < 0. \end{cases}$$

Several properties of absolute value can be used to manipulate and simplify expressions.

Properties of Absolute Value

For any real numbers a and b:

1. $|a| \geq 0$.
2. $|ab| = |a| \cdot |b|$.
3. $\left|\frac{a}{b}\right| = \frac{|a|}{|b|}$ $(b \neq 0)$.
4. $|-a| = |a|$.

We can use a grapher to confirm the first property. We might need to enter $y = |x|$ as $y = \text{abs } x$.

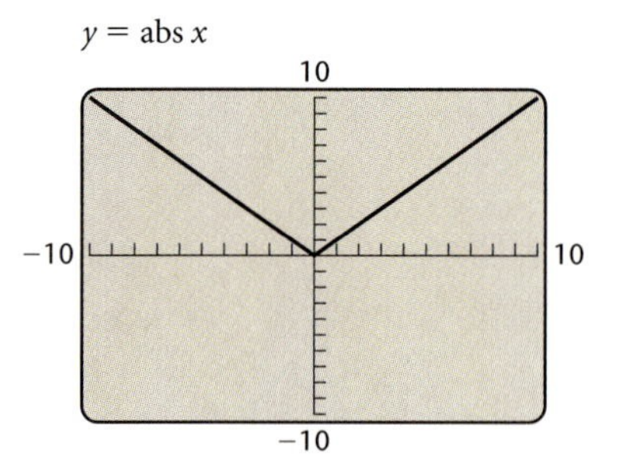

We see from the graph that all y-values are nonnegative.

Absolute value can be used to find the distance between two points on the number line.

Distance Between Two Points on the Number Line

For any real numbers a and b, the distance between a and b is $|a - b|$, or, equivalently, $|b - a|$.

To see that $|a - b|$ and $|b - a|$ are equivalent, note that by the fourth property of absolute value above, $|a - b| = |-(a - b)| = |b - a|$. We can also use a grapher to demonstrate this equivalence by graphing $y_1 = |x - a|$ and $y_2 = |a - x|$ for any nonzero real number a. For example, let's graph $y_1 = |x - 4|$ and $y_2 = |4 - x|$.

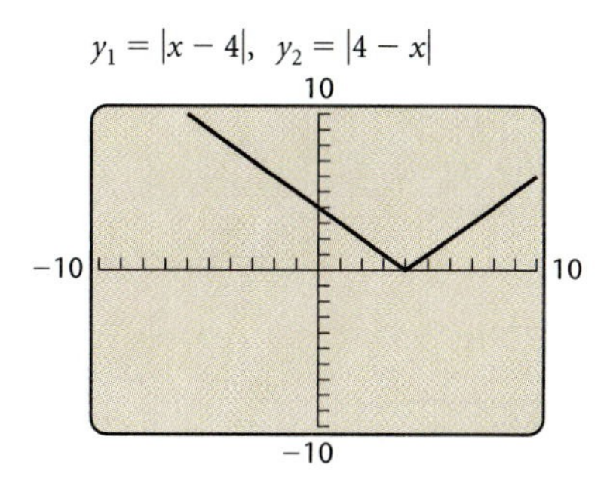

The graphs appear to coincide. Try this for several other values of a.

Example 1 Find the distance between -2 and 3.

SOLUTION

$$|-2 - 3| = |-5| = 5, \quad \text{or, equivalently,}$$
$$|3 - (-2)| = |3 + 2| = |5| = 5.$$

Properties of the Real Numbers

The following properties can be used to manipulate algebraic expressions as well as real numbers.

Properties of the Real Numbers

For any real numbers a, b, and c:

$a + b = b + a$ and $ab = ba$	Commutative properties of addition and multiplication
$a + (b + c) = (a + b) + c$ and $a(bc) = (ab)c$	Associative properties of addition and multiplication
$a + 0 = 0 + a = a$	Additive identity property
$-a + a = a + (-a) = 0$	Additive inverse property
$a \cdot 1 = 1 \cdot a = a$	Multiplicative identity property
$a \cdot \frac{1}{a} = 1 \ (a \neq 0)$	Multiplicative inverse property
$a(b + c) = ab + ac$	Distributive property

Note that the distributive property is also true for subtraction since $a(b - c) = a[b + (-c)] = ab + a(-c) = ab - ac$.

Example 2 State the property being illustrated in each sentence.

a) $8 \cdot 5 = 5 \cdot 8$

b) $5 + (m + n) = (5 + m) + n$

c) $x + 0 = x$

d) $14 + (-14) = 0$

e) $2(a - b) = 2a - 2b$

SOLUTION

SENTENCE	PROPERTY
a) $8 \cdot 5 = 5 \cdot 8$	Commutative property of multiplication: $ab = ba$
b) $5 + (m + n) = (5 + m) + n$	Associative property of addition: $a + (b + c) = (a + b) + c$
c) $x + 0 = x$	Additive identity property: $a + 0 = a$
d) $14 + (-14) = 0$	Additive inverse property: $a + (-a) = 0$
e) $2(a - b) = 2a - 2b$	Distributive property: $a(b + c) = ab + ac$

R.1 Exercise Set

In Exercises 1–6, consider the numbers -12, $\sqrt{7}$, $5.\overline{3}$, $-\frac{7}{3}$, $\sqrt[3]{8}$, 0, $5.242242224\ldots$, $-\sqrt{14}$, $\sqrt[5]{5}$, -1.96, 9, $4\frac{2}{3}$, $\sqrt{25}$, $\sqrt[3]{4}$, $\frac{5}{7}$.

1. Which are whole numbers?

2. Which are integers?

3. Which are irrational numbers?

4. Which are natural numbers?

5. Which are rational numbers?

6. Which are real numbers?

Find an example of each of the following. Answers may vary.

7. A rational number that is not an integer

8. A real number that is not a rational number

9. An integer that is not a whole number

10. A real number that is not an irrational number

In Exercises 11–28, the following notation is used: $\mathbb{N}$ *= the set of natural numbers,* $\mathbb{W}$ *= the set of whole numbers,* $\mathbb{Z}$ *= the set of integers,* $\mathbb{Q}$ *= the set of rational numbers,* $\mathbb{I}$ *= the set of irrational numbers, and* $\mathbb{R}$ *= the set of real numbers. Classify each statement as true or false.*

11. $6 \in \mathbb{N}$

12. $0 \notin \mathbb{N}$

13. $3.2 \in \mathbb{Z}$

14. $-10.\overline{1} \in \mathbb{R}$

15. $-\dfrac{11}{5} \in \mathbb{Q}$

16. $-\sqrt{6} \in \mathbb{Q}$

17. $\sqrt{11} \notin \mathbb{R}$

18. $-1 \in \mathbb{W}$

19. $24 \notin \mathbb{W}$

20. $1 \in \mathbb{Z}$

21. $1.089 \notin \mathbb{I}$

22. $\mathbb{N} \subseteq \mathbb{W}$

23. $\mathbb{W} \subseteq \mathbb{Z}$

24. $\mathbb{Z} \subseteq \mathbb{N}$

25. $\mathbb{Q} \subseteq \mathbb{R}$

26. $\mathbb{Z} \subseteq \mathbb{Q}$

27. $\mathbb{R} \subseteq \mathbb{Z}$

28. $\mathbb{Q} \subseteq \mathbb{I}$

Simplify.

29. $|-7.1|$

30. $|0|$

31. $\left|\dfrac{5}{4}\right|$

32. $|-\sqrt{3}|$

33. $|-8b|$

34. $|-13.8a|$

35. $|-5xz|$

36. $\left|-\dfrac{3}{4}ab\right|$

37. $\left|\dfrac{0.02x}{y}\right|$

38. $\left|\dfrac{6a}{b}\right|$

Find the distance between each pair of points on the number line.

39. $-5, 6$

40. $0, -2.5$

41. $-2, -8$

42. $\frac{15}{8}, \frac{23}{12}$

43. 12.1, 6.7

44. $-3, -14$

Name the property illustrated by each sentence.

45. $6x = x6$

46. $3 + (x + y) = (3 + x) + y$

47. $-3 \cdot 1 = -3$

48. $x + 4 = 4 + x$

49. $5(ab) = (5a)b$

50. $4(y - z) = 4y - 4z$

51. $2(a + b) = (a + b)2$

52. $-7 + 7 = 0$

53. $-6(m + n) = -6(n + m)$

54. $t + 0 = t$

55. $8 \cdot \frac{1}{8} = 1$

56. $9x + 9y = 9(x + y)$

Determine graphically whether each of the following appears to be an identity. (See the Introduction to Graphs and Graphers.)

57. $|x| = |-x|$

58. $|x^2| = x^2$

59. $|-5a| = 5a$

60. $\left|\frac{x}{3}\right| = \frac{|x|}{3}$

61. $x + 7 = 7 + x$

62. $8(x + 1) = 8x + 1$

Synthesis

To the student and the instructor: The synthesis exercises found at the end of every exercise set challenge students to combine concepts or skills studied in that section or in preceding parts of the text. Writing exercises, denoted by the ◈ icon, are meant to be answered with one or more sentences.

63. ◈ How would you convince a classmate that division is not associative?

64. ◈ Under what circumstances is $\sqrt{a}$ a rational number?

Between any two (different) real numbers there are many other real numbers. Find each of the following. Answers may vary.

65. An irrational number between 0.124 and 0.125

66. A rational number between $-\sqrt{2.01}$ and $-\sqrt{2}$

67. A rational number between $-\frac{1}{101}$ and $-\frac{1}{100}$

68. An irrational number between $\sqrt{5.99}$ and $\sqrt{6}$

69. The hypotenuse of an isosceles right triangle with legs of length 1 unit can be used to "measure" a value for $\sqrt{2}$ by using the Pythagorean theorem, as shown.

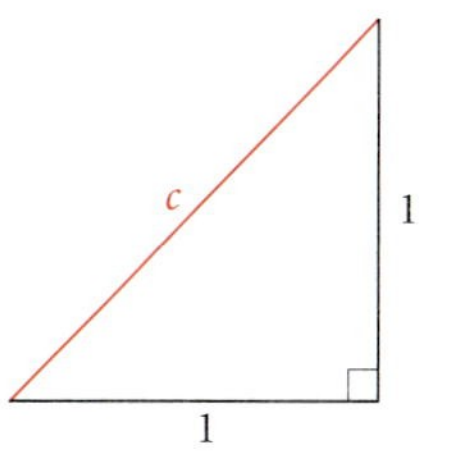

$$c^2 = 1^2 + 1^2$$
$$c^2 = 2$$
$$c = \sqrt{2}$$

Draw a right triangle that could be used to "measure" $\sqrt{10}$ units.

R.2 Integer Exponents, Scientific Notation, and Order of Operations

- *Simplify expressions with integer exponents.*
- *Solve problems using scientific notation.*
- *Use the rules for order of operations.*

Integers as Exponents

When a positive integer is used as an *exponent*, it indicates the number of times a factor appears in a product. For example, $7 \cdot 7 \cdot 7$ can be written as 7^3.

For any positive integer n,

$$a^n = \underbrace{a \cdot a \cdot a \cdots a}_{n \text{ factors}},$$

where a is the *base* and n is the *exponent.*

Zero and negative-integer exponents are defined as follows.

For any nonzero real number a and any integer n,

$$a^0 = 1 \quad \text{and} \quad a^{-n} = \frac{1}{a^n}.$$

Example 1 Simplify each of the following.

a) 6^0 b) $(-3.4)^0$ c) 4^{-5} d) $\dfrac{1}{(0.82)^{-7}}$

SOLUTION

a) $6^0 = 1$

b) $(-3.4)^0 = 1$

c) $4^{-5} = \dfrac{1}{4^5}$

d) $\dfrac{1}{(0.82)^{-7}} = (0.82)^{-(-7)} = (0.82)^7$

The following properties of exponents can be used to simplify expressions.

Properties of Exponents

For any real numbers a and b and any integers m and n, assuming 0 is not raised to a nonpositive power:

$a^m \cdot a^n = a^{m+n}$	Product rule
$\dfrac{a^m}{a^n} = a^{m-n} \quad (a \neq 0)$	Quotient rule
$(a^m)^n = a^{mn}$	Power rule
$(ab)^m = a^m b^m$	Raising a product to a power
$\left(\dfrac{a}{b}\right)^m = \dfrac{a^m}{b^m} \quad (b \neq 0)$	Raising a quotient to a power

We can use a grapher for partial confirmation of the first three properties. In the case of the product rule, for instance, we graph $y_1 = x^3 \cdot x^2$ and $y_2 = x^5$.

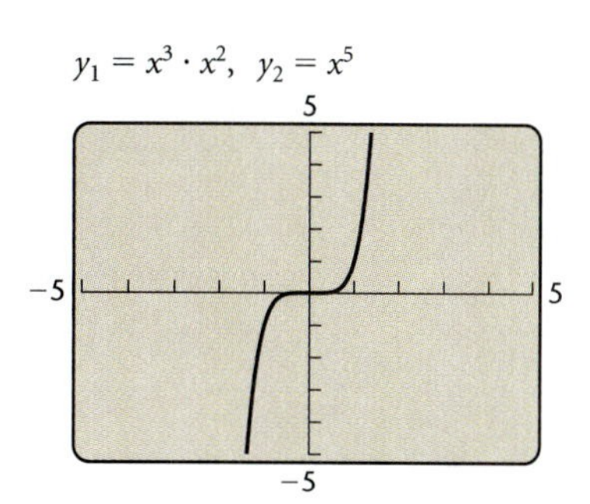

The graphs appear to coincide. We can perform similar confirmations of the quotient rule and the power rule.

Example 2 Simplify each of the following.

a) $y^{-5} \cdot y^3$ **b)** $\dfrac{48x^{12}}{16x^4}$ **c)** $(t^{-3})^5$

d) $(2s^{-2})^5$ **e)** $\left(\dfrac{45x^{-4}y^2}{9z^{-8}}\right)^{-3}$

SOLUTION

a) $y^{-5} \cdot y^3 = y^{-5+3} = y^{-2}$, or $\dfrac{1}{y^2}$

b) $\dfrac{48x^{12}}{16x^4} = \dfrac{48}{16}x^{12-4} = 3x^8$

c) $(t^{-3})^5 = t^{-3\cdot5} = t^{-15}$, or $\dfrac{1}{t^{15}}$

d) $(2s^{-2})^5 = 2^5(s^{-2})^5 = 32s^{-10}$, or $\dfrac{32}{s^{10}}$

e) $\left(\dfrac{45x^{-4}y^2}{9z^{-8}}\right)^{-3} = \left(\dfrac{5x^{-4}y^2}{z^{-8}}\right)^{-3} = \dfrac{5^{-3}x^{12}y^{-6}}{z^{24}} = \dfrac{x^{12}y^{-6}}{125z^{24}}$, or $\dfrac{x^{12}}{125y^6z^{24}}$

Scientific Notation

We can use exponential, or scientific, notation to name very large and very small positive numbers and to perform computations.

Scientific Notation

Scientific notation for a number is an expression of the type

$$N \times 10^n,$$

where $1 \le N < 10$, N is in decimal notation, and n is an integer.

Keep in mind that in scientific notation, positive exponents are used for numbers greater than 10 and negative exponents for numbers between 0 and 1.

Example 3 *National Debt.* In a recent month, the national debt increased by \$38,500,000,000. Convert this number to scientific notation. (*Source*: U.S. Treasury Department)

SOLUTION We want the decimal point to be positioned between the 3 and the 8, so we move it 10 places to the left. Since the number to be converted is greater than 10, the exponent must be positive.

$$38{,}500{,}000{,}000 = 3.85 \times 10^{10}$$

Example 4 *Mass of a Neutron.* The mass of a neutron is about 0.00000000000000000000000000167 kg. Convert this number to scientific notation.

SOLUTION We want the decimal point to be positioned between the 1 and the 6, so we move it 27 places to the right. Since the number to be converted is between 0 and 1, the exponent must be negative.

$$0.00000000000000000000000000167 = 1.67 \times 10^{-27}$$

Example 5 Convert each of the following to decimal notation.

a) 7.632×10^{-4} **b)** 9.4×10^{5}

SOLUTION

a) The exponent is negative, so the number is between 0 and 1. We move the decimal point 4 places to the left.

$$7.632 \times 10^{-4} = 0.0007632$$

b) The exponent is positive, so the number is greater than 10. We move the decimal point 5 places to the right.

$$9.4 \times 10^{5} = 940{,}000$$

Most calculators make use of scientific notation. For example, a number like 48,000,000,000,000 might be expressed as

4.8 E 13 or 4.8 13

Example 6 *Distance to a Star.* Alpha Centauri is about 4.3 light-years from the earth. One **light-year** is the distance that light travels in one year and is about 5.88×10^{12} miles. How many miles is it from earth to Alpha Centauri? Express your answer in scientific notation.

SOLUTION

$$\begin{aligned} 4.3 \times (5.88 \times 10^{12}) &= (4.3 \times 5.88) \times 10^{12} \\ &= 25.284 \times 10^{12} && \text{This is not scientific notation.} \\ &= (2.5284 \times 10^{1}) \times 10^{12} \\ &= 2.5284 \times 10^{13} \text{ miles} && \text{Writing scientific notation} \end{aligned}$$

Order of Operations

Recall that to simplify the expression $3 + 4 \cdot 5$, we multiply 4 and 5 first to get 20 and then add 3 to get 23. Mathematicians have agreed on the following procedure, or rules for order of operations.

Rules for Order of Operations

1. Do all calculations within grouping symbols before operations outside. When nested grouping symbols are present, work from the inside out.
2. Evaluate all exponential expressions.
3. Do all multiplications and divisions in order from left to right.
4. Do all additions and subtractions in order from left to right.

Most graphers use the order of operations above. Thus it is essential to remember these rules when entering a computation.

Example 7 Simplify each of the following.

a) $8(5 - 3)^3 - 20$

b) $\dfrac{10 \div (8 - 6) + 9 \cdot 4}{2^5 + 3^2}$

SOLUTION

a)

$$\begin{aligned} 8(5 - 3)^3 - 20 &= 8 \cdot 2^3 - 20 && \text{Doing the calculation within parentheses} \\ &= 8 \cdot 8 - 20 && \text{Evaluating the exponential expression} \\ &= 64 - 20 && \text{Multiplying} \\ &= 44 && \text{Subtracting} \end{aligned}$$

On a grapher this computation would be entered using the scientific keys, as follows.

```
8(5 – 3)^3 – 20          44
```

b) $\dfrac{10 \div (8 - 6) + 9 \cdot 4}{2^5 + 3^2} = \dfrac{10 \div 2 + 9 \cdot 4}{32 + 9} = \dfrac{5 + 36}{41} = \dfrac{41}{41} = 1$

Note that fraction bars act as grouping symbols. That is, the given expression is equivalent to $[10 \div (8 - 6) + 9 \cdot 4] \div (2^5 + 3^2)$. On a grapher we would enter the following.

```
(10/(8 – 6) + 9*4)/(2^5 + 3^2)
```

To confirm that parentheses are essential in Example 6(b), enter the computation on a grapher without using parentheses. What is the result?

Example 8 *Compound Interest.* If a principal P is invested at an interest rate i, compounded n times per year, in t years it will grow to an amount A given by

$$A = P\left(1 + \frac{i}{n}\right)^{nt}.$$

Thus when \$1250 is invested at 4.6% interest, compounded quarterly, the amount in the account at the end of 8 years is given by

$$A = 1250\left(1 + \frac{0.046}{4}\right)^{32}.$$

Evaluate this expression.

SOLUTION

$$\begin{aligned} A &= 1250\left(1 + \frac{0.046}{4}\right)^{32} \\ &= 1250(1 + 0.0115)^{32} && \textbf{Dividing} \\ &= 1250(1.0115)^{32} && \textbf{Adding} \\ &\approx 1250(1.441811175) && \textbf{Evaluating the exponential expression} \\ &\approx 1802.263969 && \textbf{Multiplying} \\ &\approx 1802.26 && \textbf{Rounding to the nearest cent} \end{aligned}$$

We can also use the scientific keys on a grapher.

```
1250(1 + 0.046/4)^32
                1802.263969
```

The amount in the account at the end of 8 years is \$1802.26.

R.2 Exercise Set

Simplify.

1. $5^8 \cdot 5^{-6}$
2. $6^2 \cdot 6^{-7}$
3. $m^{-5} \cdot m^5$
4. $n^9 \cdot n^{-9}$
5. $7^3 \cdot 7^{-5} \cdot 7$
6. $3^6 \cdot 3^{-5} \cdot 3^4$
7. $2x^3 \cdot 3x^2$
8. $3y^4 \cdot 4y^3$
9. $(5a^2b)(3a^{-3}b^4)$
10. $(4xy^2)(3x^{-4}y^5)$
11. $(2x)^3(3x)^2$
12. $(4y)^2(3y)^3$
13. $\frac{b^{40}}{b^{37}}$
14. $\frac{a^{39}}{a^{32}}$
15. $\frac{x^2y^{-2}}{x^{-1}y}$
16. $\frac{x^3y^{-3}}{x^{-1}y^2}$
17. $\frac{32x^{-4}y^3}{4x^{-5}y^8}$
18. $\frac{20a^5b^{-2}}{5a^7b^{-3}}$
19. $(2ab^2)^3$
20. $(4xy^3)^2$
21. $(-2x^3)^4$
22. $(-3x^2)^4$
23. $(-5c^{-1}d^{-2})^{-2}$
24. $(-4x^{-5}z^{-2})^{-3}$
25. $\left(\frac{24a^{10}b^{-8}c^7}{3a^6b^{-3}c^5}\right)^5$
26. $\left(\frac{125p^{12}q^{-14}r^{22}}{25p^8q^6r^{-15}}\right)^{-4}$

Convert to scientific notation.

27. 405,000

28. 1,670,000

29. 0.00000039

30. 0.00092

31. One cubic inch is approximately equal to $0.000016\ m^3$.

32. The average *Fortune 500* company has an annual telephone bill of \$34,000,000. (*Source:* Gallup survey of *Fortune 500* companies for Pitney Bowes)

Convert to decimal notation.

33. 8.3×10^{-5}

34. 4.1×10^{6}

35. 2.07×10^{7}

36. 3.15×10^{-6}

37. In 1994, the national debt was $\$4.69 \times 10^{12}$. (*Source:* U.S. Treasury Department)

38. The mass of a proton is about 1.67×10^{-24} g.

Compute. Write scientific notation for the answer.

39. $(3.1 \times 10^5)(4.5 \times 10^{-3})$

40. $(9.1 \times 10^{-17})(8.2 \times 10^3)$

41. $\dfrac{6.4 \times 10^{-7}}{8.0 \times 10^{6}}$

42. $\dfrac{1.1 \times 10^{-40}}{2.0 \times 10^{-71}}$

Solve. Write the answers using scientific notation.

43. *Nuclear Disintegration.* One gram of radium produces 37 billion disintegrations per second. How many disintegrations are produced in 1 hr?

44. *Length of Earth's Orbit.* The average distance from the earth to the sun is 93 million mi. About how far does the earth travel in a yearly orbit? (Assume a circular orbit. Use 3.14 for π.)

45. *Chesapeake Bay Bridge-Tunnel.* The 17.6-mile-long Chesapeake Bay Bridge-Tunnel was completed in 1964. Construction costs were \$210 million. Find the average cost per mile.

46. *Retail Hardware Sales.* In a recent year, America's largest hardware retailer had about \$7.2 billion in sales in its 240 stores. Find the average sales per store.

Calculate. Use a grapher to check each answer.

47. $3 \cdot 2 - 4 \cdot 2^2 + 6(3 - 1)$

48. $3[(2 + 4 \cdot 2^2) - 6(3 - 1)]$

49. $16 \div 4 \cdot 4 \div 2 \cdot 256$

50. $2^6 \cdot 2^{-3} \div 2^{10} \div 2^{-8}$

51. $\dfrac{4(8 - 6)^2 - 4 \cdot 3 + 2 \cdot 8}{3^1 + 19^0}$

52. $\dfrac{[4(8 - 6)^2 + 4](3 - 2 \cdot 8)}{2^2(2^3 + 5)}$

Compound Interest. *Use the compound interest formula in Exercises 53–56. Round to the nearest cent.*

53. Suppose \$2125 is invested at 6.2%, compounded semiannually. How much is in the account at the end of 5 yr?

54. Suppose \$9550 is invested at 5.4%, compounded semiannually. How much is in the account at the end of 7 yr?

55. Suppose \$6700 is invested at 4.5%, compounded quarterly. How much is in the account at the end of 6 yr?

56. Suppose \$4875 is invested at 5.8%, compounded quarterly. How much is in the account at the end of 9 yr?

Synthesis

57. ◈ Are parentheses necessary in the expression $4 \cdot 25 \div (10 - 5)$? Why or why not?

58. ◈ Is $x^{-2} < x^{-1}$ for any negative value(s) of x? Why or why not?

Simplify. Assume that all exponents are integers, all denominators are nonzero, and zero is not raised to a nonpositive power.

59. $(x^t \cdot x^{3t})^2$

60. $(x^y \cdot x^{-y})^3$

61. $(t^{a+x} \cdot t^{x-a})^4$

62. $(m^{x-b} \cdot n^{x+b})^x(m^b n^{-b})^x$

63. $\left[\dfrac{(3x^a y^b)^3}{(-3x^a y^b)^2}\right]^2$

64. $\left[\left(\dfrac{x^r}{y^t}\right)^2\left(\dfrac{x^{2r}}{y^{4t}}\right)^{-2}\right]^{-3}$

65. *Mortgage Payments.* The formula

$$M = P\left[\frac{\dfrac{i}{12}\left(1 + \dfrac{i}{12}\right)^n}{\left(1 + \dfrac{i}{12}\right)^n - 1}\right]$$

gives the monthly mortgage payment M on a home loan of P dollars at interest rate i, where n is the total number of payments (12 times the number of years).

The cost of a house is \$98,000. The down payment is \$16,000, the interest rate is $8\frac{1}{2}\%$, and the loan period is 25 yr. What is the monthly payment?

Use a grapher to graph $y_1 = x^2$, $y_2 = x^3$, and $y_3 = x^4$. Then do Exercises 66 and 67.

66. For what values of x is $x^2 > x^3$?

67. For what values of x is $x^4 > x^2$?

R.3 Addition, Subtraction, and Multiplication of Polynomials

- *Identify the terms, coefficients, and degree of a polynomial.*
- *Add, subtract, and multiply polynomials.*

Polynomials

Polynomials are a type of algebraic expression that you will often encounter in your study of algebra. Some examples of polynomials are

$$3x - 4y, \quad 5y^3 - \tfrac{7}{3}y^2 + 3y - 2, \quad 0, \quad -2.3a^4, \quad \text{and} \quad z^6 - \sqrt{5}.$$

All but the first are polynomials in one variable.

Polynomials in One Variable

A *polynomial in one variable* is any expression of the type

$$a_n x^n + a_{n-1}x^{n-1} + \cdots + a_2 x^2 + a_1 x + a_0,$$

where n is a nonnegative integer, $a_n, \ldots, a_0$ are real numbers, called *coefficients*, and $a_n \neq 0$. The parts of a polynomial separated by plus signs are called *terms*. The *degree* of the polynomial is n, the *leading coefficient* is a_n, and the *constant term* is a_0. The polynomial is said to be written in *descending order*, because the exponents decrease from left to right.

Example 1 Identify the terms of the polynomial

$$2x^4 - 7.5x^3 + x - 12.$$

SOLUTION

$$2x^4 - 7.5x^3 + x - 12 = 2x^4 + (-7.5x^3) + x + (-12),$$

so the terms are

$$2x^4, \quad -7.5x^3, \quad x, \quad \text{and} \quad -12.$$

A polynomial consisting of only a nonzero constant term, like 23, has degree 0. It is agreed that the polynomial consisting only of 0 has *no* degree.

Example 2 Find the degree of each polynomial.

a) $2x^3 - 9$ **b)** $y^2 - \frac{3}{2} + 5y^4$ **c)** 7

SOLUTION

POLYNOMIAL	DEGREE
a) $2x^3 - 9$	3
b) $y^2 - \frac{3}{2} + 5y^4 = 5y^4 + y^2 - \frac{3}{2}$	4
c) $7 = 7x^0$	0

Algebraic expressions like $3ab^3 - 8$ and $5x^4y^2 - 3x^3y^8 + 7xy^2 + 6$ are **polynomials in several variables.** The **degree of a term** is the sum of the exponents of the variables in that term. The **degree of a polynomial** is the degree of the term of highest degree.

Example 3 Find the degree of the polynomial $7ab^3 - 11a^2b^4 + 8$.

SOLUTION The degrees of the terms of $7ab^3 - 11a^2b^4 + 8$ are 4, 6, and 0, respectively, so the degree of the polynomial is 6.

A polynomial with just one term, like $-9y^6$, is a *monomial.* If a polynomial has two terms, like $x^2 + 4$, it is a *binomial.* A polynomial with three terms, like $4x^2 - 4xy + 1$, is a *trinomial.*

Expressions like

$$2x^2 - 5x + \frac{3}{x}, \qquad 9 - \sqrt{x}, \quad \text{and} \quad \frac{x+1}{x^4+5}$$

are not polynomials, because they cannot be written in the form $a_nx^n + a_{n-1}x^{n-1} + \cdots + a_1x + a_0$, where the exponents are all nonnegative integers and the coefficients are all real numbers.

Addition and Subtraction

If two terms of an expression have the same variables raised to the same powers, they are called **like terms,** or **similar terms.** We can **combine,** or **collect like terms** using the distributive property. For example, $3y^2$ and $5y^2$ are like terms and

$$3y^2 + 5y^2 = (3 + 5)y^2 = 8y^2.$$

We add or subtract polynomials by combining like terms.

Example 4 Add or subtract each of the following.

a) $(-5x^3 + 3x^2 - x) + (12x^3 - 7x^2 + 3)$

b) $(6x^2y^3 - 9xy) - (5x^2y^3 - 4xy)$

SOLUTION

a) $(-5x^3 + 3x^2 - x) + (12x^3 - 7x^2 + 3)$

$= (-5x^3 + 12x^3) + (3x^2 - 7x^2) - x + 3$ — Rearranging using the commutative and associative properties

$= (-5 + 12)x^3 + (3 - 7)x^2 - x + 3$ — Using the distributive property

$= 7x^3 - 4x^2 - x + 3$

b) We can subtract by adding an opposite:

$(6x^2y^3 - 9xy) - (5x^2y^3 - 4xy)$

$= (6x^2y^3 - 9xy) + (-5x^2y^3 + 4xy)$ — Adding the opposite of $5x^2y^3 - 4xy$

$= 6x^2y^3 - 9xy - 5x^2y^3 + 4xy$

$= x^2y^3 - 5xy.$ — Combining like terms

Multiplication

Multiplication of polynomials is based on the distributive property. For example,

$$(x + 4)(x + 3) = (x + 4)x + (x + 4)3$$ Using the distributive property

$$= x^2 + 4x + 3x + 12$$ Using the distributive property two more times

$$= x^2 + 7x + 12.$$ Combining like terms

We can find the product of two binomials, as we did above, by multiplying the **F**irst terms, then the **O**uter terms, then the **I**nner terms, then the **L**ast terms. Then we collect like terms, if possible. This procedure is sometimes called **FOIL**.

Example 5 Multiply: $(2x - 7)(3x + 4)$.

SOLUTION

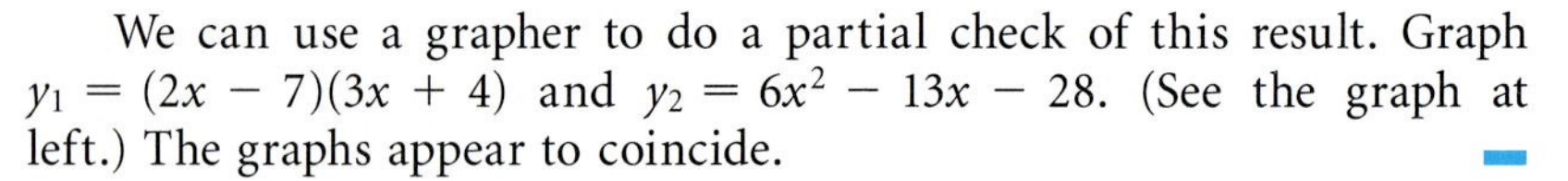

$$(2x - 7)(3x + 4) = 6x^2 + 8x - 21x - 28$$
$$= 6x^2 - 13x - 28$$

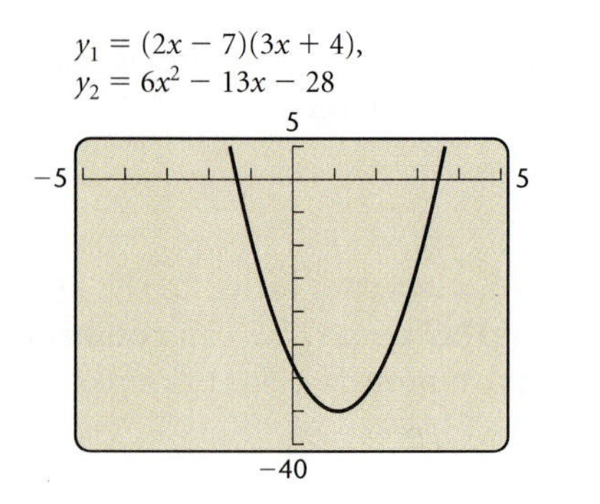

We can use a grapher to do a partial check of this result. Graph $y_1 = (2x - 7)(3x + 4)$ and $y_2 = 6x^2 - 13x - 28$. (See the graph at left.) The graphs appear to coincide.

In general, to multiply two polynomials, we multiply each term of one by each term of the other and add the products. One way to do this is to use columns.

Example 6 Multiply: $(4x^4y - 7x^2y + 3y)(2y - 3x^2y)$.

SOLUTION We can use columns to organize our work.

$$\begin{array}{rrrrl} & 4x^4y & -\ 7x^2y & +\ 3y & \\ & & 2y & -\ 3x^2y & \\ \hline -12x^6y^2 & +\ 21x^4y^2 & -\ \ 9x^2y^2 & & \text{Multiplying by } -3x^2y \\ & 8x^4y^2 & -\ 14x^2y^2 & +\ 6y^2 & \text{Multiplying by } 2y \\ \hline -12x^6y^2 & +\ 29x^4y^2 & -\ 23x^2y^2 & +\ 6y^2 & \text{Adding} \end{array}$$

Interactive Discovery

Graph $y_1 = (x + 3)^2$, $y_2 = x^2 + 9$, $y_3 = x^2 - 9$, $y_4 = x^2 + 6x + 9$, and $y_5 = (x - 3)(x + 3)$. Can you discover any identities?

We can use FOIL to find some special products.

Special Products of Binomials

$(A + B)^2 = A^2 + 2AB + B^2$	Square of a sum
$(A - B)^2 = A^2 - 2AB + B^2$	Square of a difference
$(A + B)(A - B) = A^2 - B^2$	Product of a sum and a difference

Example 7 Multiply each of the following.

a) $(4x + 1)^2$ b) $(3y^2 - 2)^2$ c) $(x^2 + 3y)(x^2 - 3y)$

SOLUTION

a) $(4x + 1)^2 = (4x)^2 + 2 \cdot 4x \cdot 1 + 1^2 = 16x^2 + 8x + 1$

b) $(3y^2 - 2)^2 = (3y^2)^2 - 2 \cdot 3y^2 \cdot 2 + 2^2 = 9y^4 - 12y^2 + 4$

c) $(x^2 + 3y)(x^2 - 3y) = (x^2)^2 - (3y)^2 = x^4 - 9y^2$

R.3 Exercise Set

Determine the terms and the degree of each polynomial.

1. $-5y^4 + 3y^3 + 7y^2 - y - 4$

2. $2m^3 - m^2 - 4m + 11$

3. $3a^4b - 7a^3b^3 + 5ab - 2$

4. $6p^3q^2 - p^2q^4 - 3pq^2 + 5$

Perform the operations indicated.

5. $(5x^2y - 2xy^2 + 3xy - 5) + (-2x^2y - 3xy^2 + 4xy + 7)$

6. $(6x^2y - 3xy^2 + 5xy - 3) + (-4x^2y - 4xy^2 + 3xy + 8)$

7. $(2x + 3y + z - 7) + (4x - 2y - z + 8) + (-3x + y - 2z - 4)$

8. $(2x^2 + 12xy - 11) + (6x^2 - 2x + 4) + (-x^2 - y - 2)$

9. $(3x^2 - 2x - x^3 + 2) - (5x^2 - 8x - x^3 + 4)$

10. $(5x^2 + 4xy - 3y^2 + 2) - (9x^2 - 4xy + 2y^2 - 1)$

11. $(x^4 - 3x^2 + 4x) - (3x^3 + x^2 - 5x + 3)$

12. $(2x^4 - 3x^2 + 7x) - (5x^3 + 2x^2 - 3x + 5)$

13. $(a - b)(2a^3 - ab + 3b^2)$

14. $(n + 1)(n^2 - 6n - 4)$

15. $(x + 5)(x - 3)$

16. $(y - 4)(y + 1)$

17. $(2a + 3)(a + 5)$

18. $(3b + 1)(b - 2)$

19. Use a grapher to check your answer to Exercise 17.

20. Use a grapher to check your answer to Exercise 18.

21. $(2x + 3y)(2x + y)$

22. $(2a - 3b)(2a - b)$

23. $(x + 5)^2$

24. $(x + 7)^2$

25. $(5y - 3)^2$

26. $(3y - 2)^2$

27. Use a grapher to check your answer to Exercise 25.

28. Use a grapher to check your answer to Exercise 26.

29. $(2x + 3y)^2$

30. $(5x + 2y)^2$

31. $(2x^2 - 3y)^2$

32. $(4x^2 - 5y)^2$

33. $(a + 3)(a - 3)$

34. $(b + 4)(b - 4)$

35. Use a grapher to check your answer to Exercise 33.

36. Use a grapher to check your answer to Exercise 34.

37. $(3x - 2y)(3x + 2y)$

38. $(3x + 5y)(3x - 5y)$

39. $(2x + 3y + 4)(2x + 3y - 4)$

40. $(5x + 2y + 3)(5x + 2y - 3)$

41. $(x + 1)(x - 1)(x^2 + 1)$

42. $(y - 2)(y + 2)(y^2 + 4)$

Synthesis

43. ◆ Is the sum of two polynomials of degree n always a polynomial of degree n? Why or why not?

44. ◆ Explain how you would convince a classmate that $(A + B)^2 \neq A^2 + B^2$.

Multiply. Assume that all exponents are natural numbers.

45. $(a^n + b^n)(a^n - b^n)$

46. $(t^a + 4)(t^a - 7)$

47. $(a^n + b^n)^2$

48. $(x^{3m} - t^{5n})^2$

49. $(x - 1)(x^2 + x + 1)(x^3 + 1)$

50. $[(2x - 1)^2 - 1]^2$

51. $(x^{a-b})^{a+b}$

52. $(t^{m+n})^{m+n} \cdot (t^{m-n})^{m-n}$

53. $(a + b + c)^2$

54. $(a + b + c)^3$

55. $(a + b)^4$

R.4 Factoring

- *Factor polynomials by removing a common factor.*
- *Factor special products of polynomials.*

To factor a polynomial, we do the reverse of multiplying; that is, we find an equivalent expression that is a product.

Terms with Common Factors

When a polynomial is to be factored, we should always look first to factor out a factor that is common to all the terms using the distributive law. We usually look for the constant common factor with the largest absolute value and for variables with the largest exponent common to all the terms. In this sense, we factor out the "largest" common factor.

Example 1 Factor each of the following.

a) $15 + 10x - 5x^2$

b) $12x^2y^2 - 20x^3y$

SOLUTION

a) $15 + 10x - 5x^2 = 5 \cdot 3 + 5 \cdot 2x - 5 \cdot x^2 = 5(3 + 2x - x^2)$

We can check by multiplying: $5(3 + 2x - x^2) = 15 + 10x - 5x^2$.

b) $12x^2y^2 - 20x^3y = 4x^2y(3y - 5x)$

Note that there are several factors common to the terms of $12x^2y^2 - 20x^3y$, but $4x^2y$ is the "largest" of these.

In some polynomials, pairs of terms have a common factor that can be removed in a process called **factoring by grouping**.

Example 2 Factor: $x^3 + 3x^2 - 5x - 15$.

SOLUTION We have

$$x^3 + 3x^2 - 5x - 15 = x^2(x + 3) - 5(x + 3)$$
$$= (x + 3)(x^2 - 5).$$

To check graphically, graph $y_1 = x^3 + 3x^2 - 5x - 15$ and $y_2 = (x + 3)(x^2 - 5)$.

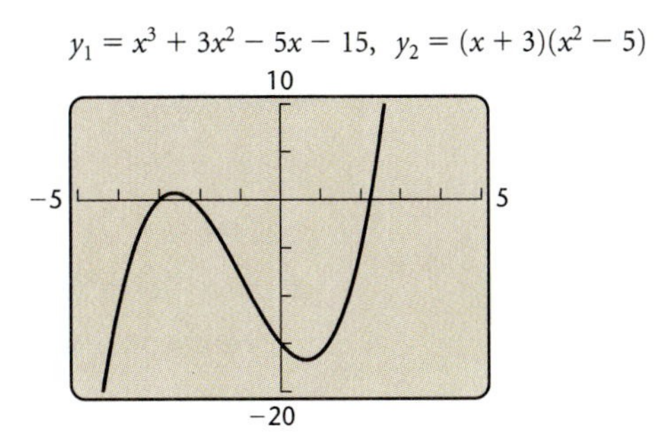

The graphs appear to coincide.

Trinomials

Some trinomials can be factored into the product of two binomials. To factor a trinomial of the form $x^2 + bx + c$, we look for two numbers with a product of c and a sum of b.

Example 3 Factor each of the following.

a) $x^2 + 5x + 6$ **b)** $y^2 - 2y - 24$

SOLUTION

a) We look for two numbers with a product of 6 and a sum of 5. They are 2 and 3. Then

$$x^2 + 5x + 6 = (x + 2)(x + 3).$$ **To check, multiply or graph.**

b) We look for two numbers with a product of -24 and a sum of -2:

$$y^2 - 2y - 24 = (y - 6)(y + 4).$$ **Note that $(-6)(4) = -24$ and $-6 + 4 = -2$.**

The next example illustrates a method for factoring a trinomial of the form $ax^2 + bx + c$, $a \neq 1$.

Example 4 Factor: $3x^2 - 10x - 8$.

SOLUTION We look for factors $px + q$ and $rx + s$ for which $px \cdot rx = 3x^2$ and $q \cdot s = -8$. When we multiply the inside terms, then the outside terms, and add, we must have $-10x$. By trial, we determine the factorization:

$$3x^2 - 10x - 8 = (3x + 2)(x - 4).$$ **To check, multiply or graph.**

Special Factorizations

We reverse the equation $(A + B)(A - B) = A^2 - B^2$ to factor a **difference of squares.**

$$A^2 - B^2 = (A + B)(A - B)$$

Example 5 Factor each of the following.

a) $x^2 - 16$ b) $9a^2 - 25$ c) $6x^4 - 6y^4$

SOLUTION

a) $x^2 - 16 = (x + 4)(x - 4)$

To check graphically, graph $y_1 = x^2 - 16$ and $y_2 = (x + 4)(x - 4)$.

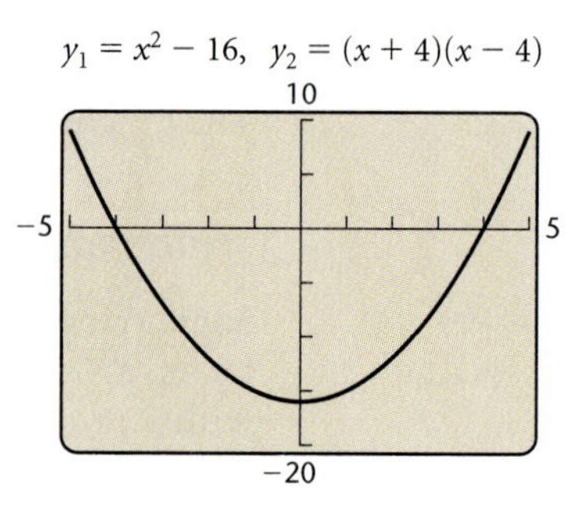

The graphs appear to coincide.

b) $9a^2 - 25 = (3a + 5)(3a - 5)$ To check, multiply or graph.

c) $6x^4 - 6y^4 = 6(x^4 - y^4)$

$= 6(x^2 + y^2)(x^2 - y^2)$

$= 6(x^2 + y^2)(x + y)(x - y)$ Because none of these factors can be factored further, we have *factored completely.*

The rules for squaring binomials can be reversed to factor trinomial squares.

$$A^2 + 2AB + B^2 = (A + B)^2$$
$$A^2 - 2AB + B^2 = (A - B)^2$$

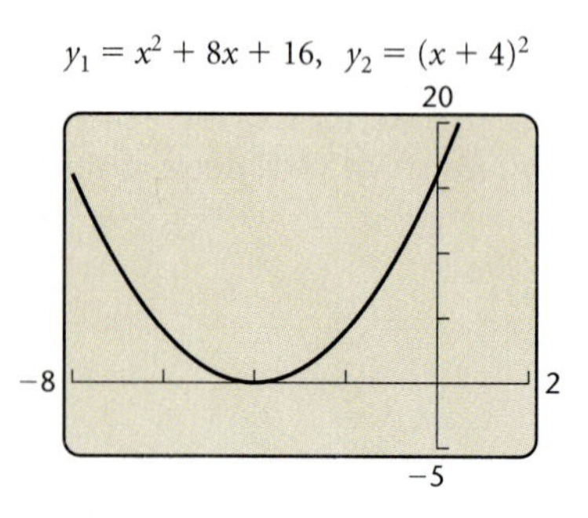

FIGURE 1

Example 6 Factor each of the following.

a) $x^2 + 8x + 16$

b) $25y^2 - 30y + 9$

SOLUTION

a) $x^2 + 8x + 16 = (x + 4)^2$ See the graphical check in Fig. 1.

b) $25y^2 - 30y + 9 = (5y - 3)^2$ To check, multiply or graph.

We can use the following rules to factor a **sum** or a **difference of cubes**:

$$A^3 + B^3 = (A + B)(A^2 - AB + B^2);$$
$$A^3 - B^3 = (A - B)(A^2 + AB + B^2).$$

These rules can be verified by multiplying.

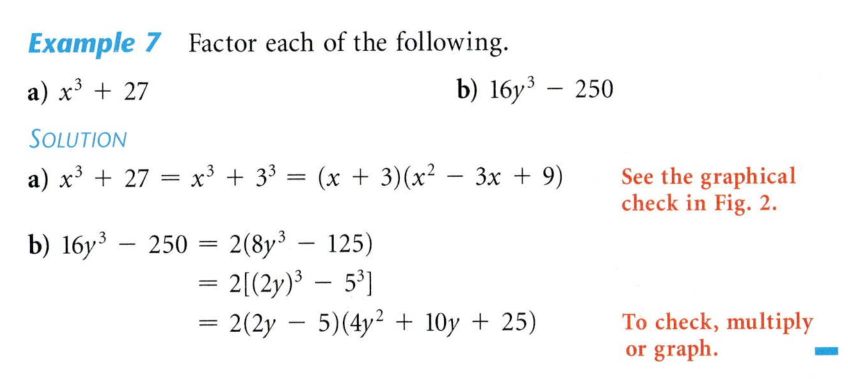

FIGURE 2

Example 7 Factor each of the following.

a) $x^3 + 27$

b) $16y^3 - 250$

SOLUTION

a) $x^3 + 27 = x^3 + 3^3 = (x + 3)(x^2 - 3x + 9)$

See the graphical check in Fig. 2.

b) $16y^3 - 250 = 2(8y^3 - 125)$
$= 2[(2y)^3 - 5^3]$
$= 2(2y - 5)(4y^2 + 10y + 25)$

To check, multiply or graph.

Not all polynomials can be factored into polynomials with integer coefficients. An example is $x^2 - x + 7$. There are no real factors of 7 whose sum is -1. In such a case we say that the polynomial is "not factorable," or **prime**.

R.4 Exercise Set

Factor out a common factor.

1. $2x - 10$

2. $7y + 42$

3. $3x^4 - 9x^2$

4. $20y^2 - 5y^5$

5. $4a^2 - 12a + 16$

6. $6n^2 + 24n - 18$

7. $a(b - 2) + c(b - 2)$

8. $a(x^2 - 3) - 2(x^2 - 3)$

Factor by grouping.

9. $x^3 + 3x^2 + 6x + 18$

10. $3x^3 + x^2 - 18x - 6$

11. $y^3 - 3y^2 - 4y + 12$

12. $p^3 - 2p^2 - 9p + 18$

13. Use a grapher to check your answer to Exercise 9.

14. Use a grapher to check your answer to Exercise 10.

Factor the trinomial.

15. $p^2 + 6p + 8$

16. $w^2 - 7w + 10$

17. $2n^2 + 9n - 56$

18. $3y^2 + 7y - 20$

19. $y^4 - 4y^2 - 21$

20. $m^4 - m^2 - 90$

21. Use a grapher to check your answer to Exercise 15.

22. Use a grapher to check your answer to Exercise 16.

Factor the difference of squares.

23. $9x^2 - 25$

24. $16x^2 - 9$

25. $4xy^4 - 4xz^2$

26. $5x^4y - 5yz^4$

Factor the trinomial square.

27. $y^2 - 6y + 9$

28. $x^2 + 8x + 16$

29. $1 - 8x + 16x^2$
30. $1 + 10x + 25x^2$
31. Use a grapher to check your answer to Exercise 29.
32. Use a grapher to check your answer to Exercise 30.

Factor the sum or difference of cubes.

33. $x^3 + 8$
34. $y^3 - 64$
35. $m^3 - 1$
36. $n^3 + 216$
37. Use a grapher to check your answer to Exercise 35.
38. Use a grapher to check your answer to Exercise 36.

Factor completely.

39. $18a^2b - 15ab^2$
40. $4x^2y + 12xy^2$
41. $x^3 - 4x^2 + 5x - 20$
42. $y^3 + 3y^2 - 3y - 9$
43. $8x^2 - 32$
44. $6y^2 - 6$
45. $4x^2 - 5$
46. $16x^2 - 7$
47. Use a grapher to check your answer to Exercise 41.
48. Use a grapher to check your answer to Exercise 42.
49. $x^2 + 9x + 20$
50. $y^2 + y - 6$
51. $y^2 - 6y + 5$
52. $x^2 - 4x - 21$
53. $2a^2 + 9a + 4$
54. $3b^2 - b - 2$
55. $6x^2 + 7x - 3$
56. $8x^2 + 2x - 15$
57. $y^2 - 18y + 81$
58. $n^2 + 2n + 1$
59. $x^2y^2 - 14xy + 49$
60. $x^2y^2 - 16xy + 64$
61. $4ax^2 + 20ax - 56a$
62. $21x^2y + 2xy - 8y$
63. $3z^3 - 24$
64. $4t^3 + 108$
65. $16a^7b + 54ab^7$
66. $24a^2x^4 - 375a^8x$
67. Use a grapher to check your answer to Exercise 55.
68. Use a grapher to check your answer to Exercise 56.

Synthesis

69. ◆ Under what circumstances can $A^2 + B^2$ be factored?
70. ◆ Explain how the rule for factoring a sum of cubes can be used to factor a difference of cubes.

Factor.

71. $y^4 - 84 + 5y^2$
72. $11x^2 + x^4 - 80$
73. $y^2 - \frac{8}{49} + \frac{2}{7}y$
74. $x^2 + \frac{3}{5}x - \frac{4}{25}$
75. $t^2 - 0.27 + 0.6t$
76. $0.4m - 0.05 + m^2$
77. $(x + h)^3 - x^3$
78. $(x + 0.01)^2 - x^2$
79. $(x + 3)^2 - 2(x + 3) - 35$
80. $(y - 4)^2 + 5(y - 4) - 24$
81. $3(a - b)^2 + 10(a - b) - 8$
82. $6(2p + q)^2 - 5(2p + q) - 25$

Factor. Assume that variables in exponents represent natural numbers.

83. $x^{2n} + 5x^n - 24$
84. $4x^{2n} - 4x^n - 3$
85. $x^2 + ax + bx + ab$
86. $bdy^2 + ady + bcy + ac$
87. $\frac{1}{4}t^2 - \frac{2}{5}t + \frac{4}{25}$
88. $\frac{4}{27}r^2 + \frac{5}{9}rs + \frac{1}{12}s^2 - \frac{1}{3}rs$
89. $25y^{2m} - (x^{2n} - 2x^n + 1)$
90. $4x^{4a} + 12x^{2a} + 10x^{2a} + 30$
91. $3x^{3n} - 24y^{3m}$
92. $x^{6a} - t^{3b}$
93. $(y - 1)^4 - (y - 1)^2$
94. $x^6 - 2x^5 + x^4 - x^2 + 2x - 1$

R.5 Rational Expressions

- *Determine the domain of a rational expression.*
- *Simplify rational expressions.*
- *Multiply, divide, add, or subtract rational expressions.*
- *Simplify complex rational expressions.*

A **rational expression** is the quotient of two polynomials. For example,

$$\frac{3}{5}, \quad \frac{2}{x - 3}, \quad \text{and} \quad \frac{x^2 - 4}{x^2 - 4x - 5}$$

are rational expressions.

The Domain of a Rational Expression

The **domain** of an algebraic expression is the set of all real numbers for which the expression is defined. Since division by zero is not defined, any number that makes the denominator zero is not in the domain of a rational expression.

Example 1 Find the domain of each of the following.

a) $\dfrac{2}{x - 3}$ b) $\dfrac{x^2 - 4}{x^2 - 4x - 5}$

SOLUTION

a) Since $x - 3 = 0$ when $x = 3$, the domain of $2/(x - 3)$ is the set of all real numbers except 3. As a partial check, graph $y = 2/(x - 3)$ using DOT mode.

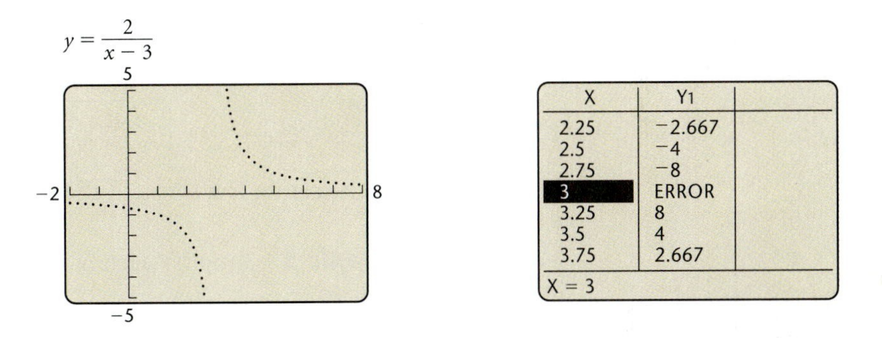

Note that there does not appear to be a point on the graph that has 3 as a first coordinate. Also note that the grapher's TABLE feature produces an error message for $x = 3$.

b) To determine the domain of $(x^2 - 4)/(x^2 - 4x - 5)$, we first factor the denominator:

$$\frac{x^2 - 4}{x^2 - 4x - 5} = \frac{x^2 - 4}{(x + 1)(x - 5)}.$$

The factor $x + 1$ is 0 when $x = -1$ and the factor $x - 5$ is 0 when $x = 5$. Since $(x + 1)(x - 5) = 0$ when $x = -1$ or $x = 5$, the domain is the set of all real numbers except -1 and 5.

We can describe the domains found in Example 1 using **set-builder notation**. For example, we write "The set of all real numbers x such that x is not equal to 3" as

$$\{x | x \text{ is a real number and } x \neq 3\}.$$

Similarly, we write "The set of all real numbers x such that x is not equal to -1 and x is not equal to 5" as

$$\{x | x \text{ is a real number and } x \neq -1 \text{ and } x \neq 5\}.$$

Simplifying, Multiplying, and Dividing Rational Expressions

To simplify rational expressions, we use the fact that

$$\frac{a \cdot c}{b \cdot c} = \frac{a}{b} \cdot \frac{c}{c} = \frac{a}{b} \cdot 1 = \frac{a}{b}.$$

Example 2 Simplify: $\dfrac{9x^2 + 6x - 3}{12x^2 - 12}$.

SOLUTION

$$\frac{9x^2 + 6x - 3}{12x^2 - 12} = \frac{3(x + 1)(3x - 1)}{3 \cdot 4(x + 1)(x - 1)}$$ Factoring the numerator and the denominator

$$= \frac{3(x + 1)}{3(x + 1)} \cdot \frac{3x - 1}{4(x - 1)}$$ Factoring the rational expression

$$= 1 \cdot \frac{3x - 1}{4(x - 1)}$$ $\dfrac{3(x + 1)}{3(x + 1)} = 1$

$$= \frac{3x - 1}{4(x - 1)}$$ Removing a factor of 1

Canceling is a shortcut that is often used to remove a factor of 1.

Example 3 Simplify each of the following.

a) $\dfrac{4x^3 + 16x^2}{2x^3 + 6x^2 - 8x}$

b) $\dfrac{2 - x}{x^2 + x - 6}$

SOLUTION

a) $$\frac{4x^3 + 16x^2}{2x^3 + 6x^2 - 8x} = \frac{2 \cdot 2 \cdot x \cdot x(x + 4)}{2 \cdot x(x + 4)(x - 1)}$$ Factoring the numerator and the denominator

$$= \frac{\not{2} \cdot 2 \cdot \not{x} \cdot x(\not{x + 4})}{\not{2} \cdot \not{x}(\not{x + 4})(x - 1)}$$ Removing a factor of 1: $\dfrac{2x(x + 4)}{2x(x + 4)} = 1$

$$= \frac{2x}{x - 1}$$

b) $$\frac{2 - x}{x^2 + x - 6} = \frac{2 - x}{(x + 3)(x - 2)}$$ Factoring the denominator

$$= \frac{-1(x - 2)}{(x + 3)(x - 2)}$$ $2 - x = -1(x - 2)$

$$= \frac{-1(\not{x - 2})}{(x + 3)(\not{x - 2})}$$ Removing a factor of 1: $\dfrac{x - 2}{x - 2} = 1$

$$= \frac{-1}{x + 3}, \text{ or } -\frac{1}{x + 3}$$

Be careful to cancel only common *factors*. Like terms that are *not* factors cannot be canceled. For example,

$$\frac{x + 4}{x} \neq 4,$$

because x is not a factor of the numerator.

To confirm this graphically, graph $y_1 = \dfrac{x + 4}{x}$ and $y_2 = 4$.

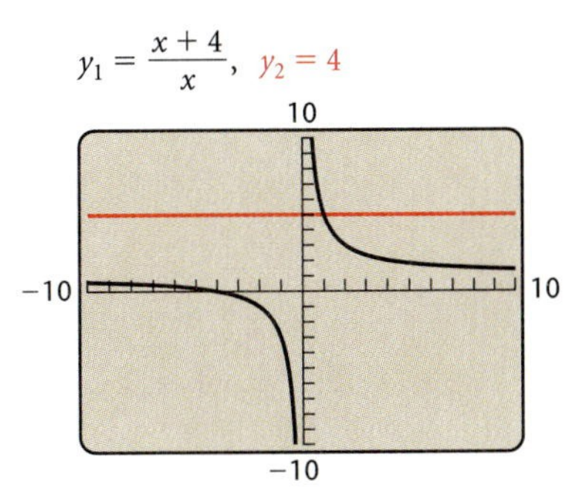

The graphs are different.

In Example 3(b), we saw that

$$\frac{2 - x}{x^2 + x - 6} \quad \text{and} \quad \frac{-1}{x + 3}$$

are **equivalent expressions**. This means that they have the same value for all numbers that are in both domains.

For

$$y_1 = \frac{2 - x}{x^2 + x - 6} \quad \text{and} \quad y_2 = \frac{-1}{x + 3},$$

a table shows that y_1 and y_2 have the same values except for $x = -3$ and $x = 2$. Note that -3 is not in the domain of *either* expression, whereas 2 is not in the domain of *both* expressions.

X	Y1	Y2
−3	ERROR	ERROR
−2	−1	−1
−1	−.5	−.5
0	−.3333	−.3333
1	−.25	−.25
2	ERROR	−.2
3	−.1667	−.1667

X = 2

To multiply rational expressions, we multiply numerators and multiply denominators and, if possible, simplify the result. To divide rational expressions, we multiply the dividend by the reciprocal of the divisor and, if possible, simplify the result.

Example 4 Multiply or divide and simplify each of the following.

a) $\dfrac{x+4}{x-3} \cdot \dfrac{x^2-9}{x^2-x-2}$ b) $\dfrac{y^3-1}{y^2-1} \div \dfrac{y^2+y+1}{y^2+2y+1}$

SOLUTION

a) $\dfrac{x+4}{x-3} \cdot \dfrac{x^2-9}{x^2-x-2} = \dfrac{(x+4)(x^2-9)}{(x-3)(x^2-x-2)}$

Multiplying the numerators and the denominators

$= \dfrac{(x+4)(x+3)\cancel{(x-3)}}{\cancel{(x-3)}(x-2)(x+1)}$ Factoring and removing a factor of 1: $\dfrac{x-3}{x-3} = 1$

$= \dfrac{(x+4)(x+3)}{(x-2)(x+1)}$

b) $\dfrac{y^3-1}{y^2-1} \div \dfrac{y^2+y+1}{y^2+2y+1} = \dfrac{y^3-1}{y^2-1} \cdot \dfrac{y^2+2y+1}{y^2+y+1}$ Multiplying by the reciprocal of the divisor

$= \dfrac{(y^3-1)(y^2+2y+1)}{(y^2-1)(y^2+y+1)}$

$= \dfrac{\cancel{(y-1)}\cancel{(y^2+y+1)}\cancel{(y+1)}(y+1)}{\cancel{(y+1)}\cancel{(y-1)}\cancel{(y^2+y+1)}}$

Factoring and removing a factor of 1

$= y+1$

Adding and Subtracting Rational Expressions

When rational expressions have the same denominator, we can add or subtract by adding or subtracting the numerators and retaining the common denominator. If the denominators differ, we must find equivalent rational expressions that have a common denominator.

It is usually most efficient to find the **least common denominator** (**LCD**) of the expressions. To do so, we factor each denominator and form the product that uses each factor the greatest number of times that it occurs in any factorization.

Example 5 Add or subtract and simplify each of the following.

a) $\dfrac{x^2-4x+4}{2x^2-3x+1} + \dfrac{x+4}{2x-2}$ b) $\dfrac{x}{x^2+11x+30} - \dfrac{5}{x^2+9x+20}$

SOLUTION

a) $\dfrac{x^2-4x+4}{2x^2-3x+1} + \dfrac{x+4}{2x-2}$

$= \dfrac{x^2-4x+4}{(2x-1)(x-1)} + \dfrac{x+4}{2(x-1)}$ Factoring the denominators

The LCD is $(2x-1)(x-1)(2)$, or $2(2x-1)(x-1)$.

$$= \frac{x^2 - 4x + 4}{(2x - 1)(x - 1)} \cdot \frac{2}{2} + \frac{x + 4}{2(x - 1)} \cdot \frac{2x - 1}{2x - 1}$$

Multiplying each term by 1 to get the LCD

$$= \frac{2x^2 - 8x + 8}{(2x - 1)(x - 1)(2)} + \frac{2x^2 + 7x - 4}{2(x - 1)(2x - 1)}$$

$$= \frac{4x^2 - x + 4}{2(x - 1)(2x - 1)}$$

Adding the numerators

b) $$\frac{x}{x^2 + 11x + 30} - \frac{5}{x^2 + 9x + 20}$$

$$= \frac{x}{(x + 5)(x + 6)} - \frac{5}{(x + 5)(x + 4)}$$

The LCD is $(x + 5)(x + 6)(x + 4)$.

$$= \frac{x}{(x + 5)(x + 6)} \cdot \frac{x + 4}{x + 4} - \frac{5}{(x + 5)(x + 4)} \cdot \frac{x + 6}{x + 6}$$

Multiplying each term by 1 to get the LCD

$$= \frac{x^2 + 4x}{(x + 5)(x + 6)(x + 4)} - \frac{5x + 30}{(x + 5)(x + 4)(x - 6)}$$

$$= \frac{x^2 + 4x - 5x - 30}{(x + 5)(x + 6)(x + 4)}$$

Be sure to change the sign of *every* term in the numerator of the expression being subtracted

$$= \frac{x^2 - x - 30}{(x + 5)(x + 6)(x + 4)}$$

$$= \frac{(x + 5)(x - 6)}{(x + 5)(x + 6)(x + 4)}$$

Removing a factor of 1: $\frac{x + 5}{x + 5} = 1$

$$= \frac{x - 6}{(x + 6)(x + 4)}$$

Complex Rational Expressions

A **complex rational expression** has rational expressions in its numerator or its denominator or both.

To simplify a complex rational expression:

Method 1. Find the LCD of all the denominators within the complex rational expression. Then multiply by 1 using that LCD as the numerator and the denominator of the expression for 1.

Method 2. First add or subtract, if necessary, to get a single rational expression in both the numerator and the denominator. Then divide by multiplying by the reciprocal of the denominator.

Example 6 Simplify: $\dfrac{\dfrac{1}{a}+\dfrac{1}{b}}{\dfrac{1}{a^3}+\dfrac{1}{b^3}}$.

SOLUTION

Method 1. The LCD of the four rational expressions in the numerator and the denominator is a^3b^3.

$$\frac{\frac{1}{a}+\frac{1}{b}}{\frac{1}{a^3}+\frac{1}{b^3}} = \frac{\frac{1}{a}+\frac{1}{b}}{\frac{1}{a^3}+\frac{1}{b^3}} \cdot \frac{a^3b^3}{a^3b^3}$$

Multiplying by 1 using $\frac{a^3b^3}{a^3b^3}$

$$= \frac{\left(\frac{1}{a}+\frac{1}{b}\right)(a^3b^3)}{\left(\frac{1}{a^3}+\frac{1}{b^3}\right)(a^3b^3)}$$

$$= \frac{a^2b^3 + a^3b^2}{b^3 + a^3}$$

$$= \frac{a^2b^2(b+a)}{(b+a)(b^2 - ba + a^2)}$$

Factoring and removing a factor of 1: $\frac{b+a}{b+a} = 1$

$$= \frac{a^2b^2}{b^2 - ab + a^2}$$

Method 2. We add in the numerator and in the denominator.

$$\frac{\frac{1}{a}+\frac{1}{b}}{\frac{1}{a^3}+\frac{1}{b^3}} = \frac{\frac{1}{a}\cdot\frac{b}{b}+\frac{1}{b}\cdot\frac{a}{a}}{\frac{1}{a^3}\cdot\frac{b^3}{b^3}+\frac{1}{b^3}\cdot\frac{a^3}{a^3}}$$

← The LCD is ab.

← The LCD is a^3b^3.

$$= \frac{\frac{b}{ab}+\frac{a}{ab}}{\frac{b^3}{a^3b^3}+\frac{a^3}{a^3b^3}}$$

$$= \frac{\frac{b+a}{ab}}{\frac{b^3+a^3}{a^3b^3}}$$

We have a single rational expression in both the numerator and the denominator.

$$= \frac{b+a}{ab} \cdot \frac{a^3b^3}{b^3+a^3}$$

Multiplying by the reciprocal of the denominator

$$= \frac{(b+a)(a)(b)(a^2b^2)}{(a)(b)(b+a)(b^2 - ba + a^2)}$$

$$= \frac{a^2b^2}{b^2 - ab + a^2}$$

R.5 Exercise Set

Find the domain of each rational expression.

1. $-\dfrac{3}{4}$

2. $\dfrac{5}{8-x}$

3. $\dfrac{3x-3}{x(x-1)}$

4. $\dfrac{(x^2-4)(x+1)}{(x+2)(x^2-1)}$

5. $\dfrac{7x^2-28x+28}{(x^2-4)(x^2+3x-10)}$

6. $\dfrac{7x^2+11x-6}{x(x^2-x-6)}$

7. Use the TABLE feature of a grapher to check your answer to Exercise 3.

8. Use the TABLE feature of a grapher to check your answer to Exercise 4.

Multiply or divide and, if possible, simplify.

9. $\dfrac{x^2-y^2}{(x-y)^2}\cdot\dfrac{1}{x+y}$

10. $\dfrac{r-s}{r+s}\cdot\dfrac{r^2-s^2}{(r-s)^2}$

11. $\dfrac{x^2-2x-35}{2x^3-3x^2}\cdot\dfrac{4x^3-9x}{7x-49}$

12. $\dfrac{x^2+2x-35}{3x^3-2x^2}\cdot\dfrac{9x^3-4x}{7x+49}$

13. $\dfrac{a^2-a-6}{a^2-7a+12}\cdot\dfrac{a^2-2a-8}{a^2-3a-10}$

14. $\dfrac{a^2-a-12}{a^2-6a+8}\cdot\dfrac{a^2+a-6}{a^2-2a-24}$

15. $\dfrac{m^2-n^2}{r+s}\div\dfrac{m-n}{r+s}$

16. $\dfrac{a^2-b^2}{x-y}\div\dfrac{a+b}{x-y}$

17. $\dfrac{3x+12}{2x-8}\div\dfrac{(x+4)^2}{(x-4)^2}$

18. $\dfrac{a^2-a-2}{a^2-a-6}\div\dfrac{a^2-2a}{2a+a^2}$

19. $\dfrac{x^2-y^2}{x^3-y^3}\cdot\dfrac{x^2+xy+y^2}{x^2+2xy+y^2}$

20. $\dfrac{c^3+8}{c^2-4}\div\dfrac{c^2-2c+4}{c^2-4c+4}$

21. $\dfrac{(x-y)^2-z^2}{(x+y)^2-z^2}\div\dfrac{x-y+z}{x+y-z}$

22. $\dfrac{(a+b)^2-9}{(a-b)^2-9}\cdot\dfrac{a-b-3}{a+b+3}$

23. Use a grapher to check your answer to Exercise 11.

24. Use a grapher to check your answer to Exercise 12.

Add or subtract and, if possible, simplify.

25. $\dfrac{3}{2a+3}+\dfrac{2a}{2a+3}$

26. $\dfrac{a-3b}{a+b}+\dfrac{a+5b}{a+b}$

27. $\dfrac{y}{y-1}+\dfrac{2}{1-y}$

28. $\dfrac{a}{a-b}+\dfrac{b}{b-a}$

29. $\dfrac{x}{2x-3y}-\dfrac{y}{3y-2x}$

30. $\dfrac{3a}{3a-2b}-\dfrac{2a}{2b-3a}$

31. $\dfrac{3}{x+2}+\dfrac{2}{x^2-4}$

32. $\dfrac{5}{a-3}-\dfrac{2}{a^2-9}$

33. $\dfrac{y}{y^2-y-20}-\dfrac{2}{y+4}$

34. $\dfrac{6}{y^2+6y+9}-\dfrac{5}{y+3}$

35. $\dfrac{3}{x+y}+\dfrac{x-5y}{x^2-y^2}$

36. $\dfrac{a^2+1}{a^2-1}-\dfrac{a-1}{a+1}$

37. $\dfrac{9x+2}{3x^2-2x-8}+\dfrac{7}{3x^2+x-4}$

38. $\dfrac{3y}{y^2-7y+10}-\dfrac{2y}{y^2-8y+15}$

39. $\dfrac{5a}{a-b}+\dfrac{ab}{a^2-b^2}+\dfrac{4b}{a+b}$

40. $\frac{6a}{a-b} - \frac{3b}{b-a} + \frac{5}{a^2-b^2}$

41. $\frac{7}{x+2} - \frac{x+8}{4-x^2} + \frac{3x-2}{4-4x+x^2}$

42. $\frac{6}{x+3} - \frac{x+4}{9-x^2} + \frac{2x-3}{9-6x+x^2}$

43. $\frac{1}{x+1} + \frac{x}{2-x} + \frac{x^2+2}{x^2-x-2}$

44. $\frac{x-1}{x-2} - \frac{x+1}{x+2} - \frac{x-6}{4-x^2}$

45. Use a grapher to check your answer to Exercise 31.

46. Use a grapher to check your answer to Exercise 32.

Simplify.

47. $\dfrac{\frac{x^2-y^2}{xy}}{\frac{x-y}{y}}$

48. $\dfrac{\frac{a-b}{b}}{\frac{a^2-b^2}{ab}}$

49. $\dfrac{a-a^{-1}}{a+a^{-1}}$

50. $\dfrac{a-\frac{a}{b}}{b-\frac{b}{a}}$

51. $\dfrac{c+\frac{8}{c^2}}{1+\frac{2}{c}}$

52. $\dfrac{x^{-1}+y^{-1}}{x^{-3}+y^{-3}}$

53. $\dfrac{x^2+xy+y^2}{\frac{x^2}{y}-\frac{y^2}{x}}$

54. $\dfrac{\frac{a^2}{b}+\frac{b^2}{a}}{a^2-ab+b^2}$

55. $\dfrac{\frac{x}{y}-\frac{y}{x}}{\frac{1}{y}+\frac{1}{x}}$

56. $\dfrac{\frac{a}{b}-\frac{b}{a}}{\frac{1}{a}-\frac{1}{b}}$

57. $\dfrac{\frac{1}{x-3}+\frac{2}{x+3}}{\frac{3}{x-1}-\frac{4}{x+2}}$

58. $\dfrac{\frac{5}{x+1}-\frac{3}{x-2}}{\frac{1}{x-5}+\frac{2}{x+2}}$

59. $\dfrac{\frac{a}{1-a}+\frac{1+a}{a}}{\frac{1-a}{a}+\frac{a}{1+a}}$

60. $\dfrac{\frac{1-x}{x}+\frac{x}{1+x}}{\frac{1+x}{x}+\frac{x}{1-x}}$

61. $\dfrac{\frac{1}{a^2}+\frac{2}{ab}+\frac{1}{b^2}}{\frac{1}{a^2}-\frac{1}{b^2}}$

62. $\dfrac{\frac{1}{x^2}-\frac{1}{y^2}}{\frac{1}{x^2}-\frac{2}{xy}+\frac{1}{y^2}}$

Synthesis

63. ◆ When adding or subtracting rational expressions, we can always find a common denominator by forming the product of all the denominators. Explain why it is usually preferable to find the least common denominator.

64. ◆ How would you determine which method to use for simplifying a particular complex rational expression?

Simplify.

65. $\dfrac{(x+h)^2-x^2}{h}$

66. $\dfrac{\frac{1}{x+h}-\frac{1}{x}}{h}$

67. $\dfrac{(x+h)^3-x^3}{h}$

68. $\dfrac{\frac{1}{(x+h)^2}-\frac{1}{x^2}}{h}$

69. $\left[\dfrac{\frac{x+1}{x-1}+1}{\frac{x+1}{x-1}-1}\right]^5$

70. $1+\cfrac{1}{1+\cfrac{1}{1+\cfrac{1}{1+\frac{1}{x}}}}$

Perform the indicated operations and, if possible, simplify.

71. $\frac{n(n+1)(n+2)}{2\cdot 3} + \frac{(n+1)(n+2)}{2}$

72. $\frac{n(n+1)(n+2)(n+3)}{2\cdot 3\cdot 4} + \frac{(n+1)(n+2)(n+3)}{2\cdot 3}$

73. $\frac{x^2-9}{x^3+27} \cdot \frac{5x^2-15x+45}{x^2-2x-3} + \frac{x^2+x}{4+2x}$

74. $\frac{x^2+2x-3}{x^2-x-12} \div \frac{x^2-1}{x^2-16} - \frac{2x+1}{x^2+2x+1}$

R.6 Radical Notation and Rational Exponents

- *Simplify radical expressions.*
- *Rationalize denominators or numerators in rational expressions.*
- *Convert between exponential and radical notation.*
- *Simplify expressions with rational exponents.*
- *Factor expressions with rational exponents.*

A number c is said to be a **square root** of a if $c^2 = a$. Thus, 3 is a square root of 9, because $3^2 = 9$, and -3 is also a square root of 9, because $(-3)^2 = 9$. Similarly, 5 is a third root (called a **cube root**) of 125, because $5^3 = 125$. The number 125 has no other real-number cube root.

nth Root

A number c is said to be an *nth root* of a if $c^n = a$.

The symbol $\sqrt{a}$ denotes the nonnegative square root of a, and the symbol $\sqrt[3]{a}$ denotes the real-number cube root of a. The symbol $\sqrt[n]{a}$ denotes the nth root of a, that is, a number whose nth power is a. The symbol $\sqrt[n]{}$ is called a **radical**, and the expression under the radical is called the **radicand**. The number n (which is omitted when it is 2) is called the **index**. Examples of roots for $n = 3$, 4, and 2, respectively, are

$$\sqrt[3]{125}, \quad \sqrt[4]{16}, \quad \text{and} \quad \sqrt{3600}.$$

Any real number has only one real-number odd root. Any positive number has two square roots, one positive and one negative. The same is true for fourth roots or roots with any even index. The positive root is called the **principal root**. When an expression such as $\sqrt{4}$ or $\sqrt[6]{23}$ is used, it is understood to represent the principal (nonnegative) root. To denote a negative root, we use $-\sqrt{4}$, $-\sqrt[6]{23}$, and so on.

Example 1 Simplify each of the following.

a) $\sqrt{36}$ b) $-\sqrt{36}$ c) $\sqrt[5]{\dfrac{32}{243}}$

d) $\sqrt[3]{-8}$ e) $\sqrt[4]{-16}$

SOLUTION

a) $\sqrt{36} = 6$, because $6^2 = 36$.

b) $-\sqrt{36} = -6$, because $6^2 = 36$ and $-(\sqrt{36}) = -(6) = -6$.

c) $\sqrt[5]{\dfrac{32}{243}} = \dfrac{2}{3}$, because $\left(\dfrac{2}{3}\right)^5 = \dfrac{2^5}{3^5} = \dfrac{32}{243}$.

d) $\sqrt[3]{-8} = -2$, because $(-2)^3 = -8$.

e) $\sqrt[4]{-16}$ is not a real number, because we cannot find a real number that can be raised to the fourth power to get -16.

We can generalize Example 1(e) and say that when a is negative and n is even, $\sqrt[n]{a}$ is not a real number. For example, $\sqrt{-4}$ and $\sqrt[4]{-81}$ are not real numbers.

Simplifying Radical Expressions

Consider the expression $\sqrt{(-3)^2}$. This is equivalent to $\sqrt{9}$, or 3. Similarly, $\sqrt{3^2} = \sqrt{9} = 3$. This illustrates the first of several properties of radicals, listed below.

Properties of Radicals

Let a and b be any real numbers or expressions for which the given roots exist. For any natural numbers m and n $(n \neq 1)$:

1. If n is even, $\sqrt[n]{a^n} = |a|$.

2. If n is odd, $\sqrt[n]{a^n} = a$.

3. $\sqrt[n]{a} \cdot \sqrt[n]{b} = \sqrt[n]{ab}$.

4. $\sqrt[n]{\dfrac{a}{b}} = \dfrac{\sqrt[n]{a}}{\sqrt[n]{b}} \quad (b \neq 0)$.

5. $\sqrt[n]{a^m} = (\sqrt[n]{a})^m$.

To illustrate the first property, graph $y_1 = \sqrt{x^2}$ and $y_2 = |x|$. We can illustrate the second property in a similar manner, graphing $y_1 = \sqrt[3]{x^3}$ and $y_2 = x$.

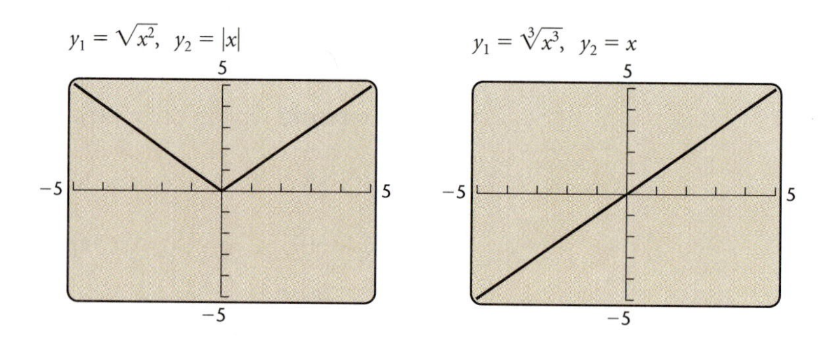

The graphs appear to coincide in each case.

Example 2 Simplify each of the following.

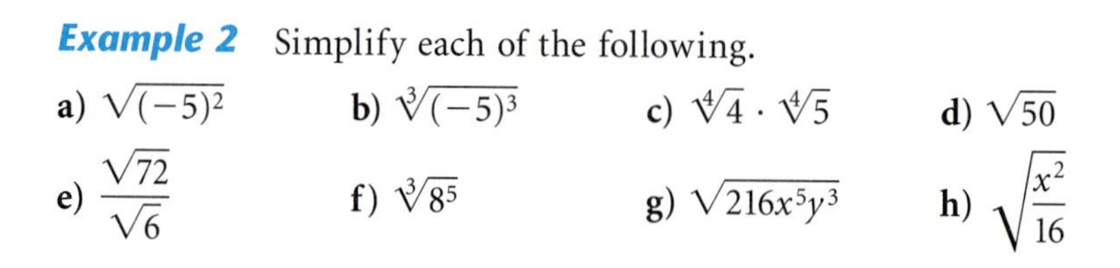

a) $\sqrt{(-5)^2}$ b) $\sqrt[3]{(-5)^3}$ c) $\sqrt[4]{4} \cdot \sqrt[4]{5}$ d) $\sqrt{50}$

e) $\dfrac{\sqrt{72}}{\sqrt{6}}$ f) $\sqrt[3]{8^5}$ g) $\sqrt{216x^5y^3}$ h) $\sqrt{\dfrac{x^2}{16}}$

SOLUTION

a) $\sqrt{(-5)^2} = |-5| = 5$ **Using Property 1**

b) $\sqrt[3]{(-5)^3} = -5$ **Using Property 2**

c) $\sqrt[4]{4} \cdot \sqrt[4]{5} = \sqrt[4]{4 \cdot 5} = \sqrt[4]{20}$ **Using Property 3**

d) $\sqrt{50} = \sqrt{25 \cdot 2} = \sqrt{25} \cdot \sqrt{2} = 5\sqrt{2}$ **Using Property 3**

e) $\dfrac{\sqrt{72}}{\sqrt{6}} = \sqrt{\dfrac{72}{6}}$ **Using Property 4**

$= \sqrt{12} = \sqrt{4 \cdot 3} = \sqrt{4}\sqrt{3}$ **Using Property 3**

$= 2\sqrt{3}$

f) $\sqrt[3]{8^5} = (\sqrt[3]{8})^5$ **Using Property 5**

$= 2^5 = 32$

g) $\sqrt{216x^5y^3} = \sqrt{36 \cdot 6 \cdot x^4 \cdot x \cdot y^2 \cdot y}$

$= \sqrt{36x^4y^2}\sqrt{6xy}$ **Using Property 3**

$= |6x^2y|\sqrt{6xy}$ **Using Property 1**

$= 6x^2|y|\sqrt{6xy}$ **$6x^2$ cannot be negative, so absolute value signs are not needed for it.**

h) $\sqrt{\dfrac{x^2}{16}} = \dfrac{\sqrt{x^2}}{\sqrt{16}}$ **Using Property 4**

$= \dfrac{|x|}{4}$ **Using Property 1**

We can check Example 2(h) graphically. Graph $y_1 = \sqrt{x^2/16}$ and $y_2 = |x|/4$.

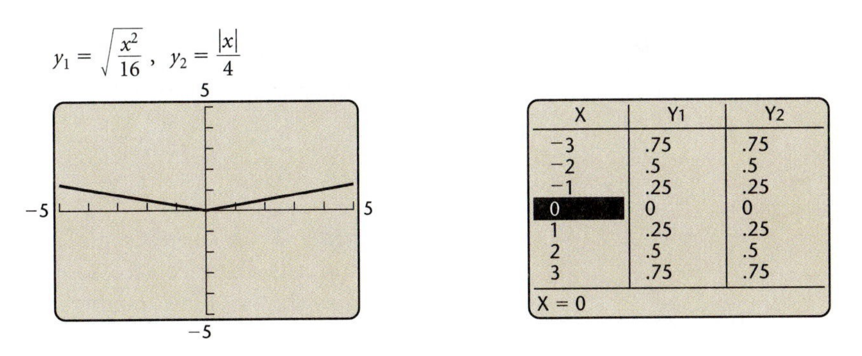

X	Y1	Y2
−3	.75	.75
−2	.5	.5
−1	.25	.25
0	0	0
1	.25	.25
2	.5	.5
3	.75	.75
X = 0		

The graphs appear to coincide. We can also use the TABLE feature to confirm that for each value of x, the y-values are the same.

In many situations, radicands are never formed by raising negative quantities to even powers. In such cases, absolute-value notation is not required. For this reason, we will henceforth assume that no radicands are formed by raising negative quantities to even powers. For example, we will write $\sqrt{x^2} = x$ and $\sqrt[4]{a^5b} = a\sqrt[4]{ab}$.

Radical expressions with the same index and the same radicand can be added or subtracted.

Example 3 Perform the operations indicated.

a) $3\sqrt{8x^2} - 5\sqrt{2x^2}$

b) $(4\sqrt{3} + \sqrt{2})(\sqrt{3} - 5\sqrt{2})$

SOLUTION

a)
$$\begin{aligned} 3\sqrt{8x^2} - 5\sqrt{2x^2} &= 3\sqrt{4x^2 \cdot 2} - 5\sqrt{x^2 \cdot 2} \\ &= 3 \cdot 2x\sqrt{2} - 5x\sqrt{2} \\ &= 6x\sqrt{2} - 5x\sqrt{2} \\ &= (6x - 5x)\sqrt{2} \qquad \text{Using the distributive property} \\ &= x\sqrt{2} \end{aligned}$$

b)
$$\begin{aligned} (4\sqrt{3} + \sqrt{2})(\sqrt{3} - 5\sqrt{2}) &= 4(\sqrt{3})^2 - 20\sqrt{6} + \sqrt{6} - 5(\sqrt{2})^2 \qquad \text{Multiplying} \\ &= 4 \cdot 3 + (-20 + 1)\sqrt{6} - 5 \cdot 2 \\ &= 12 - 19\sqrt{6} - 10 \\ &= 2 - 19\sqrt{6} \end{aligned}$$

An Application

The Pythagoren theorem relates the lengths of the sides of a right triangle. The side opposite the triangle's right angle is called the **hypotenuse**. The other sides are the **legs**.

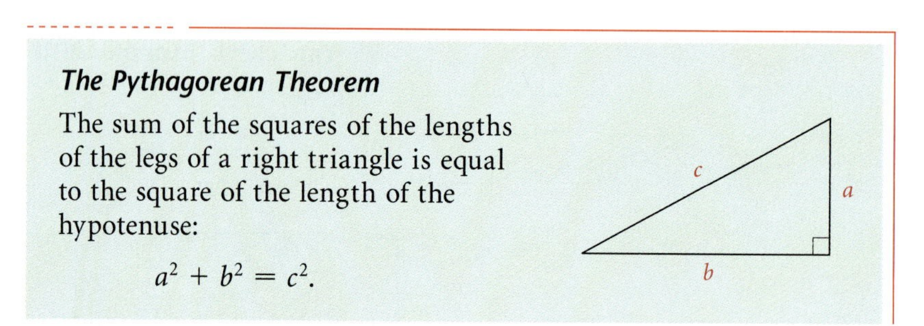

The Pythagorean Theorem

The sum of the squares of the lengths of the legs of a right triangle is equal to the square of the length of the hypotenuse:

$$a^2 + b^2 = c^2.$$

Example 4 A surveyor places poles at points A, B, and C in order to measure the distance across a pond. The distances AC and BC are measured as shown. Find the distance AB across the pond.

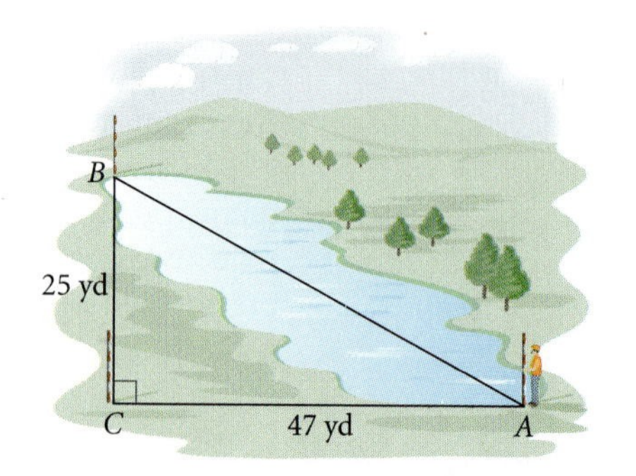

SOLUTION We see that the lengths of the legs of a right triangle are given. Thus we use the Pythagorean theorem to find the length of the hypotenuse AB:

$$AB^2 = AC^2 + BC^2$$

so

$$\begin{aligned} AB &= \sqrt{AC^2 + BC^2} \\ &= \sqrt{47^2 + 25^2} \\ &= \sqrt{2209 + 625} \\ &= \sqrt{2834} \\ &\approx 53.2. \end{aligned}$$

The distance across the pond is about 53.2 yd.

Rationalizing Denominators or Numerators

There are times when we need to remove the radicals in a denominator or a numerator. This is called **rationalizing the denominator** or **rationalizing the numerator**. It is done by multiplying by 1 in such a way as to obtain a perfect nth power.

Example 5 Rationalize the denominator of each of the following.

a) $\sqrt{\dfrac{3}{2}}$ **b)** $\dfrac{\sqrt[3]{7}}{\sqrt[3]{9}}$

SOLUTION

a) $\sqrt{\dfrac{3}{2}} = \sqrt{\dfrac{3}{2} \cdot \dfrac{2}{2}} = \sqrt{\dfrac{6}{4}} = \dfrac{\sqrt{6}}{\sqrt{4}} = \dfrac{\sqrt{6}}{2}$

b) $\dfrac{\sqrt[3]{7}}{\sqrt[3]{9}} = \dfrac{\sqrt[3]{7}}{\sqrt[3]{9}} \cdot \dfrac{\sqrt[3]{3}}{\sqrt[3]{3}} = \dfrac{\sqrt[3]{21}}{\sqrt[3]{27}} = \dfrac{\sqrt[3]{21}}{3}$

Pairs of expressions of the form $a\sqrt{b} + c\sqrt{d}$ and $a\sqrt{b} - c\sqrt{d}$ are called **conjugates**. The product of such a pair contains no radicals and can be used to rationalize a denominator or a numerator.

Example 6 Rationalize the numerator: $\dfrac{\sqrt{x} - \sqrt{y}}{5}$.

SOLUTION

$$\begin{aligned} \frac{\sqrt{x} - \sqrt{y}}{5} &= \frac{\sqrt{x} - \sqrt{y}}{5} \cdot \frac{\sqrt{x} + \sqrt{y}}{\sqrt{x} + \sqrt{y}} \\ &= \frac{(\sqrt{x})^2 - (\sqrt{y})^2}{5\sqrt{x} + 5\sqrt{y}} \\ &= \frac{x - y}{5\sqrt{x} + 5\sqrt{y}} \end{aligned}$$

The conjugate of $\sqrt{x} - \sqrt{y}$ is $\sqrt{x} + \sqrt{y}$.

Rational Exponents

We are motivated to define *rational exponents* so that the rules, or laws, for integer exponents hold for them. For example, we must have

$$a^{1/2} \cdot a^{1/2} = a^{1/2+1/2} = a^1 = a.$$

Thus we are led to define $a^{1/2}$ to mean $\sqrt{a}$. Similarly, $a^{1/n}$ would mean $\sqrt[n]{a}$. Again, if the laws of exponents are to hold, we must have

$$(a^{1/n})^m = (a^m)^{1/n} = a^{m/n}.$$

Thus we are led to define $a^{m/n}$ to mean $(\sqrt[n]{a})^m$, or, equivalently, $\sqrt[n]{a^m}$.

Rational Exponents

For any real number a and any natural numbers m and n for which $\sqrt[n]{a}$ exists,

$$a^{1/n} = \sqrt[n]{a},$$

$$a^{m/n} = \sqrt[n]{a^m} = (\sqrt[n]{a})^m, \quad \text{and}$$

$$a^{-m/n} = \frac{1}{a^{m/n}}.$$

We can use the definition of rational exponents to convert between radical and exponential notation.

Example 7 Convert to radical notation and, if possible, simplify each of the following.

a) $7^{3/4}$

b) $8^{-5/3}$

c) $m^{1/6}$

d) $(-32)^{2/5}$

SOLUTION

a) $7^{3/4} = \sqrt[4]{7^3}$, or $(\sqrt[4]{7})^3$

b) $8^{-5/3} = \dfrac{1}{8^{5/3}} = \dfrac{1}{(8^{1/3})^5} = \dfrac{1}{2^5} = \dfrac{1}{32}$

c) $m^{1/6} = \sqrt[6]{m}$

d) $(-32)^{2/5} = \sqrt[5]{(-32)^2} = \sqrt[5]{1024} = 4$, or

$(-32)^{2/5} = (\sqrt[5]{-32})^2 = (-2)^2 = 4$

Interactive Discovery

Enter each of the following expressions on a grapher and compare the results with the result of Example 7(d):

$$(-32)^{2/5}, \qquad ((-32)^2)^{1/5}, \qquad ((-32)^{1/5})^2.$$

Next, use the definition of rational exponents to find $(-8)^{5/3}$. Then enter $(-8)^{5/3}$, $((-8)^5)^{1/3}$, and $((-8)^{1/3})^5$ on a grapher and compare your results.

What do you observe about how to use a grapher to evaluate $a^{m/n}$ when a is negative and m is greater than 1?

X	Y1	Y2
−4	ERROR	2.5198
−3	ERROR	2.0801
−2	ERROR	1.5874
−1	ERROR	1
0	0	0
1	1	1
2	1.5874	1.5874
X = −1		

When using a grapher to find rational roots of the form $a^{m/n}$, with $a < 0$ and $m > 1$, we must enter the expression in the form $((a)^m)^{1/n}$ or $((a)^{1/n})^m$. Similarly, to graph $y = x^{m/n}$, where $m > 1$, we must enter $y = (x^m)^{1/n}$ or $y = (x^{1/n})^m$ in order to see the graph for negative values of x. The TABLE feature will confirm this. For example, enter $y_1 = x^{2/3}$ and $y_2 = (x^2)^{1/3}$. The table (shown at left) gives an ERROR message for values of y_1 when x is negative. Some graphers will give an ERROR message for negative values of x regardless of the form in which $x^{m/n}$, $m > 1$, is entered.

Example 8 Convert to exponential notation and, if possible, simplify each of the following.

a) $(\sqrt[4]{7xy})^5$ b) $\sqrt[6]{x^3}$ c) $\sqrt[3]{\sqrt{7}}$

SOLUTION

a) $(\sqrt[4]{7xy})^5 = (7xy)^{5/4}$

b) $\sqrt[6]{x^3} = x^{3/6} = x^{1/2}$

c) $\sqrt[3]{\sqrt{7}} = \sqrt[3]{7^{1/2}} = (7^{1/2})^{1/3} = 7^{1/6}$

We can use the laws of exponents to simplify exponential and radical expressions.

Example 9 Simplify and then write radical notation, if appropriate, for each of the following.

a) $x^{5/6} \cdot x^{2/3}$ b) $(x+3)^{5/2}(x+3)^{-1/2}$

SOLUTION

a) $x^{5/6} \cdot x^{2/3} = x^{5/6+2/3} = x^{9/6} = x^{3/2} = \sqrt{x^3} = \sqrt{x^2}\sqrt{x} = x\sqrt{x}$

b) $(x+3)^{5/2}(x+3)^{-1/2} = (x+3)^{5/2-1/2} = (x+3)^2$

Example 10 Write an expression containing a single radical: $a^{1/2}b^{5/6}$.

SOLUTION

$$a^{1/2}b^{5/6} = a^{3/6}b^{5/6} = (a^3b^5)^{1/6} = \sqrt[6]{a^3b^5}$$

Factoring Expressions with Rational Exponents

To factor expressions containing negative exponents, factor out the greatest common numerical factor and common variable factors with the smallest exponents.

Example 11 Factor: $3a^{1/2}b^{-3/4} - a^{-1/2}b^{1/4}$.

SOLUTION Note that $a^{-1/2}$ has a smaller exponent than $a^{1/2}$ and $b^{-3/4}$ has a smaller exponent than $b^{1/4}$.

$3a^{1/2}b^{-3/4} - a^{-1/2}b^{1/4}$

$= 3 \cdot a \cdot a^{-1/2}b^{-3/4} - a^{-1/2} \cdot b \cdot b^{-3/4}$ **$a^{1/2} = a \cdot a^{-1/2}$, $b^{1/4} = b \cdot b^{-3/4}$**

$= a^{-1/2}b^{-3/4}(3a - b)$ **Factoring out $a^{-1/2}b^{-3/4}$**

R.6 Exercise Set

Simplify. Assume that variables can represent any real number.

1. $\sqrt{(-11)^2}$
2. $\sqrt{(-1)^2}$
3. $\sqrt{16x^2}$
4. $\sqrt{36t^2}$
5. $\sqrt{(b+1)^2}$
6. $\sqrt{(2c-3)^2}$
7. $\sqrt[3]{-27x^3}$
8. $\sqrt[3]{-8y^3}$
9. $\sqrt{x^2-4x+4}$
10. $\sqrt{y^2+16y+64}$
11. $\sqrt[5]{32}$
12. $\sqrt[5]{-32}$
13. $\sqrt{180}$
14. $\sqrt{48}$
15. $\sqrt[3]{54}$
16. $\sqrt[3]{135}$
17. $\sqrt{128c^2d^4}$
18. $\sqrt{162c^4d^6}$
19. Use the TABLE feature of a grapher to check your answer to Exercise 9.
20. Use the TABLE feature of a grapher to check your answer to Exercise 10.

Simplify. Assume that no radicands were formed by raising negative quantities to even powers.

21. $\sqrt{2x^3y}\sqrt{12xy}$
22. $\sqrt{3y^4z}\sqrt{20z}$
23. $\sqrt[3]{3x^2y}\sqrt[3]{36x}$
24. $\sqrt[5]{8x^3y^4}\sqrt[5]{4x^4y}$
25. $\sqrt[3]{2(x+4)}\sqrt[3]{4(x+4)^4}$
26. $\sqrt[3]{4(x+1)^2}\sqrt[3]{18(x+1)^2}$
27. $\sqrt[6]{\dfrac{m^{12}n^{24}}{64}}$
28. $\sqrt[8]{\dfrac{m^{16}n^{24}}{2^8}}$
29. $\dfrac{\sqrt[3]{40m}}{\sqrt[3]{5m}}$
30. $\dfrac{\sqrt{40xy}}{\sqrt{8x}}$
31. $\dfrac{\sqrt[3]{3x^2}}{\sqrt[3]{24x^5}}$
32. $\dfrac{\sqrt{128a^2b^4}}{\sqrt{16ab}}$
33. $\sqrt[3]{\dfrac{64a^4}{27b^3}}$
34. $\sqrt{\dfrac{9x^7}{16y^8}}$
35. $\sqrt{\dfrac{7x^3}{36y^6}}$
36. $\sqrt[3]{\dfrac{2yz}{250z^4}}$
37. $9\sqrt{50}+6\sqrt{2}$
38. $11\sqrt{27}-4\sqrt{3}$
39. $8\sqrt{2x^2}-6\sqrt{20x}-5\sqrt{8x^2}$
40. $2\sqrt[3]{8x^2}+5\sqrt[3]{27x^2}-3\sqrt{x^3}$
41. $(\sqrt{3}-\sqrt{2})(\sqrt{3}+\sqrt{2})$
42. $(\sqrt{8}+2\sqrt{5})(\sqrt{8}-2\sqrt{5})$
43. $(1+\sqrt{3})^2$
44. $(\sqrt{2}-5)^2$
45. Use a grapher to check your answer to Exercise 39.
46. Use a grapher to check your answer to Exercise 40.
47. An airplane is flying at an altitude of 3700 ft. The slanted distance directly to the airport is 14,200 ft. How far horizontally is the airplane from the airport?

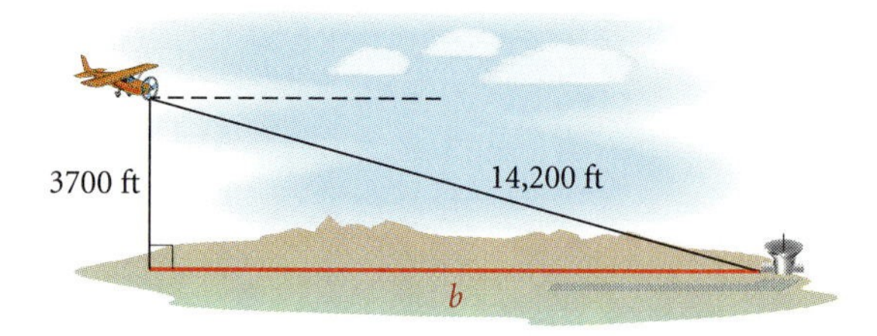

48. During a summer heat wave, a 2-mi bridge expands 2 ft in length. Assuming that the bulge occurs straight up the middle, estimate the height of the bulge. (In reality, bridges are built with expansion joints to control such buckling.)

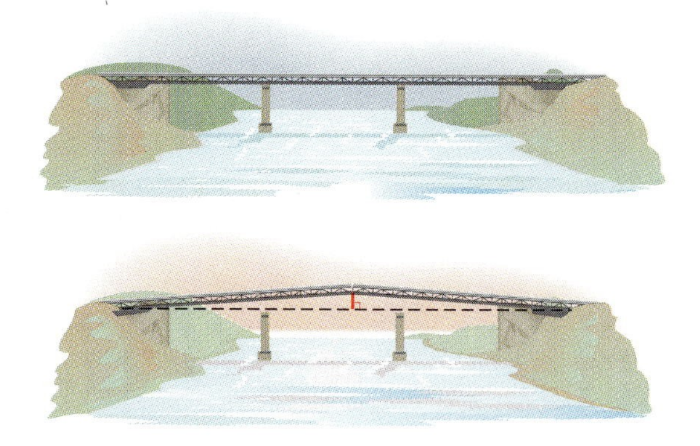

49. An *equilateral triangle* is shown below.

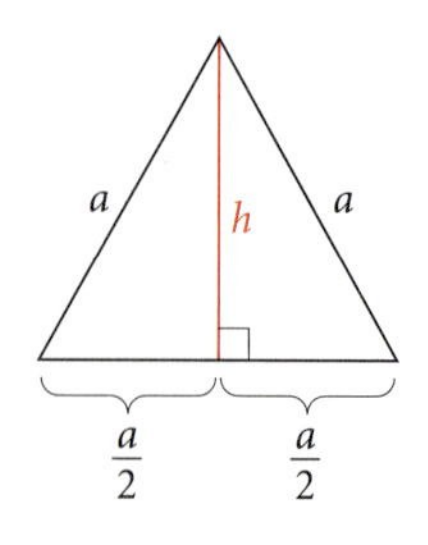

a) Find an expression for its height h in terms of a.
b) Find an expression for its area A in terms of a.

50. An isosceles right triangle has legs of length s. Find an expression for the length of the hypotenuse in terms of s.

51. The diagonal of a square has length $8\sqrt{2}$. Find the length of a side of the square.

52. The area of square $PQRS$ is 100 ft^2, and A, B, C, and D are the midpoints of the sides. Find the area of square $ABCD$.

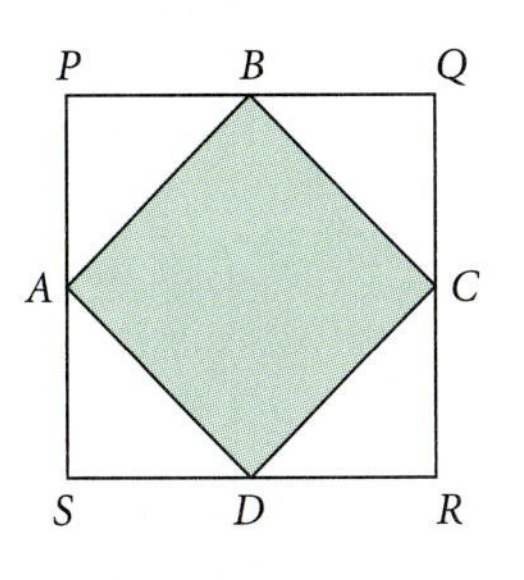

Rationalize the denominator.

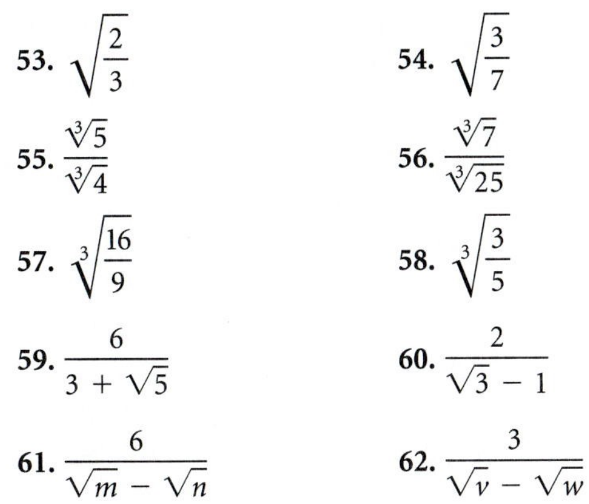

53. $\sqrt{\dfrac{2}{3}}$ **54.** $\sqrt{\dfrac{3}{7}}$

55. $\dfrac{\sqrt[3]{5}}{\sqrt[3]{4}}$ **56.** $\dfrac{\sqrt[3]{7}}{\sqrt[3]{25}}$

57. $\sqrt[3]{\dfrac{16}{9}}$ **58.** $\sqrt[3]{\dfrac{3}{5}}$

59. $\dfrac{6}{3 + \sqrt{5}}$ **60.** $\dfrac{2}{\sqrt{3} - 1}$

61. $\dfrac{6}{\sqrt{m} - \sqrt{n}}$ **62.** $\dfrac{3}{\sqrt{v} - \sqrt{w}}$

Rationalize the numerator.

63. $\dfrac{\sqrt{12}}{5}$ **64.** $\dfrac{\sqrt{50}}{3}$

65. $\sqrt[3]{\dfrac{7}{2}}$ **66.** $\sqrt[3]{\dfrac{2}{5}}$

67. $\dfrac{\sqrt{11}}{\sqrt{3}}$ **68.** $\dfrac{\sqrt{5}}{\sqrt{2}}$

69. $\dfrac{9 - \sqrt{5}}{3 - \sqrt{3}}$ **70.** $\dfrac{8 - \sqrt{6}}{5 - \sqrt{2}}$

71. $\dfrac{\sqrt{a} + \sqrt{b}}{3a}$ **72.** $\dfrac{\sqrt{p} - \sqrt{q}}{1 + \sqrt{q}}$

Convert to radical notation and simplify.

73. $x^{3/4}$ **74.** $y^{2/5}$

75. $16^{3/4}$ **76.** $4^{7/2}$

77. $125^{-1/3}$ **78.** $32^{-4/5}$

79. $a^{5/4}b^{-3/4}$ **80.** $x^{2/5}y^{-1/5}$

Convert to exponential notation and simplify.

81. $(\sqrt[4]{13})^5$ **82.** $\sqrt[5]{17^3}$

83. $\sqrt[3]{20^2}$ **84.** $(\sqrt[5]{12})^4$

85. $\sqrt[3]{\sqrt{11}}$ **86.** $\sqrt[3]{\sqrt[4]{7}}$

87. $\sqrt{5}\sqrt[3]{5}$ **88.** $\sqrt[3]{2}\sqrt{2}$

89. $\sqrt[5]{32^2}$ **90.** $\sqrt[3]{64^2}$

Simplify and then write radical notation, if appropriate.

91. $(2a^{3/2})(4a^{1/2})$ **92.** $(3a^{5/6})(8a^{2/3})$

93. $\left(\dfrac{x^6}{9b^{-4}}\right)^{1/2}$ **94.** $\left(\dfrac{x^{2/3}}{4y^{-2}}\right)^{1/2}$

95. $\dfrac{x^{2/3}y^{5/6}}{x^{-1/3}y^{1/2}}$ **96.** $\dfrac{a^{1/2}b^{5/8}}{a^{1/4}b^{3/8}}$

97. Use a grapher to check your answer to Exercise 91.

98. Use a grapher to check your answer to Exercise 92.

Write an expression containing a single radical and simplify.

99. $\sqrt[3]{6}\sqrt{2}$ **100.** $\sqrt{2}\sqrt[4]{8}$

101. $\sqrt[4]{xy}\sqrt[3]{x^2y}$ **102.** $\sqrt[3]{ab^2}\sqrt{ab}$

103. $\sqrt[3]{a^4\sqrt{a^3}}$ **104.** $\sqrt{a^3\sqrt[3]{a^2}}$

105. $\dfrac{\sqrt{(a + x)^3}\sqrt[3]{(a + x)^2}}{\sqrt[4]{a + x}}$

106. $\dfrac{\sqrt[4]{(x + y)^2}\sqrt[3]{x + y}}{\sqrt{(x + y)^3}}$

Factor and simplify.

107. $a^{-2}b^5 - a^3b^{-5}$

108. $p^8q^{-2} + p^{-5}q^4$

109. $x^{-1/3}y^{3/4} - x^{2/3}y^{-1/4}$

110. $4a^{1/2}b^{-3/4} - 6a^{-1/2}b^{1/4}$

111. $(2x - 3)^{-3}(x + 1)^{5/4} + (2x - 3)^{-2}(x + 1)^{1/4}$

112. $2x(5x + 3)^{2/3} + 3x^2(5x + 3)^{-1/3}$

Synthesis

113. ◈ Explain how you would convince a classmate that $\sqrt{a + b}$ is not equivalent to $\sqrt{a} + \sqrt{b}$, for positive real numbers a and b. Give both an algebraic explanation and a graphical explanation.

114. ◈ Explain how you would determine whether $10\sqrt{26} - 50$ is positive or negative without carrying out the actual computation.

Simplify.

115. $\sqrt{1 + x^2} + \dfrac{1}{\sqrt{1 + x^2}}$

116. $\sqrt{1 - x^2} - \dfrac{x^2}{2\sqrt{1 - x^2}}$

117. $(\sqrt{a\sqrt{a}})^{\sqrt{a}}$

118. $(2a^3b^{5/4}c^{1/7})^4 \div (54a^{-2}b^{2/3}c^{6/5})^{-1/3}$

119. Use a grapher to determine all values of x for which:

a) $x^{1/2} = x^{1/3}$
b) $x^{1/2} > x^{1/3}$
c) $x^{1/2} < x^{1/3}$

R.7 Solving Equations

- *Solve linear, quadratic, rational, and radical equations and equations with absolute value.*
- *Solve a formula for a given variable.*

An **equation** is a statement that two expressions are equal. To **solve** an equation in one variable is to find all the values of the variable that make the equation true. Each of these numbers is a **solution** of the equation. The set of all solutions of an equation is its **solution set**. Equations that have the same solution set are called **equivalent equations**.

Linear and Quadratic Equations

A *linear equation in one variable* is an equation that is equivalent to one of the form $ax + b = 0$, where a and b are real numbers and $a \neq 0$.

A *quadratic equation* is an equation that is equivalent to one of the form $ax^2 + bx + c = 0$, where a, b, and c are real numbers and $a \neq 0$.

The following principles allow us to solve many linear and quadratic equations.

Equation-Solving Principles

For any real numbers a, b, and c:

The Addition Principle: If $a = b$ is true, then $a + c = b + c$ is true.

The Multiplication Principle: If $a = b$ is true, then $ac = bc$ is true.

The Principle of Zero Products: If $ab = 0$ is true, then $a = 0$ or $b = 0$, and if $a = 0$ or $b = 0$, then $ab = 0$.

The Principle of Square Roots: If $x^2 = k$, then $x = \sqrt{k}$ or $x = -\sqrt{k}$.

Example 1 Solve: $3(7 - 2x) = 14 - 8(x - 1)$.

ALGEBRAIC SOLUTION

We have

$3(7 - 2x) = 14 - 8(x - 1)$	
$21 - 6x = 14 - 8x + 8$	Using the distributive property
$21 - 6x = 22 - 8x$	Combining like terms
$21 + 2x = 22$	Using the addition principle to add $8x$ on both sides
$2x = 1$	Using the addition principle to add -21 or subtract 21 on both sides
$x = \frac{1}{2}$.	Using the multiplication principle to multiply by $\frac{1}{2}$ or divide by 2 on both sides

CHECK:

$3(7 - 2x)$	$= 14 - 8(x - 1)$	
$3\left(7 - 2 \cdot \frac{1}{2}\right)$	? $14 - 8\left(\frac{1}{2} - 1\right)$	Substituting $\frac{1}{2}$ for x
$3(7 - 1)$	$14 - 8\left(-\frac{1}{2}\right)$	
$3 \cdot 6$	$14 + 4$	
18	18	TRUE

The solution is $\frac{1}{2}$. The set of all solutions, the solution set, is $\left\{\frac{1}{2}\right\}$.

We can also use the TABLE feature of a grapher—set in ASK mode, if available— to check the answer. We let $y_1 = 3(7 - 2x)$ and $y_2 = 14 - 8(x - 1)$. When we enter .5 for x, we see that Y_1 and Y_2 are both 18.

X	Y1	Y2
.5	18	18

X = .5

GRAPHICAL SOLUTION

We can use a grapher to solve this equation. Graph $y_1 = 3(7 - 2x)$ and $y_2 = 14 - 8(x - 1)$.

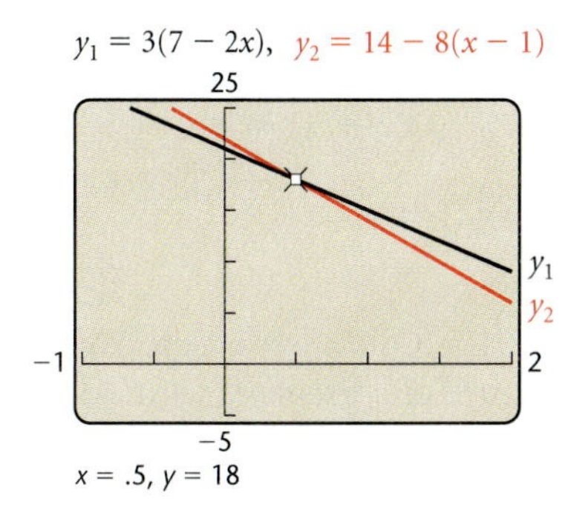

Using the TRACE and ZOOM features or the INTERSECT feature, we find that the first coordinate of the point of intersection is .5, or $\frac{1}{2}$. The solution set is $\left\{\frac{1}{2}\right\}$.

We can also use the SOLVE feature. It might be necessary to first write the equation as

$$3(7 - 2x) - (14 - 8(x - 1)) = 0.$$

The grapher returns the value .5, or $\frac{1}{2}$.

Note that the equation in Example 1 is not an identity, because it is not true for *all* values of x in its domain. In fact, we see from the graph above that $y_1 = 3(7 - 2x)$ and $y_2 = 14 - 8(x - 1)$ coincide, or intersect, only for $x = \frac{1}{2}$.

Example 2 Solve each of the following.

a) $2x^2 - x = 3$

b) $2x^2 - 10 = 0$

SOLUTION

a) We solve $2x^2 - x = 3$ both algebraically and graphically.

ALGEBRAIC SOLUTION

We have

$$\begin{aligned} 2x^2 - x &= 3 \\ 2x^2 - x - 3 &= 0 \quad \text{Subtracting 3 on both sides} \\ (2x - 3)(x + 1) &= 0 \quad \text{Factoring} \end{aligned}$$

$$2x - 3 = 0 \quad or \quad x + 1 = 0 \quad \text{Using the principle of zero products}$$

$$x = \tfrac{3}{2} \quad or \quad x = -1.$$

Both numbers check. We can use the TABLE feature of a grapher, set in ASK mode, to confirm this. Let $y = 2x^2 - x$. The y-values in the table should be 3 for both $x = 1.5$ and $x = -1$.

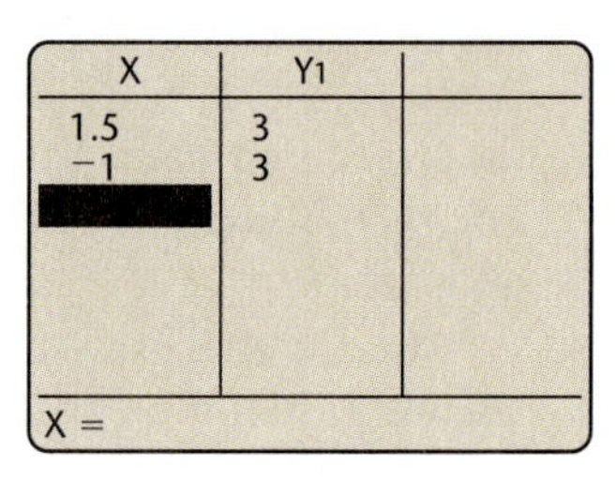

X	Y1	
1.5	3	
−1	3	

X =

The solution set is $\left\{\frac{3}{2}, -1\right\}$.

GRAPHICAL SOLUTION

Graph $y_1 = 2x^2 - x$ and $y_2 = 3$.

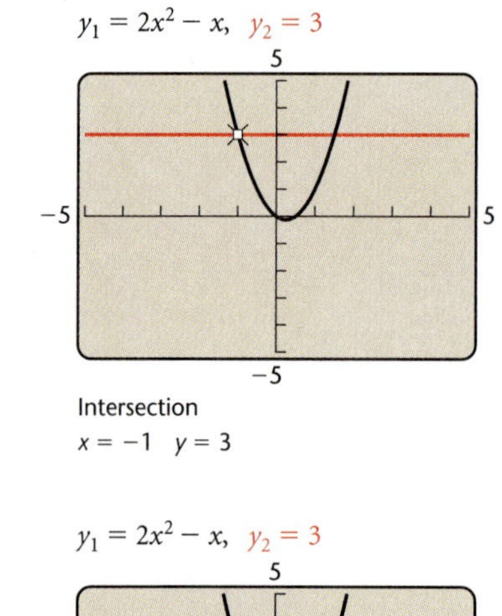

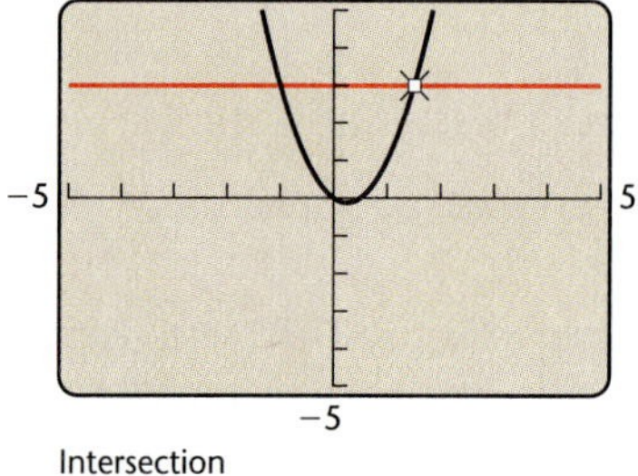

We use the TRACE and ZOOM features or the INTERSECT feature and find that the solutions are -1 and 1.5. We can also use the SOLVE feature to get this result. The solution set is $\{-1, 1.5\}$.

b) We solve $2x^2 - 10 = 0$ algebraically:

$$\begin{aligned} 2x^2 - 10 &= 0 \\ 2x^2 &= 10 \quad \text{Adding 10 on both sides} \\ x^2 &= 5 \quad \text{Dividing by 2 on both sides} \\ x = \sqrt{5} \quad or \quad x &= -\sqrt{5}. \quad \text{Using the principle of square roots} \end{aligned}$$

Both numbers check. The solution set is $\{\sqrt{5}, -\sqrt{5}\}$, or $\{\pm\sqrt{5}\}$. Use the TABLE feature of a grapher, set in ASK mode, to confirm this. Using a grapher, we find that $x \approx 2.236$ or $x \approx -2.236$.

Rational and Radical Equations

Equations containing rational expressions are called **rational equations**. Solving such equations involves multiplying on both sides by the LCD.

Example 3 Solve: $\dfrac{x^2}{x-3} = \dfrac{9}{x-3}$.

SOLUTION The LCD is $x - 3$.

$$(x-3) \cdot \frac{x^2}{x-3} = (x-3) \cdot \frac{9}{x-3}$$

$$x^2 = 9$$

$$x = -3 \quad or \quad x = 3 \qquad \text{Using the principle of square roots}$$

The possible solutions are 3 and -3. We check.

CHECK: For 3:

$$\frac{x^2}{x-3} = \frac{9}{x-3}$$

$$\frac{3^2}{3-3} \;?\; \frac{9}{3-3}$$

$$\frac{9}{0} \;\Big|\; \frac{9}{0} \qquad \text{UNDEFINED}$$

For -3:

$$\frac{x^2}{x-3} = \frac{9}{x-3}$$

$$\frac{(-3)^2}{-3-3} \;?\; \frac{9}{-3-3}$$

$$\frac{9}{-6} \;\Big|\; \frac{9}{-6} \qquad \text{TRUE}$$

Since division by 0 is undefined, 3 is not a solution. Note that 3 is not in the domain of $x^2/(x-3)$ or $9/(x-3)$. (See the table below.) The number -3 checks, so it is a solution. The solution set is $\{-3\}$.

$y_1 = \dfrac{x^2}{x-3}$, $y_2 = \dfrac{9}{x-3}$

X	Y1	Y2
0	0	−3
1	−.5	−4.5
2	−4	−9
3	ERROR	ERROR
4	16	9
5	12.5	4.5
6	12	3

X = 3

When we use the multiplication principle to multiply (or divide) both sides of an equation by an expression with a variable, we might not obtain an equivalent equation. We must check possible solutions by substituting in the original equation.

A **radical equation** is an equation in which variables appear in one or more radicands. For example,

$$\sqrt{2x-5} - \sqrt{x-3} = 1$$

is a radical equation. The following principle is needed to solve such equations.

The Principle of Powers

For any positive integer n:

If $a = b$ is true, then $a^n = b^n$ is true.

Example 4 Solve: $5 + \sqrt{x + 7} = x$.

ALGEBRAIC SOLUTION

We first isolate the radical and then use the principle of powers.

$$5 + \sqrt{x + 7} = x$$
$$\sqrt{x + 7} = x - 5 \quad \text{Subtracting 5 on both sides}$$
$$(\sqrt{x + 7})^2 = (x - 5)^2 \quad \text{Using the principle of powers; squaring both sides}$$
$$x + 7 = x^2 - 10x + 25$$
$$0 = x^2 - 11x + 18$$
$$0 = (x - 9)(x - 2)$$
$$x - 9 = 0 \quad or \quad x - 2 = 0$$
$$x = 9 \quad or \quad x = 2$$

The possible solutions are 9 and 2.

CHECK: For 9:

$$\frac{5 + \sqrt{x + 7} = x}{5 + \sqrt{9 + 7} \ ? \ 9}$$
$$5 + \sqrt{16}$$
$$5 + 4$$
$$9 \mid 9 \quad \text{TRUE}$$

For 2:

$$\frac{5 + \sqrt{x + 7} = x}{5 + \sqrt{2 + 7} \ ? \ 2}$$
$$5 + \sqrt{9}$$
$$5 + 3$$
$$8 \mid 2 \quad \text{FALSE}$$

Since 9 checks but 2 does not, the only solution is 9. The solution set is $\{9\}$.

GRAPHICAL SOLUTION

Graph $y_1 = 5 + \sqrt{x + 7}$ and $y_2 = x$.

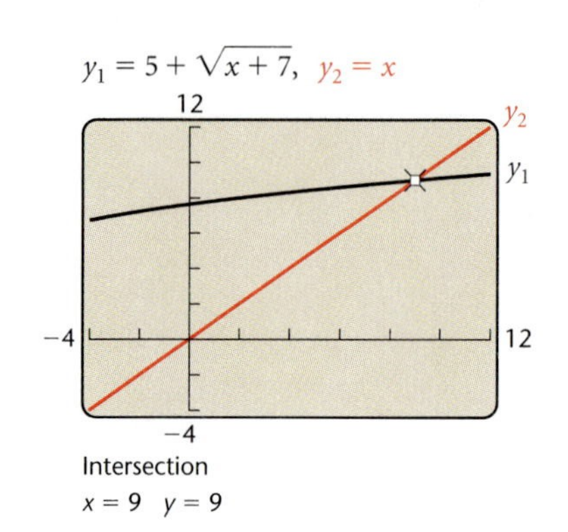

Using the TRACE and ZOOM features or the INTERSECT feature, we see that the only solution is 9. The solution set is $\{9\}$.

We can also use the SOLVE feature to get this result. It might be necessary, however, to first write the equation as

$$5 + \sqrt{x + 7} - x = 0.$$

When we raise both sides of an equation to an even power, the resulting equation can have solutions that the original equation does not. This is because the converse of the principle of powers is not necessarily true. That is, if $a^n = b^n$ is true, we do not know that $a = b$ is true. For example, $(-2)^2 = 2^2$, but $-2 \neq 2$. Thus, as we see in Example 4, it is necessary to check the possible solutions in the original equation when using the principle of powers to raise both sides of an equation to an even power.

When a radical equation has two radical terms on one side, we isolate one of them and then use the principle of powers. If, after doing so, a radical term remains, we repeat these steps.

Example 5 Solve: $\sqrt{x - 3} + \sqrt{x + 5} = 4$.

SOLUTION

$$\sqrt{x - 3} = 4 - \sqrt{x + 5} \quad \text{Isolating one radical}$$
$$(\sqrt{x - 3})^2 = (4 - \sqrt{x + 5})^2 \quad \text{Using the principle of powers; squaring both sides}$$

$$x - 3 = 16 - 8\sqrt{x + 5} + (x + 5)$$

$$x - 3 = 21 - 8\sqrt{x + 5} + x \quad \text{Combining like terms}$$

$$-24 = -8\sqrt{x + 5} \quad \text{Isolating the remaining radical}$$

$$3 = \sqrt{x + 5} \quad \text{Dividing by } -8 \text{ on both sides}$$

$$3^2 = (\sqrt{x + 5})^2 \quad \text{Using the principle of powers; squaring both sides}$$

$$9 = x + 5$$

$$4 = x$$

The number 4 checks and is the solution. Use graphs or the TABLE feature of a grapher to confirm this. The solution set is $\{4\}$.

Equations with Absolute Value

Recall that the absolute value of a number is its distance from 0 on the number line. We use this concept to solve equations with absolute value.

For $a > 0$:

$|x| = a$ is equivalent to $x = -a$ or $x = a$.

Example 6 Solve each of the following.

a) $|x| = 5$

b) $|x - 3| = 2$

SOLUTION

a) We solve $|x| = 5$ both algebraically and graphically.

ALGEBRAIC SOLUTION

We have

$$|x| = 5$$

$$x = -5 \quad or \quad x = 5.$$

Writing an equivalent statement

The solution set is $\{-5, 5\}$. Use the TABLE feature of a grapher to confirm this.

GRAPHICAL SOLUTION

Graph $y_1 = |x|$ and $y_2 = 5$ and find the first coordinates of the points of intersection using TRACE and ZOOM or INTERSECT. We can also use the SOLVE feature to get this result.

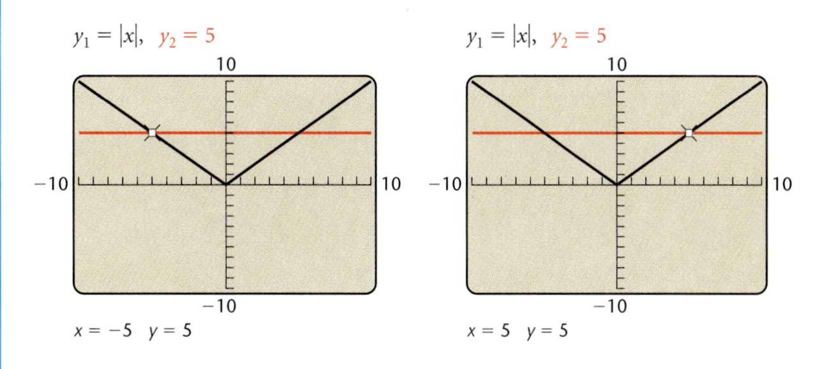

The solution set is $\{-5, 5\}$.

b) We solve $|x - 3| = 2$ algebraically:

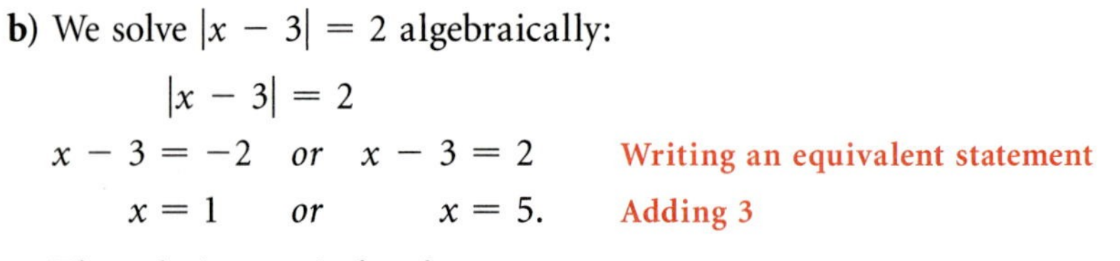

$$|x - 3| = 2$$

$$x - 3 = -2 \quad or \quad x - 3 = 2 \qquad \text{Writing an equivalent statement}$$

$$x = 1 \quad or \quad x = 5. \qquad \text{Adding 3}$$

The solution set is $\{1, 5\}$.

When $a = 0$, $|x| = a$ is equivalent to $x = 0$. Note that for $a < 0$, $|x| = a$ has no solution, because the absolute value of an expression is never negative. We can use a graph to illustrate the last statement for a specific value of a. For example, let $a = -3$. Graph $y_1 = |x|$ and $y_2 = -3$.

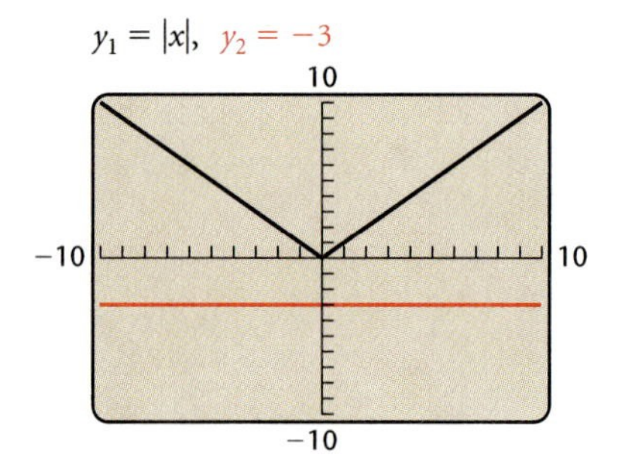

The graphs do not intersect. Thus the equation $|x| = -3$ has no solution. The solution set is the **empty set**, denoted $\varnothing$.

Formulas

A **formula** is an equation that can be used to *model* a situation. For example, the formula $P = 2l + 2w$ gives the perimeter of a rectangle with length l and width w. The equation-solving principles presented earlier can be used to solve a formula for a given variable.

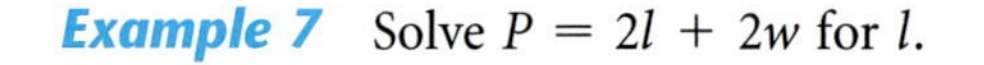

Example 7 Solve $P = 2l + 2w$ for l.

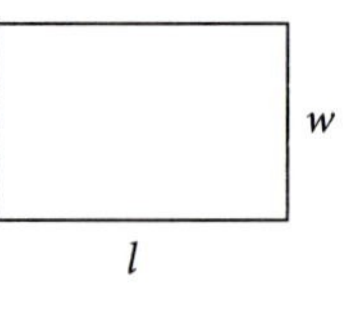

SOLUTION

$$P = 2l + 2w$$

$$P - 2w = 2l \qquad \text{Subtracting } 2w \text{ on both sides}$$

$$\frac{P - 2w}{2} = l \qquad \text{Dividing by 2 on both sides}$$

The formula $l = \dfrac{P - 2w}{2}$ can be used to determine a rectangle's length if we are given its perimeter and its width.

R.7 Exercise Set

Solve using a grapher.

1. $4x + 5 = 21$

2. $2y - 1 = 3$

3. $y + 1 = 2y - 7$

4. $5 - 4x = x - 13$

Solve algebraically. Check your answers on a grapher.

5. $5x - 2 + 3x = 2x + 6 - 4x$

6. $5x - 17 - 2x = 6x - 1 - x$

7. $7(3x + 6) = 11 - (x + 2)$

8. $4(5y + 3) = 3(2y - 5)$

9. $(2x - 3)(3x - 2) = 0$

10. $(5x - 2)(2x + 3) = 0$

11. $3x^2 + x - 2 = 0$

12. $10x^2 - 16x + 6 = 0$

13. $2x^2 = 6x$

14. $18x + 9x^2 = 0$

15. $3y^3 - 5y^2 - 2y = 0$

16. $3t^3 + 2t = 5t^2$

17. $7x^3 + x^2 - 7x - 1 = 0$
(*Hint*: Factor by grouping.)

18. $3x^3 + x^2 - 12x - 4 = 0$
(*Hint*: Factor by grouping.)

19. $\frac{1}{4} + \frac{1}{5} = \frac{1}{t}$

20. $\frac{1}{3} - \frac{5}{6} = \frac{1}{x}$

21. $\frac{x + 2}{4} - \frac{x - 1}{5} = 15$

22. $\frac{t + 1}{3} - \frac{t - 1}{2} = 1$

23. $\frac{1}{2} + \frac{2}{x} = \frac{1}{3} + \frac{3}{x}$

24. $\frac{1}{t} + \frac{1}{2t} + \frac{1}{3t} = 5$

25. $\frac{3x}{x + 2} + \frac{6}{x} = \frac{12}{x^2 + 2x}$

26. $\frac{5x}{x - 4} - \frac{20}{x} = \frac{80}{x^2 - 4x}$

27. $\frac{4}{x^2 - 1} - \frac{2}{x - 1} = \frac{3}{x + 1}$

28. $\frac{3y + 5}{y^2 + 5y} + \frac{y + 4}{y + 5} = \frac{y + 1}{y}$

29. $\frac{490}{x^2 - 49} = \frac{5x}{x - 7} - \frac{35}{x + 7}$

30. $\frac{3}{m + 2} + \frac{2}{m} = \frac{4m - 4}{m^2 - 4}$

31. $\frac{1}{x - 6} - \frac{1}{x} = \frac{6}{x^2 - 6x}$

32. $\frac{8}{x^2 - 4} = \frac{x}{x - 2} - \frac{2}{x + 2}$

33. $\frac{8}{x^2 - 2x + 4} = \frac{x}{x + 2} + \frac{24}{x^3 + 8}$

34. $\frac{18}{x^2 - 3x + 9} - \frac{x}{x + 3} = \frac{81}{x^3 + 27}$

35. $\sqrt{3x - 4} = 1$

36. $\sqrt[3]{2x + 1} = -5$

37. $\sqrt[4]{x^2 - 1} = 1$

38. $\sqrt{m + 1} - 5 = 8$

39. $\sqrt{y - 1} + 4 = 0$

40. $\sqrt[5]{3x + 4} = 2$

41. $\sqrt[3]{6x + 9} + 8 = 5$

42. $\sqrt{6x + 7} = x + 2$

43. $\sqrt{x - 3} + \sqrt{x + 2} = 5$

44. $\sqrt{x} - \sqrt{x - 5} = 1$

45. $\sqrt{3x - 5} + \sqrt{2x + 3} + 1 = 0$

46. $\sqrt{2m - 3} = \sqrt{m + 7} - 2$

47. $\sqrt{x} - \sqrt{3x - 3} = 1$

48. $\sqrt{2x + 1} - \sqrt{x} = 1$

49. $\sqrt{2y - 5} - \sqrt{y - 3} = 1$

50. $\sqrt{4p + 5} + \sqrt{p + 5} = 3$

51. $x^{1/3} = -2$

52. $t^{1/5} = 2$

53. $t^{1/4} = 3$

54. $m^{1/2} = -7$

Solve.

55. $|x| = 7$

56. $|x| = 4.5$

57. $|x| = -10.7$

58. $|x| = -\frac{3}{5}$

59. $|x - 1| = 4$

60. $|x - 7| = 5$

61. $|3x| = 1$

62. $|5x| = 4$

63. $|x| = 0$

64. $|6x| = 0$

65. $|3x + 2| = 1$

66. $|7x - 4| = 8$

67. $\left|\frac{1}{2}x - 5\right| = 17$

68. $\left|\frac{1}{3}x - 4\right| = 13$

69. $|x - 1| + 3 = 6$

70. $|x + 2| - 5 = 9$

Solve.

71. $A = \frac{1}{2}bh$, for b
(Area of a triangle)

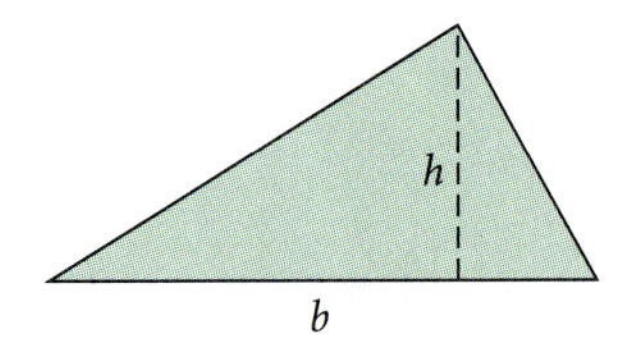

72. $A = \pi r^2$, for π
(Area of a circle)

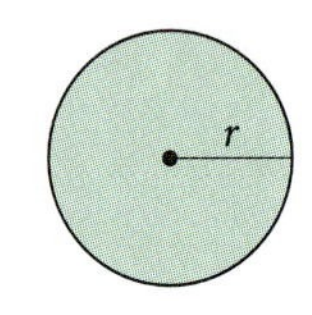

73. $P = 2l + 2w$, for w
(Perimeter of a rectangle)

74. $A = P + Prt$, for P
(Simple interest)

75. $\frac{P_1V_1}{T_1} = \frac{P_2V_2}{T_2}$, for T_1
(A chemistry formula for gases)

76. $A = \frac{1}{2}h(b_1 + b_2)$, for h
(Area of a trapezoid)

77. $\frac{1}{R} = \frac{1}{R_1} + \frac{1}{R_2}$, for R_2
(Resistance)

78. $A = P(1 + i)^2$, for i
(Compound interest)

79. $\frac{1}{F} = \frac{1}{m} + \frac{1}{p}$, for p
(A formula from optics)

80. $\frac{1}{F} = \frac{1}{m} + \frac{1}{p}$, for F
(A formula from optics)

Synthesis

81. ◈ Explain why it is necessary to check the possible solutions of a rational equation.

82. ◈ Explain in your own words why it is necessary to check the possible solutions when the principle of powers is used to solve an equation.

83. ◈ Use a graphical argument to explain why the equation $x^2 - 6x + 9 = 1 - x^4$ has no solution.

Solve.

84. $(x + 1)^3 = (x - 1)^3 + 26$

85. $(x - 2)^3 = x^3 - 2$

86. $\frac{x + 3}{x + 2} - \frac{x + 4}{x + 3} = \frac{x + 5}{x + 4} - \frac{x + 6}{x + 5}$

87. $(x - 3)^{2/3} = 2$

88. $\sqrt{15 + \sqrt{2x + 80}} = 5$

89. $\sqrt{x + 5} + 1 = \frac{6}{\sqrt{x + 5}}$

90. $x^{2/3} = x + 1$

R.8 Solving Inequalities

- *Solve linear inequalities, using interval notation to express solution sets.*
- *Solve compound inequalities.*
- *Solve inequalities with absolute value.*

An **inequality** is a sentence with $<$, $>$, $\leq$, or $\geq$ as its verb. An example is $3x - 5 < 6 - 2x$. To **solve** an inequality is to find all values of the variable that make the inequality true. Each of these numbers is a **solu-**

tion of the inequality, and the set of all such solutions is its **solution set**. Inequalities that have the same solution set are called **equivalent inequalities**.

Linear Inequalities

The principles for solving inequalities are similar to those for solving equations.

Principles for Solving Inequalities

For any real numbers a, b, and c:

The Addition Principle for Inequalities: If $a < b$ is true, then $a + c < b + c$ is true.

The Multiplication Principle for Inequalities: If $a < b$ and $c > 0$ are true, then $ac < bc$ is true. If $a < b$ and $c < 0$ are true, then $ac > bc$ is true. Similar statements hold for $a \leq b$.

Note that when both sides of an inequality are multiplied by a negative number, we must reverse the inequality sign.

First-degree inequalities with one variable, like those in Example 1 below, are **linear inequalities**.

Example 1 Solve each of the following.

a) $3x - 5 < 6 - 2x$ b) $13 - 7y \geq 10y - 4$

SOLUTION

a) $3x - 5 < 6 - 2x$

$5x - 5 < 6$ **Using the addition principle for inequalities; adding $2x$ on both sides**

$5x < 11$ **Using the addition principle for inequalities; adding 5 on both sides**

$x < \frac{11}{5}$ **Using the multiplication principle for inequalities; multiplying by $\frac{1}{5}$ or dividing by 5 on both sides**

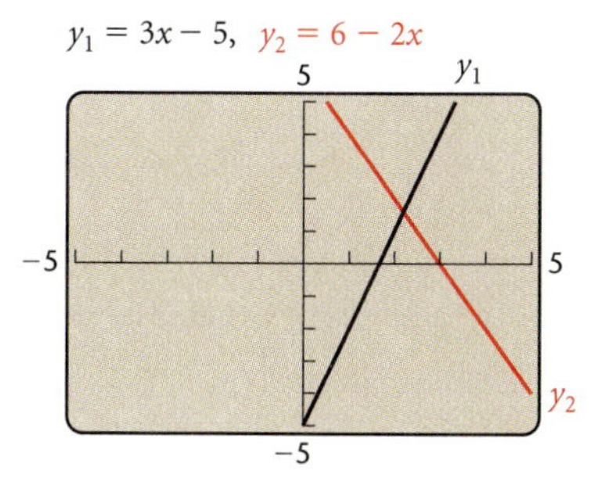

Any number less than $\frac{11}{5}$ is a solution. To check graphically, graph $y_1 = 3x - 5$ and $y_2 = 6 - 2x$. The graph (shown at left) confirms that for $x < \frac{11}{5}$, $y_1 < y_2$. The solution set is $\left\{x \middle| x < \frac{11}{5}\right\}$.

b) $13 - 7y \geq 10y - 4$

$13 - 17y \geq -4$ **Subtracting $10y$ on both sides**

$-17y \geq -17$ **Subtracting 13 on both sides**

$y \leq 1$ **Dividing by -17 on both sides and reversing the inequality sign**

The solution set is $\{y | y \leq 1\}$.

Interval Notation

Solutions of inequalities can also be expressed using **interval notation**. For example, for real numbers a and b such that $a < b$, the **open interval** (a, b) is the set of real numbers between, but not including, a and b. That is,

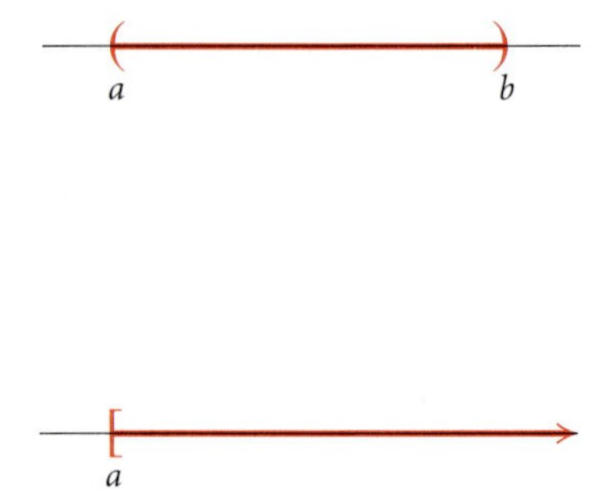

$$(a, b) = \{x \mid a < x < b\}.$$

The points a and b are the **endpoints** of the interval. The parentheses indicate that the endpoints are not included in the interval.

Some intervals extend without bound in one or both directions. The interval $[a, \infty)$, for example, begins at a and extends to the right without bound. That is,

$$[a, \infty) = \{x \mid x \geq a\}.$$

The bracket indicates that a is included in the interval.

The various types of intervals are listed below.

Intervals: Types, Notation, and Graphs

TYPE	INTERVAL NOTATION	SET NOTATION	GRAPH
Open	(a, b)	$\{x \mid a < x < b\}$	
Closed	$[a, b]$	$\{x \mid a \leq x \leq b\}$	
Half-open	$[a, b)$	$\{x \mid a \leq x < b\}$	
Half-open	$(a, b]$	$\{x \mid a < x \leq b\}$	
Open	(a, ∞)	$\{x \mid x > a\}$	
Half-open	$[a, \infty)$	$\{x \mid x \geq a\}$	
Open	$(-\infty, b)$	$\{x \mid x < b\}$	
Half-open	$(-\infty, b]$	$\{x \mid x \leq b\}$	

The interval $(-\infty, \infty)$ names the set of all real numbers, $\mathbb{R}$.

We see that we can express the solutions in Example 1 using interval notation:

$$\left\{x \mid x < \frac{11}{5}\right\} = \left(-\infty, \frac{11}{5}\right) \quad \text{and} \quad \{y \mid y \leq 1\} = (-\infty, 1].$$

Example 2 Write interval notation for each set.

a) $\{x \mid -4 < x < 5\}$

b) $\{x \mid x \geq 1.7\}$

c) $\{x \mid -10 < x \leq -5\}$

d) $\{x \mid x < \sqrt{5}\}$

SOLUTION

a) $\{x \mid -4 < x < 5\} = (-4, 5)$

b) $\{x \mid x \geq 1.7\} = [1.7, \infty)$

c) $\{x \mid -10 < x \leq -5\} = (-10, -5]$

d) $\{x \mid x < \sqrt{5}\} = (-\infty, \sqrt{5})$

Compound Inequalities

When two inequalities are joined by the word *and* or the word *or*, a **compound inequality** is formed. A compound inequality like $-3 < 2x + 5$ *and* $2x + 5 \leq 7$ is called a **conjunction**, because it uses the word *and*. The sentence $-3 < 2x + 5 \leq 7$ is an abbreviation for the preceding conjunction.

Compound inequalities can be solved using the addition and multiplication principles for inequalities.

Example 3 Solve $-3 < 2x + 5 \leq 7$. Then graph the solution set.

SOLUTION We have

$$-3 < 2x + 5 \leq 7$$

$$-8 < 2x \leq 2 \qquad \text{Subtracting 5}$$

$$-4 < x \leq 1. \qquad \text{Dividing by 2}$$

The solution set is $\{x \mid -4 < x \leq 1\}$, or $(-4, 1]$. The graph of the solution set is as shown at left.

To check, graph $y_1 = -3$, $y_2 = 2x + 5$, and $y_3 = 7$.

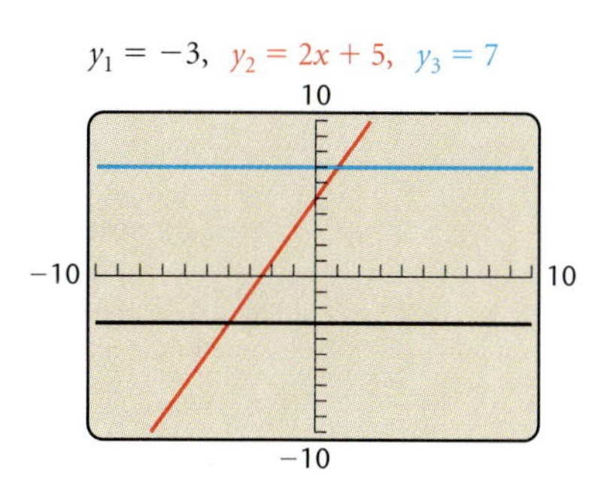

Note that $y_1 < y_2 \leq y_3$ for $-4 < x \leq 1$.

A compound inequality like $2x - 5 \leq -7$ *or* $2x - 5 > 1$ is called a **disjunction**, because it contains the word *or*. Unlike some conjunctions, it cannot be abbreviated; that is, it cannot be written without the word *or*.

Example 4 Solve $2x - 5 \leq -7$ *or* $2x - 5 > 1$. Then graph the solution set.

SOLUTION We have

$$
\begin{aligned}
2x - 5 \leq -7 \quad &or \quad 2x - 5 > 1 \\
2x \leq -2 \quad &or \quad 2x > 6 \qquad \text{Adding 5} \\
x \leq -1 \quad &or \quad x > 3. \qquad \text{Dividing by 2}
\end{aligned}
$$

The solution set is $\{x \mid x \leq -1 \text{ or } x > 3\}$. We can also write the solution using interval notation and the symbol $\cup$ for the **union** or inclusion of both sets: $(-\infty, -1] \cup (3, \infty)$. The graph of the solution set is as shown at left.

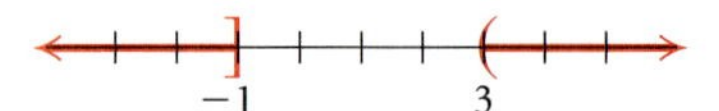

Inequalities with Absolute Value

Inequalities sometimes contain absolute value notation. The following properties are used to solve them.

For $a > 0$:

$|x| < a$ is equivalent to $-a < x < a$.

$|x| > a$ is equivalent to $x < -a$ *or* $x > a$.

Similar statements hold for $|x| \leq a$ and $|x| \geq a$.

Example 5 Solve each of the following.

a) $|3x + 2| < 5$ **b)** $|5 - 2x| \geq 1$

SOLUTION

a) $|3x + 2| < 5$

$$
\begin{aligned}
-5 < 3x + 2 < 5 \qquad &\text{Writing an equivalent inequality} \\
-7 < 3x < 3 \qquad &\text{Subtracting 2} \\
-\tfrac{7}{3} < x < 1 \qquad &\text{Dividing by 3}
\end{aligned}
$$

The solution is $\left\{x \mid -\frac{7}{3} < x < 1\right\}$, or $\left(-\frac{7}{3}, 1\right)$.

b) $|5 - 2x| \geq 1$

$$
\begin{aligned}
5 - 2x \leq -1 \quad &or \quad 5 - 2x \geq 1 \qquad \text{Writing an equivalent inequality} \\
-2x \leq -6 \quad &or \quad -2x \geq -4 \qquad \text{Subtracting 5} \\
x \geq 3 \quad &or \quad x \leq 2 \qquad \text{Dividing by } -2 \text{ and reversing the inequality signs}
\end{aligned}
$$

The solution is $\{x \mid x \leq 2 \text{ or } x \geq 3\}$, or $(-\infty, 2] \cup [3, \infty)$.

R.8 Exercise Set

Solve.

1. $x + 6 < 5x - 6$
2. $3 - x < 4x + 7$
3. $3x - 3 + 2x \geq 1 - 7x - 9$
4. $5y - 5 + y \leq 2 - 6y - 8$
5. $14 - 5y \leq 8y - 8$
6. $8x - 7 < 6x + 3$
7. $-\frac{3}{4}x \geq -\frac{5}{8} + \frac{2}{3}x$
8. $-\frac{5}{6}x \leq \frac{3}{4} + \frac{8}{3}x$
9. $4x(x - 2) < 2(2x - 1)(x - 3)$
10. $(x + 1)(x + 2) > x(x + 1)$

11. Use a grapher to check your answer to Exercise 9.

12. Use a grapher to check your answer to Exercise 10.

Write interval notation.

13. $\{x \mid -3 \leq x \leq 3\}$ **14.** $\{x \mid -4 < x < 4\}$

15. $\{x \mid -14 \leq x < -11\}$ **16.** $\{x \mid 6 < x \leq 20\}$

17. $\{x \mid x \leq -4\}$ **18.** $\{x \mid x > -5\}$

19. $\{x \mid x < 3.8\}$ **20.** $\{x \mid x \geq \sqrt{3}\}$

21. $\{x \mid x \neq 7\}$ **22.** $\{x \mid x \neq -3\}$

Write interval notation for each graph.

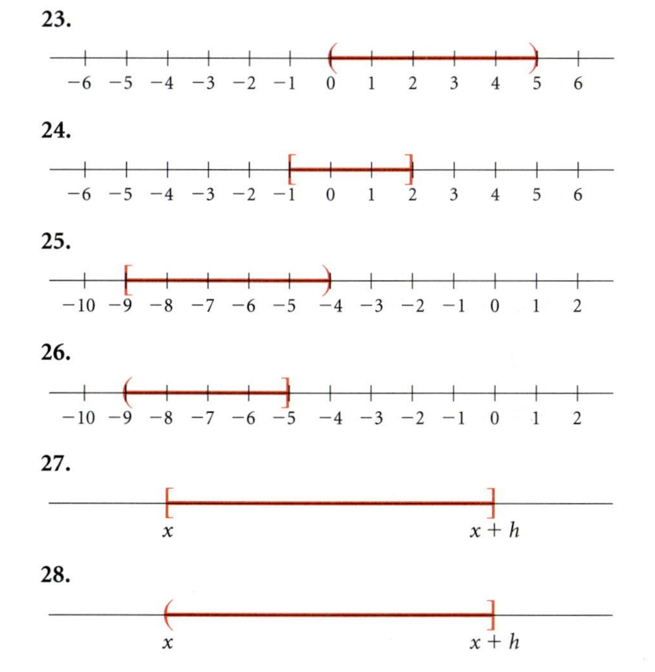

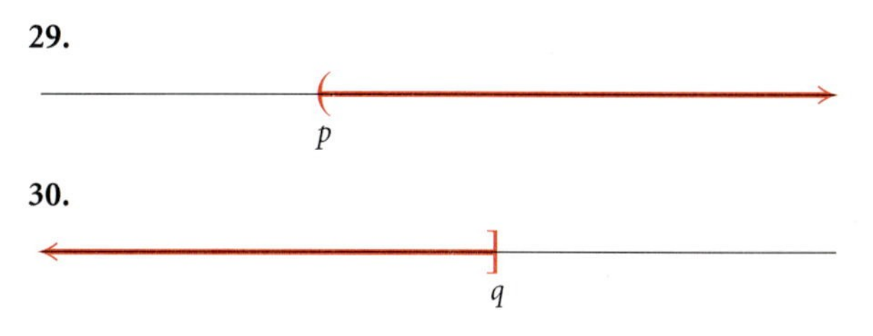

Solve. Write interval notation.

31. $-2 \leq x + 1 < 4$ **32.** $-3 < x + 2 \leq 5$

33. $5 \leq x - 3 \leq 7$ **34.** $-1 < x - 4 < 7$

35. $-3 \leq x + 4 \leq 3$ **36.** $-5 < x + 2 < 15$

37. $-2 < 2x + 1 < 5$ **38.** $-3 \leq 5x + 1 \leq 3$

39. $-4 \leq 6 - 2x < 4$ **40.** $-3 < 1 - 2x \leq 3$

41. $-5 < \frac{1}{2}(3x + 1) \leq 7$ **42.** $\frac{2}{3} \leq -\frac{4}{5}(x - 3) < 1$

43. $3x \leq -6$ *or* $x - 1 > 0$ **44.** $2x < 8$ *or* $x + 3 \geq 10$

45. $2x + 3 \leq -4$ *or* $2x + 3 \geq 4$

46. $3x - 1 < -5$ *or* $3x - 1 > 5$

47. $2x - 20 < -0.8$ *or* $2x - 20 > 0.8$

48. $5x + 11 \leq -4$ *or* $5x + 11 \geq 4$

49. $x + 14 \leq -\frac{1}{4}$ *or* $x + 14 \geq \frac{1}{4}$

50. $x - 9 < -\frac{1}{2}$ *or* $x - 9 > \frac{1}{2}$

51. Use a grapher to check your answer to Exercise 39.

52. Use a grapher to check your answer to Exercise 40.

Solve and graph the solution set.

53. $|x| < 7$ **54.** $|x| \leq 4.5$

55. $|x| \geq 4.5$ **56.** $|x| > 7$

Solve.

57. $|x + 8| < 9$ **58.** $|x + 6| \leq 10$

59. $|x + 8| \geq 9$ **60.** $|x + 6| > 10$

61. $\left|x - \frac{1}{4}\right| < \frac{1}{2}$ **62.** $|x - 0.5| \leq 0.2$

63. $|3x| < 1$ **64.** $|5x| \leq 4$

65. $|2x + 3| \leq 9$ **66.** $|3x + 4| < 13$

67. $|x - 5| > 0.1$ **68.** $|x - 7| \geq 0.4$

69. $|6 - 4x| \leq 8$ **70.** $|5 - 2x| > 10$

71. $\left|x + \frac{2}{3}\right| \leq \frac{5}{3}$ **72.** $\left|x + \frac{3}{4}\right| < \frac{1}{4}$

73. $\left|\frac{2x + 1}{3}\right| > 5$ **74.** $\left|\frac{2x - 1}{3}\right| \geq \frac{5}{6}$

75. $|2x - 4| < -5$

76. $|3x + 5| < 0$

77. Use a grapher to check your answer to Exercise 59.

78. Use a grapher to check your answer to Exercise 60.

Synthesis

79. ◆ Explain why $|x| < p$ has no solution for $p \leq 0$.

80. ◆ Explain why all real numbers are solutions of $|x| > p$, for $p < 0$.

81. ◆ Give a graphical explanation why

$$-\tfrac{1}{2}x + 1 > |x - 5|$$

has no solution.

Solve.

82. $x \leq 3x - 2 \leq 2 - x$

83. $2x \leq 5 - 7x < 7 + x$

84. $|x + 2| \leq |x - 5|$

85. $|3x - 1| > 5x - 2$

86. $|x| + |x + 1| < 10$

87. $(x - 3)^{2/3} = 4^{1/3}$

88. $|x - 3| + |2x + 5| > 6$

89. $|p - 4| + |p + 4| < 8$

R.9 Modeling and Applications

- *Solve applied problems.*

Mathematical techniques are used to answer questions arising from real-world situations. Equations and inequalities *model* many of these applications.

Solving Applied Problems

Although we will not use it in all of our problem solving, the following strategy is of great assistance.

Five Steps for Problem Solving

1. **Familiarize** yourself with the problem situation. If the problem is presented in words, then, of course, this means to read carefully. Some or all of the following can also be helpful.
   a) Make a drawing, if it makes sense to do so.
   b) Make a written list of the known facts and a list of what you wish to find out.
   c) Assign variables to represent unknown quantities.
   d) Organize the information in a chart or a table.
   e) Find further information. Look up a formula or consult a reference book or an expert in the field.
   f) Guess or estimate the answer and check your guess or estimate.

2. **Translate** the problem situation to mathematical language or symbolism. For most of the problems you will encounter in algebra, this means to write one or more equations, but sometimes an inequality or some other mathematical symbolism may be appropriate.
3. **Carry out** some type of mathematical manipulation. Use your mathematical skills to find a possible solution. In algebra, this usually means to solve an equation, an inequality, or a system of equations. The solution might also be done using a grapher.
4. **Check** to see whether your possible solution actually fits the problem situation and is thus really a solution of the problem. You might be able to solve an equation, but the solution(s) of the equation might or might not be solution(s) of the original problem.
5. **State** the answer clearly using a complete sentence.

Example 1 *U.S. versus Australian Vacation Time.* Employees of U.S. businesses receive shorter vacation times than their counterparts in other countries. For example, after one year of service, workers in Australia receive on average twice as many vacation days as U.S. workers. Together, an Australian worker and a U.S. worker have a total of 30 vacation days per year. Find the number of days each employee has. (*Source*: *Business Week*, September 4, 1995)

SOLUTION

1. **Familiarize.** Suppose the U.S. worker has 8 vacation days per year. Then the Australian worker would have $2 \cdot 8$, or 16 days, and together they have $8 + 16$, or 24 days. This estimate is too low.

   If the U.S. worker has 12 vacation days per year, then the Australian worker has $2 \cdot 12$, or 24 days, and together they have $12 + 24$, or 36 vacation days. This estimate is too high. However, we can determine from these estimates that the correct number of days is between 8 and 12 for the U.S. worker and between 16 and 24 for the Australian worker.

   We let $x =$ the number of vacation days that the U.S. employee has. Then $2x =$ the number of vacation days that the Australian employee has.
2. **Translate.** We translate as follows:

$$\underbrace{\text{The number of the U.S. employee's vacation days}}_{x} + \underbrace{\text{The number of the Australian employee's vacation days}}_{2x} = \underbrace{\text{30 days.}}_{30}$$

3. Carry out.

ALGEBRAIC SOLUTION

We solve the equation:

$$x + 2x = 30$$
$$3x = 30$$
$$x = 10.$$

GRAPHICAL SOLUTION

Graph $y_1 = x + 2x$ and $y_2 = 30$ and find the first coordinate of the point of intersection of the graphs.

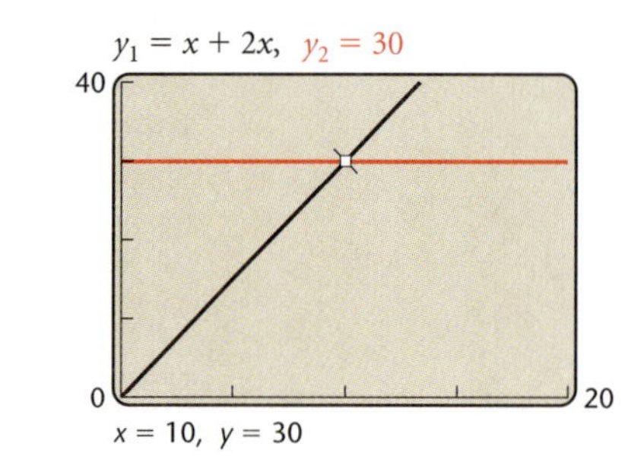

4. Check. When the U.S. employee has 10 vacation days per year, the Australian employee has $2 \cdot 10$, or 20 days. Together they have $10 + 20$, or 30 days. The answer checks.

5. State. After one year of service, U.S. employees get an average of 10 vacation days and Australian employees get an average of 20 vacation days per year.

In some applications we need to use a formula that describes the relationship between variables. When a situation involves distance, speed, and time, for example, we need to recall the **distance formula**:

$d = rt$, where d = distance, r = rate (or speed), and t = time.

Example 2 *Speed.* A 1996 BMW M3 leaves a town on the Autobahn traveling at its top speed of 237 km/h. Fifteen minutes later, a 1995 Aston Martin DB7 leaves the same town and follows the same route at its top speed of 266 km/h. How long will it take the Aston Martin to overtake the BMW? (*Sources*: *Car and Driver*, August 1995 and February 1995)

SOLUTION

1. Familiarize. We make a drawing showing both the known and the unknown information. We let t = the time, in hours, that the BMW

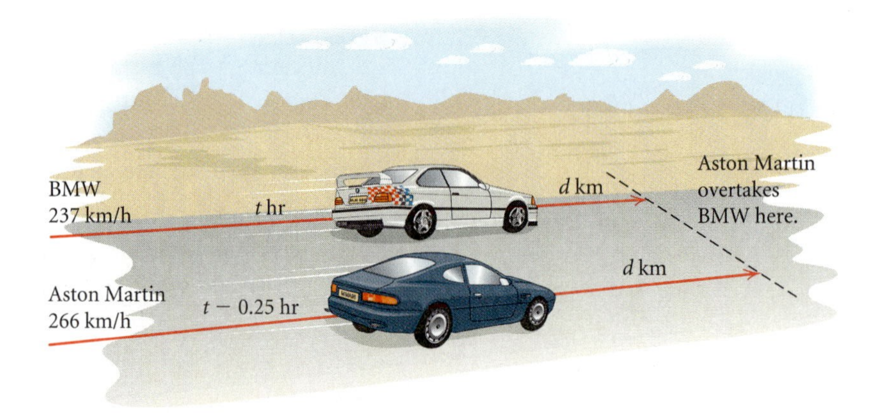

travels before being overtaken by the Aston Martin. Since the Aston Martin leaves 15 min, or 0.25 hr, after the BMW, it will travel for $t - 0.25$ hours before overtaking the BMW. The two cars have traveled the same distance, d, when one overtakes the other.

We can also organize the information in a table.

$d = r \cdot t$

	DISTANCE	RATE	TIME	
BMW	d	237	t	$\longrightarrow d = 237t$
ASTON MARTIN	d	266	$t - 0.25$	$\longrightarrow d = 266(t - 0.25)$

2. **Translate.** Using the formula $d = rt$ in each row of the table, we get two expressions for d:

$$d = 237t \quad \text{and} \quad d = 266(t - 0.25).$$

Thus we have the equation

$$237t = 266(t - 0.25).$$

3. **Carry out.**

ALGEBRAIC SOLUTION

We solve the equation:

$237t = 266(t - 0.25)$

$237t = 266t - 66.5$ Using the distributive property

$-29t = -66.5$ Subtracting $266t$ on both sides

$t \approx 2.3.$ Dividing by -29

GRAPHICAL SOLUTION

We replace t with x and d with y, graph $y_1 = 237x$ and $y_2 = 266(x - 0.25)$, and find the point of intersection of the graphs.

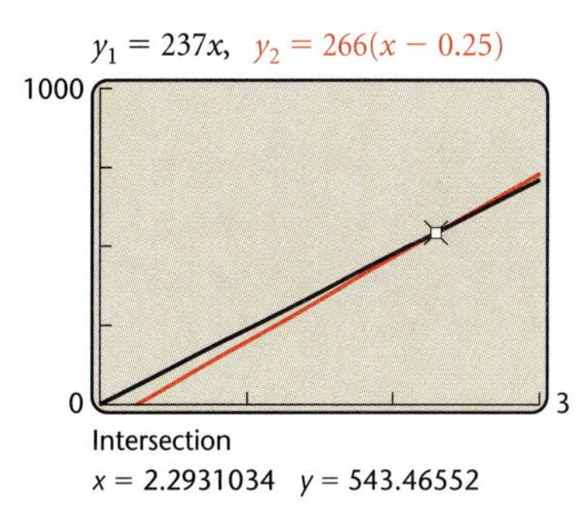

4. **Check.** In 2.3 hr, the BMW travels 237(2.3), or 545.1 km. In $2.3 - 0.25$, or 2.05 hr, the Aston Martin travels 266(2.05), or 545.3 km. Since 545.1 km $\approx$ 545.3 km, our answer checks. (The slight difference occurs because we rounded the value of t.)

5. **State.** It will take the Aston Martin about 2.3 hr to overtake the BMW.

Example 3 *Test Scores.* Kecia is taking an economics course in which there are to be four tests. To get a B or better, she must average at least 80 on the four tests. She has scores of 85, 76, and 74 on the first three tests. What scores on the last test will give her a B or better?

SOLUTION

1. **Familiarize.** Suppose Kecia gets an 82 on the fourth test. The average

of the four scores is their sum divided by the number of tests, 4:

$$\frac{85 + 76 + 74 + 82}{4} = \frac{317}{4} = 79.25.$$

Since the average must be *at least* 80 (that is, greater than or equal to 80), a score of 82 on the fourth test will not give Kecia a B.

Let's make another estimate. If Kecia gets an 88 on the fourth test, her average will be

$$\frac{85 + 76 + 74 + 88}{4} = \frac{323}{4} = 80.75.$$

This score will give her a B.

These two estimates tell us that the lowest score Kecia can get on the fourth test to ensure a grade of B or better is some number between 82 and 88. In order to find *all* scores that will give her a B or better, we translate to an inequality. We let s = Kecia's score on the fourth test.

2. **Translate.** Since the average must be *at least* 80, we translate to the inequality

$$\frac{85 + 76 + 74 + s}{4} \geq 80.$$

3. **Carry out.**

ALGEBRAIC SOLUTION

We solve the inequality:

$$4\left(\frac{85 + 76 + 74 + s}{4}\right) \geq 4 \cdot 80$$ **Multiplying by 4 to clear the fraction**

$$85 + 76 + 74 + s \geq 320$$

$$235 + s \geq 320$$ **Combining like terms**

$$s \geq 85.$$ **Subtracting 235 on both sides**

GRAPHICAL SOLUTION

Graph

$$y_1 = \frac{85 + 76 + 74 + x}{4} \quad \text{and} \quad y_2 = 80$$

and determine the values of x for which $y_1 \geq y_2$.

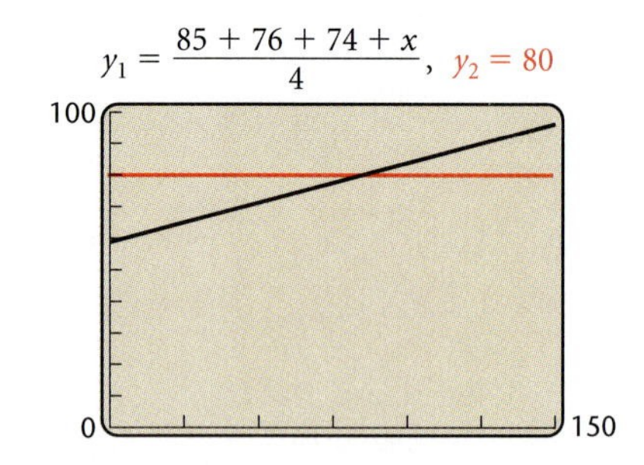

Note that $y_1 \geq y_2$ for $x \geq 85$.

4. **Check.** We can obtain a partial check by substituting a number less than 85 and then a number greater than 85 in the original inequality. We did this in the *Familiarize* step. Our answer appears to be correct.

5. **State.** A score of 85 or higher will give Kecia a B or better.

R.9 Exercise Set

Solve.

1. *Dow Chemical.* In 1994, the Dow Chemical Company spent a total of \$4.071 billion on wages and salaries and on employee benefits. The amount spent on wages and salaries was \$2.407 billion more than the amount spent on employee benefits. Find the amount spent in each category. (*Source*: Dow Chemical Company, 1994 Annual Report)

2. *Corporate Fax Expense.* The average *Fortune 500* company spends \$13 million annually for local and domestic long-distance fax service. The amount spent for local faxes is about one third the amount spent for domestic long-distance faxes. Find the amount spent for each type of fax service. (*Source*: Gallup survey of *Fortune 500* companies for Pitney Bowes)

3. *Test Plot Dimensions.* A flower seed company has a rectangular test plot with a perimeter of 322 m. The length is 25 m more than the width. Find the dimensions of the plot.

4. *Garden Dimensions.* The children at Tiny Tots Day Care plant a rectangular vegetable garden with a perimeter of 39 m. The length is twice the width. Find the dimensions of the garden.

Time of a Free Fall. The formula $s = 16t^2$ is used to approximate the distance s, in feet, that an object falls freely from rest in t seconds.

5. The Warszawa Radio Mast in Poland, at 2120 ft, is the world's tallest structure. How long would it take an object falling freely from the top to reach the ground? (*Source*: David Crystal, ed., *The Cambridge Fact Finder.* Cambridge: Cambridge University Press, p. 526)

6. The tallest structure in the United States, at 2063 ft, is the KTHI-TV tower in North Dakota. How long would it take an object falling freely from the top to reach the ground? (*Source*: David Crystal, ed., *The Cambridge Fact Finder.* Cambridge: Cambridge University Press, p. 526)

7. *Box Construction.* An open box is made from a 10-cm by 20-cm piece of tin by cutting a square from each corner and folding up the edges. The area of the resulting base is 96 cm². What is the length of the sides of the squares?

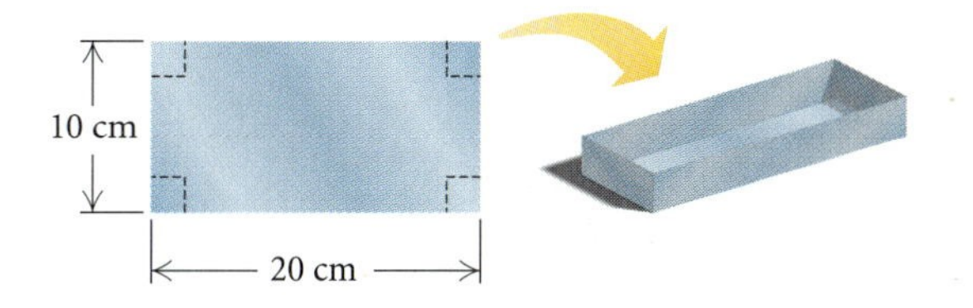

8. *Picture Frame Dimensions.* The frame of a picture is 28 cm by 32 cm outside and is of uniform width. What is the width of the frame if 192 cm² of the picture shows?

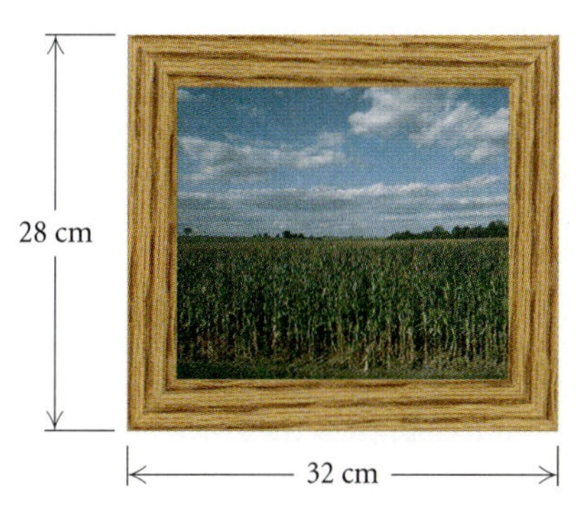

9. *Angle Measure.* In triangle ABC, angle B is five times as large as angle A. The measure of angle C is 2° less than that of angle A. Find the measures of the angles. (*Hint:* The sum of the angle measures is 180°.)

10. *Angle Measure.* In triangle ABC, angle B is twice as large as angle A. Angle C measures 20° more than angle A. Find the measures of the angles.

11. *HMO Enrollment.* The number of enrollees in health maintenance organizations (HMOs) in Kentucky in a recent year was 416,300. This was a 48.4% increase over the previous year. What was the former number of enrollees? (*Source:* Marion Merrell Dow Managed Care Department)

12. *Moving Costs.* The average cost of moving a distance of 1255 mi is $1035 if you handle the move yourself. This is about 52% of the cost of hiring a professional moving firm. What is the cost of the professionally handled move? (*Source:* American Movers Conference)

13. *Bicycling Speed.* Castulo and Linette leave a campsite, Castulo biking due north and Linette biking due east. Castulo bikes 7 km/h slower than Linette. After 4 hr, they are 68 km apart. Find the speed of each bicyclist.

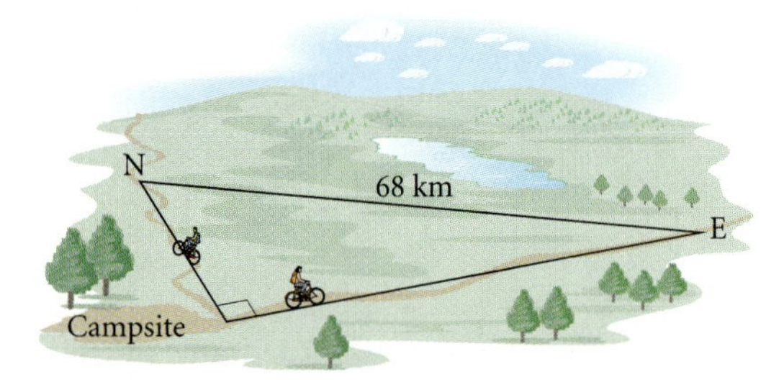

14. *Boat Speed.* A Bayline Cruises sightseeing boat travels 36 mi upriver and 36 mi back in 5 hr. The speed of the river's current is 3 mph. Find the speed of the boat in still water.

15. *Train Speeds.* The speed of an Amtrak passenger train is 14 mph faster than the speed of a Central Railway freight train. The passenger train travels 400 mi in the same time it takes the freight train to travel 330 mi. Find the speed of each train.

16. *Truck Speed.* For the first 80 mi of a trip, Micah's pickup truck traveled 10 mph slower than it did for the remaining 50 mi. The total time for the trip was 3 hr. Find the speed of the truck on each part of the trip.

17. *Distance Traveled.* An airplane leaves Chicago for Cleveland at a speed of 475 mph. Twenty minutes later, a plane going to Chicago leaves Cleveland, which is 350 mi from Chicago, at a speed of 500 mph. When they meet, how far are they from Cleveland?

18. *Distance Traveled.* A private airplane leaves Midway airport and flies due east at a speed of 180 km/h. Two hours later, a jet leaves Midway and flies due east at a speed of 900 km/h. How far from the airport will the jet overtake the private plane?

19. *Compound Interest.* An investment is made at 8%, compounded annually. It grows to $702 at the end of 1 yr. How much was originally invested?

20. *Compound Interest.* An investment made at 9%, compounded annually, grows to $926.50 at the end of 1 yr. How much was originally invested?

21. *Test Scores.* Jason is taking a literature course in which there are to be four tests. He has scores of 82, 73, and 79 on the first three tests. To get a B or better, he must average at least 80 on the four tests. What scores on the last test will give Jason a B or better?

22. *Test Scores.* Monica is taking an accounting course in which there are to be five tests, each worth 100 points. She has scores of 93, 89, and 95 on the first three tests. To get an A, she must score a total of at least 450 points. What scores on the fourth test will keep her eligible for an A?

23. *Car Rental.* A car can be rented for $35 per day with unlimited mileage or for $30 per day plus 15¢ per mile. For what daily mileages would the unlimited mileage plan cost less?

24. *Car Rental.* A car can be rented for $28 per day plus 18¢ per mile. A salesperson is on a daily car-rental budget of $91. What mileages will allow her to stay within the budget?

25. *Cost of a Service Call.* Dennis Appliance Service charges $50 plus $20 per hour for a service call. Economy Appliance charges $45 per hour for a service call. For what lengths of time is Economy Appliance more expensive?

26. *Cost of a Taxi Ride.* A taxi ride with Mason's Cab costs $1.50 for the first mile and 45¢ for each additional mile. Bell's Taxi charges 60¢ per mile. For what distances is Mason's Cab less expensive?

27. *Salary Plans.* On your new job, you can choose to be paid in one of two ways:

    Plan A: A salary of $1600 per month plus a commission of 5% of gross sales;

    Plan B: A salary of $1800 per month plus a commission of 7% of gross sales over $8000.

    For what gross sales is plan B better than plan A, assuming that gross sales are always more than $8000?

28. *Insurance Plans.* An insurance company offers two plans. With the first plan, the employee pays the first $500 of medical bills and the insurance company pays 80% of the rest. With the second plan, the employee pays the first $900 of medical bills and the insurance company pays 90% of the rest. For what amount of medical bills will the second plan cost the employee less?

Synthesis

29. ◈ Write a problem for a classmate to solve. Devise the problem so that the solution is "The dimensions of the rectangle are 36 in. by 18 in."

30. ◈ Explain why, when solving a problem, it is necessary to check your solution in the original problem situation.

31. *Speed.* A compact car is driven 144 mi. If it had traveled 4 mph faster, it could have made the trip in $\frac{1}{2}$ hr less time. What was the speed of the car?

32. *Speed.* A pickup truck is driven 280 mi. If it had traveled 5 mph faster, it could have made the trip in 1 hr less time. What was the speed of the truck?

33. *Average Speed.* Krista drove 3 hr on a freeway at a speed of 55 mph and then drove 10 mi in the city at 35 mph. What was the average speed? (*Average speed* is defined as total distance divided by total time.)

34. *Average Speed.* For the first 100 km of a 200-km trip, Bert drove at a speed of 40 km/h. For the second half of the trip, he drove at 60 km/h. What was the average speed for the entire trip? (See Exercise 33.)

35. *Average Speed.* Luke drove half the distance of a trip at a speed of 40 mph. At what speed would he have to drive for the rest of the distance so that the average speed for the entire trip would be 45 mph and the trip would be completed in 1 hr? (See Exercise 33.)

36. *Distance Traveled.* Madison drives to work at a speed of 45 mph and arrives 1 min early. At 40 mph, she would arrive 1 min late. How long is her trip to work?

37. *Distribution of Money.* Suppose your father gives half of all the money in his pockets to your mother. Then he gives one fourth of what is left to your sister, and one third of what is left after that to your brother. He then gives you half of what is left, which happens to be \$2. How much was in his pocket at the outset?

38. *Pie Sales.* Jared walks into a bakery and says to the owner, "I will buy half of all the pies in the store, plus half a pie." The sale is made. Phyllis then comes into the store and makes the same statement and purchase. Then Dexter does likewise. The owner then has exactly one pie left. How many pies did the owner have in the store at the outset?

CHAPTER R Summary and Review

Important Properties and Formulas

Properties of Absolute Value

$|a| \geq 0$

$|ab| = |a| \cdot |b|$

$\left|\frac{a}{b}\right| = \frac{|a|}{|b|} \quad (b \neq 0)$

$|-a| = |a|$

Pythagorean Theorem

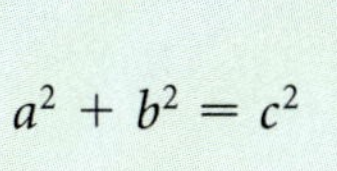

$a^2 + b^2 = c^2$

Properties of the Real Numbers

Commutative: $a + b = b + a$; $ab = ba$

Associative: $a + (b + c) = (a + b) + c$; $a(bc) = (ab)c$

Additive Identity: $a + 0 = 0 + a = a$

Additive Inverse: $-a + a = a + (-a) = 0$

Multiplicative Identity: $a \cdot 1 = 1 \cdot a = a$

Multiplicative Inverse: $a \cdot \frac{1}{a} = 1 \quad (a \neq 0)$

Distributive: $a(b + c) = ab + ac$

(continued)

Properties of Exponents

For any real numbers a and b and any integers m and n, assuming 0 is not raised to a nonpositive power:

The Product Rule: $a^m \cdot a^n = a^{m+n}$

The Quotient Rule: $\frac{a^m}{a^n} = a^{m-n} \quad (a \neq 0)$

The Power Rule: $(a^m)^n = a^{mn}$

Raising a Product to a Power:
$(ab)^m = a^m b^m$

Raising a Quotient to a Power:
$\left(\frac{a}{b}\right)^m = \frac{a^m}{b^m} \quad (b \neq 0)$

Special Products of Binomials

$(A + B)^2 = A^2 + 2AB + B^2$

$(A - B)^2 = A^2 - 2AB + B^2$

$(A + B)(A - B) = A^2 - B^2$

Sum or Difference of Cubes

$A^3 + B^3 = (A + B)(A^2 - AB + B^2)$

$A^3 - B^3 = (A - B)(A^2 + AB + B^2)$

Compound Interest Formula

$$A = P\left(1 + \frac{i}{n}\right)^{nt}$$

Properties of Radicals

Let a and b be any real numbers or expressions for which the given roots exist. For any natural numbers m and n ($n \neq 1$):

If n is even, $\sqrt[n]{a^n} = |a|$.

If n is odd, $\sqrt[n]{a^n} = a$.

$\sqrt[n]{a} \cdot \sqrt[n]{b} = \sqrt[n]{ab}$.

$\sqrt[n]{\frac{a}{b}} = \frac{\sqrt[n]{a}}{\sqrt[n]{b}} \quad (b \neq 0)$.

$\sqrt[n]{a^m} = (\sqrt[n]{a})^m$.

Five Steps for Problem Solving

1. Familiarize.
2. Translate.
3. Carry out.
4. Check.
5. State.

Equation-Solving Principles

The Addition Principle: If $a = b$ is true, then $a + c = b + c$ is true.

The Multiplication Principle: If $a = b$ is true, then $ac = bc$ is true.

The Principle of Zero Products: If $ab = 0$ is true, then $a = 0$ or $b = 0$, and if $a = 0$ or $b = 0$, then $ab = 0$.

The Principle of Square Roots: If $x^2 = k$, then $x = \sqrt{k}$ or $x = -\sqrt{k}$.

The Principle of Powers: For any positive integer n, if $a = b$ is true, then $a^n = b^n$ is true.

Principles for Solving Inequalities

The Addition Principle for Inequalities: If $a < b$ is true, then $a + c < b + c$ is true.

The Multiplication Principle for Inequalities: If $a < b$ and $c > 0$ are true, then $ac < bc$ is true.
If $a < b$ and $c < 0$ are true, then $ac > bc$ is true.

Similar statements hold for $\leq$.

Equations and Inequalities with Absolute Value

For $a > 0$,

$|x| = a \longrightarrow x = -a \quad \text{or} \quad x = a$

$|x| < a \longrightarrow -a < x < a$

$|x| > a \longrightarrow x < -a \quad \text{or} \quad x > a$

REVIEW EXERCISES

Consider the numbers -43.89, 12, -3, $-\frac{1}{5}$, $\sqrt{7}$, $\sqrt[3]{10}$, -1, $-\frac{4}{3}$, $7\frac{2}{3}$, -19, 31, 0.

1. Which are integers?
2. Which are natural numbers?
3. Which are rational numbers?
4. Which are real numbers?
5. Which are irrational numbers?
6. Which are whole numbers?

Simplify.

7. $|-3.5|$
8. $\left|-\frac{5}{6}a^2b\right|$
9. Find the distance between -7 and 3 on the number line.
10. Determine graphically whether $|x^3| = x^3$ appears to be an identity.

Compute.

11. $5^3 - [2(4^2 - 3^2 - 6)]^3$
12. $\dfrac{3^4 - (6 - 7)^4}{2^3 - 2^4}$

Convert to decimal notation.

13. 3.261×10^6
14. 4.1×10^{-4}

Convert to scientific notation.

15. 0.01432
16. $43{,}210$

Compute. Write scientific notation for the answer.

17. $\dfrac{2.5 \times 10^{-8}}{3.2 \times 10^{13}}$
18. $(8.4 \times 10^{-17})(6.5 \times 10^{-16})$

Simplify.

19. $(7a^2b^4)(-2a^{-4}b^3)$
20. $\dfrac{54x^6y^{-4}z^2}{9x^{-3}y^2z^{-4}}$
21. $\sqrt[4]{81}$
22. $\sqrt[5]{-32}$
23. $\dfrac{b - a^{-1}}{a - b^{-1}}$
24. $\dfrac{\dfrac{x^2}{y} + \dfrac{y^2}{x}}{y^2 - xy + x^2}$
25. $(\sqrt{3} - \sqrt{7})(\sqrt{3} + \sqrt{7})$
26. $(5x^2 - \sqrt{2})^2$
27. $8\sqrt{5} + \dfrac{25}{\sqrt{5}}$
28. $(x + t)(x^2 - xt + t^2)$
29. $(5a + 4b)^3$
30. $(5xy^4 - 7xy^2 + 4x^2 - 3) - (-3xy^4 + 2xy^2 - 2y + 4)$

Factor.

31. $x^3 + 2x^2 - 3x - 6$
32. $12a^3 - 27ab^4$
33. $24x + 144 + x^2$
34. $9x^3 + 35x^2 - 4x$
35. $8x^3 - 1$
36. $27x^6 + 125y^6$
37. $6x^3 + 48$
38. $4x^3 - 4x^2 - 9x + 9$
39. $9x^2 - 30x + 25$
40. $18x^2 - 3x + 6$
41. $9x^2 + 6xy + y^2 + 9x + 3y - 4$
42. Divide and simplify:
$$\frac{3x^2 - 12}{x^2 + 4x + 4} \div \frac{x - 2}{x + 2}.$$
43. Subtract and simplify:
$$\frac{x}{x^2 + 9x + 20} - \frac{4}{x^2 + 7x + 12}.$$

Write an expression containing a single radical.

44. $\sqrt{y^5}\sqrt[3]{y^2}$
45. $\dfrac{\sqrt{(a + b)^3}\sqrt[3]{a + b}}{\sqrt[6]{(a + b)^7}}$
46. Convert to radical notation: $b^{7/5}$.
47. Convert to exponential notation and simplify:
$$\sqrt[8]{\frac{m^{32}n^{16}}{3^8}}.$$
48. Rationalize the numerator:
$$\frac{\sqrt{x} - \sqrt{y}}{\sqrt{x} + \sqrt{y}}.$$

Factor and simplify.

49. $2x^{1/2}y^{-3/4} - 3x^{-1/2}y^{1/4}$
50. $(x - 2)^{-3/4}(3x + 5)^{5/2} + (x - 2)^{1/4}(3x + 5)^{3/2}$
51. How long is a guy wire reaching from the top of a 17-ft pole to a point on the ground 8 ft from the pole?

Solve.

52. $4y - 5 = 11$

53. $5(3x + 1) = 2(x - 4)$

54. $(2y + 5)(3y - 1) = 0$

55. $3x^2 + 2x = 8$

56. $\frac{5}{2x + 3} + \frac{1}{x - 6} = 0$

57. $\frac{3}{8x + 1} + \frac{8}{2x + 5} = 1$

58. $\sqrt{5x + 1} - 1 = \sqrt{3x}$

59. $\sqrt{x - 1} - \sqrt{x - 4} = 1$

60. $|x - 4| = 3$

61. $|2y + 7| = 9$

62. $-3 \leq 3x + 1 < 5$

63. $-2 < 5x - 4 \leq 6$

64. $2x < -1$ *or* $x + 3 > 0$

65. $3x + 7 \leq 2$ *or* $2x + 3 \geq 5$

66. $|6x - 1| < 5$

67. $|x + 4| \geq 2$

68. Solve $v = \sqrt{2gh}$ for h.

69. Solve $\frac{1}{a} + \frac{1}{b} = \frac{1}{t}$ for t.

70. *Legs of a Right Triangle.* The hypotenuse of a triangle is 50 ft. One leg is 10 ft longer than the other. What are the lengths of the legs?

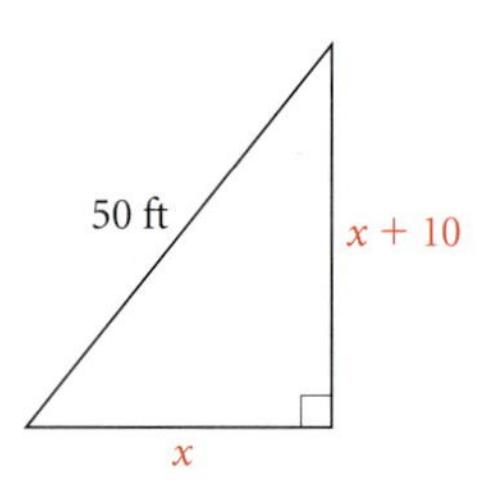

71. *Test Scores.* Jeremiah scores 73 and 79 on two tests. If the third test counts as though it were two tests, what score must he make on the third test so the average will be 85?

72. *Motion.* A Riverboat Cruise Line boat travels 8 mi upstream and 8 mi downstream. The total time for both parts of the trip is 3 hr. The speed of the stream is 2 mph. What is the speed of the boat in still water?

73. *Motion.* Two freight trains leave the same city at right angles. The first train travels at a speed of 60 km/h. In 1 hr, the trains are 100 km apart. How fast is the second train traveling?

74. *Car Rental.* A car can be rented for $32 per day with unlimited mileage or for $28 per day plus 16¢ per mile. For what daily mileages would the unlimited mileage plan cost less?

75. *Cost of a Service Call.* Harris Heating and Cooling charges $40 plus $25 per hour for a service call. Quality Cool charges $45 per hour for a service call. For what lengths of time is Quality Cool less expensive?

Synthesis

76. ◈ Write a problem for a classmate to solve. Devise the problem so that the solution is "Nisha must get a score of at least 84 on the fourth test in order to get a B or better in the course."

77. ◈ Explain the difference between equivalent expressions and equivalent equations.

Multiply. Assume that all exponents are integers.

78. $(x^n + 10)(x^n - 4)$

79. $(t^a + t^{-a})^2$

80. $(y^b - z^c)(y^b + z^c)$

81. $(a^n - b^n)^3$

Factor.

82. $y^{2n} + 16y^n + 64$

83. $x^{2t} - 3x^t - 28$

84. $m^{6n} - m^{3n}$

85. Solve: $\sqrt{\sqrt{\sqrt{x}}} = 2$.

1 Graphs, Functions, and Models

APPLICATION

During a thunderstorm, it is possible to calculate how far away, y (in miles), lightning is when the sound of thunder arrives x seconds after the lightning has been sighted. The relationship between x and y can be described by a function, $y = f(x) = \frac{1}{5}x$.

This weather application is just one example of the widespread use of graphs in today's society. In this chapter, we will focus attention on functions and equations with graphs that can be used to model real-life situations mathematically. Such models are invaluable when doing analyses and making predictions in fields ranging from astronomy to zoology.

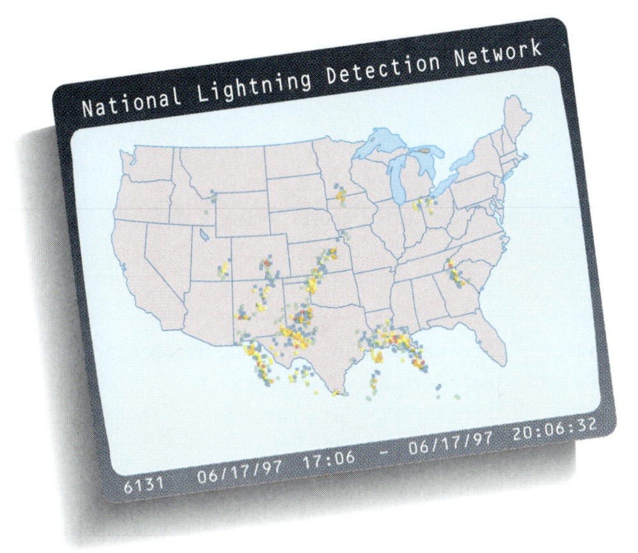

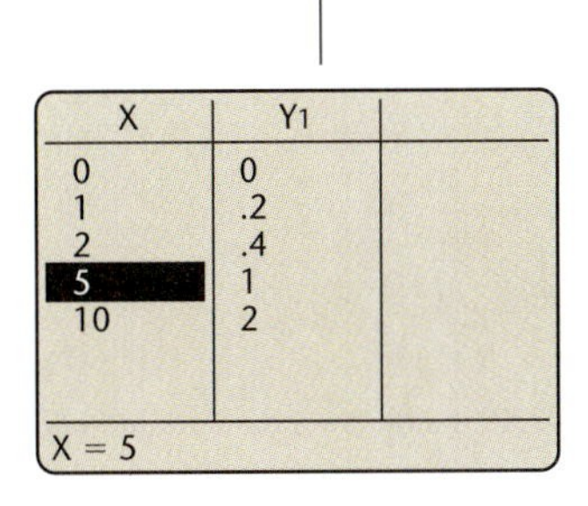

X	Y1	
0	0	
1	.2	
2	.4	
5	1	
10	2	

X = 5

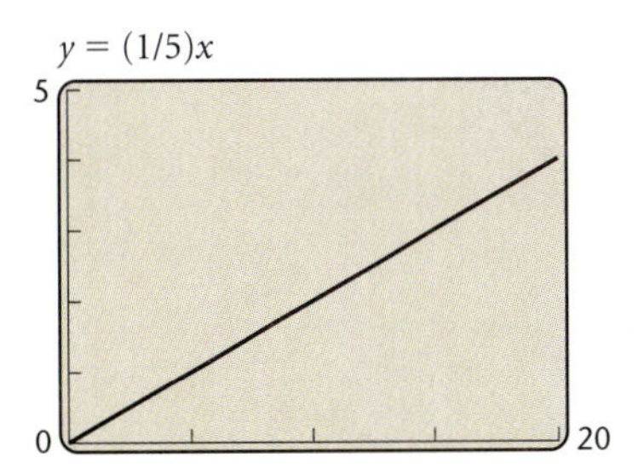

1.1 Functions, Graphs, and Graphers
1.2 Functions and Applications
1.3 Linear Functions and Applications
1.4 Data Analysis, Curve Fitting, and Linear Regression
1.5 Distance, Midpoints, and Circles
1.6 Symmetry
1.7 Transformations of Functions
1.8 The Algebra of Functions
SUMMARY AND REVIEW

1.1 Functions, Graphs, and Graphers

- *Determine whether a correspondence or a relation is a function.*
- *Find function values, or outputs, using a formula.*
- *Find the domain and the range of a function.*
- *Determine whether a graph is that of a function.*
- *Solve application problems using functions.*

We now focus attention on a concept that is fundamental to many areas of mathematics—the idea of a *function*.

Functions

We first consider an application.

Lightning–Time–Thunder Distance. During a thunderstorm, it is possible to calculate how far away, y (in miles), lightning is when the sound of thunder arrives x seconds after the lightning has been sighted. We can examine the relationship between x and y in several ways:

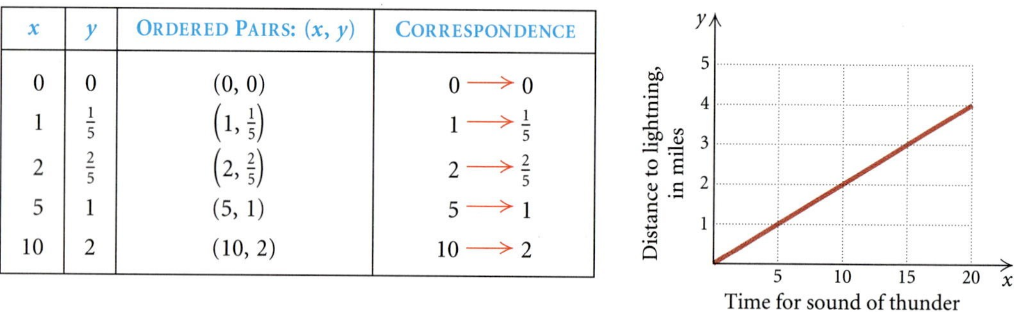

x	y	ORDERED PAIRS: (x, y)	CORRESPONDENCE
0	0	$(0, 0)$	$0 \longrightarrow 0$
1	$\frac{1}{5}$	$\left(1, \frac{1}{5}\right)$	$1 \longrightarrow \frac{1}{5}$
2	$\frac{2}{5}$	$\left(2, \frac{2}{5}\right)$	$2 \longrightarrow \frac{2}{5}$
5	1	$(5, 1)$	$5 \longrightarrow 1$
10	2	$(10, 2)$	$10 \longrightarrow 2$

The ordered pairs imply a relationship, or correspondence, between the first and second coordinates. We can see this relationship in the graph as well. There is also an equation that describes the correspondence. It is

$$y = \frac{1}{5}x.$$

This is an example of a *function*.

Let's consider some other functions before giving a definition.

DOMAIN		RANGE
To each registered student	there corresponds	an I. D. number.
To each mountain bike sold	there corresponds	its price.
To each number between -3 and 3	there corresponds	the square of that number.

In each example, the first set is called the **domain** and the second set is called the **range**. For each member, or **element**, in the domain, there is *exactly one* member of the range to which it corresponds. Thus each registered student has exactly *one* I. D. number, each mountain bike has exactly *one* price, and each number between -3 and 3 has exactly *one* square. Each correspondence is a *function*.

Function

A *function* is a correspondence between a first set, called the *domain*, and a second set, called the *range*, such that each member of the domain corresponds to *exactly one* member of the range.

It is important to note that not every correspondence between two sets is a function.

Example 1 Determine whether each of the following correspondences is a function.

a) $-3 \to 9$
   $3 \to 9$
   $2 \to 4$

b) Juan → Casandra
   Boris → Rebecca
   Boris → Helga
   Nelson → Natasha
   Bernie → Natasha

SOLUTION

a) This correspondence *is* a function because each member of the domain corresponds to exactly one member of the range.

b) This correspondence *is not* a function because there is a member of the domain (Boris) that is paired with two different members of the range.

Example 2 Determine whether each of the following correspondences is a function.

	DOMAIN	CORRESPONDENCE	RANGE
a)	Years in which a president is elected	The person elected	A set of presidents
b)	The integers	Each number's cube root	A set of real numbers
c)	All states in the United States	A senator from that state	The set of all U.S. senators

SOLUTION

a) This correspondence *is* a function, because in each presidential election *exactly one* president is elected.

b) This correspondence *is* a function, because each integer has *exactly one* cube root.

c) This correspondence *is not* a function, because each state can be paired with *two* different senators.

When a correspondence between two sets is not a function, it is still an example of a **relation**.

Relation

A *relation* is a correspondence between a first set, called the *domain*, and a second set, called the *range*, such that each member of the domain corresponds to *at least one* member of the range.

All the correspondences in Examples 1 and 2 are relations, but, as we have seen, not all are functions. Relations are sometimes written as sets of ordered pairs in which elements of the domain are the first part (coordinate) of each ordered pair and elements of the range are the second part (coordinate). For example, instead of writing $-3 \longrightarrow 9$, we write the ordered pair $(-3, 9)$.

Example 3 Determine whether each of the following relations is a function. Identify the domain and the range.

a) $\{(-2, 5), (5, 7), (0, 1), (4, -2)\}$

b) $\{(9, -5), (9, 5), (2, 4)\}$

SOLUTION

a) The relation *is* a function because no two ordered pairs have the same first coordinate and different second coordinates.

The domain is the set of all first coordinates: $\{-2, 5, 0, 4\}$.

The range is the set of all second coordinates: $\{5, 7, 1, -2\}$.

b) The relation *is not* a function because the ordered pairs $(9, -5)$ and $(9, 5)$ have the same first coordinate and different second coordinates.

The domain is the set of all first coordinates: $\{9, 2\}$.

The range is the set of all second coordinates: $\{-5, 5, 4\}$.

Notation for Functions

Functions used in mathematics are often given by equations. They generally require that certain calculations be performed in order to determine which member of the range is paired with each member of the domain. For example, in the Introduction to Graphs and Graphers, we graphed the function $y = x^2 - 5$ by doing calculations like the following:

$$\text{for } x = 3,\ y = 3^2 - 5 = 4,$$
$$\text{for } x = 2,\ y = 2^2 - 5 = -1,$$
$$\text{for } x = 1,\ y = 1^2 - 5 = -4, \quad \text{and so on.}$$

A more concise notation is often used. For $y = x^2 - 5$, the **inputs** (members of the domain) are values of x substituted into the equation. The **outputs** (members of the range) are the resulting values of y. If we call the function f, we can use x to represent an arbitrary *input* and $f(x)$—read "f of x", or "f at x", or "the value of f at x"—to represent the corresponding *output*. In this notation, the function given by $y = x^2 - 5$ is written as $f(x) = x^2 - 5$ and the above calculations would be

$$f(3) = 3^2 - 5 = 4,$$
$$f(2) = 2^2 - 5 = -1,$$
$$f(1) = 1^2 - 5 = -4.$$

Keep in mind that $f(x)$ *does not* mean $f \cdot x$.

Thus, instead of writing "when $x = 3$, the value of y is 4," we can simply write "$f(3) = 4$," which can also be read as "f of 3 is 4" or "for the input 3, the output of f is 4."

Example 4 A function f is given by $f(x) = 2x^2 - x + 3$. Find each of the following.

a) $f(0)$ **b)** $f(-7)$

c) $f(5a)$ **d)** $f(a - 4)$

SOLUTION One way to find function values when a formula is given is to think of the formula as follows:

$$f(\blacksquare) = 2(\blacksquare)^2 - (\blacksquare) + 3.$$

Then to find an output for a given input we think: "Whatever goes in the blank on the left goes in the blank(s) on the right." This gives us a "recipe" for finding outputs. We can now solve the example.

a) $f(0) = 2 \cdot (0)^2 - 0 + 3 = 0 - 0 + 3 = 3$

b) $f(-7) = 2(-7)^2 - (-7) + 3 = 2 \cdot 49 + 7 + 3 = 108$

c) $f(5a) = 2(5a)^2 - 5a + 3 = 2 \cdot 25a^2 - 5a + 3 = 50a^2 - 5a + 3$

d) $f(a - 4) = 2(a - 4)^2 - (a - 4) + 3 = 2(a^2 - 8a + 16) - a + 4 + 3$
$= 2a^2 - 16a + 32 - a + 4 + 3$
$= 2a^2 - 17a + 39$

It is important for students preparing for calculus to be able to simplify rational expressions like

$$\frac{f(a + h) - f(a)}{h}.$$

Example 5 For the function f given by $f(x) = 2x^2 - x - 3$, construct and simplify the expression

$$\frac{f(a + h) - f(a)}{h}.$$

SOLUTION We find $f(a + h)$ and $f(a)$:

$$\begin{aligned} f(a + h) &= 2(a + h)^2 - (a + h) - 3 \\ &= 2[a^2 + 2ah + h^2] - a - h - 3 \\ &= 2a^2 + 4ah + 2h^2 - a - h - 3, \end{aligned}$$

$$f(a) = 2a^2 - a - 3.$$

Then

$$\frac{f(a + h) - f(a)}{h}$$

$$= \frac{[2a^2 + 4ah + 2h^2 - a - h - 3] - [2a^2 - a - 3]}{h}$$

$$= \frac{2a^2 + 4ah + 2h^2 - a - h - 3 - 2a^2 + a + 3}{h} = \frac{4ah + 2h^2 - h}{h}$$

$$= \frac{h(4a + 2h - 1)}{h \cdot 1} = \frac{h}{h} \cdot \frac{4a + 2h - 1}{1} = 4a + 2h - 1.$$

Finding Domains of Functions

When a function is given by a formula, we find function values by making substitutions of numbers or inputs into the formula. When a function f, whose inputs and outputs are real numbers, is given by a formula, the *domain* is understood to be the set of all inputs for which the expression is defined as a real number. When a substitution results in an expression that is not defined as a real number, we say that the function value *does not exist* and that the number being substituted *is not* in the domain of the function.

Example 6 Find the indicated function values. Simplify, if possible.

a) $f(1)$ and $f(3)$, for $f(x) = \dfrac{1}{x - 3}$

b) $g(16)$ and $g(-7)$, for $g(x) = \sqrt{x} + 5$

SOLUTION

a) $f(1) = \dfrac{1}{1 - 3} = \dfrac{1}{-2} = -\dfrac{1}{2}$;

$f(3) = \dfrac{1}{3 - 3} = \dfrac{1}{0}$

Since division by 0 is not defined, the number 3 is not in the domain of f. Thus, $f(3)$ does not exist.

b) $g(16) = \sqrt{16} + 5 = 4 + 5 = 9$;
$g(-7) = \sqrt{-7} + 5$
Since $\sqrt{-7}$ is not defined as a real number, the number -7 is not in the domain of g. Thus, $g(-7)$ does not exist.

We can see from Example 7 that inputs that make a denominator 0 or a square-root radicand negative are not in the domain of a function.

Example 7 Find the domain of each of the following functions.

a) $f(x) = \dfrac{1}{x - 3}$

b) $g(x) = \sqrt{x} + 5$

c) $h(x) = \dfrac{3x^2 - x + 7}{x^2 + 2x - 3}$

d) $p(x) = \sqrt{4 - x^2}$

e) $F(x) = x^3 + |x|$

SOLUTION

a) The input 3 results in a denominator of 0. The domain is $\{x \mid x \neq 3\}$, or $(-\infty, 3) \cup (3, \infty)$.

b) We can substitute any number for which the radicand is nonnegative, that is, for which $x \geq 0$. Thus the domain is $\{x \mid x \geq 0\}$, or the interval $[0, \infty)$.

c) Although we can substitute any real number in the numerator, we must avoid inputs that make the denominator 0. To find those inputs, we solve $x^2 + 2x - 3 = 0$, or $(x + 3)(x - 1) = 0$. Thus the domain consists of the set of all real numbers except -3 and 1, or $\{x \mid x \neq -3$ *and* $x \neq 1\}$, or $(-\infty, -3) \cup (-3, 1) \cup (1, \infty)$.

d) We must avoid inputs for which the radicand is negative. Thus the domain is all real numbers for which $4 - x^2 \geq 0$, or $x^2 \leq 4$. These inputs are all numbers in the interval $[-2, 2]$. You can check this by making some substitutions or, on a grapher, by looking at the graphs of $y_1 = x^2$ and $y_2 = 4$.

e) All substitutions are suitable. The domain is the set of all real numbers, $\mathbb{R}$.

Function Values and Tables on a Grapher

A grapher can be used in many different ways to find function values. We can simply do a calculation. The process is the same as that described in the Introduction to Graphs and Graphers.

With a TABLE feature, function values can be displayed for a string of numbers like $x = -3, -2.75, -2.5, -2.25, -2, -1.75$, and so on. Consider the function $f(x) = \sqrt{4 - x^2}$. Shown at left is part of a table for this function. Note that we entered the function as $y = \sqrt{}(4 - x^2)$, or $(4 - x\wedge 2)\wedge 0.5$. An ERROR message tells us when a number is not in the domain.

X	Y1	
−3	ERROR	
−2.75	ERROR	
−2.5	ERROR	
−2.25	ERROR	
−2	0	
−1.75	.96825	
−1.50	1.3229	

X = −3

It is not easy, maybe not even possible, to determine a domain using a table, but it may point out certain numbers that are not in the domain.

Graphs of Functions

Most of the functions we study in this course and in calculus are given by formulas rather than sets of ordered pairs. We graph functions in much the same way as we do equations. We find ordered pairs (x, y) or $(x, f(x))$, look for patterns, and complete the graph.

Example 8 Graph each of the following functions.

a) $f(x) = x^2 - 5$ b) $f(x) = x^3 - x$ c) $f(x) = \sqrt{x + 4}$

SOLUTION Most graphers do not use function notation "$f(x) = \ldots$" to enter a function formula. Instead, we must enter the function using "$y = \ldots$". The graphs follow.

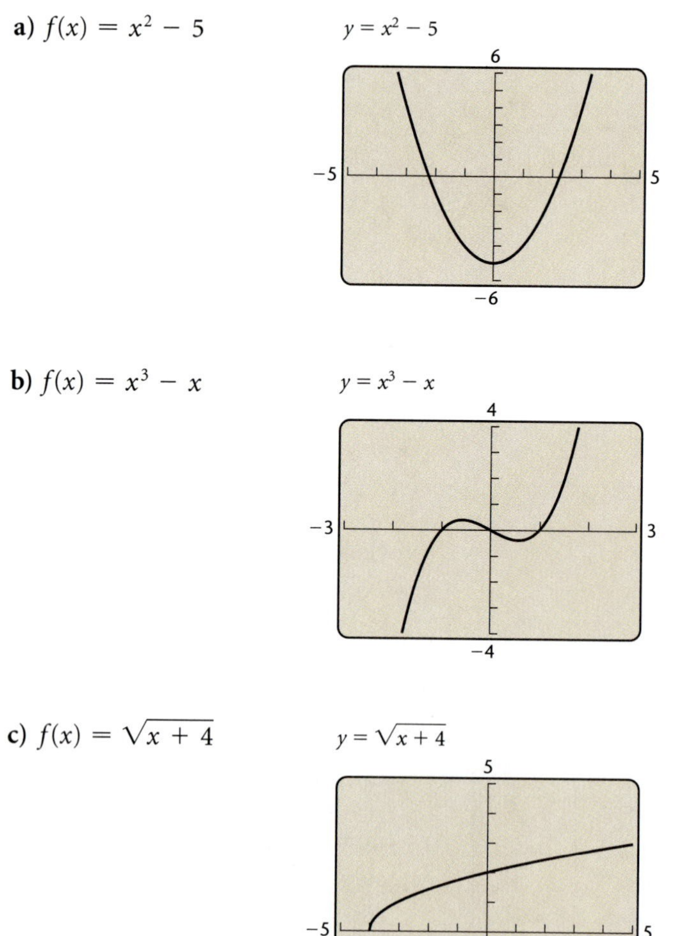

To find a function value, like $f(3)$, from a graph, we locate the input 3 on the horizontal axis, move vertically to the graph of the function, and then horizontally to find the output on the vertical axis. See the graph on the left below.

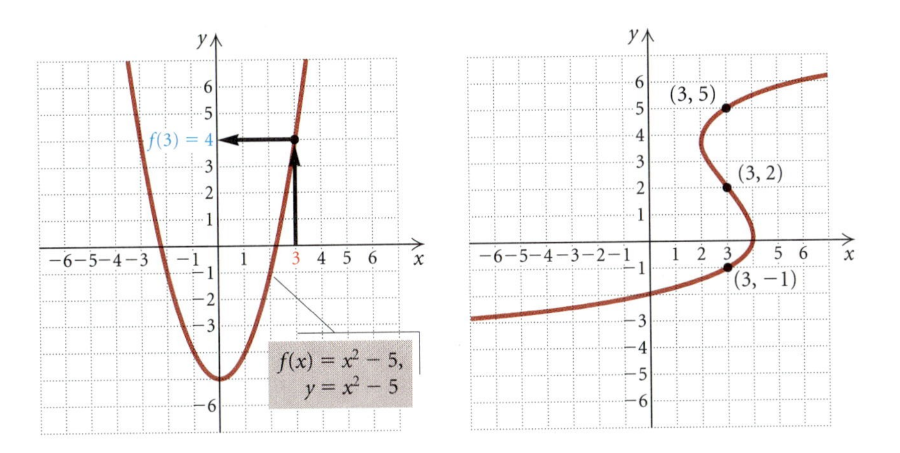

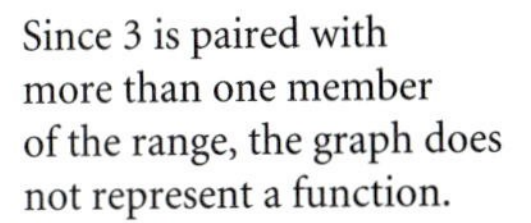
Since 3 is paired with more than one member of the range, the graph does not represent a function.

When one member of the domain is paired with two or more different members of the range, the correspondence *is not* a function. Thus, when a graph contains two or more different points with the same first coordinate, the graph cannot represent a function (see the graph on the right above). Points sharing a common first coordinate are vertically above or below each other. This leads us to the *vertical-line test.*

The Vertical-Line Test

If it is possible for a vertical line to cross a graph more than once, then the graph is not the graph of a function.

To apply the vertical-line test, we try to find a vertical line that crosses the graph more than once. If we do, then the graph is not that of a function. If we do not, then the graph is that of a function.

Example 9 Which of graphs (a) through (f) (in red) are graphs of functions?

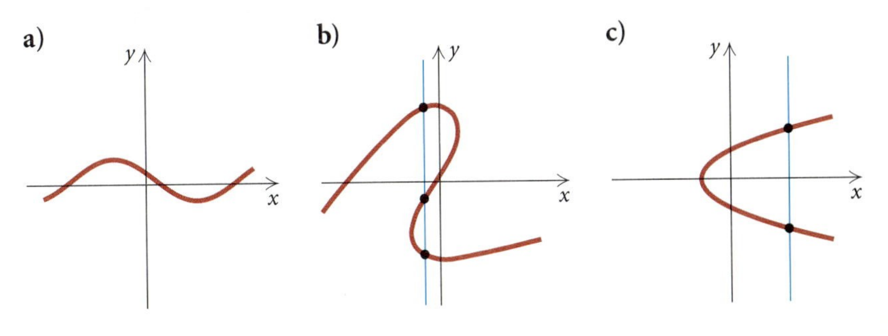

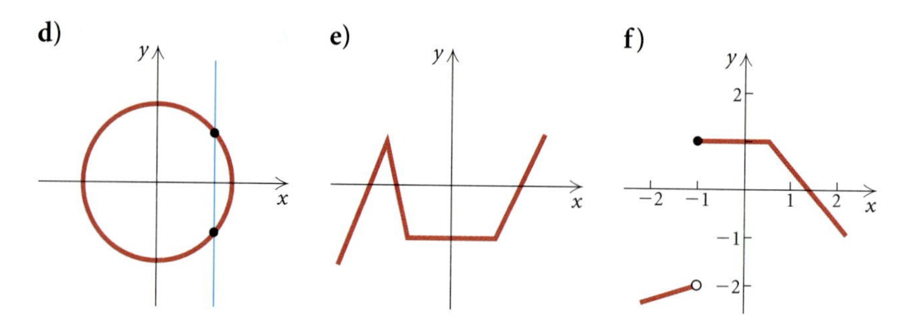

In graph (f), the solid dot shows that $(-1, 1)$ belongs to the graph. The open circle shows that $(-1, -2)$ does *not* belong to the graph.

SOLUTION Graphs (a), (e), and (f) are graphs of functions because we cannot find a vertical line that crosses any of them more than once. In (b) the vertical line crosses the graph in three points and so it is not a function. Also, in (c) and (d), we can find a vertical line that crosses the graph more than once.

The Domain and the Range Using a Grapher

Keep the following in mind regarding the *graph* of a function:

Domain = the set of a function's inputs, found on the horizontal axis;

Range = the set of a function's outputs, found on the vertical axis.

By carefully examining the graph of a function, we may be able to determine the function's domain as well as its range. Consider the graph of $f(x) = \sqrt{4 - (x - 1)^2}$, shown below. We look for the inputs on the x-axis that correspond to a point on the graph. We see that they extend from -1 to 3, inclusive. Thus the domain is $\{x \mid -1 \leq x \leq 3\}$, or the interval $[-1, 3]$.

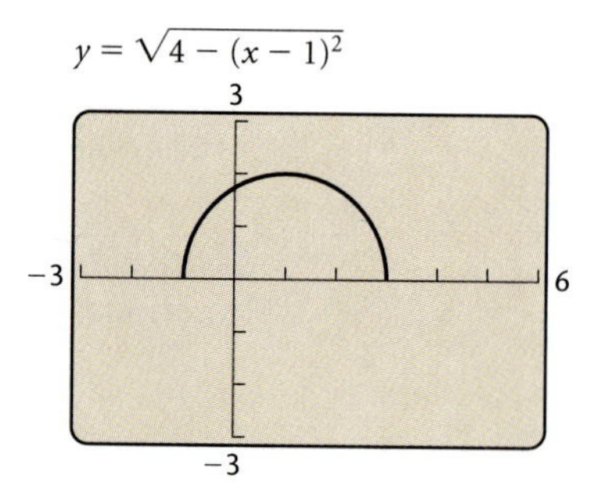

To find the range, we look for the outputs on the y-axis. We see that they extend from 0 to 2, inclusive. Thus the range of this function is $\{y \mid 0 \leq y \leq 2\}$, or the interval $[0, 2]$. We can confirm our results using the TRACE feature, moving the cursor from left to right along the curve. We can also confirm our results using a TABLE feature. (See Exercise 63 in Exercise Set 1.1.)

Example 10 Use a grapher to graph each of the following functions. Then estimate the domain and the range of each.

a) $f(x) = \sqrt{x + 4}$ b) $f(x) = x^3 - x$

c) $f(x) = \dfrac{12}{x}$ d) $f(x) = x^4 - 2x^2 - 3$

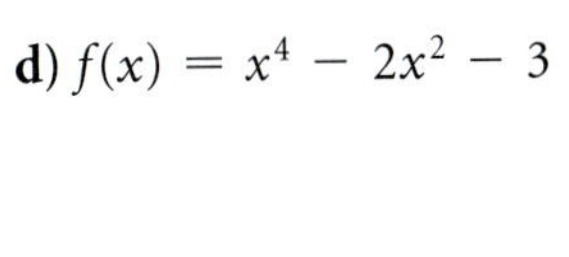

SOLUTION

a)

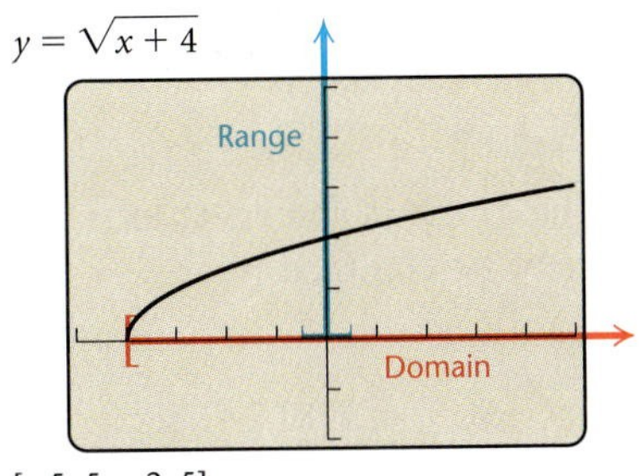

Domain $= [-4, \infty)$;
range $= [0, \infty)$

b)

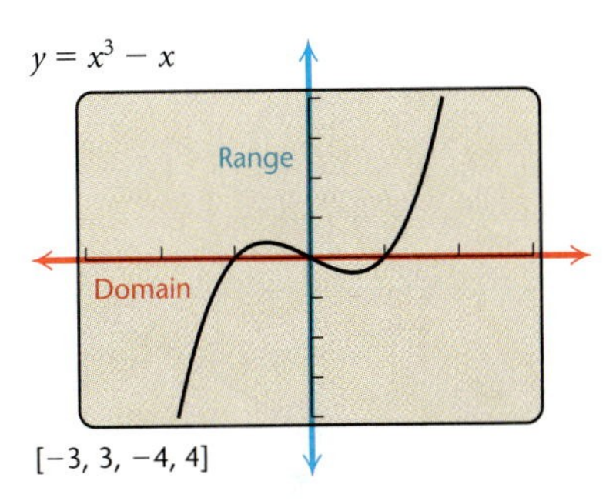

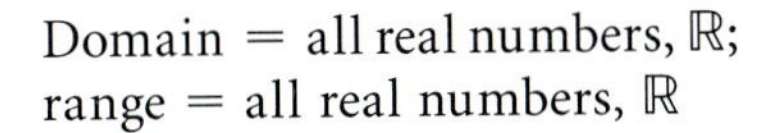

Domain = all real numbers, $\mathbb{R}$;
range = all real numbers, $\mathbb{R}$

c)

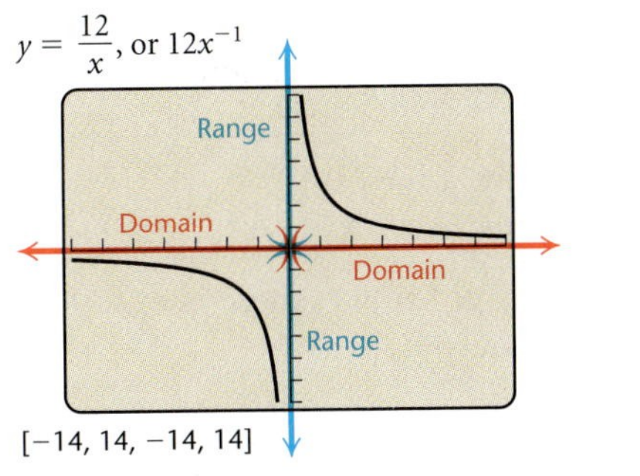

Since the graph does not cross the y-axis, 0 is excluded as an input.
Domain $= (-\infty, 0) \cup (0, \infty)$;
range $= (-\infty, 0) \cup (0, \infty)$

d)

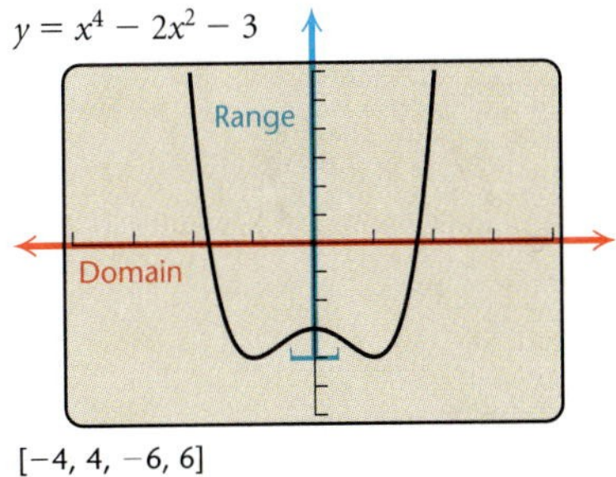

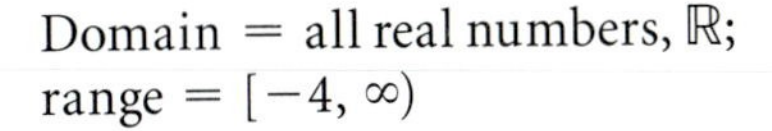

Domain = all real numbers, $\mathbb{R}$;
range $= [-4, \infty)$

Always consider adding the reasoning of Example 7 to a graphical analysis. Think, "What can I substitute?" to find the domain. Think, "What do I get out?" to find the range. Thus, in Example 10(d), it might not look like the domain is all real numbers because the graph rises steeply, but by reexamining the formula we see that we can indeed substitute any real number.

Applications of Functions

Example 11 *Speed of Sound in Air.* The speed S of sound in air is a function of the temperature t, in degrees Fahrenheit, and is given by

$$S(t) = 1087.7\sqrt{\frac{5t + 2457}{2457}},$$

where S is in feet per second.

a) Graph the function using the viewing window $[-50, 150, 1000, 1250]$ with Xscl $= 50$ and Yscl $= 50$.

b) Find the speed of sound in air when the temperature is 0°, 32°, and 70° Fahrenheit.

SOLUTION

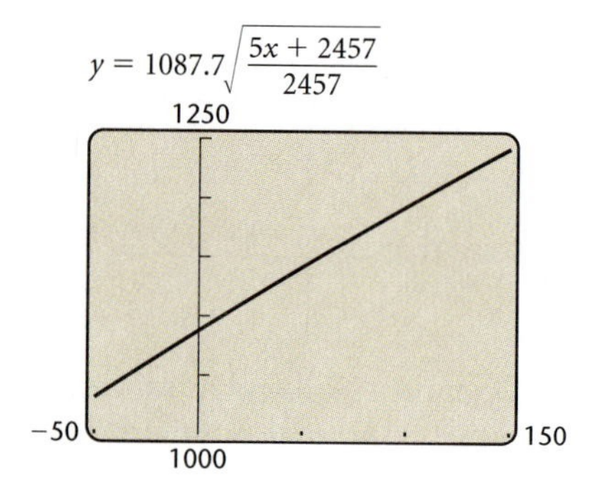

a) The graph is shown at left. Note that $S(t)$ must be changed to y and t must be changed to x.

b) We use a grapher to compute the function values. We find that

$$S(0) = 1087.7 \text{ ft/sec},$$
$$S(32) \approx 1122.6 \text{ ft/sec}, \quad \text{and}$$
$$S(70) \approx 1162.6 \text{ ft/sec}.$$

We can also use the TRACE feature to approximate these function values, or we can use the TABLE feature, set in ASK mode, to find them.

The following is a review of several of the function concepts considered in this section.

FUNCTION CONCEPTS	GRAPH
Formula for f: $f(x) = 5 + 2x^2 - x^4$. For every input, there is exactly one output. (1, 6) is on the graph. 1 is an input. 6 is an output. $f(1) = 6$ Domain: set of all inputs $= \mathbb{R}$. Range: set of all outputs $= (-\infty, 6]$.	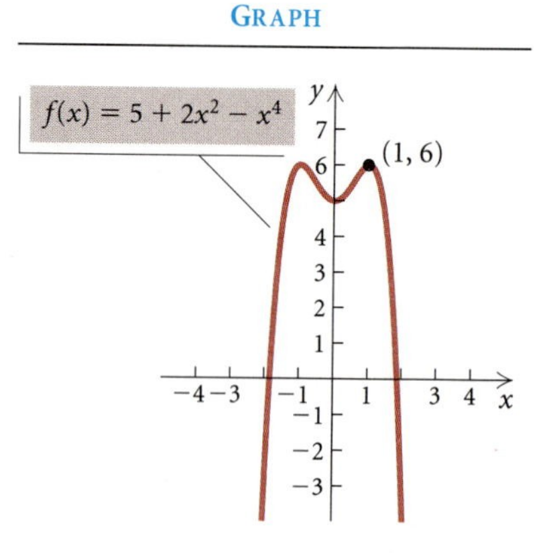

1.1 Exercise Set

Determine whether each correspondence is a function.

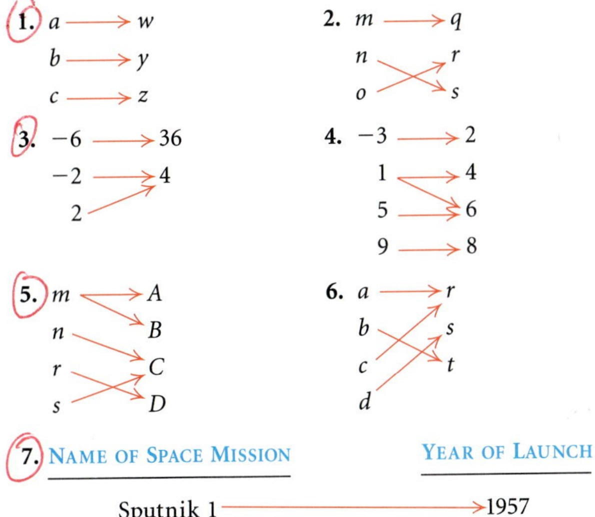

7.

NAME OF SPACE MISSION	YEAR OF LAUNCH
Sputnik 1	1957
Mercury	1962
Mariner 2	1962
Apollo 11	1969

(*Source: Cambridge Factfinder*, 1993)

8.

ANIMAL	SPEED ON THE GROUND (IN KILOMETERS/HOUR)
Cheetah	110
Lion	80
Giraffe	50
Ostrich	50

(*Source: Cambridge Factfinder*, 1993)

	DOMAIN	CORRESPONDENCE	RANGE
9.	A set of cars in a parking lot	Each car's license number	A set of numbers
10.	A set of people in a town	A doctor a person uses	A set of doctors
11.	A set of members of a family	Each person's eye color	A set of colors
12.	A set of members of a rock band	An instrument each person plays	A set of instruments
13.	A set of students in a class	A student sitting in a neighboring seat	A set of students
14.	A set of bags of chips on a shelf	Each bag's weight	A set of weights

Determine whether each relation is a function. Identify the domain and the range.

15. $\{(2, 10), (3, 15), (4, 20)\}$

16. $\{(3, 1), (5, 1), (7, 1)\}$

17. $\{(-7, 3), (-2, 1), (-2, 4), (0, 7)\}$

18. $\{(1, 3), (1, 5), (1, 7), (1, 9)\}$

19. $\{(-2, 1), (0, 1), (2, 1), (4, 1), (-3, 1)\}$

20. $\{(5, 0), (3, -1), (0, 0), (5, -1), (3, -2)\}$

21. A graph of a function f is shown below. Find $f(-1)$, $f(0)$, and $f(1)$.

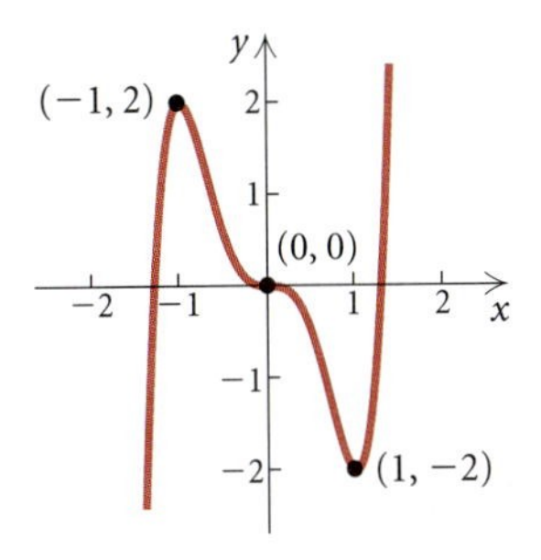

22. Given that $f(x) = 5x^2 + 4x$, find:

a) $f(0)$;
b) $f(-1)$;
c) $f(3)$;
d) $f(t)$;
e) $f(t - 1)$;
f) $\dfrac{f(a + h) - f(a)}{h}$.

23. Given that $g(x) = 3x^2 - 2x + 1$, find:

a) $g(0)$;
b) $g(-1)$;
c) $g(3)$;
d) $g(-x)$;
e) $g(1 - t)$;
f) $\dfrac{g(a + h) - g(a)}{h}$.

24. Given that $f(x) = 2|x| + 3x$, find:

a) $f(1)$;
b) $f(-2)$;
c) $f(-x)$;
d) $f(2y)$;
e) $f(2 - h)$;
f) $\dfrac{f(a + h) - f(a)}{h}$.

25. Given that $g(x) = x^3$, find:

a) $g(2)$;
b) $g(-2)$;
c) $g(-x)$;
d) $g(3y)$;
e) $g(2 + h)$;
f) $\dfrac{g(a + h) - g(a)}{h}$.

26. Given that $f(x) = \dfrac{x}{2 - x}$, find:

a) $f(2)$; b) $f(1)$;
c) $f(-16)$; d) $f(-x)$;
e) $f\left(-\dfrac{2}{3}\right)$ f) $\dfrac{f(x + h) - f(x)}{h}$.

27. Given that $g(x) = \dfrac{x - 4}{x + 3}$, find:

a) $g(5)$; b) $g(4)$;
c) $g(-3)$; d) $g(-16.25)$;
e) $g(x + h)$; f) $\dfrac{g(x + h) - g(x)}{h}$.

28. Given that $f(x) = x^2$, find

$$\frac{f(a + h) - f(a)}{h}.$$

29. Find $g(0)$, $g(-1)$, $g(5)$, and $g\left(\frac{1}{2}\right)$ for

$$g(x) = \frac{x}{\sqrt{1 - x^2}}.$$

30. Find $h(0)$, $h(2)$, and $h(-x)$ for

$$h(x) = x + \sqrt{x^2 - 1}.$$

Find the domain of each function. Do not use a grapher.

31. $f(x) = 7x + 4$
32. $f(x) = |3x - 2|$
33. $f(x) = 4 - \dfrac{2}{x}$
34. $f(x) = \dfrac{1}{x^4}$
35. $f(x) = \sqrt{7x + 4}$
36. $f(x) = \sqrt{x - 3}$
37. $f(x) = \dfrac{1}{x^2 - 4x - 5}$
38. $f(x) = \dfrac{x^4 - 2x^3 + 7}{3x^2 - 10x - 8}$
39. $f(x) = \sqrt{9 - x^2}$
40. $f(x) = \dfrac{6x}{\sqrt{9 - x^2}}$

Use a grapher to graph each function. Then visually estimate the domain and the range.

41. $f(x) = |x|$
42. $f(x) = |x| - 10.3$
43. $f(x) = \sqrt{9 - x^2}$
44. $f(x) = -\sqrt{25 - x^2}$
45. $f(x) = (x - 1)^3 + 2$
46. $f(x) = (x - 2)^4 + 1$
47. $f(x) = \sqrt{7 - x}$
48. $f(x) = \sqrt{x + 8}$
49. $f(x) = \dfrac{18}{x}$
50. $f(x) = 2x^2 - x^4 + 5$

Determine whether each graph is that of a function. An open dot indicates that the point does not belong to the graph.

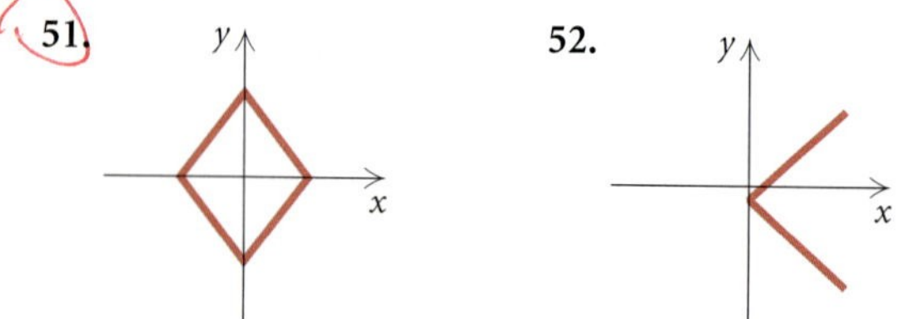

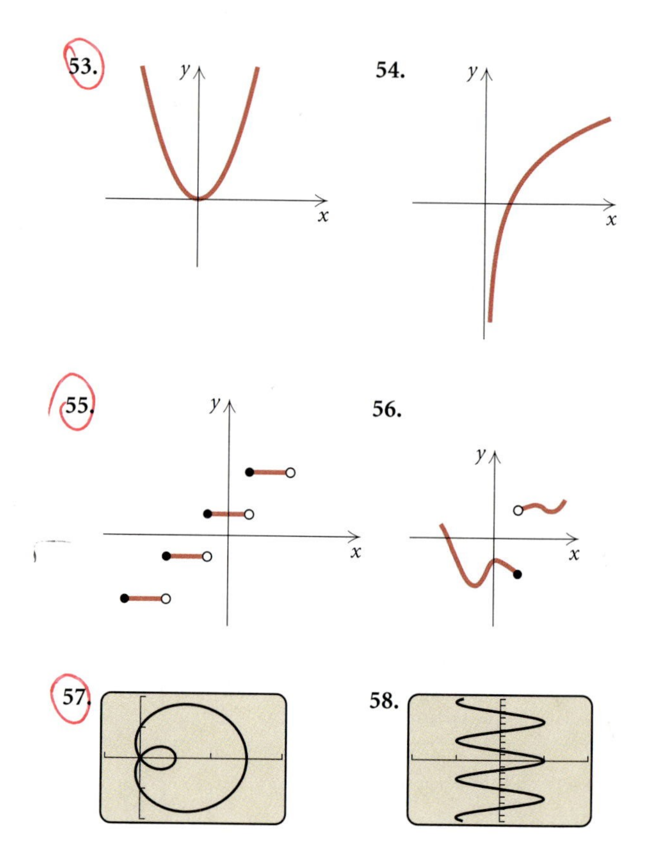

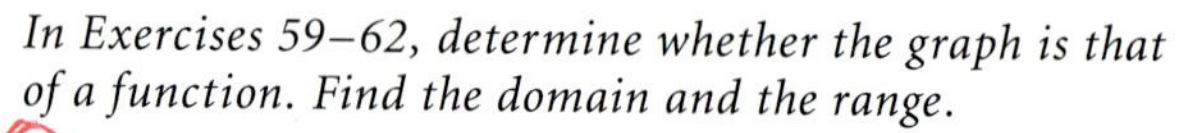
In Exercises 59–62, determine whether the graph is that of a function. Find the domain and the range.

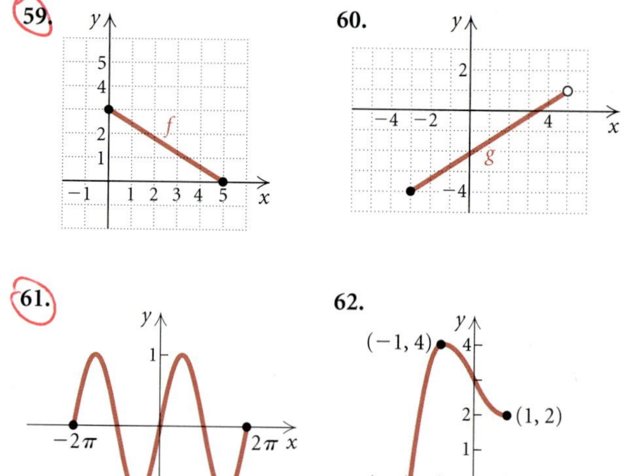

63. Use the TABLE feature to complete the input–output table for the equations

$$y_1 = \sqrt{4 - (x - 1)^2} \quad \text{and} \quad y_2 = \frac{23x}{x - 3.2}.$$

Then use the TABLE feature to consider inputs from −2.0 to 4.8.

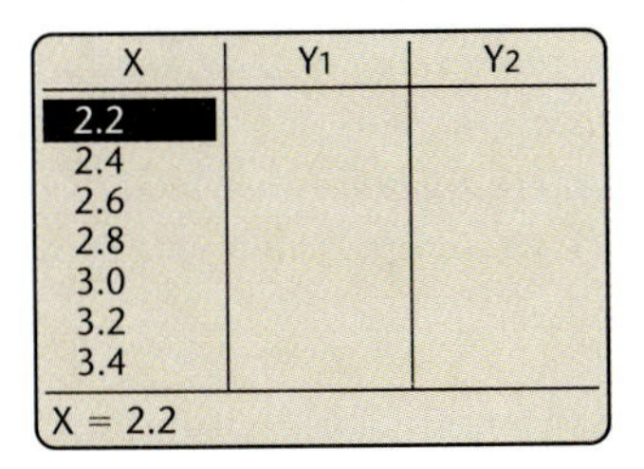

X	Y1	Y2
2.2		
2.4		
2.6		
2.8		
3.0		
3.2		
3.4		
X = 2.2		

a) What does the table tell you about the domain of each function? the range?

b) What seems to be the largest output of each function? the smallest?

64. *Average Price of a Movie Ticket.* The average price of a movie ticket, in dollars, can be estimated by the function P given by

where x = the year. Thus, $P(1998)$ is the average price of a movie ticket in 1998. The price is lower than what might be expected due to lower prices for matinees, senior citizens' discounts, and so on.

a) Use the function to predict the average price in 1998, 2000, and 2010.

b) When will the average price be $8.00?

65. *Boiling Point and Elevation.* The elevation E, in meters, above sea level at which the boiling point of water is t degrees Celsius is given by the function

$$E(t) = 1000(100 - t) + 580(100 - t)^2.$$

At what elevation is the boiling point 99.5°? 100°?

66. *Territorial Area of an Animal.* The territorial area of an animal is defined to be its defended, or exclusive region. For example, a lion has a certain region over which it is considered ruler. It has been shown that the territorial area I, in acres, of predatory animals is a function of body weight w, in pounds, and is given by the function

$$T(w) = w^{1.31}.$$

Find the territorial area of animals whose body weights are 0.5 lb, 10 lb, 20 lb, 100 lb, and 200 lb.

Skill Maintenance

To the student and the instructor: The skill maintenance exercises review skills covered previously in the text and anticipate the learning in the next section. You can expect these kinds of exercises in almost every exercise set.

Solve.

67. $\frac{2}{3}x + 7 = 12$

68. $\frac{4}{5}x + 1 = -\frac{2}{3}x + 7$

69. $\sqrt{2x - 4} = 7$

70. $3x^2 - 13x = 10$

Synthesis

To the student and the instructor: The synthesis exercises, found at the end of every exercise set, challenge students to combine concepts or skills developed in that section or in preceding parts of the text. Writing Exercises, denoted by the ◈ icon, are meant to be answered with one or more sentences.

71. ◈ Explain in your own words what a function is.

72. ◈ Explain in your own words the difference between the domain of a function and the range.

73. Construct

$$\frac{f(x + h) - f(x)}{h}$$

for $f(x) = \sqrt{x}$ and rationalize the numerator.

74. Make a hand-drawn graph of a function for which the domain is $[-4, 4]$ and the range is $[1, 2] \cup [3, 5]$. Answers may vary.

75. Give an example of two different functions that have the same domain and the same range, but have no pairs in common. Answers may vary.

76. Make a hand-drawn graph of a function for which the domain is $[-3, -1] \cup [1, 5]$ and the range is $\{1, 2, 3, 4\}$. Answers may vary.

77. Suppose that for some function f, $f(x - 1) = 5x$. Find $f(6)$.

1.2 Functions and Applications

- *Find zeros of functions.*
- *Graph functions, looking for intervals on which the function is increasing, decreasing, or constant, and determine relative maxima and minima.*
- *Graph functions defined piecewise.*
- *Given an application, find a function formula that models the application; find the domain of the function and function values, and then graph the function.*

Because functions occur in so many real-world situations, it is important to be able to analyze them carefully. In this section, we examine a variety of functions and formulate some mathematical models.

Finding Zeros on a Grapher

An input for which a function's output is 0 is called a **zero** of the function.

Zeros of Functions

An input c of a function f is called a *zero* of the function, if the output for c is 0. That is, $f(c) = 0$.

For the function given by $f(x) = x^4 - 4x^2 + 3$, the zeros are $-\sqrt{3}$, -1, 1, and $\sqrt{3}$.

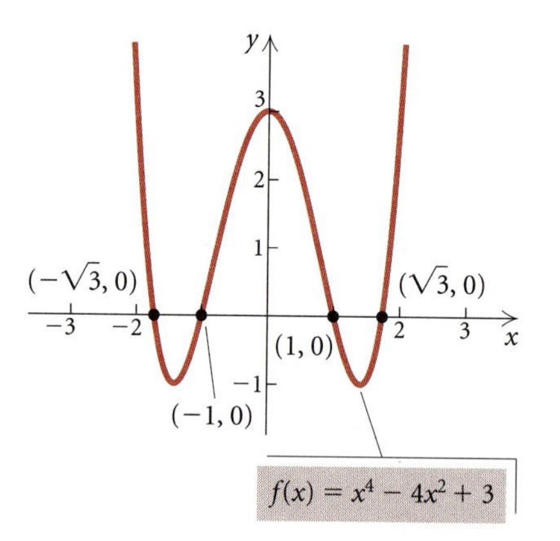

Note that these numbers are the first coordinates of the x-intercepts, $(-\sqrt{3}, 0)$, $(-1, 0)$, $(1, 0)$, and $(\sqrt{3}, 0)$. The **x-intercepts** are the points where the graph crosses the x-axis.

Finding exact values of the zeros of a function can be difficult. We can find approximations using a grapher in much the same way that we solve equations.

Example 1 Find the zeros of the function f given by

$$f(x) = 0.1x^3 - 0.6x^2 - 0.1x + 2.$$

Approximate the zeros to three decimal places.

ALGEBRAIC SOLUTION

We have not developed an algebraic procedure that will yield the zeros. We will learn some procedures in Chapter 2.

GRAPHICAL SOLUTION

We use a grapher, trying to create a graph that clearly shows the curvature. Then we look for points where the graph crosses the x-axis. It appears that there are three zeros, one near -2, one near 2, and one near 6. We use TRACE and ZOOM to approximate these zeros, or we can use a SOLVE or ROOT feature.

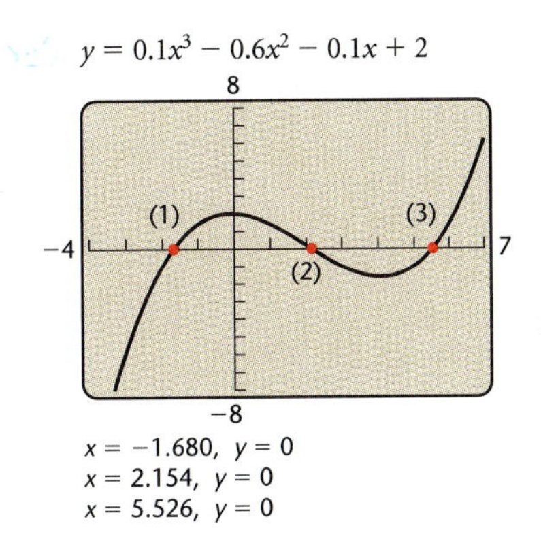

The zeros are about -1.680, 2.154, and 5.526.

Increasing, Decreasing, and Constant Functions

On a given interval, if the graph of a function rises from left to right, it is said to be **increasing** on that interval. If the graph drops from left to right, it is said to be **decreasing**. If the function stays the same from left to right, it is said to be **constant**.

Interactive Discovery

Graph each of the following functions on a grapher using the given viewing window:

$$y = |x + 1| + |x - 2| - 5, \quad [-6, 6, -6, 6];$$
$$y = 5 - (x + 2)^2, \quad [-2, 4, -10, 6];$$
$$y = 4, \quad [-20, 20, -3, 6].$$

Then using the TRACE feature, move the cursor along the graph from left to right, observing what happens to the second coordinate. Confirm the results with a TABLE feature.

We are led to the following definitions.

Increasing, Decreasing, and Constant Functions

A function f is said to be *increasing* on an interval if for all a and b in that interval, $a < b$ implies $f(a) < f(b)$.

A function f is said to be *decreasing* on an interval if for all a and b in that interval, $a < b$ implies $f(a) > f(b)$.

A function f is said to be *constant* on an interval if for all a and b in that interval, $f(a) = f(b)$.

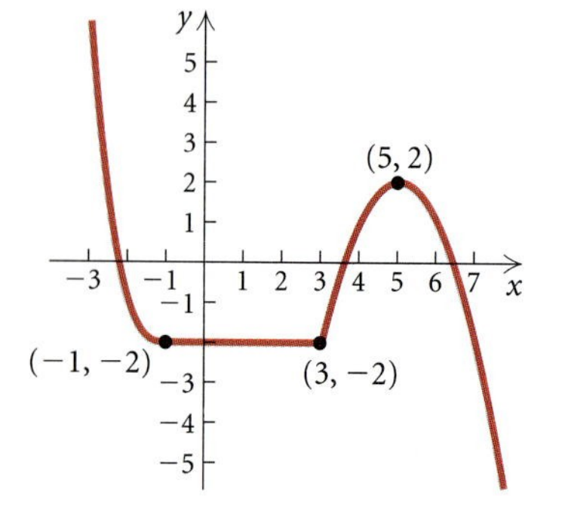

Example 2 Determine the intervals on which the function is each of the following.

a) Increasing **b)** Decreasing **c)** Constant

SOLUTION

a) The function is increasing on the interval $[3, 5]$.

b) The function is decreasing on the intervals $(-\infty, -1]$ and $[5, \infty)$.

c) The function is constant on the interval $[-1, 3]$.

Relative Maximum and Minimum Values

Consider the graph shown below. Note the "peaks" and "valleys" at the points c_1, c_2, and c_3. The function value $f(c_2)$ is called a **relative maximum** (plural, **maxima**). Each of the function values $f(c_1)$ and $f(c_3)$ is called a **relative minimum** (plural, **minima**).

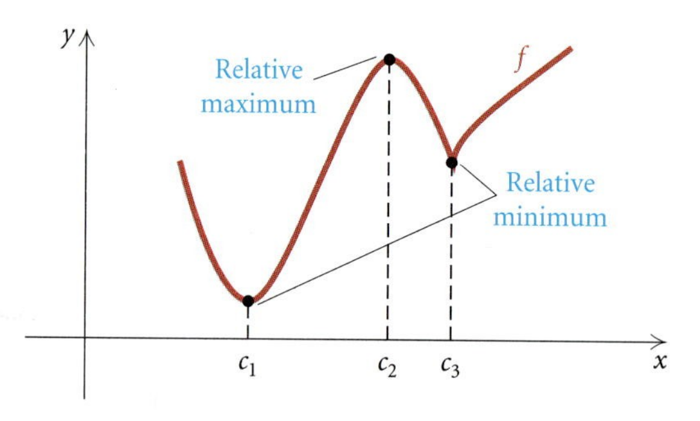

Relative Maxima and Minima

Suppose that f is a function for which $f(c)$ exists for some c in the domain of f. Then:

$f(c)$ is a *relative maximum* if there exists an open interval I containing c in the domain such that $f(c) \geq f(x)$, for all x in I; and

$f(c)$ is a *relative minimum* if there exists an open interval I containing c in the domain such that $f(c) \leq f(x)$, for all x in I.

Simply stated, $f(c)$ is a relative maximum if $f(c)$ is a "high point" in some interval, and $f(c)$ is a relative minimum if $f(c)$ is a "low point" in some interval.

We can use a grapher to approximate relative maxima or minima. On some graphers, this is done using the TRACE and ZOOM features. On other graphers, there may be a MAX or MIN feature that can be used. If you take a calculus course, you will learn a method for determining exact values of relative maxima and minima.

Example 3 Use a grapher to determine any relative maxima or minima of the function $f(x) = 0.1x^3 - 0.6x^2 - 0.1x + 2$, and to determine intervals on which the function is increasing or decreasing.

SOLUTION We first graph the function, experimenting with the window as needed. The curvature is seen fairly well with $[-4, 6, -3, 3]$.

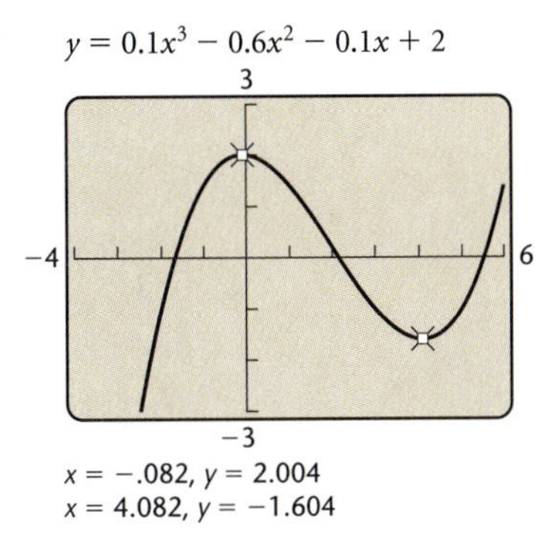

The graph starts rising, or increasing, from the left. We use the TRACE feature and move the cursor from left to right, noting whether the second coordinate increases. We note that the graph tends to stop increasing when x is somewhere around 0. Using the ZOOM feature, we get a good approximation of the relative maximum and where it occurs: 2.004 at $x = -0.082$.

From this relative maximum, the graph decreases to a point near $x = 4$ before it starts increasing again. We zoom in and get a good approximation of the relative minimum: -1.604 at $x = 4.082$.

The function is increasing on the intervals $(-\infty, -0.082]$ and $[4.082, \infty)$. It is decreasing on the interval $[-0.082, 4.082]$.

Interactive Discovery

If your grapher has a TABLE feature, try to discover a way to use it to check the results of Example 3. Consider a small interval around each input and a small step value. Are there any difficulties?

Functions Defined Piecewise

Sometimes functions are defined **piecewise** using different output formulas for different parts of the domain.

Example 4 Make a hand-drawn graph of the function defined as

$$f(x) = \begin{cases} 4, & \text{for } x \leq 0, \\ 4 - x^2, & \text{for } 0 < x \leq 2, \\ 2x - 6, & \text{for } x > 2. \end{cases}$$

SOLUTION We create the graph in three parts, as shown and described below.

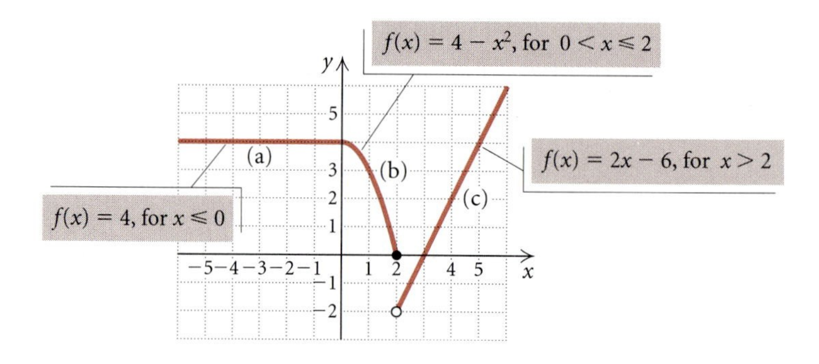

a) We graph $f(x) = 4$ *only* for inputs x less than or equal to 0 (that is, on the interval $(-\infty, 0]$).

b) We graph $f(x) = 4 - x^2$ *only* for inputs x greater than 0 and less than or equal to 2 (that is, on the interval $(0, 2]$).

c) We graph $f(x) = 2x - 6$ *only* for inputs x greater than 2 (that is, on the interval $(2, \infty)$).

To graph a function defined piecewise on a grapher, consult your manual. For piecewise-defined functions, you should select the DOT feature. Although some do not have the capability, you might try the following way to enter the function formula for Example 4. It incorporates parenthetic descriptions of the intervals:

$$f(x) = 4\ (x \leq 0) + (4 - x^2)\ (x > 0)(x \leq 2) + (2x - 6)\ (x > 2).$$

If you wish, you might enter this equation as three separate functions:

$$y_1 = 4\ (x \leq 0), \qquad y_2 = (4 - x^2)\ (x > 0)(x \leq 2),$$

$$y_3 = (2x - 6)\ (x > 2).$$

Example 5 Make a hand-drawn graph of the function defined as

$$f(x) = \begin{cases} \dfrac{x^2 - 4}{x + 2}, & \text{for } x \neq -2, \\ 3, & \text{for } x = -2. \end{cases}$$

SOLUTION When $x \neq -2$, the denominator of $(x^2 - 4)/(x + 2)$ is nonzero, so we can simplify:

$$\frac{x^2 - 4}{x + 2} = \frac{(x + 2)(x - 2)}{x + 2} = x - 2.$$

Thus,

$$f(x) = x - 2, \quad \text{for } x \neq -2.$$

The graph of this part of the function consists of a line with a "hole" at the point $(-2, -4)$, indicated by the open dot. At $x = -2$, we have $f(-2) = 3$, so the point $(-2, 3)$ is plotted above the open dot.

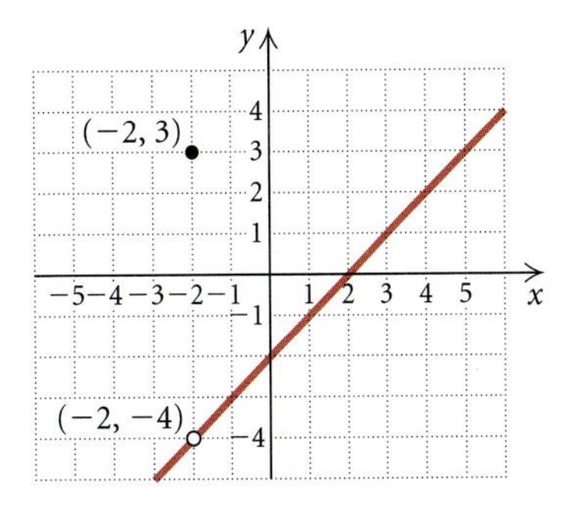

When $y = (x^2 - 4)/(x + 2)$ is graphed using a grapher, the hole may not be visible. If a TABLE feature is available, the following table can be created. The ERROR message indicates that -2 is not in the domain of the function g given by $g(x) = (x^2 - 4)/(x + 2)$. However, -2 *is* in the domain of f because $f(-2)$ is defined to be 3.

X	Y1
−2.3	−4.3
−2.2	−4.2
−2.1	−4.1
−2	ERROR
−1.9	−3.9
−1.8	−3.8
−1.7	−3.7

X = −2

A function with importance in calculus and computer programming is the *greatest integer function*, f, denoted as $f(x) = \text{INT}(x)$, or $[\![x]\!]$.

Greatest Integer Function

$f(x) = \text{INT}(x) =$ the greatest integer less than or equal to x.

The greatest integer function pairs each input with the greatest integer less than or equal to that input. To check this, note that 1, 1.2, and 1.9 are all paired with the y-value 1. Similarly, 2, 2.3, and 2.8 are all paired with 2, and -3.4, -3.06, and -3.99027 are all paired with -4.

Example 6 Graph: $f(x) = \text{INT}(x)$.

SOLUTION The greatest integer function can also be defined by a piecewise function with an infinite number of statements:

$$f(x) = \text{INT}(x) = \begin{cases} \vdots & \\ -3 & \text{for } -3 \le x < -2 \\ -2 & \text{for } -2 \le x < -1 \\ -1 & \text{for } -1 \le x < 0 \\ 0 & \text{for } 0 \le x < 1 \\ 1 & \text{for } 1 \le x < 2 \\ 2 & \text{for } 2 \le x < 3 \\ 3 & \text{for } 3 \le x < 4 \\ \vdots & \end{cases}$$

$f(x) = \text{INT}(x)$

On a grapher, we would see the graph on the left below if we used CONNECTED mode, which connects points (or rectangles called **pixels** on a grapher) with line segments. We would see the graph on the right if we used DOT mode. The DOT mode is preferable, though even it may not show the open dots at the endpoints of the segments.

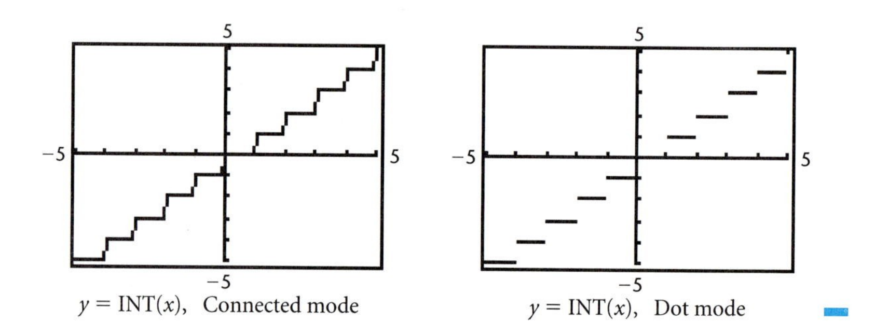

$y = \text{INT}(x)$, Connected mode $\qquad$ $y = \text{INT}(x)$, Dot mode

Applications of Functions

Many real-world situations can be modeled by functions.

Example 7 *Car Distance.* Jamie and Ruben drive away from a restaurant at right angles to each other. Jamie's speed is 65 mph and Ruben's is 55 mph.

a) Express the distance between the cars as a function of time.

b) Find the domain of the function.

c) Graph the function.

SOLUTION

a) Suppose 1 hr goes by. At that time, Jamie has traveled 65 mi and Ruben has traveled 55 mi. We can use the Pythagorean theorem then to find the distance between them. This distance would be the length of the hypotenuse of a triangle with legs measuring 65 mi and 55 mi. After 2 hr, the triangle's legs would measure 130 mi and 110 mi.

Observing that the distances will always be changing, we make a sketch and let t = the time, in hours, that Jamie and Ruben have been driving since leaving the restaurant.

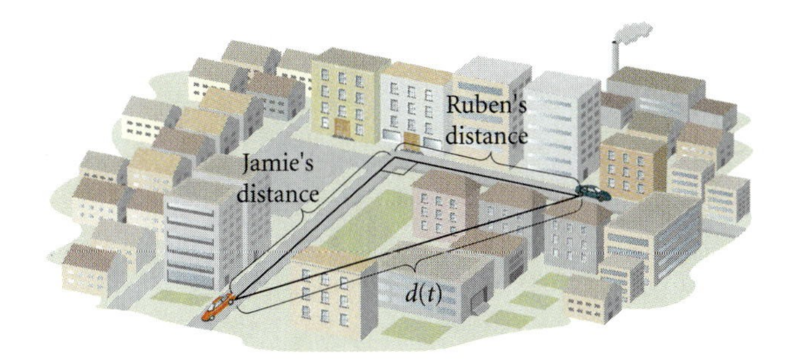

After t hours, Jamie has traveled $65t$ miles and Ruben $55t$ miles. We can use the Pythagorean theorem again:

$$[d(t)]^2 = (65t)^2 + (55t)^2.$$

Because distance must be nonnegative, we need only consider the positive square root when solving for $d(t)$:

$$\begin{aligned} d(t) &= \sqrt{(65t)^2 + (55t)^2} \\ &= \sqrt{4225t^2 + 3025t^2} \\ &= \sqrt{7250t^2} \\ &= \sqrt{7250}\,\sqrt{t^2} \\ &\approx 85.15|t| && \text{Approximating the root to two decimal places} \\ &\approx 85.15t. && \text{Since } t \geq 0,\ |t| = t. \end{aligned}$$

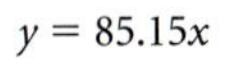
$y = 85.15x$

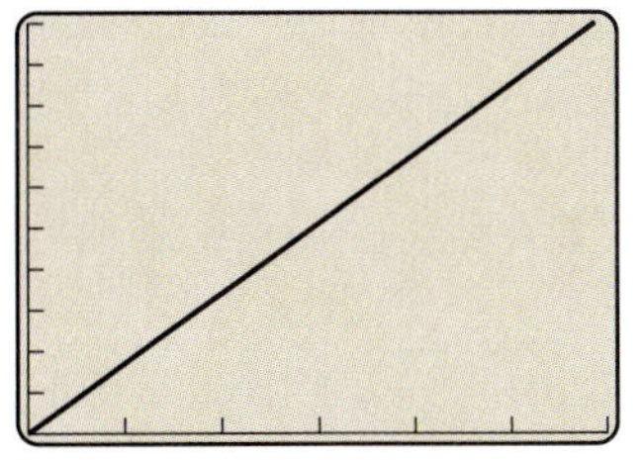
[0, 6, 0, 500]

Thus, $d(t) = 85.15t,\ t \geq 0$.

b) Since the time traveled must be nonnegative, the domain is the set of nonnegative real numbers $[0, \infty)$.

c) Because of the ease of using a grapher, we can almost always visualize a problem by making a graph. Making such graphs should become a habit as you do applications and problem solving.

Example 8 *Storage Area.* The Sound Shop has 20 ft of dividers with which to set off a rectangular area for the storage of overstock. If a corner of the store is used, the partition need only form two sides of a rectangle.

a) Express the floor area as a function of the length of the partition.

b) Find the domain of the function.

c) Graph the function.

d) Find the dimensions that maximize the area.

SOLUTION

a) Note that the dividers will form two sides of a rectangle. If, for example, 14 ft of dividers are used for the length of the rectangle, that would leave 20 − 14, or 6 ft of dividers for the width. Thus if x = the length, in feet, of the rectangle, then $20 - x$ = the width. We represent this information in a sketch.

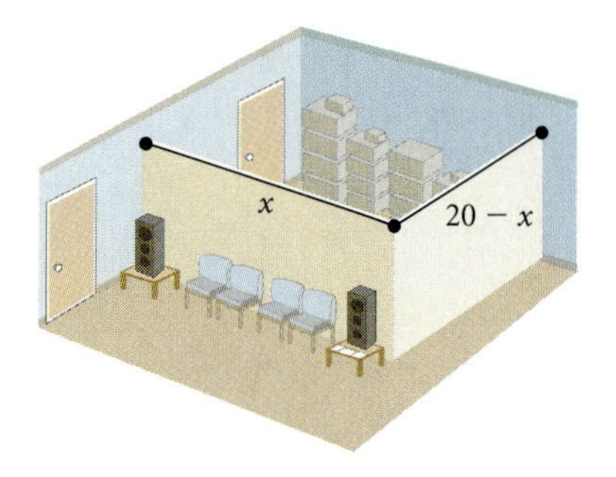

The area, $A(x)$, is given by

$$A(x) = x(20 - x) \qquad \text{Area = length} \cdot \text{width}$$
$$= 20x - x^2.$$

The function $A(x) = 20x - x^2$ can be used to express the rectangle's area as a function of the length.

b) Because the rectangle's length must be positive and less than 20 ft (why?), we restrict the domain of A to $\{x \mid 0 < x < 20\}$, that is, the interval $(0, 20)$.

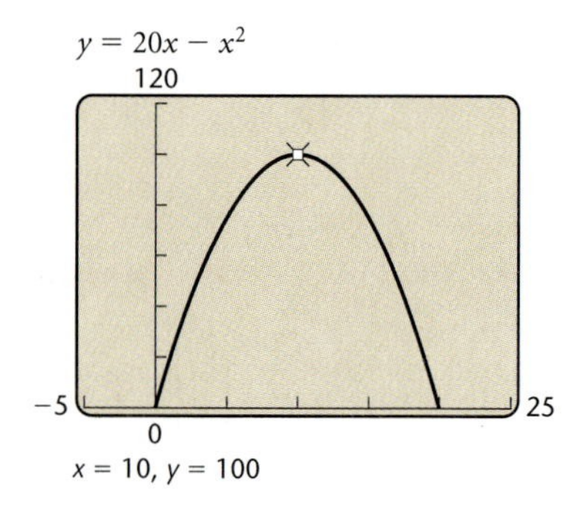

c) The graph is shown at left with the viewing window $[-5, 25, 0, 120]$.

d) We use the TRACE and ZOOM features or the MAX–MIN feature as in Example 3. The maximum value (overall largest value) of the area function is 100 when $x = 10$. Thus the dimensions that maximize the area are

Length $= x = 10$ ft and

Width $= 20 - x = 20 - 10 = 10$ ft.

Summary

Let's summarize our analysis of functions by considering the function f given by $f(x) = x^4 - 6x^3 - 4x^2 + 53.2x - 42.6$.

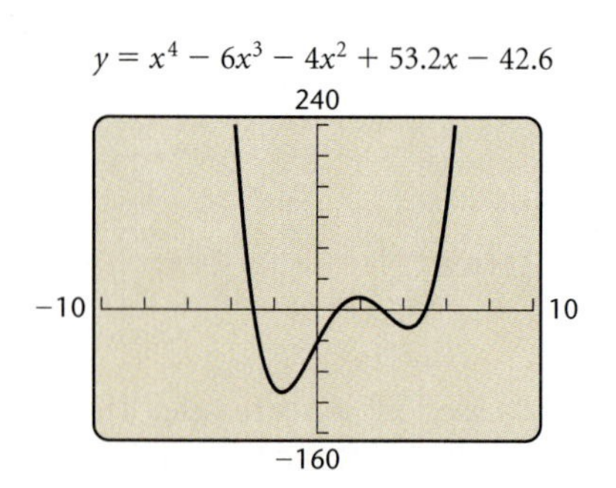

Function:	$f(x) = x^4 - 6x^3 - 4x^2 + 53.2x - 42.6$
Zeros:	-2.975, 0.950, 3, 5.025
Relative maximum:	15.921 at $x = 1.914$
Relative minima:	-106.907 at $x = -1.643$, -23.101 at $x = 4.229$
Increasing on:	$[-1.643, 1.914]$, $[4.229, \infty)$
Decreasing on:	$(-\infty, -1.643]$, $[1.914, 4.229]$
Domain:	All real numbers, $\mathbb{R}$
Range:	$[-106.907, \infty)$

1.2 Exercise Set

For each function, find the zeros.

1. $f(x) = x^3 - 3x - 1$
2. $f(x) = x^3 + 3x^2 - 9x - 13$
3. $f(x) = x^4 - 2x^2$
4. $f(x) = x^4 - 2x^3 - 5.6$
5. $f(x) = x^3 - x$
6. $f(x) = 2x^3 - x^2 - 14x - 10$
7. $f(x) = \frac{1}{2}(|x - 4| + |x - 7|) - 4$
8. $f(x) = \sqrt{7 - x^2}$
9. $f(x) = |x + 1| + |x - 2| - 5$
10. $f(x) = |x + 1| + |x - 2|$
11. $f(x) = |x + 1| + |x - 2| - 3$
12. $f(x) = x^8 + 8x^7 - 28x^6 - 56x^5 + 70x^4 + 56x^3 - 28x^2 - 8x + 1$

Determine intervals on which each function is (a) increasing, (b) decreasing, and (c) constant.

13. 14. 15. 16.

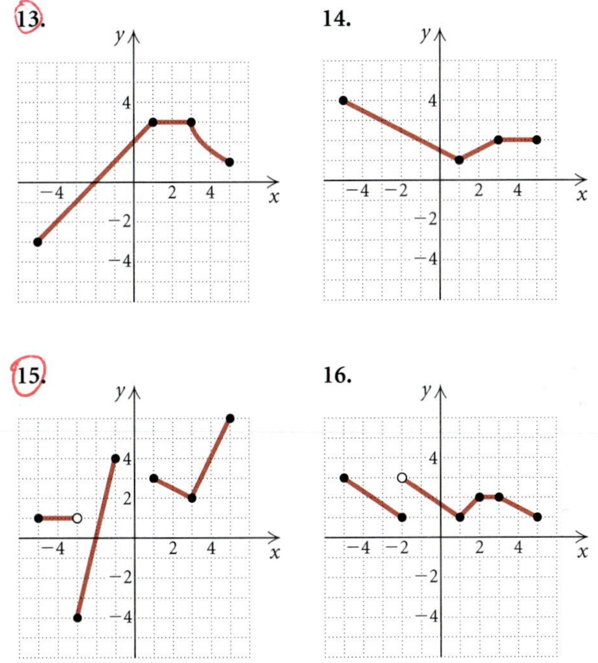

Graph each function using the given viewing window. Find where the function is increasing or decreasing and any relative maxima or minima. Change viewing windows if it seems appropriate for further analysis.

17. $f(x) = x^2$, $[-4, 4, -1, 10]$
18. $f(x) = 4 - x^2$, $[-5, 5, -5, 5]$
19. $f(x) = 5 - |x|$, $[-10, 10, -10, 10]$
20. $f(x) = |x + 3| - 5$, $[-10, 10, -10, 10]$
21. $f(x) = x^2 - 6x + 10$, $[-4, 10, -1, 20]$
22. $f(x) = -x^2 - 8x - 9$, $[-10, 2, -10, 10]$
23. $f(x) = -x^3 + 6x^2 - 9x - 4$, $[-3, 7, -20, 15]$
24. $f(x) = 0.2x^3 - 0.2x^2 - 5x - 4$, $[-10, 10, -30, 20]$
25. $f(x) = 1.1x^4 - 5.3x^2 + 4.07$, $[-4, 4, -4, 8]$
26. $f(x) = 1.2(x + 3)^4 + 10.3(x + 3)^2 + 9.78$, $[-9, 3, -40, 100]$

27. *Temperature During an Illness.* The temperature of a patient during an illness is given by the function
$$T(t) = -0.1t^2 + 1.2t + 98.6, \quad 0 \le t \le 12,$$
where T = the temperature, in degrees Fahrenheit, at time t, in days, after the onset of the illness.
   a) Graph the function using a grapher.
   b) Find the relative maximum.
   c) At what time was the patient's temperature the highest? What was the highest temperature?

28. *Advertising Effect.* A software firm estimates that it will sell N units of a new CD-ROM video game after spending a dollars on advertising, where
$$N(a) = -a^2 + 300a + 6, \quad 0 \le a \le 300,$$
and a is measured in thousands of dollars.
   a) Graph the function using a grapher.
   b) Find the relative maximum.
   c) For what advertising expenditure will the greatest number of games be sold? How many games will be sold for that amount?

Use a grapher. Find where each function is increasing or decreasing. Consider the entire set of real numbers if no domain is given.

29. $f(x) = \dfrac{8x}{x^2 + 1}$
30. $f(x) = \dfrac{-4}{x^2 + 1}$
31. $f(x) = x\sqrt{4 - x^2}$, for $-2 \le x \le 2$
32. $f(x) = -0.8x\sqrt{9 - x^2}$, for $-3 \le x \le 3$

Make a hand-drawn graph of each of the following. Check your results on a grapher.

33. $f(x) = \begin{cases} \frac{1}{2}x, & \text{for } x < 0, \\ x + 3, & \text{for } x \ge 0 \end{cases}$

34. $f(x) = \begin{cases} -\frac{1}{3}x + 2, & \text{for } x \leq 0, \\ x - 5, & \text{for } x > 0 \end{cases}$

35. $f(x) = \begin{cases} -\frac{3}{4}x + 2, & \text{for } x < 4, \\ -1, & \text{for } x \geq 4 \end{cases}$

36. $f(x) = \begin{cases} 4, & \text{for } x \leq -2, \\ x + 1, & \text{for } -2 < x < 3, \\ -x, & \text{for } x \geq -3 \end{cases}$

37. $f(x) = \begin{cases} x + 1, & \text{for } x \leq -3, \\ -1, & \text{for } -3 < x < 4, \\ \frac{1}{2}x, & \text{for } x \geq 4 \end{cases}$

38. $f(x) = \begin{cases} \dfrac{x^2 - 9}{x + 3}, & \text{for } x \neq -3, \\ 5, & \text{for } x = -3 \end{cases}$

39. $f(x) = \begin{cases} 2, & \text{for } x = 5, \\ \dfrac{x^2 - 25}{x - 5}, & \text{for } x \neq 5 \end{cases}$

40. $f(x) = \begin{cases} \dfrac{x^2 + 3x + 2}{x + 1}, & \text{for } x \neq -1, \\ 7, & \text{for } x = -1 \end{cases}$

41. $f(x) = \text{INT}(x)$

42. $f(x) = 1 + \text{INT}(x)$

43. $f(x) = \text{INT}(x) - 1$

44. $f(x) = [\![x]\!] - 2$

Graph each of the following with a grapher, if possible.

45. $f(x) = \begin{cases} \sqrt[3]{x}, & \text{for } x \leq -1, \\ x^2 - 3x, & \text{for } -1 < x < 4, \\ \sqrt{x - 4}, & \text{for } x \geq 4 \end{cases}$

46. $f(x) = \begin{cases} \left|\dfrac{x}{3} + 2\right|, & \text{for } x < 5, \\ \sqrt[3]{x - 8}, & \text{for } x \geq 5 \end{cases}$

47. $f(x) = \begin{cases} \sqrt[3]{x + 27}, & \text{for } x < 1, \\ \left|2 - \dfrac{x}{5}\right|, & \text{for } x \geq 1 \end{cases}$

48. *Airplane Distance.* An airplane is flying at an altitude of 3700 ft. The slanted distance directly to the airport is d feet. Express the horizontal distance h as a function of d.

49. *Rising Balloon.* A hot-air balloon rises straight up from the ground at a rate of 120 ft/min. The balloon is tracked from a rangefinder at point P, which is 400 ft from the release point Q of the balloon. Let d = the distance from the balloon to the rangefinder at point P and t = the time, in minutes, since the balloon was released. Express d as a function of t.

50. *Triangular Flag.* A scout troop is designing a triangular flag so that the length of its base, in inches, is 7 less than twice the height, h. Express the area of the flag as a function of the height.

51. *Garden Area.* Yardbird Landscaping has 48 m of fencing with which to enclose a rectangular garden. If the garden is x meters long, express the garden's area as a function of the length.

52. *Tablecloth Area.* A tailor uses 16 ft of lace to trim the edges of a rectangular tablecloth. If the tablecloth is w feet wide, express its area as a function of the width.

53. *Inscribed Rhombus.* A rhombus is inscribed in a rectangle that is w meters wide with a perimeter of 40 m. Each vertex of the rhombus is a midpoint of a side of the rectangle. Express the area of the rhombus as a function of the rectangle's width. (*Hint:* Consider the area of the rectangle.)

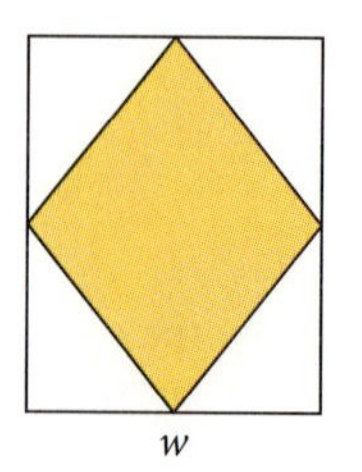

54. *Gas Tank Volume.* A gas tank has ends that are hemispheres of radius r feet. The cylindrical midsection is 6 ft long. Express the volume of the tank as a function of r.

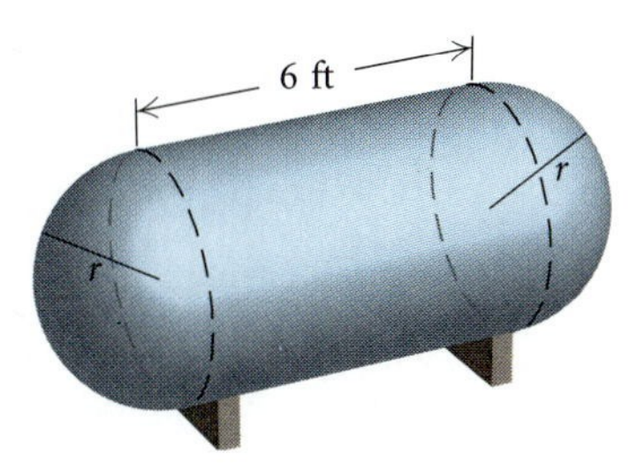

55. *Golf Distance Finder.* A device used in golf to estimate the distance d, in yards, to a hole measures the size s, in inches, that the 7-ft pin appears to be in a viewfinder. Express the distance d as a function of s.

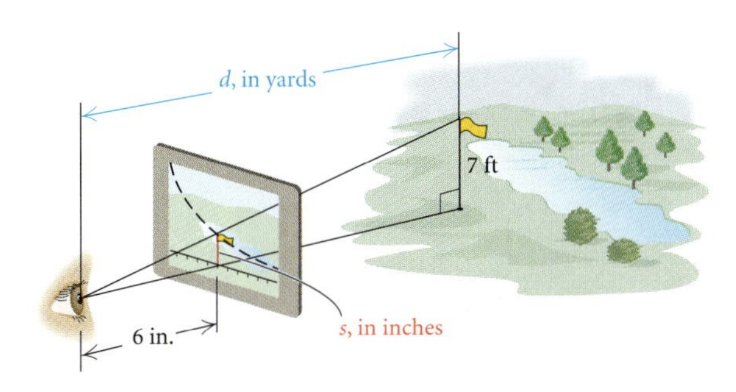

56. *Play Space.* A daycare center has 30 ft of dividers with which to enclose a rectangular play space in a corner of a large room. The sides against the wall require no partition. Suppose the play space is x feet long.

a) Express the area of the play space as a function of x.
b) Find the domain of the function.
c) Graph the function.
d) What dimensions yield the maximum area?

57. *Volume of a Box.* From a 12-cm by 12-cm piece of cardboard, square corners are cut out so that the sides can be folded up to make a box.

a) Express the volume of the box as a function of the length, x, in centimeters, of a cut-out square.
b) Find the domain of the function.
c) Graph the function.
d) What dimensions yield the maximum volume?

58. *Cost of Material.* A rectangular box with volume 320 ft^3 is built with a square base and top. The cost is $1.50/$ft^2$ for the bottom, $2.50/$ft^2$ for the sides, and $1/$ft^2$ for the top. Let $x =$ the length of the base, in feet.

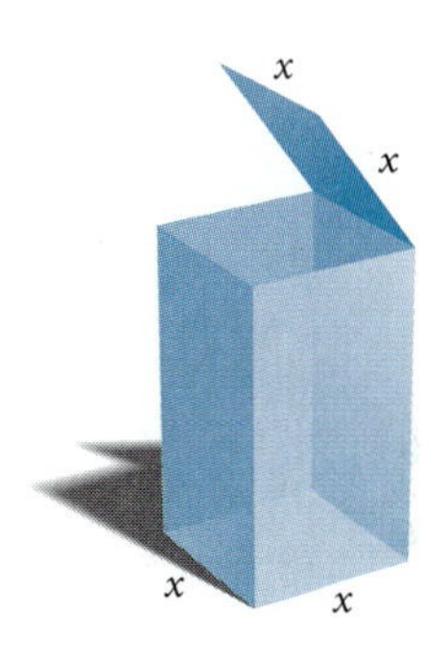

a) Express the cost of the box as a function of x.
b) Find the domain of the function.
c) Graph the function.
d) What dimensions minimize the cost of the box?

59. *Area of an Inscribed Rectangle.* A rectangle that is x feet wide is inscribed in a circle of radius 8 ft.

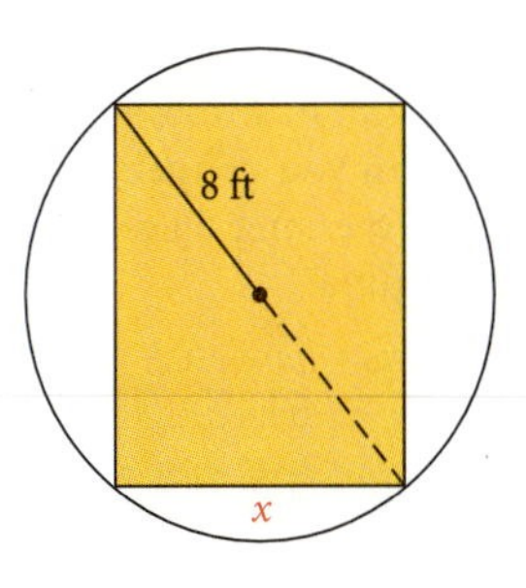

a) Express the area of the rectangle as a function of x.
b) Find the domain of the function.
c) Graph the function.
d) What dimensions maximize the area of the rectangle?

60. *Area of an Inscribed Rectangle.* A rectangle that is x meters wide is inscribed in a circle of diameter 20 m.

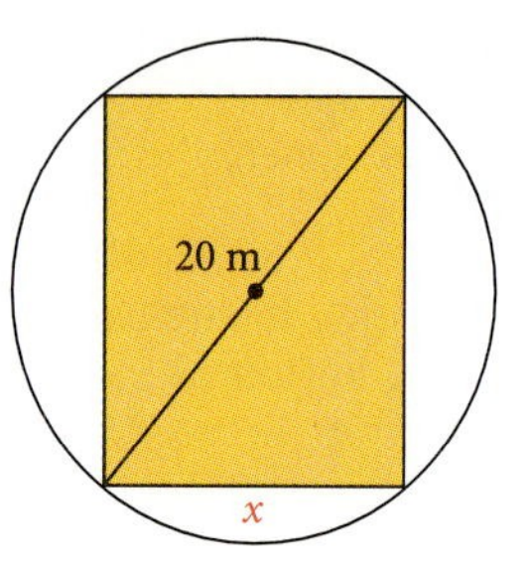

a) Express the area of the rectangle as a function of x.
b) Find the domain of the function.
c) Graph the function.
d) What dimensions maximize the area of the rectangle?

Skill Maintenance

If $f(x) = 7x + 2$, *find each of the following.*

61. $\dfrac{f(b) - f(a)}{b - a}$

62. $\dfrac{f(a + h) - f(a)}{h}$

Synthesis

63. ◈ Describe a real-world situation that could be modeled by a function that is, in turn, increasing, then constant, and finally decreasing.

64. ◈ Simply stated, a *continuous function* is a function whose graph can be drawn without lifting the pencil from the paper. Examine several functions in this exercise set to see if they are continuous. Then explore the continuous functions to find the relative maxima and minima. For continuous functions, how can you connect the ideas of increasing and decreasing on an interval to relative maxima and minima?

Use a grapher. Estimate where each function is increasing or decreasing and any relative maxima or minima. Consider the entire set of real numbers if no domain is given.

65. $f(x) = 20.17x^3 - 3.24x^5$

66. $f(x) = 3.22x^5 - 5.208x^3 - 11$

67. $f(x) = x^4 + 4x^3 - 36x^2 - 160x + 400$

68. $f(x) = -x^6 - 4x^5 + 54x^4 + 160x^3 - 641x^2 - 828x + 1200$

69. *Parking Costs.* A parking garage charges \$2 for up to (but not including) 1 hr of parking, \$4 for up to 2 hr of parking, \$6 for up to 3 hr of parking, and so on. Let $C(t)$ = the cost of parking for t hours.

a) Graph the function.
b) Write an equation for $C(t)$ using the greatest integer notation INT.

70. If $\text{INT}(x + 2) = -3$, what are the possible inputs for x?

71. If $[\text{INT}(x)]^2 = 25$, what are the possible inputs for x?

72. *Minimizing Power Line Costs.* A power line is constructed from a power station at point A to an island at point C, which is 1 mi directly out in the water from a point B on the shore. Point B is 4 mi downshore from the power station at A. It costs \$5000 per mile to lay the power line under water and \$3000 per mile to lay the power line under ground. The line comes to the shore at point S downshore from A. Note that S could very well be A or B. Let x = the distance from B to S.

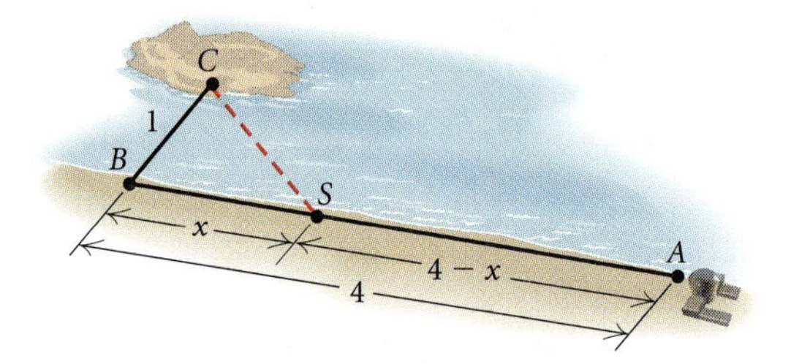

a) Express the cost C of laying the line as a function of x.
b) At what distance x should the line come to shore in order to minimize cost?

73. *Volume of an Inscribed Cylinder.* A right circular cylinder of height h and radius r is inscribed in a right circular cone with a height of 10 ft and a base with a radius of 6 ft.

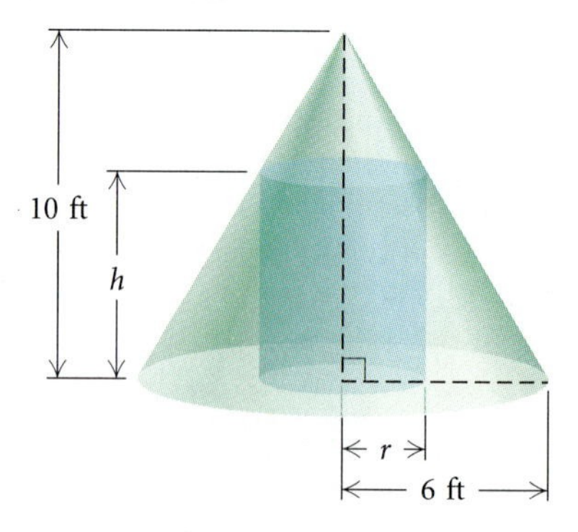

a) Express the height h of the cylinder as a function of r.
b) Express the volume V as a function of r.
c) Express the volume V as a function of h.

1.3 Linear Functions and Applications

- *Graph linear functions and equations, finding the slope and the y-intercept.*
- *Determine equations of lines.*
- *Solve applied problems involving linear functions.*

In real-life situations, we often need to make decisions on the basis of limited information. When the given information is used to formulate an equation or inequality that at least approximates the situation mathematically, we have created a **model**. The most frequently used mathematical models are *linear*—the graphs of these models are straight lines.

Linear Functions

Let's begin to examine the connections among equations, functions, and graphs that are straight lines.

Interactive Discovery

Graph each of the following equations on a grapher. Which have graphs that are lines? Look for patterns.

$$4x + 5y = 20, \qquad y = 5 - x^2,$$

$$y = \frac{x}{2} + 12, \qquad y = 1.5x - 0.05,$$

$$y = -2x - 5, \qquad y = \frac{1}{x},$$

$$x = -3, \qquad y = 6.2,$$

$$y = \sqrt{x}, \qquad x = 4.9.$$

(Some graphers are not able to graph equations of the type $x = a$. Consult your manual.) Which of these equations have graphs that are lines and are also functions?

We have the following results and related terminology.

Linear Functions

A function f is a *linear function* if it is given by $f(x) = mx + b$, where m and b are constants.

If $m = 0$, the function simplifies to the *constant function* $f(x) = b$. If $m = 1$ and $b = 0$, the function simplifies to the *identity function* $f(x) = x$.

Horizontal and Vertical Lines

Horizontal lines are given by equations of the type $y = b$ or $f(x) = b$. (They are functions.)

Vertical lines are given by equations of the type $x = a$. (They are not functions.)

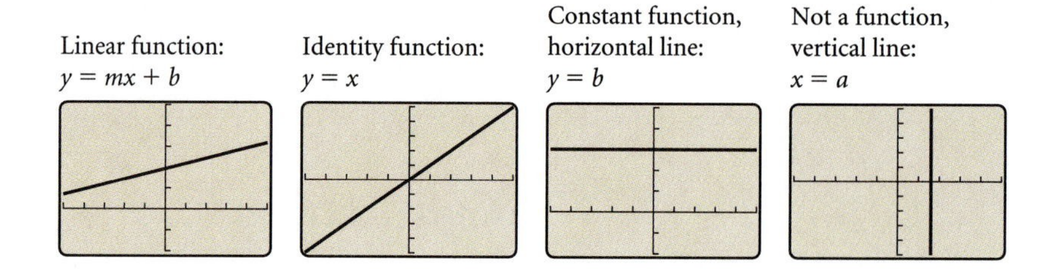

The Linear Function $f(x) = mx + b$ and Slope

To attach meaning to the constant m in the equation $f(x) = mx + b$, we first consider an application. FaxMax is an office machine business. Its total costs for two different time periods are given by two functions shown in the tables and graphs that follow. The variable x represents time, in months. The variable y represents total costs, in thousands of dollars, over that amount of time. Look for a pattern.

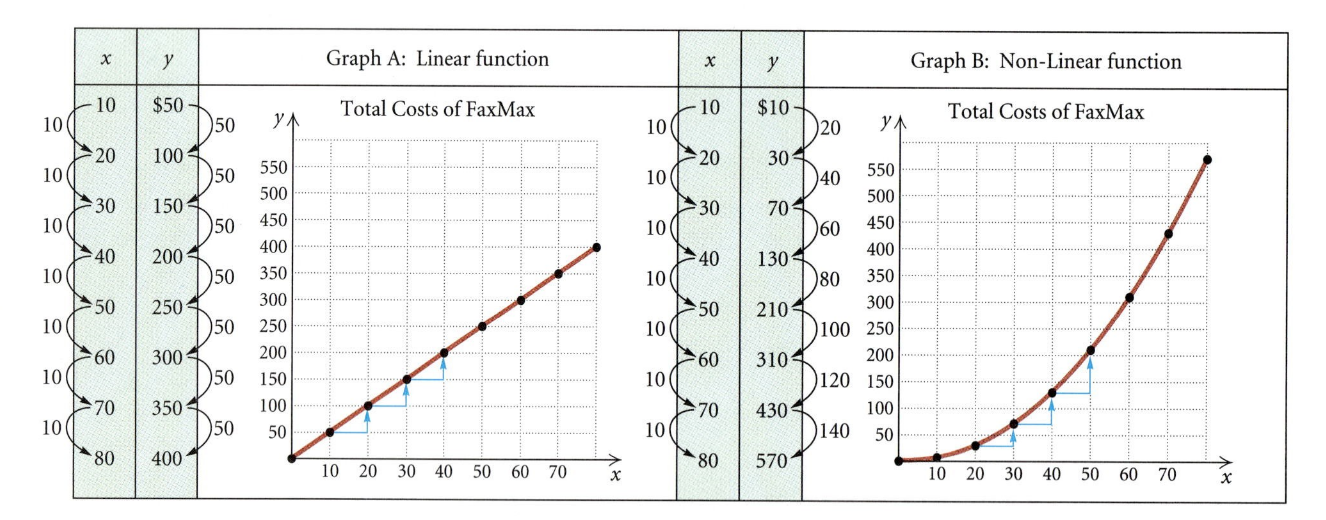

x	y
10	$50
20	100
30	150
40	200
50	250
60	300
70	350
80	400

x	y
10	$10
20	30
30	70
40	130
50	210
60	310
70	430
80	570

We see in graph A that every change of 10 months results in a $50 thousand change in total costs. But in graph B, changes of 10 months do not result in constant changes in total costs. This is a way to distinguish linear from nonlinear functions. The rate at which a linear function changes, or the steepness of its graph, is constant.

Mathematically, we define a line's steepness, or **slope**, as the ratio of its vertical change (rise) to the corresponding horizontal change (run).

Slope

The *slope m* of a line containing points (x_1, y_1) and (x_2, y_2) is given by

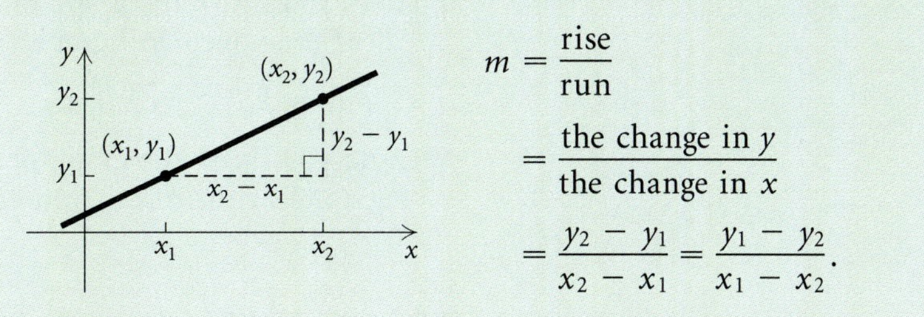

$$m = \frac{\text{rise}}{\text{run}}$$

$$= \frac{\text{the change in } y}{\text{the change in } x}$$

$$= \frac{y_2 - y_1}{x_2 - x_1} = \frac{y_1 - y_2}{x_1 - x_2}.$$

Example 1 Make a hand-drawn graph of the function $f(x) = -\frac{2}{3}x + 1$ and determine its slope.

SOLUTION Since the equation for f is in the form $f(x) = mx + b$, we know it is a linear function. We can graph it by connecting two points on the graph with a straight line. We calculate two ordered pairs, plot the points, graph the function, and determine the slope:

$$f(3) = -\frac{2}{3} \cdot 3 + 1 = -1;$$

$$f(9) = -\frac{2}{3} \cdot 9 + 1 = -5;$$

Pairs: $(3, -1)$, $(9, -5)$;

$$\text{Slope} = m = \frac{y_2 - y_1}{x_2 - x_1}$$

$$= \frac{-5 - (-1)}{9 - 3}$$

$$= \frac{-4}{6} = -\frac{2}{3}.$$

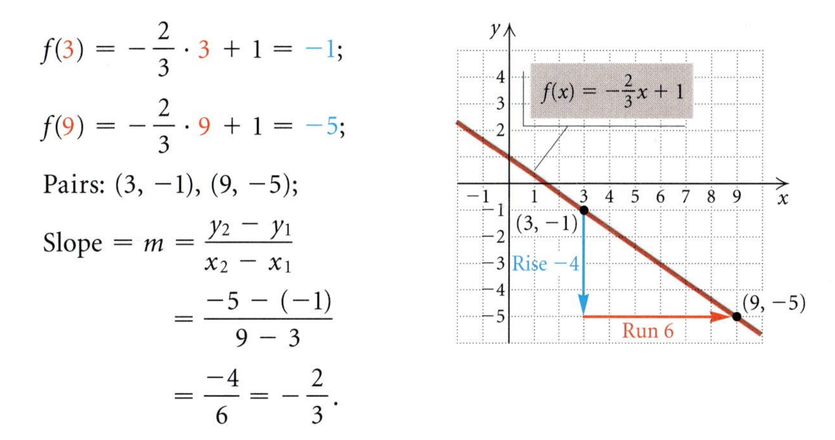

The slope is the same for any two points on a line. Thus, to check our work, note that $f(6) = -\frac{2}{3} \cdot 6 + 1 = -3$. Using the points $(6, -3)$ and $(3, -1)$, we have

$$m = \frac{-3 - (-1)}{6 - 3} = \frac{-2}{3} = -\frac{2}{3}.$$

We can also use the points in the opposite order when computing slope, so long as we are consistent. Note also that the slope of the line is indeed the number m in the equation for the function $f(x) = -\frac{2}{3}x + 1$.

Let's explore the effect of the slope m in linear equations of the type $f(x) = mx$.

Interactive Discovery

With a square viewing window (see the Introduction to Graphs and Graphers), graph the following equations:

$$y_1 = x, \quad y_2 = 2x, \quad y_3 = 5x, \quad \text{and} \quad y_4 = 10x.$$

What do you think the graph of $y = 128x$ will look like?

Clear the screen and graph the following equations:

$$y_1 = x, \quad y_2 = \tfrac{3}{4}x, \quad y_3 = 0.48x, \quad \text{and} \quad y_4 = \tfrac{3}{25}x.$$

What do you think the graph of $y = 0.000029x$ will look like?

Again clear the screen and graph each set of equations:

$$y_1 = -x, \quad y_2 = -2x, \quad y_3 = -4x, \quad \text{and} \quad y_4 = -10x$$

and $\quad y_1 = -x, \quad y_2 = -\tfrac{2}{3}x, \quad y_3 = -0.35x, \quad \text{and} \quad y_4 = -\tfrac{1}{10}x.$

From your observations, what do you think the graphs of $y = -200x$ and $y = -0.000017x$ will look like?

If a line slants up from left to right, the change in x and the change in y have the same sign, so the line has a positive slope. The larger the slope is, the steeper the line. If a line slants down from left to right, the change in x and the change in y are of opposite signs, so the line has a negative slope. The larger the absolute value of the slope, the steeper the line.

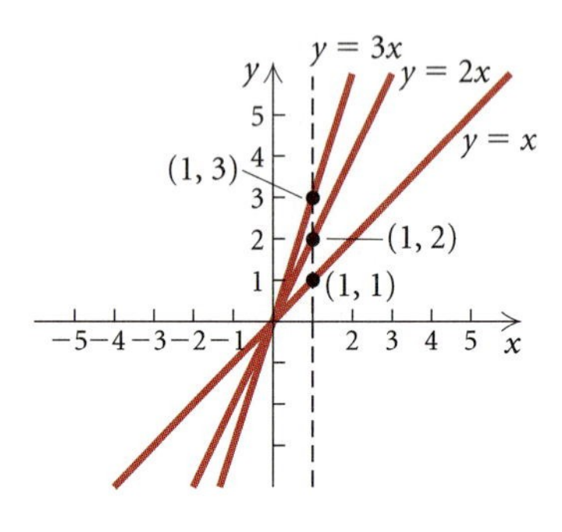

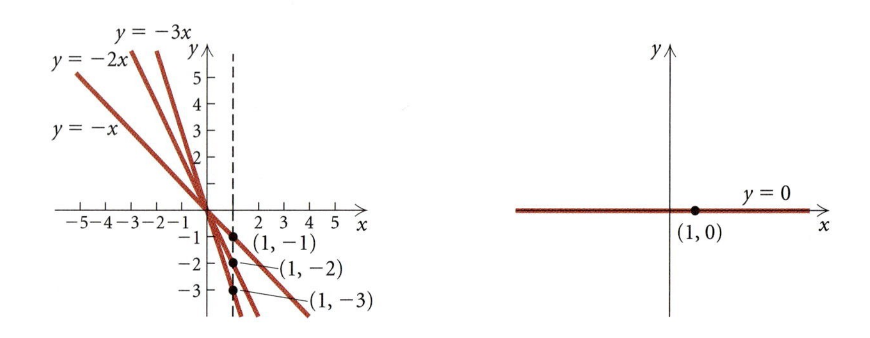

When $m = 0$, $y = 0x$, or $y = 0$, as shown in the graph on the right above. Note that this is both the x-axis and a horizontal line.

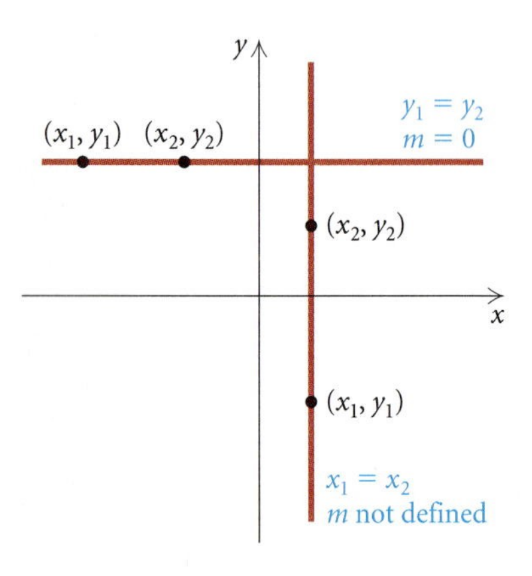

> ***Horizontal and Vertical Lines***
>
> If a line is horizontal, the change in y for any two points is 0. Thus a horizontal line has slope 0.
>
> If a line is vertical, the change in x for any two points is 0. Thus the slope is *not defined* because we cannot divide by 0.

Note that zero slope and an undefined slope are two very different concepts.

Applications of Slope

Slope has many real-world applications. For example, numbers like 2%, 4%, and 7% are often used to represent the **grade** of a road. Such a number is meant to tell how steep a road is on a hill or mountain. For example, a 4% grade means that the road rises 4 ft for every horizontal distance of 100 ft if a vehicle is going up; and −4% means that the road is dropping 4 ft for every 100 ft, if the vehicle is going down.

The concept of grade is also used in cardiology when a person runs on a treadmill. A physician may change the slope, or grade, of a treadmill to measure its effect on heart rate. Another example occurs in hydrology. When a river flows, the strength or force of the river depends on how far the river falls vertically compared to how far it flows horizontally.

Example 2 *Ramps for the Handicapped.* Construction laws regarding access ramps for the handicapped state that every vertical rise of 1 ft requires a horizontal run of 12 ft. What is the grade, or slope, of such a ramp?

SOLUTION The grade, or slope, is given by

$$m = \tfrac{1}{12} \approx 0.083 \approx 8.3\%.$$

Slope–Intercept Equations of Lines

Let's explore the effect of the constant b in linear equations of the type $f(x) = mx + b$.

Interactive Discovery

Begin with the graph of $y = x$ and a square viewing window. Now graph the lines $y = x + 3$ and $y = x - 4$ in the same viewing window. How do the lines differ from $y = x$? What do you think the line $y = x - 6$ will look like?

Try graphing $y = -0.5x$, $y = -0.5x - 4$, and $y = -0.5x + 3$ in the same viewing window. Describe what happens to the graph of $y = -0.5x$ when a number b is added.

Compare the graphs of the equations

$$y = 3x \quad \text{and} \quad y = 3x - 2.$$

Note that the graph of $y = 3x - 2$ is a shift down of the graph of $y = 3x$, and that $y = 3x - 2$ has y-intercept $(0, -2)$. That is, the graph crosses the y-axis at $(0, -2)$.

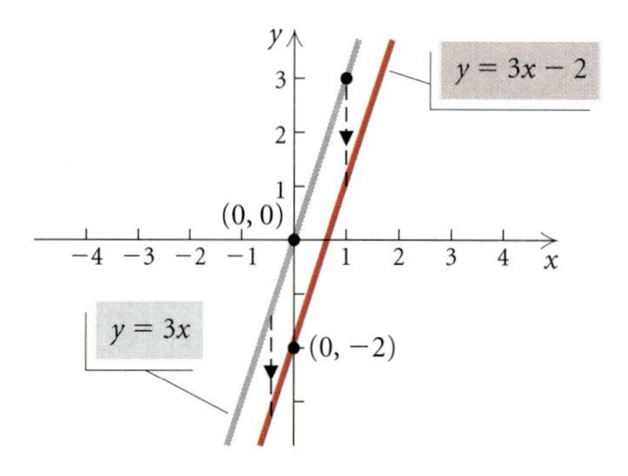

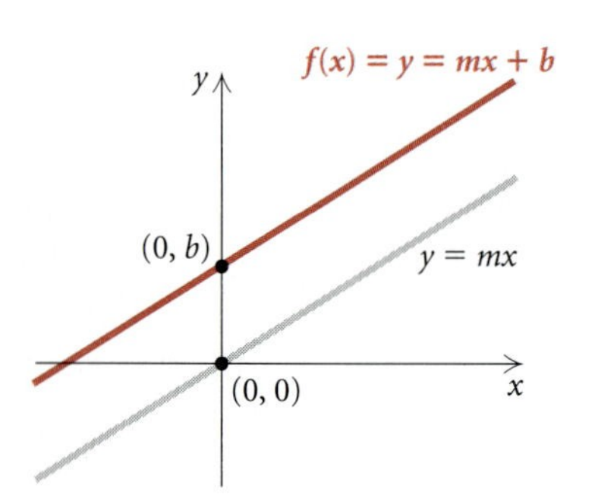

The Slope–Intercept Equation

The linear function f given by

$$f(x) = mx + b$$

has a graph that is a straight line parallel to $y = mx$. The constant m is called the slope, and the y-intercept is $(0, b)$, or b.

We know that a nonvertical line has slope m and y-intercept $(0, b)$ and that its equation is given by $y = mx + b$. The advantage of $y = mx + b$ is that we can read the slope m and the y-intercept b directly from the equation.

Example 3 Find the slope and the y-intercept of the line with equation $y = -0.25x - 3.8$.

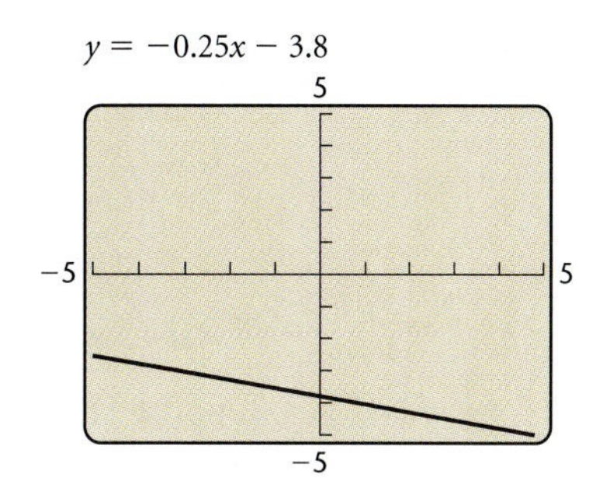

SOLUTION

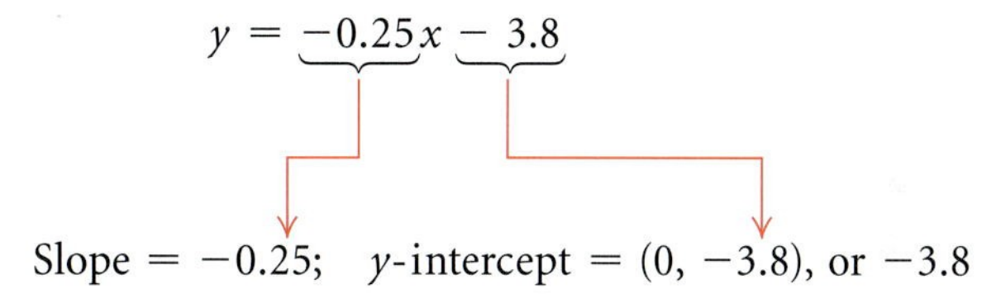

$$y = \underbrace{-0.25x}\ \underbrace{-\ 3.8}$$

Slope $= -0.25$; y-intercept $= (0, -3.8)$, or -3.8

Example 4 Find the slope and the y-intercept of the line with equation $y = 8$.

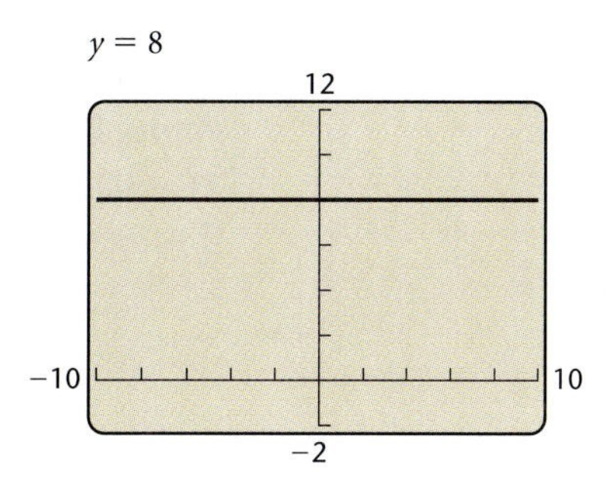

SOLUTION We can rewrite this equation as $y = 0x + 8$. We then see that the slope is 0 and the y-intercept is $(0, 8)$, or 8. The graph is a horizontal line.

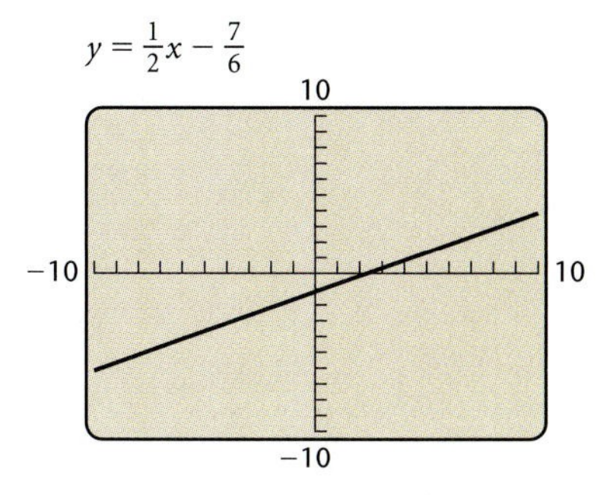

Any equation whose graph is a straight line is a **linear equation.** To find the slope and the y-intercept of the graph of a linear equation, we can solve for y, and then read the information from the equation.

Example 5 Find the slope and the y-intercept of the line with equation $3x - 6y - 7 = 0$.

SOLUTION We solve for y, obtaining $y = \frac{1}{2}x - \frac{7}{6}$. Thus the slope is $\frac{1}{2}$ and the y-intercept is $\left(0, -\frac{7}{6}\right)$, or $-\frac{7}{6}$.

Example 6 A line has slope $-\frac{7}{9}$ and y-intercept (0, 16). Find an equation of the line.

SOLUTION We use the slope–intercept equation and substitute $-\frac{7}{9}$ for m and 16 for b:

$$y = mx + b$$
$$y = -\tfrac{7}{9}x + 16.$$

Point–Slope Equations of Lines

Suppose that we have a nonvertical line and that the coordinates of point P_1 are (x_1, y_1). We can think of P_1 as fixed and imagine a movable point P on the line with coordinates (x, y). Thus the slope is given by

$$\frac{y - y_1}{x - x_1} = m.$$

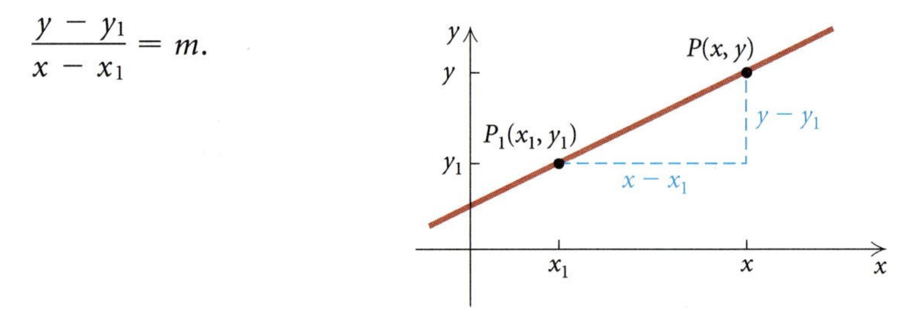

Multiplying on both sides by $x - x_1$, we get the *point–slope equation.*

Point–Slope Equation

The *point–slope equation* of the line with slope m passing through (x_1, y_1) is

$$y - y_1 = m(x - x_1).$$

Thus if we know the slope of a line and the coordinates of one point on the line, we can find an equation of the line.

Example 7 Find an equation of the line containing the point $\left(\frac{1}{2}, -1\right)$ and with slope 5.

SOLUTION If we substitute in $y - y_1 = m(x - x_1)$, we get

$$y - (-1) = 5\left(x - \tfrac{1}{2}\right),$$

which simplifies as follows:

$$y + 1 = 5x - \tfrac{5}{2}$$
$$y = 5x - \tfrac{5}{2} - 1$$
$$y = 5x - \tfrac{7}{2}. \quad \text{Slope–intercept equation}$$

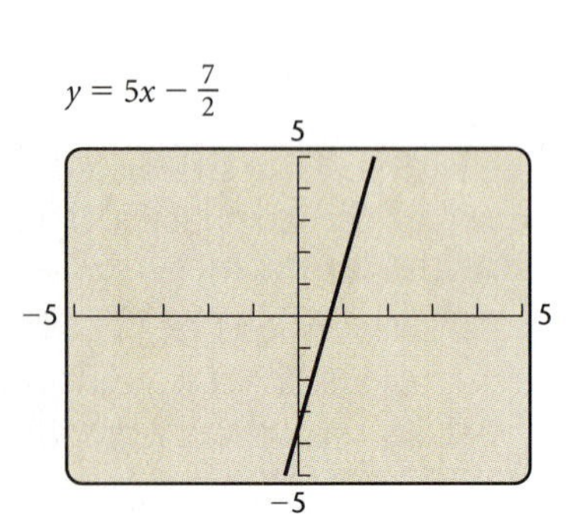

The advantage of having a grapher when doing a problem like the one in Example 7 is that you have a quick, visual way of seeing your result.

Two-Point Equations of Lines

Suppose that a nonvertical line contains the points $P_1(x_1, y_1)$ and $P_2(x_2, y_2)$. The slope of the line is

$$m = \frac{y_2 - y_1}{x_2 - x_1}.$$

If we substitute $\frac{y_2 - y_1}{x_2 - x_1}$ for m in the point–slope equation,

$$y - y_1 = m(x - x_1),$$

we get the *two-point equation.*

Two-Point Equation

The *two-point equation* of the line passing through the points (x_1, y_1) and (x_2, y_2) is

$$y - y_1 = \frac{y_2 - y_1}{x_2 - x_1}(x - x_1).$$

Example 8 Find an equation of the line containing the points (2, 3) and (1, −4).

SOLUTION If we take (2, 3) as P_1 and (1, −4) as P_2 and use the two-point equation, we get

$$y - 3 = \frac{-4 - 3}{1 - 2}(x - 2),$$

which simplifies to $y = 7x - 11$.

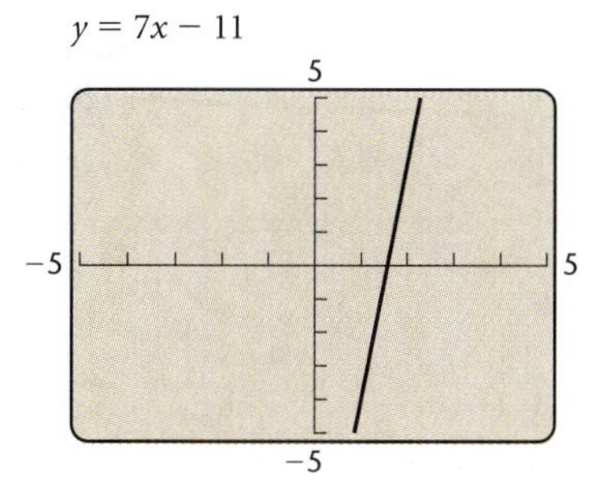

Parallel Lines

How can we examine a pair of equations to determine whether their graphs are parallel—that is, they do not intersect? We can explore this with a grapher.

Interactive Discovery

Graph each of the following pairs of lines. Try to determine whether they are parallel. You will probably need to change viewing windows and either zoom in or zoom out to decide for sure. Complete the table and look for a pattern.

PAIRS OF EQUATIONS	SLOPES		PARALLEL?
	m_1	m_2	
$y = -0.38x + 4.2$, $y = -0.37x - 5.1$			
$y = -0.38x + 4.2$, $50y + 19x = 21$			
$2x + 1 = y$, $y + 3x = 4$			
$y = 8.02x - 11.3$, $y = 7.98x - 11.3$			
$y = -3$, $y = 4.2$			
$x = -5$, $x = 2$			

Can you find a way to know whether two lines are parallel without graphing?

If two different lines are vertical, then they are parallel. Thus two equations such as $x = a_1$, $x = a_2$, where $a_1 \neq a_2$, have graphs that are *parallel lines*. Two nonvertical lines such as $y = mx + b_1$, $y = mx + b_2$, where $b_1 \neq b_2$, also have graphs that are *parallel lines*.

Parallel Lines

Vertical lines are *parallel*. Nonvertical lines are *parallel* if and only if they have the same slope and different y-intercepts.

Perpendicular Lines

How can we examine a pair of equations to determine whether they are perpendicular? Let's explore this with a grapher.

Interactive Discovery

Graph each of the following pairs of lines (see the table at the top of the following page). Try to determine whether they are perpendicular. You will probably need to change viewing windows and either zoom in or zoom out to decide for sure. However, you must use square viewing windows so that you can see or measure whether angles between lines seem to be 90°. Complete the table and look for a pattern.

(*continued*)

PAIRS OF EQUATIONS	SLOPES m_1	m_2	PRODUCT OF SLOPES $m_1 \cdot m_2$	PERPENDICULAR?
$y = -\frac{2}{5}x + 3,\ y = \frac{5}{2}x + 3$				
$y = 0.1875x - 2,\ y = -\frac{16}{3}x - 5.1$				
$3y + 4x = -21,\ 4y - 3x = 8$				
$y = -3x + 4,\ y = 2x - 7$				
$x = -5,\ y = 2$				

Can you find a way to determine whether two lines are perpendicular without graphing?

Perpendicular Lines

Two lines with slopes m_1 and m_2 are *perpendicular* if and only if the product of their slopes is -1:

$$m_1 m_2 = -1.$$

Lines are also *perpendicular* if one is vertical ($x = a$) and the other is horizontal ($y = b$).

If a line has slope m_1, the slope m_2 of a line perpendicular to it is $-1/m_1$ (the slope of one line is the opposite of the reciprocal of the other).

Example 9 Determine whether each of the following pairs of lines is parallel, perpendicular, or neither.

a) $y + 2 = 5x,\ 5y + x = -15$

b) $2y + 4x = 8,\ 5 + 2x = -y$

c) $2x + 1 = y,\ y + 3x = 4$

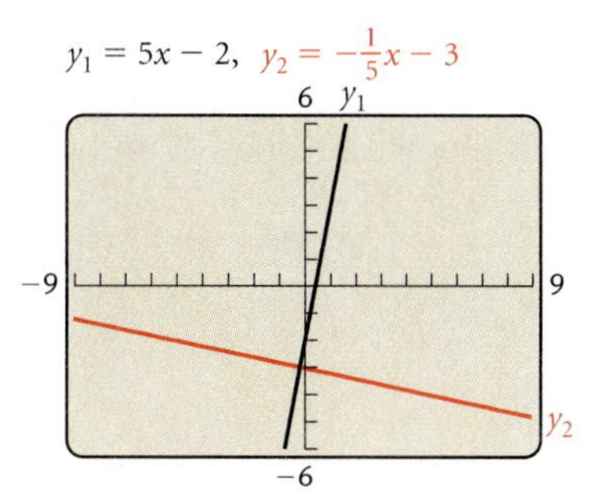

SOLUTION We use an algebraic procedure to know for sure. We can create graphs on a grapher as a check.

a) We solve each equation for y:

$$y = 5x - 2, \qquad y = -\frac{1}{5}x - 3.$$

The slopes are 5 and $-\frac{1}{5}$. Their product is -1, so the lines are perpendicular (see the figure at left).

b) Solving each equation for y, we get

$$y = -2x + 4, \qquad y = -2x - 5.$$

We see that $m_1 = -2$ and $m_2 = -2$. Since the slopes are the same and the y-intercepts, 4 and -5, are different, the lines are parallel (see the figure on the left below).

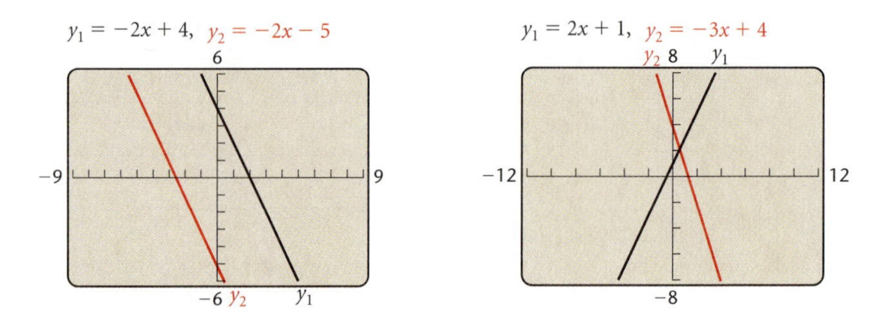

c) Solving each equation for y, we get

$$y = 2x + 1, \qquad y = -3x + 4.$$

Since $m_1 = 2$ and $m_2 = -3$, we know that $m_1 \neq m_2$ and that $m_1 m_2 \neq -1$. It follows that the lines are neither perpendicular nor parallel (see the figure on the right above).

Example 10 Write equations of the lines (a) parallel and (b) perpendicular to the graph of the line $4y - x = 20$ and both containing the point $(2, -3)$.

SOLUTION We first solve $4y - x = 20$ for y to get $y = \frac{1}{4}x + 5$. Thus the slope is $\frac{1}{4}$.

a) The line parallel to the given line will have slope $\frac{1}{4}$. We use the point–slope equation for a line with slope $\frac{1}{4}$ and containing the point $(2, -3)$:

$$y - y_1 = m(x - x_1)$$
$$y - (-3) = \tfrac{1}{4}(x - 2)$$
$$y = \tfrac{1}{4}x - \tfrac{7}{2}.$$

b) The slope of the perpendicular line is -4. Now we use the point–slope equation to write an equation for a line with slope -4 and containing the point $(2, -3)$:

$$y - y_1 = m(x - x_1)$$
$$y - (-3) = -4(x - 2)$$
$$y = -4x + 5.$$

We can visualize our work on a grapher.

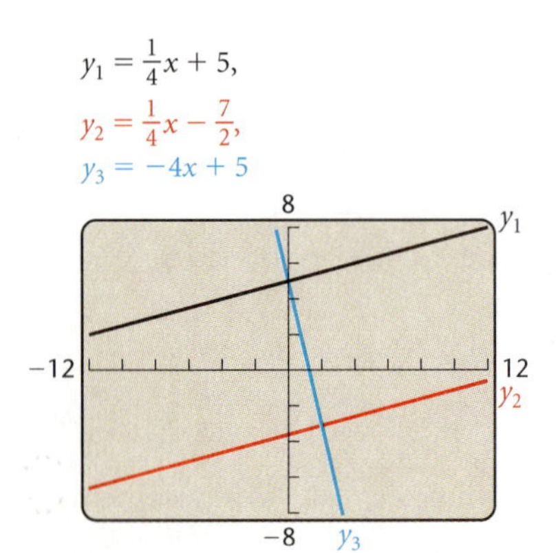

Summary of Terminology About Lines

TERMINOLOGY	MATHEMATICAL INTERPRETATION
Slope	$m = \dfrac{y_2 - y_1}{x_2 - x_1}$
Slope–intercept equation	$y = mx + b$
Point–slope equation	$y - y_1 = m(x - x_1)$
Two-point equation	$y - y_1 = \dfrac{y_2 - y_1}{x_2 - x_1}(x - x_1)$
Horizontal lines	$y = b$
Vertical lines	$x = a$
Parallel lines	$m_1 = m_2,\ b_1 \neq b_2$
Perpendicular lines	$m_1 m_2 = -1$

Applications of Linear Functions

We now consider an application of linear functions.

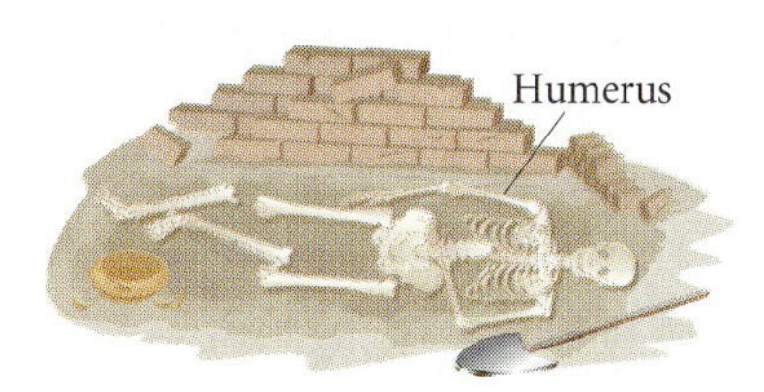

Example 11 *Anthropology Estimates.* An anthropologist can use certain linear functions to estimate the height of a male or a female, given the length of certain bones. The humerus is the bone from the elbow to the shoulder. Let $x =$ the length of the humerus, in centimeters. Then the height, in centimeters, of an adult male with a humerus of length x is given by the function

$$M(x) = 2.89x + 70.64.$$

The height, in centimeters, of an adult female with a humerus of length x is given by the function

$$F(x) = 2.75x + 71.48.$$

a) A 26-cm humerus was uncovered in a ruins. Assuming it was from a female, how tall was she?

b) Graph F.

c) Find the domain of F.

SOLUTION

a) We substitute into the function:

$$F(26) = 2.75(26) + 71.48 = 142.98.$$

Thus the female was 142.98 cm tall.

b) The graph is shown at left.

$y = 2.75x + 71.48$

1000

0

0

200

c) Theoretically, the domain of the function is the set of all real numbers. However, the context of the problem dictates a different domain. One could not find a bone with a length of 0 or less. Thus the domain consists of positive real numbers, that is, the interval $(0, \infty)$.

1.3 Exercise Set

Each table of data contains input–output values for a function. Answer the following questions for each table.

a) *Is the change in the inputs x the same?*
b) *Is the change in the outputs y the same?*
c) *Is the function linear?*

1.

x	y
−3	7
−2	10
−1	13
0	16
1	19
2	22
3	25

2.

x	y
20	12.4
30	24.8
40	49.6
50	99.2
60	198.4
70	396.8
80	793.6

3.

x	y
11	3.2
26	5.7
41	8.2
56	9.3
71	11.3
86	13.7
101	19.1

4.

x	y
2	−8
4	−12
6	−16
8	−20
10	−24
12	−28
14	−36

Find the slope of the line containing the given points.

5. **6.**

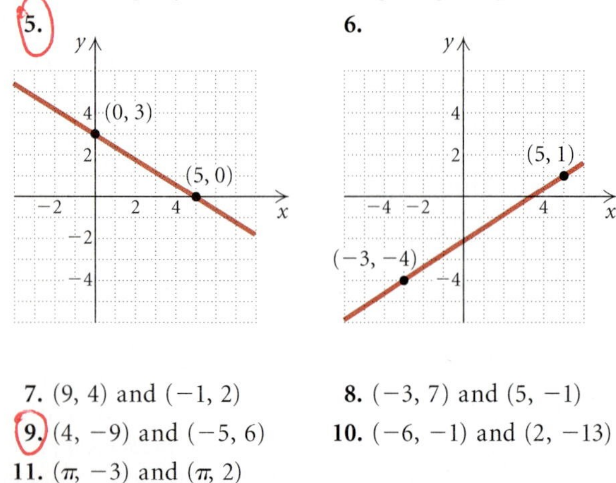

7. $(9, 4)$ and $(-1, 2)$

8. $(-3, 7)$ and $(5, -1)$

9. $(4, -9)$ and $(-5, 6)$

10. $(-6, -1)$ and $(2, -13)$

11. $(\pi, -3)$ and $(\pi, 2)$

12. $(\sqrt{2}, -4)$ and $(0.56, -4)$

13. (a, a^2) and $(a + h, (a + h)^2)$

14. $(a, 3a + 1)$ and $(a + h, 3(a + h) + 1)$

15. *Road Grade.* Using the following figure, find the road grade and an equation giving the height y as a function of the horizontal distance x.

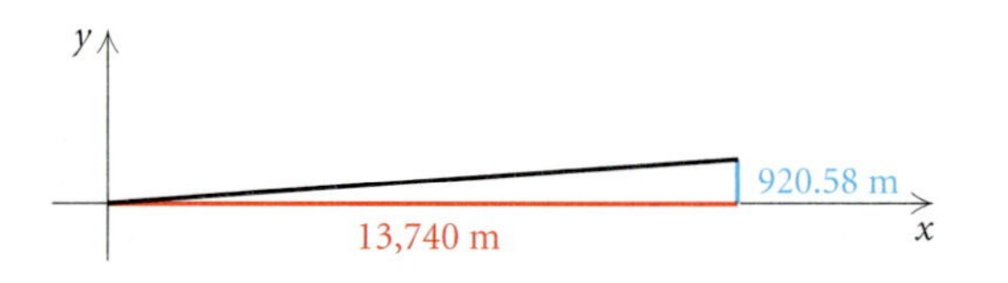

16. *Heart Arrhythmia.* A treadmill is 5 ft long and is set at an 8% grade when a heart arrhythmia occurs. How high is the end of the treadmill?

Find the slope and the y-intercept of the equation.

17. $y = \frac{3}{5}x - 7$

18. $f(x) = -2x + 3$

19. $f(x) = 5 - \frac{1}{2}x$

20. $y = 2 + \frac{3}{7}x$

21. $3x + 2y = 10$

22. $2x - 3y = 12$

23. $4x - 3f(x) - 15 = 6$

24. $9 = 3 + 5x - 2f(x)$

Write a slope–intercept equation for a line with the given characteristics.

25. $m = \frac{2}{9}$, y-intercept $(0, 4)$

26. $m = -\frac{3}{8}$, y-intercept $(0, 5)$

27. $m = -4$, y-intercept $(0, -7)$

28. $m = \frac{2}{7}$, y-intercept $(0, -6)$

29. $m = -4.2$, y-intercept $(0, \frac{3}{4})$

30. $m = -4$, y-intercept $(0, -\frac{3}{2})$

31. $m = \frac{2}{9}$, passes through $(3, 7)$

32. $m = -\frac{3}{8}$, passes through $(5, 6)$

33. $m = 3$, passes through $(1, -2)$

34. $m = -2$, passes through $(-5, 1)$

35. $m = -\frac{3}{5}$, passes through $(-4, -1)$

36. $m = \frac{2}{3}$, passes through $(-4, -5)$

37. Passes through $(-1, 5)$ and $(2, -4)$

38. Passes through $(2, -1)$ and $(7, -11)$

39. Passes through $(7, 0)$ and $(-1, 4)$

40. Passes through $(-3, 7)$ and $(-1, -5)$

Determine whether each of the following pairs of lines is parallel, perpendicular, or neither.

41. $x + 2y = 5$,
$2x + 4y = 8$

42. $2x - 5y = -3$,
$2x + 5y = 4$

43. $y = 4x - 5$,
$4y = 8 - x$

44. $y = 7 - x$,
$y = x + 3$

Write a slope–intercept equation for a line passing through the given point that is parallel to the given line. Then write a second equation for a line passing through the given point that is perpendicular to the given line.

45. $(3, 5),\ y = \frac{2}{7}x + 1$

46. $(-1, 6),\ f(x) = 2x + 9$

47. $(-7, 0),\ y = -0.3x + 4.3$

48. $(-4, -5),\ 2x + y = -4$

49. $(3, -2),\ 3x + 4y = 5$

50. $(8, -2),\ y = 4.2(x - 3) + 1$

51. $(3, -3),\ x = -1$

52. $(4, -5),\ y = -1$

53. *Ideal Weight.* One formula to estimate the ideal weight of a woman is to multiply her height by 3.5 and subtract 110. Let $W =$ the ideal weight, in pounds, and $h =$ height, in inches.

a) Express W as a linear function of h.
b) Find the ideal weight of a woman whose height is 62 in.
c) Graph W.
d) Find the domain of the function.

54. *Pressure at Sea Depth.* The function P, given by

$$P(d) = \frac{1}{33}d + 1,$$

gives the pressure, in atmospheres (atm), at a depth d, in feet, under the sea.

a) Find $P(0)$, $P(5)$, $P(10)$, $P(33)$, and $P(200)$.
b) Graph P.
c) Find the domain of the function.

55. *Stopping Distance on Glare Ice.* The stopping distance (at some fixed speed) of regular tires on glare ice is a function of the air temperature F, in degrees Fahrenheit. This function is estimated by

$$D(F) = 2F + 115,$$

where $D(F) =$ the stopping distance, in feet, when the air temperature is F, in degrees Fahrenheit.

a) Find $D(0°)$, $D(-20°)$, $D(10°)$, and $D(32°)$.
b) Graph D.
c) Explain why the domain should be restricted to $[-57.5°, 32°]$.

56. *Anthropology Estimates.* Consider Example 10 and the function

$$M(x) = 2.89x + 70.64$$

for estimating the height of a male.

a) Find the height of a male if a 26-cm humerus is found in an archeological dig.
b) Graph M.
c) Find the domain of M.

57. *Reaction Time.* While driving a car, you suddenly see a school crossing guard. Your brain registers the information and sends a signal to your foot to hit the brake. The car travels a distance D, in feet, during this time, where D is a function of the speed r, in miles per hour, that the car is traveling when you see the crossing guard. That reaction distance is a linear function given by

$$D(r) = \frac{11r + 5}{10}.$$

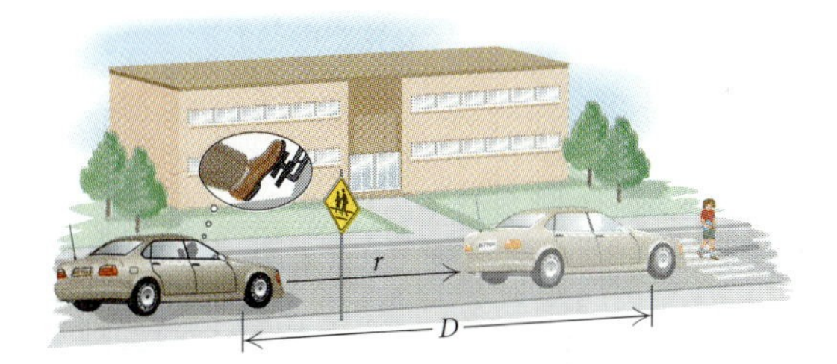

a) Find the slope of this line and interpret its meaning in this application.
b) Find $D(5)$, $D(10)$, $D(20)$, $D(50)$, and $D(65)$.
c) Graph D.
d) What is the domain of this function? Explain.

58. *Straight-line Depreciation.* A company buys a new color printer for \$5200 to print banners for a sales campaign. The printer is purchased on January 1 and is expected to last 8 yr, at the end of which time its *trade-in*, or *salvage value*, will be \$1100. If the company figures the decline or depreciation in value to be the same each year, then the salvage value V, after t years, is given by the linear function

$$V(t) = \$5200 - \$512.50t, \quad \text{for } 0 \le t \le 8.$$

a) Find $V(0)$, $V(1)$, $V(2)$, $V(3)$, and $V(8)$.
b) Graph V.
c) Find the domain and the range of this function.

For Exercises 59 and 60, express the slope as a ratio of two quantities.

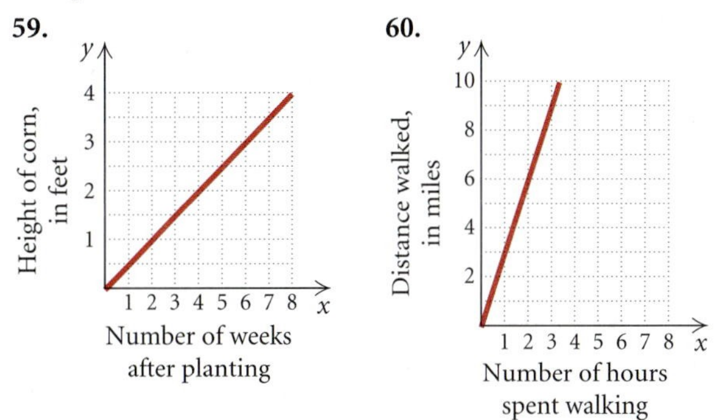

61. *Total Cost.* The Cellular Connection charges \$60 for a phone and \$40 per month under its economy plan. Write and graph an equation that can be used to determine the total cost, $C(t)$, of operating a Cellular Connection phone for t months.

62. *Total Cost.* Twin Cities Cable Television charges a \$35 installation fee and \$30 per month for "deluxe" service. Write and graph an equation that can be used to determine the total cost, $C(t)$, for t months of deluxe cable television service. Find the total cost for 8 months of service.

In Exercises 63 and 64, the term ***fixed costs*** *refers to the start-up costs of operating a business. This includes machinery and building costs. The term* ***variable costs*** *refers to what it costs a business to produce or service one item.*

63. Kara's Custom Tees experienced fixed costs of \$800 and variable costs of \$3 per shirt. Write an equation that can be used to determine the total expenses encountered by Kara's Custom Tees. Then graph the equation.

64. It's My Racquet experienced fixed costs of \$500 and variable costs of \$2 for each tennis racquet that is restrung. Write an equation that can be used to determine the total expenses encountered by It's My Racquet. Then graph the equation.

Direct Variation. Many applications can be modeled by linear functions like

$$y = kx, \quad k > 0,$$

where the variables are nonnegative: These are functions of ***direct variation.*** *The constant k is called a* ***variation constant.*** *Note that the function is increasing over the interval* $[0, \infty)$.

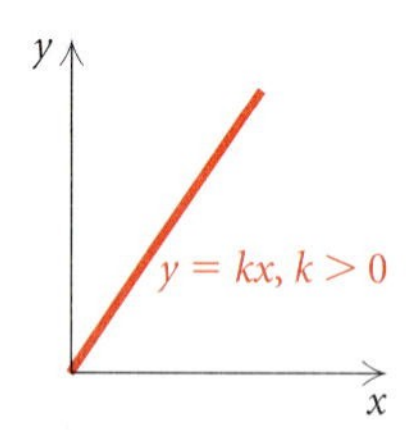

65. *House of Representatives.* The number of representatives N that each state has varies directly as the number of people P living in the state. If New York, with 18,119,416 residents, has 31 representatives, how many representatives does Michigan, with a population of 9,436,628, have?

66. *Hooke's Law.* Hooke's law states that the distance d that a spring will stretch varies directly as the mass m of an object hanging from the spring. Suppose that a 3-kg mass stretches a spring 40 cm. Find a function of direct variation. Then use the equation to predict how far the spring will stretch with a 5-kg mass.

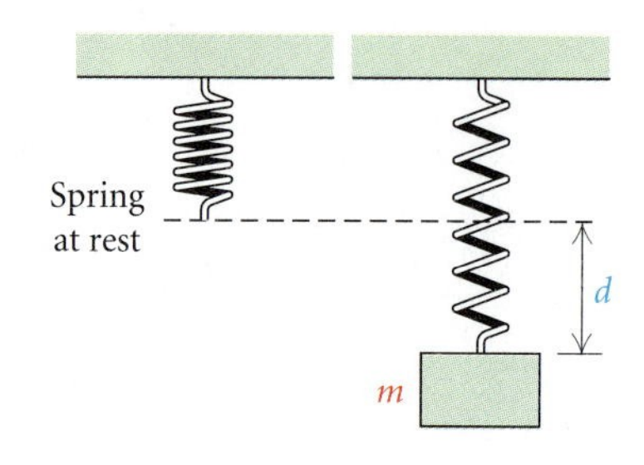

Skill Maintenance

If $f(x) = x^2 - 3x$, *find each of the following.*

67. $f(-5)$

68. $f(5)$

69. $f(-a)$

70. $f(a + h)$

Synthesis

71. ◈ Discuss why the graph of a vertical line $x = a$ cannot be that of a function.

72. ◈ Explain as you would to a fellow student how the idea of slope can be used to describe the slant and the steepness of a line using a number.

73. Find k so that the line containing the points $(-3, k)$ and $(4, 8)$ is parallel to the line containing the points $(5, 3)$ and $(1, -6)$.

74. Find an equation of the line passing through the point $(4, 5)$ perpendicular to the line passing through the points $(-1, 3)$ and $(2, 9)$.

75. *Fahrenheit and Celsius Temperatures.* Fahrenheit temperature F is a linear function of Celsius temperature C. When C is 0, F is 32; and when C is 100, F is 212. Use these data to express C as a function of F and to express F as a function of C.

Suppose that f is a linear function. Then $f(x) = mx + b$. *Determine whether each of the following is true or false.*

76. $f(c + d) = f(c) + f(d)$

77. $f(cd) = f(c)f(d)$

78. $f(kx) = kf(x)$

79. $f(c - d) = f(c) - f(d)$

Let $f(x) = mx + b$. *Find a formula for* $f(x)$ *given each of the following.*

80. $f(3x) = 3f(x)$

81. $f(x + 2) = f(x) + 2$

1.4 Data Analysis, Curve Fitting, and Linear Regression

- *Analyze a set of data to determine whether it can be modeled by a linear function.*
- *Fit a regression line to a set of data; then use the linear model to make predictions.*

Mathematical Models

When a real-world problem can be described in mathematical language, we have a **mathematical model**. For example, the natural numbers constitute a mathematical model for situations in which counting is essential. Situations in which algebra can be brought to bear often require the use of functions.

Mathematical models are abstracted from real-world situations. Procedures within the mathematical model then give results that allow one to predict what will happen in that real-world situation. If the predictions are inaccurate or the results of experimentation do not conform to the model, the model needs to be changed or discarded.

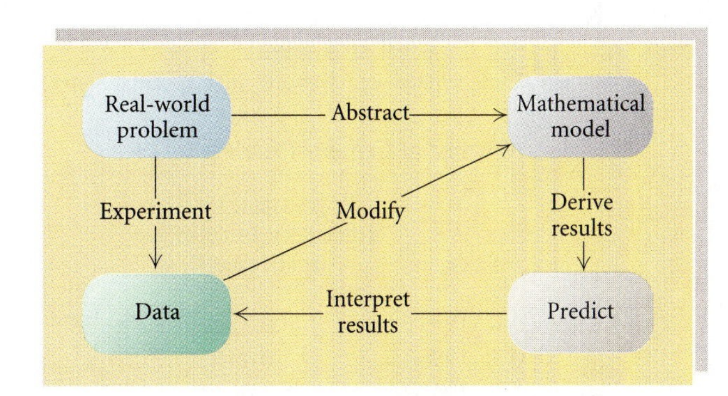

Mathematical modeling can be an ongoing process. For example, finding a mathematical model that will enable an accurate prediction of population growth is not a simple problem. Any population model that one might devise would need to be reshaped as further information is acquired.

Curve Fitting

We will develop and use many kinds of mathematical models in this text. In this chapter, we have considered many functions that can be used as models. Let's look at four of them.

Constant function: $y_1 = b$

Linear function: $y_2 = mx + b$

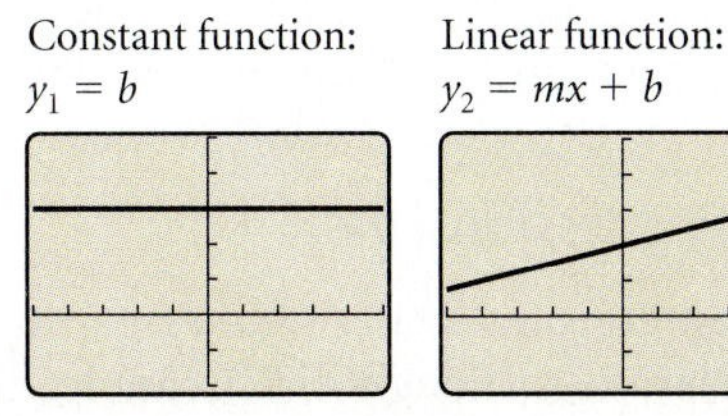

Squaring function: $y_3 = x^2$

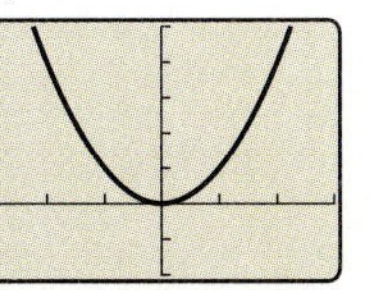

Quadratic function: $y_4 = ax^2 + bx + c, a > 0$

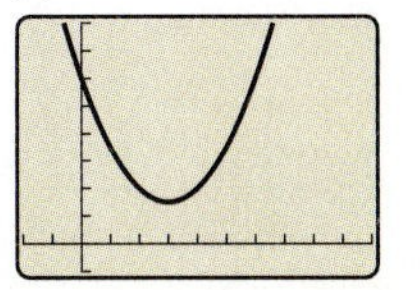

In general, we try to find a function that fits, as well as possible, observations (data), theoretical reasoning, and common sense. We call this **curve fitting**; it is one aspect of mathematical modeling.

Let's first look at some data and related graphs or scatterplots. Do any of the above functions seem to fit either set of data?

Life Expectancy of Women

YEAR, x	LIFE EXPECTANCY, y (IN YEARS)
0. 1900	49.1
1. 1910	53.7
2. 1920	56.3
3. 1930	61.4
4. 1940	65.3
5. 1950	70.9
6. 1960	73.2
7. 1970	74.8
8. 1980	77.5
9. 1990	78.6

SCATTERPLOT

Source: Statistical Abstract of the United States.

Number of Cellular Phones

YEAR, x	NUMBER OF CELLULAR PHONES, y (IN MILLIONS)
0. 1985	0.2
1. 1986	0.4
2. 1987	1.0
3. 1988	1.8
4. 1989	2.8
5. 1990	4.1
6. 1991	6.2
7. 1992	8.6
8. 1993	13.0
9. 1994	19.3

SCATTERPLOT

Source: Cellular Telecommunications Industry Association.

Looking at the scatterplots, we see that the life expectancy data seem to be rising in a manner to suggest that a linear function might fit, although a "perfect" straight line cannot be drawn through the data points. A linear function does not seem to fit the cellular phone data. Part of a quadratic function might be close. In fact, it can be modeled by an exponential function that we will consider in Chapter 3.

The Regression Line

We now consider a method to fit a linear function to a set of data: **linear regression**. Although discussion leading to a complete understanding of this method belongs in a statistics course, we present the procedure here because we can carry it out easily using technology. The grapher gives us the powerful capability to find linear models and to make predictions using them.

Example 1 *Predicting the Life Expectancy of Women.* Consider the data just presented on the life expectancy of women.

a) Fit a regression line to the data using a REGRESSION feature on a grapher.

b) Use the linear model to predict the life expectancy of women in the year 2000.

SOLUTION

a) Using linear regression, we fit a linear equation to the data. We use a grapher and enter the data into a list, like the following. We select the L_1 list to be the independent variable (x) and the L_2 list to be the dependent variable (y). The grapher can then create a scatterplot of the data, as shown on the right.

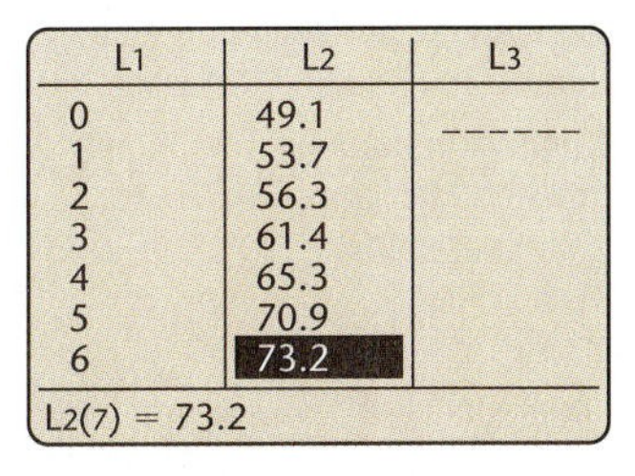

L1	L2	L3
0	49.1	------
1	53.7	
2	56.3	
3	61.4	
4	65.3	
5	70.9	
6	73.2	

L2(7) = 73.2

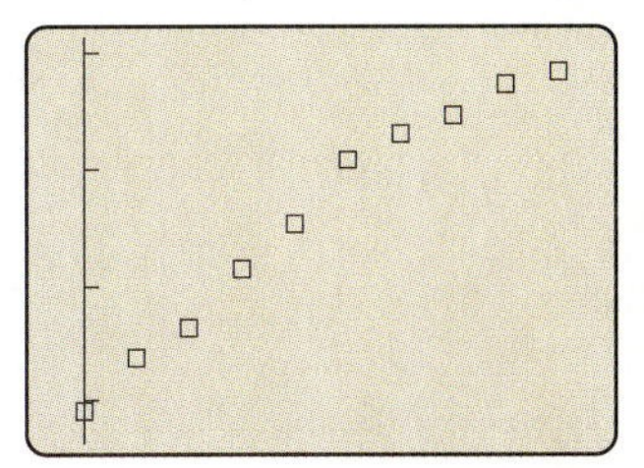

Consider the data points (0, 49.1), (1, 53.7), (2, 56.3), . . . , (9, 78.6) on the graph. We want to fit a **regression line**,

$$y = mx + b,$$

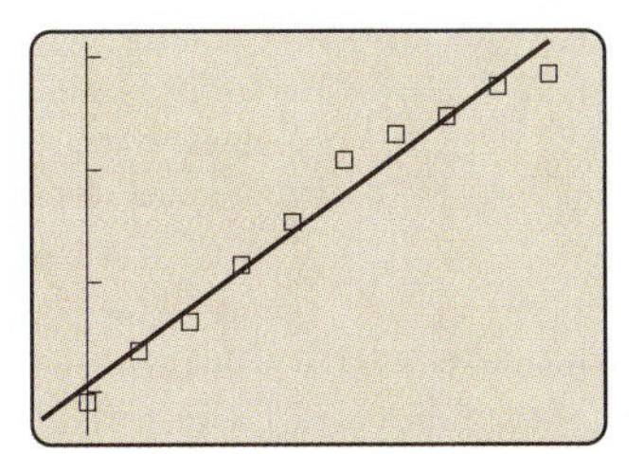

to these data points. We use the REGRESSION feature, LINREG(AX + B), to find the regression line. The result is

$$y = 3.4x + 50.7.$$ **Regression line**

We can then graph the regression line on the same graph as the scatterplot.

b) To predict the life expectancy of women in the year 2000, we substitute the corresponding x-value, $x = 10$, into the formula for the regression function:

$$y = 3.4(10) + 50.7 = 84.7.$$ **In 1980, $x = 8$; in 1990, $x = 9$. Thus in 2000, $x = 10$.**

We can also use the grapher to do the calculation. Thus we estimate the life expectancy of women in the year 2000 to be 84.7 yr.

The understanding behind the development of the regression line is best left to a course in calculus and/or statistics.

The Correlation Coefficient

On some graphers, a constant r in $[-1, 1]$, called the **coefficient of linear correlation**, appears with the equation of the regression line. Though we cannot develop a formula for calculating r in this text, keep in mind that r is a number for which $-1 \le r \le 1$. It is used to describe the strength of the linear relationship between x and y. The closer $|r|$ is to 1, the better the correlation. A positive value of r also indicates that a positive slope and an increasing function exist. A negative value of r indicates a negative slope and a decreasing function. For the life expectancy data just discussed, $r = 0.9861$, which indicates a very good linear correlation. The following scatterplots summarize the interpretation of a correlation coefficient.

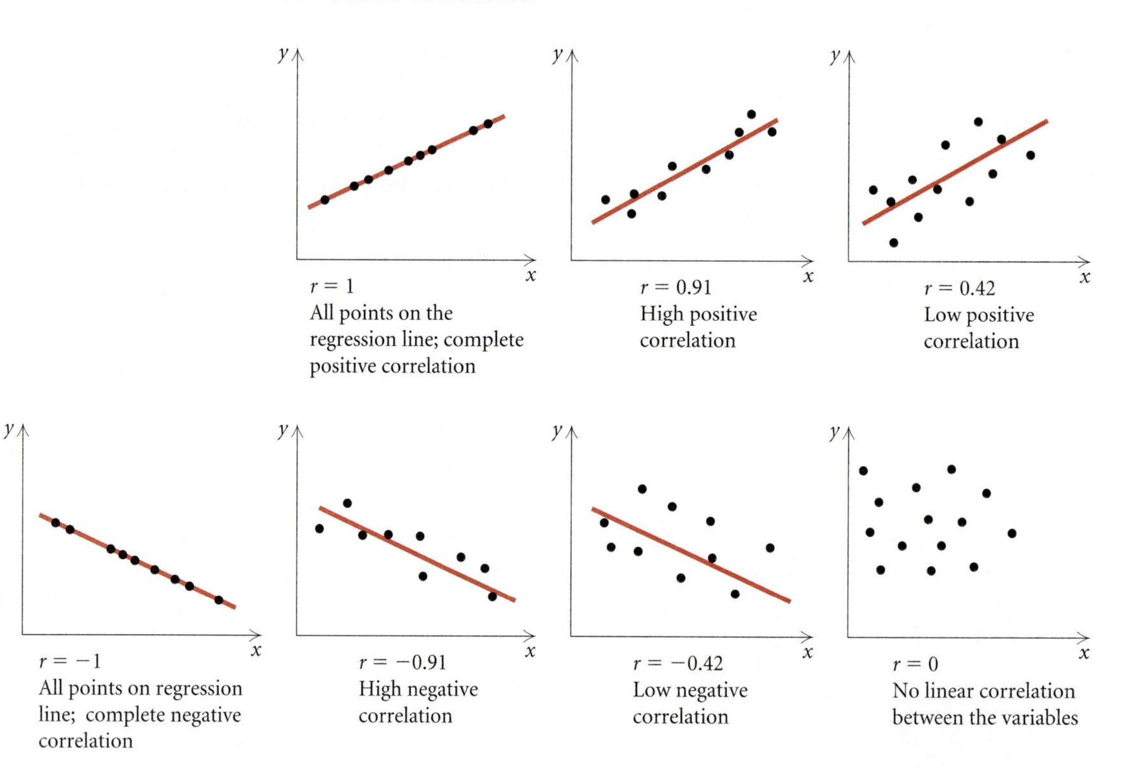

$r = 1$
All points on the regression line; complete positive correlation

$r = 0.91$
High positive correlation

$r = 0.42$
Low positive correlation

$r = -1$
All points on regression line; complete negative correlation

$r = -0.91$
High negative correlation

$r = -0.42$
Low negative correlation

$r = 0$
No linear correlation between the variables

Keep in mind that a high linear correlation coefficient does not necessarily indicate a "cause-and-effect" connection between the variables. We might be able to calculate a high positive correlation between stock prices and rainfall in India, but before we place our life savings in the market, further analysis and common sense should be applied!

1.4 Exercise Set

In Exercises 1–4, determine whether the graph might be modeled by a linear function.

1.

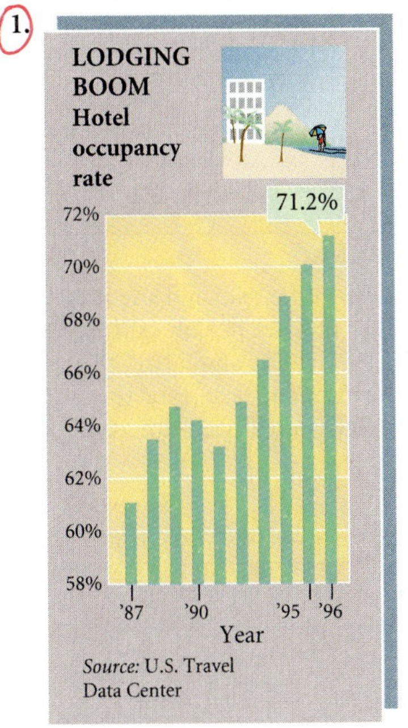

2.

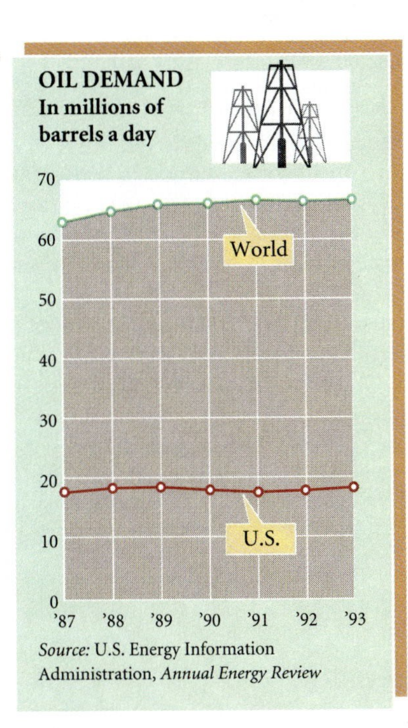

3.

THE SENIOR PROFILE

The population over 65

Since 1900, the percentage of Americans over 65 has more than tripled (4.1% in 1900 to 12.7% in 1993). The most rapid increase is expected from 2010 to 2030, when the baby boom generation reaches 65.

80 70 60 50 40 30 20 10 0

70.2 million

U.S. population over 65

3.1 million

1900 1920 1940 1960 1980 1993 2000 2010 2020 2030

Year

Source: "A Profile of Older Americans" by the American Association of Retired Persons and the Administration on Aging; and the U.S. Bureau of the Census

4.

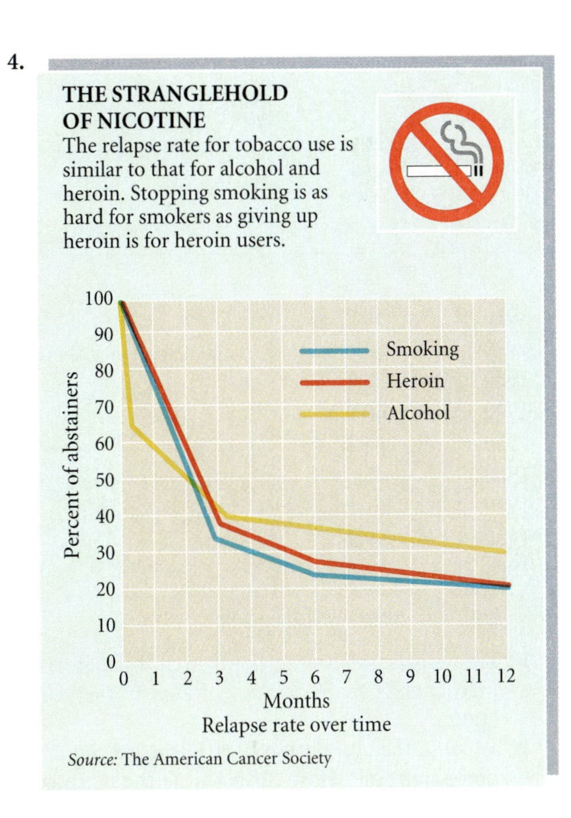

In Exercises 5–8, determine whether the scatterplot might be fit by a linear model.

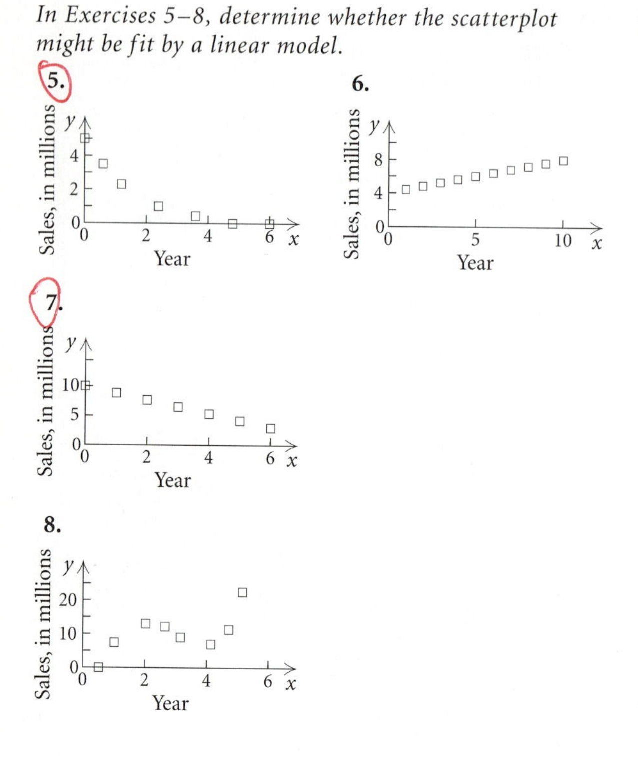

Exercises 9–14 involve creating regression lines on a grapher.

9. ***Life Expectancy of Men***

Year, x	Life Expectancy, y (in years)	Scatterplot
0. 1900	49.6	Life expectancy of men
1. 1910	50.2	
2. 1920	54.6	
3. 1930	58.0	
4. 1940	60.9	
5. 1950	65.3	
6. 1960	66.6	
7. 1970	67.1	
8. 1980	69.9	
9. 1990	71.8	

Source: Statistical Abstract of the United States.

a) Fit a regression line to the data and use it to predict the life expectancy of men in the year 2000.

b) What is the correlation coefficient for the regression line? How close a fit is the regression line?

10. ***Cases of Skin Cancers in the United States***

Year, x	Number of Cases, y (in thousands)	Scatterplot
0. 1988	5800	Skin cancer on the rise
1. 1989	6000	
2. 1990	6300	
3. 1991	6500	
4. 1992	6700	

Source: American Cancer Society.

a) Fit a regression line to the data and use it to predict the number of cases of skin cancer in the United States in 1998, 2000, and 2010.

b) What is the correlation coefficient for the regression line? How good a predictor is the regression line?

11. ***The Cost of Tuition at State Universities***

College Year	Estimated Tuition
0. (1997–1998)	$ 9,838
1. (1998–1999)	10,424
2. (1999–2000)	11,054
3. (2000–2001)	11,717
4. (2001–2002)	12,420

Source: College Board, Senate Labor Committee.

a) Fit a regression line to the data and use it to predict the cost of college tuition in 2004–2005, 2006–2007, and 2010–2011.

b) What is the correlation coefficient for the regression line? How close a fit is the regression line?

c) A college president asserts that in the college year 2006–2007, the cost of tuition will be $16,619. How well does your estimate compare?

12. ***Book Buying Growth in the United States***

Year, x	Book Sales, y (in billions)
0. 1992	$21
1. 1993	23
2. 1994	24
3. 1995	25
4. 1996	26

Source: Book Industry Trends 1995.

a) Fit a regression line to the data and use it to predict book sales in 1998, 2000, and 2010.
b) What is the correlation coefficient for the regression line? How close a fit is the regression line?
c) A bookstore owner asserts that in the year 2000, book sales will be $45 billion. How well does your estimate stack up to this?

13. *Maximum Heart Rate.* A person who is exercising should not exceed his or her maximum heart rate, which is determined on the basis of that person's sex, age, and resting heart rate. The following table relates resting heart rate and maximum heart rate for a 20-yr-old man.

RESTING HEART RATE, H (IN BEATS PER MINUTE)	MAXIMUM HEART RATE, M (IN BEATS PER MINUTE)
50	166
60	168
70	170
80	172

Source: American Heart Association.

a) Fit a regression line $M = mH + b$ to the data.
b) Predict a maximum heart rate if the resting heart rate is 40, 65, 76, and 84.
c) What is the correlation coefficient? How confident are you about using the regression line as a predictor?

EX 14. *Study Time versus Grades.* A math instructor asked her students to keep track of how much time each spent studying a chapter on functions in her algebra–trigonometry course. She collected the information together with test scores from that chapter's test. The data are listed in the table below.

STUDY TIME, x (IN HOURS)	TEST GRADE, y (IN PERCENT)
23	81
15	85
17	80
9	75
21	86
13	80
16	85
11	93

a) Fit a regression line to the data.
b) Predict a student's score if he or she studies 24 hr, 6 hr, and 18 hr.
c) What is the correlation coefficient? How confident are you about using the regression line as a predictor?

Skill Maintenance

15. Given $f(x) = 4x^3 - 5x$, find each of the following.
a) $f(2)$ b) $f(-2)$ c) $f(a)$ d) $f(-a)$

16. Given $f(x) = 5x^2 - 7$, find each of the following.
a) $f(-3)$ b) $f(3)$ c) $f(a)$ d) $f(-a)$

Synthesis

17. ◆ Why does a correlation coefficient of 1 not guarantee the existence of a cause-and-effect relationship?

18. ◆ Conduct your own experiment. Use some kind of measuring device and gather data relating body height to arm length. Fit a regression line to the data and find the correlation coefficient. How confident would you be about using these data to predict Shaquille O'Neal's arm length, knowing that his height is 7′0″?

19. *Planet Diameters versus Distance from the Sun.* The following table lists the diameters of the planets and their distances from the sun.

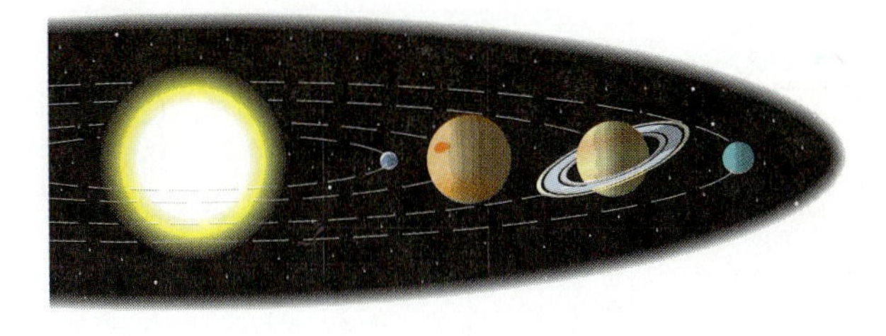

PLANET	DIAMETER, d (IN KILOMETERS)	DISTANCE FROM SUN, D (IN ASTRONOMICAL UNITS)*
Mercury	5,000	0.4
Venus	12,000	0.7
Earth	13,000	1.0
Mars	7,000	1.6
Jupiter	143,000	5.2
Saturn	120,000	10.0
Uranus	51,000	19.6
Pluto	5,000	30.8
Neptune	49,000	40.8

*1 astronomical unit (A.U.) = 150 million km.
Source: Encyclopedia Britannica.

a) Find a regression line $D = md + b$, expressing D as a function of d. What is the correlation coefficient?

b) Using a space probe, astronomers discover a new planet that has a diameter of 180,000 km. How confident would you be of using the function of part (a) to predict the distance of this planet from the sun?

20. *Ted Williams and the War Years.* Ted Williams played for the Boston Red Sox from 1939–1960. Many credit him with being the greatest hitter of all time. Unfortunately, his career totals are somewhat less impressive than others because his career was interrupted from 1943–1945 for World War II and from 1952–1953 for the Korean War. Some assert that if he had been playing during the war years, he would have broken Hank Aaron's home run record of 755 and runs batted in (RBI) record of 2297. Below are Williams' statistics.

YEAR, x	HOME RUNS, H	RBIS, R
1939	31	145
1940	23	113
1941	37	120
1942	36	137
1943	0*	0*
1944	0*	0*
1945	0*	0*
1946	38	123
1947	32	114
1948	25	127
1949	43	159
1950	28	97
1951	30	126
1952	1†	3†
1953	13†	34†
1954	29	89
1955	28	83
1956	24	82
1957	38	87
1958	26	85
1959	10	43
1960	29	72

*World War II
†Korean War

a) Excluding all data from the war years, fit a linear regression equation $H = mx + b$ to the data regarding the number of home runs. Then use the equation to predict how many home runs Williams would have hit in each of the war years. What is the correlation coefficient? How confident are you of your prediction?

b) Would Williams have broken Aaron's home run record?

c) Excluding all data from the war years, fit a linear regression equation $R = mx + b$ to the data regarding the number of RBIs. Then use the equation to predict how many RBIs Williams would have had in each of the war years. What is the correlation coefficient? How confident are you of your prediction?

d) Would Williams have broken Aaron's RBI record?

1.5 Distance, Midpoints, and Circles

- *Find the distance between two points in the plane and the midpoint of a segment.*
- *Find an equation of a circle with a given center and radius, and given an equation of a circle, find the center and the radius.*
- *Graph equations of circles.*

We have seen that graphs can provide a useful way of modeling real-world situations. In carpentry, surveying, engineering, and other fields, it is often necessary to determine distances and midpoints and to produce accurately drawn circles.

The Distance Formula

Suppose that a conservationist needs to determine the distance across an irregularly shaped pond. One way in which the conservationist might proceed is to measure two legs of a right triangle that is situated as shown below. The Pythagorean theorem, $a^2 + b^2 = c^2$, can then be used to find the length of the hypotenuse.

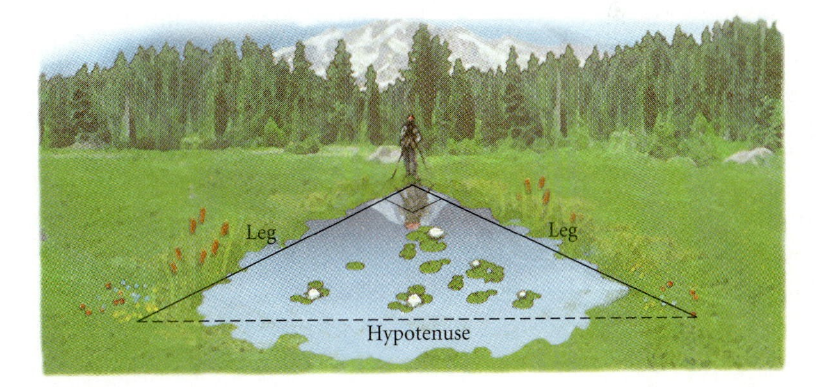

A similar strategy is used to find the distance between two points in a plane. For two points (x_1, y_1) and (x_2, y_2), we can draw a right triangle in which the legs have lengths $|x_2 - x_1|$ and $|y_2 - y_1|$.

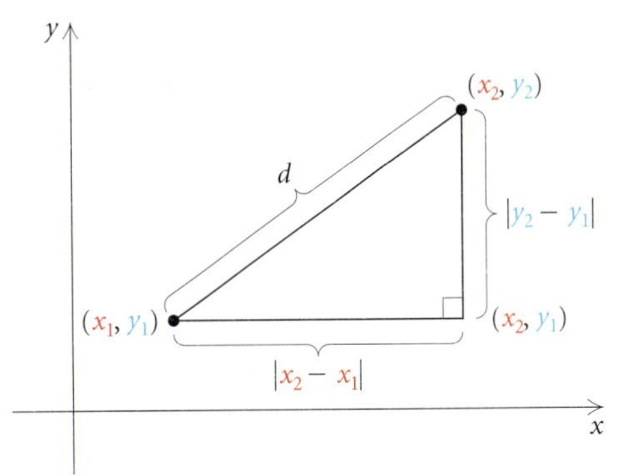

Using the Pythagorean theorem, we have

$$d^2 = |x_2 - x_1|^2 + |y_2 - y_1|^2.$$

Because we are squaring, parentheses can replace the absolute value symbols:

$$d^2 = (x_2 - x_1)^2 + (y_2 - y_1)^2.$$

Taking the principal square root, we obtain the distance formula.

The Distance Formula

The *distance d* between any two points (x_1, y_1) and (x_2, y_2) is given by

$$d = \sqrt{(x_2 - x_1)^2 + (y_2 - y_1)^2}.$$

The subtraction of the x-coordinates can be done in any order, as can the subtraction of the y-coordinates. Although we derived the distance formula by considering two points not on a horizontal or a vertical line, the distance formula holds for *any* two points.

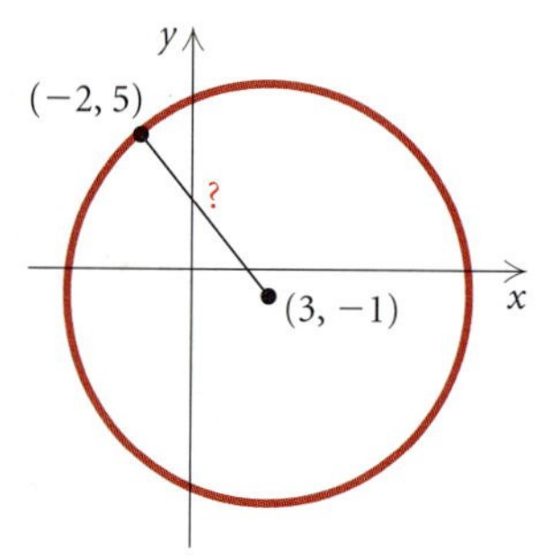

Example 1 The point $(-2, 5)$ is on a circle that has $(3, -1)$ as its center. Find the length of the radius of the circle.

SOLUTION Since the length of the radius is the distance from the center to a point on the circle, we substitute into the distance formula:

$$r = \sqrt{[3 - (-2)]^2 + [-1 - 5]^2} \qquad \text{Either point can serve as } (x_1, y_1).$$

$$= \sqrt{5^2 + (-6)^2} = \sqrt{61} \approx 7.810. \qquad \text{Rounded to the nearest thousandth}$$

The circle's radius is approximately 7.8.

Midpoints of Segments

The distance formula can be used to develop a way of determining the *midpoint* of a segment when the endpoints are known. We state the formula and leave its proof to the exercises.

The Midpoint Formula

If the endpoints of a segment are (x_1, y_1) and (x_2, y_2), then the coordinates of the *midpoint* are

$$\left(\frac{x_1 + x_2}{2}, \frac{y_1 + y_2}{2}\right).$$

Note that we obtain the coordinates of the midpoint by averaging the coordinates of the endpoints. This is an easy way to remember the midpoint formula.

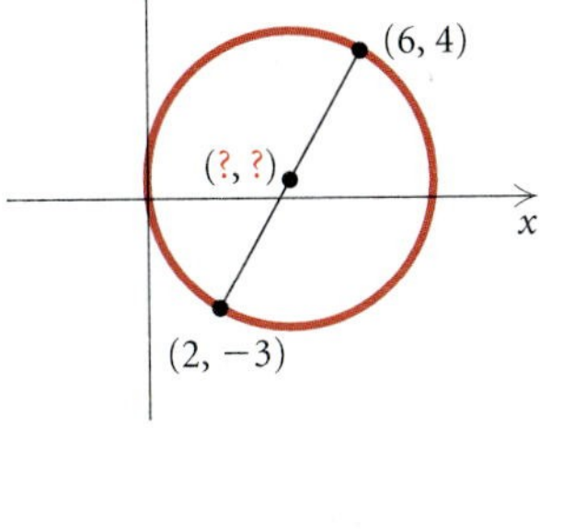

Example 2 The diameter of a circle connects two points $(2, -3)$ and $(6, 4)$ on the circle. Find the coordinates of the center of the circle.

SOLUTION Using the midpoint formula, we obtain

$$\left(\frac{2+6}{2}, \frac{-3+4}{2}\right), \quad \text{or} \quad \left(\frac{8}{2}, \frac{1}{2}\right), \quad \text{or} \quad \left(4, \frac{1}{2}\right).$$

The coordinates of the center are $\left(4, \frac{1}{2}\right)$.

Circles

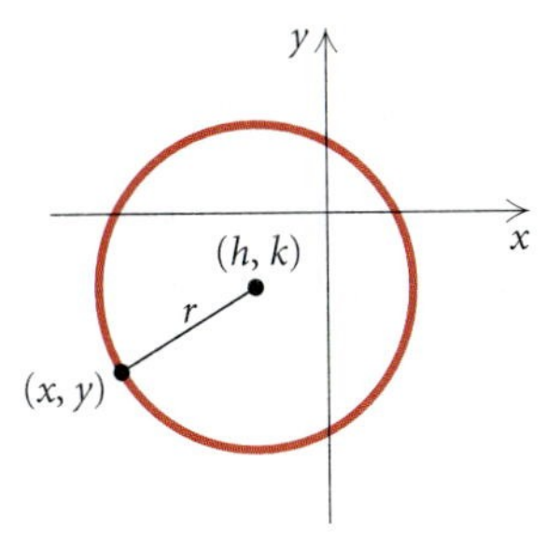

A **circle** is the set of all points in a plane that are a fixed distance r, the radius, from a center (h, k). Thus if a point (x, y) is to be r units from the center, we must have

$$r = \sqrt{(x-h)^2 + (y-k)^2}. \qquad \textbf{Using the distance formula}$$

Squaring both sides gives an equation of a circle.

The Equation of a Circle

The equation, in standard form, of a circle with center (h, k) and radius r is

$$(x-h)^2 + (y-k)^2 = r^2.$$

Example 3 Find an equation of the circle having radius 5 and center $(3, -7)$.

SOLUTION Using the standard form, we have

$$[x-3]^2 + [y-(-7)]^2 = 5^2 \qquad \textbf{Substituting}$$
$$(x-3)^2 + (y+7)^2 = 25.$$

Example 4 Graph the circle: $(x+5)^2 + (y-2)^2 = 16$.

SOLUTION We write the equation in standard form to determine the center and the radius:

$$[x-(-5)]^2 + [y-2]^2 = 4^2.$$

The center is $(-5, 2)$ and the radius is 4. We locate the center and draw the circle using a compass.

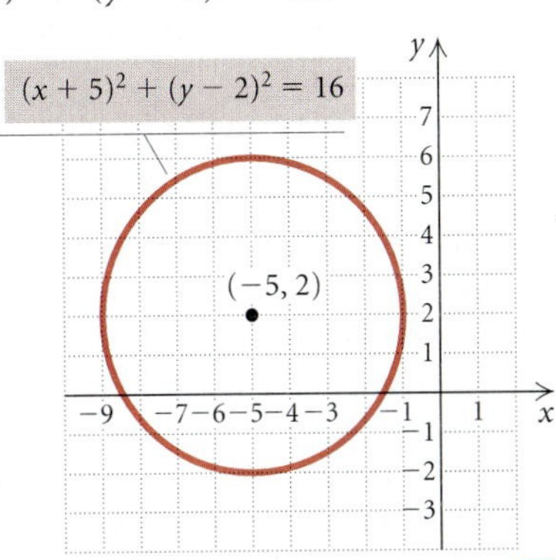

Circles can also be graphed on a grapher.

Example 5 Graph the circle: $x^2 + y^2 = 16$.

SOLUTION We first solve $x^2 + y^2 = 16$ for y:

$$x^2 + y^2 = 16$$
$$y^2 = 16 - x^2 \quad \text{Solving for } y^2$$
$$y = \pm\sqrt{16 - x^2}. \quad \text{Solving for } y$$

$x^2 + y^2 = 16$

$y_1 = \sqrt{16 - x^2}$ and $y_2 = -\sqrt{16 - x^2}$

6

−9 9

−6

We graph both equations $y_1 = \sqrt{16 - x^2}$ and $y_2 = -\sqrt{16 - x^2}$ using the same set of axes and a square viewing window. There are graphers that can draw graphs with a given center and radius directly from the DRAW menu. A square window will still be necessary. Some graphers have the ability to graph an equation like $x^2 + y^2 = 16$ without first solving for y.

1.5 Exercise Set

Find the distance between each pair of points. Give an exact answer and, where appropriate, an approximation to three decimal places.

1. (4, 6) and (5, 9)

2. (−3, 7) and (2, 11)

3. (6, −1) and (9, 5)

4. (−4, −7) and (−1, 3)

5. $(\sqrt{3}, -\sqrt{5})$ and $(-\sqrt{6}, 0)$

6. $(-\sqrt{2}, 1)$ and $(0, \sqrt{7})$

7. The points (−3, −1) and (9, 4) are the endpoints of the diameter of a circle. Find the length of the radius of the circle.

8. The point (0, 1) is on a circle that has center (−3, 5). Find the length of the diameter of the circle.

Use the distance formula and the Pythagorean theorem to determine whether each set of points could be vertices of a right triangle.

9. (−4, 5), (6, 1), and (−8, −5)

10. (−3, 1), (2, −1), and (6, 9)

11. The points (−3, 4), (0, 5), and (3, −4) are all on the circle $x^2 + y^2 = 25$. Show that these three points are vertices of a right triangle.

12. The points (−3, 4), (2, −1), (5, 2), and (0, 7) are vertices of a quadrilateral. Show that the quadrilateral is a rectangle. (*Hint:* Show that the quadrilateral's opposite sides are the same length and that the two diagonals are the same length.)

13–18. Find the midpoint of each segment having the endpoints given in Exercises 1–6, respectively.

19. Graph the rectangle described in Exercise 12. Then determine the coordinates of the midpoint for each of the four sides. Are the midpoints vertices of a rectangle?

20. Graph the square with vertices (−5, −1), (7, −6), (12, 6), and (0, 11). Then determine the midpoint for each of the four sides. Are the midpoints vertices of a square?

21. The points $(\sqrt{7}, -4)$ and $(\sqrt{2}, 3)$ are endpoints of the diameter of a circle. Determine the center of the circle.

22. The points $(-3, \sqrt{5})$ and $(1, \sqrt{2})$ are endpoints of the diagonal of a square. Determine the center of the square.

In Exercises 23 and 24, how would you change the window so the circle is not distorted? Answers may vary.

23. **24.**

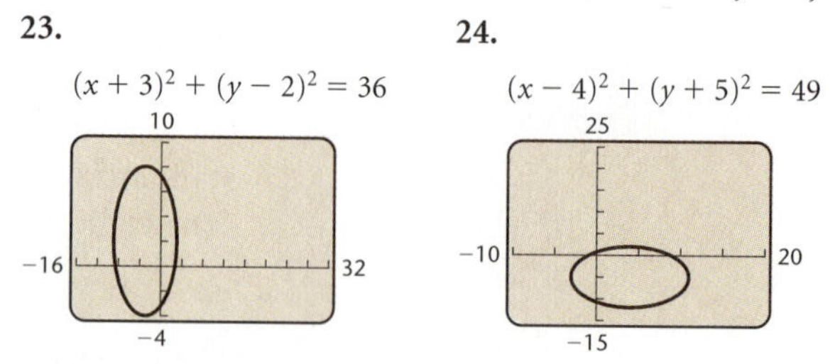

Find an equation for a circle satisfying the given conditions.

25. Center (2, 3), radius of length $\frac{5}{3}$

26. Center (4, 5), diameter of length 8.2

27. Center (−1, 4), passes through (3, 7)

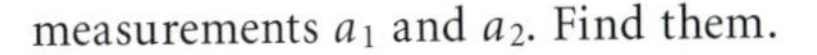

28. Center $(6, -5)$, passes through $(1, 7)$

29. The points $(7, 13)$ and $(-3, -11)$ are at either end of a diameter.

30. The points $(-9, 4)$, $(-2, 5)$, $(-8, -3)$, and $(-1, -2)$ are vertices of an inscribed square.

31. Center $(-2, 3)$, tangent (touching at one point) to the y-axis

32. Center $(4, -5)$, tangent to the x-axis

Find the center and the radius of each circle. Then graph each circle using a square viewing rectangle.

33. $x^2 + y^2 = 4$

34. $x^2 + y^2 = 81$

35. $x^2 + (y - 3)^2 = 16$
(Hi... get $y = 3 \pm \sqrt{16 - x^2}$.)

Test 447-8
445 - 9, 11
5, 8
CR: 449-50 1-9,
17-24 30-36
40, 44-45
10 Rev.

6

5

= 5 and $y = -4$.

-2.

7.

pair of points and find
ng the given points as

an theorem is used to
in standard form.
l the coordinates of a
A to point B.

...ing the given
conditions.

49. Center $(2, -7)$ with an area of 36π square units

50. Center $(-5, 8)$ with a circumference of 10π units

51. *Swimming Pool.* A swimming pool is being constructed in the corner of a yard, as shown. Before installation, the contractor needs to know measurements a_1 and a_2. Find them.

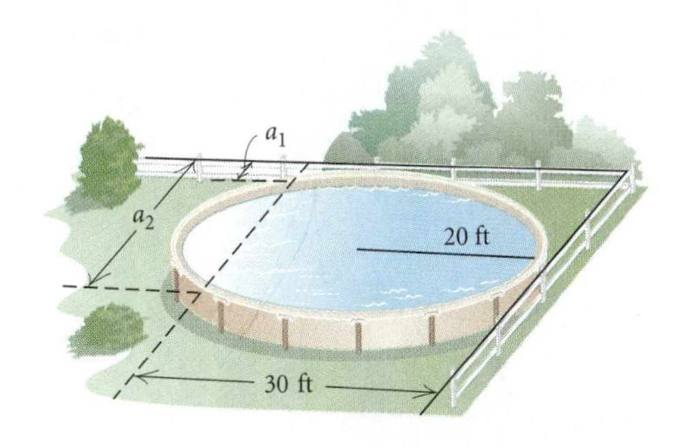

52. *An Arch of a Circle in Carpentry.* Ace Carpentry needs to cut an arch for the top of an entranceway. The arch needs to be 8 ft wide and 2 ft high. To draw the arch, the carpenters will use a stretched string with chalk attached at an end as a compass.

a) Using a coordinate system, locate the center of the circle.

b) What radius should the carpenters use to draw the arch?

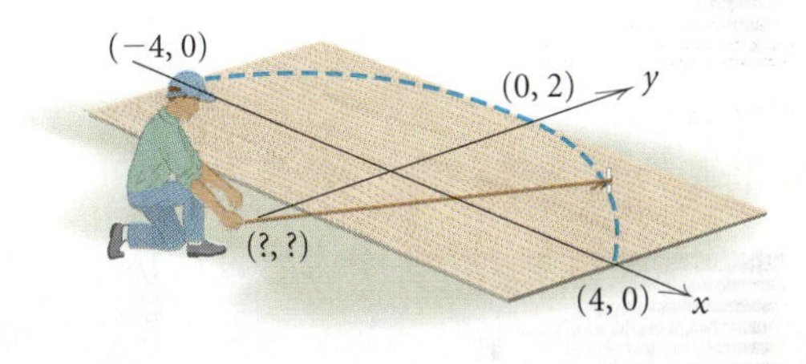

Determine whether each of the following points lies on the ***unit circle***, $x^2 + y^2 = 1$.

53. $\left(\frac{\sqrt{3}}{2}, -\frac{1}{2}\right)$

54. $(0, -1)$

55. $\left(-\frac{\sqrt{2}}{2}, \frac{\sqrt{2}}{2}\right)$

56. $\left(\frac{1}{2}, -\frac{\sqrt{3}}{2}\right)$

57. Find the point on the y-axis that is equidistant from the points $(-2, 0)$ and $(4, 6)$.

58. Consider any right triangle with base b and height h, situated as shown. Show that the midpoint of the hypotenuse P is equidistant from the three vertices of the triangle.

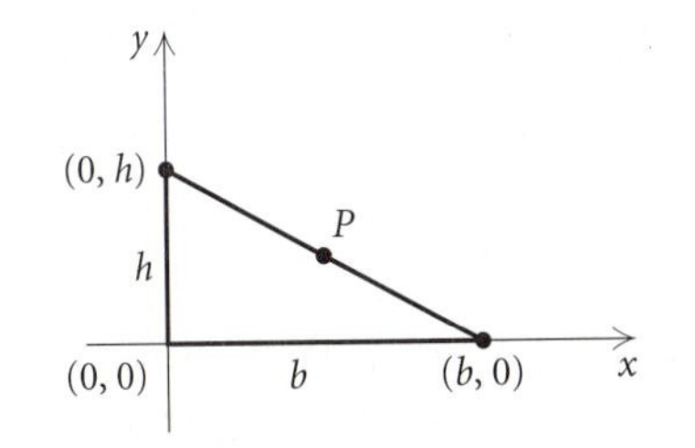

59. Prove the midpoint formula by showing that:

a) $\left(\frac{x_1 + x_2}{2}, \frac{y_1 + y_2}{2}\right)$ is equidistant from the points (x_1, y_1) and (x_2, y_2); and

b) the distance from (x_1, y_1) to the midpoint plus the distance from (x_2, y_2) to the midpoint equals the distance from (x_1, y_1) to (x_2, y_2).

1.6 Symmetry

- *Determine whether a graph is symmetric with respect to the x-axis, the y-axis, and the origin.*
- *Determine whether a function is even, odd, or neither even nor odd.*

Symmetry

Symmetry occurs often in nature and in art. For example, when viewed from the front, the bodies of most animals are at least approximately symmetric. This means that each eye is the same distance from the center of the bridge of the nose, each shoulder is the same distance from the center of the chest, and so on. Architects have used symmetry for thousands of years to enhance the beauty of buildings.

M. Hyett/VIREO

Although symmetry has important uses throughout mathematics, our present discussion is included to better enable us to graph and analyze equations and functions.

Consider the points (4, 2) and (4, −2) that appear in the graph of $x = y^2$.

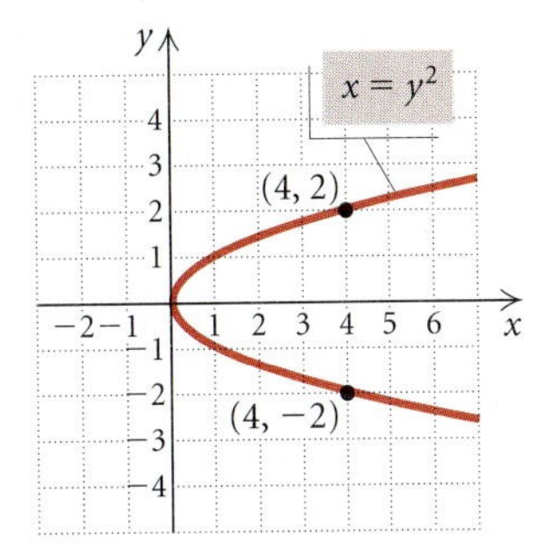

Points like these have the same x-value but opposite y-values, and are **reflections** of each other across the x-axis. Any graph that contains the reflection of every point across the x-axis is said to be **symmetric with respect to the x-axis**. If we fold the graph on the x-axis, the parts above and below the x-axis will coincide.

Consider the points (3, 4) and (−3, 4) that appear in the graph of $y = x^2 - 5$.

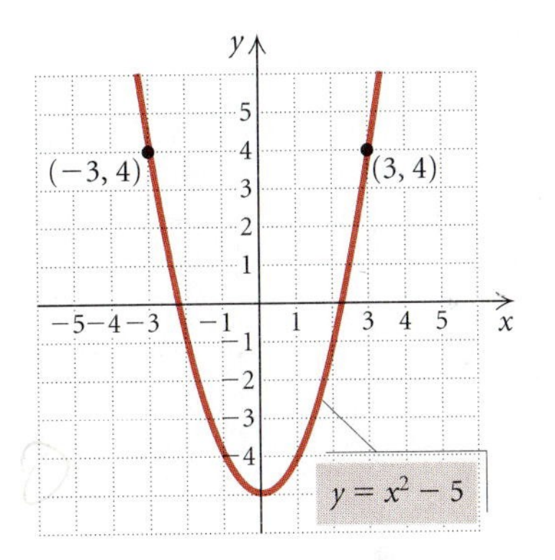

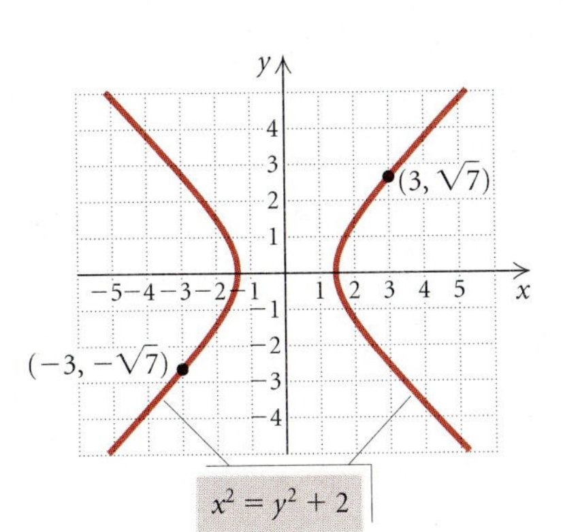

Points like these have the same y-value but opposite x-values, and are **reflections** of each other across the y-axis. Any graph that contains the reflection of every point across the y-axis is said to be **symmetric with respect to the y-axis**. If we fold the graph on the y-axis, the parts to the left and right of the y-axis will coincide.

Consider the points $(3, \sqrt{7})$ and $(-3, -\sqrt{7})$ that appear in the graph of $x^2 = y^2 + 2$. (You can create such a graph on a grapher by solving for y, obtaining $y_1 = \sqrt{x^2 - 2}$ and $y_2 = -\sqrt{x^2 - 2}$.) Note that if we take the opposites of the coordinates of one pair, we get the other pair. If for any point (x, y) on a graph the point $(-x, -y)$ is also on the graph, then the graph is said to be **symmetric with respect to the origin**. Visually, if we rotate the graph 180° about the origin, the resulting figure coincides with the original.

Algebraic Tests of Symmetry

x-axis: If replacing y by $-y$ produces an equivalent equation, then the graph is *symmetric with respect to the x-axis.*

y-axis: If replacing x by $-x$ produces an equivalent equation, then the graph is *symmetric with respect to the y-axis.*

Origin: If replacing x by $-x$ and y by $-y$ produces an equivalent equation, then the graph is *symmetric with respect to the origin.*

Example 1 Test $y = x^2 + 2$ for symmetry with respect to the x-axis, the y-axis, and the origin.

ALGEBRAIC SOLUTION

X-AXIS

Replace y by $-y$:

$$y = x^2 + 2$$
$$-y = x^2 + 2$$
$$y = -x^2 - 2. \quad \text{Multiplying both sides by } -1$$

Is the resulting equation equivalent to the original? The answer is no, so the graph *is not* symmetric with respect to the x-axis.

Y-AXIS

Replace x by $-x$:

$$y = x^2 + 2$$
$$y = (-x)^2 + 2$$
$$y = x^2 + 2. \quad \text{Simplifying}$$

Is the resulting equation equivalent to the original? The answer is yes, so the graph *is* symmetric with respect to the y-axis.

ORIGIN

Replace x by $-x$ and y by $-y$:

$$y = x^2 + 2$$
$$-y = (-x)^2 + 2$$
$$-y = x^2 + 2 \quad \text{Simplifying}$$
$$y = -x^2 - 2.$$

Is the resulting equation equivalent to the original? The answer is no, so the graph *is not* symmetric with respect to the origin.

GRAPHICAL SOLUTION

We use a grapher to graph the equation. Note that if the graph were folded on the x-axis, the parts above and below the x-axis would not coincide. If it were folded on the y-axis, the parts to the left and right of the y-axis would coincide. If we rotated it 180°, the resulting graph would not coincide with the original graph.

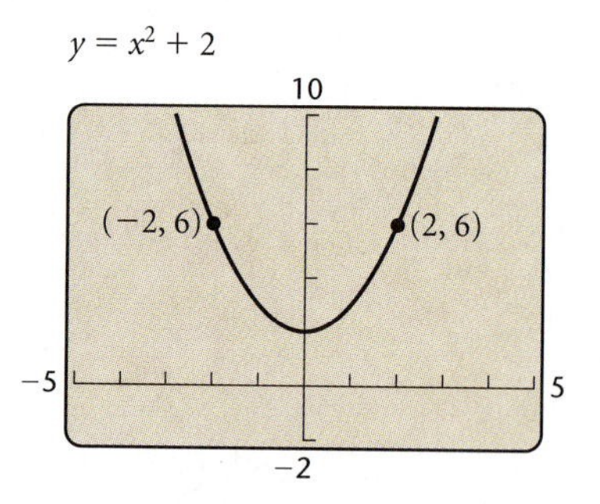

Thus we see that the graph *is not* symmetric with respect to the x-axis or the origin. The graph *is* symmetric with respect to the y-axis.

The algebraic method is often easier to apply than the graphical, especially with equations that we may not be able to graph easily.

Example 2 Test $x^2 + y^4 = 5$ for symmetry with respect to the x-axis, the y-axis, and the origin.

ALGEBRAIC SOLUTION

X-AXIS

Replace y by $-y$:

$$x^2 + y^4 = 5$$

$$x^2 + (-y)^4 = 5$$

$$x^2 + y^4 = 5.$$

The resulting equation is equivalent to the original equation. Thus the graph is symmetric with respect to the x-axis.

We leave it to the student to verify algebraically that the graph is also symmetric with respect to the y-axis and the origin.

GRAPHICAL SOLUTION

With a grapher, we see symmetry with respect to both axes and with respect to the origin.

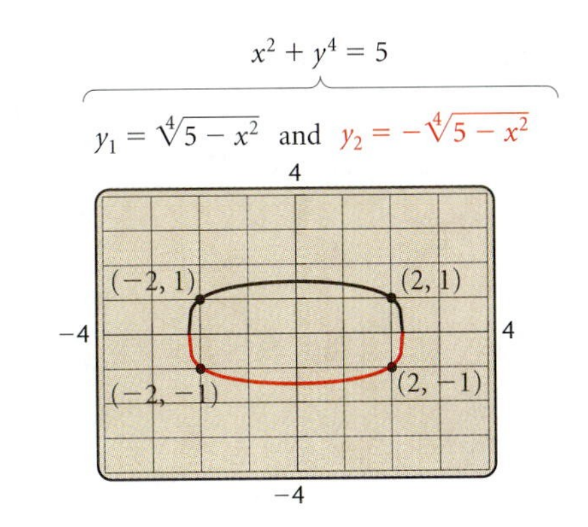

Even and Odd Functions

Now we relate symmetry to graphs of functions.

Interactive Discovery

Consider the function f given by $f(x) = x^2$. Find and graph

$$y_1 = f(x) = x^2, \quad y_2 = f(-x) = (-x)^2, \quad \text{and} \quad y_3 = -f(x) = -x^2.$$

Use the graphs to determine whether $f(x) = f(-x)$ is an identity. Use the graphs to determine whether $f(x) = -f(x)$ is an identity.

Finding $f(-x)$ is the same as replacing x with $-x$ in its equation, and finding $-f(x)$ is the same as replacing y with $-y$ in its equation. Can you make a connection between symmetry and the properties $f(x) = f(-x)$ and $f(-x) = -f(x)$?

Repeat this procedure for $g(x) = x^3$.

The preceding Interactive Discovery leads us to the following definitions and procedure.

Even and Odd Functions

If the graph of a function f is symmetric with respect to the y-axis, we say that it is an *even function*. That is, for each x in the domain of f, $f(x) = f(-x)$.

If the graph of a function f is symmetric with respect to the origin, we say that it is an *odd function*. That is, for each x in the domain of f, $f(-x) = -f(x)$.

Algebraic Procedure for Determining Even and Odd Functions

Given the function $f(x)$:

1. Find $f(-x)$ and simplify.
2. Find $-f(x)$ and simplify.
3. If $f(x) = f(-x)$, then f is even. If $f(-x) = -f(x)$, then f is odd.

Except for the function $f(x) = 0$, a function cannot be *both* even and odd. Thus if we see in step (1) that $f(x) = f(-x)$ (that is, f is even), we need not complete the process.

Example 3 Determine whether each of the following functions is even, odd, or neither.

a) $f(x) = 5x^7 - 6x^3 - 2x$ b) $g(x) = x^4 - 2x$

c) $h(x) = 5x^6 - 3x^2 - 7$

a)

ALGEBRAIC SOLUTION

$f(x) = 5x^7 - 6x^3 - 2x$

1. $f(-x) = 5(-x)^7 - 6(-x)^3 - 2(-x)$
$= 5(-x^7) - 6(-x^3) + 2x$
$(-x)^7 = (-1 \cdot x)^7 = (-1)^7x^7 = -x^7$
$= -5x^7 + 6x^3 + 2x$

2. $-f(x) = -(5x^7 - 6x^3 - 2x)$
$= -5x^7 + 6x^3 + 2x$

3. We see that $f(x) \neq f(-x)$. Thus, f is *not* even. We see that $f(-x) = -f(x)$. Thus, f is odd.

GRAPHICAL SOLUTION

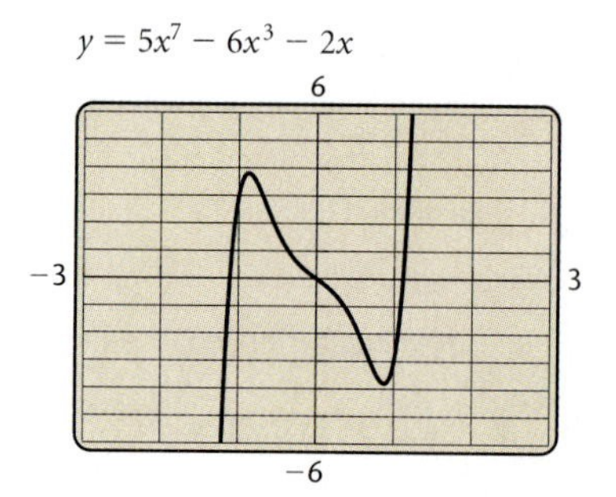

We see visually that the graph is symmetric with respect to the origin. Thus the function is odd.

b)

ALGEBRAIC SOLUTION

$g(x) = x^4 - 2x$

1. $g(-x) = (-x)^4 - 2(-x)$
$= x^4 + 2x$

2. $-g(x) = -(x^4 - 2x)$
$= -x^4 + 2x$

3. We see that $g(x) \neq g(-x)$. The function is *not* even. We see that $g(-x) \neq -g(x)$. The function is *not* odd. Thus g is neither even nor odd.

GRAPHICAL SOLUTION

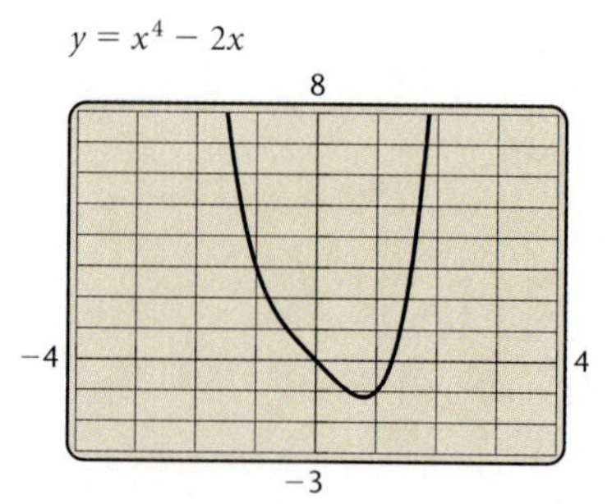

We see visually that the graph is symmetric with respect to neither the y-axis nor the origin. Thus the function is neither even nor odd.

c)

ALGEBRAIC SOLUTION

$h(x) = 5x^6 - 3x^2 - 7$

1. $h(-x) = 5(-x)^6 - 3(-x)^2 - 7$
$= 5x^6 - 3x^2 - 7$

We see that $h(x) = h(-x)$. Thus the function is even.

GRAPHICAL SOLUTION

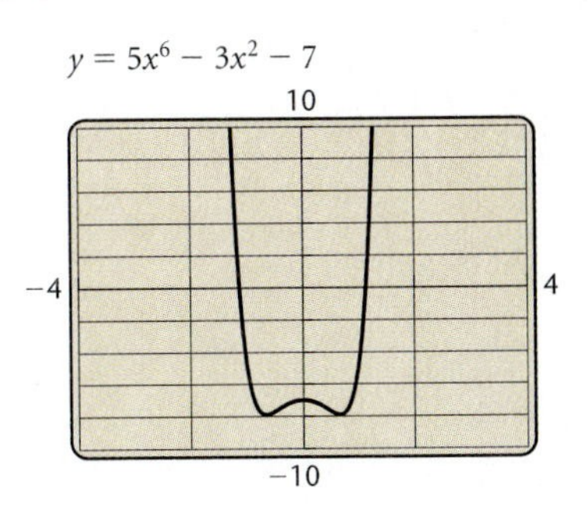

We see visually that the graph is symmetric with respect to the y-axis. Thus the function is even.

1.6 Exercise Set

Determine visually whether each graph is symmetric with respect to the x-axis, the y-axis, and the origin.

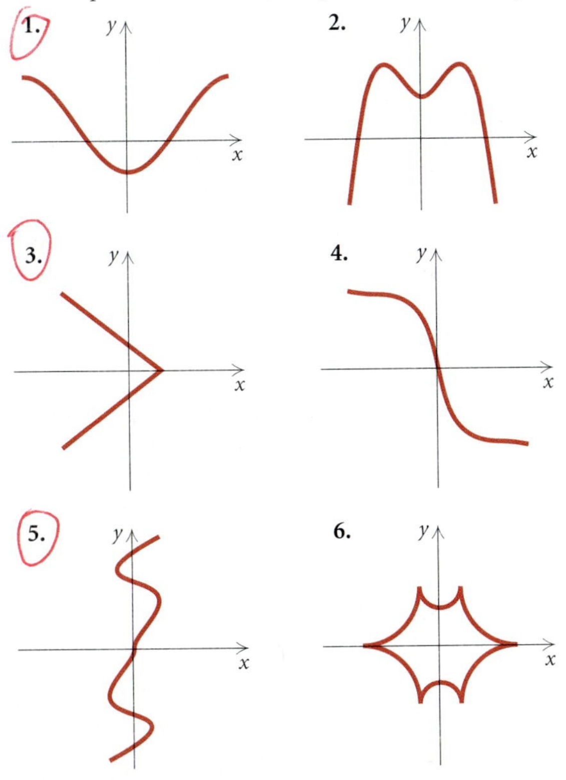

First, graph each equation and determine visually whether it is symmetric with respect to the x-axis, the y-axis, and the origin. Then verify your assertion algebraically.

7. $y = |x| - 2$

8. $y = |x + 5|$

9. $5y = 4x + 5$

10. $2x - 5 = 3y$

11. $5y = 2x^2 - 3$

12. $x^2 + 4 = 3y$

13. $y = \frac{1}{x}$

14. $y = -\frac{4}{x}$

Test algebraically whether each graph is symmetric with respect to the x-axis, the y-axis, and the origin. Then check your work graphically, if possible, using a grapher.

15. $5x - 5y = 0$

16. $6x + 7y = 0$

17. $3x^2 - 2y^2 = 3$

18. $5y = 7x^2 - 2x$

19. $y = |2x|$

20. $y^3 = 2x^2$

21. $2x^4 + 3 = y^2$

22. $2y^2 = 5x^2 + 12$

23. $3y^3 = 4x^3 + 2$

24. $3x = |y|$

25. $xy = 12$

26. $xy - x^2 = 3$

Determine visually whether each function is even, odd, or neither even nor odd.

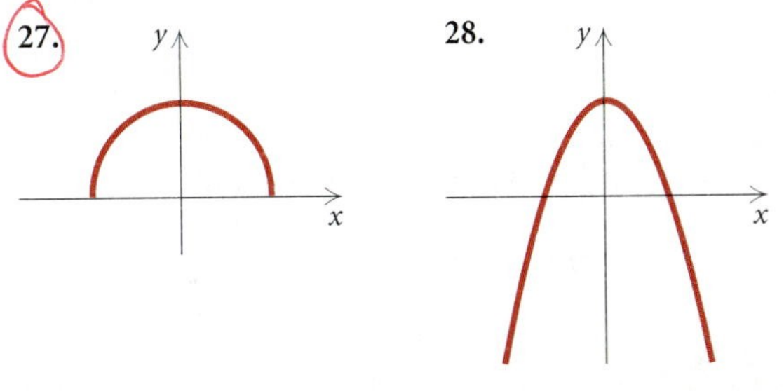

29. 30. 31. 32.

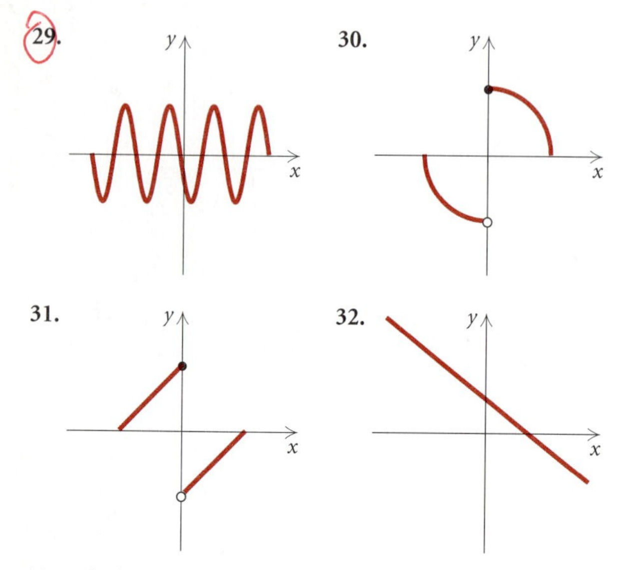

Test algebraically whether each function is even, odd, or neither even nor odd. Then check your work graphically, where possible, using a grapher.

33. $f(x) = -3x^3 + 2x$

34. $f(x) = 2x^2 + 4x$

35. $f(x) = 5x^2 + 2x^4 - 1$

36. $f(x) = 3x^4 - 4x^2$

37. $f(x) = 4x$

38. $f(x) = |3x|$

39. $f(x) = x^{24}$

40. $f(x) = 7x^3 + 4x - 2$

41. $f(x) = x^{17}$

42. $f(x) = x + \dfrac{1}{x}$

43. $f(x) = x - |x|$

44. $f(x) = \sqrt{x}$

45. $f(x) = \sqrt[3]{x}$

46. $f(x) = 8$

47. $f(x) = \sqrt[3]{x - 2}$

48. $f(x) = -\dfrac{2}{x}$

49. $f(x) = \dfrac{1}{x^2}$

50. $f(x) = \sqrt{x^2 + 1}$

Skill Maintenance

Given that $f(x) = 3x^2 - 2x$, find and simplify each of the following.

51. $f(a - 5)$

52. $f(a + 1)$

53. $f(a + 4)$

54. $f(a - 1)$

Synthesis

55. ◆ Consider the constant function $f(x) = 0$. Determine whether the graph of this function is symmetric with respect to the x-axis, the y-axis, and the origin. Determine whether this function is even or odd. In general, can a function be symmetric with respect to the x-axis? Explain.

56. ◆ Describe conditions under which you would know whether a polynomial function

$$f(x) = a_n x^n + a_{n-1}x^{n-1} + \cdots + a_2 x^2 + a_1 x + a_0$$

is even or odd without using an algebraic or graphical procedure. Explain.

Determine whether each function is even, odd, or neither even nor odd. Use any method.

57. $f(x) = x|x^3|$

58. $f(x) = x^2(5 - |x|)$

59. $f(x) = \dfrac{1 - x^4}{x^3 + 1}$

60. $f(x) = \dfrac{x^2 + 1}{x^3 - 1}$

61. $f(x) = x\sqrt{10 - x^2}$

62. $f(x) = \dfrac{-8x}{x^2 + 1}$

Determine whether each graph is symmetric with respect to the x-axis, the y-axis, and the origin. Use any method.

63. $x^4 = y^2 + x^6$

64. $x^3 = y^2(2 - x)$

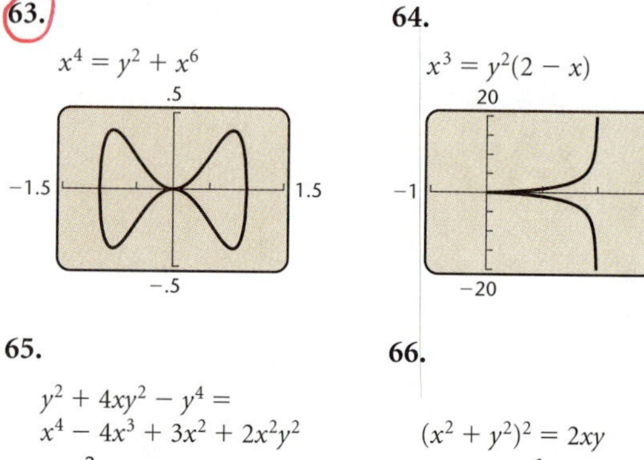

65. $y^2 + 4xy^2 - y^4 = x^4 - 4x^3 + 3x^2 + 2x^2y^2$

66. $(x^2 + y^2)^2 = 2xy$

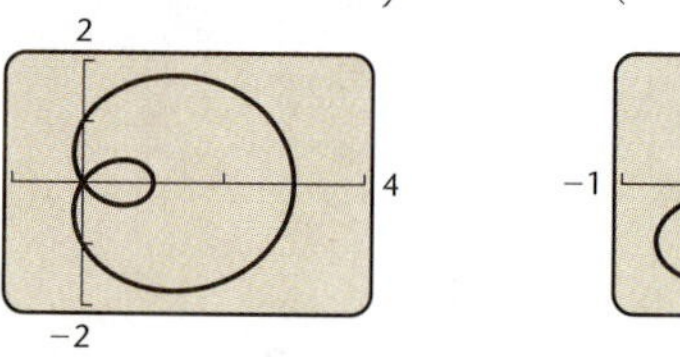

67. Given $f(x) = \frac{1}{2}x^4 - 5x^3 + 2$, let $g(x) = f(-x)$. Graph both f and g using the same set of axes. How do the graphs compare?

68. Consider symmetries with respect to the x-axis, the y-axis, and the origin. Prove that symmetry with respect to any two of these implies symmetry with respect to the other.

1.7 Transformations of Functions

- *Given the graph of a function, graph its transformation under translations, reflections, stretchings, and shrinkings.*

Throughout this chapter, we have considered many kinds of functions. Let's review some of them below.

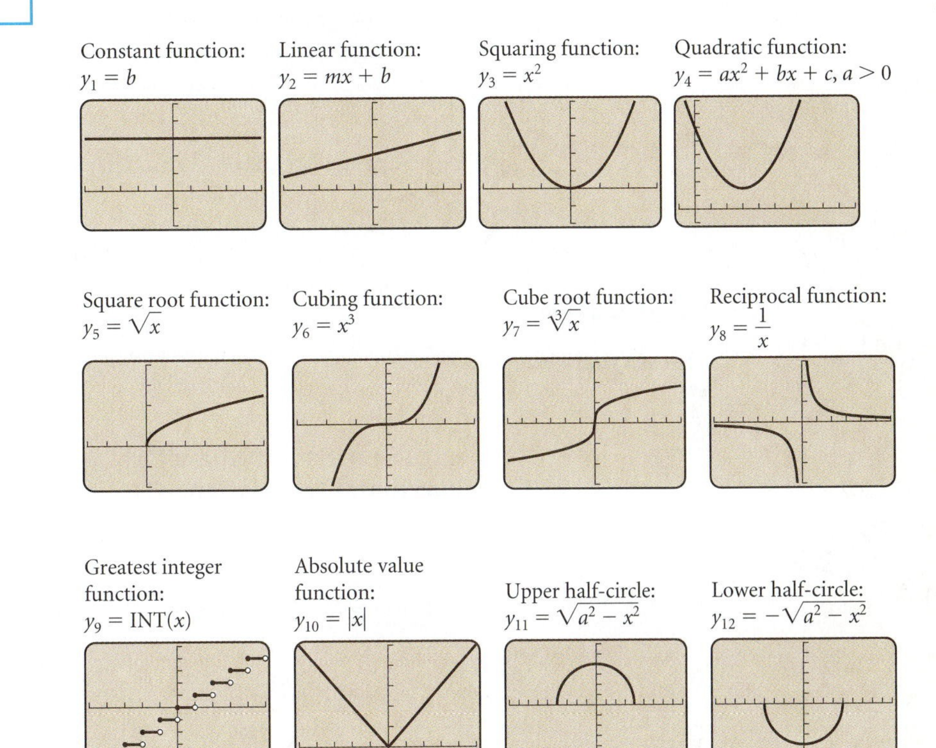

These functions can be considered building blocks for many other functions. We can create graphs of new functions by shifting them horizontally or vertically, stretching or shrinking them, and reflecting them across an axis. In this section, we consider these *transformations.*

Vertical and Horizontal Translations

Suppose we have a function given by $y_1 = f(x)$. Let's explore the graph of a new function $y_2 = f(x) + b$ or $y_3 = f(x) - b$, for $b > 0$.

Interactive Discovery

Consider the function $y_1 = \frac{1}{5}x^4$. Graph it and the functions $y_2 = \frac{1}{5}x^4 - 5$ and $y_3 = \frac{1}{5}x^4 + 3$ using the same viewing window, $[-10, 10, -8, 10]$. What pattern do you see? Test it with some other graphs.

The effect is a shift of $f(x)$ up or down. Such a shift is called a **vertical translation**.

Vertical Translation

For $b > 0$,

the graph of $y_2 = f(x) + b$ is the graph of $y_1 = f(x)$ shifted *up* b units;

the graph of $y_3 = f(x) - b$ is the graph of $y_1 = f(x)$ shifted *down* b units.

Suppose we have a function given by $y_1 = f(x)$. Let's explore the graph of a new function $y_2 = f(x + d)$ or $y_3 = f(x - d)$, for $d > 0$.

Interactive Discovery

Consider the function $y_1 = \frac{1}{5}x^4$. Graph it and the functions $y_2 = \frac{1}{5}(x - 3)^4$ and $y_3 = \frac{1}{5}(x + 7)^4$ using the same viewing window, $[-10, 10, -2, 10]$. What pattern do you see? Test it with some other graphs.

The effect is a shift of $f(x)$ to the right or left. Such a shift is called a **horizontal translation.**

Horizontal Translation

For $d > 0$;

the graph of $y_2 = f(x - d)$ is the graph of $y_1 = f(x)$ shifted *right* d units;

the graph of $y_3 = f(x + d)$ is the graph of $y_1 = f(x)$ shifted *left* d units.

Example 1 Graph each of the following. Before doing so, describe how each graph can be obtained from one of the basic graphs shown on the preceding page.

a) $g(x) = x^2 - 6$

b) $g(x) = \sqrt{x + 2}$

c) $g(x) = \dfrac{1}{x} + 2$

d) $g(x) = |x - 4|$

e) $h(x) = \sqrt{x + 2} - 3$

SOLUTION

a) To graph $g(x) = x^2 - 6$, think of the graph of $f(x) = x^2$. Since $g(x) = f(x) - 6$, the graph of $g(x) = x^2 - 6$ is the graph of $f(x) = x^2$, shifted, or translated, *down* 6 units. (See the figure at left.)

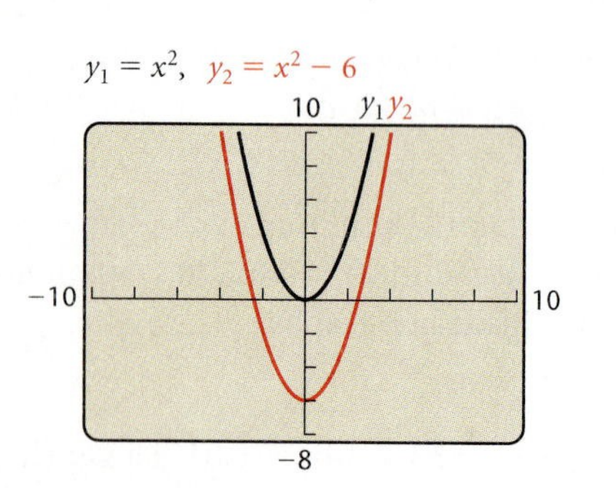

b) To graph $g(x) = \sqrt{x + 2}$, think of the graph of $f(x) = \sqrt{x}$. Since $g(x) = f(x + 2)$, the graph of $g(x) = \sqrt{x + 2}$ is the graph of $f(x) = \sqrt{x}$, shifted *left* 2 units.

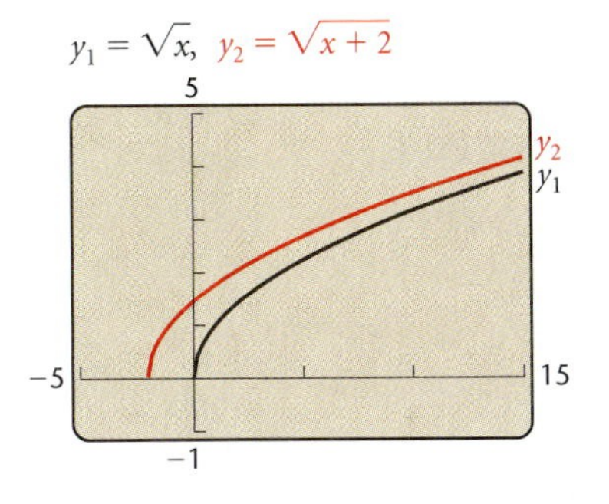

c) To graph $g(x) = 1/x + 2$, think of the graph of $f(x) = 1/x$. Since $g(x) = f(x) + 2$, the graph of $g(x) = 1/x + 2$ is the graph of $f(x) = 1/x$, shifted *up* 2 units. (See the figure on the left below.)

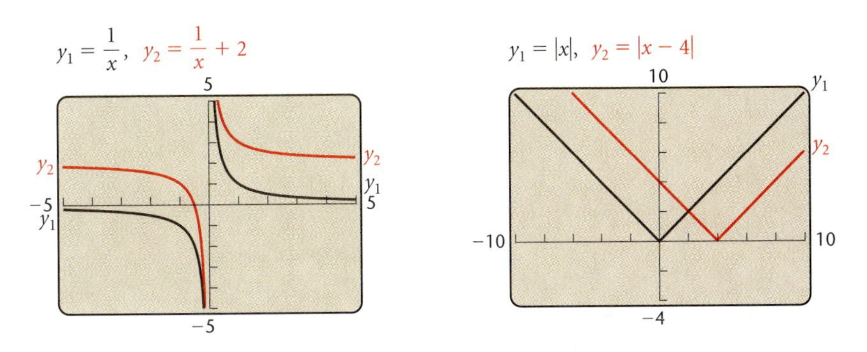

d) To graph $g(x) = |x - 4|$, think of the graph of $f(x) = |x|$. Since $g(x) = f(x - 4)$, the graph of $g(x) = |x - 4|$ is the graph of $f(x) = |x|$ shifted *right* 4 units. (See the figure on the right above.)

e) To graph $h(x) = \sqrt{x + 2} - 3$, think of the graph of $f(x) = \sqrt{x}$. In part (b), we found that the graph of $g(x) = \sqrt{x + 2}$ is the graph of $f(x) = \sqrt{x}$ shifted left 2 units. Since $h(x) = g(x) - 3$, we shift the graph of $g(x) = \sqrt{x + 2}$ *down* 3 units. Together, the graph of $f(x) = \sqrt{x}$ is shifted *left* 2 units and *down* 3 units.

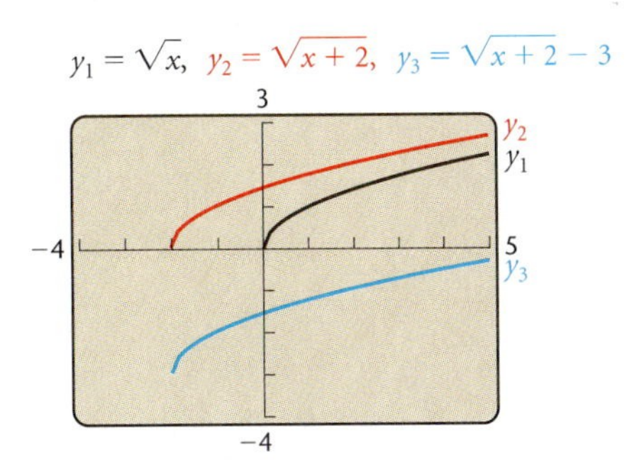

Reflections

Suppose we have a function given by $y = f(x)$. Let's explore the graph of a new function $g(x) = -f(x)$ or $g(x) = f(-x)$.

Interactive Discovery

Consider the functions $y_1 = \frac{1}{5}x^4$ and $y_2 = -\frac{1}{5}x^4$. Graph y_1 in the viewing window $[-10, 10, -10, 10]$. Then graph y_2 in the same viewing window. What pattern do you see? Test it with some other functions in which $y_2 = -y_1$.

Consider the functions $y_1 = 2x^3 - x^4 + 5$ and $y_2 = 2(-x)^3 - (-x)^4 + 5$. Graph y_1 in the viewing window $[-4, 4, -10, 10]$. Then graph y_2 in the same viewing window. What pattern do you see? Test it with some other functions in which x is replaced with $-x$.

Given the graph of $y_1 = f(x)$, we can reflect each point across the x-axis to obtain the graph of $y_2 = -f(x)$. We can reflect each point of y_1 across the y-axis to obtain the graph of $y_2 = f(-x)$. The new graphs are called **reflections** of $f(x)$.

Reflections

The graph of $y_2 = -f(x)$ is the *reflection* of the graph of $y_1 = f(x)$ across the x-axis.

The graph of $y_2 = f(-x)$ is the *reflection* of the graph of $y_1 = f(x)$ across the y-axis.

Example 2 Graph each of the following. Before doing so, describe how each graph can be obtained from the graph of $f(x) = x^3 - 4x^2$.

a) $g(x) = (-x)^3 - 4(-x)^2$ **b)** $g(x) = 4x^2 - x^3$

SOLUTION

a) We first note that

$$f(-x) = (-x)^3 - 4(-x)^2 = g(x).$$

Thus the graph of g is a reflection of the graph of f across the y-axis.

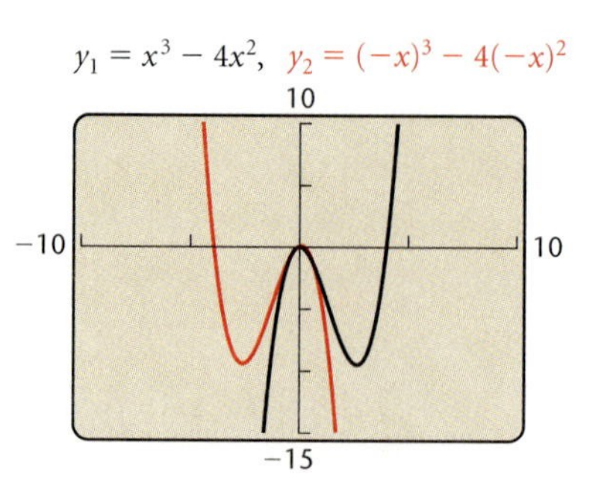

b) We first note that

$$\begin{aligned} -f(x) &= -(x^3 - 4x^2) \\ &= -x^3 + 4x^2 \\ &= 4x^2 - x^3 \\ &= g(x). \end{aligned}$$

Thus the graph of g is a reflection of the graph of f across the x-axis.

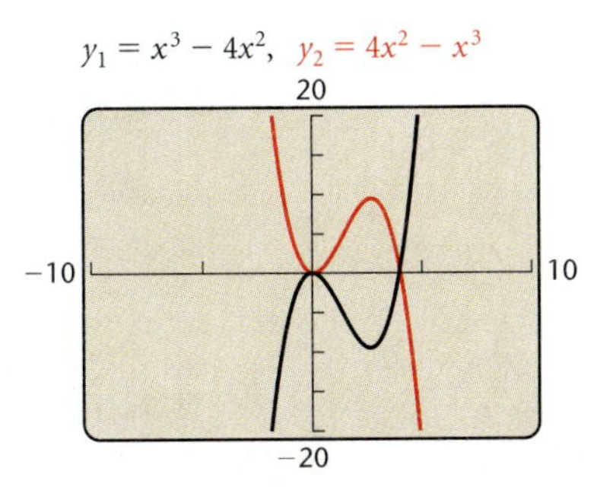

Vertical and Horizontal Stretchings and Shrinkings

Suppose we have a function given by $y_1 = f(x)$. Let's explore the graph of a new function $y_2 = af(x)$ or $y_3 = f(cx)$.

Interactive Discovery

Graph the function $y_1 = x^3 - x$ using the viewing window $[-3, 3, -1, 1]$. Then graph $y_2 = \frac{1}{10}(x^3 - x)$ using the same viewing window. Clear y_2 and graph $y_3 = 2(x^3 - x)$. Then clear y_3 and graph $y_4 = -2(x^3 - x)$. What pattern do you see? Test it with some other graphs.

Now graph y_1 and $y_2 = (2x)^3 - (2x)$ using the same viewing window. Clear y_2 and graph $y_3 = \left(\frac{1}{2}x\right)^3 - \left(\frac{1}{2}x\right)$. Then clear y_3 and graph $y_4 = \left(-\frac{1}{2}x\right)^3 - \left(-\frac{1}{2}x\right)$. What pattern do you see? Test it with some other graphs.

Consider any function f given by $y = f(x)$. Multiplying on the right by any constant a, where $|a| > 1$, to obtain $g(x) = af(x)$ will *stretch* the graph vertically away from the x-axis. If $0 < |a| < 1$, then the graph will be flattened or *shrunk* vertically toward the x-axis. If $a < 0$, the graph is also reflected across the x-axis.

Vertical Stretching and Shrinking

The graph of $y_2 = af(x)$ can be obtained from the graph of $y_1 = f(x)$ by

stretching vertically for $|a| > 1$, or
shrinking vertically for $0 < |a| < 1$.

For $a < 0$, the graph is also reflected across the x-axis.

The constant c in the equation $g(x) = f(cx)$ will *stretch* the graph of $y = f(x)$ horizontally away from the y-axis if $0 < |c| < 1$. If $|c| > 1$, the graph will be *shrunk* horizontally toward the y-axis. If $c < 0$, the graph is also reflected across the y-axis.

Horizontal Stretching and Shrinking

The graph of $y_2 = f(cx)$ can be obtained from the graph of $y_1 = f(x)$ by

shrinking horizontally for $|c| > 1$, or
stretching horizontally for $0 < |c| < 1$.

For $c < 0$, the graph is also reflected across the y-axis.

It is instructive to use these concepts now to create hand-drawn graphs from a given graph.

Example 3 Shown below is a graph of $y = f(x)$ for some function f. No formula for f is given. Make a hand-drawn graph of each of the following.

a) $g(x) = f(2x)$
b) $h(x) = f\left(\frac{1}{2}x\right)$
c) $t(x) = f\left(-\frac{1}{2}x\right)$

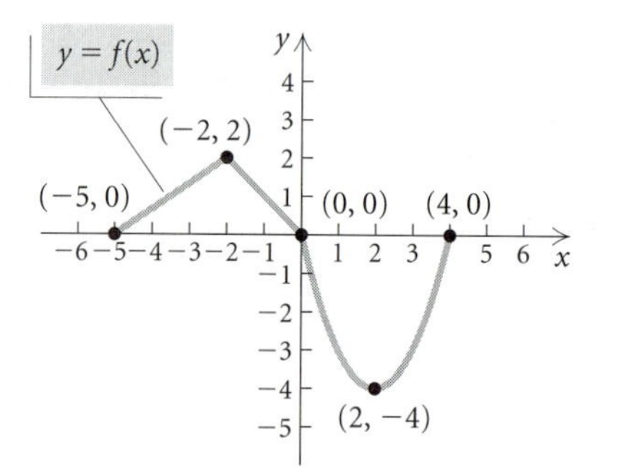

SOLUTION

a) Since $|2| > 1$, the graph of $g(x) = f(2x)$ is a horizontal shrinking of the graph of $y = f(x)$. We can consider the key points $(-5, 0)$, $(-2, 2)$, $(0, 0)$, $(2, -4)$, and $(4, 0)$. The transformation divides each x-coordinate by 2 to obtain the key points $(-2.5, 0)$, $(-1, 2)$, $(0, 0)$, $(1, -4)$, and $(2, 0)$ of the graph of $g(x) = f(2x)$. The graph is shown at the top of the following page.

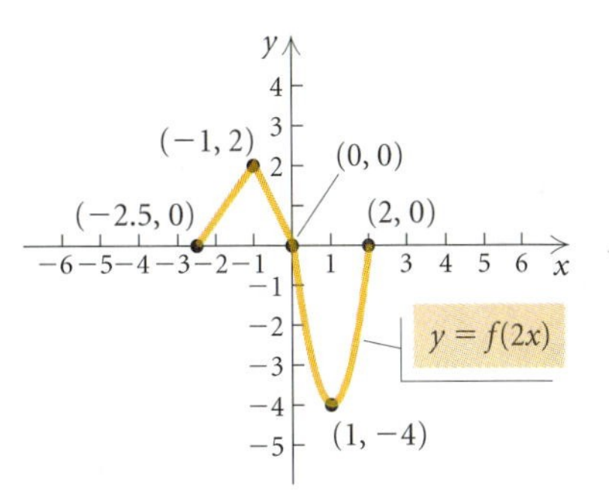

b) Since $\left|\frac{1}{2}\right| < 1$, the graph of $h(x) = f\left(\frac{1}{2}x\right)$ is a horizontal stretching of the graph of $y = f(x)$. We can consider the key points $(-5, 0)$, $(-2, 2)$, $(0, 0)$, $(2, -4)$, and $(4, 0)$. The transformation divides each x-coordinate by $\frac{1}{2}$ (which is the same as multiplying by 2) to obtain the key points $(-10, 0)$, $(-4, 2)$, $(0, 0)$, $(4, -4)$, and $(8, 0)$ of the graph of $h(x) = f\left(\frac{1}{2}x\right)$. The graph is shown below.

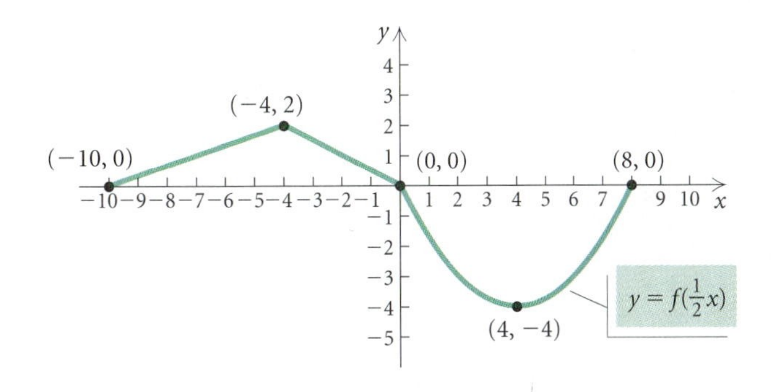

c) The graph of $t(x) = f\left(-\frac{1}{2}x\right)$ can be obtained by reflecting the graph in part (b) across the y-axis. It is shown below.

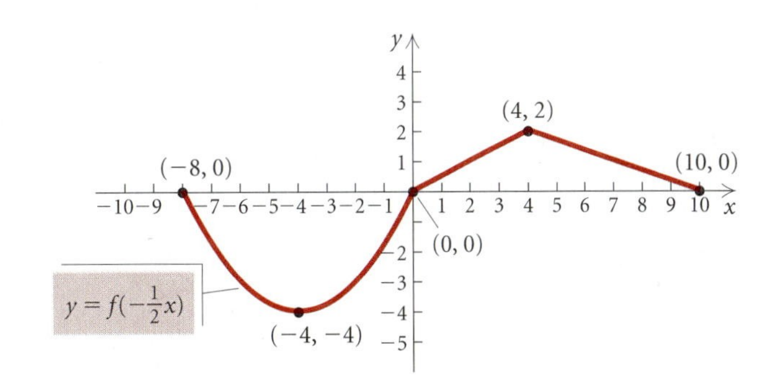

Example 4 Use the graph of $y = f(x)$ given in Example 3. Make a hand-drawn graph of

$$g(x) = -2f(x - 3) + 1.$$

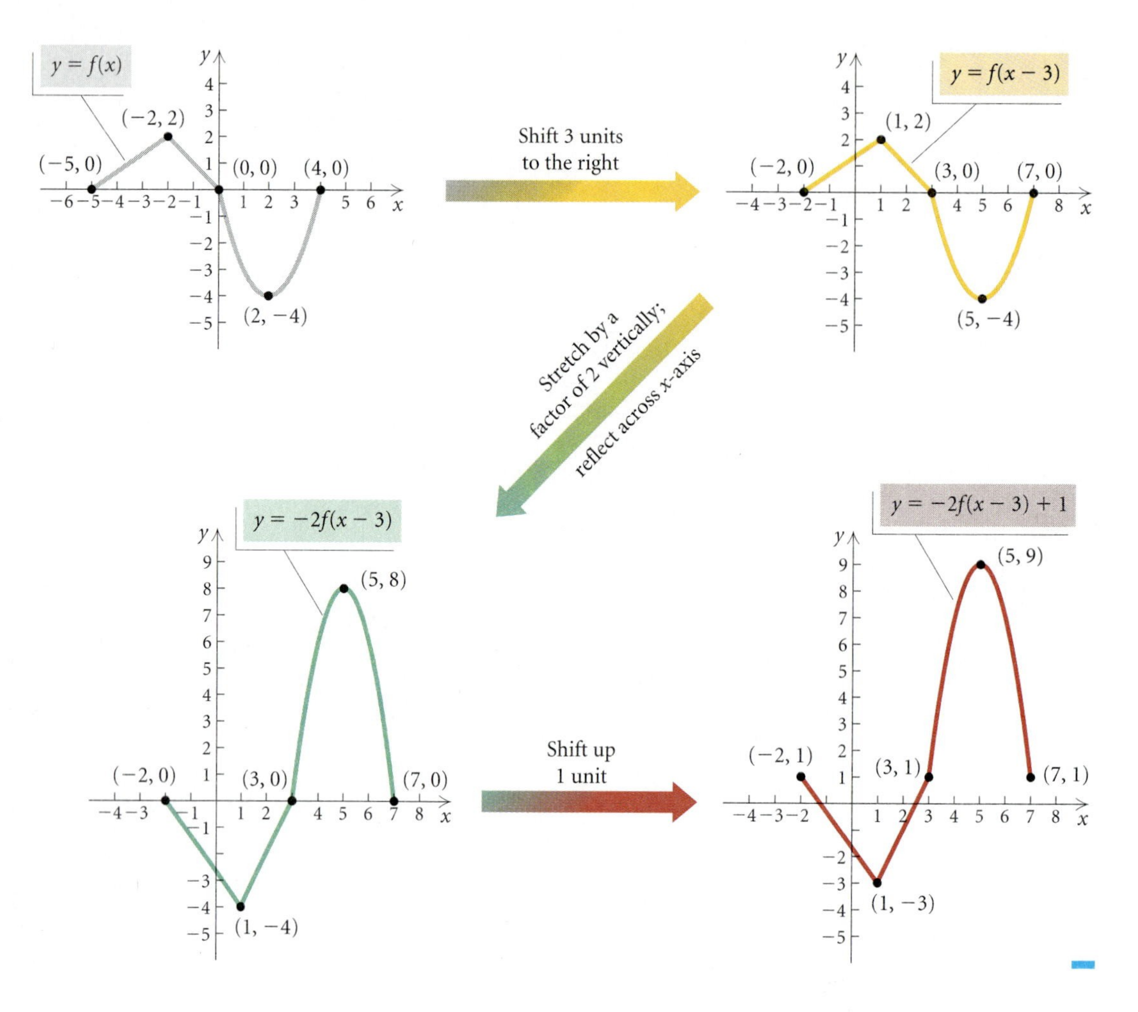

Vertical or Horizontal Translation

For $b > 0$,

the graph of $y_2 = f(x) + b$ is the graph of $y_1 = f(x)$ shifted *up* b units;

the graph of $y_2 = f(x) - b$ is the graph of $y_1 = f(x)$ shifted *down* b units.

For $d > 0$,

the graph of $y_2 = f(x - d)$ is the graph of $y_1 = f(x)$ shifted *right* d units;

the graph of $y_2 = f(x + d)$ is the graph of $y_1 = f(x)$ shifted *left* d units.

Reflections

The graph of $y_2 = -f(x)$ is the reflection of the graph of $y_1 = f(x)$ across the x-axis.

The graph of $y_2 = f(-x)$ is the reflection of the graph of $y_1 = f(x)$ across the y-axis.

Vertical Stretching or Shrinking

The graph of $y_2 = af(x)$ can be obtained from the graph of $y_1 = f(x)$ by

stretching vertically for $|a| > 1$, or
shrinking vertically for $0 < |a| < 1$.

For $a < 0$, the graph is also reflected across the x-axis.

Horizontal Stretching or Shrinking

The graph of $y_2 = f(cx)$ can be obtained from the graph of $y_1 = f(x)$ by

shrinking horizontally for $|c| > 1$, or
stretching horizontally for $0 < |c| < 1$.

For $c < 0$, the graph is also reflected across the y-axis.

1.7 Exercise Set

Graph each of the following on a grapher. Before doing so, describe how each graph can be obtained from one of the basic graphs at the beginning of this section.

1. $f(x) = x^2 + 1$
2. $f(x) = x^3 - 4$
3. $f(x) = \dfrac{1}{x} + 3$
4. $f(x) = \sqrt{x} + 2$
5. $f(x) = \sqrt{x} - 5$
6. $f(x) = -\sqrt{x}$
7. $f(x) = -x^2$
8. $f(x) = (x - 4)^2$
9. $f(x) = |x - 3|$
10. $g(x) = |3x|$
11. $g(x) = (x + 5)^3$
12. $g(x) = \dfrac{1}{x - 5}$
13. $g(x) = (x + 1)^2$
14. $f(x) = 2x^2$
15. $f(x) = \frac{1}{2}\sqrt{x}$
16. $f(x) = \frac{1}{2}x^3$
17. $g(x) = \dfrac{2}{x}$
18. $f(x) = |x - 3| - 4$
19. $f(x) = 3\sqrt{x} - 5$
20. $f(x) = 5 - \dfrac{1}{x}$
21. $f(x) = \frac{1}{2}(x - 3)^2$
22. $f(x) = \sqrt{x - 3} + 2$
23. $g(x) = \left|\frac{1}{3}x\right| - 4$
24. $f(x) = \frac{2}{3}x^3 - 4$
25. $f(x) = (x + 5)^2 - 4$
26. $f(x) = (-x)^3 - 5$
27. $f(x) = -\frac{1}{4}(x - 5)^2$
28. $g(x) = \sqrt{-x} - 2$
29. $f(x) = \dfrac{1}{x + 3} + 2$
30. $g(x) = (x - 2)^3 - 5$
31. $f(x) = (x + 4)^3 + 3$
32. $f(x) = \dfrac{1}{-x} + 2$
33. $f(x) = \sqrt{-x} + 5$
34. $f(x) = \frac{4}{5}(x - 4)^2 - 5$
35. $f(x) = 3(x + 4)^2 - 3$
36. $g(x) = 2(x + 1)^3 + 4$
37. $f(x) = 0.43(x - 3)^3 + 2.4$
38. $f(x) = 0.3|x - 5.2| + 2.8$
39. $g(x) = 2.8(x + 5.2)^2 + 1.1$
40. $f(x) = 1.8\sqrt{x - 3.4} - 4.8$

Write an equation for a function that has a graph with the given characteristics. Check your answer on a grapher.

41. The shape of $y = x^2$, but upside-down and shifted right 8 units
42. The shape of $y = \sqrt{x}$, but shifted left 6 units and down 5 units

43. The shape of $y = |x|$, but shifted left 7 units and up 2 units

44. The shape of $y = x^3$, but upside-down and shifted right 5 units

45. The shape of $y = 1/x$, but shrunk vertically by a factor of $\frac{1}{2}$ and shifted down 3 units

46. The shape of $y = x^2$, but shifted right 6 units and up 2 units

47. The shape of $y = x^2$, but upside-down and shifted right 3 units and up 4 units

48. The shape of $y = |x|$, but stretched horizontally by a factor of 2 and shifted down 5 units

49. The shape of $y = \sqrt{x}$, but reflected across the y-axis and shifted left 2 units and down 1 unit

50. The shape of $y = 1/x$, but reflected across the x-axis and shifted up 1 unit

51. The shape of $y = x^3$, but shifted left 4 units and shrunk vertically by a factor of 0.83.

A graph of $y = f(x)$ follows. No formula for f is given. Make a hand-drawn graph of each of the following.

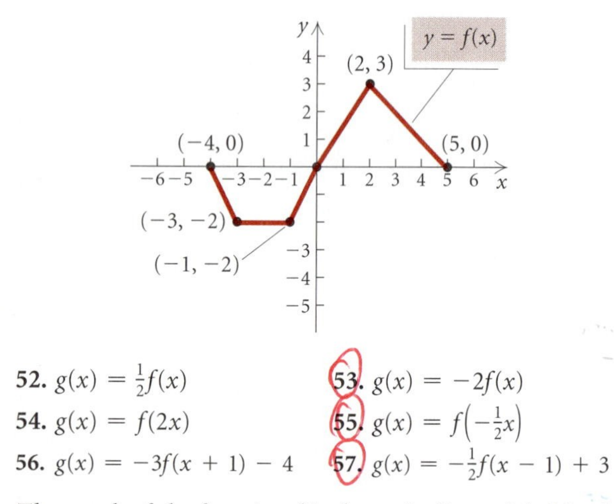

52. $g(x) = \frac{1}{2}f(x)$

53. $g(x) = -2f(x)$

54. $g(x) = f(2x)$

55. $g(x) = f\left(-\frac{1}{2}x\right)$

56. $g(x) = -3f(x + 1) - 4$

57. $g(x) = -\frac{1}{2}f(x - 1) + 3$

The graph of the function f is shown in figure (a). Match each function g in Exercises 58–65 with the appropriate graph from (a)–(h). Some graphs may be used more than once.

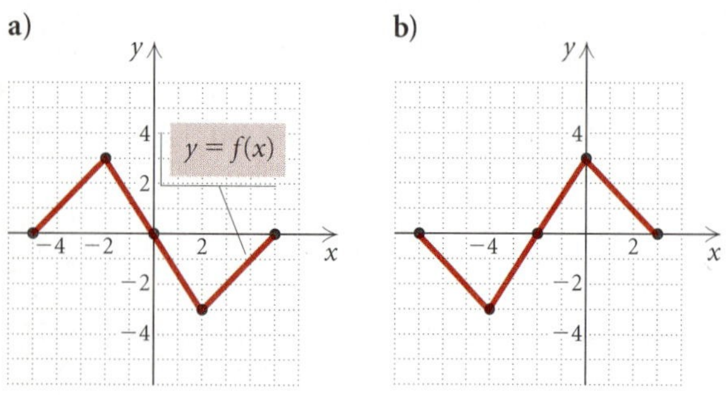

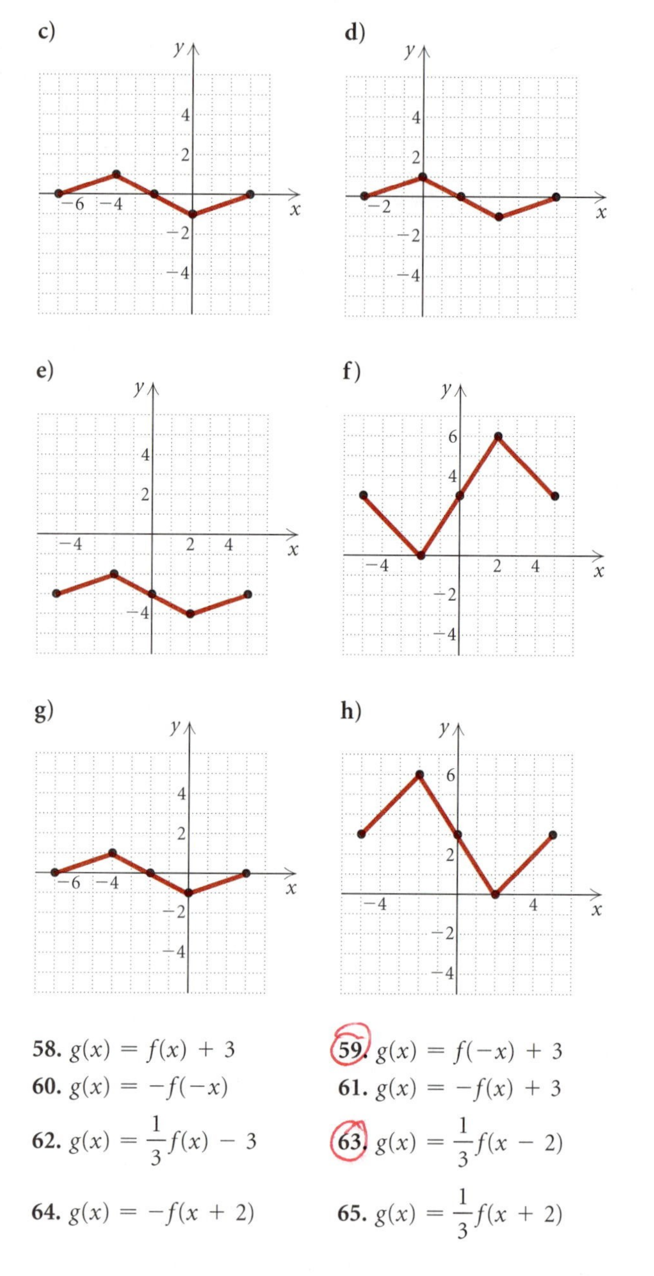

58. $g(x) = f(x) + 3$

59. $g(x) = f(-x) + 3$

60. $g(x) = -f(-x)$

61. $g(x) = -f(x) + 3$

62. $g(x) = \dfrac{1}{3}f(x) - 3$

63. $g(x) = \dfrac{1}{3}f(x - 2)$

64. $g(x) = -f(x + 2)$

65. $g(x) = \dfrac{1}{3}f(x + 2)$

Use a grapher to graph each set of functions using a SIMULTANEOUS *feature. (With this feature, you cannot tell which function is being graphed first.) After the curves are displayed, match each curve with the appropriate function. Check your answers by setting the* WINDOW *format to LabelOn. Then when you use the* TRACE *feature, the label of the curve being traced will appear.*

66. $y_1 = (x + 1)^2$,
$y_2 = (x - 3)^2 + 5$,
$y_3 = (x + 3)^2 - 5$,
$y_4 = x^2 + 1$

67. $y_1 = (x - 5)^3 + 4$,
$y_2 = (x + 5)^3 - 4$,
$y_3 = (x - 2)^3$,
$y_4 = x^3 - 2$

68. $y_1 = \dfrac{1}{x+5}$,

$y_2 = \dfrac{1}{x-3} + 4$,

$y_3 = \dfrac{1}{x+3} - 4$,

$y_4 = \dfrac{1}{x} + 5$

69. $y_1 = \sqrt{9 - x^2}$,

$y_2 = -\sqrt{9 - x^2}$,

$y_3 = \dfrac{2}{3}\sqrt{9 - x^2}$,

$y_4 = 4 - \sqrt{9 - x^2}$

70. Using a $[-10, 10, -200, 800]$ window with Yscl $= 100$, show that

$$y_1 = x^4 - 12x^3 + 34x^2 + 12x - 35$$

and

$$y_2 = x^4 + 12x^3 + 34x^2 - 12x - 35$$

are reflections of each other across the y-axis. Verify this algebraically.

For each pair of functions, determine if $g(x) = f(-x)$ using algebra. Then, using the TABLE *feature on a grapher, check your answers by looking at y_1 and y_2 for x-values near 0. You can then check graphically, but be careful to use a suitable window.*

71. $f(x) = 2x^4 - 35x^3 + 3x - 5$,
$g(x) = 2x^4 + 35x^3 - 3x - 5$

72. $f(x) = \frac{1}{4}x^4 + \frac{1}{5}x^3 - 81x^2 - 17$,
$g(x) = \frac{1}{4}x^4 + \frac{1}{5}x^3 + 81x^2 - 17$

A graph of the function $f(x) = x^3 - 3x^2$ is shown below. Exercises 73–76 show graphs of functions transformed from this one. Find a formula for each function.

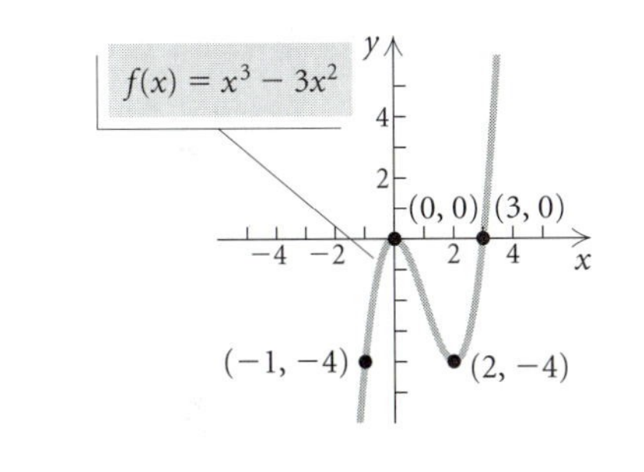

73.

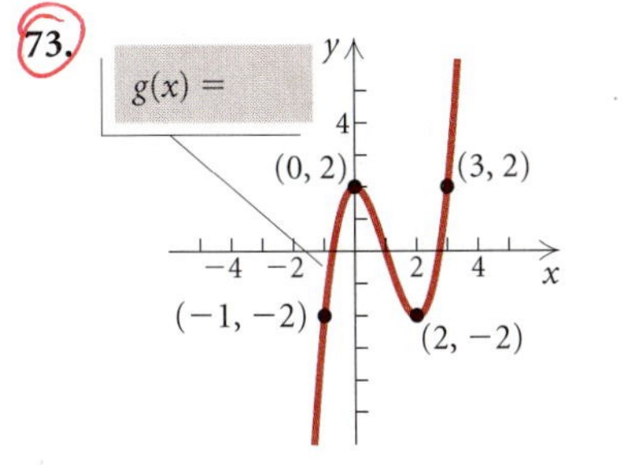

74.

75.

76.

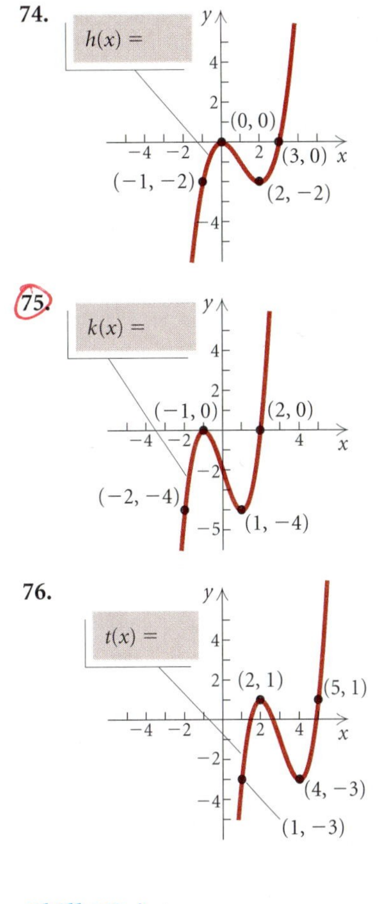

Skill Maintenance

Given that $f(x) = x^3 - 7x$ and $g(x) = x^2 + 1$, find each of the following.

77. $f(3)$

78. $g(3)$

79. $f(3) + g(3)$

80. $f(4) - g(4)$

Synthesis

81. ◈ Explain in your own words why the graph of $y = f(-x)$ is a reflection of the graph of $y = f(x)$ across the y-axis.

82. ◈ Without drawing the graph, describe what the graph of $f(x) = |x^2 - 9|$ looks like.

83. The graph of $f(x) = \sqrt{x}$ passes through the points $(0, 0)$, $(1, 1)$, and $(4, 2)$. Transform this function to one whose graph passes through the points $(4, 7)$, $(5, 6)$, and $(8, 5)$. Check your answer on a grapher.

84. The graph of $f(x) = |x|$ passes through the points $(-3, 3)$, $(0, 0)$, and $(3, 3)$. Transform this function to one whose graph passes through the points $(5, 1)$, $(8, 4)$, and $(11, 1)$. Check your answer on a grapher.

Graph each of the following on a grapher. Before doing so, describe how each graph can be obtained from a more basic graph. Give the domain and the range of each function.

85. $f(x) = \text{INT}\left(x - \frac{1}{2}\right)$

86. $g(x) = \sqrt{7 - x}$

87. $f(x) = 3 \cdot \text{INT}(x + 2) + 1$

88. $f(x) = |x^2 - 4|$

89. $g(x) = \left|\dfrac{1}{x}\right|$

90. $f(x) = |\sqrt{x} - 1|$

91. $f(x) = \big||x| - 5\big|$

92. Find the roots, or zeros, of $f(x) = 3x^5 - 20x^3$. Then without using a grapher, state the roots of $f(x - 3)$ and $f(x + 8)$.

93. If (3, 4) is a point on the graph of $y = f(x)$, what point do you know is on the graph of $y = 2f(x)$? of $y = 2 + f(x)$? of $y = f(2x)$?

94. If (−1, 5) is a point on the graph of $y = f(x)$, find b such that $(2, b)$ is on the graph of $y = f(x - 3)$.

1.8 The Algebra of Functions

- *Compute function values for the sum, the difference, the product, and the quotient of two functions, and determine the domains.*
- *Find the composition of two functions and the domain of the composition; and decompose a function as a composition of two functions.*

We now consider five methods of combining two functions to obtain a new function. Among these are addition, subtraction, multiplication, and division.

Sums, Differences, Products, and Quotients

Consider the following two functions f and g:

$$f(x) = x + 2 \quad \text{and} \quad g(x) = x^2 + 1.$$

Since $f(3) = 3 + 2 = 5$ and $g(3) = 3^2 + 1 = 10$, we have

$$f(3) + g(3) = 5 + 10 = 15, \qquad f(3) - g(3) = 5 - 10 = -5,$$

$$f(3) \cdot g(3) = 5 \cdot 10 = 50, \quad \text{and} \quad \frac{f(3)}{g(3)} = \frac{5}{10} = \frac{1}{2}.$$

In fact, so long as x is in the domain of both f and g, we can easily compute $f(x) + g(x)$, $f(x) - g(x)$, $f(x) \cdot g(x)$, and, assuming $g(x) \neq 0$, $f(x)/g(x)$. Notation has been developed to facilitate this work.

Sums, Differences, Products, and Quotients of Functions

If f and g are functions and x is in the domain of each function, then

$(f + g)(x) = f(x) + g(x)$, $\quad (f - g)(x) = f(x) - g(x)$,

$(fg)(x) = f(x) \cdot g(x)$, $\quad (f/g)(x) = f(x)/g(x)$, provided $g(x) \neq 0$.

Example 1 Given that $f(x) = x + 1$ and $g(x) = \sqrt{x + 3}$, find each of the following.

a) $(f + g)(x)$ **b)** $(f + g)(6)$ **c)** $(f + g)(-4)$

SOLUTION

a) $(f + g)(x) = f(x) + g(x)$
$= (x + 1) + \sqrt{x + 3}$ **This cannot be simplified.**

b) We can find $(f + g)(6)$ provided 6 is in the domain of each function. The domain of f is all real numbers. The domain of g is all real numbers x for which $x + 3 \geq 0$, or $x \geq -3$. This is the interval $[-3, \infty)$. Because 6 is in both domains, we have

$$f(6) = 6 + 1 = 7, \qquad g(6) = \sqrt{6 + 3} = \sqrt{9} = 3,$$
$$(f + g)(6) = f(6) + g(6) = 7 + 3 = 10.$$

Another method is to use the formula found in part (a):

$$(f + g)(6) = (6 + 1) + \sqrt{6 + 3} = 7 + \sqrt{9} = 7 + 3 = 10.$$

c) To find $(f + g)(-4)$, we must first determine whether -4 is in the domain of each function. We note that -4 is not in the domain of g, $[-3, \infty)$. That is, $\sqrt{-4 + 3}$ is not a real number. Thus, $(f + g)(-4)$ does not exist.

It is useful to view the concept of the sum of two functions graphically. In the graph on the left below, we see the graphs of two functions f and g and their sum, $f + g$. Consider finding $(f + g)(4) = f(4) + g(4)$. We can locate $g(4)$ on the graph of g and use a compass to measure it. Then we move that setting on top of $f(4)$ and add. The sum gives us $(f + g)(4)$.

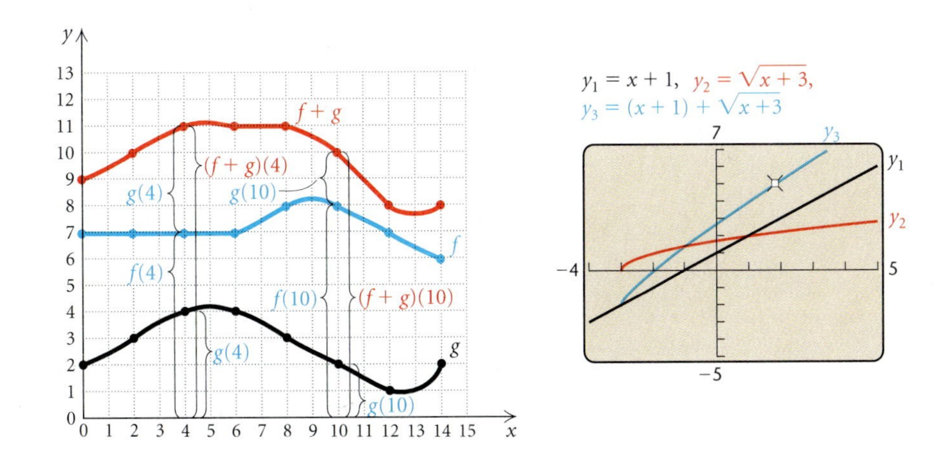

With this in mind, let's view Example 1 from a graphical perspective. See the graph on the right above. Use a grapher to graph

$$y_1 = x + 1, \qquad y_2 = \sqrt{x + 3}, \quad \text{and} \quad y_3 = (x + 1) + \sqrt{x + 3}.$$

(On some graphers, you can enter y_1 and y_2 and then key the grapher to enter $y_3 = y_1 + y_2$ directly.) Note that because the domain of y_2 is $[-3, \infty)$, the graph of the sum exists only over $[-3, \infty)$. By using the TRACE feature and jumping the cursor vertically among the three graphs, we can confirm that the y-coordinates of the graph of $(f + g)(x)$ are the sums of the corresponding y-coordinates of the graphs of $f(x)$ and $g(x)$.

Example 2 Given that $f(x) = x^2 - 4$ and $g(x) = x + 2$, find each of the following.

a) $(f + g)(x)$ **b)** $(f - g)(x)$ **c)** $(fg)(x)$

d) $(f/g)(x)$ **e)** $(gg)(x)$

f) The domain of $f + g$, $f - g$, fg, and f/g

SOLUTION

a) $(f + g)(x) = f(x) + g(x) = (x^2 - 4) + (x + 2) = x^2 + x - 2$

b) $(f - g)(x) = f(x) - g(x) = (x^2 - 4) - (x + 2) = x^2 - x - 6$

c) $(fg)(x) = f(x) \cdot g(x) = (x^2 - 4)(x + 2) = x^3 + 2x^2 - 4x - 8$

d)
$$\begin{aligned}(f/g)(x) &= \frac{f(x)}{g(x)} \\ &= \frac{x^2 - 4}{x + 2} && \text{Note that } x \neq -2. \\ &= \frac{(x + 2)(x - 2)}{x + 2} && \text{Factoring} \\ &= x - 2 && \text{Removing a factor of 1: } \frac{x+2}{x+2} = 1\end{aligned}$$

Thus, $(f/g)(x) = x - 2$ with the added stipulation that $x \neq -2$.

e) $(gg)(x) = [g(x)]^2 = (x + 2)^2 = x^2 + 4x + 4$

f) The domain of f is the set of all real numbers. The domain of g is the set of all real numbers. The domain of $f + g$, $f - g$, and fg is the set of numbers in the intersection of the domains—that is, the set of numbers in both domains, which is again the set of real numbers. For f/g, we must exclude -2, since $g(-2) = 0$. Thus the domain of f/g is the set of real numbers excluding -2, or $(-\infty, -2) \cup (-2, \infty)$.

The Composition of Functions

In the real world, it is not uncommon for a function's output to depend on some input that is itself an output of some function. For instance, the amount a person pays as state income tax usually depends on the amount of adjusted gross income on the person's federal tax return, which, in turn, depends on his or her annual earnings. Such functions are called *composite functions*.

To illustrate how composite functions work, suppose a chemistry student needs a formula to convert Fahrenheit temperatures to Kelvin units. The formula $c(t) = \frac{5}{9}(t - 32)$ gives the Celsius temperature $c(t)$ that corresponds to the Fahrenheit temperature t. The formula $k(c) = c + 273$ gives the Kelvin temperature $k(c)$ that corresponds to the Celsius temperature c. Thus, 50° Fahrenheit corresponds to

$$c(50) = \tfrac{5}{9}(50 - 32) = \tfrac{5}{9}(18) = 10° \text{ Celsius}$$

and since 10° Celsius corresponds to

$$k(10) = 10 + 273 = 283° \text{ Kelvin},$$

we see that 50° Fahrenheit is the same as 283° Kelvin. This two-step procedure can be used to convert any Fahrenheit temperature to Kelvin units.

X	Y1	Y2
50	10	283
59	15	288
68	20	293
77	25	298
86	30	303
95	35	308
104	40	313
X = 50		

= ?

In the table shown at left, we use a grapher to convert Fahrenheit temperatures, x, to Celsius temperatures, y_1, using $y_1 = \frac{5}{9}(x - 32)$. We also convert Celsius temperatures to Kelvin units, y_2, using $y_2 = y_1 + 273$.

A student making numerous conversions might look for a formula that converts directly from Fahrenheit to Kelvin. Such a formula can be found by substitution:

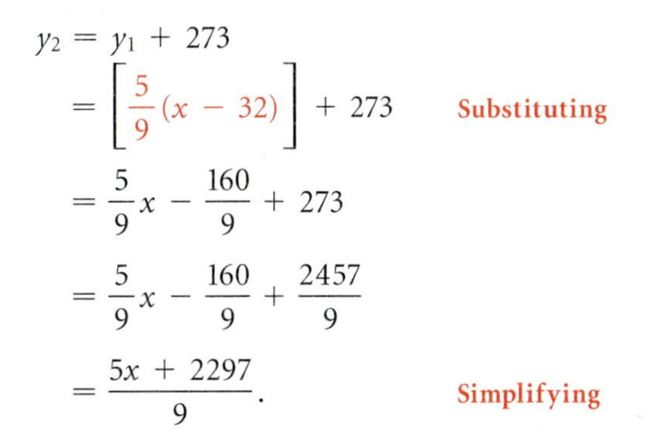

$$\begin{aligned} y_2 &= y_1 + 273 \\ &= \left[\frac{5}{9}(x - 32)\right] + 273 && \textbf{Substituting} \\ &= \frac{5}{9}x - \frac{160}{9} + 273 \\ &= \frac{5}{9}x - \frac{160}{9} + \frac{2457}{9} \\ &= \frac{5x + 2297}{9}. && \textbf{Simplifying} \end{aligned}$$

We can show on a grapher that the same values that appear in the table for y_2 will appear when y_2 is entered as

$$y_2 = \frac{5x + 2297}{9}.$$

In the more commonly used function notation, we have

$$\begin{aligned} k(c(t)) &= c(t) + 273 \\ &= \frac{5}{9}(t - 32) + 273 && \textbf{Substituting} \\ &= \frac{5t + 2297}{9}. && \textbf{Simplifying as above} \end{aligned}$$

Since the last equation expresses the Kelvin temperature as a new function, K, of the Fahrenheit temperature, t, we can write

$$K(t) = \frac{5t + 2297}{9},$$

where $K(t) =$ the Kelvin temperature corresponding to the Fahrenheit temperature, t. Here we have $K(t) = k(c(t))$. The new function K is called the **composition** of k and c and can be denoted $k \circ c$ (read "k composed with c," "the composition of k and c," or "k circle c").

Composition of Functions

The *composite function* $f \circ g$, the *composition* of f and g, is defined as

$$(f \circ g)(x) = f(g(x)),$$

where x is in the domain of g and $g(x)$ is in the domain of f.

Example 3 Given that $f(x) = 2x - 5$ and $g(x) = x^2 - 3x + 8$, find each of the following.

a) $(f \circ g)(7)$ and $(g \circ f)(7)$ **b)** $(f \circ g)(x)$ and $(g \circ f)(x)$

SOLUTION Consider each function separately:

$f(x) = 2x - 5$ **This function multiplies each input by 2 and subtracts 5.**

and

$g(x) = x^2 - 3x + 8.$ **This function squares an input, subtracts 3 times the input from the result, and then adds 8.**

a) To find $(f \circ g)(7)$, we first find $g(7)$. Then we use $g(7)$ as an input for f:

$$\begin{aligned}(f \circ g)(7) = f(g(7)) &= f(7^2 - 3 \cdot 7 + 8)\\ &= f(36) = 2 \cdot 36 - 5\\ &= 67.\end{aligned}$$

To find $(g \circ f)(7)$, we first find $f(7)$. Then we use $f(7)$ as an input for g:

$$\begin{aligned}(g \circ f)(7) &= g(f(7)) = g(2 \cdot 7 - 5)\\ &= g(9) = 9^2 - 3 \cdot 9 + 8\\ &= 62.\end{aligned}$$

b) To find $(f \circ g)(x)$, we substitute $g(x)$ for x in the equation for $f(x)$:

$$(f \circ g)(x) = f(g(x)) = f(x^2 - 3x + 8)$$

Substituting $x^2 - 3x + 8$ for $g(x)$

$$\begin{aligned}&= 2(x^2 - 3x + 8) - 5\\ &= 2x^2 - 6x + 16 - 5\\ &= 2x^2 - 6x + 11.\end{aligned}$$

To find $(g \circ f)(x)$, we substitute $f(x)$ for x in the equation for $g(x)$:

$(g \circ f)(x) = g(f(x)) = g(2x - 5)$ **Substituting $2x - 5$ for $f(x)$**

$$\begin{aligned}&= (2x - 5)^2 - 3(2x - 5) + 8\\ &= 4x^2 - 20x + 25 - 6x + 15 + 8\\ &= 4x^2 - 26x + 48.\end{aligned}$$

Note in Example 3 that, as a rule, $(f \circ g)(x) \neq (g \circ f)(x)$. We can check this graphically.

Example 4 Given that $f(x) = \sqrt{x}$ and $g(x) = x - 3$:

a) Find $h(x)$ and $k(x)$ if $h = f \circ g$ and $k = g \circ f$.

b) Graph h and k.

c) Find the domains of h and k.

SOLUTION

a) $h(x) = (f \circ g)(x) = f(g(x)) = f(x - 3) = \sqrt{x - 3}$

$k(x) = (g \circ f)(x) = g(f(x)) = g(\sqrt{x}) = \sqrt{x} - 3$

b) We actually used function composition when we worked with transformations in Section 1.7. For example, we know that the graph of $h(x) = \sqrt{x - 3}$ has the same shape as $y = \sqrt{x}$ shifted right 3 units. This occurs because g subtracts 3 units from each input before f takes the square root. When this sequence is reversed, as in the graph of $k(x) = \sqrt{x} - 3$, the subtraction of 3 occurs *after* the square root is taken. Thus the graph of k has the same shape as the graph of $y = \sqrt{x}$ shifted *down* 3 units.

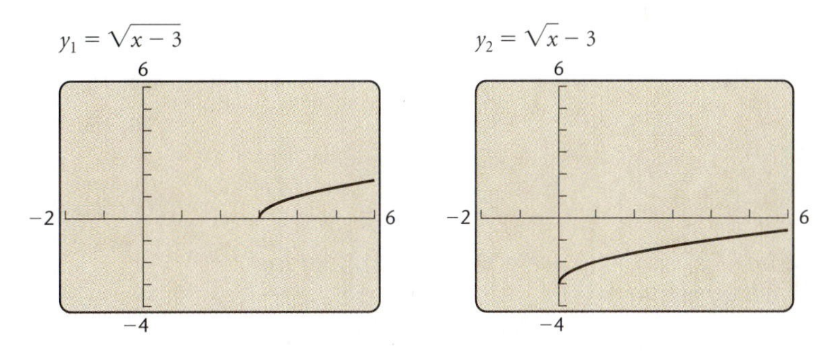

c) The domain of f is $\{x|x \geq 0\}$, or the interval $[0, \infty)$. The domain of g is $\mathbb{R}$.

To find the domain of $h = f \circ g$, we must consider that the outputs for g will serve as inputs for f. Since the inputs for f cannot be negative, we must have

$$g(x) = x - 3, \quad \text{where } x - 3 \geq 0,$$
$$\text{or} \quad g(x) = x - 3, \quad \text{where } x \geq 3.$$

Thus the domain of $h = \{x|x \geq 3\}$, or the interval $[3, \infty)$, as the graph confirms.

To find the domain of $k = g \circ f$, we must consider that the outputs for f will serve as inputs for g. Since g can accept *any* real number as an input, any output from f is acceptable. Thus the domain of $k = \{x|x \geq 0\}$, as the graph confirms.

Decomposing a Function as a Composition

In calculus, one needs to recognize how a function can be expressed as a composition. In this way, we are "decomposing" the function.

Example 5 Find $f(x)$ and $g(x)$ such that $h(x) = (f \circ g)(x)$: $h(x) = (2x - 3)^5$.

SOLUTION We have expressed $h(x)$ as $(2x - 3)$ to the 5th power. Two functions that can be used for the composition are

$$f(x) = x^5 \quad \text{and} \quad g(x) = 2x - 3.$$

We can check by forming the composition:

$$h(x) = (f \circ g)(x) = f(g(x)) = f(2x - 3) = (2x - 3)^5.$$

This is the most "obvious" answer to the question. There can be other less obvious answers. For example, if

$$f(x) = (x + 7)^5 \quad \text{and} \quad g(x) = 2x - 10,$$

then

$$h(x) = (f \circ g)(x) = f(g(x))$$
$$= f(2x - 10) = [(2x - 10) + 7]^5 = (2x - 3)^5.$$

1.8 Exercise Set

Given that $f(x) = x^2 - 3$ and $g(x) = 2x + 1$, find each of the following, if it exists.

1. $(f + g)(5)$
2. $(f - g)(3)$
3. $(f \circ g)(4)$
4. $(g \circ f)(4)$
5. $(fg)(2)$
6. $(fg)(-2)$
7. $(f/g)(-1)$
8. $(f/g)\left(-\frac{1}{2}\right)$
9. $(f/g)\left(-\frac{1}{2}\right)$
10. $(f/g)(-\sqrt{3})$
11. $(f/g)(\sqrt{3})$
12. $(f - g)(0)$

For each pair of functions in Exercises 13–16:

a) Find the domain of f, g, $f + g$, $f - g$, fg, ff, f/g, g/f, $f \circ g$, and $g \circ f$.

b) Find $(f + g)(x)$, $(f - g)(x)$, $(fg)(x)$, $(ff)(x)$, $(f/g)(x)$, $(g/f)(x)$, $(f \circ g)(x)$, and $(g \circ f)(x)$.

13. $f(x) = x - 3$, $g(x) = \sqrt{x + 4}$
14. $f(x) = x^2 - 1$, $g(x) = 2x + 5$
15. $f(x) = x^3$, $g(x) = 2x^2 + 5x - 3$
16. $f(x) = x^2$, $g(x) = \sqrt{x}$

For each pair of functions in Exercises 17–20, find $(f + g)(x)$, $(f - g)(x)$, $(fg)(x)$, $(f/g)(x)$, and $(f \circ g)(x)$.

17. $f(x) = x^2 - 4$, $g(x) = x^2 + 2$
18. $f(x) = x^2 + 2$, $g(x) = 4x - 7$
19. $f(x) = \sqrt{x - 7}$, $g(x) = x^2 - 25$
20. $f(x) = \sqrt{x - 1}$, $g(x) = \sqrt{3x - 8}$

For Exercises 21–26, consider the functions F and G as shown in the following graph.

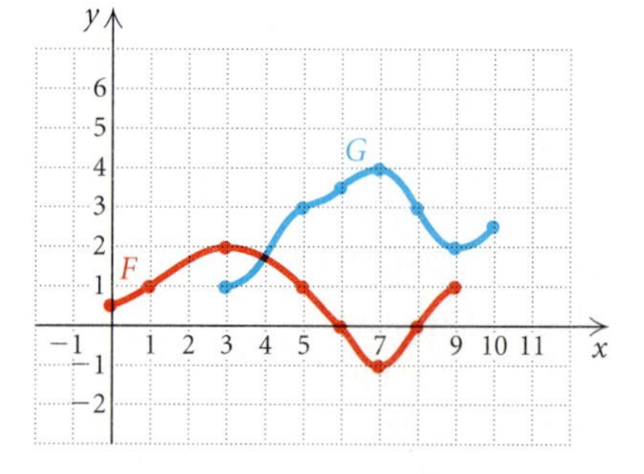

21. Find the domain of F, the domain of G, and the domain of $F + G$.
22. Find the domain of $F - G$, FG, and F/G.
23. Find the domain of G/F.
24. Graph $F + G$.
25. Graph $G - F$.
26. Graph $F - G$.

Find $(f \circ g)(x)$ and $(g \circ f)(x)$.

27. $f(x) = x + 3$, $g(x) = x - 3$
28. $f(x) = \frac{4}{5}x$, $g(x) = \frac{5}{4}x$

29. $f(x) = 3x - 7, \ g(x) = \dfrac{x + 7}{3}$

30. $f(x) = \frac{2}{3}x - \frac{4}{5}, \ g(x) = 1.5x + 1.2$

31. $f(x) = 20, \ g(x) = 0.05$

32. $f(x) = x^4, \ g(x) = \sqrt[4]{x}$

33. $f(x) = \sqrt{x + 5}, \ g(x) = x^2 - 5$

34. $f(x) = x^5 - 2, \ g(x) = \sqrt[5]{x + 2}$

35. $f(x) = \dfrac{1 - x}{x}, \ g(x) = \dfrac{1}{1 + x}$

36. $f(x) = \dfrac{x^2 - 1}{x^2 + 1}, \ g(x) = \dfrac{3x - 4}{5x - 2}$

37. $f(x) = x^3 - 5x^2 + 3x + 7, \ g(x) = x + 1$

38. $f(x) = x - 1, \ g(x) = x^3 + 2x^2 - 3x - 9$

Find $f(x)$ and $g(x)$ such that $h(x) = (f \circ g)(x)$. Answers may vary, but try to select the most obvious answer.

39. $h(x) = (4 + 3x)^5$

40. $h(x) = \sqrt[3]{x^2 - 8}$

41. $h(x) = \dfrac{1}{(x - 2)^4}$

42. $h(x) = \dfrac{1}{\sqrt{3x + 7}}$

43. $h(x) = \dfrac{x^3 - 1}{x^3 + 1}$

44. $h(x) = |9x^2 - 4|$

45. $h(x) = \left(\dfrac{2 + x^3}{2 - x^3}\right)^6$

46. $h(x) = (\sqrt{x} - 3)^4$

47. $h(x) = \sqrt{\dfrac{x - 5}{x + 2}}$

48. $h(x) = \sqrt{1 + \sqrt{1 + x}}$

49. $h(x) = (x + 2)^3 - 5(x + 2)^2 + 3(x + 2) - 1$

50. $h(x) = 2(x - 1)^{5/3} + 5(x - 1)^{2/3}$

51. *Total Cost, Revenue, and Profit.* In economics, functions that involve revenue, cost, and profit are used. For example, suppose that $R(x)$ and $C(x)$ denote the total revenue and the total cost, respectively, of producing a new kind of tool for King Hardware Stores. Then the difference

$$P(x) = R(x) - C(x)$$

represents the total profit for producing x tools. Given

$$R(x) = 60x - 0.4x^2 \quad \text{and} \quad C(x) = 3x + 13,$$

find each of the following.

a) $P(x)$

b) $R(100)$, $C(100)$, and $P(100)$

c) Use a grapher and graph the three functions using the viewing window $[0, 160, 0, 3000]$.

52. *Dress Sizes.* A dress that is size x in France is size $s(x)$ in the United States, where $s(x) = x - 32$. A dress that is size x in the United States is size $y(x)$ in Italy, where $y(x) = 2(x + 12)$. Find a function that will convert French dress sizes to Italian dress sizes.

53. *Ripple Spread.* A stone is thrown into a pond. A circular ripple is spreading over the pond in such a way that the radius is increasing at the rate of 3 ft/sec.

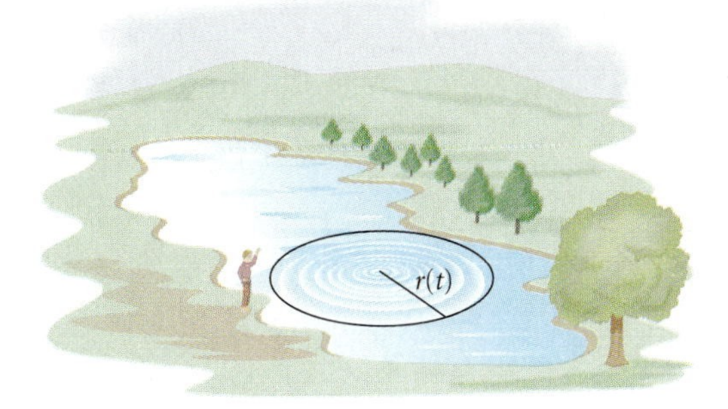

a) Find a function $r(t)$ for the radius in terms of t.

b) Find a function $A(r)$ for the area of the ripple in terms of the radius r.

c) Find $(A \circ r)(t)$. Explain the meaning of this function.

54. *Airplane Distance.* An airplane is 300 ft from the control tower at the end of the runway. It takes off at a speed of 250 mph.

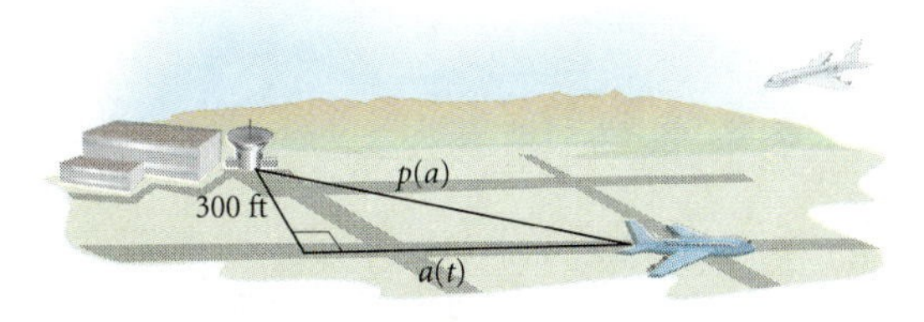

a) Let $a(t)$ = the distance that the airplane travels down the runway in time t. Find a formula for the function.

b) Let P = the distance of the airplane from the control tower. Find a formula for P in terms of the distance a—that is, find an expression for $P(a)$.

c) Find $(P \circ a)(t)$. Explain the meaning of this function.

Skill Maintenance

Refer to the graph accompanying Exercises 21–26.

55. Find the range of F.

56. Find the range of G.

57. Evaluate $5x^3 - 7x^2 + 2x - 7$ for $x = -1$.

58. Factor: $9x^5 - 64x^4 + 72x^3$.

Synthesis

59. ◈ Does $(f + g)(x) = (f \circ g)(x)$ for any functions f and g? Why or why not?

60. ◈ Explain how the function f given by $f(x) = (x - 5)^2 + 3$ can be regarded as a composition of two functions, each of which is a translation.

In Exercises 61–64, use a grapher to graph f, g, and f + g using the same viewing window.

61. $f(x) = x^2,\; g(x) = 3 - 2x$

62. $f(x) = \dfrac{2}{x},\; g(x) = x$

63. $f(x) = 5 - x^2,\; g(x) = x^2 - 5$

64. $f(x) = |x|,\; g(x) = \text{INT}(x)$

Refer to the graph accompanying Exercises 21–26.

65. Find the domain of $F \circ G$, $G \circ G$, and $F \circ F$.

66. Find the range of $F \circ G$.

67. Consider $f(x) = 3x + b$ and $g(x) = 2x - 1$. Find b such that $(f \circ g)(x) = (g \circ f)(x)$ for all real numbers x.

68. Consider $f(x) = 3x - 4$ and $g(x) = mx + b$. Find m and b such that

$$(f \circ g)(x) = (g \circ f)(x) = x$$

for all real numbers x.

69. For $f(x) = 1/(1 - x)$, find $(f \circ f)(x)$ and $(f \circ (f \circ f))(x)$.

State whether each of the following is true or false.

70. The composition of two even functions is even.

71. The sum of two even functions is even.

72. The product of two odd functions is odd.

73. The composition of two odd functions is odd.

74. The product of an even function and an odd function is odd.

75. Show that if f is *any* function, then the function E defined by

$$E(x) = \frac{f(x) + f(-x)}{2}$$

is even.

76. Show that if f is *any* function, then the function O defined by

$$O(x) = \frac{f(x) - f(-x)}{2}$$

is odd.

77. Consider the functions E and O of Exercises 75 and 76.

a) Show that $f(x) = E(x) + O(x)$. This means that every function can be expressed as the sum of an even and an odd function.

b) Let $f(x) = 4x^3 - 11x^2 + \sqrt{x} - 10$. Express f as a sum of an even function and an odd function.

CHAPTER 1 Summary and Review

Important Properties and Formulas

Terminology about Lines

Slope: $m = \dfrac{y_2 - y_1}{x_2 - x_1}$

The Slope–intercept Equation:
$y = mx + b$

The Point–slope Equation:
$y - y_1 = m(x - x_1)$

The Two-point Equation:

$$y - y_1 = \frac{y_2 - y_1}{x_2 - x_1}(x - x_1)$$

Horizontal Lines: $y = b$

Vertical Lines: $x = a$

Parallel Lines: $m_1 = m_2,\ b_1 \neq b_2$

Perpendicular Lines:
$m_1 m_2 = -1$, or
$x = a,\ y = b$

The Distance Formula

$$d = \sqrt{(x_2 - x_1)^2 + (y_2 - y_1)^2}$$

The Midpoint Formula

$$\left(\frac{x_1 + x_2}{2}, \frac{y_1 + y_2}{2}\right)$$

Equation of a Circle

$$(x - h)^2 + (y - k)^2 = r^2$$

Tests for Symmetry

x-axis: If replacing y by $-y$ produces an equivalent equation, then the graph is symmetric with respect to the x-axis.

y-axis: If replacing x by $-x$ produces an equivalent equation, then the graph is symmetric with respect to the y-axis.

Origin: If replacing x by $-x$ and y by $-y$ produces an equivalent equation, then the graph is symmetric with respect to the origin.

Even Function: $f(-x) = f(x)$

Odd Function: $f(-x) = -f(x)$

Transformations

Vertical Translation: $y = f(x) \pm b$

Horizontal Translation: $y = f(x \pm d)$

Reflection across the x-axis: $y = -f(x)$

Reflection across the y-axis: $y = f(-x)$

Vertical Stretching or Shrinking:
$y = af(x)$

Horizontal Stretching or Shrinking:
$y = f(ax)$

The Algebra of Functions

The Sum of Two Functions:
$(f + g)(x) = f(x) + g(x)$

The Difference of Two Functions:
$(f - g)(x) = f(x) - g(x)$

The Product of Two Functions:
$(fg)(x) = f(x) \cdot g(x)$

The Quotient of Two Functions:
$(f/g)(x) = f(x)/g(x),\ g(x) \neq 0$

The Composition of Two Functions:
$(f \circ g)(x) = f(g(x))$

REVIEW EXERCISES

Determine whether each relation is a function. Find the domain and the range.

1. $\{(3, 1), (5, 3), (7, 7), (3, 5)\}$

2. $\{(2, 7), (-2, -7), (7, -2), (0, 2), (1, -4)\}$

Determine whether each graph is that of a function

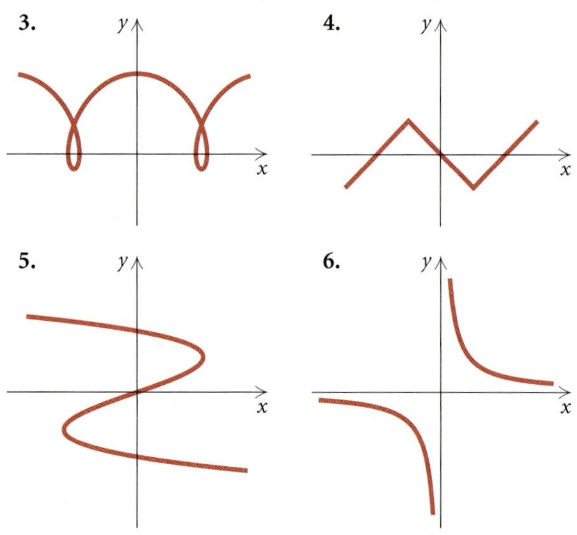

Find the domain of each function. Do not use a grapher.

7. $f(x) = 4 - 5x + x^2$

8. $f(x) = \sqrt{7 - 3x}$

9. $f(x) = \dfrac{1}{x^2 - 6x + 5}$

10. $f(x) = \dfrac{-5x}{|16 - x^2|}$

Use a grapher to graph each function. Then visually estimate the domain and the range.

11. $f(x) = \sqrt{16 - (x + 1)^2}$

12. $f(x) = |x - 3| + |x + 4| - 12$

13. $f(x) = x^3 - 5x^2 + 2x - 7$

14. $f(x) = x^4 - 8x^2 - 3$

15. Given that $f(x) = x^2 - x - 3$, find:

a) $f(0)$; **b)** $f(-3)$;

c) $f(a - 1)$; **d)** $\dfrac{f(x + h) - f(x)}{h}$.

Use a grapher to determine graphically whether each of the following seems to be an identity.

16. $\sqrt{x + 9} = \sqrt{x} + 3$

17. $x^2 \cdot x^3 = x^5$

18. $x^2 + x^3 = x^5$

19. $x + 3 = 3 + x$

Make a hand-drawn graph of each of the following. Check your results on a grapher.

20. $f(x) = \begin{cases} x^3, & \text{for } x < -2, \\ |x|, & \text{for } -2 \le x \le 2, \\ \sqrt{x - 1}, & \text{for } x > 2 \end{cases}$

21. $f(x) = \begin{cases} \dfrac{x^2 - 1}{x + 1}, & \text{for } x \ne -1, \\ 3, & \text{for } x = -1 \end{cases}$

22. $f(x) = \text{INT}(x)$

23. $f(x) = \text{INT}(x - 3)$

24. Use the TABLE feature to complete the following input–output table for the equations

$$y_1 = \sqrt{16 - (x + 1)^2} \quad \text{and} \quad y_2 = \frac{10x}{x^2 + 1}.$$

X	Y1	Y2
2.7		
2.8		
2.9		
3.0		
3.1		
3.2		
3.3		
X = 2.7		

Then use the TABLE feature to consider inputs from −5.2 to −4.6.

a) What does the table tell you about the domain of each function? the range?

b) What seems to be the largest output of each function? the smallest?

Use a grapher. For each function in Exercises 25–27, estimate:

a) *the roots, or zeros;*

b) *where the function is increasing or decreasing;*

c) *any relative maxima or minima.*

25. $f(x) = 1.1x^3 - 9.04x^2 - 18x + 702$

26. $f(x) = \dfrac{20}{(x + 4)^2 + 1} - 5$

27. $f(x) = 0.8x\sqrt{16 - (x + 1)^2}$

28. *Inscribed Rectangle.* A rectangle is inscribed in a semicircle of radius 2, as shown. The variable x = half the length of the rectangle. Express the area of the rectangle as a function of x.

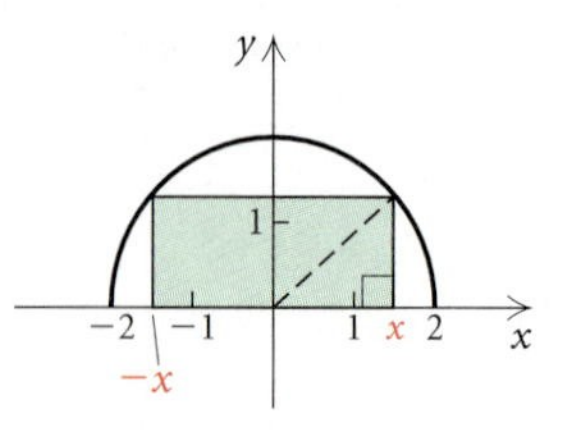

29. *Minimizing Surface Area.* A container firm is designing an open-top rectangular box, with a square base, that will hold 108 in^3. Let $x =$ the length of a side of the base.

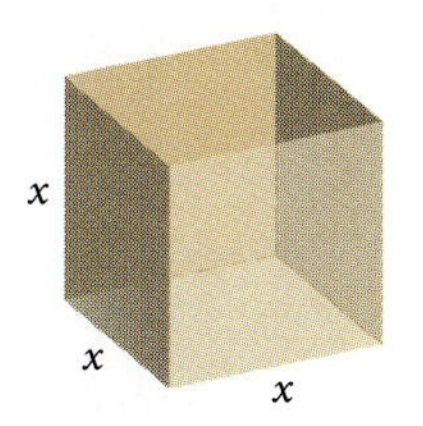

a) Express the surface area as a function of x.
b) Find the domain of the function.
c) Graph the function.
d) What dimensions minimize the surface area of the box?

Each table of data contains input–output values for a function. Answer the following questions:

a) *Is the change in the inputs, x, the same?*
b) *Is the change in the outputs, y, the same?*
c) *Is the function linear?*

30.

x	y
−3	8
−2	11
−1	14
0	17
1	20
2	22
3	26

31.

x	y
20	11.8
30	24.2
40	36.6
50	49.0
60	61.4
70	73.8
80	86.2

32. Find the slope and the y-intercept of the graph of $-2x - y = 7$.

Write a point–slope equation for a line with the following characteristics.

33. $m = -\frac{2}{3}$, y-intercept $(0, -4)$

34. $m = 3$, passes through $(-2, -1)$

35. Passes through $(4, 1)$ and $(-2, -1)$

Determine whether the lines are parallel, perpendicular, or neither.

36. $3x - 2y = 8$,
$6x - 4y = 2$

37. $y - 2x = 4$,
$2y - 3x = -7$

38. $y = \frac{3}{2}x + 7$,
$y = -\frac{2}{3}x - 4$

Given the point $(1, -1)$ *and the line* $2x + 3y = 4$:

39. Find an equation of the line containing the given point and parallel to the given line.

40. Find an equation of the line containing the given point and perpendicular to the given line.

41. *Total Cost.* Clear County Cable Television charges a $25 installation fee and $20 per month for "basic" service. Write and graph an equation that can be used to determine the total cost, $C(t)$, of t months of basic cable television service. Find the total cost of 6 months of service.

42. *Temperature and Depth of the Earth.* The function T given by

$$T(d) = 10d + 20$$

can be used to determine the temperature T, in degrees Celsius, at a depth d, in kilometers, inside the earth.

a) Find $T(5)$, $T(20)$, and $T(1000)$.
b) Graph T.
c) The radius of the earth is about 5600 km. Use this fact to determine the domain of the function.

For Exercise 43, use a grapher with a regression feature.

43. *Total Revenue of Pepsico.* Pepsico is a company that owns many fast-food chains, such as Taco Bell and Pizza Hut. It sells other food items through subsidiaries such as Frito Lay and Pepsi-Cola. Pepsico has experienced great sales growth during the past few years, as shown in the following table.

YEAR, x	TOTAL REVENUE, y (IN BILLIONS)
0. 1990	$17.8
1. 1991	19.6
2. 1992	22.0
3. 1993	25.0
4. 1994	??

Source: Pepsico Annual Report

a) Fit a regression line to the data and use it to predict total revenue in 1998, 2000, and 2010.
b) What is the correlation coefficient for the regression line? How close a fit is the regression line?

44. Find the distance between $(3, 7)$ and $(-2, 4)$.

45. Find the midpoint of the segment with endpoints $(3, 7)$ and $(-2, 4)$.

46. Find an equation of the circle with center $(-2, 6)$ and radius $\sqrt{13}$.

47. Find the center and radius of the circle

$$(x + 1)^2 + (y - 3)^2 = 16.$$

48. Find an equation of the circle having a diameter with endpoints $(-3, 5)$ and $(7, 3)$.

First, graph each equation and determine visually whether it is symmetric with respect to the x-axis, the y-axis, and the origin. Then verify your assertion algebraically.

49. $x^2 + y^2 = 4$ **50.** $y^2 = x^2 + 3$

51. $x + y = 3$ **52.** $y = x^2$

53. $y = x^3$ **54.** $y = x^4 - x^2$

Determine visually whether each function is even, odd, or neither even nor odd.

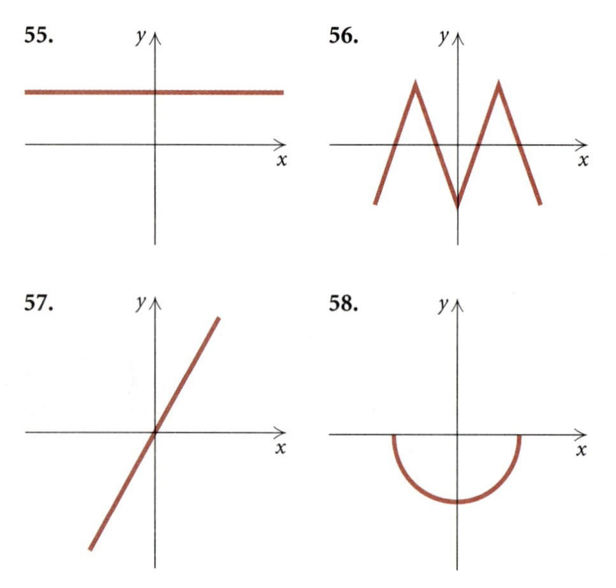

Test algebraically whether each function is even, odd, or neither even nor odd. Then check your work graphically, where possible, using a grapher.

59. $f(x) = 9 - x^2$ **60.** $f(x) = x^3 - 2x + 4$

61. $f(x) = x^7 - x^5$ **62.** $f(x) = |x|$

63. $f(x) = \sqrt{16 - x^2}$ **64.** $f(x) = \dfrac{10x}{x^2 + 1}$

Write an equation for a function that has a graph with the given characteristics. Check your answer on a grapher.

65. The shape of $y = x^2$, but shifted left 3 units

66. The shape of $y = \sqrt{x}$, but upside down and shifted right 3 units and up 4 units

67. The shape of $y = |x|$, but stretched vertically by a factor of 2 and shifted right 3 units

A graph of $y = f(x)$ is shown. No formula for f is given. Make a hand-drawn graph of each of the following.

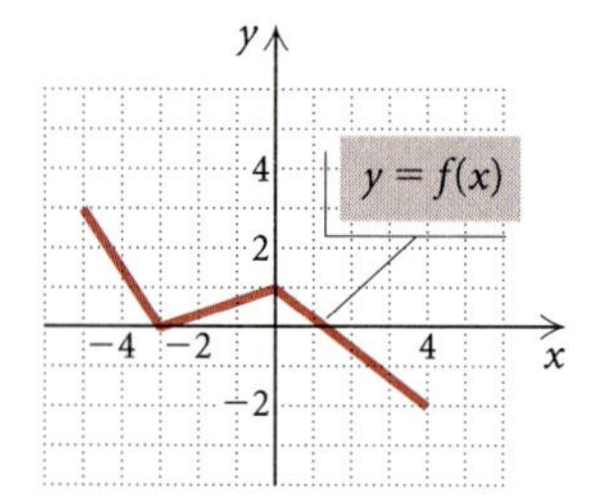

68. $y = f(x - 1)$ **69.** $y = f(2x)$

70. $y = -2f(x)$ **71.** $y = 3 + f(x)$

For each pair of functions:

a) *Find the domain of f, g, $f + g$, $f - g$, fg, f/g, $f \circ g$, and $g \circ f$.*

b) *Find $(f + g)(x)$, $(f - g)(x)$, $fg(x)$, $(f/g)(x)$, $(f \circ g)(x)$, and $(g \circ f)(x)$.*

72. $f(x) = \dfrac{4}{x^2}$; $g(x) = 3 - 2x$

73. $f(x) = 3x^2 + 4x$; $g(x) = 2x - 1$

74. Given the total-revenue and total-cost functions

$R(x) = 120x - 0.5x^2$ and $C(x) = 15x + 6$,

find total profit $P(x)$.

Find $f(x)$ and $g(x)$ such that $h(x) = (f \circ g)x$:

75. $h(x) = \sqrt{5x + 2}$ **76.** $h(x) = 4(5x - 1)^2 + 9$

Synthesis

77. ◆ Given that $f(x) = 4x^3 - 2x + 7$, find

a) $f(x) + 2$; **b)** $f(x + 2)$; **c)** $f(x) + f(2)$.

Then discuss how each expression differs from the other.

78. ◆ Given the graph of $y = f(x)$, explain and contrast the effect of the constant c on the graphs of $y = f(cx)$ and $y = cf(x)$.

79. ◆ **a)** Graph several functions of the type $y_1 = f(x)$ and $y_2 = |f(x)|$ on a grapher. Describe a procedure, involving transformations, for creating hand-drawn graphs of y_2 from y_1.

b) Describe a procedure, involving transformations, for creating hand-drawn graphs of $y_2 = f(|x|)$ from $y_1 = f(x)$.

Find the domain.

80. $f(x) = \dfrac{\sqrt{1 - x}}{x - |x|}$ **81.** $f(x) = (x - 9x^{-1})^{-1}$

82. Prove that the sum of two odd functions is odd.

83. Describe how the graph of $y = -f(-x)$ is obtained from the graph of $y = f(x)$.

Polynomial and Rational Functions 2

APPLICATION

Ibuprofen is a medication used to relieve pain. The fourth-degree, or quartic, polynomial function
$M(t) = 0.5t^4 + 3.45t^3 - 96.65t^2 + 347.7t,$
$0 \le t \le 6$
can be used to estimate the number of milligrams of ibuprofen in the bloodstream t hours after 400 mg of the medication has been swallowed.

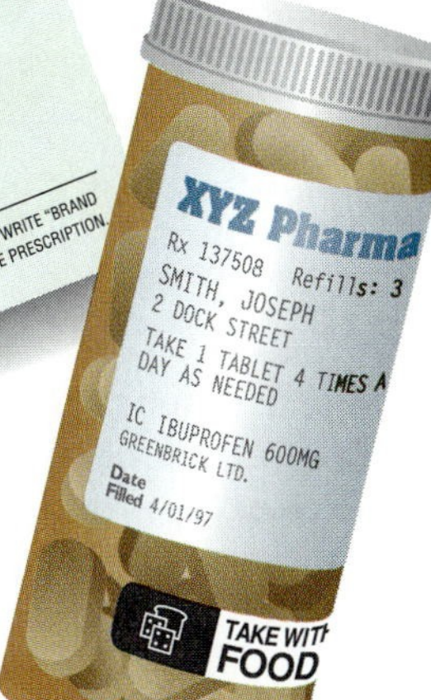

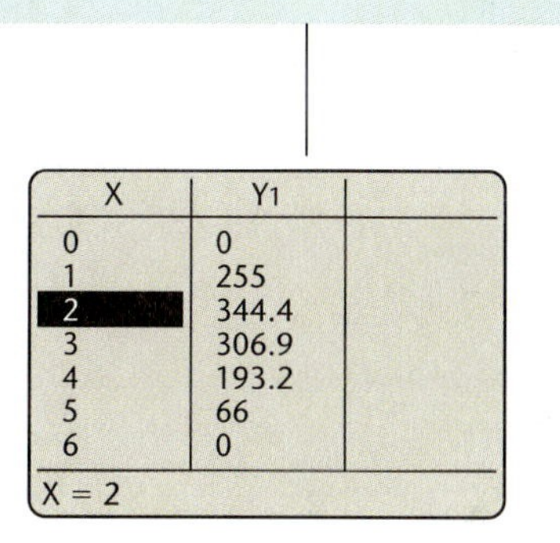

X	Y1
0	0
1	255
2	344.4
3	306.9
4	193.2
5	66
6	0

X = 2

$y = 0.5x^4 + 3.45x^3 - 96.65x^2 + 347.7x$

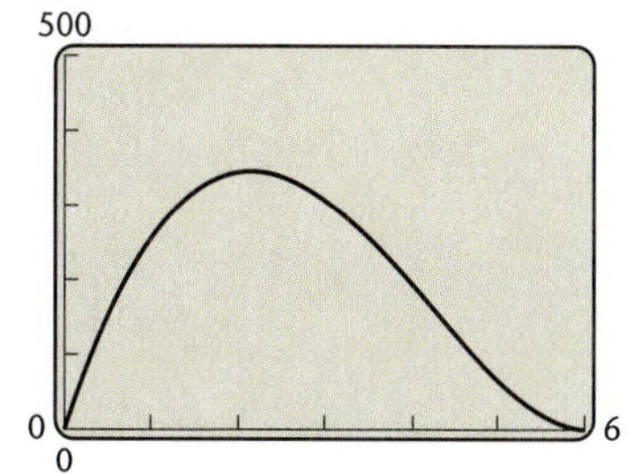

A *polynomial function* is a function that can be defined by a polynomial expression. A *rational function* is a function that can be defined as a quotient of two polynomials. We will study both kinds of functions and examine zeros of polynomials in greater depth. We will also model real data with quadratic, cubic, and quartic polynomial functions and use these functions to make predictions. In addition, we will study the graphs of rational functions and use the graphs of both polynomial and rational functions to solve related inequalities.

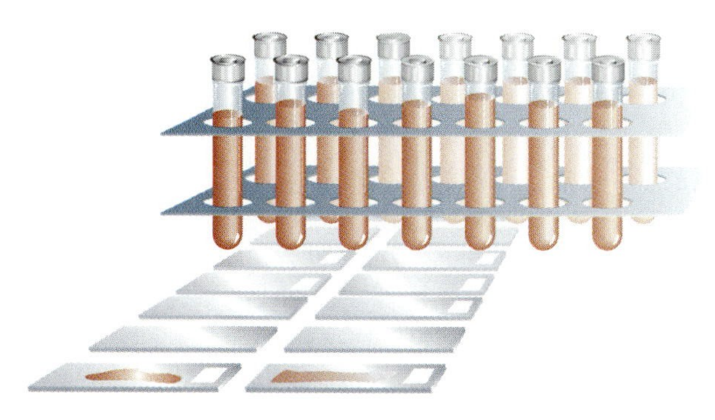

2.1 Introduction to Polynomial Functions and Complex Numbers
2.2 Quadratic Equations and Functions
2.3 Polynomial Models and Curve Fitting
2.4 Polynomial Division; The Remainder and Factor Theorems
2.5 Theorems about Zeros of Polynomial Functions
2.6 Rational Functions
2.7 Polynomial and Rational Inequalities
SUMMARY AND REVIEW

2.1 Introduction to Polynomial Functions and Complex Numbers

- *Use a grapher to graph a polynomial function and find its real-number zeros, relative maximum and minimum values, and domain and range.*
- *Perform computations involving complex numbers.*

Because of the grapher's capabilities, we have already studied many aspects of polynomial functions in Chapter 1. In this section, we begin a more rigorous study of how to find the zeros of these functions.

Polynomial Function

A *polynomial function P* is given by

$$P(x) = a_nx^n + a_{n-1}x^{n-1} + a_{n-2}x^{n-2} + \cdots + a_1x + a_0,$$

where the coefficients $a_n, a_{n-1}, \ldots, a_1, a_0$ are real numbers and the exponents are whole numbers.

In descending order, the first nonzero coefficient is assumed to be a_n and is called the **leading coefficient**. The **degree** of the polynomial function is n. Some examples of types of polynomial functions and their names are as follows.

POLYNOMIAL FUNCTION	DEGREE	EXAMPLE
Constant	0	$f(x) = 3$
Linear	1	$f(x) = \frac{2}{3}x + 1$
Quadratic	2	$f(x) = 2x^2 - x + 3$
Cubic	3	$f(x) = x^3 + 2x^2 + x - 5$
Quartic	4	$f(x) = -x^4 - 1.1x^3 + 0.3x^2 - 2.8x - 1.7$

The constant function $f(x) = 0$ is said to have no degree because there is no nonzero coefficient. This function can be described in many ways: $f(x) = 0 = 0x^2 = 0x^3 = 0x^{53}$, and so on.

From our study of functions in Chapter 1, we know how to use a grapher to at least estimate many characteristics of a polynomial function. Let's consider an example for review.

Function: $f(x) = x^4 - 6x^3 - 4x^2 + 53.2x - 42.6$

Zeros: $-2.975, 0.950, 3, 5.025$

Relative maximum: 15.921 at $x = 1.914$

Relative minima: -106.907 at $x = -1.643$, -23.101 at $x = 4.229$

Domain: All real numbers, $\mathbb{R}$

Range: $[-106.907, \infty)$

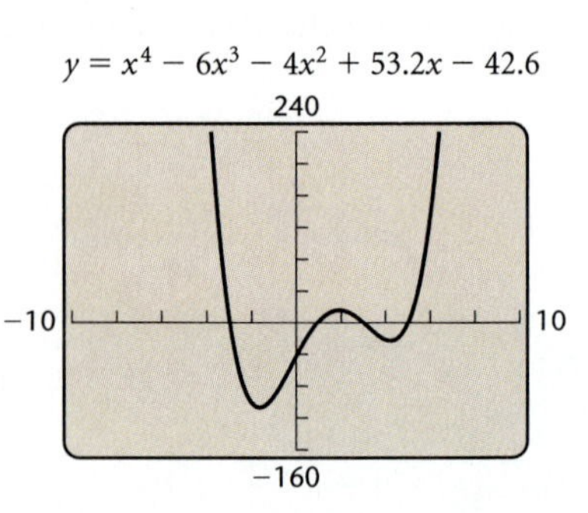

With the techniques presented in this chapter, we will be able to predict certain characteristics of the graph of a polynomial by examining its formula. This will help us to find exact, rather than approximate, values in certain cases and to have greater insight into the creation of graphs. Before proceeding, we need to expand the real-number system to one called the **complex-number system** in which certain polynomial functions have zeros that may not be real numbers.

The Complex Numbers

Let's begin by considering the polynomial function $f(x) = x^2 - 5$.

To find the zeros of the function, we solve the equation $x^2 - 5 = 0$.

Example 1 Solve: $x^2 - 5 = 0$.

ALGEBRAIC SOLUTION

We use the principle of square roots:

$$x^2 - 5 = 0$$
$$x^2 = 5 \qquad \text{Adding 5}$$
$$x = \pm\sqrt{5}. \qquad \text{Using the principle of square roots}$$

The solutions of the equation are $-\sqrt{5}$ and $\sqrt{5}$. The solution set is $\{-\sqrt{5}, \sqrt{5}\}$. Note that $\pm\sqrt{5} \approx \pm 2.236$.

GRAPHICAL SOLUTION

To solve $x^2 - 5 = 0$, we first graph the function $f(x) = x^2 - 5$ and estimate its zeros.

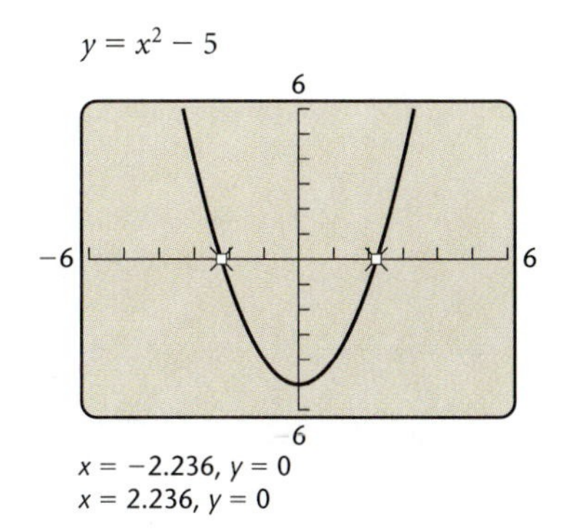

We see that there are zeros near -2 and 2. Using TRACE and ZOOM to approximate the zeros, we get the approximations -2.236 and 2.236. The zeros of the function are the solutions of the equation $x^2 - 5 = 0$. The solution set is $\{-2.236, 2.236\}$.

A faster method might be to use SOLVE. On some graphers, a POLY feature allows you to enter a polynomial and gives all the zeros directly.

Note that the real solutions found with the grapher were approximations. Now let's consider an equation that has no real-number solutions.

Example 2 Solve: $x^2 + 1 = 0$.

ALGEBRAIC SOLUTION

$$x^2 + 1 = 0$$

$$x^2 = -1 \quad \text{Subtracting 1}$$

$$x = \pm\sqrt{-1} \quad \text{Using the principle of square roots}$$

There are no real-number square roots of negative numbers. Thus the equation $x^2 + 1 = 0$ has no real-number solutions.

GRAPHICAL SOLUTION

We first graph the function $f(x) = x^2 + 1$.

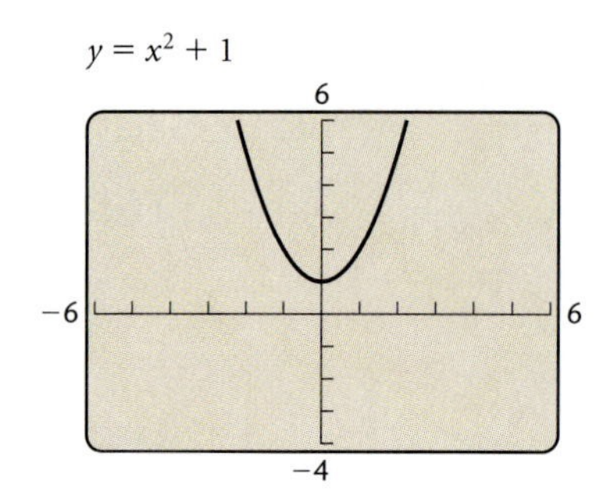

We see that since the graph does not cross the x-axis, it has no x-intercepts. This means that the function $f(x) = x^2 + 1$ has no real-number zeros. Thus there are no real-number solutions of the equation $x^2 + 1 = 0$.

A new kind of number, called *imaginary*, was defined (invented) so that negative numbers would have square roots and equations like the one in Example 2 would have solutions. These new numbers make use of a number named i, defined so that $i = \sqrt{-1}$ and $i^2 = -1$.

The Number *i*

The number i is defined such that

$$i = \sqrt{-1} \quad \text{and} \quad i^2 = -1.$$

To express roots of negative numbers in terms of i, we can use the fact that $\sqrt{-p} = \sqrt{-1}\sqrt{p}$ when p is a positive real number.

Example 3 Express each number in terms of i.

a) $\sqrt{-7}$

b) $\sqrt{-16}$

c) $-\sqrt{-13}$

d) $-\sqrt{-64}$

e) $\sqrt{-48}$

SOLUTION

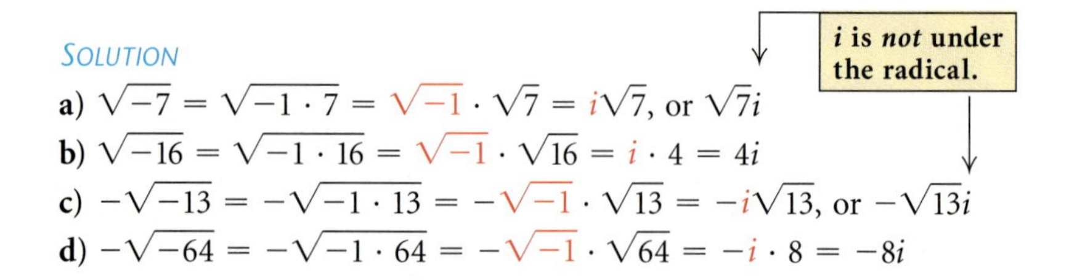

a) $\sqrt{-7} = \sqrt{-1 \cdot 7} = \sqrt{-1} \cdot \sqrt{7} = i\sqrt{7}$, or $\sqrt{7}i$

b) $\sqrt{-16} = \sqrt{-1 \cdot 16} = \sqrt{-1} \cdot \sqrt{16} = i \cdot 4 = 4i$

c) $-\sqrt{-13} = -\sqrt{-1 \cdot 13} = -\sqrt{-1} \cdot \sqrt{13} = -i\sqrt{13}$, or $-\sqrt{13}i$

d) $-\sqrt{-64} = -\sqrt{-1 \cdot 64} = -\sqrt{-1} \cdot \sqrt{64} = -i \cdot 8 = -8i$

e) $\sqrt{-48} = \sqrt{-1 \cdot 48} = \sqrt{-1} \cdot \sqrt{48}$
$= i\sqrt{48} = i \cdot 4\sqrt{3} = 4\sqrt{3}i, \quad \text{or} \quad 4i\sqrt{3}$

Imaginary Number

An *imaginary number* is a number that can be written $a + bi$, where a and b are real numbers and $b \neq 0$.

Don't let the word "imaginary" mislead you. Imaginary numbers have very "real" uses in engineering and science. The following are examples of imaginary numbers:

$$\left.\begin{array}{r} -6 + 2i \\ \frac{3}{5} - \frac{1}{4}i \\ \pi - 3.7i \\ 3 + i\sqrt{2} \end{array}\right\} \quad \text{(here } a \neq 0,\ b \neq 0\text{);}$$

$$17i \qquad \text{(here } a = 0,\ b \neq 0\text{).}$$

The number i and the numbers in Example 3 are also examples of *imaginary numbers* with $a = 0$.

When a and b are real numbers and b could be 0, the number $a + bi$ is said to be **complex**. The real number a is said to be the **real part** of $a + bi$, and the real number b is said to be the **imaginary part**.

Complex Number

A *complex number* is any number that can be written $a + bi$, where a and b are any real numbers. (Note that a and b both can be 0.)

The following are examples of complex numbers:

$7 + 3i$ (here $a \neq 0$, $b \neq 0$);
8 (here $a \neq 0$, $b = 0$);
$4i$ (here $a = 0$, $b \neq 0$);
0 (here $a = 0$, $b = 0$).

Note that when $b = 0$, $a + 0i = a$, so every real number is a complex number. Complex numbers like $17i$ or $4i$, in which $a = 0$ and $b \neq 0$, are imaginary numbers with no real part. Such numbers are called **pure imaginary numbers**. The relationships among various types of complex numbers are shown below.

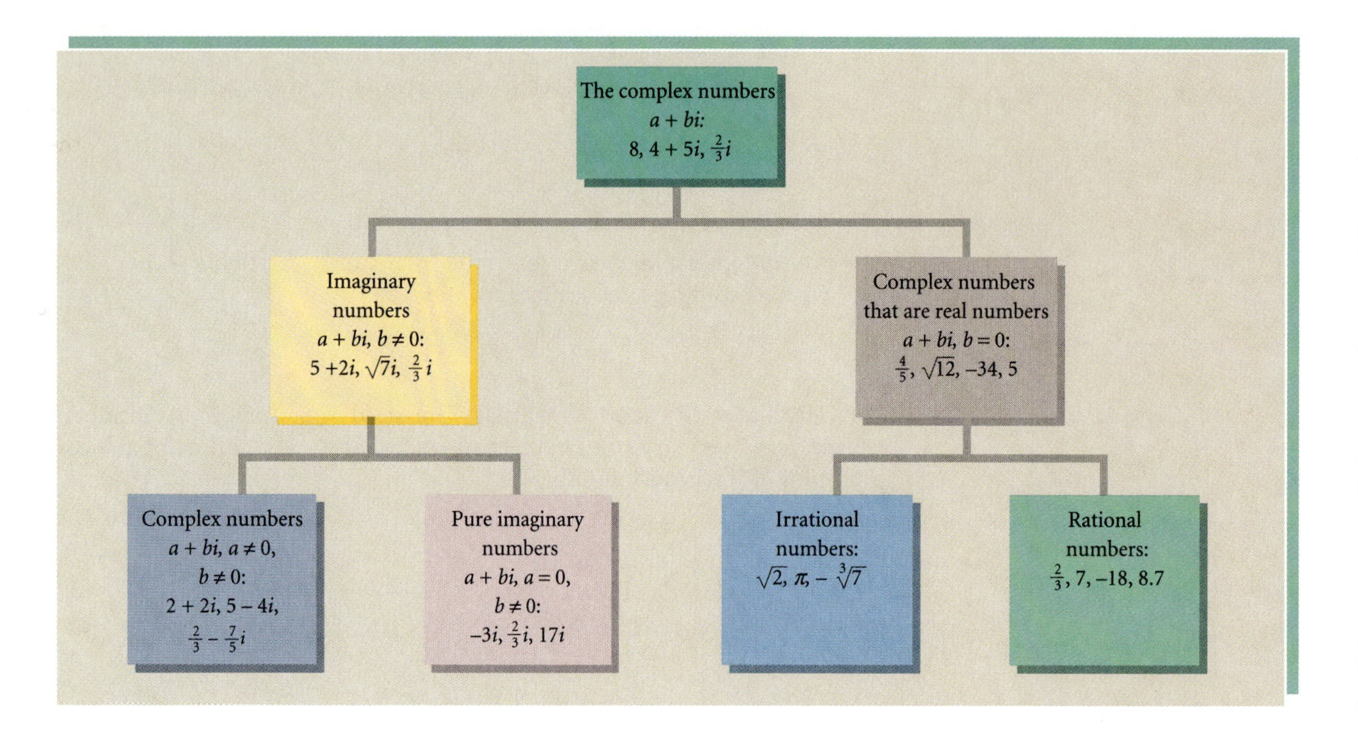

Addition and Subtraction

The complex numbers obey the commutative, associative, and distributive laws. Thus we can add and subtract them as we do binomials.

Example 4 Add or subtract and simplify each of the following.

a) $(8 + 6i) + (3 + 2i)$ **b)** $(4 + 5i) - (6 - 3i)$

SOLUTION

a) $(8 + 6i) + (3 + 2i) = (8 + 3) + (6i + 2i)$

Collecting the real parts and the imaginary parts

$= 11 + (6 + 2)i = 11 + 8i$

b) $(4 + 5i) - (6 - 3i) = (4 - 6) + [5i - (-3i)]$

Note that the 6 and the $-3i$ are both being subtracted.

$= -2 + 8i$

Multiplication

For complex numbers, the property $\sqrt{a}\sqrt{b} = \sqrt{ab}$ does *not* hold in general, but it does hold when $a = -1$ and b is a nonnegative number. To multiply square roots of negative real numbers, we first express them in terms of i. For example,

$$\sqrt{-2} \cdot \sqrt{-5} = \sqrt{-1} \cdot \sqrt{2} \cdot \sqrt{-1} \cdot \sqrt{5} = i\sqrt{2} \cdot i\sqrt{5}$$
$$= i^2\sqrt{10} = -1\sqrt{10} = -\sqrt{10} \quad \text{is correct!}$$

But

$$\sqrt{-2} \cdot \sqrt{-5} = \sqrt{(-2)(-5)} = \sqrt{10} \quad \text{is wrong!}$$

Keeping this and the fact that $i^2 = -1$ in mind, we multiply in much the same way that we do with real numbers.

Example 5 Multiply and simplify each of the following.

a) $\sqrt{-16} \cdot \sqrt{-25}$ **b)** $\sqrt{-5} \cdot \sqrt{-7}$ **c)** $-4i(3 - 5i)$

d) $(1 + 2i)(1 + 3i)$ **e)** $(3 - 7i)^2$

SOLUTION

a) $\sqrt{-16} \cdot \sqrt{-25} = \sqrt{-1} \cdot \sqrt{16} \cdot \sqrt{-1} \cdot \sqrt{25}$
$= i \cdot 4 \cdot i \cdot 5$
$= i^2 \cdot 20$
$= -1 \cdot 20$ $\quad i^2 = -1$
$= -20$

b) $\sqrt{-5} \cdot \sqrt{-7} = \sqrt{-1} \cdot \sqrt{5} \cdot \sqrt{-1} \cdot \sqrt{7}$
$= i \cdot \sqrt{5} \cdot i \cdot \sqrt{7}$
$= i^2 \cdot \sqrt{35}$
$= -1 \cdot \sqrt{35}$ $\quad i^2 = -1$
$= -\sqrt{35}$

c) $-4i(3 - 5i) = -4i \cdot 3 + (-4i)(-5i)$ **Using the distributive law**
$= -12i + 20i^2$
$= -12i - 20$ $\quad i^2 = -1$
$= -20 - 12i$ **Writing in the form $a + bi$**

d) $(1 + 2i)(1 + 3i) = 1 + 3i + 2i + 6i^2$ **Multiplying each term of one number by every term of the other (FOIL)**
$= 1 + 3i + 2i - 6$ $\quad i^2 = -1$
$= -5 + 5i$ **Collecting like terms**

e) $(3 - 7i)^2 = 3^2 - 2 \cdot 3 \cdot 7i + (7i)^2$ **Recall that $(A - B)^2 = A^2 - 2AB + B^2$**
$= 9 - 42i + 49i^2$
$= 9 - 42i - 49$ $\quad i^2 = -1$
$= -40 - 42i$

Recall that -1 raised to an *even* power is 1, and -1 raised to an *odd* power is -1. Simplifying powers of i can then be done by using the fact that $i^2 = -1$ and expressing the given power of i in terms of i^2. Consider the following:

$i = \sqrt{-1}$, $\quad i^4 = (i^2)^2 = (-1)^2 = 1$,

$i^2 = -1$, $\quad i^5 = i^4 \cdot i = (i^2)^2 \cdot i = (-1)^2 \cdot i = i$,

$i^3 = i^2 \cdot i = (-1)i = -i$, $\quad i^6 = (i^2)^3 = (-1)^3 = -1$.

Note that the powers of i cycle themselves through the values i, -1, $-i$, and 1.

Example 6 Simplify each of the following.

a) i^{37} **b)** i^{58} **c)** i^{75} **d)** i^{80}

SOLUTION

a) $i^{37} = i^{36} \cdot i = (i^2)^{18} \cdot i = (-1)^{18} \cdot i = 1 \cdot i = i$

b) $i^{58} = (i^2)^{29} = (-1)^{29} = -1$

c) $i^{75} = i^{74} \cdot i = (i^2)^{37} \cdot i = (-1)^{37} \cdot i = -1 \cdot i = -i$

d) $i^{80} = (i^2)^{40} = (-1)^{40} = 1$

Conjugates and Division

Conjugates of complex numbers are defined as follows.

Conjugate of a Complex Number

The *conjugate* of a complex number $a + bi$ is $a - bi$, and the *conjugate* of $a - bi$ is $a + bi$.

Example 7 Find the conjugate of each of the following.

a) $-3 + 7i$ **b)** $14 - 5i$ **c)** $4i$

SOLUTION

a) The conjugate of $-3 + 7i$ is $-3 - 7i$.

b) The conjugate of $14 - 5i$ is $14 + 5i$.

c) The conjugate of $4i$ is $-4i$.

The product of a complex number and its conjugate is a real number.

Example 8 Multiply each of the following.

a) $(5 + 7i)(5 - 7i)$ **b)** $(8i)(-8i)$

SOLUTION

a)
$$\begin{aligned}(5 + 7i)(5 - 7i) &= 5^2 - (7i)^2 \quad \text{Using } (A + B)(A - B) = A^2 - B^2\\ &= 25 - 49i^2\\ &= 25 - 49(-1)\\ &= 74\end{aligned}$$

b)
$$\begin{aligned}(8i)(-8i) &= -64i^2\\ &= -64(-1)\\ &= 64\end{aligned}$$

When we are dividing complex numbers, conjugates are used.

Example 9 Divide $2 - 5i$ by $1 - 6i$.

SOLUTION We write fractional notation and then multiply by 1:

$$\frac{2-5i}{1-6i} = \frac{2-5i}{1-6i}\cdot\frac{1+6i}{1+6i}$$ **Note that $1 + 6i$ is the conjugate of the divisor.**

$$= \frac{(2-5i)(1+6i)}{(1-6i)(1+6i)}$$

$$= \frac{2+7i-30i^2}{1-36i^2}$$

$$= \frac{32+7i}{1+36}$$ $i^2 = -1$

$$= \frac{32}{37} + \frac{7}{37}i.$$

2.1 Exercise Set

Use a grapher to graph each of the following polynomial functions. Then estimate the function's (a) real-number zeros; (b) relative maxima; (c) relative minima; (d) domain and range. (Review Sections 1.1 and 1.2, if needed.)

1. $f(x) = x^2 - x - 6$
2. $f(x) = 20 - x - x^2$
3. $f(x) = x^3 + 5x^2 - x - 5$
4. $f(x) = x^4 - 29x + 100$
5. $f(x) = -2x^3 + x^2 + 14x + 10$
6. $f(x) = 1.1x^4 + 4.2x^3 - 35.08x^2 - 158.8x + 294.7$
7. $f(x) = 0.001x^3 - 0.023x$
8. $f(x) = -215.3x^4 + 1647.1x^2 - 2800$

Use a grapher to show that there are no real-number zeros for each of the following polynomial functions.

9. $f(x) = x^2 + 7$
10. $f(x) = x^2 + 2$
11. $f(x) = x^2 - 6x + 10$
12. $f(x) = x^2 + 8x + 17$
13. $f(x) = x^4 - 4x^3 + 6x^2 - 4x + 3$
14. $f(x) = \frac{2}{3}(x^4 + 4x^3 + 6x^2 + 4x + 3)$

Express in terms of i.

15. $\sqrt{-17}$
16. $\sqrt{-25}$
17. $\sqrt{-49}$
18. $\sqrt{-13}$
19. $-\sqrt{-81}$
20. $-\sqrt{-20}$
21. $6 - \sqrt{-84}$
22. $7 - \sqrt{-60}$
23. $-\sqrt{76} + \sqrt{-125}$
24. $\sqrt{-4} + \sqrt{-12}$
25. $\sqrt{-5}\sqrt{-11}$
26. $-\sqrt{-9}\sqrt{-7}$
27. $\dfrac{-\sqrt{25}}{\sqrt{-16}}$
28. $\dfrac{-\sqrt{-36}}{\sqrt{-9}}$

Simplify. Write answers in the form $a + bi$, where a and b are real numbers.

29. $(-5 + 3i) + (7 + 8i)$
30. $(-6 - 5i) + (9 + 2i)$
31. $(4 - 9i) + (1 - 3i)$
32. $(7 - 2i) + (4 - 5i)$
33. $(3 + \sqrt{-16}) + (2 + \sqrt{-25})$
34. $(7 - \sqrt{-36}) + (2 + \sqrt{-9})$
35. $(10 + 7i) - (5 + 3i)$
36. $(-3 - 4i) - (8 - i)$
37. $(13 + 9i) - (8 + 2i)$
38. $(-7 + 12i) - (3 - 6i)$
39. $(1 + 3i)(1 - 4i)$
40. $(1 - 2i)(1 + 3i)$
41. $(2 + 3i)(2 + 5i)$
42. $(3 - 5i)(8 - 2i)$
43. $7i(2 - 5i)$
44. $3i(6 + 4i)$
45. $(3 + \sqrt{-16})(2 + \sqrt{-25})$
46. $(7 - \sqrt{-16})(2 + \sqrt{-9})$
47. $(5 - 4i)(5 + 4i)$
48. $(5 + 9i)(5 - 9i)$
49. $(4 + 2i)^2$
50. $(5 - 4i)^2$

51. $(-2 + 7i)^2$

52. $(-3 + 2i)^2$

53. $\dfrac{2 + \sqrt{3}i}{5 - 4i}$

54. $\dfrac{\sqrt{5} + 3i}{1 - i}$

55. $\dfrac{4 + i}{-3 - 2i}$

56. $\dfrac{5 - i}{-7 + 2i}$

57. $\dfrac{3}{5 - 11i}$

58. $\dfrac{i}{2 + i}$

59. $\dfrac{1 + i}{(1 - i)^2}$

60. $\dfrac{1 - i}{(1 + i)^2}$

61. $\dfrac{4 - 2i}{1 + i} + \dfrac{2 - 5i}{1 + i}$

62. $\dfrac{3 + 2i}{1 - i} + \dfrac{6 + 2i}{1 - i}$

Simplify.

63. i^{11}

64. i^7

65. i^{35}

66. i^{24}

67. i^{64}

68. i^{42}

69. $(-i)^{71}$

70. $(-i)^6$

71. $(5i)^4$

72. $(2i)^5$

Skill Maintenance

Factor.

73. $x^2 + 8x + 16$

74. $y^2 + 20y + 100$

75. $x^2 - 10x + 25$

76. $x^2 - 18x + 81$

Synthesis

77. ◆ Is the sum of two imaginary numbers always an imaginary number? Why or why not? What about the product?

78. ◆ Is the domain of every polynomial function $\mathbb{R}$? Why or why not?

Solve.

79. $(2 + i)x - i = 5 + i$

80. $5i = (3 + i)x + i$

81. $5ix + 3 + 2i = (3 - 2i)x + 3i$

82. $(2 + 3i)x - 2i = 2ix + 5 - 4i$

Determine whether each of the following is true or false.

83. The sum of two numbers that are conjugates of each other is always a real number.

84. The conjugate of a sum is the sum of the conjugates of the individual complex numbers.

85. The conjugate of a product is the product of the conjugates of the individual complex numbers.

Let $z = a + bi$ and $\bar{z} = a - bi$.

86. Find a general expression for $1/z$.

87. Find a general expression for $z\bar{z}$.

88. Solve $z + 6\bar{z} = 7$ for z.

2.2 Quadratic Equations and Functions

- *Solve quadratic equations by completing the square and using the quadratic formula.*
- *Solve equations that are reducible to quadratic equations.*
- *Find the vertex, the line of symmetry, and the maxima or minima of the graph of a quadratic function using the method of completing the square.*
- *Graph quadratic functions.*

In this section, we develop a method for solving any quadratic equation as well as a method for graphing quadratic functions. We define quadratic equations and functions as follows.

Quadratic Equations

A *quadratic equation* is an equation equivalent to

$$ax^2 + bx + c = 0, \quad a \neq 0,$$

where the coefficients a, b, and c are real numbers.

Quadratic Functions

A *quadratic function* f is a second-degree polynomial function

$$f(x) = ax^2 + bx + c, \quad a \neq 0,$$

where a, b, and c are real numbers.

The zeros of a quadratic function $f(x) = ax^2 + bx + c$ are the solutions of the quadratic equation $ax^2 + bx + c = 0$. We have already solved certain quadratic equations and graphed some quadratic functions. In this section, we develop methods for finding exact solutions of *any* quadratic equation and for analyzing the graph of any quadratic function. These techniques will allow us to find the exact value of a maximum or a minimum.

Completing the Square

Quadratic equations like $x^2 - 5x - 6 = 0$ and $6x^2 + 19x - 7 = 0$ can be solved by factoring (see Section R.7):

$$\begin{aligned} x^2 - 5x - 6 &= 0 \\ (x - 6)(x + 1) &= 0 \\ x - 6 = 0 \;&or\; x + 1 = 0 \\ x = 6 \;&or\; x = -1 \end{aligned}$$

The solutions are -1 and 6.

$$\begin{aligned} 6x^2 + 19x - 7 &= 0 \\ (3x - 1)(2x + 7) &= 0 \\ 3x - 1 = 0 \;&or\; 2x + 7 = 0 \\ x = \tfrac{1}{3} \;&or\; x = -\tfrac{7}{2} \end{aligned}$$

The solutions are $\frac{1}{3}$ and $-\frac{7}{2}$.

The principle of square roots can be used to solve quadratic equations like $x^2 - 9 = 0$ and $x^2 + 13 = 0$:

$$\begin{aligned} x^2 - 9 &= 0 \\ x^2 &= 9 \\ x &= \pm 3 \end{aligned}$$

The solutions are -3 and 3.

$$\begin{aligned} x^2 + 13 &= 0 \\ x^2 &= -13 \\ x &= \pm\sqrt{13}i \end{aligned}$$

The solutions are $-\sqrt{13}i$ and $\sqrt{13}i$.

We can find solutions for the first three equations graphically. A graphical approach would not yield the solution to $x^2 + 13 = 0$. Some graphers can find complex solutions, but the procedures are not graphical. None of the preceding methods would yield the exact solutions to an equation like $x^2 - 6x - 10 = 0$. If we wish to do that, we can use a procedure called *completing the square* and then use the principle of square roots.

Example 1 Solve

$$x^2 - 6x - 10 = 0$$

by completing the square.

SOLUTION Our goal is to find an equivalent polynomial equation of the form $x^2 + bx + c = d$ in which $x^2 + bx + c$ is a perfect square. This is accomplished as follows:

$x^2 - 6x - 10 = 0$

$x^2 - 6x = 10$ — **Adding 10**

$x^2 - 6x + 9 = 10 + 9$ — **Adding 9 to complete the square: $\frac{1}{2}(-6) = -3$ and $(-3)^2 = 9$**

$x^2 - 6x + 9 = 19.$

Because $x^2 - 6x + 9$ is a perfect square, we are able to write it as the square of a binomial. We can then use the principle of square roots to finish the solution:

$(x - 3)^2 = 19$ — **Factoring**

$x - 3 = \pm\sqrt{19}$ — **Using the principle of square roots**

$x = 3 \pm \sqrt{19}.$ — **Adding 3**

The solutions are $3 + \sqrt{19}$ and $3 - \sqrt{19}$, or simply $3 \pm \sqrt{19}$ (read "three plus or minus $\sqrt{19}$"). The solution set is $\{3 + \sqrt{19}, 3 - \sqrt{19}\}$, or simply $\{3 \pm \sqrt{19}\}$.

Decimal approximations for $3 \pm \sqrt{19}$ could be found on a grapher, that is, $3 + \sqrt{19} \approx 7.359$ and $3 - \sqrt{19} \approx -1.359$. The solutions could also be found using a graphical approach.

Example 2 Solve: $2x^2 - 1 = 3x$.

SOLUTION

$2x^2 - 1 = 3x$

$2x^2 - 3x - 1 = 0$ — **Subtracting $3x$; we are unable to factor the result.**

$2x^2 - 3x = 1$ — **Adding 1**

$x^2 - \frac{3}{2}x = \frac{1}{2}$ — **Dividing by 2**

$x^2 - \frac{3}{2}x + \frac{9}{16} = \frac{1}{2} + \frac{9}{16}$ — **Completing the square: $\frac{1}{2}\left(-\frac{3}{2}\right) = -\frac{3}{4}$ and $\left(-\frac{3}{4}\right)^2 = \frac{9}{16}$**

$\left(x - \frac{3}{4}\right)^2 = \frac{17}{16}$ — **Factoring and simplifying**

$x - \frac{3}{4} = \pm\frac{\sqrt{17}}{4}$ — **Using the principle of square roots and the quotient rule for radicals**

$x = \frac{3 \pm \sqrt{17}}{4}$ — **Adding $\frac{3}{4}$**

The solution set is $\left\{\frac{3 \pm \sqrt{17}}{4}\right\}$.

You should find approximate solutions using scientific keys on a grapher and check the solutions graphically.

To solve a quadratic equation by completing the square:

1. Isolate the terms with variables on one side of the equation and arrange them in descending order.
2. Divide by the coefficient of the squared term if that coefficient is not 1.
3. Complete the square by taking half the coefficient of the first-degree term and adding its square on both sides of the equation.
4. Express one side of the equation as the square of a binomial.
5. Use the principle of square roots.
6. Solve for the variable.

Using the Quadratic Formula

Because completing the square works for *any* quadratic equation, it can be used to solve the general quadratic equation $ax^2 + bx + c = 0$ for x. The result will be a formula that can be used to solve any quadratic equation quickly.

Consider any quadratic equation in standard form:

$$ax^2 + bx + c = 0, \quad a > 0.$$

If a is negative, we first multiply on both sides by -1. Then we solve by completing the square. As we carry out the steps, compare them with those of Example 2.

$$ax^2 + bx = -c \qquad \text{Adding } -c$$

$$x^2 + \frac{b}{a}x = -\frac{c}{a} \qquad \text{Dividing by } a\text{, since } a > 0$$

Half of $\frac{b}{a}$ is $\frac{b}{2a}$ and $\left(\frac{b}{2a}\right)^2 = \frac{b^2}{4a^2}$. Thus we add $\frac{b^2}{4a^2}$:

$$x^2 + \frac{b}{a}x + \frac{b^2}{4a^2} = -\frac{c}{a} + \frac{b^2}{4a^2} \qquad \text{Adding } \frac{b^2}{4a^2} \text{ to complete the square}$$

$$\left(x + \frac{b}{2a}\right)^2 = -\frac{4ac}{4a^2} + \frac{b^2}{4a^2} \qquad \text{Factoring and finding a common denominator: } -\frac{c}{a} = -\frac{4a}{4a}\cdot\frac{c}{a} = -\frac{4ac}{4a^2}$$

$$\left(x + \frac{b}{2a}\right)^2 = \frac{b^2 - 4ac}{4a^2}$$

$$x + \frac{b}{2a} = \pm\frac{\sqrt{b^2 - 4ac}}{2a} \qquad \text{Using the principle of square roots and the quotient rule for radicals; since } a > 0,\ \sqrt{4a^2} = 2a.$$

$$x = \frac{-b \pm \sqrt{b^2 - 4ac}}{2a}. \qquad \text{Adding } -\frac{b}{2a}$$

Program

QUADFORM: This program solves a quadratic equation, giving real and complex solutions. (See the Graphing Calculator Manual that accompanies this text.)

The Quadratic Formula

The solutions of $ax^2 + bx + c = 0$, $a \neq 0$, are given by

$$x = \frac{-b \pm \sqrt{b^2 - 4ac}}{2a}.$$

Example 3 Solve $3x^2 + 2x = 7$. Find exact solutions and approximate solutions rounded to the nearest thousandth.

We show both algebraic and graphical solutions. Note that only the algebraic approach yields the exact solutions.

ALGEBRAIC SOLUTION

After finding standard form, we are unable to factor, so we identify a, b, and c in order to use the quadratic formula:

$$3x^2 + 2x - 7 = 0;$$
$$a = 3, \quad b = 2, \quad c = -7.$$

We then use the quadratic formula:

$$\begin{aligned} x &= \frac{-b \pm \sqrt{b^2 - 4ac}}{2a} \\ &= \frac{-2 \pm \sqrt{2^2 - 4(3)(-7)}}{2(3)} \quad \text{Substituting} \\ &= \frac{-2 \pm \sqrt{4 + 84}}{6} = \frac{-2 \pm \sqrt{88}}{6} \\ &= \frac{-2 \pm \sqrt{4 \cdot 22}}{6} = \frac{-2 \pm 2\sqrt{22}}{6} \\ &= \frac{2}{2} \cdot \frac{-1 \pm \sqrt{22}}{3} = \frac{-1 \pm \sqrt{22}}{3}. \end{aligned}$$

The exact solutions are

$$\frac{-1 - \sqrt{22}}{3} \quad \text{and} \quad \frac{-1 + \sqrt{22}}{3}.$$

Using the scientific keys on a grapher, we approximate the solutions to be -1.897 and 1.230.

GRAPHICAL SOLUTION

To solve $3x^2 + 2x = 7$, or $3x^2 + 2x - 7 = 0$, we first graph the function $f(x) = 3x^2 + 2x - 7$. Then we look for points where the graph crosses the x-axis. It appears that there are two possible zeros, one near -2 and one near 1. We can use TRACE and ZOOM to approximate these zeros, or we can use a SOLVE or POLY feature.

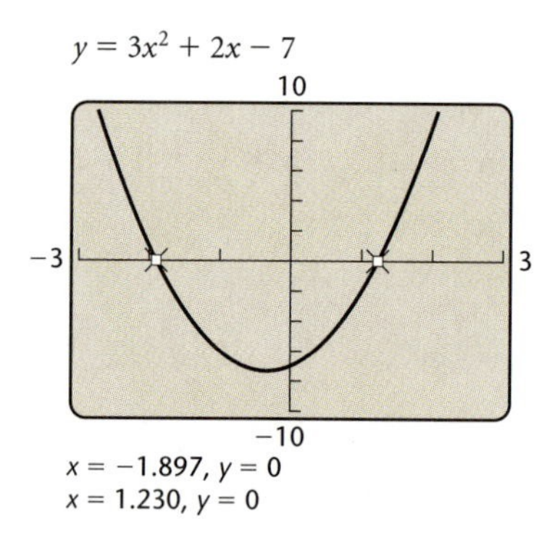

We get the approximate zeros -1.896805 and 1.2301386, or -1.897 and 1.230, rounded to three decimal places. The zeros of the function are the solutions of the equation $3x^2 + 2x = 7$.

Not all quadratic equations can be solved graphically.

Example 4 Solve: $x^2 + 5x + 8 = 0$.

ALGEBRAIC SOLUTION

To find the solutions, we use the quadratic formula. Here

$$a = 1, \quad b = 5, \quad c = 8;$$

$$\begin{aligned} x &= \frac{-b \pm \sqrt{b^2 - 4ac}}{2a} \\ &= \frac{-5 \pm \sqrt{5^2 - 4(1)(8)}}{2 \cdot 1} && \text{Substituting} \\ &= \frac{-5 \pm \sqrt{-7}}{2} && \text{Simplifying} \\ &= \frac{-5 \pm \sqrt{7}i}{2}. \end{aligned}$$

The solutions are $-\frac{5}{2} - \frac{\sqrt{7}}{2}i$ and $-\frac{5}{2} + \frac{\sqrt{7}}{2}i$.

GRAPHICAL SOLUTION

The graph of the function $f(x) = x^2 + 5x + 8$ shows no x-intercepts.

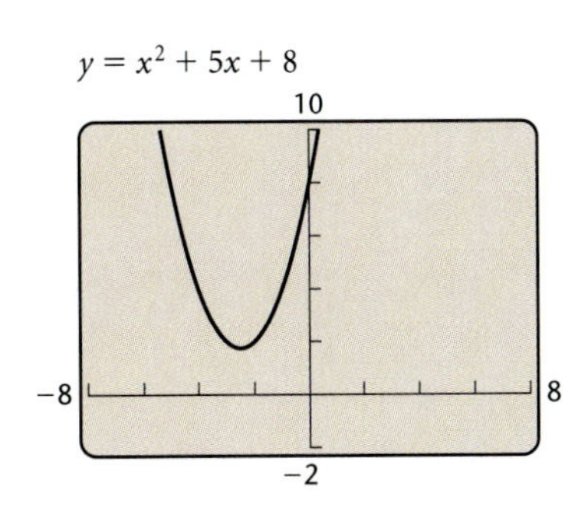

Thus there are no real-number solutions of the equation $x^2 + 5x + 8 = 0$. This is a quadratic equation that cannot be solved graphically.

The Discriminant

From the quadratic formula, we know that the solutions x_1 and x_2 of a quadratic equation are given by

$$x_1 = \frac{-b + \sqrt{b^2 - 4ac}}{2a} \quad \text{and} \quad x_2 = \frac{-b - \sqrt{b^2 - 4ac}}{2a}.$$

The expression $b^2 - 4ac$ shows the nature of the solutions. This expression is called the **discriminant**. If it is 0, then it doesn't matter whether we choose the plus or the minus sign in the formula. There is just one solution, and we sometimes say that there is one repeated real solution. If the discriminant is positive, there will be two real solutions. If it is negative, we will be taking the square root of a negative number; hence there will be two nonreal solutions, and they will be complex conjugates.

Discriminant

For $ax^2 + bx + c = 0$:

$b^2 - 4ac = 0 \longrightarrow$ One real-number solution;

$b^2 - 4ac > 0 \longrightarrow$ Two different real-number solutions;

$b^2 - 4ac < 0 \longrightarrow$ Two different imaginary-number solutions, complex conjugates.

In Example 3, the discriminant, 88, is positive, indicating that there are two different real-number solutions. If the discriminant is negative, as it is in Example 4, we know that there are two different imaginary-number solutions.

Equations Reducible to Quadratic

Some equations can be treated as quadratic, provided that we make a suitable substitution. For example, consider the following:

$$(x^2 - x)^2 - 14(x^2 - x) + 24 = 0$$

$$u^2 - 14u + 24 = 0. \quad \text{Substituting } u \text{ for } x^2 - x$$

The equation $u^2 - 14u + 24 = 0$ can be solved by factoring or using the quadratic formula. Once we have solved for u, we can reverse our substitution and solve for x. Equations that can be solved in this manner are said to be **reducible to quadratic**, or in **quadratic form**.

Example 5 Solve: $(x^2 - x)^2 - 14(x^2 - x) + 24 = 0$.

ALGEBRAIC SOLUTION

We let $u = x^2 - x$ and substitute:

$$u^2 - 14u + 24 = 0 \quad \text{Substituting } u \text{ for } x^2 - x$$

$$(u - 12)(u - 2) = 0 \quad \text{Factoring}$$

$$u - 12 = 0 \quad or \quad u - 2 = 0 \quad \text{Principle of zero products}$$

$$u = 12 \quad or \quad u = 2.$$

A common error is to stop here, having solved for u, and forget to solve for x. Remember that you must find values for the *original* variable! We now substitute $x^2 - x$ for u and solve for x:

$$x^2 - x = 12 \quad or \quad x^2 - x = 2 \quad \text{Substituting } x^2 - x \text{ for } u$$

$$x^2 - x - 12 = 0 \quad or \quad x^2 - x - 2 = 0$$

$$(x + 3)(x - 4) = 0 \quad or \quad (x + 1)(x - 2) = 0 \quad \text{Factoring}$$

$$x + 3 = 0 \text{ or } x - 4 = 0 \text{ or } x + 1 = 0 \text{ or } x - 2 = 0$$

$$x = -3 \text{ or } x = 4 \text{ or } x = -1 \text{ or } x = 2.$$

The solutions are -3, 4, -1, and 2. The solution set is $\{-3, -1, 2, 4\}$.

GRAPHICAL SOLUTION

We can obtain a graphical solution using procedures discussed earlier. We need not substitute or simplify.

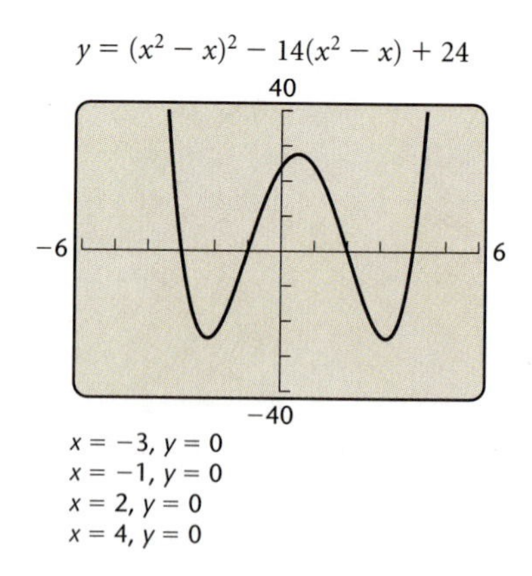

The solutions appear to be -3, -1, 2, and 4. The grapher also provides a quick visual check of the algebraic solution.

Interactive Discovery

See if you can find a way to use the TABLE feature to check the solutions in Example 5.

Example 6 Solve: $t^{2/5} - t^{1/5} - 2 = 0$.

ALGEBRAIC SOLUTION

Note that since $t^{2/5}$ can be rewritten as $(t^{1/5})^2$, the equation can be rewritten as $(t^{1/5})^2 - t^{1/5} - 2 = 0$. We let $u = t^{1/5}$ and solve the following equation:

$$u^2 - u - 2 = 0 \quad \text{Substituting } u \text{ for } t^{1/5}$$
$$(u - 2)(u + 1) = 0$$
$$u - 2 = 0 \quad or \quad u + 1 = 0$$
$$u = 2 \quad or \quad u = -1.$$

Now we substitute $t^{1/5}$ for u and solve:

$$t^{1/5} = 2 \quad or \quad t^{1/5} = -1$$
$$(t^{1/5})^5 = 2^5 \quad or \quad (t^{1/5})^5 = (-1)^5 \quad \text{Principle of powers; raising to the fifth power}$$
$$t = 32 \quad or \quad t = -1.$$

CHECK: For 32:

$$\begin{array}{r|l} t^{2/5} - t^{1/5} - 2 & = 0 \\ \hline 32^{2/5} - 32^{1/5} - 2 & ?\ 0 \\ (32^{1/5})^2 - 32^{1/5} - 2 & \\ 2^2 - 2 - 2 & \\ 0 & 0 \quad \text{TRUE} \end{array}$$

For -1:

$$\begin{array}{r|l} t^{2/5} - t^{1/5} - 2 & = 0 \\ \hline (-1)^{2/5} - (-1)^{1/5} - 2 & ?\ 0 \\ [(-1)^{1/5}]^2 - (-1)^{1/5} - 2 & \\ (-1)^2 - (-1) - 2 & \\ 0 & 0 \quad \text{TRUE} \end{array}$$

Both numbers check. The solution set is $\{-1, 32\}$.

GRAPHICAL SOLUTION

We graph the equation $y = x^{2/5} - x^{1/5} - 2$ using a grapher, entering the equation as y = (x²)^(1/5) − x^(1/5) − 2, or as (x^(1/5))² − x^(1/5) − 2.

If entered as y = x^(2/5) − x^(1/5) − 2, most graphers will omit the portion of the graph to the left of the y-axis. Those graphers reject negative inputs even though they are in the domain of the function.

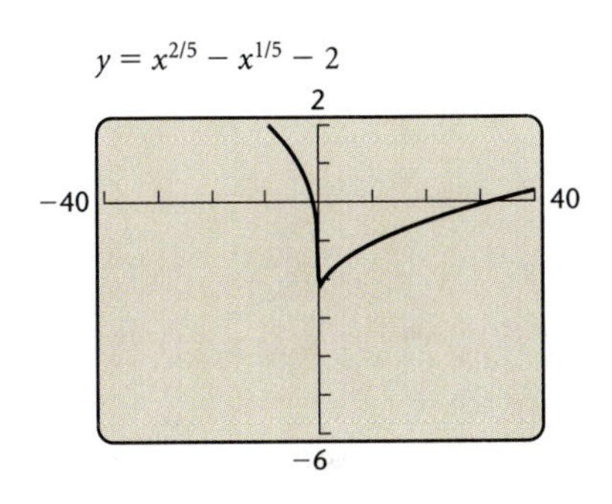

Then using graphical solving procedures as we have before, we get the solutions -1 and 32.

Example 7 Solve: $x^4 - 6x^2 + 7 = 0$.

In this case, we can find an approximate solution using a grapher, but the algebraic method yields exact solutions.

ALGEBRAIC SOLUTION

Letting $u = x^2$, we substitute u for x^2 and u^2 for x^4:

$$x^4 - 6x^2 + 7 = 0$$
$$(x^2)^2 - 6x^2 + 7 = 0$$
$$u^2 - 6u + 7 = 0.$$

To solve for u, we use the quadratic formula with $a = 1$, $b = -6$, and $c = 7$:

$$u = \frac{-b \pm \sqrt{b^2 - 4ac}}{2a}$$
$$= \frac{-(-6) \pm \sqrt{(-6)^2 - 4 \cdot 1 \cdot 7}}{2 \cdot 1}$$
$$= \frac{6 \pm \sqrt{8}}{2}$$
$$= \frac{2 \cdot 3 \pm 2\sqrt{2}}{2 \cdot 1}$$
$$= \frac{2}{2} \cdot \frac{3 \pm \sqrt{2}}{1}$$
$$= 3 \pm \sqrt{2}.$$

We have now solved for u. Next, we substitute x^2 for u and solve for x:

$$x^2 = 3 + \sqrt{2} \quad or \quad x^2 = 3 - \sqrt{2}$$
$$x = \pm\sqrt{3 + \sqrt{2}} \quad or \quad x = \pm\sqrt{3 - \sqrt{2}}.$$

Thus the solutions are

$$-\sqrt{3 + \sqrt{2}}, \quad \sqrt{3 + \sqrt{2}}, \quad -\sqrt{3 - \sqrt{2}}, \quad \text{and} \quad \sqrt{3 - \sqrt{2}},$$

or approximately

$$-2.101, \quad 2.101, \quad -1.259, \quad \text{and} \quad 1.259.$$

GRAPHICAL SOLUTION

We graph $y = x^4 - 6x^2 + 7$.

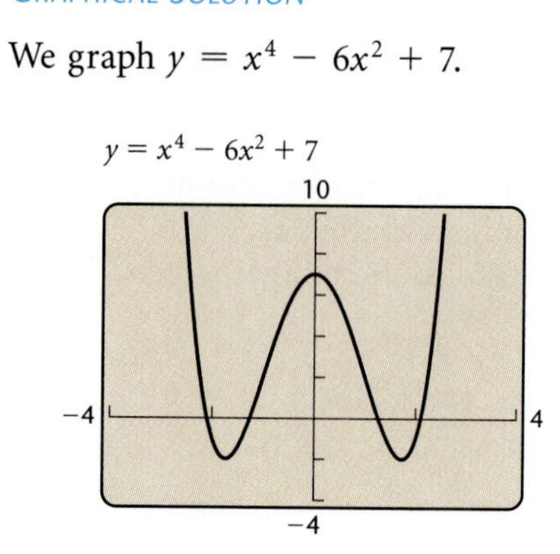

Then using graphical solving procedures as we have before, we get the approximate solutions -2.101, -1.259, 1.259, and 2.101.

Graphing Quadratic Functions of the Type $f(x) = a(x - h)^2 + k$

The graph of a quadratic function is called a **parabola**. We will see that the graph of every parabola evolves from the graph of the squaring function $f(x) = x^2$ using the transformations that we discussed in Section 1.7.

Interactive Discovery

Think of transformations and look for patterns. Consider the following functions:

$$y_1 = x^2, \qquad y_2 = -0.4x^2,$$
$$y_3 = -0.4(x - 2)^2, \qquad y_4 = -0.4(x - 2)^2 + 3.$$

Graph y_1 and y_2. How do you get from the graph of y_1 to y_2?
Graph y_2 and y_3. How do you get from the graph of y_2 to y_3?
Graph y_3 and y_4. How do you get from the graph of y_3 to y_4?

Consider the following functions:

$$y_1 = x^2, \qquad y_2 = 2x^2,$$
$$y_3 = 2(x + 3)^2, \qquad y_4 = 2(x + 3)^2 - 5.$$

Graph y_1 and y_2. How do you get from the graph of y_1 to y_2?
Graph y_2 and y_3. How do you get from the graph of y_2 to y_3?
Graph y_3 and y_4. How do you get from the graph of y_3 to y_4?

We get the graph of $f(x) = a(x - h)^2 + k$ from the graph of $f(x) = x^2$ as follows:

$f(x) = x^2$
↓
$f(x) = ax^2$ — **Vertical stretching or shrinking with a reflection across the x-axis if $a < 0$**
↓
$f(x) = a(x - h)^2$ — **Horizontal translation**
↓
$f(x) = a(x - h)^2 + k.$ — **Vertical translation**

Consider the following graphs of the form $f(x) = a(x - h)^2 + k$. The point at which the graph turns is called the **vertex**. If $f(x)$ has a maximum or a minimum value, it occurs at the vertex. Each graph has a line $x = h$ that is called the **line of symmetry**.

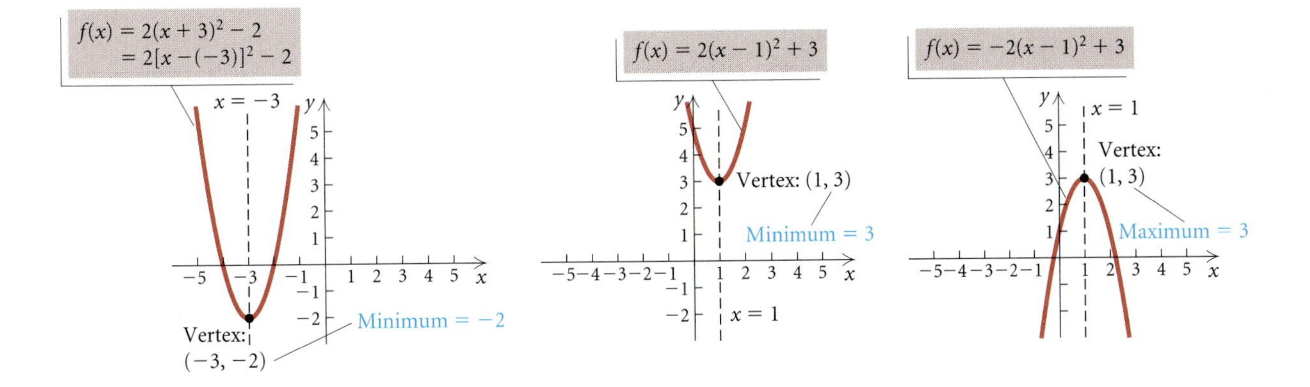

We summarize as follows.

> The graph of $f(x) = a(x - h)^2 + k$:
>
> opens up if $a > 0$ and down if $a < 0$:
>
> has (h, k) as a vertex;
>
> has $x = h$ as the line of symmetry;
>
> has k as a minimum value (output) if $a > 0$ and has k as a maximum value if $a < 0$.

If the equation is in the form $f(x) = a(x - h)^2 + k$, we can determine a great deal of information about the graph without graphing.

FUNCTION	$f(x) = 3\left(x - \frac{1}{4}\right)^2 - 2$ $= 3\left(x - \frac{1}{4}\right)^2 + (-2)$	$g(x) = -3(x + 5)^2 + 7$ $= -3[x - (-5)]^2 + 7$
VERTEX	$\left(\frac{1}{4}, -2\right)$	$(-5, 7)$
LINE OF SYMMETRY	$x = \frac{1}{4}$	$x = -5$
MAXIMUM	No: $3 > 0$, graph opens up.	Yes, 7: $-3 < 0$, graph opens down.
MINIMUM	Yes, -2; $3 > 0$, graph opens up.	No: $-3 < 0$, graph opens down.

Note that the vertex (h, k) is used to find the maximum or the minimum value of the function. The maximum or minimum is the number k, *not* the ordered pair (h, k).

Graphing Quadratic Functions of the Type $f(x) = ax^2 + bx + c,\ a \neq 0$

We now use a modification of the method of completing the square as an aid in graphing and analyzing quadratic functions of the form $f(x) = ax^2 + bx + c$, $a \neq 0$.

Example 8 Complete the square to find the vertex, the line of symmetry, and the maximum or minimum value of $f(x) = x^2 + 10x + 23$. Then graph the function.

SOLUTION To express $f(x) = x^2 + 10x + 23$ in the form $f(x) = a(x - h)^2 + k$, we proceed as follows. We construct a trinomial square. To do so, we take half the coefficient of x and square it. The number is $(10/2)^2$, or 25. We now add and subtract that number on the right-hand side. We can think of this as adding $25 - 25$, which is 0.

$$
\begin{aligned}
f(x) &= x^2 + 10x + 23 && \text{Note that 25 completes the square for } x^2 + 10x. \\
&= x^2 + 10x + 25 - 25 + 23 && \text{Adding } 25 - 25 \text{, or 0, to the right side} \\
&= (x^2 + 10x + 25) - 25 + 23 \\
&= (x + 5)^2 - 2 && \text{Factoring and simplifying} \\
&= [x - (-5)]^2 + (-2) && \text{Writing in the form } f(x) = a(x - h)^2 + k
\end{aligned}
$$

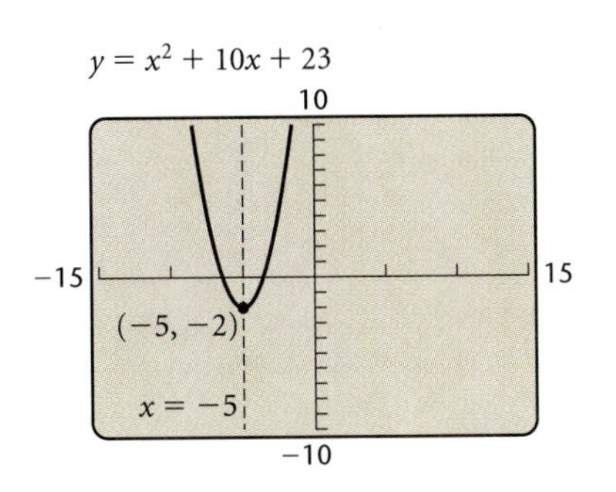

From this form of the function, we know the following:

Vertex: $(-5, -2)$;

Line of symmetry: $x = -5$;

Minimum $= -2$.

The graph of f is a shift of the graph of $y = x^2$ left 5 units and down 2 units.

Keep in mind that a line of symmetry is not part of the graph; it is a characteristic of the graph. If you fold the graph on its line of symmetry, the two halves of the graph will coincide.

Example 9 Complete the square to find the vertex, the line of symmetry, and the maximum or minimum value of $g(x) = x^2/2 - 4x + 8$. Then graph the function.

SOLUTION To complete the square, we factor $\frac{1}{2}$ out of the first two terms. This makes the coefficient of x^2 inside the parentheses 1:

$$
\begin{aligned}
g(x) &= \frac{x^2}{2} - 4x + 8 \\
&= \frac{1}{2}(x^2 - 8x) + 8. && \text{Factoring } \tfrac{1}{2} \text{ out of the first two terms}
\end{aligned}
$$

Now we complete the square inside the parentheses: Half of -8 is -4, and $(-4)^2 = 16$. We add and subtract 16 inside the parentheses:

$$
\begin{aligned}
g(x) &= \tfrac{1}{2}(x^2 - 8x + 16 - 16) + 8 \\
&= \tfrac{1}{2}(x^2 - 8x + 16) - \tfrac{1}{2} \cdot 16 + 8 && \text{Using the distributive law to remove } -16 \text{ from within the parentheses} \\
&= \tfrac{1}{2}(x - 4)^2. && \text{Factoring and simplifying}
\end{aligned}
$$

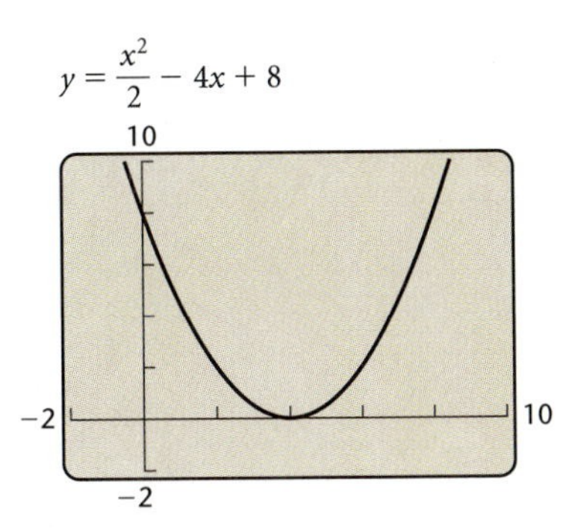

We know the following:

Vertex: $(4, 0)$;

Line of symmetry: $x = 4$;

Minimum $= 0$.

The graph of g is a shrinking of the graph of $y = x^2$ vertically followed by a shift to the right 4 units.

Example 10 Complete the square to find the vertex, the line of symmetry, and the maximum or minimum value of $f(x) = -2x^2 + 10x - \frac{23}{2}$. Then graph the function.

SOLUTION

$$\begin{aligned} f(x) &= -2x^2 + 10x - \tfrac{23}{2} \\ &= -2(x^2 - 5x) - \tfrac{23}{2} && \text{Factoring } -2 \text{ out of the first two terms} \\ &= -2\left(x^2 - 5x + \tfrac{25}{4} - \tfrac{25}{4}\right) - \tfrac{23}{2} && \text{Completing the square inside the parentheses} \\ &= -2\left(x^2 - 5x + \tfrac{25}{4}\right) - 2\left(-\tfrac{25}{4}\right) - \tfrac{23}{2} && \text{Removing } -\tfrac{25}{4} \text{ from within the parentheses} \\ &= -2\left(x - \tfrac{5}{2}\right)^2 + \tfrac{25}{2} - \tfrac{23}{2} \\ &= -2\left(x - \tfrac{5}{2}\right)^2 + 1. \end{aligned}$$

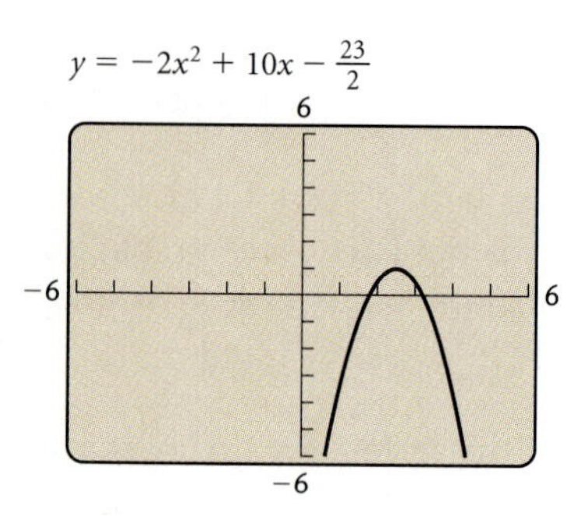

This form of the function yields the following:

Vertex: $\left(\frac{5}{2}, 1\right)$;
Line of symmetry: $x = \frac{5}{2}$;
Maximum $= 1$.

The graph is found by stretching the graph of $f(x) = x^2$ vertically, reflecting the graph across the x-axis, shifting the graph to the right $\frac{5}{2}$ units, and shifting that curve up 1 unit.

In many situations, we want to find the coordinates of the vertex directly from the equation $f(x) = ax^2 + bx + c$ using a formula. One way to develop such a formula is to consider the x-coordinate of the vertex as centered between the x-intercepts, or zeros, of the function. By averaging the two solutions of $ax^2 + bx + c = 0$, we find a formula for the x-coordinate of the vertex:

$$x\text{-coordinate of vertex} = \frac{\dfrac{-b - \sqrt{b^2 - 4ac}}{2a} + \dfrac{-b + \sqrt{b^2 - 4ac}}{2a}}{2}$$

$$= \frac{\dfrac{-2b}{2a}}{2} = \frac{-\dfrac{b}{a}}{2} = -\frac{b}{2a}.$$

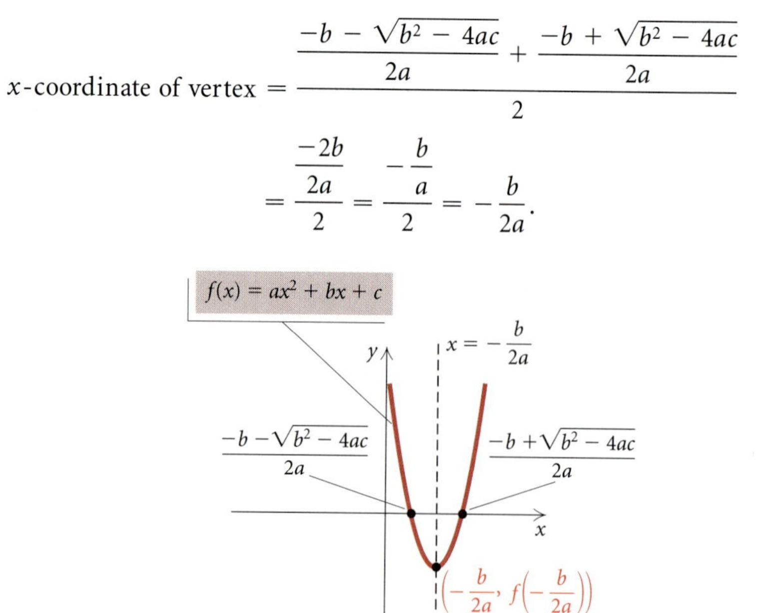

We use this value of x to find the y-coordinate of the vertex, $f\left(-\frac{b}{2a}\right)$.

The Vertex of a Parabola

The *vertex* of the graph of $f(x) = ax^2 + bx + c$ is

$$\left(-\frac{b}{2a}, f\left(-\frac{b}{2a}\right)\right).$$

We calculate the x-coordinate. — We substitute to find the y-coordinate.

Example 11 For the function $f(x) = -x^2 + 14x - 47$:

a) Find the vertex.

b) Determine whether there is a maximum or minimum value and find that value.

c) Find the range.

d) On what intervals is the function increasing? decreasing?

SOLUTION There is no need to graph the function.

a) The x-coordinate of the vertex is

$$\frac{-b}{2a}, \quad \text{or} \quad \frac{-14}{2(-1)}, \quad \text{or} \quad 7.$$

Since

$$f(7) = -7^2 + 14 \cdot 7 - 47 = 2,$$

the vertex is $(7, 2)$.

b) Since $a = -1$ and is negative, the graph opens down so the second coordinate of the vertex, 2, is the maximum value of the function.

c) The range is $(-\infty, 2]$.

d) Since the graph opens down, function values increase to the left of the vertex and decrease to the right of the vertex. Thus the function is increasing on the interval $(-\infty, 7]$ and decreasing on $[7, \infty)$.

You should visually check the results of Example 11 with a grapher.

2.2 Exercise Set

Solve by factoring.

1. $x^2 - 8x + 12 = 0$
2. $5x^2 + 42x + 16 = 0$
3. $7x^2 + 6 = 43x$
4. $23x = 12 + 10x^2$

Solve using the principle of square roots.

5. $x^2 = 81$
6. $x^2 - 100 = 0$
7. $9x^2 - 100 = 0$
8. $4x^2 - 25 = 0$
9. $x^2 + 10 = 0$
10. $x^2 = -14$
11. $(x + 3)^2 = 81$
12. $(x - 5)^2 = 64$
13. $x^2 - 6x + 9 = 1$
14. $x^2 + 8x + 16 = 9$

Solve by completing the square to obtain exact solutions. Check your work on a grapher, if appropriate.

15. $x^2 + 6x = 7$
16. $x^2 + 8x = -15$
17. $x^2 = 8x - 9$
18. $x^2 = 22 + 10x$
19. $x^2 + 8x + 25 = 0$
20. $x^2 + 6x + 13 = 0$
21. $3x^2 + 5x - 2 = 0$
22. $2x^2 - 5x - 3 = 0$
23. $6x + 1 = 4x^2$
24. $3x^2 + 5x = 3$
25. $2x^2 - 4 = 5x$
26. $4x^2 - 2 = 3x$

Solve. Use any method, but obtain exact solutions. Check your work on a grapher, if appropriate.

27. $x^2 - 2x = 15$
28. $x^2 + 4x = 5$
29. $5m^2 + 3m = 2$
30. $2y^2 - 3y - 2 = 0$
31. $3x^2 + 6 = 10x$
32. $3t^2 + 8t + 3 = 0$
33. $x^2 + x + 2 = 0$
34. $x^2 + 1 = x$
35. $5t^2 - 8t = 3$
36. $5x^2 + 2 = x$
37. $3x^2 + 4 = 5x$
38. $2t^2 - 5t = 1$

For each of the following, consider only $b^2 - 4ac$, the discriminant of the quadratic formula, to determine whether imaginary solutions exist.

39. $4x^2 = 8x + 5$
40. $4x^2 - 12x + 9 = 0$
41. $x^2 + 3x + 4 = 0$
42. $x^2 - 2x + 4 = 0$
43. $5t^2 - 7t = 0$
44. $5t^2 - 4t = 11$

Use a grapher to solve graphically. Find solutions to the nearest thousandth.

45. $x^2 - 5x - 4 = 0$
46. $x^2 + 7x - 6 = 0$
47. $3x^2 + 2x - 4 = 0$
48. $5.02x^2 - 4.19x - 2.057 = 0$

Solve. Find exact solutions. Check your work on a grapher, if possible.

49. $(2x - 3)^2 - 5(2x - 3) + 6 = 0$
50. $(3x + 2)^2 + 7(3x + 2) - 8 = 0$
51. $m^{2/3} - 2m^{1/3} - 8 = 0$
52. $t^{2/3} + t^{1/3} - 6 = 0$
53. $x^4 - 5x^2 + 4 = 0$
54. $x^4 + 3 = 4x^2$
55. $(2t^2 + t)^2 - 4(2t^2 + t) + 3 = 0$
56. $12 = (m^2 - 5m)^2 + (m^2 - 5m)$
57. $x - 3\sqrt{x} - 4 = 0$
58. $2x - 9\sqrt{x} + 4 = 0$

Complete the square to:

a) *find the vertex;*
b) *find the line of symmetry;*
c) *determine whether there is a maximum or minimum value and find that value.*

Then check your answers with a grapher.

59. $f(x) = x^2 - 8x + 12$
60. $g(x) = x^2 + 7x - 8$
61. $f(x) = x^2 - 7x + 12$
62. $g(x) = x^2 - 5x + 6$
63. $g(x) = 2x^2 + 6x + 8$
64. $f(x) = 2x^2 - 10x + 14$
65. $g(x) = -2x^2 + 2x + 1$
66. $f(x) = -3x^2 - 3x + 1$

In Exercises 67–74, match the equation with figures (a)–(h). Use a grapher only as a check.

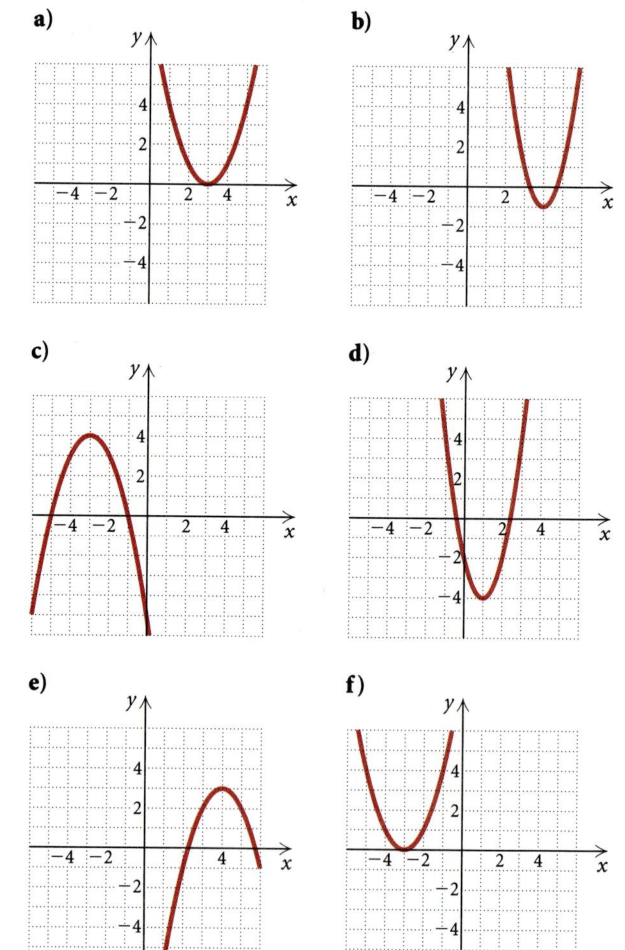

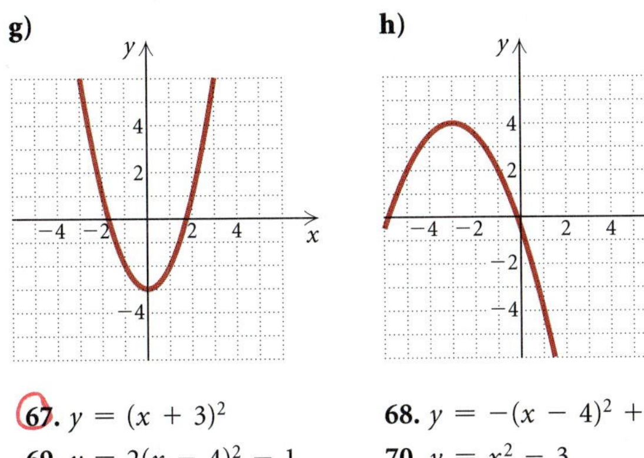

67. $y = (x + 3)^2$

68. $y = -(x - 4)^2 + 3$

69. $y = 2(x - 4)^2 - 1$

70. $y = x^2 - 3$

71. $y = -\frac{1}{2}(x + 3)^2 + 4$

72. $y = (x - 3)^2$

73. $y = -(x + 3)^2 + 4$

74. $y = 2(x - 1)^2 - 4$

In Exercises 75–82:

a) *Find the vertex.*
b) *Determine whether there is a maximum or minimum value and find that value.*
c) *Find the range.*
d) *Find the intervals on which the function is increasing and the function is decreasing.*

Then check your answers with a grapher.

75. $f(x) = x^2 - 6x + 5$

76. $f(x) = x^2 + 4x - 5$

77. $f(x) = 2x^2 + 4x - 16$

78. $f(x) = \frac{1}{2}x^2 - 3x + \frac{5}{2}$

79. $f(x) = -\frac{1}{2}x^2 + 5x - 8$

80. $f(x) = -2x^2 - 24x - 64$

81. $f(x) = 3x^2 + 6x + 5$

82. $f(x) = -3x^2 + 24x - 49$

Skill Maintenance

Solve.

83. $\dfrac{5}{x + 4} + \dfrac{14}{x + 7} = 4$

84. $\dfrac{5}{x + 2} + \dfrac{3x}{x + 6} = 2$

85. Claudia jogs at a rate of 8 mph for 45 min. How far does she run?

Synthesis

86. ◈ Is it possible for a quadratic function to have one real zero and one imaginary zero? Why or why not?

87. ◈ The graph of a quadratic function can have 0, 1, or 2 x-intercepts. How can you predict the number of x-intercepts without drawing the graph or (completely) solving an equation?

Solve.

88. $x^2 + \sqrt{5}x - \sqrt{3} = 0$

89. $x^2 + x - \sqrt{2} = 0$

90. $9t(t + 2) - 3t(t - 2) = 2(t + 4)(t + 6)$

91. $2t^2 + (t - 4)^2 = 5t(t - 4) + 24$

92. $(x - 3)^{2/3} = 2$

93. $\sqrt[4]{x + 2} = \sqrt{3x + 1}$

94. $x^6 - 28x^3 + 27 = 0$

95. $\sqrt{x - 3} - \sqrt[4]{x - 3} = 2$

96. $x^2 + 3x + 1 - \sqrt{x^2 + 3x + 1} = 8$

97. $\left(y + \dfrac{2}{y}\right)^2 + 3y + \dfrac{6}{y} = 4$

For each equation in Exercises 98–101, under the given condition:

a) *Find k.*
b) *Find a second solution.*

98. $kx^2 - 2x + k = 0$; one solution is -3

99. $kx^2 - 17x + 33 = 0$; one solution is 3

100. $x^2 - (6 + 3i)x + k = 0$; one solution is 3

101. $x^2 - kx + 2 = 0$; one solution is $1 + i$

102. Find b such that $f(x) = -4x^2 + bx + 3$ has a maximum value of 50.

103. Find c such that $f(x) = -0.2x^2 - 3x + c$ has a maximum value of -225.

104. Find a quadratic function that has $(4, -5)$ as a vertex and contains the point $(-3, 1)$.

105. Graph: $f(x) = (|x| - 5)^2 - 3$.

106. Solve $\frac{1}{2}at + v_0t + x_0 = 0$ for t.

2.3 Polynomial Models and Curve Fitting

- *Solve applications using polynomial models.*
- *Fit quadratic, cubic, and quartic polynomial functions to data and make predictions.*

Polynomial functions have many uses as models in science, engineering, and business. In this section, we examine some of these models. Keep in mind that almost any model has its limitations.

The simplest use of polynomial functions in applications occurs when we merely evaluate a polynomial function. In such cases, a model has already been developed.

Example 1 *Threshold Weight.* In a study performed by Alvin Shemesh, it was found that the **threshold weight** W, defined as the weight above which the risk of death rises dramatically, is given by

$$W(h) = \left(\frac{h}{12.3}\right)^3,$$

where W is in pounds and h is a person's height, in inches. Find the threshold weight of a person who is 5 ft, 7 in. tall.

SOLUTION Since the model has already been developed, we need only evaluate. Note that the function W expresses weight as a function of height, in inches. The domain is determined from the context of the application to be $(0, \infty)$. We convert 5 ft, 7 in. to inches,

$$5 \text{ ft}, 7 \text{ in.} = 5(12 \text{ in.}) + 7 \text{ in.} = 67 \text{ in.},$$

and evaluate the function:

$$W(67) = \left(\frac{67}{12.3}\right)^3 \approx 162.$$

A 5 ft, 7 in. person has a threshold weight of about 162 lb.

Example 2 *Ibuprofen in the Bloodstream.* Ibuprofen is a pain relief medication. The polynomial function

$$M(t) = 0.5t^4 + 3.45t^3 - 96.65t^2 + 347.7t, \; 0 \le t \le 6,$$

can be used to estimate the number of milligrams of ibuprofen in the bloodstream t hours after 400 mg of the medication has been swallowed.

a) Graph the function using the viewing windows $[-24, 15, -14{,}400, 8000]$ with Xscl = 3 and Yscl = 800 and $[0, 6, -200, 500]$ with Xscl = 1 and Yscl = 100.

b) Discuss the limitations of the model.

c) Find the domain, the relative maximum, and the range.

d) Find the number of milligrams in the bloodstream at $t = 0, 0.5, 1, 1.5$, and so on, up to 3 hr.

SOLUTION

a) The graphs using the two viewing windows are shown at the top of the following page.

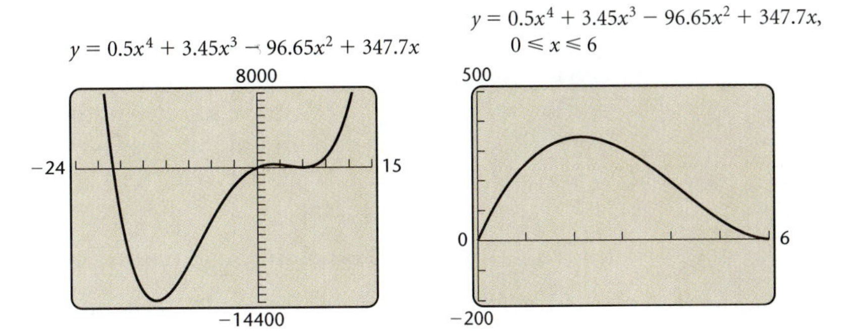

b) The domain of a polynomial function, unless restricted by a statement of the function, is $\mathbb{R}$. The graph on the left above gives a fairly complete picture of the curvature of the polynomial M. The implications of the application restrict the domain of the function. Presuming a patient had not taken any of the medication before, it seems reasonable that $M(0) = 0$; that is, at time 0, there are 0 mg of the medication in the bloodstream. After the medication has been taken, $M(t)$ will be positive for a period of time and eventually dissipate back to 0 and certainly not increase (unless another dose is taken). Thus the graph on the right above suggests that a better model includes a restriction on t,

$$M(t) = 0.5t^4 + 3.45t^3 - 96.65t^2 + 347.7t, \quad 0 \le t \le 6.$$

X	Y1
0	0
.5	150.15
1	255
1.5	318.26
2	344.4
2.5	338.625
3	306.9

X = 0

c) Given the restriction, the domain is $[0, 6]$. To find the range, we need to find the relative maximum. Some graphers can do this directly if we first enter an interval and a guess. For others, some use of TRACE and ZOOM will yield an estimate of the relative maximum (see Section 1.4). The maximum is about 345.76. Thus the range is about $[0, 345.76]$.

d) We can evaluate the function in a number of ways. To use a grapher, see your manual. If the grapher has a TABLE feature, you can start at 0 and use a step-value of 0.5.

We considered problems involving maxima or minima in Section 1.2. In those problems and in Example 2 above, the results were usually approximations. With the concepts developed in Section 2.2, we can get exact results when the function is quadratic. We can then check these results on a grapher.

Example 3 *Maximizing Area.* A stone mason has enough stones to enclose a rectangular patio with 60 ft of stone wall. If the attached house forms one side of the rectangle, what is the maximum area that the mason can enclose? What should the dimensions of the patio be in order to yield this area?

SOLUTION In this case, it is helpful to make use of the five-step problem-solving strategy that we introduced in Section R.9.

1. **Familiarize.** Suppose the patio were 10 ft wide. It would then be $60 - 2 \cdot 10 = 40$ ft long. The area would be $(10 \text{ ft})(40 \text{ ft}) = 400 \text{ ft}^2$.

To see if this is the maximum area possible, we would need to check other possible combinations of length and width. Instead, we will try to find an area function and determine where a maximum occurs.

We make a sketch of the situation, using w to represent the patio's width, in feet. Since only 60 ft of stone wall is available and the house serves as one side, then $(60 - 2w)$ feet of stone is available for the length.

2. **Translate.** Since the area of a rectangle is length times width, we have

$$A(w) = (60 - 2w)w$$
$$= -2w^2 + 60w,$$

where $A(w) =$ the area of the patio, in square feet, as a function of the width, w.

3. **Carry out.** To solve this problem, we need to determine the maximum value of $A(w)$ and find the dimensions for which that maximum occurs. Since A is a quadratic function and w^2 has a negative coefficient, we know that the function has a maximum value that occurs at the vertex of the graph of the function. The first coordinate of the vertex is

$$w = -\frac{b}{2a} = -\frac{60}{2(-2)} = 15 \text{ ft.}$$

Thus, if $w = 15$ ft, then the length $l = 60 - 2 \cdot 15 = 30$ ft; and the area is $15 \cdot 30$, or 450 ft^2.

4. **Check.** As a partial check, we note that $450 \text{ ft}^2 > 400 \text{ ft}^2$, which is the area we found in a guess in the familiarize step. A more complete check, assuming that the function $A(w)$ is correct, would examine a table of values for $A(w) = (60 - 2w)w$ and/or examine its graph.

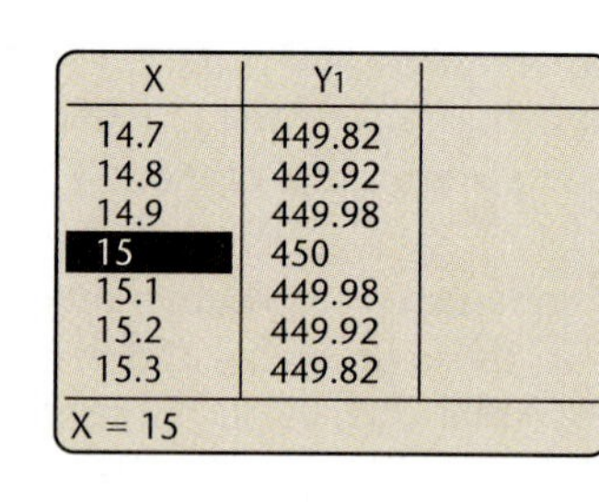

X	Y1
14.7	449.82
14.8	449.92
14.9	449.98
15	450
15.1	449.98
15.2	449.92
15.3	449.82

X = 15

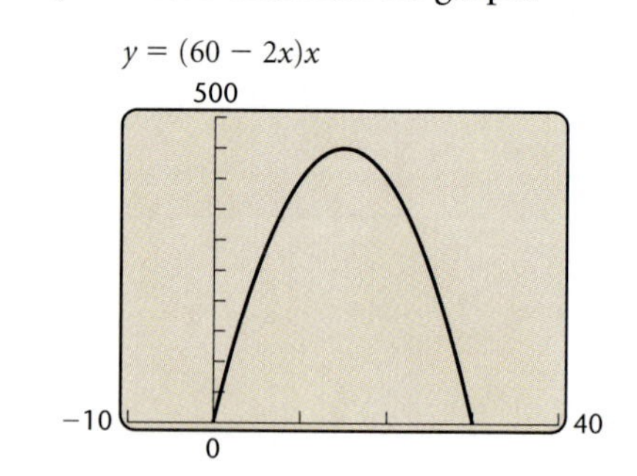

5. **State.** A maximum area of 450 ft^2 will occur if the patio is 15 ft wide and 30 ft long.

Example 4 *Finding the Depth of a Well.* A chlorine tablet is dropped into a well. Two seconds later, the sound of the splash is heard at the top of the well. The speed of the sound is 1100 ft/sec. How far is the top of the well from the water?

SOLUTION

1. **Familiarize.** We first make a drawing and label it with known and

unknown information. We let s = the depth of the well, in feet, t_1 = the time, in seconds, it takes for the tablet to hit the water, and t_2 = the time, in seconds, it takes for the sound to reach the top of the well.

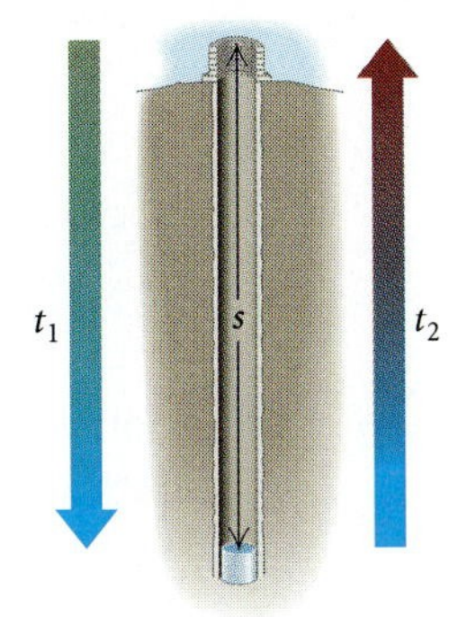

This gives us the equation

$$t_1 + t_2 = 2. \tag{1}$$

2. **Translate**. Can we find any relationship between the two times and the distance s? Often in problem solving you may need to look up related formulas in a physics book, another mathematics book, or maybe an encyclopedia. It turns out that the formula

$$s = 16t^2$$

gives the distance, in feet, that a dropped object falls in t seconds. The time t_1 that it takes the tablet to hit the water can be found as follows:

$$s = 16t_1^2, \quad \text{or} \quad t_1 = \frac{\sqrt{s}}{4}. \tag{2}$$

To find an expression for t_2, the time it takes the sound to travel to the top of the well, recall that *Distance* = *Rate* · *Time*. Thus,

$$s = 1100t_2, \quad \text{or} \quad t_2 = \frac{s}{1100}. \tag{3}$$

We now have expressions for t_1 and t_2, both in terms of s. Substituting into Equation (1), we obtain

$$t_1 + t_2 = 2, \quad \text{or} \quad \frac{\sqrt{s}}{4} + \frac{s}{1100} = 2. \tag{4}$$

3. **Carry out**. We solve Equation (4) for s. Multiplying by 1100, we get

$$275\sqrt{s} + s = 2200,$$

or

$$s + 275\sqrt{s} - 2200 = 0.$$

This equation is reducible to quadratic with $u = \sqrt{s}$. Substituting, we get

$$u^2 + 275u - 2200 = 0.$$

Using the quadratic formula, we can solve for u:

$$\begin{aligned} u &= \frac{-b \pm \sqrt{b^2 - 4ac}}{2a} \\ &= \frac{-275 + \sqrt{275^2 - 4 \cdot 1 \cdot (-2200)}}{2 \cdot 1} \\ &= \frac{-275 + \sqrt{84{,}425}}{2} \\ &\approx 7.78. \end{aligned}$$

We want only the positive solution.

Since $u \approx 7.78$, we have $\sqrt{s} \approx 7.78$, so $s \approx 60.5$.

4. **Check**. To check, we can substitute 60.5 for s in Equations (2) and (3) and see that $t_1 + t_2 \approx 2$. We leave the mathematics for the student.
5. **State**. The top of the well is about 60.5 ft above the water.

Interactive Discovery

Check the solution of Equation (4) in Example 4 with a grapher. Find the intersection of the curves $y_1 = \sqrt{x}/4 + x/1100$ and $y_2 = 2$. If necessary, review the Introduction to Graphs and Graphers for details.

Polynomial Curve Fitting

By looking at an input–output table, we can tell whether the data fit a polynomial function. Let's consider an example:

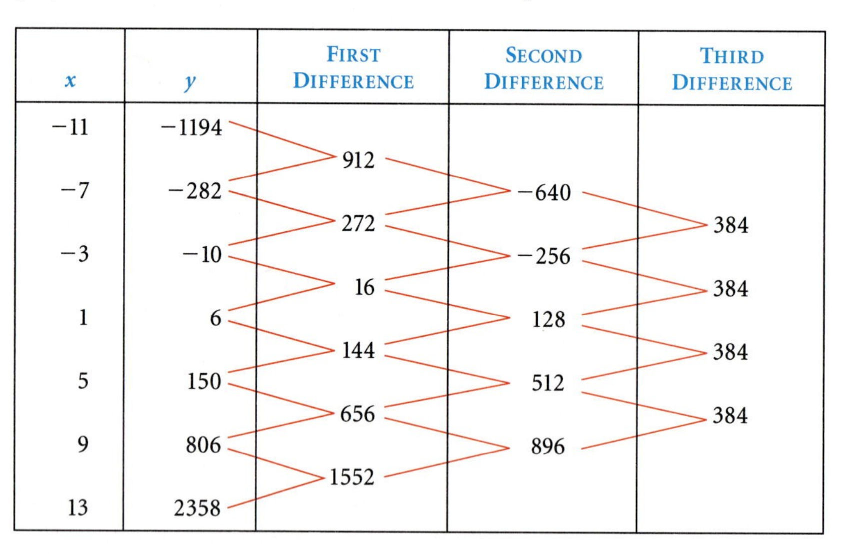

x	y	FIRST DIFFERENCE	SECOND DIFFERENCE	THIRD DIFFERENCE
−11	−1194			
		912		
−7	−282		−640	
		272		384
−3	−10		−256	
		16		384
1	6		128	
		144		384
5	150		512	
		656		384
9	806		896	
		1552		
13	2358			

In the first column, we see that the differences of the x-values are always the same, 4. Next, we take the first differences of the y-values, but these are not constant. We continue, taking the second differences; these are still not constant. The third differences, however, are all constant—the number 384. The following theorem tells us that the data in this table fit a third-degree, or cubic, polynomial.

The Polynomial Difference Theorem

A function f is a polynomial function of degree n if and only if for any set of x-values that differ by the same number, the nth differences of the corresponding y-values are the same nonzero constant.

Thus we can look at an input–output table and know for sure whether the data fit a polynomial function. Unfortunately, real-world data do not always make such a perfect fit. We can still find quadratic, cubic, and quartic polynomials that fit the data approximately. The remainder of this section presumes the use of a grapher that does quadratic, cubic, and quartic regression.

As we move through this text, we develop a "stable" of functions. These can serve as models for many applications. Let's look at some that we have considered thus far.

Linear Function:
$y_1 = mx + b$

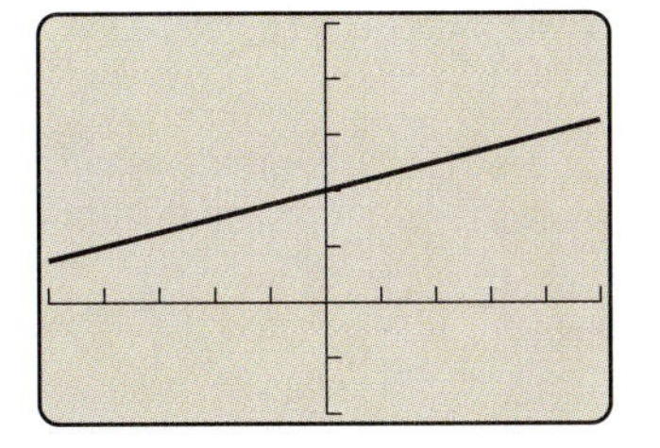

Quadratic Function:
$y_2 = ax^2 + bx + c, a > 0$

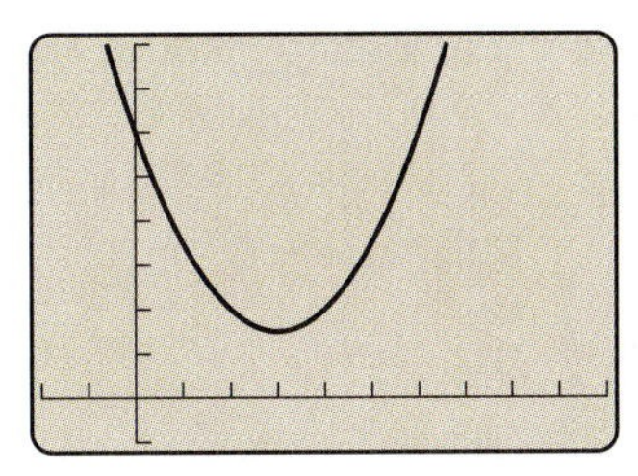

Quadratic Function:
$y_3 = ax^2 + bx + c, a < 0$

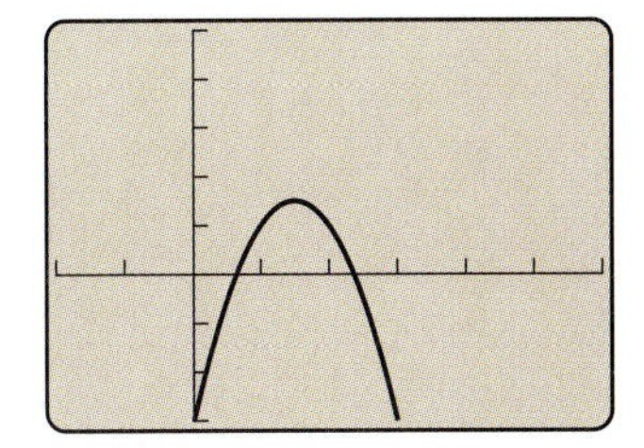

Cubic Function:
$y_4 = ax^3 + bx^2 + cx + d$

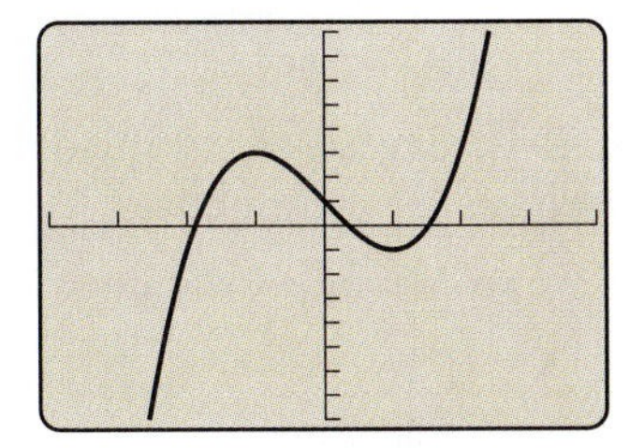

Quartic Function:
$y_5 = ax^4 + bx^3 + cx^2 + dx + e$

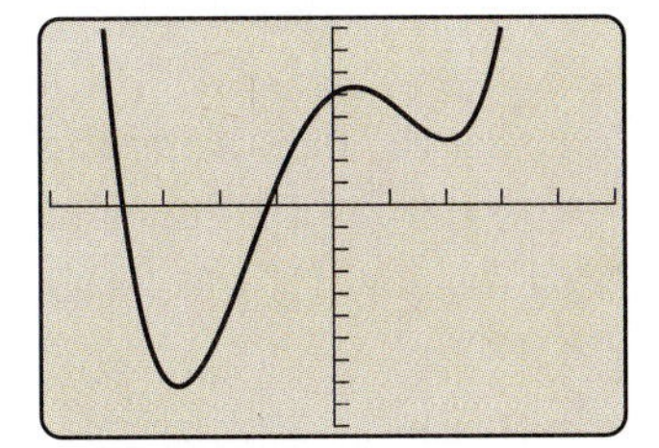

Now let's consider some real-world data. How can we decide which function might fit the data? Our simple way is to graph the data and look over the result. Simply stated, we "eyeball" the situation. Then we use a grapher to find a regression equation and make predictions.

Example 5 *Hours of Sleep versus Death Rate.* In a study by Dr. Harold J. Morowitz of Yale University, data were gathered that showed the relationship between the death rate of men and the average number of hours per day that the men slept.

Death Rate Related to Sleep

AVERAGE NUMBER OF HOURS OF SLEEP, x	DEATH RATE PER 100,000 MALES, y
5	1121
6	805
7	626
8	813
9	967

Source: "Hiding in the Hammond Report," *Hospital Practice*, by Harold J. Morowitz.

a) Make a scatterplot of the data.

b) Determine which, if any, of the functions seems to fit the data.

c) Use a grapher to fit the function to the data. Graph the equation using the same axes as the scatterplot.

d) Predict the death rate of males who sleep 4 hr, 5.5 hr, 7.5 hr, and 10 hr.

SOLUTION

a) We make a scatterplot of the data as shown on the left below.

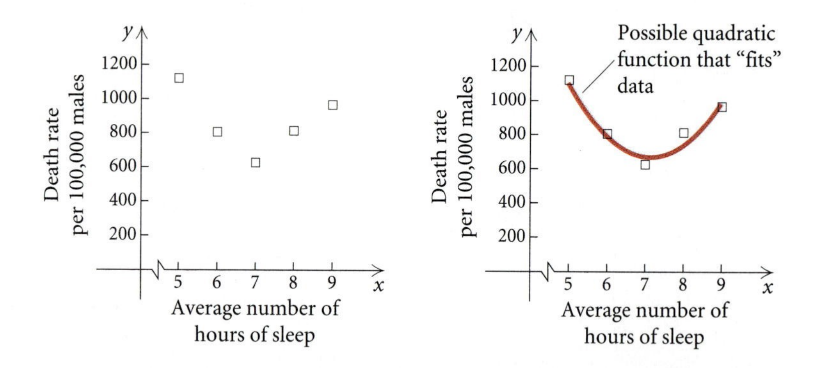

b) Note that the rate drops and then rises. This suggests that a quadratic function might fit the data. See the graph on the right above.

c) Using the quadratic REGRESSION feature, we get the following quadratic function:

$$f(x) = y = 93.28571429x^2 - 1336x + 5460.828571.$$

Depending on the accuracy required, we might round these coefficients, though it is easy to keep this equation in the grapher.

d) We then compute the outputs:

$$f(4) \approx 1609.4, \quad f(7.5) \approx 688.1,$$
$$f(5.5) \approx 934.7, \quad f(10) \approx 1429.4.$$

These can also be found with the TABLE feature set in ASK mode, after copying the quadratic function to the $y=$ screen. The results assert, for instance, that the death rate is about 688 per 100,000 for males who sleep 7.5 hr per night.

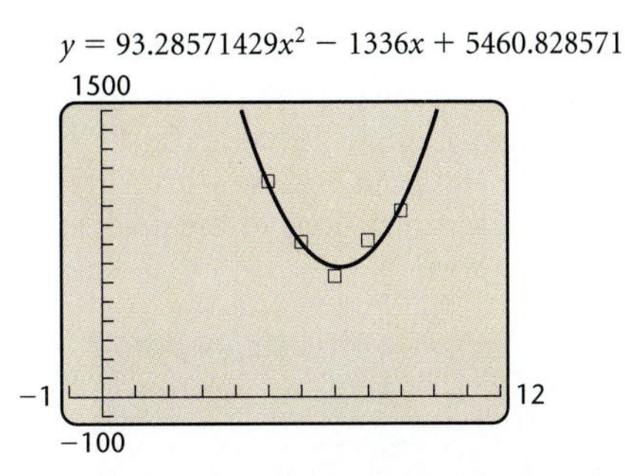

One must always be aware that a model such as the one in Example 5 has its limitations.

Interactive Discovery

In the model of Example 5, find and interpret $f(0)$ and $f(24)$. Argue whether such computations have meaning in the real world. What about $f(-1)$ and $f(25)$?

2.3 Exercise Set

1. *Vertical Leap.* A formula relating an athlete's vertical leap V, in inches, to hang time T, in seconds, is

$$V = 48T^2.$$

Anfernee Hardaway of the Orlando Magic has a vertical leap of 36 in. What is his hang time? (*Source:* National Basketball Association)

2. *Projectile Motion.* A stone thrown downward with an initial velocity of 34.3 m/sec will travel a distance of s meters, where

$$s(t) = 4.9t^2 + 34.3t$$

and t is in seconds. If a stone is thrown downward at 34.3 m/sec from a height of 294 m, how long will it take the stone to hit the ground?

3. *Games in a Sports League.* If there are x teams in a sports league and all the teams play each other twice, a total of $N(x)$ games are played, where

$$N(x) = x^2 - x.$$

A softball league has 9 teams, each of which plays the others twice. If the league pays \$45 per game for the field and umpires, how much will it cost to play the entire schedule?

4. *Windmill Power.* Under certain conditions, the power P, in watts per hour, generated by a windmill with winds blowing v miles per hour is given by

$$P(v) = 0.015v^3.$$

a) Find the power generated by 15-mph winds.

b) How fast must the wind blow in order to generate 120 watts of power in 1 hr?

5. *Interest Compounded Annually.* When P dollars is invested at interest rate i, compounded annually, for t years, the investment grows to A dollars, where

$$A = P(1 + i)^t.$$

a) Find the interest rate i if \$2560 grows to \$3610 in 2 yr.

b) Find the interest rate i if $10,000 grows to $13,310 in 3 yr.

6. *Maximizing Area.* A fourth-grade class decides to enclose a rectangular garden, using the side of the school as one side of the rectangle. What is the maximum area that the class can enclose with 32 ft of fence? What should the dimensions of the garden be in order to yield this area?

7. *Maximizing Volume.* A plastics manufacturer plans to produce a one-compartment vertical file by bending the long side of an 8-in. by 14-in. sheet of plastic along two lines to form a U-shape. How tall should the file be in order to maximize the volume that the file can hold?

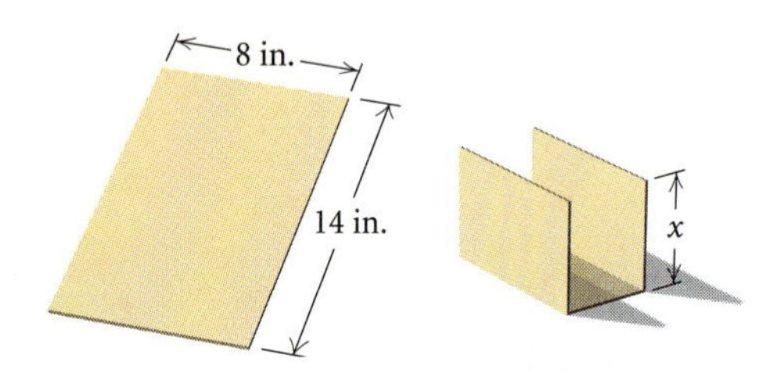

8. *Maximizing Area.* The sum of the base and the height of a parallelogram is 69 cm. Find the dimensions for which the area is a maximum.

9. *Maximizing Area.* The sum of the base and the height of a triangle is 20 cm. Find the dimensions for which the area is a maximum.

10. *Finding the Height of a Cliff.* A water balloon is dropped from a cliff. Exactly 3 sec later, the sound of the balloon hitting the ground reaches the top of the cliff. How high is the cliff? (*Hint:* See Example 4.)

11. *Finding the Height of an Elevator Shaft.* Jenelle dropped a screwdriver from the top of an elevator shaft. Exactly 5 sec later, she hears the sound of the screwdriver hitting the bottom of the shaft. How tall is the elevator shaft? (*Hint:* See Example 4.)

12. *Minimizing Cost.* Aki's Bicycle Designs has determined that when x hundred bicycles are built, the average cost per bicycle is given by

$$C(x) = 0.1x^2 - 0.7x + 2.425,$$

where $C(x)$ is in hundreds of dollars. How many bicycles should be built in order to minimize the average cost per bicycle?

Maximizing Profit. In business, profit is the difference between revenue and cost, that is,

$$\textit{Total profit} = \textit{Total revenue} - \textit{Total cost},$$
$$P(x) = R(x) - C(x),$$

where x is the number of units. Find the maximum profit and the number of units that must be sold in order to yield the maximum profit for each of the following.

13. $R(x) = 50x - 0.5x^2$, $C(x) = 10x + 3$

14. $R(x) = 5x$, $C(x) = 0.001x^2 + 1.2x + 60$

15. *Maximizing Area.* A rancher needs to enclose two adjacent rectangular corrals, one for sheep and one for cattle. If a river forms one side of the corrals and 240 yd of fencing is available, what is the largest total area that can be enclosed?

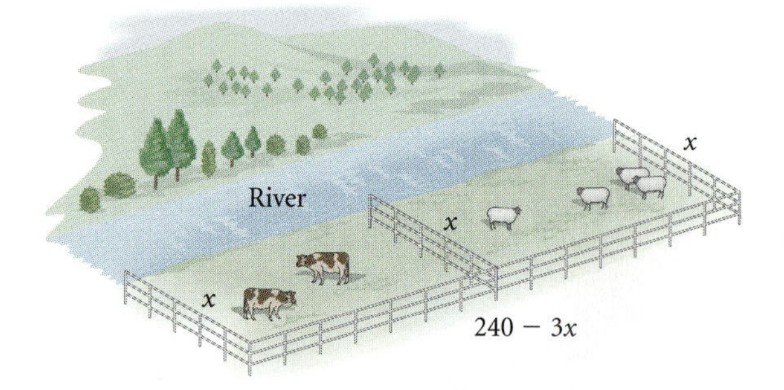

16. *Norman Window.* A Norman window is a rectangle with a semicircle on top. Sky Blue Windows is designing a Norman window that will require 24 ft of trim. What dimensions will allow the maximum amount of light to enter a house?

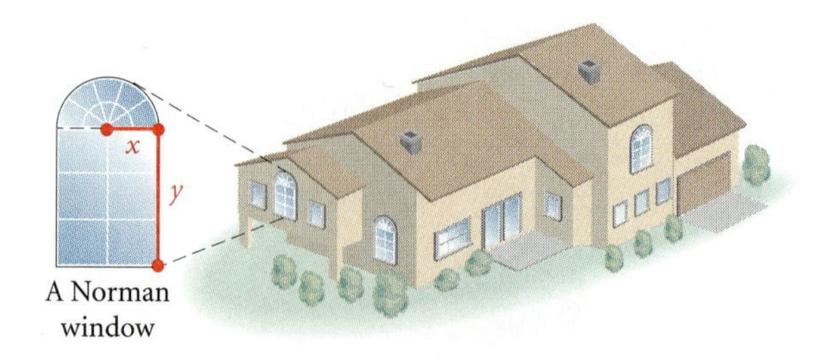

For the scatterplots and graphs in Exercises 17–25, determine which, if any, of the following functions might be used as a model for the data.

a) *Linear,* $f(x) = mx + b$

b) *Quadratic,* $f(x) = ax^2 + bx + c,\ a > 0$

c) *Quadratic,* $f(x) = ax^2 + bx + c,\ a < 0$

d) *Polynomial, not quadratic or linear*

17. **18.**

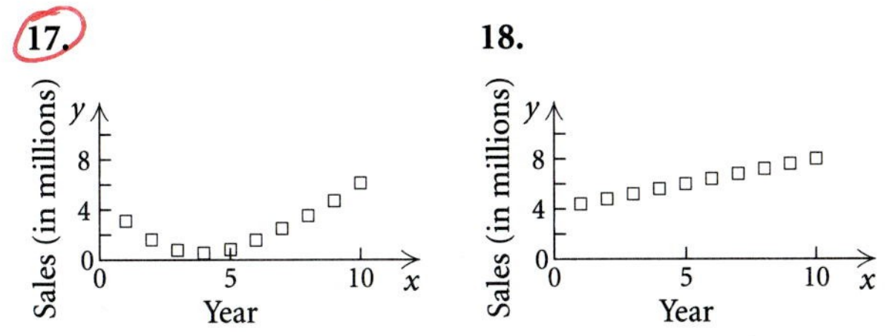

19. **20.**

21. **22.**

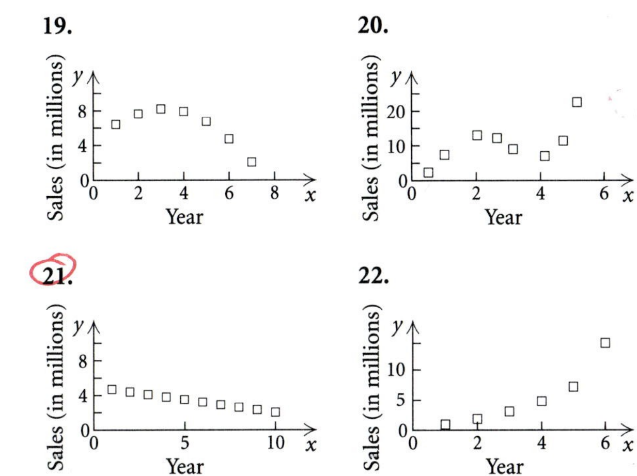

23.

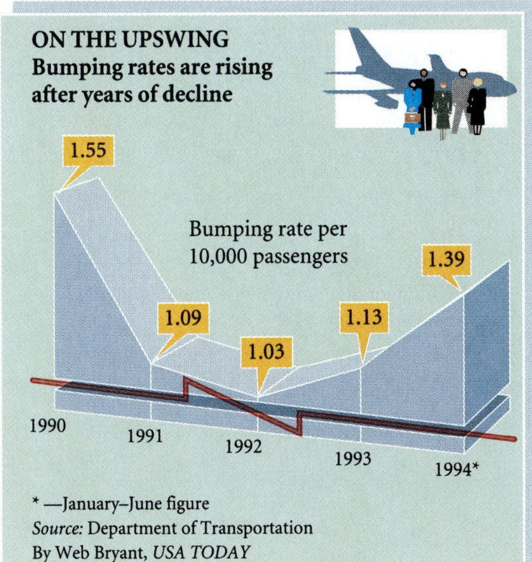

24.

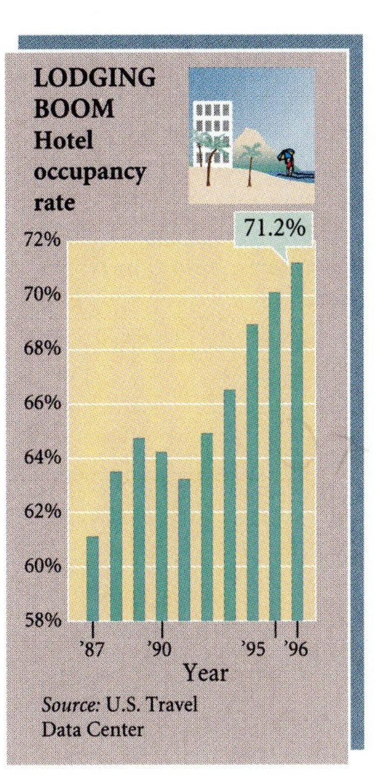

25.

DRIVER FATALITIES BY AGE

Number of licensed drivers per 100,000 who died in motor vehicle accidents in 1990. The fatality rates for both the 70–79 group and 80+ age group were lower than for the 15- to 24-year-olds.

Ages	Number of driver deaths per 100,000
15–24	28
25–39	15
40–69	10
70–79	15
80+	25

Source: National Highway Traffic Administration

For each table of data in Exercises 26–29, answer the following questions.

a) *Is the change in the inputs x the same?*
b) *Find the first differences of the y-values. Is the change in the outputs y the same?*
c) *Find the second, third, and fourth differences, as needed. Is any set of differences nonzero and constant?*
d) *Does a quadratic, cubic, or quartic function fit the data?*
e) *If so, fit a function to the data using the* REGRESSION *feature of a grapher.*
f) *Then find the missing data.*

26.

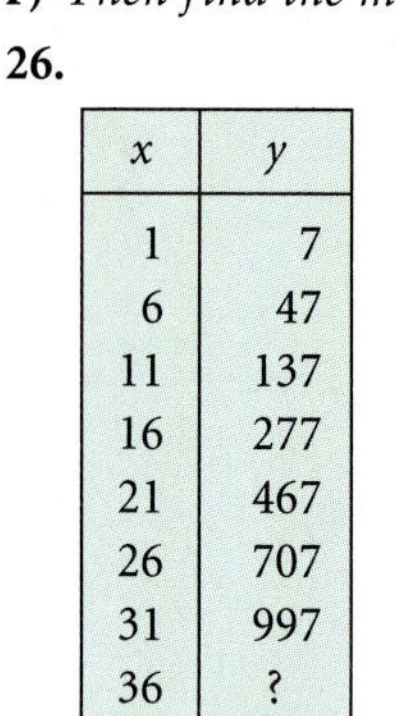

x	y
1	7
6	47
11	137
16	277
21	467
26	707
31	997
36	?
?	2657

27.

x	y
−11	1204
−7	292
−3	20
1	4
5	−140
9	−796
13	−2348
17	?
21	?

28.

x	y
10	37
12	42.6
14	47.4
16	51.4
18	54.6
20	57
22	58.6
24	?
?	45.1

29.

x	y
−8	3447
−5	369
−2	−27
1	−9
4	99
7	1917
10	9009
12	?
?	−19.94

30. *Airline Passenger Bumping.* Use the data in the graph of Exercise 23.

a) Use the REGRESSION feature of a grapher to fit a quadratic function to the data.
b) Predict the rate of passenger bumping in 1998, 2000, and 2005.

Shoe Size and Life Expectancy. The data in the following table are the result of a Swedish study relating one's life expectancy to the size of one's feet. Use these data in Exercises 31 and 32.

Shoe Size	Life Expectancy for Women (in Years)	Shoe Size	Life Expectancy for Men (in Years)
4	69	5	66
5	76	6	69
6	82	7	69
7	84	8	72
8	75	9	75
9	72	10	77
10	70	11	82
11	69	12	79
		13	72
		14	69

Source: Orthopedic Quarterly.

31. a) Use the REGRESSION feature of a grapher to fit a quadratic function to the data for women.
b) Estimate the life expectancy of women with shoe sizes of $5\frac{1}{2}$ and $8\frac{1}{2}$.

32. a) Use the REGRESSION feature of a grapher and fit a quadratic function to the data for men.
b) Estimate the life expectancy of men with shoe sizes of $10\frac{1}{2}$ and $15\frac{1}{2}$.

33. *Prices of Personal Computers.* The price P of a personal computer has varied greatly in recent years. The data in the table relate price P, in dollars, to time t, in years, where $t = 1$ corresponds to 1981.

Year, t	Average Price P of a Personal Computer
1981 ($t = 1$)	$2290
1982 ($t = 2$)	1500
1983 ($t = 3$)	1400
1984 ($t = 4$)	1850
1985 ($t = 5$)	2540
1986 ($t = 6$)	2400
1987 ($t = 7$)	1720
1988 ($t = 8$)	1860
1989 ($t = 9$)	1930
1990 ($t = 10$)	2100
1991 ($t = 11$)	1820
1992 ($t = 12$)	1640

a) Find a cubic function $y = ax^3 + bx^2 + cx + d$ that fits the data.

b) Graph the cubic function of part (a).
c) Use the cubic function to predict the price of a personal computer in 1998.
d) Find a quartic polynomial function $y = ax^4 + bx^3 + bx^2 + cx + d$ that fits the data.
e) Graph the quartic function of part (d).
f) Use the quartic function to predict the price of a personal computer in 1998.

34. *Damage to the Ozone Layer.* The concentration of chlorine compounds in the stratosphere serves as an indicator of damage to the ozone layer. The data in the following table show the relationship of estimated chlorine concentration in the atmosphere, in parts per billion (ppb), to the year.

YEAR	CHLORINE CONCENTRATION (IN PARTS PER BILLION)
1985	2.5
1995	3.3
2010	3.9
2035	3.7

Source: Adapted by U.S. EPA by NRDC Earth Action Guide, "Saving the Ozone."

a) Use the REGRESSION feature on a grapher to find a linear, a quadratic, a cubic, and a quartic function to fit the data.
b) It is estimated that in 2060 the chlorine concentration will be 3.3 ppb. Which function in part (a) would best make this prediction?
c) Use the answer to part (b) to predict the chlorine concentration in 2085.

Skill Maintenance

Simplify.

35. $\dfrac{5x^7 + 25x^6 + 10x^4 - 20x^3}{5x^3}$

36. $\dfrac{8a^5 - 10a^4 + 12a^3}{2a^3}$

37. Multiply: $(5x^3 - 2x^2 + 5x - 1)(x - 3)$.

38. Subtract: $(8x^2 + 5x - 7) - (3x^2 - 6x + 9)$.

Synthesis

39. ◈ Regarding Exercise 33, discuss the merits of which function to use to predict the price of a personal computer in 1998. How could you test your assertions?

40. ◈ Write a problem for a classmate to solve. Design the problem so that it is a maximum or minimum problem using a quadratic function.

41. In early 1995, \$2000 was deposited at a certain interest rate. One year later, \$1200 was deposited in another account at the same rate. At the end of that year, there was a total of \$3573.80 in both accounts. What is the annual interest rate?

42. *Minimizing Area.* A 24-in. piece of string is cut into two pieces. One piece is used to form a circle while the other is used to form a square. How should the string be cut so that the sum of the areas is a minimum?

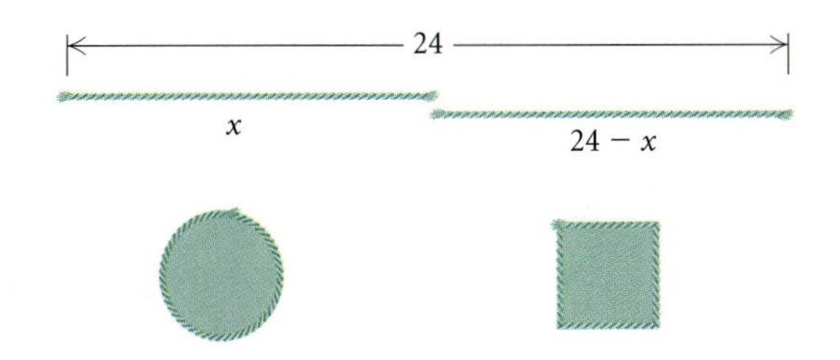

For each of the two input–output tables in Exercises 43 and 44:

a) *Fit a cubic function to the existing data. Then use that equation to find the missing values.*
b) *Fit a quartic function to the existing data. Then use that equation to find the missing values.*

43.

x	y
−4	?
−3	7
−2	11
−1	33
0	45
1	19
?	23
3	46
4	?

44.

x	y
20	12.4
?	24.3
30	28.5
40	19.6
?	39.2
50	78.4
60	196.8
70	793.6
80	?

45. *The Effect of Advertising on Movie Revenue.* The following table contains actual data pertaining to the box office revenue of certain movies together with the amount of money that was spent to advertise each movie. We want to explore whether box office revenue is directly related to advertising expenditures.

MOVIE	ADVERTISING BUDGET (IN MILLIONS)	BOX OFFICE REVENUE (IN MILLIONS)
The Lion King	\$23.3	\$300.4
Forrest Gump	25.0	298.5
True Lies	20.5	146.3
The Santa Clause	19.4	137.8
The Flintstones	10.6	130.6
Clear & Present Danger	17.9	121.8
Speed	17.2	121.2
The Mask	11.2	118.8
Maverick	13.8	101.6
Interview with the Vampire	15.4	100.7

Source: The Hollywood Reporter.

a) Using a ZOOMSTAT feature, make a scatterplot of the data.

b) Analyze the scatterplot and try to decide whether a linear, quadratic, cubic, or quartic equation might fit the data.

c) Use the REGRESSION feature on a grapher to find a quadratic, a cubic, and a quartic function to fit the data. Assume that advertising is the independent variable.

d) Plot each regression equation with the scatterplot. Examine the results and try to decide which function fits best.

e) You are about to make a blockbuster movie called *A Day in the Life of a Math Professor.* You decide to spend \$30 million to advertise the film. Use each of the three functions found in part (c) to predict the box office revenue from the movie.

2.4 Polynomial Division; The Remainder and Factor Theorems

- *Do long division with polynomials and determine whether one polynomial is a factor of another.*
- *Use synthetic division to divide a polynomial by $x - c$.*
- *Use the remainder theorem to find a function value $f(c)$.*
- *Use the factor theorem to determine whether $x - c$ is a factor of $f(x)$.*

In general, finding exact zeros of polynomial functions is neither easy nor straightforward. In this section and the one that follows, we develop concepts that help us find exact zeros of polynomial functions with degree 3 or greater.

We will now allow the coefficients of a polynomial to be complex numbers. In certain cases, we will restrict the coefficients to be real numbers, rational numbers, or integers, as shown in the following examples.

POLYNOMIAL	TYPE OF COEFFICIENT
$5x^3 - 3x^2 + (2 + 4i)x + i$	Complex
$5x^3 - 3x^2 + \sqrt{2}x - \pi$	Real
$5x^3 - 3x^2 + \frac{2}{3}x - \frac{7}{4}$	Rational
$5x^3 - 3x^2 + 8x - 11$	Integer

Let's review the meaning of a zero of a function.

Example 1 Consider $P(x) = x^3 + 2x^2 - 5x - 6$. Determine whether each of the numbers 3 and -1 is a zero of $P(x)$.

SOLUTION We have

$$P(3) = (3)^3 + 2(3)^2 - 5(3) - 6 = 24.$$ **Substituting 3 into the polynomial**

Since $P(3) \neq 0$, 3 is *not* a zero of the polynomial:

$$P(-1) = (-1)^3 + 2(-1)^2 - 5(-1) - 6 = 0.$$ **Substituting -1 into the polynomial**

Since $P(-1) = 0$, we know that -1 is a zero of $P(x)$.

Division and Factors

When we divide one polynomial by another, we obtain a quotient and a remainder. If the remainder is 0, then the divisor is a **factor** of the dividend.

Example 2 Divide to determine whether $x + 1$ is a factor of

$$x^3 + 2x^2 - 5x - 6.$$

SOLUTION

Quotient

$$\begin{array}{r|l}
 & x^2 + x - 6 \\
\hline
x + 1 & x^3 + 2x^2 - 5x - 6 \\
 & \underline{x^3 + x^2} \\
 & \quad x^2 - 5x \\
 & \quad \underline{x^2 + x} \\
 & \qquad -6x - 6 \\
 & \qquad \underline{-6x - 6} \\
 & \qquad\qquad 0
\end{array}$$

Divisor ($x + 1$) **Remainder** (0)

Since the remainder is 0, we know that $x + 1$ is a factor of $x^3 + 2x^2 - 5x - 6$.

Example 3 Divide to determine whether $x^2 + 3x - 1$ is a factor of $x^4 - 81$.

SOLUTION

$$\begin{array}{r|l}
 & x^2 - 3x + 10 \\
\hline
x^2 + 3x - 1 & x^4 \qquad\qquad\qquad\quad - 81 \\
 & \underline{x^4 + 3x^3 - x^2} \\
 & \quad -3x^3 + x^2 \\
 & \quad \underline{-3x^3 - 9x^2 + 3x} \\
 & \qquad\quad 10x^2 - 3x - 81 \\
 & \qquad\quad \underline{10x^2 + 30x - 10} \\
 & \qquad\qquad\quad -33x - 71
\end{array}$$

Note that spaces have been left for missing terms.

Since the remainder is not 0, we know that $x^2 + 3x - 1$ is not a factor of $x^4 - 81$.

When we divide a polynomial $P(x)$ by a divisor $d(x)$, a polynomial $Q(x)$ is the quotient and a polynomial $R(x)$ is the remainder. The remainder must either be 0 or have degree less than that of $d(x)$. To check a division, we multiply the quotient by the divisor and add the remainder, to see if we get the dividend. Thus these polynomials are related as follows:

$$P(x) = d(x) \cdot Q(x) + R(x)$$

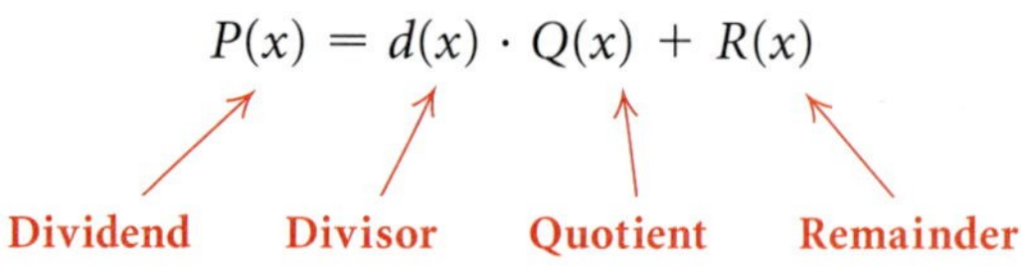

For example, if $P(x) = x^3 + 2x^2 - 5x - 6$ and $d(x) = x + 1$, then $Q(x) = x^2 + x - 6$ and $R(x) = 0$, and

$$\underbrace{x^3 + 2x^2 - 5x - 6}_{P(x)} = \underbrace{(x + 1)}_{d(x)} \cdot \underbrace{(x^2 + x - 6)}_{Q(x)} + \underbrace{0}_{R(x).}$$

If $P(x) = x^4 - 81$ and $d(x) = x^2 + 3x - 1$, then $Q(x) = x^2 - 3x + 10$ and $R(x) = -33x - 71$, and

$$\underbrace{x^4 - 81}_{P(x)} = \underbrace{(x^2 + 3x - 1)}_{d(x)} \cdot \underbrace{(x^2 - 3x + 10)}_{Q(x)} + \underbrace{(-33x - 71)}_{R(x).}$$

The Remainder Theorem and Synthetic Division

The Remainder Theorem

If a number c is substituted for x in the polynomial $f(x)$, then the result $f(c)$ is the remainder that would be obtained by dividing $f(x)$ by $x - c$. That is, if $f(x) = (x - c) \cdot Q(x) + R$, then $f(c) = R$.

Proof: The equation $f(x) = d(x) \cdot Q(x) + R(x)$, where $d(x) = x - c$, is the basis of this proof. If we divide $f(x)$ by $x - c$, we obtain a quotient $Q(x)$ and a remainder $R(x)$ related as follows:

$$f(x) = (x - c) \cdot Q(x) + R(x).$$

The remainder $R(x)$ must either be 0 or have degree less than $x - c$. Thus, $R(x)$ must be a constant. Let's call this constant R. In the expression above, we get a true sentence whenever we replace x with any number. Let's replace x with c. We get

$$\begin{aligned} f(c) &= (c - c) \cdot Q(c) + R \\ &= 0 \cdot Q(c) + R \\ &= R. \end{aligned}$$

This tells us that the function value $f(c)$ is the remainder obtained when we divide $f(x)$ by $x - c$.

The remainder theorem motivates us to find a rapid way of dividing by $x - c$, in order to find function values. To streamline division, we can arrange the work so that duplicate and unnecessary writing is avoided. Consider the following.

A.

$$\begin{array}{r}
4x^2 + 5x + 11 \\
x - 2\overline{)4x^3 - 3x^2 + x + 7} \\
\underline{4x^3 - 8x^2} \phantom{+ x + 7} \\
5x^2 + x \\
\underline{5x^2 - 10x} \\
11x + 7 \\
\underline{11x - 22} \\
29
\end{array}$$

B.

$$\begin{array}{r}
4 \quad 5 \quad 11 \\
1 - 2\overline{)4 - 3 + 1 + 7} \\
\underline{4 - 8} \\
5 + 1 \\
\underline{5 - 10} \\
11 + 7 \\
\underline{11 - 22} \\
29
\end{array}$$

The division in (B) is the same as that in (A), but we wrote only the coefficients. The color numerals are duplicated, so we look for an arrangement in which they are not duplicated. We can also simplify things by using the opposite of -2 and then adding instead of subtracting. When things are thus "collapsed," we have the algorithm known as **synthetic division.**

C. *Synthetic Division*

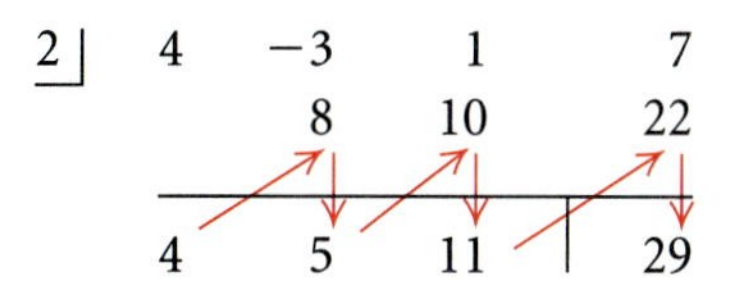

We "bring down" the 4. Then we multiply it by the 2 to get 8 and add to get 5. We then multiply 5 by 2 to get 10, add, and so on. The last number, 29, is the remainder. The others, 4, 5, and 11, are the coefficients of the quotient.

We write a 0 in the synthetic division for a missing term in the dividend.

Program

SYNDIV: This program does synthetic division.

Example 4 Use synthetic division to find the quotient and the remainder:

$$(2x^3 + 7x^2 - 5) \div (x + 3).$$

SOLUTION First we note that $x + 3 = x - (-3)$.

$$\begin{array}{r|rrrr}
-3 & 2 & 7 & 0 & -5 \\
 & & -6 & -3 & 9 \\
\hline
 & 2 & 1 & -3 & 4
\end{array}$$

Note: **We must write 0's for missing terms.**

The quotient is $2x^2 + x - 3$. The remainder is 4.

Values of Polynomial Functions

We can now apply synthetic division to find polynomial function values.

Example 5 Given that $f(x) = 2x^5 - 3x^4 + x^3 - 2x^2 + x - 8$, find $f(10)$.

SOLUTION By the remainder theorem, $f(10)$ is the remainder when $f(x)$ is divided by $x - 10$. We use synthetic division to find that remainder.

$$\begin{array}{r|rrrrrr} 10 & 2 & -3 & 1 & -2 & 1 & -8 \\ & & 20 & 170 & 1710 & 17{,}080 & 170{,}810 \\ \hline & 2 & 17 & 171 & 1708 & 17{,}081 & 170{,}802 \end{array}$$

Thus, $f(10) = 170{,}802$.

Compare the computations in Example 5 with those in a direct substitution:

$$f(10) = 2(10)^5 - 3(10)^4 + (10)^3 - 2(10)^2 + 10 - 8.$$

The computations in synthetic division are less complicated.

Example 6 Determine whether -4 is a zero of $f(x)$, where $f(x) = x^3 + 8x^2 + 8x - 32$.

SOLUTION We use synthetic division and the remainder theorem to find $f(-4)$.

$$\begin{array}{r|rrrr} -4 & 1 & 8 & 8 & -32 \\ & & -4 & -16 & 32 \\ \hline & 1 & 4 & -8 & 0 \end{array}$$

Since $f(-4) = 0$, the number -4 is a zero of $f(x)$.

Finding Factors of Polynomials

We now consider the following useful corollary of the remainder theorem.

The Factor Theorem

For a polynomial $f(x)$, if $f(c) = 0$, then $x - c$ is a factor of $f(x)$.

Proof: If we divide $f(x)$ by $x - c$, we obtain a quotient and a remainder, related as follows:

$$f(x) = (x - c) \cdot Q(x) + f(c).$$

Then if $f(c) = 0$, we have

$$f(x) = (x - c) \cdot Q(x),$$

so $x - c$ is a factor of $f(x)$.

The factor theorem is very useful in factoring polynomials, and hence in solving equations.

Example 7 Let $f(x) = x^3 + 2x^2 - 5x - 6$. Factor $f(x)$ and solve the equation $f(x) = 0$.

SOLUTION We look for linear factors of the form $x - c$. Let's try $x - 1$. We use synthetic division to see whether $f(1) = 0$.

$$\begin{array}{r|rrrr} 1 & 1 & 2 & -5 & -6 \\ & & 1 & 3 & -2 \\ \hline & 1 & 3 & -2 & -8 \end{array}$$

Since $f(1) \neq 0$, we know that $x - 1$ is not a factor of $f(x)$. We try $x + 1$ or $x - (-1)$ in the form $x - c$.

$$\begin{array}{r|rrrr} -1 & 1 & 2 & -5 & -6 \\ & & -1 & -1 & 6 \\ \hline & 1 & 1 & -6 & 0 \end{array}$$

Since $f(-1) = 0$, we know that $x + 1$ is one factor and the quotient, $x^2 + x - 6$, is another. Thus,

$$f(x) = (x + 1)(x^2 + x - 6).$$

The trinomial is easily factored, so we have

$$f(x) = (x + 1)(x + 3)(x - 2).$$

Our goal is to solve the equation $f(x) = 0$. To do so, we use the principle of zero products. The solutions are -1, -3, and 2. Thus the solution set is $\{-1, -3, 2\}$. Check these solutions on a grapher.

2.4 Exercise Set

1. Determine whether 4, 5, and -2 are zeros of
$$f(x) = x^3 - 9x^2 + 14x + 24.$$
2. Determine whether 2, 3, and -1 are zeros of
$$f(x) = 2x^3 - 3x^2 + x - 1.$$
3. For $f(x)$ in Exercise 1, use long division to determine which of the following are factors of $f(x)$.
 a) $x - 4$ **b)** $x - 5$ **c)** $x + 2$
4. For $f(x)$ in Exercise 2, use long division to determine which of the following are factors of $f(x)$.
 a) $x - 2$ **b)** $x - 3$ **c)** $x + 1$

In each of the following, a polynomial $P(x)$ and a divisor $d(x)$ are given. Find the quotient $Q(x)$ and the remainder $R(x)$ when $P(x)$ is divided by $d(x)$, and express $P(x)$ in the form $d(x) \cdot Q(x) + R(x)$.

5. $P(x) = x^3 + 6x^2 - x - 30$,
 $d(x) = x - 2$
6. $P(x) = 2x^3 - 3x^2 + x - 1$,
 $d(x) = x - 2$
7. $P(x) = x^3 + 6x^2 - x - 30$,
 $d(x) = x - 3$
8. $P(x) = 2x^3 - 3x^2 + x - 1$,
 $d(x) = x - 3$
9. $P(x) = x^3 - 8$,
 $d(x) = x + 2$
10. $P(x) = x^3 + 27$,
 $d(x) = x + 1$
11. $P(x) = x^4 + 9x^2 + 20$,
 $d(x) = x^2 + 4$
12. $P(x) = x^4 + x^2 + 2$,
 $d(x) = x^2 + x + 1$
13. $P(x) = 5x^7 - 3x^4 + 2x^2 - 3$,
 $d(x) = 2x^2 - x + 1$
14. $P(x) = 6x^5 + 4x^4 - 3x^2 + x - 2$,
 $d(x) = 3x^2 + 2x - 1$

Use synthetic division to find the quotient and the remainder.

15. $(2x^4 + 7x^3 + x - 12) \div (x + 3)$
16. $(x^3 - 7x^2 + 13x + 3) \div (x - 2)$
17. $(x^3 - 2x^2 - 8) \div (x + 2)$
18. $(x^3 - 3x + 10) \div (x - 2)$
19. $(x^4 - 1) \div (x - 1)$

20. $(x^5 + 32) \div (x + 2)$

21. $(2x^4 + 3x^2 - 1) \div \left(x - \frac{1}{2}\right)$

22. $(3x^4 - 2x^2 + 2) \div \left(x - \frac{1}{4}\right)$

23. $(x^4 - y^4) \div (x - y)$

24. $(x^3 + 3ix^2 - 4ix - 2) \div (x + i)$

Use synthetic division to find the function values.

25. $f(x) = x^3 - 6x^2 + 11x - 6$; find $f(1)$, $f(-2)$, and $f(3)$.

26. $f(x) = x^3 + 7x^2 - 12x - 3$; find $f(-3)$, $f(-2)$, and $f(1)$.

27. $f(x) = 2x^5 - 3x^4 + 2x^3 - x + 8$; find $f(20)$ and $f(-3)$.

28. $f(x) = x^5 - 10x^4 + 20x^3 - 5x - 100$; find $f(-10)$ and $f(5)$.

29. $f(x) = x^4 - 16$; find $f(2)$, $f(-2)$, $f(3)$, and $f(1 - \sqrt{2})$.

30. $f(x) = x^5 + 32$; find $f(2)$, $f(-2)$, $f(3)$, and $f(2 + 3i)$.

Using synthetic division, determine whether the numbers are zeros of the polynomials.

31. $-3, 2$; $f(x) = 3x^3 + 5x^2 - 6x + 18$

32. $-4, 2$; $f(x) = 3x^3 + 11x^2 - 2x + 8$

33. $-3, \frac{1}{2}$; $f(x) = x^3 - \frac{7}{2}x^2 + x - \frac{3}{2}$

34. $i, -i, -2$; $f(x) = x^3 + 2x^2 + x + 2$

Factor the polynomial $f(x)$. Then solve the equation $f(x) = 0$.

35. $f(x) = x^3 + 4x^2 + x - 6$

36. $f(x) = x^3 + 5x^2 - 2x - 24$

37. $f(x) = x^3 - 6x^2 + 3x + 10$

38. $f(x) = x^3 + 2x^2 - 13x + 10$

39. $f(x) = x^3 - x^2 - 14x + 24$

40. $f(x) = x^3 - 3x^2 - 10x + 24$

41. $f(x) = x^4 - x^3 - 19x^2 + 49x - 30$

42. $f(x) = x^4 + 11x^3 + 41x^2 + 61x + 30$

Skill Maintenance

Find an equation for a line with the given slope that passes through the specified point.

43. $m = \frac{3}{4}$; $(-7, 0)$

44. $m = -\frac{2}{7}$; $(5, 0)$

Find an equation for a line passing through the given points.

45. $(-1, 5)$, $(6, 3)$

46. $(1, -2)$, $(4, 7)$

Synthesis

47. ◈ Is synthetic division always the fastest way to evaluate a polynomial function? What about using a grapher? Why or why not?

48. ◈ Can an nth-degree polynomial have more than n roots? Why or why not?

In Exercises 49 and 50, a graph of a polynomial function is given. On the basis of the graph:

a) *Find as many factors as you can of the polynomial.*

b) *Construct a polynomial function with the zeros shown in the graph.*

c) *Can you find any other polynomial functions with the given zeros?*

d) *Can you find any other polynomial functions with the given zeros and the same graph?*

49. **50.**

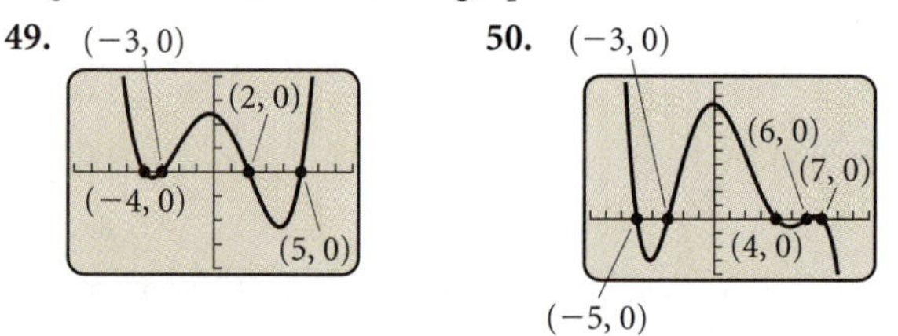

51. Find k so that $x + 2$ is a factor of $x^3 - kx^2 + 3x + 7k$.

52. For what values of k will the remainder be the same when $x^2 + kx + 4$ is divided by $x - 1$ or $x + 1$?

53. *Beam Deflection.* A beam rests at two points A and B and has a concentrated load applied to its center. Let y = the deflection, in feet, of the beam at a distance of x feet from A. Under certain conditions, this deflection is given by

$$y = \frac{1}{13}x^3 - \frac{1}{14}x.$$

Find the zeros of the polynomial in the interval $[0, 2]$.

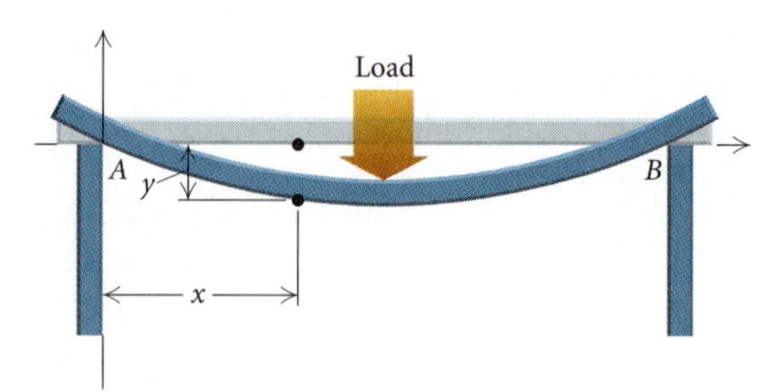

Solve.

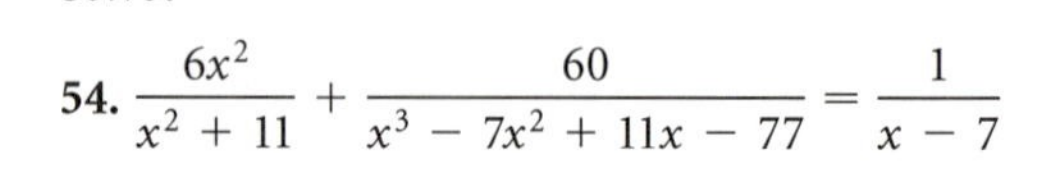

54. $\dfrac{6x^2}{x^2 + 11} + \dfrac{60}{x^3 - 7x^2 + 11x - 77} = \dfrac{1}{x - 7}$

55. $\frac{2x^2}{x^2 - 1} + \frac{4}{x + 3} = \frac{32}{x^3 + 3x^2 - x - 3}$

56. Use a grapher to graph $f(x) = x^3 + kx^2 - 19x - 35$ for $k = 15$, 19, 20, and 24. Compare the graphs. Do the zeros change when you change a coefficient?

57. Find a 15th-degree polynomial for which $x - 1$ is a factor. Answers may vary.

Use synthetic division to divide.

58. $(x^2 - 4x - 2) \div [x - (3 + 2i)]$

59. $(x^2 - 3x + 7) \div (x - i)$

2.5 Theorems about Zeros of Polynomial Functions

- *Factor polynomial functions and find the zeros and their multiplicities.*
- *Find a polynomial with specified zeros.*
- *For a polynomial function with integer coefficients, find the rational zeros and the other zeros, if possible.*
- *For a polynomial function with rational coefficients, find the rational zeros and the other zeros, if possible.*

The Fundamental Theorem of Algebra

A linear, or first-degree, polynomial function $f(x) = mx + b$ (where $m \neq 0$, of course) has just one zero, $-b/m$. It can be shown that any quadratic polynomial function with complex numbers for coefficients has at least one, and at most two, complex zeros. The following theorem is a generalization. No proof is given.

The Fundamental Theorem of Algebra

Every polynomial function of degree n, $n \geq 1$, with complex coefficients, has at least one zero in the system of complex numbers.

Note that although the fundamental theorem of algebra guarantees that a zero exists, it does not tell how to find it. Recall that the zeros of a polynomial function $f(x)$ are the solutions of the polynomial equation $f(x) = 0$. We now develop some concepts that can help in finding zeros. First, we consider a corollary of the fundamental theorem of algebra.

Every polynomial function f of degree n, where $n \geq 1$, with complex coefficients, can be factored into n linear factors (not necessarily unique); that is, $f(x) = a_n(x - c_1)(x - c_2)\cdots(x - c_n)$.

Finding Zeros of Factored Polynomial Functions

When a polynomial function is factored into a product of linear factors, it is easy to find the zeros by solving the equation $f(x) = 0$ using the principle of zero products.

Example 1 Find the zeros of

$$f(x) = (x - 3)(x + 4)(x + 1)(x - 1).$$

SOLUTION To solve the equation $f(x) = 0$, we use the principle of zero products. The zeros of $f(x)$ are 3, -4, -1, and 1.

Example 2 Find the zeros of

$$f(x) = 5(x - 2)(x - 2)(x - 2)(x + 1).$$

SOLUTION To solve the equation $f(x) = 0$, we use the principle of zero products. The zeros of $f(x)$ are 2 and -1.

In Example 2, the factor $x - 2$ occurs three times. In a case like this, we sometimes say that the zero we obtain from this factor, 2, has a **multiplicity** of 3. If we multiply out the right side, we obtain

$$f(x) = 5x^4 - 25x^3 + 30x^2 + 20x - 40.$$

Had we started with this form of the function, we might have had trouble finding the zeros. Some polynomials, however, can be factored using techniques we already know, such as factoring by grouping.

Example 3 Find the zeros of

$$f(x) = x^3 - 2x^2 - 9x + 18.$$

SOLUTION We factor by grouping, as follows:

$$\begin{aligned} f(x) &= x^3 - 2x^2 - 9x + 18 \\ &= x^2(x - 2) - 9(x - 2) \\ &= (x^2 - 9)(x - 2) \\ &= (x + 3)(x - 3)(x - 2). \end{aligned}$$

Then by the principle of zero products, the solutions of the equation $f(x) = 0$ are -3, 3, and 2. These are the zeros of $f(x)$.

Other factoring techniques can also be used.

Example 4 Find the zeros of

$$f(x) = x^4 + 4x^2 - 45.$$

SOLUTION We factor as follows:

$$\begin{aligned} f(x) &= x^4 + 4x^2 - 45 \\ &= (x^2 - 5)(x^2 + 9) \\ &= (x - \sqrt{5})(x + \sqrt{5})(x - 3i)(x + 3i). \end{aligned}$$

Then by the principle of zero products, the solutions of the equation $f(x) = 0$ are $\pm\sqrt{5}$ and $\pm 3i$. These are the zeros of $f(x)$.

If we tried to use a grapher to find the zeros of f in Example 4, the nonreal solutions $3i$ and $-3i$ could not be seen.

> Every polynomial of degree n, where $n \geq 1$, has at least one zero and at most n zeros.

This is often stated as follows: "Every polynomial of degree n, where $n \geq 1$, has *exactly* n zeros." This statement is not incompatible with the preceding statement, as it would first seem, because one must take multiplicities into account.

Finding Polynomials with Given Zeros

Interactive Discovery

Graph each of the following:

$$y_1 = (x + 2)(x - 1)\left(x - \tfrac{5}{2}\right),$$
$$y_2 = 2(x + 2)(x - 1)\left(x - \tfrac{5}{2}\right),$$
$$y_3 = -\tfrac{1}{2}(x + 2)(x - 1)\left(x - \tfrac{5}{2}\right).$$

How do the graphs compare? How do the zeros compare?

Given several numbers, we can find a polynomial with those numbers as its zeros.

Example 5 Find a polynomial function of degree 3, having the zeros -2, 1, and $3i$.

SOLUTION Such a polynomial has factors $x + 2$, $x - 1$, and $x - 3i$, so we have

$$f(x) = a_n(x + 2)(x - 1)(x - 3i).$$

The number a_n can be any nonzero number. The simplest polynomial will be obtained if we let it be 1. If we then multiply the factors, we obtain

$$f(x) = x^3 + (1 - 3i)x^2 + (-2 - 3i)x + 6i.$$

Example 6 Find a polynomial function of degree 5 with -1 as a zero of multiplicity 3, 4 as a zero of multiplicity 1, and 0 as a zero of multiplicity 1.

SOLUTION Proceeding as in Example 5, letting $a_n = 1$, we obtain

$$\begin{aligned} f(x) &= (x + 1)^3(x - 4)(x - 0) \\ &= x^5 - x^4 - 9x^3 - 11x^2 - 4x. \end{aligned}$$

Zeros of Polynomial Functions with Real Coefficients

Consider the quadratic equation $x^2 - 2x + 2 = 0$, with real coefficients. Its solutions are $1 + i$ and $1 - i$. Note that they are complex conjugates. This generalizes to any polynomial with real coefficients.

> If a complex number $a + bi$, $b \neq 0$, is a zero of a polynomial function $f(x)$ with real coefficients, then its conjugate, $a - bi$, is also a zero. (Nonreal zeros occur in conjugate pairs.)

For the preceding to be true, it is essential that the coefficients be real numbers. We see this in Example 5, where the root $3i$ occurs, but its conjugate does not. This occurs because some of the coefficients of the polynomial are not real.

Rational Coefficients

When a polynomial has rational numbers for coefficients, certain irrational zeros also occur in pairs, as described in the following theorem.

> Suppose that $f(x)$ is a polynomial with rational coefficients. Then if either of the following is a zero, so is the other: $a + c\sqrt{b}$, $a - c\sqrt{b}$, a and c rational, b not a square.

Example 7 Suppose that a polynomial function of degree 6 with rational coefficients has $-2 + 5i$, $-2i$, and $1 - \sqrt{3}$ as some of its zeros. Find the other zeros.

SOLUTION The other zeros are -2, $-5i$, $2i$, and $1 + \sqrt{3}$. There are no other zeros since the degree is 6.

Example 8 Find a polynomial function of lowest degree with rational coefficients that has $1 - \sqrt{2}$ and $1 + 2i$ as some of its zeros.

SOLUTION The function must also have the zeros $1 + \sqrt{2}$ and $1 - 2i$. Thus the polynomial function is

$$\begin{aligned} f(x) &= [x - (1 - \sqrt{2})][x - (1 + \sqrt{2})][x - (1 + 2i)][x - (1 - 2i)] \\ &= (x^2 - 2x - 1)(x^2 - 2x + 5) \\ &= x^4 - 4x^3 + 8x^2 - 8x - 5. \end{aligned}$$

Integer Coefficients and the Rational Zeros Theorem

It is not always easy to find the zeros of a polynomial function. However, if a polynomial function has integer coefficients, there is a procedure that will yield all the rational zeros.

Example 9 Let $f(x) = x^4 - 5x^3 + 10x^2 - 20x + 24$. Find the other zeros of $f(x)$, given that $2i$ is a zero.

SOLUTION Since $2i$ is a zero, we know that $-2i$ is also a zero. Thus,

$$f(x) = (x - 2i)(x + 2i) \cdot Q(x)$$

for some $Q(x)$. Since $(x - 2i)(x + 2i) = x^2 + 4$, we know that

$$f(x) = (x^2 + 4) \cdot Q(x).$$

Using division, we find that $Q(x) = x^2 - 5x + 6$, and since we can factor $x^2 - 5x + 6$, we get

$$f(x) = (x^2 + 4)(x - 2)(x - 3).$$

Thus the other zeros are $-2i$, 2, and 3.

The Rational Zeros Theorem

Let

$$P(x) = a_n x^n + a_{n-1}x^{n-1} + \cdots + a_1 x + a_0,$$

where all the coefficients are integers. Consider a rational number denoted by p/q, where p and q are relatively prime (having no common factor besides -1 and 1). If p/q is a zero of $P(x)$, then p is a factor of a_0 and q is a factor of a_n.

Example 10 Given $f(x) = 2x^5 - x^4 - 4x^3 + 2x^2 - 30x + 15$:

a) Find the rational zeros, and then the other zeros.

b) Solve the equation $f(x) = 0$.

c) Factor $f(x)$ into linear factors.

SOLUTION

a) According to the rational zeros theorem, any rational zero of f must be of the form p/q, where p is a factor of 15 and q is a factor of 2. The possibilities are

$$\frac{\textit{Possibilities for } p}{\textit{Possibilities for } q}: \quad \frac{\pm 1, \pm 3, \pm 5, \pm 15}{\pm 1, \pm 2};$$

Possibilities for p/q: $1, -1, 3, -3, 5, -5, 15, -15, \frac{1}{2}, -\frac{1}{2}, \frac{3}{2}, -\frac{3}{2}, \frac{5}{2}, -\frac{5}{2}, \frac{15}{2}, -\frac{15}{2}$.

Rather than use synthetic division to check each of these possibilities, we graph $y_1 = 2x^5 - x^4 - 4x^3 + 2x^2 - 30x + 15$ (see the figure at left). We can then inspect the graph for zeros that appear to be near any of the possible rational zeros.

$y_1 = 2x^5 - x^4 - 4x^3 + 2x^2 - 30x + 15$

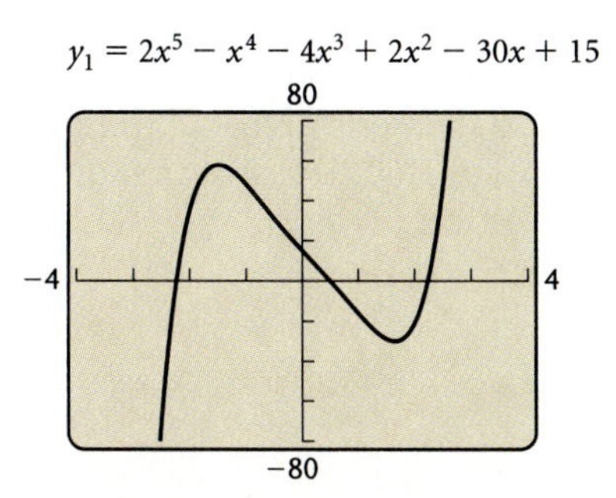

From the graph, we see that of the possibilities in the list, only the numbers $-\frac{5}{2}, \frac{1}{2}$, and $\frac{5}{2}$ might be rational zeros. By synthetic division, we see that only $\frac{1}{2}$ is actually a rational zero.

$$\begin{array}{r|rrrrrr} \frac{1}{2} & 2 & -1 & -4 & 2 & -30 & 15 \\ & & 1 & 0 & -2 & 0 & -15 \\ \hline & 2 & 0 & -4 & 0 & -30 & 0 \end{array}$$

This means that $x - \frac{1}{2}$ is a factor of $f(x)$. To find the other zeros, we write the factorization and try to factor further:

$$\begin{aligned} f(x) &= \left(x - \tfrac{1}{2}\right)(2x^4 - 4x^2 - 30) \\ &= \left(x - \tfrac{1}{2}\right) \cdot 2 \cdot (x^4 - 2x^2 - 15) && \text{Factoring out the 2} \\ &= \left(x - \tfrac{1}{2}\right) \cdot 2 \cdot (x^2 - 5)(x^2 + 3). && \text{Factoring the trinomial} \end{aligned}$$

We now use the principle of zero products to determine the zeros:

$$\begin{aligned} x - \tfrac{1}{2} &= 0 \quad \text{or} \quad x^2 - 5 = 0 \quad \text{or} \quad x^2 + 3 = 0 \\ x &= \tfrac{1}{2} \quad \text{or} \quad x^2 = 5 \quad \text{or} \quad x^2 = -3 \\ x &= \tfrac{1}{2} \quad \text{or} \quad x = \pm\sqrt{5} \quad \text{or} \quad x = \pm\sqrt{3}\,i. \end{aligned}$$

There is only one rational zero, $\frac{1}{2}$. The other zeros are $\pm\sqrt{5}$ and $\pm\sqrt{3}i$.

b) The solutions of $f(x) = 0$ are $\frac{1}{2}$, $\pm\sqrt{5}$, and $\pm\sqrt{3}\,i$.

c) The factorization into linear factors is

$$f(x) = 2\left(x - \tfrac{1}{2}\right)(x + \sqrt{5})(x - \sqrt{5})(x + \sqrt{3}\,i)(x - \sqrt{3}\,i).$$

Example 11 Given $f(x) = 3x^4 - 11x^3 + 10x - 4$:

a) Find the rational zeros, and then the other zeros.

b) Solve the equation $f(x) = 0$.

c) Factor $f(x)$ into linear factors.

SOLUTION

a) Because the degree of $f(x)$ is 4, there are at most 4 distinct zeros. Although a grapher could be useful, remember that it can only *approximate* real solutions that are not integers.

The rational zeros theorem says that if p/q is a root of $f(x)$, then p must be a factor of -4 and q must be a factor of 3. Thus the possibilities for p/q are

$$\frac{\textit{Possibilities for } p}{\textit{Possibilities for } q}: \quad \frac{\pm 1, \pm 2, \pm 4}{\pm 1, \pm 3}.$$

If we list all possible quotients, we find the following possibilities for

$$p/q: \quad 1, -1, 2, -2, 4, -4, \tfrac{1}{3}, -\tfrac{1}{3}, \tfrac{2}{3}, -\tfrac{2}{3}, \tfrac{4}{3}, -\tfrac{4}{3}.$$

We could substitute each of these possibilities to see if any are zeros, that is, if $f(c) = 0$. We could also use a TABLE feature or some other way to find function values. But, if we use synthetic division, the quotient polynomial becomes a beneficial by-product if a zero is found.

We try 1.

$$\begin{array}{r|rrrrr} 1 & 3 & -11 & 0 & 10 & -4 \\ & & 3 & -8 & -8 & 2 \\ \hline & 3 & -8 & -8 & 2 & -2 \end{array}$$

We try -1.

$$\begin{array}{r|rrrrr} -1 & 3 & -11 & 0 & 10 & -4 \\ & & -3 & 14 & -14 & 4 \\ \hline & 3 & -14 & 14 & -4 & 0 \end{array}$$

Since $f(1) = -2$, 1 is not a zero; but $f(-1) = 0$, so -1 is a zero. Using the results of the synthetic division, we can express $f(x)$ as follows:

$$f(x) = (x + 1)(3x^3 - 14x^2 + 14x - 4).$$

We now use $3x^3 - 14x^2 + 14x - 4$ and check the other possible zeros. We use synthetic division again to see whether -1 is a double zero.

$$\begin{array}{r|rrrr} -1 & 3 & -14 & 14 & -4 \\ & & -3 & 17 & -31 \\ \hline & 3 & -17 & 31 & -35 \end{array}$$

It is not. The student can check that none of the remaining integers (2, -2, 4, or -4) are zeros. Let's now try $\frac{2}{3}$.

$$\begin{array}{r|rrrr} 2/3 & 3 & -14 & 14 & -4 \\ & & 2 & -8 & 4 \\ \hline & 3 & -12 & 6 & 0 \end{array}$$

Since $f\left(\frac{2}{3}\right) = 0$, $\frac{2}{3}$ is also a zero.

Using the results of synthetic division, we can factor further:

$$f(x) = (x + 1)\left(x - \tfrac{2}{3}\right)(3x^2 - 12x + 6) \quad \text{Using the results of the last synthetic division}$$

$$= (x + 1)\left(x - \tfrac{2}{3}\right) \cdot 3 \cdot (x^2 - 4x + 2). \quad \text{Removing a factor of 3}$$

The quadratic formula can be used to find the zeros of $x^2 - 4x + 2$:

$$x = \frac{-b \pm \sqrt{b^2 - 4ac}}{2a}$$

$$= \frac{-(-4) \pm \sqrt{(-4)^2 - 4 \cdot 1 \cdot 2}}{2 \cdot 1}$$

$$= \frac{4 \pm \sqrt{8}}{2}$$

$$= 2 \pm \sqrt{2}.$$

The rational zeros are -1 and $\frac{2}{3}$. The other zeros are $2 \pm \sqrt{2}$.

b) The solutions of $f(x) = 0$ are -1, $\frac{2}{3}$, and $2 \pm \sqrt{2}$.

c) The complete factorization of $f(x)$ is

$$f(x) = (x + 1)(3x - 2)[x - (2 - \sqrt{2})][x - (2 + \sqrt{2})]$$

$$= (x + 1)(3x - 2)(x - 2 + \sqrt{2})(x - 2 - \sqrt{2}).$$

The Intermediate Value Theorem

Some functions have graphs that are curves with no breaks or holes. Such functions are called **continuous functions**. Loosely speaking, the graph of a continuous function can be drawn without lifting the pencil from the paper. The function f, sketched as follows, is continuous, but g and h are not.

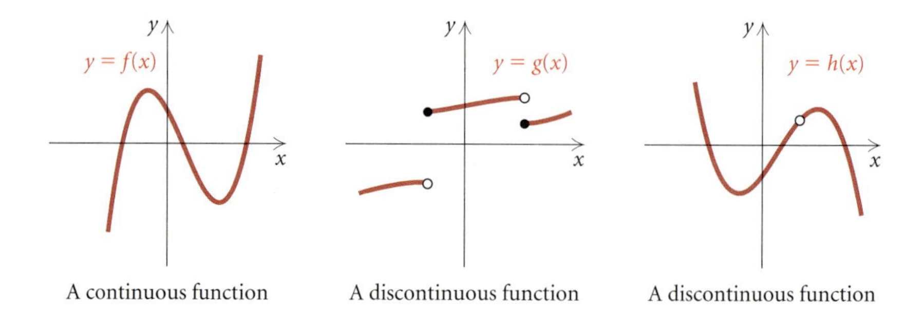

Interactive Discovery

Consider the following polynomial function, with the two given inputs.

$$P(x) = 12x^3 - 5x^2 - 11x + 6; \quad a = 0.7, b = 0.8$$

Graph this function using the viewing windows $[-1, 2, -2, 10]$ and then $[0.6, 0.92, -0.1, 0.15]$. Does there seem to be a zero between 0.7 and 0.8? Find $P(0.7)$ and $P(0.8)$. Do these function values have opposite signs?

Polynomial functions P are continuous, hence their graphs are unbroken. All polynomial functions have $\mathbb{R}$ as the domain. Suppose we have two function values $P(a)$ and $P(b)$ that have opposite signs. Since P is continuous, its graph must be a curve from $(a, P(a))$ to $(b, P(b))$ without interruption. It follows that the line must cross the x-axis somewhere between a and b.

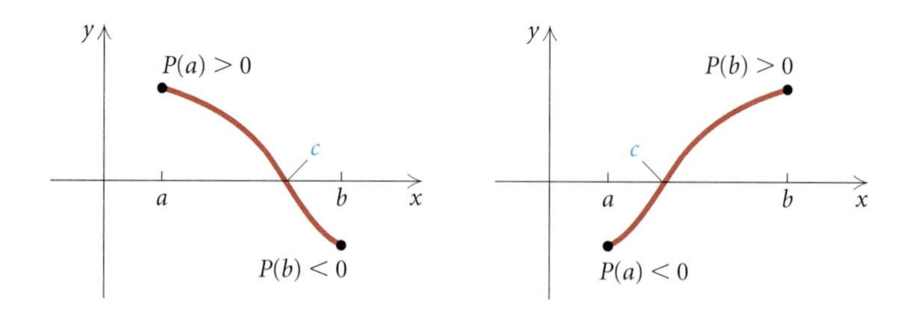

The Intermediate Value Theorem

For any polynomial function $P(x)$ with real coefficients, suppose that for $a \neq b$, $P(a)$ and $P(b)$ are of opposite signs. Then the function has a real zero between a and b.

2.5 Exercise Set

Find the zeros of the polynomial function and state the multiplicity of each.

1. $f(x) = (x + 3)^2(x - 1)$

2. $f(x) = -8(x - 3)^2(x + 4)^3x^4$

3. $f(x) = x^3(x - 1)^2(x + 4)$

4. $f(x) = (x^2 - 5x + 6)^2$

5. $f(x) = x^4 - 4x^2 + 3$

6. $f(x) = x^4 - 10x^2 + 9$

7. $f(x) = x^3 + 3x^2 - x - 3$

8. $f(x) = x^3 - x^2 - 2x + 2$

Find a polynomial function of degree 3 with the given numbers as zeros.

9. $-2, 3, 5$

10. $2, i, -i$

11. $-3, 2i, -2i$

12. $1 + 4i, 1 - 4i, -1$

13. $\sqrt{2}, -\sqrt{2}, \sqrt{3}$

14. Find a polynomial function of degree 4 with -2 as a zero of multiplicity 1, 3 as a zero of multiplicity 2, and -1 as a zero of multiplicity 1.

Suppose that a polynomial function of degree 5 with rational coefficients has the given numbers as zeros. Find the other zeros.

15. $6, -3 + 4i, 4 - \sqrt{5}$

16. $-2, 3, 4, 1 - i$

Find a polynomial function of lowest degree with rational coefficients that has the given numbers as some of its zeros.

17. $1 + i, 2$

18. $2 - i, -1$

19. $-4i, 5$

20. $2 - \sqrt{3}, 1 + i$

21. $\sqrt{5}, -3i$

22. $-\sqrt{2}, 4i$

Given that the polynomial function has the given zero, find the other zeros.

23. $f(x) = x^4 - 5x^3 + 7x^2 - 5x + 6; \quad -i$

24. $f(x) = x^4 - 16 = 0; \quad 2i$

25. $f(x) = x^3 - 6x^2 + 13x - 20; \quad 4$

26. $f(x) = x^3 - 8; \quad 2$

List all possible rational zeros.

27. $f(x) = x^5 - 3x^2 + 1$

28. $f(x) = x^7 + 37x^5 - 6x^2 + 12$

29. $f(x) = 15x^6 + 47x^2 + 2$

30. $f(x) = 10x^{25} + 3x^{17} - 35x + 6$

For each polynomial function:

a) *Find the rational zeros, and then the other zeros.*
b) *Solve the equation $f(x) = 0$.*
c) *Factor $f(x)$ into linear factors.*

31. $f(x) = x^3 + 3x^2 - 2x - 6$

32. $f(x) = x^3 - x^2 - 3x + 3$

33. $f(x) = x^3 - 3x + 2$

34. $f(x) = x^3 - 2x + 4$

35. $f(x) = x^3 - 5x^2 + 11x + 17$

36. $f(x) = 2x^3 + 7x^2 + 2x - 8$

37. $f(x) = 5x^4 - 4x^3 + 19x^2 - 16x - 4$

38. $f(x) = 3x^4 - 4x^3 + x^2 + 6x - 2$

39. $f(x) = x^4 - 3x^3 - 20x^2 - 24x - 8$

40. $f(x) = x^4 + 5x^3 - 27x^2 + 31x - 10$

41. $f(x) = x^3 - 4x^2 + 2x + 4$

42. $f(x) = x^3 - 8x^2 + 17x - 4$

43. $f(x) = x^3 + 8$

44. $f(x) = x^3 - 8$

45. $f(x) = \frac{1}{3}x^3 - \frac{1}{2}x^2 - \frac{1}{6}x + \frac{1}{6}$

46. $f(x) = \frac{2}{3}x^3 - \frac{1}{2}x^2 + \frac{2}{3}x - \frac{1}{2}$

Find only the rational zeros.

47. $f(x) = x^4 + 32$

48. $f(x) = x^6 + 8$

49. $f(x) = x^3 - x^2 - 4x + 3$

50. $f(x) = 2x^3 + 3x^2 + 2x + 3$

51. $f(x) = x^4 + 2x^3 + 2x^2 - 4x - 8$

52. $f(x) = x^4 + 6x^3 + 17x^2 + 36x + 66$

53. $f(x) = x^5 - 5x^4 + 5x^3 + 15x^2 - 36x + 20$

54. $f(x) = x^5 - 3x^4 - 3x^3 + 9x^2 - 4x + 12$

Show that the function f has a zero between a and b using the intermediate value theorem.

55. $f(x) = x^3 + 3x^2 - 9x - 13; \ a = -5, b = -4$

56. $f(x) = x^3 + 3x^2 - 9x - 13; \ a = 2, b = 3$

Skill Maintenance

Solve.

57. $\dfrac{x^2 - 4}{x^2 - 5x + 6} = 0$

58. $\dfrac{2}{x - 2} + \dfrac{1}{x + 2} = \dfrac{3}{x^2 - 4}$

Simplify.

59. $\dfrac{\dfrac{1}{x} + \dfrac{3}{5x}}{\dfrac{2}{x} - \dfrac{4}{15x}}$

60. $\dfrac{\dfrac{5}{x^2} - \dfrac{7}{3x}}{\dfrac{3}{2x^2} + \dfrac{5}{x}}$

Synthesis

61. ◈ Is it possible for a third-degree polynomial with rational coefficients to have no real zeros? Why or why not?

62. ◈ If $Q(x) = -P(x)$, do $P(x)$ and $Q(x)$ have the same zeros? Why or why not?

63. Consider $f(x) = 2x^3 - 5x^2 - 4x + 3$. Find the solutions of each equation.

a) $f(x) = 0$ b) $f(x - 1) = 0$

c) $f(x + 2) = 0$ d) $f(2x) = 0$

64. Use a grapher to find the points of intersection of the graphs of

$y_1 = 3x^5 + 3x^4 + 54x^3 - 54x^2 + 243x + 243$ and $y_2 = (x + 3)^5$.

65. Use the rational zeros theorem and the equation $x^2 - 5 = 0$ to show that $\sqrt{5}$ is not a rational number.

66. Use the rational zeros theorem and the equation $x^4 - 12 = 0$ to show that $\sqrt[4]{12}$ is irrational.

Find the rational zeros.

67. $P(x) = 2x^5 - 33x^4 - 84x^3 + 2203x^2 - 3348x - 10{,}080$

68. $P(x) = x^6 - 6x^5 - 72x^4 - 81x^2 + 486x + 5832$

2.6 Rational Functions

- *Graph a rational function, identifying all asymptotes.*
- *Solve applications involving rational functions.*

The sum, difference, or product of two polynomials is a polynomial. Now we turn our attention to functions that represent the quotient of two polynomials. In general, the quotient of two polynomials is *not* itself a polynomial.

The Domain of a Rational Function

Here are some examples of rational functions:

$$f(x) = \frac{1}{x - 3}, \qquad f(x) = \frac{8x^3 - 5x^2 + x - 1}{x^2 - 4},$$

$$f(x) = \frac{2x + 5}{3x^2 - 4x + 7}.$$

Rational Function

A *rational function* is a function f that is a quotient of two polynomials; that is,

$$f(x) = \frac{P(x)}{Q(x)},$$

where $P(x)$ and $Q(x)$ are polynomials with no common factor other than -1 and 1 and where $Q(x)$ is not the zero polynomial. The domain of f consists of all inputs x for which $Q(x) \neq 0$.

Example 1 Consider

$$f(x) = \frac{1}{x - 3}.$$

Find the domain and graph f.

SOLUTION The input 3 results in a denominator of 0. Thus the domain is

$$\{x \mid x \neq 3\}, \text{ or } (-\infty, 3) \cup (3, \infty).$$

The graph of this function is the graph of $y = 1/x$ translated right 3 units. Two versions of the graph are shown below.

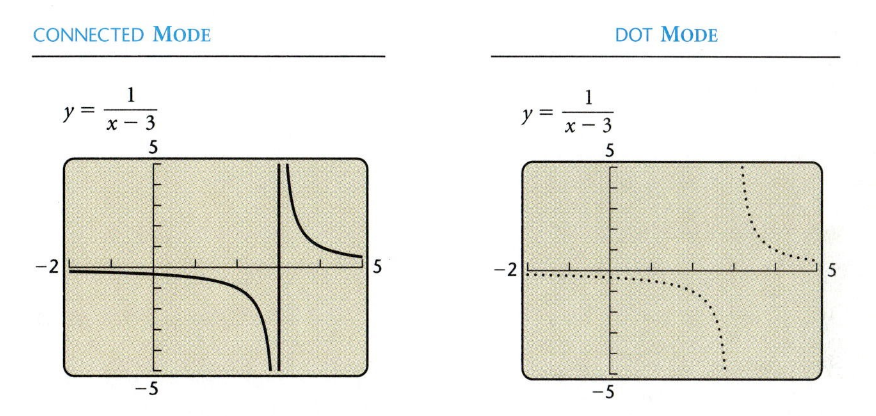

In the Introduction to Graphs and Graphers, we discussed a disadvantage of the grapher that can occur when graphing rational functions. CONNECTED mode can lead to an incorrect graph like the one on the left. Because graphs in CONNECTED mode connect plotted points with line segments, it appears as though the vertical line $x = 3$ is part of the graph. We will see later in this section that a line like $x = 3$, though important in the construction of the graph, is not part of the graph. Graphs in DOT mode, like the one on the right above, simply plot dots representing coordinates of points. If you have a choice when graphing rational functions, use DOT mode.

Because of the possible deficiencies of graphers when graphing rational functions and because we want you to understand the graphs, we will emphasize the creation of hand-drawn graphs in this section.

Asymptotes

Referring to Example 1, let's explore what happens as x becomes very large positively without bound and becomes very small negatively without bound.

Interactive Discovery

Consider $y = 1/(x - 3)$. Complete the following tables and look for a pattern. What happens to the y-values as the x-values become larger without bound? become smaller without bound? Then do the same for smaller and smaller x-values.

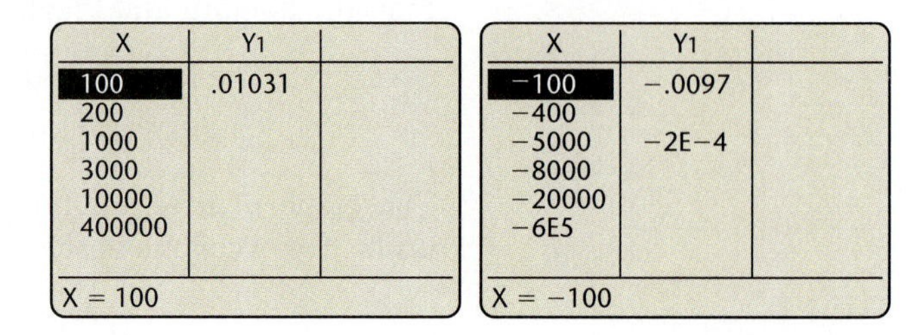

X	Y1	
100	.01031	
200		
1000		
3000		
10000		
400000		

X = 100

X	Y1	
−100	−.0097	
−400		
−5000	−2E−4	
−8000		
−20000		
−6E5		

X = −100

We see from the table on the left that "as x goes to infinity, y goes to 0." We write this as $y \to 0$ as $x \to \infty$. Using the table on the right, complete:

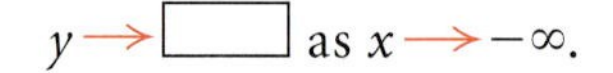

$$y \to \boxed{} \text{ as } x \to -\infty.$$

We see that

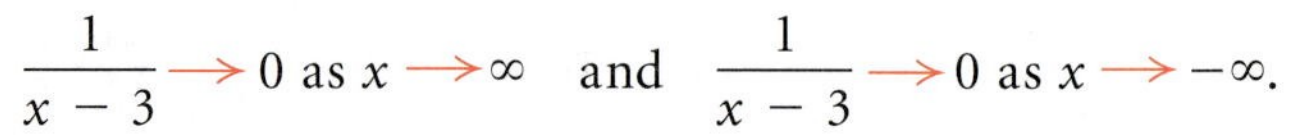

$$\frac{1}{x-3} \to 0 \text{ as } x \to \infty \quad \text{and} \quad \frac{1}{x-3} \to 0 \text{ as } x \to -\infty.$$

From these facts, we say that the curve approaches the x-axis *asymptotically* and that the x-axis is a **horizontal asymptote** for the curve.

Horizontal Asymptote

The line $y = b$ is a *horizontal asymptote* for the graph of f if either or both of the following are true:

$$f(x) \to b \text{ as } x \to \infty \quad \text{or} \quad f(x) \to b \text{ as } x \to -\infty.$$

The following figures illustrate two ways in which horizontal asymptotes can occur. In each case, the curve gets close to the line $y = b$ either as $x \to \infty$ or as $x \to -\infty$. Keep in mind that the symbols ∞ and $-\infty$ convey the idea of increasing positively without bound and decreasing negatively without bound, respectively.

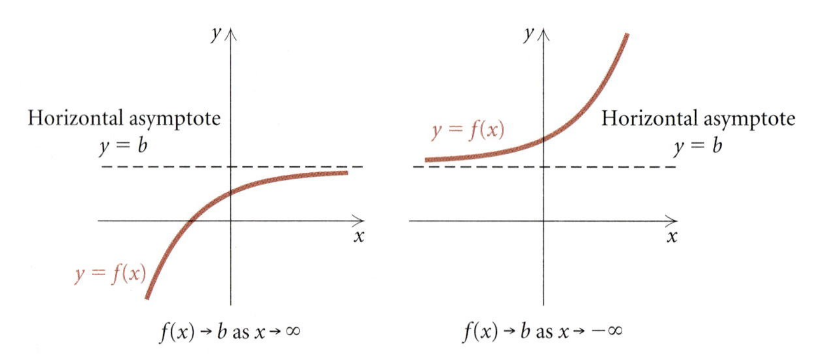

Look again at the graph in Example 1. Let's explore what happens as x-values get closer and closer to 3 from the left. We then explore what happens as x-values get closer and closer to 3 from the right.

Interactive Discovery

Consider $y = 1/(x - 3)$. Using a TABLE feature, complete the following input–output tables and look for a pattern. Trace along the curve from left to right toward an x-value of 3. Then trace from right to left toward 3. What happens to y in each case?

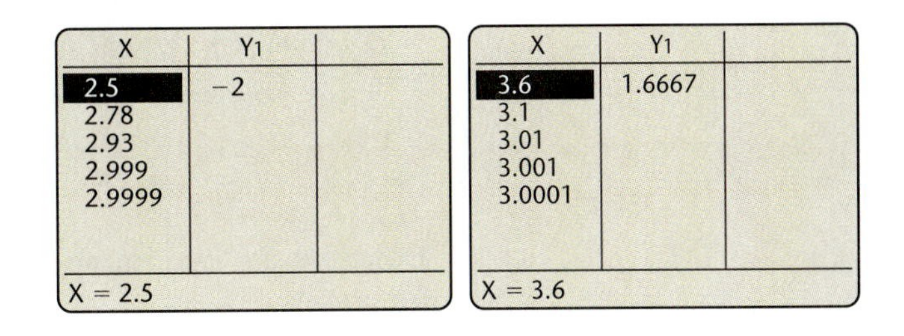

Complete:

$y \rightarrow \square$ as $x \rightarrow 3$ from the left;

$y \rightarrow \square$ as $x \rightarrow 3$ from the right.

We see that as x-values get closer and closer to 3 from the left, the function values (y-values) decrease negatively without bound. In fact, by selecting x close enough to 3, we can get the function values as small negatively as we wish. Similarly, as the x-values approach 3 from the right, the function values increase positively without bound. We write this as

$$f(x) \rightarrow -\infty \text{ as } x \rightarrow 3^- \quad \text{and} \quad f(x) \rightarrow \infty \text{ as } x \rightarrow 3^+.$$

We read "$f(x) \rightarrow -\infty$ as $x \rightarrow 3^-$" as "$f(x)$ decreases negatively without bound as x approaches 3 from the left." We read "$f(x) \rightarrow \infty$ as $x \rightarrow 3^+$" as "$f(x)$ increases positively without bound as x approaches 3 from the right." The vertical line $x = 3$ is said to be a **vertical asymptote** to this curve.

Vertical Asymptote

The line $x = a$ is a *vertical asymptote* for the graph of f if any of the following is true:

$$f(x) \rightarrow \infty \text{ as } x \rightarrow a^- \quad \text{or} \quad f(x) \rightarrow -\infty \text{ as } x \rightarrow a^-, \quad \text{or}$$
$$f(x) \rightarrow \infty \text{ as } x \rightarrow a^+ \quad \text{or} \quad f(x) \rightarrow -\infty \text{ as } x \rightarrow a^+.$$

The following figures show the four ways in which a vertical asymptote can occur.

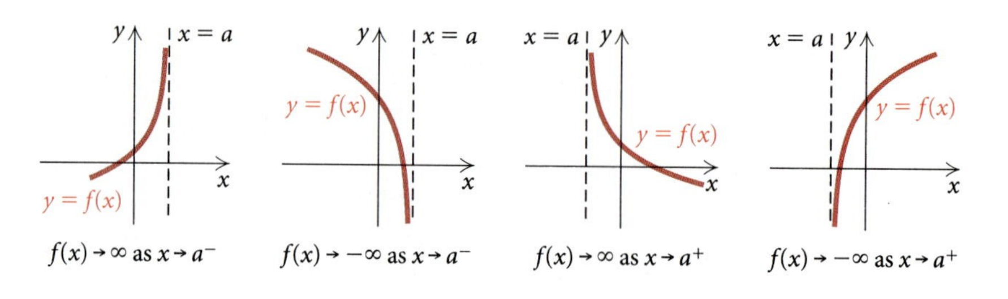

Determining Vertical Asymptotes

If a is a zero of the denominator of a rational function, then the line $x = a$ is a vertical asymptote.

Example 2 Determine the vertical asymptotes of each of the following functions.

a) $f(x) = \dfrac{3x - 2}{x^3 - 2x^2 - 15x}$ **b)** $g(x) = \dfrac{x - 2}{x^3 - 5x}$

SOLUTION

a) We factor to find the zeros of the denominator.

$$x^3 - 2x^2 - 15x = x(x - 5)(x + 3)$$

The zeros of the denominator are 0, 5, and -3. Thus the vertical asymptotes are the lines $x = 0$, $x = 5$, and $x = -3$.

b) We factor to find the zeros of the denominator.

$$x^3 - 5x = x(x^2 - 5) = x(x - \sqrt{5})(x + \sqrt{5})$$

The zeros of the denominator are 0, $\sqrt{5}$, and $-\sqrt{5}$. Thus the vertical asymptotes are the lines $x = 0$, $x = \sqrt{5}$, and $x = -\sqrt{5}$.

How can we determine a horizontal asymptote? As x gets very large or very small, the value of the polynomial function $P(x)$ is dominated by the function's leading term. Because of this, if $P(x)$ and $Q(x)$ have the *same* degree, the value of $P(x)/Q(x)$ as $|x| \longrightarrow \infty$ is dominated by the ratio of the numerator's leading coefficient to the denominator's leading coefficient. Let's explore this further.

Interactive Discovery

Consider

$$f(x) = \frac{3x^2 + 2x - 4}{2x^2 - x + 1}.$$

Let $y_1 = 3x^2 + 2x - 4$, $y_2 = 2x^2 - x + 1$, and $y_3 = y_1/y_2$. Using a TABLE feature, complete the following input–output tables and look for a pattern. Check the pattern with Example 2. Trace along the curve for larger and larger values of x. Then trace for smaller and smaller negative values. What happens to y in each case?

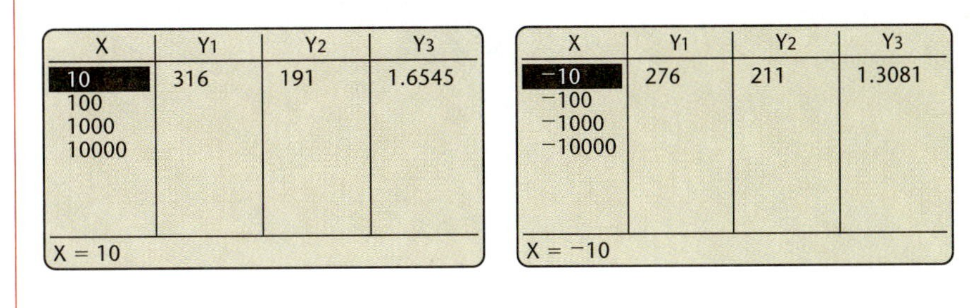

X	Y1	Y2	Y3
10	316	191	1.6545
100			
1000			
10000			

X = 10

X	Y1	Y2	Y3
−10	276	211	1.3081
−100			
−1000			
−10000			

X = −10

Complete:

$$y_3 \rightarrow \boxed{} \text{ as } x \rightarrow \infty; \qquad y_3 \rightarrow \boxed{} \text{ as } x \rightarrow -\infty.$$

It follows that when the numerator and the denominator of a rational function have the same degree, the line $y = a/b$ is the horizontal asymptote, where a and b are the leading coefficients of the numerator and the denominator, respectively.

Example 3 Find the horizontal asymptote: $f(x) = \dfrac{-7x^4 - 10x^2 + 1}{11x^4 + x - 2}$.

SOLUTION The numerator and the denominator have the same degree, so the line $y = -\frac{7}{11}$ is a horizontal asymptote.

To check Example 3, we could examine a table of values or the graph. A third check, one that is useful in calculus, is to multiply by 1, using $(1/x^4)/(1/x^4)$:

$$f(x) = \frac{-7x^4 - 10x^2 + 1}{11x^4 + x - 2} \cdot \frac{\frac{1}{x^4}}{\frac{1}{x^4}}$$

$$= \frac{-7 - \dfrac{10}{x^2} + \dfrac{1}{x^4}}{11 + \dfrac{1}{x^3} - \dfrac{2}{x^4}}.$$

As $|x|$ becomes very large, each expression with x in the denominator tends toward zero. Specifically, as $|x| \rightarrow \infty$, we have

$$f(x) \rightarrow \frac{-7 - 0 + 0}{11 + 0 - 0}, \quad \text{or} \quad f(x) \rightarrow -\frac{7}{11}.$$

The horizontal asymptote is $y = -\frac{7}{11}$.

Example 4 Find the horizontal asymptote: $f(x) = \dfrac{2x + 3}{x^3 - 2x^2 + 4}$.

SOLUTION We let $y_1 = 2x + 3$, $y_2 = x^3 - 2x^2 + 4$, and $y_3 = y_1/y_2$, and form a table of values as follows.

X	Y1	Y2	Y3
10	23	804	.02861
100	203	980004	2.1E−4
1000	2003	9.98E8	2E−6
10000	20003	1E12	2E−8

X = 10

X	Y1	Y2	Y3
−10	−17	−1196	.01421
−100	−197	−1E6	1.9E−4
−1000	−1997	−1E9	2E−6
−10000	−19997	−1E12	2E−8

X = −10

Note that as x grows large, the value of y_2 grows much faster than the value of y_1. Because of this, the ratio of y_1/y_2 shrinks toward 0. As $x \to -\infty$, the ratio y_1/y_2 behaves in a similar manner. The horizontal asymptote is $y = 0$, that is, the x-axis.

The following statements describe the two ways in which a horizontal asymptote occurs.

Determining a Horizontal Asymptote

When the numerator and the denominator of a rational function have the same degree, the line $y = a/b$ is a horizontal asymptote, where a and b are the leading coefficients of the numerator and the denominator, respectively.

When the degree of the numerator of a rational function is less than the degree of the denominator, the x-axis, or $y = 0$, is a horizontal asymptote.

Remember that the preceding statements on finding horizontal and vertical asymptotes apply to rational functions, which are defined such that the numerator and the denominator have no common factor other than -1 or 1.

The following statements are also true.

The graph of a rational function never crosses a vertical asymptote.

The graph of a rational function may or may not cross a horizontal asymptote.

Example 5 Make a hand-drawn graph of

$$g(x) = \frac{2x^2 + 1}{x^2}.$$

Include and label all asymptotes.

SOLUTION Since 0 is a zero of the denominator, the y-axis, $x = 0$, is a vertical asymptote. Note also that the degree of the numerator is the same as the degree of the denominator. Thus, $y = 2/1$, or 2, is the horizontal asymptote.

To complete the graph, we draw the asymptotes with dashed lines. Then we compute some additional ordered pairs and draw the curves.

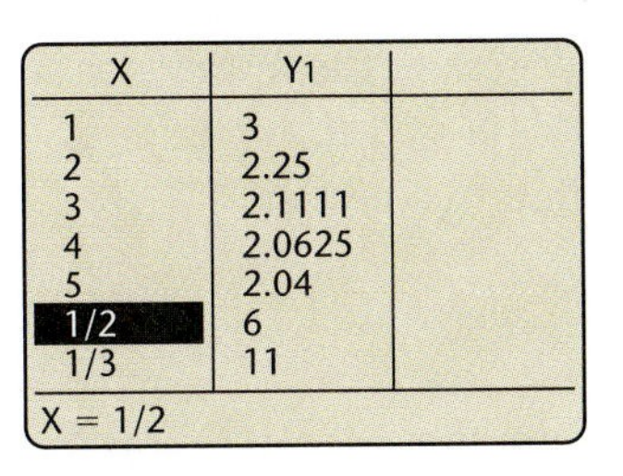

X	Y1
1	3
2	2.25
3	2.1111
4	2.0625
5	2.04
1/2	6
1/3	11

X = 1/2

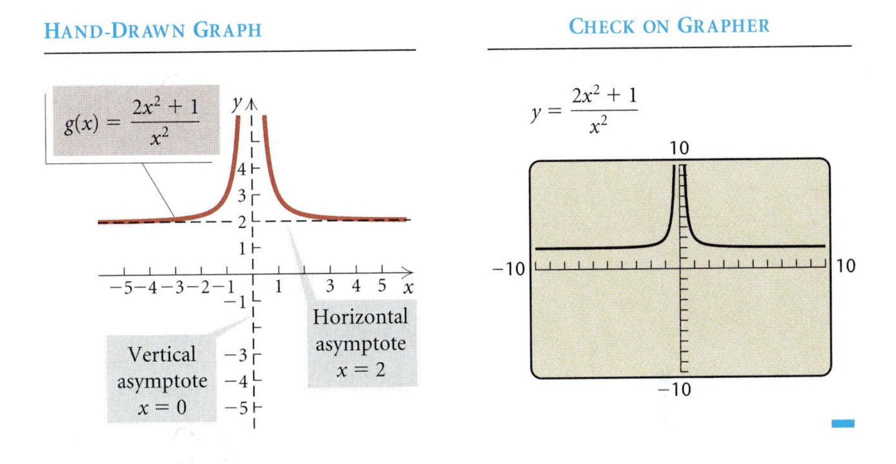

Sometimes a line that is neither horizontal nor vertical is an asymptote. Such a line is called an **oblique asymptote**, or a **slant asymptote**.

Example 6 Find all the asymptotes of

$$f(x) = \frac{2x^2 - 3x - 1}{x - 2}.$$

SOLUTION The line $x = 2$ is a vertical asymptote because 2 is a zero of the denominator. There is no horizontal asymptote. When the degree of the numerator is 1 greater than the degree of the denominator, we divide to find an equivalent expression:

$$\frac{2x^2 - 3x - 1}{x - 2} = (2x + 1) + \frac{1}{x - 2}.$$

$$\begin{array}{r} 2x \;\; + 1 \\ x - 2\overline{)2x^2 - 3x - 1} \\ \underline{2x^2 - 4x} \\ x - 1 \\ \underline{x - 2} \\ 1 \end{array}$$

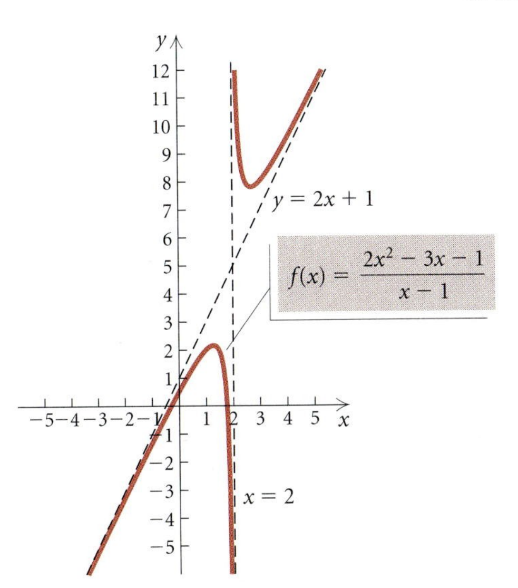

Now we see that when $|x| \to \infty$, $1/(x - 2) \to 0$ and the value of $f(x) \to 2x + 1$. This means that as $|x|$ becomes very large, the graph of $f(x)$ gets very close to the graph of $y = 2x + 1$. Thus the line $y = 2x + 1$ is the oblique asymptote.

Occurrence of Lines as Asymptotes

Vertical asymptotes occur at any x-values that make the denominator 0.

The x-axis is the horizontal asymptote when the degree of the numerator is less than the degree of the denominator.

A horizontal asymptote other than the x-axis occurs when the numerator and the denominator have the same degree.

An oblique asymptote occurs when the degree of the numerator is 1 greater than the degree of the denominator.

There can be only one horizontal or oblique asymptote.

An asymptote is *not* part of the graph.

The following is an outline of a procedure that we can follow to create accurate hand-drawn graphs of rational functions.

To graph a rational function:

1. Find the real zeros of the denominator. Determine the domain of the function and sketch the vertical asymptotes.
2. Find the horizontal or the oblique asymptote, if any, and sketch it.
3. Find the real zeros of the numerator. These are the x-intercepts of the function.
4. Find $f(0)$. This gives the y-intercept of the function.
5. Find other function values to determine the general shape. Then draw the graph.

Example 7 Graph: $f(x) = \dfrac{2x + 3}{3x^2 + 7x - 6}$.

SOLUTION

1. We find the zeros of the denominator by solving $3x^2 + 7x - 6 = 0$. Since
$$3x^2 + 7x - 6 = (3x - 2)(x + 3),$$
the zeros are $\frac{2}{3}$ and -3. Thus the domain excludes $\frac{2}{3}$ and -3 and is $(-\infty, -3) \cup \left(-3, \frac{2}{3}\right) \cup \left(\frac{2}{3}, \infty\right)$. The graph has vertical asymptotes $x = -3$ and $x = \frac{2}{3}$. We sketch these as dashed lines.
2. Because the degree of the numerator is less than the degree of the denominator, the x-axis is the horizontal asymptote. There is no oblique asymptote.
3. To find the zeros of the numerator, we solve $2x + 3 = 0$ and get $x = -\frac{3}{2}$. Thus the pair $\left(-\frac{3}{2}, 0\right)$ is the x-intercept.

4. We find $f(0)$:

$$f(0) = \frac{2 \cdot 0 + 3}{3 \cdot 0^2 + 7 \cdot 0 - 6} = \frac{3}{-6} = -\frac{1}{2}.$$

Thus, $\left(0, -\frac{1}{2}\right)$ is the y-intercept.

5. We find other function values to determine the general shape and then draw the graph. Although vertical asymptotes can never be touched by the graph of a rational function, horizontal asymptotes can be.

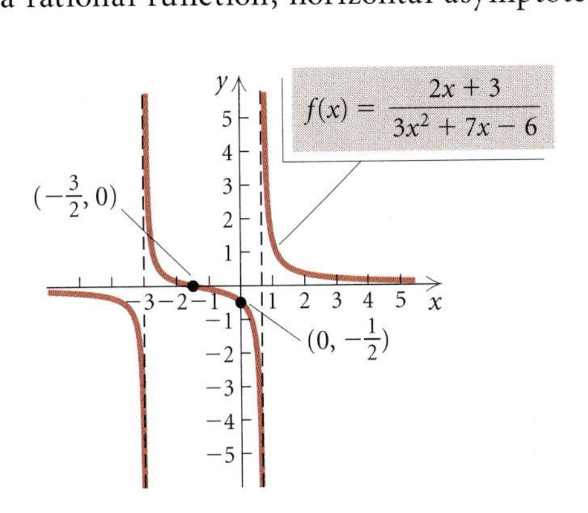

In Example 7, it is somewhat challenging to show on some graphers that $\frac{2}{3}$ is not in the domain of f. We know that the denominator, $3x^2 + 7x - 6$, is 0 when $x = \frac{2}{3}$. However, when $3(2/3)^2 + 7(2/3) - 6$ is entered, the result on a TI-82 is 1 E $^-$13, or 1×10^{-13}. This phenomenon occurs because the grapher approximates $\frac{2}{3}$ as a decimal before squaring or multiplying and this introduces an error. When the factorization $(3x - 2)(x + 3)$ is evaluated for $x = \frac{2}{3}$, the correct result, 0, is displayed. Thus a table of values for

$$y_1 = \frac{2x + 3}{3x^2 + 7x - 6} \quad \text{and} \quad y_2 = \frac{2x + 3}{(3x - 2)(x + 3)}$$

will reveal a discrepancy at $x = \frac{2}{3}$ (listed as .66667 in the x-column). Note that the ERROR message in the y_2-column is correct.

X	Y1	Y2
0	−.5	−.5
.16667	−.7018	−.7018
.33333	−1.1	−1.1
.5	−2.286	−2.286
.66667	4.3E13	ERROR
.83333	2.4348	2.4348
1	1.25	1.25

X = .666666666666...

Example 8 Graph: $g(x) = \dfrac{x^2 - 1}{x^2 + x - 6}$.

SOLUTION

1. We factor the denominator:

$$x^2 + x - 6 = (x + 3)(x - 2).$$

The domain excludes the x-values -3 and 2 and is $(-\infty, -3) \cup (-3, 2) \cup (2, \infty)$. The graph has vertical asymptotes $x = -3$ and $x = 2$. We sketch these as dashed lines.

2. The numerator and the denominator have the same degree, so the horizontal asymptote is determined by the ratio of the leading coeffi-

cients: 1/1, or 1. Thus, $y = 1$ is the horizontal asymptote. We sketch it with a dashed line. There is no oblique asymptote.

3. To find the zeros of the numerator, we solve $x^2 - 1 = 0$. The solutions are -1 and 1. Thus the pairs $(-1, 0)$ and $(1, 0)$ are the x-intercepts.
4. We find $g(0)$:

$$g(0) = \frac{0^2 - 1}{0^2 + 0 - 6} = \frac{-1}{-6} = \frac{1}{6}.$$

Thus, $\left(0, \frac{1}{6}\right)$ is the y-intercept.

5. We find other function values to determine the general shape and then draw the graph.

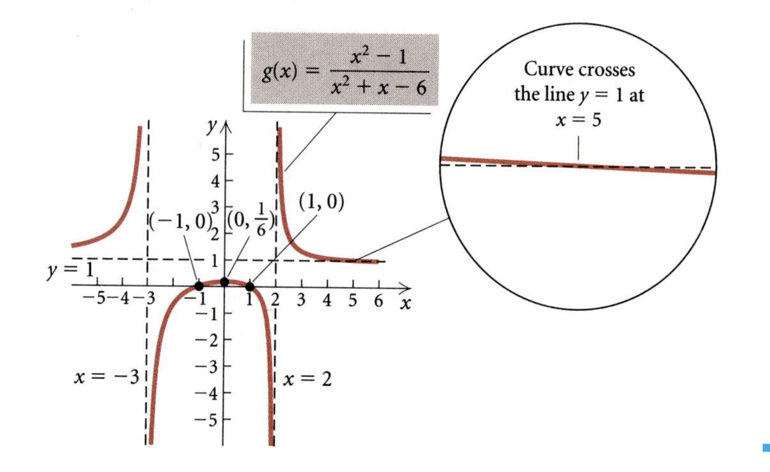

The magnified portion of the graph in Example 8 shows another situation in which a graph can cross its horizontal asymptote.

The graph of

$$f(x) = \frac{2x^3}{x^2 + 1}$$

shown below crosses its oblique asymptote $y = 2x$. Remember, graphs can cross horizontal or oblique asymptotes, but they cannot cross vertical asymptotes.

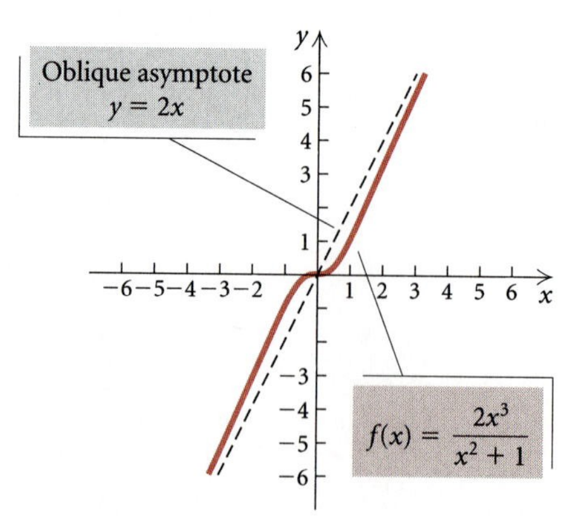

Applications

Example 9 *Temperature During an Illness.* The temperature T, in degrees Fahrenheit, of a person during an illness is given by the function

$$T(t) = \frac{4t}{t^2 + 1} + 98.6,$$

where time t is given in hours since the onset of the illness.

a) Graph the function over the interval $[0, \infty)$.

b) Find the temperature at $t = 0$, 1, 2, and 5 hr.

c) Find the horizontal asymptote. Complete:

$$T(t) \rightarrow \boxed{} \text{ as } t \rightarrow \infty.$$

d) Give the meaning of the answer to part (c) in terms of the application.

e) Find the maximum temperature during the illness.

SOLUTION

a) The graph is shown below.

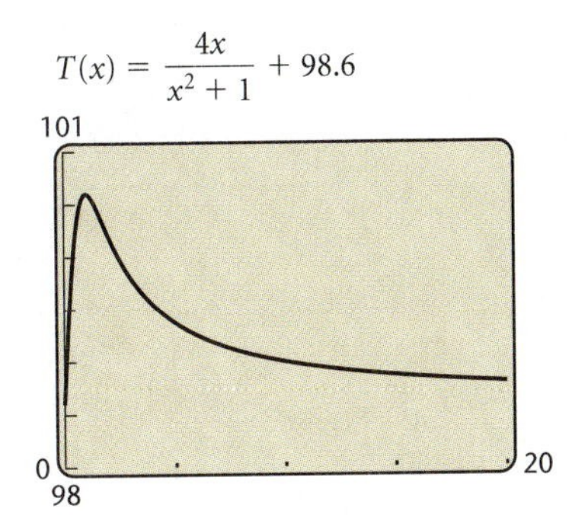

b) We have

$$T(0) = 98.6, \quad T(1) = 100.6, \quad T(2) = 100.2, \quad \text{and}$$
$$T(5) = 99.369.$$

c) Since

$$\begin{aligned} T(t) &= \frac{4t}{t^2 + 1} + 98.6 \\ &= \frac{98.6t^2 + 4t + 98.6}{t^2 + 1}, \end{aligned}$$

the horizontal asymptote is $y = 98.6$. Then it follows that $T(t) \rightarrow 98.6$ as $t \rightarrow \infty$. We can check this using a table of values.

d) As time goes on, the temperature returns to "normal," which is 98.6°.

e) Using a feature on the grapher for finding maximum values or by using TRACE and ZOOM, we find that the maximum temperature is 100.6° at $t = 1$ hr.

2.6 Exercise Set

In Exercises 1–6, use a grapher to match the equation with figures (a)–(f), which follow. List all asymptotes.

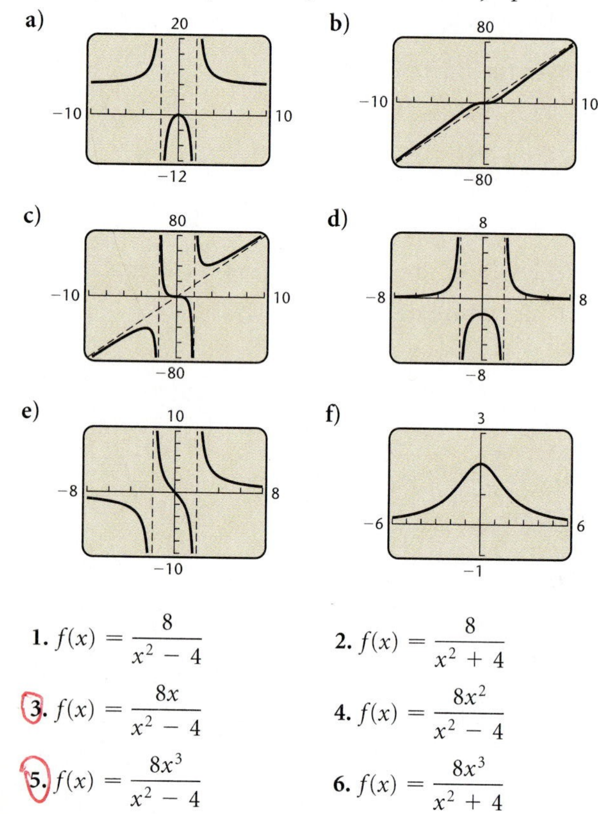

1. $f(x) = \dfrac{8}{x^2 - 4}$
2. $f(x) = \dfrac{8}{x^2 + 4}$
3. $f(x) = \dfrac{8x}{x^2 - 4}$
4. $f(x) = \dfrac{8x^2}{x^2 - 4}$
5. $f(x) = \dfrac{8x^3}{x^2 - 4}$
6. $f(x) = \dfrac{8x^3}{x^2 + 4}$

Make a hand-drawn graph for each of the following. Be sure to label all the asymptotes. Check your work on a grapher.

7. $f(x) = \dfrac{1}{x + 3}$
8. $f(x) = \dfrac{1}{x - 5}$
9. $f(x) = \dfrac{-2}{x - 5}$
10. $f(x) = \dfrac{3}{3 - x}$
11. $f(x) = \dfrac{2x + 1}{x}$
12. $f(x) = \dfrac{3x - 1}{x}$
13. $f(x) = \dfrac{1}{(x - 2)^2}$
14. $f(x) = \dfrac{-2}{(x - 3)^2}$
15. $f(x) = \dfrac{1}{x^2}$
16. $f(x) = \dfrac{1}{3x^2}$
17. $f(x) = \dfrac{1}{x^2 + 3}$
18. $f(x) = \dfrac{-1}{x^2 + 2}$
19. $f(x) = \dfrac{x - 1}{x + 2}$
20. $f(x) = \dfrac{x - 2}{x + 1}$
21. $f(x) = \dfrac{x + 3}{2x^2 - 5x - 3}$
22. $f(x) = \dfrac{3x}{x^2 + 5x + 4}$
23. $f(x) = \dfrac{x^2 - 9}{x + 1}$
24. $f(x) = \dfrac{x^2 - 4}{x - 1}$
25. $f(x) = \dfrac{x^2 + x - 2}{2x^2 + 1}$
26. $f(x) = \dfrac{x^2 - 2x - 3}{3x^2 + 2}$
27. $f(x) = \dfrac{x - 1}{x^2 - 2x - 3}$
28. $f(x) = \dfrac{x + 2}{x^2 + 2x - 15}$
29. $f(x) = \dfrac{x - 3}{(x + 1)^3}$
30. $f(x) = \dfrac{x + 2}{(x - 1)^3}$
31. $f(x) = \dfrac{x^3 + 1}{x}$
32. $f(x) = \dfrac{x^3 - 1}{x}$
33. $f(x) = \dfrac{x^3 + 2x^2 - 15x}{x^2 - 5x - 14}$
34. $f(x) = \dfrac{x^3 + 2x^2 - 3x}{x^2 - 25}$
35. $f(x) = \dfrac{5x^4}{x^4 + 1}$
36. $f(x) = \dfrac{x + 1}{x^2 + x - 6}$
37. $f(x) = \dfrac{x^2}{x^2 - x - 2}$
38. $f(x) = \dfrac{x^2 - x - 2}{x + 2}$

Find a rational function that satisfies the given conditions for each of the following. Answers may vary, but try to give the simplest answer possible.

39. Vertical asymptotes $x = -4$, $x = 5$
40. Vertical asymptotes $x = -4$, $x = 5$; x-intercept $(-2, 0)$
41. Vertical asymptotes $x = -4$, $x = 5$; horizontal asymptote $y = \frac{3}{2}$; x-intercept $(-2, 0)$
42. Oblique asymptote $y = x - 1$
43. *Medical Dosage.* The function

$$N(t) = \frac{0.8t + 1000}{5t + 4}, \quad t \geq 15$$

gives the body concentration $N(t)$, in parts per million, of a certain dosage of medication after time t, in hours.

a) Graph the function on the interval $[15, \infty)$ and complete the following:

$$N(t) \to \square \text{ as } t \to \infty.$$

b) Explain the meaning of the answer to part (a) in terms of the application.

44. *Average Cost.* The average cost per tape, in dollars, for a company to produce x videotapes on exercising is given by the function

$$A(x) = \frac{13x + 100}{x}.$$

a) Graph the function on the interval $(0, \infty)$ and complete the following:

$$A(x) \rightarrow \boxed{} \text{ as } x \rightarrow \infty.$$

b) Explain the meaning of the answer to part (a) in terms of the application.

45. *Population Growth.* The population P, in thousands, of Lordsburg is given by

$$P(t) = \frac{500t}{2t^2 + 9},$$

where t = time, in months.

a) Graph the function over the interval $[0, \infty)$.
b) Find the population at $t = 0, 1, 3,$ and 8 months.
c) Find the horizontal asymptote. Complete:

$$P(t) \rightarrow \boxed{} \text{ as } t \rightarrow \infty.$$

d) Give the meaning of the answer to part (c).
e) Find the maximum population.

46. *Minimizing Surface Area.* The Hold-It Container Co. is designing an open-top rectangular box, with a square base, that will hold 108 cubic centimeters (cc).

a) Express the surface area S as a function of the length x of a side of the base.
b) Graph the function over the interval $(0, \infty)$.
c) Estimate the minimum surface area and the value of x that will yield it.

Inverse Variation. *Rational functions like*

$$y = \frac{k}{x} \quad \text{and} \quad y = \frac{k}{x^2},$$

where the variables x, y, and k are positive, represent equations of ***inverse variation*** *in applications. The constant k is called the* ***variation constant***. *Note that the function is decreasing over the interval* $(0, \infty)$.

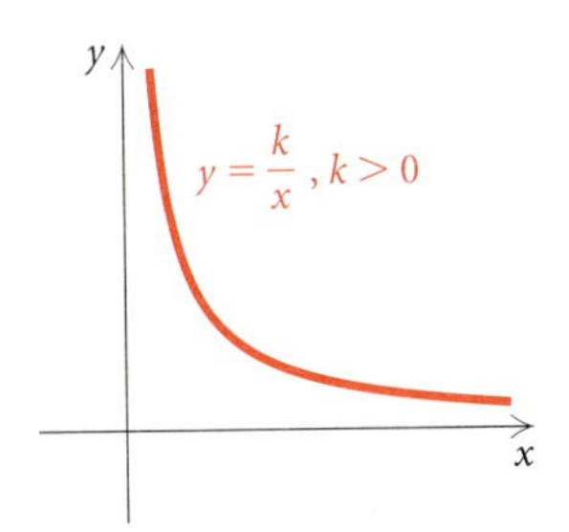

47. *Intensity of Light.* The intensity of light I from a lightbulb varies inversely as the square of the distance d from the bulb. Suppose that $I = 90\ \text{W/m}^2$ (watts per square meter) when the distance is 5 m. Find the intensity at a distance of 10 m.

48. *Stocks and Gold.* It is theorized by some economists that stock prices S vary inversely as the price G of gold. One day the Dow Jones Industrial Average was 4040 and the price of gold was \$348 per ounce. What will the Dow Jones Industrial Average be if the price of gold rises to \$440 per ounce?

Skill Maintenance

Solve.

49. $3x - 7 > 2$

50. $4x - 9 \leq 6x + 5$

51. $\dfrac{3x - 9}{x^2 + 1} = 0$

52. $\dfrac{1}{x} + \dfrac{3}{2x} = 4$

Synthesis

53. ◈ Under what circumstances will a rational function have a domain consisting of all real numbers?

54. ◈ Explain and contrast the three kinds of asymptotes of rational functions.

55. ◈ Graph

$$y_1 = \frac{x^3 + 4}{x} \quad \text{and} \quad y_2 = x^2$$

using the same viewing window. Explain how the parabola $y_2 = x^2$ can be thought of as a nonlinear asymptote for y_1.

Find the nonlinear asymptotes of each function.

56. $f(x) = \dfrac{x^4 + 3x^2}{x^2 + 1}$

57. $f(x) = \dfrac{x^5 + 2x^3 + 4x^2}{x^2 + 2}$

Graph each function. Check on a grapher.

58. $f(x) = \dfrac{x^3 + 4x^2 + x - 6}{x^2 - x - 2}$

59. $f(x) = \dfrac{2x^3 + x^2 - 8x - 4}{x^3 + x^2 - 9x - 9}$

60. $f(x) = \dfrac{x^4 + 3x^3 + 21x^2 - 50x + 80}{x^4 + 8x^3 - x^2 + 20x - 10}$

Holes. *Suppose we violate our restriction that rational functions have no common factor in the numerator and the denominator other than −1 or 1. If there is some other common factor, we can get a so-called hole in the graph where this factor is 0. Make a hand-drawn graph of each of the following, showing any holes.*

61. $f(x) = \dfrac{x^2 - 4}{x - 2}$

62. $f(x) = \dfrac{x^2 - 9}{x + 3}$

63. $g(x) = \dfrac{3x^2 - x - 2}{x - 1}$

64. $f(x) = \dfrac{2x + 1}{2x^2 - 5x - 3}$

65. ◆ Which procedure for finding asymptotes can no longer be used if we allow rational functions to have common factors other than -1 or 1 in the numerator and denominator? (See Exercises 61–64.)

Find the domain.

66. $f(x) = \sqrt{x^2 - 4x - 21}$

67. $f(x) = \sqrt{\dfrac{72}{x^2 - 4x - 21}}$

2.7 Polynomial and Rational Inequalities

• *Solve polynomial and rational inequalities.*

We will use a combination of algebraic and graphical methods to solve polynomial and rational inequalities.

Polynomial Inequalities

Just as a quadratic equation can be written in the form $ax^2 + bx + c = 0$, a **quadratic inequality** can be written in the form $ax^2 + bx + c \;\blacksquare\; 0$, where $\blacksquare$ is $<$, $>$, $\le$, or $\ge$. Here are some examples of quadratic inequalities:

$$3x^2 - 2x - 5 > 0, \qquad -\tfrac{1}{2}x^2 + 4x - 7 \le 0.$$

Quadratic inequalities are one type of **polynomial inequality**. Other examples are

$$-2x^4 + x^2 - 3 < 0, \qquad \tfrac{2}{3}x + 4 \ge 0, \quad \text{and} \quad 4x^3 - 2x^2 > 5x + 7.$$

When the inequality symbol in a polynomial inequality is replaced with an equality symbol, a **related equation** is formed. Polynomial inequalities can be easily solved once the related equation has been solved.

Example 1 Solve: $x^3 - x > 0$.

SOLUTION We are asked to find all x-values for which $x^3 - x > 0$. To locate these values, we graph $f(x) = x^3 - x$. Then we note that whenever the function changes sign, its graph passes through an x-intercept. Thus to solve $x^3 - x > 0$, we first solve the related equation to locate any x-intercepts:

$$\begin{aligned} x^3 - x &= 0 \\ x(x^2 - 1) &= 0 \\ x(x + 1)(x - 1) &= 0. \end{aligned}$$

The x-intercepts are -1, 0, and 1, as shown on the graph.

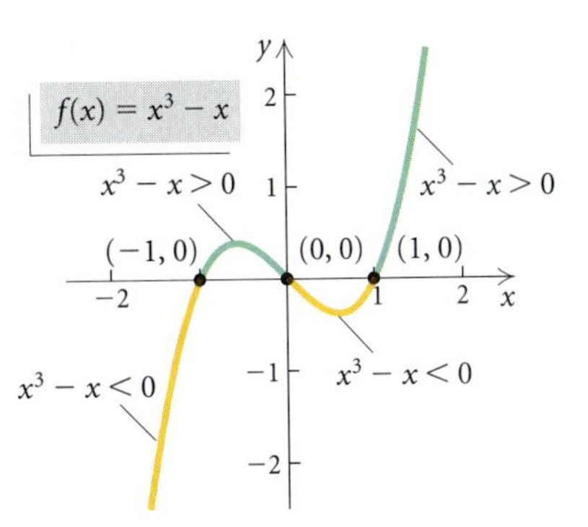

The x-intercepts divide the x-axis into four intervals: $(-\infty, -1)$, $(-1, 0)$, $(0, 1)$, and $(1, \infty)$. For x-values within each of these intervals, the sign of $x^3 - x$ must be either uniformly positive or uniformly negative. To determine which, we choose a test value for x from each interval and find $f(x)$. This can be done by substitution or by using the TABLE feature set in ASK mode. We can also simply look at the graph.

INTERVAL	TEST VALUE	SIGN OF $f(x)$
$(-\infty, -1)$	$f(-2) = -6$	Negative
$(-1, 0)$	$f(-0.5) = 0.375$	Positive
$(0, 1)$	$f(0.5) = -0.375$	Negative
$(1, \infty)$	$f(2) = 6$	Positive

Since we are solving $x^3 - x > 0$, the solution set consists of only two of the four intervals, those where the sign of $f(x)$ is *positive*. We see that the solution set is $(-1, 0) \cup (1, \infty)$, or $\{x \mid -1 < x < 0 \text{ or } x > 1\}$.

The following is a method for solving polynomial inequalities.

To solve a polynomial inequality:

1. Find an equivalent inequality with 0 on one side.
2. Solve the related polynomial equation.
3. Use the solutions to divide the x-axis into intervals. Then select a test value from each interval and determine the polynomial's sign on the interval.
4. Determine the intervals for which the inequality is satisfied and write interval notation or set-builder notation for the solution set. Include the endpoints of the intervals in the solution set if the inequality symbol is $\leq$ or $\geq$.

Example 2 Solve: $3x^4 + 10x \leq 11x^3 + 4$.

By subtracting $11x^3 + 4$, we form the equivalent inequality

$$3x^4 - 11x^3 + 10x - 4 \leq 0.$$

Algebraic Solution

To solve the related equation

$$3x^4 - 11x^3 + 10x - 4 = 0,$$

we need to use a grapher and/or the theorems of Section 2.5 (see Example 11 in Section 2.5). The solutions are

$$-1, \quad 2 - \sqrt{2}, \quad \tfrac{2}{3}, \quad \text{and} \quad 2 + \sqrt{2}.$$

These numbers divide the x-axis into five intervals.

We then let $f(x) = 3x^4 - 11x^3 + 10x - 4$ and, using test values for $f(x)$, determine the sign of $f(x)$ in each interval:

INTERVAL	TEST VALUE	SIGN OF $f(x)$
$(-\infty, -1)$	$f(-2) = 112$	Positive
$(-1, 2 - \sqrt{2})$	$f(0) = -4$	Negative
$\left(2 - \sqrt{2}, \frac{2}{3}\right)$	$f(0.6) = 0.0128$	Positive
$\left(\frac{2}{3}, 2 + \sqrt{2}\right)$	$f(1) = -2$	Negative
$(2 + \sqrt{2}, \infty)$	$f(4) = 100$	Positive

Function values are negative in the intervals $(-1, 2 - \sqrt{2})$ and $\left(\frac{2}{3}, 2 + \sqrt{2}\right)$. Since the inequality sign is $\leq$, we include the endpoints of the intervals in the solution set. The solution set is

$[-1, 2 - \sqrt{2}] \cup \left[\frac{2}{3}, 2 + \sqrt{2}\right]$, or
$\left\{x \mid -1 \leq x \leq 2 - \sqrt{2} \text{ or } \frac{2}{3} \leq x \leq 2 + \sqrt{2}\right\}$.

Graphical Solution

We graph $y = 3x^4 - 11x^3 + 10x - 4$ using a viewing window that reveals the curvature of the graph.

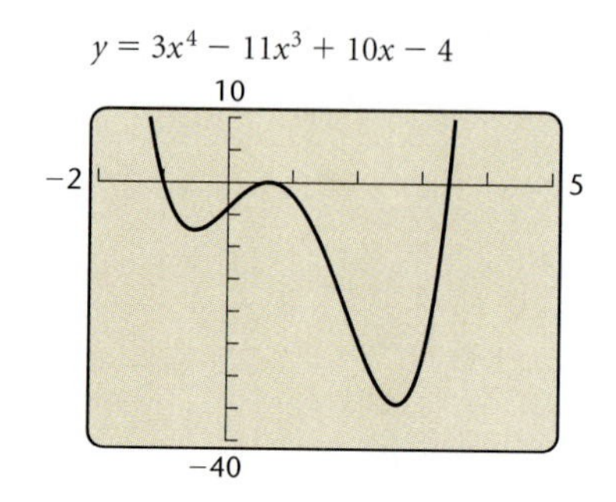

Although this window reveals the curvature, it leaves us uncertain about the number of zeros of the function in the interval [0, 1]. By using TRACE and ZOOM or the SOLVE feature (see also Example 10 in the Introduction to Graphs and Graphers), we see that the first zero is -1. The following window shows another view of the zeros in the interval [0, 1].

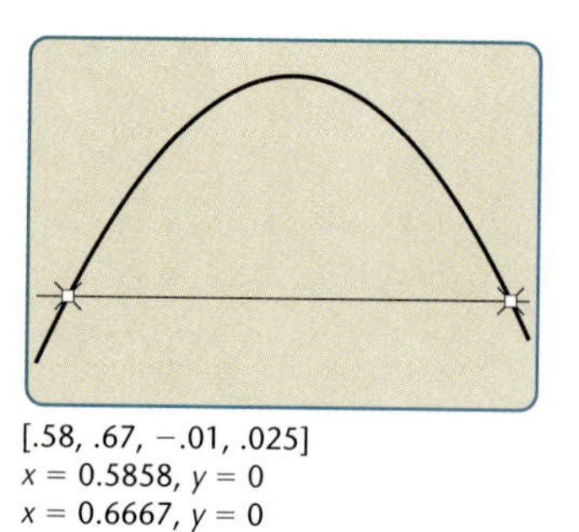

[.58, .67, −.01, .025]
$x = 0.5858$, $y = 0$
$x = 0.6667$, $y = 0$

The zeros are about -1, 0.586, 0.667, and 3.414.

The intervals to be considered are $(-\infty, -1)$, $(-1, 0.586)$, $(0.586, 0.667)$, $(0.667, 3.414)$, and $(3.414, \infty)$. We note on the graph where the function is negative. Then including appropriate endpoints, we find that the solution set is approximately

$[-1, 0.586] \cup [0.667, 3.414]$, or
$\{x \mid -1 \leq x \leq 5.586 \text{ or } 0.667 \leq x \leq 3.414\}$.

Interactive Discovery

Referring to Example 2, use the TRACE feature to confirm that the sign of $3x^4 - 11x^3 + 10x - 4$ changes as the cursor moves through each of the x-intercepts. Then let $y_1 = 3x^4 + 10x$ and $y_2 = 11x^3 + 4$. Show that the graphs of y_1 and y_2 intersect at those values for which $3x^4 - 11x^3 + 10x - 4 = 0$. How can the graphs of y_1 and y_2 be used to solve Example 2? We see from this Interactive Discovery that a grapher can be used to solve a polynomial inequality when one side is not 0.

Rational Inequalities

Some inequalities involve rational expressions and functions. These are called **rational inequalities**. To solve rational inequalities, we need to make some adjustments to the preceding method.

Example 3 Solve: $\dfrac{x-3}{x+4} \geq \dfrac{x+2}{x-5}$.

We first subtract $(x + 2)/(x - 5)$ in order to find an equivalent inequality with 0 on one side:

$$\frac{x-3}{x+4} - \frac{x+2}{x-5} \geq 0.$$

ALGEBRAIC SOLUTION

We look for all values of x for which the related function

$$f(x) = \frac{x-3}{x+4} - \frac{x+2}{x-5}$$

is not defined or is 0. These are called **critical values**.

A look at the denominators shows that $f(x)$ is not defined for $x = -4$ and $x = 5$. Next, we solve $f(x) = 0$:

$$\frac{x-3}{x+4} - \frac{x+2}{x-5} = 0$$

$$(x+4)(x-5)\left(\frac{x-3}{x+4} - \frac{x+2}{x-5}\right) = (x+4)(x-5)\cdot 0$$

$$(x-5)(x-3) - (x+4)(x+2) = 0$$

$$(x^2 - 8x + 15) - (x^2 + 6x + 8) = 0$$

$$-14x + 7 = 0$$

$$x = \tfrac{1}{2}.$$

The critical values are -4, $\frac{1}{2}$, and 5. The critical values divide the x-axis into four intervals.

We then use a test value to determine the sign of $f(x)$ in each interval:

GRAPHICAL SOLUTION

We graph

$$y = \frac{x-3}{x+4} - \frac{x+2}{x-5}$$

in a standard window, which shows its curvature.

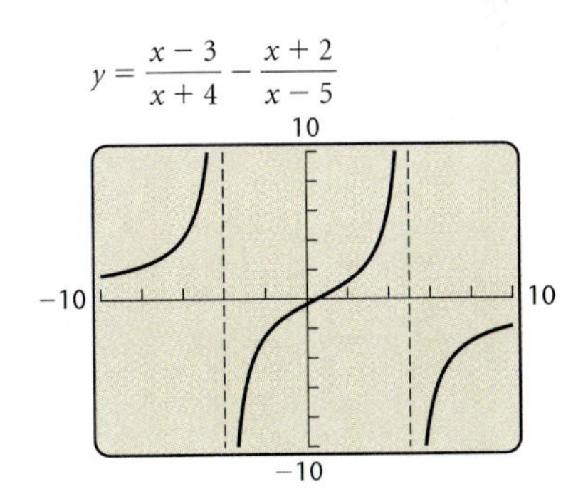

By using TRACE and ZOOM or the SOLVE feature, we find that 0.5 is a zero and that no other numbers are zeros.

We then look for values where the function is not defined. We can determine them quickly by examining the denominators $x + 4$ and $x - 5$, but it is instructive to use CONNECTED

(continued)

INTERVAL	TEST VALUE	SIGN OF $f(x)$
$(-\infty, -4)$	$f(-5) = 7.7$	Positive
$\left(-4, \frac{1}{2}\right)$	$f(-2) = -2.5$	Negative
$\left(\frac{1}{2}, 5\right)$	$f(3) = 2.5$	Positive
$(5, \infty)$	$f(6) = -7.7$	Negative

Function values are positive in the intervals $(-\infty, -4)$ and $\left(\frac{1}{2}, 5\right)$. Note that since $f\left(\frac{1}{2}\right) = 0$, $\frac{1}{2}$ must be in the solution set. Note that since neither -4 nor 5 is in the domain of f, they cannot be part of the solution set.

The solution set is

$$(-\infty, -4) \cup \left[\frac{1}{2}, 5\right).$$

mode and zoom in near the x-value -4. Then use the TRACE feature to observe how the y-values jump from very positive on the left of -4 to very negative on the right of -4. These y-values will actually be offscreen. The same thing occurs near 5.

The **critical values**, where y is either not defined or 0, are -4, 0.5, and 5.

The graph shows where y is positive and where it is negative. Note that the x-values of the asymptotes cannot be in the solution set since y is not defined for these values.

The solution set is

$$(-\infty, -4) \cup \left[\frac{1}{2}, 5\right).$$

The following is a method for solving rational inequalities.

To solve a rational inequality:

1. Find an equivalent inequality with 0 on one side.
2. Change the inequality symbol to an equals sign and solve the related equation.
3. Find x-values for which the related rational function is not defined.
4. The numbers found in steps (2) and (3) are called critical values. Use the critical values to divide the x-axis into intervals. Then test an x-value from each interval to determine the function's sign on that interval.
5. Select the intervals for which the inequality is satisfied and write interval notation or set-builder notation for the solution set. If the inequality symbol is $\leq$ or $\geq$, then the solutions to step (2) should be included in the solution set.

It works well to use a combination of algebraic and graphical methods to solve polynomial and rational inequalities. The algebraic methods give exact numbers for the critical values, and the graphical methods allow us to see easily what intervals satisfy the inequality.

2.7 Exercise Set

In Exercises 1–4, a related function is graphed. Solve the given inequality.

1. $x^3 + 6x^2 < x + 30$

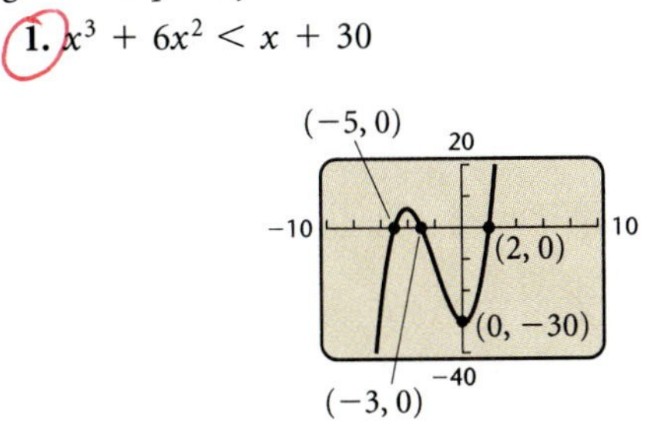

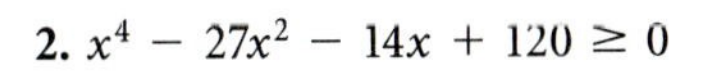

2. $x^4 - 27x^2 - 14x + 120 \geq 0$

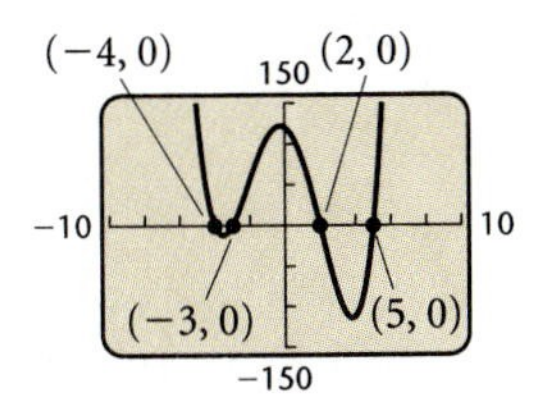

3. $\dfrac{8x}{x^2 - 4} \geq 0$

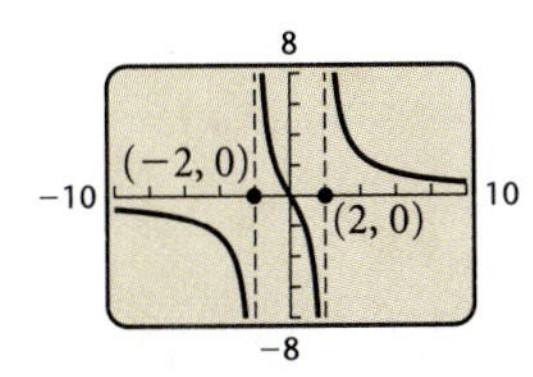

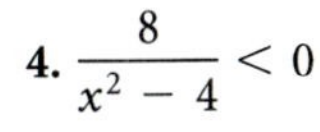

4. $\dfrac{8}{x^2 - 4} < 0$

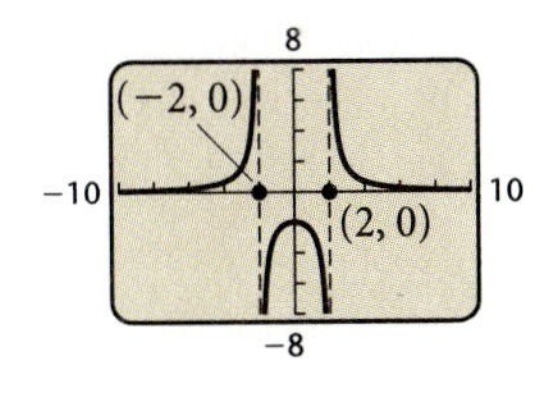

Solve.

5. $(x - 1)(x + 4) < 0$

6. $(x + 3)(x - 5) < 0$

7. $(x - 4)(x + 2) \geq 0$

8. $(x - 2)(x + 1) \geq 0$

9. $x^2 + x - 2 > 0$

10. $x^2 - x - 6 > 0$

11. $x^2 > 25$

12. $x^2 \leq 1$

13. $4 - x^2 \leq 0$

14. $11 - x^2 \geq 0$

15. $6x - 9 - x^2 < 0$

16. $x^2 + 2x + 1 \leq 0$

17. $x^2 + 12 < 4x$

18. $x^2 - 8 > 6x$

19. $4x^3 - 7x^2 \leq 15x$

20. $2x^3 - x^2 < 5$

21. $x^3 + 3x^2 - x - 3 \geq 0$

22. $x^3 + x^2 - 4x - 4 \geq 0$

23. $x^3 - 2x^2 < 5x - 6$

24. $x^3 + x \leq 6 - 4x^2$

25. $x^5 + x^2 \geq 2x^3 + 2$

26. $x^5 + 24 > 3x^3 + 8x^2$

27. $2x^3 + 6 \leq 5x^2 + x$

28. $2x^3 + x^2 < 10 + 11x$

29. $x^3 + 5x^2 - 25x \leq 125$

30. $x^3 - 9x + 27 \geq 3x^2$

31. $0.1x^3 - 0.6x^2 - 0.1x + 2 < 0$

32. $19.2x^3 + 12.8x^2 + 144 \geq 172.8x + 3.2x^4$

33. $\dfrac{1}{x + 4} > 0$

34. $\dfrac{1}{x - 3} \leq 0$

35. $\dfrac{-4}{2x + 5} < 0$

36. $\dfrac{-2}{5 - x} \geq 0$

37. $\dfrac{x - 4}{x + 3} - \dfrac{x + 2}{x - 1} \leq 0$

38. $\dfrac{x + 1}{x - 2} + \dfrac{x - 3}{x - 1} < 0$

39. $\dfrac{2x - 1}{x + 3} \geq \dfrac{x + 1}{3x + 1}$

40. $\dfrac{x + 5}{x - 4} > \dfrac{3x + 2}{2x + 1}$

41. $\dfrac{x + 1}{x - 2} \geq 3$

42. $\dfrac{x}{x - 5} < 2$

43. $x - 2 > \dfrac{1}{x}$

44. $4 \geq \dfrac{4}{x} + x$

45. $\dfrac{2}{x^2 - 4x + 3} \leq \dfrac{5}{x^2 - 9}$

46. $\dfrac{3}{x^2 - 4} \leq \dfrac{5}{x^2 + 7x + 10}$

47. $\dfrac{3}{x^2 + 1} \geq \dfrac{6}{5x^2 + 2}$

48. $\dfrac{4}{x^2 - 9} < \dfrac{3}{x^2 - 25}$

49. $\dfrac{5}{x^2 + 3x} < \dfrac{3}{2x + 1}$

50. $\dfrac{2}{x^2 + 3} > \dfrac{3}{5 + 4x^2}$

51. $\dfrac{5x}{7x - 2} > \dfrac{x}{x + 1}$

52. $\dfrac{x^2 - x - 2}{x^2 + 5x + 6} < 0$

53. $\dfrac{x}{x^2 + 4x - 5} + \dfrac{3}{x^2 - 25} \leq \dfrac{2x}{x^2 - 6x + 5}$

54. $\dfrac{2x}{x^2 - 9} + \dfrac{x}{x^2 + x - 12} \geq \dfrac{3x}{x^2 + 7x + 12}$

55. *Temperature During an Illness.* The temperature T, in degrees Fahrenheit, of a person during an illness is given by the function

$$T(t) = \frac{4t}{t^2 + 1} + 98.6,$$

where t = time, in hours. Find the interval over which the temperature was over 100°. (See Example 9 in Section 2.6.)

56. *Population Growth.* The population P, in thousands, of Lordsburg is given by

$$P(t) = \frac{500t}{2t^2 + 9},$$

where t = time, in months. Find the interval over which the population was 40 thousand or greater. (See Exercise 45 in Exercise Set 2.6.)

57. *Total Profit.* Flexl, Inc., determines that its total profit is given by the function

$$P(x) = -3x^2 + 630x - 6000.$$

a) Flexl makes a profit for those nonnegative values of x for which $P(x) > 0$. Find the values of x for which Flexl makes a profit.

b) Flexl loses money for those nonnegative values of x for which $P(x) < 0$. Find the values of x for which Flexl loses money.

58. *Height of a Thrown Object.* The function

$$S(t) = -16t^2 + 32t + 1920$$

gives the height S, in feet, of an object thrown from a cliff that is 1920 ft high. Here t is the time, in seconds, that the object is in the air.

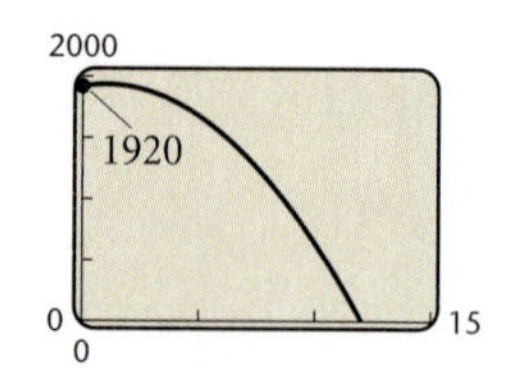

a) For what times is the height greater than 1920 ft?

b) For what times is the height less than 640 ft?

59. *Number of Diagonals.* A polygon with n sides has D diagonals, where D is given by the function

$$D(n) = \frac{n(n - 3)}{2}.$$

Find the number of sides n if

$$27 \leq D \leq 230.$$

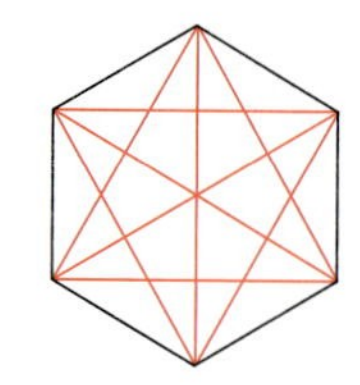

60. *Number of Handshakes.* There are n people in a room. The number N of possible handshakes by all the people in the room is given by the function

$$N(n) = \frac{n(n - 1)}{2}.$$

For what number n of people is

$$66 \leq N \leq 300?$$

Skill Maintenance

61. Solve for b: $a = 3b - 7$.

62. Solve for y: $4x - 5y = 9$.

Find $(f \circ g)(2)$ for each of the following.

63. $f(x) = 3x - 7$, $g(x) = x^3 - 1$

64. $f(x) = \sqrt{x} + 1$, $g(x) = 5x^2 - 4$

Synthesis

65. ◈ Under what circumstances would a quadratic inequality have a solution set that is a closed interval?

66. ◈ Under what circumstances would a quadratic inequality have an empty solution set?

Solve.

67. $x^2 + 9 \leq 6x$

68. $x^4 - 6x^2 + 5 > 0$

69. $x^4 + 3x^2 > 4x - 15$

70. $\left|\dfrac{x + 3}{x - 4}\right| < 2$

71. $|x^2 - 5| = 5 - x^2$

72. $(7 - x)^{-2} < 0$

73. $2|x|^2 - |x| + 2 \leq 5$

74. $|x|^2 - 4|x| + 4 \geq 9$

75. $\left|1 + \dfrac{1}{x}\right| < 3$

76. $\left|2 - \dfrac{1}{x}\right| \leq 2 + \left|\dfrac{1}{x}\right|$

77. $|x^2 + 3x - 1| < 3$

78. $|1 + 5x - x^2| \geq 5$

79. Write a quadratic inequality for which the solution set is $(-4, 3)$.

80. Write a polynomial inequality for which the solution set is $[-4, 3] \cup [7, \infty)$.

CHAPTER

2 Summary and Review

Important Properties and Formulas

Polynomial Function:

$$P(x) = a_n x^n + a_{n-1}x^{n-1} + a_{n-2}x^{n-2} + \cdots + a_1 x + a_0$$

Complex Number: $a + bi$, a, b real, $i^2 = -1$

Imaginary Number: $a + bi$, $b \neq 0$

Complex Conjugates: $a + bi$, $a - bi$

Quadratic Equation:

$$ax^2 + bx + c = 0,\ a \neq 0,\ a, b, c \text{ real}$$

Quadratic Function:

$$f(x) = ax^2 + bx + c,\ a \neq 0,\ a, b, c \text{ real}$$

Quadratic Formula:

$$x = \frac{-b \pm \sqrt{b^2 - 4ac}}{2a}$$

Polynomial Divison:

$$P(x) = D(x) \cdot Q(x) + R(x)$$

Dividend ($P(x)$) Divisor ($D(x)$) Quotient ($Q(x)$) Remainder ($R(x)$)

The Remainder Theorem: The remainder found by dividing $P(x)$ by $x - c$ is $P(c)$.

The Factor Theorem: $P(c) = 0 \leftrightarrow x - c$ is a factor of $P(x)$.

The Fundamental Theorem of Algebra: Every polynomial of degree n, $n \geq 1$, with complex coefficients has at least one complex-number zero.

The Rational Zeros Theorem: Consider the polynomial equation

$$a_n x^n + a_{n-1}x^{n-1} + a_{n-2}x^{n-2} + \cdots + a_1 x + a_0 = 0,$$

where all the coefficients are integers and $n \geq 1$. Also, consider a rational number p/q, where p and q have no common factor other than -1 and 1. If p/q is a zero of the polynomial equation, then p is a factor of a_0 and q is a factor of a_n.

The Intermediate Value Theorem: For any polynomial P with real coefficients, suppose that $P(a) \neq P(b)$ for $a \neq b$ and that $P(a)$ and $P(b)$ are of opposite signs. Then the polynomial has a real zero between a and b.

Rational Function:

$$f(x) = \frac{P(x)}{Q(x)},$$

where $P(x)$ and $Q(x)$ are polynomials with no common factor other than -1 and 1 and where $Q(x)$ is not the zero polynomial, and the domain consists of all x for which $Q(x) \neq 0$.

Horizontal Asymptote: $y = b$, where

$f(x) \to b$ as $x \to \infty$, or $f(x) \to b$ as $x \to -\infty$.

Vertical Asymptote: $x = a$ if any of the following is true:

$f(x) \to \infty$ as $x \to a^-$, or
$f(x) \to -\infty$ as $x \to a^-$, or
$f(x) \to \infty$ as $x \to a^+$, or
$f(x) \to -\infty$ as $x \to a^+$.

REVIEW EXERCISES

Use a grapher to graph each of the following polynomial functions. Then estimate the function's (a) zeros; (b) relative maxima; (c) relative minima; (d) domain and range.

1. $f(x) = -2x^2 - 3x + 6$

2. $f(x) = x^3 + 3x^2 - 2x - 6$

3. $f(x) = x^4 - 3x^3 + 2x^2$

Express in terms of i.

4. $-\sqrt{-40}$

5. $\sqrt{-12} \cdot \sqrt{-20}$

6. $\dfrac{\sqrt{-49}}{-\sqrt{-64}}$

Simplify each of the following. Leave answers in the form $a + bi$, where a and b are real numbers.

7. $(6 + 2i)(-4 - 3i)$

8. $\dfrac{2 - 3i}{1 - 3i}$

9. $(3 - 5i) - (2 - i)$

10. $(6 + 2i) + (-4 - 3i)$

11. i^{23}

12. $(-3i)^{28}$

Solve by completing the square to obtain exact solutions. Show your work. Check your work on a grapher.

13. $x^2 - 3x = 18$

14. $3x^2 - 12x - 6 = 0$

Solve. Use any method, but obtain exact solutions. Check your work on a grapher, if possible.

15. $3x^2 + 2x = 8$

16. $r^2 - 2r + 10 = 0$

17. $x^2 = 18 + 3x$

18. $x = 2\sqrt{x} - 1$

19. $y^4 - 3y^2 + 1 = 0$

20. $(x^2 - 1)^2 - (x^2 - 1) - 2 = 0$

21. $(p - 3)(3p + 2)(p + 2) = 0$

22. $x^3 + 5x^2 - 4x - 20 = 0$

In Exercises 23 and 24, complete the square to:

a) *find the vertex;*

b) *find the line of symmetry;*

c) *determine whether there is a maximum or minimum value and find that value;*

d) *find the range.*

Then check your answers using a grapher.

23. $f(x) = -4x^2 + 3x - 1$

24. $f(x) = 5x^2 - 10x + 3$

In Exercises 25–28, match the equation with figures (a)–(d), which follow. Do not use a grapher except as a check.

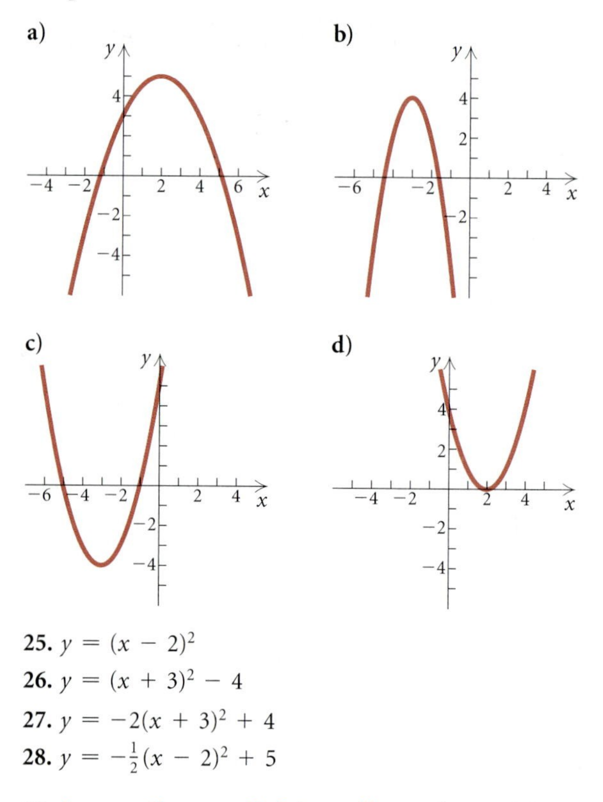

25. $y = (x - 2)^2$

26. $y = (x + 3)^2 - 4$

27. $y = -2(x + 3)^2 + 4$

28. $y = -\frac{1}{2}(x - 2)^2 + 5$

29. *Interest Compounded Annually.* When P dollars is invested at interest rate i, compounded annually, for t years, the investment grows to A dollars, where

$$A = P(1 + i)^t.$$

a) Find the interest rate i if \$6250 grows to \$6760 in 2 yr.

b) Find the interest rate i if \$1,000,000 grows to \$1,215,506.25 in 4 yr.

30. *Sidewalk Width.* A 60-ft by 80-ft parking lot is torn up to install a sidewalk of uniform width around its perimeter. The new area of the parking lot is two thirds of the old area. How wide is the sidewalk?

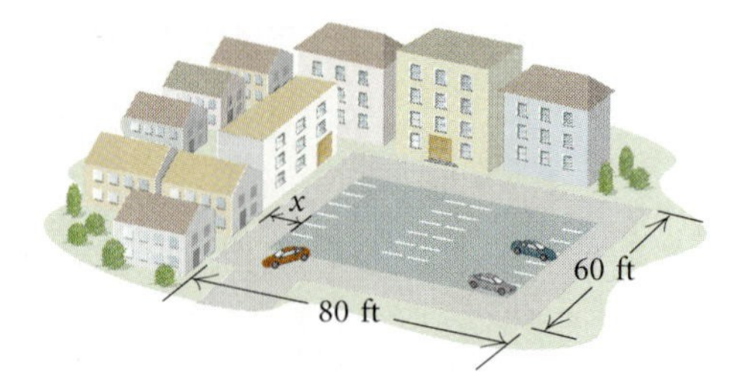

31. *Maximizing Volume.* The Berniers have 24 ft of flexible fencing with which to build a rectangular "toy corral" for a backyard. If the fencing is 2 ft high, what dimensions should the corral have in order to maximize its volume?

32. *Dimensions of a Box.* An open box is made from a 10-cm by 20-cm piece of aluminum by cutting a square from each corner and folding up the edges. The area of the resulting base is 90 cm². What is the length of the sides of the squares?

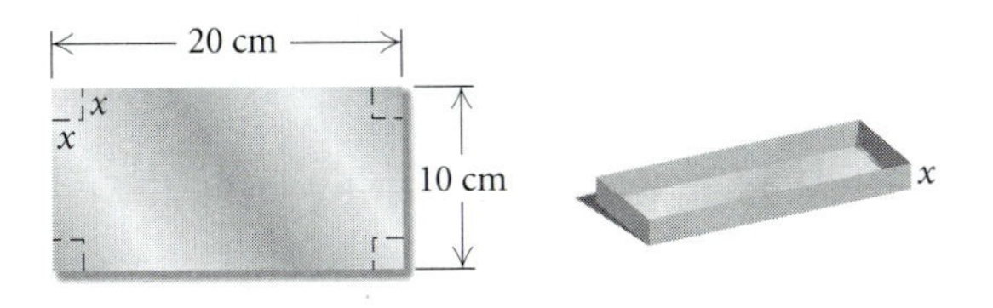

33. *Maximizing Box Dimensions.* An open box is to be made from a 10-cm by 20-cm piece of aluminum by cutting a square from each corner and folding up the edges. What should the length of the sides of the squares be in order to maximize the volume?

34. *Cholesterol Level and the Risk of Heart Attack.* The data in the following table show the relationship of cholesterol level in men to the risk of a heart attack.

CHOLESTEROL LEVEL	NUMBER OF MEN, PER 10,000, WHO SUFFER A HEART ATTACK
100	30
200	65
250	100
275	130

Source: Nutrition Action Healthletter.

a) Use the REGRESSION feature on a grapher to find a linear, a quadratic, and a cubic function to fit the data.

b) It is also known that 180 of 10,000 men have a heart attack with a cholesterol level of 300. Which function in part (a) would best make this prediction?

c) Use the answer to part (b) to predict the heart attack rate for men with a cholesterol level of 350; with one of 400.

Do the long division. Check by multiplying or a grapher.

35. $(20x^3 - 6x^2 - 9x + 10) \div (5x + 2)$

36. $(2x^5 - 3x^3 + x^2 + 4) \div (x^2 - 2)$

In Exercises 37 and 38, a polynomial equation is given along with information about factors. Use the information to solve each equation.

37. $x^3 + 2x^2 - 13x + 10 = 0$;
$x + 5$ is a factor of $x^3 + 2x^2 - 13x + 10$

38. $x^4 + 5x^3 - 23x^2 - 87x + 140 = 0$;
$x^2 + x - 20$ is a factor of
$x^4 + 5x^3 - 23x^2 - 87x + 140$

Use synthetic division to find the quotient and the remainder.

39. $(x^3 + 2x^2 - 13x + 10) \div (x - 5)$

40. $(x^4 + 3x^3 + 3x^2 + 3x + 2) \div (x + 2)$

Use synthetic division to find the indicated function value.

41. $P(x) = x^3 + 2x^2 - 13x + 10$; $P(5)$

42. $P(x) = x^4 - 16$; $P(-2)$

Determine whether the given values are zeros of $P(x)$. Then factor and find any other zeros that exist.

43. $P(x) = x^3 + 7x^2 + 7x - 15$; -1, 2, and -3

44. $P(x) = x^4 + 9x^3 + 19x^2 + 9x - 2$; 1, -4, and -2

Find a polynomial equation of lowest degree with integer coefficients and the following as some of its zeros. Use a grapher as a partial check.

45. 4, -1, -2

46. -4, $-\sqrt{3}$, $\sqrt{3}$, 1

47. 3, $2 - 3i$, $4 + \sqrt{5}$, -2

Find an equation with rational coefficients that has a graph that passes through the given points. Use a grapher as a partial check.

48. $(-2, 0)$, $(-1, 0)$, $(0, 5)$, $(4, 0)$

49. $(-\sqrt{8}, 0)$, $(-3, 0)$, $(0, 12)$, $(1, 0)$, $(\sqrt{8}, 0)$

50. $(-5, 0)$, $(1 - \sqrt{3}, 0)$, $(0, 8)$, $(1 + \sqrt{3}, 0)$

Use the rational zeros theorem and/or factoring to solve each equation.

51. $x^3 - 4x^2 + x + 6 = 0$

52. $x^3 + 7x^2 + 7x = 15$

53. $x^4 + 36x + 63 = 4x^3 + 16x^2$

For each polynomial function:

a) Solve $P(x) = 0$.

b) Graph $y = P(x)$.

c) Express $P(x)$ as a product of linear factors.

54. $P(x) = x^3 + 8x^2 - 19x + 10$

55. $P(x) = x^6 + x^5 - 28x^4 - 16x^3 + 192x^2$

Make a hand-drawn graph for each of the following. Be sure to label all the asymptotes.

56. $f(x) = \dfrac{x^2 - 5}{x + 2}$

57. $f(x) = \dfrac{5}{(x - 2)^2}$

58. $f(x) = \dfrac{x^2 + x - 6}{x^2 - x - 20}$

59. $f(x) = \dfrac{x - 2}{x^2 - 2x - 15}$

In Exercises 60 and 61, find a rational function that satisfies the given conditions. Answers may vary, but try to give the simplest answer possible.

60. Vertical asymptotes $x = -2$, $x = 3$

61. Vertical asymptotes $x = -2$, $x = 3$; horizontal asymptote $y = 4$; x-intercept $(-3, 0)$

62. *Medical Dosage.* The function

$$N(t) = \frac{0.7t + 2000}{8t + 9}, \quad t \geq 5$$

gives the body concentration $N(t)$, in parts per million, of a certain dosage of medication after time t, in hours.

a) Graph the function on the interval $[5, \infty)$ and complete the following:

$$N(t) \longrightarrow \boxed{} \text{ as } t \longrightarrow \infty.$$

b) Explain the meaning of the answer to part (a) in terms of the application.

Solve.

63. $x^2 - 9 < 0$

64. $2x^2 > 3x + 2$

65. $(1 - x)(x + 4)(x - 2) < 0$

66. $\dfrac{x - 2}{x + 3} < 4$

67. $\dfrac{x - 3}{x^2 + x - 20} \geq \dfrac{4}{x^2 - 4}$

68. *Height of a Thrown Object.* The function

$$S(t) = -16t^2 + 80t + 224$$

gives the height S, in feet, of a model rocket launched from a hill that is 224 ft high with a velocity of 80 ft/sec, where t = time, in seconds.

a) Find the maximum height of the rocket and when that height is attained.

b) Determine when the object reaches the ground.

c) Over what interval is the height greater than 320 ft?

69. *Population Growth.* The population P, in thousands, of Novi is given by

$$P(t) = \frac{8000t}{4t^2 + 10},$$

where t = time, in months. Find the interval over which the population was 400 or greater.

Synthesis

70. ◆ Explain the difference between a polynomial function and a rational function.

71. ◆ Explain and contrast the three types of asymptotes considered for rational functions.

72. Determine whether the following is an identity:

$$4(x^2 + x + 1)^3 - 27x^2(x + 1)^2 = (x - 1)^2(2x + 1)^2(x + 2)^2.$$

73. *Interest Rate.* In early 1995, \$3500 was deposited at a certain interest rate. One year later, \$4000 was deposited in another account at the same rate. At the end of that year, there was a total of \$8518.35 in both accounts. What is the annual interest rate?

Solve.

74. $x^2 \geq 5 - 2x$

75. $\left|1 - \dfrac{1}{x^2}\right| < 3$

76. $x^4 - 2x^3 + 3x^2 - 2x + 2 = 0$

77. $(x - 2)^{-3} < 0$

78. Express $x^3 - 1$ as a product of linear factors.

79. Find k so that $x + 3$ is a factor of $x^3 + kx^2 + kx - 15$.

80. When $x^2 - 4x + 3k$ is divided by $x + 5$, the remainder is 33. Find the value of k.

Find the domain of each function.

81. $f(x) = \sqrt{x^2 + 3x - 10}$

82. $f(x) = \sqrt{x^2 - 3.1x + 2.2} + 1.75$

83. $f(x) = \dfrac{1}{\sqrt{5 - |7x + 2|}}$

Exponential and Logarithmic Functions 3

APPLICATION

The number of cellular phones in this country is modeled by an exponential function, where y = the number of telephones, in millions, in the year x. Here $x = 0$ corresponds to 1985. (*Source:* Cellular Telecommunications Industry Association)

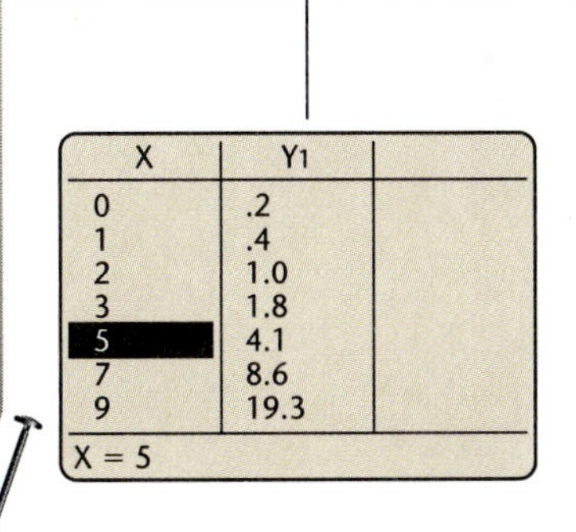

X	Y1
0	.2
1	.4
2	1.0
3	1.8
5	4.1
7	8.6
9	19.3

X = 5

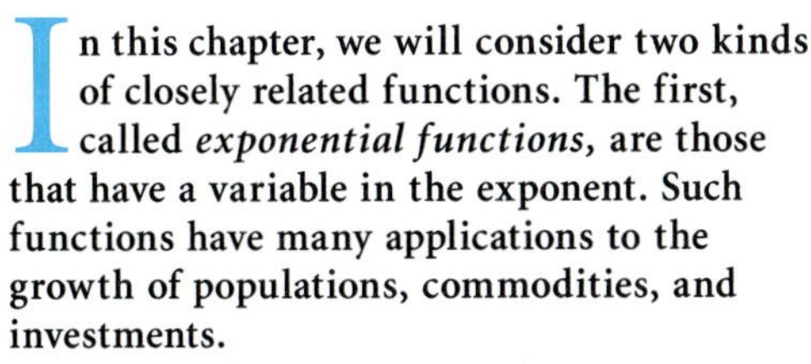

In this chapter, we will consider two kinds of closely related functions. The first, called *exponential functions,* are those that have a variable in the exponent. Such functions have many applications to the growth of populations, commodities, and investments.

Recall that a function takes an input to an output. Suppose we can reverse the process and take the output back to an input. That process produces what we call the *inverse* of the original function. Functions that are inverses of each other are closely related. The inverses of exponential functions, called *logarithmic functions,* or *logarithm functions,* are also important in many applications such as earthquake magnitude, sound level, and chemical pH.

3.1 Inverse Functions
3.2 Exponential Functions and Graphs
3.3 Logarithmic Functions and Graphs
3.4 Properties of Logarithmic Functions
3.5 Solving Exponential and Logarithmic Equations
3.6 Applications and Models: Growth and Decay
SUMMARY AND REVIEW

3.1 Inverse Functions

- *Determine whether a function is one-to-one, and if it is, find a formula for its inverse.*
- *Simplify expressions of the type $(f \circ f^{-1})(x)$ and $(f^{-1} \circ f)(x)$.*

When we go from an output of a function back to its input or inputs, we get an inverse relation. When that relation is a function, we have an inverse function. We now study inverse functions and how to find their formulas when the original function has a formula. We do this to understand the relationship between exponential and logarithmic functions.

Inverses

Consider the relation h given as follows:

$$h = \{(-8, 5), (4, -2), (-7, 1), (3.8, 6.2)\}.$$

Suppose we *interchange* the first and second coordinates. The relation we obtain is called the **inverse** of the relation h and is given as follows:

$$\text{Inverse of } h = \{(5, -8), (-2, 4), (1, -7), (6.2, 3.8)\}.$$

Inverse Relation

Interchanging the first and second coordinates of each ordered pair in a relation produces the *inverse relation.*

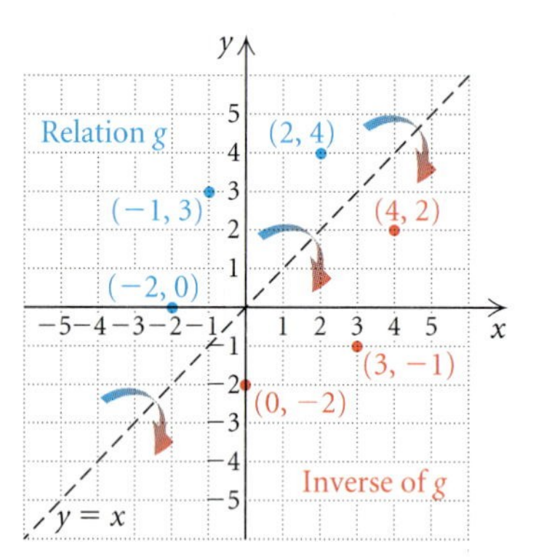

Example 1 Consider the relation g given by

$$g = \{(2, 4), (-1, 3), (-2, 0)\}.$$

Graph the relation in blue. Find the inverse and graph it in red.

SOLUTION The relation g is shown in blue in the figure at left. The inverse of the relation is

$$\{(4, 2), (3, -1), (0, -2)\}$$

and is shown in red. The pairs in the inverse are reflections across the line $y = x$.

Inverse Relation

If a relation is defined by an equation, interchanging the variables produces an equation of the *inverse relation.*

Example 2 Find an equation for the inverse of the relation:

$$y = x^2 - 5x.$$

SOLUTION We interchange x and y and obtain an equation of the inverse:

$$x = y^2 - 5y.$$

Interactive Discovery

Graph each of the following relations:

$$y = 3x + 2, \qquad y = x, \qquad xy = 2,$$
$$x^2 + 3y^2 = 4, \qquad y^2 = 4x - 5.$$

Then find the inverse of each and graph it using the same set of axes. What pattern do you see? Be sure to use a square viewing window.

If a relation is given by an equation, the solutions of the inverse can be found from those of the original equation by interchanging the first and second coordinates of each ordered pair. Thus the graphs of a relation and its inverse are always reflections of each other across the line $y - x$.

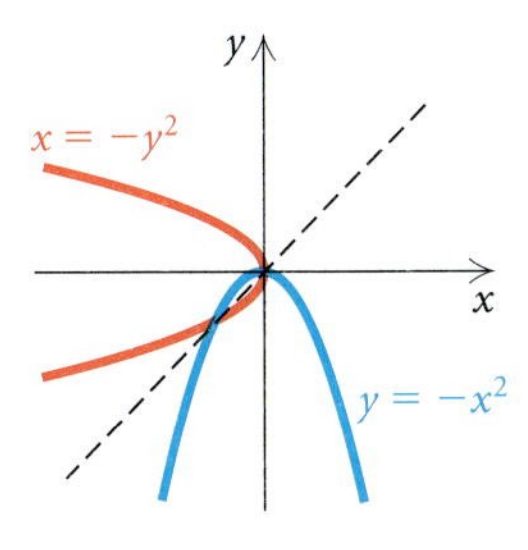

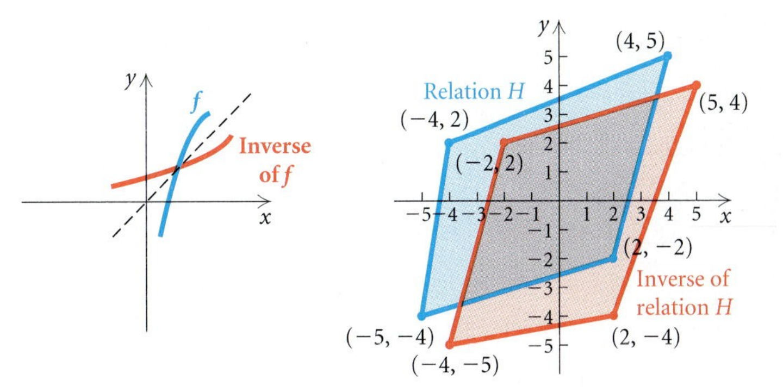

Inverses and One-to-One Functions

Let's consider the following two functions.

NUMBER (DOMAIN)	CUBING (RANGE)
−3 →	−27
−2 →	−8
−1 →	−1
0 →	0
1 →	1
2 →	8
3 →	27

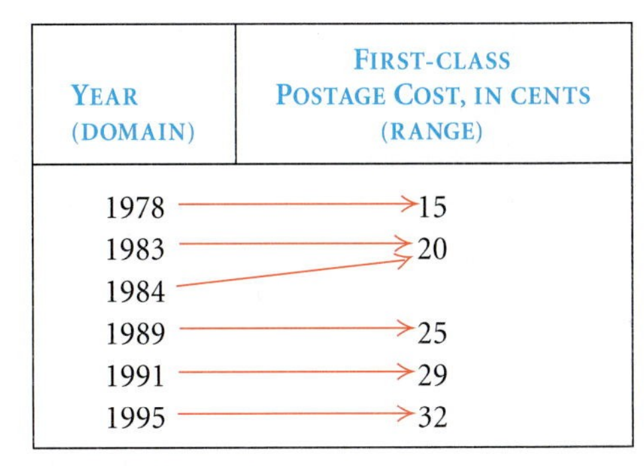

YEAR (DOMAIN)	FIRST-CLASS POSTAGE COST, IN CENTS (RANGE)
1978 →	15
1983 →	20
1984 →	20
1989 →	25
1991 →	29
1995 →	32

Source: U.S. Postal Service.

Suppose we reverse the arrows. Are these inverse relations functions?

NUMBER (RANGE)		CUBING (DOMAIN)
−3	←	−27
−2	←	−8
−1	←	−1
0	←	0
1	←	1
2	←	8
3	←	27

YEAR (RANGE)		FIRST-CLASS POSTAGE COST, IN CENTS (DOMAIN)
1978	←	15
1983	←	20
1984	← (from 20)	
1989	←	25
1991	←	29
1995	←	32

We see that the inverse of the cubing function is a function, but the inverse of the postage function is not a function. Like all functions, each input in the postage function has exactly one output. However, the output for both 1983 and 1984 is the same number, 20. Thus in the inverse of the postage function, the input 20 has *two* outputs, 1983 and 1984. When the same output comes from two or more different inputs, the inverse cannot be a function. In the cubing function, each output corresponds to exactly one input, so its inverse is also a function. The cubing function is an example of a **one-to-one function**. If the inverse of a function f is also a function, it is named f^{-1} (read "f-inverse").

The -1 in f^{-1} is *not* an exponent!

One-to-One Function and Inverses

A function f is *one-to-one* if different inputs have different outputs—that is,

if $a \neq b$, then $f(a) \neq f(b)$. Or,

A function f is *one-to-one* if when the outputs are the same, the inputs are the same—that is,

if $f(a) = f(b)$, then $a = b$.

If a function is one-to-one, then its inverse is a function.

The domain of a one-to-one function f is the range of the inverse f^{-1}.

The range of a one-to-one function f is the domain of the inverse f^{-1}.

Example 3 Given the function f described by $f(x) = 2x - 3$, prove that f is one-to-one (that is, it has an inverse that is a function).

SOLUTION To show that f is one-to-one, we show that if $f(a) = f(b)$, then $a = b$. Assume that $f(a) = f(b)$ for any numbers a and b in the domain of f. Then

$$\begin{aligned} 2a - 3 &= 2b - 3 \\ 2a &= 2b \qquad \text{Adding 3} \\ a &= b. \qquad \text{Dividing by 2} \end{aligned}$$

Thus, if $f(a) = f(b)$, then $a = b$. This shows that f is one-to-one.

Example 4 Given the function g described by $g(x) = x^2$, prove that g is not one-to-one.

SOLUTION To prove that g is not one-to-one, we need to find two numbers a and b for which $a \neq b$ and $g(a) = g(b)$. Two such numbers are -3 and 3, because $-3 \neq 3$ and $g(-3) = g(3) = 9$. Thus g is not one-to-one.

The graph below shows a function, in blue, and its inverse, in red. To determine whether the inverse is a function, we can apply the vertical-line test to its graph. By reflecting each such vertical line back across the line $y = x$, we obtain an equivalent **horizontal-line test** for the original function.

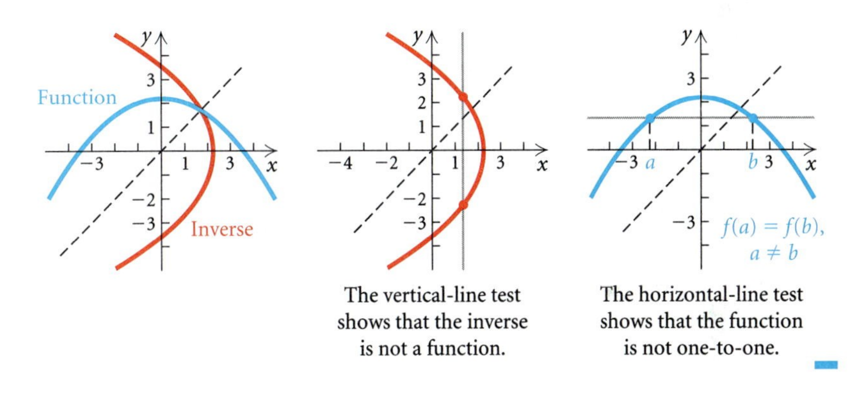

The vertical-line test shows that the inverse is not a function.

The horizontal-line test shows that the function is not one-to-one.

Horizontal-Line Test

If it is possible for a horizontal line to intersect the graph of a function more than once, then the function is *not* one-to-one and its inverse is *not* a function.

Example 5 Graph each of the following functions. Then determine whether each is one-to-one and thus has an inverse that is a function.

a) $f(x) = 4 - x$

b) $f(x) = x^2$

c) $f(x) = \sqrt[3]{x + 2} + 3$

d) $f(x) = 3x^5 - 20x^3$

SOLUTION We graph each function using a grapher. Then we apply the horizontal-line test.

a) $y = 4 - x$

8, −8, 8, −8

b) $y = x^2$

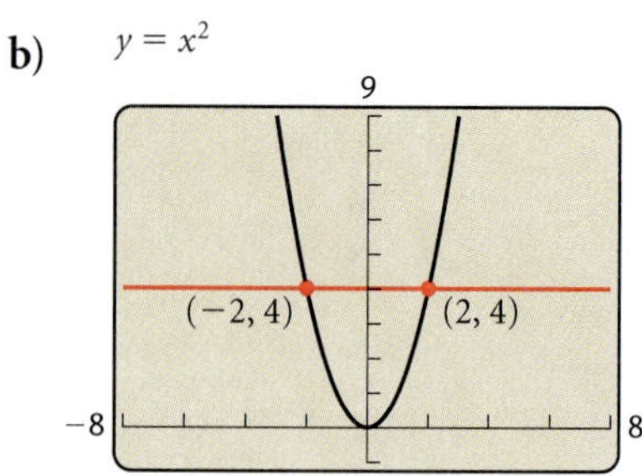

c) $y = \sqrt[3]{x + 2} + 3$

6, −12, 10, −1

d) $y = 3x^5 - 20x^3$

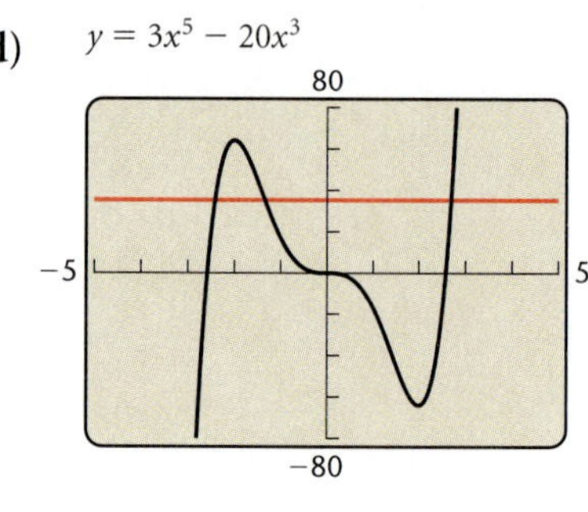

RESULT	REASON
a) One-to-one; inverse is a function	No horizontal line crosses the graph more than once.
b) Not one-to-one; inverse is not a function	There are many horizontal lines that cross the graph more than once. In particular, the line $y = 4$ does so. Note that where the line crosses, the first coordinates are -2 and 2. Although these are different inputs, they have the same output—that is, $-2 \neq 2$, but $f(-2) = (-2)^2 = 4 = 2^2 = f(2)$.
c) One-to-one; inverse is a function	No horizontal line crosses the graph more than once.
d) Not one-to-one; inverse is not a function	There are many horizontal lines that cross the graph more than once.

Finding Formulas for Inverses

Suppose that a function is described by a formula. If it has an inverse that is a function, we proceed as follows to find a formula for f^{-1}.

Obtaining a Formula for an Inverse

If a function f is one-to-one, a formula for its inverse can generally be found as follows:

1. Replace $f(x)$ with y.
2. Interchange x and y.
3. Solve for y.
4. Replace y with $f^{-1}(x)$.

Example 6 Determine whether the function $f(x) = 2x - 3$ is one-to-one, and if it is, find a formula for $f^{-1}(x)$.

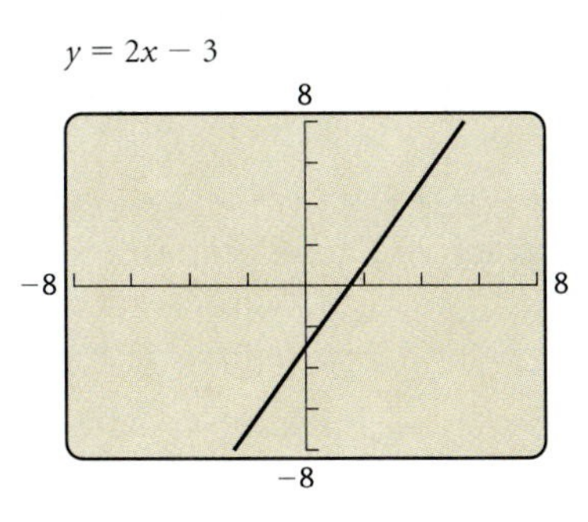

SOLUTION The graph of f is shown at left. It passes the horizontal-line test. Thus it is one-to-one and its inverse is a function.

We also proved that f is one-to-one in Example 3.

1. Replace $f(x)$ with y: $\quad y = 2x - 3$
2. Interchange x and y: $\quad x = 2y - 3$
3. Solve for y: $\quad x + 3 = 2y$

$$\frac{x + 3}{2} = y$$

4. Replace y with $f^{-1}(x)$: $\quad f^{-1}(x) = \dfrac{x + 3}{2}$.

Consider

$$f(x) = 2x - 3 \quad \text{and} \quad f^{-1}(x) = \frac{x + 3}{2}$$

from Example 6. For the input 5, we have

$$f(5) = 2 \cdot 5 - 3 = 10 - 3 = 7.$$

The output is 7. Now we use 7 for the input in the inverse:

$$f^{-1}(7) = \frac{7 + 3}{2} = \frac{10}{2} = 5.$$

The function f takes the number 5 to 7. The inverse function f^{-1} takes the number 7 back to 5.

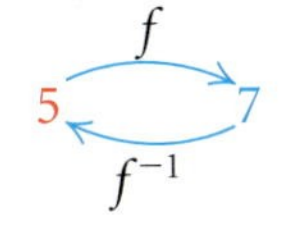

Example 7 Graph

$$f(x) = 2x - 3 \quad \text{and} \quad f^{-1}(x) = \frac{x + 3}{2}$$

using the same set of axes. Then compare the two graphs.

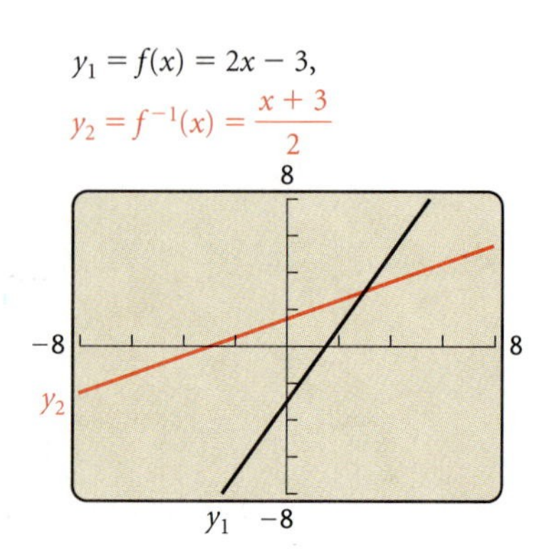

SOLUTION

Method 1. The graphs of f and f^{-1} are shown at left. Note that the graph of f^{-1} can be drawn by reflecting the graph of f across the line $y = x$. That is, if we were to graph $f(x) = 2x - 3$ in wet ink and fold along the line $y = x$, the graph of $f^{-1}(x) = (x + 3)/2$ would be formed by the ink transferred from f.

When we interchange x and y in finding a formula for the inverse of $f(x) = 2x - 3$, we are in effect reflecting the graph of that function across the line $y = x$. For example, when the coordinates of the y-intercept of the graph of f, $(0, -3)$, are reversed, we get the x-intercept of the graph of f^{-1}, $(-3, 0)$.

Method 2. Some graphers have a feature that graphs inverses automatically. If we were to start with $y_1 = 2x - 3$, the graphs of both y_1 and its inverse $y_2 = (x + 3)/2$ would be drawn. Be sure to square the viewing window. Consult your manual.

The graph of f^{-1} is a reflection of the graph of f across the line $y = x$.

Example 8 Consider $g(x) = x^3 + 2$.

a) Determine whether the function is one-to-one.

b) If it is one-to-one, find a formula for its inverse.

c) Graph the function and its inverse. Use a square viewing window.

SOLUTION

a) The graph of $g(x) = x^3 + 2$ is shown below. It passes the horizontal-line test and thus has an inverse that is a function.

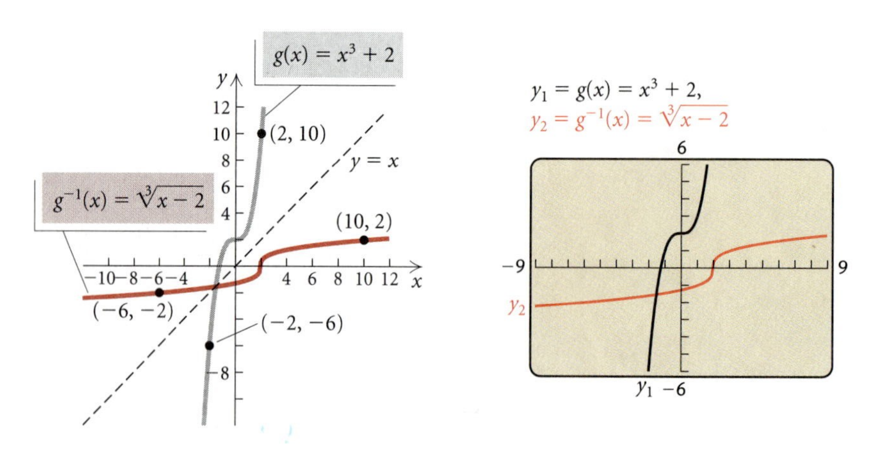

b) We follow the procedure for finding an inverse.

1. Replace $g(x)$ with y: $\quad y = x^3 + 2$

2. Interchange x and y: $\quad x = y^3 + 2$

3. Solve for y: $$x - 2 = y^3$$ $$\sqrt[3]{x - 2} = y$$

4. Replace y with $g^{-1}(x)$: $g^{-1}(x) = \sqrt[3]{x - 2}$.

c) To find the graph, we reflect the graph of $g(x) = x^3 + 2$ across the line $y = x$. This can be done by using a grapher or by plotting points. See the graph on the left above. Be sure to square the viewing window.

Inverse Functions and Composition

Suppose that we were to use some input a for a one-to-one function f and find its output, $f(a)$. The function f^{-1} would then take that output back to a. Similarly, if we began with an input b for the function f^{-1} and found its output, $f^{-1}(b)$, the original function f would then take that output back to b. This is summarized as follows.

If a function f is one-to-one, then f^{-1} is the unique function such that each of the following holds:

$(f^{-1} \circ f)(x) = f^{-1}(f(x)) = x$, for each x in the domain of f (range of f^{-1}), and

$(f \circ f^{-1})(x) = f(f^{-1}(x)) = x$, for each x in the domain of f^{-1} (range of f).

Example 9 Given that $f(x) = 5x + 8$, use composition of functions to show that $f^{-1}(x) = (x - 8)/5$.

SOLUTION We find $(f^{-1} \circ f)(x)$ and $(f \circ f^{-1})(x)$ and check to see that each is x:

$$(f^{-1} \circ f)(x) = f^{-1}(f(x)) = f^{-1}(5x + 8) = \frac{(5x + 8) - 8}{5} = \frac{5x}{5} = x;$$

$$\begin{aligned}(f \circ f^{-1})(x) &= f(f^{-1}(x)) \\ &= f\left(\frac{x - 8}{5}\right) \\ &= 5\left(\frac{x - 8}{5}\right) + 8 \\ &= x - 8 + 8 = x.\end{aligned}$$

Restricting a Domain

In the case in which the inverse of a function is not a function, the domain can be restricted to allow the inverse to be a function. We saw in Example 5 that $f(x) = x^2$ is not one-to-one. The graph is shown at the top of the following page.

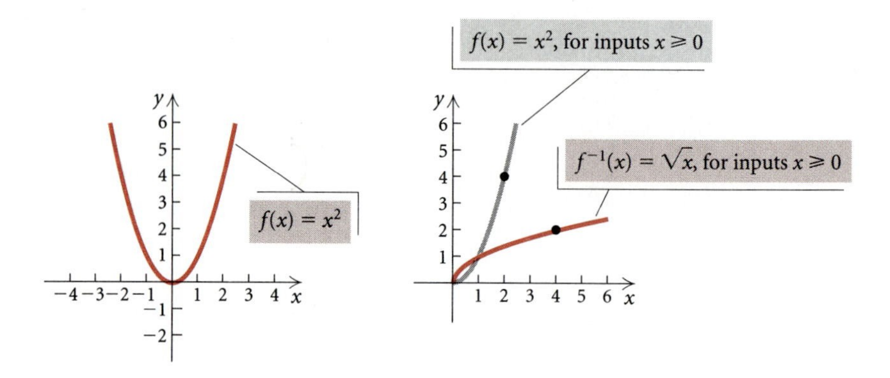

Suppose we had not known this and had tried to find a formula for the inverse as follows:

$$y = x^2 \qquad \text{Replacing } f(x) \text{ with } y$$

$$x = y^2. \qquad \text{Interchanging } x \text{ and } y$$

We cannot solve for y and get only one value, since most real numbers have two square roots:

$$\pm\sqrt{x} = y.$$

This is not the equation of a function. An input of, say, 4 would yield two outputs, -2 and 2. In such cases, it is convenient to consider "part" of the function by restricting the domain of $f(x)$. For example, if we restrict the domain of $f(x) = x^2$ to nonnegative numbers, then its inverse is a function. See the graphs of $f(x) = x^2$, $x \geq 0$, and $f^{-1}(x) = \sqrt{x}$, $x \geq 0$ on the right above.

3.1 Exercise Set

Find the inverse of each relation.

1. $\{(7, 8), (-2, 8), (3, -4), (8, -8)\}$
2. $\{(0, 1), (5, 6), (-2, -4)\}$
3. $\{(-1, -1), (-3, 4)\}$
4. $\{(-1, 3), (2, 5), (-3, 5), (2, 0)\}$

Find an equation of each inverse relation.

5. $y = 4x - 5$
6. $2x^2 + 5y^2 = 4$
7. $x^3y = -5$
8. $y = 3x^2 - 5x + 9$

Graph each equation by substituting and plotting points. Then reflect the graph across the line $y = x$ to obtain the graph of its inverse.

9. $x = y^2 - 3$
10. $y = x^2 + 1$
11. $y = |x|$
12. $x = |y|$

Using the horizontal-line test, determine whether each function is one-to-one.

13. $f(x) = 2.7^x$

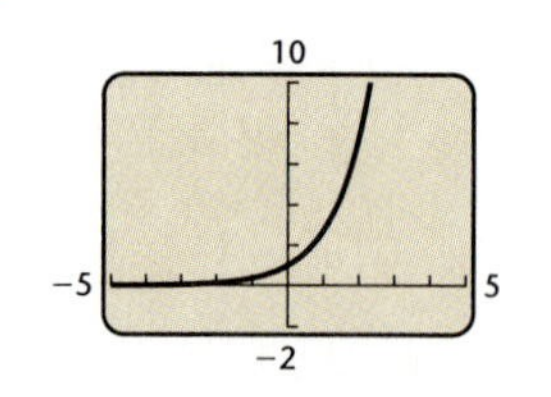

14. $f(x) = 2^{-x}$

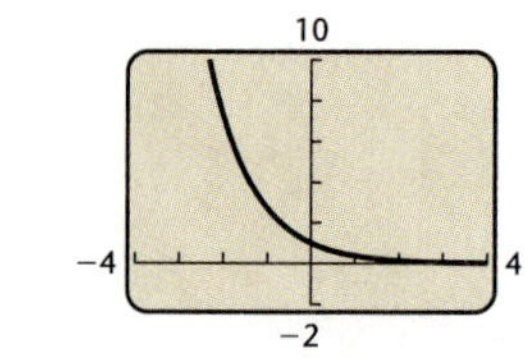

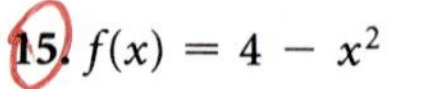

15. $f(x) = 4 - x^2$

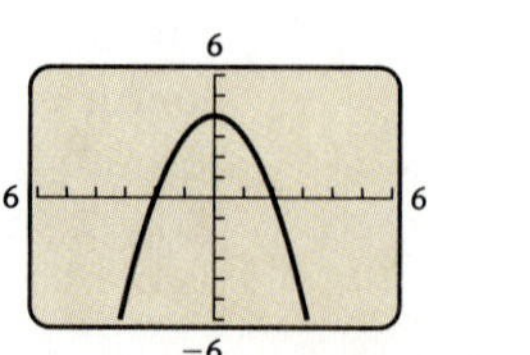

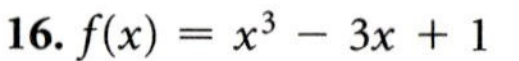

16. $f(x) = x^3 - 3x + 1$

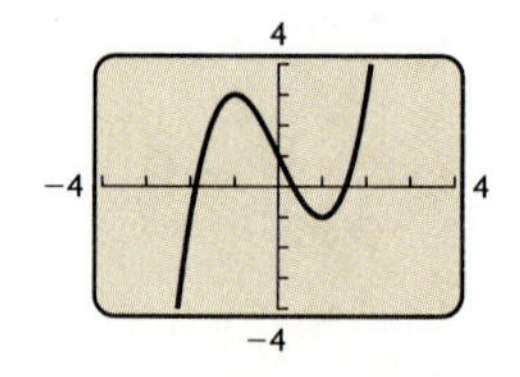

17. $f(x) = \dfrac{8}{x^2 - 4}$

18. $f(x) = \sqrt{\dfrac{10}{4 + x}}$

19. $f(x) = \sqrt[3]{x + 2} - 2$

20. $f(x) = \dfrac{8}{x}$

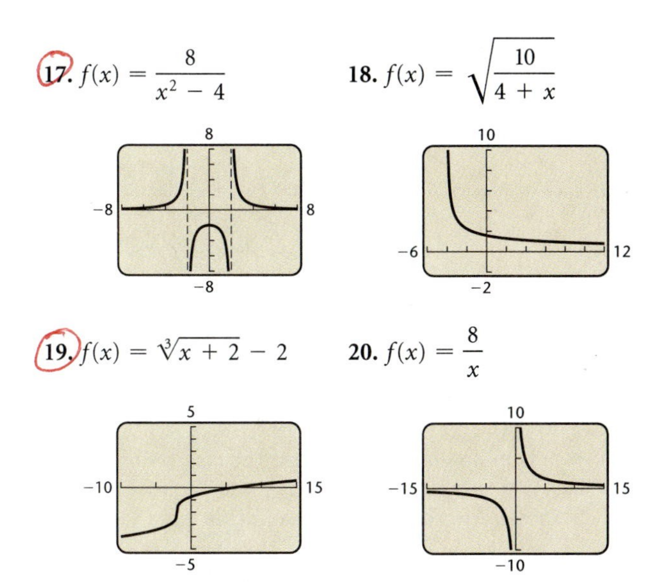

Determine whether each of the following is one-to-one. Use a grapher and the horizontal-line test or use an algebraic procedure.

21. $f(x) = 5x - 8$

22. $f(x) = 3 + 4x$

23. $f(x) = 1 - x^2$

24. $f(x) = |x| - 2$

25. $f(x) = |x + 2|$

26. $f(x) = -0.8$

27. $f(x) = -\dfrac{4}{x}$

28. $f(x) = \dfrac{2}{x + 3}$

Graph each function and its inverse using a grapher. Use an inverse drawing feature, if available. Find the domain and the range of f. Find the domain and the range of the inverse f^{-1}.

29. $f(x) = 0.8x + 1.7$

30. $f(x) = 2.7 - 1.08x$

31. $f(x) = \frac{1}{2}x - 4$

32. $f(x) = x^3 - 1$

33. $f(x) = \sqrt{x - 3}$

34. $f(x) = -\dfrac{2}{x}$

35. $f(x) = x^2 - 4,\ x \geq 0$

36. $f(x) = 3 - x^2,\ x \geq 0$

37. $f(x) = (3x - 9)^3$

38. $f(x) = \sqrt[3]{\dfrac{x - 3.2}{1.4}}$

Given each function:

a) *Determine whether it is one-to-one, using a grapher if desired.*

b) *If it is one-to-one, find a formula for the inverse.*

39. $f(x) = x + 4$

40. $f(x) = 7 - x$

41. $f(x) = 2x$

42. $f(x) = 5x + 8$

43. $f(x) = \dfrac{4}{x + 7}$

44. $f(x) = -\dfrac{3}{x}$

45. $f(x) = \dfrac{x + 4}{x - 3}$

46. $f(x) = \dfrac{5x - 3}{2x + 1}$

47. $f(x) = x^3 - 1$

48. $f(x) = (x + 5)^3$

49. $f(x) = x\sqrt{4 - x^2}$

50. $f(x) = 4x^5 - 20x^3 + 2x^2 - 5x + 1$

51. $f(x) = 5x^2 - 2,\ x \geq 2$

52. $f(x) = 4x^2 + 3,\ x \geq 0$

53. $f(x) = \sqrt{x + 1}$

54. $f(x) = \sqrt[3]{x - 8}$

Find each inverse by thinking about the operations of the function and then reversing, or undoing, them. Check your work algebraically.

	FUNCTION	INVERSE
55.	$f(x) = 3x$	$f^{-1}(x) =$
56.	$f(x) = \frac{1}{4}x + 7$	$f^{-1}(x) =$
57.	$f(x) = -x$	$f^{-1}(x) =$
58.	$f(x) = \sqrt[3]{x} - 5$	$f^{-1}(x) =$
59.	$f(x) = \sqrt[3]{x - 5}$	$f^{-1}(x) =$
60.	$f(x) = x^{-1}$	$f^{-1}(x) =$

For each function f, use composition of functions to show that f^{-1} is as given.

61. $f(x) = \dfrac{7}{8}x,\ f^{-1}(x) = \dfrac{8}{7}x$

62. $f(x) = \dfrac{(x + 5)}{4},\ f^{-1}(x) = 4x - 5$

63. $f(x) = \dfrac{(1 - x)}{x},\ f^{-1}(x) = \dfrac{1}{x + 1}$

64. $f(x) = \sqrt[3]{x + 4},\ f^{-1}(x) = x^3 - 4$

65. Find $f(f^{-1}(5))$ and $f^{-1}(f(a))$:

$$f(x) = x^3 - 4.$$

66. Find $f^{-1}(f(p))$ and $f(f^{-1}(1253))$:

$$f(x) = \sqrt[5]{\frac{2x - 7}{3x + 4}}.$$

67. *Dress Sizes in the United States and Italy.* A function that will convert dress sizes in the United States to those in Italy is

$$g(x) = 2(x + 12).$$

a) Find the dress sizes in Italy that correspond to sizes 6, 8, 10, 14, and 18 in the United States.

b) Find a formula for the inverse of the function.

c) Use the inverse function to find the dress sizes in the United States that correspond to 36, 40, 44, 52, and 60 in Italy.

68. *Bus Chartering.* An organization determines that the cost per person of chartering a bus is given by the formula

$$C(x) = \frac{100 + 5x}{x},$$

where x = the number of people in the group and $C(x)$ is in dollars. Determine $C^{-1}(x)$ and explain what it represents.

69. *Reaction Distance.* You are driving a car when a deer suddenly darts across the road in front of you. Your brain registers the emergency and sends a signal to your foot to hit the brake. The car travels a distance D, in feet, during this time, where D is a function of the speed r, in miles per hour, that the car is traveling when you see the deer. That reaction distance D is a linear function given by

$$D(r) = \frac{11r + 5}{10}.$$

a) Find $D(0)$, $D(10)$, $D(20)$, $D(50)$, and $D(65)$.
b) Graph $D(r)$.
c) Find $D^{-1}(r)$ and explain what it represents.
d) Graph the inverse.

70. *Bread Consumption.* The number N of 1-lb loaves of bread consumed per person per year t years after 1995 is given by the function

$$N(t) = 0.6514t + 53.1599.$$

(*Source:* Department of Agriculture)

a) Find the consumption of bread per person in 1998 and 2000.
b) Use a grapher to graph the function and its inverse.
c) Explain what the inverse represents.

Skill Maintenance

Graph each of the following.

71. $y = x^3 - x$

72. $x = y^3 - y$

73. $f(x) = \sqrt[3]{x}$

74. $f(x) = \dfrac{8}{x^2 - 4}$

Synthesis

75. ◈ Suppose that you have graphed a function using a grapher and you see that it is one-to-one. How could you then use the TRACE feature to make a hand-drawn graph of the inverse?

76. ◈ The following formulas for the conversion between Fahrenheit and Celsius temperatures have been considered several times in this text:

$$C = \tfrac{5}{9}(F - 32)$$

and

$$F = \tfrac{9}{5}C + 32.$$

Discuss these formulas from the standpoint of inverses.

Using only a grapher, determine whether the functions are inverses of each other.

77. $f(x) = \sqrt[3]{\dfrac{x - 3.2}{1.4}}$, $g(x) = 1.4x^3 + 3.2$

78. $f(x) = \dfrac{2x - 5}{4x + 7}$, $g(x) = \dfrac{7x - 4}{5x + 2}$

79. $f(x) = \dfrac{2}{3}$, $g(x) = \dfrac{3}{2}$

80. $f(x) = x^4$, $x \geq 0$; $g(x) = \sqrt[4]{x}$

81. Find three examples of functions that are their own inverses, that is, $f = f^{-1}$.

82. Consider the function f given by

$$f(x) = \begin{cases} x^3 + 2, & x \leq -1, \\ x^2, & -1 < x < 1, \\ x + 1, & x \geq 1. \end{cases}$$

Does f have an inverse that is a function? Why or why not?

3.2 Exponential Functions and Graphs

- *Graph exponential equations and functions.*
- *Solve problems involving applications of exponential functions and their graphs.*

We now turn our attention to the study of a set of functions very rich in application. Consider the following graphs.

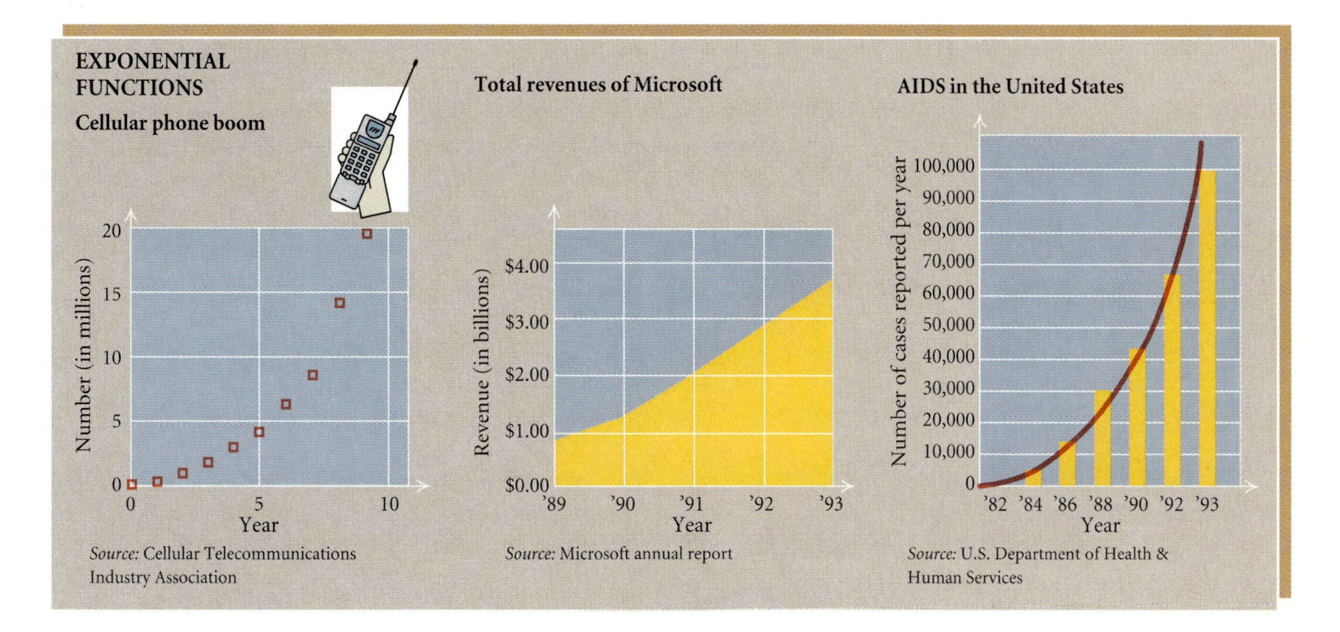

Each of these graphs illustrates the idea of an *exponential function*. The graph on the right shows the numbers of cases of AIDS reported in various years. The curve drawn along the graph approximates an exponential function. In this section, we consider such graphs, their inverses, and some important applications of exponential functions.

Graphing Exponential Functions

We now define exponential functions. We assume that a^x has meaning for any real number x and any positive real number a and that the laws of exponents still hold, though we will not prove them here.

Exponential Function

The function $f(x) = a^x$, where x is a real number, $a > 0$ and $a \neq 1$, is called the *exponential function*, base a.

We require the **base** to be positive in order to avoid the complex numbers that would occur by taking even roots of negative numbers—

an example is the square root of -1, $(-1)^{1/2}$, which is not a real number. The restriction $a \neq 1$ is made to exclude the constant function $f(x) = 1^x = 1$, which does not have an inverse because it is not one-to-one.

The following are examples of exponential functions:

$$f(x) = 2^x, \qquad f(x) = \left(\frac{1}{2}\right)^x, \qquad f(x) = (3.57)^x.$$

Note that, in contrast to functions like $f(x) = x^5$ and $f(x) = x^{1/2}$, the variable in an exponential function is *in the exponent.* Let's now consider graphs of exponential functions.

Example 1 Graph the exponential function: $y = f(x) = 2^x$.

SOLUTION We compute some function values and list the results in a table (at left).

x	y $y = f(x) = 2^x$	(x, y)
0	1	(0, 1)
1	2	(1, 2)
2	4	(2, 4)
3	8	(3, 8)
−1	$\frac{1}{2}$	$\left(-1, \frac{1}{2}\right)$
−2	$\frac{1}{4}$	$\left(-2, \frac{1}{4}\right)$
−3	$\frac{1}{8}$	$\left(-3, \frac{1}{8}\right)$

$$f(0) = 2^0 = 1; \qquad f(-1) = 2^{-1} = \frac{1}{2^1} = \frac{1}{2};$$

$$f(1) = 2^1 = 2; \qquad f(-2) = 2^{-2} = \frac{1}{2^2} = \frac{1}{4};$$

$$f(2) = 2^2 = 4; \qquad f(-3) = 2^{-3} = \frac{1}{2^3} = \frac{1}{8}$$

$$f(3) = 2^3 = 8;$$

Next, we plot these points and connect them with a smooth curve. Be sure to plot enough points to determine how steeply the curve rises.

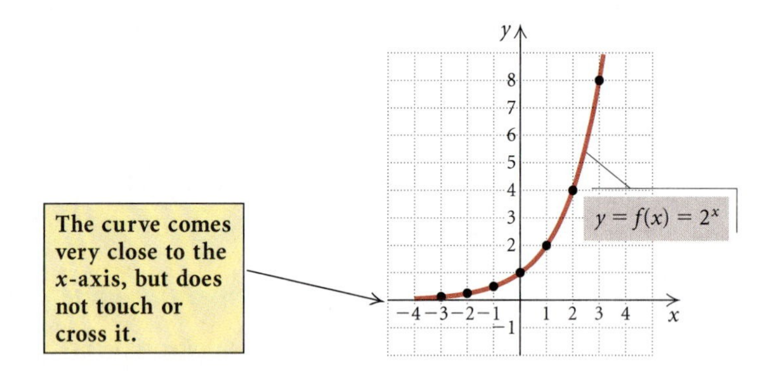

Note that as x increases, the function values increase without bound. As x decreases, the function values decrease, getting close to 0. That is, as $x \to -\infty$, $y \to 0$. The x-axis, or the line $y = 0$, is a horizontal asymptote. As the x-inputs decrease, the curve gets closer and closer to this line, but does not cross it.

Check the graph of $y = f(x) = 2^x$ on a grapher. In general, this function is entered as $y = 2\wedge x$ or $f(x) = 2\wedge x$. Then graph $y = f(x) = 3^x$. Use the TRACE and ZOOM features to confirm that the graph never touches the x-axis.

Example 2 Graph the exponential function: $y = f(x) = \left(\frac{1}{2}\right)^x$.

x	y $y = f(x) = 2^{-x}$	(x, y)
0	1	$(0, 1)$
1	$\frac{1}{2}$	$\left(1, \frac{1}{2}\right)$
2	$\frac{1}{4}$	$\left(2, \frac{1}{4}\right)$
3	$\frac{1}{8}$	$\left(3, \frac{1}{8}\right)$
-1	2	$(-1, 2)$
-2	4	$(-2, 4)$
-3	8	$(-3, 8)$

SOLUTION We compute some function values and list the results in a table (at left). Before we plot these points and draw the curve, note that

$$y = f(x) = \left(\frac{1}{2}\right)^x = (2^{-1})^x = 2^{-x}.$$

This tells us, before we begin graphing, that this graph is a reflection of the graph of $y = 2^x$ across the y-axis.

$$f(0) = 2^{-0} = 1; \qquad f(-1) = 2^{-(-1)} = 2^1 = 2;$$

$$f(1) = 2^{-1} = \frac{1}{2^1} = \frac{1}{2}; \qquad f(-2) = 2^{-(-2)} = 2^2 = 4;$$

$$f(2) = 2^{-2} = \frac{1}{2^2} = \frac{1}{4}; \qquad f(-3) = 2^{-(-3)} = 2^3 = 8$$

$$f(3) = 2^{-3} = \frac{1}{2^3} = \frac{1}{8};$$

Next, we plot these points and connect them with a smooth curve.

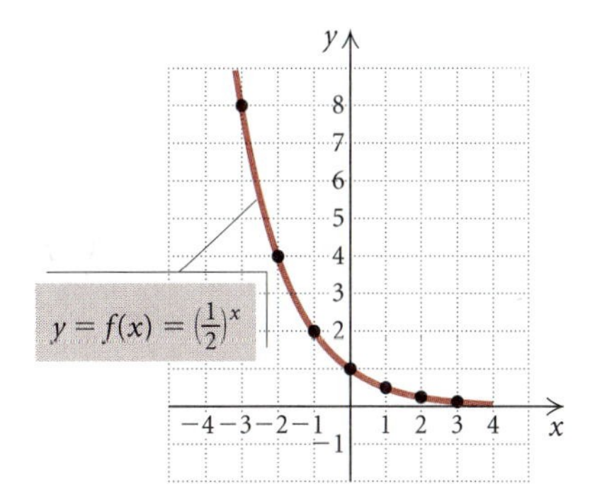

Check the graph in Example 2 on a grapher. Then graph $y = f(x) = \left(\frac{1}{3}\right)^x$. Also, try using the TABLE feature to make a table simultaneously for each of the functions $f(x) = 3^x$ and $g(x) = 3^{-x}$.

Interactive Discovery

Graph each of the following functions using the same set of axes and a viewing window of $[-10, 10, -10, 10]$. Look for patterns in the graphs.

$$f(x) = 1.1^x, \qquad f(x) = 2.7^x, \qquad f(x) = 4^x$$

Next, clear the screen and graph each of the following functions using the same set of axes and viewing window. Again, look for patterns in the graphs.

$$f(x) = 0.1^x, \qquad f(x) = 0.6^x, \qquad f(x) = 0.23^x$$

What relationship do you see between the base a and the shape of the resulting graph of a^x? What do all the graphs have in common? How do they differ?

The preceding examples and interactive discovery illustrate exponential functions with various bases. Let's list some characteristics, keeping in mind that the definition of an exponential function, $f(x) = a^x$, requires that a be positive and different from 1.

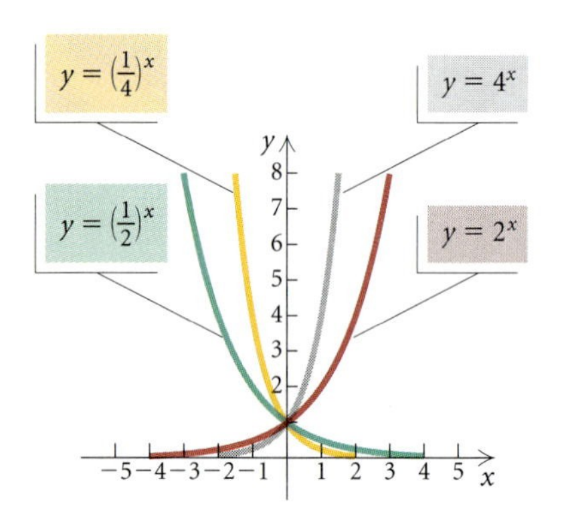

Properties of Exponential Functions

$f(x) = a^x$, $a > 1$:

Continuous

One-to-one

Domain: All real numbers, $\mathbb{R}$

Range: All positive real numbers, $(0, \infty)$

Increasing

Horizontal asymptote is x-axis: $(a^x \to 0$ as $x \to -\infty)$

y-intercept: $(0, 1)$

$f(x) = a^x$ for $0 < a < 1$ or, equivalently, $f(x) = a^{-x}$, $a > 1$:

Continuous

One-to-one

Domain: All real numbers, $\mathbb{R}$

Range: All positive real numbers, $(0, \infty)$

Decreasing

Horizontal asymptote is x-axis: $(a^x \to 0$ as $x \to \infty)$

y-intercept: $(0, 1)$

To graph other types of exponential functions, keep in mind the ideas of translation, stretching, reflection, and combinations of these ideas. All these concepts allow us to visualize the graph before drawing it.

Example 3 Graph each of the following. Before doing so, describe how each graph can be obtained from $f(x) = 2^x$.

a) $f(x) = 2^{x-2}$ **b)** $f(x) = 2^x - 4$ **c)** $f(x) = 5 - 2^{-x}$

SOLUTION

a) The graph of this function is the graph of $y = 2^x$ shifted *right* 2 units.

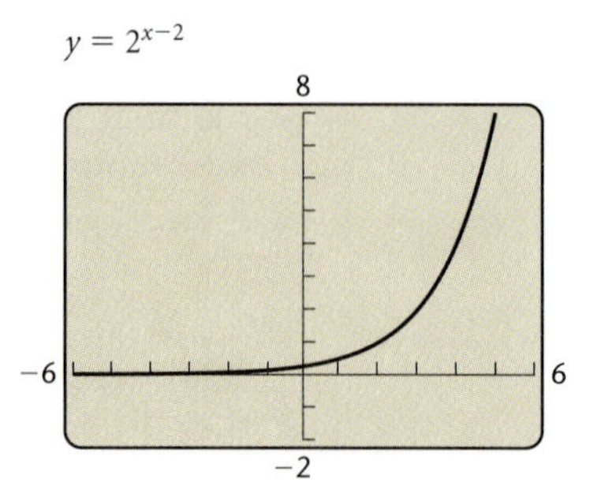

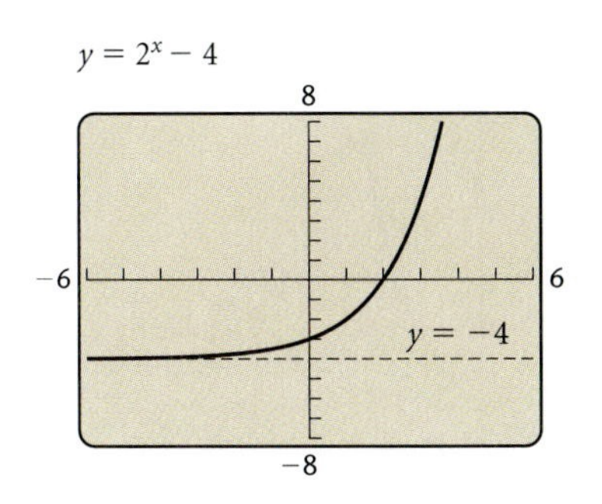

b) The graph is the graph of $y = 2^x$ shifted *down* 4 units (see the graph at left).

c) The graph is a reflection of the graph of $y = 2^x$ across the y-axis, followed by a reflection across the x-axis and then a shift *up* 5 units.

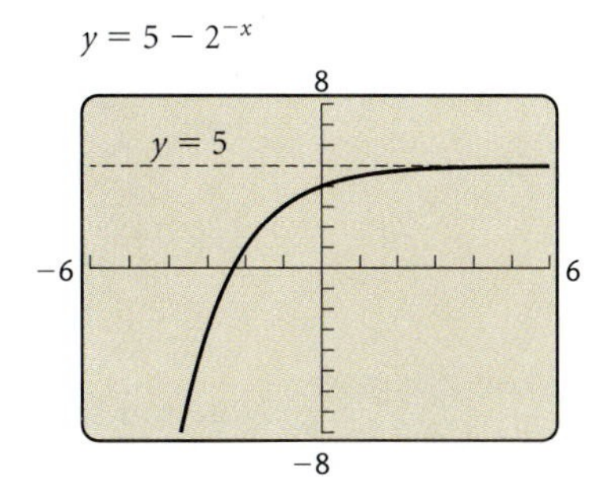

Graphs of Inverses of Exponential Functions

We have noted that every exponential function (with $a > 0$ and $a \neq 1$) is one-to-one. Thus each such function has an inverse that is a function. In the next section, we will name these inverse functions and use them in applications. For now, we draw their graphs by interchanging x and y.

Example 4 Graph: $x = 2^y$.

SOLUTION Note that x is alone on one side of the equation. We can find ordered pairs that are solutions by choosing values for y and then computing the corresponding x-values.

For $y = 0$, $x = 2^0 = 1$.

For $y = 1$, $x = 2^1 = 2$.

For $y = 2$, $x = 2^2 = 4$.

For $y = 3$, $x = 2^3 = 8$.

For $y = -1$, $x = 2^{-1} = \frac{1}{2^1} = \frac{1}{2}$.

For $y = -2$, $x = 2^{-2} = \frac{1}{2^2} = \frac{1}{4}$.

For $y = -3$, $x = 2^{-3} = \frac{1}{2^3} = \frac{1}{8}$.

x		
$x = 2^y$	y	x, y
1	0	(1, 0)
2	1	(2, 1)
4	2	(4, 2)
8	3	(8, 3)
$\frac{1}{2}$	−1	$\left(\frac{1}{2}, -1\right)$
$\frac{1}{4}$	−2	$\left(\frac{1}{4}, -2\right)$
$\frac{1}{8}$	−3	$\left(\frac{1}{8}, -3\right)$

(1) Choose values for y.
(2) Compute values for x.

We plot the points and connect them with a smooth curve. Note that the curve does not touch or cross the y-axis. The y-axis is a vertical asymptote.

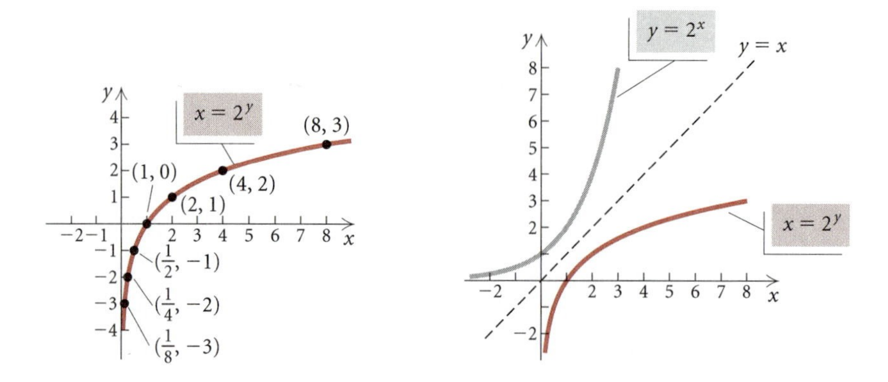

Note too that this curve looks just like the graph of $y = 2^x$, except that it is reflected across the line $y = x$, as we would expect for an inverse. The inverse of $y = 2^x$ is $x = 2^y$. We will explore the inverses of exponential functions further in the next section.

Applications

Graphers are especially helpful when working with exponential functions. They not only facilitate computations but they also allow us to visualize the function. It is worthwhile to get in the habit of creating such a graph, and perhaps looking at a table, even if an exercise or application may not specifically request a graph or table.

Example 5 *Compound Interest.* The amount of money A that a principal P will be worth after t years at interest rate i, compounded n times per year, is given by the formula

$$A = P\left(1 + \frac{i}{n}\right)^{nt}.$$

Suppose that \$100,000 is invested at 8% interest, compounded semi-annually.

a) Find a function for the amount of money after t years.

b) Find the amount of money in the account at $t = 0, 4, 8,$ and 10 yr.

c) Graph the function.

d) When will the amount of money in the account reach \$400,000?

SOLUTION

a) Since $P = \$100{,}000$, $i = 8\% = 0.08$, and $n = 2$, we can substitute these values and form the following function:

$$\begin{aligned} A(t) &= 100{,}000\left(1 + \frac{0.08}{2}\right)^{2 \cdot t} \\ &= 100{,}000(1.04)^{2t}. \end{aligned}$$

b) We can enter the function into a grapher, using a TABLE feature set in ASK mode, to compute function values. We can also calculate the values directly on a grapher by substituting in the expression for $A(t)$.

$$A(0) = 100{,}000(1.04)^{2\cdot 0} = 100{,}000;$$
$$A(4) = 100{,}000(1.04)^{2\cdot 4} \approx 136{,}856.91;$$
$$A(8) = 100{,}000(1.04)^{2\cdot 8} \approx 187{,}298.12;$$
$$A(10) = 100{,}000(1.04)^{2\cdot 10} \approx 219{,}112.31.$$

c) For the graph, we use the viewing window [0, 20, 0, 500,000] because of the large numbers and the fact that negative time values and amounts of money have no meaning in this application.

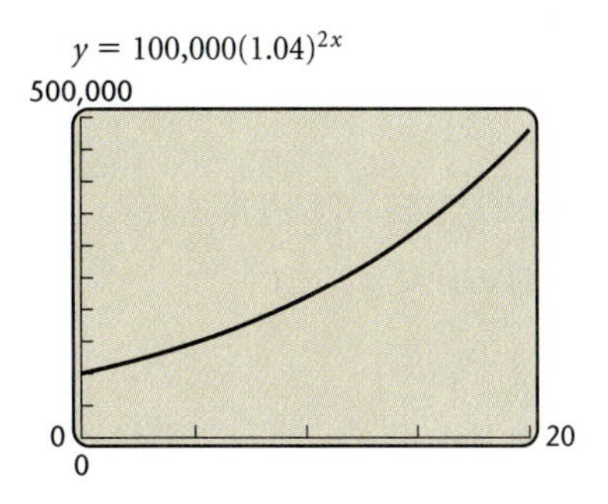

d) To find the amount of time it takes for the account to grow to $400,000, we set

$$100{,}000(1.04)^{2t} = 400{,}000$$

and solve. One way we can do this is by graphing the equations

$$y_1 = 100{,}000(1.04)^{2x} \quad \text{and} \quad y_2 = 400{,}000.$$

Then we can use the TRACE and ZOOM features or the INTERSECT feature as discussed in the Introduction to Graphs and Graphers to estimate the point of intersection.

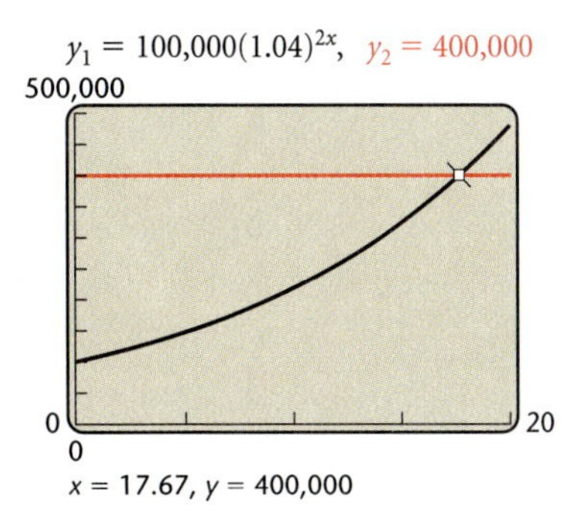

We can also use the SOLVE feature on a grapher. It might be necessary to write the equation as $100{,}000(1.04)^{2x} - 400{,}000 = 0$. In Section 3.5, we will see how to solve equations such as this algebraically.

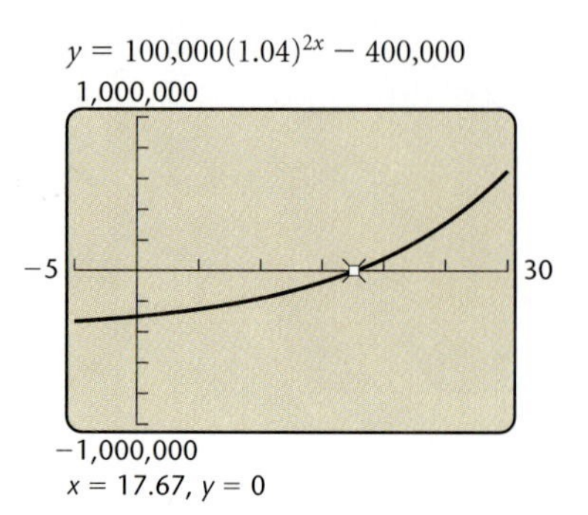

Regardless of the method we use to find the solution, we see that the account grows to \$400,000 after about 17.67 yr.

Interactive Discovery

Take a sheet of $8\frac{1}{2}$-in. by 11-in. paper and cut it into two equal pieces. Then cut each of these in half. Next, cut each of the four pieces in half again. Continue this process.

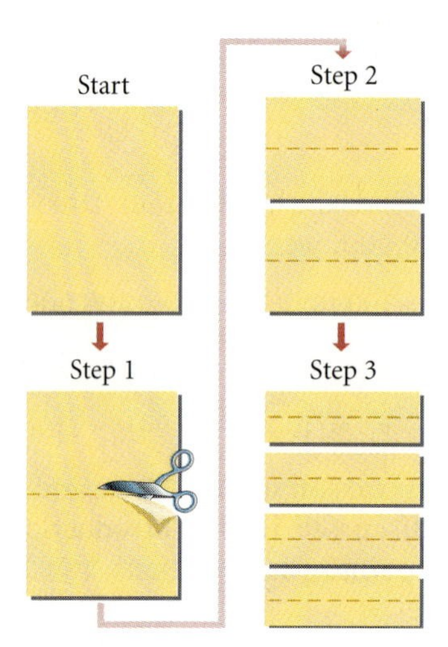

	t	$f(t) = 0.004 \cdot 2^t$
Start	0	$0.004 \cdot 2^0$, or 0.004
Step 1	1	$0.004 \cdot 2^1$, or 0.008
Step 2	2	$0.004 \cdot 2^2$, or 0.016
Step 3	3	
Step 4	4	
Step 5	5	

a) Place all the pieces in a stack and measure the thickness with a micrometer or other measuring device such as a dial caliper.

b) A piece of paper is typically 0.004 in. thick. Check the calculation in part (a) by completing the table shown here.

c) Graph the function $f(t) = 0.004(2)^t$.

d) Compute the thickness of the paper (in miles) after 25 steps.

The Number e = 2.7182818284...

We now consider a very special number in mathematics. Though you may not have encountered it before, you will see here and in future mathematics that it has many important applications. To derive this number, we use the compound interest formula $A = P(1 + i/n)^{nt}$ in Example 5. Suppose that \$1 is an initial investment at 100% interest for

1 yr (no bank would pay this). The formula above becomes a function A defined in terms of the number of compounding periods n:

$$A(n) = \left(1 + \frac{1}{n}\right)^n.$$

Let's find some function values using the scientific keys on a grapher. Rounding to six decimal places gives us the following table.

n	$A(n) = \left(1 + \frac{1}{n}\right)^n$
1 (compounded annually)	\$2.00
2 (compounded semiannually)	\$2.25
3	\$2.370370
4 (compounded quarterly)	\$2.441406
5	\$2.488320
100	\$2.704814
365 (compounded daily)	\$2.714567
8760 (compounded hourly)	\$2.718127

As the values of n get larger and larger, the function values get closer and closer to a number in mathematics, called e. Its decimal representation does not terminate or repeat; it is irrational. In 1741, Leonhard Euler named this number e.

$$e = 2.7182818284\ldots$$

Interactive Discovery

Graph the function $y = (1 + 1/x)^x$. Consider the graph for larger and larger values of x. Does this function have a horizontal asymptote? Explore with a grapher to determine the asymptote, if it exists. You might also try to use the TABLE feature.

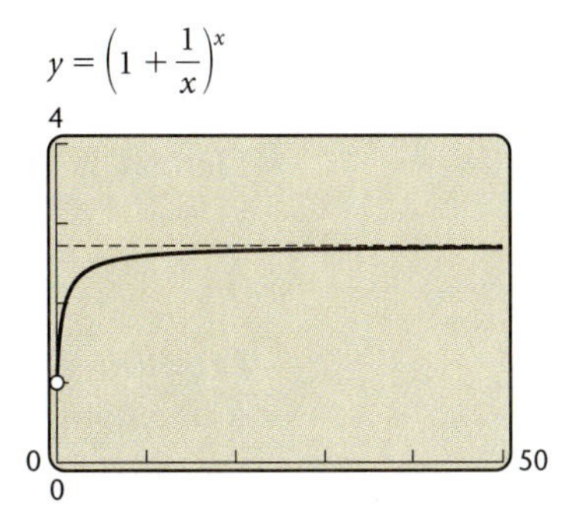

We can find values of the exponential function $f(x) = e^x$ using the scientific $\boxed{e^x}$ key on a grapher.

Example 6 Find each value of e^x, to four decimal places, on a grapher.

a) e^3 **b)** $e^{-0.23}$ **c)** e^0 **d)** $100e^{5.8}$ **e)** e^1

SOLUTION

FUNCTION VALUE	READOUT	ROUNDED
a) e^3	20.08553692	20.0855
b) $e^{-0.23}$	0.7945336025	0.7945
c) e^0	1	1
d) $100e^{5.8}$	33029.95599	33029.9560
e) e^1	2.718281828	2.7183

Graphs of Exponential Functions, Base e

We demonstrate ways in which to graph exponential functions.

Example 7 Graph $f(x) = e^x$ and $g(x) = e^{-x}$.

SOLUTION

Method 1. We can compute points for each equation using the e^x key on a grapher. Then we plot these points and draw the graphs of the functions.

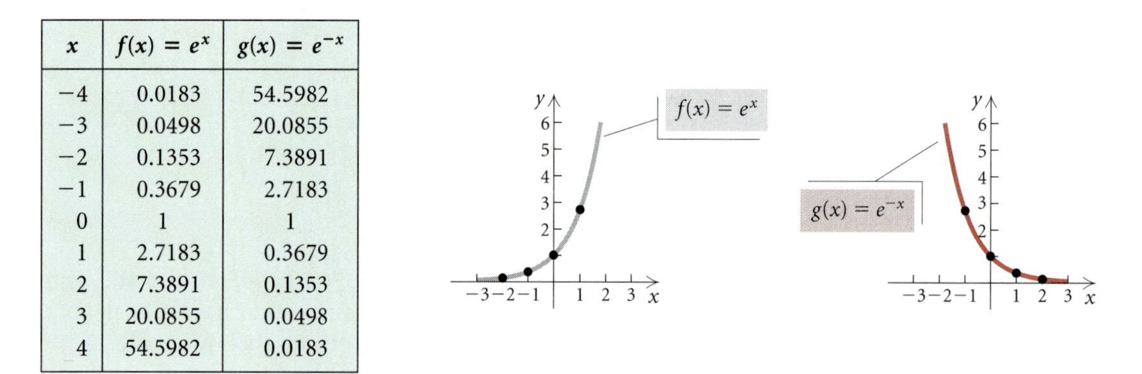

x	$f(x) = e^x$	$g(x) = e^{-x}$
−4	0.0183	54.5982
−3	0.0498	20.0855
−2	0.1353	7.3891
−1	0.3679	2.7183
0	1	1
1	2.7183	0.3679
2	7.3891	0.1353
3	20.0855	0.0498
4	54.5982	0.0183

Note that the graph of g is a reflection of the graph of f across the y-axis.

Method 2. On a grapher, we simply enter the equations $y_1 = e^x$ and $y_2 = e^{-x}$. We may need to enter these equations as $y_1 = \exp(x)$ and $y_2 = \exp(-x)$.

Example 8 Graph each of the following using a grapher. Briefly describe how each graph can be obtained from $y = e^x$.

a) $f(x) = e^{-0.5x}$

b) $f(x) = 1 - e^{-2x}$

c) $f(x) = e^{x+3}$

SOLUTION

a) We note that the graph of this function is a horizontal stretching of the graph of $y = e^x$ followed by a reflection across the y-axis.

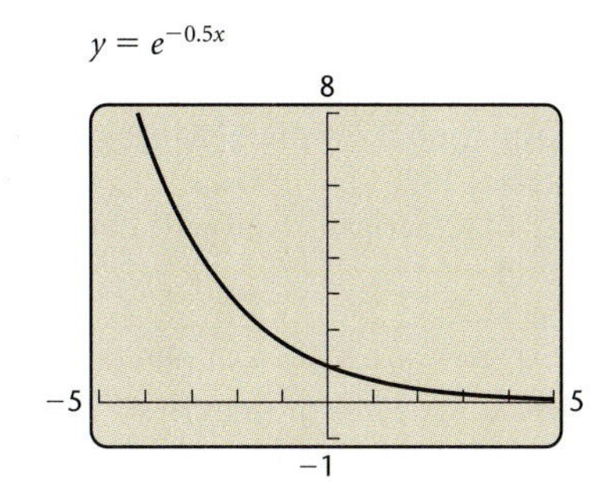

b) The graph is a horizontal shrinking of the graph of $y = e^x$, followed by a reflection across the y-axis, then across the x-axis, followed by a translation up 1 unit. (See the figure on the left below.)

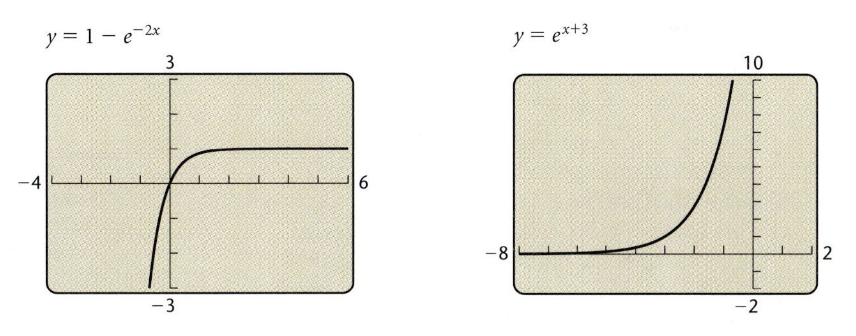

c) The graph is a translation of the graph of $y = e^x$ to the left 3 units. (See the figure on the right above.)

3.2 Exercise Set

Find each of the following, to four decimal places, on a grapher.

1. e^4

2. e^{10}

3. $e^{-2.458}$

4. $e^{1.0345}$

Make a hand-drawn graph of each function. Then check your work using a grapher, if possible.

5. $f(x) = 3^x$

6. $f(x) = 5^x$

7. $f(x) = 6^x$

8. $f(x) = 3^{-x}$

9. $f(x) = \left(\frac{1}{4}\right)^x$

10. $f(x) = \left(\frac{2}{3}\right)^x$

11. $x = 3^y$

12. $x = 4^y$

13. $x = \left(\frac{1}{2}\right)^y$

14. $x = \left(\frac{4}{3}\right)^y$

15. $y = \frac{1}{4}e^x$

16. $y = 2e^{-x}$

17. $f(x) = 1 - e^{-x}$

18. $f(x) = e^x - 2$

Graph each function using a grapher. Describe how each graph can be obtained from a basic exponential function.

19. $f(x) = 2^{x+1}$

20. $f(x) = 2^{x-1}$

21. $f(x) = 2^x - 3$

22. $f(x) = 2^x + 1$

23. $f(x) = 4 - 3^{-x}$

24. $f(x) = 2^{x-1} - 3$

25. $f(x) = \left(\frac{3}{2}\right)^{x-1}$

26. $f(x) = 3^{4-x}$

27. $f(x) = 2^{x+3} - 5$

28. $f(x) = -3^{x-2}$

29. $f(x) = e^{2x}$

30. $f(x) = e^{-0.2x}$

31. $y = e^{-x+1}$

32. $y = e^{2x} + 1$

33. $f(x) = 2(1 - e^{-x})$

34. $f(x) = 1 - e^{-0.01x}$

35. *Growth of AIDS.* The total number of Americans who have contracted AIDS is approximated by the exponential function

$$N(t) = 100{,}000(1.4)^t,$$

where $t = 0$ corresponds to 1989. (*Source:* U.S. Department of Health & Human Services)

a) According to this function, how many Americans had been infected as of 1997?
b) Predict the number of Americans who will have been infected by 2001.
c) Graph the function.
d) Include the graph of the equation $y = 1{,}000{,}000$ with the graph in part (c). Find the length of time it took for the number of Americans who contracted AIDS to reach 1 million.

36. *Growth of Bacteria Escherichia coli.* The bacteria *Escherichia coli* are commonly found in the human bladder. Suppose that 3000 of the bacteria are present at time $t = 0$. Then under certain conditions, t minutes later, the number of bacteria present is

$$N(t) = 3000(2)^{t/20}.$$

a) How many bacteria will be present after 10 min? 20 min? 30 min? 40 min? 60 min?
b) Graph the function.
c) This bacteria can cause bladder infections in humans when the number of bacteria reaches 100,000,000. Use a grapher to include the graph of the equation $y = 100{,}000{,}000$ with the graph in part (b). Find the length of time it takes for a bladder infection to be possible.

37. *Recycling Aluminum Cans.* It is estimated that two thirds of all aluminum cans distributed will be recycled each year (*Source:* Alcoa Corporation). A beverage company distributes 350,000 cans. The number still in use after time t, in years, is given by the exponential function

$$N(t) = 350{,}000\left(\tfrac{2}{3}\right)^t.$$

a) How many cans are still in use after 0 yr? 1 yr? 4 yr? 10 yr?
b) Graph the function.
c) After how long will 2000 of the cans still be in use?

38. *Interest in a College Trust Fund.* Following the birth of a child, Juan deposits $10,000 in a college trust fund where interest is 6.4%, compounded semiannually.

a) Find a function for the amount in the account after t years.
b) Find the amount of money in the account at $t = 0$, 4, 8, 10, and 18 yr.
c) Graph the function.
d) After how long will the account contain $100,000?

39. *Salvage Value.* A top-quality fax–copying machine is purchased for $5800. Its value each year is about 80% of the value of the preceding year. After t years, its value, in dollars, is given by the exponential function

$$V(t) = 5800(0.8)^t.$$

a) Find the value of the machine after 0 yr, 1 yr, 2 yr, 5 yr, and 10 yr.
b) Graph the function.
c) The company decides to replace the machine when its value has declined to $500. After how long will the machine be replaced?

40. *Revenues of Packard Bell.* Sales of Packard Bell computers have grown exponentially in recent years. The total revenue R, in billions, is given by

$$R(t) = 0.518(1.42)^t,$$

where $t =$ the number of years since 1990.

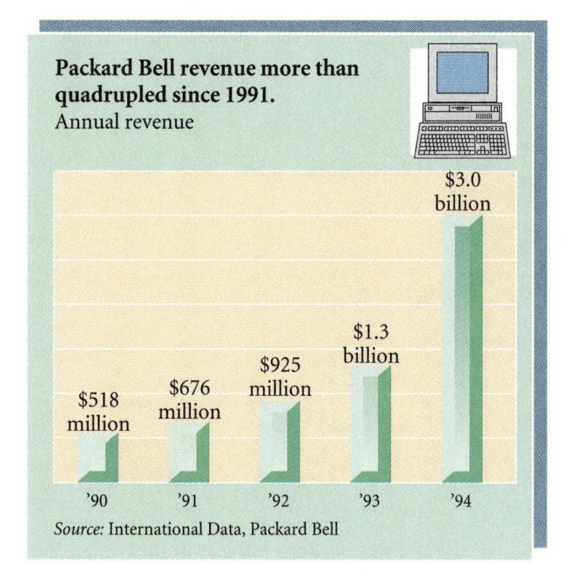

a) Find the total revenue in 1990, 1994, 1998, and 2004.
b) Graph the function.
c) When will revenues be $8 billion?

41. *Timber Demand.* World demand for timber is increasing exponentially. The demand N, in billions of cubic feet, purchased is given by

$$N(t) = 46.6(1.018)^t,$$

where $t =$ the number of years since 1981. (*Source:* U.N. Food and Agricultural Organization, American Forest and Paper Association)

a) Find the demand for timber in 1985, 1997, and 2010.
b) Graph the function.
c) After how many years will the demand for timber be 93.4 billion cubic feet?

42. *Typing Speed.* Sarah is taking typing in college. After she practices for t hours, her speed, in words per minute, is given by the function

$$S(t) = 200[1 - (0.86)^t].$$

a) What is Sarah's speed after practicing for 10 hr? 20 hr? 40 hr? 100 hr?
b) Graph the function.
c) How much time passes before Sarah's speed is 100 words per minute?
d) Does this graph have an asymptote? If so, what is it, and what is its significance to Sarah's learning?

43. *Advertising.* A company begins a radio advertising campaign in New York City to market a new CD-ROM video game. The percentage of the target market that buys a game is generally a function of the length of the advertising campaign. The estimated percentage is given by

$$f(t) = 100(1 - e^{-0.04t}),$$

where $t =$ the number of days of the campaign.

a) Find $f(25)$, the percentage of the target market that has bought the product after a 25-day advertising campaign.
b) Graph the function.
c) After how long will 90% of the target market have bought the product?

44. *Growth of a Stock.* The value of a stock is given by the function

$$V(t) = 58(1 - e^{-1.1t}) + 20,$$

where $V =$ the value of the stock after time t, in months.

a) Find $V(1)$, $V(2)$, $V(4)$, $V(6)$, and $V(12)$.
b) Graph the function.
c) After how long will the value of the stock be $75?

In Exercises 45–58, use a grapher to match the equation with one of figures (a)–(n), which follow.

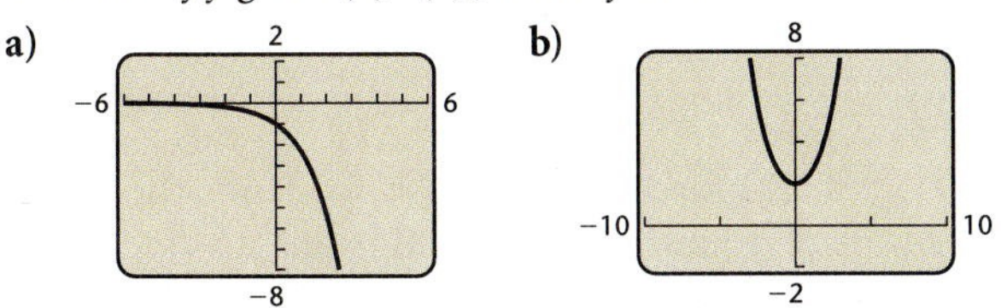

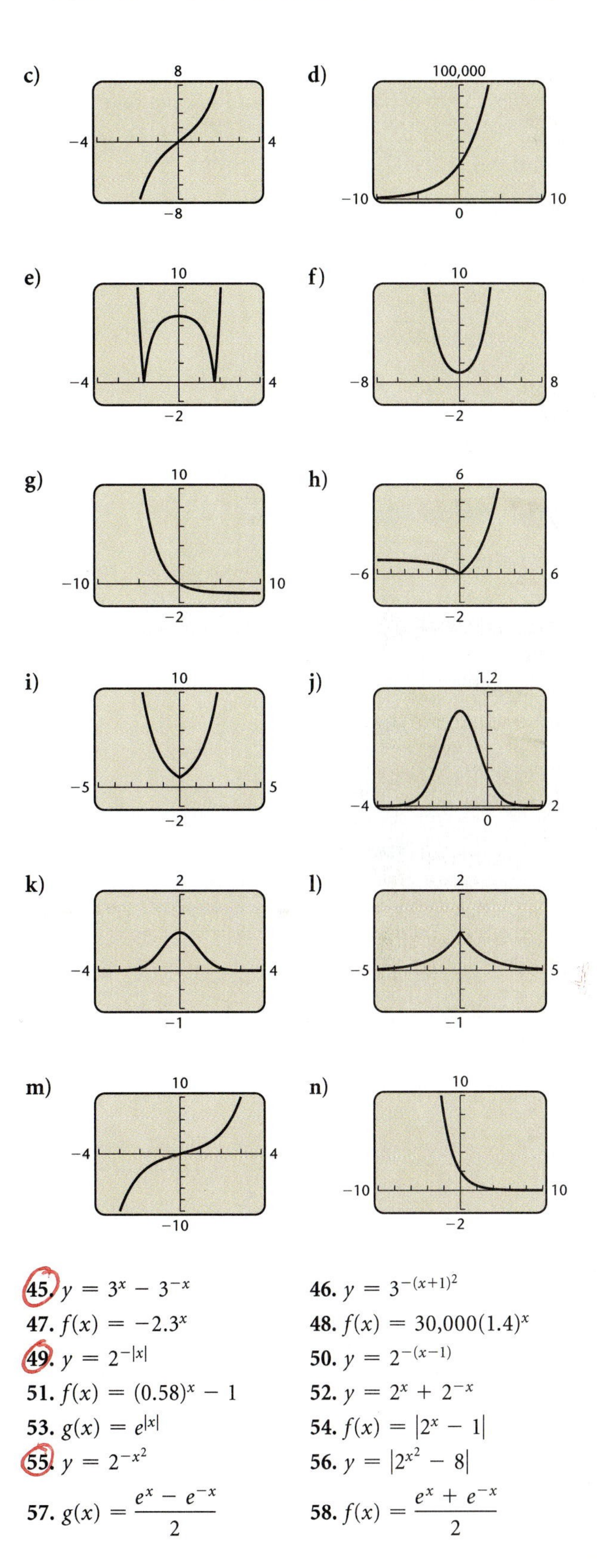

45. $y = 3^x - 3^{-x}$

46. $y = 3^{-(x+1)^2}$

47. $f(x) = -2.3^x$

48. $f(x) = 30{,}000(1.4)^x$

49. $y = 2^{-|x|}$

50. $y = 2^{-(x-1)}$

51. $f(x) = (0.58)^x - 1$

52. $y = 2^x + 2^{-x}$

53. $g(x) = e^{|x|}$

54. $f(x) = |2^x - 1|$

55. $y = 2^{-x^2}$

56. $y = |2^{x^2} - 8|$

57. $g(x) = \dfrac{e^x - e^{-x}}{2}$

58. $f(x) = \dfrac{e^x + e^{-x}}{2}$

Use a grapher to find the point(s) of intersection of the graphs of each of the following pairs of equations.

59. $y = |1 - 3^x|$, $y = 4 + 3^{-x^2}$

60. $y = 4^x + 4^{-x}$, $y = 8 - 2x - x^2$

61. $y = 2e^x - 3$, $y = \dfrac{e^x}{x}$

62. $y = \dfrac{1}{e^x + 1}$, $y = 0.3x + \dfrac{7}{9}$

Use a grapher. Solve graphically.

63. $5.3^x - 4.2^x = 1073$

64. $e^x = x^3$

65. $2^x > 1$

66. $3^x \leq 1$

67. $2^x + 3^x = x^2 + x^3$

68. $31{,}245e^{-3x} = 523{,}467$

Skill Maintenance

Simplify.

69. a^0, $a \neq 0$

70. a^1

71. 10^3

72. 10^{-2}

73. To what power must you raise 5 in order to get 25?

74. To what power must you raise 10 in order to get 0.001?

Synthesis

75. ◈ Describe the differences between the graphs of $f(x) = x^3$ and $g(x) = 3^x$.

76. ◈ Suppose that \$10,000 is invested for 8 yr at 6.4% interest, compounded annually. In what year will the most interest be earned? Why?

77. ◈ Graph each pair of equations using the same set of axes. Then compare the results in parts (a) and (b).

a) $y = 3^x$, $x = 3^y$

b) $y = 1^x$, $x = 1^y$

78. Which is larger, 7^π or π^7?

79. Graph $f(x) = x^{1/(x-1)}$. Use a grapher and the TABLE feature to identify the horizontal asymptote.

In Exercises 80 and 81:

a) *Graph using a grapher.*

b) *Approximate the zeros.*

c) *Approximate the relative maximum and minimum values. If your grapher has a* MAX–MIN *feature, use it.*

80. $f(x) = x^2e^{-x}$

81. $f(x) = e^{-x^2}$

82. Consider each of the following functions:

$$y_1 = e^x, \qquad y_2 = 1 + x + \frac{x^2}{2} + \frac{x^3}{6} + \frac{x^4}{24}.$$

a) Use a grapher to graph both functions using the viewing window $[0, 1, -1, 4]$, with Xscl $= 1$ and Yscl $= 1$.

b) On the basis of your graphs, would you consider

$$e^x = 1 + x + \frac{x^2}{2} + \frac{x^3}{6} + \frac{x^4}{24}$$

an identity?

c) See if you can prove the equation in part (b) to be an identity. Substitute $x = 1$ in each expression. What is the result?

d) Now go back to the original equations. Change the viewing window, use TRACE and ZOOM and the TABLE feature to examine the graphs in more detail. What do you discover about the graphs?

e) What caution must you be aware of using a grapher to determine whether an equation is an identity?

3.3 Logarithmic Functions and Graphs

- *Graph logarithmic functions.*
- *Convert between exponential and logarithmic equations.*
- *Find common and natural logarithms on a grapher.*

We now consider *logarithmic*, or *logarithm, functions.* These functions are inverses of exponential functions and have many applications.

Logarithmic Functions

Consider the function $f(x) = 2^x$ graphed below. We see from the graph that this function passes the horizontal-line test and is one-to-one. Thus f has an inverse that is a function.

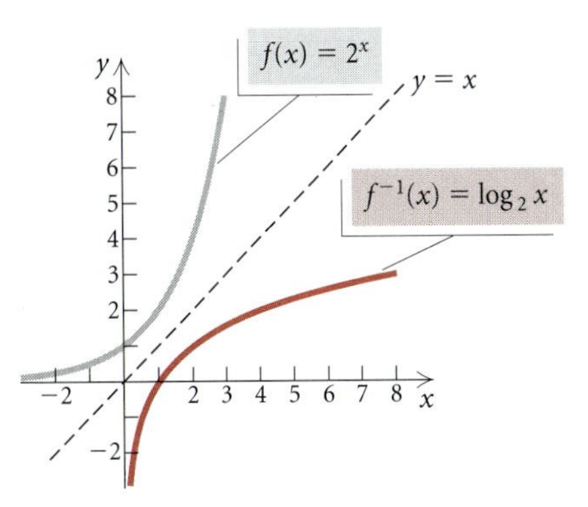

To find a formula for f^{-1} when $f(x) = 2^x$, we try to use the method of Section 3.1:

1. Replace $f(x)$ with y:	$y = 2^x$
2. Interchange x and y:	$x = 2^y$
3. Solve for y:	$y =$ the power to which we raise 2 to get x
4. Replace y with $f^{-1}(x)$:	$f^{-1}(x) =$ the power to which we raise 2 to get x.

Mathematicians have defined a new symbol to replace the words "the power to which we raise 2 to get x." That symbol is "$\log_2 x$," read "the logarithm, base 2, of x."

> ***Logarithmic Function, Base 2***
>
> "$\log_2 x$," read "the logarithm, base 2, of x," means "the power to which we raise 2 to get x."

Thus if $f(x) = 2^x$, then $f^{-1}(x) = \log_2 x$. For example,

$$f^{-1}(8) = \log_2 8$$
$$= 3,$$

because

3 is the power to which we raise 2 to get 8.

Similarly, $\log_2 13$ is the power to which we raise 2 to get 13. As yet, we have no simpler way to say this other than "$\log_2 13$ is the power to which we raise 2 to get 13." Later, however, we will learn how to approximate this expression using a grapher.

For any exponential function $f(x) = a^x$, its inverse is called a **logarithmic function, base *a*.** The graph of the inverse can be obtained by reflecting the graph of $y = a^x$ across the line $y = x$, to obtain $x = a^y$. Then $x = a^y$ is equivalent to $y = \log_a x$. We read $\log_a x$ as "the logarithm, base a, of x."

The inverse of $f(x) = a^x$ is given by $f^{-1}(x) = \log_a x$.

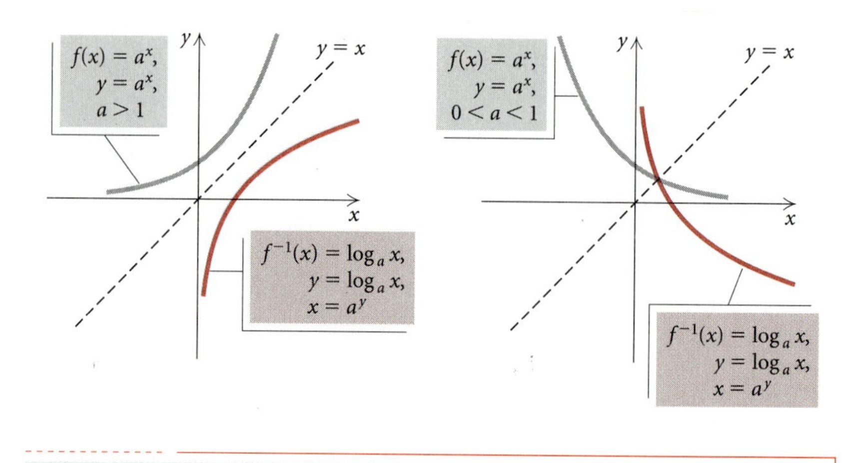

Logarithmic Function, Base a

We define $y = \log_a x$ as that number y such that $x = a^y$, where $x > 0$ and a is a positive constant other than 1.

In the following table, we compare exponential and logarithmic functions. In general, we use a number a that is greater than 1 for the logarithm base.

EXPONENTIAL FUNCTION	LOGARITHMIC FUNCTION
$y = a^x$ $f(x) = a^x$ $a > 1$ Continuous One-to-one Domain: All real numbers, $\mathbb{R}$ Range: All positive real numbers, $(0, \infty)$ Increasing Horizontal asymptote is x-axis: $(a^x \to 0$ as $x \to -\infty)$ y-intercept: (0, 1) There is no x-intercept.	$x = a^y$ $f^{-1}(x) = \log_a x$ $a > 1$ Continuous One-to-one Domain: All positive real numbers, $(0, \infty)$ Range: All real numbers, $\mathbb{R}$ Increasing Vertical asymptote is y-axis: $(\log_a x \to -\infty$ as $x \to 0^+)$ x-intercept: (1, 0) There is no y-intercept.

Converting Between Exponential and Logarithmic Equations

It is helpful in dealing with logarithmic functions to remember that a logarithm of a number is an *exponent*. It is the exponent y in $x = a^y$. You might think to yourself, "the logarithm, base a, of a number x is the power to which a must be raised to get x."

We are led to the following. (The symbol $\longleftrightarrow$ means that the two statements are equivalent; that is, when one is true, the other is true. The words "if and only if" can be used in place of $\longleftrightarrow$.)

$\log_a x = y \longleftrightarrow x = a^y$ A logarithm is an exponent!

Be sure to memorize this relationship! It is probably the most important definition in the chapter. Many times this definition will be justification for a proof or a procedure.

Example 1 Convert each of the following to a logarithmic equation.

a) $16 = 2^x$ b) $10^{-3} = 0.001$ c) $e^t = 70$

SOLUTION

The exponent is the logarithm.

a) $16 = 2^x$ $\log_2 16 = x$

The base remains the same.

b) $10^{-3} = 0.001 \longrightarrow \log_{10} 0.001 = -3$

c) $e^t = 70 \longrightarrow \log_e 70 = t$

Example 2 Convert each of the following to an exponential equation.

a) $\log_2 32 = 5$ b) $\log_a Q = 8$ c) $x = \log_t M$

SOLUTION

The logarithm is the exponent.

a) $\log_2 32 = 5$ $2^5 = 32$

The base remains the same.

b) $\log_a Q = 8 \longrightarrow a^8 = Q$

c) $x = \log_t M \longrightarrow t^x = M$

Finding Certain Logarithms

Let's use the definition of logarithms to find some logarithmic values.

Example 3 Find each of the following logarithms.

a) $\log_{10} 10{,}000$ b) $\log_{10} 0.01$

c) $\log_2 8$ d) $\log_9 3$

e) $\log_6 1$ f) $\log_8 8$

SOLUTION

RESULT	REASON
a) $\log_{10} 10{,}000 = 4$	Think of the meaning of $\log_{10} 10{,}000$. It is the exponent to which we raise 10 to get 10,000. That exponent is 4.

b) $\log_{10} 0.01 = -2$ — We have $0.01 = \frac{1}{100} = \frac{1}{10^2} = 10^{-2}$. The exponent to which we raise 10 to get 0.01 is -2.

c) $\log_2 8 = 3$ — $8 = 2^3$. The exponent to which we raise 2 to get 8 is 3.

d) $\log_9 3 = \frac{1}{2}$ — $3 = \sqrt{9} = 9^{1/2}$. The exponent to which we raise 9 to get 3 is $\frac{1}{2}$.

e) $\log_6 1 = 0$ — $1 = 6^0$. The exponent to which we raise 6 to get 1 is 0.

f) $\log_8 8 = 1$ — $8 = 8^1$. The exponent to which we raise 8 to get 8 is 1.

Examples 3(e) and 3(f) illustrate two important properties of logarithms.

$\log_a 1 = 0$ and $\log_a a = 1$, for any logarithmic base a.

$\log_a 1 = 0$ follows from the fact that $a^0 = 1$. Thus, $\log_7 1 = 0$, $\log_{10} 1 = 0$, and so on. $\log_a a = 1$ follows from the fact that $a^1 = a$. Thus, $\log_5 5 = 1$, $\log_{10} 10 = 1$, and so on.

Finding Logarithms on a Grapher

Before calculators became so widely available, base-10, or **common logarithms**, were used extensively to simplify complicated calculations. In fact, that is why logarithms were invented. The abbreviation **log**, with no base written, is used to represent the common logarithms, or base 10 logarithms. Thus,

$$\log 29 \quad \text{means} \quad \log_{10} 29.$$

We can approximate log 29 as follows:

$$\left.\begin{aligned} \log 100 = \log_{10} 100 &= 2 \\ \log 29 &= ? \\ \log 10 = \log_{10} 10 &= 1 \end{aligned}\right\}$$

It seems reasonable that log 29 is between 1 and 2.

On a grapher, the scientific key for common logarithms is generally marked LOG. Using that key, we find that

$$\log 29 \approx 1.462397998 \approx 1.4624$$

rounded to four decimal places. This also tells us that $10^{1.4624} \approx 29$.

Interactive Discovery

Find log 1000 and log 10,000 without using a grapher. Between what two whole numbers is log 9874? Now find log 9874, rounded to four decimal places, on a grapher.

Example 4 Find each of the following common logarithms on a grapher. Round to four decimal places.

a) log 645,778 **b)** log 0.0000239 **c)** log (-3)

SOLUTION

FUNCTION VALUE	READOUT	ROUNDED
a) log 645,778	5.810083246	5.8101
b) log 0.0000239	-4.621602099	-4.6216
c) log (-3)	ERROR*	Does not exist

In Example 4(c), log (-3) does not exist as a real number because there is no real-number power to which we can raise 10 to get -3. The number 10 raised to any real-number power is positive. The common logarithm of a negative number does not exist as a real number.

Natural Logarithms

Logarithms, base e, are called **natural logarithms**. The abbreviation "ln" is generally used for natural logarithms. Thus

$$\ln 53 \quad \text{means} \quad \log_e 53.$$

On a grapher, the scientific key for natural logarithms is generally marked LN. Using that key, we find that

$$\ln 53 \approx 3.970291914 \approx 3.9703$$

rounded to four decimal places.

Example 5 Find each of the following natural logarithms on a grapher. Round to four decimal places.

a) ln 645,778 **b)** ln 0.0000239 **c)** ln (-5)
d) ln e **e)** ln 1

SOLUTION

FUNCTION VALUE	READOUT	ROUNDED
a) ln 645,778	13.37821107	13.3782
b) ln 0.0000239	-10.64163210	-10.6416
c) ln (-5)	ERROR	Does not exist
d) ln e	1	1
e) ln 1	0	0

Note that $\ln e = \log_e e = 1$ and $\ln 1 = \log_e 1 = 0$.

*On some graphers, it would be expressed as a complex number—about $0.477 + 1.364i$.

Interactive Discovery

In some textbooks and computer applications, log x is used to represent $\log_e x$ rather than $\log_{10} x$. Explore some methods you might use with a grapher to discover what the base actually is.

Changing Logarithmic Bases

Most graphers give the values of both common logarithms and natural logarithms. To find a logarithm with a base other than 10 or e, we can use the following conversion formula.

The Change-of-Base Formula

For any logarithmic bases a and b, and any positive number M,

$$\log_b M = \frac{\log_a M}{\log_a b}.$$

We will prove this result in the next section.

Program

LOGBASE: This program provides a formula for changing the base of logarithms.

Example 6 Find $\log_5 8$ using common logarithms.

SOLUTION First we let $a = 10$, $b = 5$, and $M = 8$. Then we substitute into the change-of-base formula:

$$\begin{aligned}\log_5 8 &= \frac{\log_{10} 8}{\log_{10} 5} \quad \text{Substituting}\\ &\approx \frac{0.903090}{0.698970}\\ &\approx 1.2920.\end{aligned}$$

To check, we use the power key to verify that $5^{1.2920} \approx 8$.

We can also use base e for a conversion.

Example 7 Find $\log_4 31$ using natural logarithms.

SOLUTION Substituting e for a, 4 for b, and 31 for M, we have

$$\log_4 31 = \frac{\log_e 31}{\log_e 4} = \frac{\ln 31}{\ln 4} \approx \frac{3.433987}{1.386294} \approx 2.4771.$$

Graphs of Logarithmic Functions

We demonstrate several ways to graph logarithmic functions.

Example 8 Graph: $g(x) = \ln x$.

SOLUTION

Method 1. To graph $y = g(x) = \ln x$, we first write its equivalent exponential equation, $x = e^y$. Then we select values for y and use a grapher to find the corresponding values of e^y. We then plot points, remembering that x still is the first coordinate.

x, or e^y	y
1	0
2.7	1
7.4	2
20	3
0.1	−2
0.4	−1

(1) Select y.
(2) Compute x.

$g(x) = \ln x$
or $x = e^y$

Connect the points with a smooth curve.

Method 2. We can also use the LN key on a grapher to find function values. Then we plot points and draw the graph.

Method 3. We use a grapher to graph $y = \ln x$.

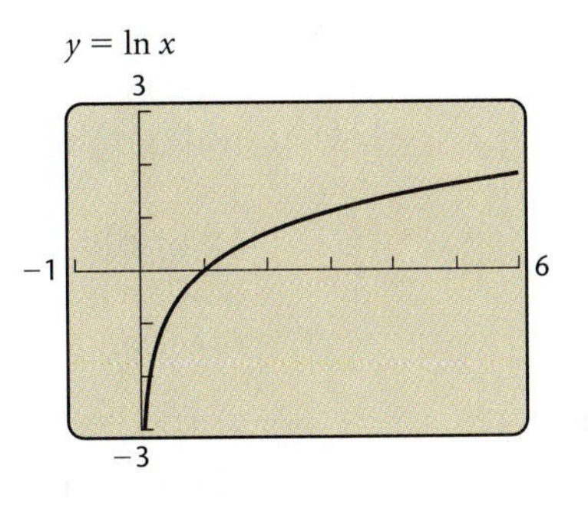

Example 9 Graph $y = f(x) = \log_5 x$.

SOLUTION

Method 1. The equation $y = \log_5 x$ is equivalent to $x = 5^y$. We can find ordered pairs that are solutions by choosing values for y and computing the x-values.

For $y = 0$, $x = 5^0 = 1$.
For $y = 1$, $x = 5^1 = 5$.
For $y = 2$, $x = 5^2 = 25$.
For $y = 3$, $x = 5^3 = 125$.
For $y = -1$, $x = 5^{-1} = \frac{1}{5}$.
For $y = -2$, $x = 5^{-2} = \frac{1}{25}$.

x, or 5^y	y
1	0
5	1
25	2
125	3
$\frac{1}{5}$	−1
$\frac{1}{25}$	−2

$y = \log_5 x$

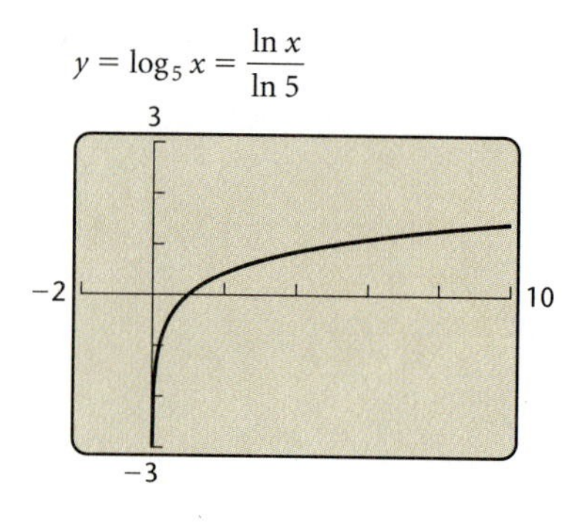

Method 2. To use a grapher, we must first change the base. Here we change from base 5 to base e:

$$y = \log_5 x = \frac{\ln x}{\ln 5}. \qquad \text{Using } \log_b M = \frac{\log_a M}{\log_a b}$$

The graph is as shown at left.

Method 3. Some graphers have a feature that graphs inverses automatically. If we begin with $y_1 = 5^x$, the graphs of both y_1 and its inverse $y_2 = \log_5 x$ will be drawn.

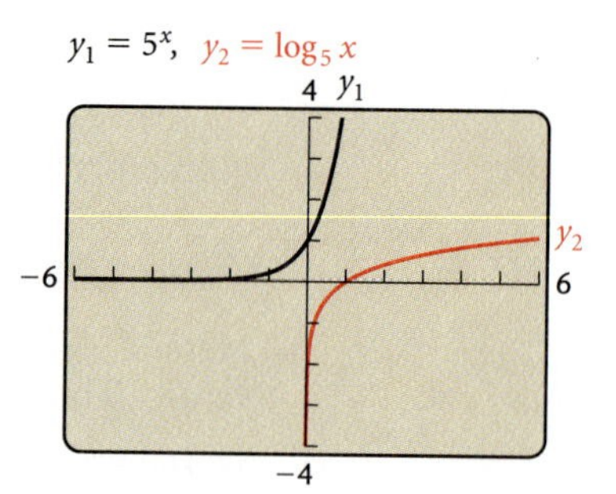

The graph of $f(x) = \log_a x$, for any base a, has the x-intercept $(1, 0)$. The domain is the set of positive real numbers. The range is the set of all real numbers.

Example 10 Graph each of the following using a grapher. Describe how each graph can be obtained from $y = \ln x$. Give the domain of each function and discuss the vertical asymptotes.

a) $f(x) = \ln (x + 3)$
b) $f(x) = 3 - \frac{1}{2} \ln x$
c) $f(x) = |\ln (x - 1)|$

SOLUTION

a) The graph is a shift of the graph of $y = \ln x$ left 3 units.

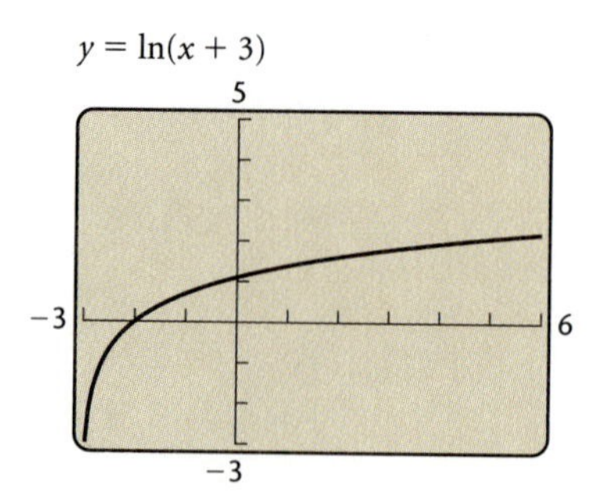

The domain is the set of all real numbers greater than -3, $(-3, \infty)$. The line $x = -3$ is a vertical asymptote.

b) The graph is a shrinking of the graph of $y = \ln x$ in the y-direction, followed by a reflection across the x-axis, and then a translation 3 units up.

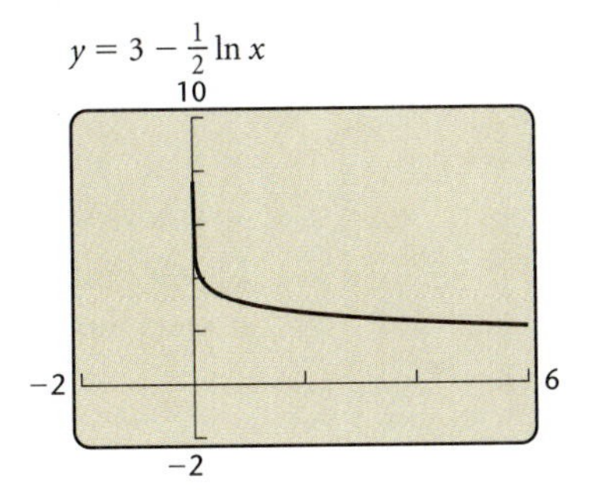

The domain is the set of all positive real numbers, $(0, \infty)$. The y-axis is a vertical asymptote.

c) The graph is a translation of the graph of $y = \ln x$, 1 unit to the right. Then the absolute value has the effect of reflecting negative outputs across the x-axis.

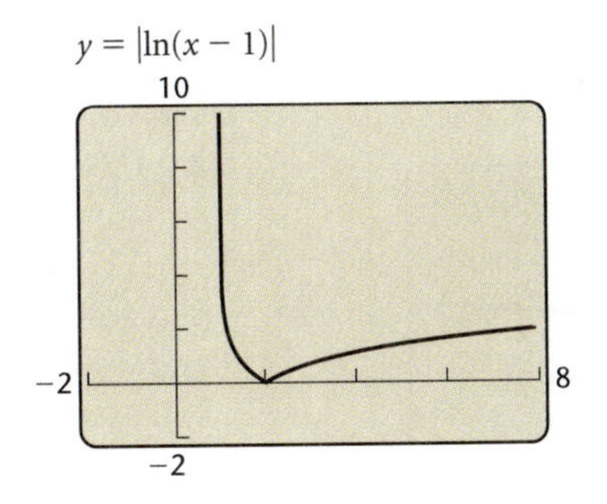

The domain is the set of all real numbers greater than 1, $(1, \infty)$. The line $x = 1$ is a vertical asymptote.

Applications

Example 11 *Walking Speed.* In a study by psychologists Bornstein and Bornstein, it was found that the average walking speed w, in feet per second, of a person living in a city of population P, in thousands, is given by the function

$$w(P) = 0.37 \ln P + 0.05.$$

(*Source: International Journal of Psychology*)

a) The population of Orlando, Florida, is 1,073,000. Find the average walking speed of people living in Orlando.

b) Graph the equation.

c) A sociologist computes the average walking speed in a city to be about 2.0 ft/sec. Use this information to estimate the population of the city.

SOLUTION

a) We substitute 1073 for P, since P is in thousands:

$$w(1073) = 0.37 \ln 1073 + 0.05 \quad \text{Substituting}$$
$$\approx 2.6 \text{ ft/sec.} \quad \text{Finding the natural logarithm and simplifying}$$

The average walking speed of people living in Orlando is about 2.6 ft/sec.

b) We graph with a viewing window of [0, 600, 0, 4] because inputs are very large and outputs are very small by comparison.

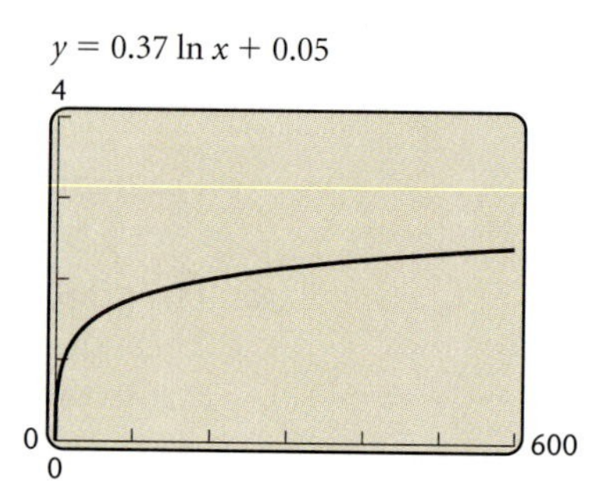

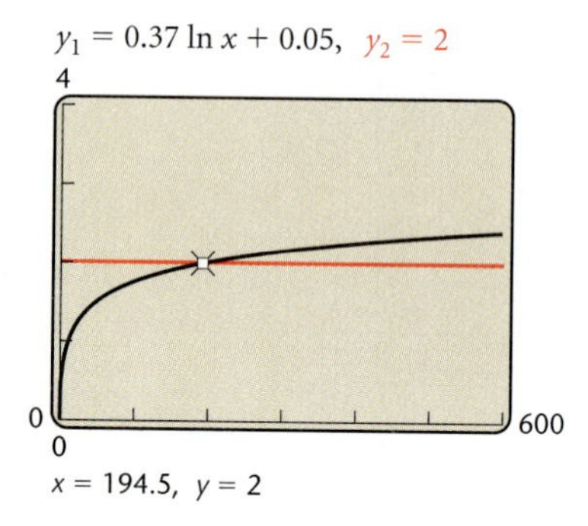

c) To find the population for which the average walking speed is 2.0 ft/sec, we set

$$2.0 = 0.37 \ln P + 0.05$$

and solve. We begin by graphing the equations $y_1 = 0.37 \ln x + 0.05$ and $y_2 = 2$. Then we can use the TRACE and ZOOM features, or the INTERSECT feature, to approximate the point of intersection. See one such TRACE at left. We see that a city with a population of about 194,500 would have an average walking speed of 2.0 ft/sec. Some graphers can solve the equation directly.

Example 12 *Earthquake Magnitude.* The magnitude R, measured on the Richter scale, of an earthquake of intensity I is defined as

$$R = \log \frac{I}{I_0},$$

where I_0 is a minimum intensity used for comparison. We can think of I_0 as a threshold intensity that is the weakest earthquake that can be recorded on a seismograph. If one earthquake is 10 times as intense as another, its magnitude on the Richter scale is 1 greater than that of the other. If one earthquake is 100 times as intense as another, its magnitude on the Richter scale is 2 higher, and so on. Thus an earthquake whose magnitude is 7 on the Richter scale is 10 times as intense as an earthquake whose magnitude is 6. Earthquakes can be interpreted as multiples of the minimum intensity I_0.

The Kobe, Japan, earthquake of 1995 had an intensity of $10^{6.8} \cdot I_0$. What was its magnitude on the Richter scale?

SOLUTION We substitute into the formula:

$$R = \log \frac{I}{I_0} = \log \frac{10^{6.8} I_0}{I_0} = \log 10^{6.8} = 6.8.$$

The magnitude of the earthquake was 6.8 on the Richter scale.

3.3 Exercise Set

Make a hand-drawn graph of each of the following. Then check your work using a grapher.

1. $y = \log_3 x$
2. $y = \log_4 x$
3. $f(x) = \log x$
4. $f(x) = \ln x$

Find each of the following. Do not use a grapher.

5. $\log_2 16$
6. $\log_3 9$
7. $\log_5 125$
8. $\log_2 64$
9. $\log 0.001$
10. $\log 100$
11. $\log_2 \frac{1}{4}$
12. $\log_8 2$
13. $\ln 1$
14. $\ln e$
15. $\log 10$
16. $\log 1$

Convert to a logarithmic equation.

17. $10^3 = 1000$
18. $5^{-3} = \frac{1}{125}$
19. $8^{1/3} = 2$
20. $10^{0.3010} = 2$
21. $e^3 = t$
22. $Q^t = x$
23. $e^2 = 7.3891$
24. $e^{-1} = 0.3679$
25. $p^k = 3$
26. $e^{-t} = 4000$

Convert to an exponential equation.

27. $\log_5 5 = 1$
28. $t = \log_4 7$
29. $\log 0.01 = -2$
30. $\log 7 = 0.845$
31. $\ln 30 = 3.4012$
32. $\ln 0.38 = -0.9676$
33. $\log_a M = -x$
34. $\log_t Q = k$
35. $\log_a T^3 = x$
36. $\ln W^5 = t$

Find each of the following on a grapher. Round to four decimal places.

37. $\log 3$
38. $\log 8$
39. $\log 532$
40. $\log 93{,}100$
41. $\log 0.57$
42. $\log 0.082$
43. $\log(-2)$
44. $\ln 50$
45. $\ln 2$
46. $\ln(-4)$
47. $\ln 809.3$
48. $\ln 0.00037$
49. $\ln(-1.32)$
50. $\ln 0$

Find the logarithm using the change-of-base formula.

51. $\log_4 100$
52. $\log_3 20$
53. $\log_{100} 0.3$
54. $\log_\pi 100$
55. $\log_{200} 50$
56. $\log_{5.3} 1700$

For each of the following functions, briefly describe how each graph can be obtained from a basic logarithmic function. Then graph the function using a grapher. Give the domain of each function and discuss the vertical asymptotes.

57. $f(x) = \log_2 (x + 3)$
58. $f(x) = \log_3 (x - 2)$
59. $y = \log_3 x - 1$
60. $y = 3 + \log_2 x$
61. $f(x) = 4 \ln x$
62. $f(x) = \frac{1}{2} \ln x$
63. $y = 2 - \ln x$
64. $y = \ln (x + 1)$

Graph each function and its inverse using the same set of axes. Use any method.

65. $f(x) = 3^x$, $f^{-1}(x) = \log_3 x$
66. $f(x) = \log_4 x$, $f^{-1}(x) = 4^x$
67. $f(x) = \log x$, $f^{-1}(x) = 10^x$
68. $f(x) = e^x$, $f^{-1}(x) = \ln x$

69. *Walking Speed.* Refer to Example 11. Various cities and their populations are given below. Find the average walking speed in each city.

a) Los Angeles, California: 3,486,000
b) Chicago, Illinois: 3,005,000
c) Tucson, Arizona: 406,000
d) Rome, New York: 50,400

70. *Earthquake Magnitude.* Refer to Example 12. Various locations of earthquakes and their intensities are given below. What was the magnitude on the Richter scale?

a) Mexico City, 1978: $10^{7.85} \cdot I_0$
b) San Francisco, 1906: $10^{8.25} \cdot I_0$
c) Chile, 1960: $10^{9.6} \cdot I_0$
d) Italy, 1980: $10^{7.85} \cdot I_0$
e) San Francisco, 1989: $10^{6.9} \cdot I_0$

71. *Forgetting.* Students in an accounting class took a final exam. They took equivalent forms of the exam in monthly intervals thereafter. The average score $S(t)$, in percent, after t months was found to be given by the function

$$S(t) = 78 - 15 \log (t + 1), \quad t \geq 0.$$

a) What was the average score when they initially took the test, $t = 0$?
b) What was the average score after 4 months? 24 months?
c) Graph the function.
d) After what time t was the average score 50?

72. *pH of Substances in Chemistry.* In chemistry, the pH of a substance is defined as

$$\text{pH} = -\log [\text{H}^+],$$

where H^+ is the hydrogen ion concentration, in moles per liter. Find the pH of each substance.

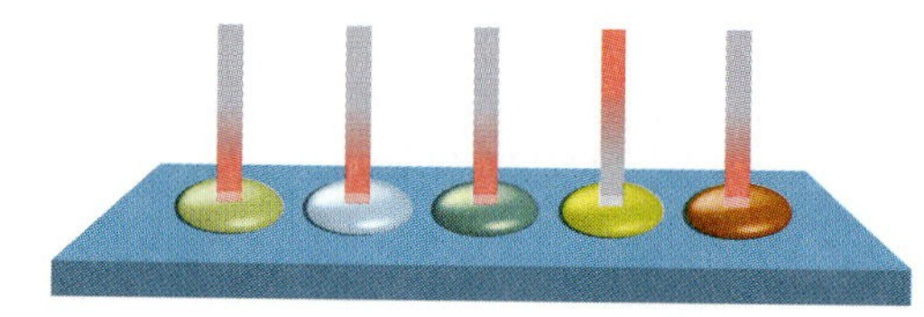

Litmus paper is used to test pH.

SUBSTANCE	HYDROGEN ION CONCENTRATION
a) Pineapple juice	1.6×10^{-4}
b) Hair rinse	0.0013
c) Mouthwash	6.3×10^{-7}
d) Eggs	1.6×10^{-8}
e) Tomatoes	6.3×10^{-5}

73. Find the hydrogen ion concentration of each substance, given the pH. Express the answer in scientific notation.

SUBSTANCE	PH
a) Tap water	7
b) Rainwater	5.4
c) Orange juice	3.2
d) Wine	4.8

74. *Advertising.* A model for advertising response is given by the function

$$N(a) = 1000 + 200 \ln a, \quad a \geq 1,$$

where $N(a)$ = the number of units sold and a = the amount spent on advertising, in thousands of dollars.

a) How many units were sold after spending $1000 ($a = 1$) on advertising?
b) How many units were sold after spending $5000?
c) Graph the function.
d) How much would have to be spent in order to sell 2000 units?

75. *Loudness of Sound.* The **loudness** L, in bels (after Alexander Graham Bell), of a sound of intensity I is defined to be

$$L = \log \frac{I}{I_0},$$

where I_0 = the minimum intensity detectable by the human ear (such as the tick of a watch at 20 ft under quiet conditions). If a sound is 10 times as intense as another, its loudness is 1 bel greater than that of the other. If a sound is 100 times as intense as another, its loudness is 2 bels greater, and so on. The bel is a large unit, so a subunit, the **decibel**, is generally used. For L, in decibels, the formula is

$$L = 10 \log \frac{I}{I_0}.$$

Find the loudness, in decibels, of each sound with the given intensity.

SOUND	INTENSITY
a) Library	$2510 \cdot I_0$
b) Dishwater	$2{,}500{,}000 \cdot I_0$
c) Conversational speech	$10^6 \cdot I_0$
d) Heavy truck	$10^9 \cdot I_0$

Skill Maintenance

76. Multiply and simplify: $x^{-5t} \cdot x^{7t}$.

77. Divide and simplify: $\dfrac{y^{-52}}{y^{-76}}$.

Simplify.

78. $(e^{-12})^{-5}$

79. $(10^4)^t$

Synthesis

80. ◆ Explain how the graph of $f(x) = \ln x$ can be used to obtain the graph of $g(x) = e^{x-2}$.

81. ◆ Explain how the graph of $f(x) = e^x$ can be used to obtain the graph of $g(x) = 3 + \ln x$.

Simplify.

82. $\dfrac{\log_3 64}{\log_3 16}$

83. $\dfrac{\log_5 8}{\log_5 2}$

Find the domain of each function.

84. $f(x) = \log_4 x^2$

85. $f(x) = \log_5 x^3$

86. $f(x) = \log (3x - 4)$

87. $f(x) = \ln |x|$

Solve.

88. $\log_2 (x - 3) \geq 4$

89. $\log_2 (2x + 5) < 0$

90. Use a grapher. Find the point(s) of intersection of the graphs.

$$y = 4 \ln x, \qquad y = \frac{4}{e^x + 1}$$

In Exercises 91–94, match the equation with one of figures (a)–(d), which follow. If needed, use a grapher.

a) b)

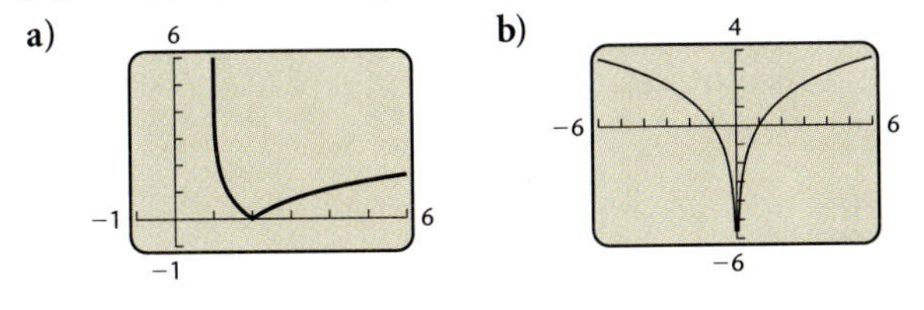

c) d)

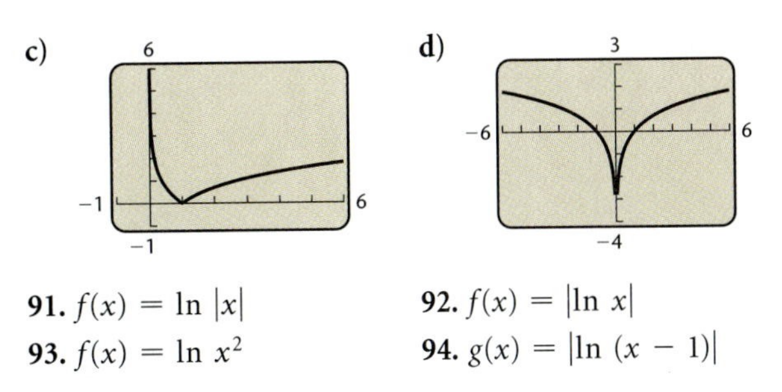

91. $f(x) = \ln |x|$

92. $f(x) = |\ln x|$

93. $f(x) = \ln x^2$

94. $g(x) = |\ln (x - 1)|$

For Exercises 95–98:

a) *Graph the function.*
b) *Estimate the zeros.*
c) *Estimate the relative maximum and the minimum values.*

95. $f(x) = x \ln x$

96. $f(x) = x^2 \ln x$

97. $f(x) = \dfrac{\ln x}{x^2}$

98. $f(x) = e^{-x} \ln x$

3.4 Properties of Logarithmic Functions

- *Convert from logarithms of products, powers, and quotients to expressions in terms of individual logarithms, and conversely.*
- *Simplify expressions of the type $\log_a a^x$ and $a^{\log_a x}$.*

We now establish some properties of logarithmic functions that are fundamental to their use here and in subsequent mathematics. These properties are based on corresponding rules for exponents.

Logarithms of Products

Interactive Discovery

Graph each of the following functions. Use the graphs to discover an identity.

$$y_1 = \log (4x^2 \cdot x^3), \quad y_2 = \log (4x^2) + \log (x^3),$$
$$y_3 = \log (4x^2 + x^3), \quad y_4 = [\log (4x^2)][\log (x^3)]$$

Test your result by using a TABLE feature to complete this table.

X	Y1	Y2	Y3	Y4
1				
10				
20				
300				

X = 1

Graph some other functions to test your discovery.

The first property of logarithms corresponds to the rule of exponents: $a^m \cdot a^n = a^{m+n}$.

The Product Rule

For any positive numbers M and N and any logarithmic base a,

$$\log_a MN = \log_a M + \log_a N.$$

(The logarithm of a product is the sum of the logarithms of the factors.)

Example 1 Express as a sum of logarithms: $\log_3 (9 \cdot 27)$.

SOLUTION We have

$$\log_3 (9 \cdot 27) = \log_3 9 + \log_3 27. \quad \text{Using the product rule}$$

As a check, note that

$$\log_3 (9 \cdot 27) = \log_3 243 = 5$$

and that

$$\log_3 9 + \log_3 27 = 2 + 3 = 5.$$

Example 2 Express as a single logarithm: $\log_2 p^3 + \log_2 q$.

SOLUTION We have

$$\log_2 p^3 + \log_2 q = \log_2 (p^3q).$$

A Proof of the Product Rule: Let $\log_a M = x$ and $\log_a N = y$. Converting to exponential equations, we have $a^x = M$ and $a^y = N$. Then

$$MN = a^x \cdot a^y$$
$$= a^{x+y}.$$

Converting back to a logarithmic equation, we get

$$\log_a MN = x + y.$$

Remembering what x and y represent, it follows that

$$\log_a MN = \log_a M + \log_a N.$$

Logarithms of Powers

Interactive Discovery

Graph each of the following functions. Use the graphs to discover an identity.

$$y_1 = \log x^5, \qquad y_2 = \log x + \log 5,$$
$$y_3 = 5 \log x, \qquad y_4 = \log (x + 5)$$

Test your result by using a TABLE feature to complete this table.

(continued)

X	Y1	Y2	Y3	Y4
1				
10				
20				
300				

X = 1

Graph some other functions to test your discovery.

The second property of logarithms corresponds to the rule of exponents: $(a^m)^n = a^{mn}$.

The Power Rule

For any positive number M, any logarithmic base a, and any real number p,

$$\log_a M^p = p \log_a M.$$

(The logarithm of a power of M is the exponent times the logarithm of M.)

Example 3 Express each of the following as a product.

a) $\log_a 11^{-3}$ **b)** $\log_a \sqrt[4]{7}$

SOLUTION

a) $\log_a 11^{-3} = -3 \log_a 11$ **Using the power rule**

b) $\log_a \sqrt[4]{7} = \log_a 7^{1/4}$ **Writing exponential notation**

$= \frac{1}{4} \log_a 7$ **Using the power rule**

A Proof of the Power Rule: Let $x = \log_a M$. The equivalent exponential equation is $a^x = M$. Raising both sides to the power p, we obtain

$$(a^x)^p = M^p,$$

or

$$a^{xp} = M^p.$$

Converting back to a logarithmic equation, we get

$$\log_a M^p = xp.$$

But $x = \log_a M$, so substituting gives us

$$\log_a M^p = (\log_a M)p$$
$$= p \log_a M.$$

Logarithms of Quotients

Interactive Discovery

Graph each of the following functions. Use the graphs to discover an identity.

$$y_1 = \log\left(\frac{4x^5}{x^2}\right), \qquad y_2 = \log(4x^5) + \log(x^2),$$

$$y_3 = \log(4x^5) - \log(x^2), \qquad y_4 = \log(4x + 5x^2)$$

Test your result by using a TABLE feature to complete this table.

X	Y1	Y2	Y3	Y4
1				
10				
20				
300				

X = 1

Graph some other functions to test your discovery.

The third property of logarithms corresponds to the rule of exponents:

$$\frac{a^m}{a^n} = a^{m-n}.$$

The Quotient Rule

For any positive numbers M and N, and any logarithmic base a,

$$\log_a \frac{M}{N} = \log_a M - \log_a N.$$

(The logarithm of a quotient is the logarithm of the numerator minus the logarithm of the denominator.)

Example 4 Express as a difference of logarithms: $\log_t \dfrac{8}{w}$.

SOLUTION

$$\log_t \frac{8}{w} = \log_t 8 - \log_t w \qquad \text{Using the quotient rule}$$

Example 5 Express as a single logarithm: $\log_b 64 - \log_b 16$.

SOLUTION

$$\log_b 64 - \log_b 16 = \log_b \frac{64}{16} = \log_b 4$$

A Proof of the Quotient Rule: The proof follows from both the product and the power rules:

$$\log_a \frac{M}{N} = \log_a MN^{-1}$$

$$= \log_a M + \log_a N^{-1}$$ Using the product rule

$$= \log_a M + (-1) \log_a N$$ Using the power rule

$$= \log_a M - \log_a N.$$

Note the following.

$\log_a MN \neq (\log_a M)(\log_a N)$	The logarithm of a product is *not* the product of the logarithms.
$\log_a (M + N) \neq \log_a M + \log_a N$	The logarithm of a sum is *not* the sum of the logarithms.
$\log_a \frac{M}{N} \neq \frac{\log_a M}{\log_a N}$	The logarithm of a quotient is *not* the quotient of the logarithms.

Using the Properties Together

Example 6 Express each of the following in terms of sums and differences of logarithms.

a) $\log_a \frac{x^2y^5}{z^4}$ b) $\log_a \sqrt[3]{\frac{a^2b}{c^5}}$ c) $\log_b \frac{ay^5}{m^3n^4}$

SOLUTION

a) $\log_a \frac{x^2y^5}{z^4} = \log_a (x^2y^5) - \log_a z^4$ Using the quotient rule

$= \log_a x^2 + \log_a y^5 - \log_a z^4$ Using the product rule

$= 2 \log_a x + 5 \log_a y - 4 \log_a z$ Using the power rule

b) $\log_a \sqrt[3]{\frac{a^2b}{c^5}} = \log_a \left(\frac{a^2b}{c^5}\right)^{1/3}$ Writing exponential notation

$= \frac{1}{3} \log_a \frac{a^2b}{c^5}$ Using the power rule

$= \frac{1}{3} (\log_a a^2b - \log_a c^5)$ Using the quotient rule

$= \frac{1}{3} (2 \log_a a + \log_a b - 5 \log_a c)$ Using the product and power rules. The parentheses are important.

$= \frac{1}{3} (2 + \log_a b - 5 \log_a c)$ $\log_a a = 1$

$= \frac{2}{3} + \frac{1}{3} \log_a b - \frac{5}{3} \log_a c$ Multiplying to remove parentheses

c) $\log_b \dfrac{ay^5}{m^3n^4} = \log_b ay^5 - \log_b m^3n^4$ — Using the quotient rule

$= (\log_b a + \log_b y^5) - (\log_b m^3 + \log_b n^4)$ — Using the product rule

$= \log_b a + \log_b y^5 - \log_b m^3 - \log_b n^4$ — Removing parentheses

$= \log_b a + 5 \log_b y - 3 \log_b m - 4 \log_b n$ — Using the power rule

Example 7 Express as a single logarithm:

$$5 \log_b x - \log_b y + \frac{1}{4} \log_b z.$$

SOLUTION

$$5 \log_b x - \log_b y + \frac{1}{4} \log_b z = \log_b x^5 - \log_b y + \log_b z^{1/4}$$

Using the power rule

$$= \log_b \frac{x^5}{y} + \log_b z^{1/4}$$

Using the quotient rule

$$= \log_b \frac{x^5 z^{1/4}}{y}, \text{ or } \log_b \frac{x^5 \sqrt[4]{z}}{y}$$

Using the product rule

Example 8 Given that $\log_a 2 = 0.301$ and $\log_a 3 = 0.477$, find each of the following.

a) $\log_a 6$

b) $\log_a \dfrac{2}{3}$

c) $\log_a 81$

d) $\log_a \sqrt{a}$

e) $\log_a 5$

f) $\dfrac{\log_a 3}{\log_a 2}$

SOLUTION

a) $\log_a 6 = \log_a (2 \cdot 3) = \log_a 2 + \log_a 3$ — Using the product rule

$= 0.301 + 0.477$

$= 0.778$

b) $\log_a \frac{2}{3} = \log_a 2 - \log_a 3$ — Using the quotient rule

$= 0.301 - 0.477 = -0.176$

c) $\log_a 81 = \log_a 3^4 = 4 \log_a 3$ — Using the power rule

$= 4(0.477) = 1.908$

d) $\log_a \sqrt{a} = \log_a a^{1/2} = \frac{1}{2} \log_a a$ — Using the power rule

$= \frac{1}{2} \cdot 1 = \frac{1}{2}$

e) $\log_a 5$ *cannot be found using these properties and the given information.*

$$(\log_a 5 \neq \log_a 2 + \log_a 3)$$

f) $\dfrac{\log_a 3}{\log_a 2} = \dfrac{0.477}{0.301} \approx 1.585$ **We simply divided, not using any of the properties.**

In Example 8, a is actually 10 so we have common logarithms. Check as many results as possible using a grapher.

Simplifying Expressions of the Type $\log_a a^x$ *and* $a^{\log_a x}$

We have two final properties to consider. The first follows from the product rule: Since $\log_a a^x = x \log_a a = x \cdot 1 = x$, we have $\log_a a^x = x$. This property also follows from the definition of a logarithm: x is the power to which we raise a in order to get a^x.

The Logarithm of a Base to a Power

For any base a and any real number x,

$$\log_a a^x = x.$$

(The logarithm, base a, of a to a power is the power.)

Example 9 Simplify each of the following.

a) $\log_a a^8$ b) $\ln e^{-t}$ c) $\log 10^{3k}$

SOLUTION

a) $\log_a a^8 = 8$ **8 is the power to which we raise a in order to get a^8.**

b) $\ln e^{-t} = \log_e e^{-t} = -t$

c) $\log 10^{3k} = \log_{10} 10^{3k} = 3k$

Let $M = \log_a x$. Then $a^M = x$. Substituting $\log_a x$ for M, we obtain $a^{\log_a x} = x$. This also follows from the definition of a logarithm: $\log_a x$ is the power to which a is raised in order to get x.

A Base to a Logarithmic Power

For any base a and any positive real number x,

$$a^{\log_a x} = x.$$

(The number a raised to the power $\log_a x$ is x.)

Example 10 Simplify each of the following.

a) $4^{\log_4 k}$ b) $e^{\ln 5}$ c) $10^{\log 7t}$

SOLUTION

a) $4^{\log_4 k} = k$

b) $e^{\ln 5} = e^{\log_e 5} = 5$

c) $10^{\log 7t} = 10^{\log_{10} 7t} = 7t$

A Proof of the Change-of-Base Formula: We close this section by proving the change-of-base formula and summarizing the properties of logarithms considered thus far in this chapter. In Section 3.3, we used the change-of-base formula,

$$\log_b M = \frac{\log_a M}{\log_a b},$$

to make base conversions in order to graph logarithmic functions on a grapher. Let $x = \log_b M$. Then by the definition of logarithms, $b^x = M$. We have established that logarithmic functions are one-to-one. This says that $M = N \longleftrightarrow \log_a M = \log_a N$ for any positive numbers M and N. Thus we can take the logarithm base a on both sides of an expression such as $b^x = M$. Then we get $\log_a b^x = \log_a M$. Using the power rule, we obtain $x \log_a b = \log_a M$. Solving for x gives us

$$x = \frac{\log_a M}{\log_a b}, \quad \text{so} \quad x = \log_b M = \frac{\log_a M}{\log_a b}.$$

Following is a summary of the properties of logarithms.

Summary of the Properties of Logarithms

The Product Rule:	$\log_a MN = \log_a M + \log_a N$
The Power Rule:	$\log_a M^p = p \log_a M$
The Quotient Rule:	$\log_a \frac{M}{N} = \log_a M - \log_a N$
The Change-of-Base Formula:	$\log_b M = \frac{\log_a M}{\log_a b}$
Other Formulas:	$\log_a a = 1$, $\log_a 1 = 0$, $\log_a a^x = x$, $a^{\log_a x} = x$

3.4 Exercise Set

Express as a sum of logarithms.

1. $\log_3 (81 \cdot 27)$
2. $\log_2 (8 \cdot 64)$
3. $\log_5 (5 \cdot 125)$
4. $\log_4 (64 \cdot 32)$
5. $\log_t 8Y$
6. $\log_e Qx$

Express as a product.

7. $\log_b t^3$
8. $\log_a x^4$
9. $\log y^8$
10. $\ln y^5$
11. $\log_c K^{-6}$
12. $\log_b Q^{-8}$

Express as a difference of logarithms.

13. $\log_t \frac{M}{8}$
14. $\log_a \frac{76}{13}$

15. $\log_a \frac{x}{y}$

16. $\log_b \frac{3}{w}$

Express in terms of sums and differences of logarithms.

17. $\log_a 6xy^5z^4$

18. $\log_a x^3y^2z$

19. $\log_b \frac{p^2q^5}{m^4b^9}$

20. $\log_b \frac{x^2y}{b^3}$

21. $\log_a \sqrt{\frac{x^6}{p^5q^8}}$

22. $\log_c \sqrt[3]{\frac{y^3z^2}{x^4}}$

23. $\log_a \sqrt[4]{\frac{m^8n^{12}}{a^3b^5}}$

24. $\log_a \sqrt{\frac{a^6b^8}{a^2b^5}}$

Express as a single logarithm and simplify, if possible.

25. $\log_a 75 + \log_a 2$

26. $\log 0.01 + \log 1000$

27. $\log 10{,}000 - \log 100$

28. $\ln 54 - \ln 6$

29. $\frac{1}{2} \log_a x + 4 \log_a y - 3 \log_a x$

30. $\frac{2}{5} \log_a x - \frac{1}{3} \log_a y$

31. $\ln x^2 - 2 \ln \sqrt{x}$

32. $\ln 2x + 3(\ln x - \ln y)$

33. $\ln (x^2 - 4) - \ln (x + 2)$

34. $\log_a \frac{a}{\sqrt{x}} - \log_a \sqrt{ax}$

35. $\ln x - 3[\ln (x - 5) + \ln (x + 5)]$

36. $\frac{2}{3}[\ln (x^2 - 9) - \ln (x + 3)] + \ln (x + y)$

37. $\frac{3}{2} \ln 4x^6 - \frac{4}{5} \ln 2y^{10}$

38. $120(\ln \sqrt[5]{x^3} + \ln \sqrt[3]{y^2} - \ln \sqrt[4]{16z^5})$

Given that $\log_b 3 = 1.0986$ *and* $\log_b 5 = 1.6094$, *find each of the following.*

39. $\log_b \frac{3}{5}$

40. $\log_b 15$

41. $\log_b \frac{1}{5}$

42. $\log_b \frac{5}{3}$

43. $\log_b \sqrt{b}$

44. $\log_b \sqrt{b^3}$

45. $\log_b 5b$

46. $\log_b 9$

47. $\log_b 75$

48. $\log_b \frac{1}{b}$

Simplify.

49. $\log_p p^3$

50. $\log_t t^{2713}$

51. $\log_e e^{|x-4|}$

52. $\log_q q^{\sqrt{3}}$

53. $3^{\log_3 4x}$

54. $5^{\log_5 (4x-3)}$

55. $10^{\log w}$

56. $e^{\ln x^3}$

57. $\ln e^{8t}$

58. $\log 10^{-k}$

Skill Maintenance

Solve.

59. $3x - 7 = 5$

60. $x^4 - 6x^2 + 7 = 0$

61. $\frac{2x - 1}{x - 4} = 9$

62. $x(x - 3) = 10$

Synthesis

63. ◆ Given that $f(x) = a^x$ and $g(x) = \log_a x$, find $(f \circ g)(x)$ and $(g \circ f)(x)$. These results are alternative proofs of what properties of logarithms already proven in this section? Explain.

64. ◆ Explain the errors, if any, in the following:

$$\log_a ab^3 = (\log_a a)(\log_a b^3) = 3 \log_a b.$$

Solve for x. Do an algebraic solution.

65. $5^{\log_5 8} = 2x$

66. $\ln e^{3x-5} = -8$

Express as a single logarithm and simplify, if possible.

67. $\log_a (x^2 + xy + y^2) + \log_a (x - y)$

68. $\log_a (a^{10} - b^{10}) - \log_a (a + b)$

Express as a sum or a difference of logarithms.

69. $\log_a \frac{x - y}{\sqrt{x^2 - y^2}}$

70. $\log_a \sqrt{9 - x^2}$

71. Given that $\log_a x = 2$, $\log_a y = 3$, and $\log_a z = 4$, find

$$\log_a \frac{\sqrt[4]{y^2z^5}}{\sqrt[4]{x^3z^{-2}}}.$$

Determine whether each of the following is true. Assume that a, x, M, and N are positive.

72. $\log_a M + \log_a N = \log_a (M + N)$

73. $\log_a M - \log_a N = \log_a \frac{M}{N}$

74. $\frac{\log_a M}{\log_a N} = \log_a M - \log_a N$

75. $\frac{\log_a M}{x} = \log_a M^{1/x}$

76. $\log_a x^3 = 3 \log_a x$

77. $\log_a 8x = \log_a x + \log_a 8$

78. $\log_N (MN)^x = x \log_N M + x$

Suppose that $\log_a x = 2$. *Find each of the following.*

79. $\log_a \left(\dfrac{1}{x}\right)$

80. $\log_{1/a} x$

81. Simplify:

$\log_{10} 11 \cdot \log_{11} 12 \cdot \log_{12} 13 \cdots \log_{998} 999 \cdot \log_{999} 1000.$

Prove each of the following for any base a and any positive number x.

82. $\log_a \left(\dfrac{1}{x}\right) = -\log_a x = \log_{1/a} x$

83. $\log_a \left(\dfrac{x + \sqrt{x^2 - 5}}{5}\right) = -\log_a (x - \sqrt{x^2 - 5})$

3.5 Solving Exponential and Logarithmic Equations

- *Solve exponential and logarithmic equations.*

Solving Exponential Equations

Equations with variables in the exponents, such as $3^x = 20$ and $2^{5x} = 64$, are called **exponential equations**. We now consider solving exponential equations.

Sometimes, as is the case with the equation $2^{5x} = 64$, we can write each side as a power of the same number:

$$2^{5x} = 2^6.$$

We can then set the exponents equal and solve:

$$5x = 6$$
$$x = \tfrac{6}{5}.$$

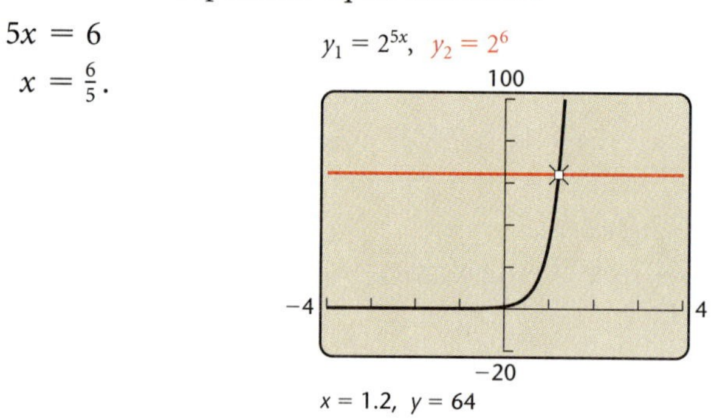

We use the following property.

Base–Exponent Property

For any $a > 0$, $a \neq 1$,

$$a^x = a^y \longleftrightarrow x = y.$$

This property follows from the fact that for any $a > 0$, $a \neq 1$, $f(x) = a^x$ is a one-to-one function. If $a^x = a^y$, then $f(x) = f(y)$. Then since f is one-to-one (see the definition in Section 3.1), it follows that $x = y$. Conversely, if $x = y$, it follows that $a^x = a^y$, since we are raising a to the same power.

Example 1 Solve: $2^{3x-7} = 32$.

ALGEBRAIC SOLUTION

Note that $32 = 2^5$. Thus we can write each side as a power of the same number:

$$2^{3x-7} = 2^5.$$

Since the bases are the same number, 2, we can use the base–exponent property and set the exponents equal:

$$3x - 7 = 5$$
$$3x = 12$$
$$x = 4$$

CHECK:

$$2^{3x-7} = 32$$
$$2^{3(4)-7} \ ? \ 32$$
$$2^{12-7}$$
$$2^5$$
$$32 \mid 32 \quad \text{TRUE}$$

The solution is 4. The solution set is $\{4\}$.

GRAPHICAL SOLUTION

To solve on a grapher, we graph the equations

$$y_1 = 2^{3x-7} \quad \text{and} \quad y_2 = 32.$$

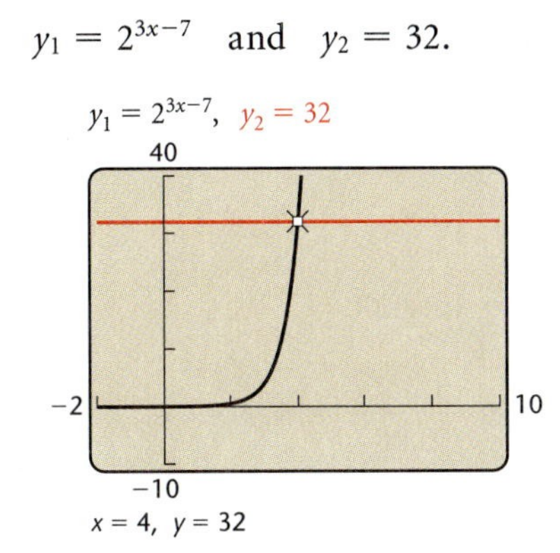

Then we can use the TRACE and ZOOM features or the INTERSECT feature to approximate the point of intersection.

We see that the solution is 4. On some graphers, we can get the solution by solving the equation $2^{3x-7} - 32 = 0$ using a SOLVE feature.

When it does not seem possible to write each side as a power of the same base, we can take the common or natural logarithm on each side and use the power rule for logarithms.

Example 2 Solve: $3^x = 20$.

ALGEBRAIC SOLUTION

We have

$$3^x = 20$$

$\log 3^x = \log 20$ — **Taking the common logarithm on both sides**

$x \log 3 = \log 20$ — **Using the power rule**

$x = \dfrac{\log 20}{\log 3}$. — **Dividing by log 3**

This is an exact answer. We cannot simplify further, but we can approximate using a grapher:

$$x = \frac{\log 20}{\log 3} \approx \frac{1.3010}{0.4771} \approx 2.7268.$$

We can check this by finding $3^{2.7268}$. The solution is about 2.727.

GRAPHICAL SOLUTION

We graph the equations

$$y_1 = 3^x \quad \text{and} \quad y_2 = 20$$

and use TRACE and ZOOM or INTERSECT to determine a point of intersection.

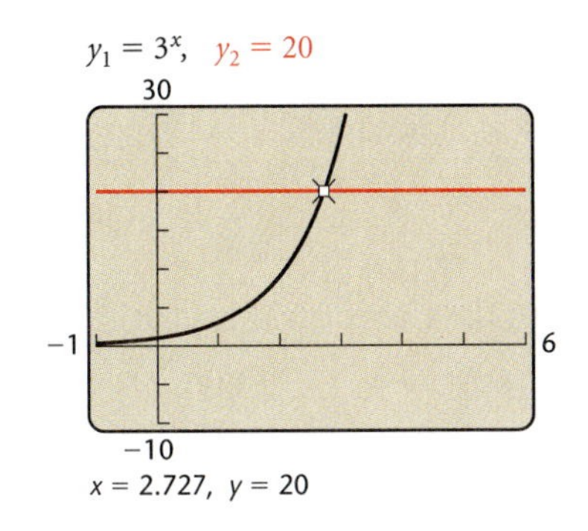

The solution is about 2.727.

It will ease our work if we take the natural logarithm when working with equations that have e as a base.

Example 3 Solve: $e^{0.08t} = 2500$.

ALGEBRAIC SOLUTION

We have

$$\begin{aligned} e^{0.08t} &= 2500 \\ \ln e^{0.08t} &= \ln 2500 && \text{Taking the natural logarithm on both sides} \\ 0.08t &= \ln 2500 && \text{Finding the logarithm of a base to a power: } \log_a a^x = x \\ t &= \frac{\ln 2500}{0.08} \approx 97.8. \end{aligned}$$

The solution is about 97.8.

GRAPHICAL SOLUTION

We graph the equations

$$y_1 = e^{0.08x} \quad \text{and} \quad y_2 = 2500$$

and determine a point of intersection.

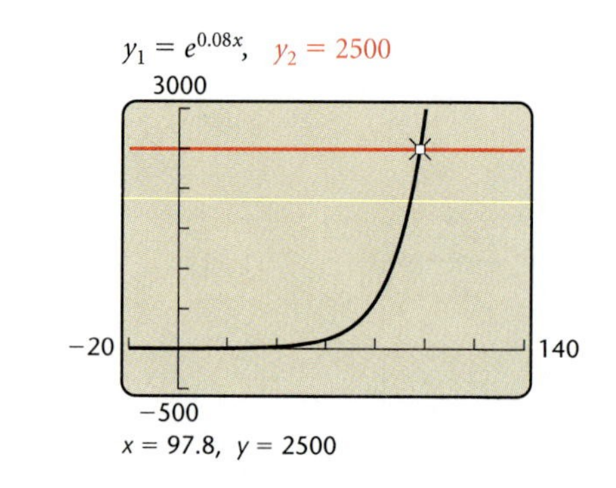

The solution is about 97.8.

Example 4 Solve: $e^x + e^{-x} - 6 = 0$.

ALGEBRAIC SOLUTION

In this case, we have more than one term with x in the exponent. To get a single expression with x in the exponent, we do the following:

$$\begin{aligned} e^x + e^{-x} - 6 &= 0 \\ e^x + \frac{1}{e^x} - 6 &= 0 && \text{Rewriting } e^{-x} \text{ with a positive exponent} \\ e^{2x} + 1 - 6e^x &= 0. && \text{Multiplying on both sides by } e^x \end{aligned}$$

This equation is reducible to quadratic with $u = e^x$:

$$u^2 - 6u + 1 = 0.$$

The coefficients of the reduced quadratic equation are $a = 1$, $b = -6$, and $c = 1$. Using the quadratic formula, we obtain

$$u = e^x = 3 \pm \sqrt{8}.$$

We now take the natural logarithm on both sides:

$$\begin{aligned} \ln e^x &= \ln (3 \pm \sqrt{8}) \\ x &= \ln (3 \pm \sqrt{8}). && \text{Using } \ln e^x = x \end{aligned}$$

Approximating each of the solutions, we obtain 1.76 and −1.76.

GRAPHICAL SOLUTION

We begin by graphing $f(x) = e^x + e^{-x} - 6$. We look for the first coordinates of the points where the graph of $y = e^x + e^{-x} - 6$ crosses the x-axis. The zeros of the function are the solutions of the equation.

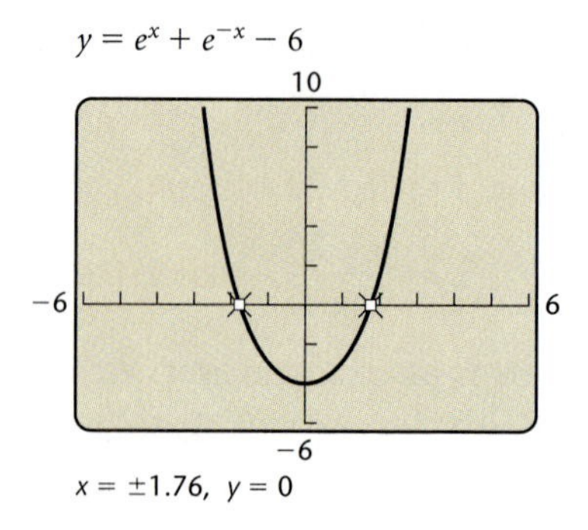

We obtain the approximate solutions −1.76 and 1.76.

It is possible that when encountering an equation like the one in Example 4, you might not recognize that it could be solved in the algebraic manner shown. This points out the value of the graphical solution.

Solving Logarithmic Equations

Equations containing variables in logarithmic expressions, such as $\log_2 x = 4$ and $\log x + \log (x + 3) = 1$, are called **logarithmic equations.**

> To solve logarithmic equations algebraically, first try to obtain a single logarithmic expression on one side and then write an equivalent exponential equation.

Example 5 Solve: $\log_3 x = -2$.

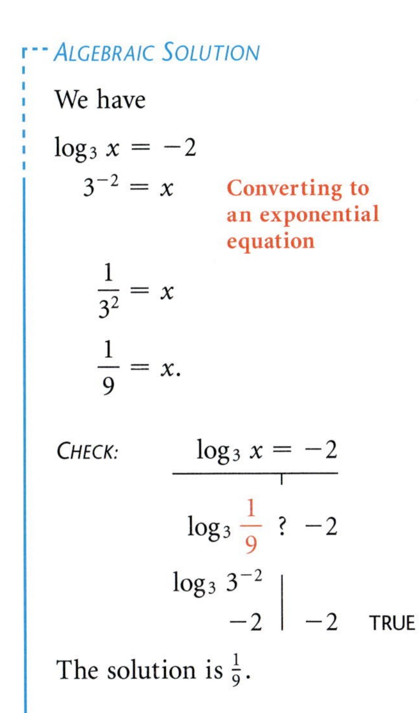

ALGEBRAIC SOLUTION

We have

$$\log_3 x = -2$$

$$3^{-2} = x \qquad \text{Converting to an exponential equation}$$

$$\frac{1}{3^2} = x$$

$$\frac{1}{9} = x.$$

CHECK:

$\log_3 x = -2$	
$\log_3 \frac{1}{9}$?	-2
$\log_3 3^{-2}$	
-2	-2 TRUE

The solution is $\frac{1}{9}$.

GRAPHICAL SOLUTION

Use the change-of-base formula and graph the equations

$$y_1 = \log_3 x = \frac{\ln x}{\ln 3}$$

and

$$y_2 = -2.$$

Then look for the point(s) of intersection.

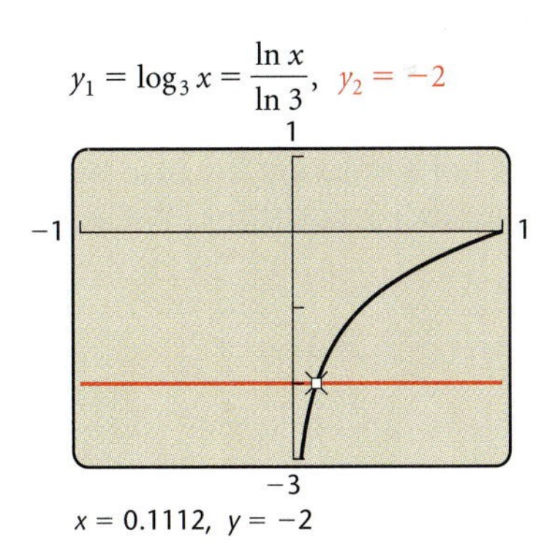

The approximate solution is 0.111, which is about $\frac{1}{9}$.

Example 6 Solve: $\log x + \log (x + 3) = 1$.

ALGEBRAIC SOLUTION

In this case, we have common logarithms. Including the base 10's will help us understand the problem:

$$\log_{10} x + \log_{10} (x + 3) = 1$$

$$\log_{10} [x(x + 3)] = 1$$ **Using the product rule to obtain a single logarithm**

$$x(x + 3) = 10^1$$ **Writing an equivalent exponential equation**

$$x^2 + 3x = 10$$

$$x^2 + 3x - 10 = 0$$

$$(x - 2)(x + 5) = 0$$ **Factoring**

$$x - 2 = 0 \quad or \quad x + 5 = 0$$

$$x = 2 \quad or \quad x = -5.$$

CHECK: For 2:

$$\log x + \log (x + 3) = 1$$

$$\log 2 + \log (2 + 3) \ ? \ 1$$

$$\log 2 + \log 5$$

$$\log 10$$

$$1 \ | \ 1 \quad \text{TRUE}$$

For -5:

$$\log x + \log (x + 3) = 1$$

$$\log (-5) + \log (-5 + 3) \ ? \ 1 \quad \text{FALSE}$$

The number -5 is not a solution because negative numbers do not have real-number logarithms. The solution is 2.

GRAPHICAL SOLUTION

We graph the equations

$$y_1 = \log x + \log (x + 3)$$

and

$$y_2 = 1$$

and find the point(s) of intersection.

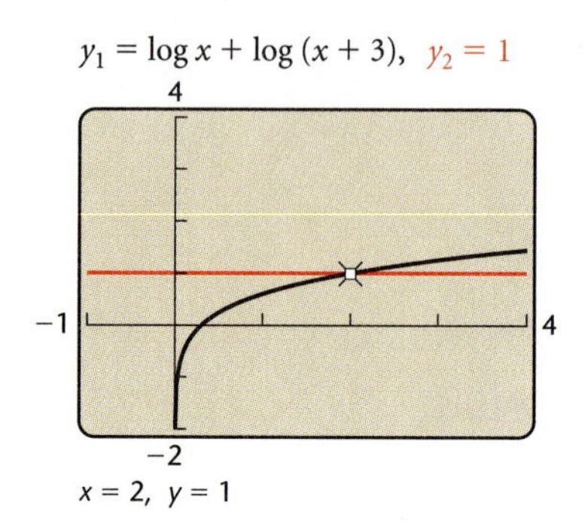

The solution is 2.

Example 7 Solve: $\log_3(2x - 1) - \log_3(x - 4) = 2$.

ALGEBRAIC SOLUTION

We have

$$\log_3(2x - 1) - \log_3(x - 4) = 2$$

$$\log_3 \frac{2x - 1}{x - 4} = 2 \quad \text{Using the quotient rule}$$

$$\frac{2x - 1}{x - 4} = 3^2 \quad \text{Writing an equivalent exponential equation}$$

$$\frac{2x - 1}{x - 4} = 9$$

$$2x - 1 = 9(x - 4) \quad \text{Multiplying by the LCD, } x - 4$$

$$2x - 1 = 9x - 36$$

$$35 = 7x$$

$$5 = x.$$

CHECK:

$\log_3(2x - 1) - \log_3(x - 4)$	$= 2$
$\log_3(2 \cdot 5 - 1) - \log_3(5 - 4)$	? 2
$\log_3 9 - \log_3 1$	
$2 - 0$	
2	2 TRUE

The solution is 5.

GRAPHICAL SOLUTION

We use the change-of-base formula and graph the equations

$$y_1 = \frac{\ln(2x - 1)}{\ln 3} - \frac{\ln(x - 4)}{\ln 3}$$

and

$$y_2 = 2.$$

Then we find the point(s) of intersection.

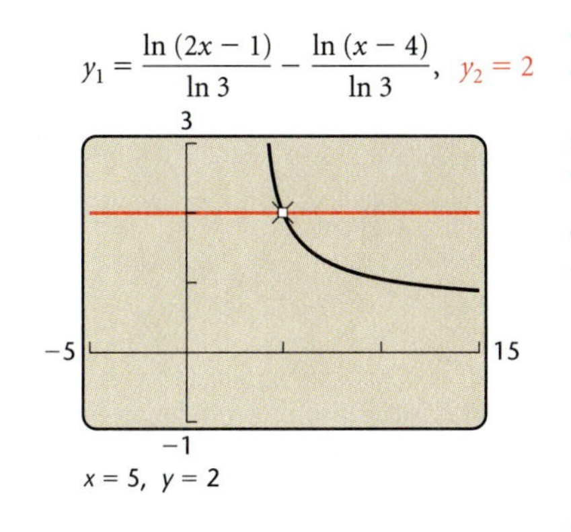

The solution is 5.

Sometimes we encounter equations for which an algebraic solution seems difficult or impossible.

Example 8 Solve: $e^{0.5x} - 7.3 = 2.08x + 6.2$.

ALGEBRAIC SOLUTION

In this case, we have an equation for which an algebraic solution seems difficult or impossible.

GRAPHICAL SOLUTION

We graph the equations

$$y_1 = e^{0.5x} - 7.3 \quad \text{and} \quad y_2 = 2.08x + 6.2$$

and look for points of intersection. (See the figure at left.)

We can also consider the equation

$$y = e^{0.5x} - 7.3 - 2.08x - 6.2, \quad \text{or} \quad y = e^{0.5x} - 2.08x - 13.5,$$

and look for zeros using TRACE and ZOOM or a SOLVE feature. The approximate solutions are −6.471 and 6.610.

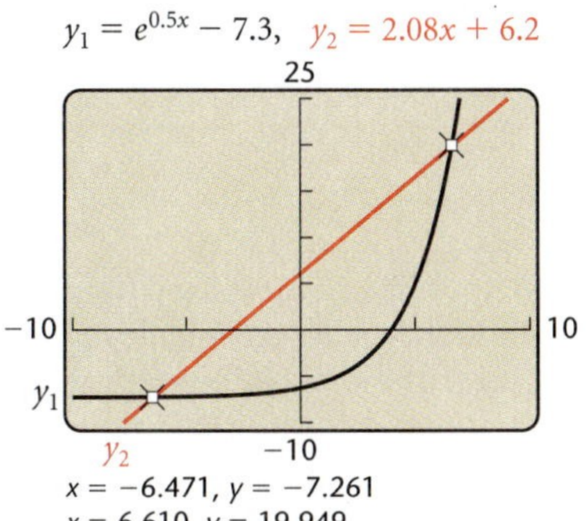

3.5 Exercise Set

Solve each exponential equation algebraically. Then check on a grapher.

1. $3^x = 81$

2. $2^x = 32$

3. $2^{2x} = 8$

4. $3^{7x} = 27$

5. $2^x = 33$

6. $2^x = 40$

7. $5^{4x-7} = 125$

8. $4^{3x-5} = 16$

9. $27 = 3^{5x} \cdot 9^{x^2}$

10. $3^{x^2+4x} = \frac{1}{27}$

11. $84^x = 70$

12. $28^x = 10^{-3x}$

13. $e^t = 1000$

14. $e^{-t} = 0.04$

15. $e^{-0.03t} = 0.08$

16. $1000e^{0.09t} = 5000$

17. $3^x = 2^{x-1}$

18. $5^{x+2} = 4^{1-x}$

19. $(3.9)^x = 48$

20. $250 - (1.87)^x = 0$

21. $e^x + e^{-x} = 5$

22. $e^x - 6e^{-x} = 1$

23. $\dfrac{e^x + e^{-x}}{e^x - e^{-x}} = 3$

24. $\dfrac{5^x - 5^{-x}}{5^x + 5^{-x}} = 8$

Solve each logarithmic equation algebraically. Then check on a grapher.

25. $\log_5 x = 4$

26. $\log_2 x = -3$

27. $\log x = -4$

28. $\log x = 1$

29. $\ln x = 1$

30. $\ln x = -2$

31. $\log_2 (10 + 3x) = 5$

32. $\log_5 (8 - 7x) = 3$

33. $\log x + \log (x - 9) = 1$

34. $\log_2 (x + 1) + \log_2 (x - 1) = 3$

35. $\log_8 (x + 1) - \log_8 x = 2$

36. $\log x - \log (x + 3) = -1$

37. $\log_4 (x + 3) + \log_4 (x - 3) = 2$

38. $\ln (x + 1) - \ln x = \ln 4$

39. $\log (2x + 1) - \log (x - 2) = 1$

40. $\log_5 (x + 4) + \log_5 (x - 4) = 2$

Use only a grapher. Find approximate solutions of each equation or approximate the point(s) of intersection of a pair of equations.

41. $e^{7.2x} = 14.009$

42. $0.082e^{0.05x} = 0.034$

43. $xe^{3x} - 1 = 3$

44. $5e^{5x} + 10 = 3x + 40$

45. $4 \ln (x + 3.4) = 2.5$

46. $\ln x^2 = -x^2$

47. $\log_8 x + \log_8 (x + 2) = 2$

48. $\log_3 x + 7 = 4 - \log_5 x$

49. $\log_5 (x + 7) - \log_5 (2x - 3) = 1$

50. $y = \ln 3x, \; y = 3x - 8$

51. $2.3x + 3.8y = 12.4, \; y = 1.1 \ln (x - 2.05)$

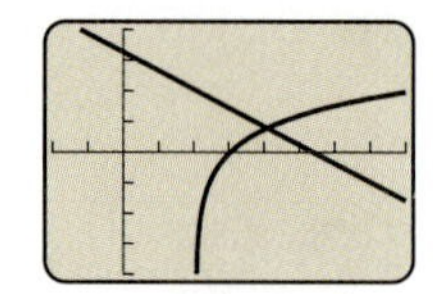

52. $y = 2.3 \ln (x + 10.7), \; y = 10e^{-0.07x^2}$

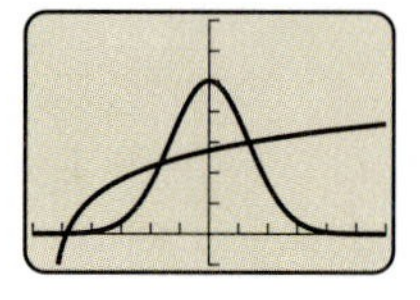

53. $y = 2.3 \ln (x + 10.7), \; y = 10e^{-0.007x^2}$

Skill Maintenance

54. Solve $K = \frac{1}{2}mv^2$ for v.

Solve.

55. $x^4 + 5x^2 = 36$

56. $t^{2/3} - 10 = 3t^{1/3}$

57. *Total Sales of Goodyear.* The following table shows factual data regarding total sales of The Goodyear Tire and Rubber Company.

YEAR, x	TOTAL SALES, y (IN MILLIONS)
1. 1991	\$10,906.8
2. 1992	11,784.9
3. 1993	11,643.4
4. 1994	12,288.2

Source: The Goodyear Tire and Rubber Company Annual Report.

a) Use linear regression on a grapher to fit an equation $y = mx + b$, where $x = 1$ corresponds to 1991, to the data points. Predict total sales in 1999.

b) Use quadratic regression on a grapher to fit a quadratic equation $y = ax^2 + bx + c$ to the data points. Predict total sales in 1999.

Synthesis

58. ◈ In Example 3, we took the natural logarithm on both sides. What would have happened had we taken the common logarithm? Explain which seems best to you and why.

59. ◈ Explain how Exercises 29 and 30 could be solved using the graph of $f(x) = \ln x$.

Solve using any method.

60. $\ln(\ln x) = 2$

61. $\ln(\log x) = 0$

62. $\ln \sqrt[4]{x} = \sqrt{\ln x}$

63. $\sqrt{\ln x} = \ln \sqrt{x}$

64. $\log_3(\log_4 x) = 0$

65. $(\log_3 x)^2 - \log_3 x^2 = 3$

66. $(\log x)^2 - \log x^2 = 3$

67. $\ln x^2 = (\ln x)^2$

68. $e^{2x} - 9 \cdot e^x + 14 = 0$

69. $5^{2x} - 3 \cdot 5^x + 2 = 0$

70. $x\left(\ln \frac{1}{6}\right) = \ln 6$

71. $\log_3 |x| = 2$

72. $x^{\log x} = \dfrac{x^3}{100}$

73. $\ln x^{\ln x} = 4$

74. $\dfrac{(e^{3x+1})^2}{e^4} = e^{10x}$

75. $\dfrac{\sqrt{(e^{2x} \cdot e^{-5x})^{-4}}}{e^x \div e^{-x}} = e^7$

76. $|\log_a x| = \log_a |x|$

77. $\ln(x - 2) > 4$

78. $e^x < \dfrac{4}{5}$

79. $|\log_5 x| + 3 \log_5 |x| = 4$

80. $|2^{x^2} - 8| = 3$

81. Given that $a = \log_8 225$ and $b = \log_2 15$, express a as a function of b.

82. Given that $a = (\log_{125} 5)^{\log_5 125}$, find the value of $\log_3 a$.

83. Given that

$$\begin{aligned}\log_2[\log_3(\log_4 x)] &= \log_3[\log_2(\log_4 y)]\\ &= \log_4[\log_3(\log_2 z)]\\ &= 0,\end{aligned}$$

find $x + y + z$.

84. Given that $f(x) = e^x - e^{-x}$, find $f^{-1}(x)$ if it exists.

3.6 Applications and Models: Growth and Decay

- *Solve applications involving exponential growth and decay.*
- *Find models involving exponential and logarithmic functions.*

Exponential and logarithmic functions with base e are rich in applications to many fields such as business, science, psychology, and sociology. In this section, we consider some basic applications and then use curve fitting to do others.

Population Growth

The function

$$P(t) = P_0 e^{kt}$$

is a model of many kinds of population growth, whether it be a population of people, bacteria, cellular phones, or money. In this function, P_0 is the population at time 0, P is the population after time t, and k is called the **exponential growth rate**. The graph of such an equation is shown at right.

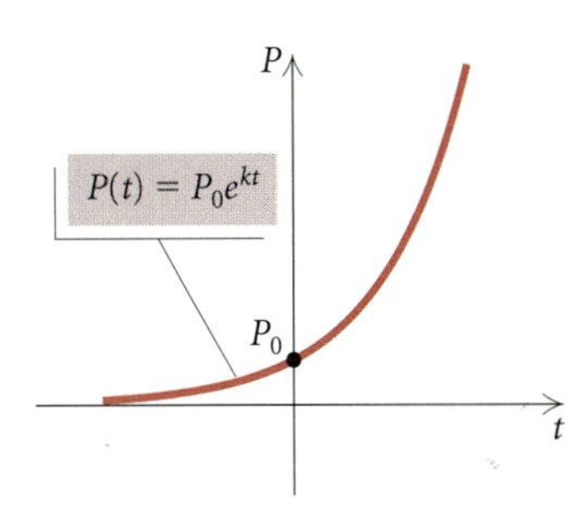

Example 1 *Population Growth of the United States.* In 1997, the population of the United States was about 266 million and the exponential growth rate was 0.9% per year. (*Source:* Statistical Abstract of the United States)

a) Find the exponential growth function.

b) What will the population be in 2002?

c) Graph the exponential growth function.

d) When will the population be double what it was in 1997?

SOLUTION

a) At $t = 0$ (1997), the population was 266 million. We substitute 266 for P_0 and 0.9%, or 0.009, for k to obtain the exponential growth function

$$P(t) = 266e^{0.009t}.$$

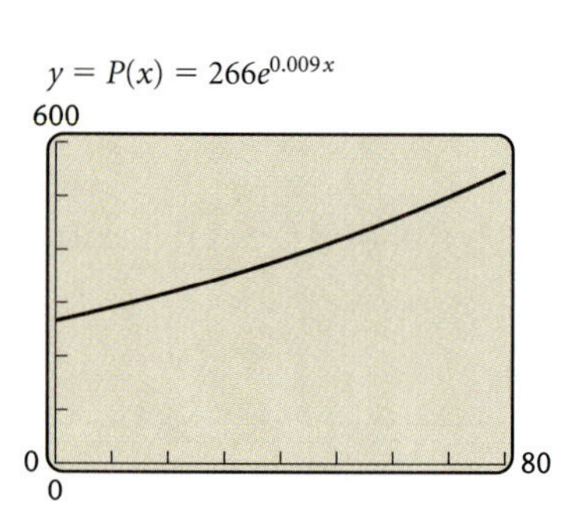

b) In 2002, $t = 5$; that is, 5 yr have passed. To find the population in 2002, we substitute 5 for t:

$$P(5) = 266e^{0.009(5)} = 266e^{0.045} \approx 278.$$

In 2002, the population of the United States will be about 278 million.

c) Using a grapher, we obtain the graph (at left) of the exponential growth function.

d) We are looking for the time T for which $P(T) = 2 \cdot 266$, or 532. The number T is called the **doubling time**. To find T, we solve the equation $532 = 266e^{0.009T}$.

ALGEBRAIC SOLUTION

We have

$$532 = 266e^{0.009T} \quad \text{Substituting 532 for } P(T)$$

$$2 = e^{0.009T} \quad \text{Dividing by 266}$$

$$\ln 2 = \ln e^{0.009T} \quad \text{Taking the natural logarithm on both sides}$$

$$\ln 2 = 0.009T \quad \ln e^x = x$$

$$\frac{\ln 2}{0.009} = T \quad \text{Dividing by 0.009}$$

$$77 \approx T.$$

The population of the United States will be double what it was in 1997 in about 77 yr.

GRAPHICAL SOLUTION

To solve on a grapher, we graph the equations

$$y_1 = 266e^{0.009x} \quad \text{and} \quad y_2 = 532$$

and find the first coordinate of their point of intersection.

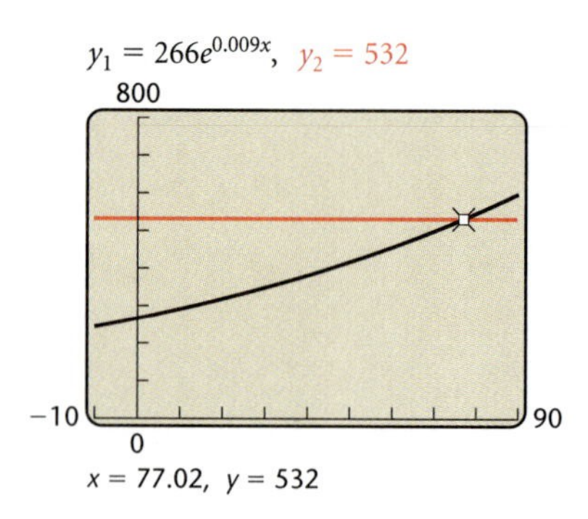

On some graphers, we can get the solution by having the grapher solve the equation directly. We see that the solution is about 77 yr.

No wonder ecologists are so concerned about population growth.

Interest Compounded Continuously

Here we explore the mathematics behind the concept of **interest compounded continuously**. Suppose that an amount P_0 is invested in a savings account at interest rate k *compounded continuously.* The amount $P(t)$ in the account after t years is given by the exponential function

$$P(t) = P_0e^{kt}.$$

Example 2 *Interest Compounded Continuously.* Suppose \$2000 is invested at interest rate k, compounded continuously, and grows to \$2983.65 in 5 yr.

a) What is the interest rate?

b) Find the exponential growth function.

c) What will the balance be after 10 yr?

d) After how long will the \$2000 have doubled?

SOLUTION

a) At $t = 0$, $P(0) = P_0 = \$2000$. Thus the exponential growth function is

$$P(t) = 2000e^{kt}.$$

We know that $P(5) = \$2983.65$. We substitute and solve for k, as shown below.

ALGEBRAIC SOLUTION

We have

$$2983.65 = 2000e^{k(5)}$$ **Substituting 2983.65 for $P(t)$**

$$2983.65 = 2000e^{5k}$$

$$\frac{2983.65}{2000} = e^{5k}$$ **Dividing by 2000**

$$\ln \frac{2983.65}{2000} = \ln e^{5k}$$ **Taking the natural logarithm**

$$\ln \frac{2983.65}{2000} = 5k$$ **Using $\ln e^x = x$**

$$\frac{\ln \frac{2983.65}{2000}}{5} = k$$

$$0.08 \approx k.$$

The interest rate is about 0.08, or 8%.

GRAPHICAL SOLUTION

We can solve by graphing the equations

$$y_1 = 2000e^{5x} \quad \text{and} \quad y_2 = 2983.65$$

on a grapher. Then we can use TRACE and ZOOM or other features of a grapher to find an approximation for the first coordinate of the point of intersection.

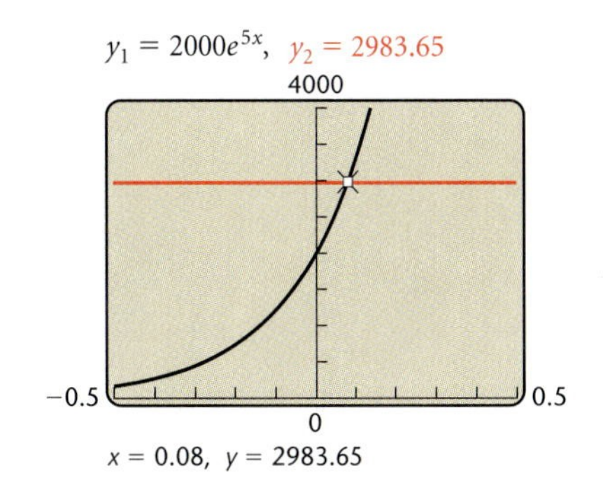

The solution is about 0.08, or 8%.

b) The exponential growth function is

$$P(t) = 2000e^{0.08t}.$$

c) The balance after 10 yr is

$$P(10) = 2000e^{0.08(10)}$$
$$= 2000e^{0.8}$$
$$\approx \$4451.08.$$

d) We solve using both the algebraic method and a grapher.

Algebraic Solution

To find the doubling time T, we set $P(T) = \$4000$ and solve for T:

$4000 = 2000e^{0.08T}$

$2 = e^{0.08T}$ **Dividing by 2000**

$\ln 2 = \ln e^{0.08T}$ **Taking the natural logarithm**

$\ln 2 = 0.08T$ **$\ln e^x = x$**

$\frac{\ln 2}{0.08} = T$ **Dividing by 0.08**

$8.7 \approx T.$

Thus the original investment of \$2000 will double in about 8.7 yr.

Graphical Solution

To solve on a grapher, we graph the equations

$$y_1 = 2000e^{0.08x} \quad \text{and} \quad y_2 = 4000$$

and find the first coordinate of their point of intersection.

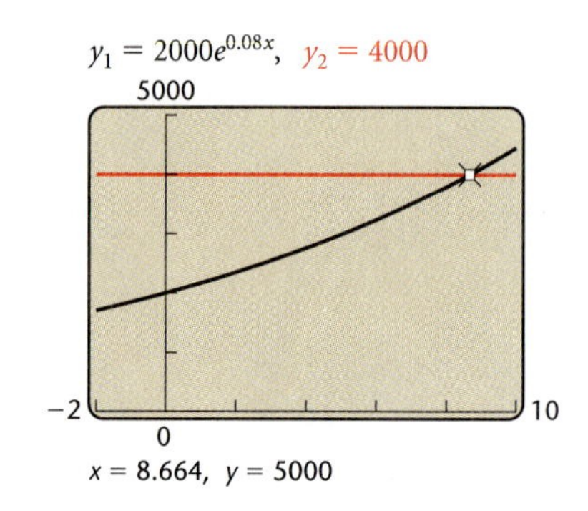

The solution is about 8.7.

We can find a general expression relating the growth rate k and the doubling time T by solving the following equation:

$2P_0 = P_0e^{kT}$ **Substituting $2P_0$ for P and T for t**

$2 = e^{kT}$ **Dividing by P_0**

$\ln 2 = \ln e^{kt}$ **Taking the natural logarithm**

$\ln 2 = kT$ **Using $\ln e^x = x$**

$\frac{\ln 2}{k} = T.$

Growth Rate and Doubling Time

The *growth rate* k and the *doubling time* T are related by

$$kT = \ln 2, \quad \text{or} \quad k = \frac{\ln 2}{T}, \quad \text{or} \quad T = \frac{\ln 2}{k}.$$

Note that the relationship between k and T does not depend on P_0.

Example 3 *World Population Growth.* The population of the world is now doubling every 24.8 yr. What is the exponential growth rate?

SOLUTION We have

$$k = \frac{\ln 2}{T} \approx \frac{0.693147}{24.8} \approx 2.8\%.$$

The growth rate of the world population is about 2.8% per year.

Models of Limited Growth

The model $P(t) = P_0e^{kt}$ has many applications involving what may seem to be unlimited population growth. There can be factors that prevent a population from exceeding some limiting value—perhaps a limitation on food, living space, or other natural resources. One model of such growth is

$$P(t) = \frac{a}{1 + be^{-kt}},$$

which is called a **logistic equation**. This function increases toward a *limiting value a* as $t \to \infty$.

Example 4 *Limited Population Growth.* A ship carrying 1000 passengers has the misfortune to be shipwrecked on a small island from which the passengers are never rescued. The natural resources of the island limit the population to 5780. The population gets closer and closer to this limiting value, but never reaches it. The population of the island after time t, in years, is given by the logistic equation

$$P(t) = \frac{5780}{1 + 4.78e^{-0.4t}}.$$

a) Find the population after 0, 1, 2, 5, 10, and 20 yr.

b) Graph the function.

SOLUTION

X	Y1	
0	1000	
1	1375	
2	1836	
5	3510	
10	5315	
20	5771	

X = 0

a) We use a grapher to find the function values, listing them in a table as shown at left. (We round to the nearest unit.) A grapher with a TABLE feature set in ASK mode can also be used. Thus the population will be 1000 after 0 yr, 1375 after 1 yr, 1836 after 2 yr, 3510 after 5 yr, 5315 after 10 yr, and 5771 after 20 yr.

b) We use a grapher to graph the function. The graph is the S-shaped curve shown below. Note that this function increases toward a limiting value of 5780.

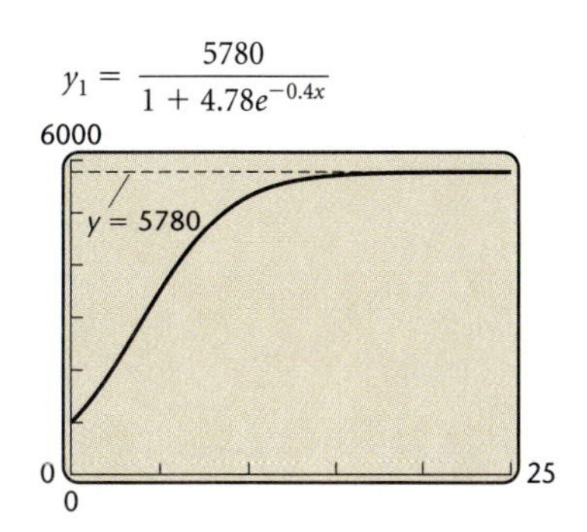

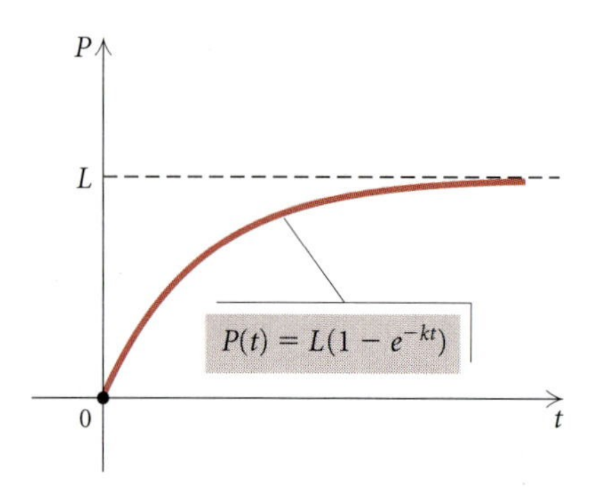

Another model of limited growth is provided by the function

$$P(t) = L(1 - e^{-kt}),$$

which is shown graphed at left. This function also increases toward a limiting value L, as $x \to \infty$.

Exponential Decay

The function

$$P(t) = P_0e^{-kt}$$

is an effective model of the decline, or decay, of a population. An example is the decay of a radioactive substance. In this case, P_0 is the amount of the substance at time $t = 0$, and $P(t)$ is the amount of the substance left after time t, where k is a positive constant that depends on the situation. The constant k is called the **decay rate**.

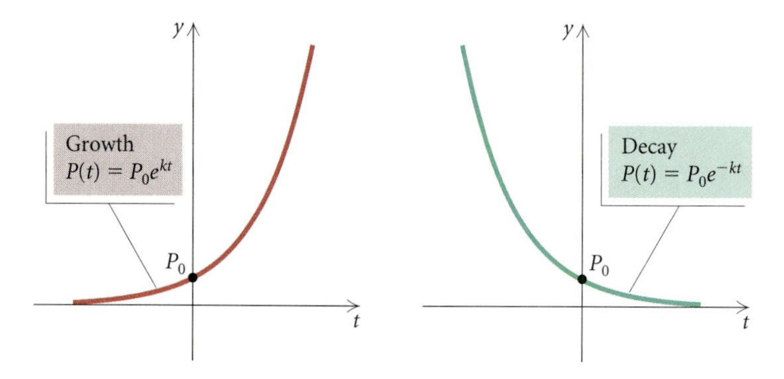

How can scientists determine that an animal bone has lost 30% of its carbon-14? The assumption is that the percentage of carbon-14 in the atmosphere and in living plants and animals is the same. When a plant or an animal dies, the amount of carbon-14 decays exponentially. The scientist burns the animal bone and uses a Geiger counter to determine the percentage of the smoke that is carbon-14. It is the amount that this varies from the percentage in the atmosphere that tells how much carbon-14 has been lost.

The process of carbon-14 dating was developed by the American chemist Willard E. Libby in 1952. It is known that the radioactivity in a living plant is 16 disintegrations per gram per minute. Since the half-life of carbon-14 is 5750 years, an object with an activity of 8 disintegrations per gram per minute is 5750 years old, one with an activity of 4 disintegrations per gram per minute is 11,500 years old, and so on. Carbon-14 dating can be used to measure the age of objects from 30,000 to 40,000 years old. Beyond such an age, it is too difficult to measure the radioactivity and some other method would have to be used.

Carbon-14 was indeed used to find the age of the Dead Sea Scrolls. It was used recently to refute the authenticity of the Shroud of Turin, presumed to have covered the body of Christ.

The **half-life** of bismuth is 5 days. This means that half of an amount of bismuth will cease to be radioactive in 5 days. The effect of half-life T is shown in the graph below for nonnegative inputs. The exponential function gets close to 0, but never reaches 0, as t gets very large. Thus, according to an exponential decay model, a radioactive substance never completely decays.

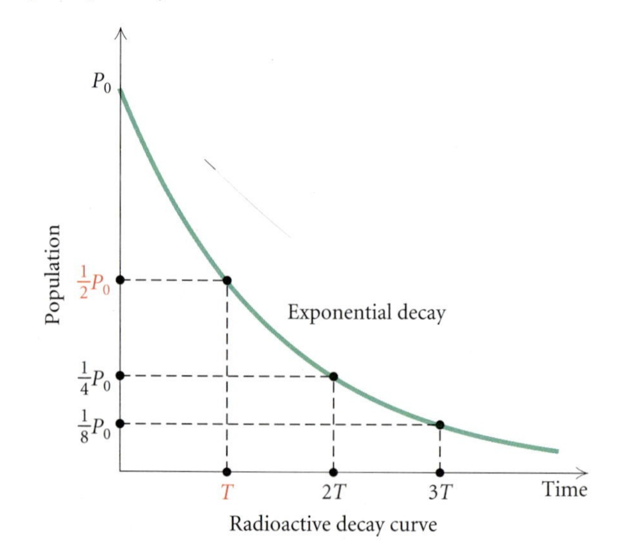

Radioactive decay curve

In 1947, a Bedouin youth looking for a stray goat climbed into a cave at Kirbet Qumran on the shores of the Dead Sea near Jericho and came upon earthenware jars containing an incalculable treasure of ancient manuscripts. Shown here are fragments of those so-called Dead Sea Scrolls, a portion of some 600 or so texts found so far and which concern the Jewish books of the Bible. Officials date them before 70 A.D., making them the oldest Biblical manuscripts by 1000 years.

Example 5 *Carbon Dating.* The radioactive element carbon-14 has a half-life of 5750 yr. The percentage of carbon-14 present in the remains of organic matter can be used to determine the age of that organic matter. Archeologists discovered that the linen wrapping from one of the Dead Sea Scrolls had lost 22.3% of its carbon-14 at the time it was found. How old was the linen wrapping?

SOLUTION We first find k. When $t = 5750$ (the half-life), $P(t)$ will be half of P_0. We substitute $\frac{1}{2}P_0$ for $P(t)$ and 5750 for t and solve for k. Then

$$\frac{1}{2}P_0 = P_0e^{-k(5750)}$$

or

$$\frac{1}{2} = e^{-5750k}.$$

We take the natural logarithm on both sides:

$$\ln \frac{1}{2} = \ln e^{-5750k}$$
$$= -5750k.$$

Then

$$k = \frac{\ln 0.5}{-5750}$$
$$\approx 0.00012.$$

We could also solve the equation $\frac{1}{2} = e^{-5750k}$ using a grapher. Now we have the function

$$P(t) = P_0e^{-0.00012t}.$$

(This equation can be used for any subsequent carbon-dating problem.) If the linen wrapping has lost 22.3% of its carbon-14 from an initial amount P_0, then $77.7\%P_0$ is the amount present. To find the age t of the wrapping, we solve the following equation for t:

$$77.7\%P_0 = P_0e^{-0.00012t} \quad \text{Substituting } 77.7\%P_0 \text{ for } P$$
$$0.777 = e^{-0.00012t}$$
$$\ln 0.777 = \ln e^{-0.00012t}$$
$$\ln 0.777 = -0.00012t \quad \ln e^x = x$$
$$\frac{\ln 0.777}{-0.00012} = t$$
$$2103 \approx t.$$

Thus the linen wrapping on the Dead Sea Scrolls is about 2103 yr old.

Exponential and Logarithmic Curve Fitting

We have added several new functions to our candidates for curve fitting. Let's review some of them.

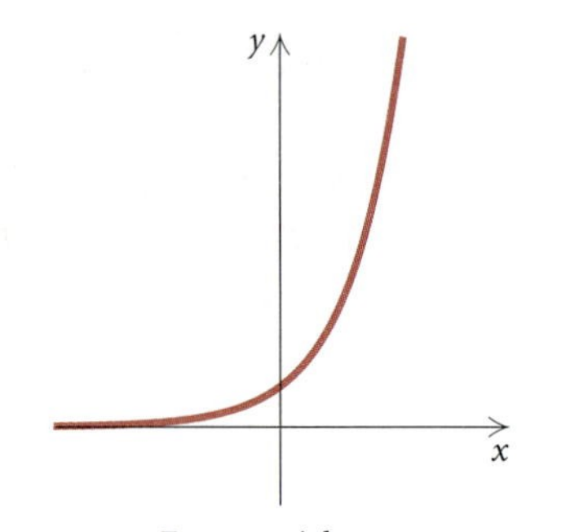

Exponential:
$f(x) = ab^x$, or ae^{kx}
$a, b > 0, k > 0$

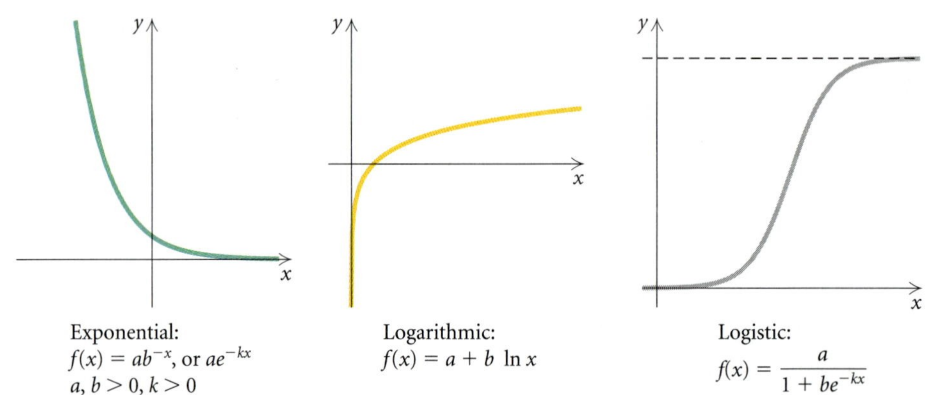

Exponential:
$f(x) = ab^{-x}$, or ae^{-kx}
$a, b > 0, k > 0$

Logarithmic:
$f(x) = a + b \ln x$

Logistic:
$$f(x) = \frac{a}{1 + be^{-kx}}$$

Now, when we analyze a set of data for curve fitting, these models can be considered as well as linear, quadratic, polynomial, and rational functions. Indeed, some graphers can use *regression* to fit an exponential, a logarithmic, or a logistic* equation to a set of data.

Example 6 *Cellular Phones.* The number of cellular phones in use has grown dramatically in recent years. Let's examine the following data.

x, YEAR	NUMBER OF CELLULAR PHONES, y (IN MILLIONS)
0. 1985	0.2
1. 1986	0.4
2. 1987	1.0
3. 1988	1.8
4. 1989	2.8
5. 1990	4.1
6. 1991	6.2
7. 1992	8.6
8. 1993	13.0
9. 1994	19.3

Source: Cellular Telecommunications Industry Association

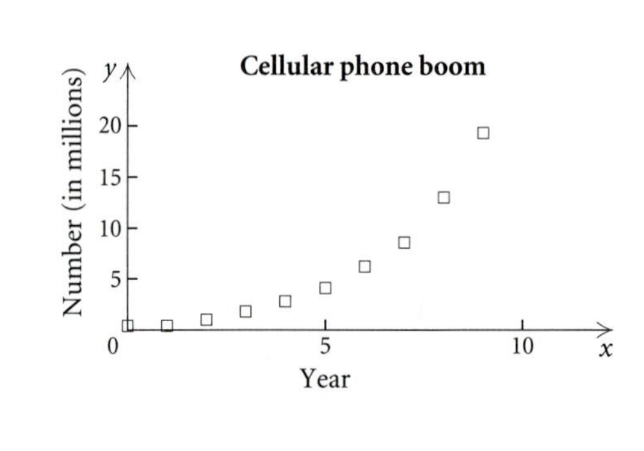

From the scatterplot on the right above, it would appear that we have exponential growth.

a) Use a grapher and regression to fit an exponential function to the data.

b) Graph the function.

c) Predict the number of cellular phones in use in the year 2000.

*The TI-83 is an example.

SOLUTION

a) We are fitting the data to an equation of the type $y = a \cdot b^x$. Entering the data into the grapher and carrying out the regression procedure, we find that

$$a = 0.3039703339,$$
$$b = 1.627328317,$$
$$r = 0.9860606791.$$

This tells us that the equation is

$$y = 0.3039703339(1.627328317)^x,$$

or about

$$y = 0.304(1.627)^x.$$

The *correlation coefficient* is very close to 1. This gives us a good indication that the data fit an exponential equation.

b) The graph is shown at left.

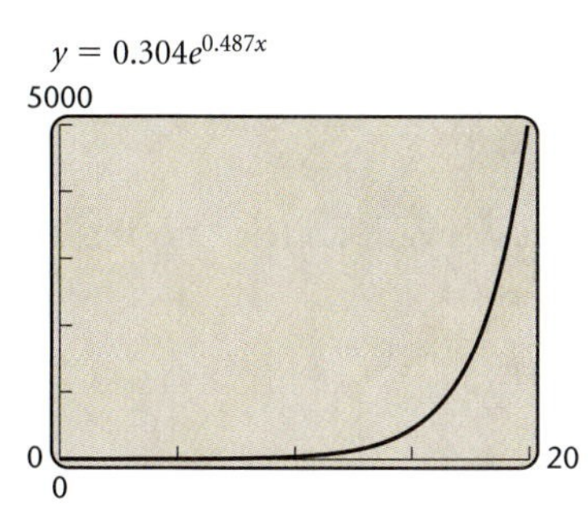

c) To predict the number of cellular phones in use in 2000, we substitute 15 for x in the exponential equation:

$$y = 0.304(1.627)^x$$
$$= 0.304(1.627)^{15} \approx 450.$$

Thus according to this model, there will be about 450 million cellular phones in use in 2000.

On some graphers, there may be a REGRESSION feature that yields an exponential function, base e. If not, and you wish to find such a function, a conversion can be done using the following.

Converting from Base b to Base e

$$b^x = e^{x(\ln b)}$$

Then, for the equation in Example 6, we have

$$y = 0.304(1.627)^x$$
$$= 0.304e^{x(\ln 1.627)} = 0.304e^{0.487x}.$$

We can prove this conversion formula using properties of logarithms, as follows:

$$e^{x(\ln b)} = e^{\ln b^x} = b^x.$$

Power Models

There are many situations in which so-called **power models** $y = ax^b$ can be fit to data using regression. Note that the constants to be determined are a and b. The base is the variable x.

Example 7 *Cholesterol Level and the Risk of Heart Attack.* The data in the following table show the relationship of cholesterol level in men to the risk of a heart attack.

CHOLESTEROL LEVEL, x	MEN, PER 10,000, WHO SUFFER A HEART ATTACK, y
100	30
200	65
250	100
275	130
300	180

Source: Nutrition Action Healthletter.

a) Use the REGRESSION feature on a grapher to find a power function to fit the data.

b) Graph the function.

c) Use the answer to part (b) to predict the heart attack rate for men with cholesterol levels of 350 and 400.

SOLUTION

a) We are fitting an equation of the type $y = ax^b$ to the data. Entering the data into the grapher and carrying out the regression procedure, we find that

$$a = 0.0241789574,$$
$$b = 1.527457172,$$
$$r = 0.9739361336.$$

This tells us that the equation is about $y = 0.024x^{1.527}$. The *correlation coefficient* of about 0.974 is very close to 1. This indicates that the data fit a power function fairly well, although an exponential function with $r = 0.996$ is a better fit. For illustrative purposes, though, we will continue with the power model.

b) The graph is shown at left.

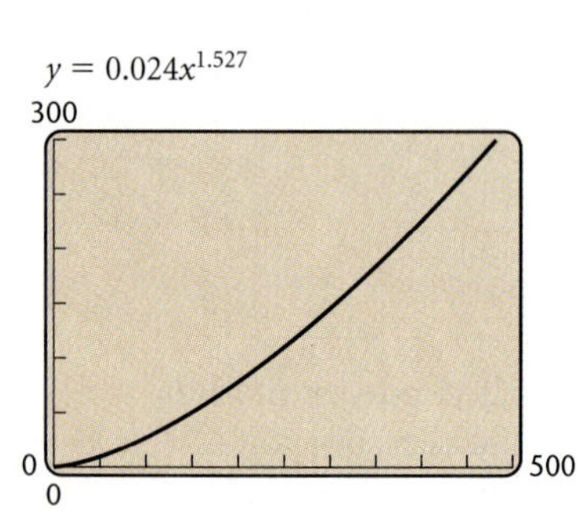

c) To predict the heart attack rate for men with cholesterol levels of 350 and 400, we substitute 350 and 400 for x in the power equation:

$$y = 0.024x^{1.527} = 0.024(350)^{1.527} \approx 184,$$
$$y = 0.024x^{1.527} = 0.024(400)^{1.527} \approx 226.$$

Thus the heart attack rate is about 184 out of 10,000 for men with a cholesterol level of 350 and about 226 out of 10,000 for men with a cholesterol level of 400.

3.6 Exercise Set

1. *World Population Growth.* In 1997, the world population was 5.8 billion. The exponential growth rate was 1.5% per year.

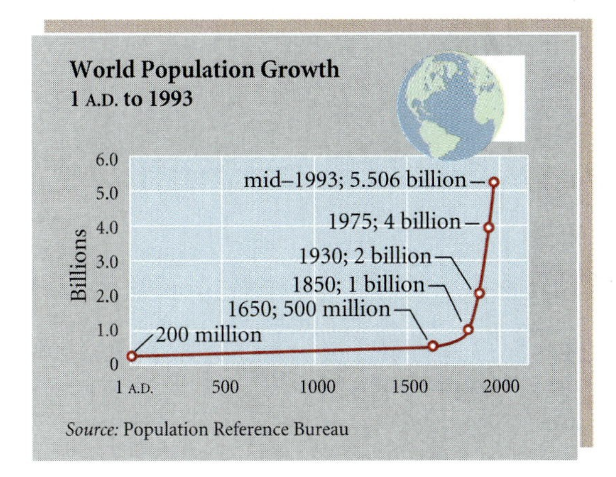

a) Find the exponential growth function.
b) Predict the population of the world in 2000 and in 2010.
c) When will the world population be 8 billion?
d) Find the doubling time.

2. *Population Growth of Rabbits.* Under ideal conditions, a population of rabbits has an exponential growth rate of 11.7% per day. Consider an initial population of 100 rabbits.

a) Find the exponential growth function.
b) What will the population be after 7 days?
c) Find the doubling time.
d) Graph the function.

3. *Population Growth.* Complete the following table.

POPULATION	GROWTH RATE, k	DOUBLING TIME, T
a) Mexico	3.5% per year	
b) Europe		69.31 yr
c) Oil reserves	10% per year	
d) Coal reserves	4% per year	
e) Alaska		24.8 yr
f) Central America		19.8 yr

4. *Female Olympic Athletes.* In 1985, the number of female athletes participating in Summer Olympic-Type Games was 500. In 1996, about 3600 participated in the Summer Olympics in Atlanta. Assuming the exponential model applies:

a) Find the value of k ($P_0 = 500$), and write the function.
b) Estimate the number of female athletes in the Summer Olympics of 2000.

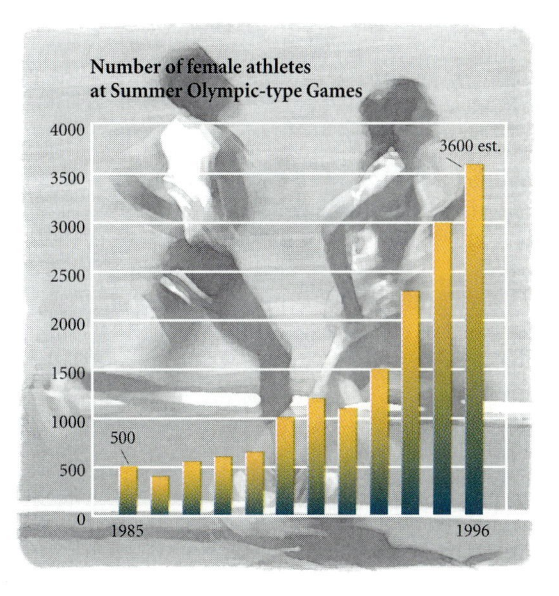

5. *Population Growth of the Virgin Islands.* The population of the U.S. Virgin Islands has a growth rate of 2.6% per year. In 1990, the population was 512,000. The land area of the Virgin Islands is 3,097,600 square yards. Assuming this growth rate continues and is exponential, after how long will there be one person for every square yard of land? (*Source:* Statistical Abstract of the United States)

6. *Value of Manhattan Island.* In 1626, Peter Minuit of the Dutch West India Company purchased Manhattan Island from the Indians for $24. Assuming an exponential rate of inflation of 8% per year, how much will Manhattan be worth in 2000?

7. *Interest Compounded Continuously.* Suppose $10,000 is invested at an interest rate of 5.4% per year, compounded continuously.

a) Find the exponential function that describes the amount in the account after time t, in years.
b) What is the balance after 1 yr? 2 yr? 5 yr? 10 yr?
c) What is the doubling time?

8. *Interest Compounded Continuously.* Complete the following table.

INITIAL INVESTMENT AT $t = 0$, P_0	INTEREST RATE, k	DOUBLING TIME, T	AMOUNT AFTER 5 YR
a) \$35,000	6.2%		
b) \$5000			\$ 7,130.90
c)	8.4%		\$11,414.71
d)		11 yr	\$17,539.32

9. *Carbon Dating.* A mummy discovered in the pyramid Khufu in Egypt has lost 46% of its carbon-14. Determine its age.

10. *Carbon Dating.* The statue of Zeus at Olympia in Greece is one of the Seven Wonders of the World. It is made of gold and ivory. The ivory was found to have lost 35% of its carbon-14. Determine the age of the statue.

11. *Radioactive Decay.* Complete the following table.

RADIOACTIVE SUBSTANCE	DECAY RATE, k	HALF-LIFE, T
a) Polonium		3 min
b) Lead		22 yr
c) Iodine-131	9.6% per day	
d) Krypton-85	6.3% per year	
e) Strontium-90		25 yr
f) Uranium-238		4560 yr
g) Plutonium		23,105 yr

12. *Decline of Long-Playing Records.* The sales S of long-playing records has declined considerably in the past 10 yr because of the emergence of the cassette tape and compact disc. In 1983, 205 million LP records were sold and in 1993, 1.2 million records were sold. (*Source:* Recording Industry Association of America) Assuming the sales are decreasing according to the exponential-decay model:

a) Find the value k, and write an exponential function that describes the number of long-playing records sold after time t, in years.

b) Estimate the sales of LP records in the year 2000.

c) In what year (theoretically) will only 1 long-playing record be sold?

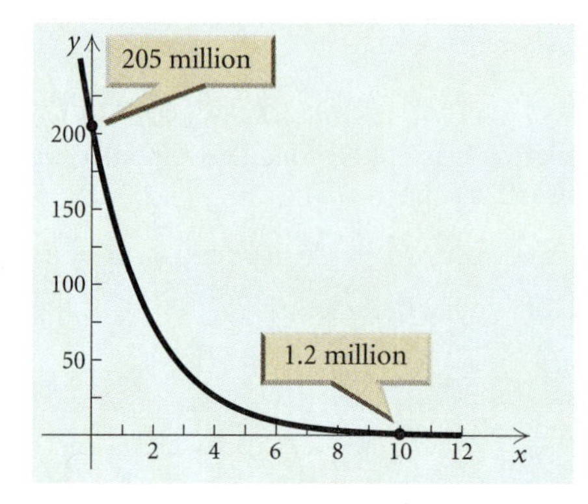

13. *Decline in Beef Consumption.* In 1985, the average annual consumption of beef B was about 80 lb per person. In 1996, it was about 67 lb per person. Assuming consumption is decreasing according to the exponential-decay model:

a) Find the value k, and write an equation that describes beef consumption after time t, in years.

b) Estimate the consumption of beef in the year 2000.

c) After how many years (theoretically) will the average annual consumption of beef be 20 lb per person?

14. *The Value of Eddie Murray's Baseball Card.* The collecting of baseball cards has become a popular hobby. The card shown here is a photograph of Eddie Murray in his rookie season of 1978.

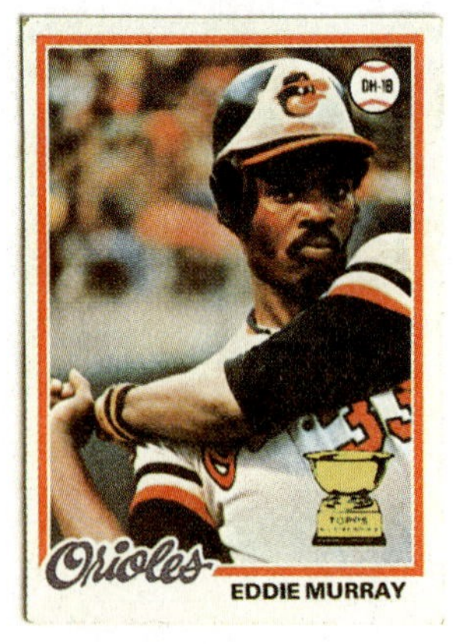

In 1983, the value of the card was \$7.75 and in 1987, its value was \$27.00. (*Source: Sport Collectors Digest*) Assuming that the value of the card has grown exponentially:

a) Find the value k and determine the exponential growth function, assuming $V_0 = 7.75$.

b) Estimate the value of the card in 1995 and in 2000. Check your answer for 1995 with a baseball card dealer.
c) What is the doubling time for the value of the card?
d) After how long will the value of the card be $2000?

15. *Spread of an Epidemic.* In a town whose population is 2000, a disease creates an epidemic. The number of people N infected t days after the disease has begun is given by the function

$$N(t) = \frac{2000}{1 + 19.9e^{-0.6t}}.$$

a) How many are initially infected with the disease ($t = 0$)?
b) Find the number infected after 2 days, 5 days, 8 days, 12 days, and 16 days.
c) Graph the function.

16. *Acceptance of Seat Belt Laws.* In recent years, many states have passed mandatory seat belt laws. The total number of states N that have passed a seat belt law t years after 1984 is given by the function

$$N(t) = \frac{50}{1 + 22e^{-0.6t}}.$$

(*Source:* National Highway Traffic Safety Administration)

a) How many states had passed the law in 1984? ($t = 0$ corresponds to 1984.)
b) Find the number of states that had passed the law by 1988, 1992, 1996, and 2002.
c) Graph the function.
d) If the function were to continue to be appropriate, would all 50 states ever pass the law? Explain.

17. *Limited Population Growth in a Lake.* A lake is stocked with 400 fish of a new variety. The size of the lake, the availability of food, and the number of other fish restrict growth in the lake to a *limiting value* of 2500. The population of fish in the lake after time t, in months, is given by the function

$$P(t) = \frac{2500}{1 + 5.25e^{-0.32t}}.$$

a) Find the population after 0, 1, 5, 10, 15, and 20 months.
b) Graph the function.

For each of the following scatterplots, determine which, if any, of these functions might be used as a model for the data.

a) *Quadratic,* $f(x) = ax^2 + bx + c$
b) *Polynomial, not quadratic*
c) *Exponential,* $f(x) = ab^x$, *or* Be^{kx}, $k > 0$
d) *Exponential,* $f(x) = ab^x$, *or* Be^{-kx}, $k > 0$
e) *Logarithmic,* $f(x) = a + b \ln x$
f) *Logistic,* $f(x) = \dfrac{a}{1 + be^{-kx}}$

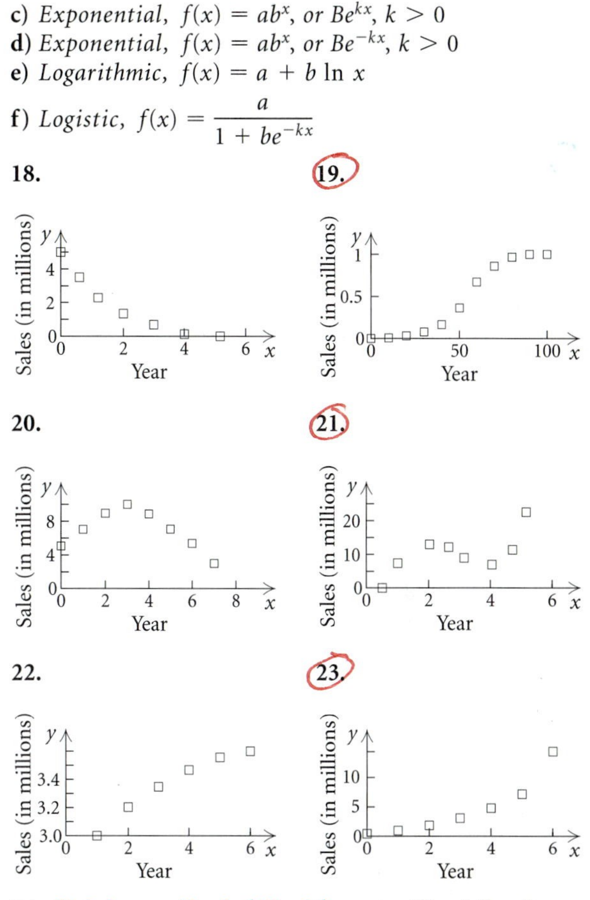

24. *Opinion on Capital Punishment.* The following table contains factual data from a recent survey of college freshmen regarding their opinion of capital punishment.

x, YEAR	PERCENTAGE OF COLLEGE FRESHMEN AGREEING THAT CAPITAL PUNISHMENT SHOULD BE ABOLISHED, y
0. 1971	56
8. 1979	32
13. 1984	24
18. 1989	21
23. 1994	18

Source: UCLA Higher Education Research Institute.

Using a grapher:

a) Create a scatterplot of the data. Determine whether an exponential function appears to fit the data.
b) Use regression to fit an exponential function $y = ab^x$, or ae^{-kx}, to the data, where $x =$ the number of years after 1971.

c) Use the function to predict the percent of college freshmen in 2000 agreeing that capital punishment should be abolished.
d) In what year would only 1% agree that capital punishment should be abolished?

25. *Total Revenue of Microsoft, Inc.*

x, YEAR	TOTAL REVENUE, y (IN BILLIONS)
0. 1989	$0.8
1. 1990	1.1
2. 1991	1.8
3. 1992	2.7
4. 1993	3.75

Source: Microsoft Annual Report.

Using a grapher:

a) Create a scatterplot of the data. Determine whether the data seem to fit an exponential function.
b) Use regression to fit an exponential function $y = ab^x$, or ae^{kx}, to the data, where $x =$ the number of years after 1989.
c) Use the function to predict the total revenue of Microsoft, Inc., in 2000.
d) In what year would the revenue be $20 billion?

26. *Forgetting.* In an art class, students were tested at the end of the course on a final exam. Then they were retested with an equivalent test at subsequent time intervals. Their scores after time t, in months, are given in the following table.

TIME, t (IN MONTHS)	SCORE, y
1	84.9%
2	84.6%
3	84.4%
4	84.2%
5	84.1%
6	83.9%

Using a grapher:

a) Use regression to fit a logarithmic function $y = a + b \ln x$ to the data.
b) Use the function to predict test scores after 8, 10, 24, and 36 months.
c) After how long will the test scores fall below 82%?

27. *Video Rentals.* Video rental spending has been on the increase. The data in the following table shows the average amount of video rental spending per family in recent years.

x, YEAR	VIDEO RENTAL SPENDING PER FAMILY, y
1. 1991	$113
2. 1992	114
3. 1993	119
4. 1994	122
5. 1995	129

Source: Media Group Research.

a) Use the REGRESSION feature on a grapher to fit both an exponential and a power function to the data, where $x = 1$ corresponds to 1991.
b) Graph each function.
c) Use each function in part (a) to predict the video rental spending in 2005.
d) Discuss the relative merits of using each function as a predictor.

28. *Effect of Advertising.* A company introduces a new software product on a trial run in a city. They advertised the product on television and found the following data relating the percentage P of people who bought the product after x ads were run.

NUMBER OF ADS, x	PERCENTAGE WHO BOUGHT, P
0	0.2
10	0.7
20	2.7
30	9.2
40	27
50	57.6
60	83.3
70	94.8
80	98.5
90	99.6

a) Use the REGRESSION feature on a grapher to fit a logistic function

$$P(x) = \frac{a}{1 + be^{-kx}}$$

to the data.
b) What percent will buy the product when 55 ads are run? 100 ads?
c) Find an asymptote for the graph. Interpret the asymptote in terms of the advertising situation.

Skill Maintenance

Find the missing lengths in each right triangle.

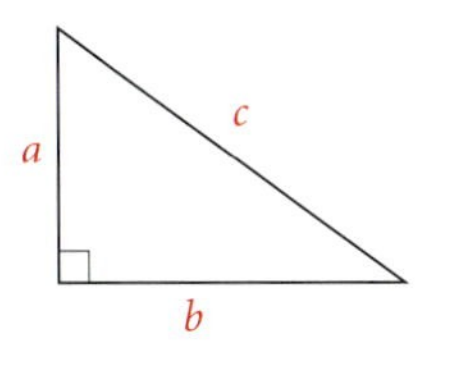

29. $b = 1,\ c = 2$

30. $a = 1,\ b = 1$

31. $a = 47,\ b = 34$

32. $b = \sqrt{13},\ c = 200$

Synthesis

33. ◈ Browse through some newspapers or magazines until you find some data and/or a graph that seem as though they can be fit to an exponential function. Make a case for why such a fit is appropriate. Then fit an exponential function to the data and make some predictions.

34. ◈ *Atmospheric Pressure.* Atmospheric pressure P at an altitude a is given by

$$P = P_0 e^{-0.00005a},$$

where P_0 = the pressure at sea level ≈ 14.7 lb/in^2 (pounds per square inch). Explain how a barometer, or some device for measuring atmospheric pressure, can be used to find the height of a skyscraper.

35. *Present Value.* Following the birth of a child, a parent wants to make an initial investment P_0 that will grow to \$50,000 for the child's education at age 18. Interest is compounded continuously at 7%. What should the initial investment be? Such an amount is called the **present value** of \$50,000 due 18 yr from now.

36. *Present Value.* Referring to Exercise 35:

a) Solve $P = P_0 e^{kt}$ for P_0.

b) Find the present value of \$50,000 due 18 yr from now at interest rate 6.4%.

37. *Supply and Demand.* The supply and demand for the sale of a certain type of VCR are given by

$$S(p) = 480e^{-0.003p} \quad \text{and} \quad D(p) = 150e^{0.004p},$$

where $S(p)$ = the number of VCRs that the company is willing to sell at price p and $D(p)$ = the quantity that the public is willing to buy at price p. Find p, called the **equilibrium price**, such that $D(p) = S(p)$.

38. *Carbon Dating.* Recently, while digging in Chaco Canyon, New Mexico, archeologists found corn pollen that was 4000 yr old. This was evidence that Native Americans had been cultivating crops in the Southwest centuries earlier than scientists had thought. What percent of the carbon-14 had been lost from the pollen? (*Source: American Anthropologist*)

39. *Newton's Law of Cooling.* Suppose a body with temperature T_1 is placed in surroundings with temperature T_0 different from that of T_1. The body will either cool or warm to temperature $T(t)$ after time t, in minutes, where

$$T(t) = T_0 + |T_1 - T_0|e^{-kt}.$$

A cup of coffee with temperature 105°F is placed in a freezer with temperature 0°F. After 5 min, the temperature of the coffee is 70°F. What will its temperature be after 10 min?

40. *When Was the Murder Committed?* The police discover the body of a math professor. Critical to solving the crime is determining when the murder was committed. The coroner arrives at the murder scene at 12:00 P.M. She immediately takes the temperature of the body and finds it to be 94.6°. She then takes the temperature 1 hr later and finds it to be 93.4°. The temperature of the room is 70°. When was the murder committed? (Use Newton's law of cooling in Exercise 39.)

41. *Electricity.* The formula

$$i = \frac{V}{R}\left[1 - e^{-(R/L)t}\right]$$

occurs in the theory of electricity. Solve for t.

42. *The Beer–Lambert Law.* A beam of light enters a medium such as water or smog with initial intensity I_0. Its intensity decreases depending on the thickness (or concentration) of the medium. The intensity I at a depth (or concentration) of x units is given by

$$I = I_0 e^{-\mu x}.$$

The constant μ (the Greek letter "mu") is called the **coefficient of absorption**, and it varies with the medium. For sea water, $\mu = 1.4$.

a) What percentage of light intensity I_0 remains at a depth of sea water that is 1 m? 3 m? 5 m? 50 m?

b) Plant life cannot exist below 10 m. What percentage of I_0 remains at 10 m?

43. Given that $y = ae^x$, take the natural logarithm on both sides. Let $Y = \ln y$. Consider Y as a function of x. What kind of function is Y?

44. Given that $y = ax^b$, take the natural logarithm on both sides. Let $Y = \ln y$ and $X = \ln x$. Consider Y as a function of X. What kind of function is Y?

CHAPTER 3 Summary and Review

Important Properties and Formulas

One-to-One Function:	$f(a) = f(b) \longrightarrow a = b$
Exponential Function:	$f(x) = a^x$
The Number $e = 2.7182818284\ldots$	
Logarithmic Function:	$f(x) = \log_a x$
A Logarithm is an Exponent:	$\log_a x = y \longleftrightarrow x = a^y$
The Change-of-Base Formula:	$\log_b M = \dfrac{\log_a M}{\log_a b}$
The Product Rule:	$\log_a MN = \log_a M + \log_a N$
The Power Rule:	$\log_a M^p = p \log_a M$
The Quotient Rule:	$\log_a \dfrac{M}{N} = \log_a M - \log_a N$
Other Properties:	$\log_a a = 1, \quad \log_a 1 = 0,$
	$\log_a a^x = x, \quad a^{\log_a x} = x$
Base–Exponent Property:	$a^x = a^y \longleftrightarrow x = y$
Exponential Growth Model:	$P(t) = P_0 e^{kt}$
Exponential Decay Model:	$P(t) = P_0 e^{-kt}$
Interest Compounded Continuously:	$P(t) = P_0 e^{kt}$
Limited Growth:	$P(t) = \dfrac{a}{1 + be^{-kt}}$

REVIEW EXERCISES

1. Find the inverse of the relation
$\{(1.3, -2.7), (8, -3), (-5, 3), (6, -3), (7, -5)\}$.

2. Find an equation of the inverse relation.

a) $y = 3x^2 + 2x - 1$
b) $0.8x^3 - 5.4y^2 = 3x$

Given each function:

a) *Determine whether it is one-to-one, using a grapher if desired.*
b) *If it is one-to-one, find a formula for the inverse.*

3. $f(x) = \sqrt{x - 6}$ **4.** $f(x) = x^3 - 8$

5. $f(x) = 3x^2 + 2x - 1$ **6.** $f(x) = e^x$

7. Find $f(f^{-1}(657))$: $f(x) = \dfrac{4x^5 - 16x^{37}}{119x}$, $x > 0$.

In Exercises 8–13, match the equation with one of figures (a)–(f), which follow. If needed, use a grapher.

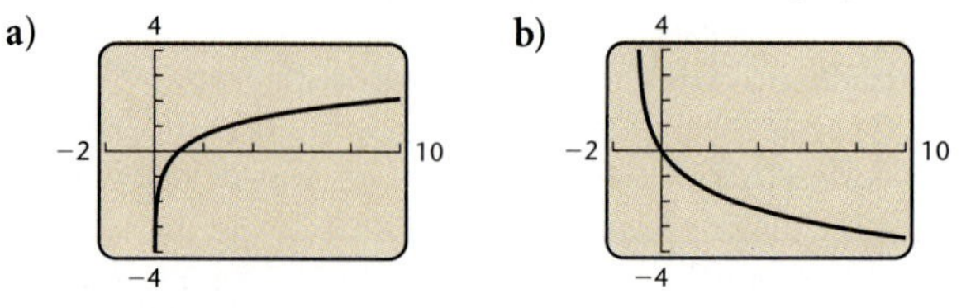

c) d)

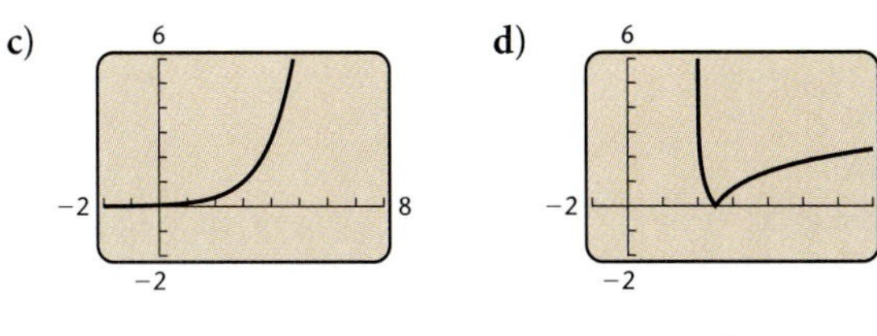

e)

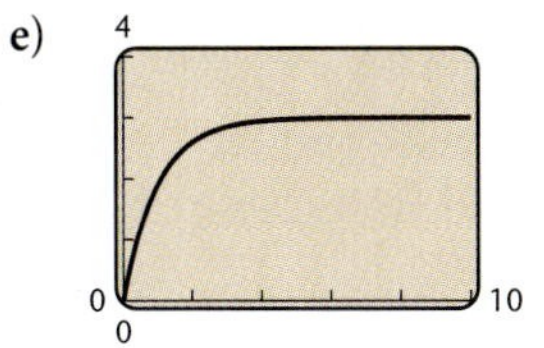

f)

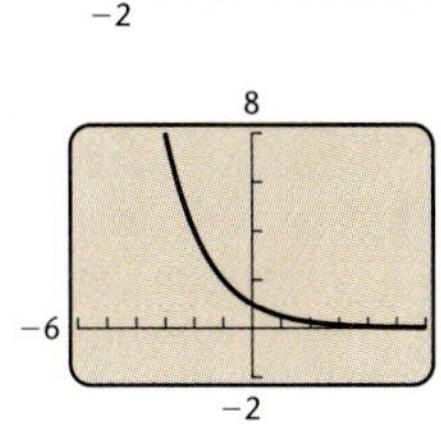

8. $f(x) = e^{x-3}$

9. $f(x) = \log_3 x$

10. $y = -\log_3 (x + 1)$

11. $y = \left(\frac{1}{2}\right)^x$

12. $f(x) = 3(1 - e^{-x}), \ x \geq 0$

13. $f(x) = |\ln (x - 4)|$

14. Convert to an exponential equation:

$$\log_4 x = 2.$$

15. Convert to a logarithmic equation:

$$e^x = 80.$$

Solve. Use any method.

16. $\log_4 x = 2$

17. $3^{1-x} = 9^{2x}$

18. $e^x = 80$

19. $4^{2x-1} - 3 = 61$

20. $\log_{16} 4 = x$

21. $\log_x 125 = 3$

22. $\log_2 x + \log_2 (x - 2) = 3$

23. $\log (x^2 - 1) - \log (x - 1) = 1$

24. $\log x^2 = \log x$

25. $e^{-x} = 0.02$

Express as a single logarithm and simplify if possible.

26. $3 \log_b x - 4 \log_b y + \frac{1}{2} \log_b z$

27. $\ln (x^3 - 8) - \ln (x^2 + 2x + 4) + \ln (x + 2)$

Express in terms of sums and differences of logarithms.

28. $\ln \sqrt[4]{wr^3}$

29. $\log \sqrt[3]{\dfrac{M^2}{N}}$

Given that $\log_a 2 = 0.301$, $\log_a 5 = 0.699$, *and* $\log_a 6 = 0.778$, *find each of the following.*

30. $\log_a 3$

31. $\log_a 50$

32. $\log_a \frac{1}{5}$

33. $\log_a \sqrt[3]{5}$

Simplify.

34. $\ln e^{-5k}$

35. $\log_5 5^{-6t}$

36. How long will it take an investment to double itself if it is invested at 5.4%, compounded continuously?

37. The population of a city doubled in 30 yr. What was the exponential growth rate?

38. How old is a skeleton that has lost 27% of its carbon-14?

39. The hydrogen ion concentration of milk is 2.3×10^{-6}. What is the pH?

40. What is the loudness, in decibels, of a sound whose intensity is $1000I_0$?

41. *The Population of Brazil.* The population of Brazil was 52 million in 1959, and the exponential growth rate was 2.8% per year. (*Source:* U.S. Bureau of the Census, World Population Profile)

a) Find the exponential growth function.
b) What will the population be in 2000? in 2020?
c) When will the population be 300 million?
d) What was the doubling time?

42. *Toll-free 800 Numbers.* The use of toll-free 800 numbers has grown exponentially. In 1967, there were 7 million such calls and in 1991, there were 10.2 billion such calls. (*Source:* Federal Communication Commission)

a) Find the exponential growth rate k.
b) Find the exponential growth function.
c) Graph the exponential growth function.
d) How many toll-free 800 number calls will be placed in 1998? in 2005?
e) In what year will 20 billion such calls be placed?

43. *Walking Speed.* The average walking speed w, in feet per second, of a person living in a city of population P, in thousands, is given by the function

$$w(P) = 0.37 \ln P + 0.05.$$

a) The population of Austin, Texas, is 466,000. Find the average walking speed.
b) A city's population has an average walking speed of 3.4 ft/sec. Find the population.

44. *Multimedia Personal Computers.* The following table contains estimated data regarding the number of multimedia personal computers installed in homes in various years.

x, YEAR	MULTIMEDIA PERSONAL COMPUTERS, y (IN MILLIONS)
0. 1992	2
1. 1993	4.0
2. 1994	8.0
3. 1995	16.0
4. 1996	22.5
5. 1997	31.4

Source: Piper Joffray Research.

Using a grapher:

a) Create a scatterplot of the data. Determine whether the data seem to fit an exponential function.

b) Use regression to fit an exponential function $y = ab^x$, or ae^{kx}, to the data, where $x =$ the number of years after 1992.

c) Use the function to predict the number of multimedia personal computers in homes in 2000.

d) In how many years will there be 100 million such computers in homes?

Synthesis

45. ◆ Suppose that you were trying to convince a fellow student that

$$\log_2 (x + 5) \neq \log_2 x + \log_2 5.$$

Give as many explanations as you can.

46. ◆ Describe the difference between $f^{-1}(x)$ and $[f(x)]^{-1}$.

Solve.

47. $|\log_4 x| = 3$

48. $\log x = \ln x$

49. $5^{\sqrt{x}} = 625$

50. a) Use a grapher to graph $f(x) = 5e^{-x} \ln x$ in the viewing window $[-1, 10, -5, 5]$.

b) Estimate the relative maximum and the minimum values. Use a MAX–MIN feature if such exists on your grapher.

51. Use only a grapher. Determine whether the following functions are inverses of each other:

$$f(x) = \frac{4 + 3x}{x - 2}, \qquad g(x) = \frac{x + 4}{x - 3}.$$

52. Find the domain: $f(x) = \log_3 (\ln x)$.

53. Find the points of intersection of the graphs of the following equations:

$$y = 5x^2e^{-x}, \qquad y = 2 - e^{-x^2}.$$

The Trigonometric Functions 4

APPLICATION

Musical instruments can generate complex sine waves. Consider two tones whose graphs are $y = 2 \sin x$ and $y = \sin 2x$. The combination of the two tones produces a new sound whose graph is $y = 2 \sin x + \sin 2x$.

X	Y1	Y2
0	0	0
.7854	1.4142	1
1.5708	2	0
2.3562	1.4142	−1
3.1416	0	0
3.927	−1.414	1
4.7124	−2	0

X = 3.1416

$y_1 = 2 \sin x$, $y_2 = \sin 2x$,
$y_3 = 2 \sin x + \sin 2x$

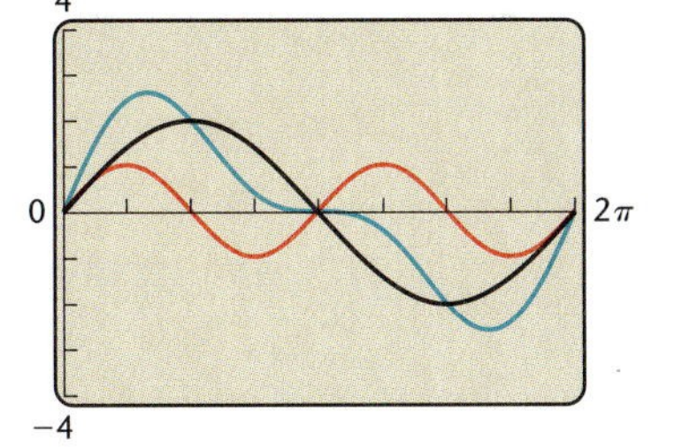

We now consider an important class of functions called *trigonometric*, or *circular*, *functions*. Historically, these functions arose from a study of triangles; hence the name trigonometric. We will begin our study with right triangles and degree measure and solve applications involving right triangles. Then we will consider trigonometric functions of angles or rotations of any size with both degree and radian measure. A circle of radius 1 (a unit circle) is then used to define the six basic trigonometric functions; hence the name circular functions. The domains and ranges of these functions consist of real numbers.

4.1 Trigonometric Functions of Acute Angles
4.2 Applications of Right Triangles
4.3 Trigonometric Functions of Any Angle
4.4 Radians, Arc Length, and Angular Speed
4.5 Circular Functions: Graphs and Properties
4.6 Graphs of Transformed Sine and Cosine Functions
SUMMARY AND REVIEW

4.1 Trigonometric Functions of Acute Angles

- *Determine the six trigonometric ratios for a given acute angle of a right triangle.*
- *Determine the trigonometric function values of 30°, 45°, and 60°.*
- *Using a grapher, find function values for any acute angle, and given a function value of an acute angle, find the angle.*
- *Given the function values of an acute angle, find the function values of its complement.*

The Trigonometric Ratios

We begin our study of trigonometry by considering right triangles and acute angles measured in degrees. Recall that an **acute angle** is an angle that measures between 0° and 90°. Greek letters such as α (alpha), β (beta), γ (gamma), θ (theta), and ϕ (phi) are used to denote an angle. Consider a right triangle with one of its acute angles labeled θ. The side opposite the right angle is called the **hypotenuse**. The other sides of the triangle are referenced by their position relative to the acute angle θ. One side is opposite θ and one is adjacent to θ.

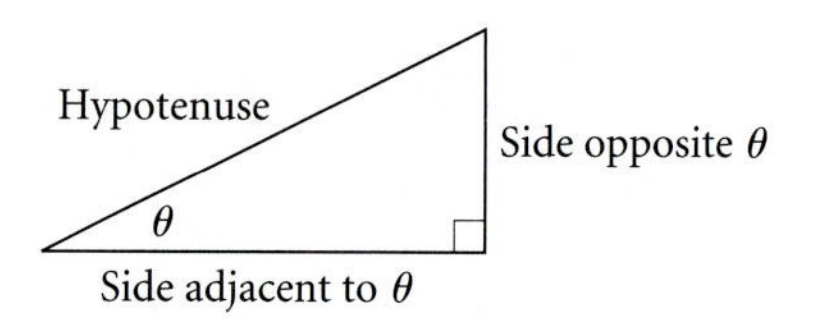

The *lengths* of the sides of the triangle are used to define the six trigonometric ratios:

sine (sin),	cosecant (csc),
cosine (cos),	secant (sec),
tangent (tan),	cotangent (cot).

The **sine of θ** is the *length* of the side opposite θ divided by the *length* of the hypotenuse:

$$\sin\theta = \frac{\text{length of side opposite } \theta}{\text{length of hypotenuse}}.$$

The ratio depends on angle θ and thus is a function of θ. The notation $\sin\theta$ actually means $\sin(\theta)$, where sin, or sine, is the name of the function.

The **cosine of θ** is the *length* of the side adjacent to θ divided by the *length* of the hypotenuse:

$$\cos\theta = \frac{\text{length of side adjacent to } \theta}{\text{length of hypotenuse}}.$$

The six trigonometric ratios, or trigonometric functions, are defined as follows.

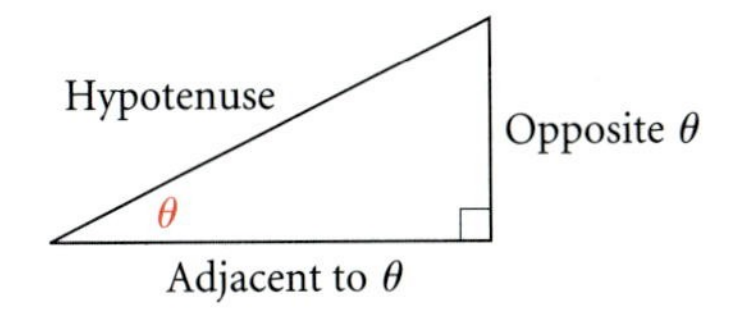

Trigonometric Function Values of an Acute Angle θ

Let θ be an acute angle of a right triangle. Then the six trigonometric functions of θ are as follows:

$$\sin\theta = \frac{\text{side opposite } \theta}{\text{hypotenuse}}, \qquad \csc\theta = \frac{\text{hypotenuse}}{\text{side opposite } \theta},$$

$$\cos\theta = \frac{\text{side adjacent to } \theta}{\text{hypotenuse}}, \qquad \sec\theta = \frac{\text{hypotenuse}}{\text{side adjacent to } \theta},$$

$$\tan\theta = \frac{\text{side opposite } \theta}{\text{side adjacent to } \theta}, \qquad \cot\theta = \frac{\text{side adjacent to } \theta}{\text{side opposite } \theta}.$$

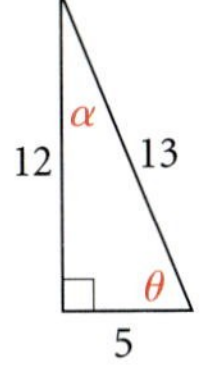

Example 1 In the right triangle shown at left, find the six trigonometric function values of (a) θ and (b) α.

SOLUTION We use the definitions.

a) $\sin\theta = \dfrac{\text{opp}}{\text{hyp}} = \dfrac{12}{13}$, $\quad \csc\theta = \dfrac{\text{hyp}}{\text{opp}} = \dfrac{13}{12}$,

$\cos\theta = \dfrac{\text{adj}}{\text{hyp}} = \dfrac{5}{13}$, $\quad \sec\theta = \dfrac{\text{hyp}}{\text{adj}} = \dfrac{13}{5}$,

$\tan\theta = \dfrac{\text{opp}}{\text{adj}} = \dfrac{12}{5}$, $\quad \cot\theta = \dfrac{\text{adj}}{\text{opp}} = \dfrac{5}{12}$

The references to opposite, adjacent, and hypotenuse are relative to θ.

b) $\sin\alpha = \dfrac{\text{opp}}{\text{hyp}} = \dfrac{5}{13}$, $\quad \csc\alpha = \dfrac{\text{hyp}}{\text{opp}} = \dfrac{13}{5}$,

$\cos\alpha = \dfrac{\text{adj}}{\text{hyp}} = \dfrac{12}{13}$, $\quad \sec\alpha = \dfrac{\text{hyp}}{\text{adj}} = \dfrac{13}{12}$,

$\tan\alpha = \dfrac{\text{opp}}{\text{adj}} = \dfrac{5}{12}$, $\quad \cot\alpha = \dfrac{\text{adj}}{\text{opp}} = \dfrac{12}{5}$

The references to opposite, adjacent, and hypotenuse are relative to α.

In Example 1(a), we note that the value of $\sin\theta$, $\frac{12}{13}$, is the reciprocal of $\frac{13}{12}$, the value of $\csc\theta$. Likewise, we see the same reciprocal relationship between the values of $\cos\theta$ and $\sec\theta$ and between the values of $\tan\theta$ and $\cot\theta$. For any angle, the cosecant, secant, and cotangent values are the reciprocals of the sine, cosine, and tangent function values, respectively.

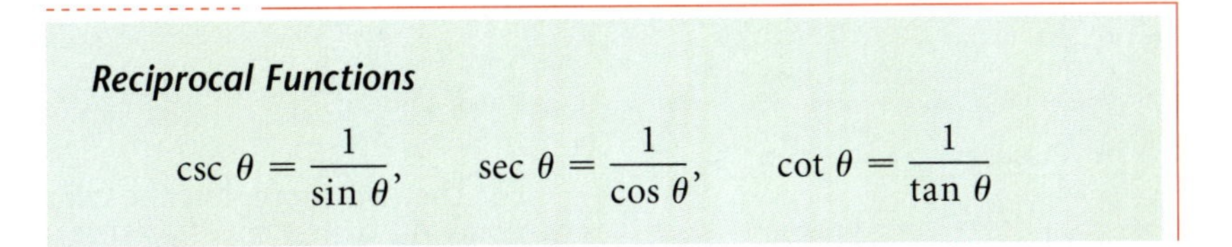

Reciprocal Functions

$$\csc\theta = \frac{1}{\sin\theta}, \qquad \sec\theta = \frac{1}{\cos\theta}, \qquad \cot\theta = \frac{1}{\tan\theta}$$

If we know the values of the sine, cosine, and tangent functions of an angle, we can use these reciprocal relationships to find the values of the cosecant, secant, and cotangent functions of that angle.

Example 2 Given that $\sin \phi = \frac{4}{5}$, $\cos \phi = \frac{3}{5}$, and $\tan \phi = \frac{4}{3}$, find $\csc \phi$, $\sec \phi$, and $\tan \phi$.

SOLUTION Using the reciprocal relationships, we have

$$\csc \phi = \frac{1}{\sin \phi} = \frac{1}{\frac{4}{5}} = \frac{5}{4}, \qquad \sec \phi = \frac{1}{\cos \phi} = \frac{1}{\frac{3}{5}} = \frac{5}{3},$$

and

$$\cot \phi = \frac{1}{\tan \phi} = \frac{1}{\frac{4}{3}} = \frac{3}{4}.$$

Triangles are said to be **similar** if their corresponding angles have the *same* measure. In similar triangles, the lengths of corresponding sides are in the same ratio. The following right triangles are similar—note that the corresponding angles are equal and the length of each side of the second triangle is four times the length of the corresponding side of the first triangle.

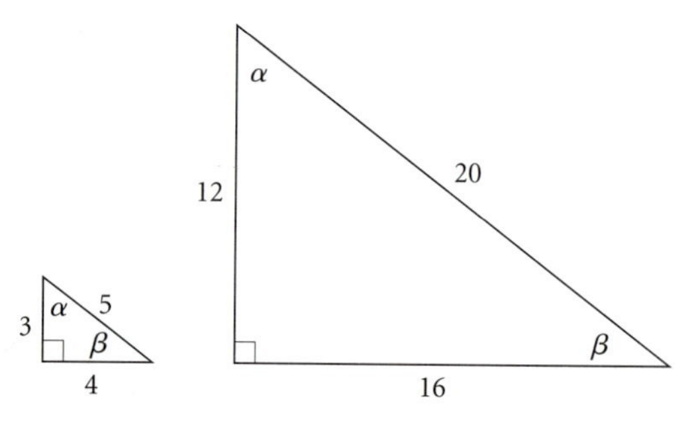

Let's observe the sine, cosine, and tangent values of β in each triangle. Can we expect these values to be the same?

FIRST TRIANGLE	SECOND TRIANGLE
$\sin \beta = \frac{3}{5}$	$\sin \beta = \frac{12}{20} = \frac{3}{5}$
$\cos \beta = \frac{4}{5}$	$\cos \beta = \frac{16}{20} = \frac{4}{5}$
$\tan \beta = \frac{3}{4}$	$\tan \beta = \frac{12}{16} = \frac{3}{4}$

For the two triangles, the corresponding values of $\sin \beta$, $\cos \beta$, and $\tan \beta$ are the same. The lengths of the sides are proportional—thus the *ratios* are the same. This must be the case because in order for the sine, cosine, and tangent to be functions, there must be only one output (the ratio) for each input (the angle β).

> The trigonometric function values of θ depend only on the size of the angle, not on the size of the triangle.

The Six Functions Related

We can find all six trigonometric function values of an acute angle when one of the function-value ratios is known.

Example 3 If $\sin \beta = \frac{6}{7}$ and β is an acute angle, find the other five trigonometric function values of β.

SOLUTION We know from the definition of the sine function that the ratio

$$\frac{6}{7} \text{ is } \frac{\text{opp}}{\text{hyp}}.$$

Using this information, let's consider a right triangle in which the hypotenuse has length 7 and the side opposite β has length 6. To find the length of the side adjacent to β, we recall the *Pythagorean theorem*:

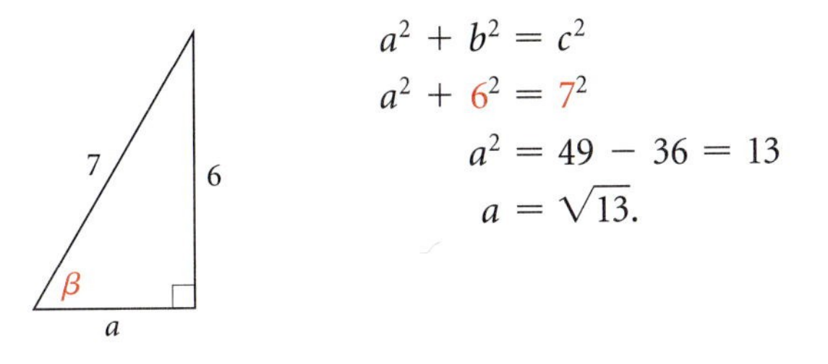

$$a^2 + b^2 = c^2$$
$$a^2 + 6^2 = 7^2$$
$$a^2 = 49 - 36 = 13$$
$$a = \sqrt{13}.$$

We now use the lengths of the three sides to find the other five ratios:

$$\sin \beta = \frac{6}{7}, \qquad \csc \beta = \frac{7}{6},$$

$$\cos \beta = \frac{\sqrt{13}}{7}, \qquad \sec \beta = \frac{7}{\sqrt{13}}, \text{ or } \frac{7\sqrt{13}}{13},$$

$$\tan \beta = \frac{6}{\sqrt{13}}, \text{ or } \frac{6\sqrt{13}}{13}, \qquad \cot \beta = \frac{\sqrt{13}}{6}.$$

Function Values of 30°, 45°, and 60°

In Examples 1 and 3, we found the trigonometric function values of an acute angle of a right triangle when the lengths of the three sides were known. In most situations, we are asked to find the function values when the measure of the acute angle is given. For certain special angles such as 30°, 45°, and 60°, which are frequently seen in applications, we can use geometry to determine the function values.

A right triangle with a 45° angle actually has two 45° angles. Thus the triangle is *isosceles*, and the legs are the same length. Let's consider such a triangle whose legs have length 1. Then if its hypotenuse has length c, we can find that length using the Pythagorean theorem as follows:

$$1^2 + 1^2 = c^2, \quad \text{or} \quad c^2 = 2, \quad \text{or} \quad c = \sqrt{2}.$$

Such a triangle is shown below. From this diagram, we can easily determine the trigonometric function values of 45°.

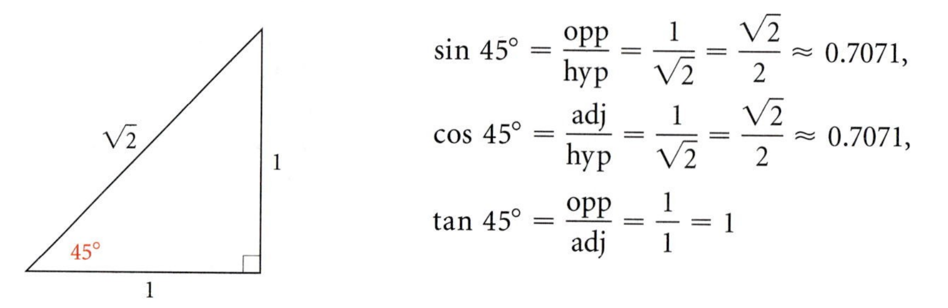

$$\sin 45^\circ = \frac{\text{opp}}{\text{hyp}} = \frac{1}{\sqrt{2}} = \frac{\sqrt{2}}{2} \approx 0.7071,$$

$$\cos 45^\circ = \frac{\text{adj}}{\text{hyp}} = \frac{1}{\sqrt{2}} = \frac{\sqrt{2}}{2} \approx 0.7071,$$

$$\tan 45^\circ = \frac{\text{opp}}{\text{adj}} = \frac{1}{1} = 1$$

It is sufficient to find only the function values of the sine, cosine, and tangent, since the others are their reciprocals.

It is also possible to determine the function values of 30° and 60°. A right triangle with 30° and 60° acute angles is half of an equilateral triangle, as shown in the following figure. Thus if we choose an equilateral triangle whose sides have length 2 and take half of it, we obtain a right triangle that has a hypotenuse of length 2 and a leg of length 1. The other leg has length a, which can be found as follows:

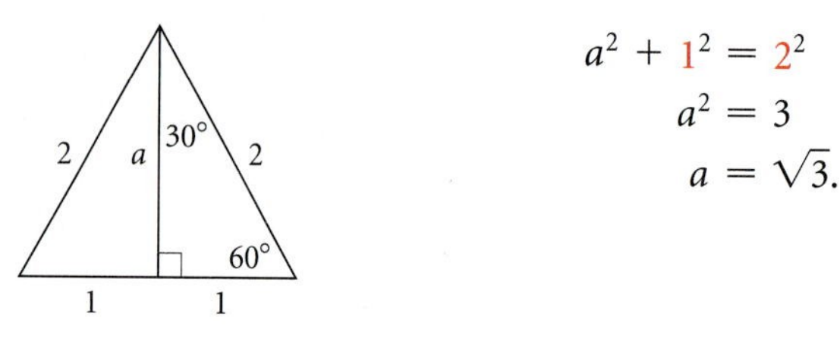

$$a^2 + 1^2 = 2^2$$
$$a^2 = 3$$
$$a = \sqrt{3}.$$

We can now determine the function values of 30° and 60°:

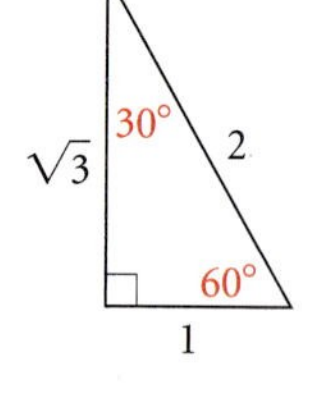

$$\sin 30^\circ = \frac{1}{2} = 0.5, \qquad \sin 60^\circ = \frac{\sqrt{3}}{2} \approx 0.8660,$$

$$\cos 30^\circ = \frac{\sqrt{3}}{2} \approx 0.8660, \qquad \cos 60^\circ = \frac{1}{2} = 0.5,$$

$$\tan 30^\circ = \frac{1}{\sqrt{3}} = \frac{\sqrt{3}}{3} \approx 0.5774, \qquad \tan 60^\circ = \frac{\sqrt{3}}{1} = \sqrt{3} \approx 1.7321$$

Since we will often use the function values of 30°, 45°, and 60°, either the triangles that yield them or the values themselves should be memorized.

Let's now use what we have learned about trigonometric functions of special angles to solve problems. We will consider such applications in greater detail in Section 4.2.

Example 4 *Height of a Hot-air Balloon.* A hot-air balloon ground crew drove 1.2 mi to an observation station as soon as the balloon began to rise. The initial observation estimated the angle between the ground and the line of sight to be 30°. Approximately how high was the balloon at that point? (We are assuming that the wind velocity was low and that the balloon rose vertically for the first few minutes.)

SOLUTION We begin with a sketch of the situation.

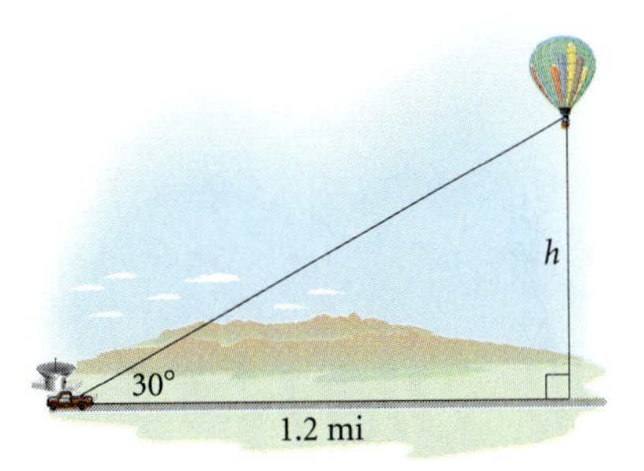

We know the measure of an acute angle and the length of its adjacent side. Since we want to determine the length of the opposite side, we can use the tangent ratio, or the cotangent ratio. Here we use the tangent ratio.

$$\tan 30° = \frac{\text{opp}}{\text{adj}} = \frac{h}{1.2}$$

$$1.2 \tan 30° = h$$

$$1.2\left(\frac{\sqrt{3}}{3}\right) = h \qquad \textbf{Substituting}$$

$$0.4\sqrt{3} = h$$

$$0.7 \approx h$$

The balloon is approximately 0.7 mi, or 3696 ft, high.

Function Values of Any Acute Angle

Historically, the measure of an angle has been expressed in degrees, minutes, and seconds. One minute, denoted 1′, is such that 60′ = 1°, or $1' = \frac{1}{60} \cdot (1°)$. One second, denoted 1″, is such that 60″ = 1′, or $1'' = \frac{1}{60} \cdot (1')$. Then 61 degrees, 27 minutes, 4 seconds could be written as 61°27′4″. This D°M′S″ form was common before the widespread use of graphers. Now the preferred notation is to express fractional parts of degrees in *decimal degree form.* Although the D°M′S″ notation is still widely used in navigation, we will most often use the decimal form in this text.

Most graphers can convert D°M′S″ notation to decimal degree notation and vice versa. This is often accomplished using a feature in the ANGLE menu. Procedures among graphers vary.

Example 5 Convert $5°42'30''$ to decimal degree notation.

SOLUTION Using a feature in the ANGLE menu on a grapher (TI-82), we enter $5°42'30''$ as $5'42'30'$. Pressing ENTER gives us

$$5°42'30'' \approx 5.71°,$$

rounded to the nearest hundredth of a degree.

Example 6 Convert $72.18°$ to D°M′S″ notation.

SOLUTION Using a grapher, we enter 72.18 and access the ►DMS feature in the ANGLE menu. The result is

$$72.18° = 72°10'48''.$$

So far we have measured angles using degrees. Another useful unit for angle measure is the radian, which we will study in Section 4.4. Graphing calculators work with either degrees or radians. Be sure to use whichever mode is appropriate. In this section, we use the degree mode.

Keep in mind the difference between an exact answer and an approximation. For example,

$$\sin 60° = \frac{\sqrt{3}}{2}. \qquad \textbf{This is exact!}$$

But on a grapher, you might get an answer like

$$\sin 60° \approx 0.8660254038. \qquad \textbf{This is approximate!}$$

Graphers usually provide values only of the sine, cosine, and tangent functions. You can find values of the cosecant, secant, and cotangent by taking reciprocals of the sine, cosine, and tangent functions, respectively.

Example 7 Find the trigonometric function value, rounded to four decimal places, of each of the following.

a) $\tan 29.7°$ b) $\sec 48°$ c) $\sin 84°10'39''$

SOLUTION

a) We check to be sure that the grapher is in degree mode. Some graphers require you to first enter 29.7 and then press the TAN key. Others require you to precede the number with the function key. The function value is

$$\tan 29.7° \approx 0.5703899297$$
$$\approx 0.5704. \qquad \text{Rounded to four decimal places}$$

b) The secant function value can be found by taking the reciprocal of the cosine function value:

$$\sec 48^\circ = \frac{1}{\cos 48^\circ} \approx 1.49447655 \approx 1.4945.$$

c) Using a feature in the ANGLE menu on a grapher, we enter sin 84°10′39″ as sin 84′10′39′. Pressing ENTER gives us

$$\sin 84^\circ 10' 39'' \approx 0.9948409474 \approx 0.9948.$$

We can use a TABLE feature on a grapher to find an angle for which we know a trigonometric function value.

Example 8 Find the acute angle, to the nearest tenth of a degree, whose sine value is approximately 0.20113.

SOLUTION With a grapher set in degree mode, we first enter the equation $y = \sin x$. With a minimum value of 0 and a step-value of 0.1, we scroll through the table of values looking for the y_1 value closest to 0.20113.

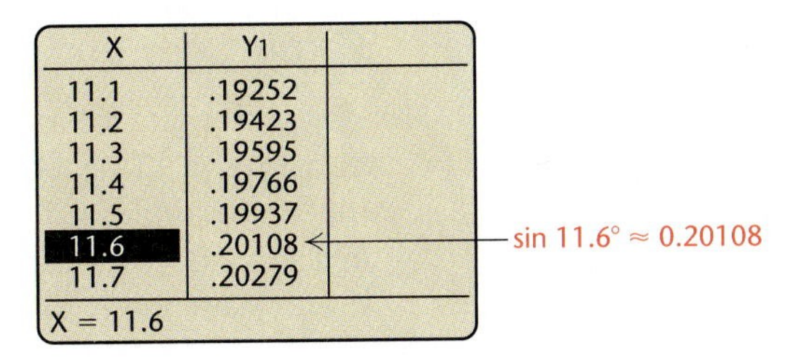

X	Y1
11.1	.19252
11.2	.19423
11.3	.19595
11.4	.19766
11.5	.19937
11.6	.20108
11.7	.20279

X = 11.6

We find that 11.6° is the angle whose sine value is about 0.20113.

The quickest way to find the angle with a grapher is to use an inverse function key. (We first studied inverse functions in Section 3.1 and will consider inverse *trigonometric* functions in Section 5.4.) Usually two keys must be pressed in sequence. For this example, if we press

2nd SIN .20113 ENTER,

we find that the acute angle whose sine is 0.20113 is approximately 11.60304613°, or 11.6°.

Example 9 *Ladder Safety.* A paint crew has purchased new 30-ft extension ladders. The manufacturer states that the safest placement on a wall is to extend the ladder to 25 ft and to position the base $6\frac{1}{2}$ ft from the wall. (*Source:* R. D. Werner Co., Inc., Corporate Headquarters: P.O. Box 580, Greenville, PA 16125) What angle does this safest position determine with the ground?

SOLUTION We draw a diagram and then use the most convenient trigonometric function.

From the definition of the cosine function, we have

$$\cos\theta = \frac{6.5\text{ ft}}{25\text{ ft}} = 0.26.$$

Using a grapher, we find that

$$\theta \approx 74.92993786°.$$

Thus the ladder is at its safest position when it makes an angle of about 75° with the ground.

Cofunctions and Complements

We recall that two angles are **complementary** whenever the sum of their measures is 90°. Each is the complement of the other. In a right triangle, the acute angles are complementary, since the sum of all three angle measures is 180° and the right angle accounts for 90° of this total. Thus if one acute angle of a right triangle is θ, the other is $90° - \theta$.

Interactive Discovery

Find the six trigonometric function values of the acute angles given in this triangle. Note that 53° and 37° are complementary since $53° + 37° = 90°$.

53°
37°

$\sin 37° = ?$ $\csc 37° = ?$

$\cos 37° = ?$ $\sec 37° = ?$

$\tan 37° = ?$ $\cot 37° = ?$

$\sin 53° = ?$ $\csc 53° = ?$

$\cos 53° = ?$ $\sec 53° = ?$

$\tan 53° = ?$ $\cot 53° = ?$

Try this with acute angles 20.3° and 69.7° as well. What pattern do you observe? Look for this same pattern in Example 1 earlier in this section.

Note that the sine of an angle is also the cosine of the angle's complement. Similarly, the tangent of an angle is the cotangent of the angle's complement, and the secant of an angle is the cosecant of the angle's complement. These pairs of functions are called **cofunctions**. A list of cofunction identities is as follows.

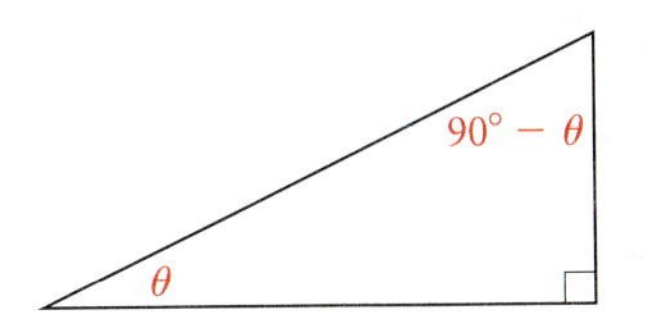

Cofunction Identities

$\sin\theta = \cos(90° - \theta)$, $\cos\theta = \sin(90° - \theta)$,
$\tan\theta = \cot(90° - \theta)$, $\cot\theta = \tan(90° - \theta)$,
$\sec\theta = \csc(90° - \theta)$, $\csc\theta = \sec(90° - \theta)$

Example 10 Given that $\sin 18° \approx 0.3090$, $\cos 18° \approx 0.9511$, and $\tan 18° \approx 0.3249$, find the six trigonometric function values of 72°.

SOLUTION Using reciprocal relationships, we know that

$$\csc 18° = \frac{1}{\sin 18°} \approx 3.2361, \qquad \sec 18° = \frac{1}{\cos 18°} \approx 1.0515,$$

and $$\cot 18° = \frac{1}{\tan 18°} \approx 3.0777.$$

Since 72° and 18° are complementary, we have

$\sin 72° = \cos 18° \approx 0.9511$, $\cos 72° = \sin 18° \approx 0.3090$,
$\tan 72° = \cot 18° \approx 3.0777$, $\cot 72° = \tan 18° \approx 0.3249$,
$\sec 72° = \csc 18° \approx 3.2361$, $\csc 72° = \sec 18° \approx 1.0515$.

4.1 Exercise Set

Find the six trigonometric function values of the specified angle.

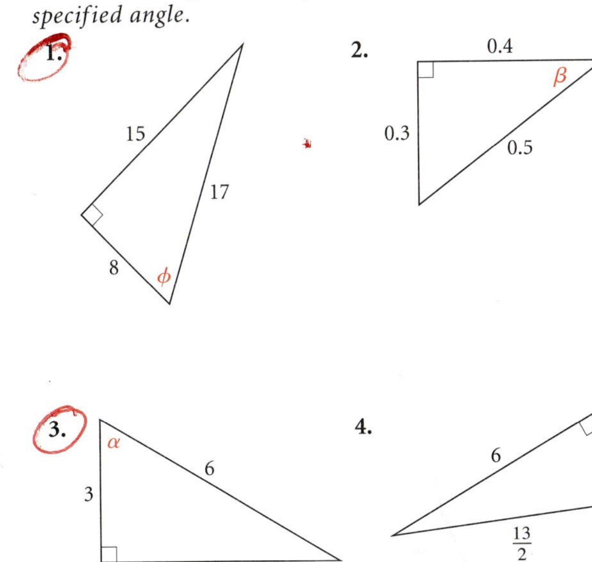

5.

6.

Given a function value of an acute angle, find the other five trigonometric function values.

7. $\sin\theta = \frac{24}{25}$

8. $\cos\theta = 0.7$

9. $\tan\phi = 2$

10. $\cot\theta = \frac{1}{3}$

11. $\csc\theta = 1.5$

12. $\sec\beta = \sqrt{17}$

Find the exact function value.

13. cos 45° **14.** tan 30°

15. sec 60° **16.** sin 45°

17. cot 60° **18.** csc 45°

19. sin 30° **20.** cos 60°

21. *Distance Between Bases.* A baseball diamond is really a square 90 ft on a side. If a line is drawn from third base to first base, then a right triangle QPH is formed, where $\angle QPH$ is 45°. Using a trigonometric function, find the distance from third base to first base.

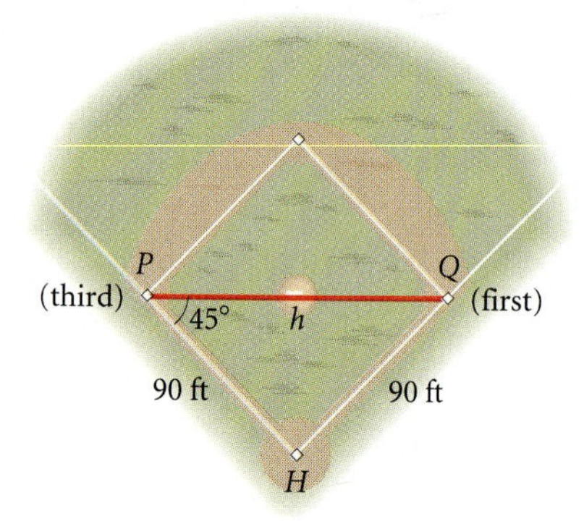

22. *Distance Across a River.* Find the distance a across the river.

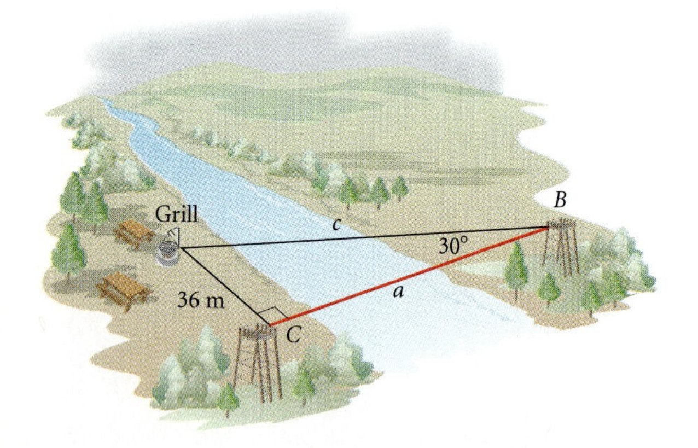

Convert to degrees, minutes, and seconds. Round to the nearest second.

23. 17.6° **24.** 20.14°

25. 83.025° **26.** 67.84°

27. 11.75° **28.** 29.8°

29. 47.8268° **30.** 0.253°

Convert to decimal degree notation. Round to two decimal places.

31. 9°43′ **32.** 52°15′

33. 35°50′ **34.** 64°53′

35. 3°2′ **36.** 19°47′23″

37. 49°38′46″ **38.** 76°11′34″

Find each function value. Round to four decimal places.

39. cos 51° **40.** cot 17°

41. tan 4°13′ **42.** sin 26.1°

43. sec 38.43° **44.** cos 74°10′40″

45. cos 40.35° **46.** csc 45.2°

47. sin 69° **48.** tan 63°48′

49. tan 85.4° **50.** cos 4°

51. csc 89.5° **52.** sec 35.28°

53. cot 30°25′6″ **54.** sin 59.2°

Using a grapher, find the acute angle θ, to the nearest tenth of a degree, for the given function value.

55. $\sin\theta = 0.5125$ **56.** $\tan\theta = 2.032$

57. $\tan\theta = 0.2226$ **58.** $\cos\theta = 0.3842$

59. $\sin\theta = 0.9022$ **60.** $\tan\theta = 3.056$

61. $\cos\theta = 0.6879$ **62.** $\sin\theta = 0.4005$

63. $\cot\theta = 2.127$ $\left(\text{Hint: } \tan\theta = \frac{1}{\cot\theta}.\right)$ **64.** $\csc\theta = 1.147$

65. $\sec\theta = 1.279$ **66.** $\cot\theta = 1.351$

Find the exact acute angle θ for the given function value.

67. $\sin\theta = \frac{\sqrt{2}}{2}$ **68.** $\cot\theta = \frac{\sqrt{3}}{3}$

69. $\cos\theta = \frac{1}{2}$ **70.** $\sin\theta = \frac{1}{2}$

71. $\tan\theta = 1$ **72.** $\cos\theta = \frac{\sqrt{3}}{2}$

Complete.

73. cos 20° = _____ 70° = $\frac{1}{\_\_\_\_\_\ 20°}$

74. sin 64° = _____ 26° = $\frac{1}{\_\_\_\_\_\ 64°}$

75. tan 52° = cot _____ = $\frac{1}{\_\_\_\_\_\ 52°}$

76. sec 13° = csc _____ = $\frac{1}{\_\_\_\_\_\ 13°}$

77. Given that

sin 65° ≈ 0.9063, cos 65° ≈ 0.4226,
tan 65° ≈ 2.145, cot 65° ≈ 0.4663,
sec 65° ≈ 2.366, csc 65° ≈ 1.103,

find the six function values of 25°.

78. Given that $\sin 38.7° \approx 0.6252$, $\cos 38.7° \approx 0.7804$, and $\tan 38.7° \approx 0.8012$, find the six function values of $51.3°$.

Skill Maintenance

Solve. Round to two decimal places.

79. $0.1284 = \dfrac{d}{11.8}$

80. $0.6045 = \dfrac{30}{h}$

81. $5(x + 20.5) = 1.8x$

82. $\dfrac{3}{4} = \dfrac{t}{0.42}$

83. Use the Pythagorean theorem to find the missing length. Round the answer to the nearest tenth.

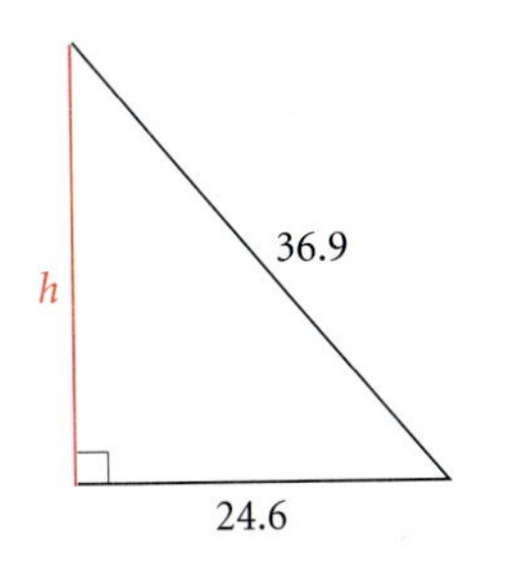

Synthesis

84. ◈ Explain the difference between two functions being reciprocal functions and two functions being cofunctions.

85. ◈ Explain why it is not necessary to memorize the function values for both 30° and 60°.

86. Given that $\cos 15.17° = 0.9651$, find $\csc 74.83°$.

87. Given that $\sec 49.2° = 1.5304$, find $\sin 40.8°$.

88. Find the six trigonometric function values of α.

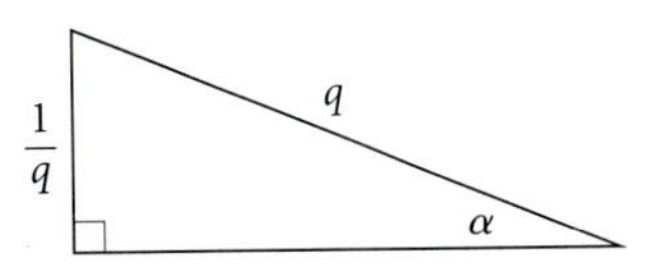

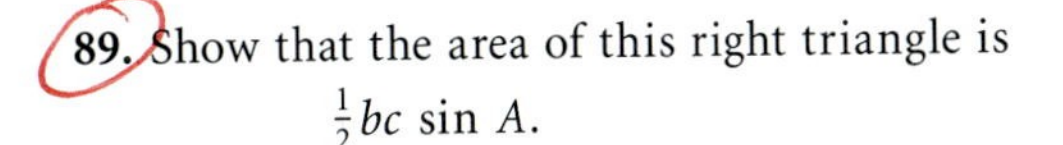

89. Show that the area of this right triangle is $\frac{1}{2}bc \sin A$.

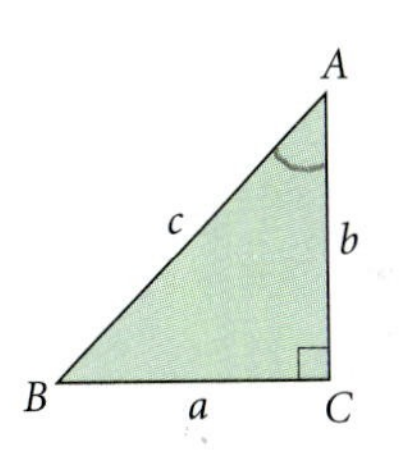

90. Show that the area of this triangle is $\frac{1}{2}ab \sin \theta$.

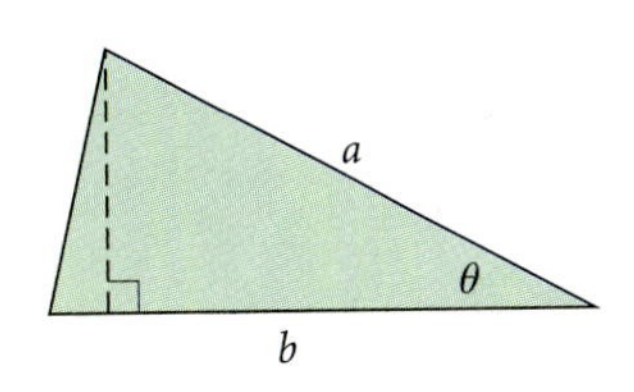

4.2 Applications of Right Triangles

- *Solve right triangles.*
- *Solve applied problems involving right triangles and trigonometric functions.*

Solving Right Triangles

Now that we can find function values for any acute angle, it is possible to *solve* right triangles. To **solve** a triangle means to find the lengths of all sides and the measures of all angles not already known. Triangles that are not right triangles will be studied later.

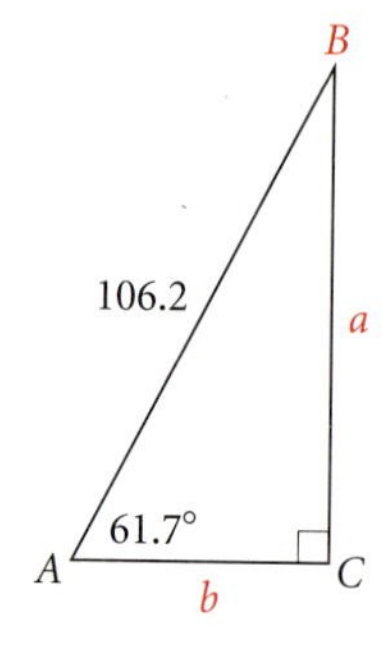

Example 1 In $\triangle ABC$ (shown at left), find a, b, and B, where a and b represent lengths of sides and B represents the measure of $\angle B$.

SOLUTION In $\triangle ABC$, we know three of the measures:

$$A = 61.7°, \quad a = ?,$$
$$B = ?, \quad b = ?,$$
$$C = 90°, \quad c = 106.2.$$

Since the sum of the angle measures of any triangle is 180° and $C = 90°$, the sum of A and B is 90°. Thus,

$$B = 90° - A = 90° - 61.7° = 28.3°.$$

We are given an acute angle and the hypotenuse. This suggests that we can use the sine and cosine ratios to find a and b, respectively:

$$\sin 61.7° = \frac{\text{opp}}{\text{hyp}} = \frac{a}{106.2} \quad \text{and} \quad \cos 61.7° = \frac{\text{adj}}{\text{hyp}} = \frac{b}{106.2}.$$

Solving for a and b, we get

$$a = 106.2 \sin 61.7° \quad \text{and} \quad b = 106.2 \cos 61.7°$$
$$a \approx 93.5 \qquad\qquad b \approx 50.3.$$

Thus,

$$A = 61.7°, \quad a \approx 93.5,$$
$$B = 28.3°, \quad b \approx 50.3,$$
$$C = 90°, \quad c = 106.2.$$

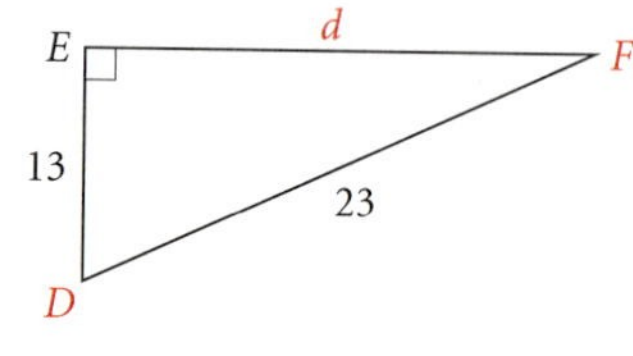

Example 2 In $\triangle DEF$ (shown at left), find D and F. Then find d.

SOLUTION In $\triangle DEF$, we know three of the measures:

$$D = ?, \quad d = ?,$$
$$E = 90°, \quad e = 23,$$
$$F = ?, \quad f = 13.$$

We know the side adjacent to D and the hypotenuse. This suggests the use of the cosine ratio:

$$\cos D = \frac{\text{adj}}{\text{hyp}} = \frac{13}{23}.$$

We now use either the TABLE feature or an inverse function key to find the angle whose cosine is $\frac{13}{23}$. To the nearest hundredth of a degree,

$$D \approx 55.58°.$$ Pressing 2nd cos (13/23) ENTER

Program

TRISOL90: This program solves right triangles.

Since the sum of D and F is 90°, we can find F by subtracting:

$$F = 90° - D \approx 90° - 55.58° \approx 34.42°.$$

To find d, we can use either the Pythagorean theorem or a trigonometric function. We could use $\cos F$, $\sin D$, or the tangent or cotangent ratios for either D or F. Let's use tan D:

$$\tan D = \frac{\text{opp}}{\text{adj}} = \frac{d}{13}, \quad \text{or} \quad \tan 55.58^\circ \approx \frac{d}{13}.$$

Then

$$d \approx 13 \tan 55.58^\circ \approx 19.$$

The six measures are

$$D \approx 55.58^\circ, \qquad d \approx 19,$$
$$E = 90^\circ, \qquad e = 23,$$
$$F \approx 34.42^\circ, \qquad f = 13.$$

Applications

Right triangles can be used to model many applications in the real world. To solve such problems, we solve, or at least partially solve, the triangle.

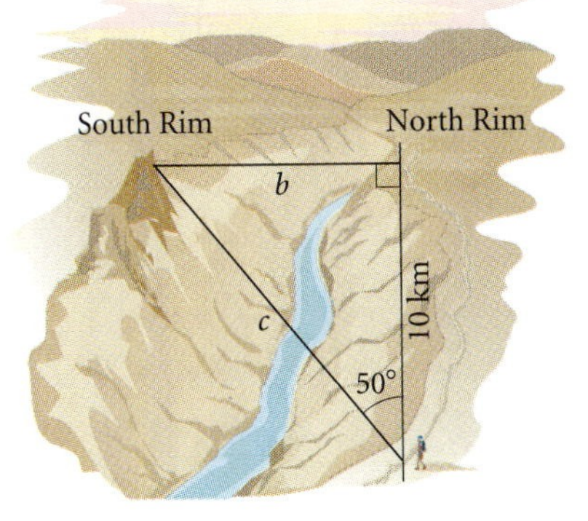

Example 3 *Hiking at the Grand Canyon.* A backpacker hiking east along the north rim of the Grand Canyon notices an unusual rock formation directly across the canyon. She decides to continue watching the landmark while hiking along the rim. In 2 hr, she has gone 10 km due east and the landmark is still visible but at approximately a 50° angle to the north rim. (See the figure at left.)

a) How many kilometers is she from the rock formation?

b) How far is it across the canyon at her starting point?

SOLUTION

a) We know the side adjacent to the 50° angle and want to find the hypotenuse. We can use the cosine ratio:

$$\cos 50^\circ = \frac{10 \text{ km}}{c}$$

$$c = \frac{10 \text{ km}}{\cos 50^\circ} \approx 15.6 \text{ km}.$$

After hiking for 10 km, she is approximately 15.6 km from the rock formation.

b) We know the side adjacent to the 50° angle and want to find the opposite side. We can use the tangent ratio:

$$\tan 50^\circ = \frac{b}{10 \text{ km}}$$

$$b = 10 \text{ km} \cdot \tan 50^\circ \approx 11.9 \text{ km}.$$

Thus it is approximately 11.9 km across the canyon at her starting point.

Many applications with right triangles involve an *angle of elevation* or an *angle of depression*. The angle between the horizontal and a line of sight above the horizontal is called an **angle of elevation**. The angle between the horizontal and a line of sight below the horizontal is called an **angle of depression**. For example, suppose that you are looking straight ahead and then you move your eyes up to look at an approaching airplane. The angle that your eyes pass through is an angle of elevation. If the pilot of the plane is looking forward and then looks down, the pilot's eyes pass through an angle of depression.

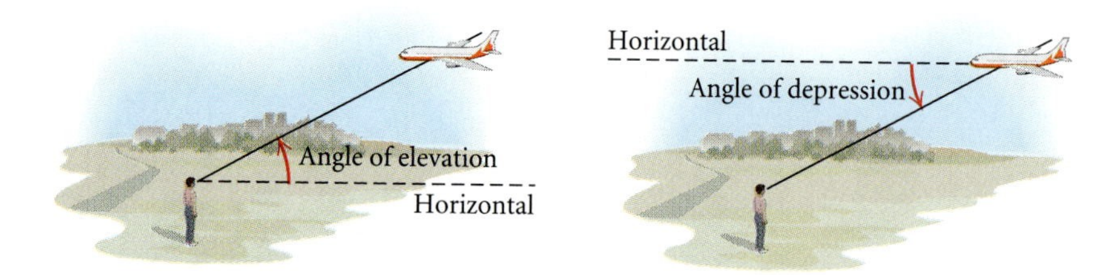

Example 4 *Aerial Photography.* An aerial photographer who photographs farm properties for a real estate company has determined from experience that the best photo is taken at a height of approximately 475 ft and a distance of 850 ft from the farmhouse. What is the angle of depression from the plane to the house?

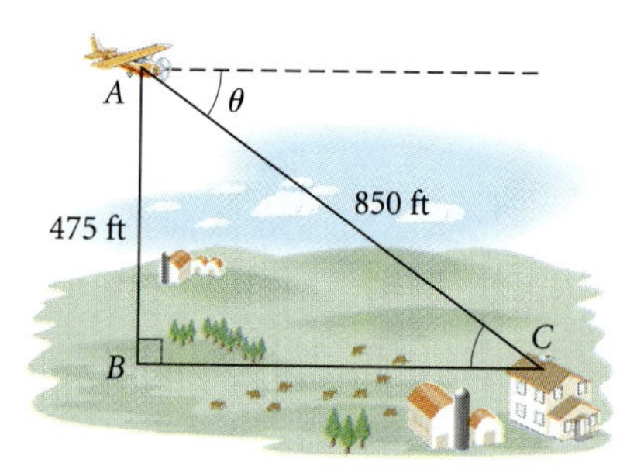

SOLUTION When parallel lines are cut by a transversal, alternate interior angles are equal. Thus the angle of depression from the plane to the house is equal to the angle of elevation from the house to the plane, so we can use the right triangle shown in the figure. Since we know the side opposite $\angle C$ and the hypotenuse, we can find θ by using the sine function:

$$\sin \theta = \sin C = \frac{475 \text{ ft}}{850 \text{ ft}}.$$

Using the inverse sine key, we find that

$$\theta \approx 34°.$$ **Pressing** 2nd sin (475/850) ENTER

Thus the angle of depression is approximately 34°.

Example 5 *Cloud Height.* To measure cloud height at night, a vertical beam of light is directed on a spot on the cloud. From a point 135 ft away from the light source, the angle of elevation to the spot is found to be 67.35°. Find the height of the cloud.

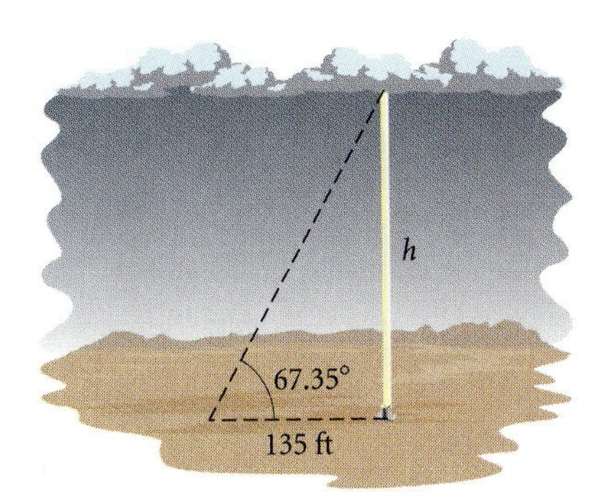

SOLUTION From the figure, we have

$$\tan 67.35° = \frac{h}{135 \text{ ft}}$$

$$h = 135 \text{ ft} \cdot \tan 67.35° \approx 324 \text{ ft}.$$

The height of the cloud is about 324 ft.

Some applications of trigonometry involve the concept of direction, or bearing. In this text we present two ways of giving direction, the first in Example 6 below and the second in Exercise Set 4.3.

BEARING: FIRST-TYPE One method of giving direction, or bearing, involves reference to a north–south line using an acute angle. For example, N55°W means 55° west of north and S67°E means 67° east of south.

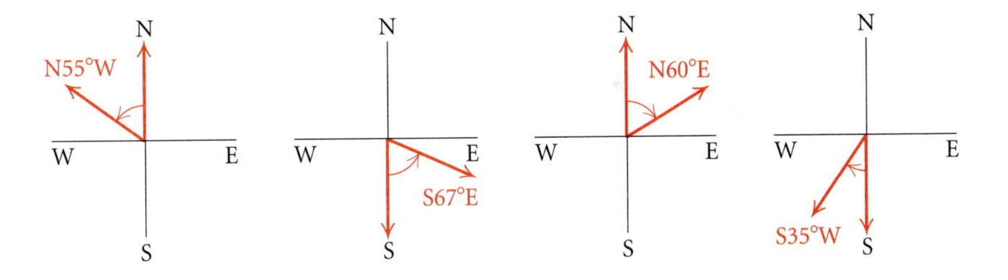

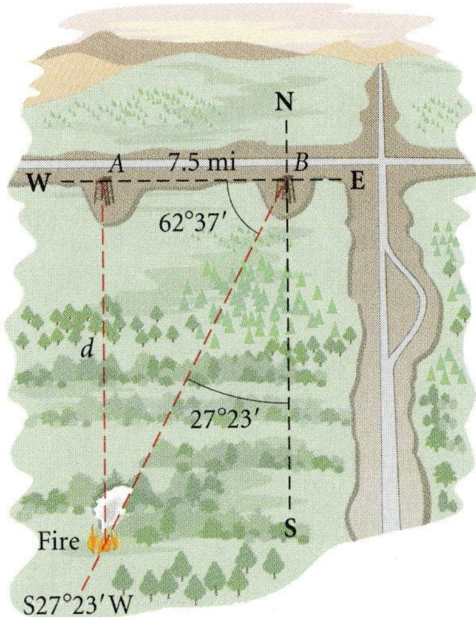

Example 6 *Distance to a Forest Fire.* A forest ranger at point A sights a fire directly south. A second ranger at point B, 7.5 mi east, sights the same fire at a bearing of S27°23′W. How far from A is the fire?

$$\begin{aligned} B &= 90° - 27°23' \\ &= 89°60' - 27°23' \\ &= 62°37' \end{aligned}$$

SOLUTION From the figure at left, we see that the desired distance d is part of a right triangle, as shown. We have

$$\frac{d}{7.5 \text{ mi}} = \tan 62°37'$$

$$d = 7.5 \text{ mi} \tan 62°37' \approx 14.5 \text{ mi}.$$

The forest ranger at A is about 14.5 mi from the fire.

Example 7 *Comiskey Park.* In the new Comiskey Park, the home of the Chicago White Sox baseball team, the first row of seats in the upper deck is farther away from home plate than the last row of seats in the old Comiskey Park. Although there is no obstructed view in the new park, some of the fans still complain about the present distance from home plate to the upper deck of seats. From a seat in the last row of the upper deck directly behind the batter, the angle of depression to home plate is 29.9° and the angle of depression to the pitcher's mound is 24.2°. Find (a) the viewing distance to home plate and (b) the viewing distance to the pitcher's mound.

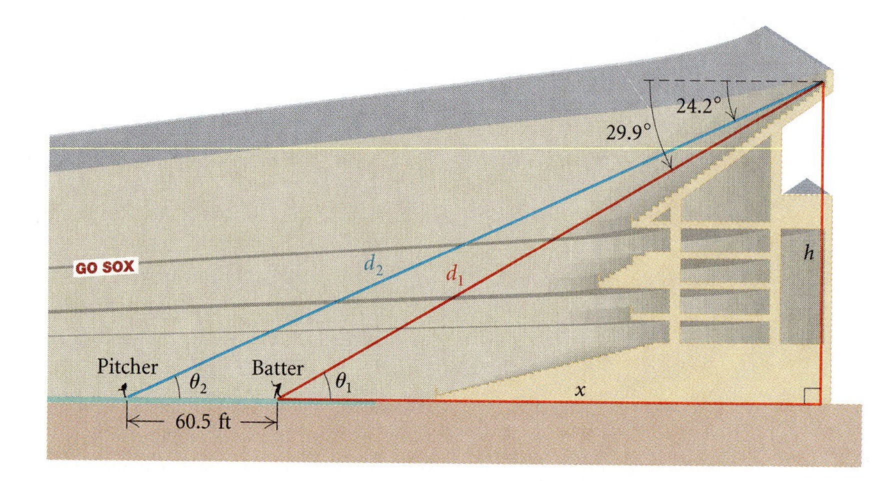

SOLUTION From geometry we know that $\theta_1 = 29.9°$ and $\theta_2 = 24.2°$. The standard distance from home plate to the pitcher's mound is 60.5 ft. In the drawing, we let d_1 be the viewing distance to home plate, d_2 the viewing distance to the pitcher, h the elevation of the last row, and x the horizontal distance from the batter to a point directly below the seat in the last row of the upper deck.

We begin by determining the distance x. We use the tangent function with $\theta_1 = 29.9°$ and $\theta_2 = 24.2°$:

$$\tan 29.9° = \frac{h}{x} \qquad \text{and} \qquad \tan 24.2° = \frac{h}{x + 60.5}$$

or

$$h = x \tan 29.9° \quad \text{and} \quad h = (x + 60.5) \tan 24.2°.$$

Then substituting $x \tan 29.9°$ for h in the second equation, we obtain

$$x \tan 29.9° = (x + 60.5) \tan 24.2°.$$

Solving for x, we get

$$\begin{aligned} x \tan 29.9° &= x \tan 24.2° + 60.5 \tan 24.2° \\ x(\tan 29.9° - \tan 24.2°) &= 60.5 \tan 24.2° \\ x &= \frac{60.5 \tan 24.2°}{\tan 29.9° - \tan 24.2°} \\ x &\approx 216.5. \end{aligned}$$

We can then find d_1 and d_2 using the cosine function:

$$\cos 29.9^\circ = \frac{216.5}{d_1} \quad \text{and} \quad \cos 24.2^\circ = \frac{216.5 + 60.5}{d_2}$$

or

$$d_1 = \frac{216.5}{\cos 29.9^\circ} \quad \text{and} \quad d_2 = \frac{277}{\cos 24.2^\circ}$$

$$d_1 \approx 249.7 \qquad\qquad d_2 \approx 303.7.$$

The distance to home plate is about 250 ft,* and the distance to the pitcher's mound is about 304 ft.

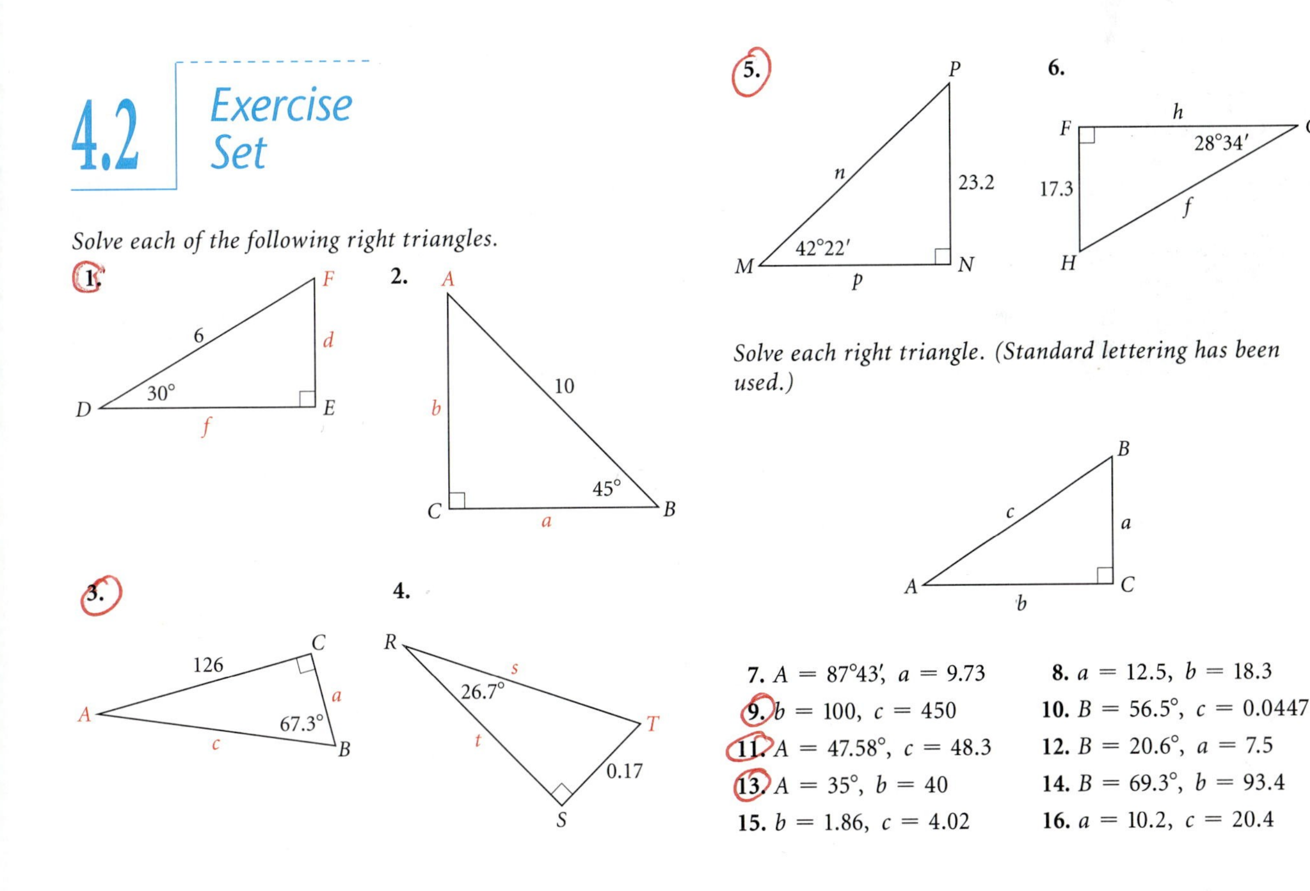

4.2 Exercise Set

Solve each of the following right triangles.

1.

2.

3.

4.

5.

6.

Solve each right triangle. (Standard lettering has been used.)

7. $A = 87°43'$, $a = 9.73$
8. $a = 12.5$, $b = 18.3$
9. $b = 100$, $c = 450$
10. $B = 56.5°$, $c = 0.0447$
11. $A = 47.58°$, $c = 48.3$
12. $B = 20.6°$, $a = 7.5$
13. $A = 35°$, $b = 40$
14. $B = 69.3°$, $b = 93.4$
15. $b = 1.86$, $c = 4.02$
16. $a = 10.2$, $c = 20.4$

*In the old Comiskey Park, the distance to home plate was only 150 ft (*Chicago Tribune*, September 19, 1993).

17. *Safety Line to Raft.* Each spring Madison uses his vacation time to ready his lake property for the summer. He wants to run a new safety line from point B on the shore to the corner of the anchored diving raft. The current safety line, which runs perpendicular to the shore line to point A, is 40 ft long. He estimates the angle from B to the corner of the raft to be 50°. Approximately how much rope does he need for the new safety line if he allows 5 ft of rope at each end to fasten the rope?

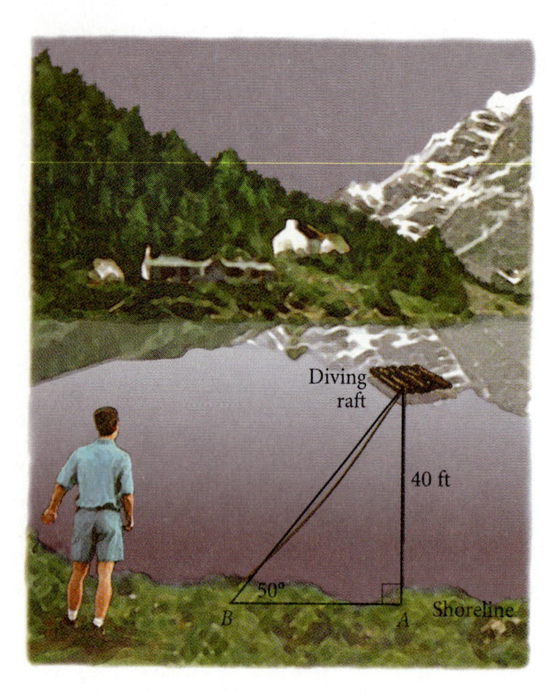

18. *Enclosing an Area.* Glynis is enclosing a triangular area in a corner of her fenced rectangular backyard for her Labrador retriever. In order for a certain tree to be included in this pen, one side needs to be 14.5 ft and make a 53° angle with the new side. How long is the new side?

19. *Easel Display.* A marketing group is designing an easel to display posters advertising their newest products. They want the easel to be 6 ft tall and fit flush against a wall. For optimal eye contact, the best angle between the front and back legs of the easel is 23°. How far from the wall should the front legs be placed in order to obtain this angle?

20. *Height of a Tree.* A supervisor must train a new team of loggers to estimate the heights of trees. As an example, she walks off 40 ft from the base of a tree and estimates the angle of elevation to the tree's peak to be 70°. Approximately how tall is the tree?

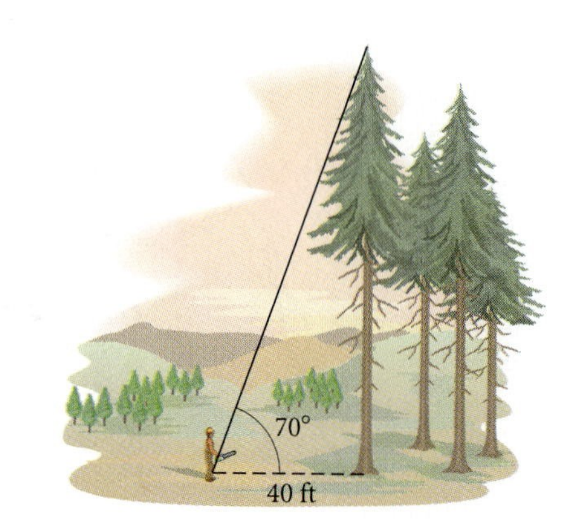

21. *Sand Dunes National Park.* While visiting the Sand Dunes National Park in Colorado, Ian approximated the angle of elevation to the top of a sand dune to be 20°. After walking 800 ft closer, he guessed that the angle of elevation had increased by 15°. Approximately how tall is the dune he was observing?

22. *Tee-Shirt Design.* A new tee-shirt design is to have a regular pentagon inscribed in a circle, as shown in the figure. Each side of the pentagon is to be 3.5 in. long. Find the radius of the circumscribed circle.

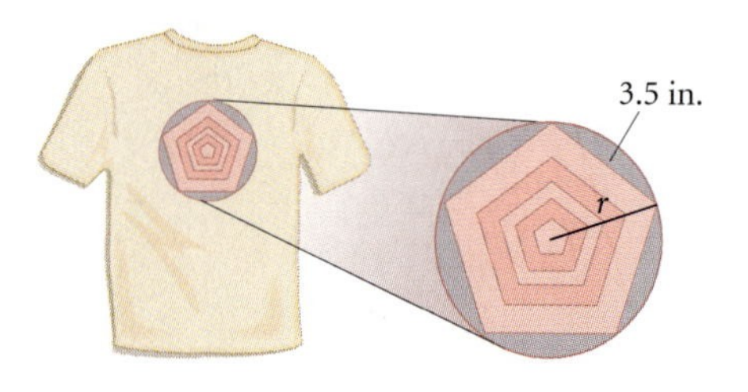

23. *Inscribed Octagon.* A regular octagon is inscribed in a circle of radius 15.8 cm. Find the perimeter of the octagon.

24. *Height of a Weather Balloon.* A weather balloon is directly west of two observing stations that are 10 mi apart. The angles of elevation of the balloon from the two stations are 17.6° and 78.2°. How high is the balloon?

25. *Height of a Kite.* For a science fair project, a group of students tested different materials used to construct kites. Their instructor provided an instrument that accurately measures the angle of elevation. In one of the tests, the angle of elevation was 63.4° with 670 ft of string out. Assuming the string was taut, how high was the kite?

26. *Height of a Building.* A window washer on a ladder looks at a nearby building 100 ft away, noting that the angle of elevation of the top of the building is 18.7° and the angle of depression of the bottom of the building is 6.5°. How tall is the nearby building?

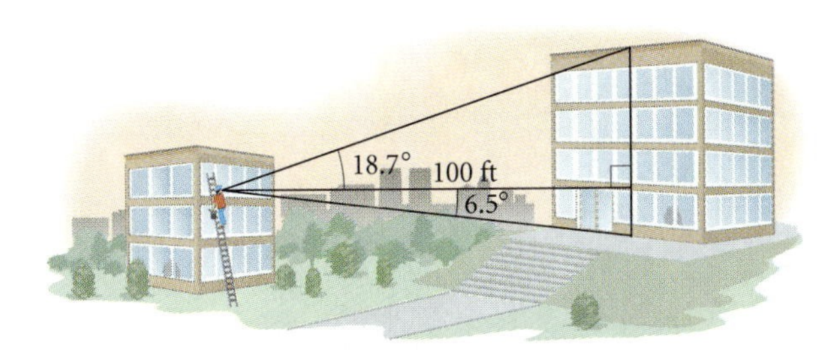

27. *Distance Between Towns.* From a hot-air balloon 2 km high, the angles of depression to two towns, in line with the balloon, are 81.2° and 13.5°. How far apart are the towns?

28. *Angle of Elevation.* What is the angle of elevation of the sun when a 35-ft mast casts a 20-ft shadow?

29. *Distance from a Lighthouse.* From the top of a lighthouse 55 ft above sea level, the angle of depression to a small boat is 11.3°. How far from the foot of the lighthouse is the boat?

30. *Lightning Detection.* In extremely large forests, it is not cost-effective to position forest rangers in towers or to use small aircraft to continually watch for fires. Since lightning is a frequent cause of fire, lightning detectors are now commonly used instead. These devices not only give a bearing on the location but also measure the intensity of the lightning. A detector at point Q is situated 15 mi west of a central fire station at point R. The bearing from Q to where lightning hits due south of R is S37.6°E. How far is the hit from point R?

31. *Lobster Boat.* A lobster boat is situated due west of a lighthouse. A barge is 12 km south of the lobster boat. From the barge, the bearing to the lighthouse is N63°20′E. How far is the lobster boat from the lighthouse?

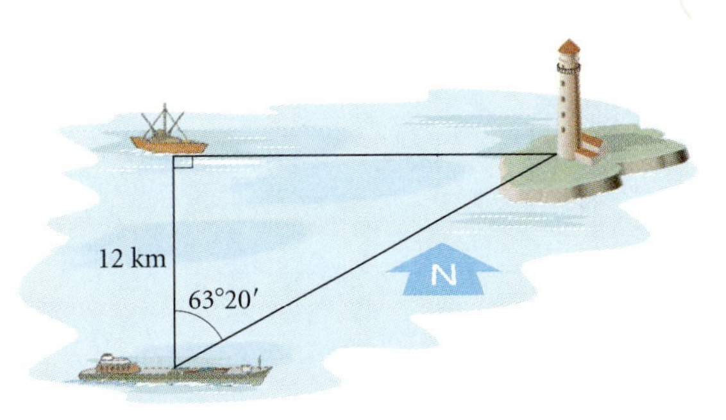

32. *Length of an Antenna.* A vertical antenna is mounted atop a 50-ft pole. From a point on the level ground 75 ft from the base of the pole, the antenna subtends an angle of 10.5°. Find the length of the antenna.

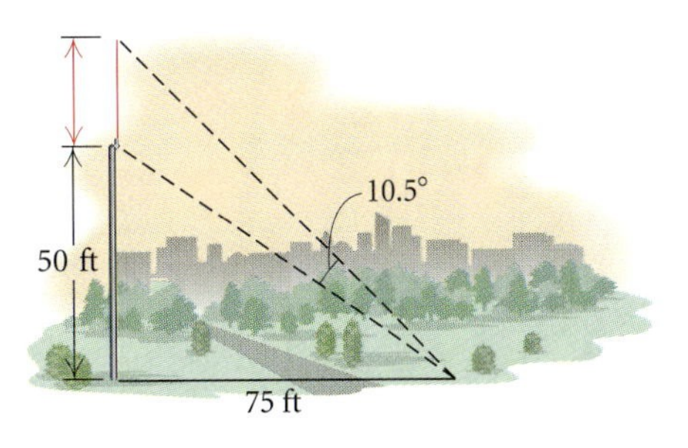

Skill Maintenance

Find the distance between the points.

33. (8, −2) and (−6, −4) **34.** (−9, 3) and (0, 0)

Evaluate the function value exactly.

35. cos 30° **36.** sin 60°

37. tan 30° **38.** sin 45°

Synthesis

39. ◈ In this section, the trigonometric functions have been defined as functions of acute angles. Thus $(0°, 90°)$ is the domain for each function. What appears to be the range for the sine, the cosine, and the tangent functions considering this limited domain?

40. ◈ Explain in your own words five ways in which length c can be determined in this triangle. Which way seems the most efficient?

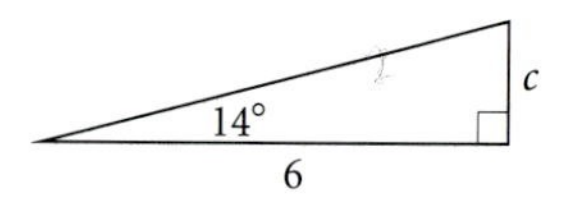

41. Find h, to the nearest tenth.

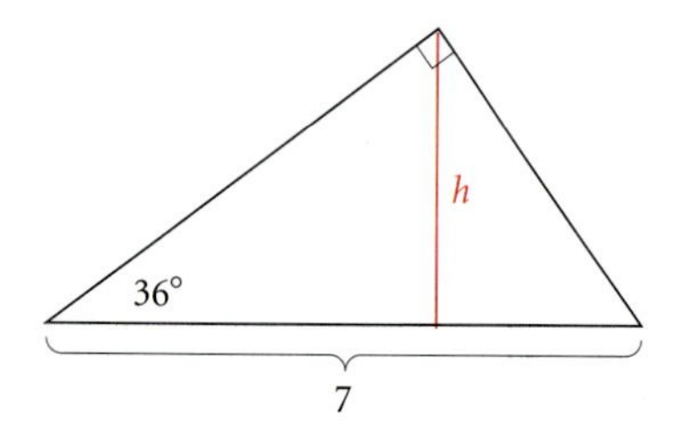

42. Find a, to the nearest tenth.

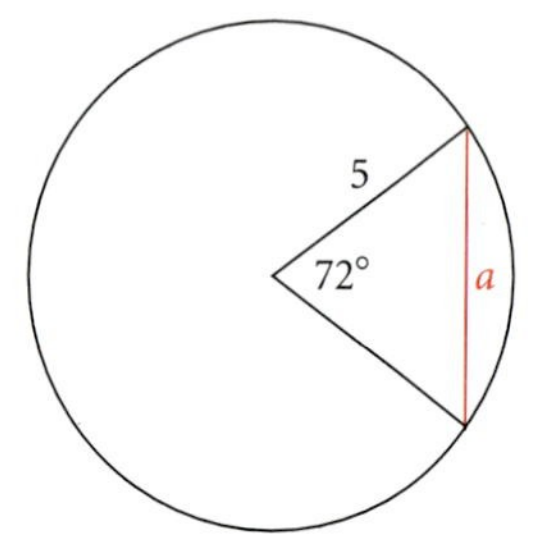

43. *Construction of Picnic Pavilions.* A construction company is mass-producing picnic pavilions for national parks, as shown in the figure. The rafter ends are to be sawed in such a way that they will be vertical when in place. The front wall is 8 ft high, the back wall is $6\frac{1}{2}$ ft high, and the distance between walls is 8 ft. At what angle should the rafters be cut?

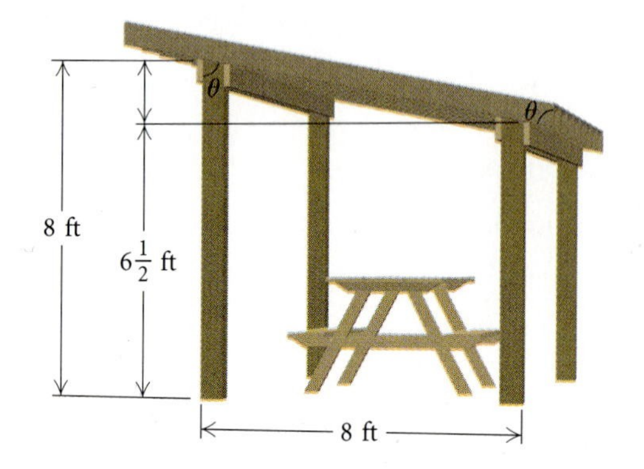

44. *Diameter of a Pipe.* A V-gauge is used to find the diameter of a pipe. The advantage of such a device is that it is rugged, it is accurate, and it has no moving parts to break down. In the figure, the measure of angle AVB is 54°. A pipe is placed in the V-shaped slot and the distance VP is used to predict the diameter.

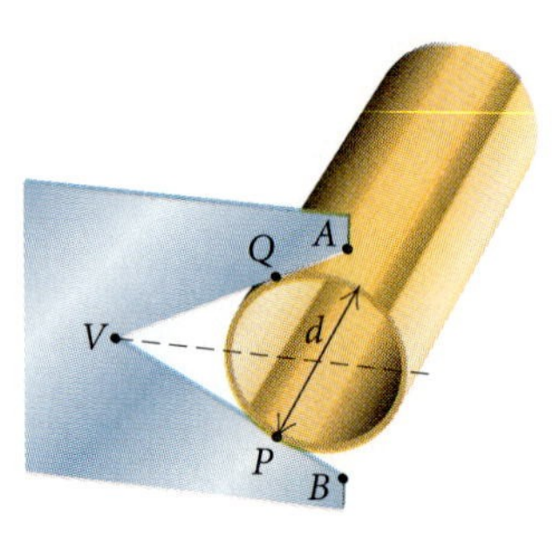

a) Suppose that the diameter of a pipe is 2 cm. What is the distance VP?
b) Suppose that the distance VP is 3.93 cm. What is the diameter of the pipe?
c) Find a formula for d in terms of VP.
d) Find a formula for VP in terms of d.

The line VP is calibrated by listing as its units the corresponding diameters. This, in effect, establishes a function between VP and d.

45. *Sound of an Airplane.* It is common experience to hear the sound of a low-flying airplane, and look at the wrong place in the sky to see the plane. Suppose that a plane is traveling directly at you at a speed of 200 mph and an altitude of 3000 ft, and you hear the sound at what seems to be an angle of inclination of 20°. At what angle θ should you actually look in order to see the plane? Consider the speed of sound to be 1100 ft/sec.

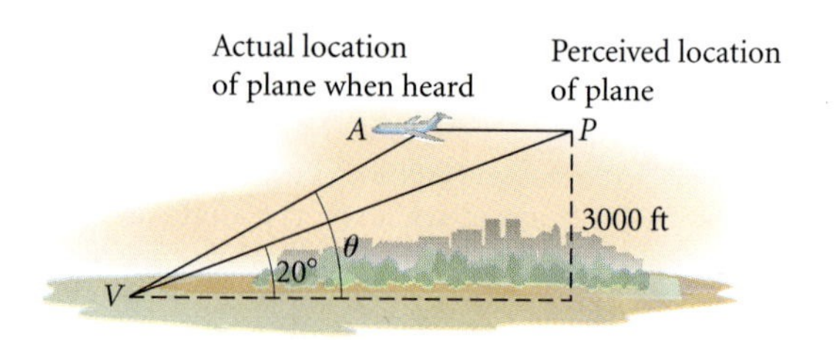

46. *Measuring the Radius of the Earth.* One way to measure the radius of the earth is to climb to the top of a mountain whose height above sea level is known and measure the angle between a vertical line to the center of the earth from the top of the mountain and a line drawn from the top of the mountain to the horizon, as shown in the figure. The height of Mt. Shasta in California is 14,162 ft. From the top of Mt. Shasta, one can see the horizon on the Pacific Ocean. The angle formed between a line to the horizon and the vertical is found to be 87°53′. Use this information to estimate the radius of the earth, in miles.

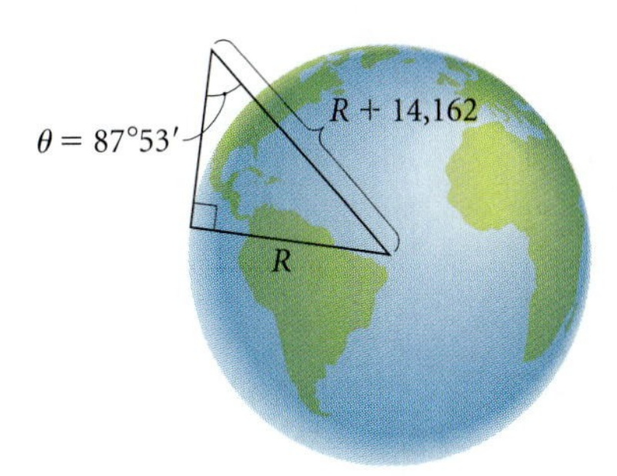

4.3 Trigonometric Functions of Any Angle

- *Find angles that are coterminal with a given angle and find the complement and the supplement of a given angle.*
- *Determine the six trigonometric function values for any angle in standard position when the coordinates of a point on the terminal side are given.*
- *Find the function values for any angle whose terminal side lies on an axis.*
- *Find the function values for an angle whose terminal side makes an angle of 30°, 45°, or 60° with the x-axis.*
- *Use a grapher to find function values and angles.*

Angles, Rotations, and Degree Measure

An *angle* is a familiar figure in the world around us.

An **angle** is the union of two rays with a common endpoint called the **vertex**. In trigonometry, we often think of an angle as a **rotation**. To do so, think of locating a ray along the positive x-axis with its endpoint at the origin. This ray is called the **initial side** of the angle. Though we leave that ray fixed, think of making a copy of it and rotating it. A rotation *counterclockwise* is a **positive rotation**, and a rotation *clockwise* is a **negative rotation**. The ray at the end of the rotation is called the

terminal side of the angle. The angle formed is said to be in **standard position**.

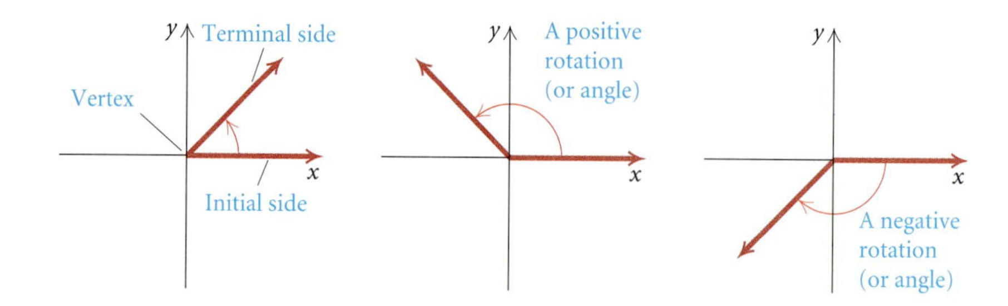

The measure of an angle or rotation may be given in degrees. The Babylonians developed the idea of dividing the circumference of a circle into 360 equal parts, or degrees. If we let the measure of one of these parts be 1°, then one complete positive revolution or rotation has a measure of 360°. One half of a revolution has a measure of 180°, one fourth of a revolution has a measure of 90°, and so on. We can also speak of an angle of measure 60°, 135°, 330°, or 420°. The terminal sides of these angles lie in quadrants I, II, IV, and I, respectively. The negative rotations −30°, −110°, and −225° represent angles with terminal sides in quadrants IV, III, and II, respectively.

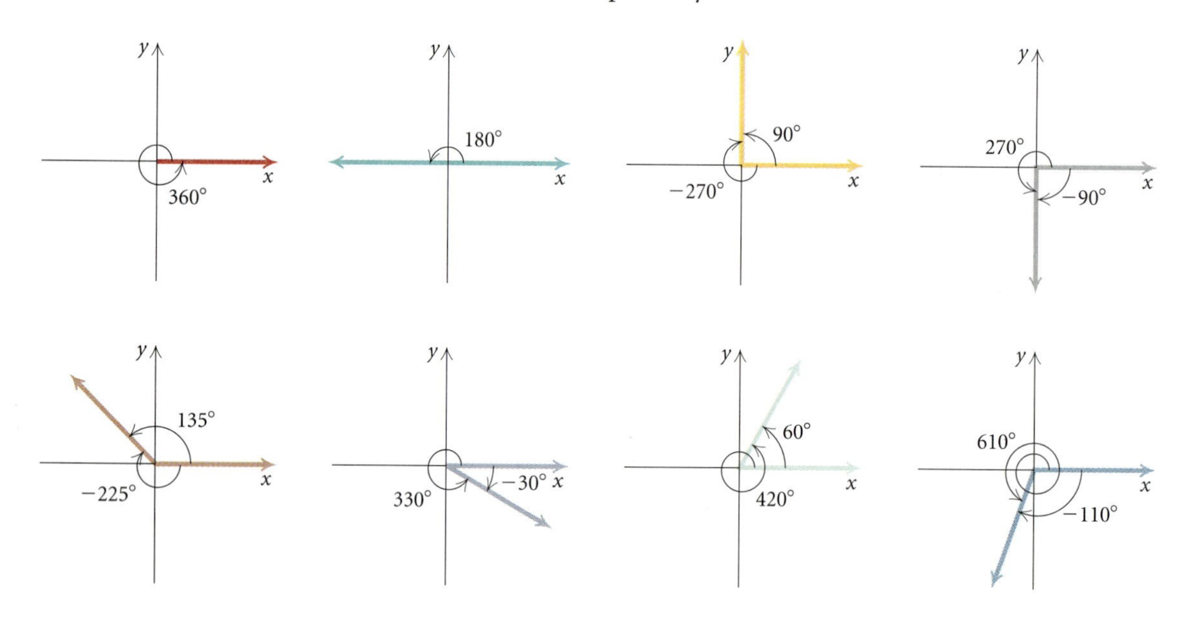

When the measure of an angle is greater than 360° or less than −360°, the rotating ray has gone through more than one complete rotation. If two or more angles have the same terminal side, the angles are said to be **coterminal**. To find angles coterminal with a given angle, we add or subtract multiples of 360°. For example, 420°, shown above, has the same terminal side as 60°, since 420° = 360° + 60°. Thus we say that angles of measure 60° and 420° are coterminal. The negative rotation that measures −300° is also coterminal with 60° because

$60° - 360° = -300°$. Other examples of coterminal angles shown above are

$90°$ and $-270°$,
$-90°$ and $270°$,
$135°$ and $-225°$,
$-30°$ and $330°$,
$-110°$ and $610°$.

Example 1 Find two positive and two negative angles that are coterminal with (a) $51°$ and (b) $-7°$.

SOLUTION

a) We add and subtract multiples of $360°$. Many answers are possible.

$$51° + 360° = 411° \qquad 51° + 3(360°) = 1131°$$
$$51° - 360° = -309° \qquad 51° - 2(360°) = -669°$$

Thus angles of measure $411°$, $1131°$, $-309°$, and $-669°$ are coterminal with $51°$.

b) We have the following.

$$-7° + 360° = 353° \qquad -7° + 2(360°) = 713°$$
$$-7° - 360° = -367° \qquad -7° - 10(360°) = -3607°$$

Thus angles of measure $353°$, $713°$, $-367°$, and $-3607°$ are coterminal with $-7°$.

Angles can also be classified by their measures, as seen in the following figure.

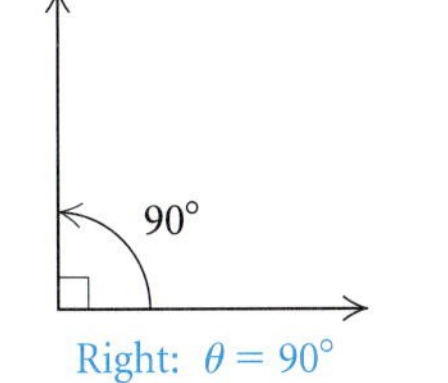

Right: $\theta = 90°$

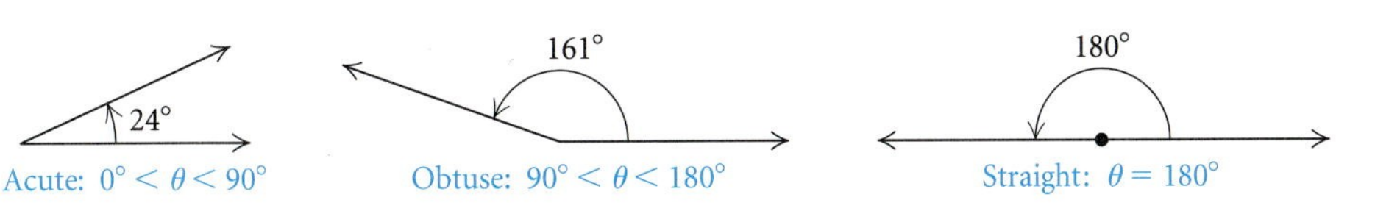

Acute: $0° < \theta < 90°$ Obtuse: $90° < \theta < 180°$ Straight: $\theta = 180°$

Two acute angles are **complementary** if their sum is $90°$. For example, angles that measure $10°$ and $80°$ are complementary because $10° + 80° = 90°$. Two positive angles are **supplementary** if their sum is $180°$. For example, angles that measure $45°$ and $135°$ are supplementary because $45° + 135° = 180°$.

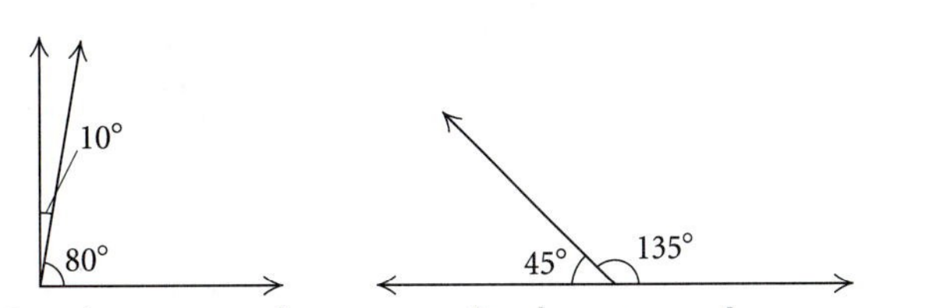

Complementary angles Supplementary angles

Example 2 Find the complement and the supplement of $71.46°$.

SOLUTION We have

$$90° - 71.46° = 18.54°,$$
$$180° - 71.46° = 108.54°.$$

Thus the complement of $71.46°$ is $18.54°$ and the supplement is $108.54°$.

Trigonometric Functions of Angles or Rotations

Many applied problems in trigonometry involve the use of angles that are not acute. Thus we need to extend the domains of the trigonometric functions defined in Section 4.1 to angles, or rotations, of any size. To do this, we first consider a right triangle with one vertex at the origin of a coordinate system and one vertex on the positive x-axis. The other vertex is at P, a point on the circle whose center is at the origin and whose radius r is the length of the hypotenuse of the triangle. This triangle is a **reference triangle** for angle θ, which is in standard position. Note that y is the length of the side opposite θ and x is the length of the side adjacent to θ.

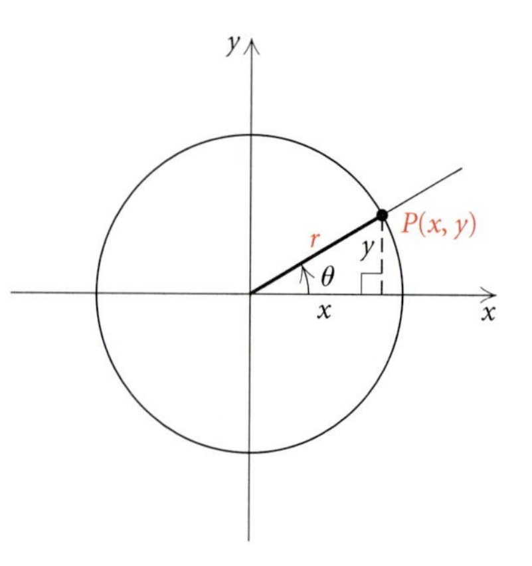

Recalling the definitions in Section 4.1, we note that three of the trigonometric functions of angle θ are defined as follows:

$$\sin\theta = \frac{\text{opp}}{\text{hyp}} = \frac{y}{r}, \qquad \cos\theta = \frac{\text{adj}}{\text{hyp}} = \frac{x}{r}, \qquad \tan\theta = \frac{\text{opp}}{\text{adj}} = \frac{y}{x}.$$

Since x and y are the coordinates of the point P and the length of the radius is the length of the hypotenuse, we can also define these functions as follows:

$$\sin\theta = \frac{y\text{-coordinate}}{\text{radius}},$$

$$\cos\theta = \frac{x\text{-coordinate}}{\text{radius}},$$

$$\tan\theta = \frac{y\text{-coordinate}}{x\text{-coordinate}}.$$

We will use these definitions for functions of angles of any measure. The following figures show angles whose terminal sides lie in quadrants II, III, and IV.

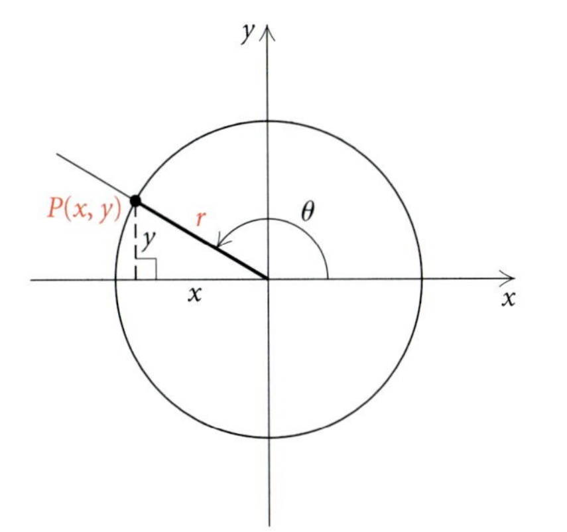

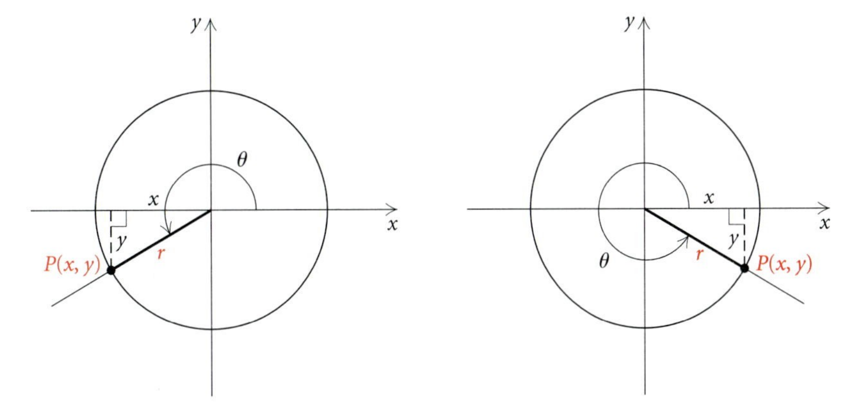

The point P, which is a point other than the vertex on the terminal side of an angle, can be anywhere on a circle of radius r. Each of its two coordinates may be positive, negative, or zero, depending on the location of the terminal side. A reference triangle can be drawn for any angle, as shown. The length of the radius, which is also the length of the hypotenuse of the reference triangle, is always considered positive. The angle is always measured from the positive half of the x-axis. Regardless of the location of P, we have the following definitions.

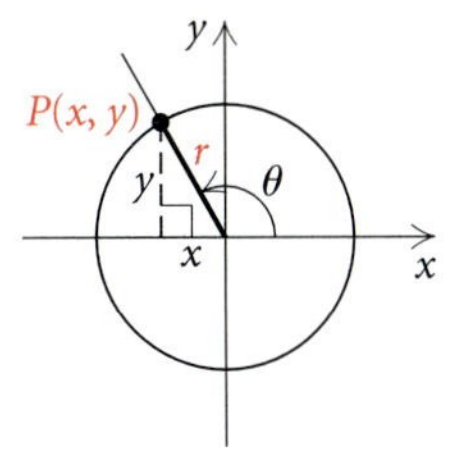

Trigonometric Functions of Any Angle θ

Suppose that $P(x, y)$ is any point on the terminal side of any angle θ in standard position, and r is the radius, or distance, from the origin of $P(x, y)$. Then the trigonometric functions are defined as follows.

$$\sin\theta = \frac{y\text{-coordinate}}{\text{radius}} = \frac{y}{r} \qquad \csc\theta = \frac{\text{radius}}{y\text{-coordinate}} = \frac{r}{y}$$

$$\cos\theta = \frac{x\text{-coordinate}}{\text{radius}} = \frac{x}{r} \qquad \sec\theta = \frac{\text{radius}}{x\text{-coordinate}} = \frac{r}{x}$$

$$\tan\theta = \frac{y\text{-coordinate}}{x\text{-coordinate}} = \frac{y}{x} \qquad \cot\theta = \frac{x\text{-coordinate}}{y\text{-coordinate}} = \frac{x}{y}$$

Function values of the trigonometric functions can be positive, negative, or zero, depending on where the terminal side of the angle lies. The length of the radius is always positive. Thus the signs of the function values depend only on the coordinates of the point P on the terminal side. In the first quadrant, all function values are positive because both coordinates are positive. In the second quadrant, first coordinates are negative and second coordinates are positive; thus only the sine and the cosecant values are positive. Similarly, we can determine the signs, + or −, of the function values in the third and fourth quadrants. Because of the reciprocal relationships, we need to learn only the signs for the sine, cosine, and tangent functions.

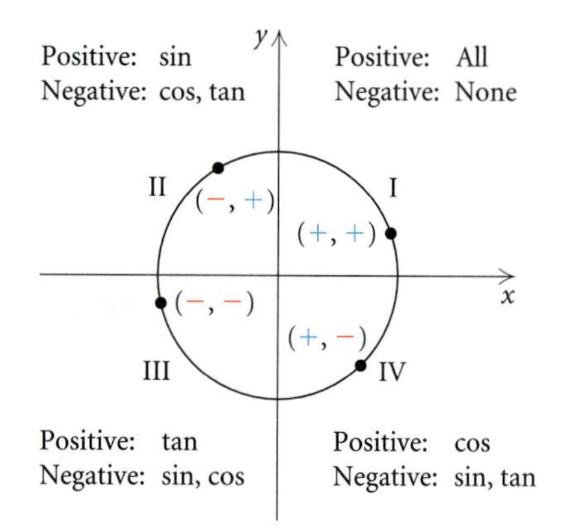

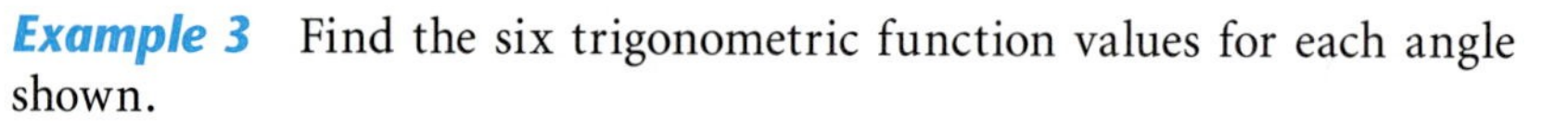

Example 3 Find the six trigonometric function values for each angle shown.

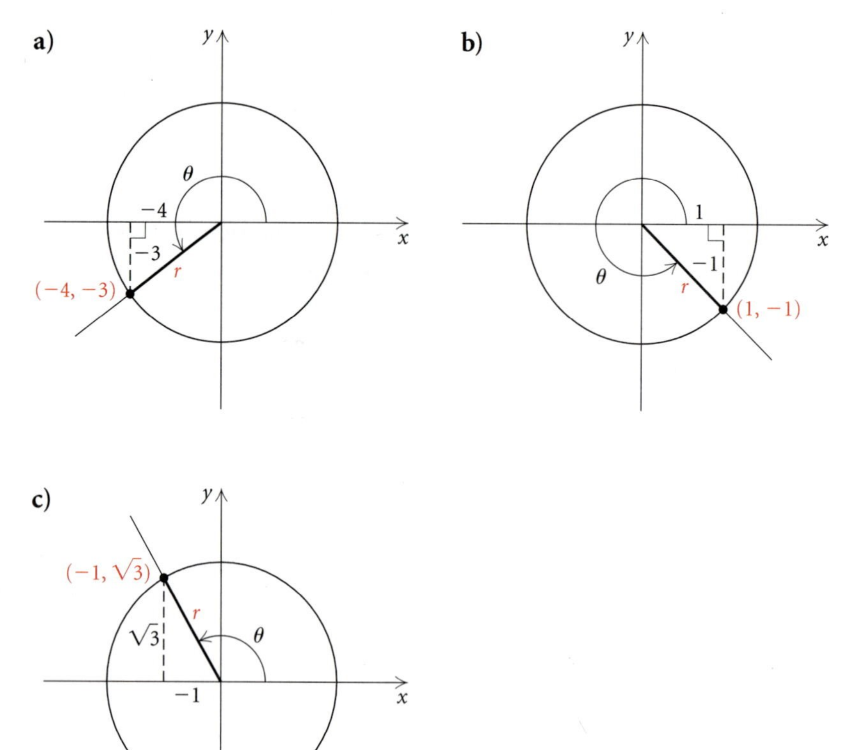

SOLUTION

a) We first determine r, the distance from the origin $(0, 0)$ to the point $(-4, -3)$. The distance between $(0, 0)$ and any point (x, y) on the terminal side is

$$\begin{aligned} r &= \sqrt{(x-0)^2 + (y-0)^2} \\ &= \sqrt{x^2 + y^2}. \end{aligned}$$

Substituting -4 for x and -3 for y, we find

$$\begin{aligned} r &= \sqrt{(-4)^2 + (-3)^2} \\ &= \sqrt{16 + 9} = \sqrt{25} = 5. \end{aligned}$$

Using the definitions of the trigonometric functions, we can now find the function values of θ. We substitute -4 for x, -3 for y, and 5 for r:

$$\sin\theta = \frac{y}{r} = \frac{-3}{5} = -0.6, \qquad \csc\theta = \frac{r}{y} = \frac{5}{-3} \approx -1.67,$$

$$\cos\theta = \frac{x}{r} = \frac{-4}{5} = -0.8, \qquad \sec\theta = \frac{r}{x} = \frac{5}{-4} = -1.25,$$

$$\tan\theta = \frac{y}{x} = \frac{-3}{-4} = 0.75, \qquad \cot\theta = \frac{x}{y} = \frac{-4}{-3} \approx 1.33.$$

As expected, the tangent and the cotangent values are positive and the other four are negative. This is true for all angles in quadrant III.

b) We first determine r, the distance from the origin to the point $(1, -1)$:

$$r = \sqrt{1^2 + (-1)^2} = \sqrt{1 + 1} = \sqrt{2}.$$

Substituting 1 for x, -1 for y, and $\sqrt{2}$ for r, we find

$$\sin\theta = \frac{y}{r} = \frac{-1}{\sqrt{2}} = -\frac{\sqrt{2}}{2}, \qquad \csc\theta = \frac{r}{y} = \frac{\sqrt{2}}{-1} = -\sqrt{2},$$

$$\cos\theta = \frac{x}{r} = \frac{1}{\sqrt{2}} = \frac{\sqrt{2}}{2}, \qquad \sec\theta = \frac{r}{x} = \frac{\sqrt{2}}{1} = \sqrt{2},$$

$$\tan\theta = \frac{y}{x} = \frac{-1}{1} = -1, \qquad \cot\theta = \frac{x}{y} = \frac{1}{-1} = -1.$$

c) We determine r, the distance from the origin to the point $(-1, \sqrt{3})$:

$$r = \sqrt{(-1)^2 + (\sqrt{3})^2} = \sqrt{1 + 3} = \sqrt{4} = 2.$$

Substituting -1 for x, $\sqrt{3}$ for y, and 2 for r, we find the trigonometric function values of θ are

$$\sin\theta = \frac{\sqrt{3}}{2}, \qquad \csc\theta = \frac{2}{\sqrt{3}} = \frac{2\sqrt{3}}{3},$$

$$\cos\theta = \frac{-1}{2} = -\frac{1}{2}, \qquad \sec\theta = \frac{2}{-1} = -2,$$

$$\tan\theta = \frac{\sqrt{3}}{-1} = -\sqrt{3}, \qquad \cot\theta = \frac{-1}{\sqrt{3}} = -\frac{\sqrt{3}}{3}.$$

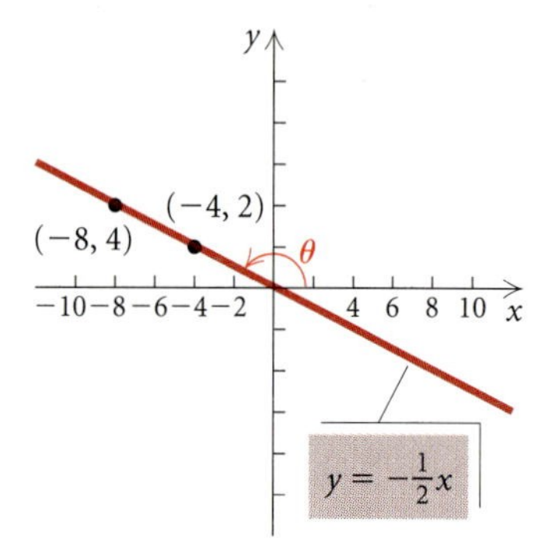

Any point other than the origin on the terminal side can be used to determine the trigonometric function values of that angle. The function values are the same regardless of which point is used. To illustrate this, let's consider an angle θ in standard position whose terminal side lies on the line $y = -\frac{1}{2}x$. We can determine two second-quadrant solutions of the equation, find the length r for each point, and then compare the sine, cosine, and tangent function values using each point.

If $x = -4$, then $y = -\frac{1}{2}(-4) = 2$.
If $x = -8$, then $y = -\frac{1}{2}(-8) = 4$.
For $(-4, 2)$, $r = \sqrt{(-4)^2 + 2^2} = \sqrt{20} = 2\sqrt{5}$.
For $(-8, 4)$, $r = \sqrt{(-8)^2 + 4^2} = \sqrt{80} = 4\sqrt{5}$.

Using $(-4, 2)$ and $r = 2\sqrt{5}$, we find that

$$\sin \theta = \frac{2}{2\sqrt{5}} = \frac{\sqrt{5}}{5}, \qquad \cos \theta = \frac{-4}{2\sqrt{5}} = -\frac{2\sqrt{5}}{5},$$

and $$\tan \theta = \frac{2}{-4} = -\frac{1}{2}.$$

Using $(-8, 4)$ and $r = 4\sqrt{5}$, we find that

$$\sin \theta = \frac{4}{4\sqrt{5}} = \frac{\sqrt{5}}{5}, \qquad \cos \theta = \frac{-8}{4\sqrt{5}} = -\frac{2\sqrt{5}}{5},$$

and $$\tan \theta = \frac{4}{-8} = -\frac{1}{2}.$$

We see that the function values are the same using either point. Any point other than the origin on the terminal side of an angle can be used to determine the trigonometric function values.

> The trigonometric function values of θ depend only on the angle, not on the choice of the point on the terminal side.

The Six Functions Related

When we know one of the function values of an angle, we can find the other five if we know the quadrant in which the terminal side lies. The idea is to sketch a reference triangle in the appropriate quadrant, use the Pythagorean theorem as needed to find the lengths of its sides, and then read off the ratios of the sides.

Example 4 Given that $\tan \theta = -\frac{2}{3}$ and θ is in the second quadrant, find the other function values.

SOLUTION We first sketch a second-quadrant angle. Since

$$\tan \theta = \frac{y}{x} = -\frac{2}{3} = \frac{2}{-3},$$

Expressing $-\frac{2}{3}$ as $\frac{2}{-3}$ since θ is in quadrant II

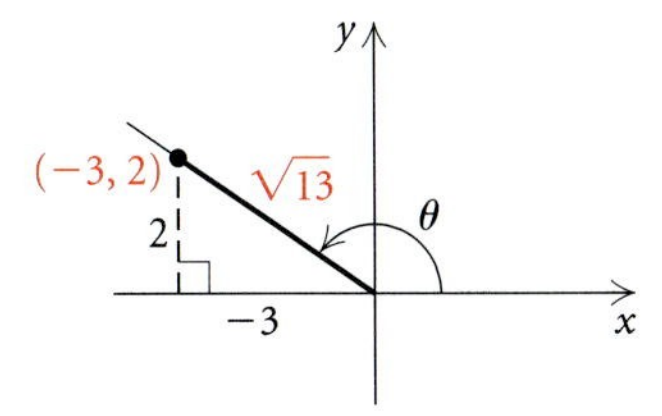

we make the legs of lengths 2 and 3. The hypotenuse must then have length $\sqrt{13}$. Now we can read off the appropriate ratios:

$$\sin\theta = \frac{2}{\sqrt{13}}, \quad \text{or} \quad \frac{2\sqrt{13}}{13}, \qquad \csc\theta = \frac{\sqrt{13}}{2},$$

$$\cos\theta = -\frac{3}{\sqrt{13}}, \quad \text{or} \quad -\frac{3\sqrt{13}}{13}, \qquad \sec\theta = -\frac{\sqrt{13}}{3},$$

$$\tan\theta = -\frac{2}{3}, \qquad \cot\theta = -\frac{3}{2}.$$

Terminal Side on an Axis

Suppose that the terminal side of an angle falls on one of the axes. Then one of the coordinates of any point on that side is 0. The definitions of the functions still apply, but in some cases, functions will not be defined because a denominator will be 0.

Example 5 Find the sine, cosine, and tangent values for 90°, 180°, 270°, and 360°.

SOLUTION We first draw a sketch of each angle in standard position and label a point on the terminal side. Since the function values are the same for all points on the terminal side, we choose (0, 1), (−1, 0), (0, −1), and (1, 0) for convenience. Note also that $r = 1$ for each choice.

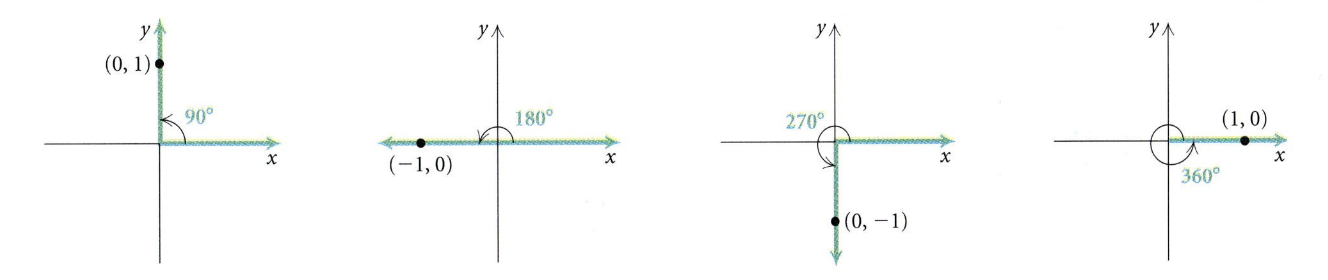

Then by the definitions we get

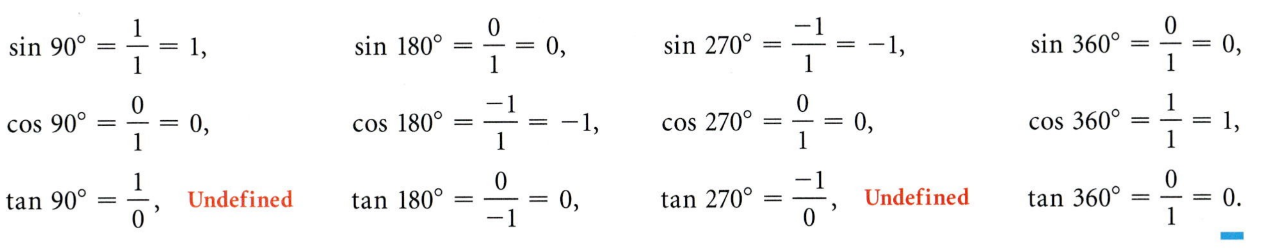

$$\sin 90° = \frac{1}{1} = 1, \qquad \sin 180° = \frac{0}{1} = 0, \qquad \sin 270° = \frac{-1}{1} = -1, \qquad \sin 360° = \frac{0}{1} = 0,$$

$$\cos 90° = \frac{0}{1} = 0, \qquad \cos 180° = \frac{-1}{1} = -1, \qquad \cos 270° = \frac{0}{1} = 0, \qquad \cos 360° = \frac{1}{1} = 1,$$

$$\tan 90° = \frac{1}{0}, \ \textbf{Undefined} \qquad \tan 180° = \frac{0}{-1} = 0, \qquad \tan 270° = \frac{-1}{0}, \ \textbf{Undefined} \qquad \tan 360° = \frac{0}{1} = 0.$$

In Example 5, all the values could be found using a grapher, but you will find that it is convenient to be able to compute them mentally. It is also helpful to note that coterminal angles have the same function values.

Example 6 Find each of the following.

a) $\sin(-90°)$ **b)** $\csc 540°$

SOLUTION

a) We observe that $-90°$ is coterminal with $270°$. Thus,

$$\sin(-90°) = \sin 270° = \frac{-1}{1} = -1.$$

b) Since $540° = 180° + 360°$, $540°$ and $180°$ are coterminal. Thus,

$$\csc 540° = \csc 180° = \frac{1}{\sin 180°} = \frac{1}{0}, \quad \text{which is undefined.}$$

Trigonometric values can always be checked with a grapher. When the value is undefined, the display will read

ERR: DOMAIN or ERR: DIVIDE BY 0

Reference Angles: 30°, 45°, and 60°

We can also mentally determine trigonometric function values whenever the terminal side makes a 30°, 45°, or 60° angle with the x-axis. Consider, for example, an angle of 150°. The terminal side makes a 30° angle with the x-axis, since $180° - 150° = 30°$. As the figure shows, $\triangle ONP$ is congruent to $\triangle ON'P'$; then the ratios of the sides of the two triangles are the same.

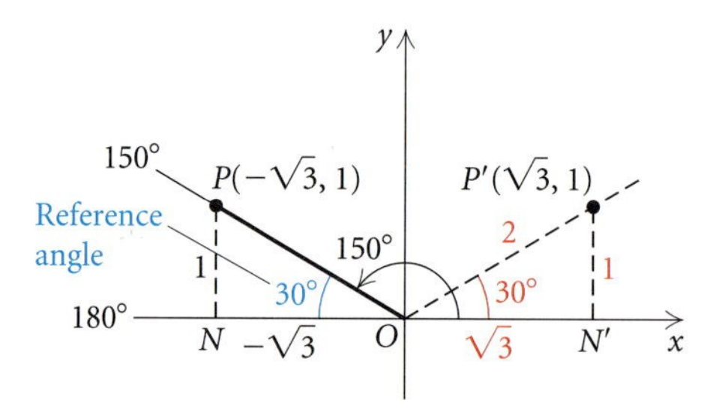

Thus the trigonometric function values are the same except perhaps for the sign. We could determine the function values directly from $\triangle ONP$, but this is not necessary. If we remember that in quadrant II, the sine is positive and the cosine and the tangent are negative, we can simply use the function values of 30° that we already know and prefix the appropriate sign. Thus,

$$\sin 150° = \sin 30° = \frac{1}{2}, \qquad \cos 150° = -\cos 30° = -\frac{\sqrt{3}}{2},$$

and $$\tan 150° = -\tan 30° = -\frac{\sqrt{3}}{3}.$$

Triangle ONP is called a *reference triangle* and its acute angle, $\angle NOP$, is called a *reference angle*.

Reference Angle

The *reference angle* for an angle is the acute angle formed by the terminal side and the x-axis.

Example 7 Find the sine, cosine, and tangent function values for each of the following.

a) 225° **b)** -780°

SOLUTION

a) We draw a figure showing the terminal side of a 225° angle. The reference angle is $225^\circ - 180^\circ$, or 45°.

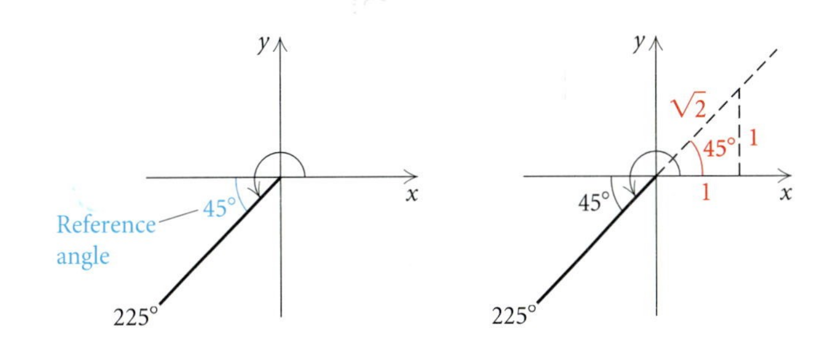

Recall from Section 4.1 that $\sin 45^\circ = \sqrt{2}/2$, $\cos 45^\circ = \sqrt{2}/2$, and $\tan 45^\circ = 1$. Also note that in the third quadrant, the sine and the cosine are negative and the tangent is positive. Thus we have

$$\sin 225^\circ = -\frac{\sqrt{2}}{2}, \qquad \cos 225^\circ = -\frac{\sqrt{2}}{2}, \quad \text{and} \quad \tan 225^\circ = 1.$$

b) We draw a figure showing the terminal side of a -780° angle. Since $-780^\circ + 2(360^\circ) = -60^\circ$, we know that -780° and -60° are coterminal.

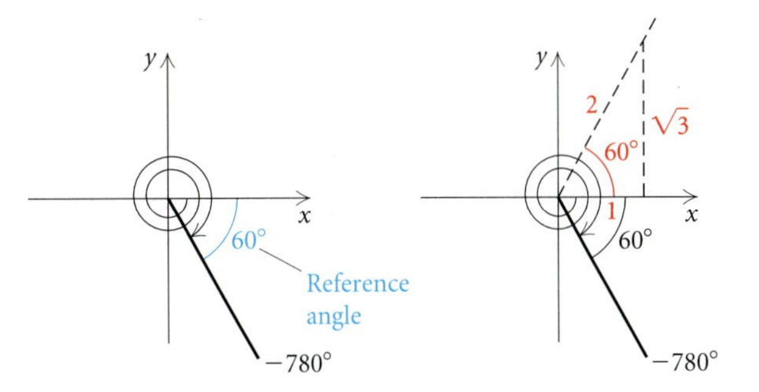

The reference angle for -60° is the acute angle formed by the terminal side and the x-axis. Thus the reference angle for -60° is 60°. We know that since -780° is a fourth-quadrant angle, the cosine is positive and the sine and the tangent are negative. Recalling that

$\sin 60° = \sqrt{3}/2$, $\cos 60° = 1/2$, and $\tan 60° = \sqrt{3}$, we have

$$\sin(-780°) = -\frac{\sqrt{3}}{2}, \qquad \cos(-780°) = \frac{1}{2},$$

and $\tan(-780°) = -\sqrt{3}$.

Function Values for Any Angle

When the terminal side of an angle falls on one of the axes or makes a 30°, 45°, or 60° angle with the x-axis, we can find exact function values without the use of a grapher. We can also find approximations of these values with a grapher. But this group is only a small subset of *all* angles. With a grapher, we can approximate the trigonometric function values of *any* angle. In fact, we can approximate function values of all angles without using a reference angle.

Example 8 Find each of the following function values.

a) $\cos 112°$ **b)** $\sec 500°$ **c)** $\tan(-83.4°)$

d) $\csc 351.75°$ **e)** $\cos 2400°$ **f)** $\tan 495°15'$

g) $\sin 175°40'9''$ **h)** $\cot(-135°)$

SOLUTION

FUNCTION VALUE	READOUT	ROUNDED
a) $\cos 112°$	-.3746065934	-0.3746
b) $\sec 500° = \dfrac{1}{\cos 500°}$	-1.305407289	-1.3054
c) $\tan(-83.4°)$	-8.642747461	-8.6427
d) $\csc 351.75° = \dfrac{1}{\sin 351.75°}$	-6.968999424	-6.9690
e) $\cos 2400°$	-.5	-0.5
f) $\tan 495°15' = \tan 495.25°$	-.9913112106	-0.9913
g) $\sin 175°40'9'' \approx \sin 175.67°$	.0755008414	0.0755
h) $\cot(-135°) = \dfrac{1}{\tan(-135°)}$	1	1

In many applications, we have a trigonometric function value and want to find the measure of an angle. To do so, we use either the TABLE feature or an inverse key on a grapher. When only acute angles are considered, there is only one output (angle) for each input (the trigonometric function value). This is not the case when we extend the domain of the trigonometric functions to the set of *all* angles. For a given function value, there is an infinite number of angles that have that function value. But there are only two such angles for each value in the range from 0° to 360°. To determine a unique answer, the quadrant in which the terminal side lies must be specified. The grapher deals with this situation by giving the reference angle as an output for each function value that is entered as an input. With this reference angle and knowing the quadrant in which the terminal side lies, we can find the specified angle.

Example 9 Given the function value and the quadrant restriction, find θ.

a) $\sin\theta = 0.2812, \quad 90° < \theta < 180°$

b) $\cot\theta = -0.1611, \quad 270° < \theta < 360°$

SOLUTION

a) We first sketch the angle in the second quadrant. We then enter 0.2812 and find the reference angle to be approximately 16.33°.

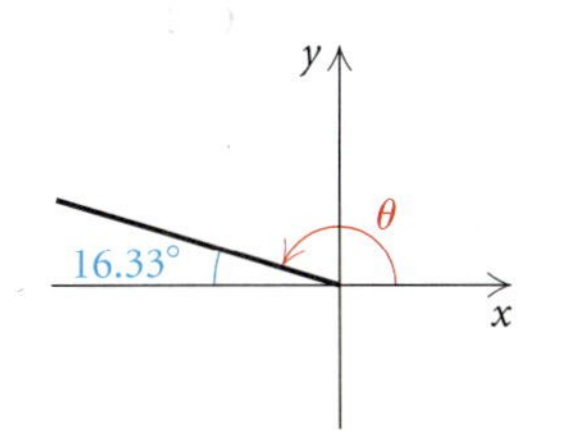

We find the angle θ by subtracting 16.33° from 180°:

$$180° - 16.33° = 163.67°.$$

Thus, $\theta \approx 163.67°$.

b) We begin by sketching the angle in the fourth quadrant. Because the cotangent value is the reciprocal of the tangent value, we know that

$$\tan\theta \approx \frac{1}{-0.1611}$$
$$\approx -6.2073.$$

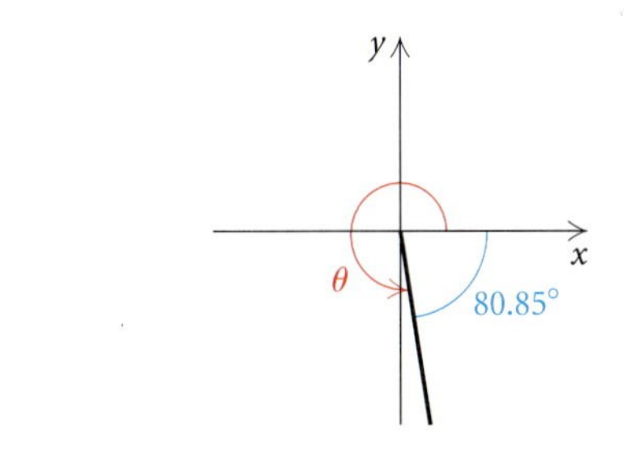

We then enter 6.2073, ignoring the fact that $\tan\theta$ is negative, and find that the reference angle is approximately 80.85°. We find angle θ by subtracting 80.85° from 360°:

$$360° - 80.85° = 279.15°.$$

Thus, $\theta \approx 279.15°$.

4.3 Exercise Set

For angles of the following measures, state in which quadrant the terminal side lies.

1. $187°$ **2.** $-14.3°$ **3.** $245°15'$
4. $-120°$ **5.** $800°$ **6.** $1075°$
7. $-460.5°$ **8.** $315°$ **9.** $-912°$
10. $13°15'60''$ **11.** $537°$ **12.** $-345.14°$

Find two positive angles and two negative angles that are coterminal with the given angle. Answers may vary.

13. $74°$ **14.** $-81°$ **15.** $115.3°$
16. $275°10'$ **17.** $-180°$ **18.** $-310°$

Find the complement and the supplement.

19. $17.11°$ **20.** $47°38'$ **21.** $12°3'14''$
22. $9.038°$ **23.** $45.2°$ **24.** $67.31°$

Find the six trigonometric function values for the angle shown.

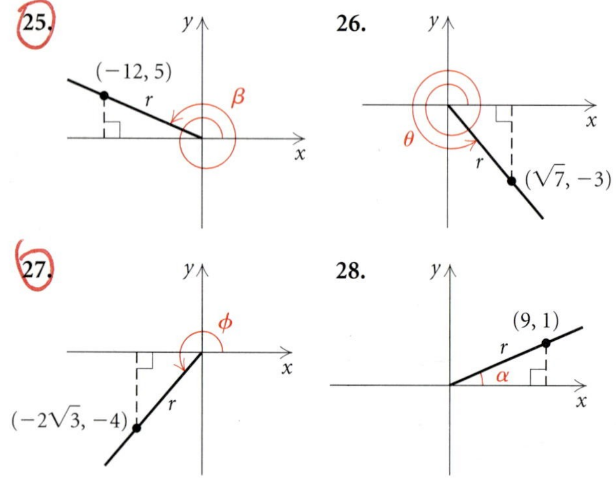

The terminal side of angle θ in standard position lies on the given line in the given quadrant. Find $\sin\theta$, $\cos\theta$, and $\tan\theta$.

29. $2x + 3y = 0$; quadrant IV
30. $4x + y = 0$; quadrant II
31. $5x - 4y = 0$; quadrant I
32. $y = 0.8x$; quadrant III

A function value and a quadrant are given. Find the other five function values.

33. $\sin\theta = -\frac{1}{3}$, quadrant III
34. $\tan\beta = 5$, quadrant I
35. $\cot\theta = -2$, quadrant IV
36. $\cos\alpha = -\frac{4}{5}$, quadrant II
37. $\cos\phi = \frac{3}{5}$, quadrant IV
38. $\sin\theta = -\frac{5}{13}$, quadrant III

Find the exact function value if it exists.

39. $\cos 150°$ **40.** $\sec(-225°)$
41. $\tan(-135°)$ **42.** $\sin(-45°)$
43. $\sin 7560°$ **44.** $\tan 270°$
45. $\cos 495°$ **46.** $\tan 675°$
47. $\csc(-210°)$ **48.** $\sin 300°$
49. $\cot 570°$ **50.** $\cos(-120°)$
51. $\tan 330°$ **52.** $\cot 855°$
53. $\sec(-90°)$ **54.** $\sin 90°$
55. $\cos(-180°)$ **56.** $\csc 90°$
57. $\tan 240°$ **58.** $\cot(-180°)$
59. $\sin 495°$ **60.** $\sin 1050°$
61. $\csc 225°$ **62.** $\sin(-450°)$
63. $\cos 0°$ **64.** $\tan 480°$
65. $\cot(-90°)$ **66.** $\sec 315°$
67. $\cos 90°$ **68.** $\sin(-135°)$
69. $\cos 270°$ **70.** $\tan 0°$

Find the signs of the six trigonometric function values for the given angles.

71. $319°$ **72.** $-57°$
73. $194°$ **74.** $-620°$
75. $-215°$ **76.** $290°$
77. $-272°$ **78.** $91°$

Use a calculator in Exercises 79–82, but do not use the trigonometric function keys.

79. Given that

$$\sin 41° = 0.6561,\quad \cos 41° = 0.7547,\quad \tan 41° = 0.8693,$$

find the trigonometric function values for $319°$.

80. Given that

$$\sin 27° = 0.4540,\quad \cos 27° = 0.8910,\quad \tan 27° = 0.5095,$$

find the trigonometric function values for $333°$.

81. Given that

$$\sin 65^\circ = 0.9063,$$
$$\cos 65^\circ = 0.4226,$$
$$\tan 65^\circ = 2.1445,$$

find the trigonometric function values for 115°.

82. Given that

$$\sin 35^\circ = 0.5736,$$
$$\cos 35^\circ = 0.8192,$$
$$\tan 35^\circ = 0.7002,$$

find the trigonometric function values for 215°.

Aerial Navigation. In aerial navigation, directions are given in degrees clockwise from north. Thus, east is 90°, south is 180°, and west is 270°. Several aerial directions or bearings are given below.

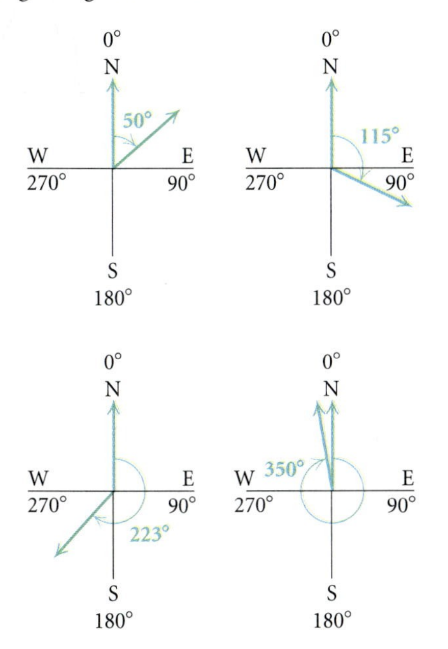

83. An airplane flies 150 km from an airport in a direction of 120°. How far east of the airport is the plane then? How far south?

84. An airplane leaves an airport and travels for 100 mi in a direction of 300°. How far north of the airport is the plane then? How far west?

85. An airplane travels at 150 km/h for 2 hr in a direction of 138° from Omaha. At the end of this time, how far south of Omaha is the plane?

86. An airplane travels at 120 km/h for 2 hr in a direction of 319° from Chicago. At the end of this time, how far north of Chicago is the plane?

Find the function value. Round to four decimal places.

87. $\tan 310.8^\circ$

88. $\cos 205.5^\circ$

89. $\cot 146.15^\circ$

90. $\sin (-16.4^\circ)$

91. $\sin 118^\circ 42'$

92. $\cos 273^\circ 45'$

93. $\cos (-295.8^\circ)$

94. $\tan 1086.2^\circ$

95. $\cos 5417^\circ$

96. $\sec 240^\circ 55'$

97. $\csc 520^\circ$

98. $\sin 3824^\circ$

Given the function value and the quadrant restriction, find θ.

FUNCTION VALUE	INTERVAL	θ
99. $\sin \theta = -0.9956$	$(270^\circ, 360^\circ)$	
100. $\tan \theta = 0.2460$	$(180^\circ, 270^\circ)$	
101. $\cos \theta = -0.9388$	$(180^\circ, 270^\circ)$	
102. $\sec \theta = -1.0485$	$(90^\circ, 180^\circ)$	
103. $\tan \theta = -3.0545$	$(270^\circ, 360^\circ)$	
104. $\sin \theta = -0.4313$	$(180^\circ, 270^\circ)$	
105. $\csc \theta = 1.0480$	$(0^\circ, 90^\circ)$	
106. $\cos \theta = -0.0990$	$(90^\circ, 180^\circ)$	

Skill Maintenance

Find decimal degree notation.

107. $44^\circ 10' 35''$

108. $382^\circ 20' 16''$

109. Convert 45 mph to ft/sec.

110. Convert 8.5 in./sec to ft/hr.

Synthesis

111. ◆ Why do the function values of θ depend only on the angle and not on the choice of a point on the terminal side?

112. ◆ Why is the domain of the tangent function different from the domains of the sine and the cosine functions?

113. *Valve Cap on a Bicycle.* The valve cap on a bicycle wheel is 12.5 in. from the center of the wheel. From the position shown, the wheel starts to roll. After the wheel has turned 390°, how far above the ground is the valve cap? Assume that the outer radius of the tire is 13.375 in.

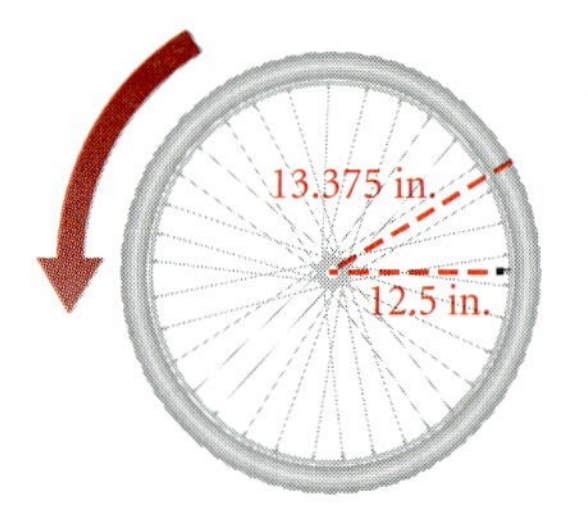

114. *Seats of a Ferris Wheel.* The seats of a ferris wheel are 35 ft from the center of the wheel. When you board the wheel, you are 5 ft above the ground. After you have rotated through an angle of 765°, how far above the ground are you?

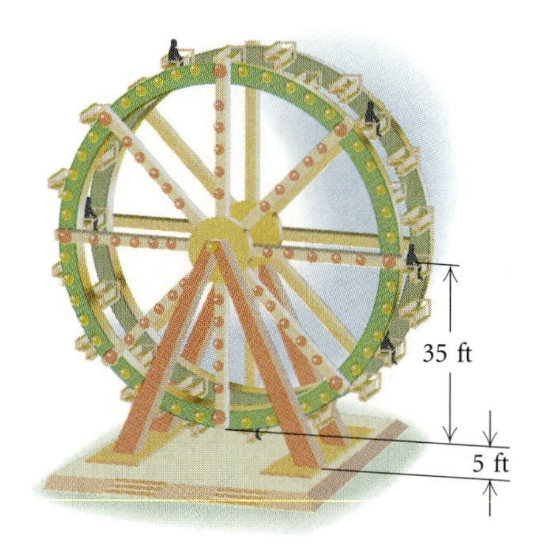

4.4 Radians, Arc Length, and Angular Speed

- *Find points on the unit circle determined by real numbers.*
- *Convert between radian and degree measure; find coterminal, complementary, and supplementary angles.*
- *Find the length of an arc of a circle; find the measure of a central angle of a circle.*
- *Convert between linear speed and angular speed.*

Another useful unit of angle measure is called a *radian*. To introduce radian measure, we use a circle centered at the origin with a radius of length 1. Such a circle is called a **unit circle**. Its equation is $x^2 + y^2 = 1$ (see Section 1.5).

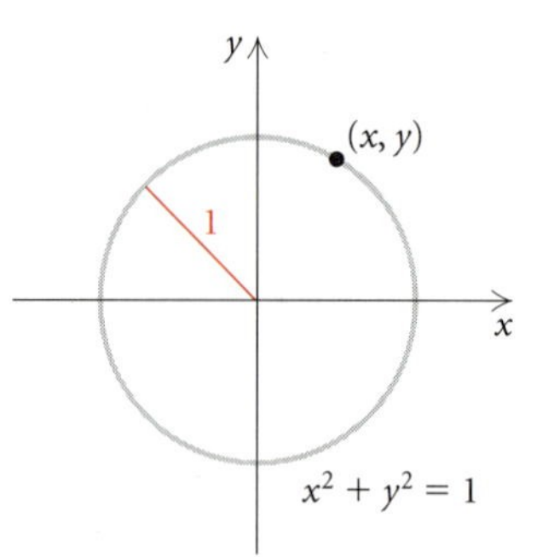

Distances on the Unit Circle

The circumference of a circle of radius r is $2\pi r$. Thus for the unit circle, where $r = 1$, the circumference is 2π. If a point starts at A and travels around the circle (Fig. 1), it will travel a distance of 2π. If it travels halfway around the circle (Fig. 2), it will travel a distance of $\frac{1}{2} \cdot 2\pi$, or π.

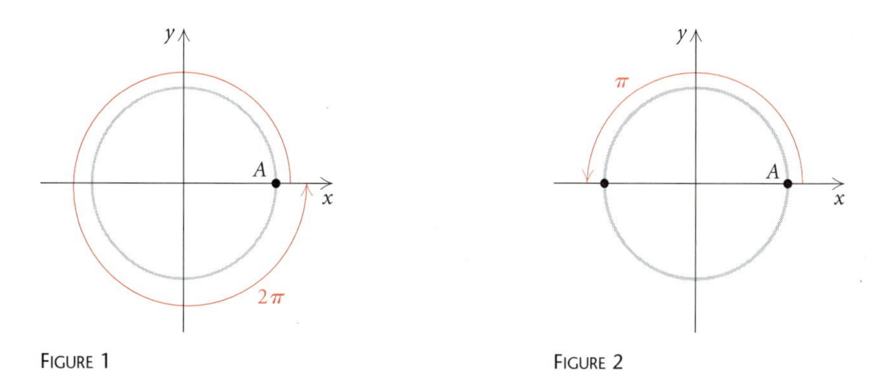

FIGURE 1

FIGURE 2

If a point C travels $\frac{1}{8}$ of the way around the circle (Fig. 3), it will travel a distance of $\frac{1}{8} \cdot 2\pi$, or $\pi/4$. Note that C is $\frac{1}{4}$ of the way from A to B. If a point D travels $\frac{1}{6}$ of the way around the circle (Fig. 4), it will travel a distance of $\frac{1}{6} \cdot 2\pi$, or $\pi/3$. Note that D is $\frac{1}{3}$ of the way from A to B.

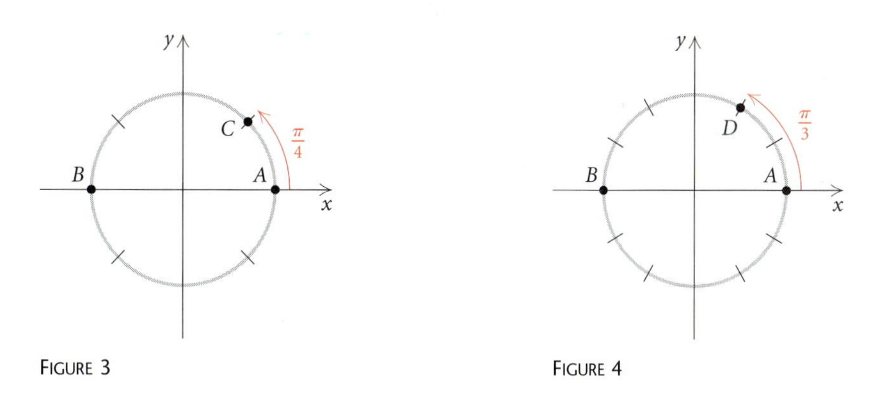

FIGURE 3

FIGURE 4

Example 1 How far will a point travel if it goes (a) $\frac{1}{4}$, (b) $\frac{1}{12}$, (c) $\frac{3}{8}$, and (d) $\frac{5}{6}$ of the way around the unit circle?

SOLUTION

a) $\frac{1}{4}$ of the total distance around the circle is $\frac{1}{4} \cdot 2\pi$, which is $\frac{1}{2} \cdot \pi$, or $\pi/2$.

b) The distance will be $\frac{1}{12} \cdot 2\pi$, which is $\frac{1}{6}\pi$, or $\pi/6$.

c) The distance will be $\frac{3}{8} \cdot 2\pi$, which is $\frac{3}{4}\pi$, or $3\pi/4$.

d) The distance will be $\frac{5}{6} \cdot 2\pi$, which is $\frac{5}{3}\pi$, or $5\pi/3$. Think of $5\pi/3$ as $\pi + \frac{2}{3}\pi$.

These distances are illustrated in the following figures.

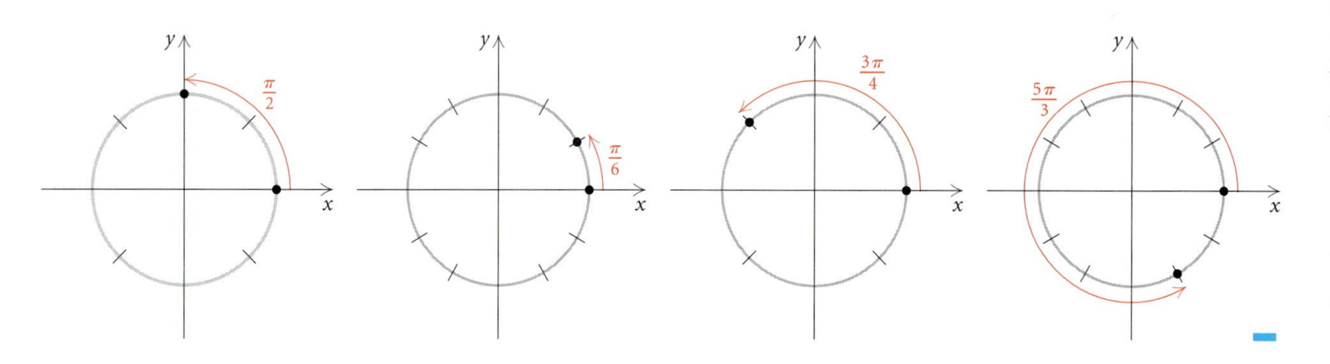

A point may travel completely around the circle and then continue. For example, if it goes around once and then continues $\frac{1}{4}$ of the way around, it will have traveled a distance of $2\pi + \frac{1}{4} \cdot 2\pi$, or $5\pi/2$ (Fig. 5). *Every* real number determines a point on the unit circle. For the positive number 10, for example, we start at A and travel counterclockwise a distance of 10. The point at which we stop is the point "determined" by the number 10 (Fig. 6).

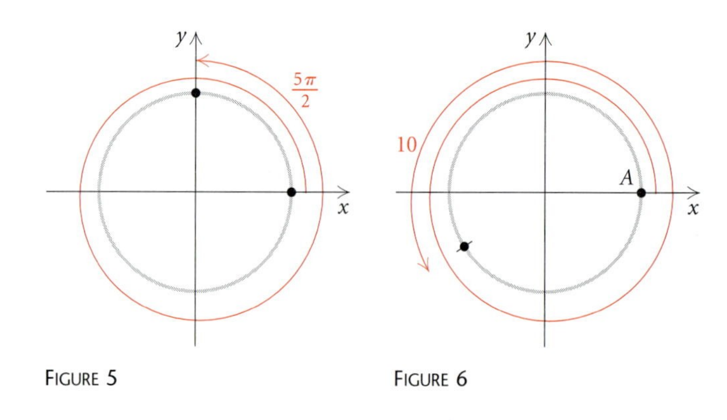

FIGURE 5

FIGURE 6

For a negative number, we move clockwise around the circle. Points for $-\pi/4$ and $-3\pi/2$ are shown in the figure below. The number 0 determines the point A.

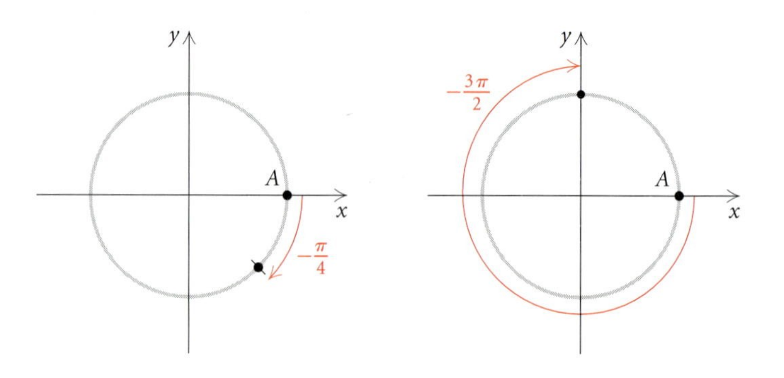

Example 2 On the unit circle, mark the point determined by each of the following real numbers.

a) $\dfrac{9\pi}{4}$ **b)** $-\dfrac{7\pi}{6}$

SOLUTION

a) Think of $9\pi/4$ as $2\pi + \frac{1}{4}\pi$ (see the figure on the left below). Since $9\pi/4 > 0$, the point moves counterclockwise. The point goes completely around once and then continues $\frac{1}{4}$ of the way from A to B.

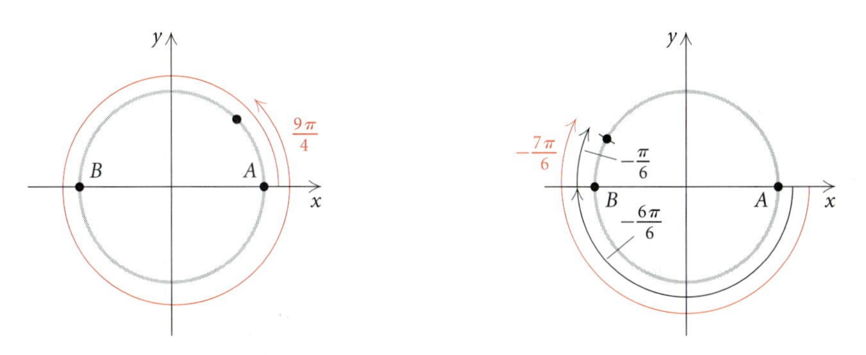

b) The number is negative, so the point moves clockwise. From A to B, the distance is π, or $\frac{6}{6}\pi$, so we need to go beyond B another distance of $\pi/6$, clockwise. (See the figure on the right above.)

Radian Measure

Degree measure is a common unit of angle measure in many everyday applications. But in many scientific fields and in mathematics (calculus, in particular), there is another commonly used unit of measure called the *radian*.

Consider the unit circle whose radius has length 1. Suppose we measure, moving counterclockwise, an arc of length 1, and mark a point T on the circle.

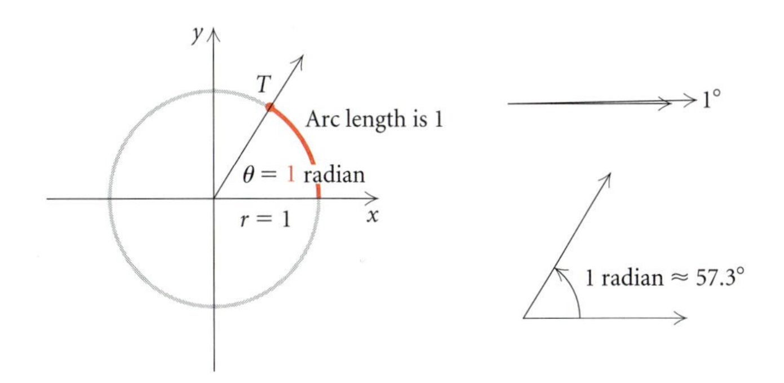

If we draw a ray from the origin through T, we have formed an angle. The measure of that angle is 1 **radian**. The word radian comes from the word *radius*. Thus measuring 1 "radius" along the circumference of the circle determines an angle whose measure is 1 *radian*. One radian is about 57.3°. Angles that measure 2 radians, 3 radians, and 6 radians are shown below.

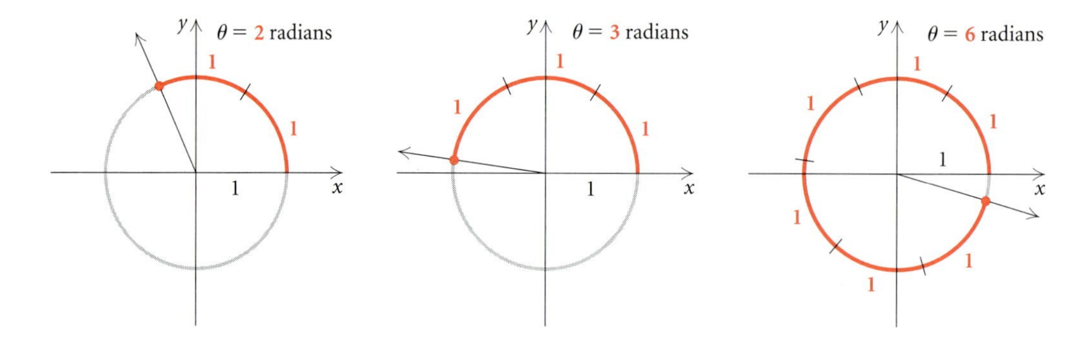

When we make a complete (counterclockwise) revolution, the terminal side coincides with the initial side on the positive x-axis. We then have an angle whose measure is 2π radians, or about 6.28 radians, which is the circumference of the circle:

$$2\pi r = 2\pi(1) = 2\pi.$$

Thus a rotation of 360° (1 revolution) has a measure of 2π radians. A half revolution is a rotation of 180°, or π radians. A quarter revolution is a rotation of 90°, or $\pi/2$ radians, and so on.

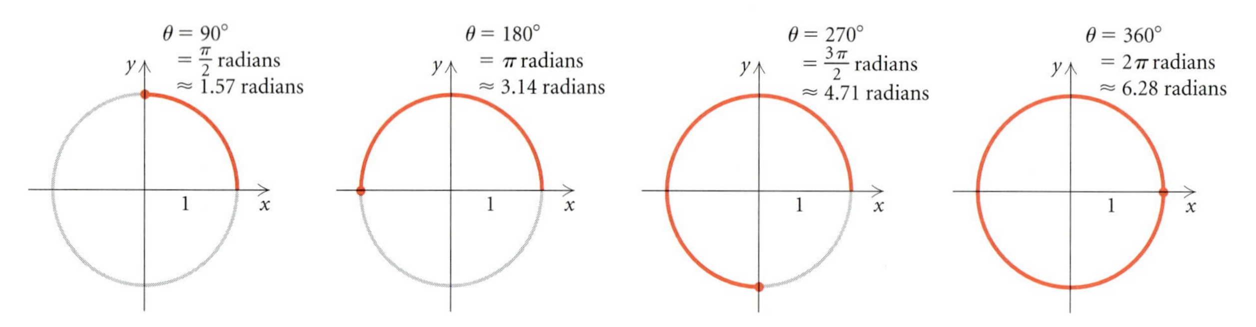

To convert between degrees and radians, we first note that

$$360° = 2\pi \text{ radians}.$$

It follows that

$$180° = \pi \text{ radians}.$$

To make conversions, we multiply by one, noting that:

$$\frac{\pi \text{ radians}}{180°} = \frac{180°}{\pi \text{ radians}} = 1.$$

Some graphers can convert directly from one measure to the other.

Example 3 Convert each of the following to radians.

a) $120°$

b) $-297.25°$

SOLUTION

a) $$120° = 120° \cdot \frac{\pi \text{ radians}}{180°}$$ **Multiplying by 1**

$$= \frac{120°}{180°}\pi \text{ radians}$$

$$= \frac{2\pi}{3} \text{ radians}, \quad \text{or about} \quad 2.09 \text{ radians}$$

b) $$-297.25° = -297.25° \cdot \frac{\pi \text{ radians}}{180°}$$

$$= -\frac{297.25°}{180°}\pi \text{ radians}$$

$$= -\frac{297.25\pi}{180} \text{ radians}$$

$$\approx -5.19 \text{ radians}$$

Example 4 Convert each of the following to degrees.

a) $\frac{3\pi}{4}$ radians

b) 8.5 radians

SOLUTION

a) $$\frac{3\pi}{4} \text{ radians} = \frac{3\pi}{4} \text{ radians} \cdot \frac{180°}{\pi \text{ radians}}$$ **Multiplying by 1**

$$= \frac{3\pi}{4\pi} \cdot 180° = \frac{3}{4} \cdot 180° = 135°$$

b) $$8.5 \text{ radians} = 8.5 \text{ radians} \cdot \frac{180°}{\pi \text{ radians}}$$

$$= \frac{8.5(180°)}{\pi} \approx 487.01°$$

The radian–degree equivalents of the most commonly used angle measures are illustrated in the following figures.

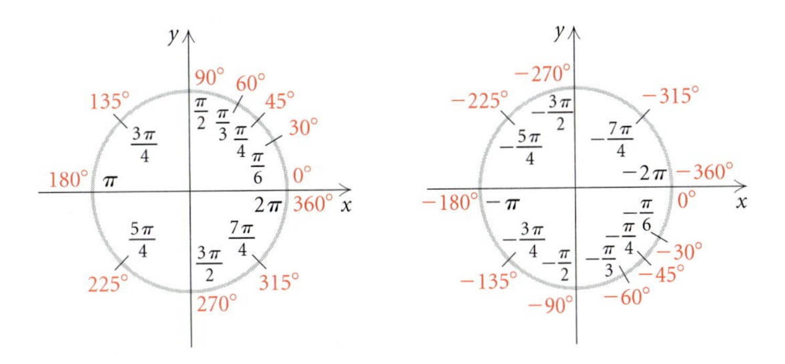

When a rotation is given in radians, the word "radians" is optional and is most often omitted. Thus if no unit is given for a rotation, the rotation is understood to be in radians.

We can also find coterminal, complementary, and supplementary angles in radian measure just as we did for degree measure in Section 4.3.

Example 5 Find a positive angle and a negative angle that are coterminal with $2\pi/3$. Many answers are possible.

SOLUTION To find angles coterminal with a given angle, we add or subtract multiples of 2π:

$$\frac{2\pi}{3} + 2\pi = \frac{2\pi}{3} + \frac{6\pi}{3} = \frac{8\pi}{3},$$

$$\frac{2\pi}{3} - 3(2\pi) = \frac{2\pi}{3} - \frac{18\pi}{3} = -\frac{16\pi}{3}.$$

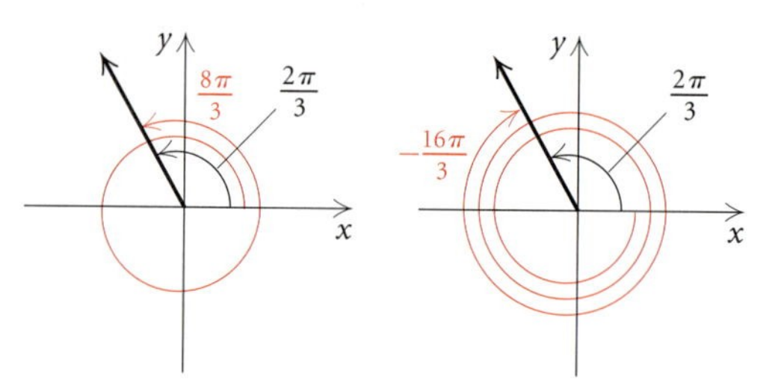

Thus, $8\pi/3$ and $-16\pi/3$ are coterminal with $2\pi/3$.

Example 6 Find the complement and the supplement of $\pi/6$.

SOLUTION Since 90° equals $\pi/2$ radians, the complement of $\pi/6$ is

$$\frac{\pi}{2} - \frac{\pi}{6} = \frac{3\pi}{6} - \frac{\pi}{6} = \frac{2\pi}{6}, \quad \text{or} \quad \frac{\pi}{3}.$$

Since 180° equals π radians, the supplement of $\pi/6$ is

$$\pi - \frac{\pi}{6} = \frac{6\pi}{6} - \frac{\pi}{6} = \frac{5\pi}{6}.$$

Thus the complement of $\pi/6$ is $\pi/3$ and the supplement is $5\pi/6$.

Arc Length and Central Angles

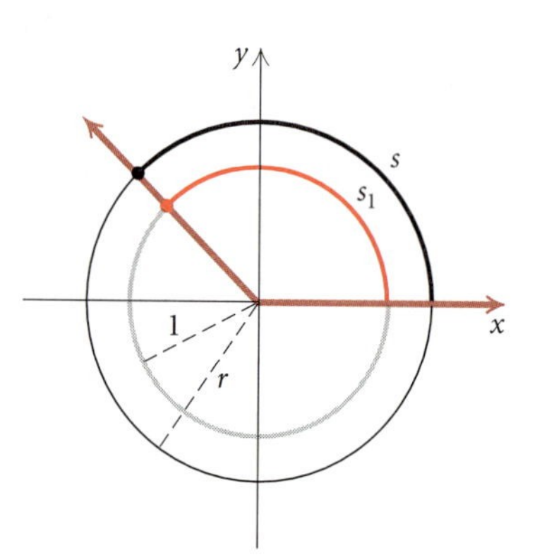

Radian measure can be determined using a circle other than a unit circle. In the figure at left, a unit circle (with radius 1) is shown along with another circle (with radius r). The angle shown is a central angle of both circles.

From geometry, we know that the arcs that the angle subtends have their lengths in the same ratio as the radii of the circles. The radii of the circles are r and 1. The corresponding arc lengths are s and s_1. Thus we have the proportion

$$\frac{s}{s_1} = \frac{r}{1},$$

which can also be written as

$$\frac{s_1}{1} = \frac{s}{r}.$$

Now s_1 is the *radian measure* of the rotation in question. It is common to use a Greek letter, such as θ, for the measure of an angle or rotation and the letter s for arc length. Adopting this convention, we rewrite the proportion above as

$$\theta = \frac{s}{r}.$$

In any circle, the measure (in radians) of a central angle, the arc length the angle subtends, and the length of the radius are related in this fashion. Or, in general, the following is true.

Radian Measure

The *radian measure* θ of a rotation is the ratio of the distance s traveled by a point at a radius r from the center of rotation, to the length of the radius r:

$$\theta = \frac{s}{r}.$$

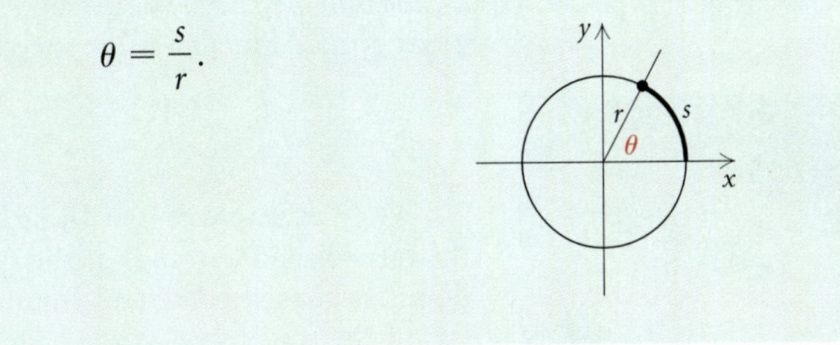

In using the above formula $\theta = s/r$, you must make sure that θ is in radians and that s and r are expressed in the same unit.

Example 7 Find the measure of a rotation in radians where a point 2 m from the center of rotation travels 4 m.

SOLUTION

$$\theta = \frac{s}{r} = \frac{4 \text{ m}}{2 \text{ m}} = 2$$ **The unit is understood to be radians.**

Example 8 Find the length of an arc of a circle of radius 5 cm associated with an angle of $\pi/3$ radians.

SOLUTION We have

$$\theta = \frac{s}{r}, \quad \text{or} \quad s = r\theta.$$

Therefore, $s = 5 \cdot \pi/3$ cm, or about 5.24 cm.

A look at Examples 7 and 8 will show why the word radian is most often omitted. In Example 7, we have the division 4 m/2 m, which simplifies to the number 2, since m/m = 1. From this point of view, it would seem preferable to omit the word radians. In Example 8, had we used the word radians all the way through, our answer would have come out to be 5.23 cm-radians. It is a distance we seek; hence we know the unit should be centimeters. Thus we must omit the word radians. Since a measure in radians is simply a real number, it is usually preferable to omit the word radians.

Linear Speed and Angular Speed

Linear speed is defined to be distance traveled per unit of time. If we use v for linear speed, s for distance, and t for time, then

$$v = \frac{s}{t}.$$

Similarly, angular speed is defined to be amount of rotation per unit of time. For example, we might speak of the angular speed of a bicycle wheel as 150 revolutions per minute or the angular speed of the earth as 2π radians per day. The Greek letter ω (omega) is generally used for angular speed. Thus angular speed is defined as

$$\omega = \frac{\theta}{t}.$$

As an example of how these definitions can be applied, let's consider the refurbished carousel at the Children's Museum in Indianapolis, Indiana. It consists of three circular rows of animals. All animals, regardless of the row, travel at the same angular speed. But the animals in the outer row travel at a greater linear speed than those in the inner rows. What is the relationship between the linear speed v and the angular speed ω?

To develop the relationship we seek, recall that $\theta = s/r$. This is equivalent to

$$s = r\theta.$$

We divide by time, t, to obtain

$$\frac{s}{t} = r\frac{\theta}{t}.$$

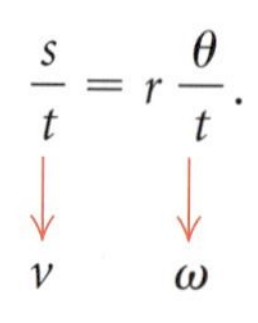

Now s/t is linear speed v and θ/t is angular speed ω. Thus we have the relationship we seek, $v = r\omega$.

Linear Speed in Terms of Angular Speed

The *linear speed* v of a point a distance r from the center of rotation is given by

$$v = r\omega,$$

where ω is the *angular speed* in radians per unit of time.

For our new formula $v = r\omega$, the units of distance for v and r must be the same, ω must be in radians per unit of time, and the units of time for v and ω must be the same.

Example 9 *Linear Speed of an Earth Satellite.* An earth satellite in circular orbit 1200 km high makes one complete revolution every 90 min. What is its linear speed? Use 6400 km for the length of a radius of the earth.

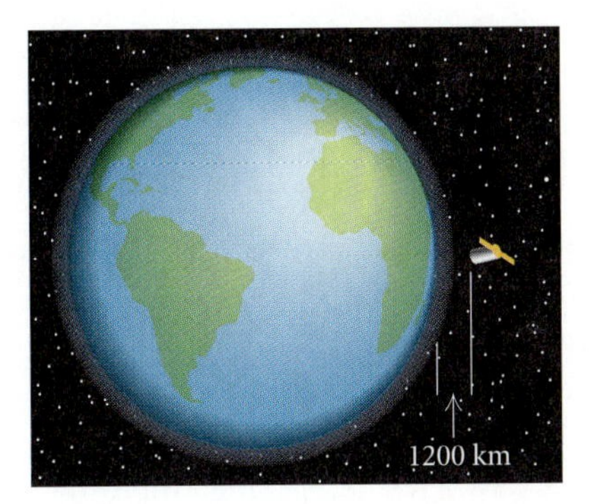

SOLUTION To use the formula $v = r\omega$, we will need to know r and ω:

$$r = 6400 \text{ km} + 1200 \text{ km}$$ **Radius of earth plus height of satellite**

$$= 7600 \text{ km},$$

$$\omega = \frac{\theta}{t} = \frac{2\pi}{90 \text{ min}} = \frac{\pi}{45 \text{ min}}.$$ **We have, as usual, omitted the word radians.**

Now, using $v = r\omega$, we have

$$v = 7600 \text{ km} \cdot \frac{\pi}{45 \text{ min}}$$

$$= \frac{7600\pi}{45} \cdot \frac{\text{km}}{\text{min}} \approx 531 \frac{\text{km}}{\text{min}}.$$

Thus the linear speed of the satellite is approximately 531 km/min.

Example 10 *Angular Speed of a Capstan.* An anchor is hoisted at a rate of 2 ft/sec as the chain is wound around a capstan with a 1.8-yd diameter. What is the angular speed of the capstan?

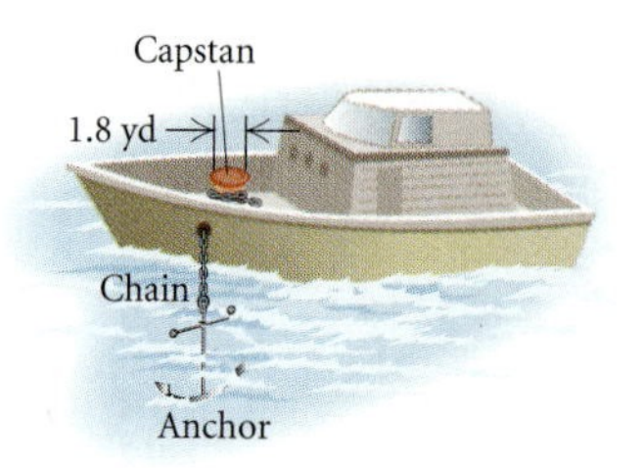

SOLUTION We will use the formula $v = r\omega$ in the form $\omega = v/r$, taking care to use the proper units. Since v is given in feet per second, we need r in feet:

$$r = \frac{d}{2} = \frac{1.8}{2}\text{ yd} \cdot \frac{3\text{ ft}}{1\text{ yd}} = 2.7\text{ ft}.$$

Then ω will be in radians per second:

$$\omega = \frac{v}{r} = \frac{2\text{ ft/sec}}{2.7\text{ ft}}$$
$$= \frac{2\text{ ft}}{\text{sec}} \cdot \frac{1}{2.7\text{ ft}} \approx 0.741/\text{sec}.$$

Thus the angular speed is approximately 0.741 radian/sec.

The formulas $\theta = \omega t$ and $v = r\omega$ can be used in combination to find distances and angles in various situations involving rotational motion.

Example 11 *Angle of Revolution.* A Porsche 911 is traveling at a speed of 65 mph. Its tires have an outside diameter of 25.086 in. Find the angle through which a tire turns in 10 sec.

SOLUTION Recall that $\omega = \theta/t$, or $\theta = \omega t$. Thus we can find θ if we know ω and t. To find ω, we use the formula $v = r\omega$. For convenience, we first convert 65 mph to feet per second:

$$v = 65\frac{\text{mi}}{\text{hr}} \cdot \frac{1\text{ hr}}{60\text{ min}} \cdot \frac{1\text{ min}}{60\text{ sec}} \cdot \frac{5280\text{ ft}}{1\text{ mi}}$$
$$\approx 95.333\,\frac{\text{ft}}{\text{sec}}.$$

Now $r = d/2 = 25.086 \text{ in.}/2 = 12.543$ in. We will convert to feet, since v is in feet per second:

$$r = 12.543 \text{ in.} \cdot \frac{1 \text{ ft}}{12 \text{ in.}}$$
$$= \frac{12.543}{12} \text{ ft} \approx 1.045 \text{ ft}.$$

Using $v = r\omega$, we have

$$95.333 \frac{\text{ft}}{\text{sec}} = 1.045 \text{ ft} \cdot \omega,$$

so

$$\omega = \frac{95.333 \text{ ft/sec}}{1.045 \text{ ft}} \approx \frac{91.228}{\text{sec}}.$$

Then in 10 sec,

$$\theta = \omega t = \frac{91.228}{\text{sec}} \cdot 10 \text{ sec} \approx 912.$$

Thus the angle, in radians, through which the tire turns in 10 sec is 912.

4.4 Exercise Set

For each of Exercises 1–4, sketch a unit circle and mark the points determined by the given real numbers.

1. a) $\frac{\pi}{4}$ **b)** $\frac{3\pi}{2}$ **c)** $\frac{3\pi}{4}$

d) π **e)** $\frac{11\pi}{4}$ **f)** $\frac{17\pi}{4}$

2. a) $\frac{\pi}{2}$ **b)** $\frac{5\pi}{4}$ **c)** 2π

d) $\frac{9\pi}{4}$ **e)** $\frac{13\pi}{4}$ **f)** $\frac{23\pi}{4}$

3. a) $\frac{\pi}{6}$ **b)** $\frac{2\pi}{3}$ **c)** $\frac{7\pi}{6}$

d) $\frac{10\pi}{6}$ **e)** $\frac{14\pi}{6}$ **f)** $\frac{23\pi}{4}$

4. a) $-\frac{\pi}{2}$ **b)** $-\frac{3\pi}{4}$ **c)** $-\frac{5\pi}{6}$

d) $-\frac{5\pi}{2}$ **e)** $-\frac{17\pi}{6}$ **f)** $-\frac{9\pi}{4}$

Find two real numbers between -2π and 2π that determine each of the points on the unit circle.

5.

6.

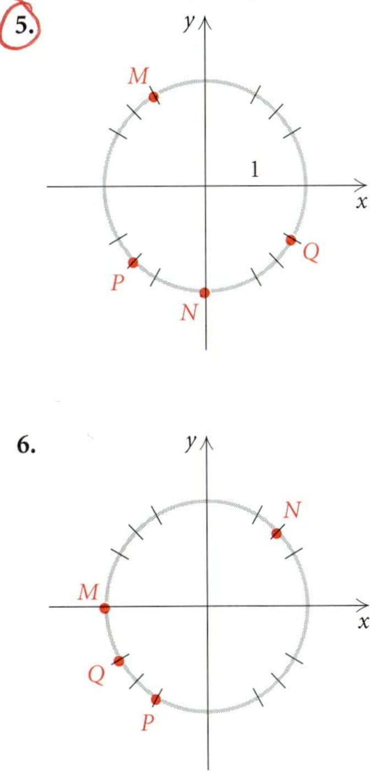

For Exercises 7 and 8, sketch a unit circle and mark the approximate location of the point determined by the given real number.

7. a) 2.4 **b)** 7.5 **c)** 32 **d)** 320

8. a) 0.25 **b)** 1.8 **c)** 47 **d)** 500

Find a positive angle and a negative angle that are coterminal with the given angle. Answers may vary.

9. $\frac{\pi}{4}$ **10.** $\frac{5\pi}{3}$ **11.** $\frac{7\pi}{6}$ **12.** π

Find the complement and the supplement.

13. $\frac{\pi}{3}$ **14.** $\frac{5\pi}{12}$ **15.** $\frac{3\pi}{8}$ **16.** $\frac{\pi}{4}$

Convert to radian measure. Leave your answer in terms of π.

17. 75° **18.** 30°

19. 200° **20.** −135°

21. −214.6° **22.** 37.71°

23. −180° **24.** 90°

Convert to radian measure. Round your answer to two decimal places.

25. 240° **26.** 15°

27. −60° **28.** 145°

29. 117.8° **30.** −231.2°

31. 1.354° **32.** 584°

Convert to degree measure. Round your answer to two decimal places.

33. $-\frac{3\pi}{4}$ **34.** $\frac{7\pi}{6}$

35. 8π **36.** $-\frac{\pi}{3}$

37. 1 **38.** −17.6

39. 2.347 **40.** 25

41. Certain positive angles are marked here in degrees. Find the corresponding radian measures.

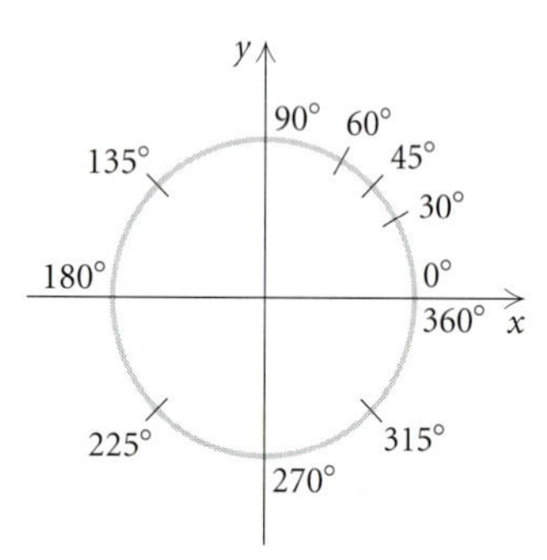

42. Certain negative angles are marked here in degrees. Find the corresponding radian measures.

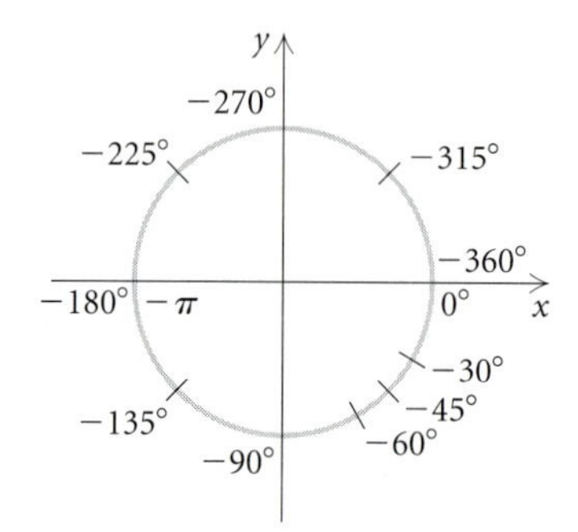

43. In a circle with a 120-cm radius, an arc 132 cm long subtends an angle of how many radians? how many degrees, to the nearest degree?

44. In a circle with a 5-m radius, how long is an arc associated with an angle of 2.1 radians?

45. *Angle of Revolution.* Through how many radians does the minute hand of a clock rotate from 12:40 P.M. to 1:30 P.M.?

46. *Angle of Revolution.* A tire on a Dodge Neon has an outside diameter of 23.468 in. Through what angle (in radians) does the tire turn while traveling 1 mi?

47. *Linear Speed.* A flywheel with a 15-cm diameter is rotating at a rate of 7 radians/sec. What is the linear speed of a point on its rim, in centimeters per minute?

48. *Linear Speed.* A wheel with a 30-cm radius is rotating at a rate of 3 radians/sec. What is the linear speed of a point on its rim, in meters per minute?

49. *Angular Speed on a Printing Press.* This text was printed on a four-color Webb heatset offset press. A cylinder on this press has a 13.37-in. diameter. The linear speed of a point on the cylinder's surface is 16.53 feet per second. What is the angular speed of the cylinder in revolutions per hour? Printers often refer to the angular speed as impressions per hour (IPH). (*Source:* Mark Krahforst, Rand Printing Company, Taunton, MA)

50. *Linear Speeds on a Carousel.* When Alicia and Zoe ride the carousel described earlier in this section, Alicia always selects a horse on the outside row, whereas Zoe prefers the row closest to the center. These rows are 19 ft 3 in. and 13 ft 11 in. from the center, respectively. The angular speed of the carousel is 2.4 revolutions per minute. What is the difference, in miles per hour, in the linear speeds of Alicia and Zoe? (*Source:* The Children's Museum, Indianapolis, IN)

51. *Linear Speed at the Equator.* The earth has a 4000-mi radius and rotates one revolution every 24 hr. What is the linear speed of a point on the equator, in miles per hour?

52. *Linear Speed of the Earth.* The earth is 93,000,000 mi from the sun and traverses its orbit, which is nearly circular, every 365.25 days. What is the linear velocity of the earth in its orbit, in miles per hour?

53. *Determining the Speed of a River.* A water wheel has a 10-ft radius. To get a good approximation of the speed of the river, you count the revolutions of the wheel and find that it makes 14 revolutions per minute (rpm). What is the speed of the river in miles per hour?

54. *The Tour de France.* Miguel Indurain won the 1995 Tour de France bicycle race. The wheel of his bicycle had a 67-cm diameter. His average linear speed during the 19th stage of the race was 48.461 km/h. What was the angular speed of the wheel in revolutions per hour? (*Source: Cycling Weekly*, July 29, 1995, p. 19, London)

55. *John Deere Tractor.* A rear wheel on a John Deere 8300 farm tractor has a 23-in. radius. Find the angle (in radians) through which a wheel rotates in 12 sec if the tractor is traveling at a speed of 22 mph.

Skill Maintenance

Graph each function. Sketch and label any vertical asymptote.

56. $f(x) = \dfrac{1}{x^2 - 25}$

57. $g(x) = x^3 - 2x + 1$

58. $f(x) = \dfrac{4}{x - 3}$

Determine the domain and the range of each function.

59. $f(x) = \dfrac{x - 4}{x + 2}$

60. $g(x) = \dfrac{x^2 - 9}{2x^2 - 7x - 15}$

Synthesis

61. ◈ Explain in your own words why it is preferable to omit the word, or unit, *radians* in radian measures.

62. ◈ In circular motion with a fixed angular speed, the length of the radius is directly proportional to the linear speed. Explain why with an example.

63. ◈ Two new cars are each driven at an average speed of 60 mph for an extended highway test drive of 2000 mi. The diameter of the wheels of the two cars are 15 in. and 16 in., respectively. If the cars use tires of equal durability and profile, differing only by the diameter, which car will probably need new tires first? Explain your result.

64. A point on the unit circle has y-coordinate $-\sqrt{21}/5$. What is its x-coordinate? Check with a grapher.

65. On the earth, one degree of latitude is how many kilometers? how many miles? (Assume that the radius of the earth is 6400 km, or 4000 mi, approximately.)

66. The **grad** is a unit of angle measure similar to a degree. A right angle has a measure of 100 grads. Convert each of the following to grads.

a) $48°$ b) $\frac{5\pi}{7}$

67. A **mil** is a unit of angle measure. A right angle has a measure of 1600 mils. Convert each of the following to degrees, minutes, and seconds.

a) 100 mils b) 350 mils

68. *Angular Speed of a Pulley.* Two pulleys, 50 cm and 30 cm in diameter, respectively, are connected by a belt. The larger pulley makes 12 revolutions per minute. Find the angular speed of the smaller pulley, in radians per second.

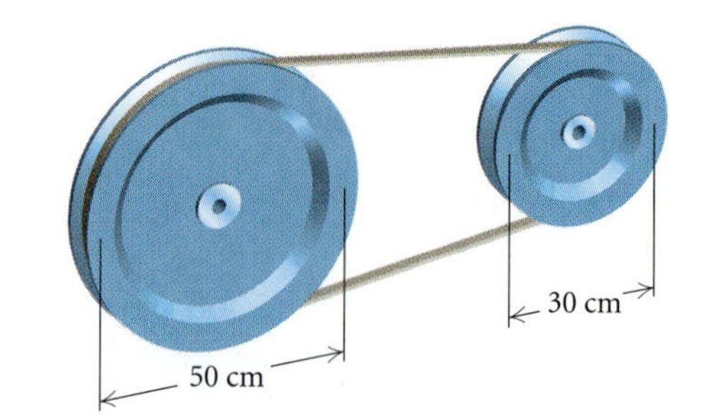

69. *Angular Speed of a Gear Wheel.* One gear wheel turns another, the teeth being on the rims. The wheels have 40-cm and 50-cm radii, and the smaller wheel rotates at 20 rpm. Find the angular speed of the larger wheel, in radians per second.

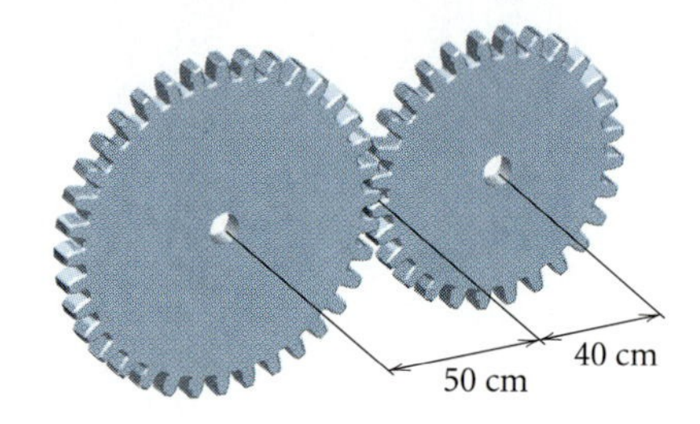

70. *Hands of a Clock.* At what times between noon and 1:00 P.M. are the hands of a clock perpendicular?

71. *Distance between Points on the Earth.* To find the distance between two points on the earth when their latitude and longitude are known, we can use a right triangle for an excellent approximation if the points are not too far apart. Point A is at latitude 38°27′30″ N, longitude 82°57′15″ W; and point B is at latitude 38°28′45″ N, longitude 82°56′30″ W. Find the distance from A to B in nautical miles (one minute of latitude is one nautical mile).

4.5 Circular Functions: Graphs and Properties

- *Given the coordinates of a point on the unit circle, find its reflections across the x-axis, the y-axis, and the origin.*
- *Determine the six trigonometric function values for a real number when the coordinates of the point on the unit circle determined by that real number are given.*
- *Find function values with a grapher for any real number.*
- *Graph the six circular functions and state their properties.*

The domains of the trigonometric functions, defined in Sections 4.1 and 4.3, have been sets of angles or rotations measured in a real number of degree units. We can also consider the domains to be sets of real numbers, or radians, introduced in Section 4.4. Many applications in calculus use the trigonometric functions referring only to radians.

Let's again consider radian measure and the unit circle. We defined radian measure for θ as

$$\theta = \frac{s}{r}.$$

When $r = 1$,

$$\theta = \frac{s}{1}, \quad \text{or} \quad \theta = s.$$

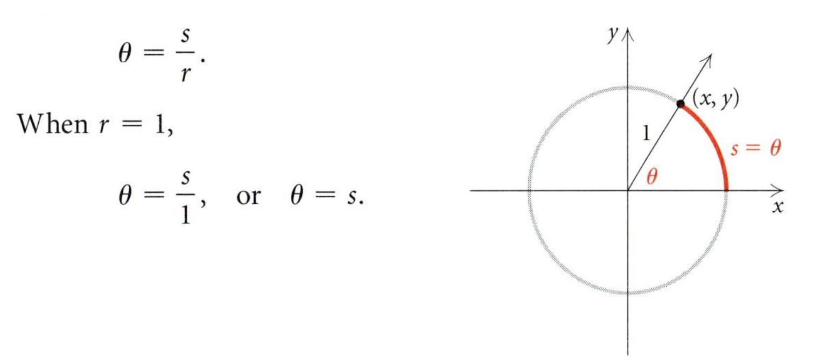

Thus the arc length s on the unit circle is the same as the radian measure of the angle θ.

In the figure above, the point (x, y) is the point where the terminal side of the angle with radian measure s intersects the unit circle. We can now extend our definitions of the trigonometric functions using domains of real numbers, or radians.

In the definitions, s can be considered the radian measure of an angle or the measure of an arc length on the unit circle. Either way, s is a real number. To each real number s, there corresponds an arc length s on the unit circle. Trigonometric functions with a domain of real numbers are called **circular functions**.

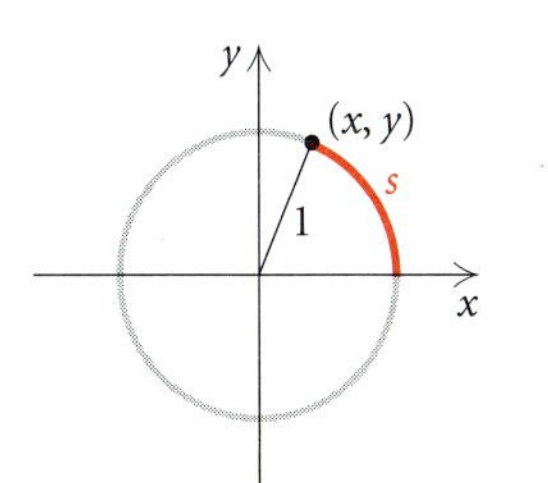

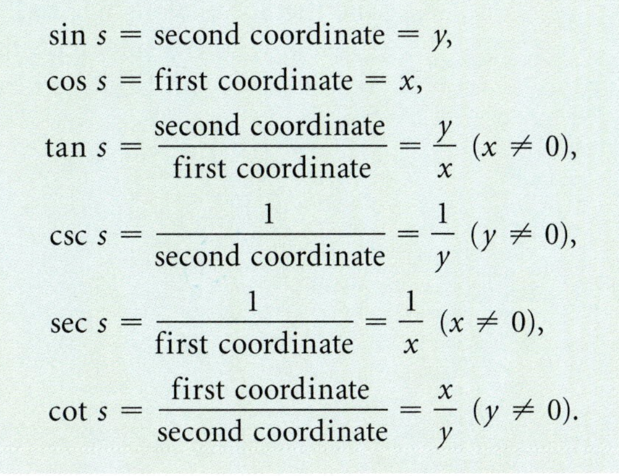

Basic Circular Functions

For a real number s that determines a point (x, y) on the unit circle:

$$\sin s = \text{second coordinate} = y,$$

$$\cos s = \text{first coordinate} = x,$$

$$\tan s = \frac{\text{second coordinate}}{\text{first coordinate}} = \frac{y}{x} \quad (x \neq 0),$$

$$\csc s = \frac{1}{\text{second coordinate}} = \frac{1}{y} \quad (y \neq 0),$$

$$\sec s = \frac{1}{\text{first coordinate}} = \frac{1}{x} \quad (x \neq 0),$$

$$\cot s = \frac{\text{first coordinate}}{\text{second coordinate}} = \frac{x}{y} \quad (y \neq 0).$$

Because of these definitions, we can consider the trigonometric functions with a domain of real numbers rather than angles. We can determine these values for a specific real number if we know the coordinates of the point on the unit circle determined by that number. As with degree measure, we can also find these function values directly with a grapher.

Reflections on the Unit Circle

Let's consider the unit circle and a few of its points. For any point (x, y) on the unit circle, with equation $x^2 + y^2 = 1$, we know that $-1 \le x \le 1$ and $-1 \le y \le 1$. If we know the x- or y-coordinate of a point on the unit circle, we can find the other coordinate. If $x = \frac{3}{5}$, then

$$\left(\tfrac{3}{5}\right)^2 + y^2 = 1$$
$$y^2 = 1 - \tfrac{9}{25} = \tfrac{16}{25}$$
$$y = \pm\tfrac{4}{5}.$$

Thus, $\left(\frac{3}{5}, \frac{4}{5}\right)$ and $\left(\frac{3}{5}, -\frac{4}{5}\right)$ are points on the unit circle. There are two points with an x-coordinate of $\frac{3}{5}$.

Now let's consider the radian measure $\pi/3$ and determine the coordinates of the point on the unit circle determined by $\pi/3$. We construct a right triangle by dropping a perpendicular segment from the point to the x-axis.

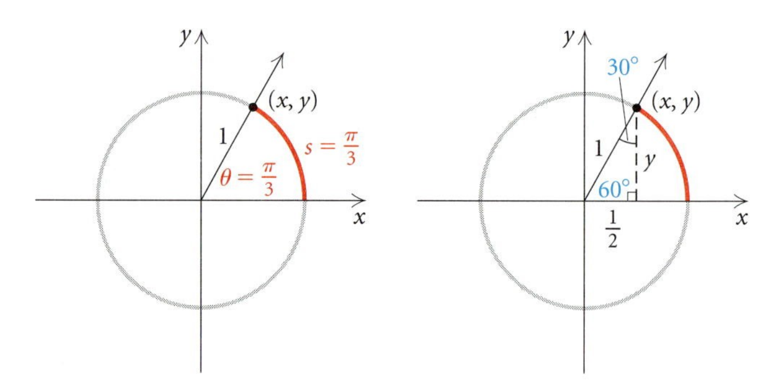

Since $\pi/3 = 60°$, we have a 30°–60° right triangle in which the side opposite the 30° angle is one half of the hypotenuse. The hypotenuse, or radius, is 1, so the side opposite the 30° angle is $\frac{1}{2}$. Using the Pythagorean theorem, we can find the other side:

$$\left(\frac{1}{2}\right)^2 + y^2 = 1$$
$$y^2 = 1 - \frac{1}{4} = \frac{3}{4}$$
$$y = \sqrt{\frac{3}{4}} = \frac{\sqrt{3}}{2}.$$

We know y is positive since the point is in the first quadrant. Thus the coordinates of the point determined by $\pi/3$ are $x = 1/2$ and $y = \sqrt{3}/2$, or $(1/2, \sqrt{3}/2)$. We can always check to see if a point is on the unit circle by substituting into the equation $x^2 + y^2 = 1$.

Because a unit circle is symmetric with respect to the x-axis, the y-axis, and the origin, we can use the coordinates of one point on the unit circle to find coordinates of its reflections.

Example 1 Each of the following points lies on the unit circle. Find their reflections across the x-axis, the y-axis, and the origin.

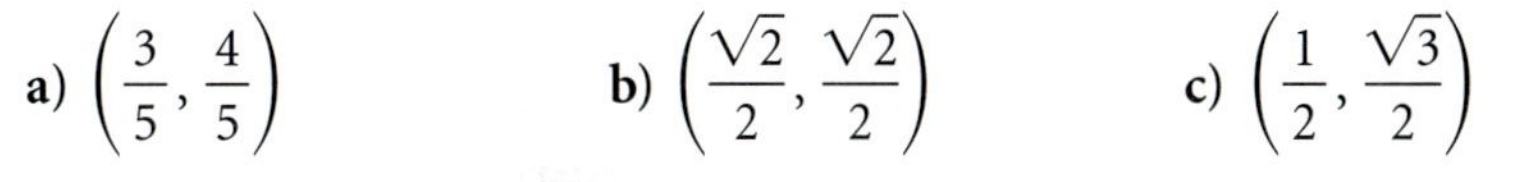

a) $\left(\frac{3}{5}, \frac{4}{5}\right)$ **b)** $\left(\frac{\sqrt{2}}{2}, \frac{\sqrt{2}}{2}\right)$ **c)** $\left(\frac{1}{2}, \frac{\sqrt{3}}{2}\right)$

SOLUTION

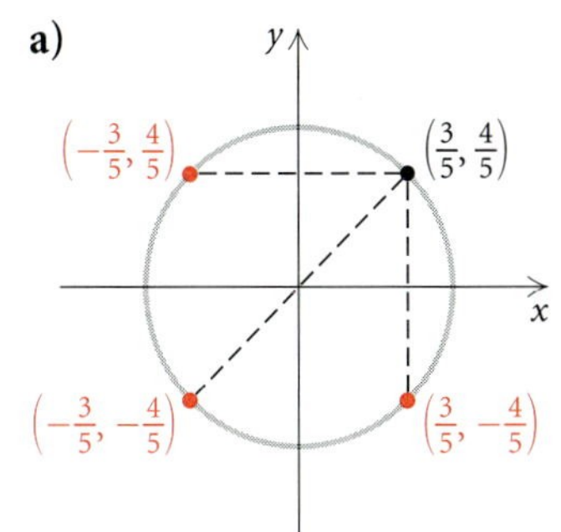

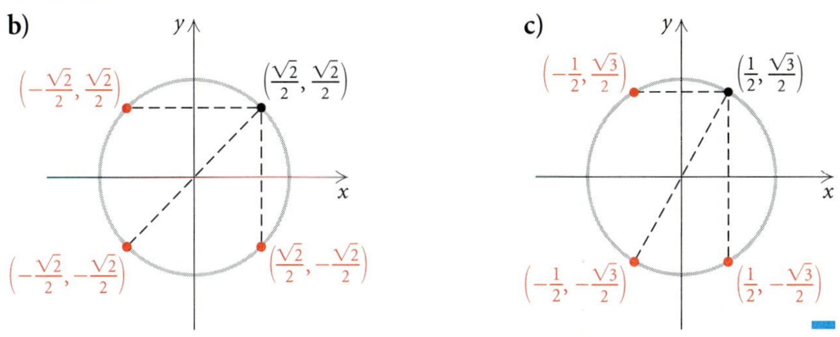

Finding Function Values

Knowing the coordinates of only a few points on the unit circle along with their reflections allows us to find trigonometric function values of the most frequently used real numbers, or radians. The coordinates on the unit circle below should be memorized.

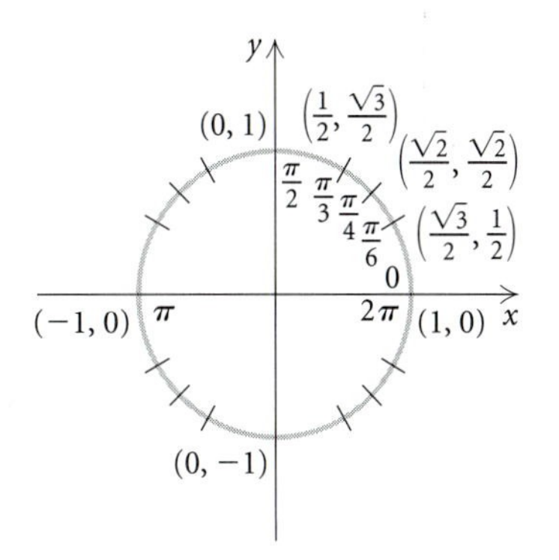

Example 2 Find each of the following function values.

a) $\tan \frac{\pi}{3}$ **b)** $\cos \frac{3\pi}{4}$ **c)** $\sin \left(-\frac{\pi}{6}\right)$

d) $\cos \dfrac{4\pi}{3}$ **e)** $\cot \pi$ **f)** $\csc\left(-\dfrac{7\pi}{2}\right)$

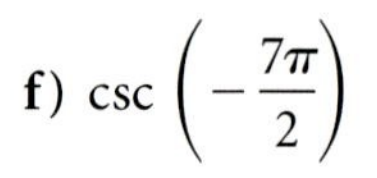

SOLUTION

a) The coordinates of the point determined by $\pi/3$ are $(1/2, \sqrt{3}/2)$. Thus,

$$\tan \frac{\pi}{3} = \frac{y}{x} = \frac{\sqrt{3}/2}{1/2} = \sqrt{3}.$$

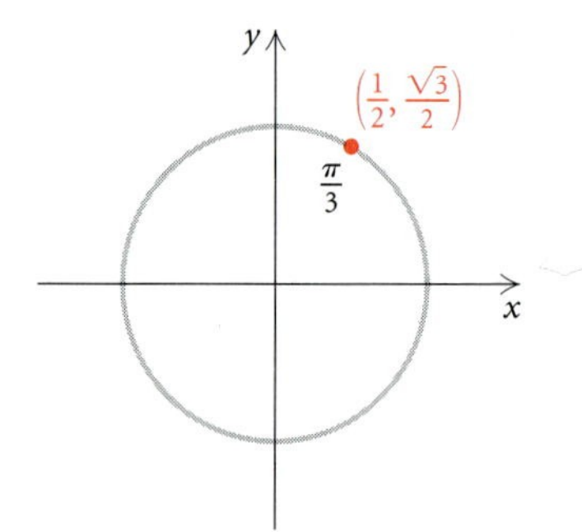

b) The reflection of $(\sqrt{2}/2, \sqrt{2}/2)$ across the y-axis is $(-\sqrt{2}/2, \sqrt{2}/2)$. Thus,

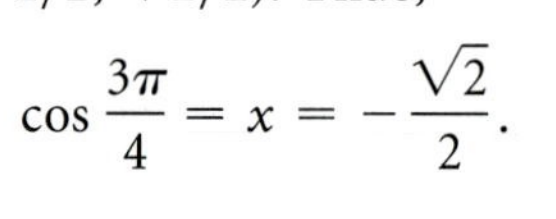

$$\cos \frac{3\pi}{4} = x = -\frac{\sqrt{2}}{2}.$$

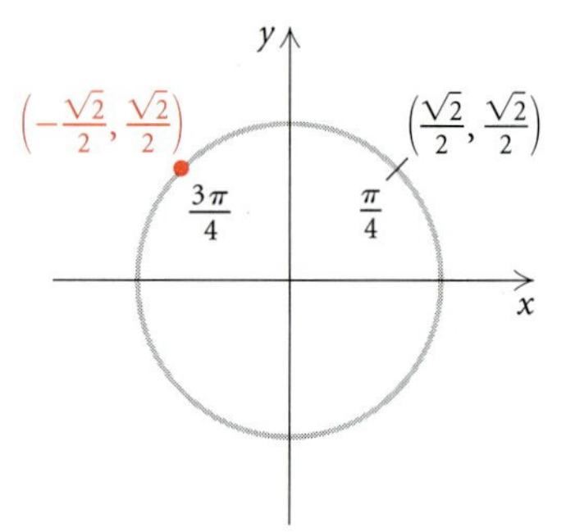

c) The reflection of $(\sqrt{3}/2, 1/2)$ across the x-axis is $(\sqrt{3}/2, -1/2)$. Thus,

$$\sin\left(-\frac{\pi}{6}\right) = y = -\frac{1}{2}.$$

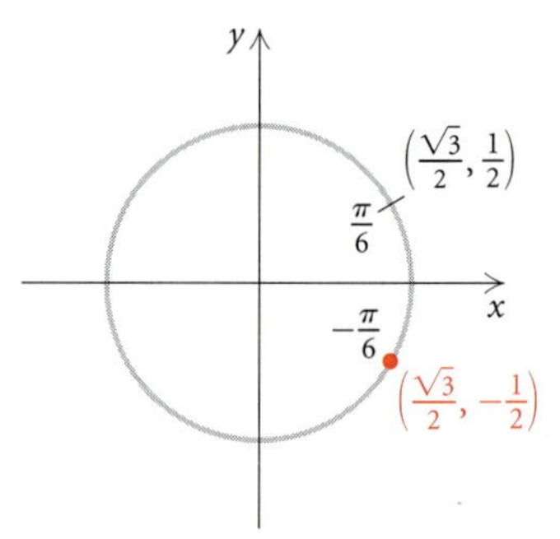

d) The reflection of $(1/2, \sqrt{3}/2)$ across the origin is $(-1/2, -\sqrt{3}/2)$. Thus,

$$\cos \frac{4\pi}{3} = x = -\frac{1}{2}.$$

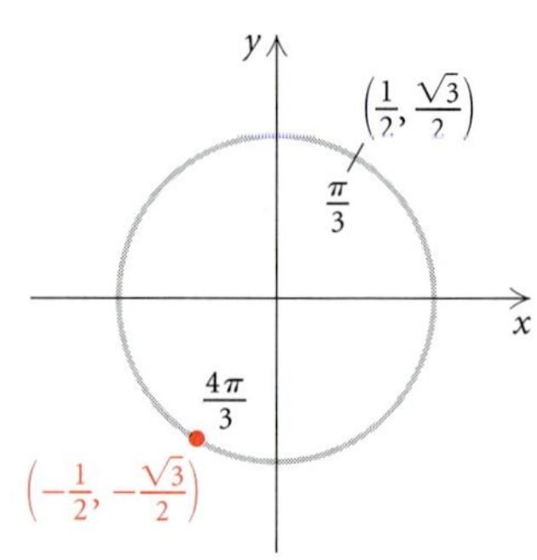

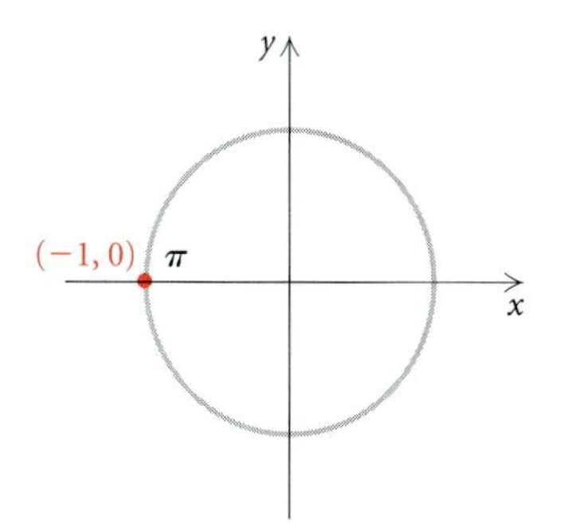

e) The coordinates of the point determined by π are $(-1, 0)$. Thus,

$$\cot \pi = \frac{x}{y} = \frac{-1}{0}, \quad \text{which is undefined.}$$

We can also think of $\cot \pi$ as the reciprocal of $\tan \pi$. Since $\tan \pi = y/x = 0/-1 = 0$ and the reciprocal of 0 is not defined, we know that $\cot \pi$ is undefined.

f) The coordinates of the point determined by $-7\pi/2$ are $(0, 1)$. Thus,

$$\csc\left(-\frac{7\pi}{2}\right) = \frac{1}{y} = \frac{1}{1} = 1.$$

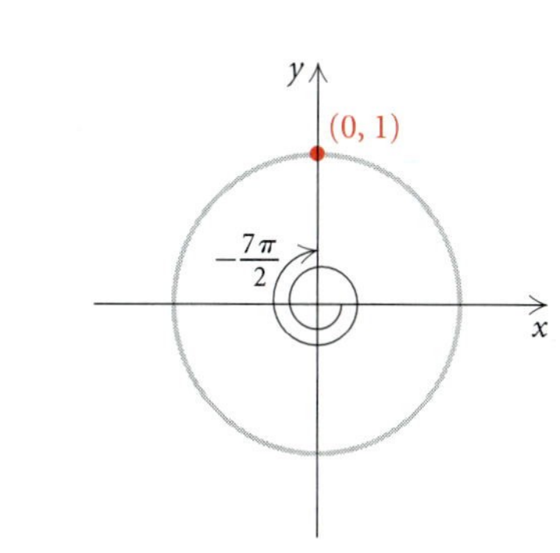

Using a grapher, we can find trigonometric function values of any real number without knowing the coordinates of the point it determines on the unit circle. Most graphers have both degree and radian modes. When finding function values of radian measures, or real numbers, we *must* set the grapher in radian mode.

Example 3 Find each of the following function values of radian measures using a grapher.

a) $\cos \frac{2\pi}{5}$ b) $\tan(-3)$ c) $\sin 24.9$ d) $\sec \frac{\pi}{7}$

SOLUTION

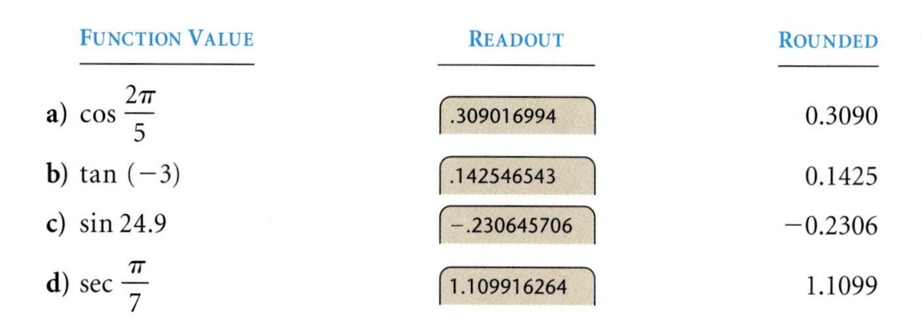

FUNCTION VALUE	READOUT	ROUNDED
a) $\cos \frac{2\pi}{5}$	.309016994	0.3090
b) $\tan(-3)$	.142546543	0.1425
c) $\sin 24.9$	−.230645706	−0.2306
d) $\sec \frac{\pi}{7}$	1.109916264	1.1099

Note in part (d) that the secant function value can be found by taking the reciprocal of the cosine value. Thus we enter $\cos \pi/7$ and use the reciprocal key.

Interactive Discovery

We can graph the unit circle on a grapher. We use parametric mode with the following window and let XIT = cos T and YIT = sin T.

WINDOW

Tmin = 0
Tmax = 2π, or 6.2831853
Tstep = $\pi/12$, or 0.2617993
Xmin = −1.5
Xmax = 1.5
Xscl = 1
Ymin = −1
Ymax = 1
Yscl = 1

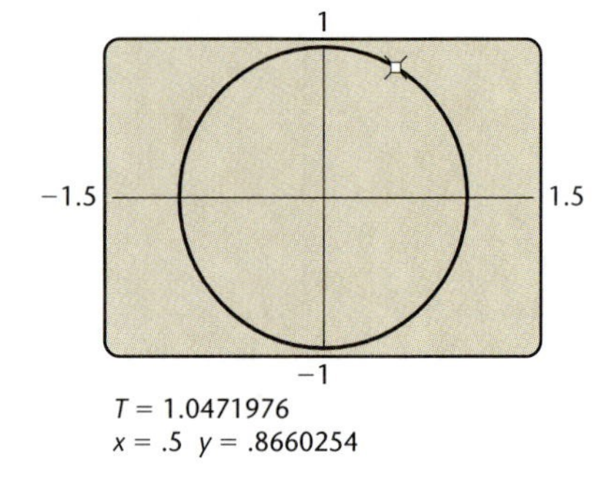

Using the trace key and an arrow key to move the cursor around the unit circle, we see the T, X, and Y values appear on the screen. What do they represent? (For more on parametric equations, see Appendix B.)

From the definitions on page 373 and from the preceding exploration, we can relabel any point (x, y) on the unit circle as $(\cos s, \sin s)$, where s is any real number.

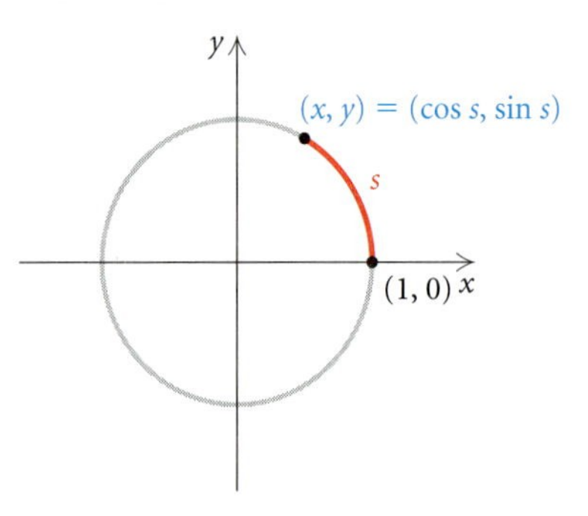

Graphs of the Sine and Cosine Functions

Properties of functions can be easily observed from their graphs. We begin by graphing the sine and cosine functions. We make a table of values, plot the points, and then connect those points with a smooth curve. It is helpful to first draw a unit circle and label a few points with coordinates.

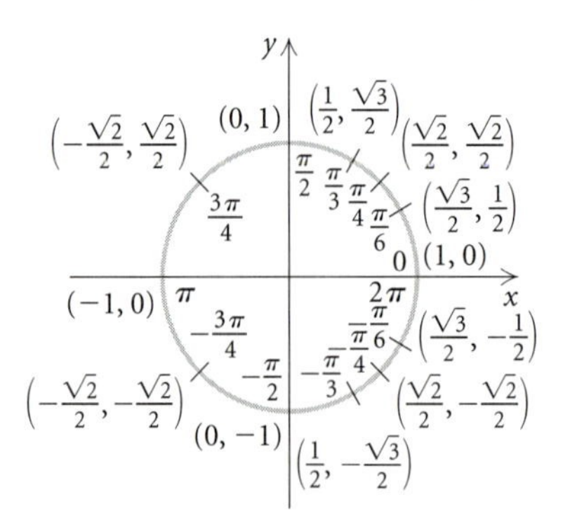

We can either use the coordinates as the function values or find approximate sine and cosine values directly with a grapher.

s	$\sin s$	$\cos s$
0	0	1
$\pi/6$	0.5	0.8660
$\pi/4$	0.7071	0.7071
$\pi/3$	0.8660	0.5
$\pi/2$	1	0
$3\pi/4$	0.7071	−0.7071
π	0	−1
$5\pi/4$	−0.7071	−0.7071
$3\pi/2$	−1	0
$7\pi/4$	−0.7071	0.7071
2π	0	1

s	$\sin s$	$\cos s$
0	0	1
$-\pi/6$	−0.5	0.8660
$-\pi/4$	−0.7071	0.7071
$-\pi/3$	−0.8660	0.5
$-\pi/2$	−1	0
$-3\pi/4$	−0.7071	−0.7071
$-\pi$	0	−1
$-5\pi/4$	0.7071	−0.7071
$-3\pi/2$	1	0
$-7\pi/4$	0.7071	0.7071
-2π	0	1

The graphs are as follows.

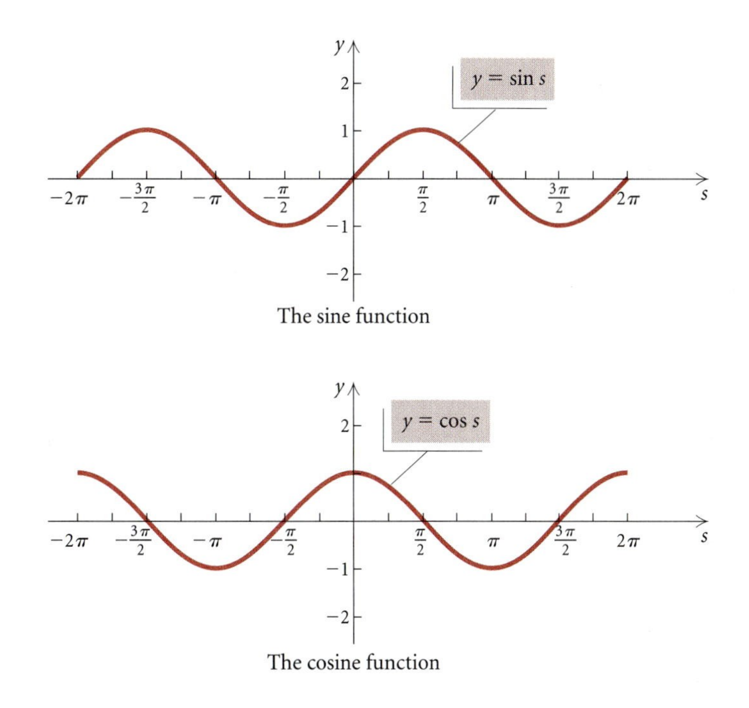

The sine function

The cosine function

We can check our graphs with a grapher. When graphing trigonometric functions, the grapher can be in radian or degree mode. Here we use radian mode.

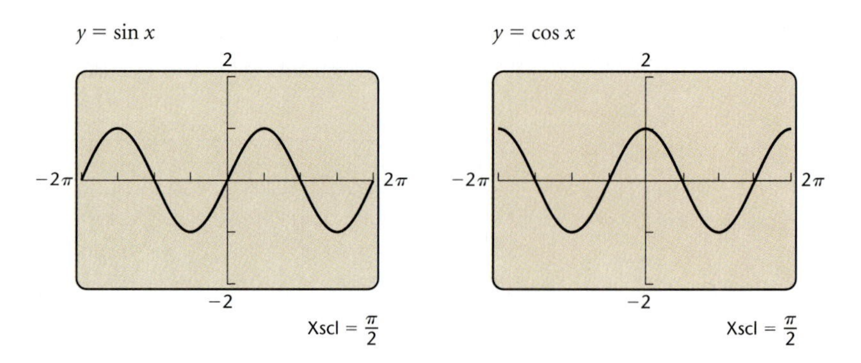

Interactive Discovery

Another way to construct the sine and cosine graphs is by considering the unit circle and transferring vertical distances for the sine function and horizontal distances for the cosine function. Using the parametric graphing feature of a grapher, we can visualize the transfer of these distances. We use PARAMETRIC mode with the following window and let $X_{1T} = \cos T - 1$ and $Y_{1T} = \sin T$ for the unit circle centered at $(-1, 0)$ and $X_{2T} = T$ and $Y_{2T} = \sin T$ for the sine curve.

Tmin = 0	Xmin = −2	Ymin = −3
Tmax = 2π	Xmax = 2π	Ymax = 3
Tstep = .1	Xscl = π/2	Yscl = 1

With the grapher set to SIMUL mode, we can actually watch the sine function "unwind" from the unit circle. In the two screens below, we partially illustrate this animated procedure for the sine function.

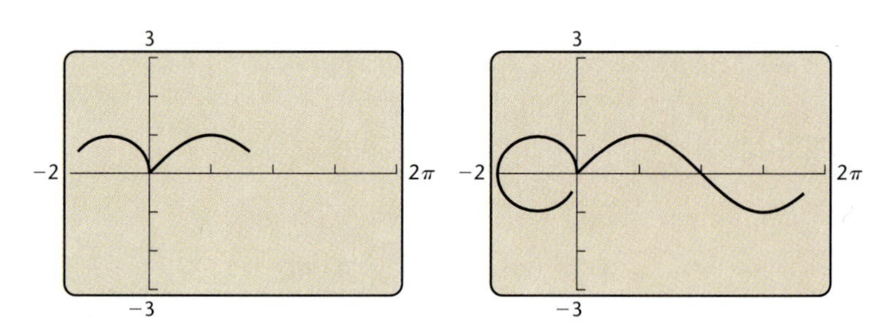

Consult your grapher's instruction manual for specific keystrokes and graph both the sine curve and the cosine curve in this manner. (For more on parametric equations, see Appendix B.)

The sine and cosine functions are continuous functions. Note in the graph of the sine function that function values increase from 0 at $s = 0$ to 1 at $s = \pi/2$, then decrease to 0 at $s = \pi$, decrease further to -1 at $s = 3\pi/2$, and increase to 0 at 2π. The reverse pattern follows when s decreases from 0 to -2π. Note in the graph of the cosine function that function values start at 1 when $s = 0$, and decrease to 0 at $s = \pi/2$. They decrease further to -1 at $s = \pi$, then increase to 0 at $s = 3\pi/2$, and increase further to 1 at $s = 2\pi$. An identical pattern follows when s decreases from 0 to -2π.

Interactive Discovery

To use the TABLE feature on a grapher to verify the y-value patterns for $y = \sin x$ and $y = \cos x$, enter $\sin x$ for Y_1 and $\cos x$ for Y_2. We then set TblMin $= 0$ and ΔTbl $= \pi/12$ and begin scrolling.

X	Y1	Y2
0	0	1
.2618	.25882	.96593
.5236	.5	.86603
.7854	.70711	.70711
1.0472	.86603	.5
1.309	.96593	.25882
1.5708	1	0
X = 0		

Can you determine the domain and the range for the sine function and for the cosine function?

From the unit circle, the graphs of the functions, and the TABLE feature on the grapher, we know that the domain of both the sine and cosine functions is the entire set of real numbers, $(-\infty, \infty)$. The range of each function is the set of all real numbers from -1 to 1, $[-1, 1]$.

A function with a repeating pattern is called **periodic**. The sine and cosine functions are examples of periodic functions. The function values of each function repeat themselves every 2π units. In other words, for any s, we have

$$\sin (s + 2\pi) = \sin s \quad \text{and} \quad \cos (s + 2\pi) = \cos s.$$

To see this another way, think of the part of the graph between 0 and 2π and note that the rest of the graph consists of copies of it. If we translate the graph of $y = \sin x$ or $y = \cos x$ to the left or right 2π units, we will obtain the original graph. We say that each of these functions has a period of 2π.

Periodic Function

A function f is said to be *periodic* if there exists a positive constant p such that

$$f(s + p) = f(s)$$

for all s in the domain of f. The smallest such positive number p is called the period of the function.

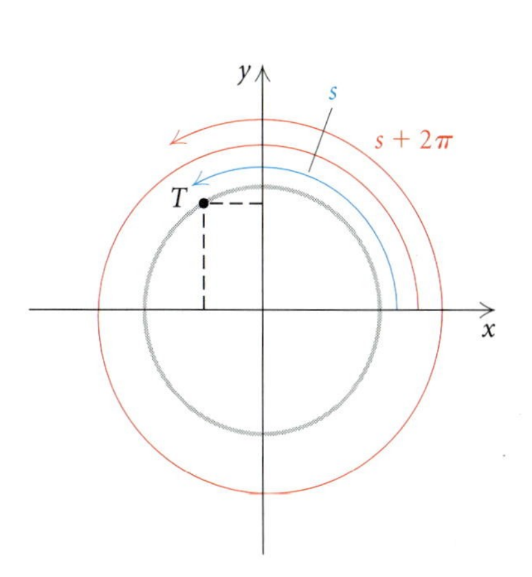

The period p can be thought of as the length of the shortest recurring interval.

We can also use the unit circle to verify that the period of the sine and cosine functions is 2π. Consider any real number s and the point T that it determines on a unit circle. If we increase s by 2π, the point determined by $s + 2\pi$ is again the point T. Hence for any real number s,

$$\sin (s + 2\pi) = \sin s \quad \text{and} \quad \cos (s + 2\pi) = \cos s.$$

Thus we have shown that for any integer k, the following equations are identities:

$$\sin [s + k(2\pi)] = \sin s \quad \text{and} \quad \cos [s + k(2\pi)] = \cos s,$$

or

$$\sin s = \sin (s + 2k\pi) \quad \text{and} \quad \cos s = \cos (s + 2k\pi).$$

The **amplitude** of a periodic function is defined as one half of the distance between its maximum and minimum function values. It is always positive. Both the graphs and the unit circle verify that the maximum value of the sine and cosine functions is 1, whereas the minimum value of each is -1. Thus,

$$\text{the amplitude of the sine function} = \tfrac{1}{2}[1 - (-1)] = 1$$

and

$$\text{the amplitude of the cosine function is } \tfrac{1}{2}[1 - (-1)] = 1.$$

Interactive Discovery

Using the TABLE feature on a grapher, compare the y-values for $y_1 = \sin x$ and $y_2 = \sin (-x)$ and for $y_3 = \cos x$ and $y_4 = \cos (-x)$. We set TblMin $= 0$ and $\triangle$Tbl $= \pi/12$.

X	Y1	Y2
0	0	0
.2618	.25882	−.2588
.5236	.5	−.5
.7854	.70711	−.7071
1.0472	.86603	−.866
1.309	.96593	−.9659
1.5708	1	−1
X = 0		

X	Y3	Y4
0	1	1
.2618	.96593	.96593
.5236	.86603	.86603
.7854	.70711	.70711
1.0472	.5	.5
1.309	.25882	.25882
1.5708	0	0
X = 0		

What appears to be the relationship between sin x and sin $(-x)$ and between cos x and cos $(-x)$?

Consider any real number s and its opposite, $-s$. These numbers determine points T and T_1 on a unit circle that are symmetric with respect to the x-axis.

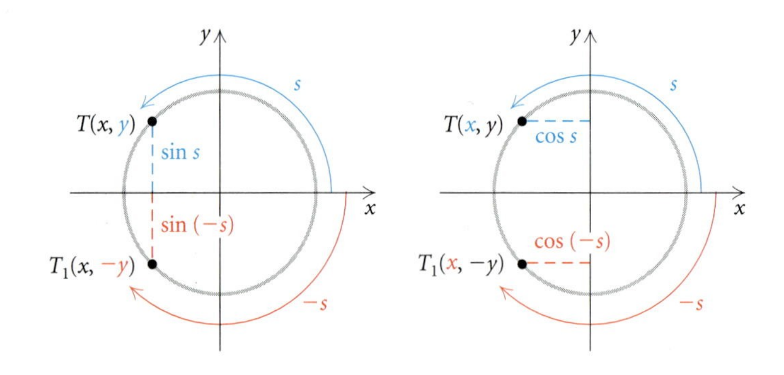

Because their second coordinates are opposites of each other, we know that for any number s,

$$\sin(-s) = -\sin s.$$

Because their first coordinates are the same, we know that for any number s,

$$\cos(-s) = \cos s.$$

Thus we have shown that the sine function is odd and the cosine function is even.

The following table is a summary of the properties of the sine and cosine functions.

SINE FUNCTION	COSINE FUNCTION
1. Continuous **2.** Period: 2π **3.** Domain: All real numbers **4.** Range: $[-1, 1]$ **5.** Amplitude: 1 **6.** Odd: $\sin(-s) = -\sin s$	**1.** Continuous **2.** Period: 2π **3.** Domain: All real numbers **4.** Range: $[-1, 1]$ **5.** Amplitude: 1 **6.** Even: $\cos(-s) = \cos s$

Graphs of the Tangent, Cotangent, Cosecant, and Secant Functions

To graph the tangent function, we could make a table of values with a grapher, but in this case it is easier to begin with the definition of tangent and the coordinates of a few points on the unit circle. We recall that

$$\tan s = \frac{y}{x} = \frac{\sin s}{\cos s}.$$

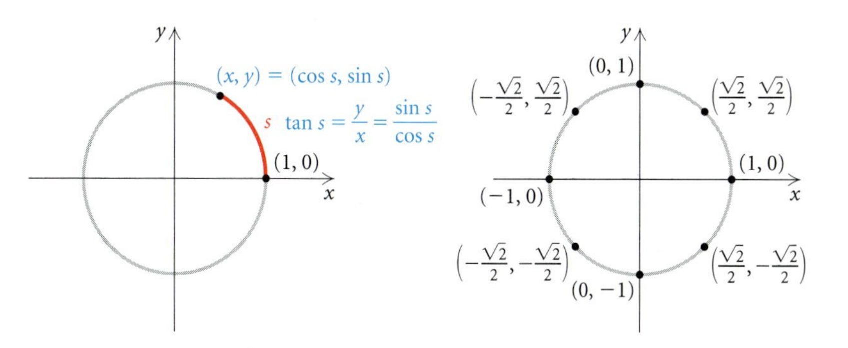

The tangent function is undefined when x, the first coordinate, is 0. That is, it is undefined for any number s whose cosine is 0:

$$s = \pm\frac{\pi}{2}, \pm\frac{3\pi}{2}, \pm\frac{5\pi}{2}, \ldots.$$

We draw vertical asymptotes at these locations (see Fig. 1 below).

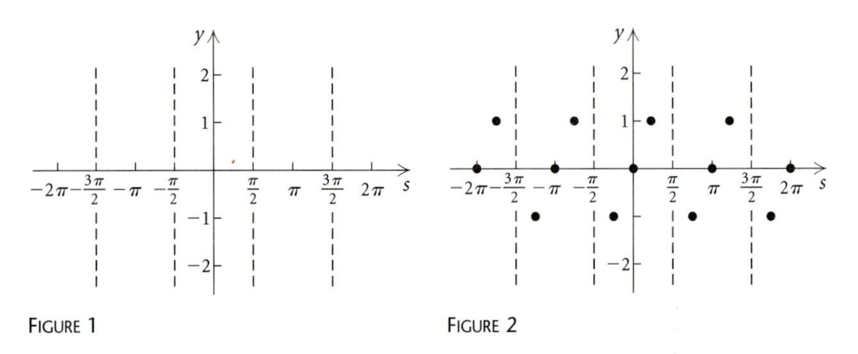

FIGURE 1

FIGURE 2

We also note that

$$\tan s = 0 \text{ at } s = 0, \pm\pi, \pm 2\pi, \pm 3\pi, \ldots,$$

$$\tan s = 1 \text{ at } s = \pm\frac{\pi}{4}, \pm\frac{5\pi}{4}, \pm\frac{9\pi}{4}, \ldots, \quad \text{and}$$

$$\tan s = -1 \text{ at } s = \pm\frac{3\pi}{4}, \pm\frac{7\pi}{4}, \pm\frac{11\pi}{4}, \ldots.$$

We can add these ordered pairs to the graph (see Fig. 2 above) and investigate the values in $(-\pi/2, \pi/2)$ with a grapher. Note that the function value is 0 when $s = 0$, and the values increase without bound as s increases toward $\pi/2$. The graph gets closer and closer to an asymptote as s gets closer to $\pi/2$, but it never touches the line. As s decreases from 0 to $-\pi/2$, the values decrease without bound. Again the graph gets closer and closer to an asymptote, but it never touches it. We now complete the graph.

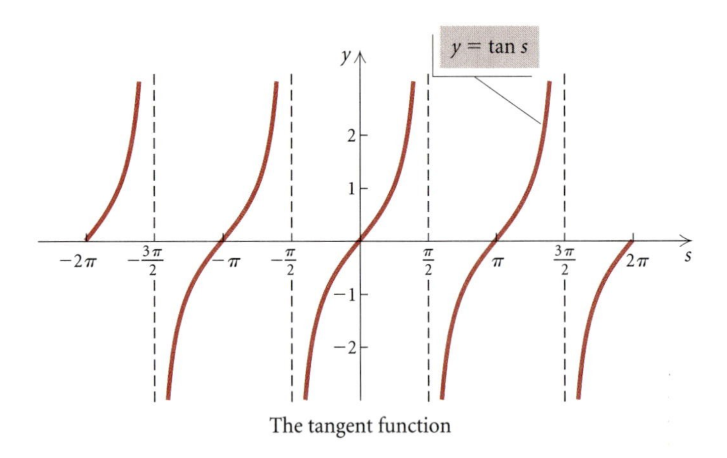

The tangent function

From the graph, we see that the tangent function is continuous except where it is not defined. Unlike the sine and cosine functions with a period of 2π, the tangent function has a period of π. Tan s is not defined where $\cos s = 0$. Thus the domain of the tangent function is the set of all real numbers except $(\pi/2) + k\pi$, where k is an integer. The range of the function is the set of all real numbers.

The cotangent function is undefined when y, the second coordinate, is 0—that is, it is undefined for any number s whose sine is 0. Thus the cotangent is undefined for $s = 0, \pm\pi, \pm 2\pi, \pm 3\pi, \ldots$. The graph of the function is shown below.

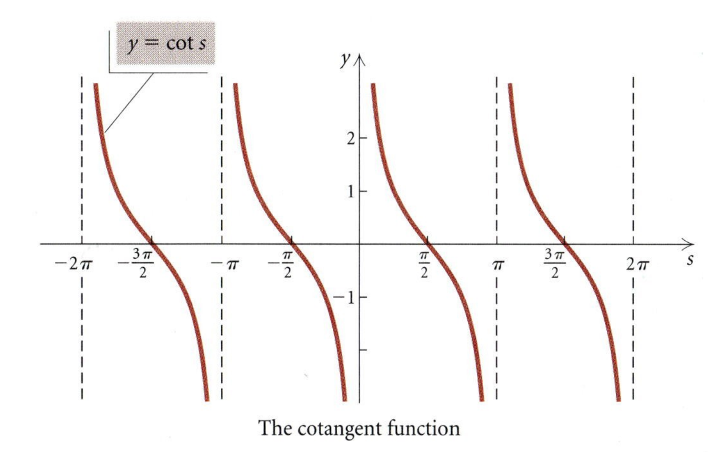

The cotangent function

The cosecant and sine functions are reciprocal functions, as are the secant and cosine functions. The graphs of the cosecant and secant functions can be constructed by finding the reciprocal of the values of the sine and cosine functions, respectively. Thus the functions will be positive together and negative together. The cosecant function is not defined for those numbers s whose sine is 0. The secant function is not defined for those numbers s whose cosine is 0. In the graphs below, the sine and cosine functions are shown for reference by the gray curves.

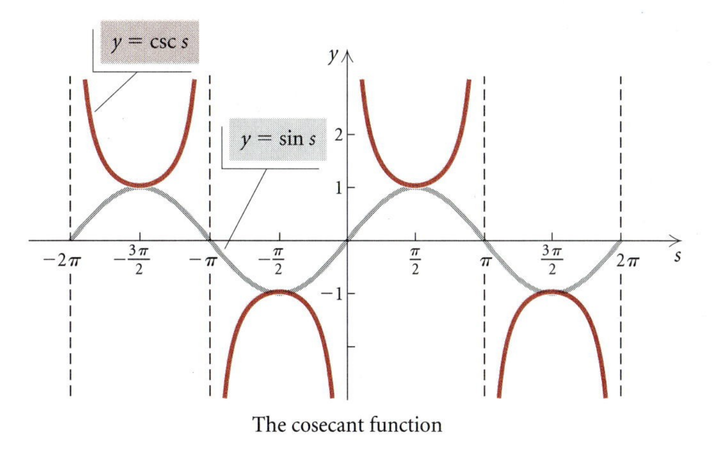

The cosecant function

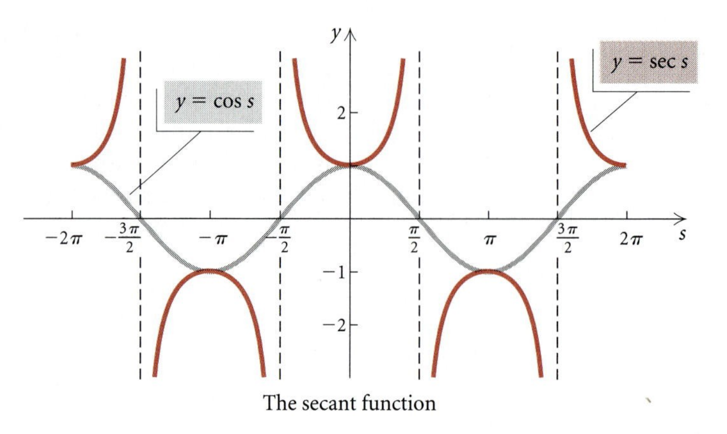

The secant function

The following is a summary of the basic properties of the tangent, cotangent, cosecant, and secant functions. These functions are continuous except where not defined.

Tangent Function

1. Period: π
2. Domain: All real numbers except $(\pi/2) + k\pi$, where k is an integer
3. Range: All real numbers

Cotangent Function

1. Period: π
2. Domain: All real numbers except $k\pi$, where k is an integer
3. Range: All real numbers

Cosecant Function

1. Period: 2π
2. Domain: All real numbers except $k\pi$, where k is an integer
3. Range: $(-\infty, -1] \cup [1, \infty)$

Secant Function

1. Period: 2π
2. Domain: All real numbers except $(\pi/2) + k\pi$, where k is an integer
3. Range: $(-\infty, -1] \cup [1, \infty)$

In this chapter, we have been using the letter s for arc length and have avoided the letters x and y, which generally represent first and second coordinates. In fact, we can represent the arc length on a unit circle by any variable, such as s, t, x, or θ. Each arc length determines a point that can be labeled with an ordered pair. The first coordinate of that ordered pair is the cosine of the arc length, and the second coordinate is the sine of the arc length. The identities we have developed hold no matter what symbols are used for variables—for example, $\cos(-s) = \cos s$, $\cos(-x) = \cos x$, $\cos(-\theta) = \cos \theta$, and $\cos(-t) = \cos t$.

4.5 Exercise Set

The following points are on the unit circle. Find the coordinates of their reflections across (a) the x-axis, (b) the y-axis, and (c) the origin.

1. $\left(-\frac{3}{4}, \frac{\sqrt{7}}{4}\right)$

2. $\left(\frac{2}{3}, \frac{\sqrt{5}}{3}\right)$

3. $\left(\frac{2}{5}, -\frac{\sqrt{21}}{5}\right)$

4. $\left(-\frac{\sqrt{3}}{2}, -\frac{1}{2}\right)$

5. The number $\pi/4$ determines a point on the unit circle with coordinates $(\sqrt{2}/2, \sqrt{2}/2)$. What are the coordinates of the point determined by $-\pi/4$?

6. A number β determines a point on the unit circle with coordinates $(-2/3, \sqrt{5}/3)$. What are the coordinates of the point determined by $-\beta$?

Find the function value using coordinates of points on the unit circle.

7. $\sin \pi$

8. $\cos\left(-\frac{\pi}{3}\right)$

9. $\cot \frac{7\pi}{6}$

10. $\tan \frac{11\pi}{4}$

11. $\sin(-3\pi)$

12. $\csc \frac{3\pi}{4}$

13. $\cos \frac{5\pi}{6}$

14. $\tan\left(-\frac{\pi}{4}\right)$

15. $\cos 10\pi$

16. $\sec \frac{\pi}{2}$

17. $\cos \frac{\pi}{6}$

18. $\sin \frac{2\pi}{3}$

19. $\sin \frac{5\pi}{4}$

20. $\cos \frac{11\pi}{6}$

21. $\sin(-5\pi)$

22. $\tan \frac{3\pi}{2}$

23. $\cot \frac{5\pi}{2}$

24. $\tan \frac{5\pi}{3}$

Find the function value using a grapher. Round the answer to four decimal places.

25. $\tan \frac{\pi}{7}$

26. $\cos\left(-\frac{2\pi}{5}\right)$

27. $\sec 37$

28. $\sin 11.7$

29. $\cot 342$

30. $\tan 1.3$

31. $\cos 6\pi$

32. $\sin \frac{\pi}{10}$

33. $\csc 4.16$

34. $\sec \frac{10\pi}{7}$

35. $\tan \frac{7\pi}{4}$

36. $\cos 2000$

37. $\sin\left(-\frac{\pi}{4}\right)$

38. $\cot 7\pi$

39. $\sin 0$

40. $\cos(-29)$

41. $\tan \frac{2\pi}{9}$

42. $\sin \frac{8\pi}{3}$

In Exercises 43–48, make hand-drawn graphs.

43. a) Sketch a graph of $y = \sin x$.
b) By reflecting the graph in part (a), sketch a graph of $y = \sin(-x)$.
c) By reflecting the graph in part (a), sketch a graph of $y = -\sin x$.
d) How do the graphs in parts (b) and (c) compare?

44. a) Sketch a graph of $y = \cos x$.
b) By reflecting the graph in part (a), sketch a graph of $y = \cos(-x)$.
c) By reflecting the graph in part (a), sketch a graph of $y = -\cos x$.
d) How do the graphs in parts (a) and (b) compare?

45. a) Sketch a graph of $y = \sin x$.
b) By translating, sketch a graph of $y = \sin(x + \pi)$.
c) By reflecting the graph of part (a), sketch a graph of $y = -\sin x$.
d) How do the graphs of parts (b) and (c) compare?

46. a) Sketch a graph of $y = \sin x$.
b) By translating, sketch a graph of $y = \sin(x - \pi)$.
c) By reflecting the graph of part (a), sketch a graph of $y = -\sin x$.
d) How do the graphs of parts (b) and (c) compare?

47. a) Sketch a graph of $y = \cos x$.
b) By translating, sketch a graph of $y = \cos(x + \pi)$.
c) By reflecting the graph of part (a), sketch a graph of $y = -\cos x$.
d) How do the graphs of parts (b) and (c) compare?

48. a) Sketch a graph of $y = \cos x$.
b) By translating, sketch a graph of $y = \cos(x - \pi)$.
c) By reflecting the graph of part (a), sketch a graph of $y = -\cos x$.
d) How do the graphs of parts (b) and (c) compare?

49. Of the six circular functions, which are even? Which are odd?

50. Of the six circular functions, which have period π? Which have period 2π?

Consider the coordinates on the unit circle for Exercises 51–54.

51. In which quadrants is the tangent function positive? negative?

52. In which quadrants is the sine function positive? negative?

53. In which quadrants is the cosine function positive? negative?

54. In which quadrants is the cosecant function positive? negative?

Skill Maintenance

Graph both functions in the same viewing window and describe how g is a transformation of f.

55. $f(x) = x^2$, $g(x) = 2x^2 - 3$

56. $f(x) = x^2$, $g(x) = (x - 2)^2$

57. $f(x) = |x|$, $g(x) = \frac{1}{2}|x - 4| + 1$

58. $f(x) = x^3$, $g(x) = -x^3$

Write an equation for a function that has a graph with the given characteristics. Check with a grapher.

59. The shape of $y = x^3$, but reflected across the x-axis, shifted right 2 units, and shifted down 1 unit

60. The shape of $y = 1/x$, but shrunk vertically by a factor of $\frac{1}{4}$ and shifted up 3 units

Synthesis

61. ◆ Describe how the graphs of the sine and cosine functions are related.

62. ◆ Explain why both the sine and cosine functions are continuous, but the tangent function, defined as sine/cosine, is not continuous.

Complete. (For example, $\sin(x + 2\pi) = \sin x$.*)*

63. $\cos(-x) =$ _______

64. $\sin(-x) =$ _______

65. $\sin(x + 2k\pi)$, $k \in \mathbb{Z} =$ _______

66. $\cos(x + 2k\pi)$, $k \in \mathbb{Z} =$ _______

67. $\sin(\pi - x) =$ _______

68. $\cos(\pi - x) =$ _______

69. $\cos(x - \pi) =$ _______

70. $\cos(x + \pi) =$ _______

71. $\sin(x + \pi) =$ _______

72. $\sin(x - \pi) =$ _______

73. Find all numbers x that satisfy the following. Check with a grapher.

a) $\sin x = 1$ b) $\cos x = -1$
c) $\sin x = 0$ d) $\sin x < \cos x$

74. Find $f \circ g$ and $g \circ f$, where $f(x) = x^2 + 2x$ and $g(x) = \cos x$.

Use a grapher to determine the domain, the range, the period, and the amplitude of each of these functions.

75. $y = (\sin x)^2$

76. $y = |\cos x| + 1$

Determine the domain of each function.

77. $f(x) = \sqrt{\cos x}$

78. $g(x) = \dfrac{1}{\sin x}$

79. $f(x) = \dfrac{\sin x}{\cos x}$

80. $g(x) = \log(\sin x)$

81. Consider $(\sin x)/x$, where x is between 0 and $\pi/2$. As x approaches 0, this function approaches a limiting value. What is it?

82. One of the motivations for developing trigonometry with a unit circle is that you can actually "see" $\sin\theta$ and $\cos\theta$ on the circle. Note in the figure that $AP = \sin\theta$ and $OA = \cos\theta$. It turns out that you can also "see" the other four trigonometric functions. Prove each of the following.

a) $BD = \tan\theta$ b) $OD = \sec\theta$
c) $OE = \csc\theta$ d) $CE = \cot\theta$

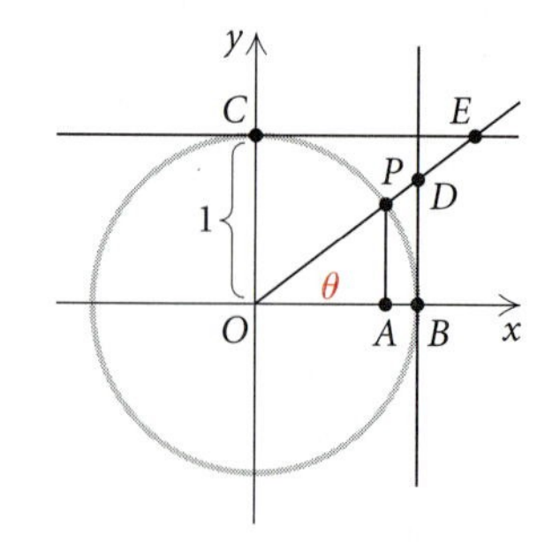

Graph.

83. $y = 3\sin x$

84. $y = \sin|x|$

85. $y = \sin x + \cos x$

86. $y = |\cos x|$

4.6 Graphs of Transformed Sine and Cosine Functions

- *Graph transformations of $y = \sin x$ and $y = \cos x$ in the form*

$$y = A \sin (Cx - D) + B$$

and

$$y = A \cos (Cx - D) + B$$

and determine the amplitude, the period, and the phase shift.
- *Graph sums of functions.*

Variations of Basic Graphs

In Section 4.5, we graphed all six trigonometric functions. In this section, we will consider variations of the sine and cosine graphs. For example, we will graph equations like the following:

$$y = 5 \sin \tfrac{1}{2}x, \qquad y = \cos (2x - \pi), \quad \text{and} \quad y = \tfrac{1}{2} \sin x - 3.$$

In particular, we are interested in graphs in the form

$$y = A \sin (Cx - D) + B$$

and

$$y = A \cos (Cx - D) + B,$$

where A, B, C, and D are constants. These constants have the effect of translating, reflecting, stretching, and shrinking the basic graphs. (It might be helpful to review Section 1.7.) Let's first examine the effect of each constant individually. Then we will consider the combined effects of more than one constant.

We first consider the effect of the constant B.

Interactive Discovery

Graph the first two functions in each group and then predict what the graph of the third function looks like. Then graph that function on a grapher.

$$y_1 = \sin x, \qquad y_2 = \sin x + 3.5, \qquad y_3 = \sin x - 4;$$

$$y_1 = \cos x, \qquad y_2 = \cos x - 2, \qquad y_3 = \cos x + 0.5.$$

What is the effect of the constant B on the graph of the basic function?

Example 1 Sketch a graph of $y = \sin x + 3$.

SOLUTION The graph of $y = \sin x + 3$ is a vertical translation of the graph of $y = \sin x$ up 3 units. One way to sketch the graph is to first consider $y = \sin x$ on an interval of length 2π, say, $[0, 2\pi]$. The zeros of the function and the maximum and minimum values can be considered key points. These are

$$(0, 0), \quad \left(\frac{\pi}{2}, 1\right), \quad (\pi, 0), \quad \left(\frac{3\pi}{2}, -1\right), \quad (2\pi, 0).$$

These key points are transformed up 3 units to obtain the key points of the graph of $y = \sin x + 3$. These are

$$(0, 3), \quad \left(\frac{\pi}{2}, 4\right), \quad (\pi, 3), \quad \left(\frac{3\pi}{2}, 2\right), \quad (2\pi, 3).$$

The graph of $y = \sin x + 3$ can be sketched over the interval $[0, 2\pi]$ and extended to obtain the rest of the graph by repeating the graph over intervals of length 2π.

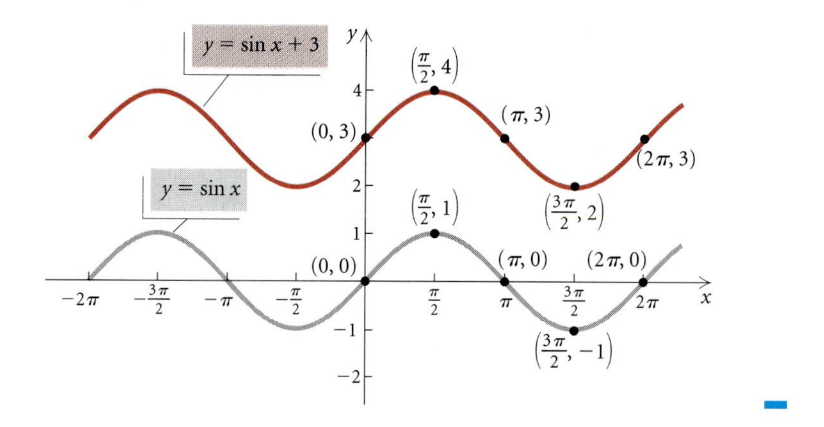

The constant B in

$$y = A \sin (Cx - D) + B \quad \text{and} \quad y = A \cos (Cx - D) + B$$

translates the graphs up B units if $B > 0$ or down $|B|$ units if $B < 0$.

Next we consider the effect of the constant A.

Interactive Discovery

Graph the first two functions in each group and then predict what the graph of the third function looks like. Then graph that function on a grapher.

$$y_1 = \sin x, \qquad y_2 = 2 \sin x, \qquad y_3 = \tfrac{1}{2} \sin x;$$
$$y_1 = \cos x, \qquad y_2 = 0.8 \cos x, \qquad y_3 = -0.8 \cos x.$$

What is the effect of the constant A on the graph of the basic function (a) when $0 < A < 1$? (b) when $A > 1$? (c) when $-1 < A < 0$? (d) when $A < -1$?

Example 2 Sketch a graph of $y = 2 \cos x$. What is the amplitude?

SOLUTION The constant 2 in $y = 2 \cos x$ has the effect of stretching the graph of $y = \cos x$ vertically by a factor of 2 units. The function values of $y = \cos x$ are such that $-1 \le \cos x \le 1$. The function values of $y = 2 \cos x$ are such that $-2 \le 2 \cos x \le 2$. The maximum value of $y = 2 \cos x$ is 2, and the minimum value is -2. Thus the amplitude, A, is $\frac{1}{2}[2 - (-2)]$, or 2.

We draw the graph of $y = \cos x$ and consider its key points,

$$(0, 1),\quad \left(\frac{\pi}{2}, 0\right),\quad (\pi, -1),\quad \left(\frac{3\pi}{2}, 0\right),\quad (2\pi, 1),$$

over the interval $[0, 2\pi]$.

We then multiply the second coordinates by 2 to obtain the key points of $y = 2 \cos x$. These are

$$(0, 2),\quad \left(\frac{\pi}{2}, 0\right),\quad (\pi, -2),\quad \left(\frac{3\pi}{2}, 0\right),\quad (2\pi, 2).$$

We plot these points and sketch the graph over the interval $[0, 2\pi]$. Then we repeat this part of the graph over adjacent intervals of length 2π.

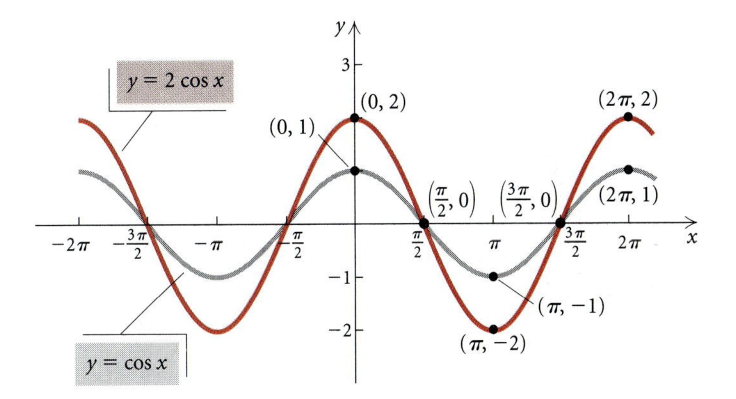

If $|A| > 1$, then there will be a vertical stretching. If $|A| < 1$, then there will be a vertical shrinking.

Amplitude

The *amplitude* of the graphs of $y = A \sin (Cx - D) + B$ and $y = A \cos (Cx - D) + B$ is $|A|$.

If $A < 0$, the graph is also reflected across the x-axis.

Example 3 Sketch a graph of $y = -\frac{1}{2} \sin x$.

SOLUTION The amplitude of the graph is $\left|-\frac{1}{2}\right|$, or $\frac{1}{2}$. The graph of $y = -\frac{1}{2} \sin x$ is a vertical shrinking and a reflection of the graph of $y = \sin x$ across the x-axis. In graphing, the key points of $y = \sin x$,

$$(0, 0),\quad \left(\frac{\pi}{2}, 1\right),\quad (\pi, 0),\quad \left(\frac{3\pi}{2}, -1\right),\quad (2\pi, 0),$$

are transformed to

$$(0, 0),\quad \left(\frac{\pi}{2}, -\frac{1}{2}\right),\quad (\pi, 0),\quad \left(\frac{3\pi}{2}, \frac{1}{2}\right),\quad (2\pi, 0).$$

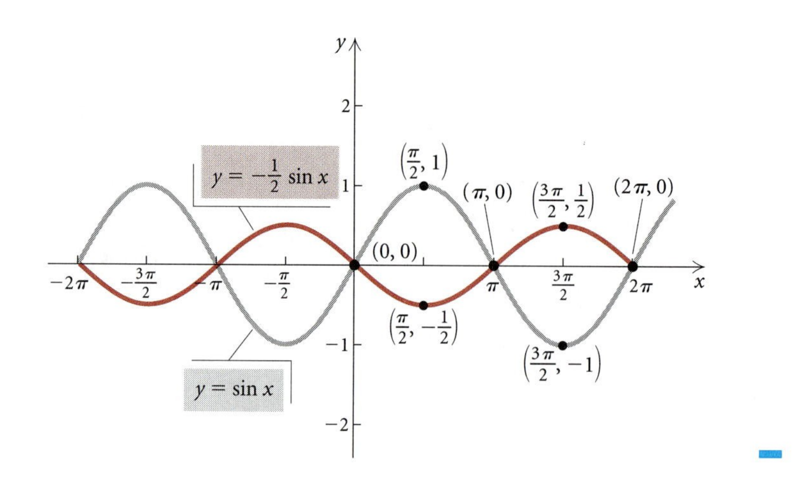

Now we consider the effect of the constant C.

Interactive Discovery

The period of the sine and cosine functions is 2π. Each of the graphs in Examples 1–3 also has a period of 2π. In general, changes in the constants A and B do not change the period. To see what effect a change in the constant C will have on the basic functions, graph

$$y_1 = \cos x,$$
$$y_2 = \cos(-2x),$$
$$y_3 = \cos\frac{1}{2}x.$$

Then visualize what the graphs of $y_4 = \cos 4x$ and $y_5 = \cos\left(-\frac{1}{2}x\right)$ will look like and check your guesses with a grapher. What is the period of each of these graphs?

Example 4 Sketch a graph of $y = \sin 4x$. What is the period?

SOLUTION The constant C has the effect of changing the period. Recall from Section 1.7 that the graph of $y = f(4x)$ is obtained from the graph of $y = f(x)$ by shrinking the graph horizontally. The new graph is obtained by dividing the first coordinate of each ordered-pair solution of $y = f(x)$ by 4. The key points of $y = \sin x$ are

$$(0, 0), \quad \left(\frac{\pi}{2}, 1\right), \quad (\pi, 0), \quad \left(\frac{3\pi}{2}, -1\right), \quad (2\pi, 0).$$

These are transformed to the key points of $y = \sin 4x$, which are

$$(0, 0), \quad \left(\frac{\pi}{8}, 1\right), \quad \left(\frac{\pi}{4}, 0\right), \quad \left(\frac{3\pi}{8}, -1\right), \quad \left(\frac{\pi}{2}, 0\right).$$

We plot these key points and sketch in the graph over the shortened interval $[0, \pi/2]$, which is of length $\pi/2$. Then we repeat the graph over other intervals of length $\pi/2$.

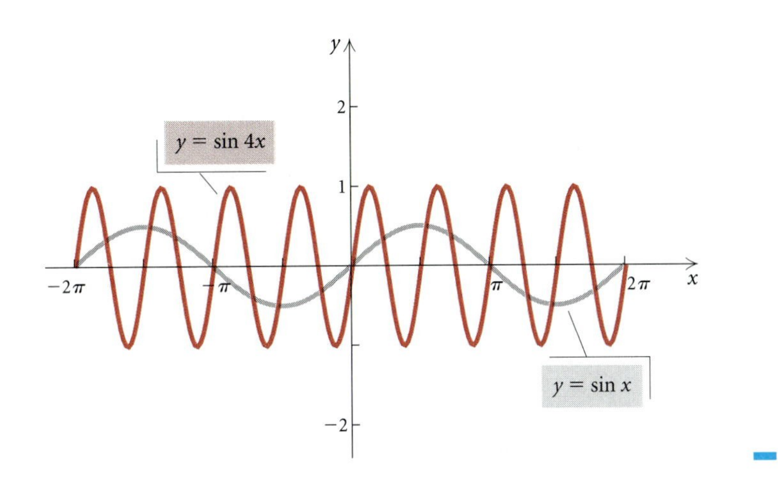

If $|C| < 1$, then there will be a horizontal stretching. If $|C| > 1$, then there will be a horizontal shrinking. If $C < 0$, the graph is also reflected across the y-axis.

Period

The *period* of the graphs of $y = A \sin (Cx - D) + B$ and $y = A \cos (Cx - D) + B$ is $\left|\frac{2\pi}{C}\right|$.

Now we examine the effect of the constant D.

Interactive Discovery

Graph

$$y_1 = \cos x,$$
$$y_2 = \cos (x - \pi),$$
$$y_3 = \cos \left(x + \frac{\pi}{2}\right),$$
$$y_4 = \cos \left(x - \frac{\pi}{4}\right).$$

Each curve has an amplitude of 1 and a period of 2π, but there are four distinct graphs. What is the effect of the constant D?

Example 5 Sketch a graph of $y = \sin \left(x - \frac{\pi}{2}\right)$.

SOLUTION The amplitude is 1, and the period is 2π. Recall from Section 1.7 that the graph of $y = f(x - d)$ is obtained from the graph of $y = f(x)$ by translating the graph horizontally—to the right d units if $d > 0$ and to the left $|d|$ units if $d < 0$. The graph of $y = \sin (x - \pi/2)$ is a translation of the graph of $y = \sin x$ to the right $\pi/2$ units. The value

$\pi/2$ is called the *phase shift*. The key points of $y = \sin x$,

$$(0, 0), \quad \left(\frac{\pi}{2}, 1\right), \quad (\pi, 0), \quad \left(\frac{3\pi}{2}, -1\right), \quad (2\pi, 0),$$

are transformed by adding $\pi/2$ to each of the first coordinates to obtain the following key points of $y = \sin(x - \pi/2)$:

$$\left(\frac{\pi}{2}, 0\right), \quad (\pi, 1), \quad \left(\frac{3\pi}{2}, 0\right), \quad (2\pi, -1), \quad \left(\frac{5\pi}{2}, 0\right).$$

We plot these key points and sketch the curve over the interval $[\pi/2, 5\pi/2]$. Then we repeat the graph over other intervals of length 2π.

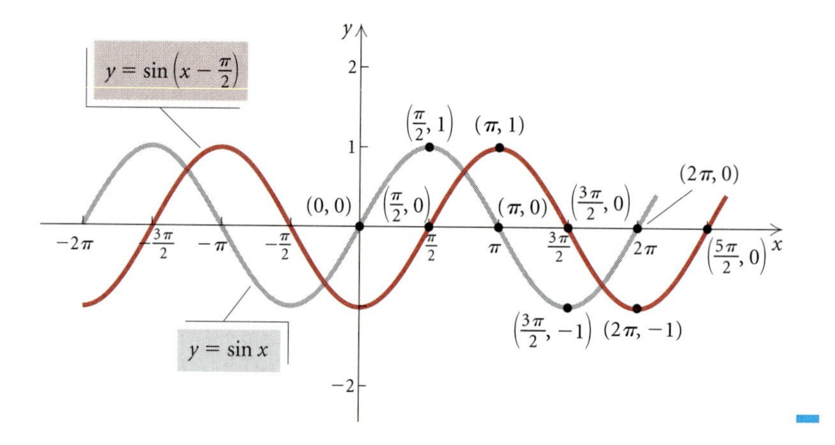

Now we consider combined transformations of graphs. It is helpful to rewrite

$$y = A \sin(Cx - D) + B \qquad \text{and} \quad y = A \cos(Cx - D) + B$$

as

$$y = A \sin\left[C\left(x - \frac{D}{C}\right)\right] + B \quad \text{and} \quad y = A \cos\left[C\left(x - \frac{D}{C}\right)\right] + B.$$

Example 6 Sketch a graph of $y = \cos(2x - \pi)$. What is the period?

SOLUTION The graph of

$$y = \cos(2x - \pi)$$

is the same as the graph of

$$y = 1 \cdot \cos\left[2\left(x - \frac{\pi}{2}\right)\right].$$

The amplitude is 1. The factor 2 shrinks the period by half, making the period $|2\pi/2|$, or π. The $\pi/2$ translates the graph of $y = \cos 2x$ to the right $\pi/2$ units. Thus to form the graph, we first graph $y = \cos x$, followed by $y = \cos 2x$ and then $y = \cos[2(x - \pi/2)]$.

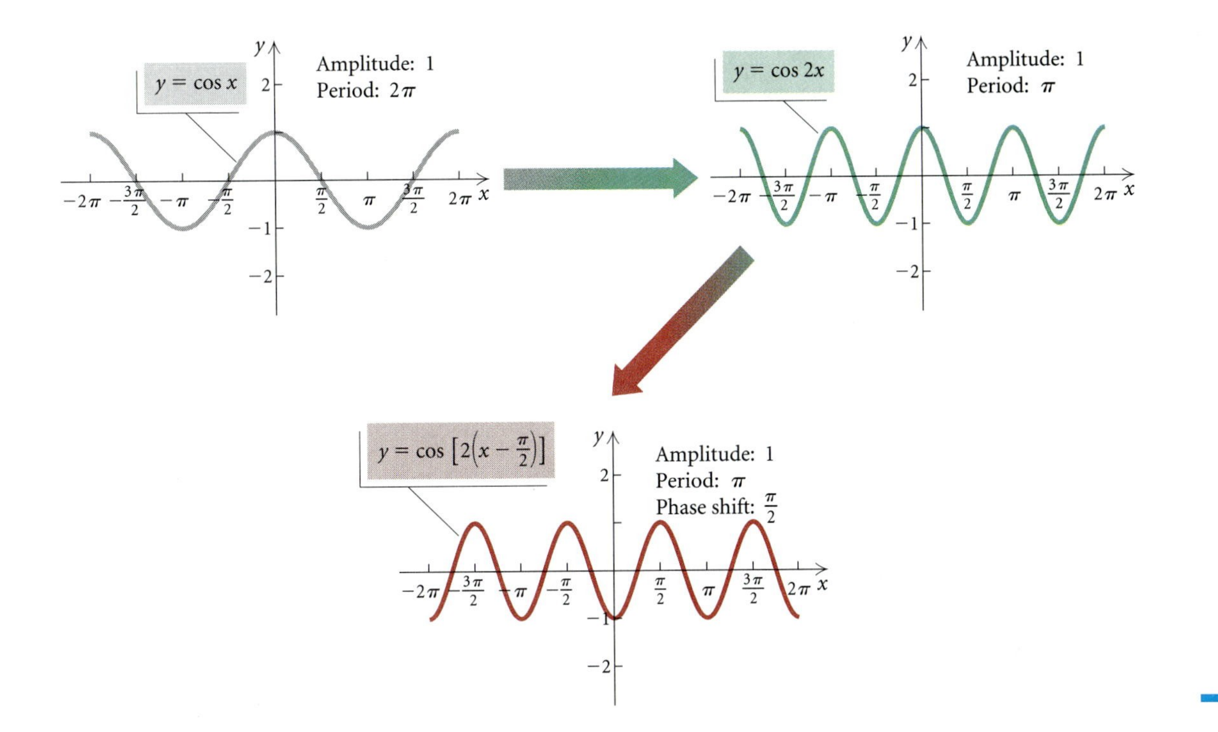

> ***Phase Shift***
>
> The *phase shift* of the graphs
>
> $$y = A \sin (Cx - D) + B = A \sin \left[C\left(x - \frac{D}{C} \right) \right] + B$$
>
> and
>
> $$y = A \cos (Cx - D) + B = A \cos \left[C\left(x - \frac{D}{C} \right) \right] + B$$
>
> is the quantity $\dfrac{D}{C}$.

If $D/C > 0$, the graph is translated to the right D/C units. If $D/C < 0$, the graph is translated to the left $|D/C|$ units. Be sure that the horizontal stretching or shrinking based on the constant C is done before the translation based on the phase shift D/C.

Let's now summarize the effect of the constants. We carry out the procedures in the order listed.

Transformations of Sine and Cosine Functions

To graph

$$y = A \sin (Cx - D) + B = A \sin \left[C\left(x - \frac{D}{C} \right) \right] + B$$

and

$$y = A \cos (Cx - D) + B = A \cos \left[C\left(x - \frac{D}{C} \right) \right] + B,$$

follow the steps listed below in the order in which they are listed.

1. Stretch or shrink the graph horizontally according to C.

$\lvert C\rvert < 1$	Stretch horizontally
$\lvert C\rvert > 1$	Shrink horizontally
$C < 0$	Reflect across the y-axis

The *period* is $\left|\frac{2\pi}{C}\right|$.

2. Stretch or shrink the graph vertically according to A.

$\lvert A\rvert < 1$	Shrink vertically
$\lvert A\rvert > 1$	Stretch vertically
$A < 0$	Reflect across the x-axis

The *amplitude* is $|A|$.

3. Translate the graph horizontally according to D/C.

$\frac{D}{C} < 0$	$\left\lvert\frac{D}{C}\right\rvert$ units to the left
$\frac{D}{C} > 0$	$\frac{D}{C}$ units to the right

The *phase shift* is $\frac{D}{C}$.

4. Translate the graph vertically according to B.

$B < 0$	$\lvert B\rvert$ units down
$B > 0$	B units up

Example 7 Sketch a graph of $y = 3 \sin (2x + \pi/2) + 1$. Find the amplitude, the period, and the phase shift.

SOLUTION We first note that

$$y = 3 \sin \left(2x + \frac{\pi}{2} \right) + 1 = 3 \sin \left[2\left(x - \left(-\frac{\pi}{4} \right) \right) \right] + 1.$$

Then we have the following:

$$\text{Amplitude} = |A| = |3| = 3,$$
$$\text{Period} = \left|\frac{2\pi}{C}\right| = \left|\frac{2\pi}{2}\right| = \pi,$$
$$\text{Phase shift} = \frac{D}{C} = -\frac{\pi}{4}.$$

To create the final graph, we begin with the basic sine curve, $y = \sin x$. Then we sketch graphs of each of the following equations in sequence.

1. $y = \sin 2x$
2. $y = 3 \sin 2x$
3. $y = 3 \sin\left[2\left(x - \left(-\frac{\pi}{4}\right)\right)\right]$
4. $y = 3 \sin\left[2\left(x - \left(-\frac{\pi}{4}\right)\right)\right] + 1$

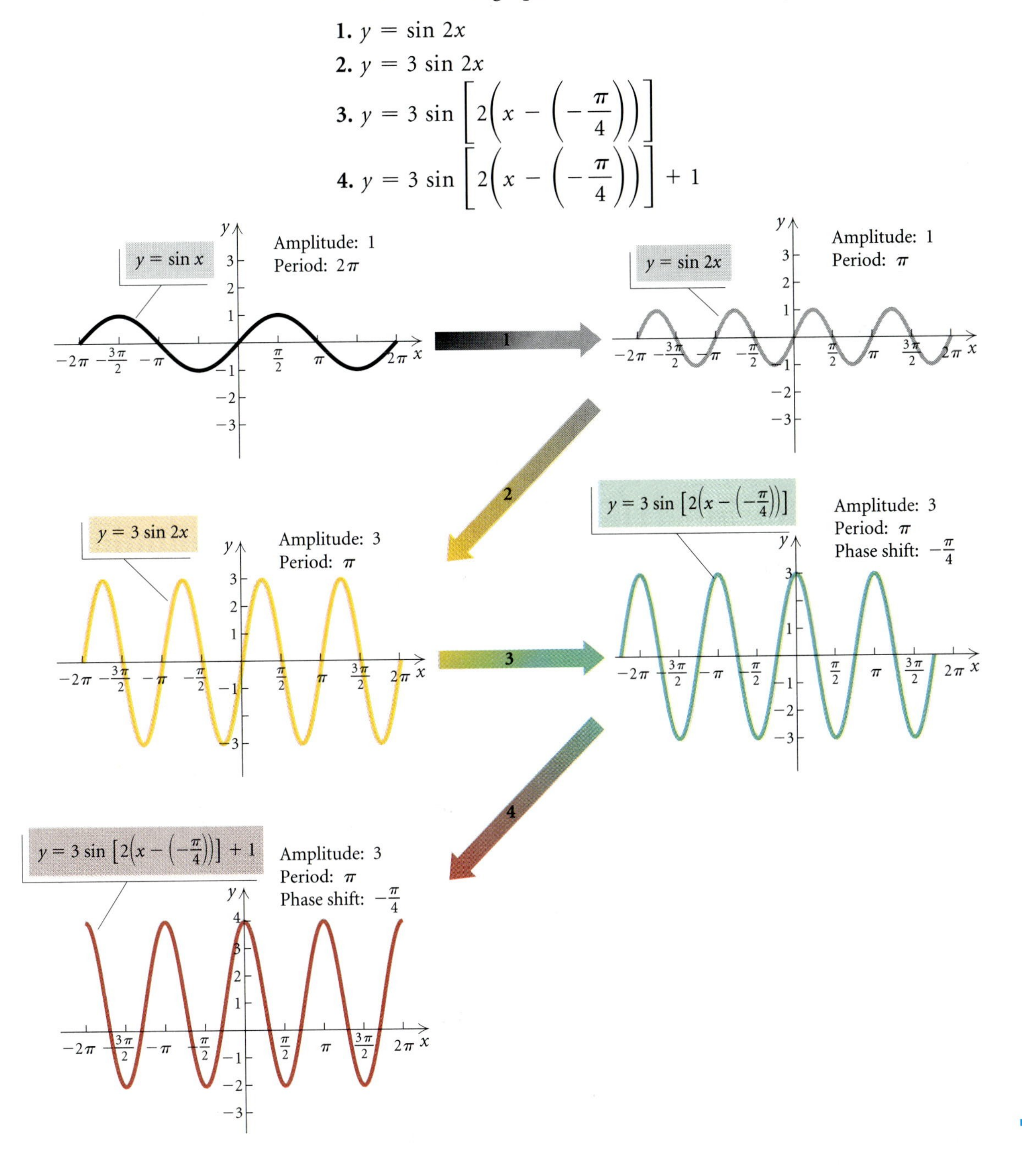

All the graphs in Examples 1–7 can be checked with a grapher. Even though it is faster and more accurate to graph with a grapher, graphing by hand gives us a greater understanding of the effect of changing the constants A, B, C, and D.

Graphers are especially convenient when a period or a phase shift is not a multiple of $\pi/4$, such as 4.3, 1, or $2\pi/3$.

Example 8 Graph $y = 3\cos(2\pi x) - 1$. Find the amplitude, the period, and the phase shift.

SOLUTION First we note the following:

$$\text{Amplitude} = |A| = |3| = 3,$$

$$\text{Period} = \left|\frac{2\pi}{C}\right| = \left|\frac{2\pi}{2\pi}\right| = |1| = 1,$$

$$\text{Phase shift} = \frac{D}{C} = \frac{0}{2\pi} = 0.$$

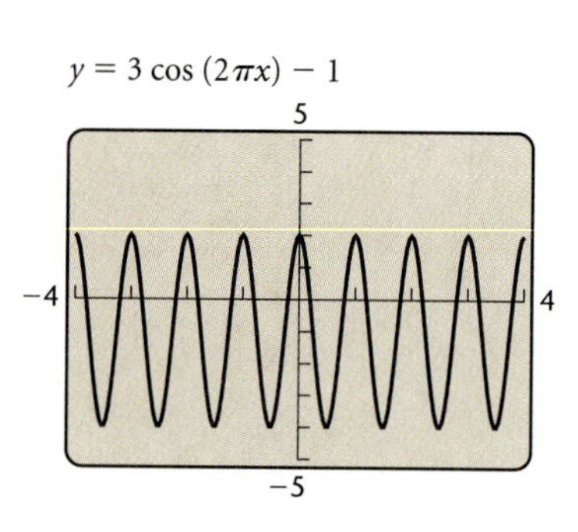

There is no phase shift in this case because the constant $D = 0$. The graph has a vertical translation of the graph of the cosine function down 1 unit, an amplitude of 3, and a period of 1, so we can use $[-4, 4, -5, 5]$ as the viewing window.

The transformational techniques that we learned in this section for graphing the sine and cosine functions can also be applied in the same manner to the other trigonometric functions. A few of these transformations are addressed in the synthesis exercises in Exercise Set 4.6.

An **oscilloscope** is an electronic device that converts electrical signals into graphs like those in the preceding examples. These graphs are often called sine waves. By manipulating the controls, we can change the amplitude, the period, and the phase of sine waves. The oscilloscope has many applications, and the trigonometric functions play a major role in many of them.

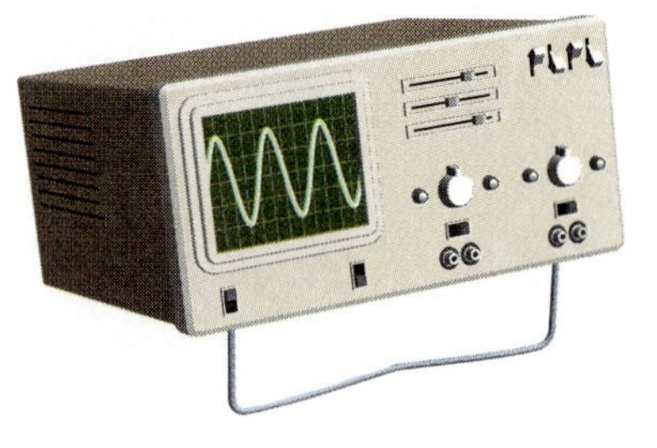

Graphs of Sums: Addition of Ordinates

The output of an electronic synthesizer used in the recording and playing of music can be converted into sine waves by an oscilloscope. The following graphs illustrate simple tones of different frequencies. The frequency of a simple tone is the number of vibrations in the signal of the tone per second. The loudness or intensity of the tone is reflected in the height of the graph (its amplitude). The three tones in the diagrams below all have the same intensity.

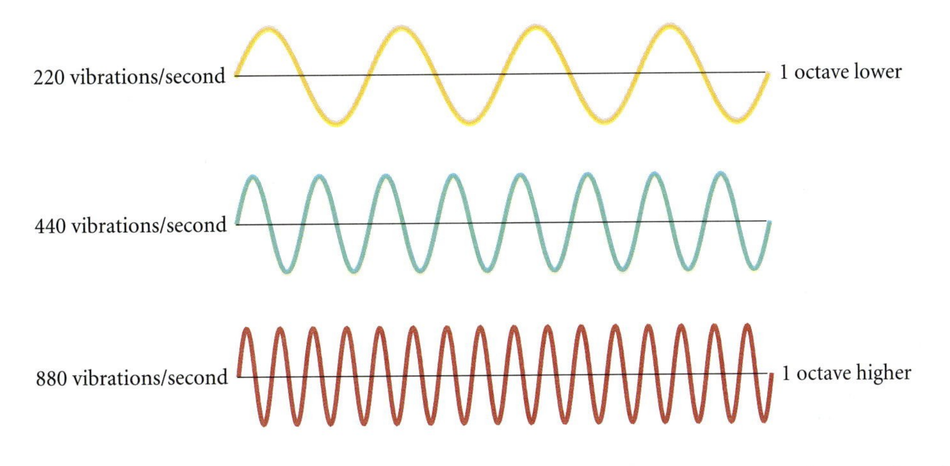

Musical instruments can generate extremely complex sine waves. On a single instrument, overtones can become superimposed on a simple tone. When multiple notes are played simultaneously, graphs become very complicated. This can happen when multiple notes are played on a single instrument or a group of instruments, or even when the same simple note is played on different instruments.

Combinations of simple tones produce interesting curves. Consider two tones whose graphs are $y_1 = 2 \sin x$ and $y_2 = \sin 2x$. The combination of the two tones produces a new sound whose graph is $y = 2 \sin x + \sin 2x$, shown in the following example.

Example 9 Graph: $y = 2 \sin x + \sin 2x$.

SOLUTION We graph $y = 2 \sin x$ and $y = \sin 2x$ using the same set of axes.

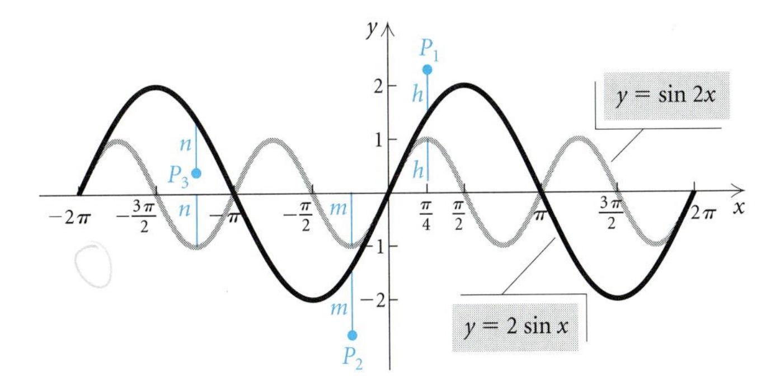

Now we graphically add some y-coordinates, or ordinates, to obtain points on the graph that we seek. At $x = \pi/4$, we transfer the distance h, which is the value of $\sin 2x$, up to add it to the value of $2 \sin x$. Point P_1 is on the graph that we seek. At $x = -\pi/4$, we use a similar procedure, but this time both ordinates are negative. Point P_2 is on the graph. At $x = -5\pi/4$, we add the negative ordinate of $\sin 2x$ to the positive ordinate of $2 \sin x$. Point P_3 is also on the graph. We continue to plot points in this fashion and then connect them to get the desired graph, shown at the top of the following page.

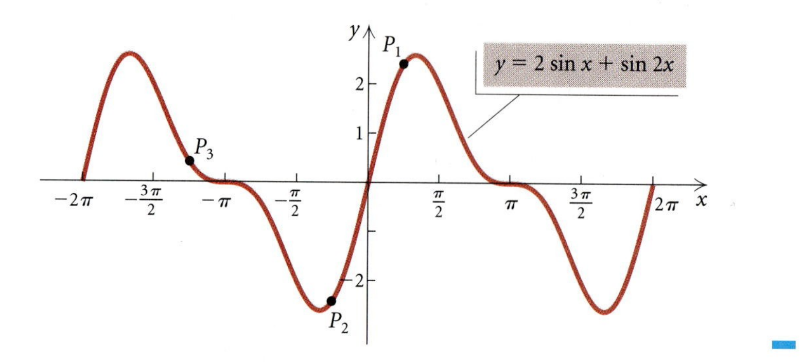

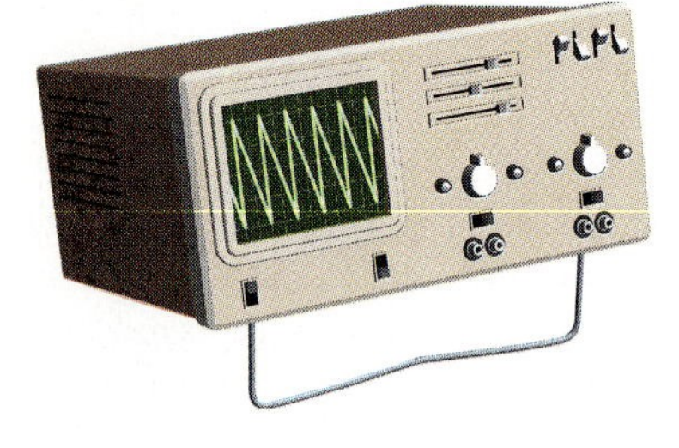

A "sawtooth function," such as the one shown on the oscilloscope at left, has numerous applications. This curve, which is actually not a function, can be approximated extremely well by adding, electronically, several sine and cosine functions.

With a grapher, we can quickly determine the period of a trigonometric function that is a combination of sine and cosine functions.

Example 10 Graph $y = 2 \cos x - \sin 3x$ and determine its period.

SOLUTION We graph $y = 2 \cos x - \sin 3x$ with appropriate dimensions. The period appears to be 2π.

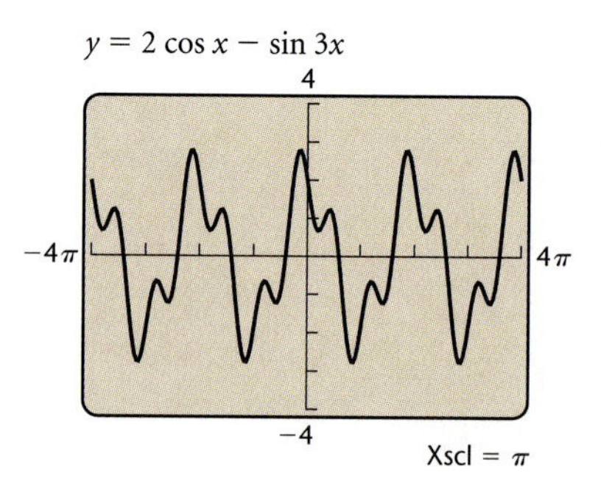

4.6 Exercise Set

Determine the amplitude, the period, and the phase shift of each function, and, without a grapher, sketch the graph of the function. Then check the graph with a grapher.

1. $y = \sin x + 1$
2. $y = \frac{1}{4} \cos x$
3. $y = -3 \cos x$
4. $y = \sin(-2x)$
5. $y = 2 \sin\left(\frac{1}{2}x\right)$
6. $y = \cos\left(x - \frac{\pi}{2}\right)$
7. $y = \frac{1}{2} \sin\left(x + \frac{\pi}{2}\right)$
8. $y = \cos x - \frac{1}{2}$
9. $y = 3 \cos(x - \pi)$
10. $y = -\sin\left(\frac{1}{4}x\right) + 1$
11. $y = \frac{1}{3} \sin x - 4$
12. $y = \cos\left(\frac{1}{2}x + \frac{\pi}{2}\right)$
13. $y = -\cos(-x) + 2$

14. $y = \frac{1}{2} \sin \left(2x - \frac{\pi}{4}\right)$

Determine the amplitude, the period, and the phase shift of each function. Then graph the function with a grapher. Try to visualize the graph before creating it.

15. $y = 2 \cos \left(\frac{1}{2}x - \frac{\pi}{2}\right)$

16. $y = 4 \sin \left(\frac{1}{4}x + \frac{\pi}{8}\right)$

17. $y = -\frac{1}{2} \sin \left(2x + \frac{\pi}{2}\right)$

18. $y = -3 \cos (4x - \pi) + 2$

19. $y = 2 + 3 \cos (\pi x - 3)$

20. $y = 5 - 2 \cos \left(\frac{\pi}{2}x + \frac{\pi}{2}\right)$

21. $y = -\frac{1}{2} \cos (2\pi x) + 2$

22. $y = -2 \sin (-2x + \pi) - 2$

23. $y = -\sin \left(\frac{1}{2}x - \frac{\pi}{2}\right) + \frac{1}{2}$

24. $y = \frac{1}{3} \cos (-3x) + 1$

25. $y = \cos (-2\pi x) + 2$

26. $y = \frac{1}{2} \sin (2\pi x + \pi)$

27. $y = -\frac{1}{4} \cos (\pi x - 4)$

28. $y = 2 \sin (2\pi x + 1)$

In Exercises 29–36, without a grapher, match each function with graphs (a)–(h), which follow. Then check your work with a grapher.

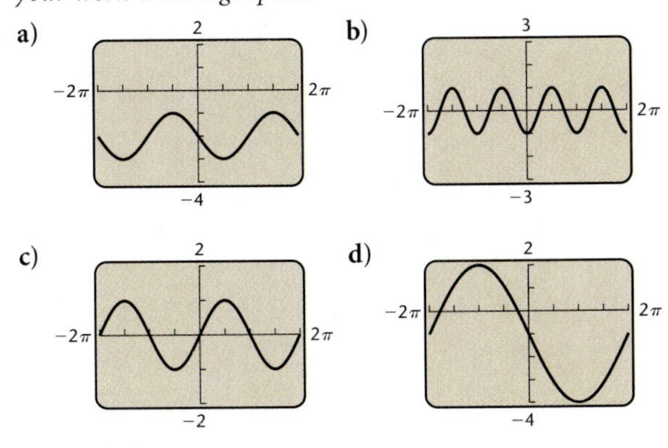

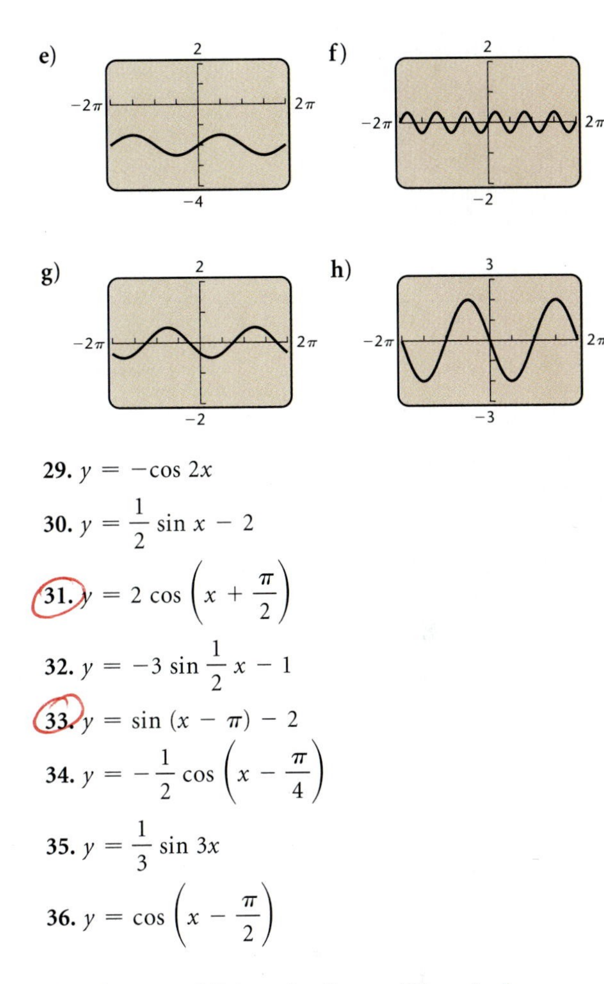

29. $y = -\cos 2x$

30. $y = \frac{1}{2} \sin x - 2$

31. $y = 2 \cos \left(x + \frac{\pi}{2}\right)$

32. $y = -3 \sin \frac{1}{2}x - 1$

33. $y = \sin (x - \pi) - 2$

34. $y = -\frac{1}{2} \cos \left(x - \frac{\pi}{4}\right)$

35. $y = \frac{1}{3} \sin 3x$

36. $y = \cos \left(x - \frac{\pi}{2}\right)$

Graph using addition of ordinates. Then check your graph with a grapher.

37. $y = 2 \cos x + \cos 2x$

38. $y = 3 \cos x + \cos 3x$

39. $y = \sin x + \cos 2x$

40. $y = 2 \sin x + \cos 2x$

41. $y = \sin x - \cos x$

42. $y = 3 \cos x - \sin x$

43. $y = 3 \cos x + \sin 2x$

44. $y = 3 \sin x - \cos 2x$

Use a grapher to graph each function.

45. $y = x + \sin x$

46. $y = \cos x - x$

47. $y = \cos 2x + 2x$

48. $y = \cos 3x + \sin 3x$

49. $y = 4 \cos 2x - 2 \sin x$

50. $y = 7.5 \cos x + \sin 2x$

Skill Maintenance

51. Add: $\dfrac{2x}{x^2 - 1} + \dfrac{4}{x - 1}$.

52. Multiply: $(x^2 - 4)(3x + 7)$.

53. Simplify: $\dfrac{2x^2 - 3x - 20}{2x^2 + 3x - 5}$.

54. Subtract: $\dfrac{1}{x - 6} - 4$.

Synthesis

55. ◈ In the equations $y = A \sin (Cx - D) + B$ and $y = A \cos (Cx - D) + B$, which constants translate the graphs and which constants stretch and shrink the graphs? Describe in your own words the effect of each constant.

56. ◈ In the transformation steps listed in this section, why must step (1) precede step (3)? Give an example that illustrates this.

Some graphers can use regression to fit a trigonometric function to a set of data.

57. *Sales.* Sales of certain products fluctuate in cycles. The data in the following table show the total sales of skis per month for a business in a northern climate.

x, MONTH	TOTAL SALES, y (IN THOUSANDS OF DOLLARS)
8. August	0
11. November	7
2. February	14
5. May	7
8. August	0

a) Using the SINE REGRESSION feature on a grapher, fit a sine function of the form $y = A \sin (Cx - D) + B$ to this set of data.

b) Approximate the total sales for December and for July.

58. *Daylight Hours.* The data in the following table give the number of daylight hours for certain days in Kajaani, Finland.

x, DAY	NUMBER OF DAYLIGHT HOURS, y
10. January 10	5.0
50. February 19	9.1
62. March 3	10.4
118. April 28	16.4
134. May 14	18.2
162. June 11	20.7
198. July 17	19.5
234. August 22	15.7
262. September 19	12.7
274. October 1	11.4
318. November 14	6.7
362. December 28	4.3

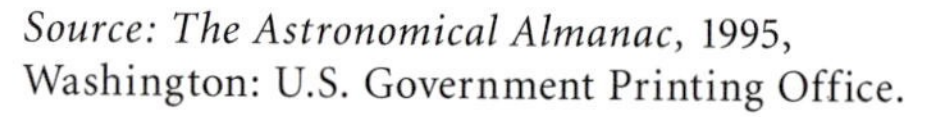

Source: The Astronomical Almanac, 1995, Washington: U.S. Government Printing Office.

a) Using the SINE REGRESSION feature on a grapher, model these data with an equation of the form $y = A \sin (Cx - D) + B$.

b) Approximate the number of daylight hours in Kajaani for April 22, July 4, and December 15.

The transformational techniques that we learned in this section for graphing the sine and cosine functions can also be applied to the other trigonometric functions. Sketch a graph of each of the following. Then check your work with a grapher.

59. $y = -\tan x$

60. $y = \tan (-x)$

61. $y = -2 + \cot x$

62. $y = -\dfrac{3}{2} \csc x$

63. $y = 2 \tan \dfrac{1}{2} x$

64. $y = \cot 2x$

65. $y = 2 \sec (x - \pi)$

66. $y = 4 \tan \left(\dfrac{1}{4}x + \dfrac{\pi}{8}\right)$

67. $y = 2 \csc \left(\dfrac{1}{2}x - \dfrac{3\pi}{4}\right)$

68. $y = 4 \sec (2x - \pi)$

Use a grapher to graph each function.

69. $f(x) = e^{-x/2} \cos x$

70. $f(x) = x \sin x$

71. $f(x) = 0.6x^2 \cos x$

72. $f(x) = 2^{-x} \cos x$

73. $f(x) = |\tan x|$

74. $f(x) = (\tan x)(\csc x)$

Use a grapher to graph each of the following over the given interval and approximate the zeros.

75. $f(x) = \dfrac{\sin x}{x}$; $[-12, 12]$

76. $f(x) = \dfrac{\cos x - 1}{x}$; $[-12, 12]$

77. $f(x) = x^3 \sin x$; $[-5, 5]$

78. $f(x) = \dfrac{(\sin x)^2}{x}$; $[-4, 4]$

79. *Temperature During an Illness.* The temperature T of a patient during a 12-day illness is given by

$$T(t) = 101.6° + 3° \sin\left(\frac{\pi}{8}t\right).$$

a) Graph the function over the interval $[0, 12]$.
b) What are the maximum and the minimum temperatures during the illness?

80. *Periodic Sales.* A company in a northern climate has sales of skis as given by

$$S(t) = 10\left(1 - \cos\frac{\pi}{6}t\right),$$

where t = time, in months ($t = 0$ corresponds to July 1), and $S(t)$ is in thousands of dollars.

a) Graph the function over a 12-month interval $[0, 12]$.
b) What is the period of the function?
c) What is the minimum amount of sales and when does it occur?
d) What is the maximum amount of sales and when does it occur?

81. *Satellite Location.* A satellite circles the earth in such a way that it is y miles from the equator (north or south, height not considered) t minutes after its launch, where

$$y(t) = 3000\left[\cos\frac{\pi}{45}(t - 10)\right].$$

a) Graph the function.
b) What are the amplitude, the period, and the phase shift?

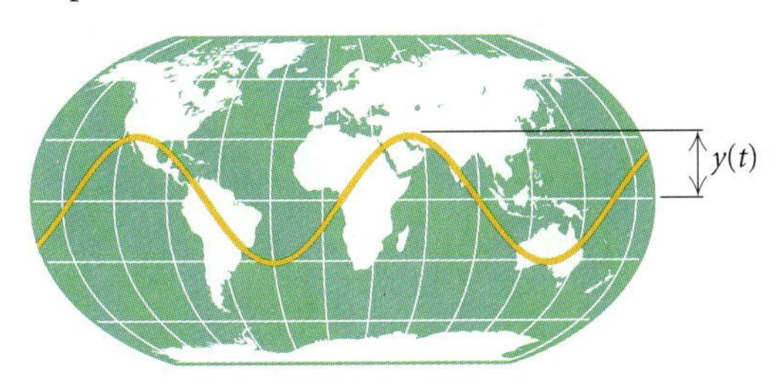

82. *Water Wave.* The cross-section of a water wave is given by

$$y = 3 \sin\left(\frac{\pi}{4}x + \frac{\pi}{4}\right),$$

where y is the vertical height of the water wave and x is the distance from the origin to the wave.

a) Graph the function.
b) What are the amplitude, the period, and the phase shift?

83. *Damped Oscillations.* Suppose that the motion of a spring is given by

$$d(t) = 6e^{-0.8t} \cos(6\pi t) + 4,$$

where d is the distance, in inches, of a weight from the point at which the spring is attached to a ceiling, after t seconds.

a) Graph the function over the interval $[0, 10]$.
b) How far do you think the spring is from the ceiling when the spring stops bobbing?

84. *Rotating Beacon.* A police car is parked 10 ft from a wall. On top of the car is a beacon rotating in such a way that the light is at a distance $d(t)$ from point Q after t seconds, where

$$d(t) = 10 \tan(2\pi t).$$

When d is positive, the light is pointing north of Q, and when d is negative, the light is pointing south of Q.

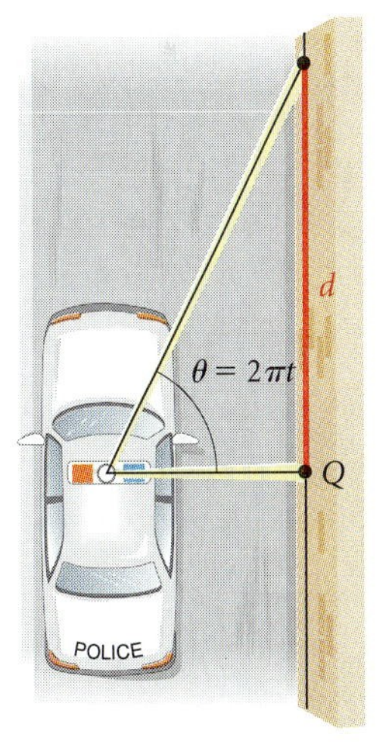

a) Graph the function over the interval $[0, 2]$.
b) Explain the meaning of the values of t for which the function is undefined.

CHAPTER 4 Summary and Review

Important Properties and Formulas

Trigonometric Function Values of an Acute Angle θ

Let θ be an acute angle of a right triangle. The six trigonometric functions of θ are as follows:

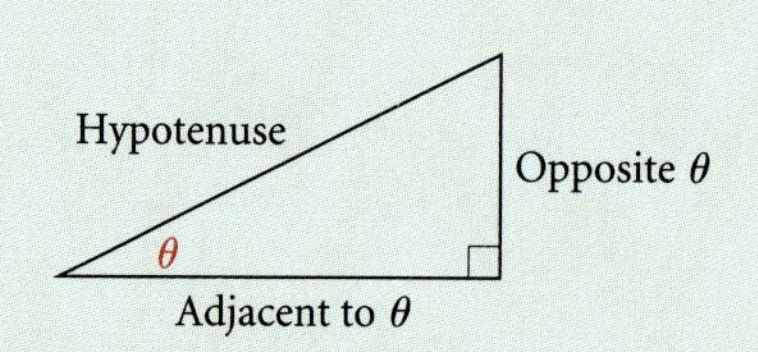

$$\sin\theta = \frac{\text{opp}}{\text{hyp}}, \qquad \cos\theta = \frac{\text{adj}}{\text{hyp}}, \qquad \tan\theta = \frac{\text{opp}}{\text{adj}}.$$

Reciprocal Functions

$$\csc\theta = \frac{1}{\sin\theta}, \qquad \sec\theta = \frac{1}{\cos\theta}, \qquad \cot\theta = \frac{1}{\tan\theta}$$

Function Values of Special Angles

	0°	30°	45°	60°	90°
sin	0	1/2	$\sqrt{2}/2$	$\sqrt{3}/2$	1
cos	1	$\sqrt{3}/2$	$\sqrt{2}/2$	1/2	0
tan	0	$\sqrt{3}/3$	1	$\sqrt{3}$	Undefined

Cofunction Identities

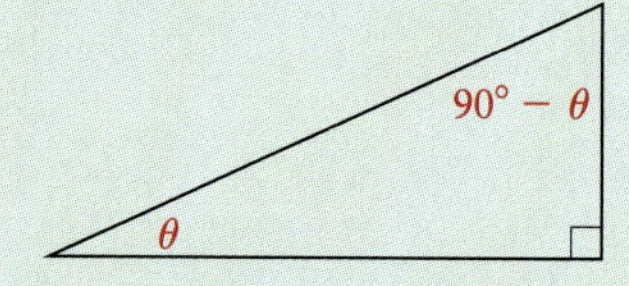

$$\sin\theta = \cos(90° - \theta) \qquad \cos\theta = \sin(90° - \theta)$$
$$\tan\theta = \cot(90° - \theta) \qquad \cot\theta = \tan(90° - \theta)$$
$$\sec\theta = \csc(90° - \theta) \qquad \csc\theta = \sec(90° - \theta)$$

Trigonometric Functions of Any Angle θ

$P(x, y)$ is any point on the terminal side of any angle θ in standard position, and r is the distance from the origin to $P(x, y)$.

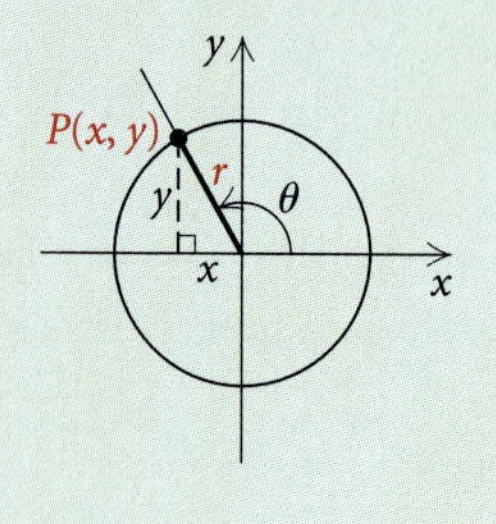

$$\sin\theta = \frac{y}{r}, \qquad \cos\theta = \frac{x}{r}, \qquad \tan\theta = \frac{y}{x}$$

(continued)

Signs of Function Values

The signs of the function values depend only on the coordinates of the point P on the terminal side.

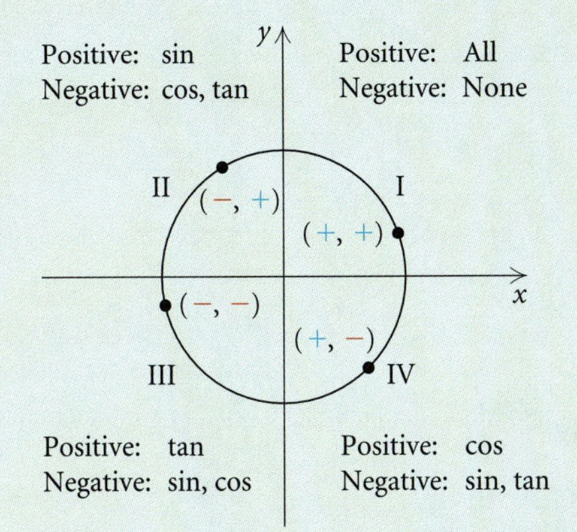

Radian–Degree Equivalents

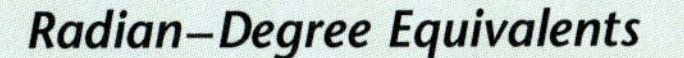

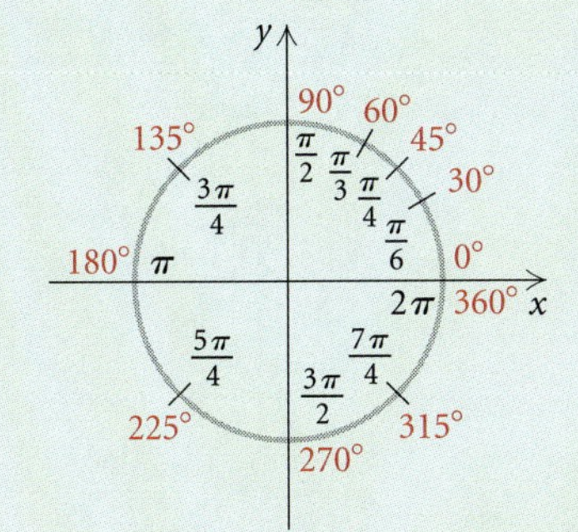

Linear Speed in Terms of Angular Speed

$$v = r\omega$$

Basic Circular Functions

For a real number s that determines a point (x, y) on the *unit circle*:

$$\sin s = y, \qquad \cos s = x, \qquad \tan s = \frac{y}{x}.$$

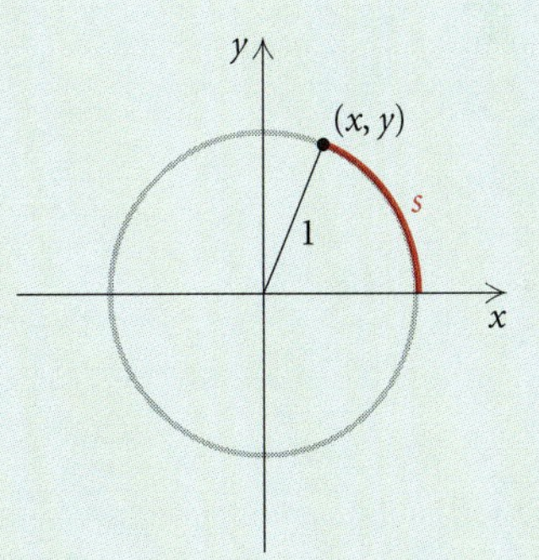

Sine (Odd Function): $\sin(-s) = -\sin s$

Cosine (Even Function): $\cos(-s) = \cos s$

Transformations of Sine and Cosine Functions

To graph $y = A \sin(Cx - D) + B$ and $y = A \cos(Cx - D) + B$:

1. Stretch or shrink the graph horizontally according to C.
2. Stretch or shrink the graph vertically according to A.
3. Translate the graph horizontally according to D/C.
4. Translate the graph vertically according to B.

REVIEW EXERCISES

1. Find the six trigonometric function values of the specified angle.

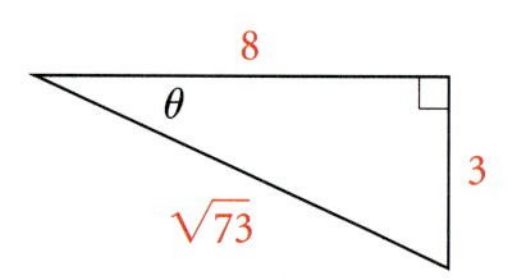

Find the exact function value, if it exists.

2. $\cos 45°$
3. $\cot 60°$
4. $\cos 495°$
5. $\sin 150°$
6. $\sec(-270°)$
7. $\tan(-600°)$

8. Convert 22.27° to degrees, minutes, and seconds. Round to the nearest second.

9. Convert 47°33′27″ to decimal degree notation. Round to two decimal places.

Find the function value. Round to four decimal places.

10. $\tan 2184°$
11. $\sec 27.9°$
12. $\cos 18°13'42''$
13. $\sin 245°24'$
14. $\cot(-33.2°)$
15. $\sin 556.13°$

Find θ in the interval indicated. Round the answer to the nearest tenth of a degree.

16. $\cos\theta = -0.9041$, $(180°, 270°)$

17. $\tan\theta = 1.0799$, $(0°, 90°)$

Find the exact acute angle θ, in degrees, given the function value.

18. $\sin\theta = \dfrac{\sqrt{3}}{2}$

19. $\tan\theta = \sqrt{3}$

20. Given that $\sin 59.1° \approx 0.8581$, $\cos 59.1° \approx 0.5135$, and $\tan 59.1° \approx 1.6709$, find the six function values for 30.9°.

Solve each of the following right triangles. Standard lettering has been used.

21. $a = 7.3$, $c = 8.6$

22. $a = 30.5$, $B = 51.17°$

23. One leg of a right triangle bears east. The hypotenuse is 734 m long and bears N57°23′E. Find the perimeter of the triangle.

24. An observer's eye is 6 ft above the floor. A mural is being viewed. The bottom of the mural is at floor level. The observer looks down 13° to see the bottom and up 17° to see the top. How tall is the mural?

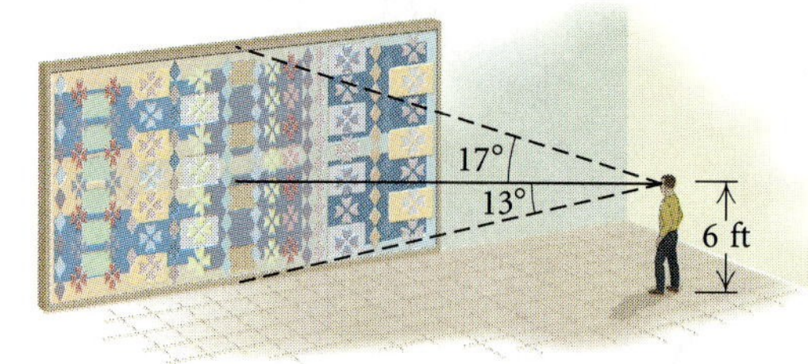

Find a positive angle and a negative angle that are coterminal with the given angle. Answers may vary.

25. $65°$
26. $\dfrac{7\pi}{3}$

Find the complement and the supplement.

27. $13.4°$
28. $\dfrac{\pi}{6}$

29. Find the six trigonometric function values for the angle θ shown.

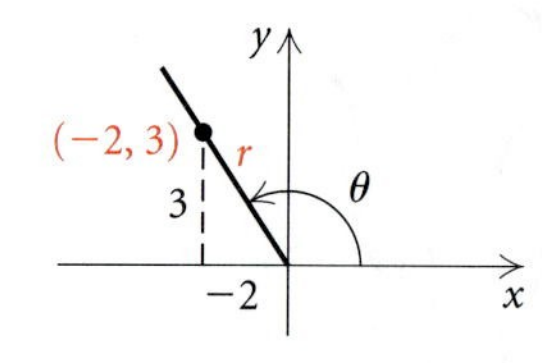

30. Given that $\tan\theta = 2/\sqrt{5}$ and that the terminal side is in quadrant III, find the other five function values.

31. An airplane travels at 530 mph for $3\frac{1}{2}$ hr in a direction of 160° from Minneapolis, Minnesota. At the end of that time, how far south of Minneapolis is the airplane?

32. On a unit circle, mark and label the points determined by $7\pi/6$, $-3\pi/4$, $-\pi/3$, and $9\pi/4$.

For angles of the following measures, state in which quadrant the terminal side lies, convert to radian measure in terms of π, and convert to radian measure not in terms of π.

33. $145.2°$
34. $-30°$

Convert to degree measure. Round the answer to two decimal places.

35. $\dfrac{3\pi}{2}$
36. 3

37. Find the length of an arc of a circle, given a central angle of $\pi/4$ and a radius of 7 cm.

38. An arc 18 m long on a circle of radius 8 m subtends an angle of how many radians? how many degrees, to the nearest degree?

39. Inside La Madeleine French Bakery and Cafe in Houston, Texas, there is one of the few remaining working watermills in the world. The 300-yr-old French-built waterwheel has a radius of 7 ft and makes one complete revolution in 70 sec. What is the linear speed in feet per minute of a point on the rim? (*Source:* La Madeleine French Bakery and Cafe, Houston, TX)

40. An automobile wheel has a diameter of 14 in. If the car travels at a speed of 55 mph, what is the angular velocity, in radians per hour, of a point on the edge of the wheel?

41. The point $\left(\frac{3}{5}, -\frac{4}{5}\right)$ is on a unit circle. Find the coordinates of its reflections across the x-axis, the y-axis, and the origin.

Find the exact function value, if it exists.

42. $\cos \pi$ **43.** $\tan \frac{5\pi}{4}$ **44.** $\sin \frac{5\pi}{3}$

45. $\sin \left(-\frac{7\pi}{6}\right)$ **46.** $\tan \frac{\pi}{6}$ **47.** $\cos (-13\pi)$

Find the function value. Round to four decimal places.

48. $\sin 24$ **49.** $\cos (-75)$

50. $\cot 16\pi$ **51.** $\tan \frac{3\pi}{7}$

52. $\sec 14.3$ **53.** $\cos \left(-\frac{\pi}{5}\right)$

54. Make a hand-drawn graph from -2π to 2π for each of the six trigonometric functions.

55. What is the period of each of the six trigonometric functions?

56. Complete the following table.

FUNCTION	DOMAIN	RANGE
sine		
cosine		
tangent		

57. Complete the following table with the sign of the specified trigonometric function value in each of the four quadrants.

FUNCTION	I	II	III	IV
sine				
cosine				
tangent				

Determine the amplitude, the period, and the phase shift of each function; and without a grapher, sketch the graph of the function. Then check the graph with a grapher.

58. $y = \sin \left(x + \frac{\pi}{2}\right)$

59. $y = 3 + \frac{1}{2} \cos \left(2x - \frac{\pi}{2}\right)$

In Exercises 60–63, without a grapher, match each function with graphs (a)–(d), which follow. Then check your work with a grapher.

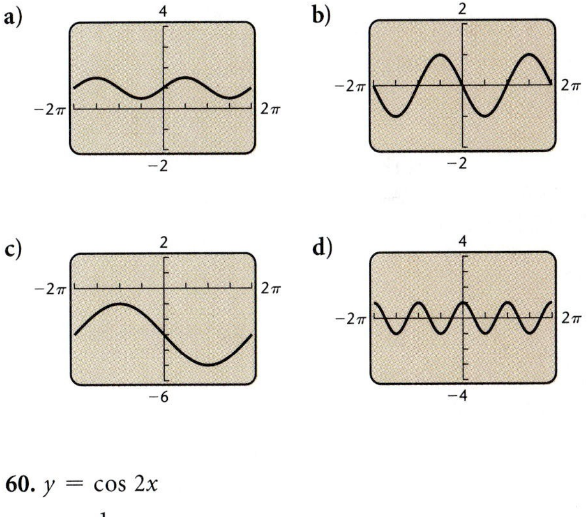

60. $y = \cos 2x$

61. $y = \frac{1}{2} \sin x + 1$

62. $y = -2 \sin \frac{1}{2} x - 3$

63. $y = -\cos \left(x - \frac{\pi}{2}\right)$

64. Sketch a graph of $y = 3 \cos x + \sin x$ for values of x between 0 and 2π.

Synthesis

65. ◈ Compare the terms radian and degree.

66. ◈ Describe the shape of the graph of the cosine function. How many maximum values are there of the cosine function? Where do they occur?

67. ◈ Explain the disadvantage of a grapher when graphing a function like

$$f(x) = \frac{\sin x}{x}.$$

68. ◈ Does $5 \sin x = 7$ have a solution for x? Why or why not?

69. For what values of x in $(0, \pi/2]$ is $\sin x < x$ true?

70. Graph $y = 3 \sin (x/2)$ and determine the domain, the range, and the period.

71. In the window below, $y_1 = \sin x$ is shown and y_2 is shown in red. Express y_2 as a transformation of the graph of y_1.

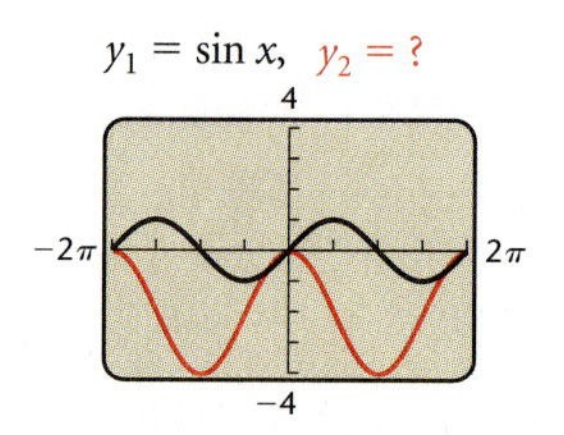

72. Find the domain of $y = \log (\cos x)$.

73. Given that $\sin x = 0.6144$ and that the terminal side is in quadrant II, find the other basic circular function values.

Trigonometric Identities, Inverse Functions, and Equations 5

APPLICATION

A satellite circles the earth in such a manner that it is y miles from the equator (north or south) t minutes after its launch, where

$$y = 5000\left[\cos \frac{\pi}{45}(t - 10)\right].$$

This trigonometric equation can be used to determine the times t in the first 4 hr that the satellite is 3000 mi north of the equator.

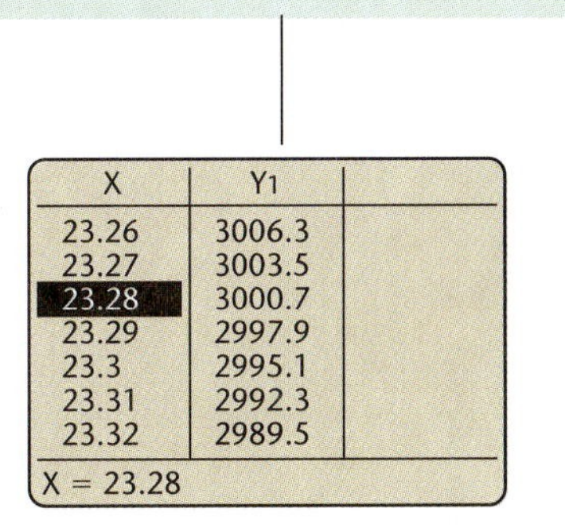

X	Y1	
23.26	3006.3	
23.27	3003.5	
23.28	3000.7	
23.29	2997.9	
23.3	2995.1	
23.31	2992.3	
23.32	2989.5	

X = 23.28

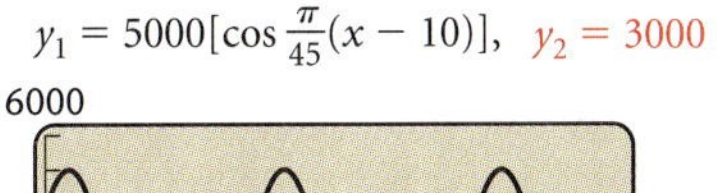

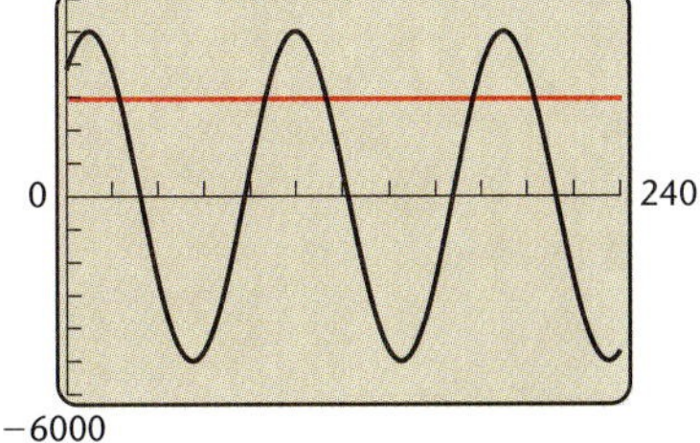

There are a number of relationships among the trigonometric functions, given by identities, that are important in algebraic and trigonometric manipulations. A large part of this chapter is devoted to those identities and their use in solving trigonometric equations. We also provide a detailed examination of inverses of the trigonometric functions. This chapter provides not only a basis for applications but also a foundation for further work in mathematics.

5.1 Identities: Pythagorean and Sum and Difference
5.2 Identities: Cofunction, Double-Angle, and Half-Angle
5.3 Proving Trigonometric Identities
5.4 Inverses of the Trigonometric Functions
5.5 Solving Trigonometric Equations
SUMMARY AND REVIEW

5.1 Identities: Pythagorean and Sum and Difference

- *State the Pythagorean identities.*
- *Simplify and manipulate expressions containing trigonometric expressions.*
- *Use the sum and difference identities to find function values.*

Recall that an identity is an equation that is true for all *possible* replacements of the variables. The following is a list of the identities studied in Chapter 4.

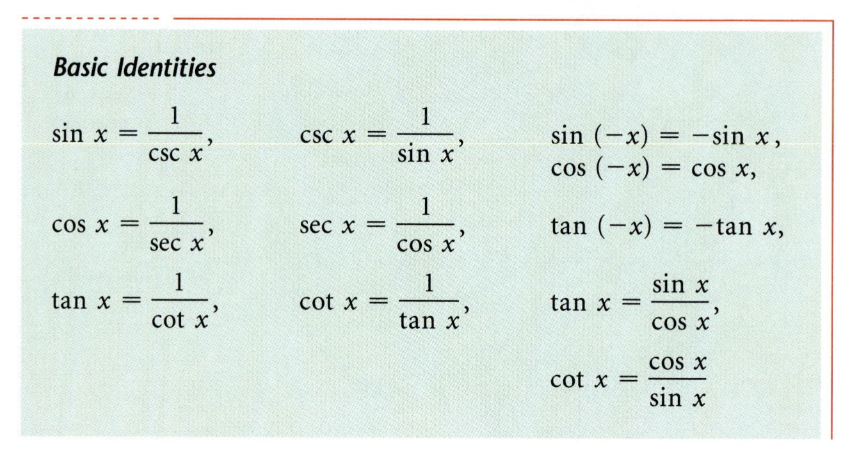

Basic Identities

$\sin x = \dfrac{1}{\csc x}$, $\csc x = \dfrac{1}{\sin x}$, $\sin(-x) = -\sin x$, $\cos(-x) = \cos x$,

$\cos x = \dfrac{1}{\sec x}$, $\sec x = \dfrac{1}{\cos x}$, $\tan(-x) = -\tan x$,

$\tan x = \dfrac{1}{\cot x}$, $\cot x = \dfrac{1}{\tan x}$, $\tan x = \dfrac{\sin x}{\cos x}$,

$\cot x = \dfrac{\cos x}{\sin x}$

In this section, we will develop some other important identities.

Pythagorean Identities

We now consider three other identities that are fundamental to a study of trigonometry. They are called the *Pythagorean identities*. Recall that the equation of a unit circle in the xy-plane is $x^2 + y^2 = 1$.

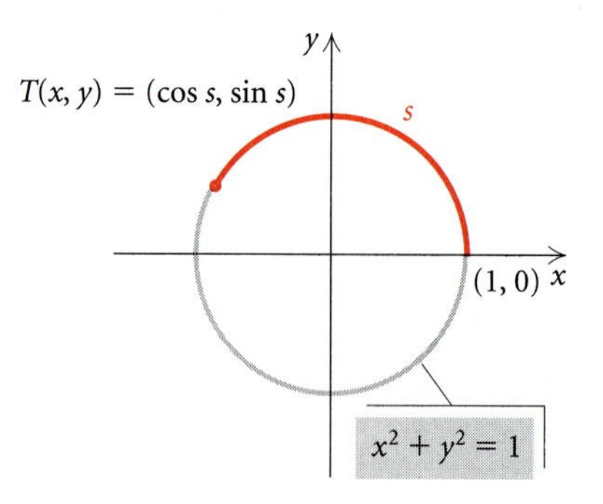

For any point on the unit circle, the coordinates x and y satisfy this equation. Suppose that a real number s determines a point T on the unit circle, with coordinates (x, y), or $(\cos s, \sin s)$. Then $x = \cos s$ and $y = \sin s$. Substituting $\cos s$ for x and $\sin s$ for y in the equation of the unit circle gives us the identity

$$(\cos s)^2 + (\sin s)^2 = 1,$$

which can be expressed as

$$\mathbf{\sin^2 s + \cos^2 s = 1.}$$

It is conventional in trigonometry to use the notation $\sin^2 s$ rather than $(\sin s)^2$, which uses parentheses. Note that $\sin^2 s \neq \sin s^2$.

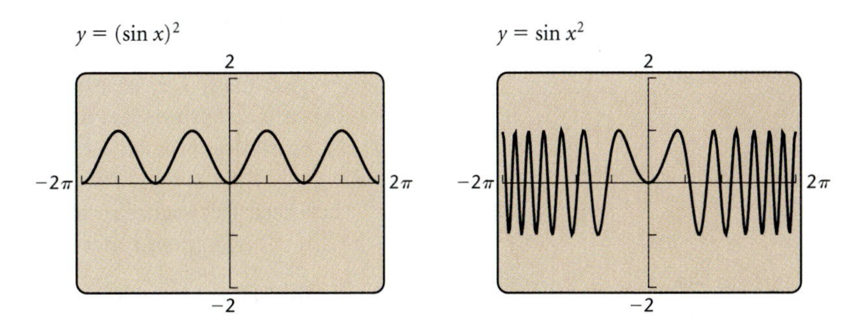

The identity $\sin^2 s + \cos^2 s = 1$ relates the sine and the cosine of any real number s. It is an important **Pythagorean identity**.

Interactive Discovery

Addition of y-values provides a unique way of developing the identity $\sin^2 x + \cos^2 x = 1$. First graph $y_1 = \sin^2 x$ and $y_2 = \cos^2 x$. By adding the y-values, visually graph the sum, $y_3 = \sin^2 x + \cos^2 x$. Then graph y_3 with a grapher and check your prediction. The resulting graph appears to be the line $y_4 = 1$, which is the graph of the expression on the right of the equals sign. These graphs do not prove the identity, but they do provide a check in the interval $[-2\pi, 2\pi]$.

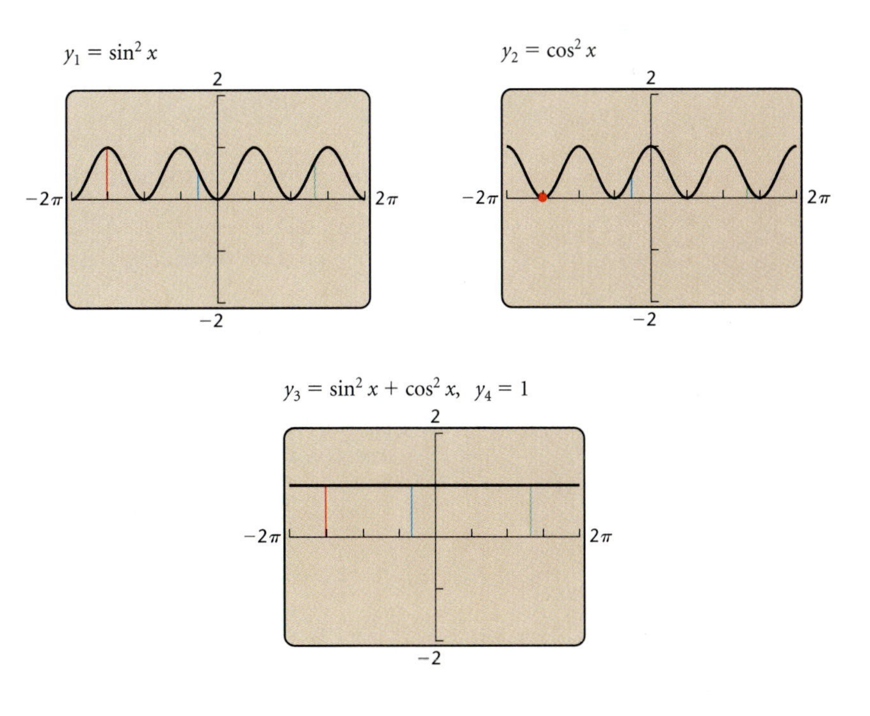

We can divide on both sides of the preceding identity by $\sin^2 s$:

$$\frac{\sin^2 s}{\sin^2 s} + \frac{\cos^2 s}{\sin^2 s} = \frac{1}{\sin^2 s}.$$

Simplifying gives us a second identity:

$$\mathbf{1 + \cot^2 s = \csc^2 s.}$$

This equation is true for any replacement of s by a real number for which $\sin^2 s \neq 0$, since we divided by $\sin^2 s$. But the numbers for which $\sin^2 s = 0$ (or $\sin s = 0$) are exactly the ones for which the cotangent and cosecant functions are undefined. Hence our new equation holds for all real numbers s for which $\cot s$ and $\csc s$ are defined and is thus an identity.

The third Pythagorean identity, obtained by dividing on both sides of the first Pythagorean identity by $\cos^2 s$, is

$$\mathbf{\tan^2 s + 1 = \sec^2 s.}$$

The identities we have developed hold no matter what symbols are used for the variables. For example, $\sin^2 s + \cos^2 s = 1$, $\sin^2 \theta + \cos^2 \theta = 1$, and $\sin^2 x + \cos^2 x = 1$.

Pythagorean Identities

$\sin^2 x + \cos^2 x = 1,$

$1 + \cot^2 x = \csc^2 x,$

$1 + \tan^2 x = \sec^2 x$

It is important to remember the Pythagorean identities and how they are developed.

Simplifying Trigonometric Expressions

We can factor, simplify, and manipulate trigonometric expressions in the same way that we manipulate strictly algebraic expressions.

Example 1 Multiply and simplify: $\cos x\,(\tan x - \sec x)$.

SOLUTION

$\cos x\,(\tan x - \sec x)$

$= \cos x \tan x - \cos x \sec x$ **Multiplying**

$= \cos x \dfrac{\sin x}{\cos x} - \cos x \dfrac{1}{\cos x}$ **Recalling the identities $\tan x = \dfrac{\sin x}{\cos x}$ and $\sec x = \dfrac{1}{\cos x}$ and substituting**

$= \sin x - 1$ **Simplifying**

There is no general procedure for creating an identity as in Example 1, but it is often helpful to write everything in terms of sines and cosines.

Example 2 Factor and simplify: $\sin^2 x \cos^2 x + \cos^4 x$.

SOLUTION

$$\sin^2 x \cos^2 x + \cos^4 x$$

$$= \cos^2 x (\sin^2 x + \cos^2 x) \qquad \text{Removing a common factor}$$

$$= \cos^2 x \qquad \text{Using } \sin^2 x + \cos^2 x = 1$$

Identities can be checked with a grapher. First graph the expression on the left side of the equals sign. Then graph the expression on the right side using the same screen. If the two graphs are indistinguishable, then we have a partial verification that the equation is an identity. Of course, we can never see the entire graph, so there can always be some doubt. Also, the graphs may not overlap precisely, but you may not be able to tell because the difference between the graphs may be less than the width of a pixel.

For example, consider the identity in Example 1:

$$\cos x (\tan x - \sec x) = \sin x - 1.$$

Graph each side, recalling that we may need to enter $\sec x$ as $1/\cos x$:

$$y_1 = \cos x (\tan x - \sec x) = \cos x [\tan x - (1/\cos x)]$$

and

$$y_2 = \sin x - 1.$$

As you can see in the screen below, the graphs appear to be identical. Thus, $\cos x (\tan x - \sec x) = \sin x - 1$ is most likely an identity.

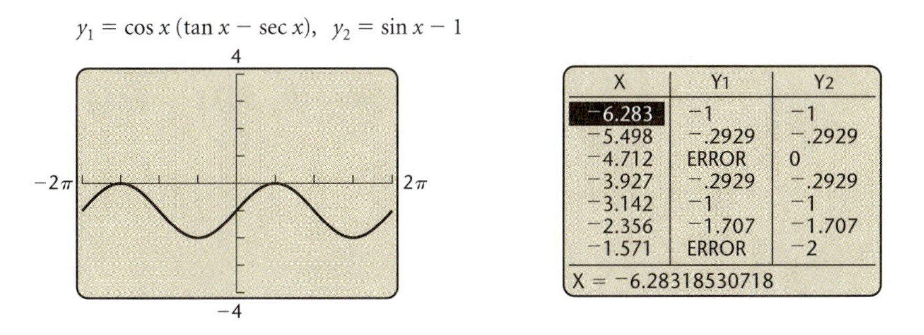

X	Y1	Y2
−6.283	−1	−1
−5.498	−.2929	−.2929
−4.712	ERROR	0
−3.927	−.2929	−.2929
−3.142	−1	−1
−2.356	−1.707	−1.707
−1.571	ERROR	−2

X = −6.28318530718

The TABLE feature can also be used to check identities. Note in the table above that the function values are the same except for those values of x for which $\cos x = 0$. The domain of y_1 excludes these values. The domain of y_2 is the set of all real numbers. Thus all real numbers except $\pm\pi/2, \pm 3\pi/2, \pm 5\pi/2, \ldots$ are possible replacements for x in the identity. Recall that an identity is an equation that is true for all *possible* replacements. Note that when the grapher graphs y_1, it simplifies the expression to $\sin x - 1$ before graphing. This is why the graph is continuous and does not show the domain restriction for y_1.

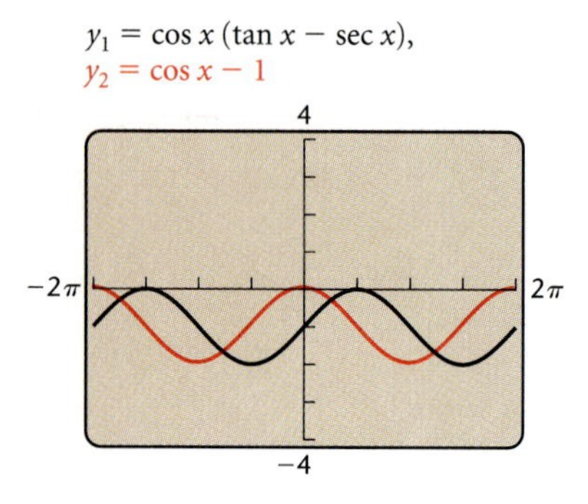

If we had simplified incorrectly in Example 1 and had gotten $\cos x - 1$ on the right instead of $\sin x - 1$, two different graphs would have appeared in the window. Thus we would have known that we did not have an identity and that $\cos x\,(\tan x - \sec x) \neq \cos x - 1$.

Example 3 Simplify each of the following trigonometric expressions.

a) $\dfrac{\cot(-\theta)}{\csc(-\theta)}$ b) $\dfrac{2\sin^2 t + \sin t - 3}{1 - \cos^2 t - \sin t}$

SOLUTION

a) $$\frac{\cot(-\theta)}{\csc(-\theta)} = \frac{\dfrac{\cos(-\theta)}{\sin(-\theta)}}{\dfrac{1}{\sin(-\theta)}} = \frac{\cos(-\theta)}{\sin(-\theta)} \cdot \sin(-\theta) = \cos(-\theta) = \cos\theta$$

b) $$\frac{2\sin^2 t + \sin t - 3}{1 - \cos^2 t - \sin t}$$

$$= \frac{2\sin^2 t + \sin t - 3}{\sin^2 t - \sin t}$$ Using $\sin^2 x + \cos^2 x = 1$, or $\sin^2 x = 1 - \cos^2 x$, and substituting $\sin^2 t$ for $1 - \cos^2 t$

$$= \frac{(2\sin t + 3)(\sin t - 1)}{\sin t\,(\sin t - 1)}$$ Factoring in both numerator and denominator

$$= \frac{2\sin t + 3}{\sin t}$$ Simplifying

Check the results on a grapher.

We can add and subtract trigonometric fractional expressions in the same way that we did algebraic expressions.

Example 4 Add and simplify: $\dfrac{\cos x}{1 + \sin x} + \tan x$.

SOLUTION

$$\frac{\cos x}{1 + \sin x} + \tan x = \frac{\cos x}{1 + \sin x} + \frac{\sin x}{\cos x}$$ Using $\tan x = \dfrac{\sin x}{\cos x}$

$$= \frac{\cos x}{1 + \sin x} \cdot \frac{\cos x}{\cos x} + \frac{\sin x}{\cos x} \cdot \frac{1 + \sin x}{1 + \sin x}$$ Multiplying by forms of 1

$$= \frac{\cos^2 x + \sin x + \sin^2 x}{\cos x\,(1 + \sin x)}$$ Adding

$$= \frac{1 + \sin x}{\cos x\,(1 + \sin x)}$$ Using $\sin^2 x + \cos^2 x = 1$

$$= \frac{1}{\cos x}, \quad \text{or} \quad \sec x$$ Simplifying

Check the result on a grapher.

When radicals occur, the use of absolute value is sometimes necessary, but it can be difficult to determine when to use it. In Examples 5 and 6, we will assume that all radicands are nonnegative. This means that the identities are meant to be confined to certain quadrants.

Example 5 Multiply and simplify: $\sqrt{\sin^3 x \cos x} \cdot \sqrt{\cos x}$.

SOLUTION

$$\begin{aligned}\sqrt{\sin^3 x \cos x} \cdot \sqrt{\cos x} &= \sqrt{\sin^3 x \cos^2 x} \\ &= \sqrt{\sin^2 x \cos^2 x \sin x} \\ &= \sin x \cos x \sqrt{\sin x}\end{aligned}$$

Example 6 Rationalize the denominator: $\sqrt{\dfrac{2}{\tan x}}$.

SOLUTION

$$\sqrt{\frac{2}{\tan x}} = \sqrt{\frac{2}{\tan x} \cdot \frac{\tan x}{\tan x}} = \sqrt{\frac{2 \tan x}{\tan^2 x}} = \frac{\sqrt{2 \tan x}}{\tan x}$$

Often in calculus, a substitution is a useful manipulation, as we show in the following example.

Example 7 Express $\sqrt{9 + x^2}$ as a trigonometric function of θ without using radicals by letting $x = 3 \tan \theta$. Assume that $0 < \theta < \pi/2$. Then find $\sin \theta$ and $\cos \theta$.

SOLUTION We have

$$\begin{aligned}\sqrt{9 + x^2} &= \sqrt{9 + (3 \tan \theta)^2} && \text{Substituting } 3 \tan \theta \text{ for } x \\ &= \sqrt{9 + 9 \tan^2 \theta} \\ &= \sqrt{9(1 + \tan^2 \theta)} && \text{Factoring} \\ &= \sqrt{9 \sec^2 \theta} && \text{Using } 1 + \tan^2 x = \sec^2 x \\ &= 3|\sec \theta| = 3 \sec \theta. && \text{For } 0 < \theta < \pi/2, \sec \theta > 0.\end{aligned}$$

We can express $\sqrt{9 + x^2} = 3 \sec \theta$ as

$$\sec \theta = \frac{\sqrt{9 + x^2}}{3}.$$

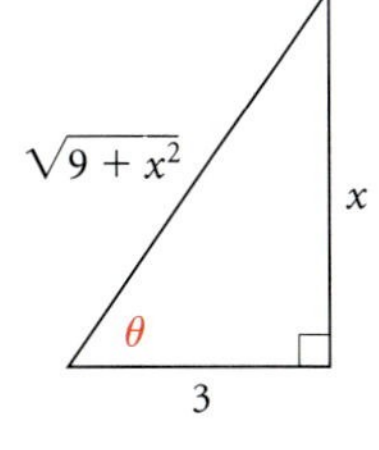

In a right triangle, we know that $\sec \theta$ is hypotenuse/adjacent, when θ is one of the acute angles. Using the Pythagorean theorem, we can determine that the side opposite θ is x. Then from the right triangle, we see that

$$\sin \theta = \frac{x}{\sqrt{9 + x^2}} \quad \text{and} \quad \cos \theta = \frac{3}{\sqrt{9 + x^2}}.$$

Sum and Difference Identities

We now develop some important identities involving sums or differences of two numbers (or angles), beginning with an identity for the cosine of the difference of two numbers. We use the letters u and v for these numbers.

Let's consider a real number u in the interval $[\pi/2, \pi]$ and a real number v in the interval $[0, \pi/2]$. These determine points A and B on the unit circle as shown. The arc length s is $u - v$, and we know that $0 \le s \le \pi$. Recall that the coordinates of A are $(\cos u, \sin u)$, and the coordinates of B are $(\cos v, \sin v)$.

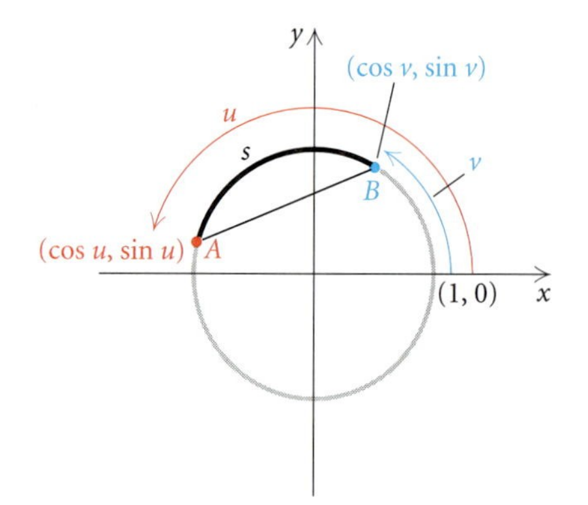

Using the distance formula, we can write an expression for the distance AB:

$$AB = \sqrt{(\cos u - \cos v)^2 + (\sin u - \sin v)^2}.$$

This can be simplified as follows:

$$\begin{aligned} AB &= \sqrt{\cos^2 u - 2\cos u \cos v + \cos^2 v + \sin^2 u - 2\sin u \sin v + \sin^2 v} \\ &= \sqrt{(\sin^2 u + \cos^2 u) + (\sin^2 v + \cos^2 v) - 2(\cos u \cos v + \sin u \sin v)} \\ &= \sqrt{2 - 2(\cos u \cos v + \sin u \sin v)}. \end{aligned}$$

Now let's imagine rotating the circle above so that point B is at $(1, 0)$. Although the coordinates of point A are now $(\cos s, \sin s)$, the distance AB has not changed.

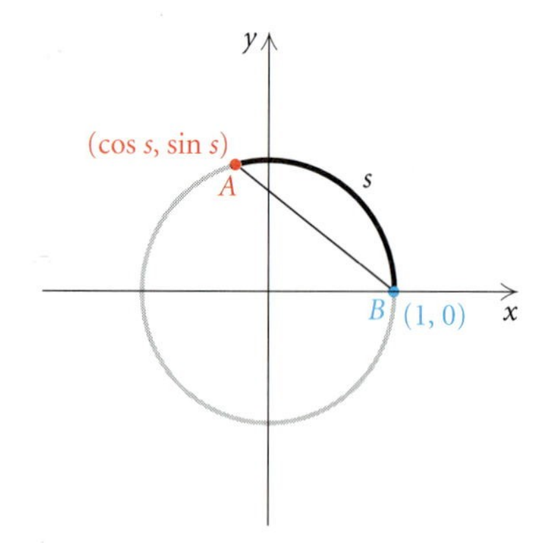

Again we use the distance formula to write an expression for the distance AB:

$$AB = \sqrt{(\cos s - 1)^2 + (\sin s - 0)^2}.$$

This simplifies as follows:

$$\begin{aligned} AB &= \sqrt{\cos^2 s - 2\cos s + 1 + \sin^2 s} \\ &= \sqrt{(\sin^2 s + \cos^2 s) + 1 - 2\cos s} \\ &= \sqrt{2 - 2\cos s}. \end{aligned}$$

Equating our two expressions for AB, we obtain

$$\sqrt{2 - 2(\cos u \cos v + \sin u \sin v)} = \sqrt{2 - 2\cos s}.$$

Solving this equation for $\cos s$ gives

$$\cos s = \cos u \cos v + \sin u \sin v. \qquad (1)$$

But $s = u - v$, so we have the equation

$$\cos (u - v) = \cos u \cos v + \sin u \sin v. \qquad (2)$$

Formula (1) above holds when s is the length of the shortest arc from A to B. Given any real numbers u and v, the length of the shortest arc from A to B is not always $u - v$. In fact, it could be $v - u$. However, since $\cos(-x) = \cos x$, we know that $\cos(v - u) = \cos(u - v)$. Thus, $\cos s$ is always equal to $\cos(u - v)$. Formula (2) holds for all real numbers u and v. That formula is thus the identity we sought:

$$\mathbf{\cos (u - v) = \cos u \cos v + \sin u \sin v.}$$

On a grapher, graph

$$y_1 = \cos (x - 3)$$

and $\quad y_2 = \cos x \cos 3 + \sin x \sin 3$

to illustrate this result.

The cosine sum formula follows easily from the one we have just derived. Let's consider $\cos(u + v)$. This is equal to $\cos[u - (-v)]$, and by the identity above, we have

$$\cos (u + v) = \cos [u - (-v)] = \cos u \cos (-v) + \sin u \sin (-v).$$

But $\cos(-v) = \cos v$ and $\sin(-v) = -\sin v$, so the identity we seek is the following:

$$\mathbf{\cos (u + v) = \cos u \cos v - \sin u \sin v.}$$

On a grapher, graph

$$y_1 = \cos (x + 2)$$

and $\quad y_2 = \cos x \cos 2 - \sin x \sin 2$

to illustrate this result.

Example 8 Find $\cos(5\pi/12)$ exactly.

SOLUTION We can express $5\pi/12$ as a difference of two numbers whose sine and cosine values are known:

$$\frac{5\pi}{12} = \frac{9\pi}{12} - \frac{4\pi}{12}, \quad \text{or} \quad \frac{3\pi}{4} - \frac{\pi}{3}.$$

Then, using $\cos(u - v) = \cos u \cos v + \sin u \sin v$, we have

$$\cos\frac{5\pi}{12} = \cos\left(\frac{3\pi}{4} - \frac{\pi}{3}\right) = \cos\frac{3\pi}{4}\cos\frac{\pi}{3} + \sin\frac{3\pi}{4}\sin\frac{\pi}{3}$$
$$= -\frac{\sqrt{2}}{2}\cdot\frac{1}{2} + \frac{\sqrt{2}}{2}\cdot\frac{\sqrt{3}}{2}$$
$$= -\frac{\sqrt{2}}{4} + \frac{\sqrt{6}}{4} = \frac{\sqrt{6} - \sqrt{2}}{4}.$$

We can check with a grapher:

$$\cos\frac{5\pi}{12} \approx 0.2588 \quad \text{and} \quad \frac{\sqrt{6} - \sqrt{2}}{4} \approx 0.2588.$$

Consider $\cos(\pi/2 - \theta)$. We can use the identity for the cosine of a difference to simplify as follows:

$$\cos\left(\frac{\pi}{2} - \theta\right) = \cos\frac{\pi}{2}\cos\theta + \sin\frac{\pi}{2}\sin\theta$$
$$= 0 \cdot \cos\theta + 1 \cdot \sin\theta = \sin\theta.$$

This identity appeared first in Section 4.1.

Thus we have developed the identity

$$\sin\theta = \cos\left(\frac{\pi}{2} - \theta\right). \tag{3}$$

This identity holds for any real number θ. From it we can obtain an identity for the sine function. We first let α be any real number. Then we replace θ in $\sin\theta = \cos(\pi/2 - \theta)$ by $\pi/2 - \alpha$. This gives us

$$\sin\left(\frac{\pi}{2} - \alpha\right) = \cos\left[\frac{\pi}{2} - \left(\frac{\pi}{2} - \alpha\right)\right] = \cos\alpha,$$

which yields the identity

$$\cos\alpha = \sin\left(\frac{\pi}{2} - \alpha\right). \tag{4}$$

Using identities (3) and (4) and the identity for the cosine of a difference, we can obtain an identity for the sine of a sum. We start with identity (3) and substitute $u + v$ for θ:

$$\sin\theta = \cos\left(\frac{\pi}{2} - \theta\right) \quad \text{Identity (3)}$$

$$\sin(u + v) = \cos\left[\frac{\pi}{2} - (u + v)\right] \quad \text{Substituting } u + v \text{ for } \theta$$

$$= \cos\left[\left(\frac{\pi}{2} - u\right) - v\right]$$

$$= \cos\left(\frac{\pi}{2} - u\right)\cos v + \sin\left(\frac{\pi}{2} - u\right)\sin v \quad \text{Using the identity for the cosine of a difference}$$

$$= \sin u \cos v + \cos u \sin v. \quad \text{Using identities (3) and (4)}$$

Thus the identity we seek is

$$\mathbf{\sin (u + v) = \sin u \cos v + \cos u \sin v.}$$

To find a formula for the sine of a difference, we can use the identity just derived, substituting $-v$ for v:

$$\sin (u + (-v)) = \sin u \cos (-v) + \cos u \sin (-v).$$

Simplifying gives us

$$\mathbf{\sin (u - v) = \sin u \cos v - \cos u \sin v.}$$

Example 9 Find sin 105° exactly.

SOLUTION We can express 105° as the sum of two measures:

$$105° = 45° + 60°.$$

Then

$$\begin{aligned} \sin 105° &= \sin (45° + 60°) \\ &= \sin 45° \cos 60° + \cos 45° \sin 60° \quad \text{Using } \sin (u + v) = \sin u \cos v + \cos u \sin v \\ &= \frac{\sqrt{2}}{2} \cdot \frac{1}{2} + \frac{\sqrt{2}}{2} \cdot \frac{\sqrt{3}}{2} \\ &= \frac{\sqrt{2} + \sqrt{6}}{4}. \end{aligned}$$

We can easily check this result on a grapher:

$$\sin 105° \approx 0.9659 \quad \text{and} \quad \frac{\sqrt{2} + \sqrt{6}}{4} \approx 0.9659.$$

Formulas for the tangent of a sum or a difference can be derived using identities already established. A summary of the sum and difference identities follows.*

Sum and Difference Identities

$$\sin (u \pm v) = \sin u \cos v \pm \cos u \sin v,$$

$$\cos (u \pm v) = \cos u \cos v \mp \sin u \sin v,$$

$$\tan (u \pm v) = \frac{\tan u \pm \tan v}{1 \mp \tan u \tan v}$$

Example 10 Find tan 15° exactly.

SOLUTION We rewrite 15° as 45° − 30° and use the identity for the tan-

*There are six identities here, half of them obtained by using the signs shown in color.

gent of a difference:

$$\begin{aligned}\tan 15° &= \tan(45° - 30°)\\ &= \frac{\tan 45° - \tan 30°}{1 + \tan 45° \tan 30°}\\ &= \frac{1 - \sqrt{3}/3}{1 + 1 \cdot \sqrt{3}/3}\\ &= \frac{3 - \sqrt{3}}{3 + \sqrt{3}}.\end{aligned}$$

Example 11 Assume that $\sin \alpha = \frac{2}{3}$ and $\sin \beta = \frac{1}{3}$ and that α and β are between 0 and $\pi/2$. Then evaluate $\sin(\alpha + \beta)$.

SOLUTION Using the identity for the sine of a sum, we have

$$\begin{aligned}\sin(\alpha + \beta) &= \sin \alpha \cos \beta + \cos \alpha \sin \beta\\ &= \tfrac{2}{3} \cos \beta + \tfrac{1}{3} \cos \alpha.\end{aligned}$$

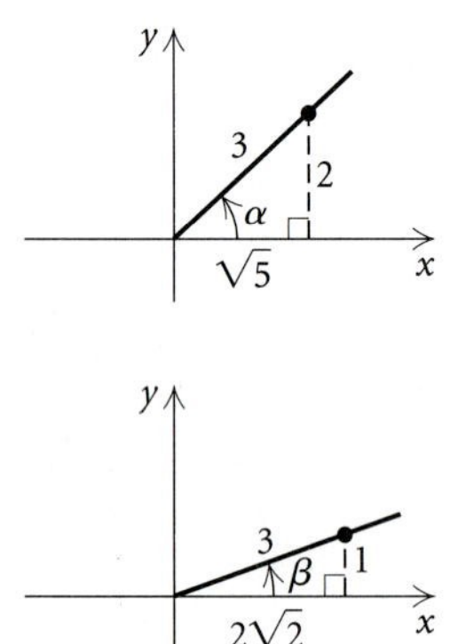

To finish, we need to know the values of $\cos \beta$ and $\cos \alpha$. Using reference triangles and the Pythagorean theorem, we can determine these values from the diagrams:

$$\cos \alpha = \frac{\sqrt{5}}{3} \quad \text{and} \quad \cos \beta = \frac{2\sqrt{2}}{3}.$$

Substituting these values gives us

$$\begin{aligned}\sin(\alpha + \beta) &= \frac{2}{3} \cdot \frac{2\sqrt{2}}{3} + \frac{1}{3} \cdot \frac{\sqrt{5}}{3}\\ &= \frac{4}{9}\sqrt{2} + \frac{1}{9}\sqrt{5}, \quad \text{or} \quad \frac{4\sqrt{2} + \sqrt{5}}{9}.\end{aligned}$$

Multiply and simplify. Check your result with a grapher.

1. $(\sin x - \cos x)(\sin x + \cos x)$
2. $\tan x (\cos x - \csc x)$
3. $\cos y \sin y (\sec y + \csc y)$
4. $(\sin x + \cos x)(\sec x + \csc x)$
5. $(\sin \phi - \cos \phi)^2$
6. $(1 + \tan x)^2$
7. $(\sin x + \csc x)(\sin^2 x + \csc^2 x - 1)$
8. $(1 - \sin t)(1 + \sin t)$

Factor and simplify. Check your result with a grapher.

9. $\sin x \cos x + \cos^2 x$
10. $\tan^2 \theta - \cot^2 \theta$
11. $\sin^4 x - \cos^4 x$
12. $4 \sin^2 y + 8 \sin y + 4$
13. $2 \cos^2 x + \cos x - 3$
14. $3 \cot^2 \beta + 6 \cot \beta + 3$
15. $\sin^3 x + 27$
16. $1 - 125 \tan^3 s$

Simplify and check with a grapher.

17. $\dfrac{\sin^2 x \cos x}{\cos^2 x \sin x}$
18. $\dfrac{30 \sin^3 x \cos x}{6 \cos^2 x \sin x}$

19. $\dfrac{\sin^2 x + 2 \sin x + 1}{\sin x + 1}$

20. $\dfrac{\cos^2 \alpha - 1}{\cos \alpha + 1}$

21. $\dfrac{4 \tan t \sec t + 2 \sec t}{6 \tan t \sec t + 2 \sec t}$

22. $\dfrac{\csc (-x)}{\cot (-x)}$

23. $\dfrac{\sin^4 x - \cos^4 x}{\sin^2 x - \cos^2 x}$

24. $\dfrac{4 \cos^3 x}{\sin^2 x} \cdot \left(\dfrac{\sin x}{4 \cos x}\right)^2$

25. $\dfrac{5 \cos \phi}{\sin^2 \phi} \cdot \dfrac{\sin^2 \phi - \sin \phi \cos \phi}{\sin^2 \phi - \cos^2 \phi}$

26. $\dfrac{\tan^2 y}{\sec y} \div \dfrac{3 \tan^3 y}{\sec y}$

27. $\dfrac{1}{\sin^2 s - \cos^2 s} - \dfrac{2}{\cos s - \sin s}$

28. $\left(\dfrac{\sin x}{\cos x}\right)^2 - \dfrac{1}{\cos^2 x}$

29. $\dfrac{\sin^2 \theta - 9}{2 \cos \theta + 1} \cdot \dfrac{10 \cos \theta + 5}{3 \sin \theta + 9}$

30. $\dfrac{9 \cos^2 \alpha - 25}{2 \cos \alpha - 2} \cdot \dfrac{\cos^2 \alpha - 1}{6 \cos \alpha - 10}$

Simplify and check with a grapher. Assume that all radicands are nonnegative.

31. $\sqrt{\sin^2 x \cos x} \cdot \sqrt{\cos x}$

32. $\sqrt{\cos^2 x \sin x} \cdot \sqrt{\sin x}$

33. $\sqrt{\cos \alpha \sin^2 \alpha} - \sqrt{\cos^3 \alpha}$

34. $\sqrt{\tan^2 x - 2 \tan x \sin x + \sin^2 x}$

35. $(1 - \sqrt{\sin y})(\sqrt{\sin y} + 1)$

36. $\sqrt{\cos \theta}(\sqrt{2 \cos \theta} + \sqrt{\sin \theta \cos \theta})$

Rationalize the denominator.

37. $\sqrt{\dfrac{\sin x}{\cos x}}$

38. $\sqrt{\dfrac{\cos x}{\tan x}}$

39. $\sqrt{\dfrac{\cos^2 y}{2 \sin^2 y}}$

40. $\sqrt{\dfrac{1 - \cos \beta}{1 + \cos \beta}}$

Rationalize the numerator.

41. $\sqrt{\dfrac{\cos x}{\sin x}}$

42. $\sqrt{\dfrac{\sin x}{\cot x}}$

43. $\sqrt{\dfrac{1 + \sin y}{1 - \sin y}}$

44. $\sqrt{\dfrac{\cos^2 x}{2 \sin^2 x}}$

Use the given substitution to express the given radical expression as a trigonometric function without radicals. Assume that $a > 0$ and $0 < \theta < \pi/2$. Then find expressions for the indicated trigonometric functions.

45. Let $x = a \sin \theta$ in $\sqrt{a^2 - x^2}$. Then find $\cos \theta$ and $\tan \theta$.

46. Let $x = 2 \tan \theta$ in $\sqrt{4 + x^2}$. Then find $\sin \theta$ and $\cos \theta$.

47. Let $x = 3 \sec \theta$ in $\sqrt{x^2 - 9}$. Then find $\sin \theta$ and $\cos \theta$.

48. Let $x = a \sec \theta$ in $\sqrt{x^2 - a^2}$. Then find $\sin \theta$ and $\cos \theta$.

Use the given substitution to express the given radical expression as a trigonometric function without radicals. Assume that $0 < \theta < \pi/2$.

49. Let $x = \sin \theta$ in $\dfrac{x^2}{\sqrt{1 - x^2}}$.

50. Let $x = 4 \sec \theta$ in $\dfrac{\sqrt{x^2 - 16}}{x^2}$.

Use the sum and difference identities to evaluate exactly. Then check with a grapher.

51. $\sin \dfrac{\pi}{12}$

52. $\cos 75°$

53. $\tan 105°$

54. $\tan \dfrac{5\pi}{12}$

55. $\cos 15°$

56. $\sin \dfrac{7\pi}{12}$

First write each of the following as a trigonometric function of a single angle; then evaluate.

57. $\sin 37° \cos 22° + \cos 37° \sin 22°$

58. $\cos 83° \cos 53° + \sin 83° \sin 53°$

59. $\dfrac{\tan 20° + \tan 32°}{1 - \tan 20° \tan 32°}$

60. $\dfrac{\tan 35° - \tan 12°}{1 + \tan 35° \tan 12°}$

Assuming that $\sin u = \frac{3}{5}$ and $\sin v = \frac{4}{5}$ and that u and v are between 0 and $\pi/2$, evaluate each of the following exactly.

61. $\cos (u + v)$

62. $\tan (u - v)$

63. $\sin (u - v)$

Assuming that $\sin \theta = 0.6249$ and $\cos \phi = 0.1102$ and that both θ and ϕ are first-quadrant angles, evaluate each of the following.

64. $\sin (\theta - \phi)$

65. $\tan (\theta + \phi)$

66. $\cos (\theta + \phi)$

Simplify.

67. $\sin (\alpha + \beta) + \sin (\alpha - \beta)$

68. $\cos (\alpha + \beta) - \cos (\alpha - \beta)$

69. $\cos (u + v) \cos v + \sin (u + v) \sin v$

70. $\sin (u - v) \cos v + \cos (u - v) \sin v$

Skill Maintenance

Solve.

71. $2x - 3 = 2\left(x - \frac{3}{2}\right)$

72. $x - 7 = x + 3.4$

Given that $\sin 31° = 0.5150$ *and* $\cos 31° = 0.8572$, *find the specified function value.*

73. $\sec 59°$

74. $\tan 59°$

Synthesis

75. ◈ What is the difference between a trigonometric equation that is an identity and a trigonometric equation that is not an identity? Give an example of each.

76. ◈ Why is it possible with a grapher to *disprove* that an equation is an identity but not to *prove* that one is?

Angles Between Lines. *One of the identities gives an easy way to find an angle formed by two lines. Consider two lines with equations* l_1: $y = m_1x + b_1$ *and* l_2: $y = m_2x + b_2$.

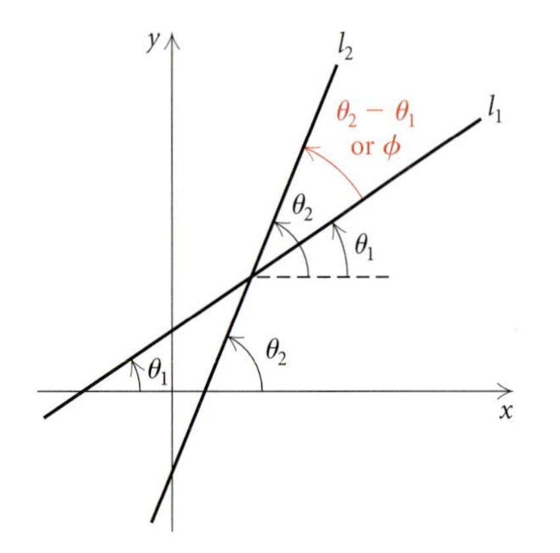

The slopes m_1 *and* m_2 *are the tangents of the angles* θ_1 *and* θ_2 *that the lines form with the positive direction of the x-axis. Thus we have* $m_1 = \tan\theta_1$ *and* $m_2 = \tan\theta_2$. *To find the measure of* $\theta_2 - \theta_1$, *or* ϕ, *we proceed as follows:*

$$\tan\phi = \tan(\theta_2 - \theta_1) = \frac{\tan\theta_2 - \tan\theta_1}{1 + \tan\theta_2\tan\theta_1} = \frac{m_2 - m_1}{1 + m_2m_1}.$$

This formula also holds when the lines are taken in the reverse order. When ϕ *is acute,* $\tan\phi$ *will be positive. When* ϕ *is obtuse,* $\tan\phi$ *will be negative.*

Find the measure of the angle from l_1 *to* l_2.

77. l_1: $2x = 3 - 2y$,
l_2: $x + y = 5$

78. l_1: $3y = \sqrt{3}x + 3$,
l_2: $y = \sqrt{3}x + 2$

79. l_1: $y = 3$,
l_2: $x + y = 5$

80. l_1: $2x + y - 4 = 0$,
l_2: $y - 2x + 5 = 0$

81. *Circus Guy Wire.* In a circus, a guy wire A is attached to the top of a 30-ft pole. Wire B is used for performers to walk up to the tight wire, 10 ft above the ground. Find the angle ϕ between the wires if they are attached to the ground 40 ft from the pole.

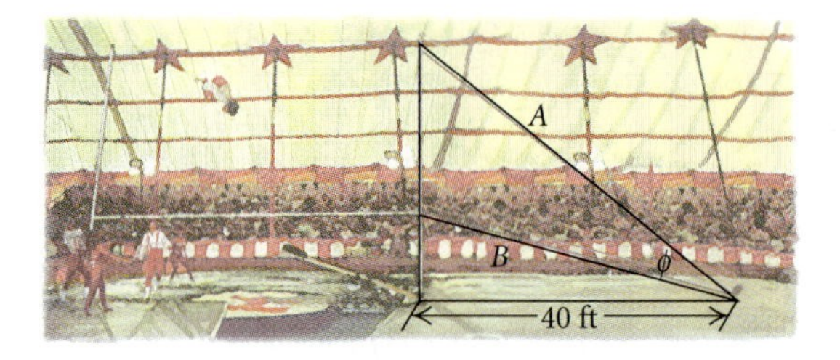

82. Given that $f(x) = \sin x$, show that

$$\frac{f(x+h) - f(x)}{h} = \sin x\left(\frac{\cos h - 1}{h}\right) + \cos x\left(\frac{\sin h}{h}\right).$$

83. Given that $f(x) = \cos x$, show that

$$\frac{f(x+h) - f(x)}{h} = \cos x\left(\frac{\cos h - 1}{h}\right) - \sin x\left(\frac{\sin h}{h}\right).$$

Show that each of the following is not an identity by first finding a replacement or replacements for which the sides of the equation do not name the same number. Then use a grapher to show that the equation is not an identity.

84. $\sqrt{\sin^2\theta} = \sin\theta$

85. $\dfrac{\sin 5x}{x} = \sin 5$

86. $\sin(-x) = \sin x$

87. $\cos(2\alpha) = 2\cos\alpha$

88. $\tan^2\theta + \cot^2\theta = 1$

89. $\dfrac{\cos 6x}{\cos x} = 6$

Find the slope of line l_1, *where* m_2 *is the slope of line* l_2 *and* ϕ *is the smallest positive angle from* l_1 *to* l_2.

90. $m_2 = \frac{4}{3}$, $\phi = 45°$

91. $m_2 = \frac{2}{3}$, $\phi = 30°$

92. Line l_1 contains the points $(-2, 4)$ and $(5, -1)$. Find the slope of line l_2 such that the angle from l_1 to l_2 is 45°.

93. Line l_1 contains the points $(-3, 7)$ and $(-3, -2)$. Line l_2 contains $(0, -4)$ and $(2, 6)$. Find the smallest positive angle from l_1 to l_2.

94. Find an identity for $\sin 2\theta$. (*Hint:* $2\theta = \theta + \theta$.)

95. Find an identity for $\cos 2\theta$. (*Hint:* $2\theta = \theta + \theta$.)

Derive each identity. Check with a grapher.

96. $\sin\left(x - \frac{3\pi}{2}\right) = \cos x$

97. $\tan\left(x + \frac{\pi}{4}\right) = \frac{1 + \tan x}{1 - \tan x}$

98. $\frac{\sin(\alpha + \beta)}{\cos(\alpha - \beta)} = \frac{\tan\alpha + \tan\beta}{1 + \tan\alpha\tan\beta}$

99. $\sin(\alpha + \beta) + \sin(\alpha - \beta) = 2\sin\alpha\cos\beta$

5.2 Identities: Cofunction, Double-Angle, and Half-Angle

- *Use cofunction identities to derive other identities.*
- *Use the double-angle identities to find function values of twice an angle when one function value is known for that angle.*
- *Use the half-angle identities to find function values of half an angle when one function value is known for that angle.*
- *Simplify trigonometric expressions using the double-angle and half-angle identities.*

Cofunction Identities

Each of the identities listed below yields a conversion to a *cofunction*. For this reason, we call them cofunction identities.

Cofunction Identities

$$\sin\left(\frac{\pi}{2} - x\right) = \cos x, \qquad \cos\left(\frac{\pi}{2} - x\right) = \sin x,$$

$$\tan\left(\frac{\pi}{2} - x\right) = \cot x, \qquad \cot\left(\frac{\pi}{2} - x\right) = \tan x,$$

$$\sec\left(\frac{\pi}{2} - x\right) = \csc x, \qquad \csc\left(\frac{\pi}{2} - x\right) = \sec x$$

We verified the first two of these identities in Section 5.1. The other four can be proved using the first two and the definitions of the trigonometric functions. These identities hold for all real numbers, and thus, for all degree measures, but if we restrict θ to values such that $0° < \theta < 90°$, then we have a special application to the acute angles of a right triangle.

Interactive Discovery

Graph $y_1 = \sin x$ and $y_2 = \sin(x + \pi/2)$ on the same screen. Note that the graph of y_2 is a translation of the graph of y_1 to the left $\pi/2$ units. Now graph $y_3 = \cos x$. We observe that the graph of y_2 is the same as the graph of y_3. This leads to a possible identity:

$$\sin\left(x + \frac{\pi}{2}\right) = \cos x.$$

(*continued*)

Repeat this exploration for

$$y = \sin\left(x - \frac{\pi}{2}\right),$$

$$y = \cos\left(x + \frac{\pi}{2}\right),$$

and $$y = \cos\left(x - \frac{\pi}{2}\right).$$

What possible identities do you observe?

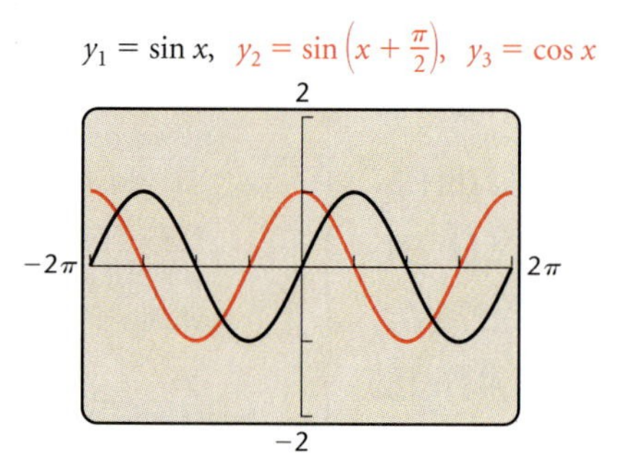

The identity $\sin(x + \pi/2) = \cos x$ can be proved using the identity for the sine of a sum developed in Section 5.1.

Example 1 Prove the identity $\sin(\theta + \pi/2) = \cos \theta$.

SOLUTION

$$\sin\left(\theta + \frac{\pi}{2}\right) = \sin\theta \cos\frac{\pi}{2} + \cos\theta \sin\frac{\pi}{2}$$ Using $\sin(u + v) = \sin u \cos v + \cos u \sin v$

$$= \sin\theta \cdot 0 + \cos\theta \cdot 1$$

$$= \cos\theta$$

We now state four more cofunction identities. These new identities that involve the sine and cosine functions can be verified using previously established identities as seen in Example 1.

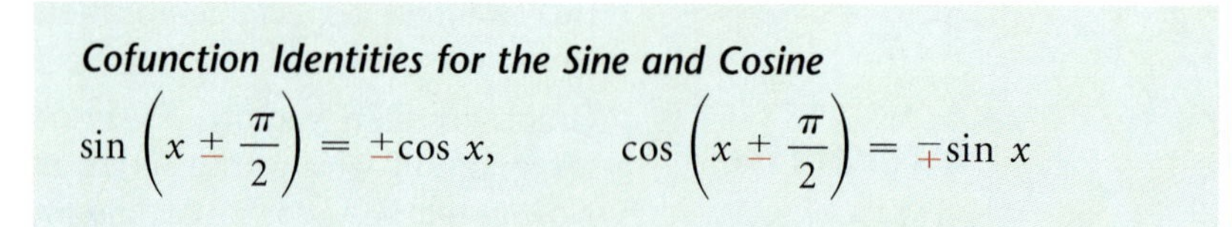

Cofunction Identities for the Sine and Cosine

$$\sin\left(x \pm \frac{\pi}{2}\right) = \pm\cos x, \qquad \cos\left(x \pm \frac{\pi}{2}\right) = \mp\sin x$$

Example 2 Find an identity for each of the following.

a) $\tan\left(x + \frac{\pi}{2}\right)$ **b)** $\sec(x - 90°)$

SOLUTION

a) We have

$$\tan\left(x + \frac{\pi}{2}\right) = \frac{\sin\left(x + \frac{\pi}{2}\right)}{\cos\left(x + \frac{\pi}{2}\right)} \qquad \text{Using } \tan x = \frac{\sin x}{\cos x}$$

$$= \frac{\cos x}{-\sin x} \qquad \text{Using cofunction identities}$$

$$= -\cot x.$$

Thus the identity we seek is

$$\tan\left(x + \frac{\pi}{2}\right) = -\cot x.$$

b) We have

$$\sec(x - 90°) = \frac{1}{\cos(x - 90°)} = \frac{1}{\sin x} = \csc x.$$

Thus, $\sec(x - 90°) = \csc x$.

Double-Angle Identities

To develop these identities, we will use the sum formulas from the preceding section. We first develop a formula for $\sin 2x$. Recall that

$$\sin(u + v) = \sin u \cos v + \cos u \sin v.$$

We will consider a number x and substitute it for both u and v in this identity. Doing so gives us

$$\sin(x + x) = \sin 2x$$
$$= \sin x \cos x + \cos x \sin x$$
$$= 2 \sin x \cos x.$$

Our first double-angle identity is thus

$$\mathbf{\sin 2x = 2 \sin x \cos x.}$$

Graphers provide visual partial checks of identities. We can graph $y_1 = \sin 2x$ and $y_2 = 2 \sin x \cos x$ and see that they appear to have the same graph in $[-2\pi, 2\pi]$.

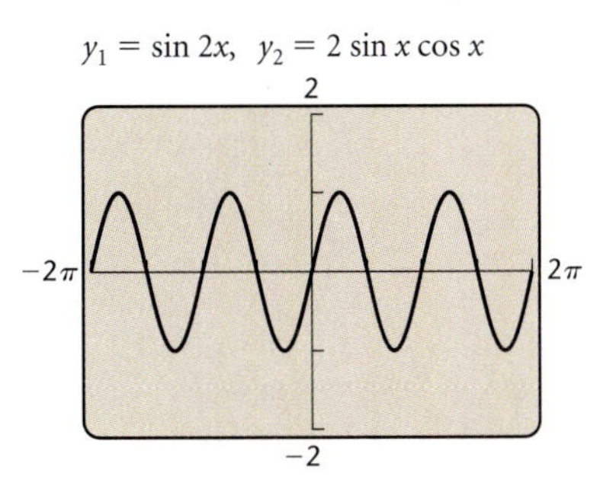

Double-angle identities for the cosine and tangent functions can be derived and checked with a grapher in much the same way as the identity above:

$$\mathbf{\cos 2x = \cos^2 x - \sin^2 x,}$$

$$\mathbf{\tan 2x = \frac{2 \tan x}{1 - \tan^2 x}.}$$

Example 3 Given that $\tan \theta = -\frac{3}{4}$ and θ is in quadrant II, find each of the following.

a) $\sin 2\theta$

b) $\cos 2\theta$

c) $\tan 2\theta$

d) The quadrant in which 2θ lies

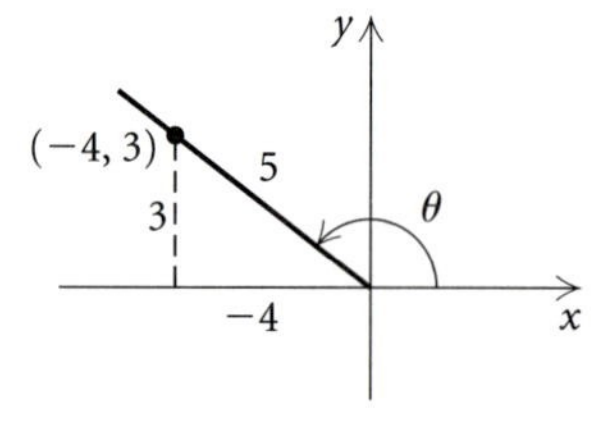

SOLUTION By drawing a diagram as shown, we find that

$$\sin \theta = \frac{3}{5} \quad \text{and} \quad \cos \theta = -\frac{4}{5}.$$

Thus we have the following.

a) $\sin 2\theta = 2 \sin \theta \cos \theta = 2 \cdot \frac{3}{5} \cdot \left(-\frac{4}{5}\right) = -\frac{24}{25}$

b) $\cos 2\theta = \cos^2 \theta - \sin^2 \theta = \left(-\frac{4}{5}\right)^2 - \left(\frac{3}{5}\right)^2 = \frac{16}{25} - \frac{9}{25} = \frac{7}{25}$

c) $\tan 2\theta = \frac{2 \tan \theta}{1 - \tan^2 \theta} = \frac{2 \cdot \left(-\frac{3}{4}\right)}{1 - \left(-\frac{3}{4}\right)^2} = \frac{-\frac{3}{2}}{1 - \frac{9}{16}} = -\frac{24}{7}$

Note that $\tan 2\theta$ could have been found more easily in this case by simply dividing:

$$\tan 2\theta = \frac{\sin 2\theta}{\cos 2\theta} = \frac{-\frac{24}{25}}{\frac{7}{25}} = -\frac{24}{7}.$$

d) Since $\sin 2\theta$ is negative and $\cos 2\theta$ is positive, we know that 2θ is in quadrant IV.

Two other useful identities for $\cos 2x$ can be derived easily, as follows.

$$\begin{aligned}\cos 2x &= \cos^2 x - \sin^2 x \\ &= (1 - \sin^2 x) - \sin^2 x \\ &= 1 - 2 \sin^2 x\end{aligned}$$

$$\begin{aligned}\cos 2x &= \cos^2 x - \sin^2 x \\ &= \cos^2 x - (1 - \cos^2 x) \\ &= 2 \cos^2 x - 1\end{aligned}$$

Double-Angle Identities

$$\sin 2x = 2 \sin x \cos x,$$

$$\tan 2x = \frac{2 \tan x}{1 - \tan^2 x},$$

$$\begin{aligned}\cos 2x &= \cos^2 x - \sin^2 x \\ &= 1 - 2 \sin^2 x \\ &= 2 \cos^2 x - 1\end{aligned}$$

Solving the last two cosine double-angle identities for $\sin^2 x$ and $\cos^2 x$, respectively, we can obtain two more identities:

$$\sin^2 x = \frac{1 - \cos 2x}{2}$$

and

$$\cos^2 x = \frac{1 + \cos 2x}{2}.$$

Using division and the last two identities gives us the following useful identity:

$$\tan^2 x = \frac{1 - \cos 2x}{1 + \cos 2x}.$$

Example 4 Find an equivalent expression for each of the following.

a) $\sin 3\theta$ in terms of function values of θ

b) $\cos^3 x$ in terms of function values of x or $2x$, raised only to the first power

SOLUTION

a)
$$\begin{aligned}\sin 3\theta &= \sin (2\theta + \theta)\\ &= \sin 2\theta \cos \theta + \cos 2\theta \sin \theta\\ &= (2 \sin \theta \cos \theta) \cos \theta + (2 \cos^2 \theta - 1) \sin \theta\end{aligned}$$

Using $\sin 2\theta = 2 \sin \theta \cos \theta$ and $\cos 2\theta = 2 \cos^2 \theta - 1$

$$\begin{aligned}&= 2 \sin \theta \cos^2 \theta + 2 \sin \theta \cos^2 \theta - \sin \theta\\ &= 4 \sin \theta \cos^2 \theta - \sin \theta\end{aligned}$$

We could also substitute $\cos^2 \theta - \sin^2 \theta$ or $1 - 2 \sin^2 \theta$ for $\cos 2\theta$. Each substitution leads to a different result, but all results are equivalent.

b)
$$\begin{aligned}\cos^3 x &= \cos^2 x \cos x\\ &= \frac{1 + \cos 2x}{2} \cos x\end{aligned}$$

Half-Angle Identities

To develop more identities, we take square roots and replace x by $x/2$. For example,

$$\sin^2 x = \frac{1 - \cos 2x}{2} \longrightarrow \left|\sin \frac{x}{2}\right| = \sqrt{\frac{1 - \cos x}{2}}.$$

The half-angle formula is the one on the right above. We can eliminate the absolute-value sign by using $\pm$ signs, with the understanding that our use of + and − depends on the quadrant in which the angle $x/2$ lies. Half-angle identities for the cosine and tangent functions can be derived in a similar manner. Two additional formulas for the half-angle tangent identity are listed on the following page.

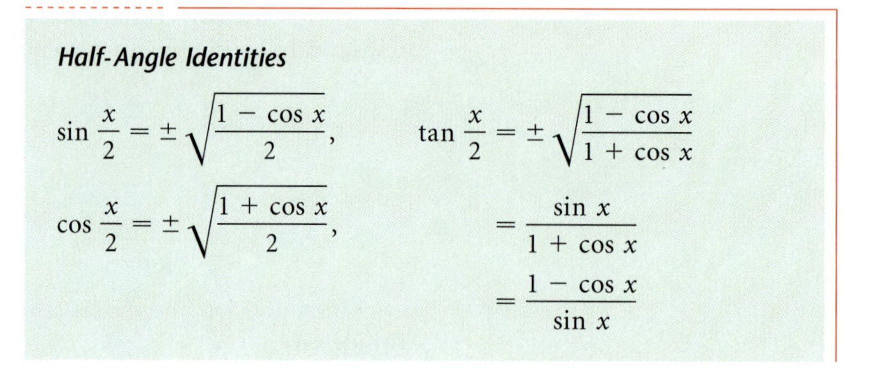

Half-Angle Identities

$$\sin \frac{x}{2} = \pm\sqrt{\frac{1 - \cos x}{2}}, \qquad \tan \frac{x}{2} = \pm\sqrt{\frac{1 - \cos x}{1 + \cos x}}$$

$$\cos \frac{x}{2} = \pm\sqrt{\frac{1 + \cos x}{2}}, \qquad = \frac{\sin x}{1 + \cos x}$$

$$= \frac{1 - \cos x}{\sin x}$$

Example 5 Find tan ($\pi/8$) exactly. Then check the answer with a grapher.

SOLUTION We have

$$\tan \frac{\pi}{8} = \tan \frac{\frac{\pi}{4}}{2} = \frac{\sin \frac{\pi}{4}}{1 + \cos \frac{\pi}{4}} = \frac{\frac{\sqrt{2}}{2}}{1 + \frac{\sqrt{2}}{2}}$$

$$= \frac{\sqrt{2}}{2 + \sqrt{2}} = \sqrt{2} - 1$$

This value checks:

$$\tan \frac{\pi}{8} \approx 0.4142 \quad \text{and} \quad \sqrt{2} - 1 \approx 0.4142.$$

The identities that we have developed are also useful for simplifying trigonometric expressions.

Example 6 Simplify each of the following.

a) $\dfrac{\sin x \cos x}{\frac{1}{2} \cos 2x}$

b) $2 \sin^2 \dfrac{x}{2} + \cos x$

SOLUTION

a) We can obtain $2 \sin x \cos x$ in the numerator by multiplying the expression by $\frac{2}{2}$:

$$\frac{\sin x \cos x}{\frac{1}{2} \cos 2x} = \frac{2}{2} \cdot \frac{\sin x \cos x}{\frac{1}{2} \cos 2x}$$

$$= \frac{2 \sin x \cos x}{\cos 2x}$$

$$= \frac{\sin 2x}{\cos 2x} \qquad \text{Using } \sin 2x = 2 \sin x \cos x$$

$$= \tan 2x.$$

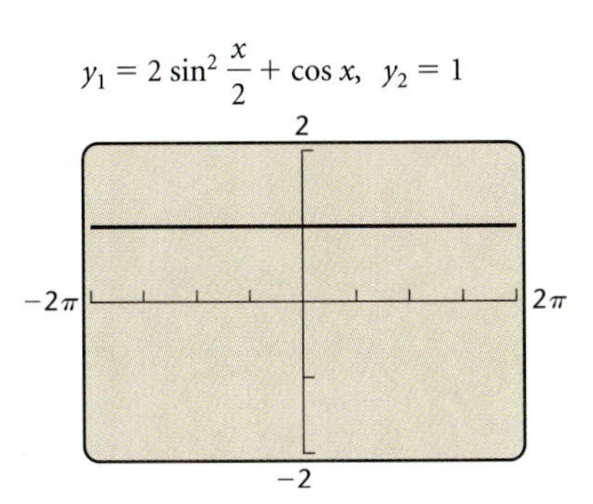

b) We have

$$2\sin^2 \frac{x}{2} + \cos x = 2\left(\frac{1 - \cos x}{2}\right) + \cos x$$

Using $\sin \frac{x}{2} = \pm\sqrt{\frac{1 - \cos x}{2}}$, or $\sin^2 \frac{x}{2} = \frac{1 - \cos x}{2}$

$$= 1 - \cos x + \cos x$$

$$= 1.$$

To check this on $[-2\pi, 2\pi]$, we graph

$$y_1 = 2\sin^2 \frac{x}{2} + \cos x \quad \text{and} \quad y_2 = 1.$$

Note that the graphs appear to be the same.

5.2 Exercise Set

1. Given that $\sin(3\pi/10) \approx 0.8090$ and $\cos(3\pi/10) \approx 0.5878$, find each of the following.
   a) The other four function values for $3\pi/10$
   b) The six function values for $\pi/5$

2. Given that
   $$\sin \frac{\pi}{12} = \frac{\sqrt{2 - \sqrt{3}}}{2} \quad \text{and} \quad \cos \frac{\pi}{12} = \frac{\sqrt{2 + \sqrt{3}}}{2},$$
   find each of the following.
   a) The other four function values for $\pi/12$
   b) The six function values for $5\pi/12$

3. Given that $\sin\theta = \frac{1}{3}$ and that the terminal side is in quadrant II, find each of the following.
   a) The other function values for θ
   b) The six function values for $\pi/2 - \theta$
   c) The six function values for $\theta - \pi/2$

4. Given that $\cos\phi = \frac{4}{5}$ and that the terminal side is in quadrant IV, find each of the following.
   a) The other function values for ϕ
   b) The six function values for $\pi/2 - \phi$
   c) The six function values for $\phi + \pi/2$

Find an equivalent expression for each of the following.

5. $\sec\left(x + \frac{\pi}{2}\right)$
6. $\cot\left(x - \frac{\pi}{2}\right)$
7. $\tan\left(x - \frac{\pi}{2}\right)$
8. $\csc\left(x + \frac{\pi}{2}\right)$

Find $\sin 2\theta$, $\cos 2\theta$, $\tan 2\theta$, *and the quadrant in which* 2θ *lies.*

9. $\sin\theta = \frac{4}{5}$, θ in quadrant I
10. $\cos\theta = \frac{5}{13}$, θ in quadrant I
11. $\cos\theta = -\frac{3}{5}$, θ in quadrant III
12. $\tan\theta = -\frac{5}{8}$, θ in quadrant II
13. $\tan\theta = -\frac{5}{12}$, θ in quadrant II
14. $\sin\theta = -\frac{\sqrt{10}}{10}$, θ in quadrant IV
15. Find an equivalent expression for $\cos 4x$ in terms of function values of x.
16. Find an equivalent expression for $\sin^4\theta$ in terms of function values of θ, 2θ, or 4θ, raised only to the first power.

Use the half-angle identities to evaluate exactly.

17. $\cos 15°$
18. $\tan 67.5°$
19. $\sin 112.5°$
20. $\cos \frac{\pi}{8}$
21. $\tan 75°$
22. $\sin \frac{5\pi}{12}$

Given that $\sin\theta = 0.3416$ *and* θ *is in quadrant I, find each of the following.*

23. $\sin 2\theta$
24. $\cos \frac{\theta}{2}$
25. $\sin \frac{\theta}{2}$
26. $\sin 4\theta$

Use a grapher to determine which of the following expressions asserts an identity. Then prove the identity algebraically.

27. $\dfrac{\cos 2x}{\cos x - \sin x} = \cdots$

a) $1 + \cos x$ b) $\cos x - \sin x$

$y = 1 + \cos x$ $y = \cos x - \sin x$

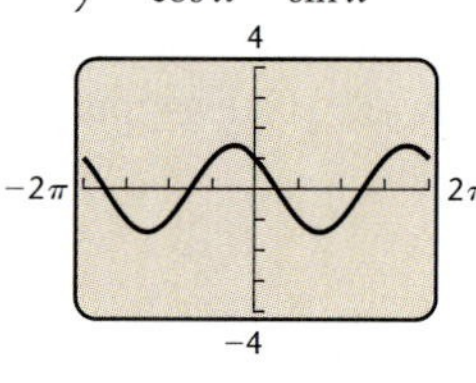

c) $-\cot x$ d) $\sin x\,(\cot x + 1)$

$y = -\cot x$ $y = \sin x\,(\cot x + 1)$

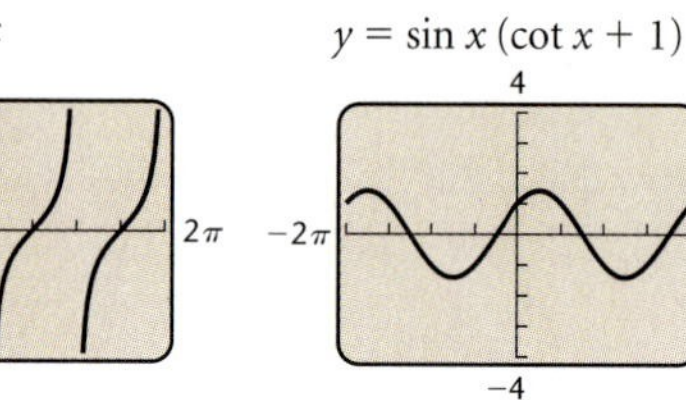

28. $2\cos^2 \dfrac{x}{2} = \cdots$

a) $\sin x\,(\csc x + \tan x)$ b) $\sin x - 2\cos x$
c) $2(\cos^2 x - \sin^2 x)$ d) $1 + \cos x$

29. $\dfrac{\sin 2x}{2\cos x} = \cdots$

a) $\cos x$ b) $\tan x$
c) $\cos x + \sin x$ d) $\sin x$

30. $2\sin\dfrac{\theta}{2}\cos\dfrac{\theta}{2} = \cdots$

a) $\cos^2 \theta$ b) $\sin \dfrac{\theta}{2}$
c) $\sin \theta$ d) $\sin\theta - \cos\theta$

Simplify. Check your results with a grapher.

31. $2\cos^2 \dfrac{x}{2} - 1$

32. $\cos^4 x - \sin^4 x$

33. $(\sin x - \cos x)^2 + \sin 2x$

34. $(\sin x + \cos x)^2$

35. $\dfrac{2 - \sec^2 x}{\sec^2 x}$

36. $\dfrac{1 + \sin 2x + \cos 2x}{1 + \sin 2x - \cos 2x}$

37. $(-4\cos x \sin x + 2\cos 2x)^2 + (2\cos 2x + 4\sin x \cos x)^2$

38. $2\sin x \cos^3 x - 2\sin^3 x \cos x$

Skill Maintenance

Complete the identity.

39. $1 - \cos^2 x =$

40. $\sec^2 x - \tan^2 x =$

41. $\sin^2 x - 1 =$

42. $1 + \cot^2 x =$

Synthesis

43. ◈ Discuss and compare the graphs of $y_1 = \sin x$, $y_2 = \sin 2x$, and $y_3 = \sin(x/2)$.

44. ◈ Find all errors in the following:

$$\begin{aligned} &2\sin^2 2x + \cos 4x \\ &= 2(2\sin x \cos x)^2 + 2\cos 2x \\ &= 8\sin^2 x \cos^2 x + 2(\cos^2 x + \sin^2 x) \\ &= 8\sin^2 x \cos^2 x + 2. \end{aligned}$$

Then verify with a grapher that

$$2\sin^2 2x + \cos 4x = 8\sin^2 x \cos^2 x + 2$$

is not an identity.

45. Given that $\cos 51° \approx 0.6293$, find the six function values for 141°.

Simplify. Check your results with a grapher.

46. $\sin\left(\dfrac{\pi}{2} - x\right)[\sec x - \cos x]$

47. $\cos(\pi - x) + \cot x \sin\left(x - \dfrac{\pi}{2}\right)$

48. $\dfrac{\cos x - \sin\left(\dfrac{\pi}{2} - x\right)\sin x}{\cos x - \cos(\pi - x)\tan x}$

49. $\dfrac{\cos^2 y \sin\left(y + \dfrac{\pi}{2}\right)}{\sin^2 y \sin\left(\dfrac{\pi}{2} - y\right)}$

Find $\sin\theta$, $\cos\theta$, and $\tan\theta$ under the given conditions.

50. $\cos 2\theta = \dfrac{7}{12}, \quad \dfrac{3\pi}{2} \le 2\theta \le 2\pi$

51. $\tan\dfrac{\theta}{2} = -\dfrac{5}{3}, \quad \pi < \theta \le \dfrac{3\pi}{2}$

52. *Nautical Mile.* *Latitude* is used to measure north–south location on the earth between the equator and the poles. For example, Chicago has latitude 42°N. (See the figure.) In Great Britain, the

nautical mile is defined as the length of a minute of arc of the earth's radius. Since the earth is flattened slightly at the poles, a British nautical mile varies with latitude. In fact, it is given, in feet, by the function

$$N(\phi) = 6066 - 31 \cos 2\phi,$$

where ϕ is the latitude in degrees.

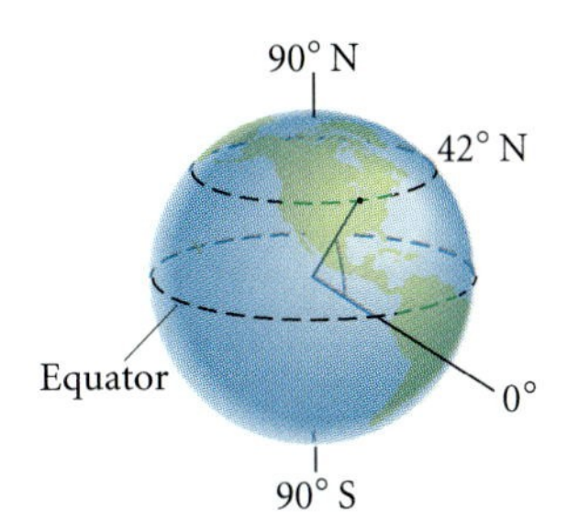

a) What is the length of a British nautical mile at Chicago?
b) What is the length of a British nautical mile at the North Pole?
c) Express $N(\phi)$ in terms of $\cos \phi$ only. That is, do not use the double angle.

53. *Acceleration Due to Gravity.* The acceleration due to gravity is often denoted by g in a formula such as $S = \frac{1}{2}gt^2$, where S is the distance that an object falls in time t. The number g relates to motion near the earth's surface and is usually considered constant. In fact, however, g is not constant, but varies slightly with latitude. If ϕ stands for latitude, in degrees, g is given with good approximation by the formula

$$g = 9.78049(1 + 0.005288 \sin^2 \phi - 0.000006 \sin^2 2\phi),$$

where g is measured in meters per second per second at sea level.

a) Chicago has latitude 42°N. Find g.
b) Philadelphia has latitude 40°N. Find g.
c) Express g in terms of $\sin \phi$ only. That is, eliminate the double angle.
d) Where on earth is g greatest? least? Use the MAX–MIN feature on a grapher.

5.3 Proving Trigonometric Identities

- *Prove identities using other identities.*

The Logic of Proving Identities

We outline two algebraic methods for proving identities.

Method 1. Start with either the left or the right side of the equation and deduce the other side. For example, suppose you are trying to prove that the equation $P = Q$ is an identity. You might try to produce a string of statements like the following, which start at P and end with Q:

$$\begin{aligned} P &= S_1 \\ &= S_2 \\ &\vdots \\ &= Q. \end{aligned}$$

Method 2. Work with each side separately until you deduce the same expression. For example, suppose you are trying to prove that $P = Q$ is an identity. You might be able to produce two strings of statements like the following, each ending with the same statement S.

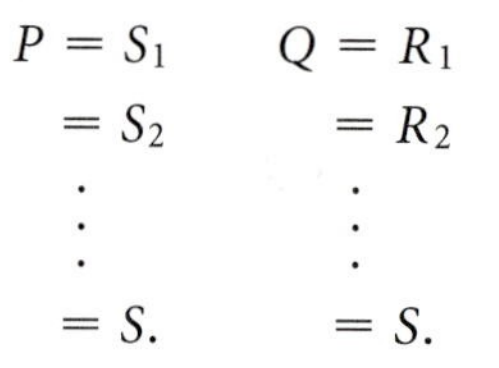

The number of steps in each string might be different, but in each case the result is S.

A first step in learning to prove identities is to have at hand a list of the identities that you have already learned. Such a list is on the inside back cover of this text. Ask your instructor which ones you are expected to memorize. The more identities you prove, the easier it will be to prove new ones. A list of helpful hints follows.

Hints for Proving Identities

1. Use methods 1 or 2 previously outlined.
2. Work with the more complex side first.
3. Carry out any algebraic manipulations, such as adding, subtracting, multiplying, or factoring.
4. Multiplying by 1 can be helpful when rational expressions are involved.
5. Converting all expressions to sines and cosines is often helpful.
6. Try something! Put your pencil to work and get involved. You will be amazed at how often this leads to success.
7. Use a grapher for a partial check of your final answer.

Proving Identities

In the work that follows, method 1 is used in Examples 1–3 and method 2 is used in Examples 4 and 5.

Example 1 Prove the identity $1 + \sin 2\theta = (\sin \theta + \cos \theta)^2$.

SOLUTION Let's use method 1. We begin with the right side and deduce the left side:

$$(\sin \theta + \cos \theta)^2 = \sin^2 \theta + 2 \sin \theta \cos \theta + \cos^2 \theta \qquad \text{Squaring}$$

$$= 1 + 2 \sin \theta \cos \theta \qquad \text{Recalling the identity } \sin^2 x + \cos^2 x = 1 \text{ and substituting}$$

$$= 1 + \sin 2\theta. \qquad \text{Using } \sin 2x = 2 \sin x \cos x$$

We could also begin with the left side and deduce the right side:

$$\begin{aligned} 1 + \sin 2\theta &= 1 + 2 \sin \theta \cos \theta && \textbf{Using } \sin 2x = 2 \sin x \cos x \\ &= \sin^2 \theta + 2 \sin \theta \cos \theta + \cos^2 \theta && \textbf{Replacing 1 with } \sin^2 \theta + \cos^2 \theta \\ &= (\sin \theta + \cos \theta)^2. && \textbf{Factoring} \end{aligned}$$

Technology allows us to do partial checks. We graph each side of the equation and look to see if the graphs appear to be the same. For instance, in Example 1 we can graph

$$y_1 = 1 + \sin 2x \quad \text{and} \quad y_2 = (\sin x + \cos x)^2$$

in the same screen and observe that the graphs appear to be identical in $[-2\pi, 2\pi]$.

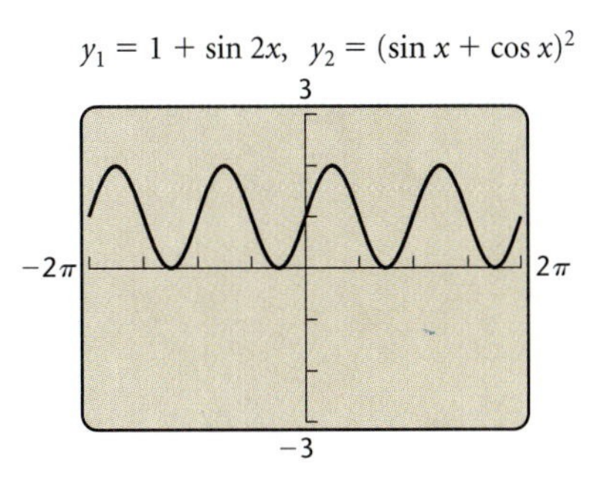

Keep in mind that this check does not *prove* an identity. But it is a useful tool for checking over specific intervals.

Example 2 Prove the identity

$$\frac{\sec t - 1}{t \sec t} = \frac{1 - \cos t}{t}.$$

SOLUTION We use method 1, starting with the left side. Note that the left side involves $\sec t$, whereas the right side involves $\cos t$, so it might be wise to make use of an identity that involves these two expressions. That basic identity is $\sec t = 1/\cos t$.

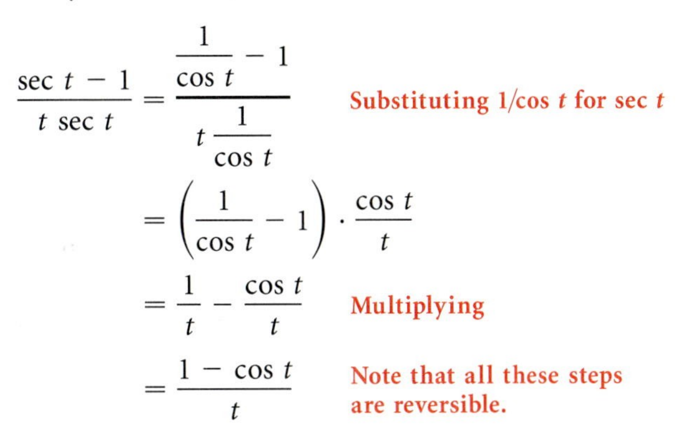

$$\begin{aligned} \frac{\sec t - 1}{t \sec t} &= \frac{\dfrac{1}{\cos t} - 1}{t \dfrac{1}{\cos t}} && \textbf{Substituting } 1/\cos t \textbf{ for } \sec t \\ &= \left(\frac{1}{\cos t} - 1\right) \cdot \frac{\cos t}{t} \\ &= \frac{1}{t} - \frac{\cos t}{t} && \textbf{Multiplying} \\ &= \frac{1 - \cos t}{t} && \textbf{Note that all these steps are reversible.} \end{aligned}$$

We started with the left side and deduced the right side, so the proof is complete. Check with a grapher.

Example 3 Prove the identity

$$\frac{\sin 2x}{\sin x} - \frac{\cos 2x}{\cos x} = \sec x.$$

SOLUTION

$$\frac{\sin 2x}{\sin x} - \frac{\cos 2x}{\cos x} = \frac{2 \sin x \cos x}{\sin x} - \frac{\cos^2 x - \sin^2 x}{\cos x}$$ Using double-angle identities

$$= 2 \cos x - \frac{\cos^2 x - \sin^2 x}{\cos x}$$ Simplifying

$$= \frac{2 \cos^2 x}{\cos x} - \frac{\cos^2 x - \sin^2 x}{\cos x}$$ Multiplying $2 \cos x$ by 1, or $\cos x/\cos x$

$$= \frac{2 \cos^2 x - \cos^2 x + \sin^2 x}{\cos x}$$ Subtracting

$$= \frac{\cos^2 x + \sin^2 x}{\cos x}$$

$$= \frac{1}{\cos x}$$ Using a Pythagorean identity

$$= \sec x$$ Recalling a basic identity

Check with a grapher.

Example 4 Prove the identity

$$\sin^2 x \tan^2 x = \tan^2 x - \sin^2 x.$$

SOLUTION For this proof, we are going to work with each side separately using method 2. We try to deduce the same expression on each side. In practice, you might work on one side for awhile, then work on the other side separately, and then go back to the other side. In other words, you work back and forth until you arrive at the same expression. Let's start with the right side:

$$\tan^2 x - \sin^2 x = \frac{\sin^2 x}{\cos^2 x} - \sin^2 x$$ Recalling the identity $\tan x = \frac{\sin x}{\cos x}$ and substituting

$$= \frac{\sin^2 x}{\cos^2 x} - \sin^2 x \cdot \frac{\cos^2 x}{\cos^2 x}$$ Multiplying by 1 in order to subtract

$$= \frac{\sin^2 x - \sin^2 x \cos^2 x}{\cos^2 x}$$ Carrying out the subtraction

$$= \frac{\sin^2 x\,(1 - \cos^2 x)}{\cos^2 x}$$ Factoring

$$= \frac{\sin^2 x \sin^2 x}{\cos^2 x}$$ Recalling the identity $\sin^2 x + \cos^2 x = 1$ or $1 - \cos^2 x = \sin^2 x$ and substituting

$$= \frac{\sin^4 x}{\cos^2 x}.$$

At this point, we stop and work with the left side, $\sin^2 x \tan^2 x$, of the original identity and try to end with the same expression that we ended with on the right side:

$$\sin^2 x \tan^2 x = \sin^2 x \frac{\sin^2 x}{\cos^2 x}$$

Recalling the identity $\tan x = \frac{\sin x}{\cos x}$ and substituting

$$= \frac{\sin^4 x}{\cos^2 x}.$$

We have deduced the same expression from each side, so the proof is complete. Check with a grapher.

Example 5 Prove the identity

$$\cot \phi + \csc \phi = \frac{\sin \phi}{1 - \cos \phi}.$$

SOLUTION We are again using method 2, beginning with the left side:

$$\cot \phi + \csc \phi = \frac{\cos \phi}{\sin \phi} + \frac{1}{\sin \phi}$$ Using basic identities

$$= \frac{1 + \cos \phi}{\sin \phi}.$$ Adding

At this point, we stop and work with the right side of the original identity:

$$\frac{\sin \phi}{1 - \cos \phi} = \frac{\sin \phi}{1 - \cos \phi} \cdot \frac{1 + \cos \phi}{1 + \cos \phi}$$ Multiplying by 1

$$= \frac{\sin \phi\,(1 + \cos \phi)}{1 - \cos^2 \phi}$$

$$= \frac{\sin \phi\,(1 + \cos \phi)}{\sin^2 \phi}$$ Using $\sin^2 x = 1 - \cos^2 x$

$$= \frac{1 + \cos \phi}{\sin \phi}.$$ Simplifying

The proof is complete since we deduced the same expression from each side. Check with a grapher.

5.3 Exercise Set

Prove each of the following identities. Check your work with a grapher.

1. $\sec x - \sin x \tan x = \cos x$

$y_1 = \sec x - \sin x \tan x,$
$y_2 = \cos x$

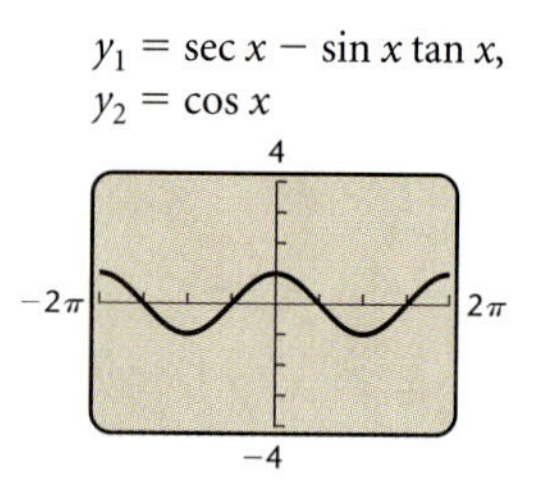

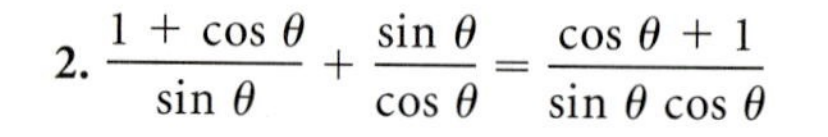

2. $\dfrac{1 + \cos\theta}{\sin\theta} + \dfrac{\sin\theta}{\cos\theta} = \dfrac{\cos\theta + 1}{\sin\theta\cos\theta}$

$y_1 = \dfrac{1 + \cos x}{\sin x} + \dfrac{\sin x}{\cos x},$

$y_2 = \dfrac{\cos x + 1}{\sin x \cos x}$

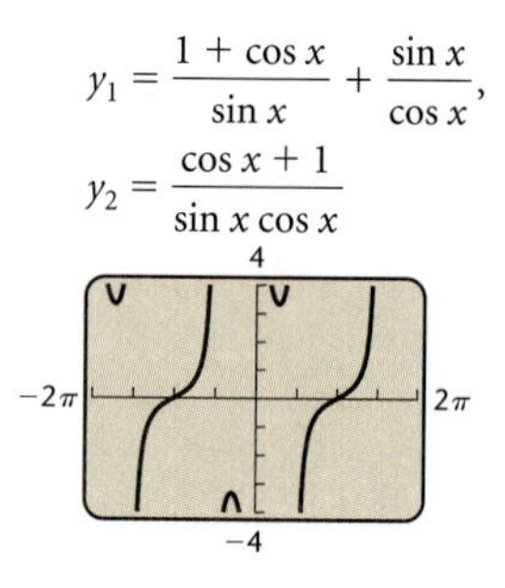

3. $\dfrac{1 - \cos x}{\sin x} = \dfrac{\sin x}{1 + \cos x}$

4. $\dfrac{1 + \tan y}{1 + \cot y} = \dfrac{\sec y}{\csc y}$

5. $\dfrac{1 + \tan\theta}{1 - \tan\theta} + \dfrac{1 + \cot\theta}{1 - \cot\theta} = 0$

6. $\dfrac{\sin x + \cos x}{\sec x + \csc x} = \dfrac{\sin x}{\sec x}$

7. $\dfrac{\cos^2\alpha + \cot\alpha}{\cos^2\alpha - \cot\alpha} = \dfrac{\cos^2\alpha \tan\alpha + 1}{\cos^2\alpha\tan\alpha - 1}$

8. $\sec 2\theta = \dfrac{\sec^2\theta}{2 - \sec^2\theta}$

9. $\dfrac{2\tan\theta}{1 + \tan^2\theta} = \sin 2\theta$

10. $\dfrac{\cos(u - v)}{\cos u \sin v} = \tan u + \cot v$

11. $1 - \cos 5\theta \cos 3\theta - \sin 5\theta \sin 3\theta = 2\sin^2\theta$

12. $\cos^4 x - \sin^4 x = \cos 2x$

13. $2\sin\theta\cos^3\theta + 2\sin^3\theta\cos\theta = \sin 2\theta$

14. $\dfrac{\tan 3t - \tan t}{1 + \tan 3t \tan t} = \dfrac{2\tan t}{1 - \tan^2 t}$

15. $\dfrac{\tan x - \sin x}{2\tan x} = \sin^2\dfrac{x}{2}$

16. $\dfrac{\cos^3\beta - \sin^3\beta}{\cos\beta - \sin\beta} = \dfrac{2 + \sin 2\beta}{2}$

17. $\sin(\alpha + \beta)\sin(\alpha - \beta) = \sin^2\alpha - \sin^2\beta$

18. $\cos^2 x(1 - \sec^2 x) = -\sin^2 x$

19. $\tan\theta(\tan\theta + \cot\theta) = \sec^2\theta$

20. $\dfrac{\cos\theta + \sin\theta}{\cos\theta} = 1 + \tan\theta$

21. $\dfrac{1 + \cos^2 x}{\sin^2 x} = 2\csc^2 x - 1$

22. $\dfrac{\tan y + \cot y}{\csc y} = \sec y$

23. $\dfrac{1 + \sin x}{1 - \sin x} + \dfrac{\sin x - 1}{1 + \sin x} = 4\sec x \tan x$

24. $\tan\theta - \cot\theta = (\sec\theta - \csc\theta)(\sin\theta + \cos\theta)$

25. $\cos^2\alpha\cot^2\alpha = \cot^2\alpha - \cos^2\alpha$

26. $\dfrac{\tan x + \cot x}{\sec x + \csc x} = \dfrac{1}{\cos x + \sin x}$

27. $2\sin^2\theta\cos^2\theta + \cos^4\theta = 1 - \sin^4\theta$

28. $\dfrac{\cot\theta}{\csc\theta - 1} = \dfrac{\csc\theta + 1}{\cot\theta}$

29. $\dfrac{1 + \sin x}{1 - \sin x} = (\sec x + \tan x)^2$

30. $\sec^4 s - \tan^2 s = \tan^4 s + \sec^2 s$

In Exercises 31–36, use a grapher to determine which expression (A)–(F) on the right can be used to complete an identity. Then try to prove that identity algebraically.

31. $\dfrac{\cos x + \cot x}{1 + \csc x}$	A. $\dfrac{\sin^3 x - \cos^3 x}{\sin x - \cos x}$
32. $\cot x + \csc x$	B. $\cos x$
33. $\sin x \cos x + 1$	C. $\tan x + \cot x$
34. $2\cos^2 x - 1$	D. $\cos^3 x + \sin^3 x$
35. $\dfrac{1}{\cot x \sin^2 x}$	E. $\dfrac{\sin x}{1 - \cos x}$
36. $(\cos x + \sin x)(1 - \sin x \cos x)$	F. $\cos^4 x - \sin^4 x$

Skill Maintenance

For each function:

a) *Graph the function.*
b) *Determine whether the function is one-to-one.*
c) *If the function is one-to-one, find an equation for its inverse.*
d) *Graph the inverse of the function.*

37. $f(x) = 3x - 2$

38. $f(x) = x^3 + 1$

39. $f(x) = x^2 - 4,\ x \geq 0$

40. $f(x) = \sqrt{x + 2}$

Synthesis

41. ◈ What restrictions must be placed on the variable in each of the following identities? Why?

a) $\sin 2x = \dfrac{2 \tan x}{1 + \tan^2 x}$

b) $\dfrac{1 - \cos x}{\sin x} = \dfrac{\sin x}{1 + \cos x}$

c) $2 \sin x \cos^3 x + 2 \sin^3 x \cos x = \sin 2x$

42. ◈ Consider each of the following functions:

$$y_1 = \frac{\pi}{2} - \frac{4}{\pi}\left(\cos x + \frac{\cos 3x}{9} + \frac{\cos 5x}{25}\right)$$

and

$$y_2 = \frac{\pi}{2} - \frac{4}{\pi}\left(\cos x + \frac{\cos 3x}{9} + \frac{\cos 5x}{25} + \frac{\cos 7x}{49}\right).$$

Use a grapher to graph both functions using the viewing window $[-10, 10, -1, 4]$, with Xscl $= 1$ and Yscl $= 1$. On the basis of your graphs, would you consider

$$\frac{\pi}{2} - \frac{4}{\pi}\left(\cos x + \frac{\cos 3x}{9} + \frac{\cos 5x}{25}\right) = \frac{\pi}{2} - \frac{4}{\pi}\left(\cos x + \frac{\cos 3x}{9} + \frac{\cos 5x}{25} + \frac{\cos 7x}{49}\right)$$

to be an identity? Now change the viewing window and use ZOOM and TRACE and the TABLE feature to examine the graphs in more detail. What do you discover about the graphs? Do we have an identity? What caution must be used in determining whether the equation is an identity considering only the graphs?

Prove each identity and verify your results with a grapher.

43. $\ln |\tan x| = -\ln |\cot x|$

44. $\ln |\sec \theta + \tan \theta| = -\ln |\sec \theta - \tan \theta|$

45. Prove the identity

$\log (\cos x - \sin x) + \log (\cos x + \sin x) = \log \cos 2x.$

46. *Mechanics.* The following equation occurs in the study of mechanics:

$$\sin \theta = \frac{I_1 \cos \phi}{\sqrt{(I_1 \cos \phi)^2 + (I_2 \sin \phi)^2}}.$$

It can happen that $I_1 = I_2$. Assuming that this happens, simplify the equation.

47. *Alternating Current.* In the theory of alternating current, the following equation occurs:

$$R = \frac{1}{\omega C(\tan \theta + \tan \phi)}.$$

Show that this equation is equivalent to

$$R = \frac{\cos \theta \cos \phi}{\omega C \sin (\theta + \phi)}.$$

48. *Electrical Theory.* In electrical theory, the following equations occur:

$$E_1 = \sqrt{2}E_t \cos\left(\theta + \frac{\pi}{P}\right)$$

and

$$E_2 = \sqrt{2}E_t \cos\left(\theta - \frac{\pi}{P}\right).$$

Assuming that these equations hold, show that

$$\frac{E_1 + E_2}{2} = \sqrt{2}E_t \cos \theta \cos \frac{\pi}{P}$$

and

$$\frac{E_1 - E_2}{2} = -\sqrt{2}E_t \sin \theta \sin \frac{\pi}{P}.$$

5.4 Inverses of the Trigonometric Functions

- *Find values of the inverse trigonometric functions.*
- *Simplify expressions such as $\sin(\sin^{-1} x)$ and $\sin^{-1}(\sin x)$.*
- *Simplify expressions involving compositions such as $\sin\left(\cos^{-1}\frac{1}{2}\right)$ without using a calculator.*
- *Simplify expressions such as $\sin \arctan(a/b)$ by making a drawing and reading off appropriate ratios.*

In this section, we develop inverse trigonometric functions. It may be helpful for you to review the material on inverse functions in Section 3.1.

The graphs of the sine, cosine, and tangent functions follow. Do these functions have inverses that are functions? They do if they are one-to-one, which means that they pass the horizontal-line test.

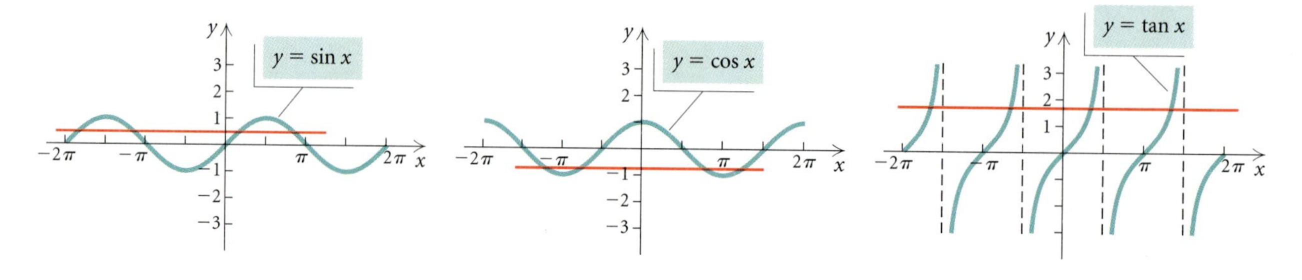

Note that each function has a horizontal line (shown in red) that crosses the graph more than once. Therefore, none of them has an inverse that is a function.

The graphs of an equation and its inverse are reflections of each other across the line $y = x$. Let's examine the inverses of each of the three functions graphed above.

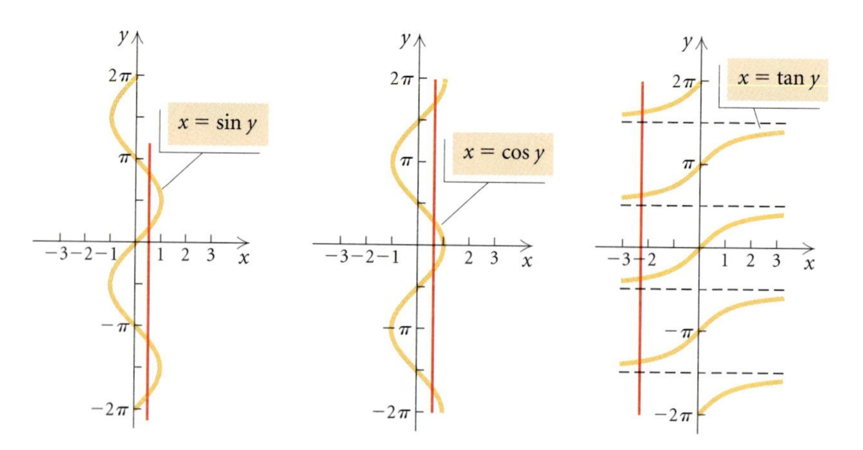

We can check again to see whether these are graphs of functions by using the vertical-line test. In each case, there is a vertical line (shown in red) that crosses the graph more than once, so each fails to be a function.

Let's look specifically at the graph of the inverse of $y = \sin x$, which is $x = \sin y$. Consider the input $x = \frac{1}{2}$. On the graph of the inverse of

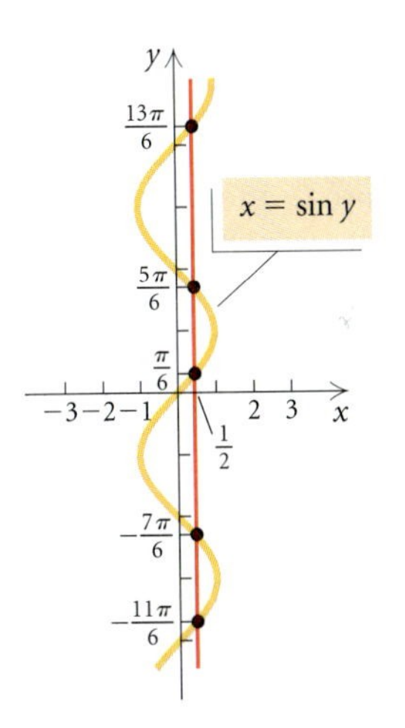

the sine functions, we draw a vertical line at $x = \frac{1}{2}$, as shown in the figure at left. The vertical line intersects the graph at points whose y-values are such that $\frac{1}{2} = \sin y$. Some numbers whose sine is $\frac{1}{2}$ are $\pi/6$, $5\pi/6$, $-7\pi/6$, and so on. The complete set of values is given by $\pi/6 + 2k\pi$ and $5\pi/6 + 2k\pi$, where k is an integer. Indeed, the vertical line crosses the curve at many points, verifying that the inverse of the sine function is not a function.

Restricting Ranges to Define Inverse Functions

Recall that a function like $f(x) = x^2$ does not have an inverse that is a function, but by restricting the domain of f to nonnegative numbers, we have a new squaring function, $f(x) = x^2$, $x \geq 0$, that has an inverse, $f^{-1}(x) = \sqrt{x}$. This is equivalent to restricting the range of the inverse relation to exclude ordered pairs that contain negative numbers.

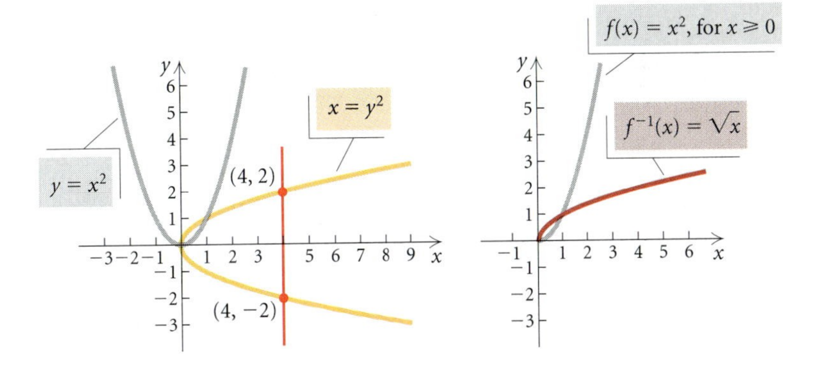

In a similar manner, we can define new trigonometric functions whose inverses are functions. We can do this by restricting either the domains of the basic trigonometric functions or the ranges of their inverse relations. This can be done in many ways, but the restrictions illustrated below with solid black curves are fairly standard in mathematics.

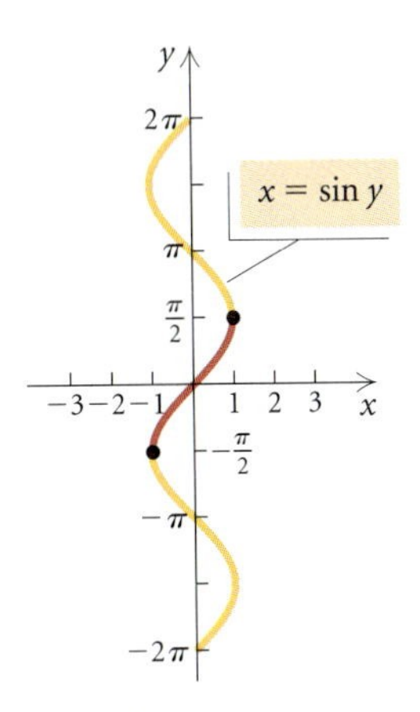

FIGURE 1

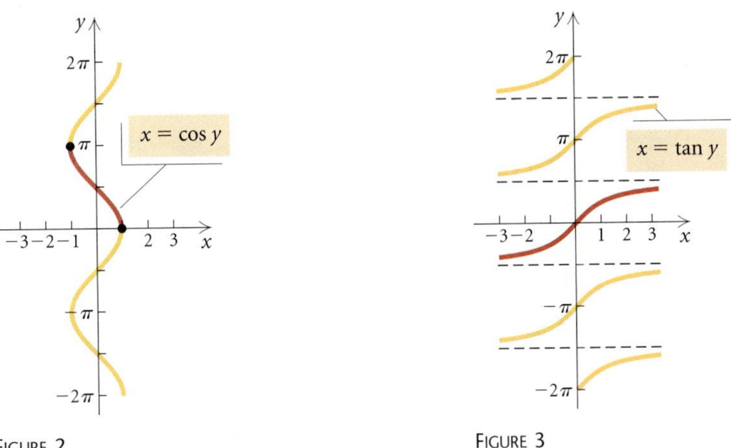

FIGURE 2

FIGURE 3

For the inverse sine function, we choose a range close to the origin that allows all inputs in the interval $[-1, 1]$ to have function values. Thus we choose the interval $[-\pi/2, \pi/2]$ for the range (Fig. 1). For the inverse cosine function, we choose a range close to the origin that allows all inputs in the interval $[-1, 1]$ to have function values. This is the interval $[0, \pi]$ (Fig. 2). For the inverse tangent function, we choose a range close to the origin that allows all real numbers to have function values. The interval $(-\pi/2, \pi/2)$ satisfies this requirement (Fig. 3).

Inverse Trigonometric Functions

FUNCTION	DOMAIN	RANGE
$y = \sin^{-1} x$ $= \arcsin x$, where $x = \sin y$	$[-1, 1]$	$[-\pi/2, \pi/2]$
$y = \cos^{-1} x$ $= \arccos x$, where $x = \cos y$	$[-1, 1]$	$[0, \pi]$
$y = \tan^{-1} x$ $= \arctan x$, where $x = \tan y$	$(-\infty, \infty)$	$(-\pi/2, \pi/2)$

The notation arcsin x arises because the function is the length of an arc on the unit circle for which the sine is x. **The notation $\sin^{-1} x$ is not exponential notation. It does not mean $1/\sin x$!** Either of the two kinds of notation above can be read "the inverse sine of x" or "the arc sine of x" or "the number (or angle) whose sine is x."

The graphs of the inverse trigonometric functions are as follows.

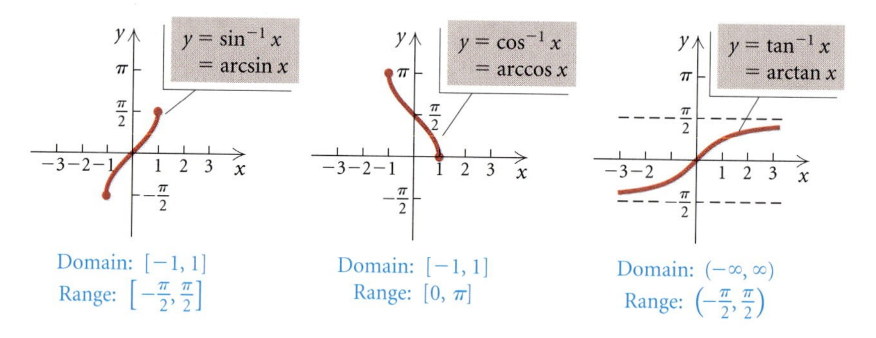

Domain: $[-1, 1]$
Range: $\left[-\frac{\pi}{2}, \frac{\pi}{2}\right]$

Domain: $[-1, 1]$
Range: $[0, \pi]$

Domain: $(-\infty, \infty)$
Range: $\left(-\frac{\pi}{2}, \frac{\pi}{2}\right)$

Interactive Discovery

Inverse trigonometric functions can be graphed on a grapher. Graph $y = \sin^{-1} x$ making sure the grapher is in RADIAN mode and changing the viewing window to $[-3, 3, -\pi, \pi]$, with Xscl $= 1$ and Yscl $= \pi/2$. Now try graphing $y = \cos^{-1} x$ and $y = \tan^{-1} x$. Then trace to confirm the domain and the range of each inverse with its graph.

The following diagrams show the restricted ranges for the inverse trigonometric functions on a unit circle. These ranges should be memorized. The missing endpoints indicate inputs that are not in the domain of the original function.

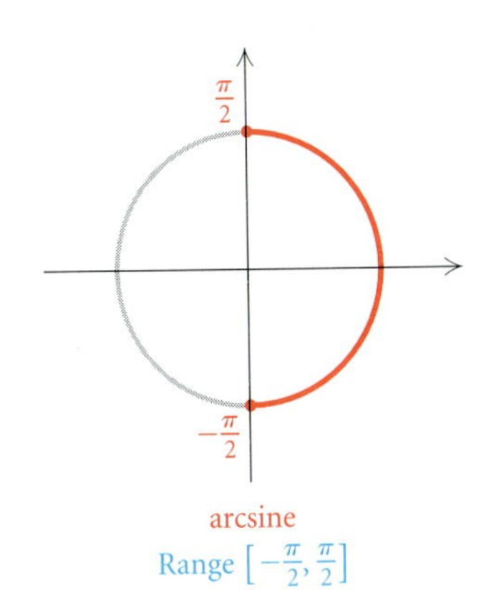

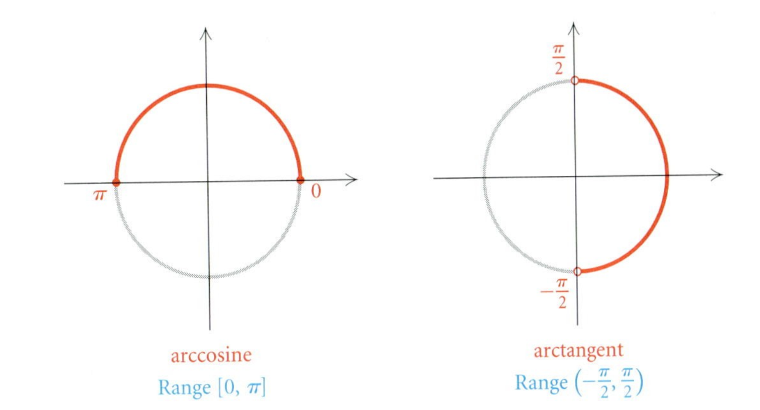

Example 1 Find each of the following function values.

a) $\arcsin \dfrac{\sqrt{2}}{2}$ b) $\cos^{-1}\left(-\dfrac{1}{2}\right)$ c) $\tan^{-1}\left(-\dfrac{\sqrt{3}}{3}\right)$

SOLUTION

a) In the restricted range $[-\pi/2, \pi/2]$, the only number with a sine of $\sqrt{2}/2$ is $\pi/4$. Thus, $\arcsin(\sqrt{2}/2) = \pi/4$, or $45°$. (See Fig. 4 at left.)

b) The only number with a cosine of $-\frac{1}{2}$ in the restricted range $[0, \pi]$ is $2\pi/3$. Thus, $\cos^{-1}\left(-\frac{1}{2}\right) = 2\pi/3$, or $120°$. (See Fig. 5 at left.)

c) The only number in the restricted range $(-\pi/2, \pi/2)$ with a tangent of $-\sqrt{3}/3$ is $-\pi/6$. Thus, $\tan^{-1}(-\sqrt{3}/3)$ is $-\pi/6$, or $-30°$. (See Fig. 6 at left.)

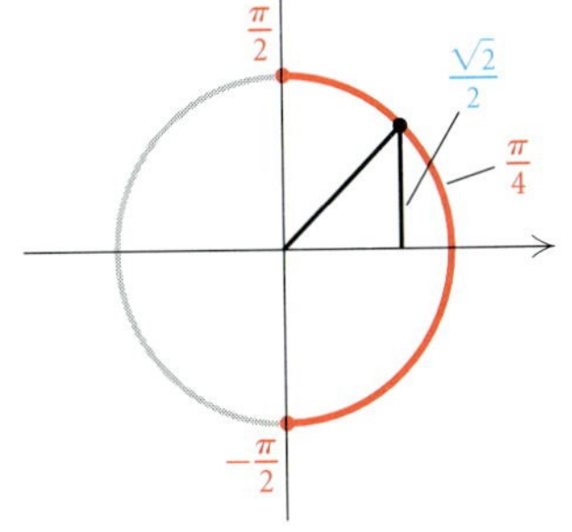

FIGURE 4

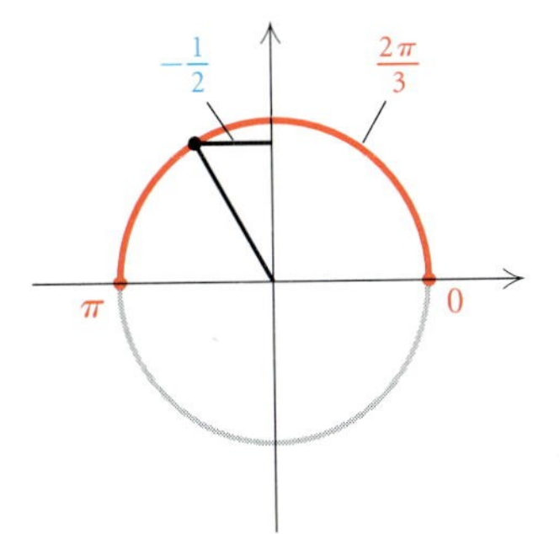

FIGURE 5

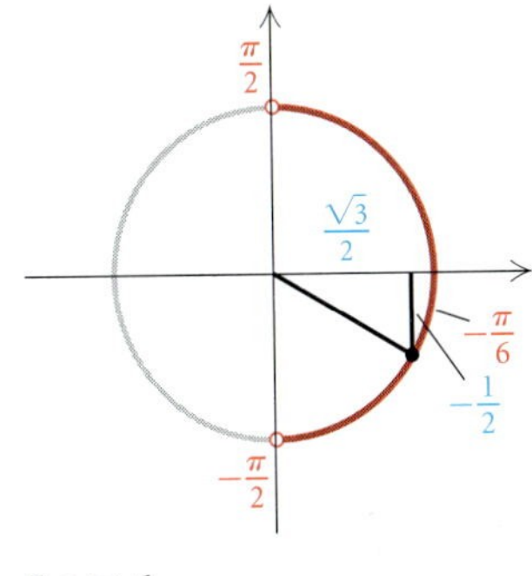

FIGURE 6

We can also use a grapher to find inverse trigonometric function values (see Section 4.1, p. 329). On most graphers, we can find inverse function values in both radians and degrees simply by selecting the appropriate mode. The key strokes involved in finding inverse function values vary with the grapher. Be sure to read the instructions for the grapher you are using.

Finding $\sin^{-1} 1$ provides a quick way to check the mode setting. If you get 1.570796327, which is about $\pi/2$, you know that the values are in radians. If you get 90, you know that the values are in degrees.

Example 2 Approximate each of the following function values in both radians and degrees.

a) $\cos^{-1}(-0.2689)$

b) $\arctan(-0.2623)$

c) $\sin^{-1} 0.20345$

d) $\arccos 1.318$

e) $\csc^{-1} 8.205$

SOLUTION

FUNCTION VALUE	MODE	READOUT	ROUNDED
a) $\cos^{-1}(-0.2689)$	Radian	1.843047111	1.8430
	Degree	105.5988209	105.6°
b) arctan (-0.2623)	Radian	-.256521214	-0.2565
	Degree	-14.69758292	$-14.7°$
c) $\sin^{-1} 0.20345$	Radian	.204880336	0.2049
	Degree	11.73877855	11.7°
d) arccos 1.318	Radian	ERR:DOMAIN	
	Degree	ERR:DOMAIN	

The value 1.318 is not in $[-1, 1]$, the domain of the arccosine function.

e) The cosecant function is the reciprocal of the sine function:

$\csc^{-1} 8.205 =$ $\sin^{-1}(1/8.205)$	Radian	.122180665	0.1222
	Degree	7.000436462	7.0°

The following is a summary of the domains and ranges of the trigonometric functions together with a summary of the domains and ranges of the inverse trigonometric functions. For completeness, we have included the arccosecant, the arcsecant, and the arccotangent, though there is a lack of uniformity in their definitions in mathematical literature.

FUNCTION	DOMAIN	RANGE	INVERSE FUNCTION	DOMAIN	RANGE
sin	All reals, $(-\infty, \infty)$	$[-1, 1]$	$\sin^{-1}$	$[-1, 1]$	$\left[-\frac{\pi}{2}, \frac{\pi}{2}\right]$
cos	All reals, $(-\infty, \infty)$	$[-1, 1]$	$\cos^{-1}$	$[-1, 1]$	$[0, \pi]$
tan	All reals except $k\pi/2$, k odd	All reals, $(-\infty, \infty)$	$\tan^{-1}$	All reals, $(-\infty, \infty)$	$\left(-\frac{\pi}{2}, \frac{\pi}{2}\right)$
csc	All reals except $k\pi$	$(-\infty, -1] \cup [1, \infty)$	$\csc^{-1}$	$(-\infty, -1] \cup [1, \infty)$	$\left[-\frac{\pi}{2}, 0\right) \cup \left(0, \frac{\pi}{2}\right]$
sec	All reals except $k\pi/2$, k odd	$(-\infty, -1] \cup [1, \infty)$	$\sec^{-1}$	$(-\infty, -1] \cup [1, \infty)$	$\left[0, \frac{\pi}{2}\right) \cup \left(\frac{\pi}{2}, \pi\right]$
cot	All reals except $k\pi$	All reals, $(-\infty, \infty)$	$\cot^{-1}$	All reals, $(-\infty, \infty)$	$\left[-\frac{\pi}{2}, 0\right) \cup \left(0, \frac{\pi}{2}\right]$

Composition of Trigonometric Functions and Their Inverses

Various compositions of trigonometric functions and their inverses often occur in practice. For example, we might want to try to simplify an expression such as

$$\sin(\sin^{-1} x) \quad \text{or} \quad \sin\left(\operatorname{arccot} \frac{x}{2}\right).$$

In the expression on the left, we are finding "the sine of a number whose sine is x." Recall from Section 3.1 that if a function f has an inverse that is also a function, then

$$f(f^{-1}(x)) = x, \quad \text{for all } x \text{ in the domain of } f^{-1},$$

and $$f^{-1}(f(x)) = x, \quad \text{for all } x \text{ in the domain of } f.$$

Thus, if $f(x) = \sin x$ and $f^{-1}(x) = \sin^{-1} x$, then

$$\mathbf{\sin(\sin^{-1} x) = x, \quad \text{for all } x \text{ in the domain of } \sin^{-1},}$$

which is any number in the interval $[-1, 1]$. Similar results hold for the other trigonometric functions.

Composition of Trigonometric Functions

$\sin(\sin^{-1} x) = x$, for all x in the domain of $\sin^{-1}$.

$\cos(\cos^{-1} x) = x$, for all x in the domain of $\cos^{-1}$.

$\tan(\tan^{-1} x) = x$, for all x in the domain of $\tan^{-1}$.

Example 3 Simplify each of the following.

a) $\cos\left(\cos^{-1} \frac{\sqrt{3}}{2}\right)$ **b)** $\sin(\sin^{-1} 1.8)$

SOLUTION

a) Since $\sqrt{3}/2$ is in the domain of $\cos^{-1}$, $[-1, 1]$, it follows that

$$\cos\left(\cos^{-1} \frac{\sqrt{3}}{2}\right) = \frac{\sqrt{3}}{2}.$$

b) Since 1.8 is not in $[-1, 1]$, the domain of $\sin^{-1}$, we cannot evaluate this expression. We know that there is no number with a sine of 1.8. Since we cannot find $\sin^{-1} 1.8$, we state that $\sin(\sin^{-1} 1.8)$ does not exist.

Now let's consider an expression like $\sin^{-1}(\sin x)$. We might also suspect that this is equal to x for any x, but this is not true unless x is in the range of the $\sin^{-1}$ function. Note that in order to define $\sin^{-1}$, we had to restrict the domain of the sine function. In doing so, we restricted the range of the inverse sine function. Thus,

$$\mathbf{\sin^{-1}(\sin x) = x, \quad \text{for all } x \text{ in the range of } \sin^{-1}.}$$

Similar results hold for the other trigonometric functions.

Special Cases

$\sin^{-1}(\sin x) = x$, for all x in the range of $\sin^{-1}$.
$\cos^{-1}(\cos x) = x$, for all x in the range of $\cos^{-1}$.
$\tan^{-1}(\tan x) = x$, for all x in the range of $\tan^{-1}$.

Example 4 Simplify each of the following.

a) $\tan^{-1}\left(\tan \frac{\pi}{6}\right)$ **b)** $\sin^{-1}\left(\sin \frac{3\pi}{4}\right)$

SOLUTION

a) Since $\pi/6$ is in $(-\pi/2, \pi/2)$, the range of the $\tan^{-1}$ function, we can use $\tan^{-1}(\tan x) = x$. Thus,

$$\tan^{-1}\left(\tan \frac{\pi}{6}\right) = \frac{\pi}{6}.$$

b) Note that $3\pi/4$ is not in $[-\pi/2, \pi/2]$, the range of the $\sin^{-1}$ function. Thus we *cannot* apply $\sin^{-1}(\sin x) = x$. Instead we first find $\sin(3\pi/4)$, which is $\sqrt{2}/2$, and substitute:

$$\sin^{-1}\left(\sin \frac{3\pi}{4}\right) = \sin^{-1}\left(\frac{\sqrt{2}}{2}\right) = \frac{\pi}{4}.$$

Now we find some other function compositions.

Example 5 Simplify each of the following.

a) $\sin[\arctan(-1)]$ **b)** $\cos^{-1}\left(\sin \frac{\pi}{2}\right)$

SOLUTION

a) $\sin[\arctan(-1)] = \sin\left[-\frac{\pi}{4}\right] = -\frac{\sqrt{2}}{2}$

b) $\cos^{-1}\left(\sin \frac{\pi}{2}\right) = \cos^{-1}(1) = 0$

Next let's consider

$$\cos\left(\arcsin \frac{3}{5}\right).$$

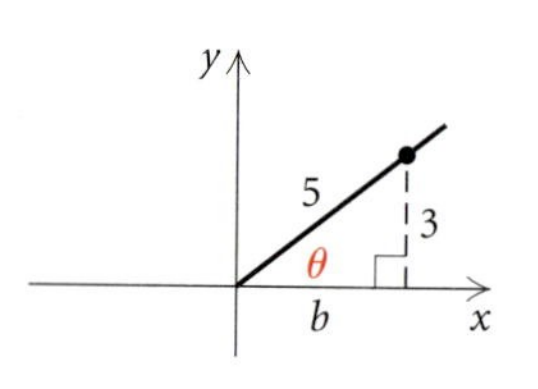

Without using a grapher, we cannot find $\arcsin \frac{3}{5}$. However, we can still evaluate the entire expression by sketching a reference triangle. We are looking for angle θ such that $\arcsin \frac{3}{5} = \theta$, or $\sin \theta = \frac{3}{5}$. Since arcsin is defined in $[-\pi/2, \pi/2]$ and $\frac{3}{5} > 0$, we know that θ is in quadrant I. We sketch a reference right triangle, as shown at left. The angle θ in this triangle is an angle whose sine is $\frac{3}{5}$. We wish to find the cosine of this angle.

Since the triangle is a right triangle, we can find the length of the base, b. It is 4. Thus we know that $\cos\theta = b/5$, or $\frac{4}{5}$. Therefore,

$$\cos\left(\arcsin\frac{3}{5}\right) = \frac{4}{5}.$$

Example 6 Find $\sin\left(\operatorname{arccot}\frac{x}{2}\right)$.

SOLUTION We draw a right triangle whose legs have lengths $|x|$ and 2, so that $\cot\theta = x/2$. If x is negative, we get the triangle in standard position shown on the left below. If x is positive, we get the triangle shown on the right.

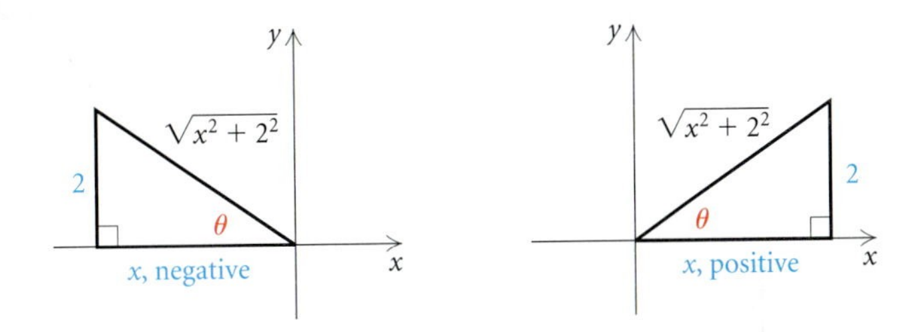

We find the length of the hypotenuse and then read off the sine ratio. In either case, we get

$$\sin\left(\operatorname{arccot}\frac{x}{2}\right) = \frac{2}{\sqrt{x^2+2^2}}, \quad \text{or} \quad \frac{2}{\sqrt{x^2+4}}.$$

In the following example, we use a sum identity to evaluate an expression.

Example 7 Evaluate: $\sin\left(\sin^{-1}\frac{1}{2} + \cos^{-1}\frac{5}{13}\right)$.

SOLUTION The expression is a sine of a sum, so we use the identity

$$\sin(u + v) = \sin u \cos v + \cos u \sin v.$$

Thus,

$$\sin\left(\sin^{-1}\frac{1}{2} + \cos^{-1}\frac{5}{13}\right)$$

$$= \sin\left(\sin^{-1}\frac{1}{2}\right)\cdot\cos\left(\cos^{-1}\frac{5}{13}\right) + \cos\left(\sin^{-1}\frac{1}{2}\right)\cdot\sin\left(\cos^{-1}\frac{5}{13}\right)$$

$$= \frac{1}{2}\cdot\frac{5}{13} + \cos\left(\sin^{-1}\frac{1}{2}\right)\cdot\sin\left(\cos^{-1}\frac{5}{13}\right)$$ **Using composition identities**

Now since $\sin^{-1}\frac{1}{2} = \pi/6$, $\cos\left(\sin^{-1}\frac{1}{2}\right)$ simplifies to $\sqrt{3}/2$. To find $\sin\left(\cos^{-1}\frac{5}{13}\right)$, we use a reference triangle in quadrant I and determine that the sine of the angle whose cosine is $\frac{5}{13}$ is $\frac{12}{13}$. Our expression now simplifies to

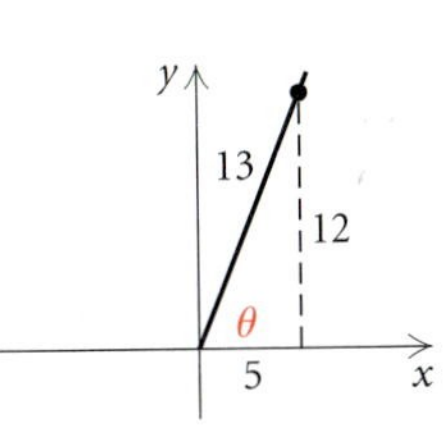

$$\frac{1}{2}\cdot\frac{5}{13} + \frac{\sqrt{3}}{2}\cdot\frac{12}{13}, \quad \text{or} \quad \frac{5+12\sqrt{3}}{26}.$$

Thus,

$$\sin\left(\sin^{-1}\frac{1}{2} + \cos^{-1}\frac{5}{13}\right) = \frac{5 + 12\sqrt{3}}{26}.$$

We can check with a grapher:

$$\sin\left(\sin^{-1}\frac{1}{2} + \cos^{-1}\frac{5}{13}\right) \approx 0.9917 \quad \text{and} \quad \frac{5 + 12\sqrt{3}}{26} \approx 0.9917.$$

5.4 Exercise Set

Find each of the following exactly in radians and degrees.

1. $\sin^{-1}\left(-\frac{\sqrt{3}}{2}\right)$
2. $\cos^{-1}\frac{1}{2}$
3. $\arctan 1$
4. $\arcsin 0$
5. $\arccos\frac{\sqrt{2}}{2}$
6. $\sec^{-1}\sqrt{2}$
7. $\tan^{-1} 0$
8. $\arctan\frac{\sqrt{3}}{3}$
9. $\cos^{-1}\frac{\sqrt{3}}{2}$
10. $\cot^{-1}\left(-\frac{\sqrt{3}}{3}\right)$
11. $\csc^{-1} 2$
12. $\sin^{-1}\frac{1}{2}$
13. $\text{arccot}\,(-\sqrt{3})$
14. $\tan^{-1}(-1)$
15. $\arcsin\left(-\frac{1}{2}\right)$
16. $\arccos\left(-\frac{\sqrt{2}}{2}\right)$
17. $\cos^{-1} 0$
18. $\sin^{-1}\frac{\sqrt{3}}{2}$
19. $\text{arcsec}\, 2$
20. $\text{arccsc}\,(-1)$

Use a grapher to find each of the following in radians, rounded to three decimal places, and in degrees, rounded to the nearest tenth of a degree.

21. $\arctan 0.3673$
22. $\cos^{-1}(-0.2935)$
23. $\sin^{-1} 0.9613$
24. $\arcsin(-0.6199)$
25. $\cos^{-1}(-0.9810)$
26. $\tan^{-1} 158$
27. $\csc^{-1}(-6.2774)$
28. $\sec^{-1} 1.1677$
29. $\tan^{-1}(1.091)$
30. $\cot^{-1} 1.265$
31. $\arcsin(-0.8192)$
32. $\arccos(-0.2716)$

33. State the domains of the inverse sine, inverse cosine, and inverse tangent functions.

34. State the ranges of the inverse sine, inverse cosine, and inverse tangent functions.

35. *Angle of Depression.* An airplane is flying at an altitude of 2000 ft toward an island. The straight-line distance from the airplane to the island is d feet. Express θ, the angle of depression, as an inverse sine function of d.

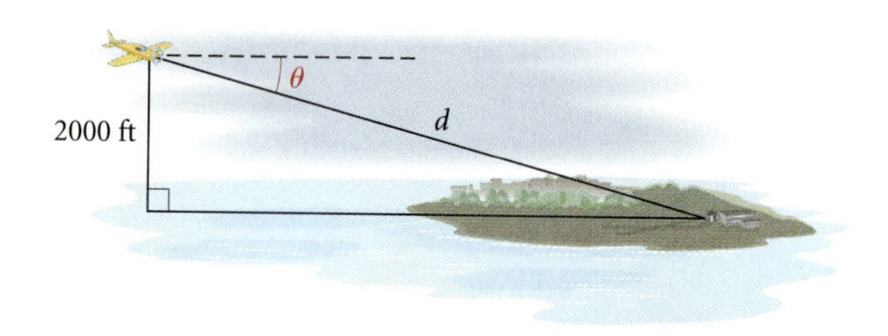

36. *Angle of Inclination.* A guy wire is attached to the top of a 50-ft pole and stretched to a point that is d feet from the bottom of the pole. Express β, the angle of inclination, as a function of d.

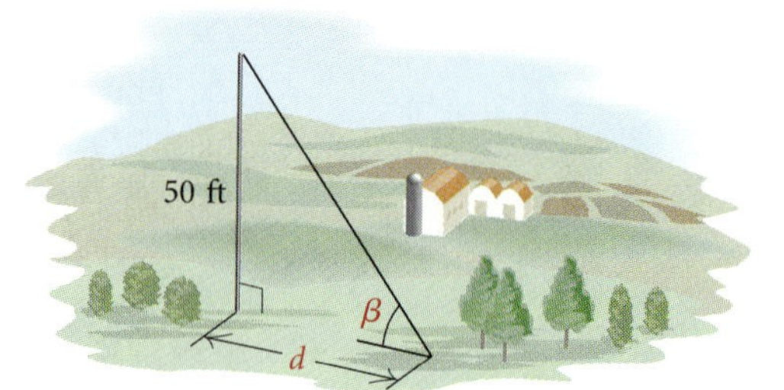

Evaluate.

37. $\sin(\arcsin 0.3)$
38. $\tan[\tan^{-1}(-4.2)]$
39. $\cos^{-1}\left[\cos\left(-\frac{\pi}{4}\right)\right]$
40. $\arcsin\left(\sin\frac{2\pi}{3}\right)$
41. $\sin^{-1}\left(\sin\frac{\pi}{5}\right)$
42. $\cot^{-1}\left(\cot\frac{2\pi}{3}\right)$
43. $\tan^{-1}\left(\tan\frac{2\pi}{3}\right)$
44. $\cos^{-1}\left(\cos\frac{\pi}{7}\right)$

45. $\sin\left(\arctan\frac{\sqrt{3}}{3}\right)$

46. $\cos\left(\arcsin\frac{\sqrt{3}}{2}\right)$

47. $\tan\left(\cos^{-1}\frac{\sqrt{2}}{2}\right)$

48. $\cos^{-1}(\sin\pi)$

49. $\arcsin\left(\cos\frac{\pi}{6}\right)$

50. $\sin^{-1}\left[\tan\left(-\frac{\pi}{4}\right)\right]$

51. $\tan(\arcsin 0.1)$

52. $\cos\left(\tan^{-1}\frac{\sqrt{3}}{4}\right)$

Find.

53. $\sin\left(\arctan\frac{a}{3}\right)$

54. $\tan\left(\cos^{-1}\frac{3}{x}\right)$

55. $\cot\left(\sin^{-1}\frac{p}{q}\right)$

56. $\sin(\cos^{-1}x)$

57. $\tan\left(\arcsin\frac{p}{\sqrt{p^2+9}}\right)$

58. $\tan\left(\frac{1}{2}\arcsin\frac{1}{2}\right)$

59. $\cos\left(\frac{1}{2}\arcsin\frac{\sqrt{3}}{2}\right)$

60. $\sin\left(2\cos^{-1}\frac{3}{5}\right)$

Evaluate.

61. $\cos\left(\sin^{-1}\frac{\sqrt{2}}{2}+\cos^{-1}\frac{3}{5}\right)$

62. $\sin\left(\sin^{-1}\frac{1}{2}+\cos^{-1}\frac{3}{5}\right)$

63. $\sin(\sin^{-1}x+\cos^{-1}y)$

64. $\cos(\sin^{-1}x-\cos^{-1}y)$

65. $\sin(\sin^{-1}0.6032+\cos^{-1}0.4621)$

66. $\cos(\sin^{-1}0.7325-\cos^{-1}0.4838)$

Skill Maintenance

Solve.

67. $2x^2 = 5x$

68. $3x^2+5x-10=18$

69. $x^4+5x^2-36=0$

70. $x^2-10x+1=0$

71. $\sqrt{x-2}=5$

72. $x=\sqrt{x+7}+5$

Synthesis

73. ◆ Explain in your own words why the ranges of the inverse trigonometric functions are restricted.

74. ◆ How does the graph of $y=\sin^{-1}x$ differ from the graph of $y=\sin x$?

75. ◆ Why is it that

$$\sin\frac{5\pi}{6}=\frac{1}{2},$$

but

$$\sin^{-1}\left(\frac{1}{2}\right)\neq\frac{5\pi}{6}?$$

76. Use a calculator to approximate the following expression:

$$16\tan^{-1}\frac{1}{5}-4\tan^{-1}\frac{1}{239}.$$

What number does this expression seem to approximate?

Prove each identity.

77. $\sin^{-1}x+\cos^{-1}x=\frac{\pi}{2}$

78. $\tan^{-1}x+\cot^{-1}x=\frac{\pi}{2}$

79. $\sin^{-1}x=\tan^{-1}\frac{x}{\sqrt{1-x^2}}$

80. $\tan^{-1}x=\sin^{-1}\frac{x}{\sqrt{x^2+1}}$

81. $\arcsin x=\arccos\sqrt{1-x^2}$, for $x\geq 0$

82. $\arccos x=\arctan\frac{\sqrt{1-x^2}}{x}$, for $x>0$

83. *Height of a Mural.* An art student's eye is at a point A, looking at a mural of height h, with the bottom of the mural y feet above the eye. The eye is x feet from the wall. Write an expression for θ in terms of x, y, and h. Then evaluate the expression when $x = 20$ ft, $y = 7$ ft, and $h = 25$ ft.

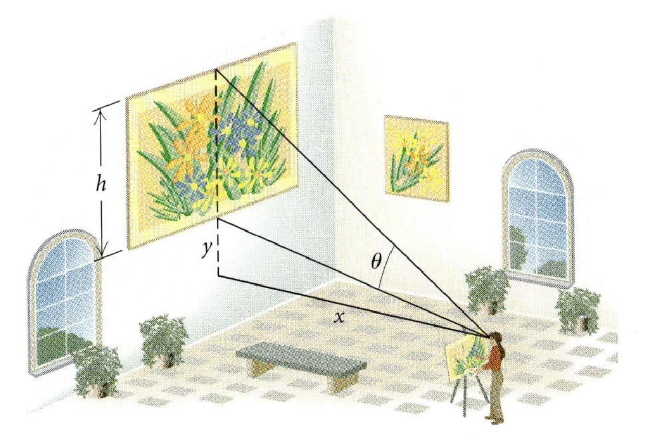

5.5 Solving Trigonometric Equations

• *Solve trigonometric equations.*

When an equation contains a trigonometric expression with a variable, such as $\cos x$, it is called a trigonometric equation. Some trigonometric equations are identities, such as $\sin^2 x + \cos^2 x = 1$. Now we consider equations, such as $2 \cos x = -1$, that are usually not identities. As we have done for other types of equations, we will solve such equations by finding all values for x that make the equation true.

Example 1 Solve: $2 \cos x = -1$.

SOLUTION We first solve for $\cos x$:

$$2 \cos x = -1$$
$$\cos x = -\frac{1}{2}.$$

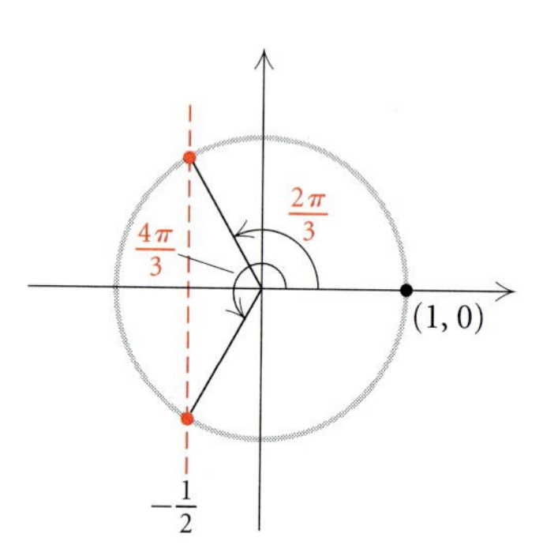

The solutions are numbers that have a cosine of $-\frac{1}{2}$. To find them, we use the unit circle (see Section 4.5).

There are just two points on it for which the cosine is $-\frac{1}{2}$, as shown in the figure at left. They are the points for $2\pi/3$ and $4\pi/3$. These numbers, plus any multiple of 2π, are the solutions:

$$\frac{2\pi}{3} + 2k\pi \quad \text{or} \quad \frac{4\pi}{3} + 2k\pi,$$

where k is any integer. In degrees, the solutions are

$$120° + k \cdot 360° \quad \text{or} \quad 240° + k \cdot 360°,$$

where k is any integer.

To check the solution to $2 \cos x = -1$, or $\cos x = -\frac{1}{2}$, we can graph $y_1 = \cos x$ and $y_2 = -\frac{1}{2}$ on the same set of axes. Using $\pi/3$ as the Xscl facilitates our reading of the solutions. First let's graph these equations in the interval from 0 to 2π as shown in the figure on the left below. The only solutions in $[0, 2\pi)$ are $2\pi/3$ and $4\pi/3$.

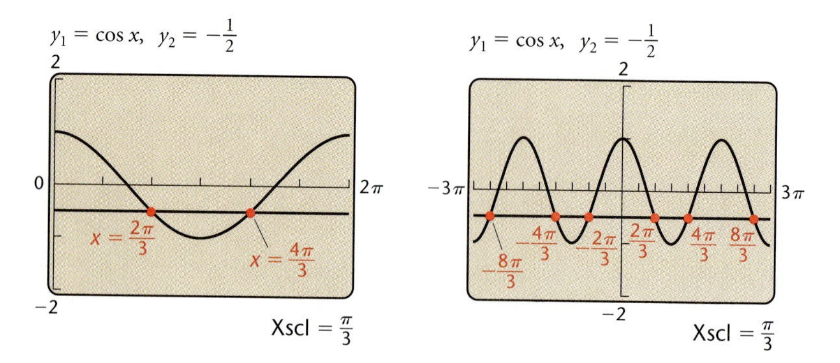

Now let's change the viewing window dimensions to $[-3\pi, 3\pi, -2, 2]$ and graph again. Since the cosine function is periodic, there is an infinite number of solutions. A few of these appear in the graph on the right above. From the graph, we see that the solutions are $2\pi/3 + 2k\pi$ and $4\pi/3 + 2k\pi$.

Example 2 Solve: $4 \sin^2 x = 1$.

SOLUTION We begin by solving for $\sin x$:

$$4 \sin^2 x = 1$$
$$\sin^2 x = \frac{1}{4}$$
$$\sin x = \pm\frac{1}{2}.$$

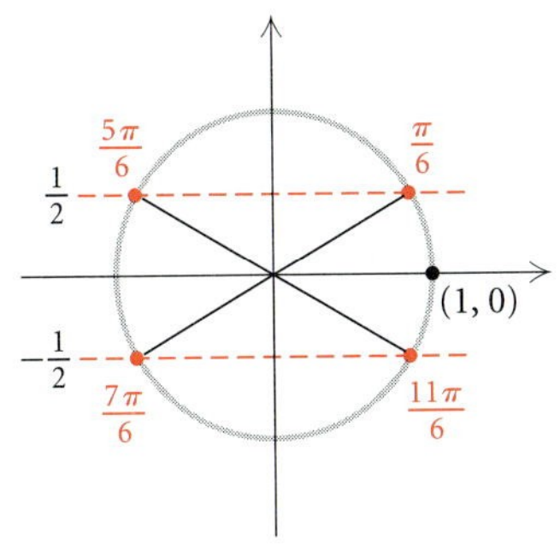

Again we use the unit circle to find those numbers having a sine of $\frac{1}{2}$ or $-\frac{1}{2}$. The solutions are

$$\frac{\pi}{6} + 2k\pi, \quad \frac{5\pi}{6} + 2k\pi, \quad \frac{7\pi}{6} + 2k\pi, \quad \text{and} \quad \frac{11\pi}{6} + 2k\pi,$$

where k is any integer. In degrees, the solutions are

$$30° + k \cdot 360°, \quad 150° + k \cdot 360°, \quad 210° + k \cdot 360°, \quad \text{and} \quad 330° + k \cdot 360°,$$

where k is any integer.

The general solutions listed above could be condensed using odd as well as even multiples of π:

$$\frac{\pi}{6} + k\pi \quad \text{and} \quad \frac{5\pi}{6} + k\pi,$$

or, in degrees,

$$30° + k \cdot 180° \quad \text{and} \quad 150° + k \cdot 180°,$$

where k is any integer.

Let's do a partial check with a grapher, checking only the solutions in $[0, 2\pi)$. We graph $y_1 = \sin x$, $y_2 = \frac{1}{2}$, and $y_3 = -\frac{1}{2}$ and observe that the solutions in $[0, 2\pi)$ are $\pi/6$, $5\pi/6$, $7\pi/6$, and $11\pi/6$.

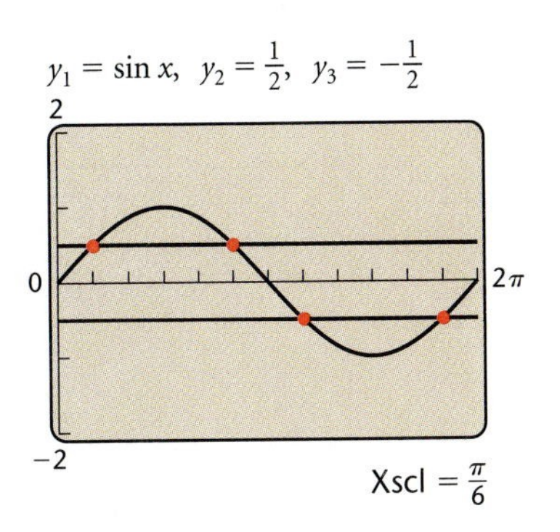

In most applications, it is sufficient to find just the solutions from 0 to 2π or from 0° to 360°. We then remember that any multiple of 2π, or 360°, can be added to obtain the rest of the solutions.

We must be careful to find all solutions in $[0, 2\pi)$ when solving trigonometric equations involving double angles.

Example 3 Solve $3 \tan 2x = -3$ in the interval $[0, 2\pi)$.

SOLUTION We first solve for $\tan 2x$:

$$3 \tan 2x = -3$$
$$\tan 2x = -1.$$

We are looking for solutions x to the equation for which

$$0 \le x < 2\pi.$$

Multiplying by 2, we get

$$0 \le 2x < 4\pi,$$

which is the interval we use when solving $\tan 2x = -1$.

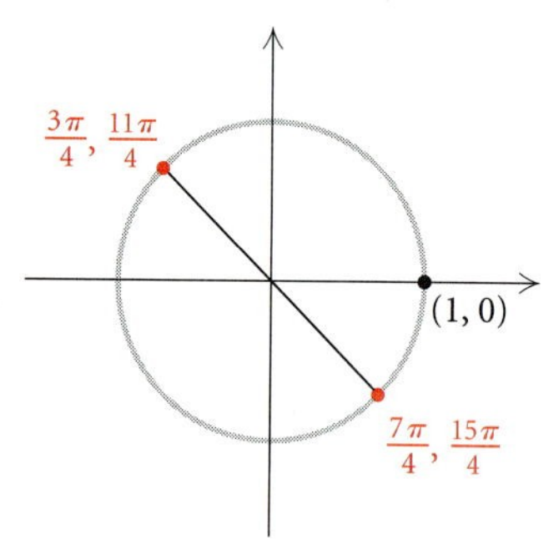

Using the unit circle, we find points $2x$ in $[0, 4\pi)$ for which $\tan 2x = -1$. These values of $2x$ are as follows:

$$2x = \frac{3\pi}{4}, \frac{7\pi}{4}, \frac{11\pi}{4}, \text{ and } \frac{15\pi}{4}.$$

Thus the desired values of x in $[0, 2\pi)$ are each of these values divided by 2. Therefore,

$$x = \frac{3\pi}{8}, \frac{7\pi}{8}, \frac{11\pi}{8}, \text{ and } \frac{15\pi}{8}.$$

Graphers are needed for some trigonometric equations. Answers can be found in radians or degrees, depending on the mode setting.

Example 4 Solve $\frac{1}{2} \cos \phi + 1 = 1.2108$ in $[0, 360°)$.

SOLUTION We have

$$\frac{1}{2} \cos \phi + 1 = 1.2108$$
$$\frac{1}{2} \cos \phi = 0.2108$$
$$\cos \phi = 0.4216.$$

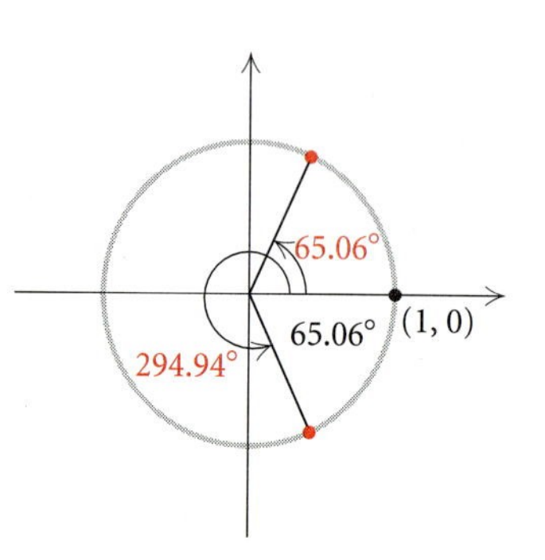

Using a grapher, we find that the reference angle, arccos 0.4216, is

$$\phi \approx 65.06°.$$

Since $\cos \phi$ is positive, the solutions are in quadrants I and IV. The solutions in $[0, 360°)$ are

$$65.06° \quad \text{and} \quad 294.94°.$$

In some cases, we may need to apply some algebra before concerning ourselves with the trigonometry.

Example 5 Solve $2\cos^2 u = 1 - \cos u$ in $[0, 2\pi)$.

SOLUTION It is instructive to begin with a grapher. We can use either of two methods to determine the solutions graphically.

Method 1. We graph the left side of the equation as one function and then the right side as another function. Then we look for points of intersection. Here it is helpful to use Xscl $= \pi/3$.

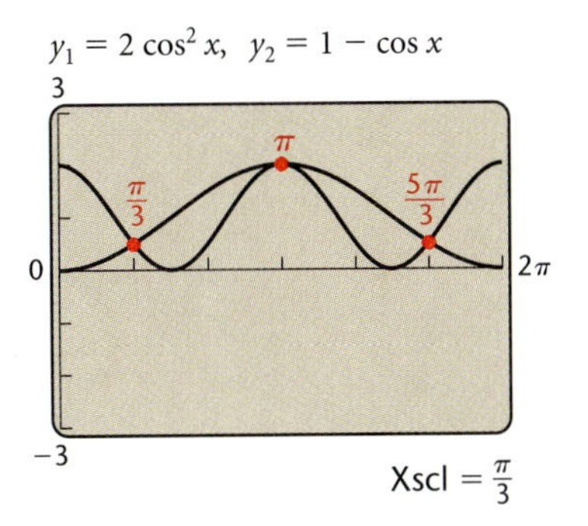

The solutions in $[0, 2\pi)$ appear to be $\pi/3$, π, and $5\pi/3$. We could use the INTERSECT feature and find approximate solutions, 1.05, 3.14, and 5.24.

Method 2. We rewrite the equation

$$2\cos^2 u = 1 - \cos u$$

as

$$2\cos^2 u + \cos u - 1 = 0.$$

Then we graph $y = 2\cos^2 u + \cos u - 1$ and determine the zeros of the function.

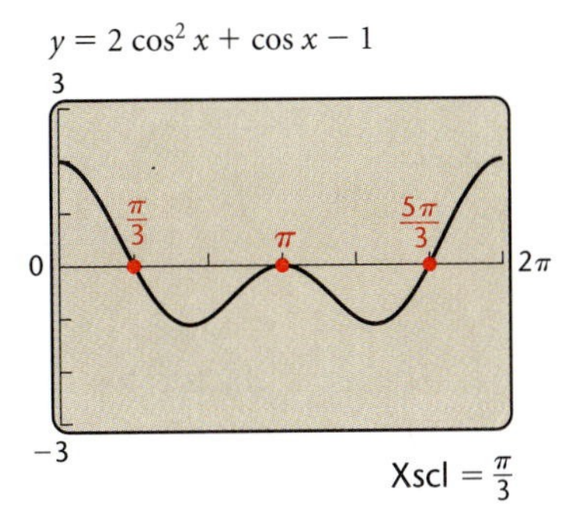

Again the solutions appear to be $\pi/3$, π, and $5\pi/3$. We could use the SOLVE feature and find approximate solutions, 1.05, 3.14, and 5.24.

To confirm the solutions algebraically, we use the principle of zero products:

$$2\cos^2 u = 1 - \cos u$$
$$2\cos^2 u + \cos u - 1 = 0$$
$$(2\cos u - 1)(\cos u + 1) = 0 \qquad \text{Factoring}$$
$$2\cos u - 1 = 0 \quad \text{or} \quad \cos u + 1 = 0 \qquad \text{Principle of zero products}$$
$$2\cos u = 1 \quad \text{or} \quad \cos u = -1$$
$$\cos u = \frac{1}{2} \quad \text{or} \quad \cos u = -1.$$

Thus,

$$u = \frac{\pi}{3}, \frac{5\pi}{3} \quad \text{or} \quad u = \pi.$$

The solutions in $[0, 2\pi)$ are $\pi/3$, π, and $5\pi/3$.

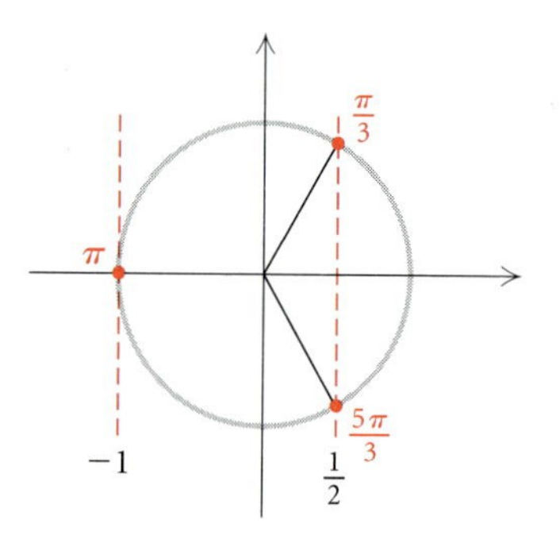

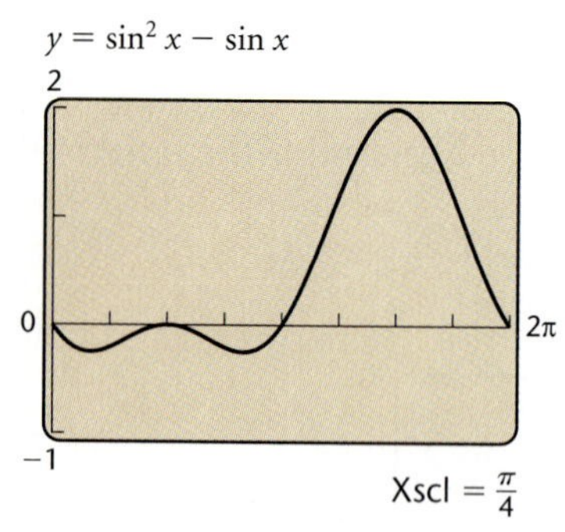

Example 6 Solve: $\sin^2 \beta - \sin \beta = 0$ in $[0, 2\pi)$.

SOLUTION We have

$$\sin^2 \beta - \sin \beta = 0$$
$$\sin \beta (\sin \beta - 1) = 0 \qquad \text{Factoring}$$
$$\sin \beta = 0 \quad \text{or} \quad \sin \beta - 1 = 0$$
$$\sin \beta = 0 \quad \text{or} \quad \sin \beta = 1$$
$$\beta = 0, \pi \quad \text{or} \quad \beta = \frac{\pi}{2}.$$

The solutions in $[0, 2\pi)$ are 0, $\pi/2$, and π.

If a trigonometric equation is quadratic but difficult or impossible to factor, we use the quadratic formula.

Example 7 Solve $10\sin^2 x - 12\sin x - 7 = 0$ in $[0°, 360°)$.

SOLUTION This equation is quadratic in $\sin x$ with $a = 10$, $b = -12$, and $c = -7$. Substituting into the quadratic formula, we get

$$\begin{aligned} \sin x &= \frac{12 \pm \sqrt{144 + 280}}{20} \quad \text{Using the quadratic formula} \\ &= \frac{12 \pm \sqrt{424}}{20} \\ &\approx \frac{12 \pm 20.5913}{20} \end{aligned}$$

$$\sin x \approx 1.6296 \quad \text{or} \quad \sin x \approx -0.4296.$$

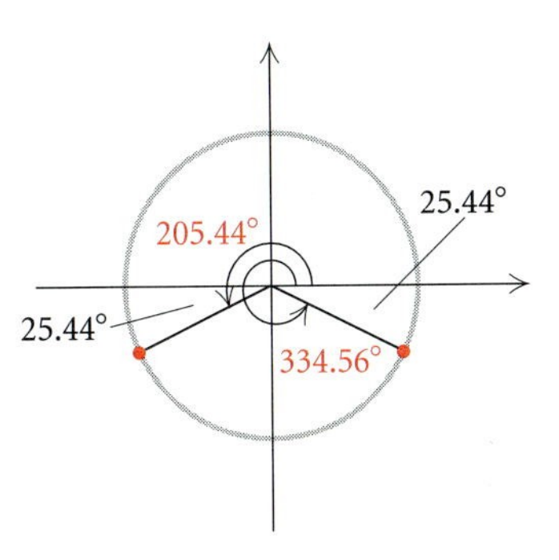

Since sine values are never greater than 1, the first of the equations has no solution. Using the other equation, we find the reference angle to be 25.44°. Since $\sin x$ is negative, the solutions are in quadrants III and IV. Thus the solutions in [0°, 360°) are

$$180° + 25.44° = 205.44° \quad \text{and} \quad 360° - 25.44° = 334.56°.$$

Trigonometric equations can involve more than one function.

Example 8 Solve $2 \cos^2 x \tan x = \tan x$ in $[0, 2\pi)$.

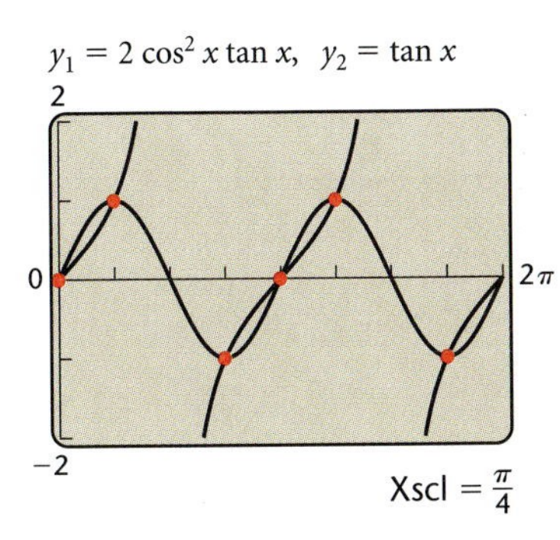

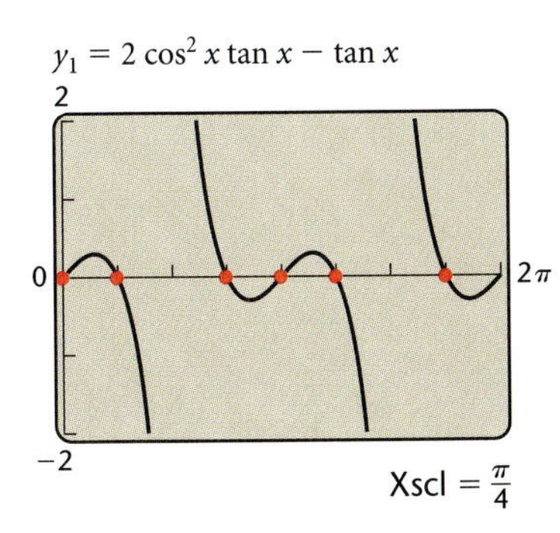

SOLUTION With a grapher, we can determine that there are six solutions. If we let Xscl $= \pi/4$, the solutions are read more easily. In the figures at left, we show two different methods of solving graphically. Each illustrates that the solutions in $[0, 2\pi)$ are

$$0, \quad \frac{\pi}{4}, \quad \frac{3\pi}{4}, \quad \pi, \quad \frac{5\pi}{4}, \quad \text{and} \quad \frac{7\pi}{4}.$$

We can verify these solutions algebraically, as follows:

$$\begin{aligned} 2\cos^2 x \tan x &= \tan x \\ 2\cos^2 x \tan x - \tan x &= 0 \\ \tan x\,(2\cos^2 x - 1) &= 0 \\ \tan x = 0 \quad \textit{or} \quad 2\cos^2 x - 1 &= 0 \\ \cos^2 x &= \frac{1}{2} \\ \cos x &= \pm\frac{\sqrt{2}}{2} \\ x = 0, \pi \quad \textit{or} \quad x &= \frac{\pi}{4}, \frac{3\pi}{4}, \frac{5\pi}{4}, \frac{7\pi}{4}. \end{aligned}$$

These solutions check with those found graphically. Thus, $x = 0$, $\pi/4$, $3\pi/4$, π, $5\pi/4$, and $7\pi/4$.

When a trigonometric equation involves more than one function, it is sometimes helpful to use identities to rewrite the equation in terms of a single function.

Example 9 Solve $\sin x + \cos x = 1$ in $[0, 2\pi)$.

ALGEBRAIC SOLUTION

We have

$$\sin x + \cos x = 1$$

$$\sin^2 x + 2 \sin x \cos x + \cos^2 x = 1$$

Squaring both sides

$$2 \sin x \cos x + 1 = 1$$

Using $\sin^2 x + \cos^2 x = 1$

$$2 \sin x \cos x = 0$$

$$\sin 2x = 0.$$

Using $2 \sin x \cos x = \sin 2x$

We are looking for solutions x to the equation for which $0 \le x < 2\pi$. Multiplying by 2, we get $0 \le 2x < 4\pi$, which is the interval we consider to solve $\sin 2x = 0$. These values of $2x$ are 0, π, 2π, and 3π. Thus the desired values of x in $[0, 2\pi)$ satisfying this equation are 0, $\pi/2$, π, and $3\pi/2$. Now we check these in the original equation:

$$\sin 0 + \cos 0 = 0 + 1 = 1,$$

$$\sin \frac{\pi}{2} + \cos \frac{\pi}{2} = 1 + 0 = 1,$$

$$\sin \pi + \cos \pi = 0 + (-1) = -1,$$

$$\sin \frac{3\pi}{2} + \cos \frac{3\pi}{2} = (-1) + 0 = -1.$$

We find that π and $3\pi/2$ do not check, but the other values do. Thus the solutions are

$$0 \quad \text{and} \quad \frac{\pi}{2}.$$

When the solution process involves squaring both sides, values are often obtained that are not solutions of the original equation. As we saw in this example, it is important to check the possible solutions.

GRAPHICAL SOLUTION

We can graph the left side and then the right side of the equation as seen in the first window below. Then we look for points of intersection. We could also rewrite the equation as $\sin x + \cos x - 1 = 0$, graph the left side, and look for the zeros of the function, as illustrated in the second window below. In each window, we see the solutions in $[0, 2\pi)$ as 0 and $\pi/2$.

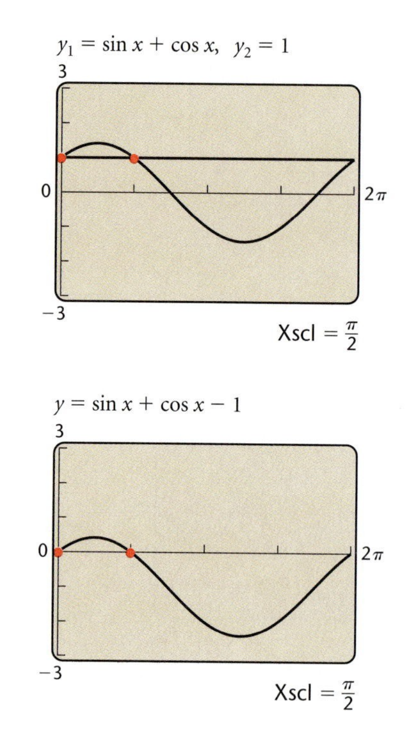

This example illustrates a valuable advantage of the grapher—that is, with a grapher, extraneous solutions do not appear.

Example 10 Solve $\cos 2x + \sin x = 1$ in $[0, 2\pi)$.

ALGEBRAIC SOLUTION

We have

$$\cos 2x + \sin x = 1$$

$$1 - 2\sin^2 x + \sin x = 1 \qquad \text{Using the identity } \cos 2x = 1 - 2\sin^2 x$$

$$-2\sin^2 x + \sin x = 0$$

$$\sin x(1 - 2\sin x) = 0 \qquad \text{Factoring}$$

$$\sin x = 0 \quad \textit{or} \quad 1 - 2\sin x = 0 \qquad \text{Principle of zero products}$$

$$\sin x = 0 \quad \textit{or} \quad \sin x = \frac{1}{2}$$

$$x = 0, \pi \quad \textit{or} \quad x = \frac{\pi}{6}, \frac{5\pi}{6}.$$

All values check. The solutions in $[0, 2\pi)$ are 0, $\pi/6$, $5\pi/6$, and π.

GRAPHICAL SOLUTION

We graph $y_1 = \cos 2x + \sin x - 1$ and look for the zeros of the function.

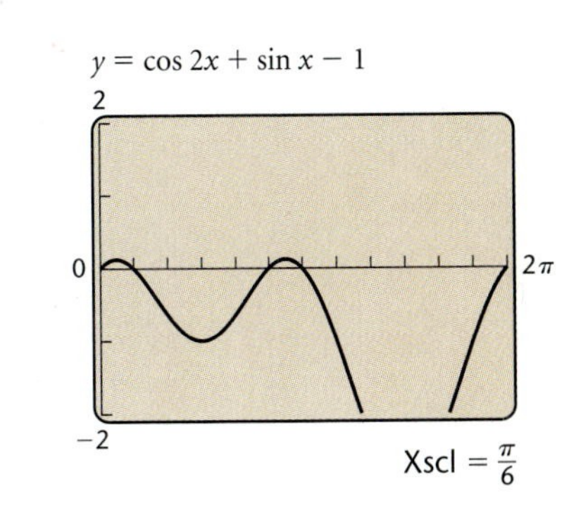

The solutions in $[0, 2\pi)$ are 0, $\pi/6$, $5\pi/6$, and π, or, approximately, 0, 0.524, 2.618, and 3.142.

Example 11 Solve $\tan^2 x + \sec x - 1 = 0$ in $[0, 2\pi)$.

ALGEBRAIC SOLUTION

We have

$$\tan^2 x + \sec x - 1 = 0$$

$$\sec^2 x - 1 + \sec x - 1 = 0 \qquad \text{Using the identity } 1 + \tan^2 x = \sec^2 x$$

$$\sec^2 x + \sec x - 2 = 0$$

$$(\sec x + 2)(\sec x - 1) = 0 \qquad \text{Factoring}$$

$$\sec x = -2 \quad \textit{or} \quad \sec x = 1 \qquad \text{Principle of zero products}$$

$$x = \frac{2\pi}{3}, \frac{4\pi}{3} \quad \textit{or} \quad x = 0$$

All these values check. The solutions in $[0, 2\pi)$ are 0, $2\pi/3$, and $4\pi/3$.

GRAPHICAL SOLUTION

We graph $y = \tan^2 x + \sec x - 1$, but we must enter this equation in the form

$$y_1 = \left(\frac{\sin x}{\cos x}\right)^2 + \left(\frac{1}{\cos x}\right) - 1$$

in some graphers. We look for the zeros of the function.

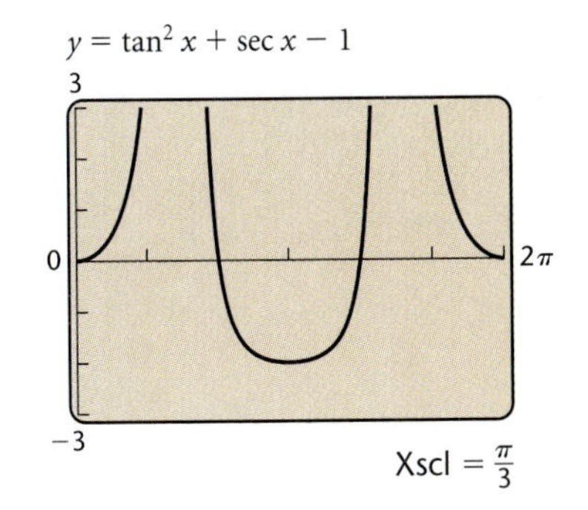

The solutions in $[0, 2\pi)$ are 0, $2\pi/3$, and $4\pi/3$.

Sometimes the only plan of action is to approximate solutions with a grapher.

Example 12 Solve each of the following in $[0, 2\pi)$.

a) $x^2 - 1.5 = \cos x$

b) $\sin x - \cos x = \cot x$

SOLUTION

a) In the screen on the left, we graph $y_1 = x^2 - 1.5$ and $y_2 = \cos x$ and look for points of intersection. In the screen on the right, we graph $y_1 = x^2 - 1.5 - \cos x$ and look for the zeros of the function.

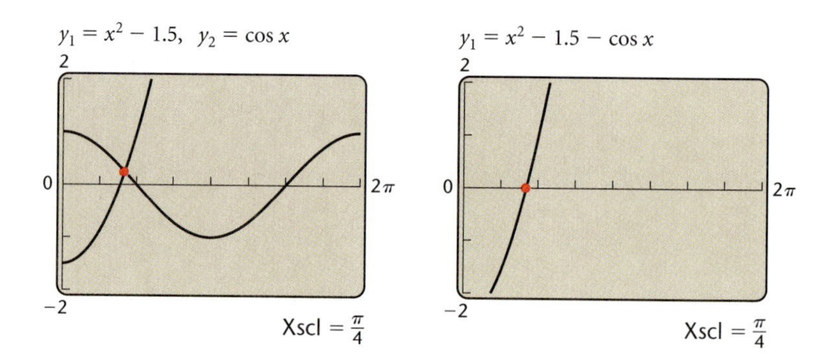

We determine the solution in $[0, 2\pi)$ to be approximately 1.32.

b) In the screen on the left, we graph $y_1 = \sin x - \cos x$ and $y_2 = \cot x$ and determine the points of intersection. In the screen on the right, we graph the function $y_1 = \sin x - \cos x - \cot x$ and determine the zeros.

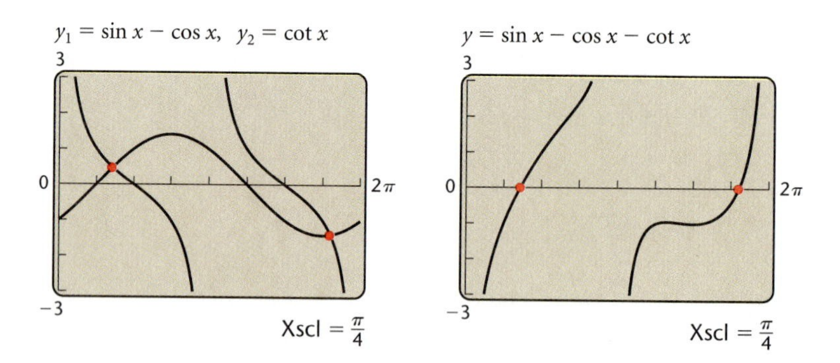

Each method leads to the approximate solutions 1.13 and 5.66 in $[0, 2\pi)$.

5.5 Exercise Set

Solve, finding all solutions. Express the solutions in both radians and degrees.

1. $\cos x = \frac{\sqrt{3}}{2}$

2. $\sin x = -\frac{\sqrt{2}}{2}$

3. $\tan x = -\sqrt{3}$

4. $\cos x = \frac{1}{2}$

Solve, finding all solutions in $[0, 2\pi)$ or $[0°, 360°)$. *Verify your answer with a grapher.*

5. $2 \cos x - 1 = -1.2814$

6. $\sin x + 3 = 2.0816$

7. $2 \sin x + \sqrt{3} = 0$

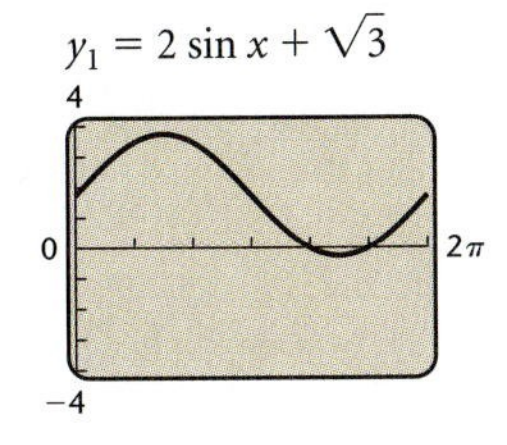

8. $2 \tan x - 4 = 1$

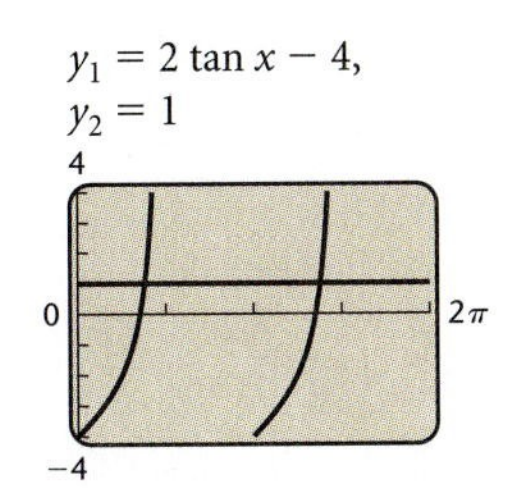

9. $2 \cos^2 x = 1$

10. $\csc^2 x - 4 = 0$

11. $2 \sin^2 x + \sin x = 1$

12. $\cos^2 x + 2 \cos x = 3$

13. $2 \cos^2 x - \sqrt{3} \cos x = 0$

14. $2 \sin^2 \theta + 7 \sin \theta = 4$

15. $6 \cos^2 \phi + 5 \cos \phi + 1 = 0$

16. $2 \sin t \cos t + 2 \sin t - \cos t - 1 = 0$

17. $\sin 2x \cos x - \sin x = 0$

18. $5 \sin^2 x - 8 \sin x = 3$

19. $\cos^2 x + 6 \cos x + 4 = 0$

20. $2 \tan^2 x = 3 \tan x + 7$

21. $7 = \cot^2 x + 4 \cot x$

22. $3 \sin^2 x = 3 \sin x + 2$

Solve, finding all solutions in $[0, 2\pi)$. *Then check each solution with a grapher.*

23. $\cos 2x - \sin x = 1$

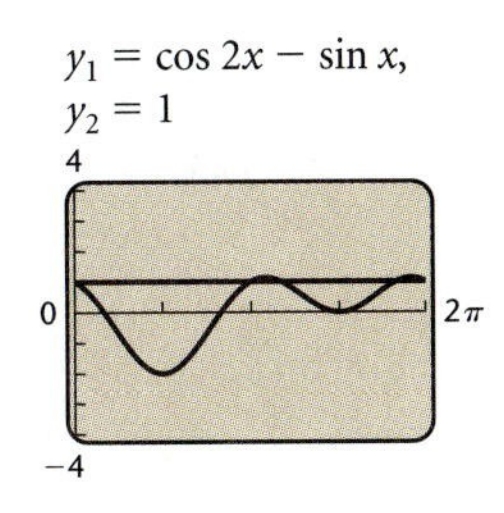

24. $2 \sin x \cos x + \sin x = 0$

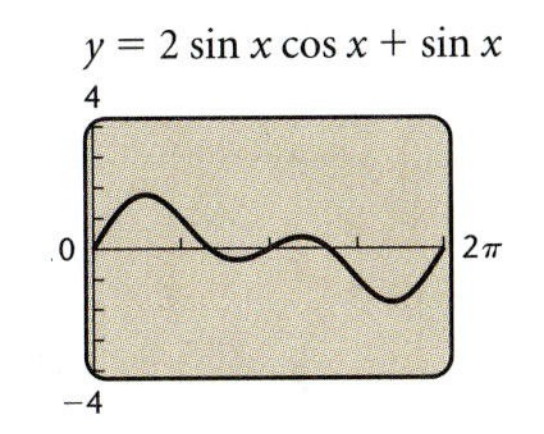

25. $\sin 4x - 2 \sin 2x = 0$

26. $\tan x \sin x - \tan x = 0$

27. $\sin 2x \cos x + \sin x = 0$

28. $\cos 2x \sin x + \sin x = 0$

29. $2 \sec x \tan x + 2 \sec x + \tan x + 1 = 0$

30. $\sin 2x \sin x - \cos 2x \cos x = -\cos x$

31. $\sin 2x + \sin x + 2 \cos x + 1 = 0$

32. $\tan^2 x + 4 = 2 \sec^2 x + \tan x$

33. $\sec^2 x - 2 \tan^2 x = 0$

34. $\cot x = \tan (2x - 3\pi)$

35. $2 \cos x + 2 \sin x = \sqrt{6}$

36. $\sqrt{3} \cos x - \sin x = 1$

37. $\sec^2 x + 2 \tan x = 6$

38. $5 \cos 2x + \sin x = 4$

39. $\cos (\pi - x) + \sin \left(x - \frac{\pi}{2}\right) = 1$

40. $\dfrac{\sin^2 x - 1}{\cos \left(\frac{\pi}{2} - x\right) + 1} = \frac{\sqrt{2}}{2} - 1$

Solve with a grapher, finding all solutions in $[0, 2\pi)$.

41. $x \sin x = 1$

42. $x^2 + 2 = \sin x$

43. $2 \cos^2 x = x + 1$

44. $x \cos x - 2 = 0$

45. $\cos x - 2 = x^2 - 3x$

46. $\sin x = \tan \frac{x}{2}$

Skill Maintenance

Solve the right triangle.

47.

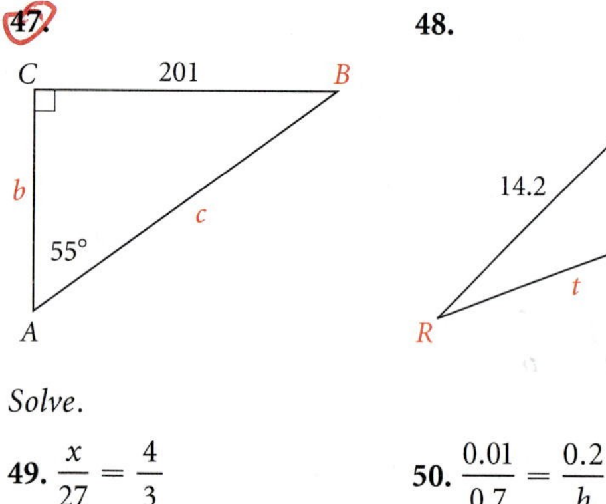

48.

T
3.8
14.2
S
t
R

Solve.

49. $\frac{x}{27} = \frac{4}{3}$

50. $\frac{0.01}{0.7} = \frac{0.2}{h}$

Synthesis

51. ◆ Jan lists her answer to a problem as $\pi/6 + k\pi$, for any integer k, while Jacob lists his answer as $\pi/6 + 2k\pi$ and $7\pi/6 + 2\pi k$, for any integer k. Are their answers equivalent? Why or why not?

52. ◆ Under what circumstances will a grapher give exact solutions of a trigonometric equation?

Solve in $[0, 2\pi)$. *Check your results with a grapher.*

53. $|\sin x| = \frac{\sqrt{3}}{2}$

54. $|\cos x| = \frac{1}{2}$

55. $\sqrt{\tan x} = \sqrt[4]{3}$

56. $12 \sin x - 7\sqrt{\sin x} + 1 = 0$

57. $\ln (\cos x) = 0$

58. $e^{\sin x} = 1$

59. $\sin (\ln x) = -1$

60. $e^{\ln (\sin x)} = 1$

61. *Temperature During an Illness.* The temperature T, in degrees Fahrenheit, of a patient t days into a 12-day illness is given by

$$T(t) = 101.6° + 3° \sin\left(\frac{\pi}{8} t\right).$$

Find the times t during the illness at which the patient's temperature was 103°.

62. *Satellite Location.* A satellite circles the earth in such a manner that it is y miles from the equator (north or south, height from the surface not considered) t minutes after its launch, where

$$y = 5000\left[\cos \frac{\pi}{45}(t - 10)\right].$$

At what times t in the interval $[0, 240]$, the first 4 hr, is the satellite 3000 mi north of the equator?

63. *Nautical Mile.* (See Exercise 52 in Exercise Set 5.2.) In Great Britain, the *nautical mile* is defined as the length of a minute of arc of the earth's radius. Since the earth is flattened at the poles, a British nautical mile varies with latitude. In fact, it is given, in feet, by the function

$$N(\phi) = 6066 - 31 \cos 2\phi,$$

where ϕ is the latitude in degrees. At what latitude north is the length of a British nautical mile found to be 6040 ft?

64. *Acceleration Due to Gravity.* (See Exercise 53 in Exercise Set 5.2.) The acceleration due to gravity is often denoted by g in a formula such as $S = \frac{1}{2}gt^2$, where $S =$ the distance that an object falls in t seconds. The number g is generally considered constant, but in fact it varies slightly with latitude. If ϕ stands for latitude, in degrees, an excellent approximation of g is given by the formula

$$g = 9.78049(1 + 0.005288 \sin^2 \phi - 0.000006 \sin^2 2\phi),$$

where g is measured in meters per second per second at sea level. At what latitude north does $g = 9.8$?

Solve.

65. $\arccos x = \arccos \frac{3}{5} - \arcsin \frac{4}{5}$

66. $\sin^{-1} x = \tan^{-1} \frac{1}{3} + \tan^{-1} \frac{1}{2}$

67. Suppose that $\sin x = 5 \cos x$. Find $\sin x \cos x$.

CHAPTER 5 Summary and Review

Important Properties and Formulas

Basic Identities

$\sin x = \dfrac{1}{\csc x}$, $\quad \sin(-x) = -\sin x$,

$\cos(-x) = \cos x$,

$\cos x = \dfrac{1}{\sec x}$, $\quad \tan(-x) = -\tan x$

$\tan x = \dfrac{1}{\cot x}$,

$\tan x = \dfrac{\sin x}{\cos x}$,

$\cot x = \dfrac{\cos x}{\sin x}$,

Pythagorean Identities

$\sin^2 x + \cos^2 x = 1$,

$1 + \cot^2 x = \csc^2 x$,

$1 + \tan^2 x = \sec^2 x$

Sum and Difference Identities

$\sin(u \pm v) = \sin u \cos v \pm \cos u \sin v$,

$\cos(u \pm v) = \cos u \cos v \mp \sin u \sin v$,

$$\tan(u \pm v) = \frac{\tan u \pm \tan v}{1 \mp \tan u \tan v}$$

Cofunction Identities

$\sin\left(\dfrac{\pi}{2} - x\right) = \cos x$,

$\tan\left(\dfrac{\pi}{2} - x\right) = \cot x$,

$\sec\left(\dfrac{\pi}{2} - x\right) = \csc x$,

$\sin\left(x \pm \dfrac{\pi}{2}\right) = \pm\cos x$,

$\cos\left(x \pm \dfrac{\pi}{2}\right) = \mp \sin x$

Double-Angle Identities

$\sin 2x = 2 \sin x \cos x$,

$$\begin{aligned}\cos 2x &= \cos^2 x - \sin^2 x \\ &= 1 - 2\sin^2 x \\ &= 2\cos^2 x - 1,\end{aligned}$$

$$\tan 2x = \frac{2 \tan x}{1 - \tan^2 x}$$

Half-Angle Identities

$\sin \dfrac{x}{2} = \pm\sqrt{\dfrac{1 - \cos x}{2}}$,

$\cos \dfrac{x}{2} = \pm\sqrt{\dfrac{1 + \cos x}{2}}$,

$$\begin{aligned}\tan \frac{x}{2} &= \pm\sqrt{\frac{1 - \cos x}{1 + \cos x}}, \\ &= \frac{\sin x}{1 + \cos x} \\ &= \frac{1 - \cos x}{\sin x}\end{aligned}$$

Inverse Trigonometric Functions

FUNCTION	DOMAIN	RANGE
$y = \sin^{-1} x$	$[-1, 1]$	$\left[-\dfrac{\pi}{2}, \dfrac{\pi}{2}\right]$
$y = \cos^{-1} x$	$[-1, 1]$	$[0, \pi]$
$y = \tan^{-1} x$	$(-\infty, \infty)$	$\left(-\dfrac{\pi}{2}, \dfrac{\pi}{2}\right)$

(continued)

Composition of Trigonometric Functions

The following are true for any x in the domain of the inverse function:

$$\sin(\sin^{-1} x) = x,$$
$$\cos(\cos^{-1} x) = x,$$
$$\tan(\tan^{-1} x) = x.$$

The following are true for any x in the range of the inverse function:

$$\sin^{-1}(\sin x) = x,$$
$$\cos^{-1}(\cos x) = x,$$
$$\tan^{-1}(\tan x) = x.$$

REVIEW EXERCISES

Complete the Pythagorean identity.

1. $1 + \cot^2 x =$
2. $\sin^2 x + \cos^2 x =$

Multiply and simplify. Check with a grapher.

3. $(\tan y - \cot y)(\tan y + \cot y)$
4. $(\cos x + \sec x)^2$

Factor and simplify. Check with a grapher.

5. $\sec x \csc x - \csc^2 x$
6. $3 \sin^2 y - 7 \sin y - 20$
7. $1000 - \cos^3 u$

Simplify and check with a grapher.

8. $\dfrac{\sec^4 x - \tan^4 x}{\sec^2 x + \tan^2 x}$
9. $\dfrac{2 \sin^2 x}{\cos^3 x} \cdot \left(\dfrac{\cos x}{2 \sin x}\right)^2$
10. $\dfrac{3 \sin x}{\cos^2 x} \cdot \dfrac{\cos^2 x + \cos x \sin x}{\sin^2 x - \cos^2 x}$
11. $\dfrac{3}{\cos y - \sin y} - \dfrac{2}{\sin^2 y - \cos^2 y}$
12. $\left(\dfrac{\cot x}{\csc x}\right)^2 + \dfrac{1}{\csc^2 x}$
13. $\dfrac{4 \sin x \cos^2 x}{16 \sin^2 x \cos x}$
14. Simplify. Assume the radicand is nonnegative.
$$\sqrt{\sin^2 x + 2 \cos x \sin x + \cos^2 x}$$
15. Rationalize the denominator: $\sqrt{\dfrac{1 + \sin x}{1 - \sin x}}$.
16. Rationalize the numerator: $\sqrt{\dfrac{\cos x}{\tan x}}$.
17. Given that $x = 3 \tan \theta$, express $\sqrt{9 + x^2}$ as a trigonometric function without radicals. Assume that $0 < \theta < \pi/2$.

Use the sum and difference formulas to write equivalent expressions. You need not simplify.

18. $\cos\left(x + \dfrac{3\pi}{2}\right)$
19. $\tan(45° - 30°)$
20. Simplify: $\cos 27° \cos 16° + \sin 27° \sin 16°$.
21. Find $\cos 165°$ exactly.
22. Given that $\tan \alpha = \sqrt{3}$ and $\sin \beta = \sqrt{2}/2$ and that α and β are between 0 and $\pi/2$, evaluate $\tan(\alpha - \beta)$ exactly.
23. Assume that $\sin \theta = 0.5812$ and $\cos \phi = 0.2341$ and that both θ and ϕ are first-quadrant angles. Evaluate $\cos(\theta + \phi)$.

Complete the cofunction identity.

24. $\cos\left(x + \dfrac{\pi}{2}\right) =$
25. $\cos\left(\dfrac{\pi}{2} - x\right) =$
26. $\sin\left(x - \dfrac{\pi}{2}\right) =$
27. Given that $\cos \alpha = -\frac{3}{5}$ and that the terminal side is in quadrant III:
    a) Find the other function values for α.
    b) Find the six function values for $\pi/2 - \alpha$.
    c) Find the six function values for $\alpha + \pi/2$.
28. Find an equivalent expression for $\csc\left(x - \dfrac{\pi}{2}\right)$.
29. Find $\tan 2\theta$, $\cos 2\theta$, and $\sin 2\theta$ and the quadrant in which 2θ lies, where $\cos \theta = -\frac{4}{5}$ and θ is in quadrant III.

30. Find $\sin \frac{\pi}{8}$ exactly.

31. Given that $\sin \beta = 0.2183$ and β is in quadrant I, find $\sin 2\beta$, $\cos \frac{\beta}{2}$, and $\cos 4\beta$.

Simplify and check with a grapher.

32. $1 - 2\sin^2 \frac{x}{2}$

33. $(\sin x + \cos x)^2 - \sin 2x$

34. $2 \sin x \cos^3 x + 2 \sin^3 x \cos x$

35. $\dfrac{2 \cot x}{\cot^2 x - 1}$

Prove each of the following identities. Check with a grapher.

36. $\dfrac{1 - \sin x}{\cos x} = \dfrac{\cos x}{1 + \sin x}$

37. $\dfrac{1 + \cos 2\theta}{\sin 2\theta} = \cot \theta$

38. $\dfrac{\tan y + \sin y}{2 \tan \theta} = \cos^2 \dfrac{y}{2}$

39. $\dfrac{\sin x - \cos x}{\cos^2 x} = \dfrac{\tan^2 x - 1}{\sin x + \cos x}$

In Exercises 40–43, use a grapher to determine which expression (A)–(D) on the right can be used to complete an identity. Then prove the identity algebraically.

40. $\csc x - \cos x \cot x$

41. $\dfrac{1}{\sin x \cos x} - \dfrac{\cos x}{\sin x}$

42. $\dfrac{\cot x - 1}{1 - \tan x}$

43. $\dfrac{\cos x + 1}{\sin x} + \dfrac{\sin x}{\cos x + 1}$

A. $\dfrac{\csc x}{\sec x}$

B. $\sin x$

C. $\dfrac{2}{\sin x}$

D. $\dfrac{\sin x \cos x}{1 - \sin^2 x}$

Find each of the following exactly in both radians and degrees.

44. $\sin^{-1}\left(-\frac{1}{2}\right)$

45. $\cos^{-1} \frac{\sqrt{3}}{2}$

46. $\arctan 1$

47. $\arcsin 0$

Use a grapher to find each of the following in radians, rounded to three decimal places, and in degrees, rounded to the nearest tenth of a degree.

48. $\arccos(-0.2194)$

49. $\cot^{-1} 2.381$

Evaluate.

50. $\cos\left(\cos^{-1} \frac{1}{2}\right)$

51. $\tan^{-1}\left(\tan \frac{\sqrt{3}}{3}\right)$

52. $\sin^{-1}\left(\sin \frac{\pi}{7}\right)$

53. $\cos\left(\arcsin \frac{\sqrt{2}}{2}\right)$

Find.

54. $\cos\left(\arctan \frac{b}{3}\right)$

55. $\cos\left(2 \sin^{-1} \frac{4}{5}\right)$

Solve, finding all solutions. Express the solutions in both radians and degrees.

56. $\cos x = -\frac{\sqrt{2}}{2}$

57. $\tan x = \sqrt{3}$

Solve, finding all solutions in $[0, 2\pi)$.

58. $4 \sin^2 x = 1$

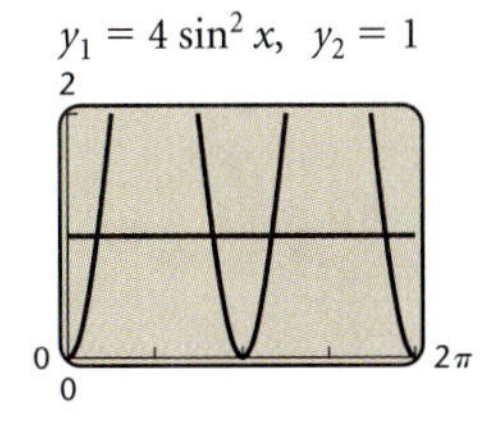

59. $\sin 2x \sin x - \cos x = 0$

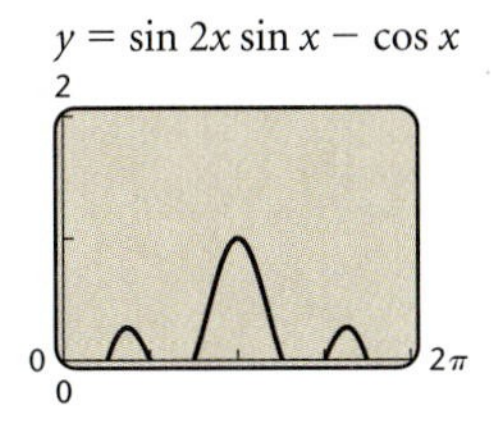

60. $2\cos^2 x + 3 \cos x = -1$

61. $\sin^2 x - 7 \sin x = 0$

62. $\csc^2 x - 2 \cot^2 x = 0$

63. $\sin 4x + 2 \sin 2x = 0$

64. $2 \cos x + 2 \sin x = \sqrt{2}$

65. $6 \tan^2 x = 5 \tan x + \sec^2 x$

Solve with a grapher, finding all solutions in $[-2\pi, 2\pi]$.

66. $x \cos x = 1$

67. $2 \sin^2 x = x + 1$

Synthesis

68. ◆ Prove the identity $2 \cos^2 x - 1 = \cos^4 x - \sin^4 x$ in three ways:

a) Start with the left side and deduce the right (method 1).

b) Start with the right side and deduce the left (method 1).

c) Work with each side separately until you deduce the same expression (method 2).

Then determine the most efficient method and explain why.

69. ◆ Why are the ranges of the inverse trigonometric functions restricted?

70. ◆ Explain why $\tan(x + 450°)$ cannot be simplified using the tangent sum formula, but can be simplified using the sine and cosine sum formulas.

71. Find the measure of the angle from l_1 to l_2:

$$l_1: x + y = 3 \qquad l_2: 2x - y = 5.$$

72. Find an identity for $\cos(u + v)$ involving only cosines.

73. Simplify: $\cos\left(\frac{\pi}{2} - x\right)[\csc x - \sin x]$.

74. Find $\sin \theta$, $\cos \theta$, and $\tan \theta$ under the given conditions:

$$\sin 2\theta = \frac{1}{5}, \quad \frac{\pi}{2} \le 2\theta < \pi.$$

75. Prove the following equation to be an identity and verify the results with a grapher:

$$\ln e^{\sin t} = \sin t.$$

76. Graph: $y = \sec^{-1} x$.

77. Show that

$$\tan^{-1} x = \frac{\sin^{-1} x}{\cos^{-1} x}$$

is *not* an identity.

78. Solve $e^{\cos x} = 1$ in $[0, 2\pi)$. Check with a grapher.

Applications of Trigonometry

6

APPLICATION

During a rescue mission, a Marine fighter pilot receives data on an unidentified aircraft from an AWACS plane and is instructed to intercept the enemy plane. The diagram shown below pops up on the screen. The law of sines,

$$\frac{x}{\sin X} = \frac{y}{\sin Y} = \frac{z}{\sin Z},$$

can be used to determine how far the pilot must fly.

Triangle trigonometry is important in applications such as large-scale construction, navigation, and surveying. In this chapter, we continue the study of triangle trigonometry that we began in Chapter 4. We will find that the trigonometric functions can also be used to solve triangles that are not right triangles.

The study of complex numbers begun in Chapter 2 is continued in this chapter. Complex numbers have applications in both electronics and engineering. We also introduce the polar coordinate system and graphs of polar equations.

This chapter also introduces the idea of a vector, which is related to the study of triangles. A vector is a quantity that has a direction. Vectors have many practical applications in the physical sciences.

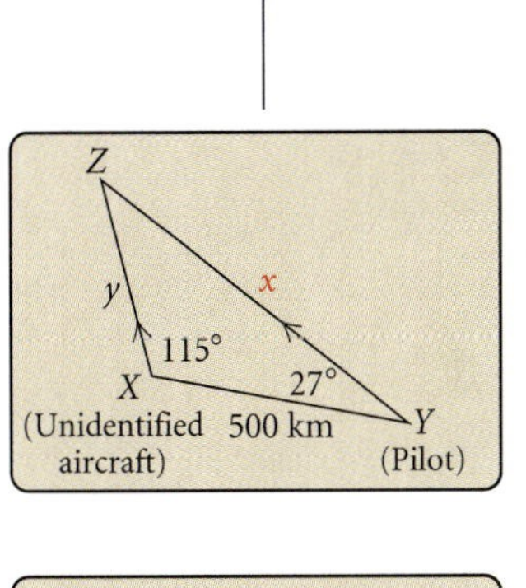

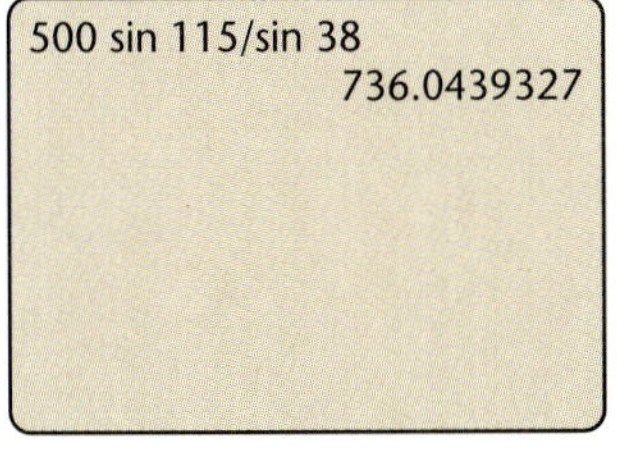

6.1 The Law of Sines
6.2 The Law of Cosines
6.3 Complex Numbers: Trigonometric Form
6.4 Polar Coordinates and Graphs
6.5 Vectors
SUMMARY AND REVIEW

6.1 The Law of Sines

- *Use the law of sines to solve triangles.*
- *Find the area of any triangle given the lengths of two sides and the measure of the included angle.*

To **solve a triangle** means to find the lengths of all its sides and the measures of all its angles. We solved right triangles in Section 4.2. For review, let's solve the right triangle shown below. We begin by listing the known measures.

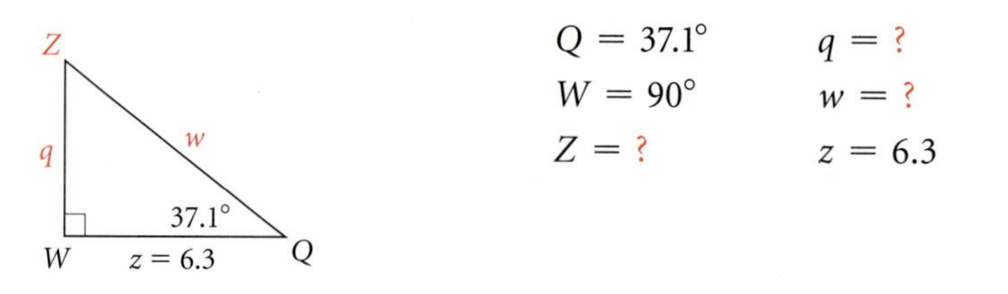

Since the sum of the three angle measures of any triangle is 180°, we can immediately find the third angle:

$$Z = 180° - (90° + 37.1°) = 52.9°.$$

Then using the tangent and cosine ratios, respectively, we can find q and w:

$$\tan 37.1° = \frac{q}{6.3}, \quad \text{or} \quad q = 6.3 \tan 37.1° \approx 4.8,$$

and

$$\cos 37.1° = \frac{6.3}{w}, \quad \text{or} \quad w = \frac{6.3}{\cos 37.1°} \approx 7.9.$$

Now all six measures are known and we have solved triangle QWZ.

$Q = 37.1°$	$q \approx 4.8$
$W = 90°$	$w \approx 7.9$
$Z = 52.9°$	$z = 6.3$

Solving Oblique Triangles

The trigonometric functions can also be used to solve triangles that are not right triangles. Such triangles are called **oblique**. Any triangle, right or oblique, can be solved *if at least one side and any other two measures are known.* The five possible situations are illustrated below.

1. AAS: Two angles of a triangle and a side opposite one of them are known.

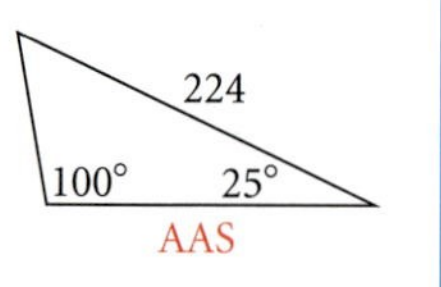

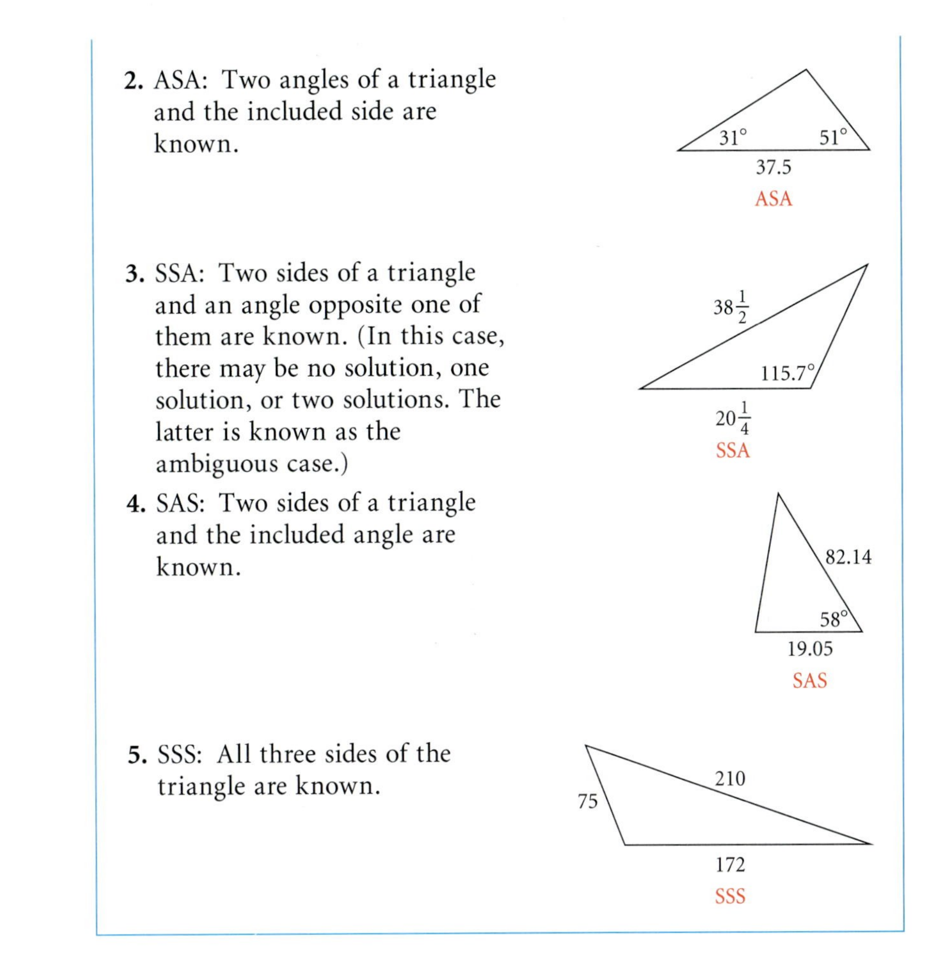

2. ASA: Two angles of a triangle and the included side are known.
3. SSA: Two sides of a triangle and an angle opposite one of them are known. (In this case, there may be no solution, one solution, or two solutions. The latter is known as the ambiguous case.)
4. SAS: Two sides of a triangle and the included angle are known.
5. SSS: All three sides of the triangle are known.

The list above does not include the situation in which only the three angle measures are given. The reason for this lies in the fact that the angle measures determine *only the shape* of the triangle and *not the size*, as shown with the following triangles.

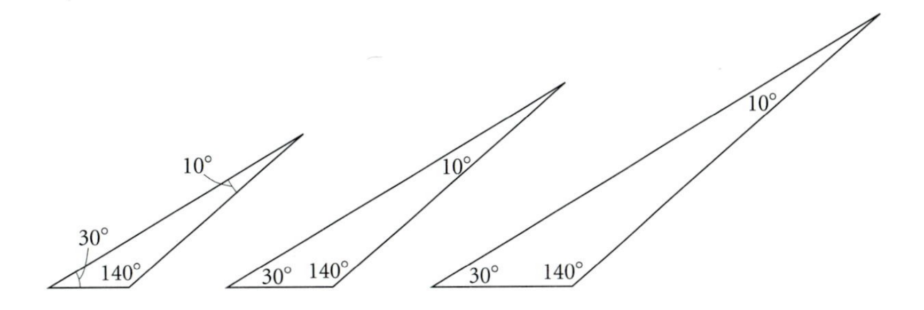

Thus we cannot solve a triangle when only the three angle measures are given.

In order to solve oblique triangles, we need to derive the *law of sines* and the *law of cosines*. The law of sines applies to the first three situations listed above. The law of cosines, which we develop in Section 6.2, applies to the last two situations.

The Law of Sines

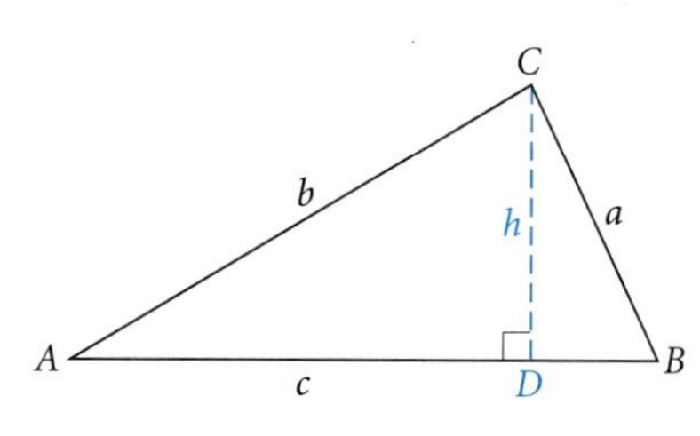

We consider any oblique triangle. It may or may not have an obtuse angle. Although we look at only the acute-triangle case, the derivation of the obtuse-triangle case is essentially the same.

In acute $\triangle ABC$ above, we have drawn an altitude from vertex C. It has length h. From $\triangle ADC$, we have

$$\sin A = \frac{h}{b}, \quad \text{or} \quad h = b \sin A.$$

From $\triangle BDC$, we have

$$\sin B = \frac{h}{a}, \quad \text{or} \quad h = a \sin B.$$

With $h = b \sin A$ and $h = a \sin B$, we now have

$$a \sin B = b \sin A$$

$$\frac{a \sin B}{\sin A \sin B} = \frac{b \sin A}{\sin A \sin B} \qquad \text{Dividing by } \sin A \sin B$$

$$\frac{a}{\sin A} = \frac{b}{\sin B}. \qquad \text{Simplifying}$$

There is no danger of dividing by 0 here because we are dealing with triangles whose angles are never 0° or 180°. Thus the sine value will never be 0.

If we were to consider altitudes from vertex A and vertex B in the triangle shown above, the same argument would give us

$$\frac{b}{\sin B} = \frac{c}{\sin C} \quad \text{and} \quad \frac{a}{\sin A} = \frac{c}{\sin C}.$$

We combine these results to obtain the law of sines.

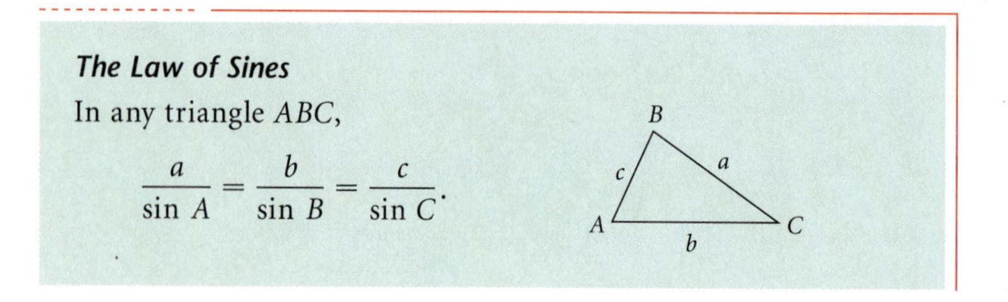

The Law of Sines

In any triangle ABC,

$$\frac{a}{\sin A} = \frac{b}{\sin B} = \frac{c}{\sin C}.$$

In any triangle, the sides are proportional to the sines of the opposite angles.

Solving Triangles (AAS and ASA)

When two angles and a side of any triangle are known, the law of sines can be used to solve the triangle.

Example 1 In $\triangle EFG$, $e = 4.56$, $E = 43°$, and $G = 57°$. Solve the triangle.

SOLUTION We first draw a sketch. We know three of the six measures.

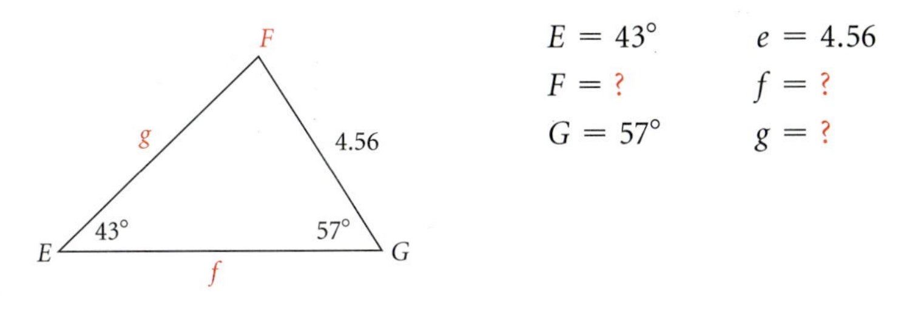

From the figure, we see that we have the AAS situation. We begin by finding F:

$$F = 180° - (43° + 57°) = 80°.$$

We can now find the other two sides, using the law of sines:

$$\frac{f}{\sin F} = \frac{e}{\sin E}$$

$$\frac{f}{\sin 80°} = \frac{4.56}{\sin 43°} \qquad \text{Substituting}$$

$$f = \frac{4.56 \sin 80°}{\sin 43°} \qquad \text{Solving for } f$$

$$f \approx 6.58;$$

$$\frac{g}{\sin G} = \frac{e}{\sin E}$$

$$\frac{g}{\sin 57°} = \frac{4.56}{\sin 43°} \qquad \text{Substituting}$$

$$g = \frac{4.56 \sin 57°}{\sin 43°} \qquad \text{Solving for } g$$

$$g \approx 5.61.$$

Thus, we have found the following to solve the triangle:

$$E = 43°, \qquad e = 4.56,$$
$$F = 80°, \qquad f \approx 6.58,$$
$$G = 57°, \qquad g \approx 5.61.$$

The law of sines is frequently used in determining distances.

Example 2 *Rescue Mission.* During a rescue mission, a Marine fighter pilot receives data on an unidentified aircraft from an AWACS plane and is instructed to intercept the aircraft. The diagram shown below pops up on the screen, but before the distance to the point of interception appears on the screen, communications are jammed. Fortunately, the pilot remembers the law of sines. How far must the pilot fly?

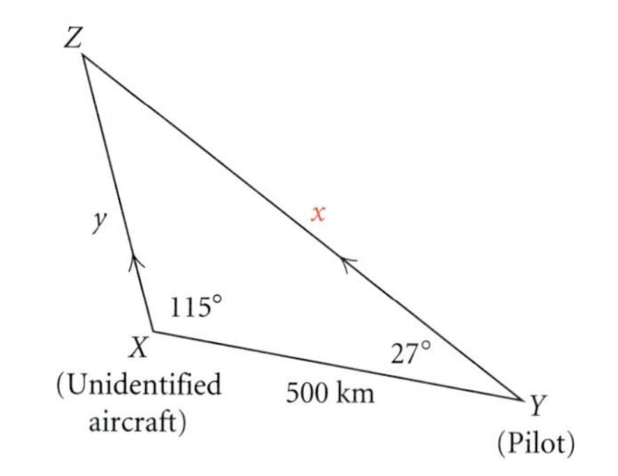

SOLUTION We let x represent the distance that the pilot must fly in order to intercept the aircraft and Z represent the point of interception. We first find angle Z:

$$\begin{aligned} Z &= 180° - (115° + 27°) \\ &= 38°. \end{aligned}$$

Because this application involves the ASA situation, we use the law of sines to determine x:

$$\frac{x}{\sin X} = \frac{z}{\sin Z}$$

$$\frac{x}{\sin 115°} = \frac{500}{\sin 38°} \quad \text{Substituting}$$

$$x = \frac{500 \sin 115°}{\sin 38°} \quad \text{Solving for } x$$

$$x \approx 736.$$

Thus the pilot must fly approximately 736 km in order to intercept the unidentified aircraft.

Solving Triangles (SSA)

When two sides of a triangle and an angle opposite one of them are known, the law of sines can be used to solve the triangle.

Suppose for $\triangle ABC$ that b, c, and B are given. The various possibilities are as shown in the eight cases below: 5 cases when B is acute and 3 cases when B is obtuse. Note that $b < c$ in cases 1, 2, 3, and 6; $b = c$ in cases 4 and 7; and $b > c$ in cases 5 and 8.

Angle B Is Acute

Case 1: No solution
$b < c$; side b is too short to reach the base. No triangle is formed.

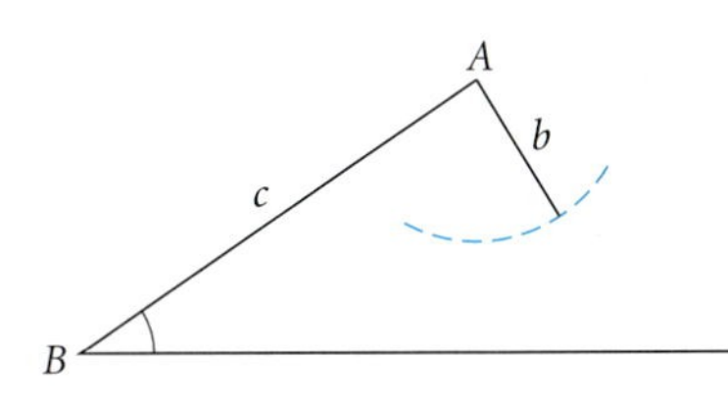

Case 2: One solution
$b < c$; side b just reaches the base and is perpendicular to it.

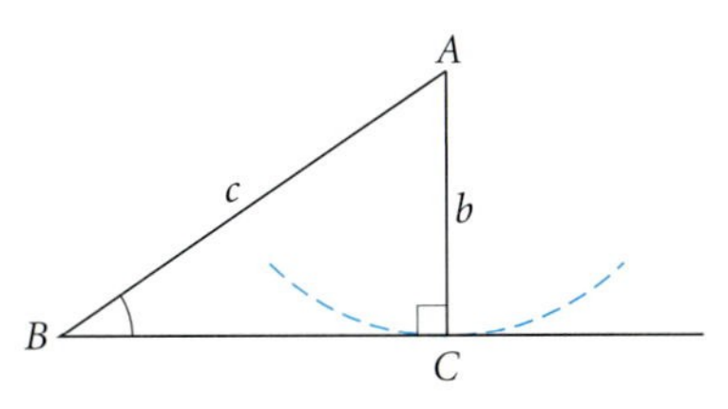

Case 3: Two solutions
$b < c$; an arc of radius b meets the base at two points. (This case is called the *ambiguous case*.)

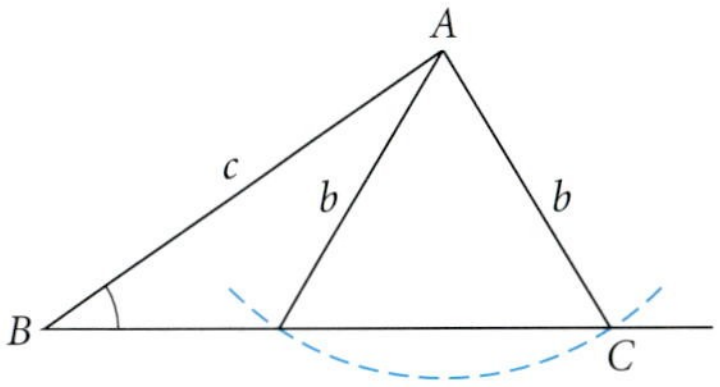

Case 4: One solution
$b = c$; an arc of radius b meets the base at just one point, other than B.

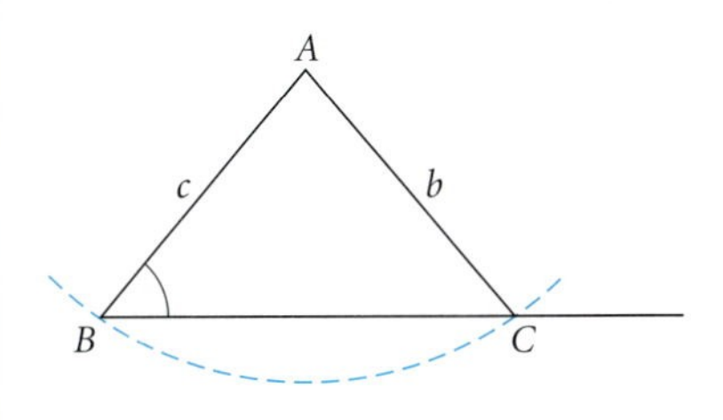

Case 5: One solution
$b > c$; an arc of radius b meets the base at just one point.

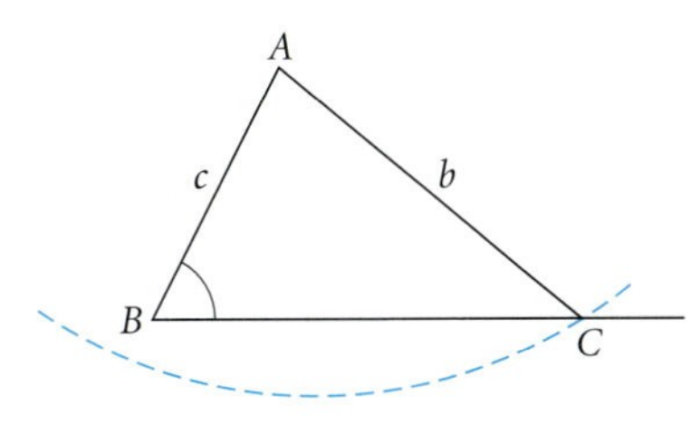

Angle B Is Obtuse

Case 6: No solution
$b < c$; side b is too short to reach the base. No triangle is formed.

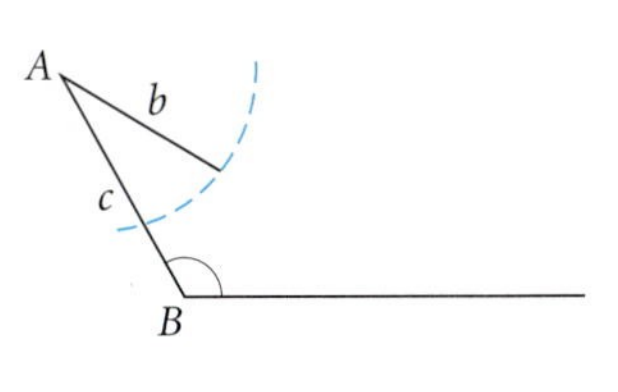

Case 7: No solution
$b = c$; an arc of radius b meets the base only at point B. No triangle is formed.

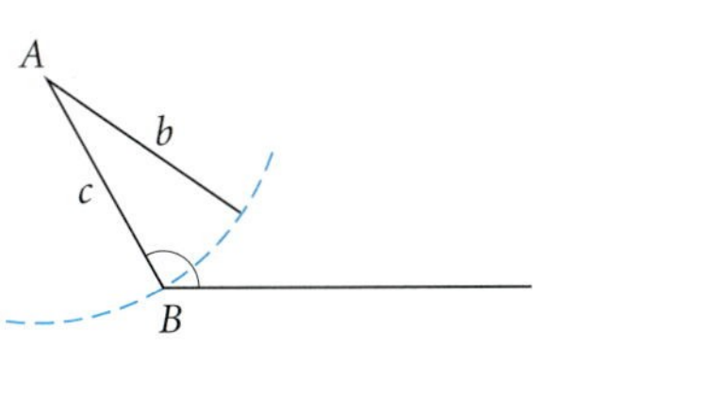

Case 8: One solution
$b > c$; an arc of radius b meets the base at just one point.

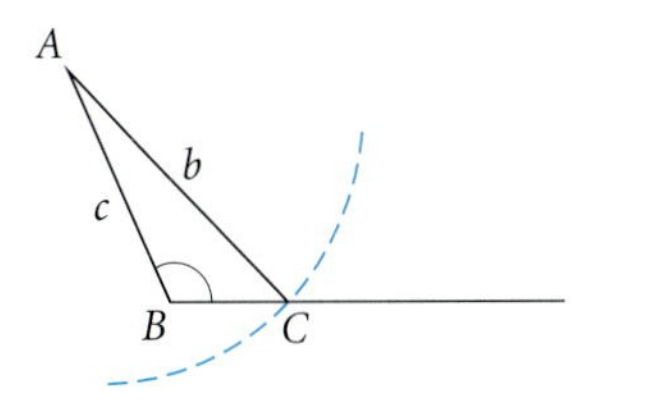

The eight cases above lead us to three possibilities in the SSA situation: *no* solution, *one* solution, or *two* solutions. Since there are two solutions in case 3, this possibility is referred to as the **ambiguous case**. Let's investigate these possibilities further, looking for ways to recognize the number of solutions.

Example 3 *No solution.* In $\triangle QRS$, $q = 15$, $r = 28$, and $Q = 43.6°$. Solve the triangle.

SOLUTION We make a drawing and list the known measures.

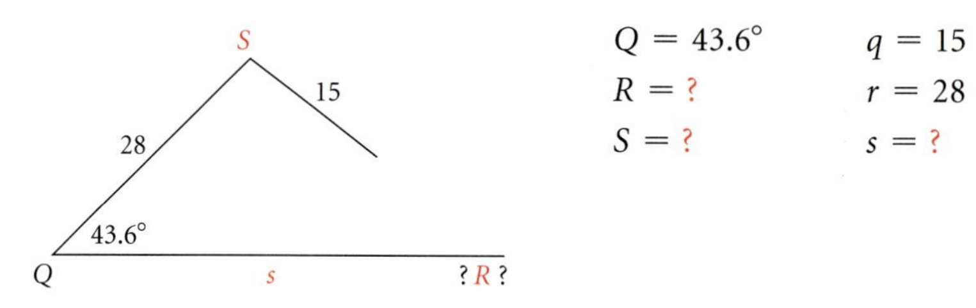

$Q = 43.6°$ $\quad q = 15$
$R = ?$ $\quad r = 28$
$S = ?$ $\quad s = ?$

We observe the SSA situation and use the law of sines to find R:

$$\frac{q}{\sin Q} = \frac{r}{\sin R}$$

$$\frac{15}{\sin 43.6°} = \frac{28}{\sin R} \qquad \text{Substituting}$$

$$\sin R = \frac{28 \sin 43.6°}{15} \qquad \text{Solving for } \sin R$$

$$\sin R \approx 1.2873.$$

Since there is no angle with a sine greater than 1, there is *no solution.*

Example 4 *One solution.* In $\triangle XYZ$, $x = 23.5$, $y = 9.8$, and $X = 39.7°$. Solve the triangle.

SOLUTION We make a drawing and organize the given information.

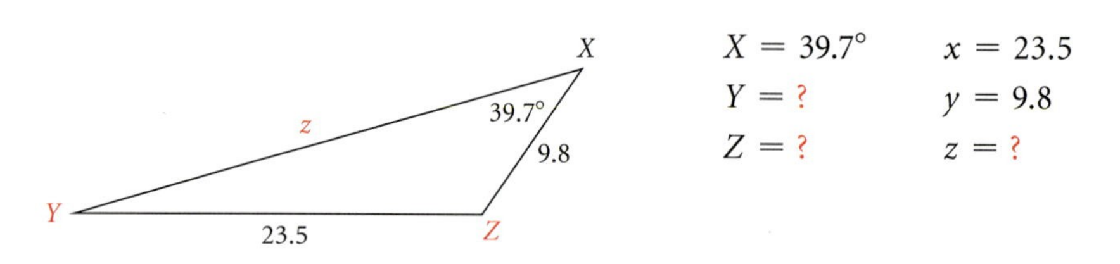

$X = 39.7°$ $\quad x = 23.5$
$Y = ?$ $\quad y = 9.8$
$Z = ?$ $\quad z = ?$

We see the SSA situation and begin by finding Y with the law of sines:

$$\frac{x}{\sin X} = \frac{y}{\sin Y}$$

$$\frac{23.5}{\sin 39.7°} = \frac{9.8}{\sin Y} \qquad \text{Substituting}$$

$$\sin Y = \frac{9.8 \sin 39.7°}{23.5} \qquad \text{Solving for } \sin Y$$

$$\sin Y \approx 0.2664.$$

Then $Y = 15.4°$ or $Y = 164.6°$, to the nearest tenth of a degree. An angle of $164.6°$ cannot be an angle of this triangle because it already has an angle of $39.7°$ and these two angles would total more than $180°$. Thus,

15.4° is the only possibility for Y. Therefore,

$$Z \approx 180^\circ - (39.7^\circ + 15.4^\circ) \approx 124.9^\circ.$$

We now find z:

$$\frac{z}{\sin Z} = \frac{x}{\sin X}$$

$$\frac{z}{\sin 124.9^\circ} = \frac{23.5}{\sin 39.7^\circ} \qquad \text{Substituting}$$

$$z = \frac{23.5 \sin 124.9^\circ}{\sin 39.7^\circ} \qquad \text{Solving for } z$$

$$z \approx 30.2.$$

We now have solved the triangle:

$$X = 39.7^\circ, \qquad x = 23.5,$$
$$Y \approx 15.4^\circ, \qquad y = 9.8,$$
$$Z \approx 124.9^\circ, \qquad z \approx 30.2.$$

In the ambiguous case, there are two possible solutions.

Example 5 *Two solutions.* In $\triangle ABC$, $b = 15$, $c = 20$, and $B = 29^\circ$. Solve the triangle.

SOLUTION We make a drawing, list the known measures, and see that we again have the SSA situation.

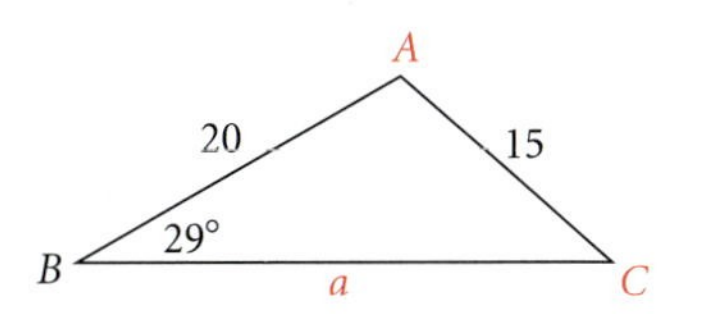

$A = ?$ $\qquad a = ?$
$B = 29^\circ$ $\qquad b = 15$
$C = ?$ $\qquad c = 20$

We first find C:

$$\frac{b}{\sin B} = \frac{c}{\sin C}$$

$$\frac{15}{\sin 29^\circ} = \frac{20}{\sin C} \qquad \text{Substituting}$$

$$\sin C = \frac{20 \sin 29^\circ}{15} \approx 0.6464. \qquad \text{Solving for } \sin C$$

There are two angles less than 180° with a sine of 0.6464. They are 40° and 140°, to the nearest degree. This gives us two possible solutions.

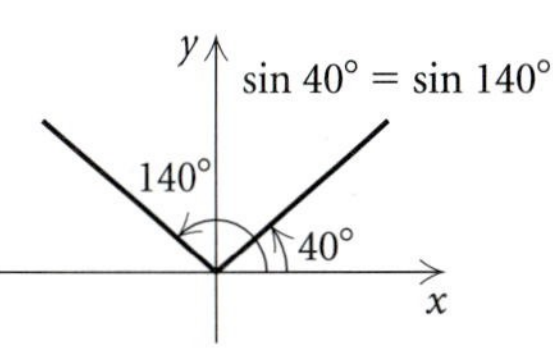

Program

TRISOL: This program solves triangles when three parts are given. It covers unsolvable triangles as well as multiple solutions.

Possible Solution I.

If $C = 40°$, then

$$A = 180° - (29° + 40°) = 111°.$$

Then we find a:

$$\frac{a}{\sin A} = \frac{b}{\sin B}$$

$$\frac{a}{\sin 111°} = \frac{15}{\sin 29°}$$

$$a = \frac{15 \sin 111°}{\sin 29°} \approx 29.$$

These measures make a triangle as shown below; thus we have a solution.

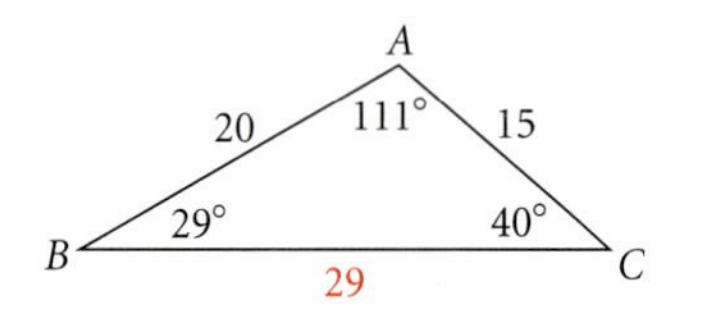

Possible Solution II.

If $C = 140°$, then

$$A = 180° - (29° + 140°) = 11°.$$

Then we find a:

$$\frac{a}{\sin A} = \frac{b}{\sin B}$$

$$\frac{a}{\sin 11°} = \frac{15}{\sin 29°}$$

$$a = \frac{15 \sin 11°}{\sin 29°} \approx 6.$$

These measures make a triangle as shown below; thus we have a second solution.

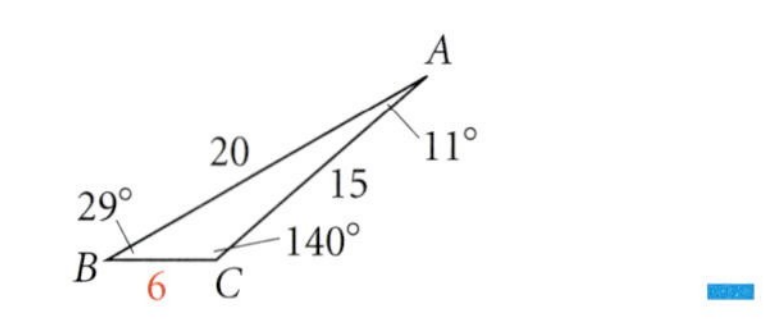

Examples 3–5 illustrate the SSA situation. Note that we need not memorize the eight cases or the procedures in finding no solution, one solution, or two solutions. When we are using the law of sines, the sine value leads us directly to the correct solution or solutions.

The Area of a Triangle

The familiar formula for the area of a triangle, $A = \frac{1}{2}bh$, can be used only when h is known. However, we can use the method used to derive the law of sines to derive an area formula that does not involve the height.

Consider a general triangle $\triangle ABC$, with area K, as shown below.

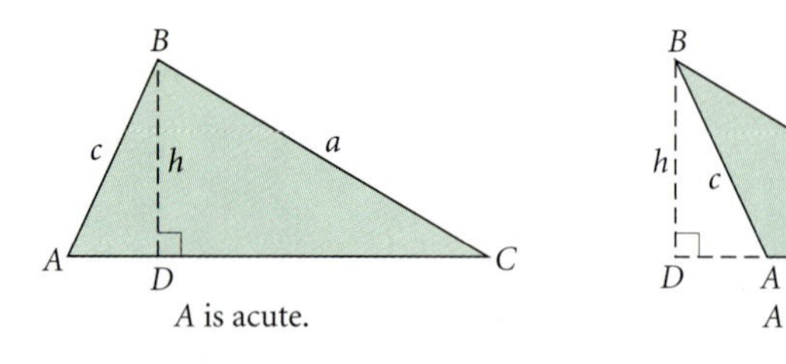

In each $\triangle ADB$,

$$\sin A = \frac{h}{c}, \quad \text{or} \quad h = c \sin A.$$

Substituting into the formula $K = \frac{1}{2}bh$, we get

$$K = \frac{1}{2}bc \sin A.$$

Any pair of sides and the included angle could have been used. Thus we also have

$$K = \frac{1}{2}ab \sin C \quad \text{and} \quad K = \frac{1}{2}ac \sin B.$$

The Area of a Triangle

The area K of any $\triangle ABC$ is one half the product of the lengths of two sides and the sine of the included angle:

$$K = \frac{1}{2}bc \sin A = \frac{1}{2}ab \sin C = \frac{1}{2}ac \sin B.$$

Example 6 *Area of the Peace Monument.* Through the Mentoring in the City Program sponsored by Marian College, Indianapolis, Indiana, children have turned a vacant downtown lot into a monument for peace.* This community project brought together neighborhood volunteers, businesses, and government in hopes of showing children how to develop positive nonviolent ways of dealing with conflict. A landscape architect† used the children's drawings and ideas to design a triangular-shaped peace garden. Two sides of the property, formed by Indiana Avenue and Senate Avenue, measure 182 ft and 230 ft, respectively, and together form a 44.7° angle. The third side of the garden, formed by an apartment building, measures 163 ft. What is the area of this property?

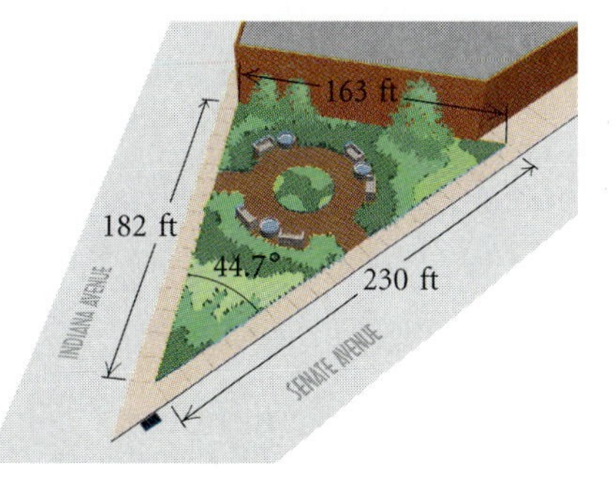

SOLUTION Since we do not know a height of the triangle, we use the area formula:

$$K = \tfrac{1}{2}bc \sin A$$
$$K = \tfrac{1}{2} \cdot 182 \text{ ft} \cdot 230 \text{ ft} \cdot \sin 44.7^\circ$$
$$K \approx 14{,}722 \text{ ft}^2.$$

The area of the property is approximately 14,722 ft^2.

**The Indianapolis Star,* August 6, 1995, p. J8.

†Alan Day, a landscape architect with Browning Day Mullins Dierdorf, Inc., donated his time to this project.

6.1 Exercise Set

Solve the triangle, if possible.

1. $B = 38°$, $C = 21°$, $b = 24$
2. $A = 131°$, $C = 23°$, $b = 10$
3. $A = 36.5°$, $a = 24$, $b = 34$
4. $B = 118.3°$, $C = 45.6°$, $b = 42.1$
5. $C = 61°10'$, $c = 30.3$, $b = 24.2$
6. $A = 126.5°$, $a = 17.2$, $c = 13.5$
7. $c = 3$ mi, $B = 37.48°$, $C = 32.16°$
8. $a = 2345$ mi, $b = 2345$ mi, $A = 124.67°$
9. $b = 56.78$ yd, $c = 56.78$ yd, $C = 83.78°$
10. $A = 129°32'$, $C = 18°28'$, $b = 1204$ in.
11. $a = 20.01$ cm, $b = 10.07$ cm, $A = 30.3°$
12. $b = 4.157$ km, $c = 3.446$ km, $C = 51°48'$
13. $A = 89°$, $a = 15.6$ in., $b = 18.4$ in.
14. $C = 46°32'$, $a = 56.2$ m, $c = 22.1$ m
15. $a = 200$ m, $A = 32.76°$, $C = 21.97°$
16. $B = 115°$, $c = 45.6$ yd, $b = 23.8$ yd

Find the area of the triangle.

17. $B = 42°$, $a = 7.2$ ft, $c = 3.4$ ft
18. $A = 17°12'$, $b = 10$ in., $c = 13$ in.
19. $C = 82°54'$, $a = 4$ yd, $b = 6$ yd
20. $C = 75.16°$, $a = 1.5$ m, $b = 2.1$ m
21. $B = 135.2°$, $a = 46.12$ ft, $c = 36.74$ ft
22. $A = 113°$, $b = 18.2$ cm, $c = 23.7$ cm

Solve. Keep in mind the two types of bearing considered in Sections 4.2 and 4.3.

23. *Area of Back Yard.* A new homeowner has a triangular-shaped back yard. Two of the three sides measure 53 ft and 42 ft and form an included angle of 135°. To determine the amount of fertilizer and grass seed to be purchased, the owner has to know, or at least approximate, the area of the yard. Find the area of the yard to the nearest square foot.

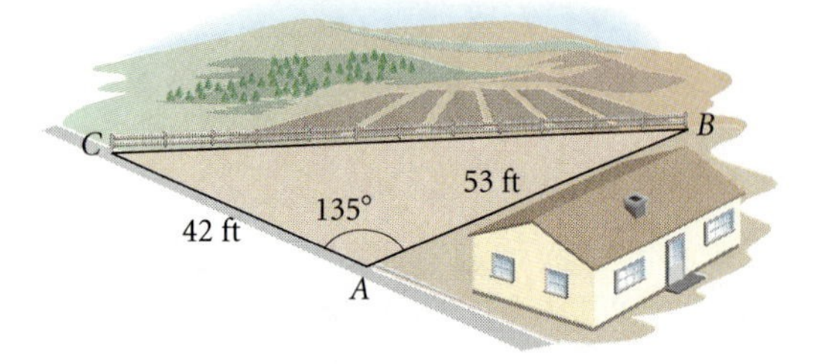

24. *Boarding Stable.* A rancher operates a boarding stable and temporarily needs to make an extra pen. He has a piece of rope 38 ft long and plans to tie the rope to one end of the barn (S) and run the rope around a tree (T) and back to the barn (Q). The tree is 21 ft from where the rope is first tied, and the rope from the barn to the tree makes an angle of 35° with the barn. Does the rancher have enough rope if he allows $4\frac{1}{2}$ ft at each end to fasten the rope?

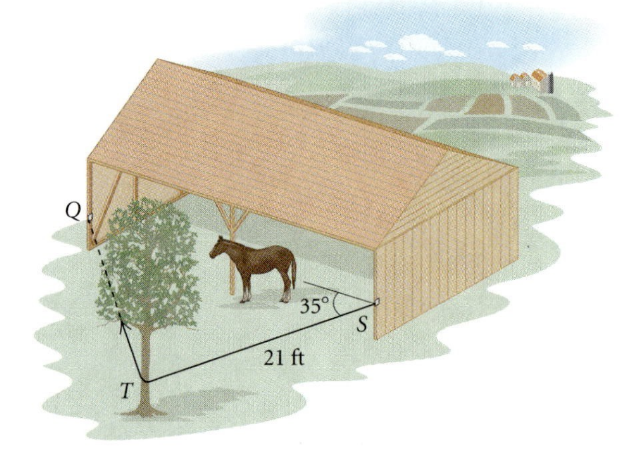

25. *Rock Concert.* In preparation for an outdoor rock concert, a stage crew must determine how far apart to place the two large speaker columns on stage. What generally works best is to place them at 50° angles to the center of the front row. The distance from the center of the front row to each of the speakers is 10 ft. How far apart does the crew need to place the speakers on stage?

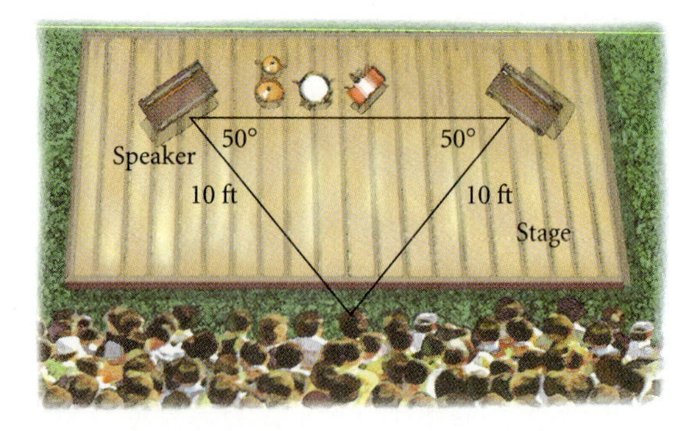

26. *Lunar Crater.* Points A and B are on opposite sides of a lunar crater. Point C is 50 m from A. The measure of $\angle BAC$ is determined to be 112° and the measure of $\angle ACB$ is determined to be 42°. What is the width of the crater?

27. *Length of Pole.* A pole leans away from the sun at an angle of 7° to the vertical. When the angle of

elevation of the sun is 51°, the pole casts a shadow 47 ft long on level ground. How long is the pole?

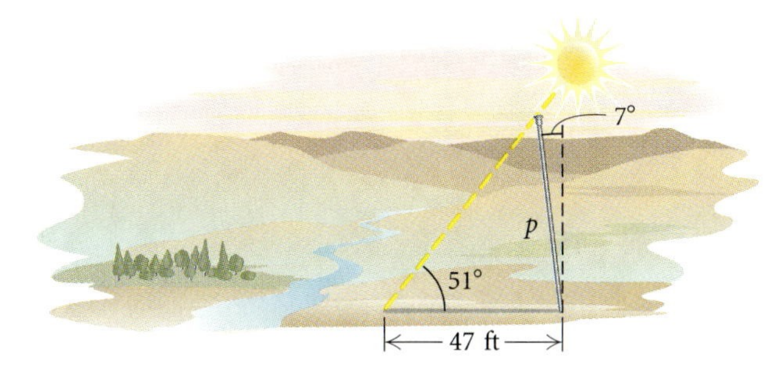

28. *Reconnaissance Airplane.* A reconnaissance airplane leaves its airport on the east coast of the United States and flies in a direction of 085°. Because of bad weather, it returns to another airport 230 km to the north of its home base. For the return trip, it flies in a direction of 283°. What is the total distance that the airplane flew?

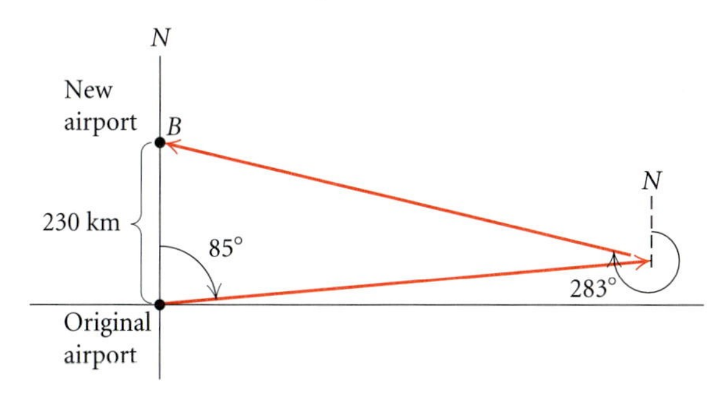

29. *Fire Tower.* A ranger in fire tower A spots a fire at a direction of 295°. A ranger in fire tower B, located 45 mi at a direction of 045° from tower A, spots the same fire at a direction of 255°. How far from tower A is the fire? from tower B?

30. *Lighthouse.* A boat leaves a lighthouse A and sails 5.1 km. At this time it is sighted from lighthouse B, 7.2 km west of A. The bearing of the boat from B is N65°10′E. How far is the boat from B?

31. *Mackinac Island.* Mackinac Island is located 35 mi N65°20′W of Cheboygan, Michigan, where the Coast Guard cutter Mackinaw is stationed. A freighter in distress radios the Coast Guard cutter for help. It radios its position as N25°40′E of Mackinac Island and N10°10′W of Cheboygan. How far is the freighter from Cheboygan?

32. *Gears.* Three gears are arranged as shown in the figure below. Find the angle ϕ.

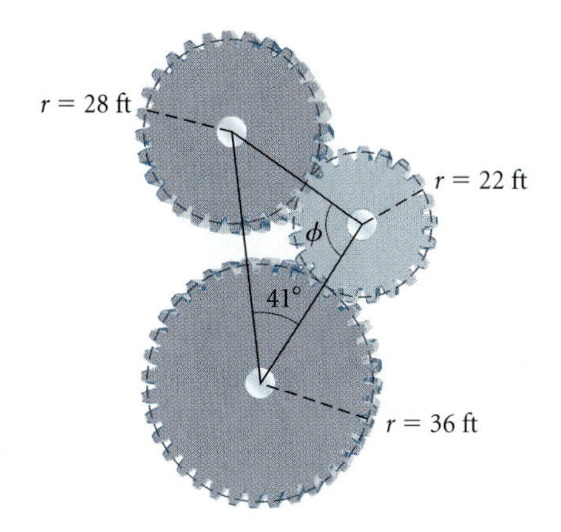

Skill Maintenance

Find the acute angle, A, in both radians and degrees, for the given function value.

33. $\cos A = 0.2213$

34. $\cos A = 1.5612$

Convert to decimal degree notation.

35. 18°14′20″

36. 125°3′42″

Synthesis

37. ◈ Explain why the law of sines cannot be used to find the first angle when solving a triangle given three sides.

38. ◈ We considered eight cases of solving triangles given two sides and an angle opposite one of them. Describe the relationship between side b and the height h in each.

39. Prove the following area formulas for a general triangle ABC with area represented by K.

$$K = \frac{a^2 \sin B \sin C}{2 \sin A} = \frac{c^2 \sin A \sin B}{2 \sin C} = \frac{b^2 \sin C \sin A}{2 \sin B}$$

40. *Area of a Parallelogram.* Prove that the area of a parallelogram is the product of two sides and the sine of the included angle.

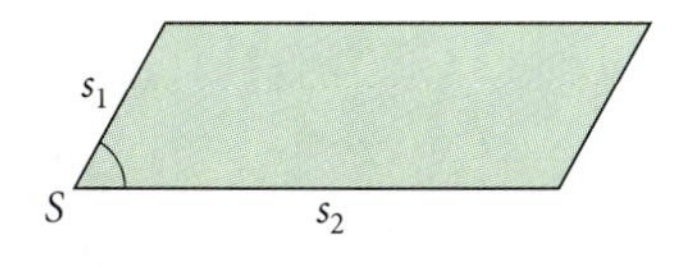

41. *Area of a Quadrilateral.* Prove that the area of a quadrilateral is one half the product of the lengths of its diagonals and the sine of the angle between the diagonals.

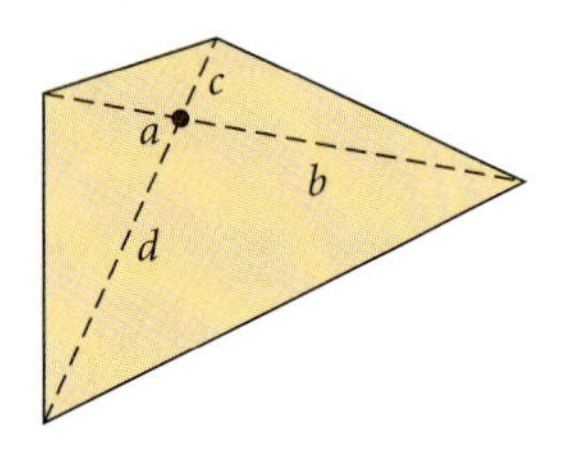

42. Find d.

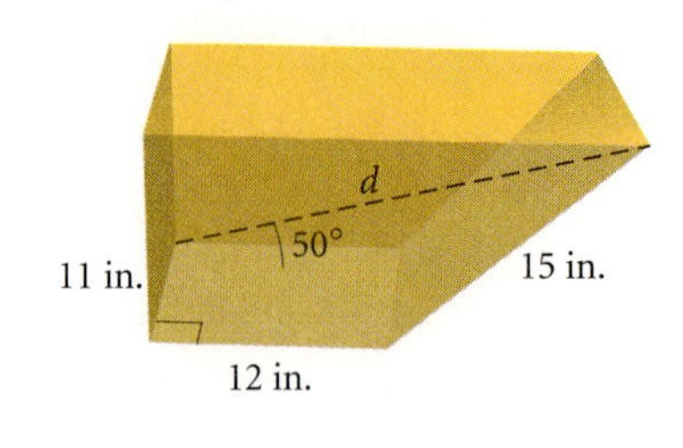

6.2 The Law of Cosines

- *Use the law of cosines to solve triangles.*
- *Determine whether the law of sines or the law of cosines should be applied to solve a triangle, and then solve the triangle.*

The law of sines is used to solve triangles given a side and two angles (AAS and ASA) or given two sides and an angle opposite one of them (SSA). A second law, called the *law of cosines,* is needed to solve triangles given two sides and the included angle (SAS) or given three sides (SSS).

The Law of Cosines

To derive this property, we consider any $\triangle ABC$ placed on a coordinate system. We position the origin at one of the vertices—say, C—and the positive half of the x-axis along one of the sides—say, CB. Let (x, y) be the coordinates of vertex A. Point B has coordinates $(a, 0)$ and point C has coordinates $(0, 0)$.

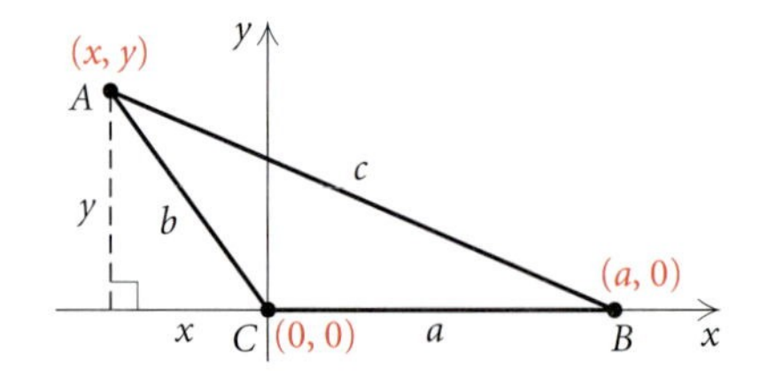

Then

$$\cos C = \frac{x}{b}, \quad \text{so} \quad x = b \cos C$$

and

$$\sin C = \frac{y}{b}, \quad \text{so} \quad y = b \sin C.$$

Thus point A has coordinates

$$(b \cos C, b \sin C).$$

Next, we use the distance formula to determine c^2:

$$c^2 = (x - a)^2 + (y - 0)^2,$$

or $$c^2 = (b \cos C - a)^2 + (b \sin C - 0)^2.$$

Now we multiply and simplify:

$$\begin{aligned} c^2 &= b^2 \cos^2 C - 2ab \cos C + a^2 + b^2 \sin^2 C \\ &= a^2 + b^2(\sin^2 C + \cos^2 C) - 2ab \cos C \\ &= a^2 + b^2 - 2ab \cos C. \end{aligned}$$

Had we placed the origin at one of the other vertices, we would have obtained

$$a^2 = b^2 + c^2 - 2bc \cos A$$

or $$b^2 = a^2 + c^2 - 2ac \cos B.$$

The Law of Cosines

In any triangle ABC,

$$a^2 = b^2 + c^2 - 2bc \cos A,$$
$$b^2 = a^2 + c^2 - 2ac \cos B,$$
or $$c^2 = a^2 + b^2 - 2ab \cos C.$$

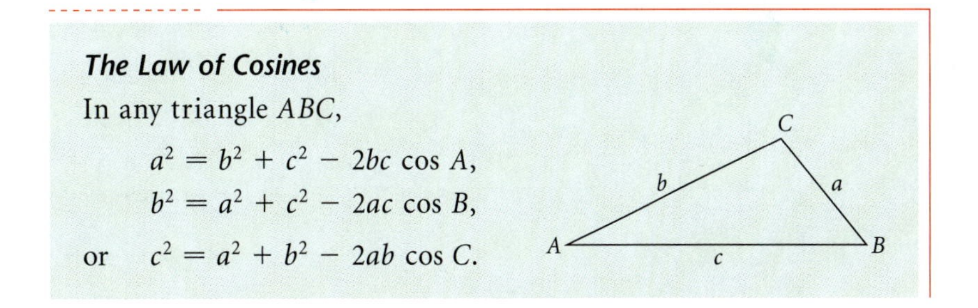

Thus, in any triangle, the square of a side is the sum of the squares of the other two sides, minus twice the product of those sides and the cosine of the included angle.

Solving Triangles (SAS)

When two sides of a triangle and the included angle are known, we can use the law of cosines to find the third side. The law of cosines or the law of sines can then be used to finish solving the triangle.

Example 1 Solve $\triangle ABC$ if $a = 32$, $c = 48$, and $B = 125.2°$.

SOLUTION We first label a triangle with the known and unknown measures.

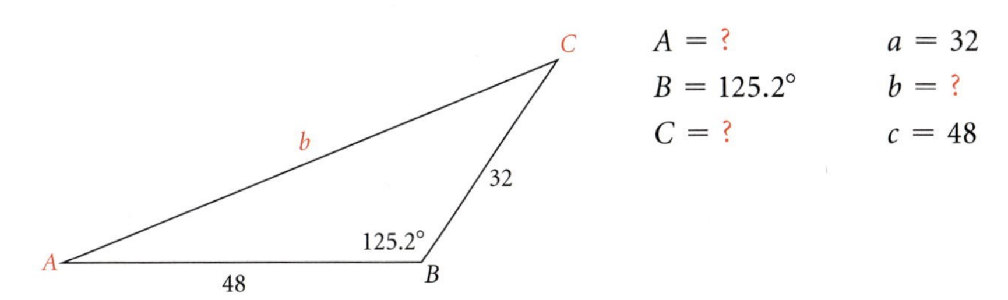

$A = ?$	$a = 32$
$B = 125.2°$	$b = ?$
$C = ?$	$c = 48$

We can find the third side using the law of cosines, as follows:

$$b^2 = a^2 + c^2 - 2ac \cos B$$
$$b^2 = 32^2 + 48^2 - 2 \cdot 32 \cdot 48 \cos 125.2^\circ \quad \textbf{Substituting}$$
$$b^2 \approx 5098.8$$
$$b \approx 71.$$

We now have $a = 32$, $b \approx 71$, and $c = 48$, and we need to find the other two angle measures. At this point, we can find them in two ways. One way uses the law of sines. The ambiguous case may arise, however, and we would have to be alert to this possibility. The advantage of using the law of cosines again is that if we solve for the cosine and find that its value is negative, then we know that the angle is obtuse. If the value of the cosine is positive, then the angle is acute. Thus we use the law of cosines to find a second angle.

Let's find angle A. We select the formula from the law of cosines that contains $\cos A$ and substitute:

$$a^2 = b^2 + c^2 - 2bc \cos A$$
$$32^2 = 71^2 + 48^2 - 2 \cdot 71 \cdot 48 \cos A \quad \textbf{Substituting}$$
$$1024 = 5041 + 2304 - 6816 \cos A$$
$$-6321 = -6816 \cos A$$
$$\cos A \approx 0.9273768$$
$$A \approx 22.0^\circ.$$

The third angle is now easy to find:

$$C \approx 180^\circ - (125.2^\circ + 22.0^\circ) \approx 32.8^\circ.$$

Thus,

$$A \approx 22.0^\circ, \qquad a = 32,$$
$$B = 125.2^\circ, \qquad b \approx 71,$$
$$C \approx 32.8^\circ, \qquad c = 48.$$

Due to errors created by rounding, answers may vary depending on the order in which they are found. Had we found the measure of angle C first in Example 1, the angle measures would have been $C \approx 34.1^\circ$ and $A \approx 20.7^\circ$. The answers at the back of the book were generated by considering alphabetical order. Variances in rounding also change the answers. Had we used 71.4 for b in Example 1, the angle measures would have been $A \approx 21.5^\circ$ and $C \approx 33.3^\circ$.

Suppose we used the law of sines at the outset in Example 1 to find b. We were given only three measures: $a = 32$, $c = 48$, and $B = 125.2^\circ$. When substituting these measures into the proportions, we see that there is not enough information to use the law of sines:

$$\frac{a}{\sin A} = \frac{b}{\sin B} \rightarrow \frac{32}{\sin A} = \frac{b}{\sin 125.2^\circ},$$
$$\frac{b}{\sin B} = \frac{c}{\sin C} \rightarrow \frac{b}{\sin 125.2^\circ} = \frac{48}{\sin C},$$

$$\frac{a}{\sin A} = \frac{c}{\sin C} \rightarrow \frac{32}{\sin A} = \frac{48}{\sin C}.$$

In all three situations, the resulting equation, after the substitutions, still has two unknowns. Thus we cannot use the law of sines to find b.

Solving Triangles (SSS)

When all three sides of a triangle are known, the law of cosines can be used to solve the triangle.

Example 2 Solve $\triangle RST$ if $r = 3.5$, $s = 4.7$, and $t = 2.8$.

SOLUTION We sketch a triangle and label it with the given measures.

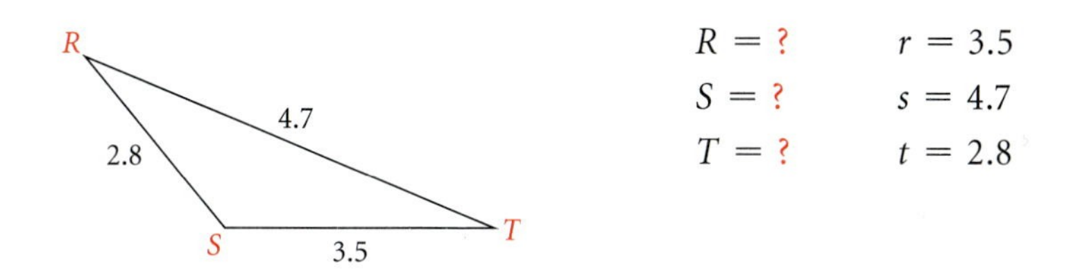

Since we do not know any of the angle measures, we cannot use the law of sines. We begin instead by finding angle S with the law of cosines. We choose the formula that contains $\cos S$:

$$s^2 = r^2 + t^2 - 2rt \cos S$$

$$(4.7)^2 = (3.5)^2 + (2.8)^2 - 2(3.5)(2.8) \cos S \qquad \text{Substituting}$$

$$\cos S = \frac{(3.5)^2 + (2.8)^2 - (4.7)^2}{2(3.5)(2.8)}$$

$$\cos S \approx -0.1020408$$

$$S \approx 95.86^\circ.$$

Similarly, we find angle R:

$$r^2 = s^2 + t^2 - 2st \cos R$$

$$(3.5)^2 = (4.7)^2 + (2.8)^2 - 2(4.7)(2.8) \cos R$$

$$\cos R = \frac{(4.7)^2 + (2.8)^2 - (3.5)^2}{2(4.7)(2.8)}$$

$$\cos R \approx 0.6717325$$

$$R \approx 47.80^\circ.$$

Then

$$T \approx 180^\circ - (95.86^\circ + 47.80^\circ) \approx 36.34^\circ.$$

Thus,

$$R \approx 47.80^\circ, \qquad r = 3.5,$$

$$S \approx 95.86^\circ, \qquad s = 4.7,$$

$$T \approx 36.34^\circ, \qquad t = 2.8.$$

Example 3 *Knife Bevel.* Knifemakers know that the *bevel* of the blade (the angle formed at the cutting edge of the blade) determines the cutting characteristics of the knife. A small bevel like that of a straight razor makes for a keen edge, but is impractical for heavy-duty cutting because the edge dulls quickly and is prone to chipping. A large bevel is suitable for heavy-duty work like chopping wood. Survival knives, being universal in application, are a compromise between small and large bevels. The diagram at left illustrates the blade of a hand-made Randall Model 18 survival knife. What is its bevel? (*Source:* Randall Made Knives, P.O. Box 1988, Orlando, FL 32802)

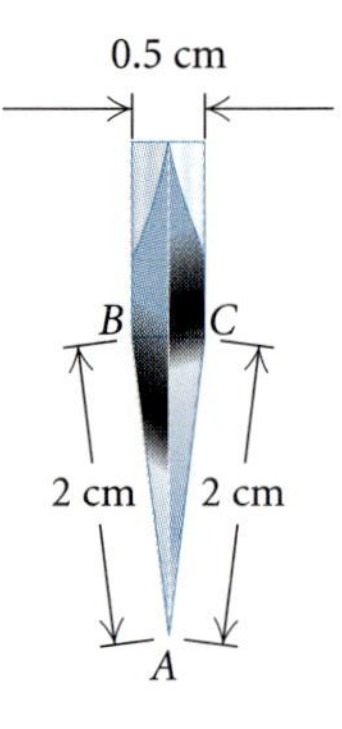

SOLUTION We know three sides of a triangle. We can use the law of cosines to find the bevel, angle A.

$$a^2 = b^2 + c^2 - 2bc \cos A$$
$$(0.5)^2 = 2^2 + 2^2 - 2 \cdot 2 \cdot 2 \cdot \cos A$$
$$\cos A = \frac{4 + 4 - 0.25}{8}$$
$$\cos A = 0.96875$$
$$A \approx 14.36^\circ.$$

Thus the bevel is approximately 14.36°.

Choosing the Appropriate Law

The following summarizes the situations in which to use the law of sines and the law of cosines.

To solve an oblique triangle:

Use the *law of sines* for:	Use the *law of cosines* for:
AAS	SAS
ASA	SSS
SSA	

The law of cosines can also be used for the SSA situation, but since the process involves solving a quadratic equation, we do not include that option in the list above.

Example 4 In $\triangle ABC$, three measures are given. Determine which law to use when solving the triangle. You need not solve the triangle.

a) $a = 14, b = 23, c = 10$

b) $a = 207, B = 43.8^\circ, C = 57.6^\circ$

c) $A = 112^\circ, C = 37^\circ, a = 84.7$

d) $B = 101^\circ, a = 960, c = 1042$

e) $b = 17.26, a = 27.29, A = 39^\circ$

f) $A = 61^\circ, B = 39^\circ, C = 80^\circ$

SOLUTION It is helpful to make a sketch of a triangle with the given information. The triangle need not be drawn to scale. The given parts are shown in color.

FIGURE	SITUATION	LAW TO USE
a)	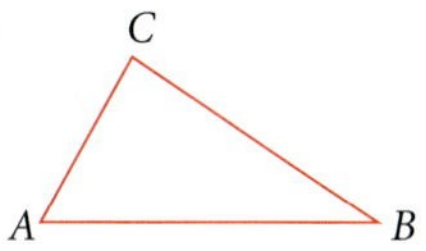 SSS	Law of Cosines
b)	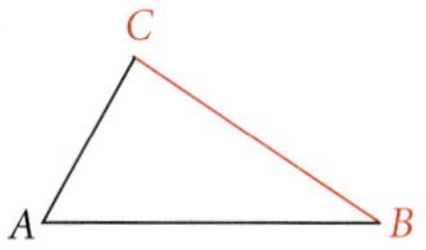 ASA	Law of Sines
c)	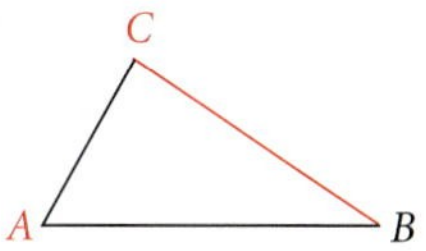 AAS	Law of Sines
d)	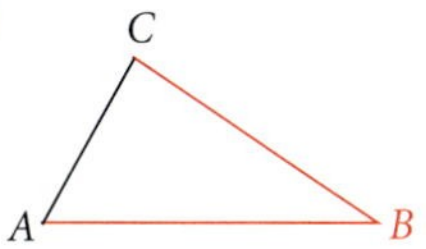 SAS	Law of Cosines
e)	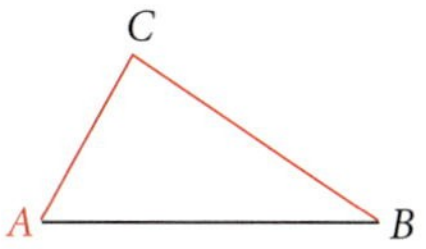 SSA	Law of Sines
f)	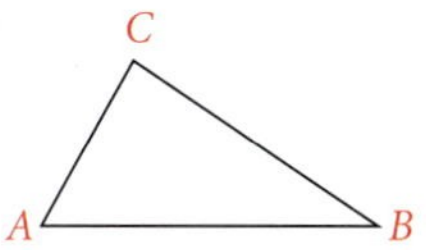 AAA	Cannot be solved

6.2 Exercise Set

Solve the triangle, if possible.

1. $A = 30°$, $b = 12$, $c = 24$
2. $B = 133°$, $a = 12$, $c = 15$
3. $a = 12$, $b = 14$, $c = 20$
4. $a = 22.3$, $b = 22.3$, $c = 36.1$
5. $B = 72°40'$, $c = 16$ m, $a = 78$ m
6. $C = 22.28°$, $a = 25.4$ cm, $b = 73.8$ cm
7. $a = 16$ m, $b = 20$ m, $c = 32$ m
8. $B = 72.66°$, $a = 23.78$ km, $c = 25.74$ km
9. $a = 2$ ft, $b = 3$ ft, $c = 8$ ft
10. $A = 96°13'$, $b = 15.8$ yd, $c = 18.4$ yd
11. $a = 26.12$ km, $b = 21.34$ km, $c = 19.25$ km
12. $C = 28°43'$, $a = 6$ mm, $b = 9$ mm
13. $a = 60.12$ mi, $b = 40.23$ mi, $C = 48.7°$
14. $a = 11.2$ cm, $b = 5.4$ cm, $c = 7$ cm
15. $b = 10.2$ in., $c = 17.3$ in., $A = 53.456°$
16. $a = 17$ yd, $b = 15.4$ yd, $c = 1.5$ yd

Determine which law applies. Then solve the triangle.

17. $A = 70°$, $B = 12°$, $b = 21.4$
18. $a = 15$, $c = 7$, $B = 62°$
19. $a = 3.3$, $b = 2.7$, $c = 2.8$
20. $a = 1.5$, $b = 2.5$, $A = 58°$

21. $A = 40.2°$, $B = 39.8°$, $C = 100°$

22. $a = 60$, $b = 40$, $C = 47°$

23. $a = 3.6$, $b = 6.2$, $c = 4.1$

24. $B = 110°30'$, $C = 8°10'$, $c = 0.912$

Solve.

25. *Poachers.* A park ranger establishes an observation post from which to watch for poachers. Despite losing her map, the ranger does have a compass and a rangefinder. She observes some poachers, and the rangefinder indicates that they are 500 ft from her position. They are headed toward big game that she knows to be 375 ft from her position. Using her compass, she finds that the poachers' azimuth (the direction measured as an angle from north) is 355° and that of the big game is 42°. What is the distance between the poachers and the game?

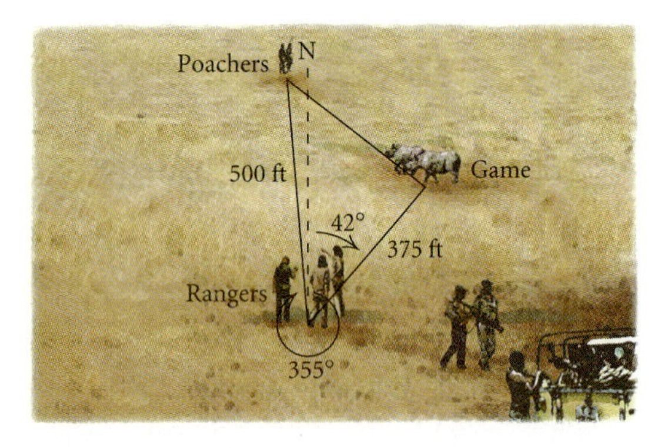

26. *Circus Highwire Act.* A circus highwire act walks up an approach wire to reach a highwire. The approach wire is 122 ft long and is currently anchored so that the angle it forms with the ground is at the maximum of 35°. A greater approach angle causes the aerialists to slip. However, the aerialists find that there is enough room to anchor the approach wire 30 ft back in order to make the approach angle less severe. When this is done, how much farther will they have to walk up the approach wire, and what will the new approach angle be?

27. *In-line Skater.* An in-line skater skates on a fitness trail along the Pacific Ocean from point A to point B. As shown below, two streets intersecting at point C also intersect the trail at A and B. In his car, the skater found the lengths of AC and BC to be approximately 0.5 mi and 1.3 mi, respectively. From a map, he estimates the included angle at C to be 110°. How far did he skate from A to B?

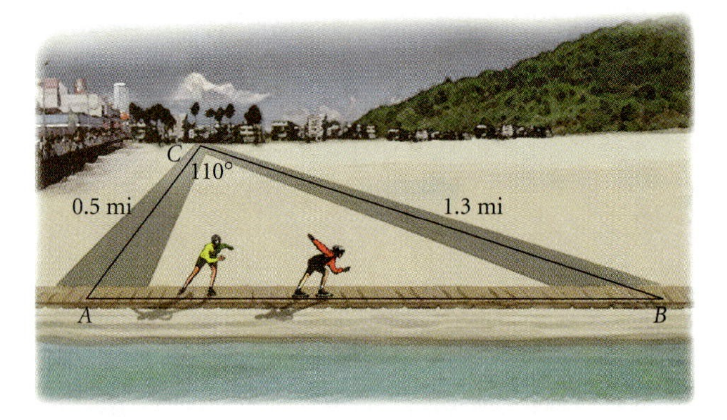

28. *Baseball Bunt.* A batter in a baseball game drops a bunt down the first-base line. It rolls 34 ft at an angle of 25° with the base path. The pitcher's mound is 60.5 ft from home plate. How far must the pitcher travel to pick up the ball? (*Hint:* A baseball diamond is a square.)

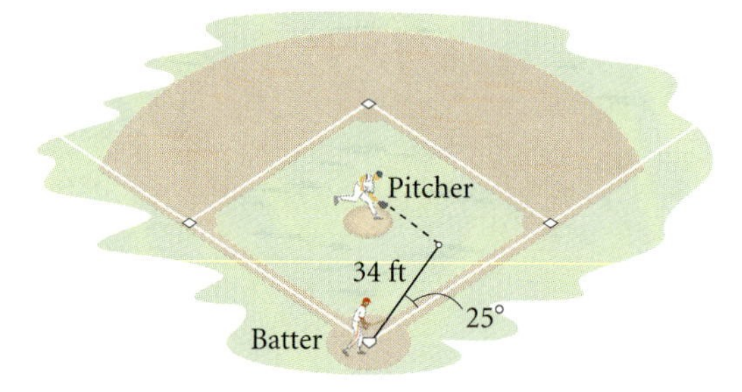

29. *Ships.* Two ships leave harbor at the same time. The first sails N15°W at 25 knots (a knot is one nautical mile per hour). The second sails N32°E at 20 knots. After 2 hr, how far apart are the ships?

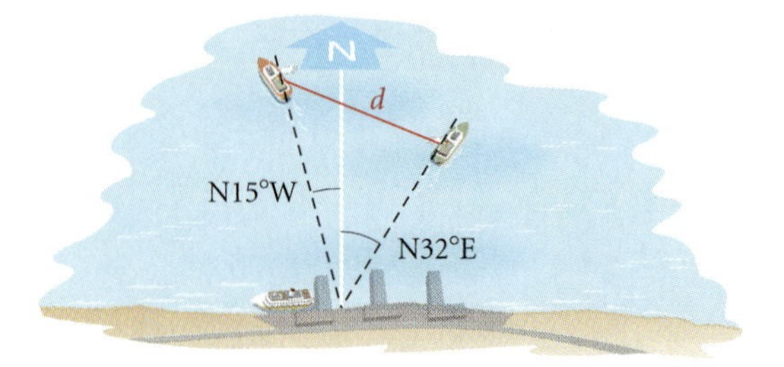

30. *Survival Trip.* A group of college students is learning to navigate for an upcoming survival trip.

On a map, they have been given three points at which they are to check in. The map also shows the distances between the points. However, to navigate they need to know the angle measurements. Calculate the angles for them.

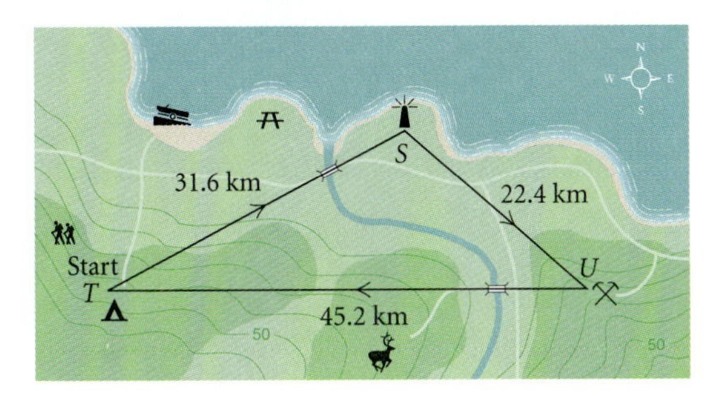

31. *Airplanes.* Two airplanes leave an airport at the same time. The first flies 150 km/h in a direction of 320°. The second flies 200 km/h in a direction of 200°. After 3 hr, how far apart are the planes?

32. *Slow-pitch Softball.* A slow-pitch softball diamond is a square 65 ft on a side. The pitcher's mound is 46 ft from home. How far is it from the pitcher's mound to first base?

33. *Isosceles Trapezoid.* The longer base of an isosceles trapezoid measures 14 ft. The nonparallel sides measure 10 ft, and the base angles measure 80°.

 a) Find the length of a diagonal.
 b) Find the area.

34. *Area of Sail.* A sail that is in the shape of an isosceles triangle has a vertex angle of 38°. The angle is included by two sides, each measuring 20 ft. Find the area of the sail.

35. Three circles are arranged as shown in the figure below. Find the length PQ.

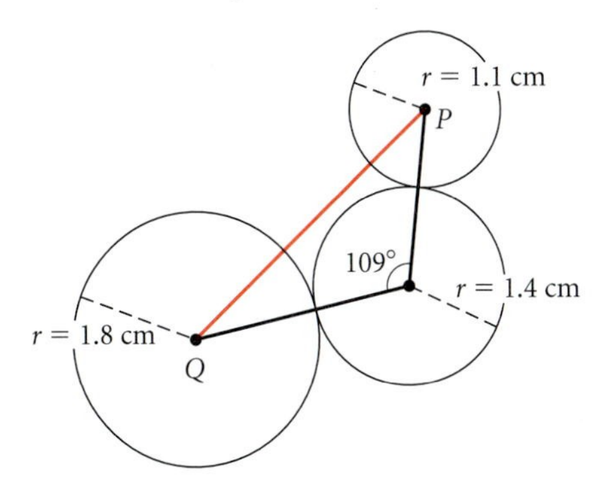

36. *Swimming Pool.* A triangular swimming pool measures 44 ft on one side and 32.8 ft on another side. These sides form an angle that measures 40.8°. How long is the other side?

Skill Maintenance

37. Find the absolute value: $|-5|$.

Find the values.

38. $\cos \dfrac{\pi}{6}$

39. $\sin 45°$

40. $\sin 300°$

41. $\cos\left(-\dfrac{2\pi}{3}\right)$

42. Multiply: $(1 - i)(1 + i)$.

Synthesis

43. ◈ Try to solve this triangle using the law of cosines. Then explain why it is easier to solve it using the law of sines.

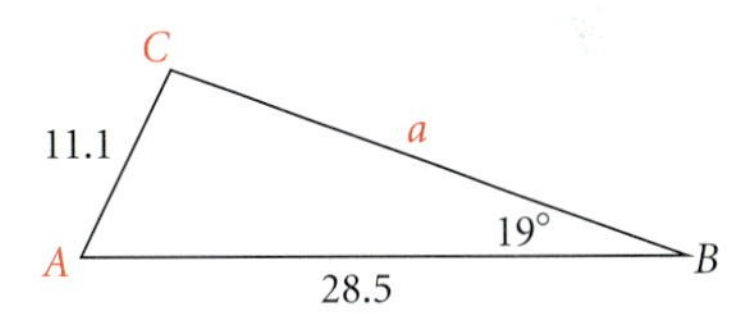

44. ◈ Explain why we cannot solve a triangle given SAS with the law of sines.

45. *Canyon Depth.* A bridge is being built across a canyon. The length of the bridge is 5045 ft. From the deepest point in the canyon, the angles of elevation of the ends of the bridge are 78° and 72°. How deep is the canyon?

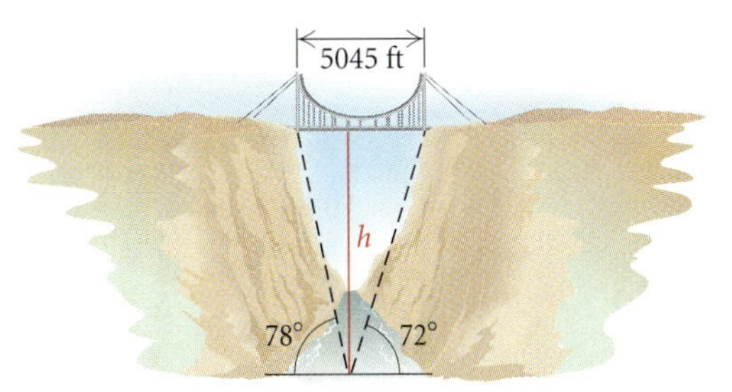

46. *Heron's Formula.* If a, b, and c are the lengths of the sides of a triangle, then the area K of the triangle is given by

$$K = \sqrt{s(s - a)(s - b)(s - c)},$$

where $s = \frac{1}{2}(a + b + c)$. The number s is called the **semiperimeter**. Prove Heron's formula. (*Hint:* Use the area formula $K = \frac{1}{2}bc \sin A$ developed in Section 6.1. Then use Heron's formula to find the area of the triangular swimming pool described in Exercise 36.)

47. *Area of Isosceles Triangle.* Find a formula for the area of an isosceles triangle in terms of the congruent sides and their included angle. Under what conditions will the area of a triangle with fixed congruent sides be maximum?

48. *Reconnaissance Plane.* A reconnaissance plane patrolling at 5000 ft sights a submarine at bearing 35° and at an angle of depression of 25°. A carrier is at bearing 105° and at an angle of depression of 60°. How far is the submarine from the carrier?

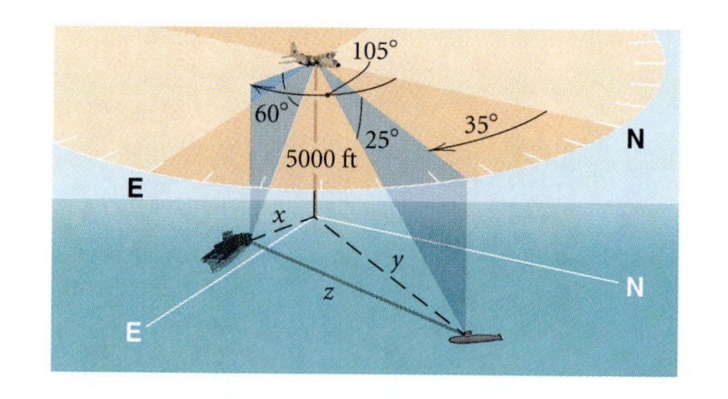

6.3 Complex Numbers: Trigonometric Form

- *Graph complex numbers.*
- *Given a complex number in standard form, find trigonometric, or polar, notation; and given a complex number in trigonometric form, find standard notation.*
- *Use trigonometric notation to multiply and divide complex numbers.*
- *Use DeMoivre's theorem to raise complex numbers to powers.*
- *Find the nth roots of a complex number.*

Graphical Representation

Just as real numbers can be graphed on a line, complex numbers can be graphed on a plane. We graph a complex number $a + bi$ in the same way that we graph an ordered pair of real numbers (a, b). However, in place of an x-axis, we have a real axis, and in place of a y-axis, we have an imaginary axis. Horizontal distances correspond to the real part of a number. Vertical distances correspond to the imaginary part.

Example 1 Graph each of the following complex numbers.

a) $3 + 2i$ b) $-4 - 5i$ c) $-3i$
d) $-1 + 3i$ e) 2

SOLUTION

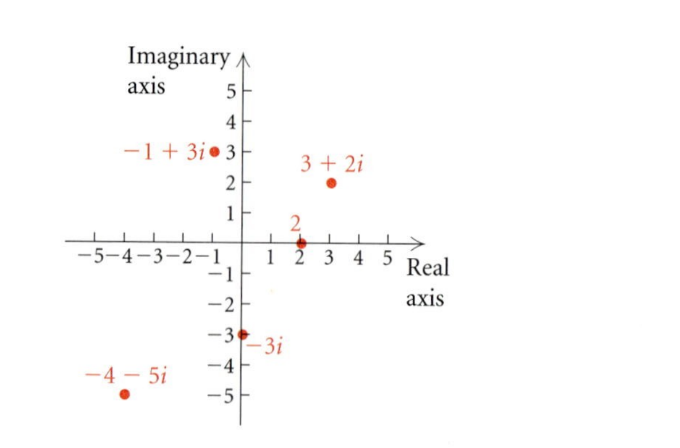

We recall that the absolute value of a real number is its distance from 0 on the number line. The absolute value of a complex number is its distance from the origin in the complex plane. For example, if $z = a + bi$, then using the distance formula, we have

$$|z| = |a + bi| = \sqrt{(a - 0)^2 + (b - 0)^2}$$
$$= \sqrt{a^2 + b^2}.$$

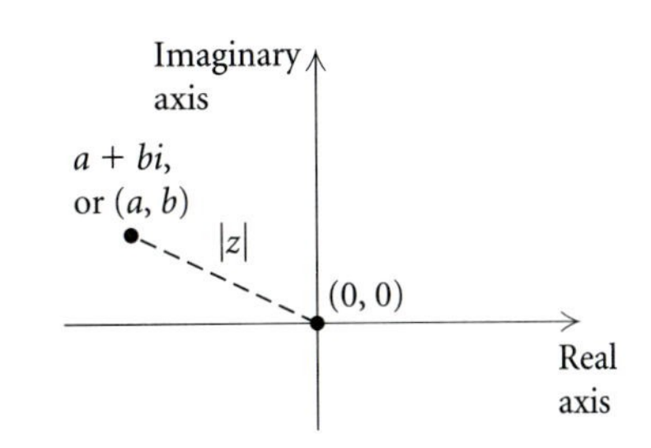

Absolute Value of a Complex Number

The absolute value of a complex number $a + bi$ is

$$|a + bi| = \sqrt{a^2 + b^2}.$$

Example 2 Find the absolute value of each of the following.

a) $3 + 4i$ b) $-2 - i$ c) $\frac{4}{5}i$

SOLUTION

a) $|3 + 4i| = \sqrt{3^2 + 4^2} = \sqrt{9 + 16} = \sqrt{25} = 5$

b) $|-2 - i| = \sqrt{(-2)^2 + (-1)^2} = \sqrt{5}$

c) $\left|\frac{4}{5}i\right| = \left|0 + \frac{4}{5}i\right| = \sqrt{\left(\frac{4}{5}\right)^2} = \frac{4}{5}$

Trigonometric Notation for Complex Numbers

Now let's consider a nonzero complex number $a + bi$. Suppose that its absolute value is r. If we let θ be an angle in standard position whose terminal side passes through the point (a, b), as shown in the figure, then

$$\cos\theta = \frac{a}{r}, \quad \text{or} \quad a = r\cos\theta$$

and

$$\sin\theta = \frac{b}{r}, \quad \text{or} \quad b = r\sin\theta.$$

Substituting these values for a and b into the $(a + bi)$ notation, we get

$$\begin{aligned} a + bi &= r \cos \theta + (r \sin \theta)i \\ &= r(\cos \theta + i \sin \theta). \end{aligned}$$

This is **trigonometric notation** for a complex number $a + bi$. The number r is called the **absolute value** of $a + bi$, and θ is called the **argument** of $a + bi$. Trigonometric notation for a complex number is also called **polar notation**.

Trigonometric Notation for Complex Numbers

$a + bi = r(\cos \theta + i \sin \theta)$

To find trigonometric notation for a complex number given in standard notation, we must find r and determine the angle θ for which $\sin \theta = b/r$ and $\cos \theta = a/r$.

Example 3 Find trigonometric notation for each of the following complex numbers.

a) $1 + i$ **b)** $\sqrt{3} - i$

SOLUTION

a) We note that $a = 1$ and $b = 1$. Then

$$r = \sqrt{a^2 + b^2} = \sqrt{1^2 + 1^2} = \sqrt{2},$$

$$\sin \theta = \frac{b}{r} = \frac{1}{\sqrt{2}}, \quad \text{or} \quad \frac{\sqrt{2}}{2},$$

and

$$\cos \theta = \frac{1}{\sqrt{2}}, \quad \text{or} \quad \frac{\sqrt{2}}{2}.$$

Thus, $\theta = \pi/4$, or $45°$, and we have

$$1 + i = \sqrt{2}\left(\cos \frac{\pi}{4} + i \sin \frac{\pi}{4}\right),$$

or $\quad 1 + i = \sqrt{2}(\cos 45° + i \sin 45°).$

b) We see that $a = \sqrt{3}$ and $b = -1$. Then

$$r = \sqrt{(\sqrt{3})^2 + (-1)^2} = 2, \qquad \sin \theta = -\frac{1}{2}, \quad \text{and} \quad \cos \theta = \frac{\sqrt{3}}{2}.$$

Thus, $\theta = 11\pi/6$, or $330°$, and we have

$$\sqrt{3} - i = 2\left(\cos \frac{11\pi}{6} + i \sin \frac{11\pi}{6}\right),$$

or $\quad \sqrt{3} - i = 2(\cos 330° + i \sin 330°).$

In changing to trigonometric notation, note that there are many angles satisfying the given conditions. We ordinarily choose the smallest positive angle.

To change from trigonometric notation to standard notation, $a + bi$, we recall that $a = r\cos\theta$ and $b = r\sin\theta$.

Example 4 Find standard notation, $a + bi$, for each of the following complex numbers.

a) $2(\cos 120^\circ + i\sin 120^\circ)$ **b)** $\sqrt{8}\left(\cos\frac{7\pi}{4} + i\sin\frac{7\pi}{4}\right)$

SOLUTION

a) Rewriting, we have

$$2(\cos 120^\circ + i\sin 120^\circ) = 2\cos 120^\circ + (2\sin 120^\circ)i.$$

Thus,

$$a = 2\cos 120^\circ = 2\cdot\left(-\frac{1}{2}\right) = -1$$

and

$$b = 2\sin 120^\circ = 2\cdot\frac{\sqrt{3}}{2} = \sqrt{3},$$

so

$$2(\cos 120^\circ + i\sin 120^\circ) = -1 + \sqrt{3}i.$$

b) Rewriting, we have

$$\sqrt{8}\left(\cos\frac{7\pi}{4} + i\sin\frac{7\pi}{4}\right) = \sqrt{8}\cos\frac{7\pi}{4} + \left(\sqrt{8}\sin\frac{7\pi}{4}\right)i.$$

Thus,

$$a = \sqrt{8}\cos\frac{7\pi}{4} = \sqrt{8}\cdot\frac{\sqrt{2}}{2} = 2$$

and

$$b = \sqrt{8}\sin\frac{7\pi}{4} = \sqrt{8}\cdot\left(-\frac{\sqrt{2}}{2}\right) = -2,$$

so

$$\sqrt{8}\left(\cos\frac{7\pi}{4} + i\sin\frac{7\pi}{4}\right) = 2 - 2i.$$

Multiplication and Division with Trigonometric Notation

Multiplication and division of complex numbers is easier to manage with trigonometric notation than with standard notation. We simply multiply the absolute values and add the arguments. Let's state this in a more formal manner.

Complex Numbers: Multiplication

For any complex numbers $r_1(\cos\theta_1 + i\sin\theta_1)$ and $r_2(\cos\theta_2 + i\sin\theta_2)$,

$$r_1(\cos\theta_1 + i\sin\theta_1)\cdot r_2(\cos\theta_2 + i\sin\theta_2) = r_1r_2[\cos(\theta_1+\theta_2) + i\sin(\theta_1+\theta_2)].$$

Proof: Let's first multiply $a_1 + b_1i$ by $a_2 + b_2i$:

$$(a_1 + b_1i)(a_2 + b_2i) = (a_1a_2 - b_1b_2) + (a_2b_1 + a_1b_2)i.$$

Recall that

$$a_1 = r_1\cos\theta_1, \qquad b_1 = r_1\sin\theta_1,$$

and

$$a_2 = r_2\cos\theta_2, \qquad b_2 = r_2\sin\theta_2.$$

We substitute these in the product above, to obtain

$$\begin{aligned} r_1(\cos\theta_1 + i\sin\theta_1)\cdot r_2(\cos\theta_2 + i\sin\theta_2) &= (r_1r_2\cos\theta_1\cos\theta_2 - r_1r_2\sin\theta_1\sin\theta_2) \\ &\quad + (r_1r_2\sin\theta_1\cos\theta_2 + r_1r_2\cos\theta_1\sin\theta_2)i. \end{aligned}$$

This simplifies to

$$r_1r_2(\cos\theta_1\cos\theta_2 - \sin\theta_1\sin\theta_2) + r_1r_2(\sin\theta_1\cos\theta_2 + \cos\theta_1\sin\theta_2)i.$$

Now, using identities for sums of angles, we simplify, obtaining

$$r_1r_2\cos(\theta_1+\theta_2) + r_1r_2\sin(\theta_1+\theta_2)i,$$

which was to be shown.

Example 5 Multiply and express the answer to each of the following in standard notation.

a) $3(\cos 40° + i\sin 40°)$ and $4(\cos 20° + i\sin 20°)$

b) $2(\cos\pi + i\sin\pi)$ and $3\left[\cos\left(-\frac{\pi}{2}\right) + i\sin\left(-\frac{\pi}{2}\right)\right]$

SOLUTION

a)
$$\begin{aligned} 3(\cos 40° + i\sin 40°)\cdot 4(\cos 20° + i\sin 20°) &= 3\cdot 4\cdot[\cos(40° + 20°) + i\sin(40° + 20°)] \\ &= 12(\cos 60° + i\sin 60°) \\ &= 12\left(\frac{1}{2} + \frac{\sqrt{3}}{2}i\right) \\ &= 6 + 6\sqrt{3}i \end{aligned}$$

b) $2(\cos \pi + i \sin \pi) \cdot 3\left[\cos\left(-\frac{\pi}{2}\right) + i \sin\left(-\frac{\pi}{2}\right)\right]$

$$= 2 \cdot 3 \cdot \left[\cos\left(\pi + \left(-\frac{\pi}{2}\right)\right) + i \sin\left(\pi + \left(-\frac{\pi}{2}\right)\right)\right]$$
$$= 6\left(\cos \frac{\pi}{2} + i \sin \frac{\pi}{2}\right)$$
$$= 6(0 + i \cdot 1)$$
$$= 6i$$

We can check by multiplying with standard notation. Let's check Example 5(b):

$$2(\cos \pi + i \sin \pi) = 2(-1 + 0) = -2$$

and

$$3\left[\cos\left(-\frac{\pi}{2}\right) + i \sin\left(-\frac{\pi}{2}\right)\right] = 3[0 + i(-1)] = -3i.$$

Thus,

$$-2 \cdot (-3i) = 6i.$$

Example 6 Convert to trigonometric notation and multiply:

$$(1 + i)(\sqrt{3} - i).$$

SOLUTION We first find trigonometric notation:

$$1 + i = \sqrt{2}(\cos 45° + i \sin 45°),$$ See Example 3(a).

$$\sqrt{3} - i = 2(\cos 330° + i \sin 330°).$$ See Example 3(b).

Then we multiply:

$$\sqrt{2}(\cos 45° + i \sin 45°) \cdot 2(\cos 330° + i \sin 330°)$$
$$= 2\sqrt{2}[\cos (45° + 330°) + i \sin (45° + 330°)]$$
$$= 2\sqrt{2}(\cos 375° + i \sin 375°).$$

To divide complex numbers, we divide the absolute values and subtract the arguments. We state this fact below, but omit the proof.

Complex Numbers: Division

For any complex numbers $r_1(\cos \theta_1 + i \sin \theta_1)$ and $r_2(\cos \theta_2 + i \sin \theta_2)$, $r_2 \neq 0$,

$$\frac{r_1(\cos \theta_1 + i \sin \theta_1)}{r_2(\cos \theta_2 + i \sin \theta_2)} = \frac{r_1}{r_2}[\cos (\theta_1 - \theta_2) + i \sin (\theta_1 - \theta_2)].$$

Example 7 Divide

$$2\left(\cos\frac{3\pi}{2} + i\sin\frac{3\pi}{2}\right) \quad \text{by} \quad 4\left(\cos\frac{\pi}{2} + i\sin\frac{\pi}{2}\right)$$

and express the solution in standard notation.

SOLUTION We have

$$\frac{2\left(\cos\frac{3\pi}{2} + i\sin\frac{3\pi}{2}\right)}{4\left(\cos\frac{\pi}{2} + i\sin\frac{\pi}{2}\right)}$$

$$= \frac{2}{4}\left[\cos\left(\frac{3\pi}{2} - \frac{\pi}{2}\right) + i\sin\left(\frac{3\pi}{2} - \frac{\pi}{2}\right)\right]$$

$$= \frac{1}{2}(\cos\pi + i\sin\pi)$$

$$= \frac{1}{2}(-1 + i\cdot 0)$$

$$= -\frac{1}{2}.$$

Example 8 Convert to trigonometric notation and divide:

$$\frac{1+i}{1-i}.$$

SOLUTION We first convert to trigonometric notation:

$$1 + i = \sqrt{2}(\cos 45° + i\sin 45°),$$ See Example 3(a).

$$1 - i = \sqrt{2}(\cos 315° + i\sin 315°).$$

We now divide:

$$\frac{\sqrt{2}(\cos 45° + i\sin 45°)}{\sqrt{2}(\cos 315° + i\sin 315°)}$$

$$= 1[\cos(45° - 315°) + i\sin(45° - 315°)]$$

$$= \cos(-270°) + i\sin(-270°)$$

$$= 0 + i\cdot 1$$

$$= i.$$

Powers of Complex Numbers

An important theorem about powers and roots of complex numbers is named for the French mathematician Abraham DeMoivre (1667–1754). Let's consider the square of a complex number $r(\cos\theta + i\sin\theta)$:

$$[r(\cos\theta + i\sin\theta)]^2 = [r(\cos\theta + i\sin\theta)]\cdot[r(\cos\theta + i\sin\theta)]$$

$$= r\cdot r\cdot[\cos(\theta+\theta) + i\sin(\theta+\theta)]$$

$$= r^2(\cos 2\theta + i\sin 2\theta).$$

Similarly, we see that

$$[r(\cos\theta + i\sin\theta)]^3$$
$$= r \cdot r \cdot r \cdot [\cos(\theta + \theta + \theta) + i\sin(\theta + \theta + \theta)]$$
$$= r^3(\cos 3\theta + i\sin 3\theta).$$

DeMoivre's theorem is the generalization of these results.

DeMoivre's Theorem

For any complex number $r(\cos\theta + i\sin\theta)$ and any natural number n,

$$[r(\cos\theta + i\sin\theta)]^n = r^n(\cos n\theta + i\sin n\theta).$$

Example 9 Find each of the following.

a) $(1 + i)^9$
b) $(\sqrt{3} - i)^{10}$

SOLUTION

a) We first find trigonometric notation:

$$1 + i = \sqrt{2}(\cos 45° + i\sin 45°).$$

Then

$$(1 + i)^9 = [\sqrt{2}(\cos 45° + i\sin 45°)]^9$$
$$= (\sqrt{2})^9[\cos(9 \cdot 45°) + i\sin(9 \cdot 45°)]$$

DeMoivre's theorem

$$= 2^{9/2}(\cos 405° + i\sin 405°)$$
$$= 16\sqrt{2}(\cos 45° + i\sin 45°)$$

405° has the same terminal side as 45°.

$$= 16\sqrt{2}\left(\frac{\sqrt{2}}{2} + i\frac{\sqrt{2}}{2}\right)$$
$$= 16 + 16i.$$

b) We first convert to trigonometric notation:

$$\sqrt{3} - i = 2(\cos 330° + i\sin 330°).$$

Then

$$(\sqrt{3} - i)^{10} = [2(\cos 330° + i\sin 330°)]^{10}$$
$$= 2^{10}(\cos 3300° + i\sin 3300°)$$
$$= 1024(\cos 60° + i\sin 60°)$$
$$= 1024\left(\frac{1}{2} + i\frac{\sqrt{3}}{2}\right)$$
$$= 512 + 512\sqrt{3}i.$$

Roots of Complex Numbers

As we will see, every nonzero complex number has two square roots. A nonzero complex number has three cube roots, four fourth roots, and so on. In general, a nonzero complex number has n different nth roots. They can be found using the formula that we now state and prove.

Roots of Complex Numbers

The nth roots of a complex number $r(\cos \theta + i \sin \theta)$, $r \neq 0$, are given by

$$r^{1/n}\left[\cos\left(\frac{\theta}{n} + k \cdot \frac{360^\circ}{n}\right) + i \sin\left(\frac{\theta}{n} + k \cdot \frac{360^\circ}{n}\right)\right],$$

where $k = 0, 1, 2, \ldots, n - 1$.

Using DeMoivre's theorem, we show that this formula gives us n different roots. First we take the expression for the nth roots and raise it to the nth power, to show that we get $r(\cos \theta + i \sin \theta)$:

$$\left[r^{1/n}\left[\cos\left(\frac{\theta}{n} + k \cdot \frac{360^\circ}{n}\right) + i \sin\left(\frac{\theta}{n} + k \cdot \frac{360^\circ}{n}\right)\right]\right]^n$$

$$= (r^{1/n})^n\left[\cos\left(n\left(\frac{\theta}{n} + k \cdot \frac{360^\circ}{n}\right)\right) + i \sin\left(n\left(\frac{\theta}{n} + k \cdot \frac{360^\circ}{n}\right)\right)\right]$$

$$= r[\cos(\theta + k \cdot 360^\circ) + i \sin(\theta + k \cdot 360^\circ)]$$

$$= r(\cos \theta + i \sin \theta).$$

Thus we know that the formula gives us nth roots for any nonnegative integer k, $k < n$.

Next, we show that there are at least n different roots. To see this, consider substituting 0, 1, 2, and so on, for k. When $k = n$, the cycle begins to repeat, but from 0 to $n - 1$, the angles obtained and their sines and cosines are all different. There cannot be more than n different nth roots. That fact follows from the fundamental theorem of algebra, considered in the study of polynomial functions (Section 2.5).

Example 10 Find the square roots of $2 + 2\sqrt{3}i$.

SOLUTION We first find trigonometric notation:

$$2 + 2\sqrt{3}i = 4(\cos 60^\circ + i \sin 60^\circ).$$

Then

$$[4(\cos 60^\circ + i \sin 60^\circ)]^{1/2}$$

$$= 4^{1/2}\left[\cos\left(\frac{60^\circ}{2} + k \cdot \frac{360^\circ}{2}\right) + i \sin\left(\frac{60^\circ}{2} + k \cdot \frac{360^\circ}{2}\right)\right], \quad k = 0, 1$$

$$= 2[\cos(30^\circ + k \cdot 180^\circ) + i \sin(30^\circ + k \cdot 180^\circ), \quad k = 0, 1.$$

Program

DEMOIVRE: This program graphs nth roots of complex numbers.

Thus the roots are

$$2(\cos 30° + i \sin 30°) \text{ for } k = 0$$

and

$$2(\cos 210° + i \sin 210°) \text{ for } k = 1,$$

or

$$\sqrt{3} + i \quad \text{and} \quad -\sqrt{3} - i.$$

In Example 10, we see that the two square roots of the number are opposites of each other. We can illustrate this graphically. We also note that the roots are equally spaced about a circle of radius r—in this case, $r = 2$. The roots are 360°/2, or 180° apart.

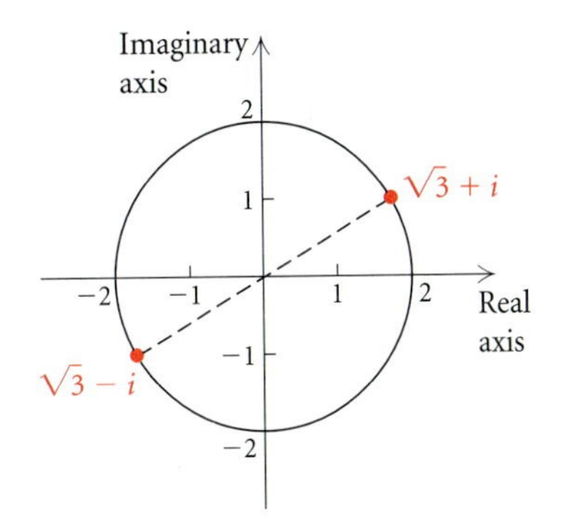

Example 11 Find the cube roots of 1. Then locate them on a graph.

SOLUTION We begin by finding trigonometric notation:

$$1 = 1(\cos 0° + i \sin 0°).$$

Then

$$[1(\cos 0° + i \sin 0°)]^{1/3}$$
$$= 1^{1/3}\left[\cos\left(\frac{0°}{3} + k \cdot \frac{360°}{3}\right) + i \sin\left(\frac{0°}{3} + k \cdot \frac{360°}{3}\right)\right], \quad k = 0, 1, 2.$$

The roots are

$$1(\cos 0° + i \sin 0°), \quad 1(\cos 120° + i \sin 120°),$$

and

$$1(\cos 240° + i \sin 240°),$$

or

$$1, \quad -\frac{1}{2} + \frac{\sqrt{3}}{2}i, \quad \text{and} \quad -\frac{1}{2} - \frac{\sqrt{3}}{2}i.$$

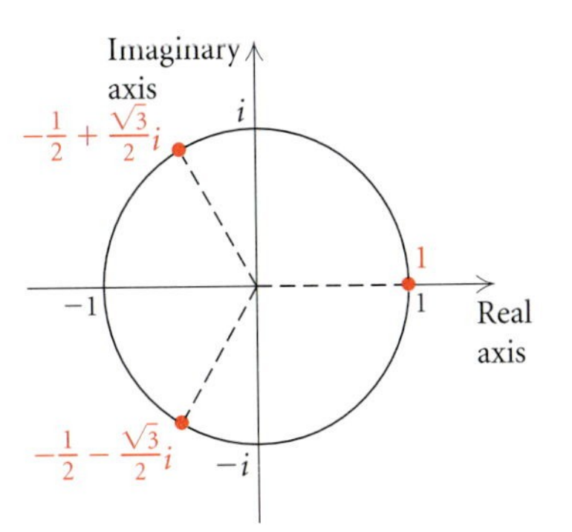

The graphs of the cube roots lie equally spaced about a circle of radius $r = 1$. The roots are 360°/3, or 120° apart.

The nth roots of 1 are often referred to as the **nth roots of unity**. In Example 11, we found the cube roots of unity.

Interactive Discovery

With a grapher in PARAMETRIC mode, we can approximate the nth roots of a number p. We use the following window and let

$$X_{1T} = (p\wedge(1/n)) \cos T \quad \text{and} \quad Y_{1T} = (p\wedge(1/n)) \sin T.$$

WINDOW

Tmin = 0

Tmax = 360, if in degree mode, or
= 2π, if in radian mode

Tstep = $360/n$, or $2\pi/n$

Xmin = −3

Xmax = 3

Xscl = 1

Ymin = −2

Ymax = 2

Yscl = 1

To find the fifth roots of 8, enter $X_{1T} = (8\wedge(1/5)) \cos T$ and $Y_{1T} = (8\wedge(1/5)) \sin T$. In this case, use DEGREE mode. After the graph has been generated, use the TRACE feature to locate the fifth roots. The T, X, and Y values appear on the screen. What do they represent?

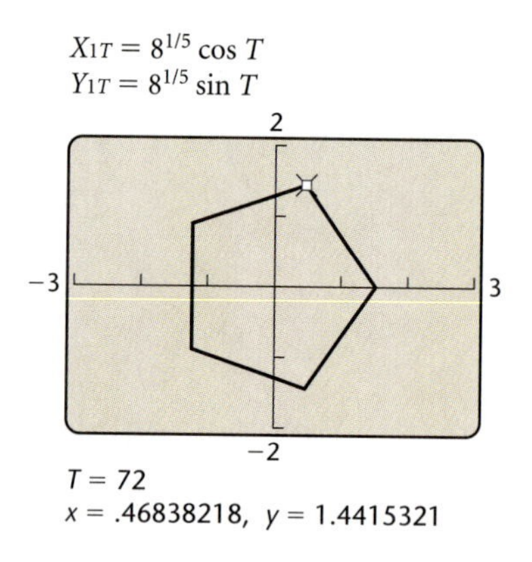

Three of the fifth roots of 8 are approximately

$$1.5157, \qquad 0.46838 + 1.44153i, \quad \text{and} \quad -1.22624 + 0.89092i.$$

Find the other two. Then use a grapher to approximate the cube roots of unity that were found in Example 11. Then approximate the fourth roots of 5 and the tenth roots of unity.

6.3 Exercise Set

Graph each complex number and find its absolute value.

1. $4 + 3i$
2. $-2 - 3i$
3. i
4. $-5 - 2i$
5. $4 - i$
6. $6 + 3i$
7. 3
8. $-2i$

Express the indicated number in both standard notation and trigonometric notation.

9.
10.
11.
12.

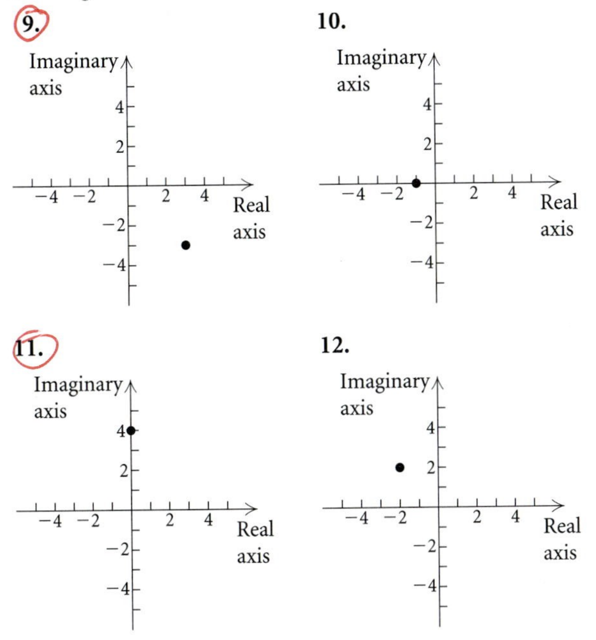

Find trigonometric notation.

13. $1 - i$
14. $-10\sqrt{3} + 10i$
15. $-3i$
16. $-5 + 5i$
17. $\sqrt{3} + i$
18. 4
19. $\frac{2}{5}$
20. $7.5i$

Find standard notation, $a + bi$.

21. $3(\cos 30° + i \sin 30°)$
22. $6(\cos 120° + i \sin 120°)$
23. $10(\cos 270° + i \sin 270°)$
24. $3(\cos 0° + i \sin 0°)$
25. $\sqrt{8}\left(\cos \frac{\pi}{4} + i \sin \frac{\pi}{4}\right)$
26. $5\left(\cos \frac{\pi}{3} + i \sin \frac{\pi}{3}\right)$
27. $2\left(\cos \frac{\pi}{2} + i \sin \frac{\pi}{2}\right)$
28. $3\left[\cos\left(-\frac{3\pi}{4}\right) + i \sin\left(-\frac{3\pi}{4}\right)\right]$

Multiply or divide and leave the answer in trigonometric notation.

29. $\dfrac{12(\cos 48° + i \sin 48°)}{3(\cos 6° + i \sin 6°)}$
30. $5\left(\cos \frac{\pi}{3} + i \sin \frac{\pi}{3}\right) \cdot 2\left(\cos \frac{\pi}{4} + i \sin \frac{\pi}{4}\right)$
31. $2.5(\cos 35° + i \sin 35°) \cdot 4.5(\cos 21° + i \sin 21°)$
32. $\dfrac{\frac{1}{2}\left(\cos \frac{2\pi}{3} + i \sin \frac{2\pi}{3}\right)}{\frac{3}{8}\left(\cos \frac{\pi}{6} + i \sin \frac{\pi}{6}\right)}$

Convert to trigonometric notation and then multiply or divide.

33. $(1 - i)(2 + 2i)$
34. $(1 + i\sqrt{3})(1 + i)$
35. $\dfrac{1 - i}{1 + i}$
36. $\dfrac{1 - i}{\sqrt{3} - i}$
37. $(3\sqrt{3} - 3i)(2i)$
38. $(2\sqrt{3} + 2i)(2i)$
39. $\dfrac{2\sqrt{3} - 2i}{1 + \sqrt{3}i}$
40. $\dfrac{3 - 3\sqrt{3}i}{\sqrt{3} - i}$

Raise the number to the given power and write trigonometric notation for the answer.

41. $\left[2\left(\cos \frac{\pi}{3} + i \sin \frac{\pi}{3}\right)\right]^3$
42. $[2(\cos 120° + i \sin 120°)]^4$
43. $(1 + i)^6$
44. $(-\sqrt{3} + i)^5$

Raise the number to the given power and write standard notation for the answer.

45. $[3(\cos 20° + i \sin 20°)]^3$
46. $[2(\cos 10° + i \sin 10°)]^9$
47. $(1 - i)^5$
48. $(2 + 2i)^4$
49. $\left(\frac{1}{\sqrt{2}} - \frac{1}{\sqrt{2}}i\right)^{12}$
50. $\left(\frac{\sqrt{3}}{2} + \frac{1}{2}i\right)^{10}$

Find the square roots of each number.

51. $-i$
52. $1 + i$
53. $2\sqrt{2} - 2\sqrt{2}i$
54. $-\sqrt{3} - i$

Find the cube roots of each number.

55. i

56. $-64i$

57. $2\sqrt{3} - 2i$

58. $1 - \sqrt{3}i$

59. Find and graph the fourth roots of 16.

60. Find and graph the fourth roots of i.

61. Find and graph the fifth roots of -1.

62. Find and graph the sixth roots of 1.

63. Find the tenth roots of 8.

64. Find the ninth roots of -4.

65. Find the sixth roots of -1.

66. Find the fourth roots of 12.

Find all the complex solutions of the equation.

67. $x^3 = 1$

68. $x^5 - 1 = 0$

69. $x^4 + i = 0$

70. $x^4 + 81 = 0$

71. $x^6 + 64 = 0$

72. $x^5 + \sqrt{3} + i = 0$

Skill Maintenance

Convert to degree measure.

73. $\frac{\pi}{12}$

74. 3π

Convert to radian measure.

75. $330°$

76. $-225°$

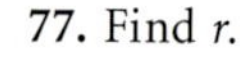

77. Find r.

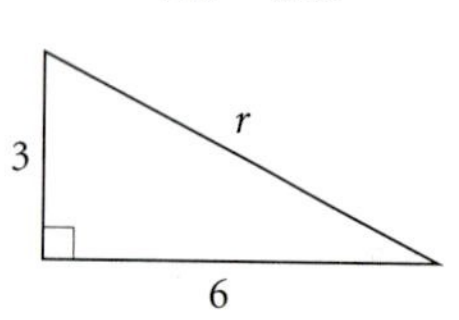

78. Graph these points in the rectangular coordinate system: $(2, -1)$, $(0, 3)$, and $\left(-\frac{1}{2}, -4\right)$.

Synthesis

79. ◈ Find and graph the square roots of $1 - i$. Explain geometrically why they are the opposites of each other.

80. ◈ Explain why trigonometric notation for a complex number is not unique, but rectangular, or standard, notation is unique.

Solve.

81. $x^2 + (1 - i)x + i = 0$

82. $3x^2 + (1 + 2i)x + 1 - i = 0$

83. Find polar notation for $(\cos\theta + i\sin\theta)^{-1}$.

84. Show that for any complex number z,

$$|z| = |-z|.$$

(*Hint:* Let $z = a + bi$.)

85. Show that for any complex number z,

$$|z| = |\bar{z}|.$$

(*Hint:* Let $z = a + bi$.)

86. Show that for any complex number z,

$$|z\bar{z}| = |z^2|.$$

87. Show that for any complex number z,

$$|z^2| = |z|^2.$$

88. Show that for any complex numbers z and w,

$$|z \cdot w| = |z| \cdot |w|.$$

(*Hint:* Let $z = r_1(\cos\theta_1 + i\sin\theta_1)$ and $w = r_2(\cos\theta_2 + i\sin\theta_2)$.)

89. Show that for any complex number z and any nonzero, complex number w,

$$\left|\frac{z}{w}\right| = \frac{|z|}{|w|}.$$ (Use the hint for Exercise 88.)

90. On a complex plane, graph $|z| = 1$.

91. On a complex plane, graph $z + \bar{z} = 3$.

6.4 Polar Coordinates and Graphs

- *Graph points given their polar coordinates.*
- *Convert from rectangular to polar coordinates and from polar to rectangular coordinates.*
- *Convert from rectangular to polar equations and from polar to rectangular equations.*
- *Graph polar equations.*

Polar Coordinates

All graphing throughout this text has been done with rectangular coordinates, (x, y), in the Cartesian coordinate system. We now introduce the polar coordinate system. As shown in the diagram below, any point P has rectangular coordinates (x, y) and polar coordinates (r, θ). On a polar graph, the origin is called the **pole** and the positive half of the x-axis is called the **polar axis**. The point P can be plotted given the directed angle θ from the polar axis to the ray OP and the directed distance r from the pole to the point. The angle θ can be expressed in degrees or radians.

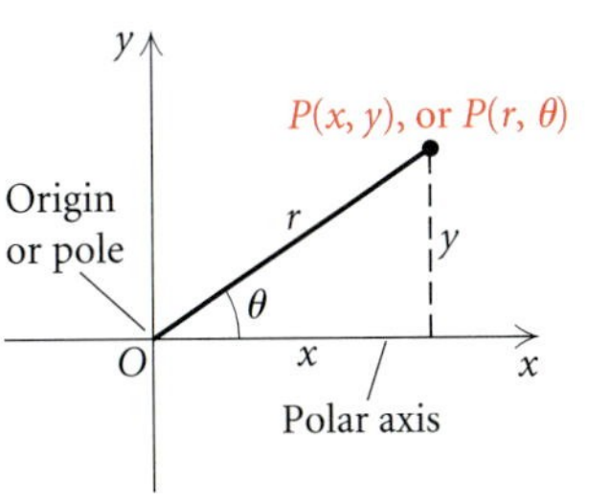

To plot points on a polar graph:

1. Locate the directed angle θ.
2. Move a directed distance r from the pole. If $r > 0$, move along ray OP. If $r < 0$, move in the opposite direction of ray OP.

Polar graph paper, shown below, facilitates plotting. Points B and G illustrate that θ may be in radians. Points E and F illustrate that the polar coordinates of a point are not unique.

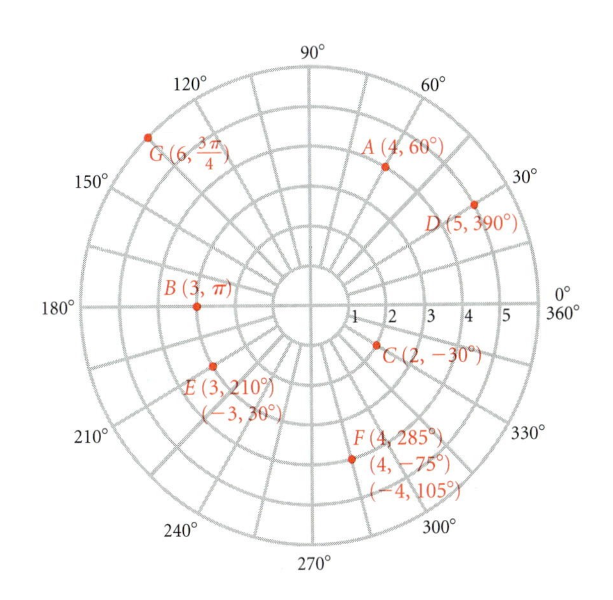

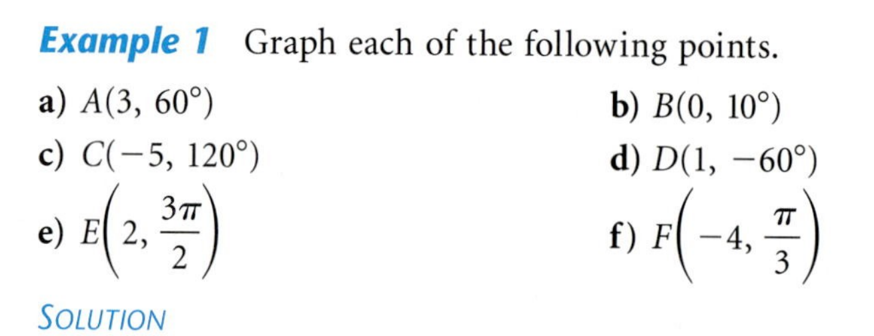

Example 1 Graph each of the following points.

a) $A(3, 60°)$ b) $B(0, 10°)$

c) $C(-5, 120°)$ d) $D(1, -60°)$

e) $E\left(2, \frac{3\pi}{2}\right)$ f) $F\left(-4, \frac{\pi}{3}\right)$

SOLUTION

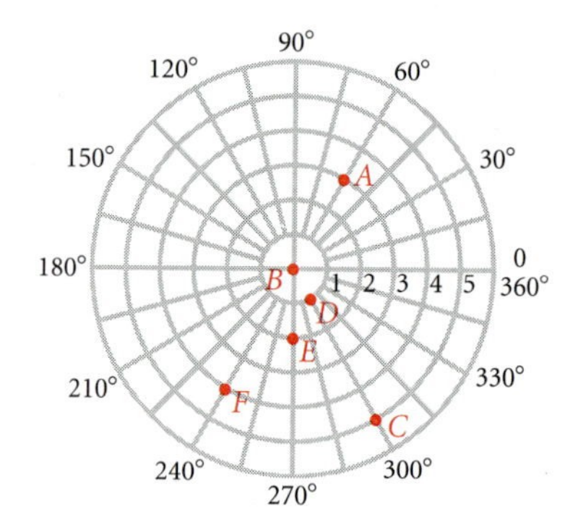

To convert from rectangular to polar coordinates and from polar to rectangular coordinates, we need to recall the following relationships.

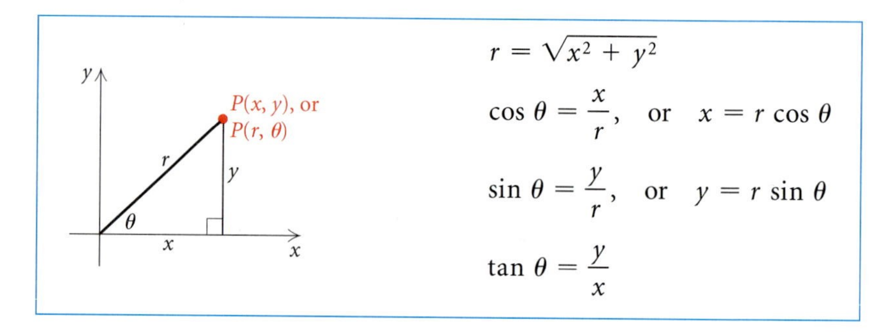

$$r = \sqrt{x^2 + y^2}$$

$$\cos\theta = \frac{x}{r}, \quad \text{or} \quad x = r\cos\theta$$

$$\sin\theta = \frac{y}{r}, \quad \text{or} \quad y = r\sin\theta$$

$$\tan\theta = \frac{y}{x}$$

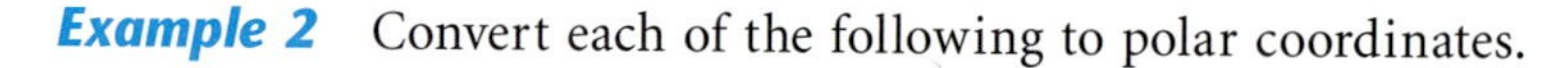

Example 2 Convert each of the following to polar coordinates.

a) $(3, 3)$

b) $(2\sqrt{3}, -2)$

SOLUTION

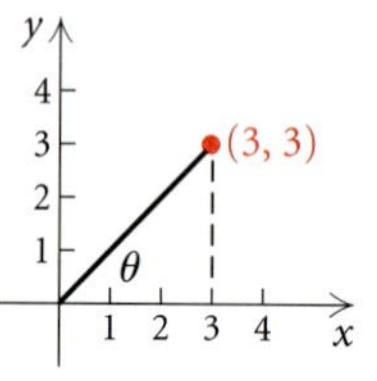

a) We first find r:

$$r = \sqrt{3^2 + 3^2}$$
$$= \sqrt{18} = 3\sqrt{2}.$$

Then we determine θ:

$$\tan\theta = \frac{3}{3} = 1; \quad \text{therefore,} \quad \theta = 45°, \text{ or } \frac{\pi}{4}.$$

We know that $\theta = \pi/4$ and not $5\pi/4$ since $(3, 3)$ is in quadrant I. Thus, $(r, \theta) = (3\sqrt{2}, 45°)$, or $(3\sqrt{2}, \pi/4)$. Other possibilities for polar coordinates include $(3\sqrt{2}, -315°)$ and $(-3\sqrt{2}, 5\pi/4)$.

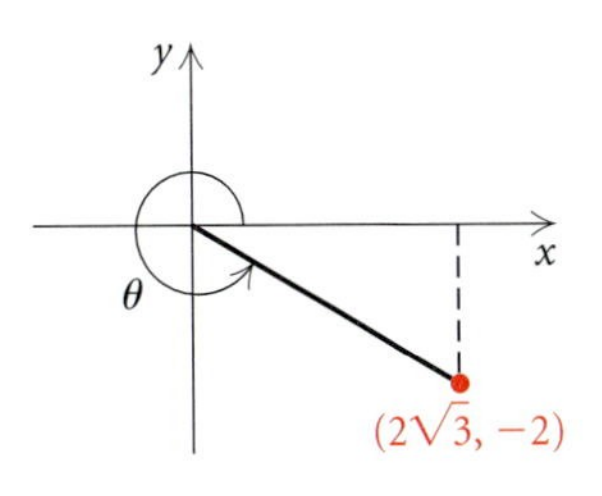

b) We first find r:

$$r = \sqrt{(2\sqrt{3})^2 + (-2)^2} = \sqrt{12 + 4} = \sqrt{16} = 4.$$

Then we determine θ:

$$\tan\theta = \frac{-2}{2\sqrt{3}} = -\frac{1}{\sqrt{3}}; \quad \text{therefore,} \quad \theta = 330°, \text{ or } \frac{11\pi}{6}.$$

Thus, $(r, \theta) = (4, 330°)$, or $(4, 11\pi/6)$. Other possibilities for polar coordinates for this point include $(4, -\pi/6)$ and $(-4, 150°)$.

It is easier to convert from polar to rectangular coordinates than from rectangular to polar coordinates.

Example 3 Convert each of the following to rectangular coordinates.

a) $\left(10, \frac{\pi}{3}\right)$ **b)** $(-5, 135°)$

SOLUTION

a) The ordered pair $(10, \pi/3)$ gives us $r = 10$ and $\theta = \pi/3$. We now find x and y:

$$x = r\cos\theta = 10\cos\frac{\pi}{3} = 10 \cdot \frac{1}{2} = 5$$

and

$$y = r\sin\theta = 10\sin\frac{\pi}{3} = 10 \cdot \frac{\sqrt{3}}{2} = 5\sqrt{3}.$$

Thus, $(x, y) = (5, 5\sqrt{3})$.

b) From the ordered pair $(-5, 135°)$, we know that $r = -5$ and $\theta = 135°$. We now find x and y:

$$x = -5\cos 135° = -5 \cdot \left(-\frac{\sqrt{2}}{2}\right) = \frac{5\sqrt{2}}{2}$$

and

$$y = -5\sin 135° = -5 \cdot \left(\frac{\sqrt{2}}{2}\right) = -\frac{5\sqrt{2}}{2}.$$

Thus, $(x, y) = (5\sqrt{2}/2, -5\sqrt{2}/2)$.

The above conversions can be easily made with some graphers.

Polar and Rectangular Equations

Some curves have simpler equations in polar coordinates than in rectangular coordinates. For others, the reverse is true.

Example 4 Convert each of the following to a polar equation.

a) $x^2 + y^2 = 25$ **b)** $2x - y = 5$

SOLUTION

a) We have

$$\begin{aligned} x^2 + y^2 &= 25 \\ (r\cos\theta)^2 + (r\sin\theta)^2 &= 25 && \textbf{Substituting for } x \textbf{ and } y \\ r^2\cos^2\theta + r^2\sin^2\theta &= 25 \\ r^2(\cos^2\theta + \sin^2\theta) &= 25 \\ r^2 &= 25 && \cos^2\theta + \sin^2\theta = 1 \\ r &= 5. \end{aligned}$$

This example illustrates that the polar equation of a circle centered at the origin is much simpler than the rectangular equation.

b) We have

$$\begin{aligned} 2x - y &= 5 \\ 2(r\cos\theta) - (r\sin\theta) &= 5 \\ r(2\cos\theta - \sin\theta) &= 5. \end{aligned}$$

In this example, we see that the rectangular equation is simpler than the polar equation.

Example 5 Convert each of the following to a rectangular equation.

a) $r = 4$

b) $r\cos\theta = 6$

c) $r = 2\cos\theta + 3\sin\theta$

SOLUTION

a) We have

$$\begin{aligned} r &= 4 \\ \sqrt{x^2 + y^2} &= 4 && \textbf{Substituting for } r \\ x^2 + y^2 &= 16. && \textbf{Squaring} \end{aligned}$$

In squaring, we must be careful not to introduce solutions of the equation that are not already present. In this case, we did not, because the graph of either equation is a circle of radius 4 centered at the origin.

b) We have

$$\begin{aligned} r\cos\theta &= 6 \\ x &= 6. && x = r\cos\theta \end{aligned}$$

The graph of $r\cos\theta = 6$, or $x = 6$, is a vertical line.

c) We have

$$\begin{aligned} r &= 2\cos\theta + 3\sin\theta \\ r^2 &= 2r\cos\theta + 3r\sin\theta && \textbf{Multiplying both sides by } r \\ x^2 + y^2 &= 2x + 3y. && \textbf{Substituting } x^2 + y^2 \textbf{ for } r^2, x \textbf{ for } r\cos\theta, \textbf{ and } y \textbf{ for } r\sin\theta. \end{aligned}$$

Graphing Polar Equations

To graph a polar equation, we can make a table of values, choosing values of θ and calculating corresponding values of r. We plot the points and complete the graph, as in the rectangular case. A difference occurs in the case of a polar equation, because as θ increases sufficiently, points may, in some cases, begin to repeat and the curve will be traced again and again. If such a point is reached, the curve is complete.

Example 6 Make a hand-drawn graph of $r = 1 - \sin\theta$.

SOLUTION We first make a table of values. The TABLE feature on a grapher is the most efficient way to create this list. Note that the points begin to repeat at $\theta = 360°$. We plot these points and draw the curve, as shown below.

θ	r
0°	1
15°	0.74118
30°	0.5
45°	0.29289
60°	0.13397
75°	0.03407
90°	0
105°	0.03407
120°	0.13397
135°	0.29289
150°	0.5
165°	0.74118
180°	1

θ	r
195°	1.2588
210°	1.5
225°	1.7071
240°	1.866
255°	1.9659
270°	2
285°	1.9659
300°	1.866
315°	1.7071
330°	1.5
345°	1.2588
360°	1
375°	0.74118
390°	0.5

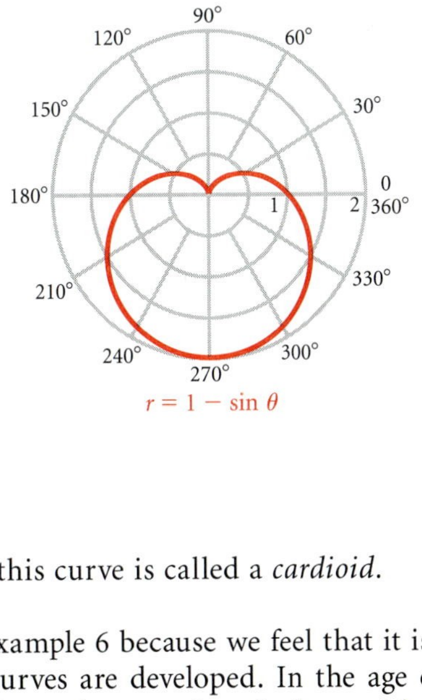

$r = 1 - \sin\theta$

Because of its heart shape, this curve is called a *cardioid*.

We plotted points in Example 6 because we feel that it is important to understand how these curves are developed. In the age of graphers, most of our work in graphing polar equations is done with technology. Nearly all graphers allow the graphing of polar equations. In general, the equation must be written first in the form $r = f(\theta)$. When graphing polar equations, it is necessary to decide on not only the best window dimensions but also the range of values for θ. Typically, we begin with a range of 0 to 2π for θ in radians and 0° to 360° for θ in degrees. Because most polar graphs are curved, it is important to square the window to minimize distortion.

Interactive Discovery

Graph $r = 4 \sin 3\theta$. Begin by setting the grapher in POLAR mode and use either of the following windows:

Window (Radians)	Window (Degrees)
θmin $= 0$	θmin $= 0$
θmax $= 2\pi$	θmax $= 360$
θstep $= \pi/24$	θstep $= 1$
Xmin $= -9$	Xmin $= -9$
Xmax $= 9$	Xmax $= 9$
Xscl $= 1$	Xscl $= 1$
Ymin $= -6$	Ymin $= -6$
Ymax $= 6$	Ymax $= 6$
Yscl $= 1$	Yscl $= 1$

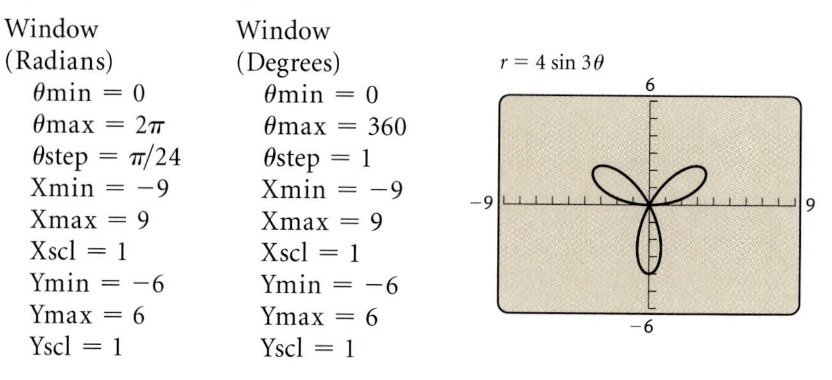

We observe the same graph in both windows. The grapher allows us to view the curve as it is formed.

Now graph each of the following equations and observe the effect of changing the coefficient of sin 3θ and the coefficient of θ:

$$r = 2 \sin 3\theta, \quad r = 6 \sin 3\theta, \quad r = 4 \sin \theta,$$
$$r = 4 \sin 5\theta, \quad r = 4 \sin 2\theta, \quad r = 4 \sin 4\theta.$$

Polar equations of the form $r = a \cos n\theta$ and $r = a \sin n\theta$ have rose-shaped curves. **The number a determines the length of the petals, and the number n determines the number of petals.** If n is odd, there are n petals. If n is even, there are $2n$ petals.

Example 7 Graph each of the following polar equations. Try to visualize the shape of the curve before graphing it.

a) $r = 3$ **b)** $r = 5 \sin \theta$ **c)** $r = \cos \theta$ **d)** $r = 2 \csc \theta$

SOLUTION

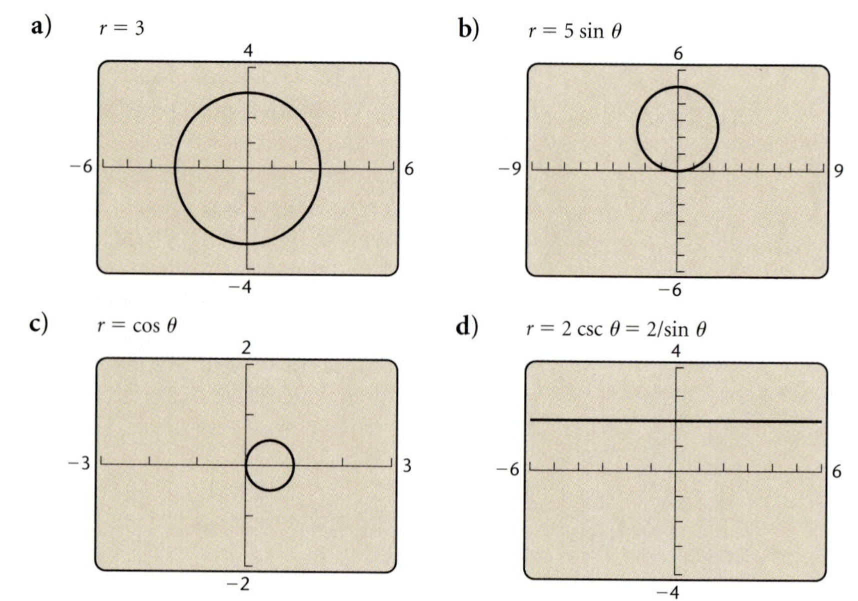

We can verify our graphs in parts (a)–(d) by converting the polar equation to the equivalent rectangular equation and visualizing the graph. For $r = 3$, we substitute $\sqrt{x^2 + y^2}$ for r and square. The resulting equation, $x^2 + y^2 = 3^2$, is the equation of a circle with radius 3 centered at the origin.

In part (d), we have

$$r = 2 \csc \theta$$

$$r = \frac{2}{\sin \theta}$$

$$r \sin \theta = 2$$

$$y = 2. \quad \text{Substituting } y \text{ for } r \sin \theta$$

The graph of $y = 2$ is a horizontal line passing through (0, 2).

Parts (b) and (c) can also be checked by rewriting the equations in rectangular form.

Example 8 Graph: $r + 1 = 2 \cos 2\theta$.

SOLUTION We first solve for r:

$$r = 2 \cos 2\theta - 1.$$

We then obtain the following graph.

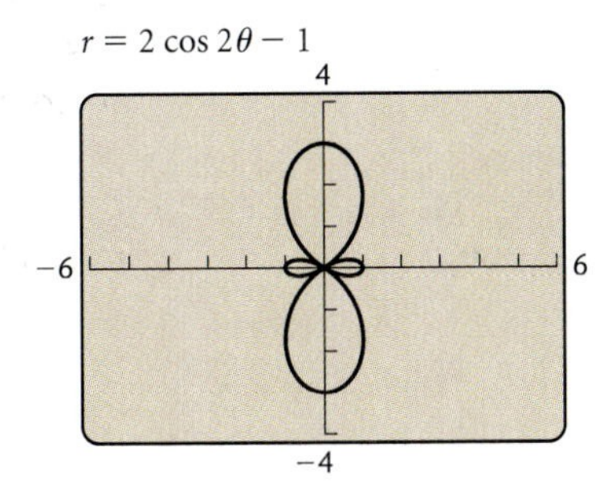

6.4 Exercise Set

Graph each point on a polar grid.

1. (2, 45°)
2. $(4, \pi)$
3. (3.5, 210°)
4. (−3, 135°)
5. $\left(1, \frac{\pi}{6}\right)$
6. (2.75, 150°)
7. $\left(-5, \frac{\pi}{2}\right)$
8. (0, 15°)
9. (3, −315°)
10. $\left(1.2, -\frac{2\pi}{3}\right)$
11. (4.3, −60°)
12. (3, 405°)

Find polar coordinates of each of these points. Give three answers for each point.

13. 14.

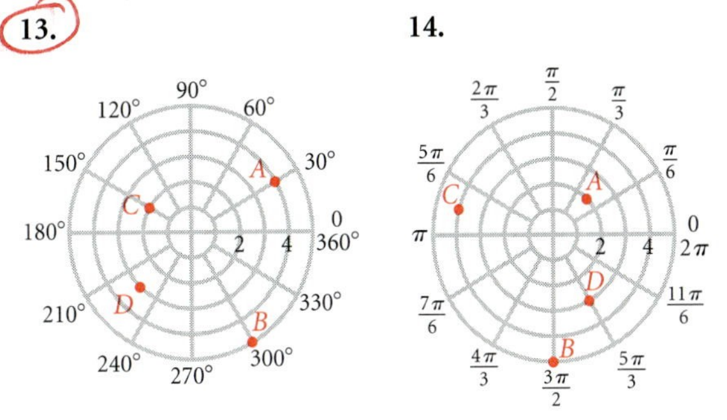

Find the polar coordinates of the point. Express the

angle in degrees and then in radians, using the smallest possible positive angle.

15. $(0, -3)$ **16.** $(-4, 4)$
17. $(3, -3\sqrt{3})$ **18.** $(-\sqrt{3}, 1)$
19. $(4\sqrt{3}, -4)$ **20.** $(2\sqrt{3}, 2)$
21. $(-\sqrt{2}, -\sqrt{2})$ **22.** $(-3, 3\sqrt{3})$

Use a grapher to convert from rectangular to polar coordinates. Express the answer in both degrees and radians, using the smallest possible positive angle.

23. $(3, 7)$ **24.** $(-2, -\sqrt{5})$
25. $(-\sqrt{10}, 3.4)$ **26.** $(0.9, -6)$

Find the rectangular coordinates of the point.

27. $(5, 60°)$ **28.** $(0, -23°)$
29. $(-3, 45°)$ **30.** $(6, 30°)$
31. $(3, -120°)$ **32.** $\left(7, \frac{\pi}{6}\right)$
33. $\left(-2, \frac{5\pi}{3}\right)$ **34.** $(1.4, 225°)$

Use a grapher to convert from polar to rectangular coordinates. Round the coordinates to the nearest hundredth.

35. $(3, -43°)$ **36.** $\left(-5, \frac{\pi}{7}\right)$
37. $\left(-4.2, \frac{3\pi}{5}\right)$ **38.** $(2.8, 166°)$

Convert to a polar equation.

39. $3x + 4y = 5$ **40.** $5x + 3y = 4$
41. $x = 5$ **42.** $y = 4$
43. $x^2 + y^2 = 36$ **44.** $x^2 - 4y^2 = 4$
45. $x^2 = 25y$ **46.** $2x - 9y + 3 = 0$
47. $y^2 - 5x - 25 = 0$ **48.** $x^2 + y^2 = 8y$

Convert to a rectangular equation.

49. $r = 5$ **50.** $\theta = \frac{3\pi}{4}$
51. $r \sin \theta = 2$ **52.** $r = -3 \sin \theta$
53. $r + r \cos \theta = 3$ **54.** $r = \frac{2}{1 - \sin \theta}$
55. $r - 9 \cos \theta = 7 \sin \theta$ **56.** $r + 5 \sin \theta = 7 \cos \theta$
57. $r = 5 \sec \theta$ **58.** $r = 3 \cos \theta$

Graph the equation by plotting points. Then check your work with a grapher.

59. $r = \sin \theta$ **60.** $r = 1 - \cos \theta$
61. $r = 4 \cos 2\theta$ **62.** $r = 1 - 2 \sin \theta$

In Exercises 63–74, use a grapher to match the equation with figures (a)–(l), which follow. Try matching the graphs mentally before using a grapher.

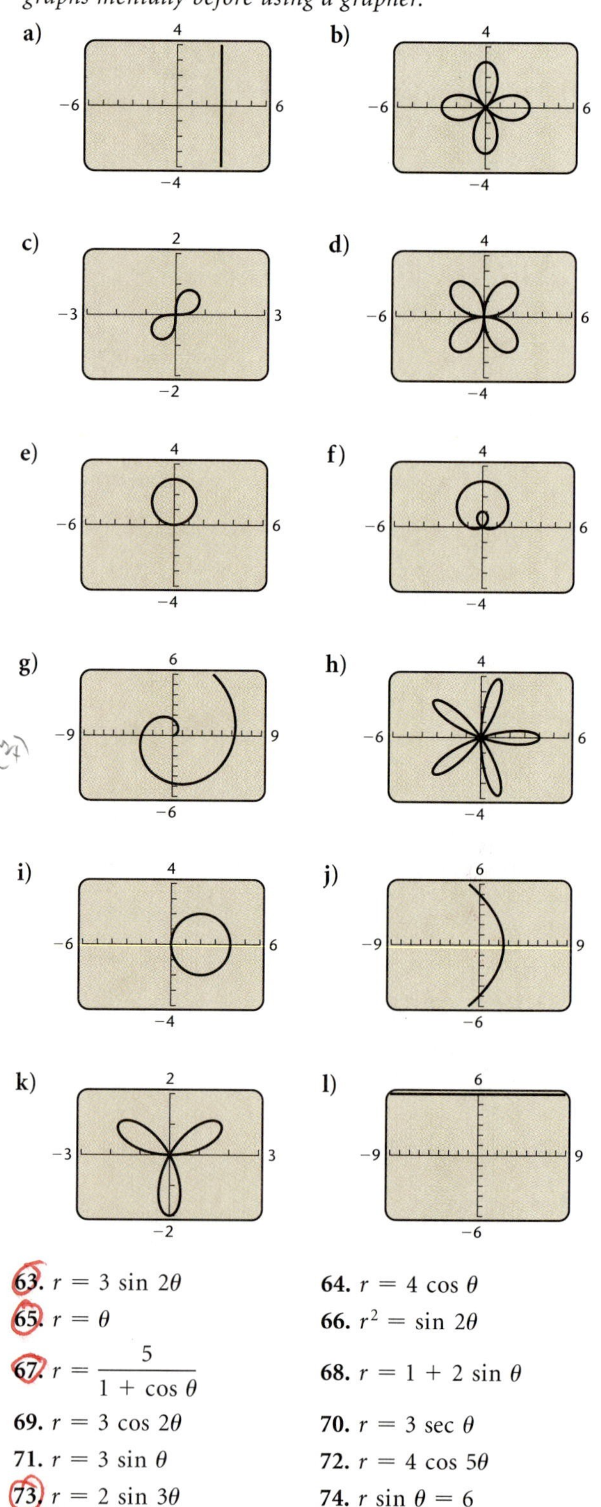

63. $r = 3 \sin 2\theta$ **64.** $r = 4 \cos \theta$
65. $r = \theta$ **66.** $r^2 = \sin 2\theta$
67. $r = \frac{5}{1 + \cos \theta}$ **68.** $r = 1 + 2 \sin \theta$
69. $r = 3 \cos 2\theta$ **70.** $r = 3 \sec \theta$
71. $r = 3 \sin \theta$ **72.** $r = 4 \cos 5\theta$
73. $r = 2 \sin 3\theta$ **74.** $r \sin \theta = 6$

Skill Maintenance

Solve.

75. $2x - 4 = x + 8$

76. $4 - 5y = 3$

Graph.

77. $y = 2x - 5$

78. $4x - y = 6$

79. $x = -3$

80. $y = 0$

Synthesis

81. ◈ Explain why the rectangular coordinates of a point are unique and the polar coordinates of a point are not unique.

82. ◈ Give an example of an equation that is easier to graph on a grapher in polar notation than in rectangular notation and explain why.

83. Convert to a rectangular equation:

$$r = \sec^2 \frac{\theta}{2}.$$

84. The center of a regular hexagon is at the origin, and one vertex is the point $(4, 0°)$. Find the coordinates of the other vertices.

Graph.

85. $r = \sin \theta \tan \theta$ (Cissoid)

86. $r = 3\theta$ (Spiral of Archimedes)

87. $r = e^{\theta/10}$ (Logarithmic spiral)

88. $r = 10^{2\theta}$ (Logarithmic spiral)

89. $r = \cos 2\theta \sec \theta$ (Strophoid)

90. $r = \cos 2\theta - 2$ (Peanut)

91. $r = \frac{1}{4} \tan^2 \theta \sec \theta$ (Semicubical parabola)

92. $r = \sin 2\theta + \cos \theta$ (Twisted sister)

6.5 Vectors

- *Perform calculations with vectors and resolve vectors into components.*
- *Solve applications involving vectors.*

In many applications, quantities occur for which a *magnitude* and a *direction* are specified. Any such quantity is called a *vector*. In this section, we consider vectors together with their properties and applications.

Vectors

We measure some quantities using only their magnitudes. For example, we describe time, length, and mass using units like seconds, feet, and kilograms, respectively. However, to measure quantities like **displacement**, **velocity**, or **force**, we need to describe a *magnitude* and a *direction*. Together these describe a **vector**. The following are some examples.

DISPLACEMENT An object moves a certain distance in a certain direction.

A surveyor steps 20 yd to the northeast.

A hiker follows a trail 5 mi to the west.

A batter hits a ball 100 m along the left-field line.

VELOCITY An object travels at a certain speed in a certain direction.

A breeze is blowing 15 mph from the northwest.

An airplane is traveling 450 km/h in a direction of 243°.

FORCE A push or pull is exerted on an object in a certain direction.

A cart is being pulled up a 30° incline, requiring an effort of 200 lb.

A 25-lb force is required to lift a box upward.

A force of 15 newtons is exerted downward on the handle of a jack. (A newton is a unit of force used in physics, abbreviated N.)

We represent vectors graphically by directed line segments, or arrows, with a tip and a tail. The length is chosen, according to some scale, to represent the **magnitude of the vector**, and the direction of the arrow represents the **direction of the vector**. For example, if we let 1 cm represent 5 km/h, then a 15-km/h wind from the northwest would be represented by an arrow 3 cm long, as shown in the figure.

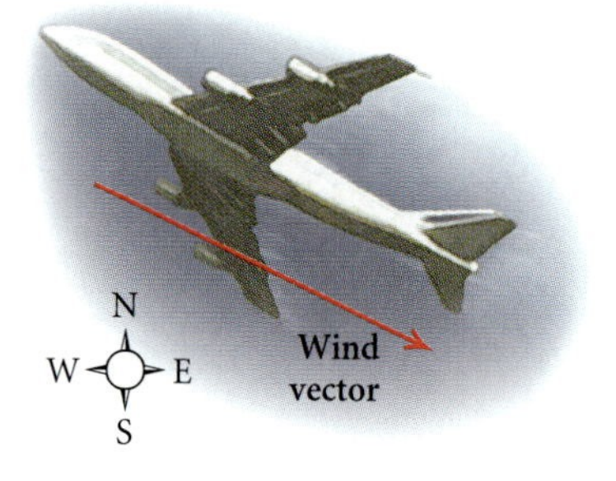

Vector

A *vector* in the plane is a directed line segment. Two vectors are *equal* or *the same* if they have the same magnitude and direction.

The vector defined by a directed line segment from point A to point B is written $\overrightarrow{AB}$ (read "vector **AB**"). The arrows we use when we draw vectors are understood to represent the same vector if they have the same length, are parallel, and point in the same direction. The four vectors in the figure below have the same length and direction. Thus they represent the *same* vector—that is,

$$\overrightarrow{AB} = \overrightarrow{CD} = \overrightarrow{OP} = \overrightarrow{EF}.$$

The vectors are all the "same" because they have the same length and direction.

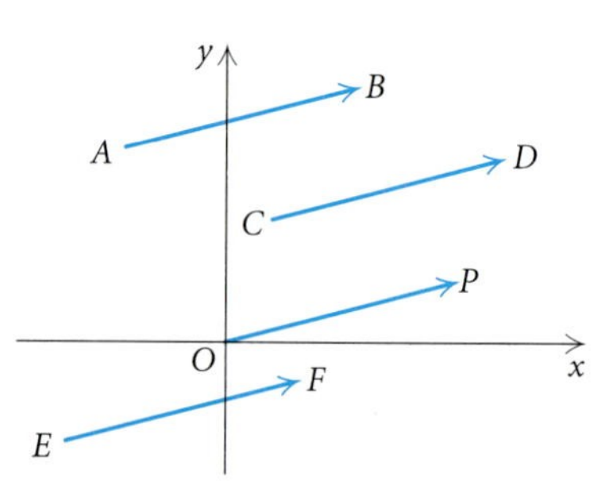

Scalars and Scalar Multiples

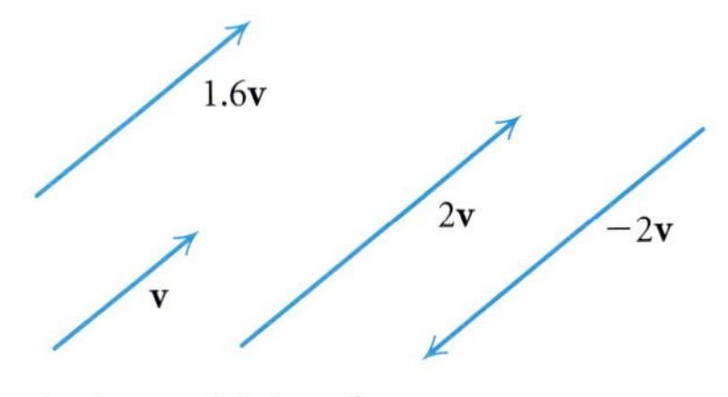

Scalar multiples of **v**

Vectors are usually written using a boldface, roman letter, such as **v** (read "vector vee"). The vectors **v**, 2**v**, 1.6**v**, and −2**v** are shown in the figure at left.

To multiply a vector **v** by a positive real number, we multiply its length by the number. Its direction stays the same. To multiply **v** by 2, we double its length and maintain its direction. To multiply a vector by 1.6, we increase its length by 60% and keep its direction. To multiply a vector by −2, we multiply by 2 and then reverse the direction. Since real numbers work like scaling factors in vector multiplication, we call them **scalars** and the products $c\mathbf{v}$ are called **scalar multiples** of **v**.

The **vector 0** has no length or direction and is understood to be the result of multiplying by the number 0.

Vector Addition

Suppose a person takes 4 steps east and then 3 steps north. He or she will then be 5 steps from the starting point in the direction shown. The **sum** of the two vectors is the vector 5 steps in magnitude and in the direction shown. The sum is also called the **resultant** of the two vectors.

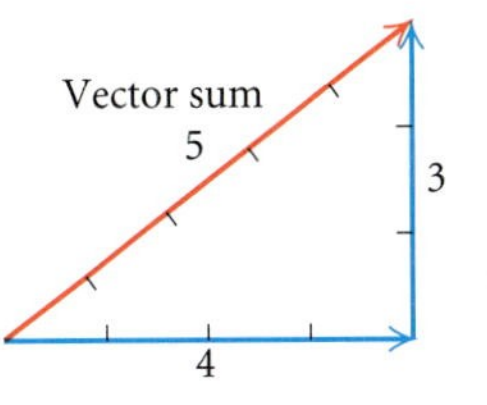

In general, two nonzero vectors $\mathbf{v}_1$ and $\mathbf{v}_2$ can be added geometrically by placing the tail of one arrow at the tip of the other and then finding the vector that forms the third side of the triangle, as shown in the following figure.

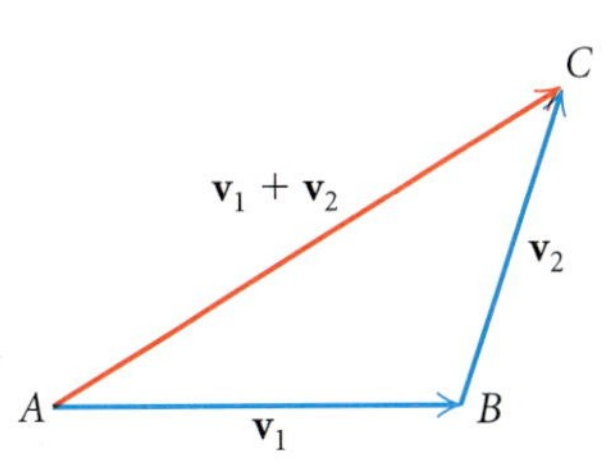

The sum $\mathbf{v}_1 + \mathbf{v}_2$ is the vector represented by the arrow from the initial point A of $\mathbf{v}_1$ to the terminal point C of $\mathbf{v}_2$. That is, if

$$\mathbf{v}_1 = \overrightarrow{AB} \quad \text{and} \quad \mathbf{v}_2 = \overrightarrow{BC},$$

then

$$\mathbf{v}_1 + \mathbf{v}_2 = \overrightarrow{AB} + \overrightarrow{BC} = \overrightarrow{AC}.$$

We can also describe vector addition by placing the tails of the arrows together, completing a parallelogram, and finding the diagonal of the parallelogram. This description of addition is sometimes called the

parallelogram law of vector addition because $\mathbf{v}_1 + \mathbf{v}_2$ is given by the diagonal of the parallelogram determined by $\mathbf{v}_1$ and $\mathbf{v}_2$.

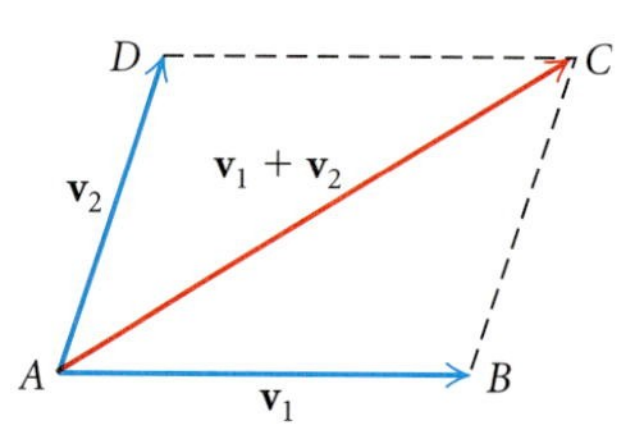

Example 1 Forces of 15 newtons and 25 newtons act on an object at right angles to each other. Find their sum, or resultant, giving the angle that it makes with the larger force.

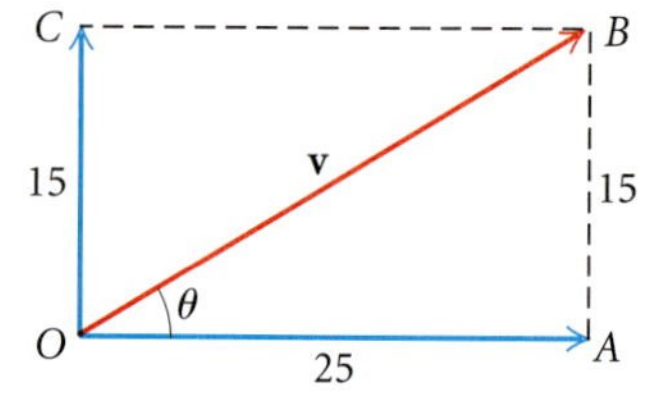

SOLUTION We make a drawing—this time, a rectangle—using $\mathbf{v}$ to represent $\overrightarrow{OB}$. To find the magnitude, we use the Pythagorean theorem:

$$|\mathbf{v}|^2 = 15^2 + 25^2$$ Here $|\mathbf{v}|$ denotes the length, or magnitude, of $\mathbf{v}$.

$$\mathbf{v} = \sqrt{15^2 + 25^2}$$

$$\mathbf{v} \approx 29.2.$$

To find the direction, we note that since OAB is a right triangle,

$$\tan\theta = \tfrac{15}{25} = 0.6.$$

From a grapher, we find that θ, the angle that the resultant makes with the larger force, is about 31°. The resultant $\overrightarrow{OB}$ has a magnitude of 29.2 and makes an angle of 31° with the larger force.

Pilots must adjust the direction of their flight when there is a crosswind. Both the wind and the aircraft velocities can be described by vectors.

Example 2 *Airplane Speed and Direction.* An airplane travels on a bearing of 100° at an airspeed of 190 km/h while a wind is blowing 48 km/h from 220°. Find the speed of the airplane over the ground and the direction of its track over the ground.

SOLUTION We first make a drawing. The wind is represented by $\overrightarrow{OC}$ and the velocity vector of the airplane by $\overrightarrow{OA}$. The resultant velocity vector is $\mathbf{v}$, the sum of the two vectors.

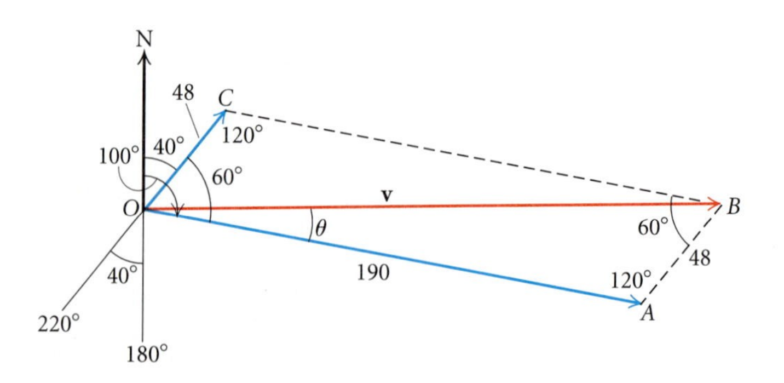

Note that the measure of $\angle CBA = 60°$. Since the sum of all the angles of the parallelogram is 360° and $\angle OCB$ and $\angle OAB$ have the same measure, each must be 120°. By the law of cosines in $\triangle OAB$, we have

$$\begin{aligned} |\mathbf{v}|^2 &= 48^2 + 190^2 - 2 \cdot 48 \cdot 190 \cos 120° \\ &= 47{,}524. \end{aligned}$$

Thus, $|\mathbf{v}|$ is 218 km/h. By the law of sines in the same triangle,

$$\frac{48}{\sin \theta} = \frac{218}{\sin 120°},$$

or

$$\begin{aligned} \sin \theta &= \frac{48 \sin 120°}{218} \\ &\approx 0.1907. \end{aligned}$$

Thus, $\theta = 11°$, to the nearest degree. The ground speed of the airplane is 218 km/h, and its track is in the direction of $100° - 11°$, or 89°.

Components

Given a vector $\mathbf{v}$, we may want to find two other vectors $\mathbf{v}_1$ and $\mathbf{v}_2$ whose sum is $\mathbf{v}$. The vectors $\mathbf{v}_1$ and $\mathbf{v}_2$ are called **components** of $\mathbf{v}$ and the process of finding them is called **resolving**, or **representing**, a vector into its components.

When we resolve a vector, we generally look for perpendicular components. Most often, one component will be parallel to the x-axis and the other will be parallel to the y-axis. For this reason, they are often called the **horizontal** and **vertical** components of a vector. In the following figure, the vector $\mathbf{v} = \overrightarrow{AC}$ is resolved as the sum of $\mathbf{v}_1 = \overrightarrow{AB}$ and $\mathbf{v}_2 = \overrightarrow{BC}$. The horizontal component of $\mathbf{v}$ is $\mathbf{v}_1$ and the vertical component is $\mathbf{v}_2$.

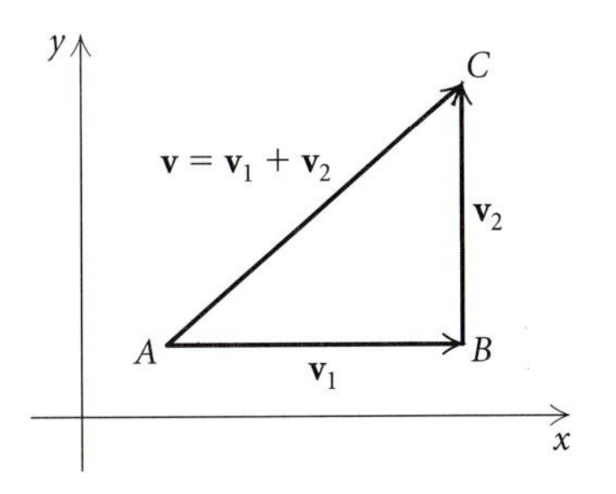

Example 3 A certain vector $\mathbf{v}$ has a magnitude of 130 and is inclined 40° with the horizontal. Find the magnitude of the horizontal and vertical components of the vector.

SOLUTION We first make a drawing showing horizontal and vertical vectors $\mathbf{v}_1$ and $\mathbf{v}_2$ whose sum is $\mathbf{v}$.

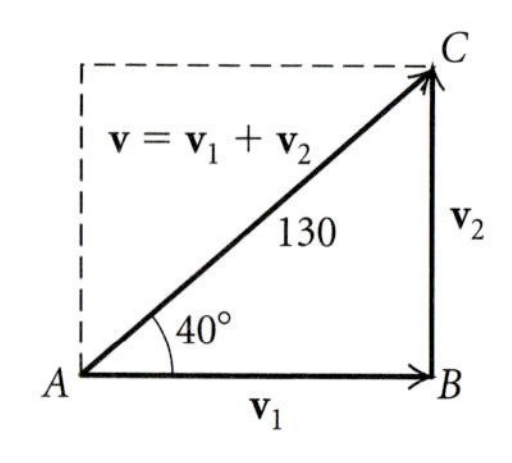

From $\triangle ABC$, we see that the magnitude of the horizontal component $\mathbf{v}_1$ is

$$130 \cos 40° = 100$$

and the magnitude of the vertical component $\mathbf{v}_2$ is

$$130 \sin 40° = 84.$$

In Example 3, we could not express $\mathbf{v}$ as a sum because 100 and 84 are scalar magnitudes, not vectors. To express $\mathbf{v}$ as a sum of vectors $\mathbf{v}_1$ and $\mathbf{v}_2$, recall that when a vector is multiplied by a positive scalar, its magnitude is changed, but its direction stays the same. If we define the vector $\mathbf{i}$ as a vector of length 1, parallel to the x-axis, and in the positive x-direction, then $\mathbf{v}_1$ can be written

$$\mathbf{v}_1 = 100\mathbf{i}.$$

Similarly, if $\mathbf{j}$ is a vector of length 1, parallel to the y-axis, and in the positive y-direction, then

$$\mathbf{v}_2 = 84\mathbf{j}$$

and

$$\mathbf{v} = 100\mathbf{i} + 84\mathbf{j}.$$

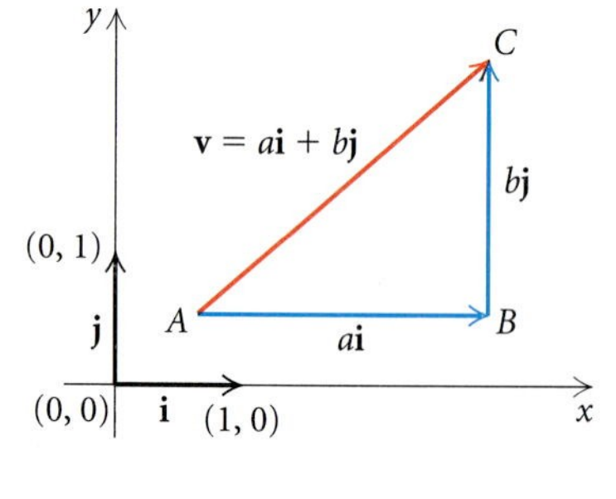

In general, a vector $\mathbf{v}$ can be resolved into a sum

$$\mathbf{v} = a\mathbf{i} + b\mathbf{j},$$

where a and b are scalars and indicate the magnitude of the horizontal and vertical components of $\mathbf{v}$.

Vector Components

If $\mathbf{v} = a\mathbf{i} + b\mathbf{j}$, the vectors $a\mathbf{i}$ and $b\mathbf{j}$ are the *vector components* of $\mathbf{v}$ in the directions of $\mathbf{i}$ and $\mathbf{j}$. The numbers a and b are the *scalar components* of $\mathbf{v}$ in the direction of $\mathbf{i}$ and $\mathbf{j}$.

Example 4 An airplane is flying at 200 km/h in a direction of 305°. Resolve the vector into its components.

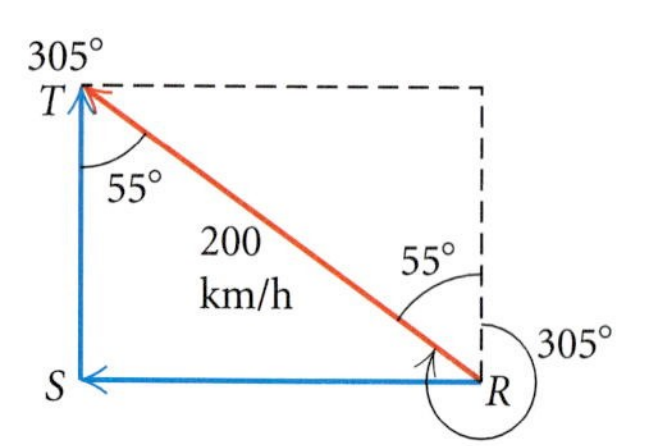

SOLUTION We first make a drawing showing westerly and northerly vectors whose sum is the given velocity. From $\triangle RST$, we see that

$$a = -200 \sin 55° \approx -164 \text{ km/h}$$

(note that a is negative because the direction is westerly) and

$$b = 200 \cos 55° \approx 115 \text{ km/h}.$$

Then

$$\mathbf{v} = -164\mathbf{i} + 115\mathbf{j}.$$

Adding and Subtracting Using Components

We can add vectors by adding their corresponding scalar components, as shown in the following figure.

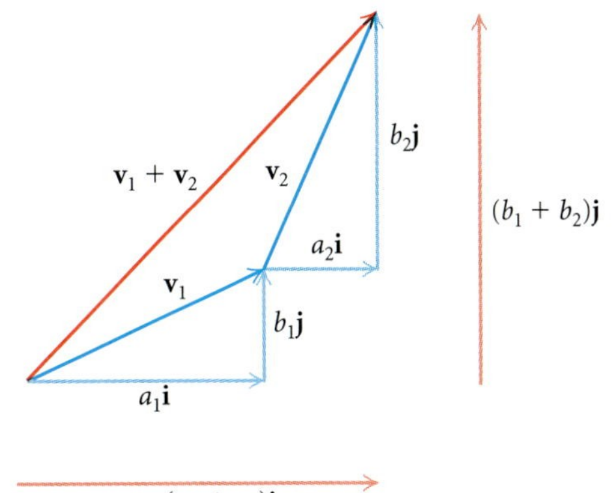

Vector Addition

If $\mathbf{v}_1 = a_1\mathbf{i} + b_1\mathbf{j}$ and $\mathbf{v}_2 = a_2\mathbf{i} + b_2\mathbf{j}$, then

$$\mathbf{v}_1 + \mathbf{v}_2 = (a_1 + a_2)\mathbf{i} + (b_1 + b_2)\mathbf{j}.$$

Example 5 Add the vectors $\mathbf{u} = 3\mathbf{i} - 7\mathbf{j}$ and $\mathbf{v} = 4\mathbf{i} + 2\mathbf{j}$.

SOLUTION We have

$$\begin{aligned}\mathbf{u} + \mathbf{v} &= (3\mathbf{i} - 7\mathbf{j}) + (4\mathbf{i} + 2\mathbf{j}) \\ &= (3 + 4)\mathbf{i} + (-7 + 2)\mathbf{j} = 7\mathbf{i} - 5\mathbf{j}.\end{aligned}$$

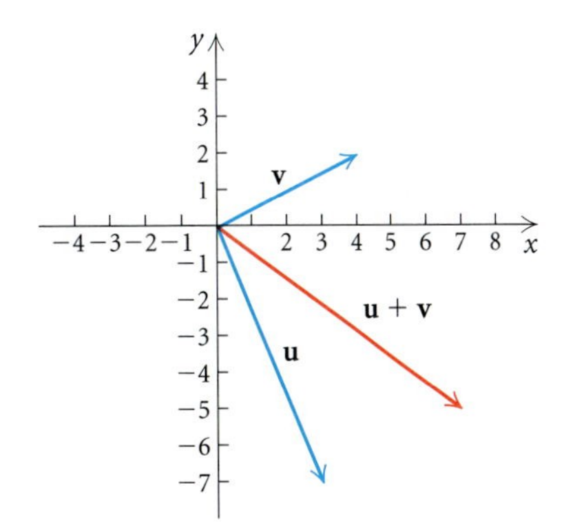

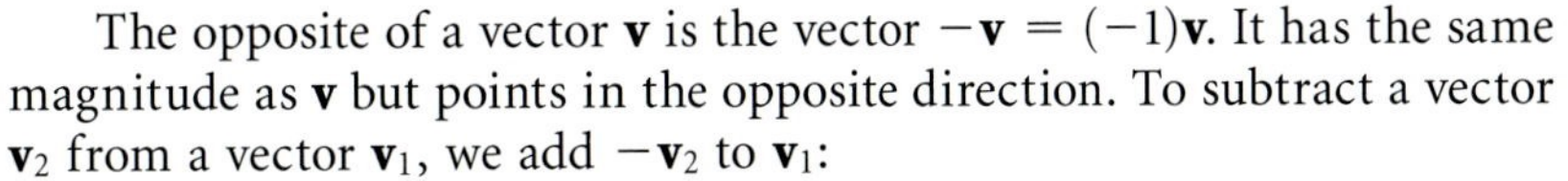

The opposite of a vector $\mathbf{v}$ is the vector $-\mathbf{v} = (-1)\mathbf{v}$. It has the same magnitude as $\mathbf{v}$ but points in the opposite direction. To subtract a vector $\mathbf{v}_2$ from a vector $\mathbf{v}_1$, we add $-\mathbf{v}_2$ to $\mathbf{v}_1$:

$$\begin{aligned}\mathbf{v}_1 - \mathbf{v}_2 &= \mathbf{v}_1 + (-\mathbf{v}_2) \\ &= \mathbf{v}_1 + (-1)\mathbf{v}_2.\end{aligned}$$

Geometrically, we can represent subtraction of vectors as shown below. We can draw $-\mathbf{v}_2$ from the tip of $\mathbf{v}_1$ and then draw the vector from the tail of $\mathbf{v}_1$ to the tip of $-\mathbf{v}_2$.

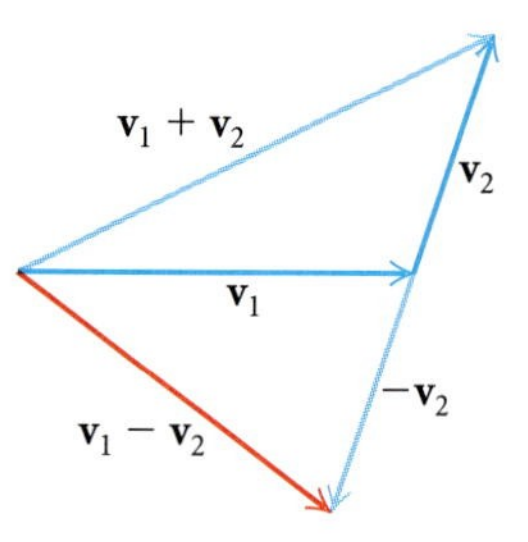

Vector Subtraction

If $\mathbf{v}_1 = a_1\mathbf{i} + b_1\mathbf{j}$ and $\mathbf{v}_2 = a_2\mathbf{i} + b_2\mathbf{j}$, then

$$\mathbf{v}_1 - \mathbf{v}_2 = (a_1 - a_2)\mathbf{i} + (b_1 - b_2)\mathbf{j}.$$

Example 6 Subtract: $(3\mathbf{i} - 9\mathbf{j}) - (7\mathbf{i} + 5\mathbf{j})$.

SOLUTION We have

$$\begin{aligned}(3\mathbf{i} - 9\mathbf{j}) - (7\mathbf{i} + 5\mathbf{j}) &= (3 - 7)\mathbf{i} + (-9 - 5)\mathbf{j} \\ &= -4\mathbf{i} - 14\mathbf{j}.\end{aligned}$$

Magnitude and Scalar Multiplication

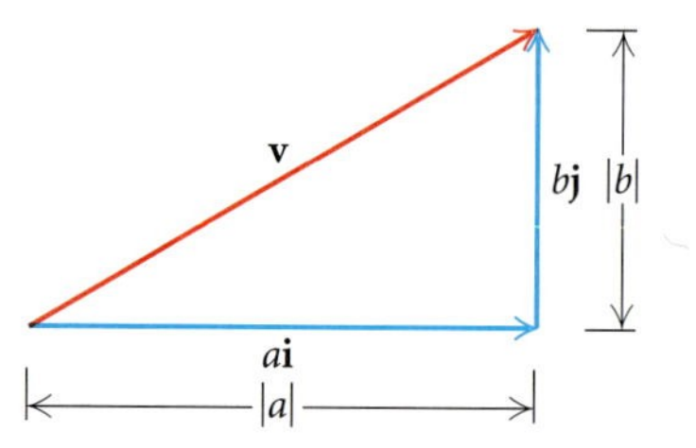

The magnitude or length of vector $\mathbf{v} = a\mathbf{i} + b\mathbf{j}$ is $|\mathbf{v}| = \sqrt{a^2 + b^2}$. This follows from the Pythagorean theorem with the right triangle formed by $\mathbf{v}$ and its two vector components, as shown in the figure at left.

When $\mathbf{v}$ is multiplied by a scalar, we can express the magnitude of the new vector in terms of $|\mathbf{v}|$.

Scalar Multiplication

If c is a scalar and $\mathbf{v} = a\mathbf{i} + b\mathbf{j}$ is a vector, then

$$c\mathbf{v} = c(a\mathbf{i} + b\mathbf{j}) = (ca)\mathbf{i} + (cb)\mathbf{j}.$$

Example 7 Do the following calculations where $\mathbf{u} = 4\mathbf{i} + 3\mathbf{j}$ and $\mathbf{v} = -6\mathbf{i} + 8\mathbf{j}$.

a) $5\mathbf{u} - 2\mathbf{v}$

b) $|5\mathbf{u} - 2\mathbf{v}|$

SOLUTION

a) $$\begin{aligned}5\mathbf{u} - 2\mathbf{v} &= 5(4\mathbf{i} + 3\mathbf{j}) - 2(-6\mathbf{i} + 8\mathbf{j}) \\ &= (20\mathbf{i} + 15\mathbf{j}) + (12\mathbf{i} - 16\mathbf{j}) = 32\mathbf{i} - \mathbf{j}\end{aligned}$$

b) Using the result of part (a), we have

$$\begin{aligned} |5\mathbf{u} - 2\mathbf{v}| &= \sqrt{(32)^2 + (-1)^2} \\ &= \sqrt{1025} \\ &\approx 32.02. \end{aligned}$$

Zero and Unit Vectors

It is convenient for work in subsequent courses, such as calculus, to consider zero vectors and unit vectors in terms of components and to have a way to express a vector so that both its magnitude and direction can be determined, or read, easily. In terms of components, the zero vector is

$$\mathbf{0} = 0\mathbf{i} + 0\mathbf{j}.$$

Any vector whose length is 1 is a **unit vector**. The vectors **i** and **j** are unit vectors:

$$\begin{aligned} |\mathbf{i}| &= |1\mathbf{i} + 0\mathbf{j}| = \sqrt{1^2 + 0^2} = 1, \\ |\mathbf{j}| &= |0\mathbf{i} + 1\mathbf{j}| = \sqrt{0^2 + 1^2} = 1. \end{aligned}$$

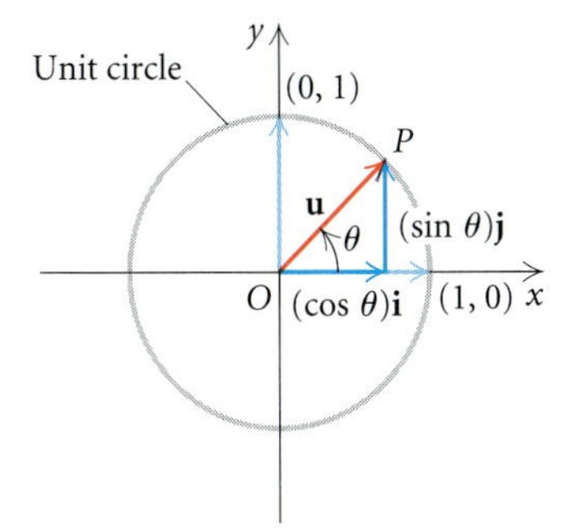

If **u** is the unit vector obtained by rotating **i** through an angle θ in the positive direction, then **u** has a horizontal component $\cos\theta$ and a vertical component $\sin\theta$, as shown in the figure at left. Thus,

$$\mathbf{u} = (\cos\theta)\mathbf{i} + (\sin\theta)\mathbf{j}. \tag{1}$$

As θ varies from 0 to 2π, the point P traces the circle $x^2 + y^2 = 1$ counterclockwise. This takes in all possible directions for unit vectors, so equation (1) describes every possible unit vector in the plane.

Example 8 Calculate and sketch the unit vector $\mathbf{u} = (\cos\theta)\mathbf{i} + (\sin\theta)\mathbf{j}$ for $\theta = 2\pi/3$. Include the unit circle in your sketch.

SOLUTION

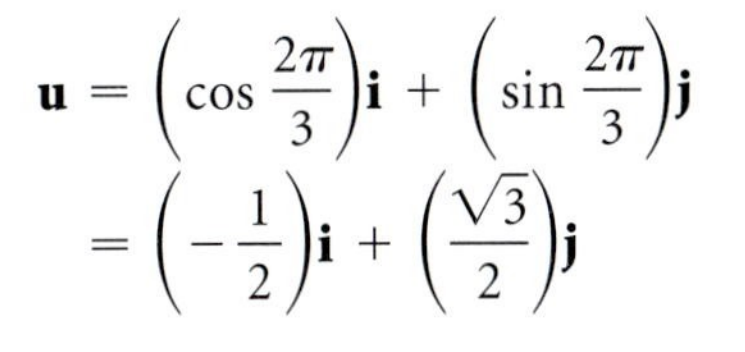

$$\begin{aligned} \mathbf{u} &= \left(\cos\frac{2\pi}{3}\right)\mathbf{i} + \left(\sin\frac{2\pi}{3}\right)\mathbf{j} \\ &= \left(-\frac{1}{2}\right)\mathbf{i} + \left(\frac{\sqrt{3}}{2}\right)\mathbf{j} \end{aligned}$$

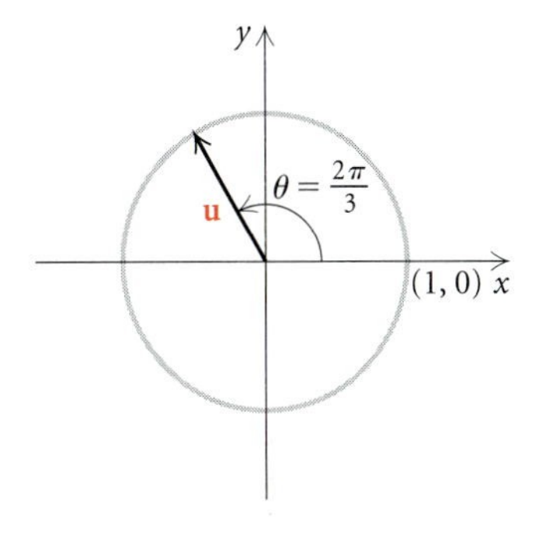

Magnitude and Direction

If $\mathbf{v} \neq \mathbf{0}$, then

$$\left|\frac{\mathbf{v}}{|\mathbf{v}|}\right| = \left|\frac{1}{|\mathbf{v}|}\mathbf{v}\right| = \frac{1}{|\mathbf{v}|}|\mathbf{v}| = 1,$$

so $\mathbf{v}/|\mathbf{v}|$ is a unit vector in the direction of $\mathbf{v}$. We can therefore express $\mathbf{v}$ in terms of its two important features, magnitude and direction, by writing $\mathbf{v} = |\mathbf{v}|(\mathbf{v}/|\mathbf{v}|)$.

> **Magnitude and Direction**
>
> If $\mathbf{v} \neq \mathbf{0}$, then:
>
> $\dfrac{\mathbf{v}}{|\mathbf{v}|}$ is a unit vector in the direction of $\mathbf{v}$, and
>
> the equation $\mathbf{v} = |\mathbf{v}|\dfrac{\mathbf{v}}{|\mathbf{v}|}$ expresses $\mathbf{v}$ in terms of its magnitude and direction.

Example 9 Express $\mathbf{v} = 3\mathbf{i} - 4\mathbf{j}$ as a product of its magnitude and direction.

SOLUTION We have

$$\text{Magnitude of } \mathbf{v}: \quad |\mathbf{v}| = \sqrt{3^2 + (-4)^2} = \sqrt{25} = 5;$$

$$\text{Direction of } \mathbf{v}: \quad \frac{\mathbf{v}}{|\mathbf{v}|} = \frac{3\mathbf{i} - 4\mathbf{j}}{5} = \frac{3}{5}\mathbf{i} - \frac{4}{5}\mathbf{j}.$$

Then

$$\mathbf{v} = 3\mathbf{i} - 4\mathbf{j} = 5\left(\frac{3}{5}\mathbf{i} - \frac{4}{5}\mathbf{j}\right).$$

Magnitude Direction

Analytic Representation of Vectors

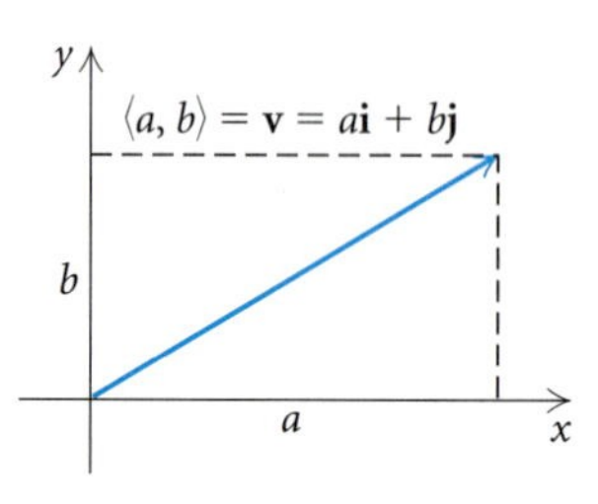

If we place a coordinate system so that the origin is at the tail of an arrow representing a vector $\mathbf{v}$, we say that the vector is in **standard position**. Then if we know the coordinates of the other end of the vector, we know the vector. The coordinates will be the scalar components of $\mathbf{v}$ in the direction of $\mathbf{i}$ and $\mathbf{j}$. Thus we can consider an ordered pair (a, b) to be a vector. To emphasize that we are thinking of a vector and to avoid the confusion of notation with ordered-pair and interval notation, we usually write $\langle a, b\rangle$. When vectors are given in this form, it is easy to add them. We simply add the respective scalar components. We do scalar multiplication by multiplying each component by the scalar.

Example 10 Do the following calculations where $\mathbf{u} = \langle 4, 3\rangle$ and $\mathbf{v} = \langle -5, 8\rangle$.

a) $\mathbf{u} + \mathbf{v}$

b) $\mathbf{u} - \mathbf{v}$

c) $-7\mathbf{v}$

d) $|3\mathbf{u} - 4\mathbf{v}|$

SOLUTION

a) $\mathbf{u} + \mathbf{v} = \langle 4, 3\rangle + \langle -5, 8\rangle = \langle -1, 11\rangle$

b) $\mathbf{u} - \mathbf{v} = \langle 4 - (-5), 3 - 8\rangle = \langle 9, -5\rangle$

c) $-7\mathbf{v} = \langle -7 \cdot (-5), -7 \cdot 8\rangle = \langle 35, -56\rangle$

d) $|3\mathbf{u} - 4\mathbf{v}| = |3\langle 4, 3\rangle - 4\langle -5, 8\rangle| = |\langle 12, 9\rangle - \langle -20, 32\rangle|$
$= |\langle 32, -23\rangle| = \sqrt{32^2 + (-23)^2} = \sqrt{1553} \approx 39.41$

6.5 Exercise Set

In Exercises 1–8, magnitudes of vectors **u** *and* **v** *and the angle* θ *between the vectors are given. Find the sum* **u** + **v**. *Give the magnitude to the nearest tenth and give the direction by specifying to the nearest degree the angle that the resultant makes with* **u**.

1. $|\mathbf{u}| = 45$, $|\mathbf{v}| = 35$, $\theta = 90°$
2. $|\mathbf{u}| = 54$, $|\mathbf{v}| = 43$, $\theta = 150°$
3. $|\mathbf{u}| = 10$, $|\mathbf{v}| = 12$, $\theta = 67°$
4. $|\mathbf{u}| = 25$, $|\mathbf{v}| = 30$, $\theta = 75°$
5. $|\mathbf{u}| = 20$, $|\mathbf{v}| = 20$, $\theta = 117°$
6. $|\mathbf{u}| = 30$, $|\mathbf{v}| = 30$, $\theta = 123°$
7. $|\mathbf{u}| = 23$, $|\mathbf{v}| = 47$, $\theta = 27°$
8. $|\mathbf{u}| = 32$, $|\mathbf{v}| = 74$, $\theta = 72°$

The vectors **u**, **v**, *and* **w** *are drawn below. Copy them on a sheet of paper. Then sketch each of the vectors in Exercises 9–12.*

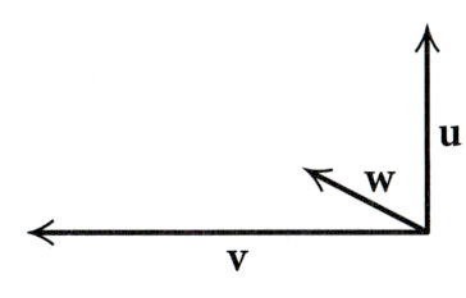

9. $\mathbf{u} + \mathbf{v}$
10. $\mathbf{u} - 2\mathbf{v}$
11. $\mathbf{u} + \mathbf{v} + \mathbf{w}$
12. $\frac{1}{2}\mathbf{u} - \mathbf{w}$

13. Vectors **u**, **v**, and **w** are determined by the sides of $\triangle ABC$ below.

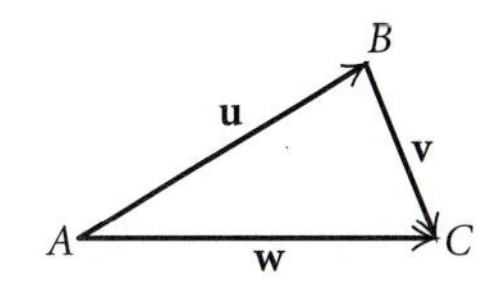

a) Find an expression for **w** in terms of **u** and **v**.
b) Find an expression for **v** in terms of **u** and **w**.

14. In $\triangle ABC$, vectors **u** and **w** are determined by the sides shown, where P is the midpoint of side BC. Find an expression for **v** in terms of **u** and **w**.

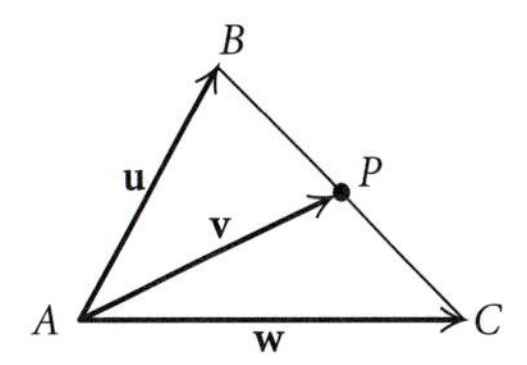

Express each vector in Exercises 15–24 in the form $a\mathbf{i} + b\mathbf{j}$. *That is, resolve each vector into scalar components.*

15. $\langle -5, -7\rangle$
16. $\langle 9, -20\rangle$
17.

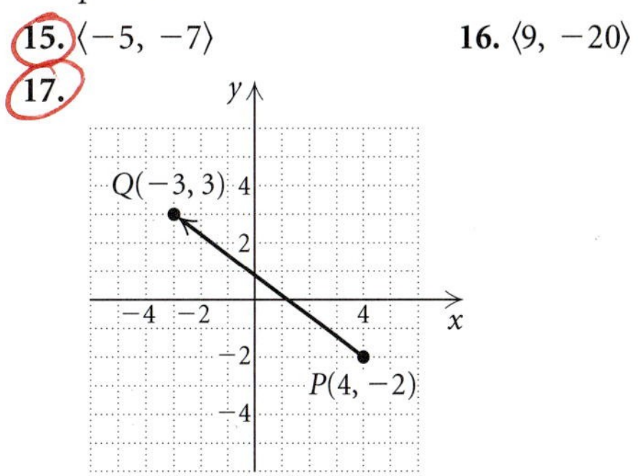

18.

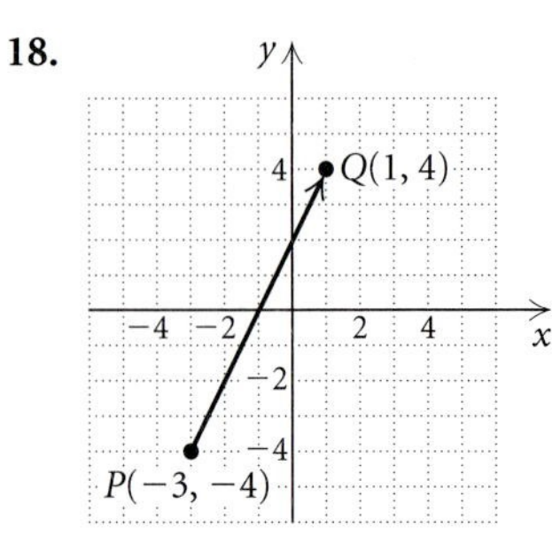

19. The sum of the vectors $\langle -2, 3\rangle$ and $\langle 10, -4\rangle$

20. $\overrightarrow{PQ}$ if P is the point (1, 2) and Q is the point (−3, 5)

21. $\overrightarrow{AB}$ if A is the point (−5, 3) and B is the point (−10, 8)

22. $\overrightarrow{CD}$ if C is the point (1, 3) and D is the midpoint of the line segment AB joining $A(2, -1)$ and $B(-4, 3)$

23. The sum of the vectors $\overrightarrow{AB}$ and $\overrightarrow{CD}$, given the four points $A(1, -1)$, $B(2, 0)$, $C(-1, 3)$, and $D(-2, 2)$

24. Vector $\mathbf{v}$ from point A to the origin, where $\overrightarrow{AB} = 4\mathbf{i} - 2\mathbf{j}$ and B is the point (−2, 5).

25. Given the vector $\overrightarrow{AB} = 3\mathbf{i} - \mathbf{j}$ and A is the point (2, 9), find the point B.

26. *Baseball.* A baseball player throws a baseball with a speed **S** of 72 mph at an angle of 45° with the horizontal. Resolve the vector **S** into components.

27. *Airplane.* An airplane takes off at a speed **S** of 225 mph at an angle of 17° with the horizontal. Resolve the vector **S** into components.

28. *Wheelbarrow.* A wheelbarrow is pushed by applying a 97-lb force **F** that makes a 38° angle with the horizontal. Resolve **F** into its horizontal and vertical components. (The horizontal component is the effective force in the direction of motion and the vertical component adds weight to the wheelbarrow.)

29. *Luggage Wagon.* A luggage wagon is being pulled with vector force **V**, which has a magnitude of 780 lb at an angle of elevation of 60°. Resolve the vector **V** into components.

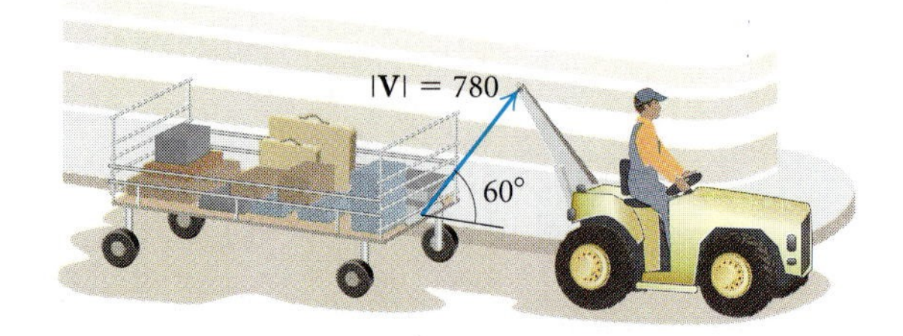

30. *Hot-air Balloon.* A hot-air balloon exerts a 1200-lb pull on a tether line at a 45° angle with the horizontal. Resolve the vector **B** into components.

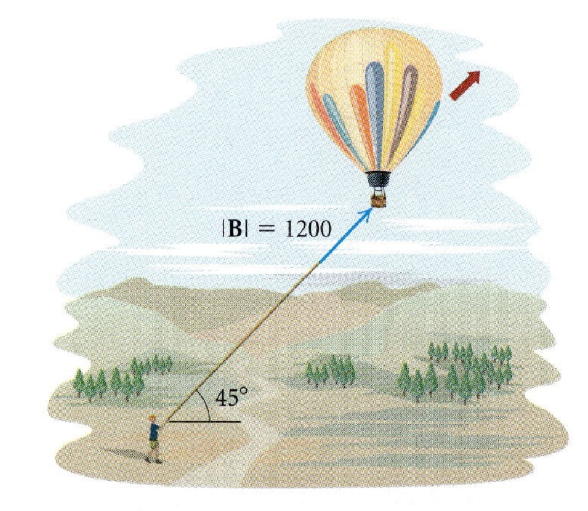

31. Two forces of 32 N (newtons) and 45 N act on an object at right angles. Find the magnitude of the resultant and the angle that it makes with the smaller force.

32. Forces of 410 N and 600 N act on an object. The angle between the forces is 47°. Find the resultant, giving the angle that it makes with the smaller force.

33. A vector of magnitude 100 points southeast. Resolve the vector into easterly and southerly components.

34. *Wind.* A wind has an easterly component (*from* the east) of 10 km/h and a southerly component (*from* the south) of 16 km/h. Find the magnitude and the direction of the wind.

35. *Hot-air Balloon.* A hot-air balloon is rising vertically 10 ft/sec while the wind is blowing horizontally 5 ft/sec. Find the speed of the balloon and the angle that it makes with the horizontal.

36. *Boat.* A boat heads 35°, propelled by a force of 750 lb. A wind from 320° exerts a force of 150 lb on

the boat. How large is the resultant force, and in what direction is the boat moving?

37. *Ship.* A ship sails first N80°E for 120 nautical mi, and then S20°W for 200 nautical mi. How far is it, then, from the starting point, and in what direction?

38. *Airplane.* An airplane flies 032° for 210 km, and then 280° for 170 km. How far is it, then, from the starting point, and in what direction?

39. *Airplane.* An airplane has an airspeed of 150 km/h. It is to make a flight in a direction of 070° while there is a 25-km/h wind from 340°. What will the airplane's actual heading be?

40. A vector $\mathbf{u}$ with a magnitude of 150 lb is inclined to the right and upward 52° from the horizontal. Resolve the vector into components.

Do the calculations in Exercises 41–50 for the vectors

$\mathbf{u} = 3\mathbf{i} + 4\mathbf{j}$, $\mathbf{v} = 5\mathbf{i} + 12\mathbf{j}$, and $\mathbf{w} = \langle -6, 8 \rangle$.

41. $3\mathbf{u} + 2\mathbf{v}$

42. $3\mathbf{u} - 2\mathbf{v}$

43. $(\mathbf{u} + \mathbf{v}) - \mathbf{w}$

44. $\mathbf{u} - (\mathbf{v} + \mathbf{w})$

45. $|\mathbf{u}| - |\mathbf{v}|$

46. $|\mathbf{u}| + |\mathbf{v}|$

47. $|\mathbf{u} + \mathbf{v}|$

48. $|\mathbf{u} - \mathbf{v}|$

49. $2|\mathbf{u}| + 2|\mathbf{v}|$

50. $2|\mathbf{u} + \mathbf{v}|$

Express each vector in Exercises 51–54 in the form $a\mathbf{i} + b\mathbf{j}$ *and sketch each as an arrow in the coordinate plane.*

51. The unit vectors $\mathbf{u} = (\cos\theta)\mathbf{i} + (\sin\theta)\mathbf{j}$ for $\theta = \pi/6$ and $\theta = 3\pi/4$. Include the unit circle $x^2 + y^2 = 1$ in your sketch.

52. The unit vectors $\mathbf{u} = (\cos\theta)\mathbf{i} + (\sin\theta)\mathbf{j}$ for $\theta = -\pi/4$ and $\theta = -3\pi/4$. Include the unit circle $x^2 + y^2 = 1$ in your sketch.

53. The unit vector obtained by rotating $\mathbf{j}$ counterclockwise $3\pi/4$ radians about the origin

54. The unit vector obtained by rotating $\mathbf{j}$ clockwise $2\pi/3$ radians about the origin

For the vectors in Exercises 55 and 56, find the unit vectors $\mathbf{u} = (\cos\theta)\mathbf{i} + (\sin\theta)\mathbf{j}$ *in the same direction.*

55. $-\mathbf{i} + 3\mathbf{j}$

56. $6\mathbf{i} - 8\mathbf{j}$

For the vectors in Exercises 57 and 58, express each vector as a product of its magnitude and its direction.

57. $2\mathbf{i} - 3\mathbf{j}$

58. $5\mathbf{i} + 12\mathbf{j}$

59. Use a sketch to show that

$$\mathbf{v} = 3\mathbf{i} - 6\mathbf{j} \quad \text{and} \quad \mathbf{u} = -\mathbf{i} + 2\mathbf{j}$$

have opposite directions.

60. Use a sketch to show that

$$\mathbf{v} = 3\mathbf{i} - 6\mathbf{j} \quad \text{and} \quad \mathbf{u} = \tfrac{1}{2}\mathbf{i} - \mathbf{j}$$

have the same directions.

Skill Maintenance

Find the point(s) of intersection of each pair of equations.

61. $y_1 = 3x + 2$, $y_2 = -\frac{1}{2}x + 1$

62. $y_1 = 1.1 + x$, $y_2 = 3.4 - 0.9x$

63. $y_1 = 5 + 2x - x^2$, $y_2 = x^2 + 4x - 7$

64. $y_1 = 5 + 2x - x^2$, $y_2 = x^4 - 7x^3 + 5x + 4$

Synthesis

65. ◈ Describe the concept of a vector as though you were explaining it to a classmate. What might an arrow shot from a bow have to do with the explanation?

66. ◈ Write a vector sum problem for a classmate for which the answer is $\mathbf{v} = 5\mathbf{i} - 8\mathbf{j}$.

67. Find all the unit vectors that are parallel to the vector $\langle 3, -4 \rangle$.

68. Find a vector of length 2 whose direction is the opposite of the direction of the vector $\mathbf{v} = -\mathbf{i} + 2\mathbf{j}$. How many such vectors are there?

69. *Eagle's Flight.* An eagle flies from its nest 7 mi in the direction northeast, where it stops to rest on a cliff. It then flies 8 mi in the direction S30°W to land on top of a tree. Place an xy-coordinate system so that the origin is the bird's nest, the x-axis points east, and the y-axis points north.

a) At what point is the cliff located?
b) At what point is the tree?

70. If $\overrightarrow{PQ}$ is any vector, what is $\overrightarrow{PQ} + \overrightarrow{QP}$?

71. The **inner product**, $\mathbf{u} \cdot \mathbf{v}$, of two vectors is a scalar, defined as follows:

$$\mathbf{u} \cdot \mathbf{v} = |\mathbf{u}||\mathbf{v}| \cos\theta,$$

where θ is the angle between the vectors. Show that vectors $\mathbf{u}$ and $\mathbf{v}$ are perpendicular if and only if $\mathbf{u} \cdot \mathbf{v} = 0$.

CHAPTER

6 Summary and Review

Important Properties and Formulas

The Law of Sines

$$\frac{a}{\sin A} = \frac{b}{\sin B} = \frac{c}{\sin C}$$

The Law of Cosines

$$a^2 = b^2 + c^2 - 2bc \cos A,$$
$$b^2 = a^2 + c^2 - 2ac \cos B,$$
$$c^2 = a^2 + b^2 - 2ab \cos C$$

The Area of a Triangle

$$K = \frac{1}{2}bc \sin A = \frac{1}{2}ab \sin C = \frac{1}{2}ac \sin B$$

Complex Numbers

Absolute Value: $|a + bi| = \sqrt{a^2 + b^2}$

Trigonometric Notation: $a + bi = r(\cos \theta + i \sin \theta)$

Multiplication: $r_1(\cos \theta_1 + i \sin \theta_1) \cdot r_2(\cos \theta_2 + i \sin \theta_2)$
$= r_1 r_2[\cos (\theta_1 + \theta_2) + i \sin (\theta_1 + \theta_2)]$

Division: $\dfrac{r_1(\cos \theta_1 + i \sin \theta_1)}{r_2(\cos \theta_2 + i \sin \theta_2)} = \dfrac{r_1}{r_2}[\cos (\theta_1 - \theta_2) + i \sin (\theta_1 - \theta_2)], \quad r_2 \neq 0$

DeMoivre's Theorem

$$[r(\cos \theta + i \sin \theta)]^n = r^n(\cos n\theta + i \sin n\theta)$$

Roots of Complex Numbers

The nth roots of $r(\cos \theta + i \sin \theta)$ are

$$r^{1/n}\left[\cos\left(\frac{\theta}{n} + k \cdot \frac{360^\circ}{n}\right) + i \sin\left(\frac{\theta}{n} + k \cdot \frac{360^\circ}{n}\right)\right], \quad r \neq 0, k = 0, 1, 2, \ldots, n - 1.$$

Vectors

If $\mathbf{v}_1 = a_1\mathbf{i} + b_1\mathbf{j}$ and $\mathbf{v}_2 = a_2\mathbf{i} + b_2\mathbf{j}$ and c is a scalar, then:

Addition: $\mathbf{v}_1 + \mathbf{v}_2 = (a_1 + a_2)\mathbf{i} + (b_1 + b_2)\mathbf{j}$

Subtraction: $\mathbf{v}_1 - \mathbf{v}_2 = (a_1 - a_2)\mathbf{i} + (b_1 - b_2)\mathbf{j}$

Scalar Multiplication: $c\mathbf{v} = c(a\mathbf{i} + b\mathbf{j}) = (ca)\mathbf{i} + (cb)\mathbf{j}$

REVIEW EXERCISES

Solve $\triangle ABC$, if possible.

1. $a = 23.4$ ft, $b = 15.7$ ft, $c = 8.3$ ft
2. $B = 27°$, $C = 35°$, $b = 19$ in.
3. $A = 133°28'$, $C = 31°42'$, $b = 890$ m
4. $B = 37°$, $b = 4$ yd, $c = 8$ yd
5. Find the area of $\triangle ABC$ if $b = 9.8$ m, $c = 7.3$ m, and $A = 67.3°$.
6. A parallelogram has sides of lengths 3.21 ft and 7.85 ft. One of its angles measures 147°. Find the area of the parallelogram.
7. *Sandbox.* A child-care center has a triangular-shaped sandbox. Two of the three sides measure 15 ft and 12.5 ft and form an included angle of 42°. To determine the amount of sand that is needed to fill the box, the director must determine the area of the triangular area. Find the area of the floor of the box to the nearest square foot.

8. *Flower Garden.* A triangular flower garden has sides of lengths 11 m, 9 m, and 6 m. Find the angles of the garden to the nearest tenth of a degree.
9. In an isosceles triangle, the base angles each measure 52.3° and the base is 513 ft long. Find the lengths of the other two sides to the nearest foot.
10. *Airplanes.* Two airplanes leave an airport at the same time. The first flies 175 km/h in a direction of 305.6°. The second flies 220 km/h in a direction of 195.5°. After 2 hr, how far apart are the planes?

Graph each complex number and find its absolute value.

11. $2 - 5i$
12. 4
13. $2i$
14. $-3 + i$

Find trigonometric notation.

15. $1 + i$
16. $-4i$
17. $-5\sqrt{3} + 5i$
18. $\frac{3}{4}$

Find standard notation, $a + bi$.

19. $4(\cos 60° + i \sin 60°)$
20. $7(\cos 0° + i \sin 0°)$
21. $5\left(\cos \frac{2\pi}{3} + i \sin \frac{2\pi}{3}\right)$
22. $2\left[\cos\left(-\frac{\pi}{3}\right) + i \sin\left(-\frac{\pi}{3}\right)\right]$

Convert to trigonometric notation and then multiply or divide, expressing the answer in standard notation.

23. $(1 + i\sqrt{3})(1 - i)$
24. $\frac{2 - 2i}{2 + 2i}$
25. $\frac{2 + 2\sqrt{3}i}{\sqrt{3} - i}$
26. $i(3 - 3\sqrt{3}i)$

Raise the number to the given power and write trigonometric notation for the answer.

27. $[2(\cos 60° + i \sin 60°)]^3$
28. $(1 - i)^4$

Raise the number to the given power and write standard notation for the answer.

29. $(1 + i)^6$
30. $\left(\frac{1}{2} + \frac{\sqrt{3}}{2}i\right)^{10}$
31. Find the square roots of $-1 + i$.
32. Find the cube roots of $3\sqrt{3} - 3i$.
33. Find and graph the fourth roots of 81.
34. Find and graph the fifth roots of 1.

Find all the complex solutions of the equation.

35. $x^4 - i = 0$
36. $x^3 + 1 = 0$
37. Find the polar coordinates of each of these points. Give three answers for each point.

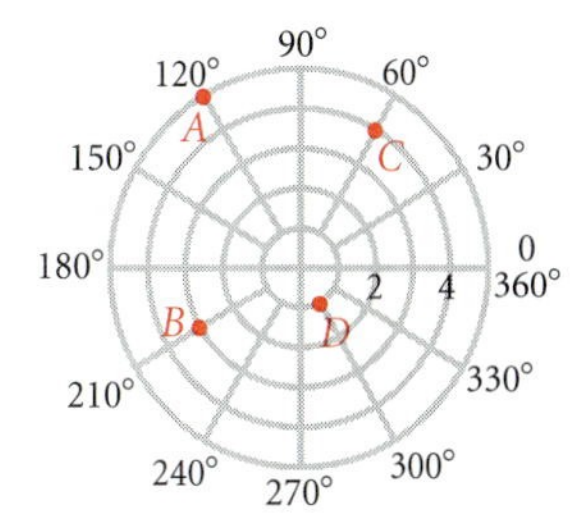

Find the polar coordinates of the point. Express the answer in degrees and then in radians.

38. $(-4\sqrt{2}, 4\sqrt{2})$
39. $(0, -5)$

Use a grapher to convert from rectangular to polar coordinates. Express the answer in degrees and then in radians.

40. $(-2, 5)$
41. $(-4.2, \sqrt{7})$

Find the rectangular coordinates of the point.

42. $\left(3, \frac{\pi}{4}\right)$

43. $(-6, -120°)$

Use a grapher to convert from polar to rectangular coordinates. Round the coordinates to the nearest hundredth.

44. $(2, -15°)$

45. $\left(-2.3, \frac{\pi}{5}\right)$

Convert to a polar equation.

46. $5x - 2y = 6$

47. $y = 3$

48. $x^2 + y^2 = 9$

49. $y^2 - 4x - 16 = 0$

Convert to a rectangular equation.

50. $r = 6$

51. $r + r \sin \theta = 1$

52. $r = \dfrac{3}{1 - \cos \theta}$

53. $r - 2 \cos \theta = 3 \sin \theta$

In Exercises 54–57, use a grapher to match the equation with figures (a)–(d), which follow. Try matching the graphs mentally before using a grapher.

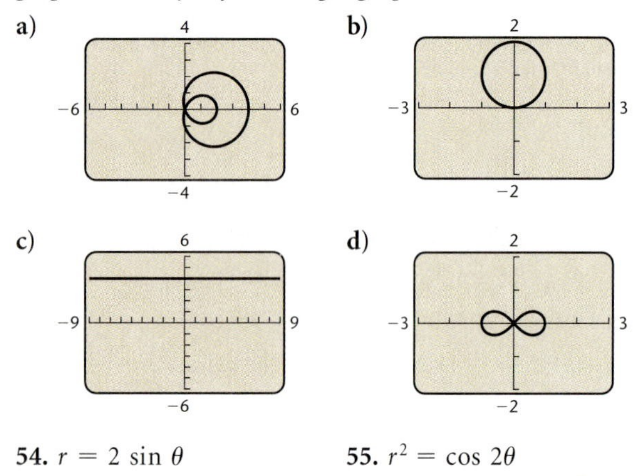

54. $r = 2 \sin \theta$

55. $r^2 = \cos 2\theta$

56. $r = 1 + 3 \cos \theta$

57. $r \sin \theta = 4$

Magnitudes of vectors **u** *and* **v** *and the angle* θ *between the vectors are given. Find the magnitude of the sum,* **u** + **v**, *to the nearest tenth and give the direction by specifying to the nearest degree the angle that it makes with the vector* **u**.

58. $|\mathbf{u}| = 12,\ |\mathbf{v}| = 15,\ \theta = 120°$

59. $|\mathbf{u}| = 41,\ |\mathbf{v}| = 60,\ \theta = 25°$

The vectors **u**, **v**, *and* **w** *are drawn below. Copy them on a sheet of paper. Then sketch each of the vectors in Exercises 60 and 61.*

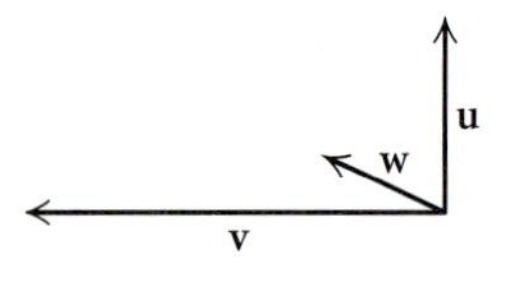

60. $\mathbf{u} - \mathbf{v}$

61. $\mathbf{u} + \frac{1}{2}\mathbf{w}$

62. Forces of 230 N and 500 N act on an object. The angle between the forces is 52°. Find the resultant, giving the angle that it makes with the smaller force.

63. *Wind.* A wind has an easterly component of 15 km/h and a southerly component of 25 km/h. Find the magnitude and the direction of the wind.

64. *Ship.* A ship sails N75°E for 90 nautical mi, and then S10°W for 100 nautical mi. How far is it, then, from the starting point, and in what direction?

Do the calculations in Exercises 65–68 for the vectors $\mathbf{u} = 2\mathbf{i} + 5\mathbf{j}$, $\mathbf{v} = -3\mathbf{i} + 10\mathbf{j}$, *and* $\mathbf{w} = 4\mathbf{i} + 7\mathbf{j}$.

65. $5\mathbf{u} - 8\mathbf{v}$

66. $\mathbf{u} - (\mathbf{v} + \mathbf{w})$

67. $|\mathbf{u} - \mathbf{v}|$

68. $3|\mathbf{w}| + |\mathbf{v}|$

69. Express the vector $\overrightarrow{PQ}$ in the form $a\mathbf{i} + b\mathbf{j}$, if P is the point $(1, -3)$ and Q is the point $(-4, 2)$.

Express each vector in Exercises 70 and 71 in the form $a\mathbf{i} + b\mathbf{j}$ *and sketch each as an arrow in the coordinate plane.*

70. The unit vectors $\mathbf{u} = (\cos \theta)\mathbf{i} + (\sin \theta)\mathbf{j}$ for $\theta = \pi/4$ and $\theta = 5\pi/4$. Include the unit circle $x^2 + y^2 = 1$ in your sketch.

71. The unit vector obtained by rotating **j** counterclockwise $2\pi/3$ radians about the origin.

72. Express the vector $3\mathbf{i} - \mathbf{j}$ as a product of its magnitude and its direction.

Synthesis

73. ◈ Summarize how you can tell algebraically when solving triangles whether there is no solution, one solution, or two solutions.

74. ◈ Explain why the law of sines cannot be used to find the first angle when solving a triangle given three sides.

75. A parallelogram has sides of lengths 3.42 and 6.97. Its area is 18.4. Find the sizes of its angles.

76. Let $\mathbf{u} = 12\mathbf{i} + 5\mathbf{j}$. Find a vector that has the same direction as **u** but has length 3.

77. Convert to a rectangular equation:

$$r = \csc^2 \frac{\theta}{2}.$$

Systems and Matrices 7

APPLICATION

The equilibrium point (x, y) for a product indicates the price y at which the amount x of the product that the seller willingly supplies is the same as the amount that the consumer willingly demands. The equilibrium point for the Great Tunes portable CD player is the solution of this system of equations, where y is the price, in dollars, and x is the number of units, in millions:

$$y = 200 - 25x,$$
$$y = 90 + 30x.$$

X	Y1	Y2
0	200	90
1	175	120
2	150	150
3	125	180
4	100	210
5	75	240
6	50	270

X = 2

$y_1 = 200 - 25x$, $y_2 = 90 + 30x$

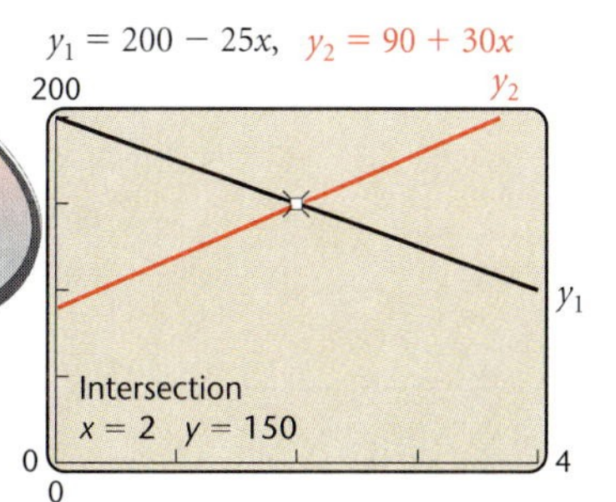

It is often desirable or even necessary to use two or more variables to model a situation in a field such as business, science, psychology, engineering, education, or sociology. When this is the case, we write and solve a *system of equations* in order to answer questions about that situation. In this chapter, we study systems of equations and several methods for solving them. We also use *systems of inequalities*, along with *linear programming*, to model applications and to find the maximum and minimum values of functions subject to a set of restrictions, or constraints.

7.1 Systems of Equations in Two Variables
7.2 Systems of Equations in Three or More Variables
7.3 Matrices and Systems of Equations
7.4 Matrix Operations
7.5 Inverses of Matrices
7.6 Systems of Inequalities and Linear Programming
7.7 Partial Fractions
SUMMARY AND REVIEW

7.1 Systems of Equations in Two Variables

- *Solve a system of two linear equations in two variables by graphing.*
- *Solve a system of two linear equations in two variables using the substitution and the elimination methods.*
- *Use systems of two linear equations to solve applied problems.*

A **system of equations** is composed of two or more equations considered simultaneously. For example,

$$x + y = 11,$$
$$3x - y = 5$$

is a system of two linear equations in two variables. The solution set of this system consists of all ordered pairs that make *both* equations true. Note that (4, 7) is a solution of the system of equations above. We can verify this by substituting 4 for x and 7 for y in *each* equation.

$x + y = 11$			$3x - y = 5$		
$4 + 7$	? 11		$3 \cdot 4 - 7$	? 5	
11	11	TRUE	$12 - 7$		
			5	5	TRUE

Interactive Discovery

Graph $y_1 = 2x - 5$ and $y_2 = -x + 1$ in the same viewing window. Use the grapher's TRACE and ZOOM features or the INTERSECT feature to find the coordinates of the point of intersection of the graphs. Do the same for $y_1 = x + 4$ and $y_2 = -1.25x - 5$ and for $y_1 = 3x - 7$ and $y_2 = -2x + \frac{1}{2}$. What does each of these ordered pairs represent?

Solving Systems of Equations Graphically

Recall that the graph of a linear equation is a line that contains all the ordered pairs in the solution set of the equation. When we graph a system of linear equations, each point at which the equations intersect is a solution of *both* equations and therefore a solution of the system of equations.

Consider the system

$$x + y = 11,$$
$$3x - y = 5.$$

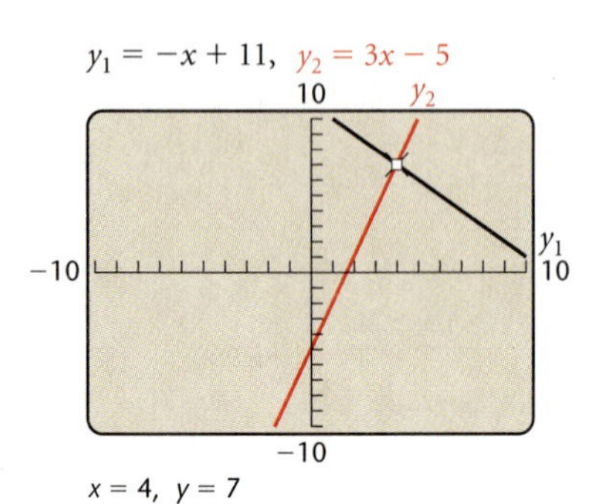

$x = 4,\ y = 7$

When we graph these equations on the same set of axes, we see that they intersect at a single point, (4, 7), so (4, 7) is a solution of the system of equations. (Note that when a grapher is used, it might be necessary to write each equation in "$Y = \cdots$" form. If so, we would graph $y_1 = -x + 11$ and $y_2 = 3x - 5$.)

To check this result, we can substitute 4 for x and 7 for y in both equations, as we did above. We can also use the TABLE feature on a grapher to do a check, observing that when $x = 4$, $y_1 = 7$ and $y_2 = 7$.

X	Y1	Y2
1	10	−2
2	9	1
3	8	4
4	7	7
5	6	10
6	5	13
7	4	16
X = 4		

The graphs of most of the systems of equations that we use to model applications intersect at a single point, like the system above. However, it is possible that the graphs will have no points in common or infinitely many points in common. Each of these possibilities is illustrated below.

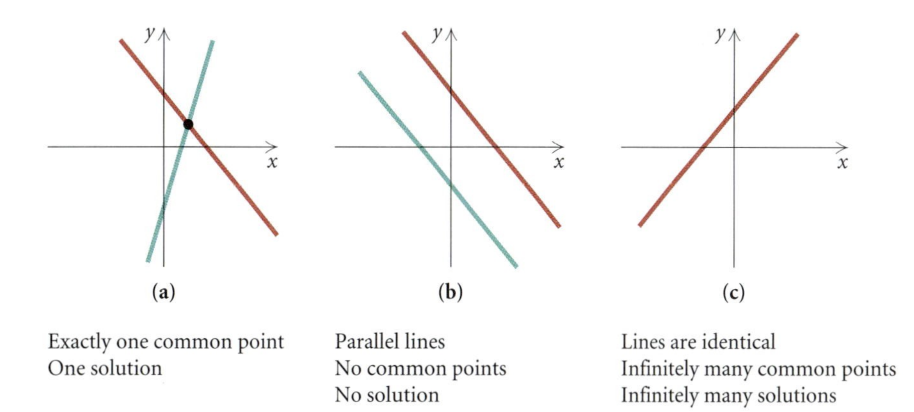

(a) Exactly one common point
One solution

(b) Parallel lines
No common points
No solution

(c) Lines are identical
Infinitely many common points
Infinitely many solutions

If a system of two linear equations in two variables has at least one solution, it is **consistent**. If the system has no solutions, it is **inconsistent**. In addition, if a system of two linear equations in two variables has an infinite number of solutions, it is **dependent**. Otherwise, it is **independent**.

The system graphed in figure (a) above is consistent and independent; the system in (b) is inconsistent and independent; and the system in (c) is consistent and dependent.

The Substitution Method

Graphing helps us picture the solution of a system of equations, but solving by graphing is not always accurate when solutions are not integers. A grapher can improve the accuracy of the graphing method, but when the solution is a pair like $\left(\frac{43}{27}, -\frac{19}{27}\right)$, the grapher generally gives only an approximate solution.

Algebraic methods for solving systems of equations, when used correctly, always give accurate results. One such technique is the **substitution method**. It is used most often when a variable is alone on one side of an equation or when it is easy to solve for a variable. To apply the substitution method, we begin by using one of the equations to express one variable in terms of the other; then we substitute that expression in the other equation of the system.

Example 1 Use the substitution method to solve the system

$$x + y = 11, \qquad (1)$$
$$3x - y = 5. \qquad (2)$$

SOLUTION First, we solve equation (1) for y. (We could just as well solve for x.)

$$y = 11 - x$$

Then we substitute $11 - x$ for y in equation (2). This gives an equation in one variable, which we know how to solve.

$$3x - (11 - x) = 5$$
$$3x - 11 + x = 5 \qquad \text{Removing parentheses}$$
$$4x - 11 = 5$$
$$4x = 16$$
$$x = 4$$

Now we substitute 4 for x in either equation (1) or (2) (this is called **back-substitution**) and solve for y. We choose equation (1):

$$4 + y = 11$$
$$y = 7.$$

We have previously checked the pair (4, 7) in both equations both algebraically and graphically. The solution of the system of equations is (4, 7).

The Elimination Method

Another algebraic technique for solving systems of equations is the **elimination method**. With this method, we eliminate a variable by adding two equations. Before we add, it might be necessary to multiply one or both equations by suitable constants in order to find two equations in which coefficients differ only by sign.

Example 2 Solve each of the following systems using the elimination method.

a) $2x + y = 2, \qquad (1)$
$x - y = 7 \qquad (2)$

b) $4x + 3y = 11, \qquad (1)$
$-5x + 2y = 15 \qquad (2)$

SOLUTION

a) Since the y-coefficients differ only by sign, we can eliminate y by adding the equations:

$$2x + y = 2 \qquad (1)$$
$$x - y = 7 \qquad (2)$$
$$3x \quad = 9 \qquad \text{Adding}$$
$$x = 3.$$

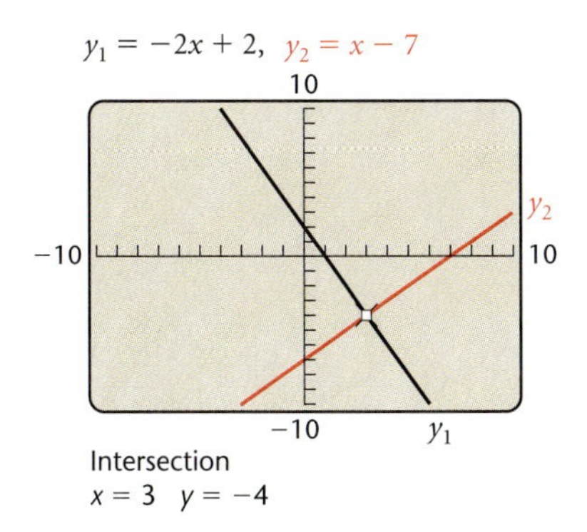

Intersection
$x = 3$ $y = -4$

We then back-substitute 3 for x in either equation and solve for y. We choose equation (1):

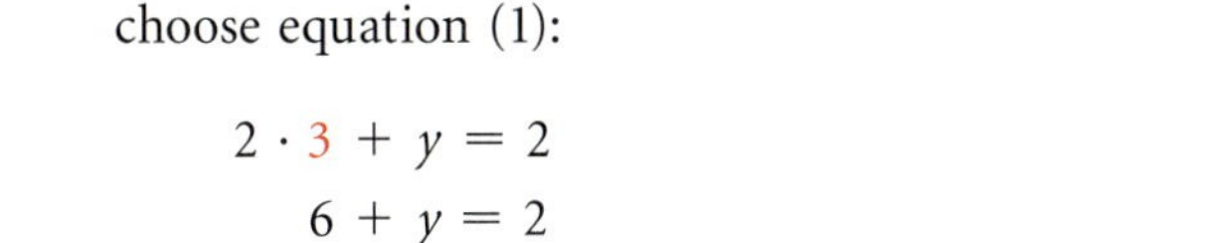

$$\begin{aligned} 2 \cdot 3 + y &= 2 \\ 6 + y &= 2 \\ y &= -4. \end{aligned}$$

We can check the pair $(3, -4)$ either algebraically by substituting in both equations or graphically using the INTERSECT feature, as shown at left. The solution is $(3, -4)$.

b) We can obtain x-coefficients that differ only by sign by multiplying the first equation by 5 and the second equation by 4:

$$\begin{aligned} 20x + 15y &= 55 && \textbf{Multiplying equation (1) by 5} \\ -20x + 8y &= 60 && \textbf{Multiplying equation (2) by 4} \\ \hline 23y &= 115 && \textbf{Adding} \\ y &= 5. \end{aligned}$$

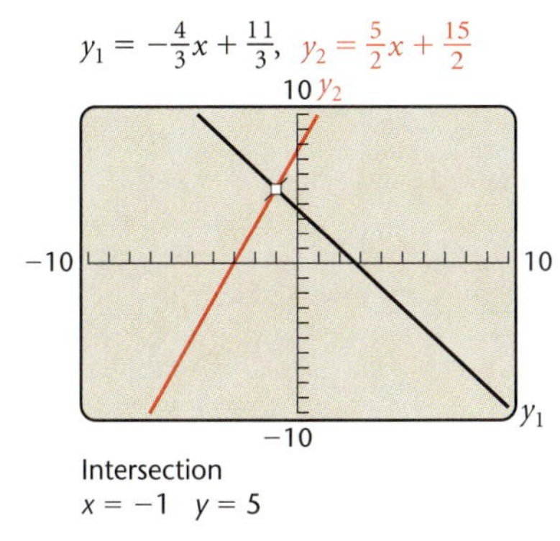

Intersection
$x = -1$ $y = 5$

We then back-substitute 5 for y in either equation (1) or (2) and solve for x. We choose equation (1):

$$\begin{aligned} 4x + 3 \cdot 5 &= 11 && \textbf{Substituting 5 for } y \textbf{ in equation (1)} \\ 4x + 15 &= 11 \\ 4x &= -4 \\ x &= -1. \end{aligned}$$

We can check the pair $(-1, 5)$ either algebraically by substituting in both equations or graphically, as shown at left. The solution is $(-1, 5)$.

In Example 2(b), the two systems

$$\begin{aligned} 4x + 3y &= 11, \\ -5x + 2y &= 15 \end{aligned} \quad \text{and} \quad \begin{aligned} 20x + 15y &= 55, \\ -20x + 8y &= 60 \end{aligned}$$

are **equivalent** because they have exactly the same solutions. When we use the elimination method, we often multiply one or both equations by constants to find equivalent equations that allow us to eliminate a variable by adding.

Example 3 Solve each of the following systems using the elimination method.

a)
$$\begin{aligned} x - 3y &= 1, && (1) \\ -2x + 6y &= 5 && (2) \end{aligned}$$

b)
$$\begin{aligned} 2x + 3y &= 6, && (1) \\ 4x + 6y &= 12 && (2) \end{aligned}$$

SOLUTION

a) We multiply equation (1) by 2 and add:

$$\begin{array}{rl} 2x - 6y = 2 & \textbf{Multiplying equation (1) by 2} \\ -2x + 6y = 5 & (2) \\ \hline 0 = 7. & \textbf{Adding} \end{array}$$

There are no values of x and y for which $0 = 7$ is true, so the system has no solution. The solution set is $\varnothing$. The system of equations is inconsistent.

The graphs of the equations are parallel lines. In fact, we see this when we write the equations in simplified "$Y = \cdots$" form to enter them on a grapher. The slopes are the same, $\left(\frac{1}{3}\right)$, but the y-intercepts are different, $\left(0, \frac{5}{6}\right)$ and $\left(0, -\frac{1}{3}\right)$.

$y_1 = \frac{1}{3}x - \frac{1}{3}$, $y_2 = \frac{1}{3}x + \frac{5}{6}$

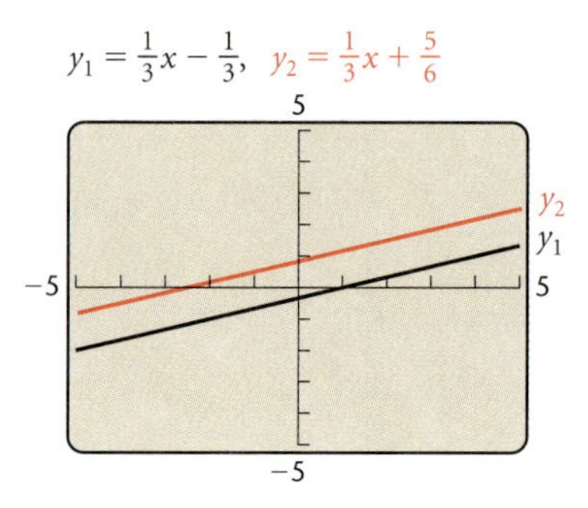

b) We multiply equation (1) by −2 and add:

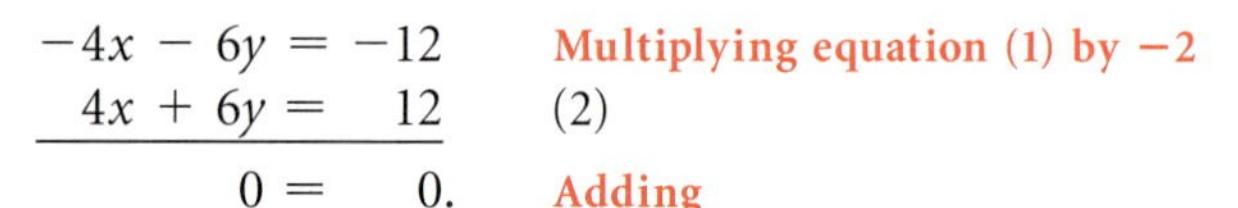

$-4x - 6y = -12$ Multiplying equation (1) by −2
$4x + 6y = 12$ (2)
$0 = 0.$ Adding

$y_1 = -\frac{2}{3}x + 2$, $y_2 = -\frac{2}{3}x + 2$

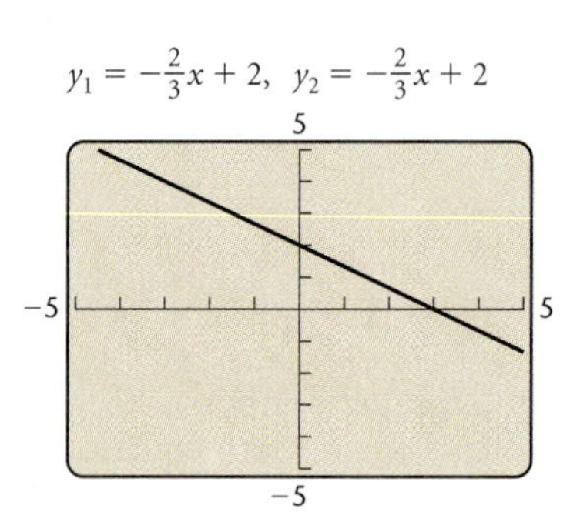

We obtain the equation $0 = 0$, which is true for all values of x and y. This tells us that the system of equations is dependent, so there are infinitely many solutions. That is, any solution of one equation of the system is also a solution of the other. The graphs of the equations are identical. (In fact, when we write the equations in simplified "$Y = \cdots$" form to enter them on a grapher, we get identical equations.)

Using either equation, we can write $y = -\frac{2}{3}x + 2$, so we can write the solutions of the system as ordered pairs of the form $\left(x, -\frac{2}{3}x + 2\right)$. Any real value that we choose for x then gives us a value for y and thus an ordered pair in the solution set. Some of the solutions are $(-3, 4)$, $(0, 2)$, and $(6, -2)$. Similarly, we can write $x = -\frac{3}{2}y + 3$, so the solutions can also be expressed as $\left(-\frac{3}{2}y + 3, y\right)$.

Applications

Frequently the most challenging and time-consuming step in the problem-solving process is translating a situation to mathematical language. However, in many cases, this task is facilitated if we translate to more than one equation in more than one variable.

Example 4 *Motion.* A Boeing 747 flies the 3000-mi distance from Los Angeles to New York, with a tail wind, in 5 hr. The return trip, against the wind, takes 6 hr. Find thc speed of the plane and the speed of the wind.

SOLUTION We will use the five-step problem-solving process. (See Section R.9.)

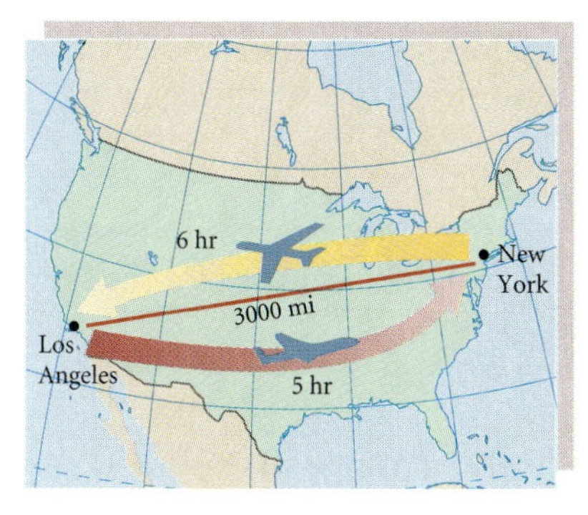

1. **Familiarize.** We first make a drawing. Let p = the speed of the plane in still air and w = the speed of the wind, both in miles per hour. The distances are the same. When the plane flies east with the wind, its speed is $p + w$. When it flies west against the wind, its speed is $p - w$. We organize this information in a table, the columns of the table coming from the motion formula $d = rt$.

	DISTANCE	SPEED	TIME	
EAST (WITH THE WIND)	3000	$p + w$	5	$\rightarrow 3000 = (p + w)5$
WEST (AGAINST THE WIND)	3000	$p - w$	6	$\rightarrow 3000 = (p - w)6$

2. **Translate.** Using $d = rt$ in each row of the table, we get an equation. Thus we have a system of equations:

$$3000 = (p + w)5 = 5p + 5w,$$
$$3000 = (p - w)6 = 6p - 6w.$$

3. **Carry out.** We solve the system

$$5p + 5w = 3000, \quad (1)$$
$$6p - 6w = 3000. \quad (2)$$

ALGEBRAIC SOLUTION

Note that we can simplify these equations by multiplying equation (1) by $\frac{1}{5}$ and equation (2) by $\frac{1}{6}$. Then we have

$$p + w = 600 \quad (1a)$$
$$p - w = 500. \quad (2a)$$

We add these equations to eliminate w:

$$\begin{aligned} p + w &= 600 \\ p - w &= 500 \\ \hline 2p \quad &= 1100 \\ p &= 550. \end{aligned}$$

Next, we back-substitute to find w:

$$550 + w = 600 \quad \textbf{Substituting in equation (1a)}$$
$$w = 50.$$

GRAPHICAL SOLUTION

We replace p with x and w with y, graph $y_1 = 600 - x$ and $y_2 = x - 500$, and find the point of intersection of the graphs.

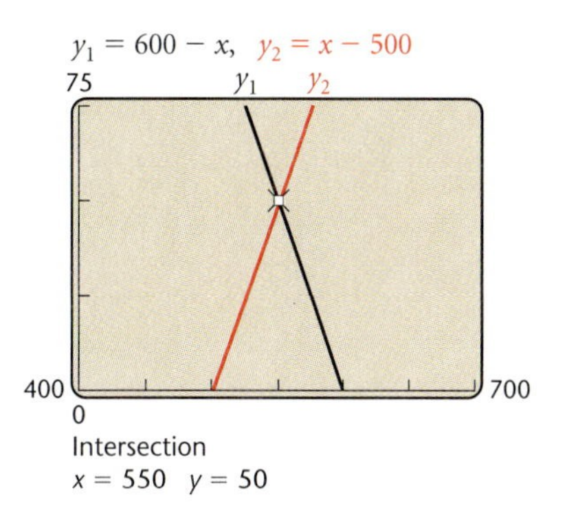

4. **Check.** When the plane is traveling east, its speed is $550 + 50$, or 600 mph, and the trip takes $3000/600$, or 5 hr. When the plane travels west, its speed is $550 - 50$, or 500 mph, and the trip takes $3000/500$, or 6 hr. Our solution checks.
5. **State.** The solution of the system of equations is (550, 50). The speed of the plane is 550 mph, and the speed of the wind is 50 mph.

Example 5 *Supply and Demand.* Suppose that the price and the supply of the Great Tunes portable CD player are related by the equation

$$y = 200 - 25x,$$

where y is the price, in dollars, at which the seller is willing to supply x million units. Also suppose that the price and the demand for the same model of CD player are related by the equation

$$y = 90 + 30x,$$

where y is the price, in dollars, at which the consumer is willing to buy x million units.

The *equilibrium point* for this product is the pair (x, y) that is a solution of both equations. The *equilibrium price* is the price at which the amount of the product that the seller is willing to supply is the same as the amount demanded by the consumer. Find the equilibrium point for this product.

SOLUTION

1., 2. **Familiarize** and **Translate.** We are given a system of equations in the statement of the problem, so no further translation is necessary.

$$y = 200 - 25x, \quad (1)$$
$$y = 90 + 30x. \quad (2)$$

We substitute some values for x in each equation to get an idea of the corresponding prices. When $x = 1$,

$$y = 200 - 25 \cdot 1 = 175, \quad \text{Substituting in equation (1)}$$
$$y = 90 + 30 \cdot 1 = 120. \quad \text{Substituting in equation (2)}$$

This indicates that the price when 1 million units are supplied is higher than the price when 1 million units are demanded.

When $x = 4$,

$$y = 200 - 25 \cdot 4 = 100, \quad \text{Substituting in equation (1)}$$
$$y = 90 + 30 \cdot 4 = 210. \quad \text{Substituting in equation (2)}$$

In this case, the price related to supply is lower than the price related to demand. It would appear that the x-value we are looking for is between 1 and 4.

3. **Carry out.** We use substitution:

$$200 - 25x = 90 + 30x \quad \text{Substituting } 200 - 25x \text{ for } y \text{ in equation (2)}$$
$$110 = 55 \quad \text{Adding } 25x \text{ and subtracting 90 on both sides}$$
$$2 = x. \quad \text{Dividing by 55 on both sides}$$

We now back-substitute 2 for x in either equation and find y:

$$y = 200 - 25 \cdot 2 \qquad \textbf{Substituting in equation (1)}$$
$$y = 150.$$

4. **Check.** We can check algebraically by substituting 2 for x and 150 for y in both equations. Since the equations were given in the statement of the problem, we can also check graphically, as shown below.

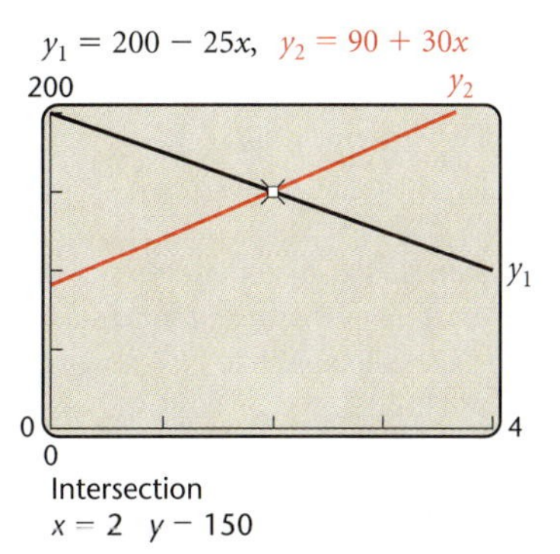

5. **State.** The equilibrium point is (2, \$150). That is, the equilibrium supply is 2 million units and the equilibrium price is \$150.

There is another algebraic method for solving systems of two linear equations in two variables known as *Cramer's rule*. It is presented in Appendix A. Systems of equations can be solved on some graphers using the SIMULT feature.

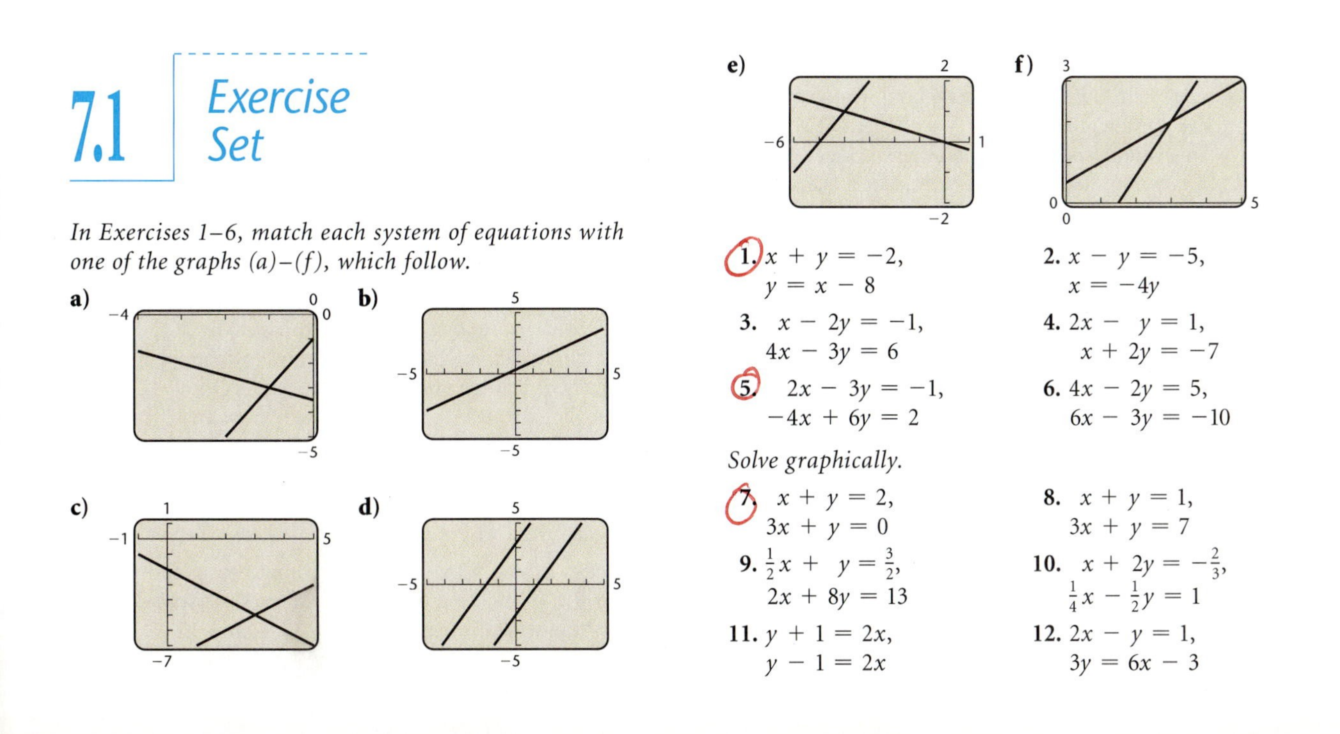

7.1 Exercise Set

In Exercises 1–6, match each system of equations with one of the graphs (a)–(f), which follow.

a) b) c) d) e) f)

1. $x + y = -2,$
   $y = x - 8$

2. $x - y = -5,$
   $x = -4y$

3. $x - 2y = -1,$
   $4x - 3y = 6$

4. $2x - y = 1,$
   $x + 2y = -7$

5. $2x - 3y = -1,$
   $-4x + 6y = 2$

6. $4x - 2y = 5,$
   $6x - 3y = -10$

Solve graphically.

7. $x + y = 2,$
   $3x + y = 0$

8. $x + y = 1,$
   $3x + y = 7$

9. $\frac{1}{2}x + y = \frac{3}{2},$
   $2x + 8y = 13$

10. $x + 2y = -\frac{2}{3},$
    $\frac{1}{4}x - \frac{1}{2}y = 1$

11. $y + 1 = 2x,$
    $y - 1 = 2x$

12. $2x - y = 1,$
    $3y = 6x - 3$

Solve using the substitution method. Use a grapher to check your answer.

13. $x + y = 9,$ $2x - 3y = -2$

14. $3x - y = 5,$ $x + y = \frac{1}{2}$

15. $x - 2y = 7,$ $x = y + 4$

16. $x + 4y = 6,$ $x = -3y + 3$

17. $y = 2x - 6,$ $5x - 3y = 16$

18. $3x + 5y = 2,$ $2x - y = -3$

19. $x - 5y = 4,$ $y = 7 - 2x$

20. $5x + 3y = -1,$ $x + y = 1$

21. $2x - 3y = 5,$ $5x + 4y = 1$

22. $3x + 4y = 6,$ $2x + 3y = 5$

Solve using the elimination method. Also determine whether each system is consistent or inconsistent and dependent or independent. Use a grapher to check your answer.

23. $x + 2y = 7,$ $x - 2y = -5$

24. $3x + 4y = -2,$ $-3x - 5y = 1$

25. $x - 3y = 2,$ $6x + 5y = -34$

26. $x + 3y = 0,$ $20x - 15y = 75$

27. $0.3x - 0.2y = -0.9,$ $0.2x - 0.3y = -0.6$

28. $0.2x - 0.3y = 0.3,$ $0.4x + 0.6y = -0.2$

29. $3x - 12y = 6,$ $2x - 8y = 4$

30. $2x + 6y = 7,$ $3x + 9y = 10$

31. $\frac{1}{5}x + \frac{1}{2}y = 6,$ $\frac{3}{5}x - \frac{1}{2}y = 2$

32. $\frac{2}{3}x + \frac{3}{5}y = -17,$ $\frac{1}{2}x - \frac{1}{3}y = -1$

33. $2x = 5 - 3y,$ $4x = 11 - 7y$

34. $7(x - y) = 14,$ $2x = y + 5$

35. *Sales Promotions.* During a one-month promotional campaign, Video Village gave either a free video rental or a 12-serving box of microwave popcorn to new members. It cost the store \$1 for each free rental and \$2 for each box of popcorn. In all, 48 new members were signed up and the store's cost for the incentives was \$86. How many of each incentive were given away?

36. *Concert Ticket Prices.* One evening 1500 concert tickets were sold for the Fairmont Summer Jazz Festival. Tickets cost \$25 for covered pavilion seats and \$15 for lawn seats. Total receipts were \$28,500. How many of each type of ticket were sold?

37. *Museum Admission Prices.* Admission to the Indianapolis Children's Museum costs twice as much for adults as for children. Admission to the museum for 4 adults and 16 children from the Tiny Tots Day Care Center cost \$72. Find the cost of each adult's admission and each child's admission.

38. *Mail-Order Business.* A mail-order lacrosse equipment business shipped 120 packages one day. Customers are charged \$3.50 for each standard-delivery package and \$7.50 for each express-delivery package. Total shipping charges for the day were \$596. How many of each kind of package were shipped?

39. *Supply and Demand.* The supply and demand for a particular model of electronic organizer are related to price by the equations

$$y = 175 - 5x,$$
$$y = 70 + 2x,$$

respectively, where y is the price, in dollars, and x is the number of units, in thousands. Find the equilibrium point for this product.

40. *Supply and Demand.* The supply and demand for a particular model of treadmill are related to price by the equations

$$y = 500 - 25x,$$
$$y = 240 + 40x,$$

respectively, where y is the price, in dollars, and x is the number of units, in thousands. Find the equilibrium point for this product.

The point at which a company's costs equal its revenues is the break-even point. In Exercises 41–44, C represents the production cost, in dollars, of x units of a product and R represents the revenue, in dollars, from the sale of x units. Find the number of units that must be produced and sold in order to break even. That is, find the value of x for which C = R.

41. $C = 14x + 350,$ $R = 16.5x$

42. $C = 8.5x + 75,$ $R = 10x$

43. $C = 15x + 12{,}000,$ $R = 18x - 6000$

44. $C = 3x + 400,$ $R = 7x - 600$

45. *Nutrition.* A one-cup serving of spaghetti with meatballs contains 260 Cal (calories) and 32 g of carbohydrates. A one-cup serving of chopped iceberg lettuce contains 5 Cal and 1 g of carbohydrates. How many servings of each would be required to obtain 400 Cal and 50 g of carbohydrates? (*Source: Home and Garden Bulletin No. 72,* U.S. Government Printing Office, Washington, D.C. 20402)

46. *Nutrition.* One serving of tomato soup contains 100 Cal and 18 g of carbohydrates. One slice of whole wheat bread contains 70 Cal and 13 g of

carbohydrates. How many servings of each would be required to obtain 230 Cal and 42 g of carbohydrates? (*Source: Home and Garden Bulletin No. 72,* U.S. Government Printing Office, Washington, D.C. 20402)

47. *Motion.* A Leisure Time Cruises riverboat travels 46 km downstream in 2 hr. It travels 51 km upstream in 3 hr. Find the speed of the boat and the speed of the stream.

48. *Motion.* A DC10 travels 3000 km with a tail wind in 3 hr. It travels 3000 km with a head wind in 4 hr. Find the speed of the plane and the speed of the wind.

49. *Investment.* Bernadette inherited \$15,000 and invested it in two municipal bonds, which pay 7% and 9% simple interest. The annual interest is \$1230. Find the amount invested at each rate.

50. *Tee Shirt Sales.* Mack's Tee Shirt Shack sold 36 shirts one day. All short-sleeved tee shirts cost \$12 and all long-sleeved tee shirts cost \$18. Total receipts for the day were \$522. How many of each kind of shirt were sold?

51. *Coffee Mixtures.* The owner of The Daily Grind coffee shop mixes French roast coffee worth \$9.00 per pound with Kenyan coffee worth \$7.50 per pound in order to get 10 lb of a mixture worth \$8.40 per pound. How much of each type of coffee was used?

52. *Snack Mixtures.* At Sunda's Snacks, caramel corn worth \$2.50 per pound is mixed with honey roasted peanuts worth \$7.50 per pound in order to get 20 lb of a mixture worth \$3 per pound. How much of each snack was used?

53. *Commissions.* Jackson Manufacturing offers its sales representatives a choice between being paid a commission of 8% of sales or being paid a monthly salary of \$1500 plus a commission of 1% of sales. For what monthly sales do the two plans pay the same amount?

54. *Motion.* Two private airplanes travel toward each other from cities that are 780 km apart at speeds of 190 km/h and 200 km/h. They left at the same time. In how many hours will they meet?

Skill Maintenance

55. Evaluate $2x - 3y$ for $x = -2$ and $y = 5$.

56. Evaluate $-x + 5y - 4z$ for $x = -1$, $y = 5$, and $z = 2$.

57. Graph: $y = 2x^2 - 3x + 4$.

58. Graph: $y = 0.4x^2 + 0.01x - 0.05$.

Synthesis

59. ◈ Write a problem for a classmate to solve that can be translated to a system of equations. Devise the problem so that the solution is "The concert promoter sold 200 reserved seat tickets and 950 general admission tickets."

60. ◈ Explain in your own words when the elimination method for solving a system of equations is preferable to the substitution method.

61. *Motion.* Nancy jogs and walks to campus each day. She averages 4 km/h walking and 8 km/h jogging. The distance from home to the campus is 6 km and she makes the trip in 1 hr. How far does she jog on each trip?

62. *Mail-Order Business.* A mail-order catalog advertises a limited-time sale, offering 1 turtleneck for \$15 and 2 turtlenecks for \$25. A total of 1250 turtlenecks are sold and \$16,750 is taken in. How many customers ordered 2 turtlenecks?

63. *Motion.* A train leaves Union Station for Central Station, 216 km away, at 9 A.M. One hour later, a train leaves Central Station for Union Station. They meet at noon. If the second train had started at 9 A.M. and the first train at 10:30 A.M., they would still have met at noon. Find the speed of each train.

64. *Antifreeze Mixtures.* An automobile radiator contains 16 L of antifreeze and water. This mixture is 30% antifreeze. How much of this mixture should be drained and replaced with pure antifreeze so that the final mixture will be 50% antifreeze?

65. *Value of a Horse.* A stablehand agreed to work for 1 yr. At the end of that year, he was to receive \$240 and one horse. After 7 months, the stablehand quit the job but still received the horse and \$100. What is the value of the horse?

66. *Ticket Line.* You are in line at a ticket window. There are 2 more people ahead of you in line than there are behind you. In the entire line, there are three times as many people as there are behind you. How many people are ahead of you?

67. Two solutions of the equation $y = mx + b$ are $(-2, 3)$ and $(4, -5)$. Find m and b.

68. Two solutions of the equation $Ax + By = 1$ are $(3, -1)$ and $(-4, -2)$. Find A and B.

69. *Gas Mileage.* A 1994 Dodge Stealth gets 18 miles per gallon (mpg) in city driving and 24 mpg in highway driving. The car is driven 465 mi on a full tank of 23 gal of gasoline. How many miles were driven in the city and how many were driven on the highway?

70. *Motion.* Heather is out hiking and is standing on a railroad bridge, as shown in the figure at right. A train is approaching from the direction shown by the arrow. If Heather runs at a speed of 10 mph toward the train, she will reach point P on the bridge at the same moment that the train does. If she runs to point Q at the other end of the bridge at a speed of 10 mph, she will reach point Q also at the same moment that the train does. How fast, in miles per hour, is the train traveling?

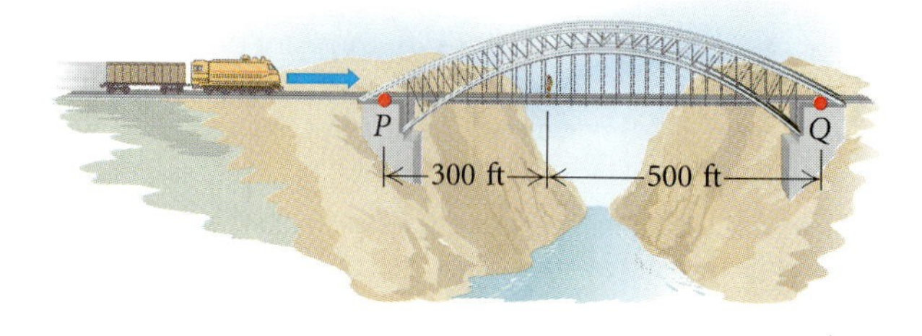

7.2 Systems of Equations in Three or More Variables

- *Solve systems of linear equations in three or more variables.*
- *Use systems of three equations to solve applied problems.*
- *Model a situation using a quadratic function.*

A **linear equation in three variables** is an equation equivalent to one of the form $Ax + By + Cz = D$, where A, B, C, and D are real numbers and A, B, and C are not all 0. A **solution of a system of three equations in three variables** is an ordered triple that makes all three equations true. For example, the triple $(2, -1, 0)$ is a solution of the system of equations

$$\begin{aligned} 4x + 2y + 5z &= 6, \\ 2x - y + z &= 5, \\ 3x + 2y - z &= 4. \end{aligned}$$

We can verify this by substituting 2 for x, -1 for y, and 0 for z in each equation.

Solving Systems of Equations in Three or More Variables

We will solve systems of equations in three or more variables using an algebraic method called **Gaussian elimination**, named for the German mathematician Karl Friedrich Gauss (1777–1855). Our goal is to transform the original system to an equivalent one of the form

$$\begin{aligned} Ax + By + Cz &= D, \\ Ey + Fz &= G, \\ Hz &= K. \end{aligned}$$

Then we solve the third equation for z and back-substitute to find y and then x.

Each of the following operations can be used to transform the original system to an equivalent system in the desired form.

1. Interchange any two equations.
2. Multiply both sides of one of the equations by a nonzero constant.
3. Add a nonzero multiple of one equation to another equation.

Example 1 Solve the following system:

$$\begin{aligned} x - 2y + 3z &= 11, && (1)\\ 4x + 2y - 3z &= 4, && (2)\\ 3x + 3y - z &= 4. && (3) \end{aligned}$$

SOLUTION We multiply equation (1) by -4 and add it to equation (2). We also multiply equation (1) by -3 and add it to equation (3). These operations produce an equivalent system of equations in which the x-terms have been eliminated from both the second and the third equations.

$$\begin{aligned} -4x + 8y - 12z &= -44\\ 4x + 2y - 3z &= 4 \quad \text{Eq. (2)}\\ \hline 10y - 15z &= -40 \end{aligned}$$

$$\begin{aligned} x - 2y + 3z &= 11,\\ 10y - 15z &= -40,\\ 9y - 10z &= -29 \end{aligned}$$

$$\begin{aligned} -3x + 6y - 9z &= -33\\ 3x + 3y - z &= 4 \quad \text{Eq. (3)}\\ \hline 9y - 10z &= -29 \end{aligned}$$

Now we multiply the last equation by 10 to make the y-coefficient a multiple of the y-coefficient in the second equation:

$$\begin{aligned} x - 2y + 3z &= 11, && (1)\\ 10y - 15z &= -40, && (4)\\ 90y - 100z &= -290. && (5) \end{aligned}$$

Next we multiply equation (4) by -9 and add it to equation (5).

$$\begin{aligned} -90y + 135z &= 360\\ 90y - 100z &= -290 \quad \text{Eq. (5)}\\ \hline 35z &= 70 \end{aligned}$$

$$\begin{aligned} x - 2y + 3z &= 11, && (1)\\ 10y - 15z &= -40, && (4)\\ 35z &= 70. && (6) \end{aligned}$$

Now we solve equation (6) for z:

$$\begin{aligned} 35z &= 70\\ z &= 2. \end{aligned}$$

Then we back-substitute 2 for z in equation (4) and solve for y:

$$\begin{aligned} 10y - 15 \cdot 2 &= -40\\ 10y - 30 &= -40\\ 10y &= -10\\ y &= -1. \end{aligned}$$

Finally, we back-substitute -1 for y and 2 for z in equation (1) and solve for x:

$$\begin{aligned} x - 2(-1) + 3 \cdot 2 &= 11\\ x + 2 + 6 &= 11\\ x &= 3. \end{aligned}$$

We can check the triple $(3, -1, 2)$ in each of the three original equations. Since it makes all three equations true, the solution is $(3, -1, 2)$.

Example 2 Solve the following system:

$$\begin{aligned} x + y + z &= 7, && (1)\\ 3x - 2y + z &= 3, && (2)\\ x + 6y + 3z &= 25. && (3) \end{aligned}$$

SOLUTION We multiply equation (1) by -3 and add it to equation (2). We also multiply equation (1) by -1 and add it to equation (3).

$$\begin{aligned} x + y + z &= 7, && (1)\\ -5y - 2z &= -18, && (4)\\ 5y + 2z &= 18 && (5) \end{aligned}$$

Now we add equation (4) to equation (5):

$$\begin{aligned} x + y + z &= 7,\\ -5y - 2z &= -18,\\ 0 &= 0. \end{aligned}$$

The equation $0 = 0$ tells us that equation (3) of the original system is dependent on the first two equations. Thus the original system is equivalent to

$$\begin{aligned} x + y + z &= 7,\\ 3x - 2y + z &= 3 \end{aligned}$$

and has infinitely many solutions. To find an expression for these solutions, we first solve equation (4) for either y or z. In this case, we choose to solve for y:

$$\begin{aligned} -5y - 2z &= -18\\ -5y &= 2z - 18\\ y &= -\tfrac{2}{5}z + \tfrac{18}{5}. \end{aligned}$$

Then we back-substitute in equation (1) to find an expression for x in terms of z:

$$\begin{aligned} x - \tfrac{2}{5}z + \tfrac{18}{5} + z &= 7\\ x + \tfrac{3}{5}z + \tfrac{18}{5} &= 7\\ x + \tfrac{3}{5}z &= \tfrac{17}{5}\\ x &= -\tfrac{3}{5}z + \tfrac{17}{5}. \end{aligned}$$

The solutions of the system of equations are ordered triples of the form $\left(-\tfrac{3}{5}z + \tfrac{17}{5}, -\tfrac{2}{5}z + \tfrac{18}{5}, z\right)$, where z can be any real number. Any real number that we use for z then gives us values for x and y and thus an ordered triple in the solution set.

If we get a false equation, such as $0 = -5$, at some stage of the elimination process, we conclude that the original system is inconsistent. That is, it has no solutions.

Just as with systems of equations in two variables, systems of three linear equations in three variables can also be solved using Cramer's rule. See Appendix A.

Program

SYSLINEQ: This program solves 2×2, 3×3, and 4×4 systems of linear equations.

Although systems of three linear equations in three variables do not lend themselves well to graphical solutions, it is of interest to picture some possible solutions. The graph of a linear equation in three variables is a plane. Thus the solution set of such a system is the intersection of three planes. Some possibilities are shown below.

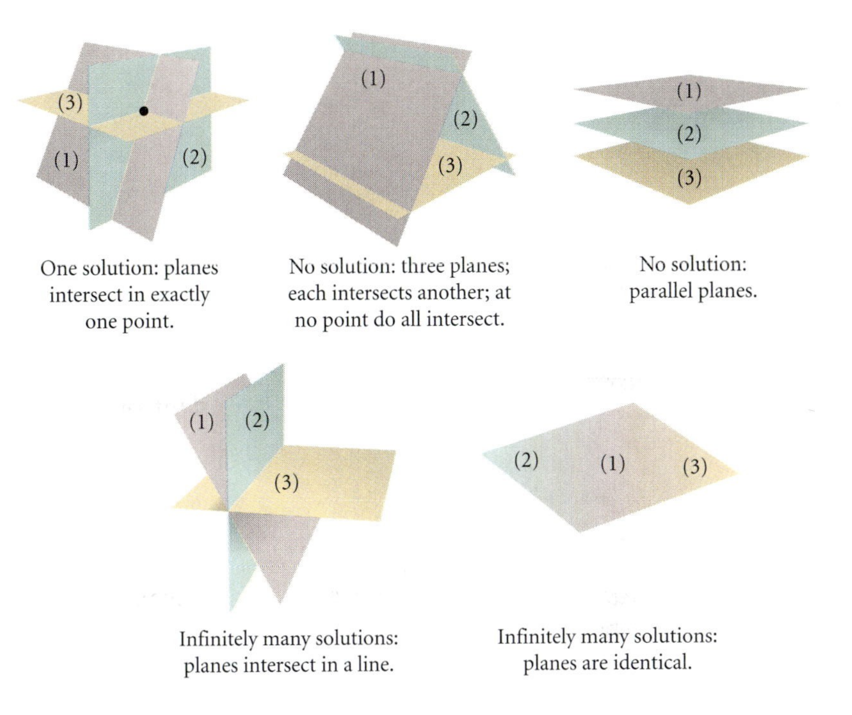

One solution: planes intersect in exactly one point.

No solution: three planes; each intersects another; at no point do all intersect.

No solution: parallel planes.

Infinitely many solutions: planes intersect in a line.

Infinitely many solutions: planes are identical.

If you have access to a grapher that graphs equations in three variables, use it to graph the systems in Examples 1 and 2.

Applications

Systems of equations in three or more variables allow us to solve many problems in fields such as business, the social and natural sciences, and engineering.

Example 3 *Investment.* Moira inherited \$15,000 and invested part of it in a money market account, part in municipal bonds, and part in a mutual fund. After 1 yr, she received a total of \$730 in simple interest from the three investments. The money market account paid 4% annually, the bonds paid 5% annually, and the mutual fund paid 6% annually. There was \$2000 more invested in the mutual fund than in bonds. Find the amount that Moira invested in each category.

SOLUTION

1. **Familiarize.** We let x, y, and z represent the amounts invested in the money market account, the bonds, and the mutual fund, respectively. Then the amounts of income produced annually by each investment are given by $4\%x$, $5\%y$, and $6\%z$, or $0.04x$, $0.05y$, and $0.06z$.
2. **Translate.** The fact that a total of \$15,000 is invested gives us one equation:

$$x + y + z = 15{,}000.$$

Since the total interest is \$730, we have a second equation:

$$0.04x + 0.05y + 0.06z = 730.$$

Another statement in the problem gives us a third equation.

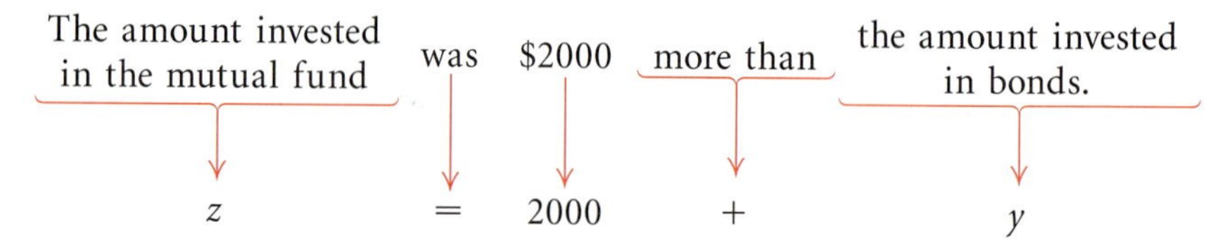

We now have a system of three equations:

$$\begin{aligned} x + y + z &= 15{,}000, \\ 0.04x + 0.05y + 0.06z &= 730, \\ z &= 2000 + y; \end{aligned} \quad \text{or} \quad \begin{aligned} x + y + z &= 15{,}000, \\ 4x + 5y + 6z &= 73{,}000, \\ -y + z &= 2000. \end{aligned}$$

3. **Carry out.** Solving the system of equations, we get (7000, 3000, 5000).
4. **Check.** The sum of the numbers is 15,000. The income produced is

$$0.04(7000) + 0.05(3000) + 0.06(5000) = 280 + 150 + 300, \quad \text{or} \quad 730.$$

Also the amount invested in the mutual fund, \$5000, is \$2000 more than the amount invested in bonds, \$3000. Our solution checks in the original problem.

5. **State.** Moira invested \$7000 in a money market account, \$3000 in municipal bonds, and \$5000 in a mutual fund.

Mathematical Models and Applications

In a situation in which a quadratic function will serve as a mathematical model, we may wish to find an equation, or formula, for the function. For a linear model, we can find an equation if we know two data points. For a quadratic function, we need three data points.

Example 4 *The Cost of Operating an Automobile at Various Speeds.* Under certain conditions, it is found that the cost of operating an automobile as a function of speed is approximated by a quadratic function. Use the data shown below to find an equation of the function. Then use the equation to determine the cost of operating the automobile at 60 mph and at 80 mph.

SPEED (IN MILES PER HOUR)	OPERATING COST PER MILE (IN CENTS)
10	22
20	20
50	20

SOLUTION Letting x = the speed and $f(x)$ = the cost, we use the three data points (10, 22), (20, 20), and (50, 20) to find a, b, and c in the equation $f(x) = ax^2 + bx + c$. First we substitute:

For (10, 22): $22 = a \cdot 10^2 + b \cdot 10 + c$;

For (20, 20): $20 = a \cdot 20^2 + b \cdot 20 + c$;

For (50, 20): $20 = a \cdot 50^2 + b \cdot 50 + c$.

We now have a system of equations in the variables a, b, and c:

$$100a + 10b + c = 22,$$
$$400a + 20b + c = 20,$$
$$2500a + 50b + c = 20.$$

Solving this system of equations, we obtain (0.005, −0.35, 25). Thus,

$$f(x) = 0.005x^2 - 0.35x + 25.$$

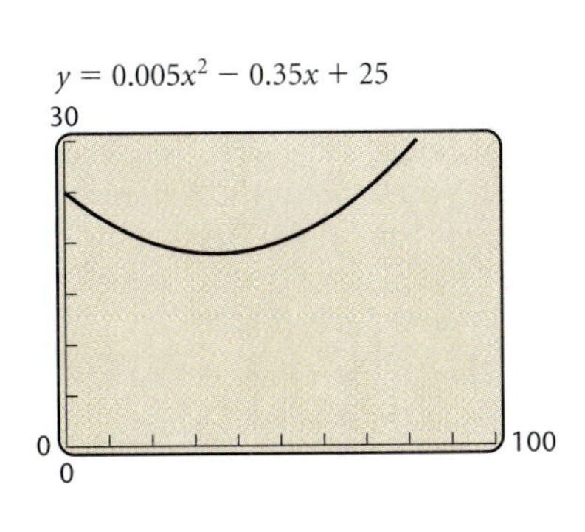

To determine the cost of operating an automobile at 60 mph, we find $f(60)$:

$$f(60) = 0.005(60)^2 - 0.35(60) + 25 = 22¢.$$

To find the cost of operating an automobile at 80 mph, we find $f(80)$:

$$f(80) = 0.005(80)^2 - 0.35(80) + 25 = 29¢.$$

The graph of the cost function is shown at left.

The function in Example 4 can also be found using the QUADRATIC REGRESSION feature in the STAT package on a grapher as we did in Section 2.3. Note that the method of Example 4 works when we have exactly three data points, whereas the QUADRATIC REGRESSION feature on a grapher can be used for three or more points.

7.2 Exercise Set

Solve each of the following systems.

1. $x + y + z = 2,$ $6x - 4y + 5z = 31,$ $5x + 2y + 2z = 13$

2. $x + 6y + 3z = 4,$ $2x + y + 2z = 3,$ $3x - 2y + z = 0$

3. $x - y + 2z = -3,$ $x + 2y + 3z = 4,$ $2x + y + z = -3$

4. $x + y + z = 6,$ $2x - y - z = -3,$ $x - 2y + 3z = 6$

5. $x + 2y - z = 5,$ $2x - 4y + z = 0,$ $3x + 2y + 2z = 3$

6. $2x + 3y - z = 1,$ $x + 2y + 5z = 4,$ $3x - y - 8z = -7$

7. $x + 2y - z = -8,$ $2x - y + z = 4,$ $8x + y + z = 2$

8. $x + 2y - z = 4,$ $4x - 3y + z = 8,$ $5x - y = 12$

9. $2x + y - 3z = 1,$
$x - 4y + z = 6,$
$4x - 7y - z = 13$

10. $x + 3y + 4z = 1,$
$3x + 4y + 5z = 3,$
$x + 8y + 11z = 2$

11. $4a + 9b = 8,$
$8a + 6c = -1,$
$6b + 6c = -1$

12. $3p + 2r = 11,$
$q - 7r = 4,$
$p - 6q = 1$

13. $w + x + y + z = 2,$
$w + 2x + 2y + 4z = 1,$
$-w + x - y - z = -6,$
$-w + 3x + y - z = -2$

14. $w + x - y + z = 0,$
$-w + 2x + 2y + z = 5,$
$-w + 3x + y - z = -4,$
$-2w + x + y - 3z = -7$

15. *Mail-Order Business.* Computer Warehouse charges \$3 for shipping orders up to 10 lb, \$5 for orders from 10 lb up to 15 lb, and \$7.50 for orders of 15 lb or more. One day shipping charges for 150 orders totaled \$680. The number of orders under 10 lb was three times the number of orders weighing 15 lb or more. Find the number of packages shipped at each rate.

16. *Mail-Order Business.* Natural Fibers Clothing charges \$4 for shipping orders of \$30 or less, \$6 for orders from \$30 to \$70, and \$7 for orders over \$70. One week shipping charges for 600 orders totaled \$3340. Eighty more orders for \$30 or less were shipped than orders for more than \$70. Find the number of orders shipped at each rate.

17. *Nutrition.* A hospital dietician must plan a lunch menu that provides 485 Cal, 41.5 g of carbohydrates, and 35 mg of calcium. A 3-oz serving of broiled ground beef contains 245 Cal, 0 g of carbohydrates, and 9 mg of calcium. One baked potato contains 145 Cal, 34 g of carbohydrates, and 8 mg of calcium. A one-cup serving of strawberries contains 45 Cal, 10 g of carbohydrates, and 21 mg of calcium. How many servings of each are required to provide the desired nutritional values? (*Source: Home and Garden Bulletin No. 72,* U.S. Government Printing Office, Washington, D.C. 20402)

18. *Nutrition.* A diabetic patient wishes to prepare a meal consisting of roasted chicken breast, mashed potatoes, and peas. A 3-oz serving of roasted, skinless chicken breast contains 140 Cal, 27 g of protein, and 64 mg of sodium. A one-cup serving of mashed potatoes contains 160 Cal, 4 g of protein, and 636 mg of sodium, and a one-cup serving of peas contains 125 Cal, 8 g of protein, and 139 mg of sodium. How many servings of each should be used if the meal is to contain 415 Cal, 50.5 g of protein, and 553 mg of sodium? (*Source: Home and Garden Bulletin No. 72,* U.S. Government Printing Office, Washington, D.C. 20402)

19. *Investment.* Jamal earns a year-end bonus of \$5000 and puts it in 3 one-year investments that pay \$302 in simple interest. Part is invested at 4%, part at 6%, and part at 7%. There is \$1500 more invested at 7% than at 4%. Find the amount invested at each rate.

20. *Investment.* Casey receives \$336 per year in simple interest from three investments. Part is invested at 8%, part at 9%, and part at 10%. There is \$500 more invested at 9% than at 8%. The amount invested at 10% is three times the amount invested at 9%. Find the amount invested at each rate.

21. *Price Increases.* Orange juice, a raisin bagel, and a cup of coffee from Kelly's Koffee Kart cost a total of \$3. Kelly posts a notice announcing that, effective next week, the price of orange juice will increase 50% and the price of bagels will increase 20%. After the increase, the same purchase will cost a total of \$3.75, and orange juice will cost twice as much as coffee. Find the price of each item before the increase.

22. *Cost of Snack Food.* Martin and Eva pool their loose change to buy snacks on their coffee break. One day, they spent \$1.85 on 1 carton of milk, 2 donuts, and 1 cup of coffee. The next day, they spent \$2.30 on 3 donuts and 2 cups of coffee. The third day, they bought 1 carton of milk, 1 donut, and 2 cups of coffee and spent \$1.75. On the fourth day, they have a total of \$1.80 left. Is this enough to buy 2 cartons of milk and 2 donuts?

23. *Passenger Transportation.* The total volume of passenger traffic on domestic airways, by bus (excluding school buses and urban transit buses), and by railroads in the United States in a recent year was 405 billion passenger-miles. (One passenger-mile is the transportation of one passenger the distance of one mile.) The volume of bus traffic was 10 billion passenger-miles more than the volume of railroad traffic. The total volume of railroad traffic was 329 billion passenger-miles less than the volume of traffic on domestic airways. What was the volume of each type of passenger traffic? (*Source: Transportation in America,* Eno Transportation Foundation, Inc., Landsdowne, VA)

24. *Cheese Consumption.* The total per capita consumption (the average amount consumed per person) of cheddar, mozzarella, and Swiss cheese in the United States in a recent year was 18.1 lb. The total consumption of mozzarella and Swiss cheese was 0.3 lb less than that of cheddar cheese. The consumption of mozzarella cheese was 6.5 lb more than that of Swiss cheese. What was the per capita consumption of each type of cheese? (*Source:* U.S. Department of Agriculture, Economic Research Service, *Food Consumption, Prices, and Expenditures*)

25. *Golf.* On an 18-hole golf course, there are par-3 holes, par-4 holes, and par-5 holes. A golfer who shoots par on every hole has a score of 72. The sum of the number of par-3 holes and the number of par-5 holes is 8. How many of each type of hole are there on the golf course?

26. *Golf.* On an 18-hole golf course, there are par-3 holes, par-4 holes, and par-5 holes. A golfer who shoots par on every hole has a score of 70. There are twice as many par-4 holes as there are par-5 holes. How many of each type of hole are there on the golf course?

27. *Cost of a New Car.* The following table shows the average new car transaction price in the United States as a percentage of disposable income for three recent years, represented as years since 1960.

YEAR	PERCENT OF DISPOSABLE INCOME
(1960) 0	43
(1973) 13	31
(1992) 32	43

Source: Kemper Reports, Fall/Winter, 1995, p. 11.

a) Find a quadratic function $f(x) = ax^2 + bx + c$ that fits the data.
b) Use the function to predict the price of a new car as a percentage of disposable income in the year 2000.

28. *Number of Marriages.* The following table shows the number of marriages, in thousands, in the mountain states in three recent years, presented as years since 1980.

YEAR	NUMBER OF MARRIAGES (IN THOUSANDS)
(1980) 0	242
(1985) 5	234
(1993) 13	255

Source: U.S. National Center for Health Statistics, *Vital Statistics of the United States*, annual.

a) Find a quadratic function $f(x) = ax^2 + bx + c$ that fits the data.
b) Use the function to predict the number of marriages in the mountain states in 2000.

29. *Milk Consumption.* The following table shows the per capita milk consumption, in pounds, in the United States in three recent years, represented as years since 1975.

YEAR	MILK CONSUMPTION (IN POUNDS)
(1975) 0	539
(1985) 10	594
(1992) 17	565

Source: U.S. Department of Agriculture, Economic Research Service, *Food Consumption, Prices, and Expenditures.*

a) Find a quadratic function $f(x) = ax^2 + bx + c$ that fits the data.
b) Use the function to predict the per capita consumption of milk in 2005.

30. *Crude Steel Production.* The following table shows the crude steel production, in millions of metric tons, in the United States in three recent years, represented as years since 1980.

YEAR	CRUDE STEEL PRODUCTION (IN MILLIONS OF METRIC TONS)
(1980) 0	80
(1988) 8	92
(1993) 13	89

Source: Statistical Office of the United Nations, New York, NY, *Statistical Yearbook* and *Industrial Commodity Statistics Yearbook*, annuals.

a) Find a quadratic function $f(x) = ax^2 + bx + c$ that fits the data.
b) Use the function to predict the crude steel production in the United States in 2000.

Skill Maintenance

For $A = 2x - 3y$ and $B = -x + 5y$, find each of the following.

31. $2A$

32. $-A$

33. $A + B$

34. $-3A + B$

Synthesis

35. Given two linear equations in three variables, $Ax + By + Cz = D$ and $Ex + Fy + Gz = H$, explain how you would find a third equation such that the system composed of the three equations is dependent.

36. Write a problem for a classmate to solve that can be translated to a system of three equations in three variables.

In Exercises 37 and 38, let u represent 1/x, v represent 1/y, and w represent 1/z. Solve first for u, v, and w. Then solve the system.

37. $\frac{2}{x} - \frac{1}{y} - \frac{3}{z} = -1,$
$\frac{2}{x} - \frac{1}{y} + \frac{1}{z} = -9,$
$\frac{1}{x} + \frac{2}{y} - \frac{4}{z} = 17$

38. $\frac{2}{x} + \frac{2}{y} - \frac{3}{z} = 3,$
$\frac{1}{x} - \frac{2}{y} - \frac{3}{z} = 9,$
$\frac{7}{x} - \frac{2}{y} + \frac{9}{z} = -39$

39. Find the sum of the angle measures at the tips of the star.

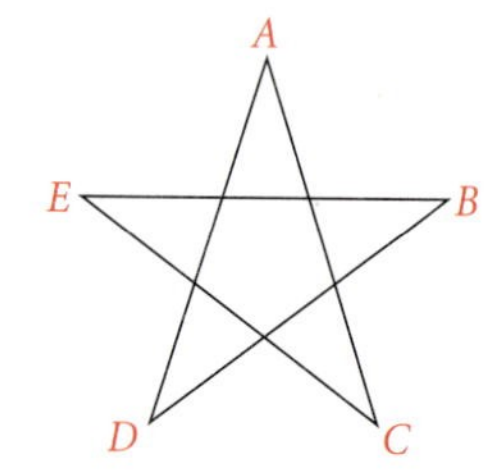

40. *Transcontinental Railroad.* Use the following facts to find the year in which the first U.S. transcontinental railroad was completed. The sum of the digits in the year is 24. The units digit is 1 more than the hundreds digit. Both the tens and the units digits are multiples of three.

In Exercises 41 and 42, three solutions of an equation are given. Use a system of three equations in three variables to find the constants and write the equation.

41. $Ax + By + Cz = 12;$
$\left(1, \frac{3}{4}, 3\right)$, $\left(\frac{4}{3}, 1, 2\right)$, and $(2, 1, 1)$

42. $y = B - Mx - Nz;$
$(1, 1, 2)$, $(3, 2, -6)$, and $\left(\frac{3}{2}, 1, 1\right)$

In Exercises 43 and 44, four solutions of the equation $y = ax^3 + bx^2 + cx + d$ are given. Use a system of four equations in four variables to find the constants and write the equation.

43. $(-2, 59)$, $(-1, 13)$, $(1, -1)$, and $(2, -17)$

44. $(-2, -39)$, $(-1, -12)$, $(1, -6)$, and $(3, 16)$

45. *Theater Attendance.* A performance at the Bingham Performing Arts Center was attended by 100 people. The audience consisted of adults, students, and children. The ticket prices were \$10 for adults, \$3 for students, and 50 cents for children. The total amount of money taken in was \$100. How many adults, students, and children were in attendance? Does there seem to be some information missing? Do some careful reasoning.

7.3 Matrices and Systems of Equations

- *Solve systems of equations using matrices.*

Matrices and Row-Equivalent Operations

In this section, we consider additional techniques for solving systems of equations. You have probably observed that when we solve a system of equations, we perform computations with the coefficients and the constants and continually rewrite the variables. We can streamline the solution process by omitting the variables until a solution is found. For example, the system

$$2x - 3y = 7,$$
$$x + 4y = -2$$

can be written more simply as

$$\left[\begin{array}{rr|r} 2 & -3 & 7 \\ 1 & 4 & -2 \end{array}\right].$$

The vertical line replaces the equals signs.

The rectangular array of numbers above is called a **matrix** (pl., **matrices**). In fact, it is called an **augmented matrix** for the given system of equations, because it contains not only the coefficients but also the constant terms. The matrix

$$\begin{bmatrix} 2 & -3 \\ 1 & 4 \end{bmatrix}$$

is called the **coefficient matrix** of the system.

The **rows** of a matrix are horizontal, and the **columns** are vertical. The augmented matrix above has 2 rows and 3 columns, and the coefficient matrix has 2 rows and 2 columns. A matrix with m rows and n columns is said to be of **order** $m \times n$. Thus the order of the augmented matrix above is 2×3, and the order of the coefficient matrix is 2×2. When $m = n$, a matrix is said to be **square**. The coefficient matrix above is a square matrix. The numbers 2 and 4 lie on the **main diagonal** of the coefficient matrix. The numbers in a matrix are called **entries**.

Gaussian Elimination with Matrices

In Section 7.2, we described a series of operations that can be used to transform a system of equations to an equivalent system. Each of these operations corresponds to one that can be used to produce *row-equivalent matrices*.

Row-Equivalent Operations

1. Interchange any two rows.
2. Multiply each entry in a row by the same nonzero constant.
3. Add a nonzero multiple of one row to another row.

We can use these operations on the augmented matrix of a system of equations to solve the system.

Example 1 Solve the following system:

$$\begin{aligned} 2x - y + 4z &= -3, \\ x - 2y - 10z &= -6, \\ 3x \qquad + 4z &= 7. \end{aligned}$$

SOLUTION First we write the augmented matrix, writing 0 for the missing y-term in the last equation:

$$\left[\begin{array}{rrr|r} 2 & -1 & 4 & -3 \\ 1 & -2 & -10 & -6 \\ 3 & 0 & 4 & 7 \end{array}\right].$$

Our goal is to find a row-equivalent matrix of the form

$$\left[\begin{array}{ccc|c} 1 & a & b & c \\ 0 & 1 & d & e \\ 0 & 0 & 1 & f \end{array}\right].$$

The variables can then be reinserted to form equations from which we can complete the solution by working from the bottom equation to the top and using back-substitution.

The first step is to multiply and/or interchange rows so that each number in the first column below the first number is a multiple of that number. In this case, we interchange the first and second rows to obtain a 1 in the upper left-hand corner.

$$\left[\begin{array}{rrr|r} 1 & -2 & -10 & -6 \\ 2 & -1 & 4 & -3 \\ 3 & 0 & 4 & 7 \end{array}\right] \quad \begin{array}{l} \text{New row 1} = \text{row 2} \\ \text{New row 2} = \text{row 1} \\ \\ \end{array}$$

Next we multiply the first row by -2 and add it to the second row. We also multiply the first row by -3 and add it to the third row.

$$\left[\begin{array}{rrr|r} 1 & -2 & -10 & -6 \\ 0 & 3 & 24 & 9 \\ 0 & 6 & 34 & 25 \end{array}\right] \quad \begin{array}{l} \\ \text{New row 2} = -2(\text{row 1}) + \text{row 2} \\ \text{New row 3} = -3(\text{row 1}) + \text{row 3} \end{array}$$

Now we multiply the second row by $\frac{1}{3}$ to get a 1 in the second row, second column.

$$\left[\begin{array}{rrr|r} 1 & -2 & -10 & -6 \\ 0 & 1 & 8 & 3 \\ 0 & 6 & 34 & 25 \end{array}\right] \quad \begin{array}{l} \\ \text{New row 2} = \frac{1}{3}(\text{row 2}) \\ \\ \end{array}$$

Then we multiply the second row by -6 and add it to the third row.

$$\left[\begin{array}{rrr|r} 1 & -2 & -10 & -6 \\ 0 & 1 & 8 & 3 \\ 0 & 0 & -14 & 7 \end{array}\right] \quad \begin{array}{l} \\ \\ \text{New row 3} = -6(\text{row 2}) + \text{row 3} \end{array}$$

Finally we multiply the third row by $-\frac{1}{14}$ to get a 1 in the third row, third column.

$$\left[\begin{array}{rrr|r} 1 & -2 & -10 & -6 \\ 0 & 1 & 8 & 3 \\ 0 & 0 & 1 & -\frac{1}{2} \end{array}\right] \qquad \text{New row 3} = -\tfrac{1}{14}(\text{row 3})$$

Now we can write the system of equations that corresponds to the last matrix above:

$$\begin{aligned} x - 2y - 10z &= -6, \quad &(1)\\ y + 8z &= 3, \quad &(2)\\ z &= -\tfrac{1}{2}. \quad &(3) \end{aligned}$$

We back-substitute $-\frac{1}{2}$ for z in equation (2) and solve for y:

$$\begin{aligned} y + 8\left(-\tfrac{1}{2}\right) &= 3\\ y - 4 &= 3\\ y &= 7. \end{aligned}$$

Next we back-substitute 7 for y and $-\frac{1}{2}$ for z in equation (1) and solve for x:

$$\begin{aligned} x - 2 \cdot 7 - 10\left(-\tfrac{1}{2}\right) &= -6\\ x - 14 + 5 &= -6\\ x - 9 &= -6\\ x &= 3. \end{aligned}$$

The triple $\left(3, 7, -\frac{1}{2}\right)$ checks in the original system of equations, so it is the solution.

The procedure followed in Example 1 is called **Gaussian elimination** for matrices. The last matrix above is in **row-echelon form**. To be in this form, a matrix must have the following properties.

Row-Echelon Form

1. If a row does not consist entirely of 0's, then the first nonzero element in the row is a 1 (called a **leading 1**).
2. For any two successive nonzero rows, the leading 1 in the lower row is farther to the right than the leading 1 in the higher row.
3. All the rows consisting entirely of 0's are at the bottom of the matrix.

If a fourth property is also satisfied, a matrix is said to be in **reduced row-echelon form**:

4. Each column that contains a leading 1 has 0's everywhere else.

Example 2 Which of the following matrices are in row-echelon form? Which, if any, are in reduced row-echelon form?

a) $\begin{bmatrix} 1 & -3 & 5 & -2 \\ 0 & 1 & -4 & 3 \\ 0 & 0 & 1 & 10 \end{bmatrix}$ b) $\begin{bmatrix} 0 & -1 & 2 \\ 0 & 1 & 5 \end{bmatrix}$ c) $\begin{bmatrix} 1 & -2 & -6 & 4 & 7 \\ 0 & 3 & 5 & -8 & -1 \\ 0 & 0 & 1 & 9 & 2 \end{bmatrix}$

d) $\begin{bmatrix} 1 & 0 & 0 & -2.4 \\ 0 & 1 & 0 & 0.8 \\ 0 & 0 & 1 & 5.6 \end{bmatrix}$ e) $\begin{bmatrix} 1 & 0 & 0 & 0 & \frac{2}{3} \\ 0 & 1 & 0 & 0 & -\frac{1}{4} \\ 0 & 0 & 1 & 0 & \frac{6}{7} \\ 0 & 0 & 0 & 0 & 0 \end{bmatrix}$ f) $\begin{bmatrix} 1 & -4 & 2 & 5 \\ 0 & 0 & 0 & 0 \\ 0 & 1 & -3 & -8 \end{bmatrix}$

SOLUTION The matrices in (a), (d), and (e) satisfy the row-echelon criteria and, thus, are in row-echelon form. In (b) and (c), the first nonzero elements of the first and second rows, respectively, are not 1. In (f), the row consisting entirely of 0's is not at the bottom of the matrix. Thus the matrices in (b), (c), and (f) are not in row-echelon form. In (d) and (e), not only are the row-echelon criteria met but each column that contains a leading 1 also has 0's elsewhere, so these matrices are in reduced row-echelon form.

Gauss–Jordan Elimination

We have seen that with Gaussian elimination we perform row-equivalent operations on a matrix to obtain a row-equivalent matrix in row-echelon form. When we continue to apply these operations until we have a matrix in *reduced* row-echelon form, we are using **Gauss–Jordan elimination**. This method is named for Karl Friedrich Gauss and Wilhelm Jordan (1842–1899).

Example 3 Use Gauss–Jordan elimination to solve the system of equations in Example 1.

SOLUTION Using Gaussian elimination in Example 1, we obtained the matrix

$$\left[\begin{array}{ccc|c} 1 & -2 & -10 & -6 \\ 0 & 1 & 8 & 3 \\ 0 & 0 & 1 & -\frac{1}{2} \end{array}\right].$$

We continue to perform row-equivalent operations until we have a matrix in reduced row-echelon form. First we multiply the third row by -8 and add it to the second row. We also multiply the third row by 10 and add it to the first row.

$$\left[\begin{array}{ccc|c} 1 & -2 & 0 & -11 \\ 0 & 1 & 0 & 7 \\ 0 & 0 & 1 & -\frac{1}{2} \end{array}\right]$$

New row 1 = 10(row 3) + row 1
New row 2 = −8(row 3) + row 2

Then we multiply the second row by 2 and add it to the first row.

$$\left[\begin{array}{ccc|c} 1 & 0 & 0 & 3 \\ 0 & 1 & 0 & 7 \\ 0 & 0 & 1 & -\frac{1}{2} \end{array}\right] \qquad \text{New row 1} = 2(\text{row 2}) + \text{row 1}$$

Writing the system of equations that corresponds to this matrix, we have

$$\begin{aligned} x \qquad &= 3, \\ y \quad &= 7, \\ z &= -\tfrac{1}{2}. \end{aligned}$$

We can actually read the solution, $\left(3, 7, -\frac{1}{2}\right)$, directly from the reduced row-echelon matrix.

Example 4 Solve the following system:

$$\begin{aligned} 3x - 4y - z &= 6, \\ 2x - y + z &= -1, \\ 4x - 7y - 3z &= 13. \end{aligned}$$

SOLUTION We write the augmented matrix and use Gauss–Jordan elimination.

$$\left[\begin{array}{ccc|c} 3 & -4 & -1 & 6 \\ 2 & -1 & 1 & -1 \\ 4 & -7 & -3 & 13 \end{array}\right]$$

We begin by multiplying the second and third rows by 3 so that each number in the first column below the first number, 3, is a multiple of that number.

$$\left[\begin{array}{ccc|c} 3 & -4 & -1 & 6 \\ 6 & -3 & 3 & -3 \\ 12 & -21 & -9 & 39 \end{array}\right] \qquad \begin{array}{l} \\ \text{New row 2} = 3(\text{row 2}) \\ \text{New row 3} = 3(\text{row 3}) \end{array}$$

Next we multiply the first row by -2 and add it to the second row. We also multiply the first row by -4 and add it to the third row.

$$\left[\begin{array}{ccc|c} 3 & -4 & -1 & 6 \\ 0 & 5 & 5 & -15 \\ 0 & -5 & -5 & 15 \end{array}\right] \qquad \begin{array}{l} \\ \text{New row 2} = -2(\text{row 1}) + \text{row 2} \\ \text{New row 3} = -4(\text{row 1}) + \text{row 3} \end{array}$$

Now we add the second row to the third row.

$$\left[\begin{array}{ccc|c} 3 & -4 & -1 & 6 \\ 0 & 5 & 5 & -15 \\ 0 & 0 & 0 & 0 \end{array}\right] \qquad \begin{array}{l} \\ \\ \text{New row 3} = \text{row 2} + \text{row 3} \end{array}$$

We can stop at this stage because we have a row consisting entirely of 0's. The last row of the matrix corresponds to the equation $0 = 0$, which is true for all values of x, y, and z. Consequently, we have a dependent system of equations that is equivalent to

$$\begin{aligned} 3x - 4y - z &= 6, \\ 5y + 5z &= -15. \end{aligned}$$

Solving the second equation for y gives us

$$y = -z - 3.$$

Substituting for y in the first equation, we get

$$3x - 4(-z - 3) - z = 6$$
$$x = -z - 2.$$

Then the solutions of this system are of the form

$$(-z - 2, -z - 3, z),$$

where z can be any real number.

Similarly, if we obtain a row whose only nonzero entry occurs in the last column, we have an inconsistent system of equations. For example, in the matrix

$$\left[\begin{array}{ccc|c} 1 & 0 & 3 & -2 \\ 0 & 1 & 5 & 4 \\ 0 & 0 & 0 & 6 \end{array}\right],$$

the last row corresponds to the false equation $0 = 6$, so we know the original system of equations has no solution.

Row-equivalent operations on matrices can be performed using the MATRIX feature on most graphers. Refer to your user's manual for instructions. Then use a grapher to perform the row-equivalent operations in Examples 1, 3, and 4.

7.3 Exercise Set

Determine the order of each matrix.

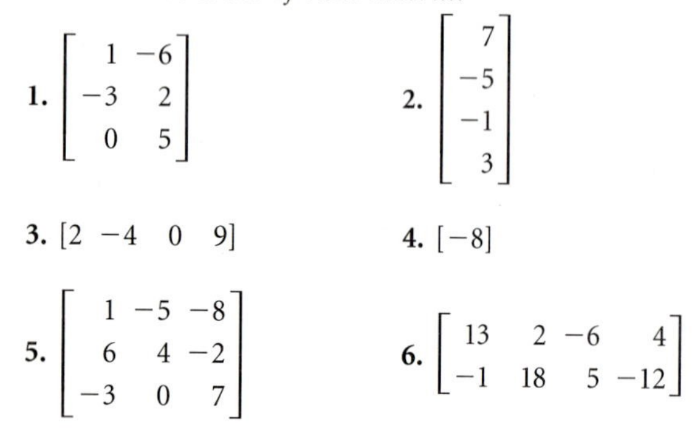

1. $\begin{bmatrix} 1 & -6 \\ -3 & 2 \\ 0 & 5 \end{bmatrix}$

2. $\begin{bmatrix} 7 \\ -5 \\ -1 \\ 3 \end{bmatrix}$

3. $[2 \quad -4 \quad 0 \quad 9]$

4. $[-8]$

5. $\begin{bmatrix} 1 & -5 & -8 \\ 6 & 4 & -2 \\ -3 & 0 & 7 \end{bmatrix}$

6. $\begin{bmatrix} 13 & 2 & -6 & 4 \\ -1 & 18 & 5 & -12 \end{bmatrix}$

Write the augmented matrix for each system of equations.

7. $\begin{aligned} 2x - y &= 7, \\ x + 4y &= -5 \end{aligned}$

8. $\begin{aligned} 3x + 2y &= 8, \\ 2x - 3y &= 15 \end{aligned}$

9. $\begin{aligned} x - 2y + 3z &= 12, \\ 2x \qquad - 4z &= 8, \\ 3y + z &= 7 \end{aligned}$

10. $\begin{aligned} x + y - z &= 7, \\ 3y + 2z &= 1, \\ -2x - 5y \qquad &= 6 \end{aligned}$

Write the system of equations that corresponds to each augmented matrix.

11. $\left[\begin{array}{cc|c} 3 & -5 & 1 \\ 1 & 4 & -2 \end{array}\right]$

12. $\left[\begin{array}{cc|c} 1 & 2 & -6 \\ 4 & 1 & -3 \end{array}\right]$

13. $\left[\begin{array}{rrr|r} 2 & 1 & -4 & 12 \\ 3 & 0 & 5 & -1 \\ 1 & -1 & 1 & 2 \end{array}\right]$

14. $\left[\begin{array}{rrr|r} -1 & -2 & 3 & 6 \\ 0 & 4 & 1 & 2 \\ 2 & -1 & 0 & 9 \end{array}\right]$

Solve each system of equations using Gaussian elimination or Gauss–Jordan elimination.

15. $4x + 2y = 11,$ $3x - y = 2$

16. $2x + y = 1,$ $3x + 2y = -2$

17. $5x - 2y = -3,$ $2x + 5y = -24$

18. $2x + y = 1,$ $3x - 6y = 4$

19. $3x + 4y = 7,$ $-5x + 2y = 10$

20. $5x - 3y = -2,$ $4x + 2y = 5$

21. $3x + 2y = 6,$ $2x - 3y = -9$

22. $x - 4y = 9,$ $2x + 5y = 5$

23. $x - 3y = 8,$ $-2x + 6y = 3$

24. $4x - 8y = 12,$ $-x + 2y = -3$

25. $-2x + 6y = 4,$ $3x - 9y = -6$

26. $6x + 2y = -10,$ $-3x - y = 6$

27. $x + 2y - 3z = 9,$ $2x - y + 2z = -8,$ $3x - y - 4z = 3$

28. $x - y + 2z = 0,$ $x - 2y + 3z = -1,$ $2x - 2y + z = -3$

29. $4x - y - 3z = 1,$ $8x + y - z = 5,$ $2x + y + 2z = 5$

30. $3x + 2y + 2z = 3,$ $x + 2y - z = 5,$ $2x - 4y + z = 0$

31. $x - 2y + 3z = -4,$ $3x + y - z = 0,$ $2x + 3y - 5z = 1$

32. $2x - 3y + 2z = 2,$ $x + 4y - z = 9,$ $-3x + y - 5z = 5$

33. $2x - 4y - 3z = 3,$ $x + 3y + z = -1,$ $5x + y - 2z = 2$

34. $x + y - 3z = 4,$ $4x + 5y + z = 1,$ $2x + 3y + 7z = -7$

35. $p + q + r = 1,$ $p + 2q + 3r = 4,$ $4p + 5q + 6r = 7$

36. $m + n + t = 9,$ $m - n - t = -15,$ $3m + n + t = 2$

37. $a + b - c = 7,$ $a - b + c = 5,$ $3a + b - c = -1$

38. $a - b + c = 3,$ $2a + b - 3c = 5,$ $4a + b - c = 11$

39. $-2w + 2x + 2y - 2z = -10,$ $w + x + y + z = -5,$ $3w + x - y + 4z = -2,$ $w + 3x - 2y + 2z = -6$

40. $-w + 2x - 3y + z = -8,$ $-w + x + y - z = -4,$ $w + x + y + z = 22,$ $-w + x - y - z = -14$

Use Gaussian elimination or Gauss–Jordan elimination in Exercises 41–44.

41. *Time of Return.* The Houlihans pay their babysitter $5 per hour before 11 P.M. and $7.50 per hour after 11 P.M. One evening they went out for 5 hr and paid the sitter $30. What time did they come home?

42. *Advertising Expense.* Cabletron Systems, Inc., spent a total of $17.5 million on advertising in fiscal years 1992, 1993, and 1994. The amount spent in 1994 was three fourths the total amount spent in 1992 and 1993. The amount spent in 1993 was $2.4 million less than the amount spent in 1994. How much was spent on advertising each year? (*Source:* Cabletron Systems Annual Report, 1994)

43. *Borrowing.* Gonzalez Manufacturing borrowed $30,000 to buy a new piece of equipment. Part of the money was borrowed at 8%, part at 10%, and part at 12%. The annual interest was $3040, and the total amount borrowed at 8% and 10% was twice the amount borrowed at 12%. How much was borrowed at each rate?

44. *Stamp Purchase.* Ricardo spent $18.30 on 32¢ and 23¢ stamps. He bought a total of 60 stamps. How many of each type did he buy?

Skill Maintenance

45. Find the opposite of -8.

46. Solve: $y + 3 = -2$.

47. Simplify: $3 - 12$.

48. Simplify: $5(-1) + 4 \cdot 3 - 8 \cdot 2$.

Synthesis

49. ◆ Solve the following system of equations using matrices and Gaussian elimination. Then solve it again using Gauss–Jordan elimination. Do you prefer one method over the other? Why or why not?

$$3x + 4y + 2z = 0,$$
$$x - y - z = 10,$$
$$2x + 3y + 3z = -10$$

50. ◆ Explain in your own words why the augmented matrix below represents a dependent system of equations.

$$\left[\begin{array}{rrr|r} 1 & -3 & 2 & -5 \\ 0 & 1 & -4 & 8 \\ 0 & 0 & 0 & 0 \end{array}\right]$$

In Exercises 51 and 52, three solutions of the equation $y = ax^2 + bx + c$ are given. Use a system of three equations in three variables and Gaussian elimination or Gauss–Jordan elimination to find the constants and write the equation.

51. $(-3, 12)$, $(-1, -7)$, and $(1, -2)$

52. $(-1, 0)$, $(1, -3)$, and $(3, -22)$

53. Find two different row-echelon forms of

$$\begin{bmatrix} 1 & 5 \\ 3 & 2 \end{bmatrix}.$$

54. Consider the system of equations

$$\begin{aligned} x - y + 3z &= -8, \\ 2x + 3y - z &= 5, \\ 3x + 2y + 2kz &= -3k. \end{aligned}$$

For what value(s) of k, if any, will the system have

a) no solution?

b) exactly one solution?

c) infinitely many solutions?

Solve using matrices.

55. $y = x + z,$
$3y + 5z = 4,$
$x + 4 = y + 3z$

56. $x + y = 2z,$
$2x - 5z = 4,$
$x - z = y + 8$

57. $x - 4y + 2z = 7,$
$3x + y + 3z = -5$

58. $x - y - 3z = 3,$
$-x + 3y + z = -7$

59. $4x + 5y = 3,$
$-2x + y = 9,$
$3x - 2y = -15$

60. $2x - 3y = -1,$
$-x + 2y = -2,$
$3x - 5y = 1$

7.4 Matrix Operations

- *Add, subtract, and multiply matrices when possible.*
- *Write a matrix equation equivalent to a system of equations.*

In Section 7.3, we used matrices to solve systems of equations. Matrices are useful in many other types of applications as well. In this section, we study matrices and some of their properties.

A capital letter is generally used to name a matrix, and lower-case letters with double subscripts generally denote its entries. For example, a_{47}, read "a sub four seven," indicates the entry in the fourth row and the seventh column. A general term is represented by a_{ij}. The notation a_{ij} indicates the entry in row i and column j. In general, we can write a matrix as

$$\mathbf{A} = [a_{ij}] = \begin{bmatrix} a_{11} & a_{12} & a_{13} & \cdots & a_{1n} \\ a_{21} & a_{22} & a_{23} & \cdots & a_{2n} \\ a_{31} & a_{32} & a_{33} & \cdots & a_{3n} \\ \vdots & \vdots & \vdots & & \vdots \\ a_{m1} & a_{m2} & a_{m3} & \cdots & a_{mn} \end{bmatrix}.$$

The matrix above has m rows and n columns. That is, its order is $m \times n$.

Two matrices are **equal** if they have the same order and corresponding entries are equal.

Interactive Discovery

Enter the following matrices on a grapher.

$$\mathbf{A} = \begin{bmatrix} 2 & -3 \\ -1 & 4 \end{bmatrix}, \qquad \mathbf{B} = \begin{bmatrix} -3 & 7 \\ 1 & -5 \end{bmatrix},$$

$$\mathbf{C} = \begin{bmatrix} 4 & 0 & -2 \\ 10 & -4 & -6 \\ -1 & 5 & 3 \end{bmatrix}, \qquad \mathbf{D} = \begin{bmatrix} 1 & -1 & 8 \\ 0 & 3 & -4 \\ 2 & -5 & 7 \end{bmatrix}$$

Use the MATRIX feature of the grapher to find $\mathbf{A} + \mathbf{B}$, $\mathbf{A} - \mathbf{B}$, $\mathbf{C} + \mathbf{D}$, and $\mathbf{C} - \mathbf{D}$. What relationship do you observe between corresponding entries of each pair of matrices and the entries of the sum? the difference?

What happens when you try to find $\mathbf{A} + \mathbf{C}$?

Matrix Addition and Subtraction

To add or subtract matrices, we add or subtract their corresponding entries. The matrices must have the same order for this to be possible.

Addition and Subtraction of Matrices

Given two $m \times n$ matrices $\mathbf{A} = [a_{ij}]$ and $\mathbf{B} = [b_{ij}]$, their sum is

$$\mathbf{A} + \mathbf{B} = [a_{ij} + b_{ij}]$$

and their difference is

$$\mathbf{A} - \mathbf{B} = [a_{ij} - b_{ij}].$$

Addition of matrices is both commutative and associative.

Example 1 Find $\mathbf{A} + \mathbf{B}$ for each of the following.

a) $\mathbf{A} = \begin{bmatrix} -5 & 0 \\ 4 & \frac{1}{2} \end{bmatrix}, \quad \mathbf{B} = \begin{bmatrix} 6 & -3 \\ 2 & 3 \end{bmatrix}$

b) $\mathbf{A} = \begin{bmatrix} 1 & 3 \\ -1 & 5 \\ 6 & 0 \end{bmatrix}, \quad \mathbf{B} = \begin{bmatrix} -1 & -2 \\ 1 & -2 \\ -3 & 1 \end{bmatrix}$

SOLUTION Each pair of matrices has the same order. We add the corresponding entries.

a) $\mathbf{A} + \mathbf{B} = \begin{bmatrix} -5 & 0 \\ 4 & \frac{1}{2} \end{bmatrix} + \begin{bmatrix} 6 & -3 \\ 2 & 3 \end{bmatrix}$

$$= \begin{bmatrix} -5 + 6 & 0 + (-3) \\ 4 + 2 & \frac{1}{2} + 3 \end{bmatrix} = \begin{bmatrix} 1 & -3 \\ 6 & 3\frac{1}{2} \end{bmatrix}$$

b) $\mathbf{A} + \mathbf{B} = \begin{bmatrix} 1 & 3 \\ -1 & 5 \\ 6 & 0 \end{bmatrix} + \begin{bmatrix} -1 & -2 \\ 1 & -2 \\ -3 & 1 \end{bmatrix}$

$$= \begin{bmatrix} 1 + (-1) & 3 + (-2) \\ -1 + 1 & 5 + (-2) \\ 6 + (-3) & 0 + 1 \end{bmatrix} = \begin{bmatrix} 0 & 1 \\ 0 & 3 \\ 3 & 1 \end{bmatrix}$$

Use a grapher to verify these results.

Example 2 Find $\mathbf{C} - \mathbf{D}$ for each of the following.

a) $\mathbf{C} = \begin{bmatrix} 1 & 2 \\ -2 & 0 \\ -3 & -1 \end{bmatrix}$, $\mathbf{D} = \begin{bmatrix} 1 & -1 \\ 1 & 3 \\ 2 & 3 \end{bmatrix}$ b) $\mathbf{C} = \begin{bmatrix} 5 & -6 \\ -3 & 4 \end{bmatrix}$, $\mathbf{D} = \begin{bmatrix} -4 \\ 1 \end{bmatrix}$

SOLUTION

a) We subtract corresponding entries:

$$\mathbf{C} - \mathbf{D} = \begin{bmatrix} 1 & 2 \\ -2 & 0 \\ -3 & -1 \end{bmatrix} - \begin{bmatrix} 1 & -1 \\ 1 & 3 \\ 2 & 3 \end{bmatrix}$$

$$= \begin{bmatrix} 1 - 1 & 2 - (-1) \\ -2 - 1 & 0 - 3 \\ -3 - 2 & -1 - 3 \end{bmatrix} = \begin{bmatrix} 0 & 3 \\ -3 & -3 \\ -5 & -4 \end{bmatrix}$$

b) The matrices do not have the same order, so we cannot subtract.

Use a grapher to verify these results.

Interactive Discovery

Enter each of the following matrices on a grapher.

$$\mathbf{A} = \begin{bmatrix} -2 & 4 \\ 1 & -5 \end{bmatrix}, \qquad \mathbf{B} = \begin{bmatrix} 2 & -4 \\ -1 & 5 \end{bmatrix},$$

$$\mathbf{C} = \begin{bmatrix} 0 & -3 & -2 \\ 8 & 4 & -4 \\ -1 & -5 & 7 \end{bmatrix}, \qquad \mathbf{D} = \begin{bmatrix} 0 & 3 & 2 \\ -8 & -4 & 4 \\ 1 & 5 & -7 \end{bmatrix}$$

What relationship do you observe between the corresponding entries of **A** and **B**? of **C** and **D**? Use the MATRIX feature of the grapher to find $\mathbf{A} + \mathbf{B}$ and $\mathbf{C} + \mathbf{D}$. What relationship do you observe between each pair of matrices and its sum?

The **opposite**, or **additive inverse**, of a matrix is obtained by replacing each entry by its opposite.

Example 3 Find $-\mathbf{A}$ and $\mathbf{A} + (-\mathbf{A})$ for

$$\mathbf{A} = \begin{bmatrix} 1 & 0 & 2 \\ 3 & -1 & 5 \end{bmatrix}.$$

SOLUTION To find $-\mathbf{A}$, we replace each entry of A by its opposite.

$$-\mathbf{A} = \begin{bmatrix} -1 & 0 & -2 \\ -3 & 1 & -5 \end{bmatrix},$$

$$\mathbf{A} + (-\mathbf{A}) = \begin{bmatrix} 1 & 0 & 2 \\ 3 & -1 & 5 \end{bmatrix} + \begin{bmatrix} -1 & 0 & -2 \\ -3 & 1 & -5 \end{bmatrix}$$

$$= \begin{bmatrix} 0 & 0 & 0 \\ 0 & 0 & 0 \end{bmatrix}$$

Use a grapher to verify these results.

Interactive Discovery

Enter each of the following matrices on a grapher.

$$\mathbf{A} = \begin{bmatrix} 1 & -3 \\ -1 & 10 \end{bmatrix}, \qquad \mathbf{B} = \begin{bmatrix} 0 & 0 \\ 0 & 0 \end{bmatrix},$$

$$\mathbf{C} = \begin{bmatrix} 3 & -3 & 4 \\ -8 & -1 & -7 \end{bmatrix}, \qquad \mathbf{D} = \begin{bmatrix} 0 & 0 & 0 \\ 0 & 0 & 0 \end{bmatrix}$$

Use the MATRIX feature of the grapher to find $\mathbf{A} + \mathbf{B}$ and $\mathbf{C} + \mathbf{D}$. What is the outcome when a matrix consisting entirely of 0's is added to another matrix of the same order?

A matrix having 0's for all its entries is called a **zero matrix**. When a zero matrix is added to a second matrix of the same order, the second matrix is unchanged. Thus a zero matrix is an **additive identity**. For example,

$$\begin{bmatrix} 0 & 0 & 0 \\ 0 & 0 & 0 \end{bmatrix}$$

is the additive identity for any 2×3 matrix.

Scalar Multiplication

When we find the product of a number and a matrix, we obtain a *scalar product*.

Scalar Product

The *scalar product* of a number k and a matrix $\mathbf{A}$ is the matrix denoted $k\mathbf{A}$, obtained by multiplying each entry of $\mathbf{A}$ by the number k. The number k is called a *scalar*.

Example 4 Find $3\mathbf{A}$ and $(-1)\mathbf{A}$ for

$$\mathbf{A} = \begin{bmatrix} -3 & 0 \\ 4 & 5 \end{bmatrix}.$$

SOLUTION We have

$$3\mathbf{A} = 3\begin{bmatrix} -3 & 0 \\ 4 & 5 \end{bmatrix} = \begin{bmatrix} 3(-3) & 3 \cdot 0 \\ 3 \cdot 4 & 3 \cdot 5 \end{bmatrix} = \begin{bmatrix} -9 & 0 \\ 12 & 15 \end{bmatrix},$$

$$(-1)\mathbf{A} = -1\begin{bmatrix} -3 & 0 \\ 4 & 5 \end{bmatrix} = \begin{bmatrix} -1(-3) & -1 \cdot 0 \\ -1 \cdot 4 & -1 \cdot 5 \end{bmatrix} = \begin{bmatrix} 3 & 0 \\ -4 & -5 \end{bmatrix}.$$

Use a grapher to verify these results.

Example 5 *Production.* Mitchell Fabricators manufactures three styles of bicycle frames in its two plants. The following table shows the number of each style produced at each plant in April.

	MOUNTAIN BIKE	RACING BIKE	TOURING BIKE
NORTH PLANT	150	120	100
SOUTH PLANT	180	90	130

a) Write a 2×3 matrix **A** that represents the information in the table.

b) The manufacturer increased production by 20% in May. Find a matrix **M** that represents the increased production figures.

c) Find the matrix $\mathbf{A} + \mathbf{M}$ and tell what it represents.

SOLUTION

a) Write the entries in the table in a 2×3 matrix **A**.

$$\mathbf{A} = \begin{bmatrix} 150 & 120 & 100 \\ 180 & 90 & 130 \end{bmatrix}$$

b) The production in May will be represented by $\mathbf{A} + 20\%\mathbf{A}$, or $\mathbf{A} + 0.2\mathbf{A}$, or $1.2\mathbf{A}$. Thus,

$$\mathbf{M} = (1.2)\begin{bmatrix} 150 & 120 & 100 \\ 180 & 90 & 130 \end{bmatrix} = \begin{bmatrix} 180 & 144 & 120 \\ 216 & 108 & 156 \end{bmatrix}.$$

c) $$\mathbf{A} + \mathbf{M} = \begin{bmatrix} 150 & 120 & 100 \\ 180 & 90 & 130 \end{bmatrix} + \begin{bmatrix} 180 & 144 & 120 \\ 216 & 108 & 156 \end{bmatrix}$$

$$= \begin{bmatrix} 330 & 264 & 220 \\ 396 & 198 & 286 \end{bmatrix}$$

The matrix $\mathbf{A} + \mathbf{M}$ represents the total production of each of the three types of frames at each plant in April and May.

The properties of matrix addition and scalar multiplication are similar to the properties of addition and multiplication of real numbers.

Properties of Matrix Addition and Scalar Multiplication

For any $m \times n$ matrices $\mathbf{A}$, $\mathbf{B}$, and $\mathbf{C}$ and any scalars k and l:

$\mathbf{A} + \mathbf{B} = \mathbf{B} + \mathbf{A}$.	*Commutative Property*
$\mathbf{A} + (\mathbf{B} + \mathbf{C}) = (\mathbf{A} + \mathbf{B}) + \mathbf{C}$.	*Associative Property of Addition*
$(kl)\mathbf{A} = k(l\mathbf{A})$.	*Associative Property of Scalar Multiplication*

There exists a unique matrix $\mathbf{0}$ such that:

$\mathbf{A} + \mathbf{0} = \mathbf{0} + \mathbf{A} = \mathbf{A}$.	*Additive Identity Property*

There exists a unique matrix $-\mathbf{A}$ such that:

$\mathbf{A} + (-\mathbf{A}) = -\mathbf{A} + \mathbf{A} = \mathbf{0}$.	*Additive Inverse Property*
$k(\mathbf{A} + \mathbf{B}) = k\mathbf{A} + k\mathbf{B}$.	*Distributive Property*
$(k + l)\mathbf{A} = k\mathbf{A} + l\mathbf{A}$.	*Distributive Property*

Products of Matrices

Matrix multiplication is defined in such a way that it can be used in solving systems of equations and in many applications.

Matrix Multiplication

For an $m \times n$ matrix $\mathbf{A} = [a_{ij}]$ and an $n \times p$ matrix $\mathbf{B} = [b_{ij}]$, the *product* $\mathbf{AB} = [c_{ij}]$ is an $m \times p$ matrix, where

$$c_{ij} = a_{i1} \cdot b_{1j} + a_{i2} \cdot b_{2j} + a_{i3} \cdot b_{3j} + \cdots + a_{in} \cdot b_{nj}.$$

In other words, the entry c_{ij} in $\mathbf{AB}$ is obtained by multiplying the entries in row i of $\mathbf{A}$ by the corresponding entries in column j of $\mathbf{B}$ and adding the results. Note that we can multiply two matrices only when the number of columns in the first matrix is equal to the number of rows in the second matrix.

Example 6 For

$$\mathbf{A} = \begin{bmatrix} 3 & 1 & -1 \\ 2 & 0 & 3 \end{bmatrix}, \quad \mathbf{B} = \begin{bmatrix} 1 & 6 \\ 3 & -5 \\ -2 & 4 \end{bmatrix}, \quad \text{and} \quad \mathbf{C} = \begin{bmatrix} 4 & -6 \\ 1 & 2 \end{bmatrix},$$

find each of the following.

a) $\mathbf{AB}$

b) $\mathbf{BA}$

c) $\mathbf{BC}$

d) $\mathbf{AC}$

SOLUTION

a) **A** is a 2×3 matrix and **B** is a 3×2 matrix, so **AB** will be a 2×2 matrix.

$$\mathbf{AB} = \begin{bmatrix} 3 & 1 & -1 \\ 2 & 0 & 3 \end{bmatrix} \begin{bmatrix} 1 & 6 \\ 3 & -5 \\ -2 & 4 \end{bmatrix}$$

$$= \begin{bmatrix} 3 \cdot 1 + 1 \cdot 3 + (-1)(-2) & 3 \cdot 6 + 1(-5) + (-1)(4) \\ 2 \cdot 1 + 0 \cdot 3 + 3(-2) & 2 \cdot 6 + 0(-5) + 3 \cdot 4 \end{bmatrix}$$

$$= \begin{bmatrix} 8 & 9 \\ -4 & 24 \end{bmatrix}$$

b) **B** is a 3×2 matrix and **A** is a 2×3 matrix, so **BA** will be a 3×3 matrix.

$$\mathbf{BA} = \begin{bmatrix} 1 & 6 \\ 3 & -5 \\ -2 & 4 \end{bmatrix} \begin{bmatrix} 3 & 1 & -1 \\ 2 & 0 & 3 \end{bmatrix}$$

$$= \begin{bmatrix} 1 \cdot 3 + 6 \cdot 2 & 1 \cdot 1 + 6 \cdot 0 & 1(-1) + 6 \cdot 3 \\ 3 \cdot 3 + (-5)(2) & 3 \cdot 1 + (-5)(0) & 3(-1) + (-5)(3) \\ -2 \cdot 3 + 4 \cdot 2 & -2 \cdot 1 + 4 \cdot 0 & -2(-1) + 4 \cdot 3 \end{bmatrix}$$

$$= \begin{bmatrix} 15 & 1 & 17 \\ -1 & 3 & -18 \\ 2 & -2 & 14 \end{bmatrix}$$

c) **B** is a 3×2 matrix and **C** is a 2×2 matrix, so **BC** will be a 3×2 matrix.

$$\mathbf{BC} = \begin{bmatrix} 1 & 6 \\ 3 & -5 \\ -2 & 4 \end{bmatrix} \begin{bmatrix} 4 & -6 \\ 1 & 2 \end{bmatrix}$$

$$= \begin{bmatrix} 1 \cdot 4 + 6 \cdot 1 & 1(-6) + 6 \cdot 2 \\ 3 \cdot 4 + (-5)(1) & 3(-6) + (-5)(2) \\ -2 \cdot 4 + 4 \cdot 1 & -2(-6) + 4 \cdot 2 \end{bmatrix}$$

$$= \begin{bmatrix} 10 & 6 \\ 7 & -28 \\ -4 & 20 \end{bmatrix}$$

d) The product **AC** is not defined because the number of columns of **A** (3) is not equal to the number of rows of **C** (2).

Use a grapher to verify these results. What message appeared on the grapher in part (d)?

Note in Example 6 that $\mathbf{AB} \neq \mathbf{BA}$. Multiplication of matrices is generally not commutative.

Example 7 *Dairy Profit.* Dalton's Dairy produces no-fat ice cream and frozen yogurt. The table below shows the number of gallons of each product that are shipped to the dairy's three retail outlets one week.

	STORE 1	STORE 2	STORE 3
NO-FAT ICE CREAM	100	80	120
FROZEN YOGURT	160	120	100

On each gallon of no-fat ice cream, the dairy's profit is \$4, and on each gallon of frozen yogurt, it is \$3. Use matrices to find the total profit on these items at each store for the given week.

SOLUTION We can write the table showing the distribution of the products as a 2×3 matrix:

$$\mathbf{D} = \begin{bmatrix} 100 & 80 & 120 \\ 160 & 120 & 100 \end{bmatrix}.$$

The profit per gallon for each product can also be written as a matrix:

$$\mathbf{P} = [4 \quad 3].$$

Then the total profit at each store is given by the matrix product **PD**:

$$\begin{aligned} \mathbf{PD} &= [4 \quad 3]\begin{bmatrix} 100 & 80 & 120 \\ 160 & 120 & 100 \end{bmatrix} \\ &= [4 \cdot 100 + 3 \cdot 160 \quad 4 \cdot 80 + 3 \cdot 120 \quad 4 \cdot 120 + 3 \cdot 100] \\ &= [880 \quad 680 \quad 780]. \end{aligned}$$

Thus the total profit on no-fat ice cream and frozen yogurt for the given week was \$880 at store 1, \$680 at store 2, and \$780 at store 3. You can verify these results on a grapher.

A matrix that consists of a single row, like **P** in Example 7, is called a **row matrix**. Similarly, a matrix that consists of a single column, like

$$\begin{bmatrix} 8 \\ -3 \\ 5 \end{bmatrix},$$

is called a **column matrix**.

We have already seen that matrix multiplication is generally not commutative. Nevertheless, matrix multiplication does have some properties that are similar to those for multiplication of real numbers.

Properties of Matrix Multiplication

For matrices **A**, **B**, and **C**, assuming that the indicated operation is possible:

$\mathbf{A}(\mathbf{BC}) = (\mathbf{AB})\mathbf{C}$.	*Associative Property of Multiplication*
$\mathbf{A}(\mathbf{B} + \mathbf{C}) = \mathbf{AB} + \mathbf{AC}$.	*Distributive Property*
$(\mathbf{B} + \mathbf{C})\mathbf{A} = \mathbf{BA} + \mathbf{CA}$.	*Distributive Property*

Matrix Equations

We can write a matrix equation equivalent to a system of equations.

Example 8 Write a matrix equation equivalent to the following system of equations:

$$\begin{aligned} 4x + 2y - z &= 3, \\ 9x \qquad + z &= 5, \\ 4x + 5y - 2z &= 1. \end{aligned}$$

SOLUTION We write the coefficients on the left in a matrix. We then write the product of that matrix and the column matrix containing the variables, and set the result equal to the column matrix containing the constants on the right:

$$\begin{bmatrix} 4 & 2 & -1 \\ 9 & 0 & 1 \\ 4 & 5 & -2 \end{bmatrix} \begin{bmatrix} x \\ y \\ z \end{bmatrix} = \begin{bmatrix} 3 \\ 5 \\ 1 \end{bmatrix}.$$

If we let

$$\mathbf{A} = \begin{bmatrix} 4 & 2 & -1 \\ 9 & 0 & 1 \\ 4 & 5 & -2 \end{bmatrix}, \quad \mathbf{X} = \begin{bmatrix} x \\ y \\ z \end{bmatrix}, \quad \text{and} \quad \mathbf{B} = \begin{bmatrix} 3 \\ 5 \\ 1 \end{bmatrix},$$

we can write this matrix equation as $\mathbf{AX} = \mathbf{B}$.

In the next section, we will solve systems of equations using a matrix equation like the one in Example 8.

7.4 Exercise Set

Find x, y, and z.

1. $[5 \quad x] = [y \quad -3]$

2. $\begin{bmatrix} 6x \\ 25 \end{bmatrix} = \begin{bmatrix} -9 \\ 5y \end{bmatrix}$

3. $\begin{bmatrix} 3 & 2x \\ y & -8 \end{bmatrix} = \begin{bmatrix} 3 & -2 \\ 1 & -8 \end{bmatrix}$

4. $\begin{bmatrix} x - 1 & 4 \\ y + 3 & -7 \end{bmatrix} = \begin{bmatrix} 0 & 4 \\ -2 & -7 \end{bmatrix}$

For Exercises 5–20, let

$$\mathbf{A} = \begin{bmatrix} 1 & 2 \\ 4 & 3 \end{bmatrix}, \quad \mathbf{B} = \begin{bmatrix} -3 & 5 \\ 2 & -1 \end{bmatrix},$$

$$\mathbf{C} = \begin{bmatrix} 1 & -1 \\ -1 & 1 \end{bmatrix}, \quad \mathbf{D} = \begin{bmatrix} 1 & 1 \\ 1 & 1 \end{bmatrix},$$

$$\mathbf{E} = \begin{bmatrix} 1 & 3 \\ 2 & 6 \end{bmatrix}, \quad \mathbf{F} = \begin{bmatrix} 3 & 3 \\ -1 & -1 \end{bmatrix},$$

$$\mathbf{O} = \begin{bmatrix} 0 & 0 \\ 0 & 0 \end{bmatrix}, \quad \mathbf{I} = \begin{bmatrix} 1 & 0 \\ 0 & 1 \end{bmatrix}.$$

Find each of the following. Use a grapher to check your answers.

5. **A** + **B**
6. **B** + **A**
7. **E** + **O**
8. 2**A**
9. 3**F**
10. (−1)**D**
11. 3**F** + 2**A**
12. **A** − **B**
13. **B** − **A**
14. **AB**
15. **BA**
16. **OF**
17. **CD**
18. **EF**
19. **AI**
20. **IA**

21. *Budget.* For the month of June, Nelia budgets $150 for food, $80 for clothes, and $40 for entertainment.
   a) Write a 1 × 3 matrix **B** that represents these items.
   b) After receiving a raise, Nelia increases the amount budgeted for each item in July by 5%. Find a matrix **R** that represents the new amounts.
   c) Find **B** + **R** and tell what the entries represent.

22. *Produce.* The produce manager at Dugan's Market orders 40 lb of tomatoes, 20 lb of zucchini, and 30 lb of onions from a local farmer one week.
   a) Write a 1 × 3 matrix **A** that represents these items.
   b) The following week the produce manager increases her order by 10%. Find a matrix **B** that represents this order.
   c) Find **A** + **B** and tell what the entries represent.

23. *Nutrition.* A 3-oz serving of roasted, skinless chicken breast contains 140 Cal, 27 g of protein, 3 g of fat, 13 mg of calcium, and 64 mg of sodium. One-half cup of potato salad contains 180 Cal, 4 g of protein, 11 g of fat, 24 mg of calcium, and 662 mg of sodium. One broccoli spear contains 50 Cal, 5 g of protein, 1 g of fat, 82 mg of calcium, and 20 mg of sodium. (*Source: Home and Garden Bulletin No. 72,* U.S. Government Printing Office, Washington, D.C. 20402)
   a) Write 1 × 5 matrices **C**, **P**, and **B** that represent the nutritional values of each food.
   b) Find **C** + 2**P** + 3**B** and tell what the entries represent.

24. *Nutrition.* One slice of cheese pizza contains 290 Cal, 15 g of protein, 9 g of fat, and 39 g of carbohydrates. One-half cup of gelatin dessert contains 70 Cal, 2 g of protein, 0 g of fat, and 17 g of carbohydrates. One cup of whole milk contains 150 Cal, 8 g of protein, 8 g of fat, and 11 g of carbohydrates. (*Source: Home and Garden Bulletin No. 72,* U.S. Government Printing Office, Washington, D.C. 20402)
   a) Write 1 × 4 matrices **P**, **G**, and **M** that represent the nutritional values of each food.
   b) Find 3**P** + 2**G** + 2**M** and tell what the entries represent.

Use a grapher to find each product, if possible.

25. $\begin{bmatrix} -1 & 0 & 7 \\ 3 & -5 & 2 \end{bmatrix} \begin{bmatrix} 6 \\ -4 \\ 1 \end{bmatrix}$

26. $\begin{bmatrix} 6 & -1 & 2 \end{bmatrix} \begin{bmatrix} 1 & 4 \\ -2 & 0 \\ 5 & -3 \end{bmatrix}$

27. $\begin{bmatrix} -2 & 4 \\ 5 & 1 \\ -1 & -3 \end{bmatrix} \begin{bmatrix} 3 & -6 \\ -1 & 4 \end{bmatrix}$

28. $\begin{bmatrix} 2 & -1 & 0 \\ 0 & 5 & 4 \end{bmatrix} \begin{bmatrix} -3 & 1 & 0 \\ 0 & 2 & -1 \\ 5 & 0 & 4 \end{bmatrix}$

29. $\begin{bmatrix} 1 \\ -5 \\ 3 \end{bmatrix} \begin{bmatrix} -6 & 5 & 8 \\ 0 & 4 & -1 \end{bmatrix}$

30. $\begin{bmatrix} 2 & 0 & 0 \\ 0 & -1 & 0 \\ 0 & 0 & 3 \end{bmatrix} \begin{bmatrix} 0 & -4 & 3 \\ 2 & 1 & 0 \\ -1 & 0 & 6 \end{bmatrix}$

31. $\begin{bmatrix} 1 & -4 & 3 \\ 0 & 8 & 0 \\ -2 & -1 & 5 \end{bmatrix} \begin{bmatrix} 3 & 0 & 0 \\ 0 & -4 & 0 \\ 0 & 0 & 1 \end{bmatrix}$

32. $\begin{bmatrix} 4 \\ -5 \end{bmatrix} \begin{bmatrix} 2 & 0 \\ 6 & -7 \\ 0 & -3 \end{bmatrix}$

33. *Food Service Management.* The food service manager at a large hospital is concerned about maintaining reasonable food costs. The table below shows the cost per serving, in cents, for items on four menus.

MENU	MEAT	POTATO	VEGETABLE	SALAD	DESSERT
1	45.29	6.63	10.94	7.42	8.01
2	53.78	4.95	9.83	6.16	12.56
3	47.13	8.47	12.66	8.29	9.43
4	51.64	7.12	11.57	9.35	10.72

On a particular day a dietician orders 65 meals from menu 1, 48 from menu 2, 93 from menu 3, and 57 from menu 4.

a) Write the information in the table as a 4×5 matrix **M**.
b) Write a row matrix **N** that represents the number of each menu ordered.
c) Find the product **NM**.
d) State what the entries of **NM** represent.

34. *Food Service Management.* A college food service manager uses a table like the one below to show the number of units of ingredients, by weight, required for various menu items.

	WHITE CAKE	BREAD	COFFEE CAKE	SUGAR COOKIES
FLOUR	1	2.5	0.75	0.5
MILK	0	0.5	0.25	0
EGGS	0.75	0.25	0.5	0.5
BUTTER	0.5	0	0.5	1

The cost per unit of each ingredient is 15 cents for flour, 28 cents for milk, 54 cents for eggs, and 83 cents for butter.

a) Write the information in the table as a 4×4 matrix **M**.
b) Write a row matrix **C** that represents the cost per unit of each ingredient.
c) Find the product **CM**.
d) State what the entries of **CM** represent.

35. *Production Cost.* Karin supplies two small campus coffee shops with homemade chocolate chip cookies, oatmeal cookies, and peanut butter cookies. The table below shows the number of each type of cookie, in dozens, that Karin sold in one week.

	MUGSY'S COFFEE SHOP	THE COFFEE CLUB
CHOCOLATE CHIP	8	15
OATMEAL	6	10
PEANUT BUTTER	4	3

Karin spends \$3 for the ingredients for one dozen chocolate chip cookies, \$1.50 for the ingredients for one dozen oatmeal cookies, and \$2 for the ingredients for one dozen peanut butter cookies.

a) Write the information in the table as a 3×2 matrix **S**.
b) Write a row matrix **C** that represents the cost, per dozen, of the ingredients for each type of cookie.
c) Find the product **CS**.
d) State what the entries of **CS** represent.

36. *Profit.* A manufacturer produces exterior plywood, interior plywood, and fiberboard, which are shipped to two distributors. The table below shows the number of units of each type of product that are shipped to each warehouse.

	DISTRIBUTOR 1	DISTRIBUTOR 2
EXTERIOR PLYWOOD	900	500
INTERIOR PLYWOOD	450	1000
FIBERBOARD	600	700

The profits from each unit of exterior plywood, interior plywood, and fiberboard are \$5, \$8, and \$4, respectively.

a) Write the information in the table as a 3×2 matrix **M**.
b) Write a row matrix **P** that represents the profit, per unit, of each type of product.
c) Find the product **PM**.
d) State what the entries of **PM** represent.

37. *Profit.* In Exercise 35, suppose that Karin's profits on one dozen chocolate chip, oatmeal, and peanut butter cookies are \$6, \$4.50, and \$5.20, respectively.

a) Write a row matrix **P** that represents this information.
b) Use the matrices **S** and **P** to find Karin's total profit from each snack bar.

38. *Production Cost.* In Exercise 36, suppose that the manufacturer's production costs for each unit of exterior plywood, interior plywood, and fiberboard are \$20, \$25, and \$15, respectively.

a) Write a row matrix **C** that represents this information.

b) Use the matrices **M** and **C** to find the total production cost for the products shipped to each distributor.

Write a matrix equation equivalent to the system of equations.

39. $2x - 3y = 7,$
$x + 5y = -6$

40. $-x + y = 3,$
$5x - 4y = 16$

41. $x + y - 2z = 6,$
$3x - y + z = 7,$
$2x + 5y - 3z = 8$

42. $3x - y + z = 1,$
$x + 2y - z = 3,$
$4x + 3y - 2z = 11$

43. $3x - 2y + 4z = 17,$
$2x + y - 5z = 13$

44. $3x + 2y + 5z = 9,$
$4x - 3y + 2z = 10$

45. $-4w + x - y + 2z = 12,$
$w + 2x - y - z = 0,$
$-w + x + 4y - 3z = 1,$
$2w + 3x + 5y - 7z = 9$

46. $12w + 2x + 4y - 5z = 2,$
$-w + 4x - y + 12z = 5,$
$2w - x + 4y = 13,$
$2x + 10y + z = 5$

Skill Maintenance

Solve.

47. $5x = 45$

48. $-13x = 52$

49. $4x = -28$

50. $ax = b$, for x

Synthesis

51. ◆ Is it true that if $\mathbf{AB} = \mathbf{0}$, for matrices **A** and **B**, then $\mathbf{A} = \mathbf{0}$ or $\mathbf{B} = \mathbf{0}$? Why or why not?

52. ◆ Explain how Karin could use the matrix products found in Exercises 35 and 37 in making business decisions.

For Exercises 53–56, let

$$\mathbf{A} = \begin{bmatrix} -1 & 0 \\ 2 & 1 \end{bmatrix} \quad \textit{and} \quad \mathbf{B} = \begin{bmatrix} 1 & -1 \\ 0 & 2 \end{bmatrix}.$$

53. Show that

$$(\mathbf{A} + \mathbf{B})(\mathbf{A} - \mathbf{B}) \neq \mathbf{A}^2 - \mathbf{B}^2,$$

where

$$\mathbf{A}^2 = \mathbf{AA} \quad \text{and} \quad \mathbf{B}^2 = \mathbf{BB}.$$

54. Show that

$$(\mathbf{A} + \mathbf{B})(\mathbf{A} + \mathbf{B}) \neq \mathbf{A}^2 + 2\mathbf{AB} + \mathbf{B}^2.$$

55. Show that

$$(\mathbf{A} + \mathbf{B})(\mathbf{A} - \mathbf{B}) = \mathbf{A}^2 + \mathbf{BA} - \mathbf{AB} - \mathbf{B}^2.$$

56. Show that

$$(\mathbf{A} + \mathbf{B})(\mathbf{A} + \mathbf{B}) = \mathbf{A}^2 + \mathbf{BA} + \mathbf{AB} + \mathbf{B}^2.$$

In Exercises 57–61, let

$$\mathbf{A} = \begin{bmatrix} a_{11} & a_{12} & a_{13} & \cdots & a_{1n} \\ a_{21} & a_{22} & a_{23} & \cdots & a_{2n} \\ a_{31} & a_{32} & a_{33} & \cdots & a_{3n} \\ \vdots & \vdots & \vdots & & \vdots \\ a_{m1} & a_{m2} & a_{m3} & \cdots & a_{mn} \end{bmatrix},$$

$$\mathbf{B} = \begin{bmatrix} b_{11} & b_{12} & b_{13} & \cdots & b_{1n} \\ b_{21} & b_{22} & b_{23} & \cdots & b_{2n} \\ b_{31} & b_{32} & b_{33} & \cdots & b_{3n} \\ \vdots & \vdots & \vdots & & \vdots \\ b_{m1} & b_{m2} & b_{m3} & \cdots & b_{mn} \end{bmatrix},$$

$$\textit{and } \mathbf{C} = \begin{bmatrix} c_{11} & c_{12} & c_{13} & \cdots & c_{1n} \\ c_{21} & c_{22} & c_{23} & \cdots & c_{2n} \\ c_{31} & c_{32} & c_{33} & \cdots & c_{3n} \\ \vdots & \vdots & \vdots & & \vdots \\ c_{m1} & c_{m2} & c_{m3} & \cdots & c_{mn} \end{bmatrix},$$

and let k and l be any scalars.

57. Prove that $\mathbf{A} + \mathbf{B} = \mathbf{B} + \mathbf{A}$.

58. Prove that $\mathbf{A} + (\mathbf{B} + \mathbf{C}) = (\mathbf{A} + \mathbf{B}) + \mathbf{C}$.

59. Prove that $(kl)\mathbf{A} = k(l\mathbf{A})$.

60. Prove that $k(\mathbf{A} + \mathbf{B}) = k\mathbf{A} + k\mathbf{B}$.

61. Prove that $(k + l)\mathbf{A} = k\mathbf{A} + l\mathbf{A}$.

7.5 Inverses of Matrices

- *Find the inverse of a square matrix, if it exists.*
- *Use inverses of matrices to solve systems of equations.*

In this section, we continue our study of matrix algebra, finding the **multiplicative inverse**, or simply **inverse**, of a square matrix, if it exists. Then we use such inverses to solve systems of equations.

Interactive Discovery

Enter the following matrices on a grapher.

$$\mathbf{A} = \begin{bmatrix} 2 & -3 \\ -1 & 4 \end{bmatrix}, \qquad \mathbf{B} = \begin{bmatrix} 1 & 0 \\ 0 & 1 \end{bmatrix},$$

$$\mathbf{C} = \begin{bmatrix} 4 & 0 & -2 \\ 10 & -4 & -6 \\ -1 & 5 & 3 \end{bmatrix}, \qquad \mathbf{D} = \begin{bmatrix} 1 & 0 & 0 \\ 0 & 1 & 0 \\ 0 & 0 & 1 \end{bmatrix}$$

Use the MATRIX feature of the grapher to find **AB**, **BA**, **CD**, and **DC**. What is the effect of multiplying by **B**? by **D**?

The Identity Matrix

Recall that, for real numbers, $a \cdot 1 = 1 \cdot a = 1$; 1 is the multiplicative identity. A multiplicative identity matrix is very similar to the number 1.

Identity Matrix

For any positive integer n, the $n \times n$ *identity matrix* is an $n \times n$ matrix with 1's on the main diagonal and 0's elsewhere and is denoted by

$$\mathbf{I} = \begin{bmatrix} 1 & 0 & 0 & \cdots & 0 \\ 0 & 1 & 0 & \cdots & 0 \\ 0 & 0 & 1 & \cdots & 0 \\ \vdots & \vdots & \vdots & & \vdots \\ 0 & 0 & 0 & \cdots & 1 \end{bmatrix}.$$

Then $\mathbf{AI} = \mathbf{IA} = \mathbf{A}$, for any $n \times n$ matrix $\mathbf{A}$.

Example 1 For

$$\mathbf{A} = \begin{bmatrix} 4 & -7 \\ -3 & 2 \end{bmatrix}$$

and $$\mathbf{I} = \begin{bmatrix} 1 & 0 \\ 0 & 1 \end{bmatrix},$$

find each of the following.

a) **AI**

b) **IA**

SOLUTION

$$\text{a) } \mathbf{AI} = \begin{bmatrix} 4 & -7 \\ -3 & 2 \end{bmatrix}\begin{bmatrix} 1 & 0 \\ 0 & 1 \end{bmatrix}$$

$$= \begin{bmatrix} 4 \cdot 1 - 7 \cdot 0 & 4 \cdot 0 - 7 \cdot 1 \\ -3 \cdot 1 + 2 \cdot 0 & -3 \cdot 0 + 2 \cdot 1 \end{bmatrix}$$

$$= \begin{bmatrix} 4 & -7 \\ -3 & 2 \end{bmatrix} = \mathbf{A}$$

$$\text{b) } \mathbf{IA} = \begin{bmatrix} 1 & 0 \\ 0 & 1 \end{bmatrix}\begin{bmatrix} 4 & -7 \\ -3 & 2 \end{bmatrix}$$

$$= \begin{bmatrix} 1 \cdot 4 + 0(-3) & 1(-7) + 0 \cdot 2 \\ 0 \cdot 4 + 1(-3) & 0(-7) + 1 \cdot 2 \end{bmatrix} = \begin{bmatrix} 4 & -7 \\ -3 & 2 \end{bmatrix} = \mathbf{A}$$

The Inverse of a Matrix

Recall that for every nonzero real number a, there is a multiplicative inverse $1/a$ such that $a(1/a) = (1/a)a = 1$. The multiplicative inverse of a matrix behaves in a similar manner.

Inverse of a Matrix

For an $n \times n$ matrix $\mathbf{A}$, if there is a matrix $\mathbf{A}^{-1}$ for which $\mathbf{A}^{-1} \cdot \mathbf{A} = \mathbf{I} = \mathbf{A} \cdot \mathbf{A}^{-1}$, then $\mathbf{A}^{-1}$ is the *inverse* of $\mathbf{A}$.

Note that $\mathbf{A}^{-1}$ is read "**A** inverse."

Example 2 Verify that

$$\mathbf{B} = \begin{bmatrix} 4 & -3 \\ 3 & -2 \end{bmatrix} \quad \text{is the inverse of} \quad \mathbf{A} = \begin{bmatrix} -2 & 3 \\ -3 & 4 \end{bmatrix}.$$

SOLUTION We show that $\mathbf{BA} = \mathbf{I} = \mathbf{AB}$.

$$\mathbf{BA} = \begin{bmatrix} 4 & -3 \\ 3 & -2 \end{bmatrix}\begin{bmatrix} -2 & 3 \\ -3 & 4 \end{bmatrix} = \begin{bmatrix} 1 & 0 \\ 0 & 1 \end{bmatrix}$$

$$\mathbf{AB} = \begin{bmatrix} -2 & 3 \\ -3 & 4 \end{bmatrix}\begin{bmatrix} 4 & -3 \\ 3 & -2 \end{bmatrix} = \begin{bmatrix} 1 & 0 \\ 0 & 1 \end{bmatrix}$$

We can find the inverse of a square matrix, if it exists, by using row-equivalent operations as in the Gauss–Jordan elimination method. For example, consider the matrix

$$\mathbf{A} = \begin{bmatrix} -2 & 3 \\ -3 & 4 \end{bmatrix}.$$

To find its inverse, we first form an **augmented matrix** consisting of **A** on the left side and the 2 × 2 identity matrix on the right side:

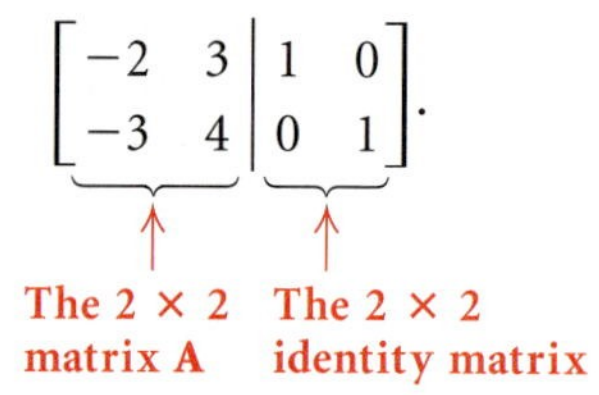

Then we attempt to transform the augmented matrix to one of the form

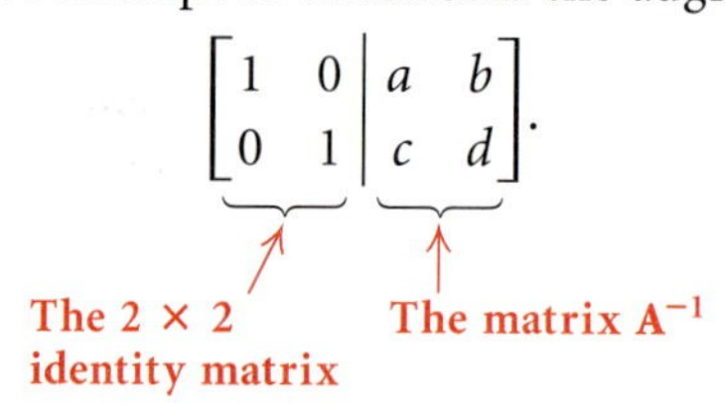

$$\left[\begin{array}{rr|rr} 1 & 0 & a & b \\ 0 & 1 & c & d \end{array}\right].$$
The 2 × 2 identity matrix
The matrix A^{-1}

The matrix on the right, $\begin{bmatrix} a & b \\ c & d \end{bmatrix}$, is $\mathbf{A}^{-1}$.

Example 3 Find $\mathbf{A}^{-1}$, where

$$\mathbf{A} = \begin{bmatrix} -2 & 3 \\ -3 & 4 \end{bmatrix}.$$

SOLUTION First we write the augmented matrix. Then we transform it to the desired form.

$$\left[\begin{array}{rr|rr} -2 & 3 & 1 & 0 \\ -3 & 4 & 0 & 1 \end{array}\right]$$

$$\left[\begin{array}{rr|rr} 1 & -\frac{3}{2} & -\frac{1}{2} & 0 \\ -3 & 4 & 0 & 1 \end{array}\right] \qquad \text{New row 1} = -\tfrac{1}{2}(\text{row 1})$$

$$\left[\begin{array}{rr|rr} 1 & -\frac{3}{2} & -\frac{1}{2} & 0 \\ 0 & -\frac{1}{2} & -\frac{3}{2} & 1 \end{array}\right] \qquad \text{New row 2} = 3(\text{row 1}) + \text{row 2}$$

$$\left[\begin{array}{rr|rr} 1 & -\frac{3}{2} & -\frac{1}{2} & 0 \\ 0 & 1 & 3 & -2 \end{array}\right] \qquad \text{New row 2} = -2(\text{row 2})$$

$$\left[\begin{array}{rr|rr} 1 & 0 & 4 & -3 \\ 0 & 1 & 3 & -2 \end{array}\right] \qquad \text{New row 1} = \tfrac{3}{2}(\text{row 2}) + \text{row 1}$$

Thus,

$$\mathbf{A}^{-1} = \begin{bmatrix} 4 & -3 \\ 3 & -2 \end{bmatrix},$$

which we verified in Example 2.

A grapher can be used to find the inverse of a matrix quickly. Check the user's manual for the procedure. Then use a grapher to verify the result of Example 3.

If a matrix has an inverse, we say that it is **invertible**, or **nonsingular**. When we cannot obtain the identity matrix on the left using the Gauss–Jordan method, then no inverse exists. This occurs when we obtain a row consisting entirely of 0's in either of the two matrices in the augmented matrix. In this case, we say that **A** is a **singular matrix**. A grapher will produce an error message similar to "ERR: SINGULAR MATRIX" in this situation.

Solving Systems of Equations

We can write a system of n linear equations in n variables as a matrix equation $\mathbf{AX} = \mathbf{B}$. (See Section 7.4.) If **A** has an inverse, then the system of equations has a unique solution that can be found by solving for **X**, as follows:

$$\begin{aligned} \mathbf{AX} &= \mathbf{B} \\ \mathbf{A}^{-1}(\mathbf{AX}) &= \mathbf{A}^{-1}\mathbf{B} && \text{Multiplying by } \mathbf{A}^{-1} \text{ on the left on both sides} \\ (\mathbf{A}^{-1}\mathbf{A})\mathbf{X} &= \mathbf{A}^{-1}\mathbf{B} && \text{Using the associative property of matrix multiplication} \\ \mathbf{IX} &= \mathbf{A}^{-1}\mathbf{B} && \mathbf{A}^{-1}\mathbf{A} = \mathbf{I} \\ \mathbf{X} &= \mathbf{A}^{-1}\mathbf{B}. && \mathbf{IX} = \mathbf{X} \end{aligned}$$

Matrix Solutions of Systems of Equations

For a system of n linear equations in n variables, $\mathbf{AX} = \mathbf{B}$, if **A** is an invertible matrix, then the unique solution of the system is given by

$$\mathbf{X} = \mathbf{A}^{-1}\mathbf{B}.$$

Since matrix multiplication is not commutative in general, care must be taken to multiply *on the left* by $\mathbf{A}^{-1}$.

Example 4 Use an inverse matrix to solve the following system of equations:

$$\begin{aligned} x + 2y - z &= -2, \\ 3x + 5y + 3z &= 3, \\ 2x + 4y + 3z &= 1. \end{aligned}$$

SOLUTION We write an equivalent matrix equation, $\mathbf{AX} = \mathbf{B}$:

$$\begin{bmatrix} 1 & 2 & -1 \\ 3 & 5 & 3 \\ 2 & 4 & 3 \end{bmatrix} \begin{bmatrix} x \\ y \\ z \end{bmatrix} = \begin{bmatrix} -2 \\ 3 \\ 1 \end{bmatrix}.$$

Then we use a grapher to find $\mathbf{A}^{-1}\mathbf{B}$. We get

$$\mathbf{X} = \mathbf{A}^{-1}\mathbf{B} = \begin{bmatrix} -0.6 & 2 & -2.2 \\ 0.6 & -1 & 1.2 \\ -0.4 & 0 & 0.2 \end{bmatrix}\begin{bmatrix} -2 \\ 3 \\ 1 \end{bmatrix} = \begin{bmatrix} 5 \\ -3 \\ 1 \end{bmatrix},$$

or

$$\begin{bmatrix} x \\ y \\ z \end{bmatrix} = \begin{bmatrix} 5 \\ -3 \\ 1 \end{bmatrix},$$

so the solution is $x = 5$, $y = -3$, and $z = 1$, or $(5, -3, 1)$.

When a grapher is used, it is not actually necessary to enter the matrix $\mathbf{A}^{-1}$. After the matrices $\mathbf{A}$ and $\mathbf{B}$ are entered on a grapher, the computation $[\mathbf{A}]^{-1} * [\mathbf{B}]$ is entered and only the result

$$\begin{bmatrix} 5 \\ -3 \\ 1 \end{bmatrix}$$

is displayed.

7.5 Exercise Set

Determine whether **B** *is the inverse of* **A**.

1. $\mathbf{A} = \begin{bmatrix} 1 & -3 \\ -2 & 7 \end{bmatrix}$, $\mathbf{B} = \begin{bmatrix} 7 & 3 \\ 2 & 1 \end{bmatrix}$

2. $\mathbf{A} = \begin{bmatrix} 3 & 2 \\ 4 & 3 \end{bmatrix}$, $\mathbf{B} = \begin{bmatrix} 3 & -2 \\ -4 & 3 \end{bmatrix}$

3. $\mathbf{A} = \begin{bmatrix} -1 & -1 & 6 \\ 1 & 0 & -2 \\ 1 & 0 & -3 \end{bmatrix}$, $\mathbf{B} = \begin{bmatrix} 2 & 3 & 2 \\ 3 & 3 & 4 \\ 1 & 1 & 1 \end{bmatrix}$

4. $\mathbf{A} = \begin{bmatrix} -2 & 0 & -3 \\ 5 & 1 & 7 \\ -3 & 0 & 4 \end{bmatrix}$, $\mathbf{B} = \begin{bmatrix} 4 & 0 & -3 \\ 1 & 1 & 1 \\ -3 & 0 & 2 \end{bmatrix}$

Use the Gauss–Jordan method to find $\mathbf{A}^{-1}$, *if it exists. Check your answers on a grapher by finding* $\mathbf{A}^{-1}\mathbf{A}$ *and* $\mathbf{A}\mathbf{A}^{-1}$.

5. $\mathbf{A} = \begin{bmatrix} 3 & 2 \\ 5 & 3 \end{bmatrix}$

6. $\mathbf{A} = \begin{bmatrix} 3 & 5 \\ 1 & 2 \end{bmatrix}$

7. $\mathbf{A} = \begin{bmatrix} 11 & 3 \\ 7 & 2 \end{bmatrix}$

8. $\mathbf{A} = \begin{bmatrix} 8 & 5 \\ 5 & 3 \end{bmatrix}$

9. $\mathbf{A} = \begin{bmatrix} 6 & 9 \\ 4 & 6 \end{bmatrix}$

10. $\mathbf{A} = \begin{bmatrix} -4 & -6 \\ 2 & 3 \end{bmatrix}$

11. $\mathbf{A} = \begin{bmatrix} 3 & 1 & 0 \\ 1 & 1 & 1 \\ 1 & -1 & 2 \end{bmatrix}$

12. $\mathbf{A} = \begin{bmatrix} 1 & 0 & 1 \\ 2 & 1 & 0 \\ 1 & -1 & 1 \end{bmatrix}$

13. $\mathbf{A} = \begin{bmatrix} 1 & -1 & 2 \\ 0 & 1 & 3 \\ 2 & 1 & -2 \end{bmatrix}$

14. $\mathbf{A} = \begin{bmatrix} 1 & -1 & 2 \\ 0 & 1 & 2 \\ 1 & -3 & -4 \end{bmatrix}$

15. $\mathbf{A} = \begin{bmatrix} 1 & -4 & 8 \\ 1 & -3 & 2 \\ 2 & -7 & 10 \end{bmatrix}$

16. $\mathbf{A} = \begin{bmatrix} -2 & 5 & 3 \\ 4 & -1 & 3 \\ 7 & -2 & 5 \end{bmatrix}$

Use a grapher to find $\mathbf{A}^{-1}$, *if it exists.*

17. $\mathbf{A} = \begin{bmatrix} 4 & -3 \\ 1 & -2 \end{bmatrix}$

18. $\mathbf{A} = \begin{bmatrix} 0 & -1 \\ 1 & 0 \end{bmatrix}$

19. $\mathbf{A} = \begin{bmatrix} 2 & 3 & 2 \\ 3 & 3 & 4 \\ -1 & -1 & -1 \end{bmatrix}$

20. $\mathbf{A} = \begin{bmatrix} 1 & 2 & 3 \\ 2 & -1 & -2 \\ -1 & 3 & 3 \end{bmatrix}$

21. $\mathbf{A} = \begin{bmatrix} 1 & 2 & -1 \\ -2 & 0 & 1 \\ 1 & -1 & 0 \end{bmatrix}$

22. $\mathbf{A} = \begin{bmatrix} 7 & -1 & -9 \\ 2 & 0 & -4 \\ -4 & 0 & 6 \end{bmatrix}$

23. $\mathbf{A} = \begin{bmatrix} 1 & 1 & 0 \\ 1 & 0 & -1 \\ 0 & -1 & 1 \end{bmatrix}$

24. $\mathbf{A} = \begin{bmatrix} 1 & 0 & 0 \\ -2 & 1 & 0 \\ 1 & -2 & 1 \end{bmatrix}$

25. $\mathbf{A} = \begin{bmatrix} 1 & 3 & -1 \\ 0 & 2 & -1 \\ 1 & 1 & 0 \end{bmatrix}$

26. $\mathbf{A} = \begin{bmatrix} -1 & 0 & -1 \\ -1 & 1 & 0 \\ 0 & 1 & 1 \end{bmatrix}$

27. $\mathbf{A} = \begin{bmatrix} 1 & 2 & 3 & 4 \\ 0 & 1 & 3 & -5 \\ 0 & 0 & 1 & -2 \\ 0 & 0 & 0 & -1 \end{bmatrix}$

28. $\mathbf{A} = \begin{bmatrix} -2 & -3 & 4 & 1 \\ 0 & 1 & 1 & 0 \\ 0 & 4 & -6 & 1 \\ -2 & -2 & 5 & 1 \end{bmatrix}$

29. $\mathbf{A} = \begin{bmatrix} 1 & -14 & 7 & 38 \\ -1 & 2 & 1 & -2 \\ 1 & 2 & -1 & -6 \\ 1 & -2 & 3 & 6 \end{bmatrix}$

30. $\mathbf{A} = \begin{bmatrix} 10 & 20 & -30 & 15 \\ 3 & -7 & 14 & -8 \\ -7 & -2 & -1 & 2 \\ 4 & 4 & -3 & 1 \end{bmatrix}$

Solve each system of equations using the inverse of the coefficient matrix of the equivalent matrix equation.

31. $4x + 3y = 2,$
$x - 2y = 6$

32. $2x - 3y = 7,$
$4x + y = -7$

33. $5x + y = 2,$
$3x - 2y = -4$

34. $x - 6y = 5,$
$-x + 4y = -5$

35. $x + y = 7,$
$2x - y = 2$

36. $2x + 5y = -3,$
$3x + 7y = -5$

37. $x + z = 1,$
$2x + y = 3,$
$x - y + z = 4$

38. $x + 2y + 3z = -1,$
$2x - 3y + 4z = 2,$
$-3x + 5y - 6z = 4$

39. $2x + 3y + 4z = 2,$
$x - 4y + 3z = 2,$
$5x + y + z = -4$

40. $x + y = 2,$
$3x + 2z = 5,$
$2x + 3y - 3z = 9$

41. $x + y + z = 6,$
$2x - y - z = 9,$
$3x - 2y + z = -5$

42. $x + y + z = 1,$
$2x - y + 3z = -2,$
$3x + 2y - 2z = 15$

43. $2w - 3x + 4y - 5z = 0,$
$3w - 2x + 7y - 3z = 2,$
$w + x - y + z = 1,$
$-w - 3x - 6y + 4z = 6$

44. $5w - 4x + 3y - 2z = -6,$
$w + 4x - 2y + 3z = -5,$
$2w - 3x + 6y - 9z = 14,$
$3w - 5x + 2y - 4z = -3$

45. $w + x + y + z = 0,$
$2w - 3x + 2y + 5z = 3,$
$w - 2x - y + 2z = 7,$
$3w - x - 2y - 7z = 5$

46. $w + x - y + 2z = -7,$
$2w - x + y - z = 7,$
$-w - x + 2y - 4z = 9,$
$3w + 2x + y + z = 0$

47. *Sales.* Stefan sold a total of 145 Italian sausages and hot dogs from his curbside pushcart and collected $242.05. He sold 45 more hot dogs than sausages. How many of each did he sell?

48. *Price of School Supplies.* Miranda bought 4 lab record books and 3 highlighters for $13.93. Victor bought 3 lab record books and 2 highlighters for $10.25. Find the price of each item.

49. *Cost.* Evergreen Landscaping bought 4 tons of topsoil, 3 tons of mulch, and 6 tons of pea gravel for $2825. The next week the firm bought 5 tons of topsoil, 2 tons of mulch, and 5 tons of pea gravel for $2663. Pea gravel costs $17 less per ton than topsoil. Find the price per ton for each item.

50. *Investment.* Selena receives $537.75 per year in simple interest from three investments totaling $8500. Part is invested at 5.15%, part at 6.05%, and the rest at 7.2%. There is $1500 more invested at 7.2% than at 5.15%. Find the amount invested at each rate.

Skill Maintenance

Graph.

51. $y = x - 3$

52. $2x + 3y = 6$

53. $y = -5$

54. $x - y = 7,$
$2x - 3y = 15$

Synthesis

55. ◈ For square matrices **A** and **B**, is it true, in general, that $(\mathbf{A} + \mathbf{B})^{-1} = \mathbf{A}^{-1} + \mathbf{B}^{-1}$? Explain.

56. ◈ For square matrices **A** and **B**, is it true, in general, that $(\mathbf{AB})^{-1} = \mathbf{A}^{-1}\mathbf{B}^{-1}$? Explain.

State the conditions under which $\mathbf{A}^{-1}$ exists. Then find a formula for $\mathbf{A}^{-1}$.

57. $\mathbf{A} = [x]$

58. $\mathbf{A} = \begin{bmatrix} x & 0 \\ 0 & y \end{bmatrix}$

59. $\mathbf{A} = \begin{bmatrix} 0 & 0 & x \\ 0 & y & 0 \\ z & 0 & 0 \end{bmatrix}$

60. $\mathbf{A} = \begin{bmatrix} x & 1 & 1 & 1 \\ 0 & y & 0 & 0 \\ 0 & 0 & z & 0 \\ 0 & 0 & 0 & w \end{bmatrix}$

7.6 Systems of Inequalities and Linear Programming

- *Graph linear inequalities.*
- *Graph systems of linear inequalities.*
- *Solve linear programming problems.*

A graph of an inequality is a drawing that represents its solutions. We have already seen that an inequality in one variable can be graphed on a number line. An inequality in two variables can be graphed on a coordinate plane.

Graphs of Linear Inequalities

A statement like $5x - 4y < 20$ is a linear inequality in two variables.

Linear Inequality in Two Variables

A *linear inequality in two variables* is an inequality that can be written in the form

$$Ax + By < C,$$

where A, B, and C are real numbers and A and B are not both zero. The symbol $<$ may be replaced by $\leq$, $>$, or $\geq$.

A solution of a linear inequality in two variables is an ordered pair (x, y) for which the inequality is true. For example, $(1, 3)$ is a solution of $5x - 4y < 20$ because $5 \cdot 1 - 4 \cdot 3 < 20$, or $-7 < 20$, is true. On the other hand, $(2, -6)$ is not a solution of $5x - 4y < 20$ because $5 \cdot 2 - 4 \cdot (-6) \not< 20$.

The **solution set** of an inequality is the set of all the ordered pairs that make it true. A **graph of an inequality** represents its solution set.

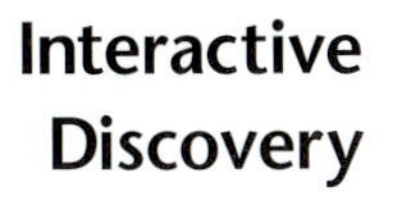

Graph $y_1 = x - 4$. Select three or four points (x, y) in the region below the graph and compare the values of y_1 and $x - 4$ for each point. Do the same for three or four points in the region above the graph. Which points satisfy the inequality $y_1 < x - 4$? Shade that region. (Most graphers are capable of shading a region above or below a graph or between two graphs. Consult your user's manual for the keystrokes to use.) What inequality is satisfied by the points in the region that is not shaded?

Example 1 Graph: $y < x + 3$.

SOLUTION We begin by graphing the related equation $y = x + 3$. We use a dashed line because the inequality symbol is $<$. This indicates that the line itself is not in the solution set of the inequality.

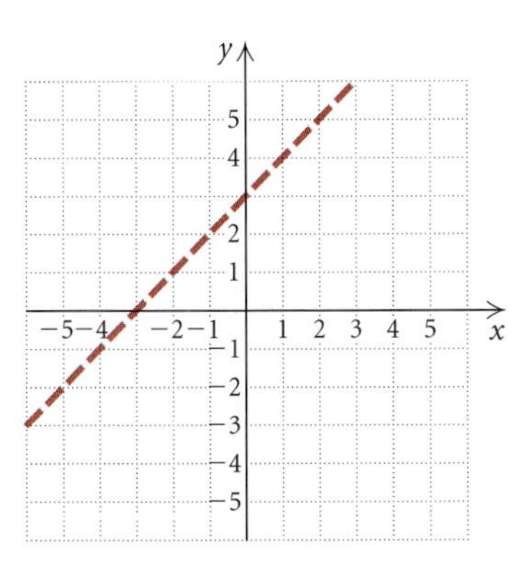

Note that the line divides the coordinate plane into two regions called **half-planes**, one of which satisfies the inequality. Either *all* points in a half-plane are in the solution set of the inequality or *none* are.

To determine which half-plane satisfies the inequality, we try a test point in either region. The point $(0, 0)$ is usually a convenient choice so long as it does not lie on the line.

$$\begin{array}{c|c} y & < x + 3 \\ \hline 0 & ?\ 0 + 3 \\ 0 & 3 \end{array} \quad \text{TRUE}$$

Since $(0, 0)$ satisfies the inequality, so do all points in the half-plane that contains $(0, 0)$. We shade this region to show the solution set of the inequality.

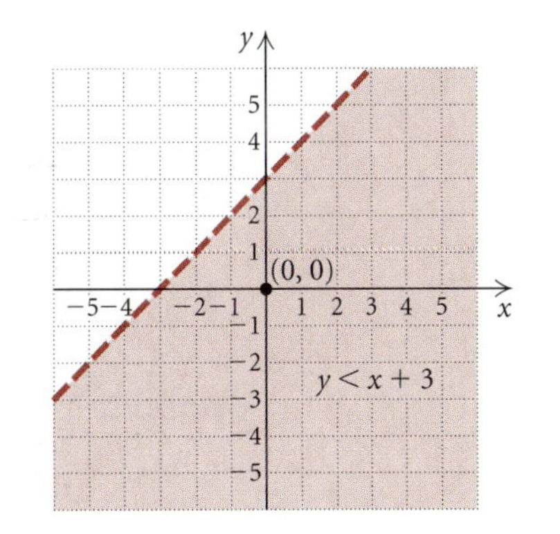

In general, we use the following procedure to graph linear inequalities in two variables.

To graph a linear inequality in two variables:

1. Replace the inequality symbol with an equals sign and graph this related equation. If the inequality symbol is $<$ or $>$, draw the line dashed. If the inequality symbol is $\leq$ or $\geq$, draw the line solid.
2. The graph consists of a half-plane on one side of the line and, if the line is solid, the line as well. To determine which half-plane to shade, test a point not on the line in the original inequality. If that point is a solution, shade the half-plane containing that point. If not, shade the opposite half-plane.

Example 2 Graph: $3x + 4y \geq 12$.

SOLUTION

1. First we graph the related equation $3x + 4y = 12$. We use a solid line because the inequality symbol is $\geq$. This indicates that the line is included in the solution set.

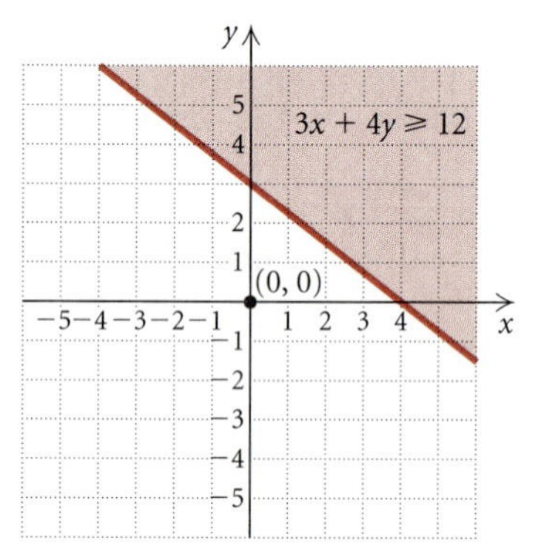

2. To determine which half-plane to shade, we test a point in either region. We choose (0, 0).

$$\begin{array}{c|c} 3x + 4y & \geq 12 \\ \hline 3 \cdot 0 + 4 \cdot 0 & ?\ 12 \\ 0 & 12 \quad \text{FALSE} \end{array}$$

Because (0, 0) is *not* a solution, all the points in the half-plane that does *not* contain (0, 0) are solutions. We shade that region.

This inequality can also be graphed using a grapher with a SHADE feature.

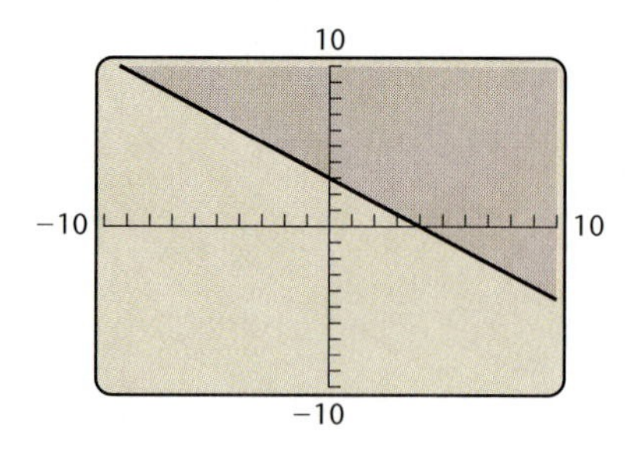

Example 3 Graph $x > -3$ on a plane.

SOLUTION

1. First we graph the related equation $x = -3$. We use a dashed line because the inequality symbol is $>$. This indicates that the line is not included in the solution set.

2. The inequality tells us that all points (x, y) for which $x > -3$ are solutions. These are the points to the right of the line. We can also use a test point to determine the solutions. We choose (5, 1).

$$x > -3$$

$$5 \ ? \ -3 \quad \text{TRUE}$$

Because (5, 1) is a solution, we shade the region containing that point—that is, the region to the right of the dashed line.

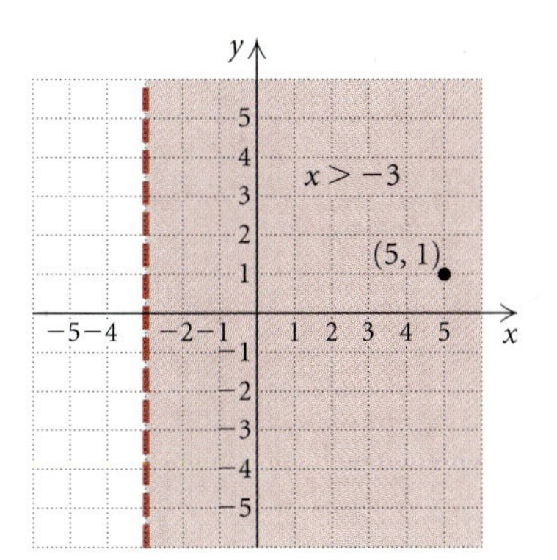

Example 4 Graph $y \leq 4$ on a plane.

SOLUTION

1. First we graph the related equation $y = 4$. We use a solid line because the inequality symbol is $\leq$.
2. The inequality tells us that all points (x, y) for which $y \leq 4$ are solutions of the inequality. These are the points below the line. We can also use a test point to determine the solutions. We choose $(-2, 5)$.

$$y \leq 4$$

$$5 \ ? \ 4 \quad \text{FALSE}$$

Because $(-2, 5)$ is not a solution, we shade the half-plane that does not contain that point.

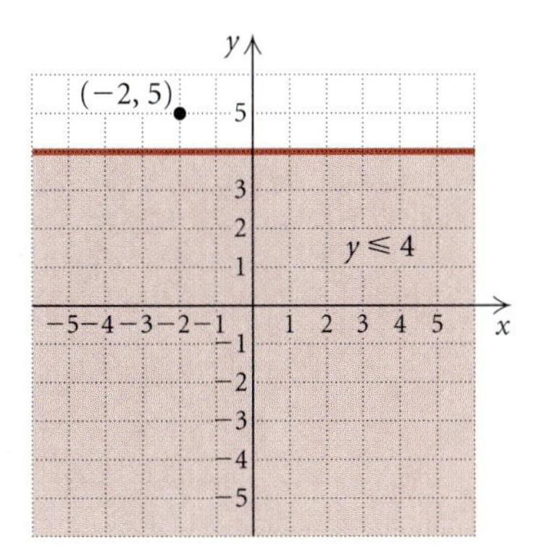

We can also use a grapher with the SHADE feature to graph this inequality.

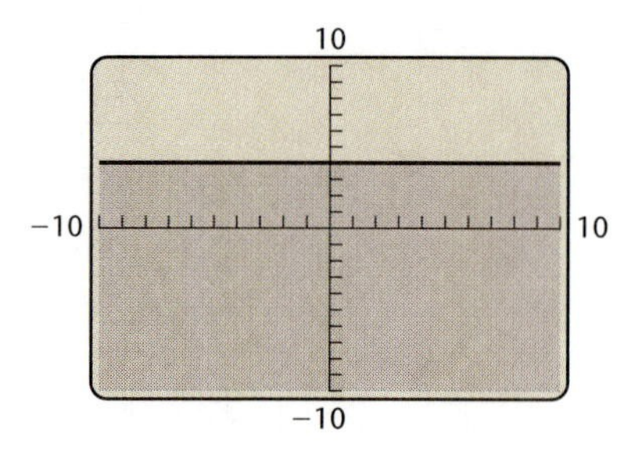

Systems of Linear Inequalities

A system of inequalities in two variables consists of two or more inequalities in two variables considered simultaneously. For example,

$$x + y \leq 4,$$
$$x - y \geq 2$$

is a system of two *linear* inequalities in two variables.

A solution of a system of inequalities is an ordered pair that is a solution of each inequality in the system. To graph a system of linear inequalities, we graph each inequality and determine the region that is common to all the solution sets.

Example 5 Graph the solution set of the system

$$x + y \leq 4,$$
$$x - y \geq 2.$$

SOLUTION We graph $x + y \leq 4$ by first graphing the equation $x + y = 4$ using a solid line. Next, we choose (0, 0) as a test point and find that it is a solution of $x + y \leq 4$, so we shade the half-plane containing (0, 0) using red. Next we graph $x - y = 2$ using a solid line. We find that (0, 0) is not a solution of $x - y \geq 2$, so we shade the half-plane that does not contain (0, 0) using blue. The arrows at the ends of each line help to indicate the half-plane that contains each solution set.

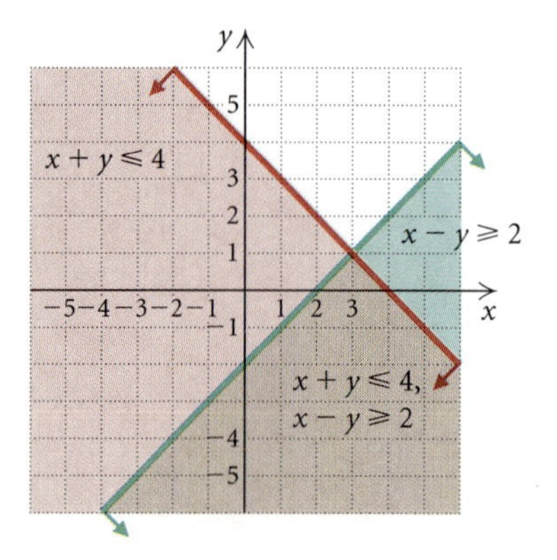

The solution set of the system of equations is the region shaded both blue and red, or purple, including parts of the lines $x + y = 4$ and $x - y = 2$.

A system of inequalities may have a graph that consists of a polygon and its interior. As we will see later in this section, it is important in many applications to be able to find the vertices of such a polygon.

Example 6 Graph the following system of inequalities and find the coordinates of any vertices formed:

$$3x - y \leq 6, \qquad (1)$$
$$y - 3 \leq 0, \qquad (2)$$
$$x + y \geq 0. \qquad (3)$$

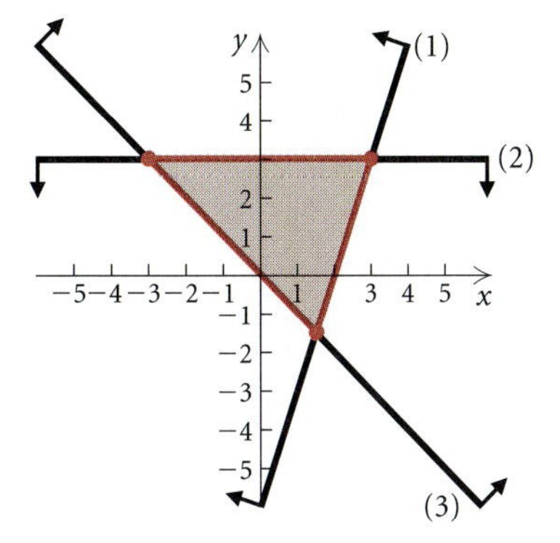

SOLUTION We graph the related equations $3x - y = 6$, $y - 3 = 0$, and $x + y = 0$ using solid lines. The half-plane containing the solution set for each inequality is indicated by the arrows at the ends of each line. We shade the region common to all three solution sets.

To find the vertices, we solve three systems of equations. The system of equations from inequalities (1) and (2) is

$$3x - y = 6,$$
$$y - 3 = 0.$$

Solving, we obtain the vertex (3, 3).

The system of equations from inequalities (1) and (3) is

$$3x - y = 6,$$
$$x + y = 0.$$

Solving, we obtain the vertex $\left(\frac{3}{2}, -\frac{3}{2}\right)$.

The system of equations from inequalities (2) and (3) is

$$y - 3 = 0,$$
$$x + y = 0.$$

Solving, we obtain the vertex $(-3, 3)$.

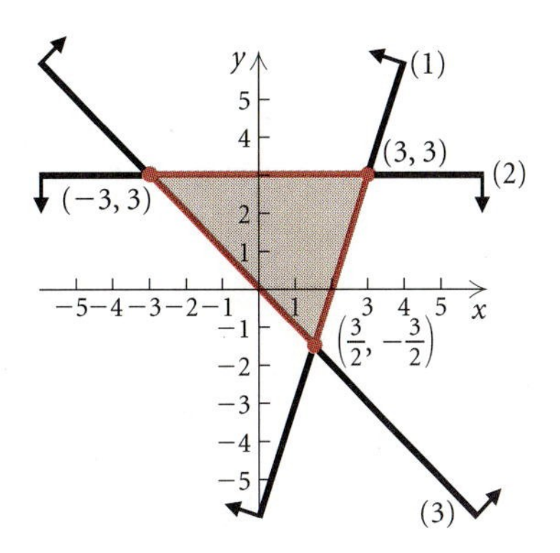

We could also graph the system of related equations on a grapher and use the INTERSECT feature or the TRACE and ZOOM features to find the vertices. However, most graphers are not capable of shading a region determined by three or more graphs like the one above.

Applications: Linear Programming

In many applications, we want to find a maximum or minimum value. In business, for example, we might want to maximize profit and minimize cost. **Linear programming** can tell us how to do this.

In our study of linear programming, we will consider linear functions of two variables that are to be maximized or minimized subject to several conditions, or **constraints**. These constraints are expressed as inequalities. The solution set of the system of inequalities made up of the constraints contains all the **feasible solutions** of a linear programming problem. The function that we want to maximize or minimize is called the **objective function**.

It can be shown that the maximum and minimum values of the objective function occur at a vertex of the region of feasible solutions. Thus we have the following procedure.

Linear Programming Procedure

To find the maximum or minimum value of a linear objective function subject to a set of constraints:

1. Graph the region of feasible solutions.
2. Determine the coordinates of the vertices of the region.
3. Evaluate the objective function at each vertex. The largest and smallest of those values are the maximum and minimum values of the function, respectively.

Example 7 *Maximizing Profit.* Dovetail Carpentry Shop makes bookcases and desks. Each bookcase requires 5 hr of woodworking and 4 hr of finishing. Each desk requires 10 hr of woodworking and 3 hr of finishing. Each month the shop has 600 hr of labor available for woodworking and 240 hr for finishing. The profit on each bookcase is \$40 and on each desk is \$75. How many of each product should be made each month in order to maximize profit?

SOLUTION We let $x =$ the number of bookcases to be produced and $y =$ the number of desks. Then the profit P is given by the function

$$P = 40x + 75y.$$

To emphasize that P is a function of two variables, we sometimes write $P(x, y) = 40x + 75y$.

We know that x bookcases require $5x$ hr of woodworking and y desks require $10y$ hr of woodworking. Since there is no more than 600 hr of labor available for woodworking, we have one constraint:

$$5x + 10y \leq 600.$$

Similarly, the bookcases and desks require $4x$ hr and $3y$ hr of finishing, respectively. There is no more than 240 hr of labor available for finishing, so we have a second constraint:

$$4x + 3y \leq 240.$$

We also know that $x \geq 0$ and $y \geq 0$ because the carpentry shop cannot make a negative number of either product.

Thus we want to maximize the objective function

$$P = 40x + 75y$$

subject to the constraints

$$\begin{aligned} 5x + 10y &\leq 600, \\ 4x + 3y &\leq 240, \\ x &\geq 0, \\ y &\geq 0. \end{aligned}$$

We graph the system of inequalities and determine the vertices.

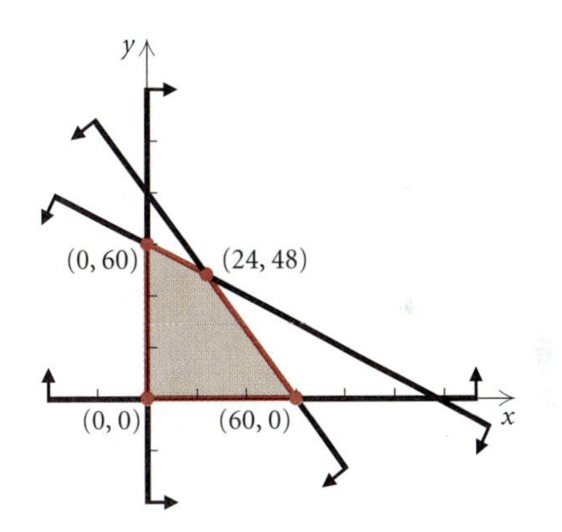

Now we evaluate the objective function P at each vertex.

VERTICES (x, y)	PROFIT $P = 40x + 75y$
(0, 0)	$P = 40 \cdot 0 + 75 \cdot 0 = 0$
(60, 0)	$P = 40 \cdot 60 + 75 \cdot 0 = 2400$
(24, 48)	$P = 40 \cdot 24 + 75 \cdot 48 = 4560$ ← Maximum
(0, 60)	$P = 40 \cdot 0 + 75 \cdot 60 = 4500$

The carpentry shop will make a maximum profit of \$4560 when 24 bookcases and 48 desks are produced.

7.6 Exercise Set

In Exercises 1–8, match each inequality with one of the graphs (a)–(h), which follow.

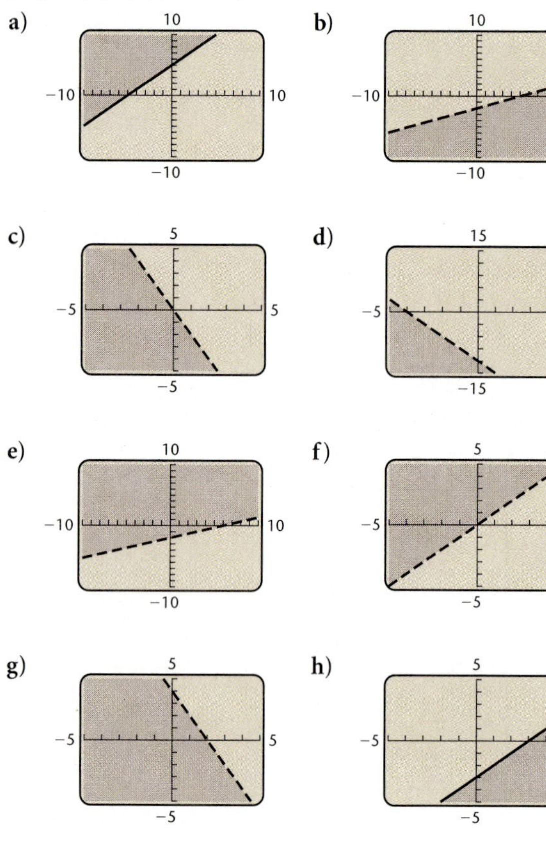

1. $y > x$

2. $y < -2x$

3. $y \leq x - 3$

4. $y \geq x + 5$

5. $2x + y < 4$

6. $3x + y < -12$

7. $2x - 5y > 10$

8. $3x - 9y < 18$

Graph.

9. $y > 2x$

10. $2y < x$

11. $y + x \geq 0$

12. $y - x < 0$

13. $y > x - 3$

14. $y \leq x + 4$

15. $x + y < 4$

16. $x - y \geq 5$

17. $3x - 2y \leq 6$

18. $2x - 5y < 10$

19. $3y + 2x \geq 6$

20. $2y + x \leq 4$

21. $3x - 2 \leq 5x + y$

22. $2x - 6y \geq 8 + 2y$

23. $x < -4$

24. $y \geq 5$

25. $y > -3$

26. $x \leq 5$

27. $-4 < y < -1$
(*Hint*: Think of this as $-4 < y$ *and* $y < -1$.)

28. $-3 < x < 3$
(*Hint*: Think of this as $-3 < x$ *and* $x < 3$.)

29. $y \geq |x|$

30. $y \leq |x + 2|$

In Exercises 31–36, match each system of inequalities with one of the graphs (a)–(f), which follow.

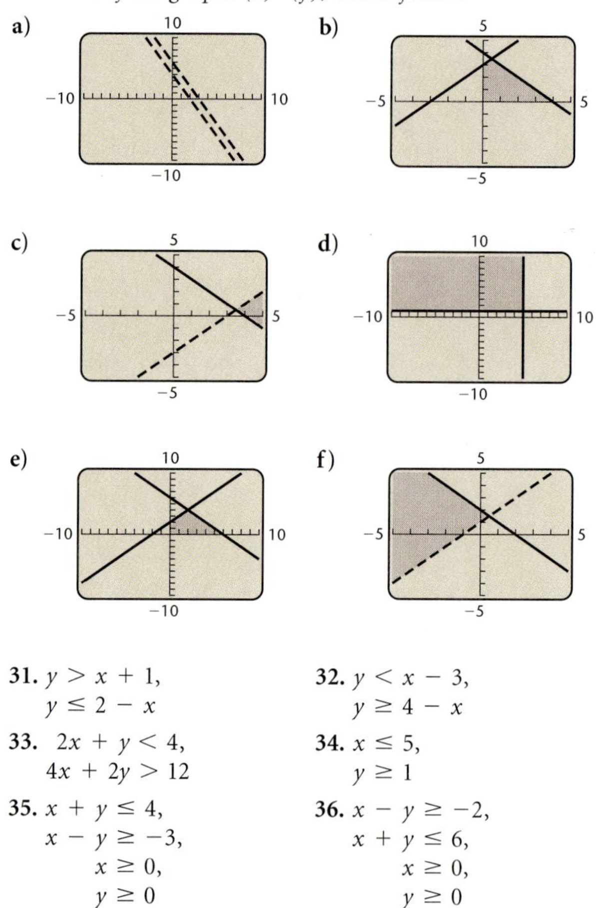

31. $y > x + 1,$
$y \leq 2 - x$

32. $y < x - 3,$
$y \geq 4 - x$

33. $2x + y < 4,$
$4x + 2y > 12$

34. $x \leq 5,$
$y \geq 1$

35. $x + y \leq 4,$
$x - y \geq -3,$
$x \geq 0,$
$y \geq 0$

36. $x - y \geq -2,$
$x + y \leq 6,$
$x \geq 0,$
$y \geq 0$

Graph the system of inequalities. Then find the coordinates of the vertices.

37. $y \leq x,$
$y \geq 3 - x$

38. $y \geq x,$
$y \leq x - 5$

39. $y \geq x,$
$y \leq x - 4$

40. $y \geq x,$
$y \leq 2 - x$

41. $y \geq -3,$
$x \geq 1$

42. $y \leq -2,$
$x \geq 2$

43. $x \leq 3,$
$y \geq 2 - 3x$

44. $x \geq -2,$
$y \leq 3 - 2x$

45. $x + y \leq 1,$
$x - y \leq 2$

46. $y + 3x \geq 0,$
$y + 3x \leq 2$

47. $2y - x \le 2,$
$y + 3x \ge -1$

48. $y \le 2x + 1,$
$y \ge -2x + 1,$
$x - 2 \le 0$

49. $x - y \le 2,$
$x + 2y \ge 8,$
$y - 4 \le 0$

50. $x + 2y \le 12,$
$2x + y \le 12,$
$x \ge 0,$
$y \ge 0$

51. $4y - 3x \ge -12,$
$4y + 3x \ge -36,$
$y \le 0,$
$x \le 0$

52. $8x + 5y \le 40,$
$x + 2y \le 8,$
$x \ge 0,$
$y \ge 0$

53. $3x + 4y \ge 12,$
$5x + 6y \le 30,$
$1 \le x \le 3$

54. $y - x \ge 1,$
$y - x \le 3,$
$2 \le x \le 5$

Find the maximum and the minimum values of the function and the values of x and y for which they occur.

55. $P = 17x - 3y + 60$, subject to

$$6x + 8y \le 48,\quad 0 \le y \le 4,\quad 0 \le x \le 7.$$

56. $Q = 28x - 4y + 72$, subject to

$$5x + 4y \ge 20,\quad 0 \le y \le 4,\quad 0 \le x \le 3.$$

57. $F = 5x + 36y$, subject to

$$5x + 3y \le 34,\quad 3x + 5y \le 30,\quad x \ge 0,\quad y \ge 0.$$

58. $G = 16x + 14y$, subject to

$$3x + 2y \le 12,\quad 7x + 5y \le 29,\quad x \ge 0,\quad y \ge 0.$$

59. *Maximizing Income.* Golden Harvest Foods makes jumbo biscuits and regular biscuits. The oven can cook at most 200 biscuits per day. Each jumbo biscuit requires 2 oz of flour, each regular biscuit requires 1 oz of flour, and there is 300 oz of flour available. The income from each jumbo biscuit is \$0.10 and from each regular biscuit is \$0.08. How many of each size biscuit should be made in order to maximize income? What is the maximum income?

60. *Maximizing Mileage.* Omar owns a car and a moped. He can afford 12 gal of gasoline to be split between the car and the moped. Omar's car gets 20 mpg and holds at most 10 gal of gas. His moped gets 100 mpg and holds at most 3 gal of gas. How many gallons of gasoline should each vehicle use if Omar wants to travel as far as possible? What is the maximum number of miles that he can travel?

61. *Maximizing Profit.* Norris Mill can convert logs into lumber and plywood. In a given week, the mill can turn out 400 units of production, of which 100 units of lumber and 150 units of plywood are required by regular customers. The profit is \$20 per unit of lumber and \$30 per unit of plywood. How many units of each should the mill produce in order to maximize the profit?

62. *Maximizing Profit.* Sunnydale Farm includes 240 acres of cropland. The farm owner wishes to plant this acreage in corn and oats. The profit per acre in corn production is \$40 and in oats is \$30. A total of 320 hr of labor is available. Each acre of corn requires 2 hr of labor, whereas each acre of oats requires 1 hr of labor. How should the land be divided between corn and oats in order to yield the maximum profit? What is the maximum profit?

63. *Minimizing Cost.* An animal feed to be mixed from soybean meal and oats must contain at least 120 lb of protein, 24 lb of fat, and 10 lb of mineral ash. Each 100-lb sack of soybean meal costs \$15 and contains 50 lb of protein, 8 lb of fat, and 5 lb of mineral ash. Each 100-lb sack of oats costs \$5 and contains 15 lb of protein, 5 lb of fat, and 1 lb of mineral ash. How many sacks of each should be used to satisfy the minimum requirements at minimum cost?

64. *Minimizing Cost.* Suppose that in the preceding problem the oats were replaced by alfalfa, which costs \$8 per 100 lb and contains 20 lb of protein, 6 lb of fat, and 8 lb of mineral ash. How much of each is now required in order to minimize the cost?

65. *Maximizing Income.* Clayton is planning to invest up to \$40,000 in corporate and municipal bonds. The least he is allowed to invest in corporate bonds is \$6000, and he does not want to invest more than \$22,000 in corporate bonds. He also does not want to invest more than \$30,000 in municipal bonds. The interest is 8% on corporate bonds and $7\frac{1}{2}$% on municipal bonds. This is simple interest for one year. How much should he invest in each type of bond in order to maximize his income? What is the maximum income?

66. *Maximizing Income.* Margaret is planning to invest up to \$22,000 in certificates of deposit at City Bank and People's Bank. She wants to invest at least \$2000 but no more than \$14,000 at City Bank. People's Bank does not insure more than a \$15,000 investment, so she will invest no more than that in People's Bank. The interest is 6% at City Bank and $6\frac{1}{2}$% at People's Bank. This is simple interest for one year. How much should she invest in each bank in order to maximize her income? What is the maximum income?

67. *Minimizing Transportation Cost.* An airline with two types of airplanes, P_1 and P_2, has contracted with a tour group to provide transportation for a minimum of 2000 first-class, 1500 tourist-class, and 2400 economy-class passengers. For a certain trip, airplane P_1 costs $12 thousand to operate and can accommodate 40 first-class, 40 tourist-class, and 120 economy-class passengers, whereas airplane P_2 costs $10 thousand to operate and can accommodate 80 first-class, 30 tourist-class, and 40 economy-class passengers. How many of each type of airplane should be used in order to minimize the operating cost?

68. *Minimizing Transportation Cost.* Suppose that in the preceding problem a new airplane P_3 becomes available, having an operating cost for the same trip of $15 thousand and accommodating 40 first-class, 40 tourist-class, and 80 economy-class passengers. If airplane P_1 were replaced by airplane P_3, how many of P_2 and P_3 should be used in order to minimize the operating cost?

69. *Maximizing Profit.* It takes Just Sew 2 hr of cutting and 4 hr of sewing to make a knit suit. It takes 4 hr of cutting and 2 hr of sewing to make a worsted suit. At most 20 hr per day is available for cutting and at most 16 hr per day is available for sewing. The profit is $34 on a knit suit and $31 on a worsted suit. How many of each kind of suit should be made each day in order to maximize profit? What is the maximum profit?

70. *Maximizing Profit.* Cambridge Metal Works manufactures two sizes of gears. The smaller gear requires 4 hr of machining and 1 hr of polishing and yields a profit of $25. The larger gear requires 1 hr of machining and 1 hr of polishing and yields a profit of $10. The firm has available at most 24 hr per day for machining and 9 hr per day for polishing. How many of each type of gear should be produced each day in order to maximize profit? What is the maximum profit?

71. *Minimizing Nutrition Cost.* Suppose it takes 12 units of carbohydrates and 6 units of protein to satisfy Jacob's minimum weekly requirements. A particular type of meat contains 2 units of carbohydrates and 2 units of protein per pound. A particular cheese contains 3 units of carbohydrates and 1 unit of protein per pound. The meat costs $3.50 per pound and the cheese costs $4.60 per pound. How many pounds of each are needed in order to minimize the cost and still meet the minimum requirements?

72. *Minimizing Salary Cost.* The Spring Hill school board is analyzing education costs for Hill Top School. It wants to hire teachers and teacher's aides to make up a faculty that satisfies its needs at minimum cost. The average annual salary for a teacher is $35,000 and for a teacher's aide is $18,000. The school building can accommodate a faculty of no more than 50 but needs at least 20 faculty members to function properly. The school must have at least 12 aides, but the number of teachers must be at least twice the number of aides in order to accommodate the expectations of the community. How many teachers and teacher's aides should be hired in order to minimize salary costs?

73. *Maximizing Animal Support in a Forest.* A certain area of forest is populated by two species of animals, which scientists refer to as A and B for simplicity. The forest supplies two kinds of food, referred to as F_1 and F_2. For one year, species A requires 1 unit of F_1 and 0.5 unit of F_2. Species B requires 0.2 unit of F_1 and 1 unit of F_2. The forest can normally supply at most 600 units of F_1 and 525 units of F_2 per year. What is the maximum total number of these animals that the forest can support?

74. *Maximizing Animal Support in a Forest.* In reference to Exercise 73, if there is a wet spring, then supplies of food increase to 1080 units of F_1 and 810 units of F_2. In this case, what is the maximum total number of these animals that the forest can support?

Skill Maintenance

Factor.

75. $x^2 - 2x - 15$

76. $2x^2 + x - 3$

Add.

77. $\frac{3}{x+1} + \frac{2}{x-4}$

78. $\frac{4}{x-2} + \frac{1}{x^2+3}$

Synthesis

79. ◆ Write an applied linear programming problem for a classmate to solve. Devise your problem so that the answer is "The bakery will make a maximum profit when 5 dozen pies and 12 dozen cookies are baked."

80. ◆ Describe how the graph of a linear inequality differs from the graph of a linear equation.

Graph the system of inequalities.

81. $y \geq x^2 - 2,$
$y \leq 2 - x^2$

82. $y < x + 1,$
$y \geq x^2$

Graph the inequality.

83. $|x + y| \leq 1$

84. $|x| + |y| \leq 1$

85. $|x| > |y|$

86. $|x - y| > 0$

87. *Allocation of Resources.* Significant Sounds manufactures two types of speaker assemblies. The less expensive assembly, which sells for $350, consists of one midrange speaker and one tweeter. The more expensive speaker assembly, which sells for $600, consists of one woofer, one midrange speaker, and two tweeters. The manufacturer has in stock 44 woofers, 60 midrange speakers, and 90 tweeters. How many of each type of speaker assembly should be made in order to maximize income? What is the maximum income?

88. *Allocation of Resources.* Sitting Pretty Furniture produces chairs and sofas. The chairs require 20 ft of wood, 1 lb of foam rubber, and 2 yd^2 of fabric. The sofas require 100 ft of wood, 50 lb of foam rubber, and 20 yd^2 of fabric. The manufacturer has in stock 1900 ft of wood, 500 lb of foam rubber, and 240 yd^2 of fabric. The chairs can be sold for $80 each and the sofas for $300 each. How many of each should be produced in order to maximize income? What is the maximum income?

7.7 Partial Fractions

- *Decompose rational expressions into partial fractions.*

There are situations in calculus in which it is useful to write a rational expression as a sum of two or more simpler rational expressions. For example, in the equation

$$\frac{4x - 13}{2x^2 + x - 6} = \frac{3}{x + 2} + \frac{-2}{2x - 3},$$

each fraction on the right side is called a **partial fraction**. The expression on the right side is the **partial fraction decomposition** of the rational expression on the left side. In this section, we learn how such decompositions are created.

Interactive Discovery

Look at the table of values for

$$y_1 = \frac{13x + 5}{3x^2 - 7x - 6} \quad \text{and} \quad y_2 = \frac{1}{3x + 2} + \frac{4}{x - 3}.$$

Then do the same for the following pairs of equations.

$$y_1 = \frac{3x^2 - 3x - 2}{(x + 1)(x - 1)^2}, \qquad y_2 = \frac{1}{x + 1} + \frac{2}{x - 1} + \frac{-1}{(x - 1)^2};$$

$$y_1 = \frac{2x^2 + 4x + 5}{(x^2 + 1)(x + 2)}, \qquad y_2 = \frac{x + 2}{x^2 + 1} + \frac{1}{x + 2}$$

What do you observe about the relationship between each pair of equations?

Partial Fraction Decompositions

The procedure for finding the partial fraction decomposition of a rational expression involves factoring its denominator into linear and quadratic factors.

Procedure for Decomposing a Rational Expression into Partial Fractions

Consider any rational expression $P(x)/Q(x)$ such that $P(x)$ and $Q(x)$ have no common factor other than 1 or -1.

1. If the degree of $P(x)$ is greater than or equal to the degree of $Q(x)$, divide to express $P(x)/Q(x)$ as a quotient + remainder/$Q(x)$ and follow steps (2)–(5) to decompose the resulting rational expression.
2. If the degree of $P(x)$ is less than the degree of $Q(x)$, factor $Q(x)$ into linear factors of the form $(px + q)^n$ and/or quadratic factors of the form $(ax^2 + bx + c)^m$. Any quadratic factor $ax^2 + bx + c$ must be *irreducible*, meaning that it cannot be factored into linear factors with real coefficients.
3. Assign to each linear factor $(px + q)^n$ the sum of n partial fractions:

$$\frac{A_1}{px + q} + \frac{A_2}{(px + q)^2} + \cdots + \frac{A_n}{(px + q)^n}.$$

4. Assign to each quadratic factor $(ax^2 + bx + c)^m$ the sum of m partial fractions:

$$\frac{B_1x + C_1}{ax^2 + bx + c} + \frac{B_2x + C_2}{(ax^2 + bx + c)^2} + \cdots + \frac{B_mx + C_m}{(ax^2 + bx + c)^m}.$$

5. Apply algebraic methods, as illustrated in the following examples, to find the constants in the numerators of the partial fractions.

Example 1 Decompose into partial fractions:

$$\frac{4x - 13}{2x^2 + x - 6}.$$

SOLUTION The degree of the numerator is less than the degree of the denominator. We begin by factoring the denominator: $(x + 2)(2x - 3)$. We know that there are constants A and B such that

$$\frac{4x - 13}{(x + 2)(2x - 3)} = \frac{A}{x + 2} + \frac{B}{2x - 3}.$$

To determine A and B, we add the expressions on the right:

$$\frac{4x - 13}{(x + 2)(2x - 3)} = \frac{A(2x - 3) + B(x + 2)}{(x + 2)(2x - 3)}.$$

Next, we equate the numerators:

$$4x - 13 = A(2x - 3) + B(x + 2).$$

Since the last equation containing A and B is true for all x, we can substitute any value of x and still have a true equation. In order to have $2x - 3 = 0$, we choose $x = \frac{3}{2}$. This gives us

$$4\left(\tfrac{3}{2}\right) - 13 = A\left(2 \cdot \tfrac{3}{2} - 3\right) + B\left(\tfrac{3}{2} + 2\right)$$
$$-7 = 0 + \tfrac{7}{2}B.$$

Solving, we obtain $B = -2$.

In order to have $x + 2 = 0$, we choose $x = -2$, which gives us

$$4(-2) - 13 = A[2(-2) - 3] + B(-2 + 2).$$

Solving, we obtain $A = 3$.

The decomposition is as follows:

$$\frac{4x - 13}{2x^2 + x - 6} = \frac{3}{x + 2} + \frac{-2}{2x - 3}, \quad \text{or} \quad \frac{3}{x + 2} - \frac{2}{2x - 3}.$$

To check, we can add to see if we get the expression on the left. We can also perform a partial check by graphing

$$y_1 = \frac{4x - 13}{2x^2 + x - 6} \quad \text{and} \quad y_2 = \frac{3}{x + 2} - \frac{2}{2x - 3}.$$

If the graphs are identical, the decomposition is probably correct. As we see in the following figure, the graphs appear to be the same.

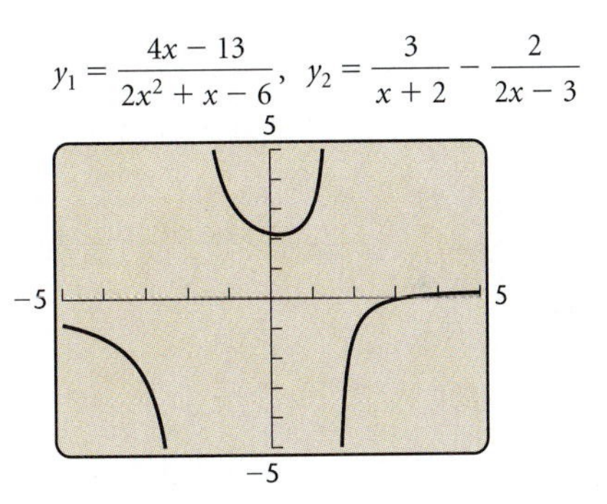

We can also use the TABLE feature, comparing values of

$$y_1 = \frac{4x - 13}{2x^2 + x - 6} \quad \text{and} \quad y_2 = \frac{3}{x + 2} - \frac{2}{2x - 3}$$

for the same values of x. Yet another check using a grapher is accomplished by graphing

$$y = \frac{4x - 13}{2x^2 + x - 6} - \left(\frac{3}{x + 2} - \frac{2}{2x - 3}\right).$$

If the partial fraction decomposition is correct, the right-hand side of the equation will be 0 and the graph will be $y = 0$, or the x-axis.

Example 2 Decompose into partial fractions:

$$\frac{7x^2 - 29x + 24}{(2x - 1)(x - 2)^2}.$$

SOLUTION The degree of the numerator is less than the degree of the denominator. The decomposition has the following form:

$$\frac{7x^2 - 29x + 24}{(2x - 1)(x - 2)^2} = \frac{A}{2x - 1} + \frac{B}{x - 2} + \frac{C}{(x - 2)^2}.$$

As in Example 1, we add and equate the numerators. This gives us

$$7x^2 - 29x + 24 = A(x - 2)^2 + B(2x - 1)(x - 2) + C(2x - 1).$$

Since the equation containing A, B, and C is true for all x, we can substitute any value of x and still have a true equation. In order to have $2x - 1 = 0$, we let $x = \frac{1}{2}$. This gives us

$$7\left(\tfrac{1}{2}\right)^2 - 29 \cdot \tfrac{1}{2} + 24 = A\left(\tfrac{1}{2} - 2\right)^2 + 0.$$

Solving we obtain $A = 5$.

In order to have $x - 2 = 0$, we let $x = 2$. Substituting gives us

$$7(2)^2 - 29(2) + 24 = 0 + C(2 \cdot 2 - 1).$$

Solving, we obtain $C = -2$.

To find B, we choose any value for x except $\frac{1}{2}$ or 2 and replace A with 5 and C with -2. We let $x = 1$:

$$\begin{aligned} 7 \cdot 1^2 - 29 \cdot 1 + 24 &= 5(1 - 2)^2 + B(2 \cdot 1 - 1)(1 - 2) \\ &\quad + (-2)(2 \cdot 1 - 1) \\ 2 &= 5 - B - 2 \\ B &= 1. \end{aligned}$$

The decomposition is as follows:

$$\frac{7x^2 - 29x + 24}{(2x - 1)(x - 2)^2} = \frac{5}{2x - 1} + \frac{1}{x - 2} - \frac{2}{(x - 2)^2}.$$

We can check the result on a grapher using one of the methods discussed in Example 1. Here we let

$$y_1 = \frac{7x^2 - 29x + 24}{(2x - 1)(x - 2)^2} \quad \text{and} \quad y_2 = \frac{5}{2x - 1} + \frac{1}{x - 2} - \frac{2}{(x - 2)^2}$$

and check a table of values.

X	Y1	Y2
−5	−.6382	−.6382
−4	−.7778	−.7778
−3	−.9943	−.9943
−2	−1.375	−1.375
−1	−2.222	−2.222
0	−6	−6
1	2	2

X = −5

Since $y_1 = y_2$ for given values of x as we scroll through the table, the decomposition appears to be correct.

Example 3 Decompose into partial fractions:

$$\frac{6x^3 + 5x^2 - 7}{3x^2 - 2x - 1}.$$

SOLUTION The degree of the numerator is greater than that of the denominator. Thus we divide and find an equivalent expression:

$$\begin{array}{r} 2x + 3 \\ 3x^2 - 2x - 1 \overline{)6x^3 + 5x^2 - 7} \\ \underline{6x^3 - 4x^2 - 2x } \\ 9x^2 + 2x - 7 \\ \underline{9x^2 - 6x - 3} \\ 8x - 4 \end{array}$$

The original expression is thus equivalent to

$$2x + 3 + \frac{8x - 4}{3x^2 - 2x - 1}.$$

We decompose the fraction to get

$$\frac{8x - 4}{(3x + 1)(x - 1)} = \frac{5}{3x + 1} + \frac{1}{x - 1}.$$

The final result is

$$2x + 3 + \frac{5}{3x + 1} + \frac{1}{x - 1}.$$

We can check the result on a grapher using one of the methods discussed in Example 1. Here we graph

$$y = \frac{6x^3 + 5x^2 - 7}{3x^2 - 2x - 1} - \left(2x + 3 + \frac{5}{3x + 1} + \frac{1}{x - 1}\right).$$

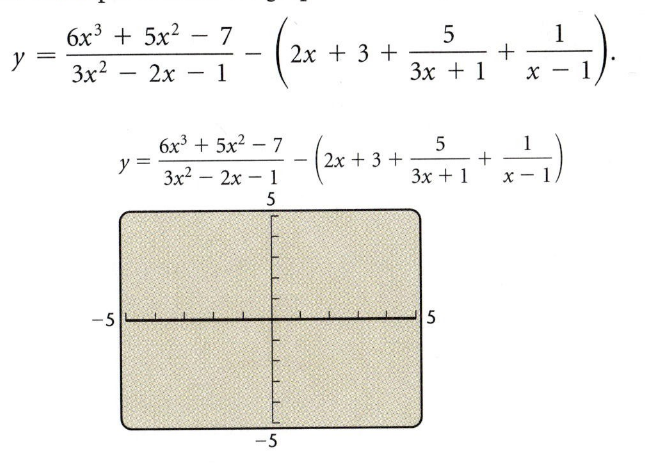

The graph appears to be $y = 0$, or the x-axis, so the decomposition is probably correct.

Systems of equations can be used to decompose rational expressions. Let's reconsider Example 2.

Example 4 Decompose into partial fractions:

$$\frac{7x^2 - 29x + 24}{(2x - 1)(x - 2)^2}.$$

SOLUTION The decomposition has the following form:

$$\frac{A}{2x - 1} + \frac{B}{x - 2} + \frac{C}{(x - 2)^2}.$$

We first add:

$$\frac{7x^2 - 29x + 24}{(2x - 1)(x - 2)^2} = \frac{A}{2x - 1} + \frac{B}{x - 2} + \frac{C}{(x - 2)^2}$$
$$= \frac{A(x - 2)^2}{(2x - 1)(x - 2)^2} + \frac{B(2x - 1)(x - 2)}{(2x - 1)(x - 2)^2}$$
$$+ \frac{C(2x - 1)}{(2x - 1)(x - 2)^2}.$$

Then we equate numerators:

$$7x^2 - 29x + 24$$
$$= A(x - 2)^2 + B(2x - 1)(x - 2) + C(2x - 1)$$
$$= A(x^2 - 4x + 4) + B(2x^2 - 5x + 2) + C(2x - 1)$$
$$= Ax^2 - 4Ax + 4A + 2Bx^2 - 5Bx + 2B + 2Cx - C,$$

or

$$7x^2 - 29x + 24$$
$$= (A + 2B)x^2 + (-4A - 5B + 2C)x + (4A + 2B - C).$$

Next, we equate corresponding coefficients:

$$7 = A + 2B,$$ The coefficients of the x^2-terms must be the same.

$$-29 = -4A - 5B + 2C,$$ The coefficients of the x-terms must be the same.

$$24 = 4A + 2B - C.$$ The constant terms must be the same.

We now have a system of three equations. You should confirm that the solution of the system is

$$A = 5, \quad B = 1, \quad \text{and} \quad C = -2.$$

The decomposition is as follows:

$$\frac{7x^2 - 29x + 24}{(2x - 1)(x - 2)^2} = \frac{5}{2x - 1} + \frac{1}{x - 2} - \frac{2}{(x - 2)^2}.$$

Example 5 Decompose into partial fractions:

$$\frac{11x^2 - 8x - 7}{(2x^2 - 1)(x - 3)}.$$

SOLUTION The decomposition has the following form:

$$\frac{11x^2 - 8x - 7}{(2x^2 - 1)(x - 3)} = \frac{Ax + B}{2x^2 - 1} + \frac{C}{x - 3}.$$

Adding and equating numerators, we get

$$\begin{aligned} 11x^2 - 8x - 7 &= (Ax + B)(x - 3) + C(2x^2 - 1) \\ &= Ax^2 - 3Ax + Bx - 3B + 2Cx^2 - C, \end{aligned}$$

or $$11x^2 - 8x - 7 = (A + 2C)x^2 + (-3A + B)x + (-3B - C).$$

We then equate corresponding coefficients:

$11 = A + 2C,$ The coefficients of the x^2-terms

$-8 = -3A + B,$ The coefficients of the x-terms

$-7 = -3B - C.$ The constant terms

We solve this system of three equations and obtain

$$A = 3, \quad B = 1, \quad \text{and} \quad C = 4.$$

The decomposition is as follows:

$$\frac{11x^2 - 8x - 7}{(2x^2 - 1)(x - 3)} = \frac{3x + 1}{2x^2 - 1} + \frac{4}{x - 3}.$$

We can check the result on a grapher.

$$y_1 = \frac{11x^2 - 8x - 7}{(2x^2 - 1)(x - 3)}, \quad y_2 = \frac{3x + 1}{2x^2 - 1} + \frac{4}{x - 3}$$

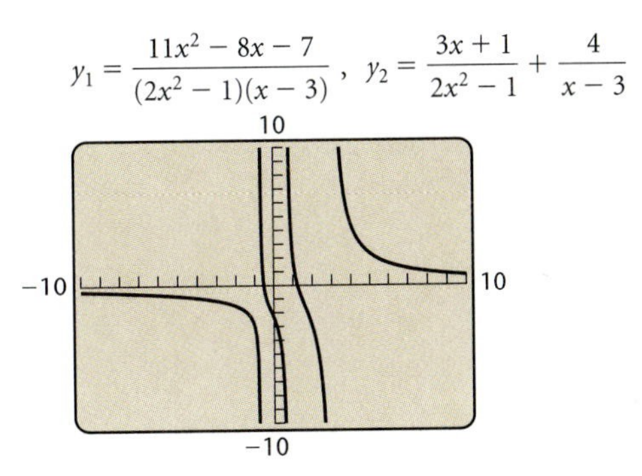

The graphs appear to be the same.

7.7 Exercise Set

Decompose into partial fractions. Check your answers on a grapher.

1. $\dfrac{x + 7}{(x - 3)(x + 2)}$

2. $\dfrac{2x}{(x + 1)(x - 1)}$

3. $\dfrac{7x - 1}{6x^2 - 5x + 1}$

4. $\dfrac{13x + 46}{12x^2 - 11x - 15}$

5. $\dfrac{3x^2 - 11x - 26}{(x^2 - 4)(x + 1)}$

6. $\dfrac{5x^2 + 9x - 56}{(x - 4)(x - 2)(x + 1)}$

7. $\dfrac{9}{(x + 2)^2(x - 1)}$

8. $\dfrac{x^2 - x - 4}{(x - 2)^3}$

9. $\dfrac{2x^2 + 3x + 1}{(x^2 - 1)(2x - 1)}$

10. $\dfrac{x^2 - 10x + 13}{(x^2 - 5x + 6)(x - 1)}$

11. $\dfrac{x^4 - 3x^3 - 3x^2 + 10}{(x + 1)^2(x - 3)}$

12. $\dfrac{10x^3 - 15x^2 - 35x}{x^2 - x - 6}$

13. $\dfrac{-x^2 + 2x - 13}{(x^2 + 2)(x - 1)}$

14. $\dfrac{26x^2 + 208x}{(x^2 + 1)(x + 5)}$

15. $\dfrac{6 + 26x - x^2}{(2x - 1)(x + 2)^2}$

16. $\dfrac{5x^3 + 6x^2 + 5x}{(x^2 - 1)(x + 1)^3}$

17. $\dfrac{6x^3 + 5x^2 + 6x - 2}{2x^2 + x - 1}$

18. $\dfrac{2x^3 + 3x^2 - 11x - 10}{x^2 + 2x - 3}$

19. $\dfrac{2x^2 - 11x + 5}{(x - 3)(x^2 + 2x - 5)}$

20. $\dfrac{3x^2 - 3x - 8}{(x - 5)(x^2 + x - 4)}$

21. $\dfrac{-4x^2 - 2x + 10}{(3x + 5)(x + 1)^2}$

22. $\dfrac{26x^2 - 36x + 22}{(x - 4)(2x - 1)^2}$

23. $\dfrac{36x + 1}{12x^2 - 7x - 10}$

24. $\dfrac{-17x + 61}{6x^2 + 39x - 21}$

25. $\dfrac{-4x^2 - 9x + 8}{(3x^2 + 1)(x - 2)}$

26. $\dfrac{11x^2 - 39x + 16}{(x^2 + 4)(x - 8)}$

Skill Maintenance

Factor.

27. $4x^2 - 25y^2$

28. $3x^2 + 10x - 8$

Complete the square and write the resulting expression in the form $(x + a)^2$.

29. $x^2 + 8x$

30. $x^2 - x$

Synthesis

31. ◆ Consider the three methods of checking the partial fraction decomposition of a rational expression in Example 1. Which do you prefer? Why?

32. ◆ What would you say to a classmate who tells you that the partial fraction decomposition of

$$\frac{3x^2 - 8x + 9}{(x + 3)(x^2 - 5x + 6)}$$

is

$$\frac{2}{x + 3} + \frac{x - 1}{x^2 - 5x + 6}?$$

Explain.

Decompose into partial fractions.

33. $\dfrac{x}{x^4 - a^4}$

34. $\dfrac{9x^3 - 24x^2 + 48x}{(x - 2)^4(x + 1)}$

[*Hint*: Let the expression equal

$$\frac{A}{x + 1} + \frac{P(x)}{(x - 2)^4}$$

and find $P(x)$.]

35. $\dfrac{1 + \ln x^2}{(\ln x + 2)(\ln x - 3)^2}$

36. $\dfrac{1}{e^{-x} + 3 + 2e^x}$

37. ◆ Explain the error in the following process.

$$\frac{x^2 + 4}{(x + 2)(x + 1)} = \frac{A}{x + 2} + \frac{B}{x + 1} = \frac{A(x + 1) + B(x + 2)}{(x + 2)(x + 1)}$$

Then

$$x^2 + 4 = A(x + 1) + B(x + 2).$$

When $x = -1$:

$$(-1)^2 + 4 = A(-1 + 1) + B(-1 + 2)$$
$$5 = B.$$

When $x = -2$:

$$(-2)^2 + 4 = A(-2 + 1) + B(-2 + 2)$$
$$8 = -A$$
$$-8 = A.$$

Thus,

$$\frac{x^2 + 4}{(x + 2)(x + 1)} = \frac{-8}{x + 2} + \frac{5}{x + 1}.$$

CHAPTER 7 Summary and Review

Important Properties and Formulas

Row-Equivalent Operations

1. Interchange any two rows.
2. Multiply each entry in a row by the same nonzero constant.
3. Add a nonzero multiple of one row to another row.

Row-Echelon Form

1. If a row does not consist entirely of 0's, then the first nonzero element in the row is a 1 (called a leading 1).
2. For any two successive nonzero rows, the leading 1 in the lower row is farther to the right than the leading 1 in the higher row.
3. All the rows consisting entirely of 0's are at the bottom of the matrix.

If a fourth property is also satisfied, a matrix is said to be in reduced row-echelon form:

4. Each column that contains a leading 1 has 0's everywhere else.

Properties of Matrix Addition and Scalar Multiplication

For any $m \times n$ matrices $\mathbf{A}$, $\mathbf{B}$, and $\mathbf{C}$ and any scalars k and l:

Commutative Property:

$\mathbf{A} + \mathbf{B} = \mathbf{B} + \mathbf{A}$.

Associative Property of Addition:

$\mathbf{A} + (\mathbf{B} + \mathbf{C}) = (\mathbf{A} + \mathbf{B}) + \mathbf{C}$.

Associative Property of Scalar Multiplication:

$(kl)\mathbf{A} = k(l\mathbf{A})$.

Additive Identity Property:

There exists a unique matrix $\mathbf{0}$ such that $\mathbf{A} + \mathbf{0} = \mathbf{0} + \mathbf{A} = \mathbf{A}$.

Additive Inverse Property:

There exists a unique matrix $-\mathbf{A}$ such that $\mathbf{A} + (-\mathbf{A}) = -\mathbf{A} + \mathbf{A} = \mathbf{0}$.

Distributive Properties:

$k(\mathbf{A} + \mathbf{B}) = k\mathbf{A} + k\mathbf{B}$,
$(k + l)\mathbf{A} = k\mathbf{A} + l\mathbf{A}$.

Properties of Matrix Multiplication

For matrices $\mathbf{A}$, $\mathbf{B}$, and $\mathbf{C}$, assuming that the indicated operation is possible:

Associative Property of Multiplication:

$\mathbf{A}(\mathbf{BC}) = (\mathbf{AB})\mathbf{C}$

Distributive Properties:

$\mathbf{A}(\mathbf{B} + \mathbf{C}) = \mathbf{AB} + \mathbf{AC}$,
$(\mathbf{B} + \mathbf{C})\mathbf{A} = \mathbf{BA} + \mathbf{CA}$

Matrix Solutions of Systems of Equations

For a system of n linear equations in n variables, $\mathbf{AX} = \mathbf{B}$, if $\mathbf{A}$ is an invertible matrix, then the unique solution of the system is given by $\mathbf{X} = \mathbf{A}^{-1}\mathbf{B}$.

To Graph a Linear Inequality in Two Variables:

1. Replace the inequality symbol with an equals sign and graph this related equation. If the inequality symbol is $<$ or $>$, draw the line dashed. If the inequality symbol is $\le$ or $\ge$, draw the line solid.
2. The graph consists of a half-plane on one side of the line and, if the line is solid, the line as well. To determine which half-plane to shade, test a point not on the line in the original inequality. If that point is a solution, shade the half-plane containing that point. If not, shade the opposite half-plane.

(continued)

Linear Programming Procedure

To find the maximum or minimum value of a linear objective function subject to a set of constraints:

1. Graph the region of feasible solutions.
2. Determine the coordinates of the vertices of the region.
3. Evaluate the objective function at each vertex. The largest and smallest of those values are the maximum and minimum values of the function, respectively.

Procedure for Decomposing a Rational Expression into Partial Fractions

Consider any rational expression $P(x)/Q(x)$ such that $P(x)$ and $Q(x)$ have no common factor other than 1 or -1.

1. If the degree of $P(x)$ is greater than or equal to the degree of $Q(x)$, divide to express $P(x)/Q(x)$ as a quotient + remainder/$Q(x)$ and follow steps (2)–(5) to decompose the resulting rational expression.
2. If the degree of $P(x)$ is less than the degree of $Q(x)$, factor $Q(x)$ into linear factors of the form $(px + q)^n$ and/or quadratic factors of the form $(ax^2 + bx + c)^m$. Any quadratic factor $ax^2 + bx + c$ must be irreducible, meaning that it cannot be factored into linear factors with real coefficients.
3. Assign to each linear factor $(px + q)^n$ the sum of n partial fractions:
$$\frac{A_1}{px + q} + \frac{A_2}{(px + q)^2} + \cdots + \frac{A_n}{(px + q)^n}.$$
4. Assign to each quadratic factor $(ax^2 + bx + c)^m$ the sum of m partial fractions:
$$\frac{B_1x + C_1}{ax^2 + bx + c} + \frac{B_2x + C_2}{(ax^2 + bx + c)^2} + \cdots + \frac{B_mx + C_m}{(ax^2 + bx + c)^m}.$$
5. Apply algebraic methods to find the constants in the numerators of the partial fractions.

REVIEW EXERCISES

Match each of the following with one of the graphs (a)–(h), which follow.

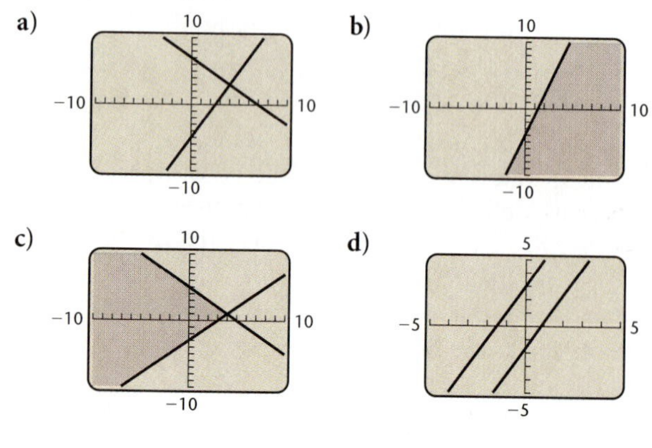

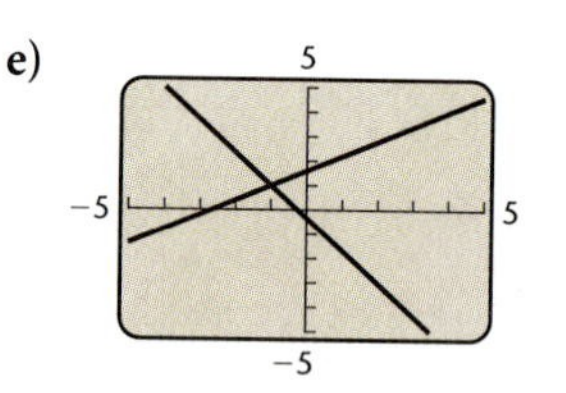

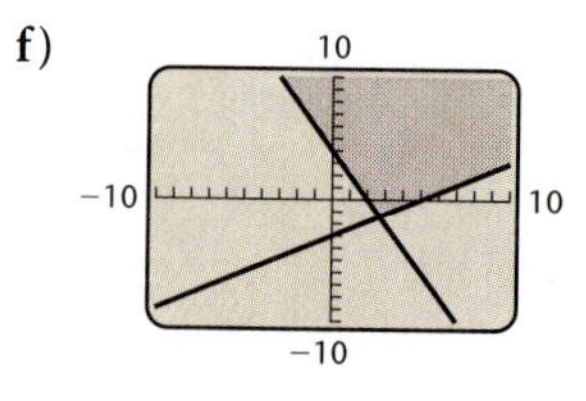

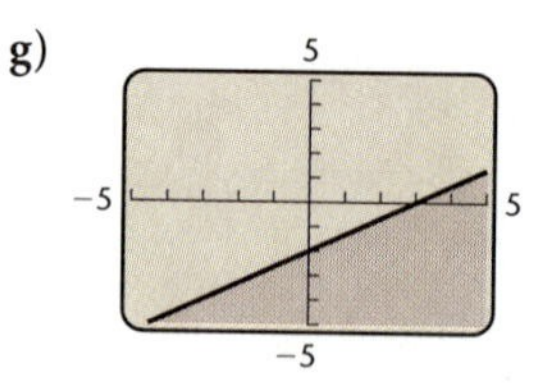

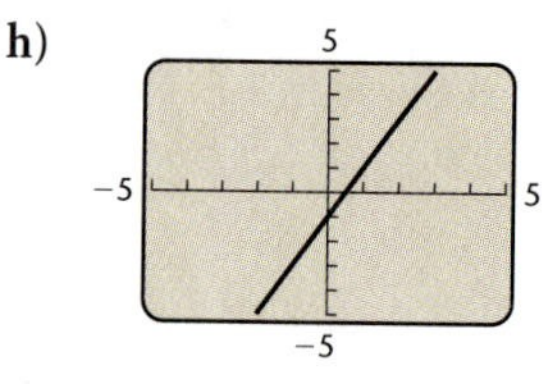

1. $x + y = 7,$
   $2x - y = 5$

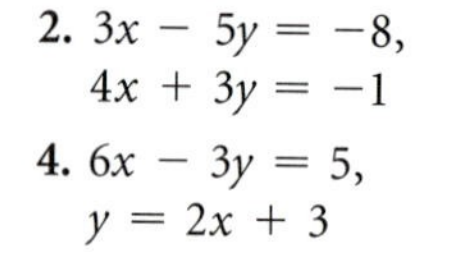

2. $3x - 5y = -8,$
   $4x + 3y = -1$

3. $y = 2x - 1,$
   $4x - 2y = 2$

4. $6x - 3y = 5,$
   $y = 2x + 3$

5. $y \leq 3x - 4$

6. $2x - 3y \geq 6$

7. $x - y \leq 3,$
$x + y \leq 5$

8. $2x + y \geq 4,$
$3x - 5y \leq 15$

Solve.

9. $5x - 3y = -4,$
$3x - y = -4$

10. $2x + 3y = 2,$
$5x - y = -29$

11. $x + 5y = 12,$
$5x + 25y = 12$

12. $x - y = -2,$
$-3x + 3y = 6$

13. $2x - 4y + 3z = -3,$
$-5x + 2y - z = 7,$
$3x + 2y - 2z = 4$

14. $x + 5y + 3z = 0,$
$3x - 2y + 4z = 0,$
$2x + 3y - z = 0$

15. $x - y = 5,$
$y - z = 6,$
$z - w = 7,$
$x + w = 8$

16. Classify each of the systems in Exercises 9–15 as consistent or inconsistent.

17. Classify each of the systems in Exercises 9–15 as dependent or independent.

Solve each system of equations using Gaussian elimination or Gauss–Jordan elimination.

18. $x + 2y = 5,$
$2x - 5y = -8$

19. $3x + 4y + 2z = 3,$
$5x - 2y - 13z = 3,$
$4x + 3y - 3z = 6$

20. $3x + 5y + z = 0,$
$2x - 4y - 3z = 0,$
$x + 3y + z = 0$

21. $w + x + y + z = -2,$
$-3w - 2x + 3y + 2z = 10,$
$2w + 3x + 2y - z = -12,$
$2w + 4x - y + z = 1$

22. *Coins.* The value of 75 coins, consisting of nickels and dimes, is \$5.95. How many of each kind are there?

23. *Investment.* The Mendez family invested \$5000, part at 10% and the remainder at 10.5%. The annual income from both investments is \$517. What is the amount invested at each rate?

24. *Nutrition.* A dietician must plan a breakfast menu that provides 460 Cal, 9 g of fat, and 55 mg of calcium. One plain bagel contains 200 Cal, 2 g of fat, and 29 mg of calcium. A one-tablespoon serving of cream cheese contains 100 Cal, 10 g of fat, and 24 mg of calcium. One banana contains 105 Cal, 1 g of fat, and 7 mg of calcium. How many servings of each are required to provide the desired nutritional values? (*Source: Home and Garden Bulletin No. 72,* U.S. Government Printing Office, Washington, D.C. 20402)

25. *Test Scores.* A student has a total of 225 on three tests. The sum of the scores on the first and second tests exceeds the score on the third test by 61. The first score exceeds the second by 6. Find the three scores.

26. *Ice Milk Consumption.* The table below shows the per capita ice milk consumption, in pounds, in the United States in three recent years, represented as years since 1985.

YEAR	ICE MILK CONSUMPTION (IN POUNDS)
(1985) 0	6.9
(1990) 5	7.7
(1992) 7	7.1

Source: U.S. Department of Agriculture, Economic Research Service, *Food Consumption, Prices, and Expenditures,* annual.

a) Find a quadratic function $f(x) = ax^2 + bx + c$ that fits the data.

b) Use the function to predict the per capita ice milk consumption in 1998.

For Exercises 26–33, let

$$\mathbf{A} = \begin{bmatrix} 1 & -1 & 0 \\ 2 & 3 & -2 \\ -2 & 0 & 1 \end{bmatrix}, \quad \mathbf{B} = \begin{bmatrix} -1 & 0 & 6 \\ 1 & -2 & 0 \\ 0 & 1 & -3 \end{bmatrix},$$

and

$$\mathbf{C} = \begin{bmatrix} -2 & 0 \\ 1 & 3 \end{bmatrix}.$$

Find each of the following, if possible.

27. $\mathbf{A} + \mathbf{B}$

28. $-3\mathbf{A}$

29. $-\mathbf{A}$

30. $\mathbf{AB}$

31. $\mathbf{B} + \mathbf{C}$

32. $\mathbf{A} - \mathbf{B}$

33. $2\mathbf{A} - \mathbf{B}$

34. $\mathbf{A} + 3\mathbf{B}$

35. *Food Service Management.* The table below shows the cost per serving, in cents, for items on four menus that are served at an elder-care facility.

MENU	MEAT	POTATO	VEGETABLE	SALAD	DESSERT
1	46.1	5.9	10.1	8.5	11.4
2	54.6	4.6	9.6	7.6	10.6
3	48.9	5.5	12.7	9.4	9.3
4	51.3	4.8	11.3	6.9	12.7

On a particular day, a dietician orders 32 meals from menu 1, 19 from menu 2, 43 from menu 3, and 38 from menu 4.

a) Write the information in the table as a 4×5 matrix **M**.
b) Write a row matrix **N** that represents the number of each menu ordered.
c) Find the product **NM**.
d) State what the entries of **NM** represent.

Find $\mathbf{A}^{-1}$, *if it exists.*

36. $\mathbf{A} = \begin{bmatrix} -2 & 0 \\ 1 & 3 \end{bmatrix}$

37. $\mathbf{A} = \begin{bmatrix} 0 & 0 & 3 \\ 0 & -2 & 0 \\ 4 & 0 & 0 \end{bmatrix}$

38. $\mathbf{A} = \begin{bmatrix} 1 & 0 & 0 & 0 \\ 0 & 4 & -5 & 0 \\ 0 & 2 & 2 & 0 \\ 0 & 0 & 0 & 1 \end{bmatrix}$

39. Write a matrix equation equivalent to this system of equations:

$$3x - 2y + 4z = 13,$$
$$x + 5y - 3z = 7,$$
$$2x - 3y + 7z = -8.$$

Solve each system of equations using the inverse of the coefficient matrix of the equivalent matrix equation.

40. $2x + 3y = 5,$
$3x + 5y = 11$

41. $5x - y + 2z = 17,$
$3x + 2y - 3z = -16,$
$4x - 3y - z = 5$

42. $w - x - y + z = -1,$
$2w + 3x - 2y - z = 2,$
$-w + 5x + 4y - 2z = 3,$
$3w - 2x + 5y + 3z = 4$

Graph.

43. $y \leq 3x + 6$

44. $4x - 3y \geq 12$

45. Graph this system of inequalities and find the coordinates of any vertices formed.

$$2x + y \geq 9,$$
$$4x + 3y \geq 23,$$
$$x + 3y \geq 8,$$
$$x \geq 0,$$
$$y \geq 0$$

46. Find the maximum and minimum values of $T = 6x + 10y$ subject to

$$x + y \leq 10,$$
$$5x + 10y \geq 50,$$
$$x \geq 2,$$
$$y \geq 0.$$

47. *Maximizing a Test Score.* Marita is taking a test that contains questions in group A worth 7 points each and questions in group B worth 12 points each. The total number of questions answered must be at least 8. If Marita knows that group A questions take 8 min each and group B questions take 10 min each and the maximum time for the test is 80 min, how many questions from each group must she answer correctly in order to maximize her score? What is the maximum score?

Decompose into partial fractions.

48. $\dfrac{5}{(x + 2)^2(x + 1)}$

49. $\dfrac{-8x + 23}{2x^2 + 5x - 12}$

Synthesis

50. ◆ Write a problem for a classmate to solve that can be translated to a system of equations. Devise the problem so that the solution is "The caterer sold 20 cheese trays and 35 seafood trays."

51. ◆ For square matrices **A** and **B**, is it true, in general, that $(\mathbf{AB})^2 = \mathbf{A}^2\mathbf{B}^2$? Explain.

52. One year, Don invested a total of \$40,000, part at 12%, part at 13%, and the rest at $14\frac{1}{2}$%. The total amount of interest received on the investments was \$5370. The interest received on the $14\frac{1}{2}$% investment was \$1050 more than the interest received on the 13% investment. How much was invested at each rate?

Solve.

53. $\dfrac{2}{3x} + \dfrac{4}{5y} = 8,$
$\dfrac{5}{4x} - \dfrac{3}{2y} = -6$

54. $\dfrac{3}{x} - \dfrac{4}{y} + \dfrac{1}{z} = -2,$
$\dfrac{5}{x} + \dfrac{1}{y} - \dfrac{2}{z} = 1,$
$\dfrac{7}{x} + \dfrac{3}{y} + \dfrac{2}{z} = 19$

Graph.

55. $|x| - |y| \leq 1$

56. $|xy| > 1$

Conic Sections 8

APPLICATION

For a student recreation building at Southport Community College, an architect wants to lay out a rectangular piece of ground that has a perimeter of 204 m and an area of 2565 m^2. The dimensions of the piece of ground are solutions of the nonlinear system of equations

$$2x + 2y = 204,$$
$$xy = 2565.$$

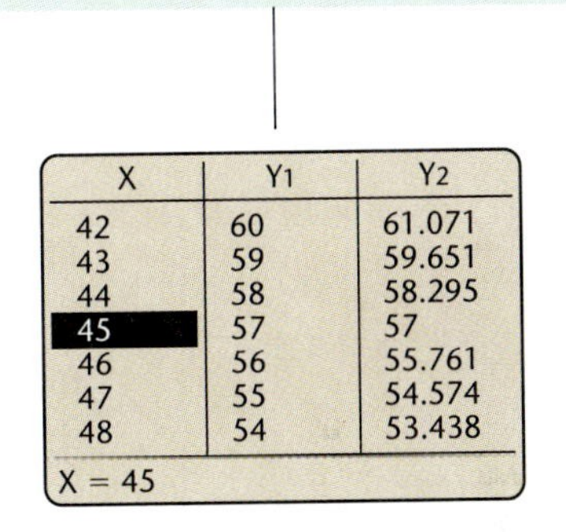

X	Y1	Y2
42	60	61.071
43	59	59.651
44	58	58.295
45	57	57
46	56	55.761
47	55	54.574
48	54	53.438

X = 45

$y_1 = (204 - 2x)/2,\ y_2 = 2565/x$

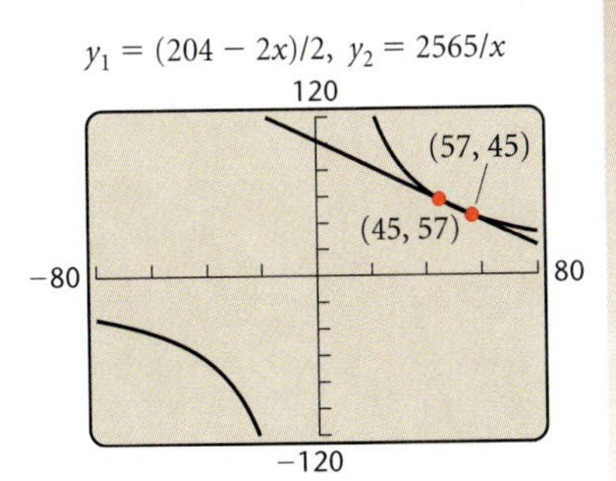

In this chapter, we study *conic sections*. These curves are formed by the intersection of a cone and a plane. Conic sections and their properties were first studied by the Greeks. Today they have many applications, as we will see.

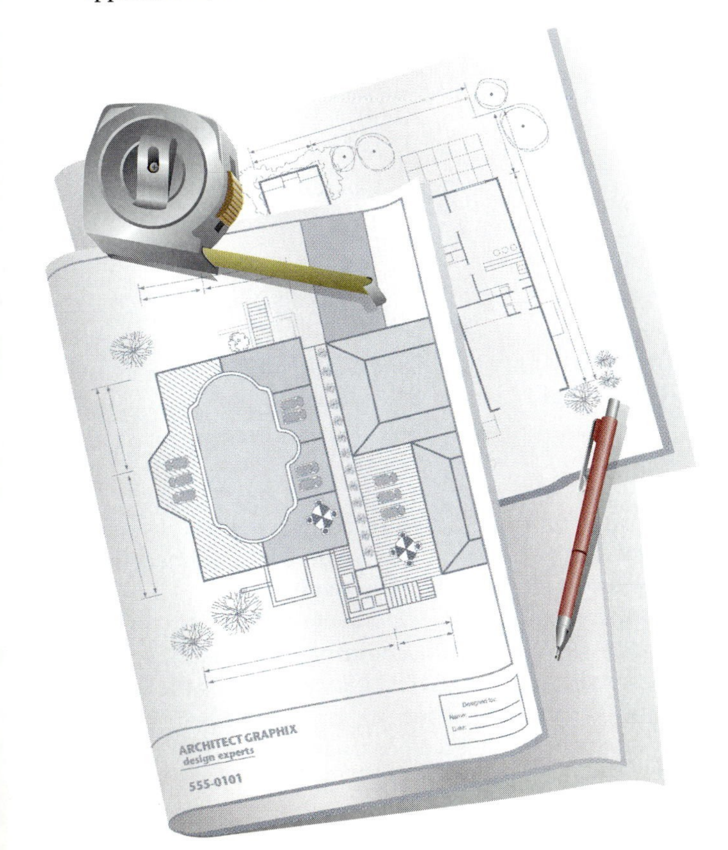

8.1 The Parabola
8.2 The Circle and the Ellipse
8.3 The Hyperbola
8.4 Nonlinear Systems of Equations
SUMMARY AND REVIEW

8.1 The Parabola

- *Given an equation of a parabola, complete the square, if necessary, and then find the vertex, the focus, and the directrix and graph the parabola.*

A **conic section** is formed when a right circular cone with two parts, called *nappes*, is intersected by a plane. One of four types of curves can be formed: a parabola, a circle, an ellipse, or a hyperbola.

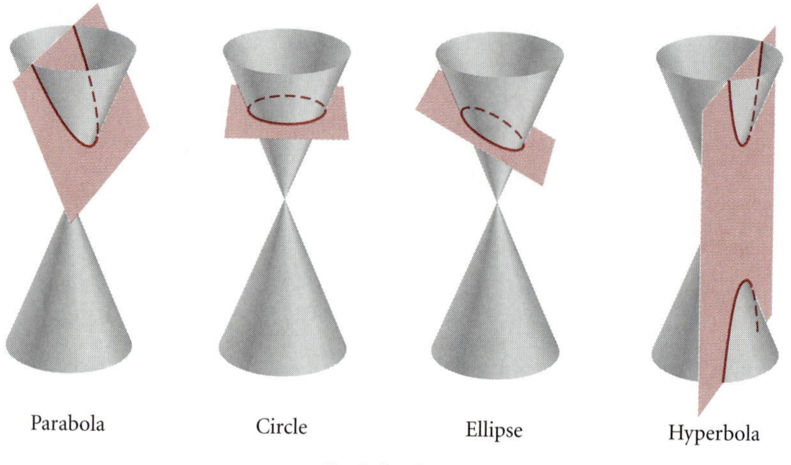

Conic Sections

Parabolas

Conic sections can be defined algebraically using second-degree equations of the form $Ax^2 + Bxy + Cy^2 + Dx + Ey + F = 0$. In addition, they can be defined geometrically as a set of points that satisfy certain conditions.

In Section 2.2, we saw that the graph of the quadratic function $f(x) = ax^2 + bx + c$, $a \neq 0$, is a parabola. A parabola can be defined geometrically.

Parabola

A *parabola* is the set of all points in a plane equidistant from a fixed line (the *directrix*) and a fixed point not on the line (the *focus*).

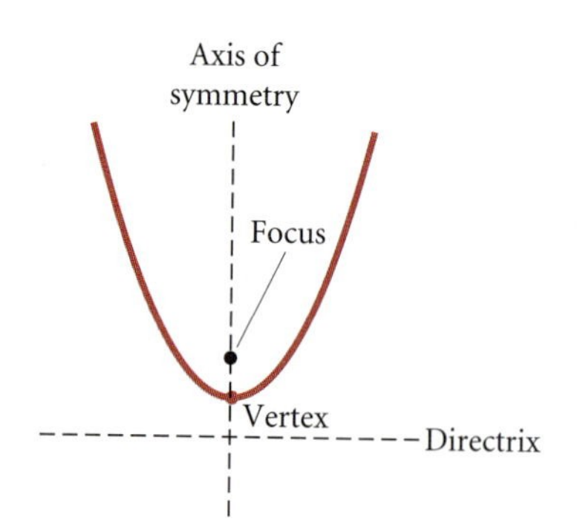

The line that is perpendicular to the directrix and contains the focus is the **axis of symmetry**. The **vertex** is the midpoint of the segment between the focus and the directrix. (See the figure at left.)

Let's derive the standard equation of a parabola with vertex $(0, 0)$ and directrix $y = -p$, $p > 0$. We place the coordinate axes as shown in the figure at the top of the following page. The y-axis contains the focus F. The distance from the focus to the vertex is the same as the distance from the vertex to the directrix. Thus the coordinates of F are $(0, p)$.

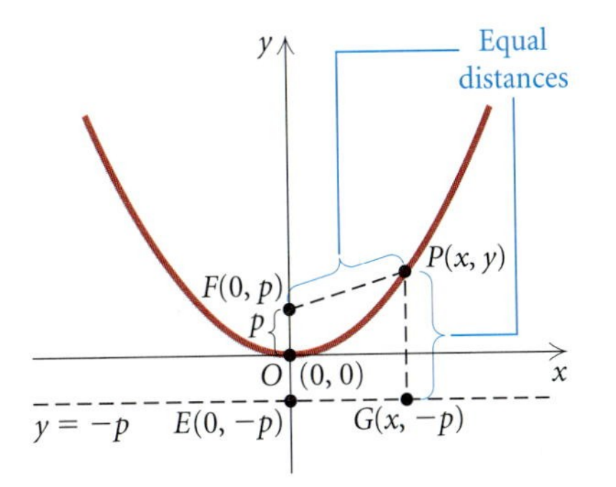

Let $P(x, y)$ be any point of the parabola and consider $\overline{PG}$ perpendicular to the line $y = -p$. The coordinates of G are $(x, -p)$. By the definition of a parabola,

$PF = PG$. **The distance from P to the focus is the same as the distance from P to the directrix.**

Then using the distance formula, we have

$$\sqrt{(x - 0)^2 + (y - p)^2} = \sqrt{(x - x)^2 + (y + p)^2}$$

$$x^2 + y^2 - 2py + p^2 = y^2 + 2py + p^2 \quad \textbf{Squaring both sides}$$

$$x^2 = 4py.$$

We have shown that if $P(x, y)$ is on the parabola shown above, then its coordinates satisfy this equation. The converse is also true, but we will not prove it here.

Note that if $p > 0$, as above, the graph opens up. If $p < 0$, the graph opens down.

The equation of a parabola with vertex $(0, 0)$ and directrix $x = -p$ is derived similarly. Such a parabola opens either right $(p > 0)$ or left $(p < 0)$.

Standard Equation of a Parabola with Vertex at the Origin

The standard equation of a parabola with vertex $(0, 0)$ and directrix $y = -p$ is

$$x^2 = 4py.$$

The focus is $(0, p)$ and the y-axis is the axis of symmetry.

The standard equation of a parabola with vertex $(0, 0)$ and directrix $x = -p$ is

$$y^2 = 4px.$$

The focus is $(p, 0)$ and the x-axis is the axis of symmetry.

Example 1 Find the focus and the directrix of the parabola $y = -\frac{1}{12}x^2$. Then graph the parabola.

SOLUTION We write $y = -\frac{1}{12}x^2$ in the form $x^2 = 4py$:

$$-\frac{1}{12}x^2 = y \qquad \text{Given equation}$$
$$x^2 = -12y \qquad \text{Multiplying both sides by } -12$$
$$x^2 = 4(-3)y. \qquad \text{Standard form}$$

Thus, $p = -3$, so the focus is $(0, p)$, or $(0, -3)$. The directrix is $y = -p = -(-3) = 3$.

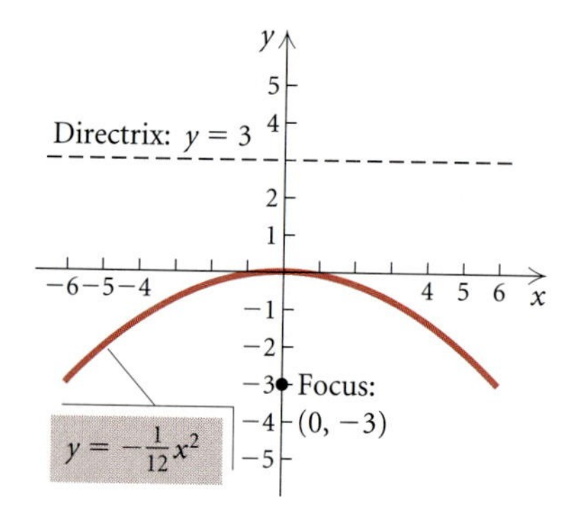

Example 2 Find an equation of the parabola with vertex (0, 0) and focus (5, 0). Then graph the parabola.

SOLUTION The focus is on the x-axis so the line of symmetry is the x-axis. Thus the equation is of the type

$$y^2 = 4px.$$

Since the focus is 5 units to the right of the vertex, $p = 5$ and the equation is

$$y^2 = 4(5)x, \quad \text{or}$$
$$y^2 = 20x.$$

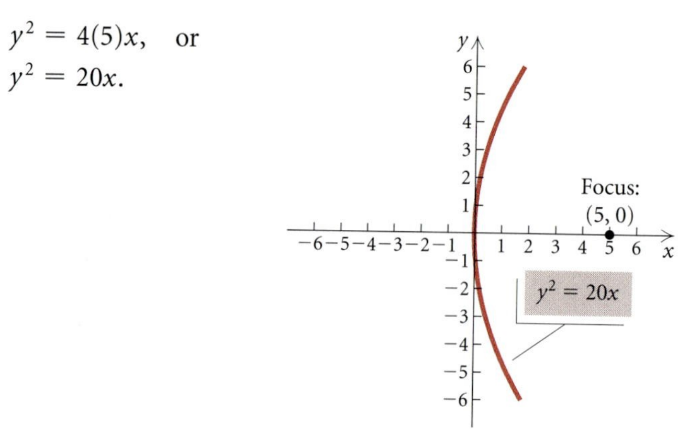

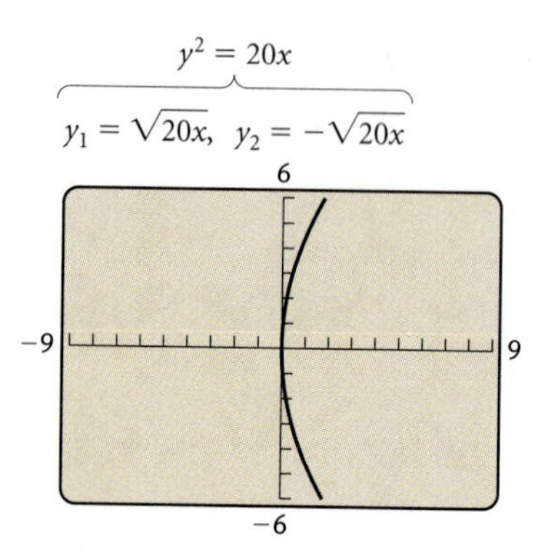

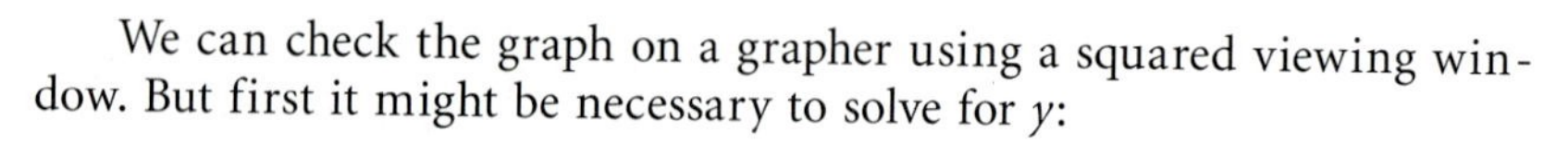

We can check the graph on a grapher using a squared viewing window. But first it might be necessary to solve for y:

$$y^2 = 20x$$
$$y = \pm\sqrt{20x}. \qquad \text{Using the principle of square roots}$$

We now graph $y_1 = \sqrt{20x}$ and $y_2 = -\sqrt{20x}$. On some graphers, it is possible to graph $y_1 = \sqrt{20x}$ and $y_2 = -y_1$ by using the Y-VARS menu.

Finding Standard Form by Completing the Square

If a parabola with vertex at the origin is translated $|h|$ units horizontally and $|k|$ units vertically, it has an equation as follows.

Standard Equation of a Parabola with Vertex (h, k) and Vertical Axis of Symmetry

The standard equation of a parabola with vertex (h, k) and vertical axis of symmetry is

$$(x - h)^2 = 4p(y - k),$$

where the vertex is (h, k), the focus is $(h, k + p)$, and the directrix is $y = k - p$.

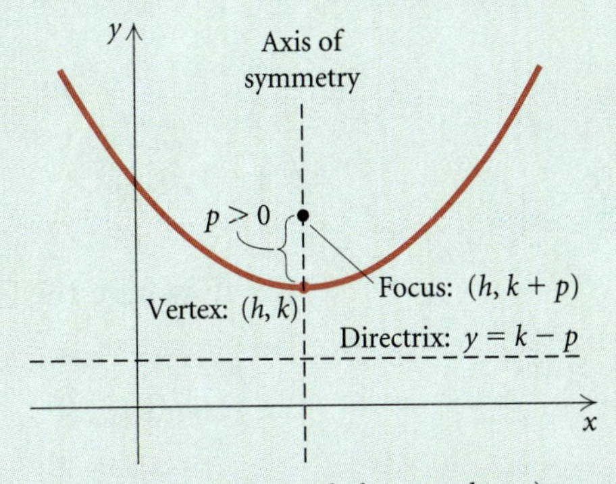

(When $p < 0$, the parabola opens down.)

Standard Equation of a Parabola with Vertex (h, k) and Horizontal Axis of Symmetry

The standard equation of a parabola with vertex (h, k) and horizontal axis of symmetry is

$$(y - k)^2 = 4p(x - h),$$

where the vertex is (h, k), the focus is $(h + p, k)$, and the directrix is $x = h - p$.

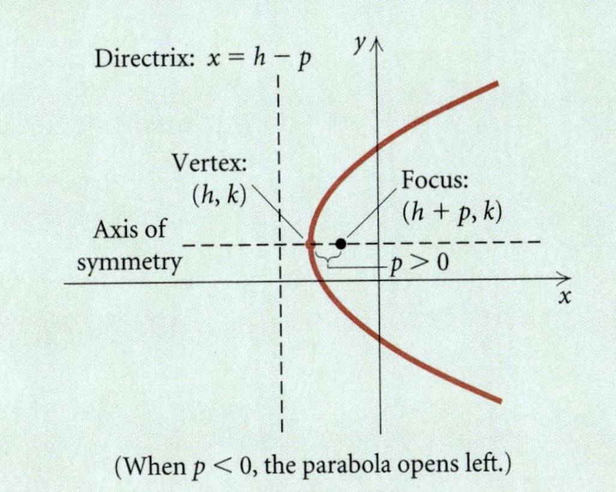

(When $p < 0$, the parabola opens left.)

We can complete the square on equations of the form

$$y = ax^2 + bx + c \quad \text{or} \quad x = ay^2 + by + c$$

in order to write them in standard form.

Example 3 For the parabola

$$x^2 + 6x + 4y + 5 = 0,$$

find the vertex, the focus, and the directrix. Then draw the graph.

SOLUTION We first complete the square:

$$x^2 + 6x + 4y + 5 = 0$$

$$x^2 + 6x = -4y - 5 \qquad \text{Subtracting } 4y \text{ and 5 on both sides}$$

$$x^2 + 6x + 9 = -4y - 5 + 9 \qquad \text{Adding 9 on both sides to complete the square on the left side}$$

$$x^2 + 6x + 9 = -4y + 4$$

$$(x + 3)^2 = -4(y - 1) \qquad \text{Factoring}$$

$$[(x - (-3)]^2 = 4(-1)(y - 1). \qquad \text{Writing standard form: } (x - h)^2 = 4p(y - k)$$

We now have the following:

Vertex (h, k): $(-3, 1)$;
Focus $(h, k + p)$: $(-3, 1 + (-1))$, or $(-3, 0)$;
Directrix $y = k - p$: $y = 1 - (-1)$, or $y = 2$.

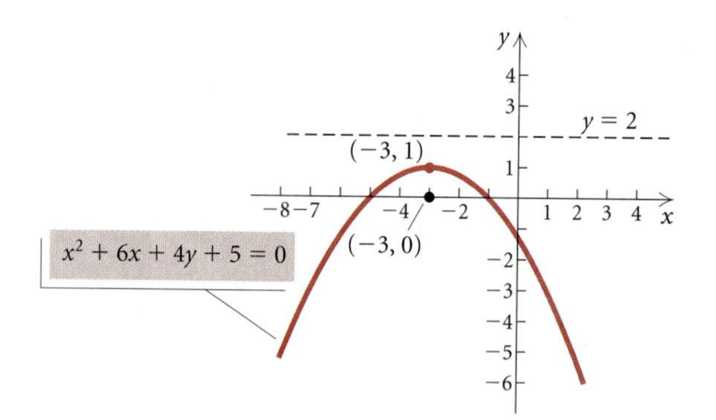

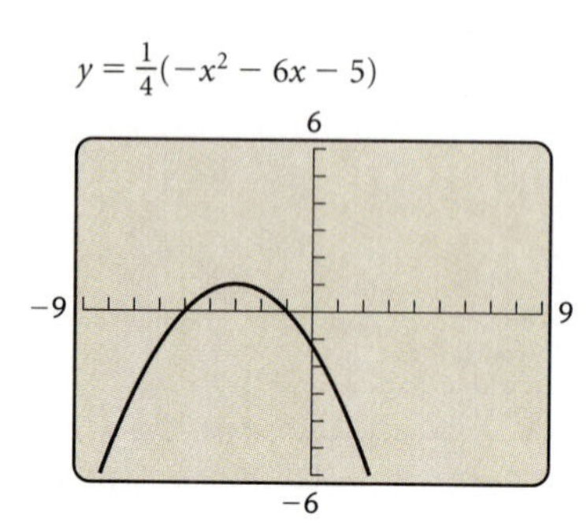

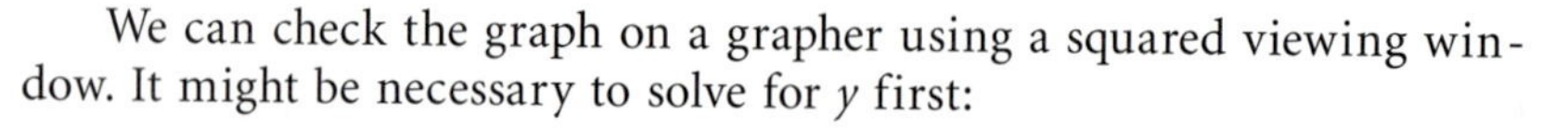

We can check the graph on a grapher using a squared viewing window. It might be necessary to solve for y first:

$$x^2 + 6x + 4y + 5 = 0$$

$$4y = -x^2 - 6x - 5$$

$$y = \tfrac{1}{4}(-x^2 - 6x - 5).$$

The hand-drawn graph appears to be correct.

Example 4 For the parabola

$$y^2 - 2y - 8x - 31 = 0,$$

find the vertex, the focus, and the directrix. Then draw the graph.

SOLUTION We first complete the square:

$$y^2 - 2y - 8x - 31 = 0$$

$$y^2 - 2y = 8x + 31 \qquad \text{Adding } 8x \text{ and } 31 \text{ on both sides}$$

$$y^2 - 2y + 1 = 8x + 31 + 1 \qquad \text{Adding 1 on both sides to complete the square on the left side}$$

$$y^2 - 2y + 1 = 8x + 32$$

$$(y - 1)^2 = 8(x + 4)$$

$$(y - 1)^2 = 4(2)[x - (-4)]. \qquad \text{Writing standard form: } (y - k)^2 = 4p(x - h)$$

We now have the following:

Vertex (h, k): $(-4, 1)$;
Focus $(h + p, k)$: $(-4 + 2, 1)$, or $(-2, 1)$;
Directrix $x = h - p$: $x = -4 - 2$, or $x = -6$.

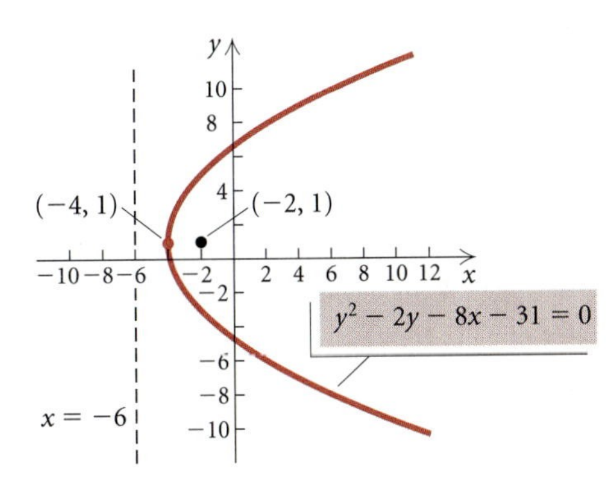

We can check the graph on a grapher using a squared window. We solve the original equation for y using the quadratic formula:

$$y^2 - 2y - 8x - 31 = 0$$

$$y^2 - 2y + (-8x - 31) = 0$$

$$a = 1, \quad b = -2, \quad c = -8x - 31$$

$$y = \frac{-(-2) \pm \sqrt{(-2)^2 - 4 \cdot 1(-8x - 31)}}{2 \cdot 1}$$

$$y = \frac{2 \pm \sqrt{32x + 128}}{2}.$$

We now graph

$$y_1 = \frac{2 + \sqrt{32x + 128}}{2} \quad \text{and} \quad y_2 = \frac{2 - \sqrt{32x + 128}}{2}.$$

The hand-drawn graph appears to be correct.

$$y^2 - 2y - 8x - 31 = 0$$

$$y_1 = \frac{2 + \sqrt{32x + 128}}{2}, \quad y_2 = \frac{2 - \sqrt{32x + 128}}{2}$$

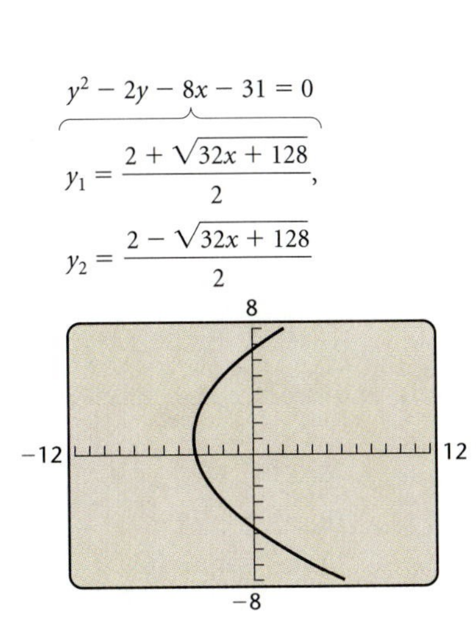

Applications

Parabolas have many applications. For example, cross sections of car headlights, flashlights, and searchlights are parabolas. The bulb is located at the focus and light from that point is reflected outward parallel to the axis of symmetry. Satellite dishes and field microphones used at sporting events often have parabolic cross sections. Incoming radio waves or sound waves, parallel to the axis, are reflected into the focus. Cables hung between structures in suspension bridges, such as the Golden Gate Bridge, form parabolas. When a cable supports only its own weight, however, it forms a curve called a *catenary* rather than a parabola.

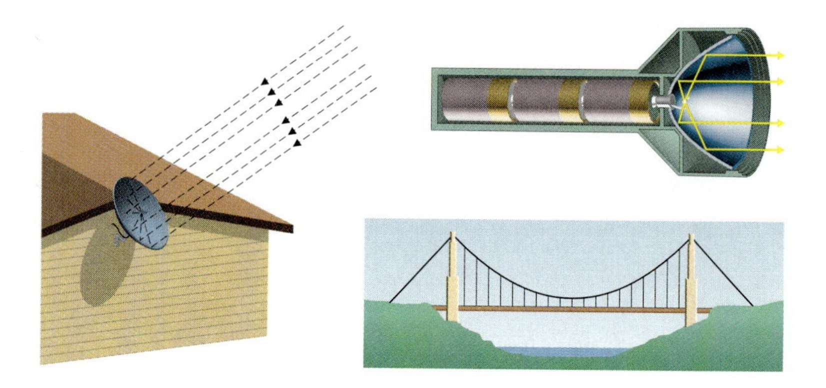

8.1 Exercise Set

In Exercises 1–6, match each equation with one of the graphs (a)–(f), which follow.

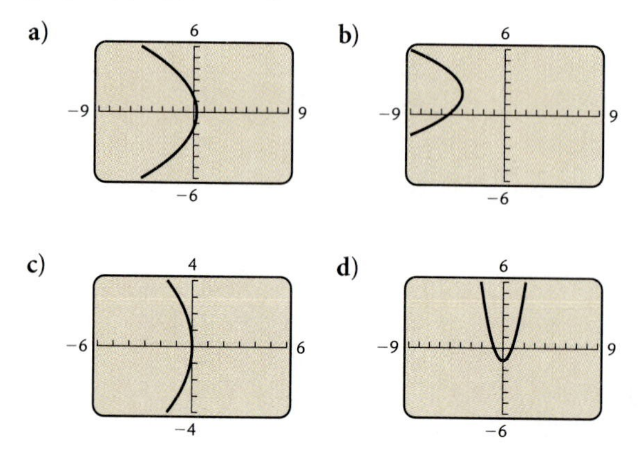

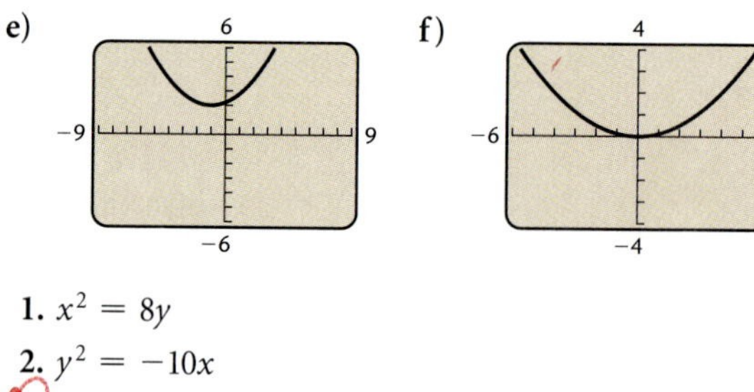

1. $x^2 = 8y$
2. $y^2 = -10x$
3. $(y - 2)^2 = -3(x + 4)$
4. $(x + 1)^2 = 5(y - 2)$
5. $13x^2 - 8y - 9 = 0$
6. $41x + 6y^2 = 12$

Find the vertex, the focus, and the directrix. Then draw the graph.

7. $x^2 = 20y$
8. $x^2 = 16y$
9. $y^2 = -6x$
10. $y^2 = -2x$
11. $x^2 - 4y = 0$
12. $y^2 + 4x = 0$
13. $x = 2y^2$
14. $y = \frac{1}{2}x^2$

Find an equation of a parabola satisfying the given conditions.

15. Focus $(4, 0)$, directrix $x = -4$
16. Focus $\left(0, \frac{1}{4}\right)$, directrix $y = -\frac{1}{4}$
17. Focus $(0, -\pi)$, directrix $y = \pi$
18. Focus $(-\sqrt{2}, 0)$, directrix $x = \sqrt{2}$
19. Focus $(3, 2)$, directrix $x = -4$
20. Focus $(-2, 3)$, directrix $y = -3$

Find the vertex, the focus, and the directrix. Then draw the graph.

21. $(x + 2)^2 = -6(y - 1)$
22. $(y - 3)^2 = -20(x + 2)$
23. $x^2 + 2x + 2y + 7 = 0$
24. $y^2 + 6y - x + 16 = 0$
25. $x^2 - y - 2 = 0$
26. $x^2 - 4x - 2y = 0$
27. $y = x^2 + 4x + 3$
28. $y = x^2 + 6x + 10$
29. $y^2 - y - x + 6 = 0$
30. $y^2 + y - x - 4 = 0$

31. *Satellite Dish.* An engineer designs a satellite dish with a parabolic cross section. The dish is 15 ft wide at the opening and the focus is placed 4 ft from the vertex.
    a) Position a coordinate system with the origin at the vertex and the x-axis on the parabola's axis of symmetry and find an equation of the parabola.
    b) Find the depth of the satellite dish at the vertex.

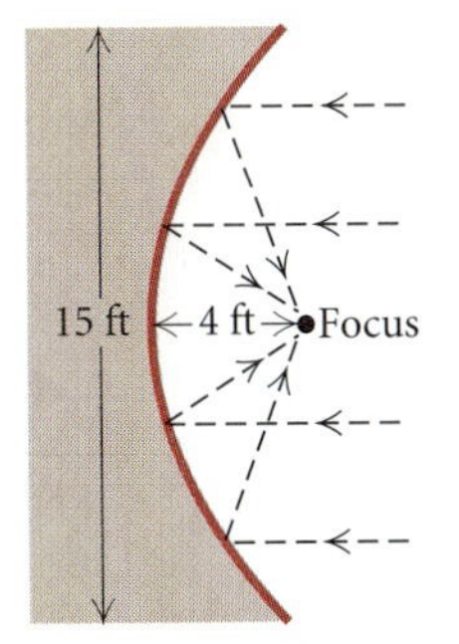

32. *Headlight Mirror.* A car headlight mirror has a parabolic cross section with diameter 6 in. and depth 1 in.
    a) Position a coordinate system with the origin at the vertex and the x-axis on the parabola's axis of symmetry and find an equation of the parabola.
    b) How far from the vertex should the bulb be positioned if it is to be placed at the focus?

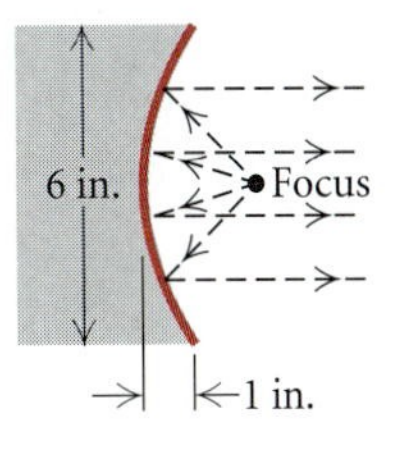

33. *Spotlight.* A spotlight has a parabolic cross section that is 4 ft wide at the opening and 1.5 ft deep at the vertex. How far from the vertex is the focus?

34. *Field Microphone.* A field microphone used at a football game has a parabolic cross section and is 18 in. deep. The focus is 4 in. from the vertex. Find the width of the microphone at the opening.

Skill Maintenance

Complete the square and write the result as the square of a binomial.

35. $x^2 + 10x$
36. $y^2 - 9y$

Find the center and the radius of each circle.

37. $(x - 1)^2 + (y + 2)^2 = 9$
38. $(x + 3)^2 + (y - 5)^2 = 36$

Synthesis

39. ◆ Is a parabola always the graph of a function? Why or why not?

40. ◆ Explain how the distance formula is used to find the standard equation of a parabola.

41. Find an equation of the parabola with a vertical axis of symmetry and vertex $(-1, 2)$ and containing the point $(-3, 1)$.

42. Find an equation of a parabola with a horizontal axis of symmetry and vertex $(-2, 1)$ and containing the point $(-3, 5)$.

Use a grapher to find the vertex, the focus, and the directrix of each of the following.

43. $4.5x^2 - 7.8x + 9.7y = 0$
44. $134.1y^2 + 43.4x - 316.6y - 122.4 = 0$

45. *Suspension Bridge.* The cables of a suspension bridge are 50 ft above the roadbed at the ends of the bridge and 10 ft above it in the center of the bridge. The roadbed is 200 ft long. Vertical cables are to be spaced every 20 ft along the bridge. Calculate the lengths of these vertical cables.

8.2 The Circle and the Ellipse

- *Given an equation of a circle, complete the square, if necessary, and then find the center and the radius and graph the circle.*
- *Given an equation of an ellipse, complete the square, if necessary, and then find the center, the vertices, and the foci and graph the ellipse.*

Circles

We can define a circle geometrically.

Circle

A *circle* is the set of all points in a plane that are at a fixed distance from a fixed point (the *center*) in the plane.

Circles were introduced in Section 1.5. Recall the standard equation of a circle with center (h, k) and radius r.

Standard Equation of a Circle

The standard equation of a circle with center (h, k) and radius r is

$$(x - h)^2 + (y - k)^2 = r^2.$$

Example 1 For the circle

$$x^2 + y^2 - 16x + 14y + 32 = 0,$$

find the center and the radius. Then graph the circle.

SOLUTION First we complete the square twice:

$$x^2 + y^2 - 16x + 14y + 32 = 0$$
$$x^2 - 16x \quad + y^2 + 14y \quad = -32$$
$$x^2 - 16x + 64 + y^2 + 14y + 49 = -32 + 64 + 49$$

Adding 64 and 49 on both sides to complete the square twice on the left side

$$(x - 8)^2 + (y + 7)^2 = 81$$
$$(x - 8)^2 + [y - (-7)]^2 = 9^2. \quad \textbf{Writing standard form}$$

The center is $(8, -7)$ and the radius is 9.

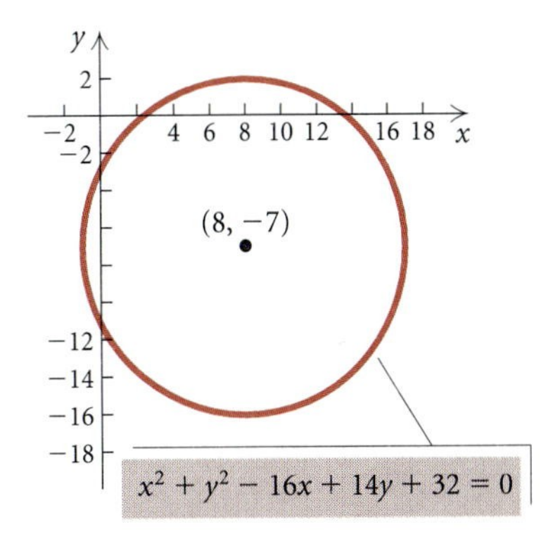

To use a grapher to graph the circle, it might be necessary to solve for y first. The original equation can be solved using the quadratic formula, or the standard form of the equation can be solved using the principle of square roots. The second alternative is illustrated here:

$$(x - 8)^2 + (y + 7)^2 = 81$$
$$(y + 7)^2 = 81 - (x - 8)^2$$
$$y + 7 = \pm\sqrt{81 - (x - 8)^2} \quad \textbf{Using the principle of square roots}$$
$$y = -7 \pm \sqrt{81 - (x - 8)^2}.$$

$(x - 8)^2 + (y + 7)^2 = 81$

$y_1 = -7 + \sqrt{81 - (x - 8)^2}$,

$y_2 = -7 - \sqrt{81 - (x - 8)^2}$

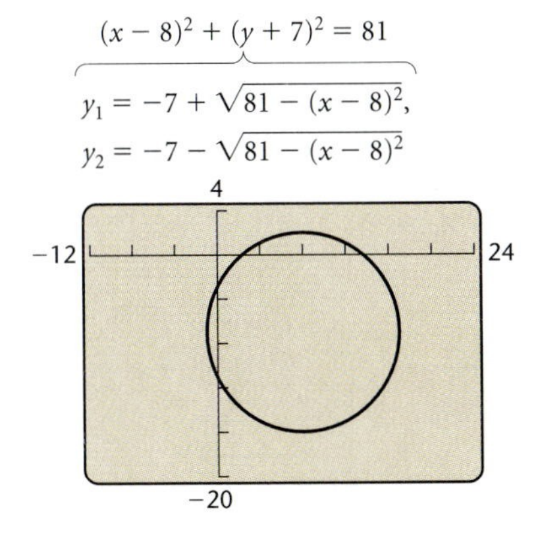

Then we graph

$$y_1 = -7 + \sqrt{81 - (x - 8)^2}$$

and

$$y_2 = -7 - \sqrt{81 - (x - 8)^2}$$

in a squared viewing window.

The hand-drawn graph appears to be correct.

Some graphers have a DRAW feature that provides a quick way to graph a circle when the center and the radius are known.

Ellipses

We have studied two conic sections, the parabola and the circle. Now we turn our attention to a third, the *ellipse*.

Ellipse

An *ellipse* is the set of all points in a plane, the sum of whose distances from two fixed points (the *foci*) is constant. The *center* of an ellipse is the midpoint of the segment between the foci.

We can draw an ellipse by first placing two thumbtacks in a piece of cardboard. These are the foci (singular, *focus*). We then attach a piece of string to the tacks. Its length is the constant sum of the distances $d_1 + d_2$ from the foci to any point on the ellipse. Next, we trace a curve with a pencil held tight against the string. The figure traced is an ellipse.

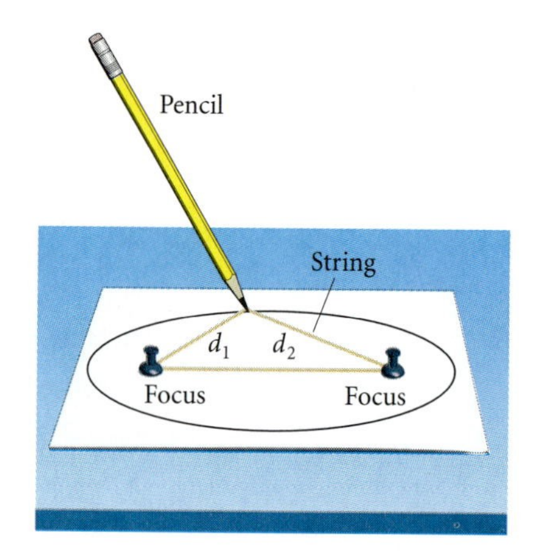

Let's first consider the ellipse shown below with center at the origin. The points F_1 and F_2 are the foci. The segment $\overline{A'A}$ is the **major axis**, and the points A' and A are the **vertices**. The segment $\overline{B'B}$ is the **minor axis**, and the points B and B' are the ***y*-intercepts**. Note that the major axis of an ellipse is always longer than the minor axis.

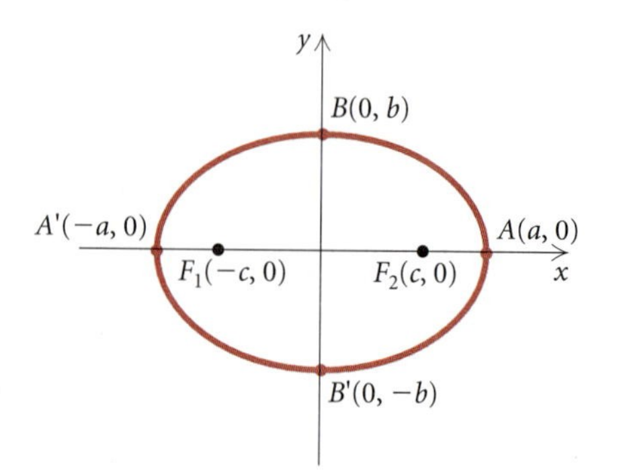

Standard Equation of an Ellipse with Center at the Origin

Major Axis Horizontal

$$\frac{x^2}{a^2} + \frac{y^2}{b^2} = 1, \; a > b > 0$$

Vertices: $(-a, 0)$, $(a, 0)$

y-intercepts: $(0, -b)$, $(0, b)$

Foci: $(-c, 0)$, $(c, 0)$, where $c^2 = a^2 - b^2$

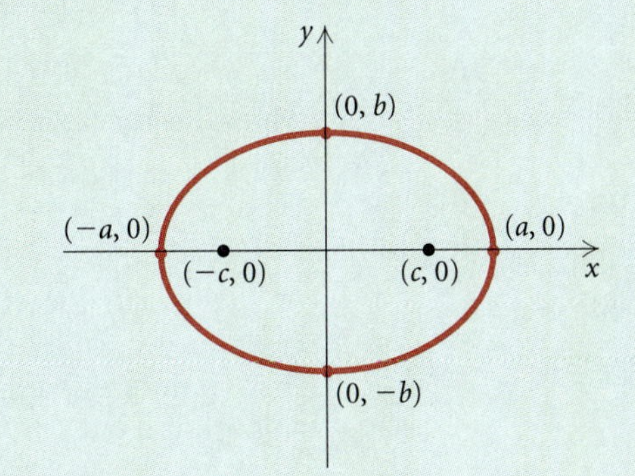

Major Axis Vertical

$$\frac{x^2}{b^2} + \frac{y^2}{a^2} = 1, \; a > b > 0$$

Vertices: $(0, -a)$, $(0, a)$

x-intercepts: $(-b, 0)$, $(b, 0)$

Foci: $(0, -c)$, $(0, c)$, where $c^2 = a^2 - b^2$

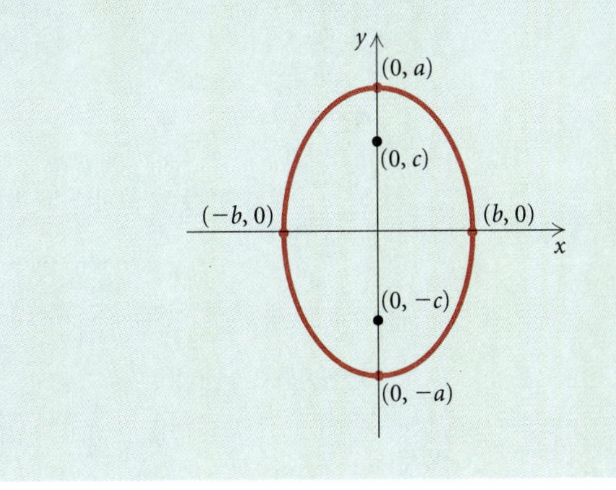

Example 2 Find the standard equation of the ellipse with vertices $(-5, 0)$ and $(5, 0)$ and foci $(-3, 0)$ and $(3, 0)$. Then graph the ellipse.

SOLUTION Since the foci are on the x-axis and the origin is the midpoint of the segment between them, the major axis is horizontal and $(0, 0)$ is the center of the ellipse. Thus the equation is of the form

$$\frac{x^2}{a^2} + \frac{y^2}{b^2} = 1.$$

Since the vertices are $(-5, 0)$ and $(5, 0)$ and the foci are $(-3, 0)$ and $(3, 0)$, we know that $a = 5$ and $c = 3$. These values can be used to find b^2:

$$\begin{aligned} c^2 &= a^2 - b^2 \\ 3^2 &= 5^2 - b^2 \\ 9 &= 25 - b^2 \\ b^2 &= 16. \end{aligned}$$

Thus the equation of the ellipse is

$$\frac{x^2}{25} + \frac{y^2}{16} = 1.$$

To graph the ellipse, we plot the vertices $(-5, 0)$ and $(5, 0)$. Since $b^2 = 16$, we know that the y-intercepts are $(0, -4)$ and $(0, 4)$. We plot these points as well and connect the four points we have plotted with a smooth curve.

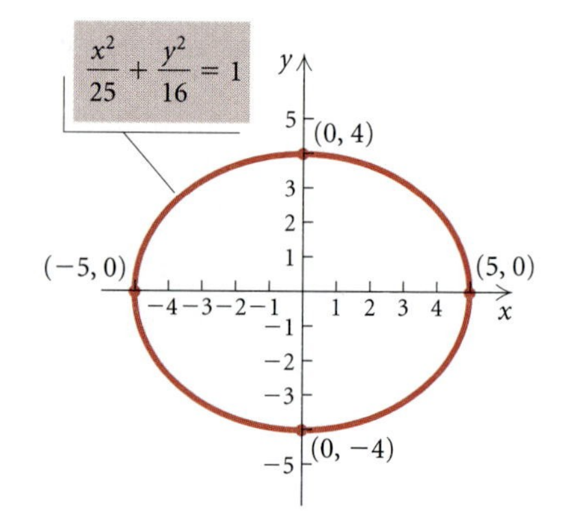

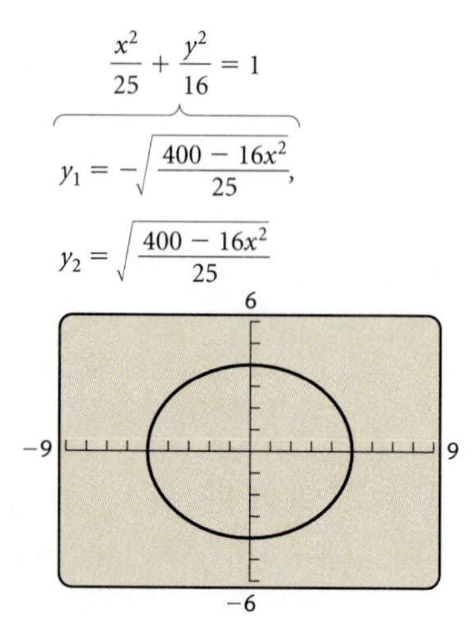

To draw the graph using a grapher, it might be necessary to solve for y first:

$$y = \pm\sqrt{\frac{400 - 16x^2}{25}}.$$

Then we graph

$$y_1 = -\sqrt{\frac{400 - 16x^2}{25}} \quad \text{and} \quad y_2 = \sqrt{\frac{400 - 16x^2}{25}}$$

or

$$y_1 = -\sqrt{\frac{400 - 16x^2}{25}} \quad \text{and} \quad y_2 = -y_1$$

in a squared viewing window.

Example 3 For the ellipse

$$9x^2 + 4y^2 = 36,$$

find the vertices and the foci. Then draw the graph.

SOLUTION We first find standard form:

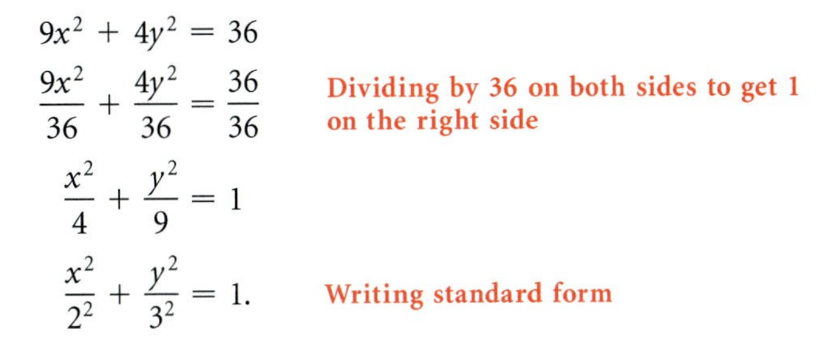

$$9x^2 + 4y^2 = 36$$

$$\frac{9x^2}{36} + \frac{4y^2}{36} = \frac{36}{36} \quad \text{Dividing by 36 on both sides to get 1 on the right side}$$

$$\frac{x^2}{4} + \frac{y^2}{9} = 1$$

$$\frac{x^2}{2^2} + \frac{y^2}{3^2} = 1. \quad \text{Writing standard form}$$

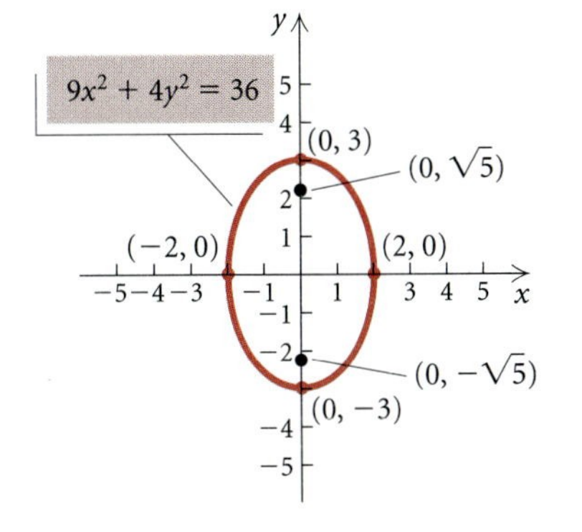

Thus, $a = 3$ and $b = 2$. The major axis is vertical, so the vertices are $(0, -3)$ and $(0, 3)$. Since we know that $c^2 = a^2 - b^2$, we have $c^2 = 9 - 4$, so $c = \sqrt{5}$ and the foci are $(0, -\sqrt{5})$ and $(0, \sqrt{5})$.

To graph the ellipse, we plot the vertices. Note also that since $b = 2$, the x-intercepts are $(-2, 0)$ and $(2, 0)$. We plot these points as well and connect the four points we have plotted with a smooth curve.

If the center of an ellipse is not at the origin but at some point (h, k), then we can think of an ellipse with center at the origin being translated $|h|$ units left or right and $|k|$ units up or down.

Standard Equation of an Ellipse with Center at (h, k)

Major Axis Horizontal

$$\frac{(x - h)^2}{a^2} + \frac{(y - k)^2}{b^2} = 1, \quad a > b > 0$$

Vertices: $(h - a, k)$, $(h + a, k)$

Length of minor axis: $2b$

Foci: $(h - c, k)$, $(h + c, k)$, where $c^2 = a^2 - b^2$

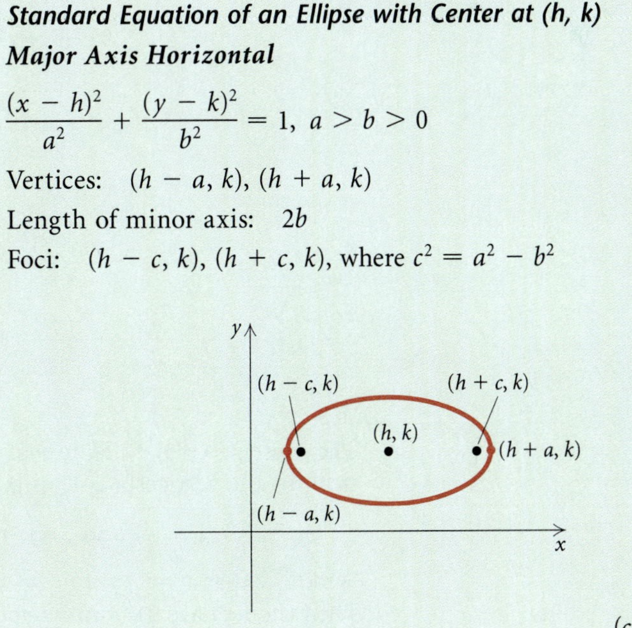

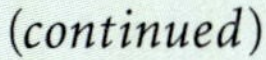

(continued)

Major Axis Vertical

$$\frac{(x-h)^2}{b^2} + \frac{(y-k)^2}{a^2} = 1,\ a > b > 0$$

Vertices: $(h, k - a), (h, k + a)$

Length of minor axis: $2b$

Foci: $(h, k - c), (h, k + c)$, where $c^2 = a^2 - b^2$

Example 4 For the ellipse

$$4x^2 + y^2 + 24x - 2y + 21 = 0,$$

find the center, the vertices, and the foci. Then draw the graph.

SOLUTION First we complete the square to get standard form:

$$4x^2 + y^2 + 24x - 2y + 21 = 0$$

$$4(x^2 + 6x \qquad) + (y^2 - 2y \qquad) = -21$$

$$4(x^2 + 6x + 9) + (y^2 - 2y + 1) = -21 + 4 \cdot 9 + 1$$

Completing the square twice by adding $4 \cdot 9$ and 1 on both sides

$$4(x + 3)^2 + (y - 1)^2 = 16$$

$$\frac{1}{16}[4(x + 3)^2 + (y - 1)^2] = \frac{1}{16} \cdot 16$$

$$\frac{(x + 3)^2}{4} + \frac{(y - 1)^2}{16} = 1$$

$$\frac{[x - (-3)]^2}{2^2} + \frac{(y - 1)^2}{4^2} = 1.$$

Writing standard form: $\frac{(x-h)^2}{b^2} + \frac{(y-k)^2}{a^2} = 1$

The center is $(-3, 1)$. Note that $a = 4$ and $b = 2$. The major axis is vertical, so the vertices are 4 units above and below the center:

$$(-3, 1 + 4) \text{ and } (-3, 1 - 4), \quad \text{or} \quad (-3, 5) \text{ and } (-3, -3).$$

We know that $c^2 = a^2 - b^2$, so $c^2 = 16 - 4 = 12$ and $c = \sqrt{12} = 2\sqrt{3}$. Then the foci are $2\sqrt{3}$ units above and below the center:

$$(-3, 1 + 2\sqrt{3}) \quad \text{and} \quad (-3, 1 - 2\sqrt{3}).$$

To graph the ellipse, we plot the vertices. Note also that since $b = 2$, two other points on the graph are the endpoints of the minor axis, 2 units right and left of the center:

$$(-3 + 2, 1) \text{ and } (-3 - 2, 1), \quad \text{or} \quad (-1, 1) \text{ and } (-5, 1).$$

We plot these points as well and connect the four points with a smooth curve.

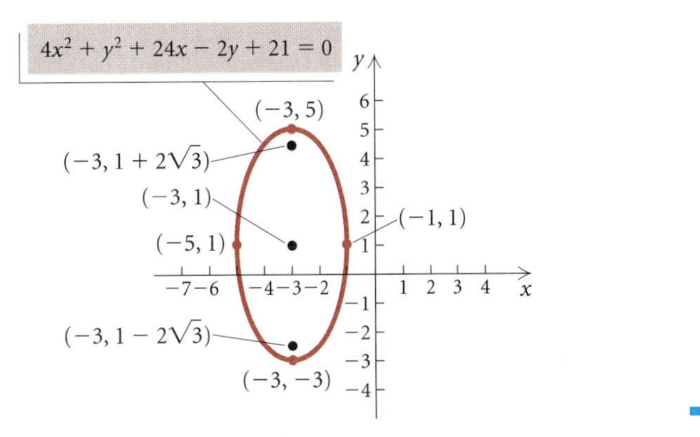

Applications

One of the most exciting recent applications of an ellipse is a medical device called a *lithotripter*. This machine uses underwater shock waves to crush kidney stones. The waves originate at one focus of an ellipse and are reflected to the kidney stone, which is positioned at the other focus. Recovery time following the use of this technique is much shorter than with conventional surgery and the mortality rate is far lower.

A room with an ellipsoidal ceiling is known as a *whispering gallery*. In such a room, a word whispered at one focus can be clearly heard at the other. Whispering galleries are found in the rotunda of the Capitol Building in Washington, D.C., and in the Mormon Tabernacle in Salt Lake City.

Ellipses have many other applications. Planets travel around the sun in elliptical orbits with the sun at one focus, for example, and satellites travel around the earth in elliptical orbits as well.

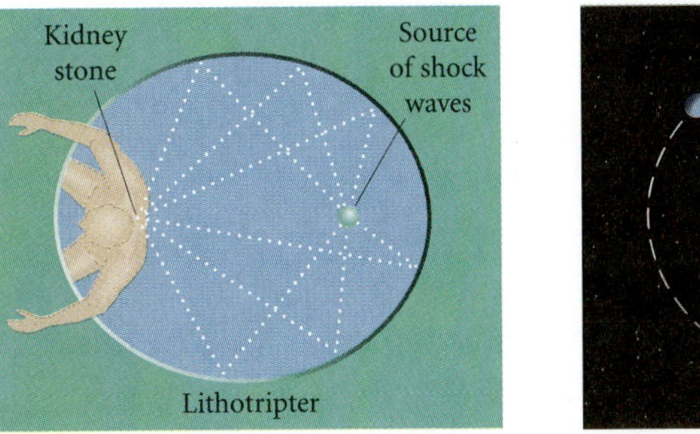

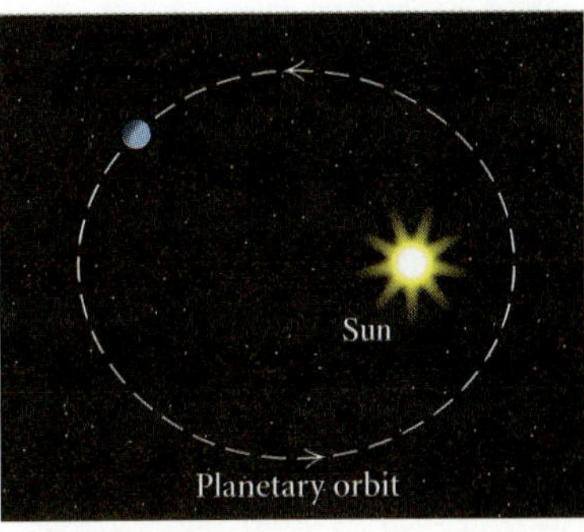

8.2 Exercise Set

In Exercises 1–6, match each equation with one of the graphs (a)–(f), which follow.

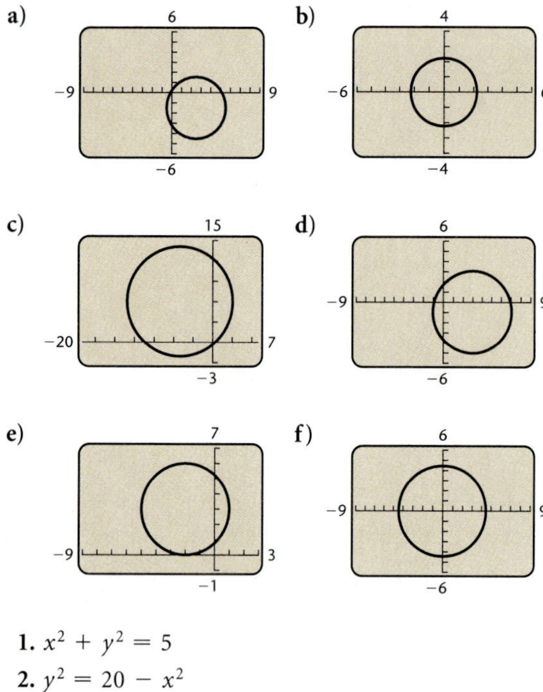

1. $x^2 + y^2 = 5$
2. $y^2 = 20 - x^2$
3. $x^2 + y^2 - 6x + 2y = 6$
4. $x^2 + y^2 + 10x - 12y = 3$
5. $x^2 + y^2 - 5x + 3y = 0$
6. $x^2 + 4x - 2 = 6y - y^2 - 6$

Find the center and the radius of the circle with the given equation. Then draw the graph.

7. $x^2 + y^2 - 14x + 4y = 11$
8. $x^2 + y^2 + 2x - 6y = -6$
9. $x^2 + y^2 + 4x - 6y - 12 = 0$
10. $x^2 + y^2 - 8x - 2y - 19 = 0$
11. $x^2 + y^2 + 6x - 10y = 0$
12. $x^2 + y^2 - 7x - 2y = 0$
13. $x^2 + y^2 - 9x = 7 - 4y$
14. $y^2 - 6y - 1 = 8x - x^2 + 3$

In Exercises 15–18, match each equation with one of the graphs (a)–(d), which follow.

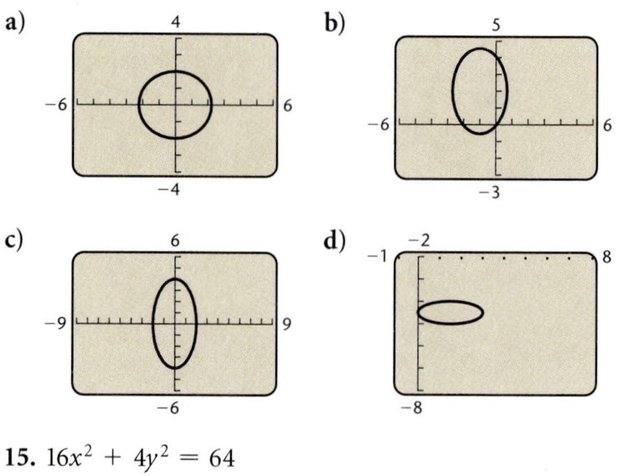

15. $16x^2 + 4y^2 = 64$
16. $4x^2 + 5y^2 = 20$
17. $x^2 + 9y^2 - 6x + 90y = -225$
18. $9x^2 + 4y^2 + 18x - 16y = 11$

Find the vertices and the foci of the ellipse with the given equation. Then draw the graph.

19. $\frac{x^2}{4} + \frac{y^2}{1} = 1$
20. $\frac{x^2}{25} + \frac{y^2}{36} = 1$
21. $16x^2 + 9y^2 = 144$
22. $9x^2 + 4y^2 = 36$
23. $2x^2 + 3y^2 = 6$
24. $5x^2 + 7y^2 = 35$
25. $4x^2 + 9y^2 = 1$
26. $25x^2 + 16y^2 = 1$

Find an equation of an ellipse satisfying the given conditions.

27. Vertices: $(-7, 0)$ and $(7, 0)$; foci: $(-3, 0)$ and $(3, 0)$
28. Vertices: $(0, -6)$ and $(0, 6)$; foci: $(0, -4)$ and $(0, 4)$
29. Vertices: $(0, -8)$ and $(0, 8)$; length of minor axis: 10
30. Vertices: $(-5, 0)$ and $(5, 0)$; length of minor axis: 6
31. Foci: $(-2, 0)$ and $(2, 0)$; length of major axis: 6
32. Foci: $(0, -3)$ and $(0, 3)$; length of major axis: 10

Find the center, the vertices, and the foci of each ellipse. Then draw the graph.

33. $\frac{(x-1)^2}{9} + \frac{(y-2)^2}{4} = 1$
34. $\frac{(x-1)^2}{1} + \frac{(y-2)^2}{4} = 1$
35. $\frac{(x+3)^2}{25} + \frac{(y-5)^2}{36} = 1$

36. $\dfrac{(x-2)^2}{16} + \dfrac{(y+3)^2}{25} = 1$

37. $3(x+2)^2 + 4(y-1)^2 = 192$

38. $4(x-5)^2 + 3(y-4)^2 = 48$

39. $4x^2 + 9y^2 - 16x + 18y - 11 = 0$

40. $x^2 + 2y^2 - 10x + 8y + 29 = 0$

41. $4x^2 + y^2 - 8x - 2y + 1 = 0$

42. $9x^2 + 4y^2 + 54x - 8y + 49 = 0$

*The **eccentricity** of an ellipse is defined as $e = c/a$. For an ellipse, $0 < c < a$, so $0 < e < 1$. When e is close to 0, an ellipse appears to be nearly circular. When e is close to 1, an ellipse is very flat.*

43. Observe the shapes of the ellipses in Examples 2 and 4. Which ellipse has the smaller eccentricity? Confirm your answer by computing the eccentricity of each ellipse.

44. Which ellipse below has the smaller eccentricity? (Assume that the coordinate systems have the same scale.)

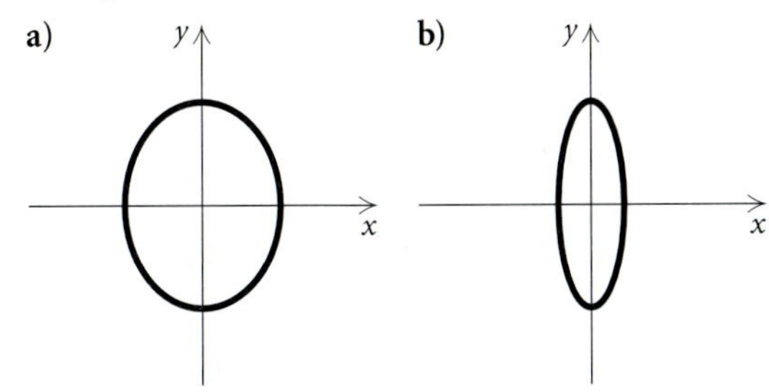

45. Find an equation of an ellipse with vertices $(0, -4)$ and $(0, 4)$ and $e = \frac{1}{4}$.

46. Find an equation of an ellipse with vertices $(-3, 0)$ and $(3, 0)$ and $e = \frac{7}{10}$.

47. *Bridge Supports.* The bridge support shown in the figure below is the top half of an ellipse. Assuming that a coordinate system is superimposed on the drawing in such a way that the center of the ellipse is at point Q, find an equation of the ellipse.

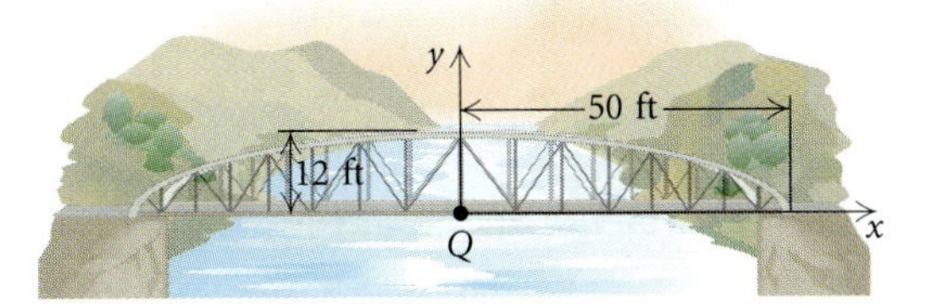

48. *The Ellipse.* In Washington, D.C., there is a large grassy area south of the White House known as the Ellipse. It is actually an ellipse with major axis of length 1048 ft and minor axis of length 898 ft. Assuming that a coordinate system is superimposed on the area in such a way that the center is at the origin and the major and minor axes are on the x- and y-axes of the coordinate system, respectively, find an equation of the ellipse.

49. *The Earth's Orbit.* The maximum distance of the earth from the sun is 9.3×10^7 miles. The minimum distance is 9.1×10^7 miles. The sun is at one focus of the elliptical orbit. Find the distance from the sun to the other focus.

50. *Carpentry.* A carpenter is cutting a 3-ft by 4-ft elliptical sign from a 3-ft by 4-ft piece of plywood. The ellipse will be drawn using a string attached to the board at the foci of the ellipse.

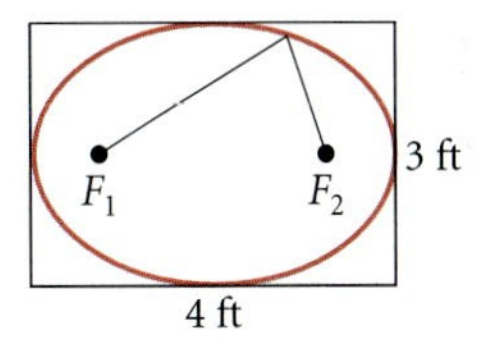

a) How far from the ends of the board should the string be attached?
b) How long should the string be?

Skill Maintenance

Graph.

51. $y = \frac{5}{2}x$

52. $y = -\frac{3}{4}x$

53. $y - 1 = \frac{1}{3}(x + 3)$

54. $y + 2 = -\frac{5}{4}(x - 5)$

Synthesis

55. ◆ Explain why function notation is not used in this section.

56. ◆ Would you prefer to graph the ellipse
$$\frac{(x-3)^2}{4} + \frac{(y+5)^2}{9} = 1$$
by hand or using a grapher? Explain why you answered as you did.

Find an equation of an ellipse satisfying the given conditions.

57. Vertices: $(3, -4)$, $(3, 6)$;
endpoints of minor axis: $(1, 1)$, $(5, 1)$

58. Vertices: $(-1, -1)$, $(-1, 5)$;
endpoints of minor axis: $(-3, 2)$, $(1, 2)$

59. Vertices: $(-3, 0)$ and $(3, 0)$; passing through $\left(2, \frac{22}{3}\right)$

60. Center: $(-2, 3)$; major axis vertical;
length of major axis: 4;
length of minor axis: 1

Use a grapher to find the center and the vertices of each of the following.

61. $4x^2 + 9y^2 - 16.025x + 18.0927y - 11.346 = 0$

62. $9x^2 + 4y^2 + 54.063x - 8.016y + 49.872 = 0$

63. *Bridge Arch.* A bridge with a semielliptical arch spans a river as shown below. What is the clearance 6 ft from the riverbank?

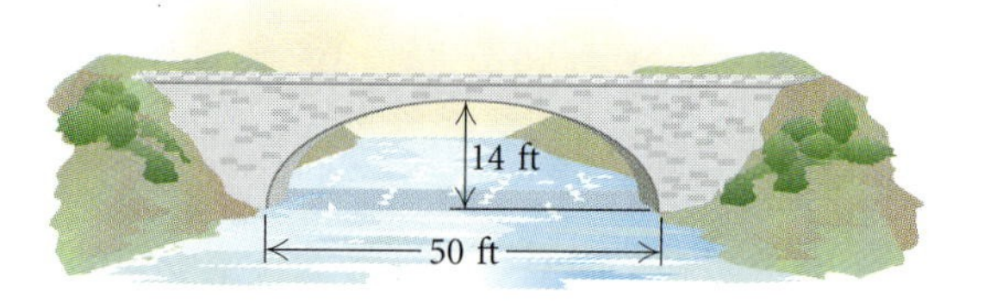

8.3 The Hyperbola

- *Given an equation of a hyperbola, complete the square, if necessary, and then find the center, the vertices, and the foci and graph the hyperbola.*

The last type of conic section that we will study is the *hyperbola*.

> **Hyperbola**
>
> A *hyperbola* is the set of all points in a plane for which the absolute value of the difference of the distances from two fixed points (the *foci*) is constant. The midpoint of the segment between the foci is the *center* of the hyperbola.

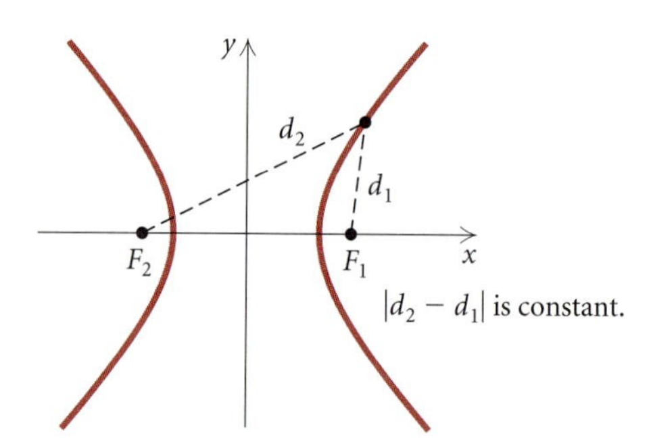

Standard Equations of Hyperbolas

We first consider the equation of a hyperbola with center at the origin. In the figure at the top of the following page, F_1 and F_2 are the foci. The segment $\overline{V_2V_1}$ is the **transverse axis** and the points V_2 and V_1 are the **vertices**.

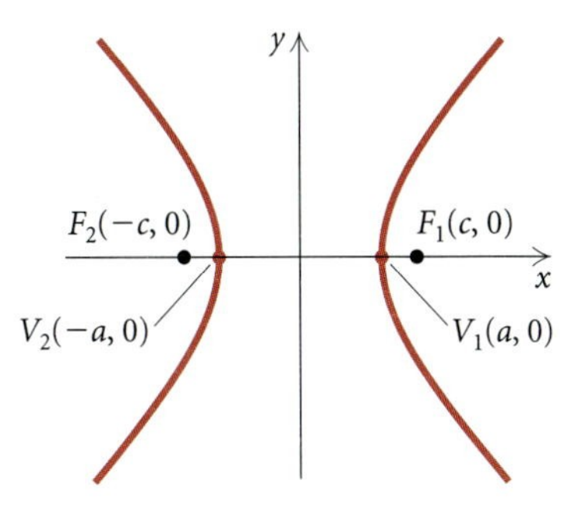

Standard Equation of a Hyperbola with Center at the Origin

Transverse Axis Horizontal

$$\frac{x^2}{a^2} - \frac{y^2}{b^2} = 1$$

Vertices: $(-a, 0)$, $(a, 0)$

Foci: $(-c, 0)$, $(c, 0)$, where $c^2 = a^2 + b^2$

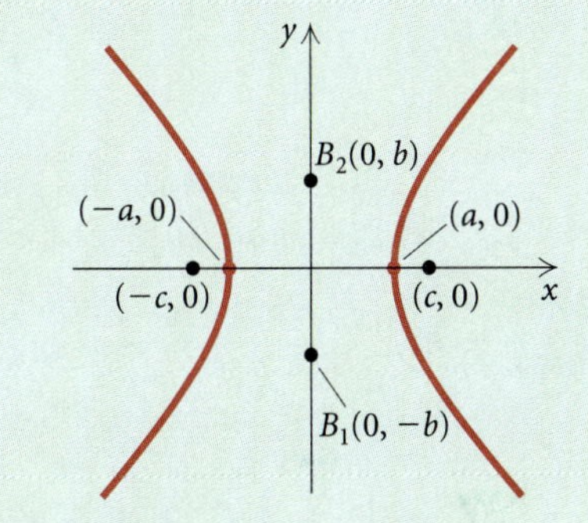

Transverse Axis Vertical

$$\frac{y^2}{a^2} - \frac{x^2}{b^2} = 1$$

Vertices: $(0, -a)$, $(0, a)$

Foci: $(0, -c)$, $(0, c)$, where $c^2 = a^2 + b^2$

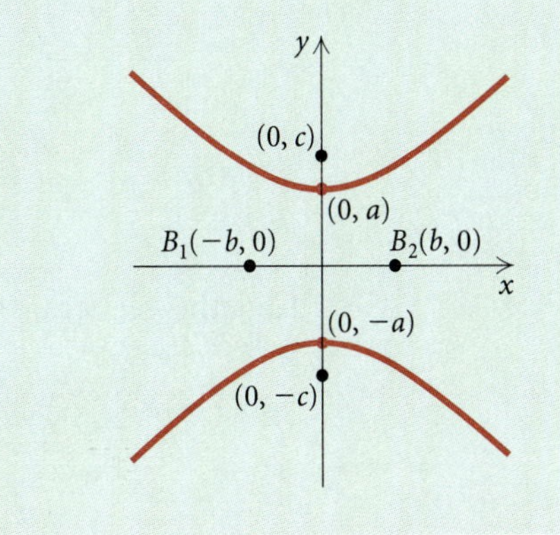

The segment $\overline{B_1B_2}$ is the **conjugate axis** of the hyperbola.

To graph a hyperbola with a horizontal transverse axis, it is helpful to begin by graphing the lines $y = -(b/a)x$ and $y = (b/a)x$. These are the **asymptotes** of the hyperbola. For a hyperbola with a vertical transverse axis, the asymptotes are $y = -(a/b)x$ and $y = (a/b)x$. As $|x|$ gets larger and larger, the graph of the hyperbola gets closer and closer to the asymptotes.

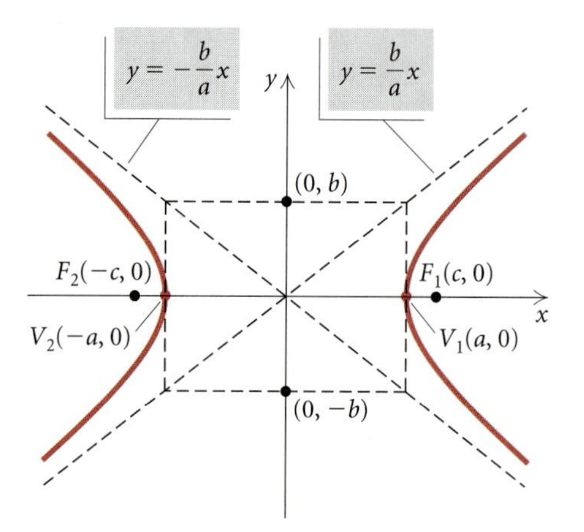

Example 1 Find an equation of the hyperbola with vertices $(0, -4)$ and $(0, 4)$ and foci $(0, -6)$ and $(0, 6)$.

SOLUTION We know that $a = 4$ and $c = 6$. We find b^2.

$$c^2 = a^2 + b^2$$
$$6^2 = 4^2 + b^2$$
$$36 = 16 + b^2$$
$$20 = b^2$$

Since the vertices and the foci are on the y-axis, we know that the transverse axis is vertical. We can now write the equation of the hyperbola:

$$\frac{y^2}{a^2} - \frac{x^2}{b^2} = 1$$
$$\frac{y^2}{16} - \frac{x^2}{20} = 1.$$

Example 2 For the hyperbola given by

$$9x^2 - 16y^2 = 144,$$

find the vertices, the foci, and the asymptotes. Then graph the hyperbola.

SOLUTION First, we find standard form:

$$9x^2 - 16y^2 = 144$$

$$\frac{1}{144}(9x^2 - 16y^2) = \frac{1}{144} \cdot 144$$ **Multiplying by $\frac{1}{144}$ to get 1 on the right side**

$$\frac{x^2}{16} - \frac{y^2}{9} = 1$$

$$\frac{x^2}{4^2} - \frac{y^2}{3^2} = 1.$$ **Writing standard form**

The hyperbola has a horizontal transverse axis, so the vertices are $(-a, 0)$ and $(a, 0)$, or $(-4, 0)$ and $(4, 0)$. From the standard form of the equation, we know that $a^2 = 4^2$, or 16, and $b^2 = 3^2$, or 9. We find the foci:

$$c^2 = a^2 + b^2$$
$$c^2 = 16 + 9$$
$$c^2 = 25$$
$$c = 5.$$

Thus the foci are $(-5, 0)$ and $(5, 0)$.

Next we find the asymptotes:

$$y = \frac{b}{a}x = \frac{3}{4}x$$

and

$$y = -\frac{b}{a}x = -\frac{3}{4}x.$$

To draw the graph, we sketch the asymptotes first. This is easily done by drawing the rectangle with horizontal sides passing through $(0, 3)$ and $(0, -3)$ and vertical sides through $(4, 0)$ and $(-4, 0)$. Then we draw and extend the diagonals of this rectangle. The two extended diagonals are the asymptotes of the hyperbola. Next, we plot the vertices and draw the branches of the hyperbola outward from the vertices toward the asymptotes.

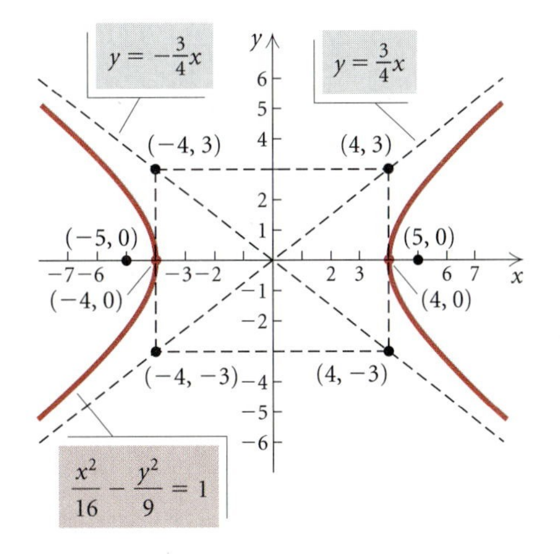

We can use a grapher as a check. It might be necessary to solve for y first and then graph the top and bottom halves of the hyperbola in the same squared viewing window.

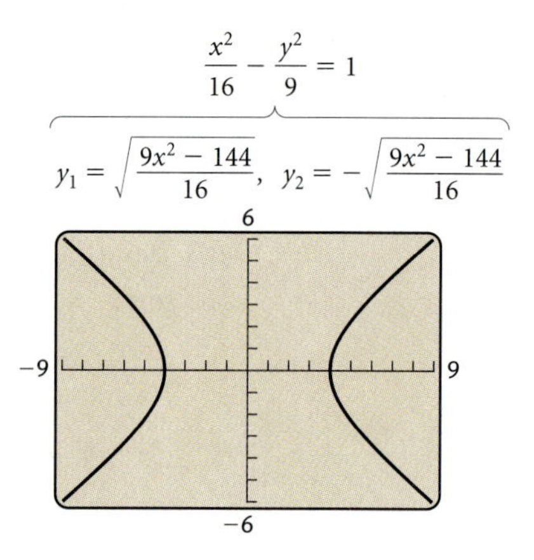

If a hyperbola with center at the origin is translated $|h|$ units left or right and $|k|$ units up or down, the center is at the point (h, k).

Standard Equation of a Hyperbola with Center (h, k)

Transverse Axis Horizontal

$$\frac{(x - h)^2}{a^2} - \frac{(y - k)^2}{b^2} = 1$$

Vertices: $(h - a, k)$, $(h + a, k)$

Asymptotes: $y - k = \frac{b}{a}(x - h)$, $y - k = -\frac{b}{a}(x - h)$

Foci: $(h - c, k)$, $(h + c, k)$, where $c^2 = a^2 + b^2$

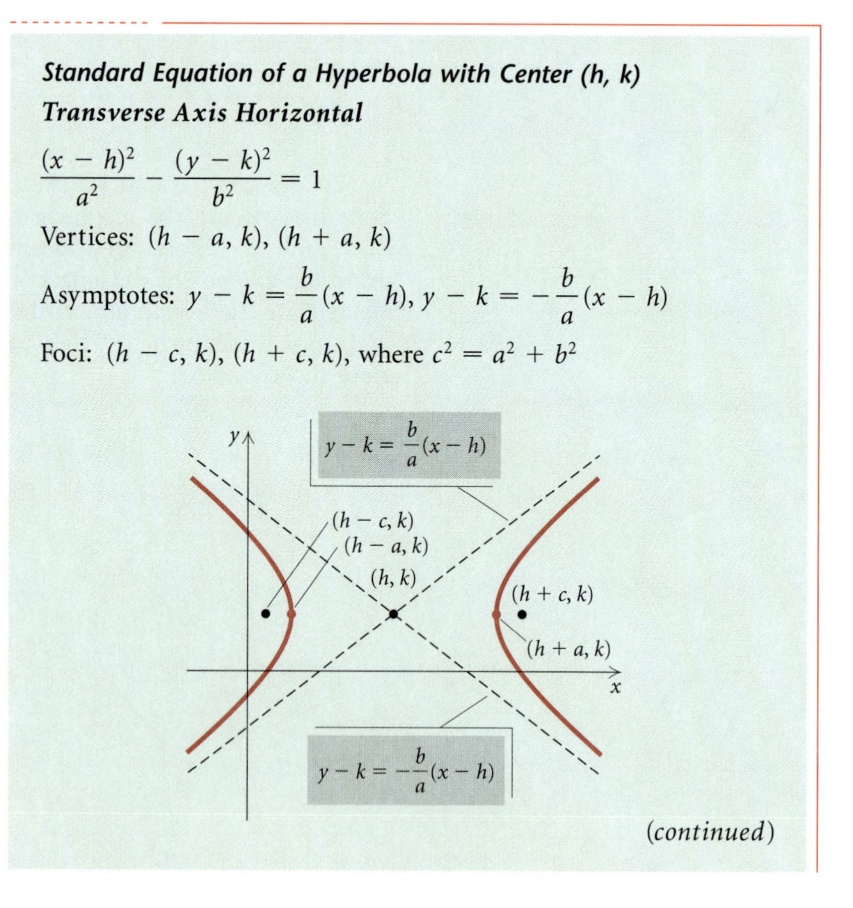

(continued)

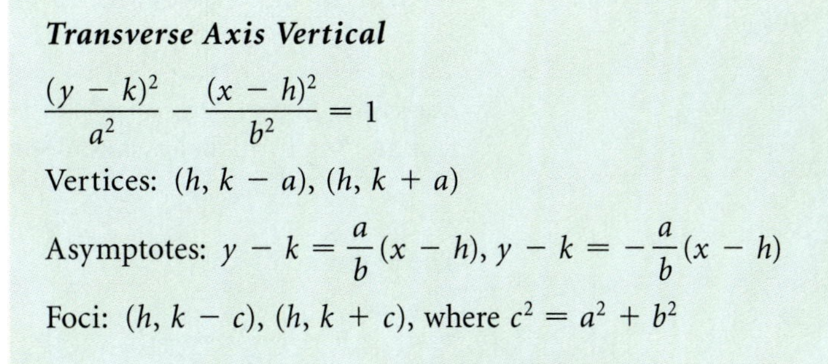

Transverse Axis Vertical

$$\frac{(y-k)^2}{a^2} - \frac{(x-h)^2}{b^2} = 1$$

Vertices: $(h, k-a)$, $(h, k+a)$

Asymptotes: $y - k = \frac{a}{b}(x-h)$, $y - k = -\frac{a}{b}(x-h)$

Foci: $(h, k-c)$, $(h, k+c)$, where $c^2 = a^2 + b^2$

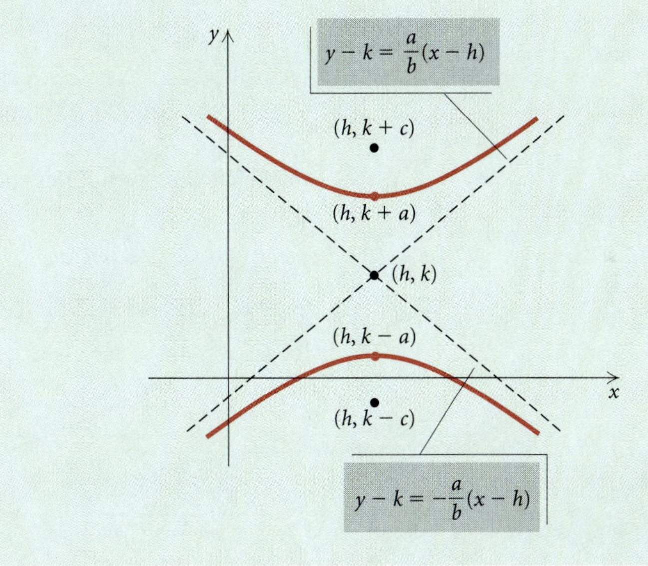

Example 3 For the hyperbola given by

$$4y^2 - x^2 + 24y + 4x + 28 = 0,$$

find the center, the vertices, and the foci. Then draw the graph.

SOLUTION First, we complete the square to get standard form:

$$\begin{aligned}
4y^2 - x^2 + 24y + 4x + 28 &= 0 \\
4(y^2 + 6y \qquad\quad) - (x^2 - 4x \qquad\quad) &= -28 \\
4(y^2 + 6y + 9 - 9) - (x^2 - 4x + 4 - 4) &= -28 \\
4(y^2 + 6y + 9) - (x^2 - 4x + 4) &= -28 + 36 - 4 \\
4(y+3)^2 - (x-2)^2 &= 4 \\
\frac{(y+3)^2}{1} - \frac{(x-2)^2}{4} &= 1 \qquad \textbf{Dividing by 4} \\
\frac{[y-(-3)]^2}{1^2} - \frac{(x-2)^2}{2^2} &= 1. \qquad \textbf{Standard form}
\end{aligned}$$

The center is $(2, -3)$. Note that $a = 1$ and $b = 2$. The transverse axis is vertical, so the vertices are 1 unit below and above the center:

$$(2, -3 - 1) \text{ and } (2, -3 + 1), \quad \text{or} \quad (2, -4) \text{ and } (2, -2).$$

We know that $c^2 = a^2 + b^2$, so $c^2 = 1 + 4 = 5$ and $c = \sqrt{5}$. Thus the foci are $\sqrt{5}$ units below and above the center:

$$(2, -3 - \sqrt{5}) \quad \text{and} \quad (2, -3 + \sqrt{5}).$$

The asymptotes are

$$y - (-3) = \frac{1}{2}(x - 2) \quad \text{and} \quad y - (-3) = -\frac{1}{2}(x - 2),$$

or

$$y + 3 = \frac{1}{2}(x - 2) \quad \text{and} \quad y + 3 = -\frac{1}{2}(x - 2).$$

We sketch the asymptotes, plot the vertices, and draw the graph.

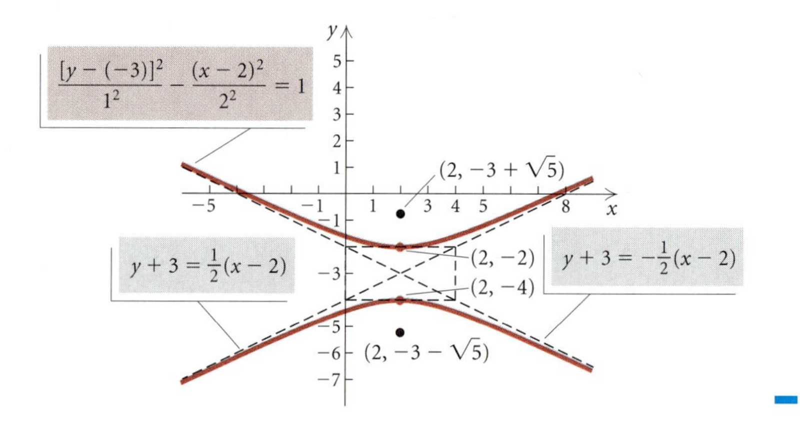

Applications

Some comets travel in hyperbolic paths with the sun at one focus. Such comets pass by the sun only one time unlike those with elliptical orbits, which reappear at intervals. A cross section of an amphitheater might be one branch of a hyperbola. A cross section of a nuclear cooling tower might also be a hyperbola.

One other important application of hyperbolas is in the long range navigation system LORAN. This system uses transmitting stations in three locations to send out simultaneous signals to a ship or aircraft. The difference in the arrival times of the signals from one pair of transmitters is recorded on the ship or aircraft. This difference is also recorded for signals from another pair of transmitters. For each pair, a computation is performed to determine the difference in the distances from each member of the pair to the ship or aircraft. If each pair of differences is kept constant, two hyperbolas can be drawn. Each has one of the pairs of transmitters as foci and the ship or aircraft lies on the intersection of two of their branches.

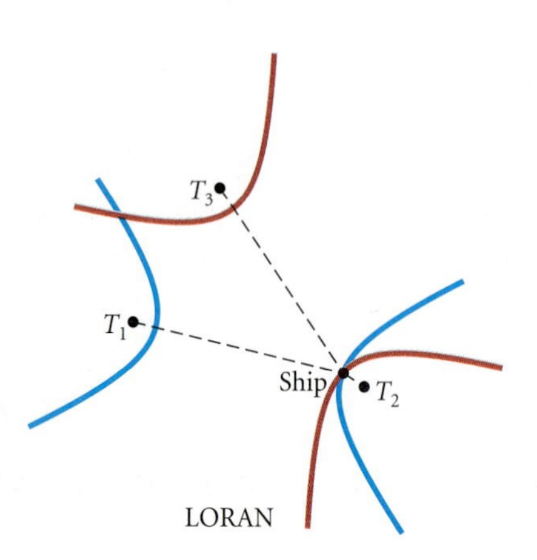

8.3 Exercise Set

In Exercises 1–6, match each equation with one of the graphs (a)–(f), which follow.

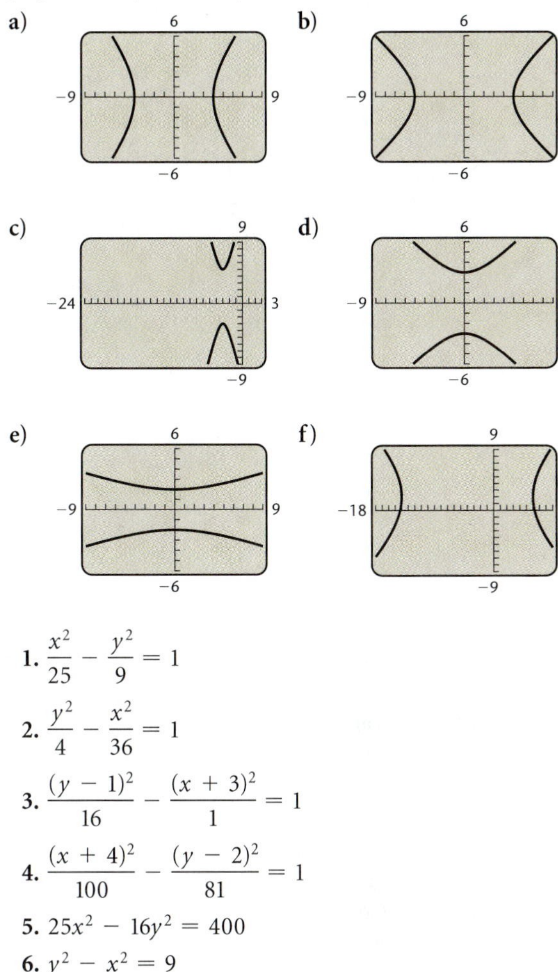

1. $\frac{x^2}{25} - \frac{y^2}{9} = 1$
2. $\frac{y^2}{4} - \frac{x^2}{36} = 1$
3. $\frac{(y-1)^2}{16} - \frac{(x+3)^2}{1} = 1$
4. $\frac{(x+4)^2}{100} - \frac{(y-2)^2}{81} = 1$
5. $25x^2 - 16y^2 = 400$
6. $y^2 - x^2 = 9$

Find an equation of a hyperbola satisfying the given conditions.

7. Vertices at $(0, 3)$ and $(0, -3)$; foci at $(0, 5)$ and $(0, -5)$
8. Vertices at $(1, 0)$ and $(-1, 0)$; foci at $(2, 0)$ and $(-2, 0)$
9. Asymptotes $y = \frac{3}{2}x$, $y = -\frac{3}{2}x$; one vertex $(2, 0)$
10. Asymptotes $y = \frac{5}{4}x$, $y = -\frac{5}{4}x$; one vertex $(0, 3)$

Find the center, the vertices, the foci, and the asymptotes. Then draw the graph.

11. $\frac{x^2}{4} - \frac{y^2}{4} = 1$
12. $\frac{x^2}{1} - \frac{y^2}{9} = 1$
13. $\frac{(x-2)^2}{9} - \frac{(y+5)^2}{1} = 1$
14. $\frac{(x-5)^2}{16} - \frac{(y+2)^2}{9} = 1$
15. $\frac{(y+3)^2}{4} - \frac{(x+1)^2}{16} = 1$
16. $\frac{(y+4)^2}{25} - \frac{(x+2)^2}{16} = 1$
17. $x^2 - 4y^2 = 4$
18. $4x^2 - y^2 = 16$
19. $9y^2 - x^2 = 81$
20. $y^2 - 4x^2 = 4$
21. $x^2 - y^2 = 2$
22. $x^2 - y^2 = 3$
23. $y^2 - x^2 = \frac{1}{4}$
24. $y^2 - x^2 = \frac{1}{9}$

Find the center, the vertices, the foci, and the asymptotes of each hyperbola. Then draw the graph.

25. $x^2 - y^2 - 2x - 4y - 4 = 0$
26. $4x^2 - y^2 + 8x - 4y - 4 = 0$
27. $36x^2 - y^2 - 24x + 6y - 41 = 0$
28. $9x^2 - 4y^2 + 54x + 8y + 41 = 0$
29. $9y^2 - 4x^2 - 18y + 24x - 63 = 0$
30. $x^2 - 25y^2 + 6x - 50y = 41$
31. $x^2 - y^2 - 2x - 4y = 4$
32. $9y^2 - 4x^2 - 54y - 8x + 41 = 0$
33. $y^2 - x^2 - 6x - 8y - 29 = 0$
34. $x^2 - y^2 = 8x - 2y - 13$

The ***eccentricity*** *of a hyperbola is defined as $e = c/a$. For a hyperbola, $c > a > 0$, so $e > 1$. When e is close to 1, a hyperbola appears to be very narrow. As the eccentricity increases, the hyperbola becomes "wider."*

35. Observe the shapes of the hyperbolas in Examples 2 and 3. Which hyperbola has the larger eccentricity? Confirm your answer by computing the eccentricity of each hyperbola.

36. Which hyperbola below has the larger eccentricity? (Assume that the coordinate systems have the same scale.)

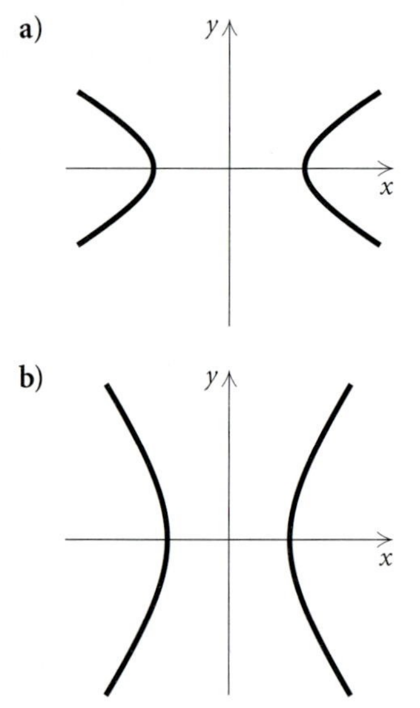

37. Find an equation of a hyperbola with vertices (3, 7) and (−3, 7) and $e = \frac{5}{3}$.

38. Find an equation of a hyperbola with vertices (−1, 3) and (−1, 7) and $e = 4$.

39. *Hyperbolic Mirror.* Certain telescopes contain both a parabolic and a hyperbolic mirror. In the telescope shown in the figure below, the parabola and the hyperbola share focus F_1, which is 14 m above the vertex of the parabola. The hyperbola's second focus F_2 is 2 m above the parabola's vertex. The vertex of the hyperbolic mirror is 1 m below F_1. Position a coordinate system with the origin at the center of the hyperbola and with the foci on the y-axis. Then find the equation of the hyperbola.

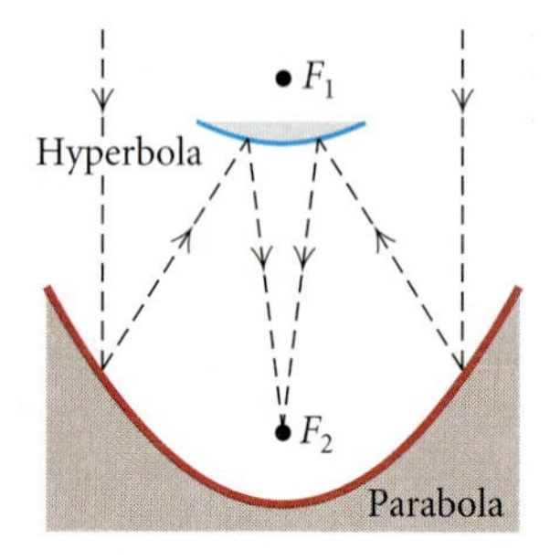

40. *Nuclear Cooling Tower.* A cross section of a nuclear cooling tower is a hyperbola with equation

$$\frac{x^2}{90^2} - \frac{y^2}{130^2} = 1.$$

The tower is 450 ft tall and the distance from the top of the tower to the center of the hyperbola is half the distance from the base of the tower to the center of the hyperbola. Find the diameter of the top and the base of the tower.

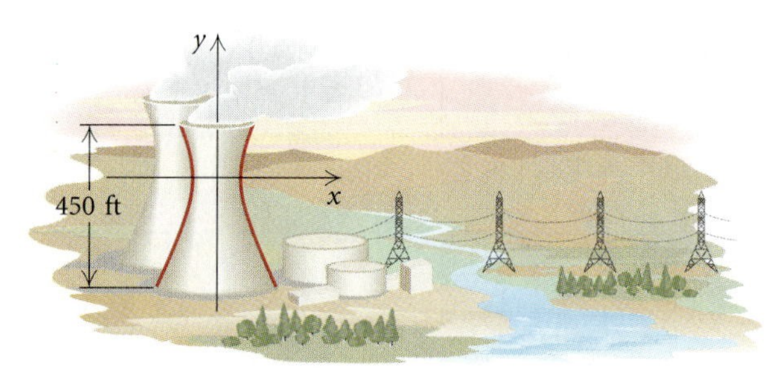

Skill Maintenance

Solve.

41. $x + y = 5,$
$x - y = 7$

42. $3x - 2y = 5,$
$5x + 2y = 3$

43. $2x - 3y = 7,$
$3x + 5y = 1$

44. $3x + 2y = -1,$
$2x + 3y = 6$

Synthesis

45. ◈ How does the graph of a parabola differ from the graph of one branch of a hyperbola?

46. ◈ Would you prefer to graph the hyperbola

$$\frac{(y + 5)^2}{16} + \frac{(x - 4)^2}{36} = 1$$

by hand or using a grapher? Explain why you answered as you did.

Find an equation of a hyperbola satisfying the given conditions.

47. Vertices at (3, −8) and (3, −2); asymptotes $y = 3x - 14$, $y = -3x + 4$

48. Vertices at (−9, 4) and (−5, 4); asymptotes $y = 3x + 25$, $y = -3x - 17$

Use a grapher to find the center, the vertices, and the asymptotes.

49. $5x^2 - 3.5y^2 + 14.6x - 6.7y + 3.4 = 0$

50. $x^2 - y^2 - 2.046x - 4.088y - 4.228 = 0$

51. *Navigation.* Two radio transmitters positioned 300 mi apart send simultaneous signals to a ship that is 200 mi offshore, sailing parallel to the shoreline. The signal from transmitter S reaches the ship 200 microseconds later than the signal from transmitter T. The signals travel at a speed of 186,000 miles per second, or 0.186 mile per microsecond. Find the equation of the hyperbola with foci S and T on which the ship is located. (*Hint:* For any point on the hyperbola, the absolute value of the difference of its distances from the foci is $2a$.)

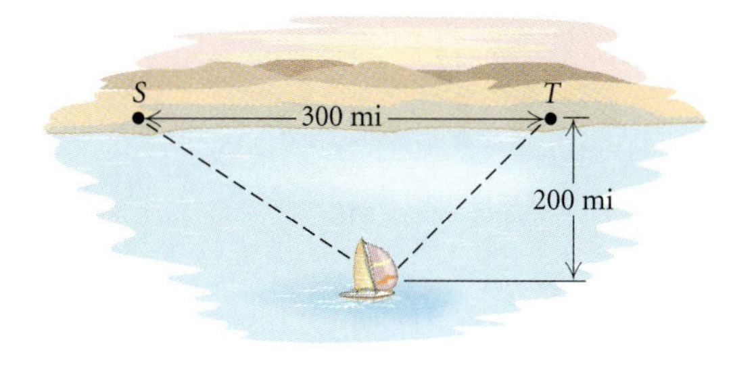

8.4 Nonlinear Systems of Equations

- *Solve a nonlinear system of equations.*
- *Use nonlinear systems of equations to solve applied problems.*

The systems of equations that we have studied so far have been composed of linear equations. Now we consider systems of two equations in two variables in which at least one equation is not linear.

Interactive Discovery

Graph the circle $x^2 + y^2 = 5$ and the line $y = x - 4$ in the window $[-6, 6, -4, 4]$. How many points of intersection are there? What conclusion can you draw about the number of real-number solutions of the system of equations composed of $x^2 + y^2 = 5$ and $y = x - 4$?

Now graph $x^2 + y^2 = 5$ and $y = -0.5x + 2.5$ in the same window and determine the number of points of intersection. What does each point of intersection represent?

Finally, graph $x^2 + y^2 = 5$ and $y = x$ in the same window. How many points of intersection are there? What does each represent?

Nonlinear Systems of Equations

The graph of a nonlinear system of equations can have no point of intersection or one or more points of intersection. The coordinates of each point of intersection are a solution of the system of equations. When no point of intersection exists, the system of equations has no real-number solution.

Real-number solutions of nonlinear systems of equations can be found algebraically, using the substitution or elimination method, or graphically. However, since a graph shows *only* real-number solutions, solutions involving imaginary numbers must always be found algebraically.

Example 1 Solve the following system of equations:

$$x^2 + y^2 = 25, \quad (1) \qquad \text{(The graph is a circle.)}$$
$$3x - 4y = 0. \quad (2) \qquad \text{(The graph is a line.)}$$

ALGEBRAIC SOLUTION

We use the substitution method. First, we solve equation (2) for x:

$$x = \tfrac{4}{3}y. \quad (3) \qquad \text{We could have solved for } y \text{ instead.}$$

Next, we substitute $\frac{4}{3}y$ for x in equation (1) and solve for y:

$$\left(\tfrac{4}{3}y\right)^2 + y^2 = 25$$
$$\tfrac{16}{9}y^2 + y^2 = 25$$
$$\tfrac{25}{9}y^2 = 25$$
$$y^2 = 9 \qquad \text{Multiplying by } \tfrac{9}{25}$$
$$y = \pm 3.$$

Now we substitute these numbers for y in equation (3) and solve for x:

$$x = \tfrac{4}{3}(3) = 4, \qquad \text{The pair (4, 3) appears to be a solution.}$$
$$x = \tfrac{4}{3}(-3) = -4. \qquad \text{The pair } (-4, -3) \text{ appears to be a solution.}$$

CHECK: For (4, 3):

$x^2 + y^2 = 25$			$3x - 4y = 0$		
$4^2 + 3^2$	? 25		$3(4) - 4(3)$	? 0	
$16 + 9$			$12 - 12$		
25	25	TRUE	0	0	TRUE

For $(-4, -3)$:

$x^2 + y^2 = 25$			$3x - 4y = 0$		
$(-4)^2 + (-3)^2$	? 25		$3(-4) - 4(-3)$	? 0	
$16 + 9$			$-12 + 12$		
25	25	TRUE	0	0	TRUE

The pairs (4, 3) and $(-4, -3)$ check, so they are the solutions.

GRAPHICAL SOLUTION

We graph both equations in the same viewing window and note that there are two points of intersection. We can find their coordinates using the INTERSECT feature or the TRACE and ZOOM features.

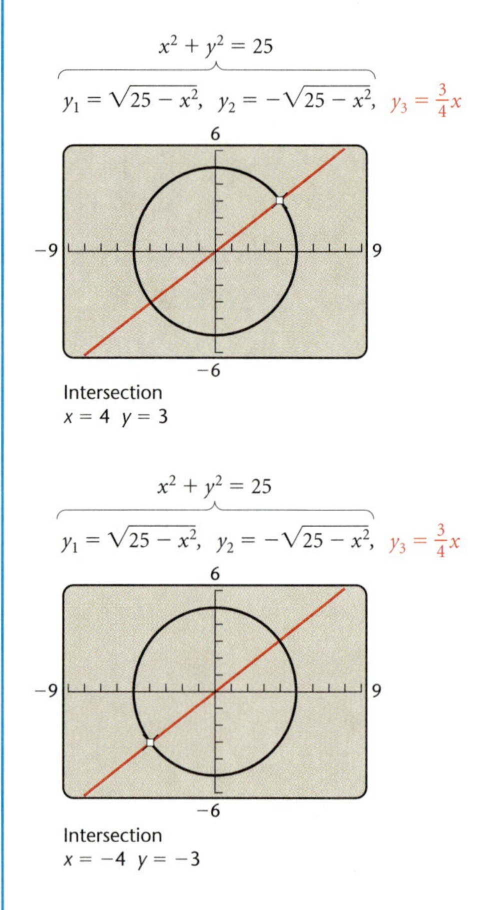

The graph can also be used to check the solutions that we found algebraically.

In the algebraic solution above, suppose that to find x, we had substituted 3 and -3 in equation (1) rather than equation (3). If $y = 3$, $y^2 = 9$, and if $y = -3$, $y^2 = 9$, so both substitutions can be performed at the same time:

$$\begin{aligned} x^2 + y^2 &= 25 \qquad (1) \\ x^2 + (\pm 3)^2 &= 25 \\ x^2 + 9 &= 25 \\ x^2 &= 16 \\ x &= \pm 4. \end{aligned}$$

Thus, if $y = 3$, $x = 4$ or $x = -4$, and if $y = -3$, $x = 4$ or $x = -4$. The possible solutions are $(4, 3)$, $(-4, 3)$, $(4, -3)$, and $(-4, -3)$. A check reveals that $(4, -3)$ and $(-4, 3)$ are not solutions of equation (2). Had we graphed the system of equations before solving algebraically, it would have been clear that there are only two real-number solutions.

Example 2 Solve the following system of equations:

$$x + y = 5, \qquad (1) \qquad \text{(The graph is a line.)}$$

$$y = 3 - x^2. \qquad (2) \qquad \text{(The graph is a parabola.)}$$

ALGEBRAIC SOLUTION

We use the substitution method, substituting $3 - x^2$ for y in the first equation:

$$\begin{aligned} x + 3 - x^2 &= 5 \\ -x^2 + x - 2 &= 0 \\ x^2 - x + 2 &= 0. \end{aligned}$$

Next, we use the quadratic formula:

$$x = \frac{-b \pm \sqrt{b^2 - 4ac}}{2a} = \frac{-(-1) \pm \sqrt{(-1)^2 - 4(1)(2)}}{2(1)}$$

$$= \frac{1 \pm \sqrt{1 - 8}}{2} = \frac{1 \pm \sqrt{-7}}{2} = \frac{1 \pm i\sqrt{7}}{2} = \frac{1}{2} \pm \frac{\sqrt{7}}{2}i.$$

Now, we substitute these values for x in equation (1) and solve for y:

$$\frac{1}{2} + \frac{\sqrt{7}}{2}i + y = 5$$

$$y = 5 - \frac{1}{2} - \frac{\sqrt{7}}{2}i = \frac{9}{2} - \frac{\sqrt{7}}{2}i$$

and $\frac{1}{2} - \frac{\sqrt{7}}{2}i + y = 5$

$$y = 5 - \frac{1}{2} + \frac{\sqrt{7}}{2}i = \frac{9}{2} + \frac{\sqrt{7}}{2}i.$$

The solutions are

$$\left(\frac{1}{2} + \frac{\sqrt{7}}{2}i, \frac{9}{2} - \frac{\sqrt{7}}{2}i\right) \text{ and } \left(\frac{1}{2} - \frac{\sqrt{7}}{2}i, \frac{9}{2} + \frac{\sqrt{7}}{2}i\right).$$

There are no real-number solutions.

GRAPHICAL SOLUTION

We graph both equations in the same viewing window.

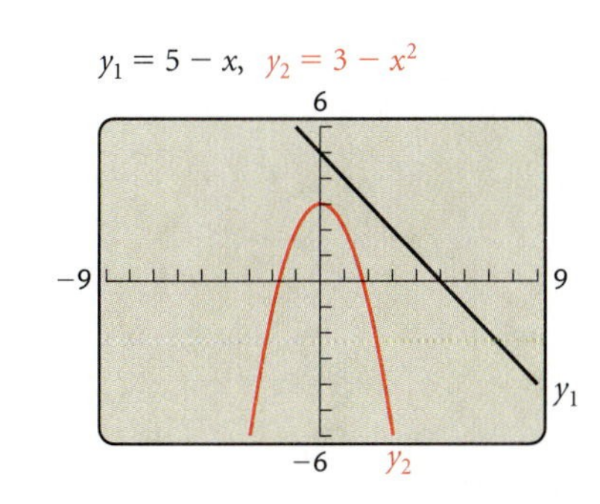

Note that there are no points of intersection. This indicates that there are no real-number solutions. Algebra must be used, as at left, to find the complex-number solutions.

Example 3 Solve the following system of equations:

$$2x^2 + 5y^2 = 39, \quad (1) \quad \text{(The graph is an ellipse.)}$$
$$3x^2 - y^2 = -1. \quad (2) \quad \text{(The graph is a hyperbola.)}$$

ALGEBRAIC SOLUTION

We use the elimination method. First we multiply equation (2) by 5 and add to eliminate the y^2-term:

$$2x^2 + 5y^2 = 39 \quad (1)$$
$$15x^2 - 5y^2 = -5 \quad \text{Multiplying (2) by 5}$$
$$17x^2 = 34 \quad \text{Adding}$$
$$x^2 = 2$$
$$x = \pm\sqrt{2}.$$

If $x = \sqrt{2}$, $x^2 = 2$, and if $x = -\sqrt{2}$, $x^2 = 2$. Thus substituting $\sqrt{2}$ or $-\sqrt{2}$ for x in equation (2) gives us

$$3(\pm\sqrt{2})^2 - y^2 = -1$$
$$3 \cdot 2 - y^2 = -1$$
$$6 - y^2 = -1$$
$$-y^2 = -7$$
$$y^2 = 7$$
$$y = \pm\sqrt{7}.$$

Thus, for $x = \sqrt{2}$, we have $y = \sqrt{7}$ or $y = -\sqrt{7}$, and for $x = -\sqrt{2}$, we have $y = \sqrt{7}$ or $y = -\sqrt{7}$. The possible solutions are $(\sqrt{2}, \sqrt{7})$, $(\sqrt{2}, -\sqrt{7})$, $(-\sqrt{2}, \sqrt{7})$, and $(-\sqrt{2}, -\sqrt{7})$. All four pairs check, so they are the solutions.

GRAPHICAL SOLUTION

We graph both equations in the same viewing window and note that there are four points of intersection. We can use the INTERSECT feature or the TRACE and ZOOM features to find their coordinates.

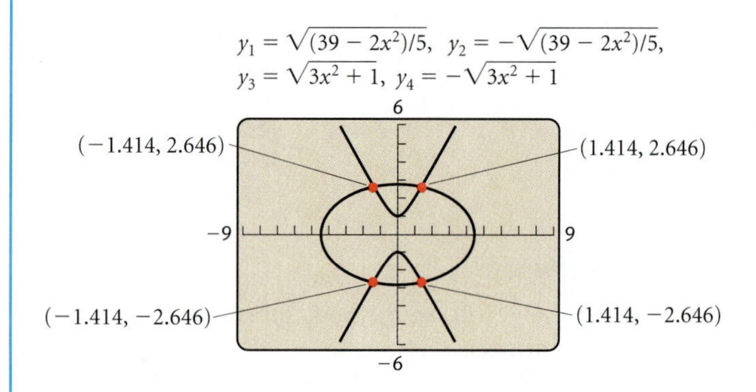

Note that the algebraic method yields exact solutions whereas the graphical method yields decimal approximations of the solutions.

Example 4 Solve the following system of equations:

$$x^2 - 3y^2 = 6, \qquad (1)$$
$$xy = 3. \qquad (2)$$

ALGEBRAIC SOLUTION

We use the substitution method. First, we solve equation (2) for y:

$$xy = 3$$
$$y = \frac{3}{x}. \qquad (3)$$

Then we substitute $3/x$ for y in equation (1) and solve for x:

$$x^2 - 3\left(\frac{3}{x}\right)^2 = 6$$
$$x^2 - 3 \cdot \frac{9}{x^2} = 6$$
$$x^2 - \frac{27}{x^2} = 6$$
$$x^4 - 27 = 6x^2 \qquad \textbf{Multiplying by } x^2$$
$$x^4 - 6x^2 - 27 = 0$$
$$u^2 - 6u - 27 = 0 \qquad \textbf{Letting } u = x^2$$
$$(u - 9)(u + 3) = 0 \qquad \textbf{Factoring}$$
$$u = 9 \quad or \quad u = -3 \qquad \textbf{Principle of zero products}$$
$$x^2 = 9 \quad or \quad x^2 = -3$$
$$x = \pm 3 \quad or \quad x = \pm i\sqrt{3}.$$

Since $y = 3/x$,

when $x = 3$, $\quad y = \frac{3}{3} = 1;$

when $x = -3$, $\quad y = \frac{3}{-3} = -1;$

when $x = i\sqrt{3}$, $\quad y = \frac{3}{i\sqrt{3}} = \frac{3}{i\sqrt{3}} \cdot \frac{-i\sqrt{3}}{-i\sqrt{3}} = -i\sqrt{3};$

when $x = -i\sqrt{3}$, $\quad y = \frac{3}{-i\sqrt{3}} = \frac{3}{-i\sqrt{3}} \cdot \frac{i\sqrt{3}}{i\sqrt{3}} = i\sqrt{3}.$

The pairs $(3, 1)$, $(-3, -1)$, $(i\sqrt{3}, -i\sqrt{3})$, and $(-i\sqrt{3}, i\sqrt{3})$ check, so they are the solutions.

GRAPHICAL SOLUTION

We graph both equations in the same viewing window and find the coordinates of their points of intersection.

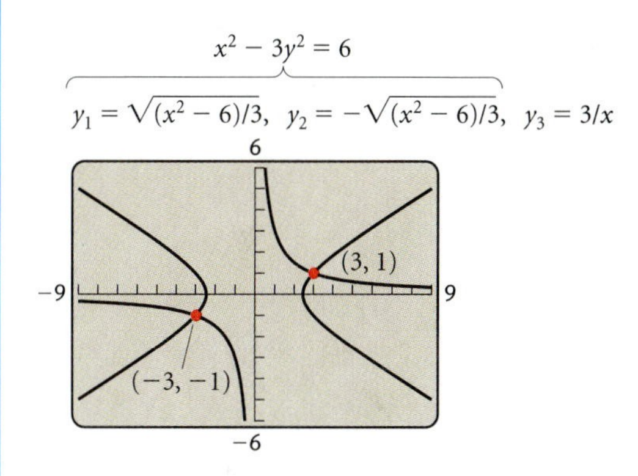

Again, note that the graphical method yields only the real-number solutions of the system of equations. The algebraic method must be used to find *all* the solutions.

Modeling and Problem Solving

Example 5 *Dimensions of a Piece of Land.* For a student recreation building at Southport Community College, an architect wants to lay out a rectangular piece of land that has a perimeter of 204 m and an area of 2565 m^2. Find the dimensions of the piece of land.

SOLUTION

1. **Familiarize.** We make a drawing and label it, letting l = the length, in meters, and w = the width, in meters.

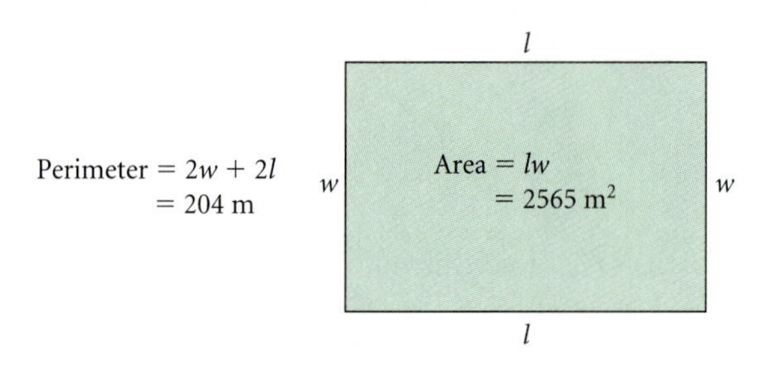

2. **Translate.** We now have the following:

$$\text{Perimeter:} \quad 2w + 2l = 204, \quad (1)$$
$$\text{Area:} \quad lw = 2565. \quad (2)$$

3. **Carry out.** We solve the system of equations both algebraically and graphically.

ALGEBRAIC SOLUTION

We solve the system of equations

$$2w + 2l = 204,$$
$$lw = 2565.$$

Solving the second equation for l gives us $l = 2565/w$. We then substitute $2565/w$ for l in the first equation and solve for w:

$$2w + 2\left(\frac{2565}{w}\right) = 204$$

$$2w^2 + 2(2565) = 204w \quad \textbf{Multiplying by } w$$

$$2w^2 - 204w + 2(2565) = 0$$

$$w^2 - 102w + 2565 = 0 \quad \textbf{Multiplying by } \tfrac{1}{2}$$

$$(w - 57)(w - 45) = 0$$

$$w = 57 \quad or \quad w = 45.$$

Principle of zero products

If $w = 57$, then $l = 2565/w = 2565/57 = 45$. If $w = 45$, then $l = 2565/w = 2565/45 = 57$. Since length is generally considered to be longer than width, we have the solution $l = 57$ and $w = 45$, or $(57, 45)$.

GRAPHICAL SOLUTION

We replace l with x and w with y, graph $y_1 = (204 - 2x)/2$ and $y_2 = 2565/x$, and find the point(s) of intersection of the graphs.

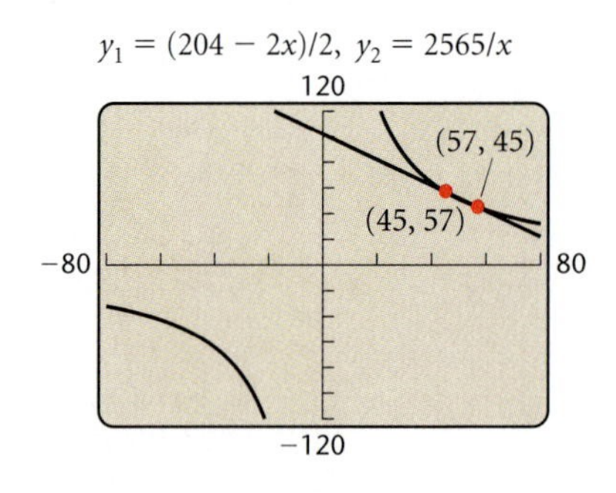

As in the algebraic solution, we have two possible solutions: $(45, 57)$ and $(57, 45)$. Since length (x) is generally considered to be longer than width (y), we have the solution $(57, 45)$.

4. **Check.** If $l = 57$ and $w = 45$, the perimeter is $2 \cdot 57 + 2 \cdot 45$, or 204. The area is $57 \cdot 45$, or 2565. The numbers check.

5. **State.** The length of the piece of land is 57 m and the width is 45 m.

8.4 Exercise Set

In Exercises 1–6, match each system of equations with one of the graphs (a)–(f), which follow.

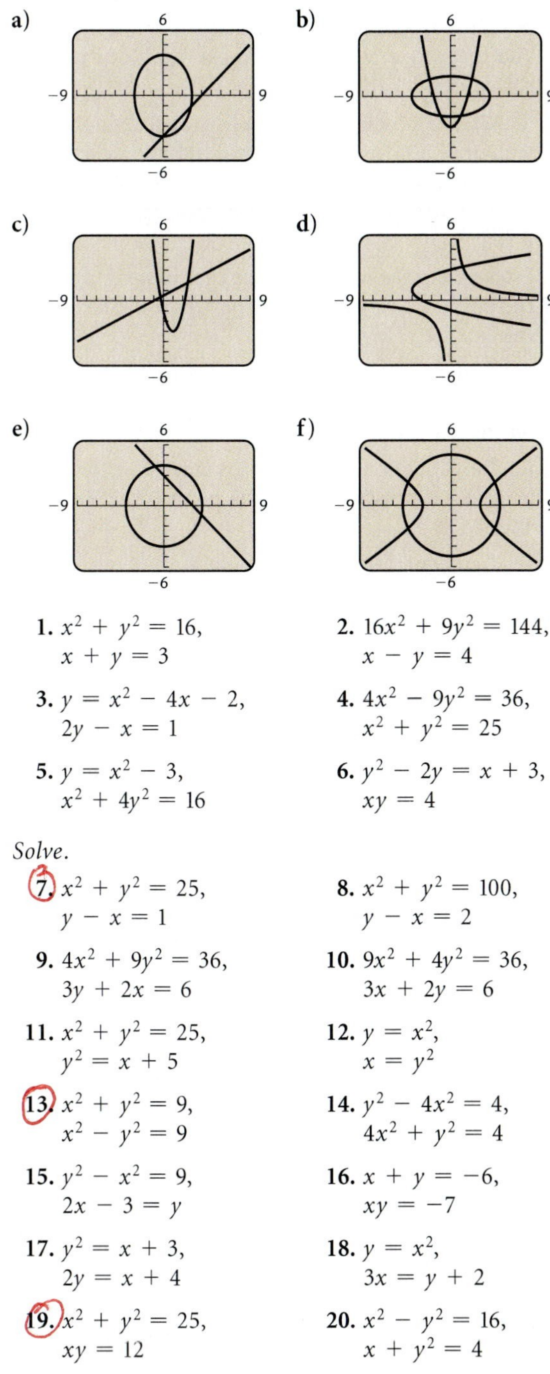

1. $x^2 + y^2 = 16,$
$x + y = 3$

2. $16x^2 + 9y^2 = 144,$
$x - y = 4$

3. $y = x^2 - 4x - 2,$
$2y - x = 1$

4. $4x^2 - 9y^2 = 36,$
$x^2 + y^2 = 25$

5. $y = x^2 - 3,$
$x^2 + 4y^2 = 16$

6. $y^2 - 2y = x + 3,$
$xy = 4$

Solve.

7. $x^2 + y^2 = 25,$
$y - x = 1$

8. $x^2 + y^2 = 100,$
$y - x = 2$

9. $4x^2 + 9y^2 = 36,$
$3y + 2x = 6$

10. $9x^2 + 4y^2 = 36,$
$3x + 2y = 6$

11. $x^2 + y^2 = 25,$
$y^2 = x + 5$

12. $y = x^2,$
$x = y^2$

13. $x^2 + y^2 = 9,$
$x^2 - y^2 = 9$

14. $y^2 - 4x^2 = 4,$
$4x^2 + y^2 = 4$

15. $y^2 - x^2 = 9,$
$2x - 3 = y$

16. $x + y = -6,$
$xy = -7$

17. $y^2 = x + 3,$
$2y = x + 4$

18. $y = x^2,$
$3x = y + 2$

19. $x^2 + y^2 = 25,$
$xy = 12$

20. $x^2 - y^2 = 16,$
$x + y^2 = 4$

21. $x^2 + y^2 = 4,$
$16x^2 + 9y^2 = 144$

22. $x^2 + y^2 = 25,$
$25x^2 + 16y^2 = 400$

23. $x^2 + 4y^2 = 25,$
$x + 2y = 7$

24. $y^2 - x^2 = 16,$
$2x - y = 1$

25. $x^2 - xy + 3y^2 = 27,$
$x - y = 2$

26. $2y^2 + xy + x^2 = 7,$
$x - 2y = 5$

27. $x^2 + y^2 = 16,$
$y^2 - 2x^2 = 10$

28. $x^2 + y^2 = 14,$
$x^2 - y^2 = 4$

29. $x^2 + y^2 = 5,$
$xy = 2$

30. $x^2 + y^2 = 20,$
$xy = 8$

31. $3x + y = 7,$
$4x^2 + 5y = 56$

32. $2y^2 + xy = 5,$
$4y + x = 7$

33. $a + b = 7,$
$ab = 4$

34. $p + q = -4,$
$pq = -5$

35. $x^2 + y^2 = 13,$
$xy = 6$

36. $x^2 + 4y^2 = 20,$
$xy = 4$

37. $x^2 + y^2 + 6y + 5 = 0,$
$x^2 + y^2 - 2x - 8 = 0$

38. $2xy + 3y^2 = 7,$
$3xy - 2y^2 = 4$

39. $2a + b = 1,$
$b = 4 - a^2$

40. $4x^2 + 9y^2 = 36,$
$x + 3y = 3$

41. $a^2 + b^2 = 89,$
$a - b = 3$

42. $xy = 4,$
$x + y = 5$

43. $xy - y^2 = 2,$
$2xy - 3y^2 = 0$

44. $4a^2 - 25b^2 = 0,$
$2a^2 - 10b^2 = 3b + 4$

45. $m^2 - 3mn + n^2 + 1 = 0,$
$3m^2 - mn + 3n^2 = 13$

46. $ab - b^2 = -4,$
$ab - 2b^2 = -6$

47. $x^2 + y^2 = 5,$
$x - y = 8$

48. $4x^2 + 9y^2 = 36,$
$y - x = 8$

49. $a^2 + b^2 = 14,$
$ab = 3\sqrt{5}$

50. $x^2 + xy = 5,$
$2x^2 + xy = 2$

51. $x^2 + y^2 = 25,$
$9x^2 + 4y^2 = 36$

52. $x^2 + y^2 = 1,$
$9x^2 - 16y^2 = 144$

53. $5y^2 - x^2 = 1,$
$xy = 2$

54. $x^2 - 7y^2 = 6,$
$xy = 1$

55. *Picture Frame Dimensions.* Frank's Frame Shop is building a frame for a rectangular oil painting with a perimeter of 28 cm and a diagonal of 10 cm. Find the dimensions of the painting.

56. *Landscaping.* Green Leaf Landscaping is planting a rectangular wildflower garden with a perimeter of 6 m and a diagonal of $\sqrt{5}$ m. Find the dimensions of the garden.

57. *Pamphlet Design.* A graphic artist is designing a rectangular advertising pamphlet that is to have an area of 20 in^2 and a perimeter of 18 in. Find the dimensions of the pamphlet.

58. *Sign Dimensions.* Peden's Advertising is building a rectangular sign with an area of 2 yd^2 and a perimeter of 6 yd. Find the dimensions of the sign.

59. *Banner Design.* A rectangular banner with an area of $\sqrt{3}$ m^2 is being designed to advertise an exhibit at the Madison Art League. The length of a diagonal is 2 m. Find the dimensions of the banner.

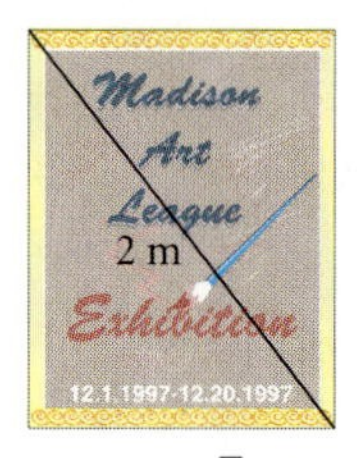

Area = $\sqrt{3}$ m^2

60. *Carpentry.* A carpenter wants to make a rectangular tabletop with an area of $\sqrt{2}$ m^2 and a diagonal of $\sqrt{3}$ m. Find the dimensions of the tabletop.

61. *Fencing.* It will take 210 yd of fencing to enclose a rectangular dog run. The area of the run is 2250 yd^2. What are the dimensions of the run?

62. *Office Dimensions.* The diagonal of the floor of a rectangular office cubicle is 1 ft longer than the length of the cubicle and 3 ft longer than twice the width. Find the dimensions of the cubicle.

63. *Seed Test Plots.* The Burton Seed Company has two square test plots. The sum of their areas is 832 ft^2 and the difference of their areas is 320 ft^2. Find the length of a side of each plot.

64. *Investment.* Jenna made an investment for 1 yr that earned \$7.50 simple interest. If the principal had been \$25 more and the interest rate 1% less, the interest would have been the same. Find the principal and the rate.

Skill Maintenance

Evaluate each expression for the given value of n.

65. $n - \dfrac{1}{n}$, for $n = 5$

66. $(3n + 1)^2$, for $n = 3$

67. $\dfrac{n+3}{n+5}$, for $n = 20$

68. $(-1)^n n^2$, for $n = 7$

Synthesis

69. ◆ What would you say to a classmate who tells you that any system of nonlinear equations can be solved graphically?

70. ◆ Write a problem that can be translated to a system of nonlinear equations, and ask a classmate to solve it. Devise the problem so that the solution is "The dimensions of the rectangle are 6 ft by 8 ft."

71. Find an equation of the circle that passes through the points (2, 4) and (3, 3) and whose center is on the line $3x - y = 3$.

72. Find an equation of the circle that passes through the points $(-2, 3)$ and $(-4, 1)$ and whose center is on the line $5x + 8y = -2$.

73. Find an equation of an ellipse centered at the origin that passes through the points $(1, \sqrt{3}/2)$ and $(\sqrt{3}, 1/2)$.

74. Find an equation of a hyperbola of the type

$$\frac{x^2}{b^2} - \frac{y^2}{a^2} = 1$$

that passes through the points $(-3, -3\sqrt{5}/2)$ and $(-3/2, 0)$.

75. Find an equation of the circle that passes through the points (4, 6), $(-6, 2)$, and $(1, -3)$.

76. Find an equation of the circle that passes through the points (2, 3), (4, 5), and $(0, -3)$.

77. Show that a hyperbola does not intersect its asymptotes. That is, solve the system of equations

$$\frac{x^2}{a^2} - \frac{y^2}{b^2} = 1,$$

$$y = \frac{b}{a}x \quad \left(\text{or } y = -\frac{b}{a}x\right).$$

78. *Numerical Relationship.* Find two numbers whose product is 2 and the sum of whose reciprocals is $\frac{33}{8}$.

79. *Numerical Relationship.* The square of a number exceeds twice the square of another number by $\frac{1}{8}$. The sum of their squares is $\frac{5}{16}$. Find the numbers.

80. *Box Dimensions.* Four squares with sides 5 in. long are cut from the corners of a rectangular metal sheet that has an area of 340 in^2. The edges are bent up to form an open box with a volume of 350 in^3. Find the dimensions of the box.

81. *Numerical Relationship.* The sum of two numbers is 1, and their product is 1. Find the sum of their cubes. There is a method to solve this problem that is easier than solving a nonlinear system of equations. Can you discover it?

82. Solve for x and y:

$$x^2 - y^2 = a^2 - b^2,$$
$$x - y = a - b.$$

Solve.

83. $x^3 + y^3 = 72,$
$x + y = 6$

84. $a + b = \frac{5}{6},$
$\frac{a}{b} + \frac{b}{a} = \frac{13}{6}$

85. $p^2 + q^2 = 13,$
$\frac{1}{pq} = -\frac{1}{6}$

86. $x^2 + y^2 = 4,$
$(x - 1)^2 + y^2 = 4$

87. $5^{x+y} = 100,$
$3^{2x-y} = 1000$

88. $e^x - e^{x+y} = 0,$
$e^y - e^{x-y} = 0$

Solve using a grapher.

89. $y - \ln x = 2,$
$y = x^2$

90. $y = \ln (x + 4),$
$x^2 + y^2 = 6$

91. $e^x - y = 1,$
$3x + y = 4$

92. $y - e^{-x} = 1,$
$y = 2x + 5$

93. $y = e^x,$
$x - y = -2$

94. $y = e^{-x},$
$x + y = 3$

95. $x^2 + y^2 = 19{,}380{,}510.36,$
$27{,}942.25x - 6.125y = 0$

96. $2x + 2y = 1660,$
$xy = 35{,}325$

97. $14.5x^2 - 13.5y^2 - 64.5 = 0,$
$5.5x - 6.3y - 12.3 = 0$

98. $13.5xy + 15.6 = 0,$
$5.6x - 6.7y - 42.3 = 0$

99. $0.319x^2 + 2688.7y^2 = 56{,}548,$
$0.306x^2 - 2688.7y^2 = 43{,}452$

100. $18.465x^2 + 788.723y^2 = 6408,$
$106.535x^2 - 788.723y^2 = 2692$

CHAPTER 8 Summary and Review

Important Properties and Formulas

Standard Equation of a Parabola with Vertex at the Origin

The standard equation of a parabola with vertex $(0, 0)$ and directrix $y = -p$ is

$$x^2 = 4py.$$

The focus is $(0, p)$ and the y-axis is the axis of symmetry.

The standard equation of a parabola with vertex $(0, 0)$ and directrix $x = -p$ is

$$y^2 = 4px.$$

The focus is $(p, 0)$ and the x-axis is the axis of symmetry.

Standard Equation of a Parabola with Vertex (h, k) and Vertical Axis of Symmetry

The standard equation of a parabola with vertex (h, k) and vertical axis of symmetry is

$$(x - h)^2 = 4p(y - k),$$

where the vertex is (h, k), the focus is $(h, k + p)$, and the directrix is $y = k - p$.

Standard Equation of a Parabola with Vertex (h, k) and Horizontal Axis of Symmetry

The standard equation of a parabola with vertex (h, k) and horizontal axis of symmetry is

$$(y - k)^2 = 4p(x - h),$$

where the vertex is (h, k), the focus is $(h + p, k)$, and the directrix is $x = h - p$.

(continued)

Standard Equation of a Circle

The standard equation of a circle with center (h, k) and radius r is

$$(x - h)^2 + (y - k)^2 = r^2.$$

Standard Equation of an Ellipse with Center at the Origin

Major axis horizontal

$$\frac{x^2}{a^2} + \frac{y^2}{b^2} = 1, a > b > 0$$

Vertices: $(-a, 0), (a, 0)$

y-intercepts: $(0, -b), (0, b)$

Foci: $(-c, 0), (c, 0)$, where $c^2 = a^2 - b^2$

Major axis vertical

$$\frac{x^2}{b^2} + \frac{y^2}{a^2} = 1, a > b > 0$$

Vertices: $(0, -a), (0, a)$

x-intercepts: $(-b, 0), (b, 0)$

Foci: $(0, -c), (0, c)$, where $c^2 = a^2 - b^2$

Standard Equation of an Ellipse with Center at (h, k)

Major axis horizontal

$$\frac{(x - h)^2}{a^2} + \frac{(y - k)^2}{b^2} = 1, a > b > 0$$

Vertices: $(h + a, k), (h - a, k)$

Length of minor axis: $2b$

Foci: $(h - c, k), (h + c, k)$, where $c^2 = a^2 - b^2$

Major axis vertical

$$\frac{(x - h)^2}{b^2} + \frac{(y - k)^2}{a^2} = 1, a > b > 0$$

Vertices: $(h, k - a), (h, k + a)$

Length of minor axis: $2b$

Foci: $(h, k - c), (h, k + c)$, where $c^2 = a^2 - b^2$

Standard Equation of a Hyperbola with Center at the Origin

Transverse axis horizontal

$$\frac{x^2}{a^2} - \frac{y^2}{b^2} = 1$$

Vertices: $(-a, 0), (a, 0)$

Foci: $(-c, 0), (c, 0)$, where $c^2 = a^2 + b^2$

Transverse axis vertical

$$\frac{y^2}{a^2} - \frac{x^2}{b^2} = 1$$

Vertices: $(0, -a), (0, a)$

Foci: $(0, -c), (0, c)$, where $c^2 = a^2 + b^2$

Standard Equation of a Hyperbola with Center at (h, k)

Transverse axis horizontal

$$\frac{(x - h)^2}{a^2} - \frac{(y - k)^2}{b^2} = 1$$

Vertices: $(h - a, k), (h + a, k)$

Asymptotes: $y - k = \frac{b}{a}(x - h)$,

$$y - k = -\frac{b}{a}(x - h)$$

Foci: $(h - c, k), (h + c, k)$, where $c^2 = a^2 + b^2$

Transverse axis vertical

$$\frac{(y - k)^2}{a^2} - \frac{(x - h)^2}{b^2} = 1$$

Vertices: $(h, k - a), (h, k + a)$

Asymptotes: $y - k = \frac{a}{b}(x - h)$,

$$y - k = -\frac{a}{b}(x - h)$$

Foci: $(h, k - c), (h, k + c)$, where $c^2 = a^2 + b^2$

REVIEW EXERCISES

In Exercises 1–8, match each equation with one of the graphs (a)–(h), which follow.

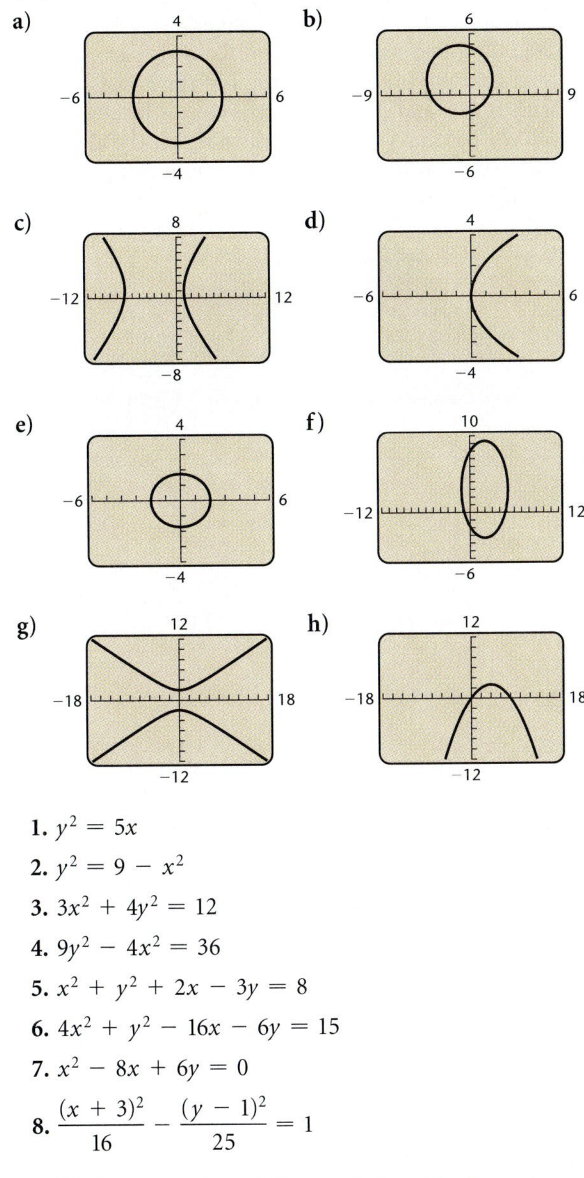

1. $y^2 = 5x$
2. $y^2 = 9 - x^2$
3. $3x^2 + 4y^2 = 12$
4. $9y^2 - 4x^2 = 36$
5. $x^2 + y^2 + 2x - 3y = 8$
6. $4x^2 + y^2 - 16x - 6y = 15$
7. $x^2 - 8x + 6y = 0$
8. $\frac{(x+3)^2}{16} - \frac{(y-1)^2}{25} = 1$

9. Find an equation of the parabola with directrix $y = \frac{3}{2}$ and focus $\left(0, -\frac{3}{2}\right)$.

10. Find the focus, the vertex, and the directrix of the parabola given by
$$y^2 = -12x.$$

11. Find the vertex, the focus, and the directrix of the parabola given by
$$x^2 + 10x + 2y + 9 = 0.$$

12. Find the center, the vertices, and the foci of the ellipse given by
$$16x^2 + 25y^2 - 64x + 50y - 311 = 0.$$
Then draw the graph.

13. Find an equation of the ellipse having vertices $(0, -4)$ and $(0, 4)$ with minor axis of length 6.

14. Find the center, the vertices, the foci, and the asymptotes of the hyperbola given by
$$x^2 - 2y^2 + 4x + y - \tfrac{1}{8} = 0.$$

15. *Spotlight.* A spotlight has a parabolic cross section that is 2 ft wide at the opening and 1.5 ft deep at the vertex. How far from the vertex is the focus?

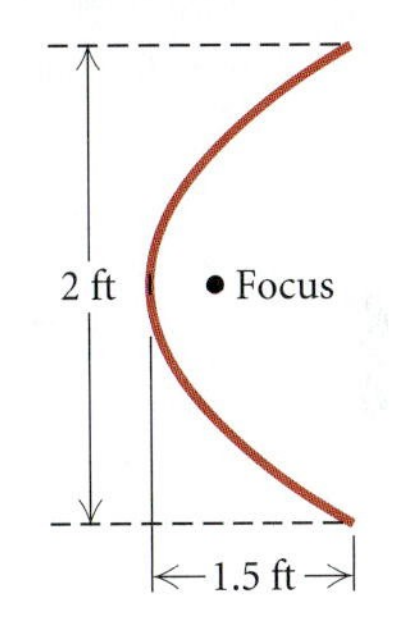

Solve.

16. $x^2 - 16y = 0,$
$x^2 - y^2 = 64$

17. $4x^2 + 4y^2 = 65,$
$6x^2 - 4y^2 = 25$

18. $x^2 - y^2 = 33,$
$x + y = 11$

19. $x^2 - 2x + 2y^2 = 8,$
$2x + y = 6$

20. $x^2 - y = 3,$
$2x - y = 3$

21. $x^2 + y^2 = 25,$
$x^2 - y^2 = 7$

22. $x^2 - y^2 = 3,$
$y = x^2 - 3$

23. $x^2 + y^2 = 18,$
$2x + y = 3$

24. $x^2 + y^2 = 100,$
$2x^2 - 3y^2 = -120$

25. $x^2 + 2y^2 = 12,$
$xy = 4$

26. *Numerical Relationship.* The sum of two numbers is 11 and the sum of their squares is 65. Find the numbers.

27. *Dimensions of a Rectangle.* A rectangle has a perimeter of 38 m and an area of 84 m^2. What are the dimensions of the rectangle?

28. *Numerical Relationship.* Find two positive integers whose sum is 12 and the sum of whose reciprocals is $\frac{3}{8}$.

29. *Perimeter.* The perimeter of a square is 12 cm more than the perimeter of another square. The area of the first square exceeds the area of the other by 39 cm^2. Find the perimeter of each square.

30. *Radius of a Circle.* The sum of the areas of two circles is 130π ft^2. The difference of the areas is 112π ft^2. Find the radius of each circle.

Synthesis

31. ◆ What would you say to a classmate who tells you that an algebraic solution of a nonlinear system of equations is always preferable to a graphical solution?

32. ◆ Is a circle a special type of ellipse? Why or why not?

33. Find an equation of the ellipse containing the point $(-1/2, 3\sqrt{3}/2)$ and with vertices $(0, -3)$ and $(0, 3)$.

34. Find two numbers whose product is 4 and the sum of whose reciprocals is $\frac{65}{56}$.

35. Find an equation of the circle that passes through the points $(10, 7)$, $(-6, 7)$, and $(-8, 1)$.

36. *Navigation.* Two radio transmitters positioned 400 mi apart send simultaneous signals to a ship that is 250 mi offshore, sailing parallel to the shoreline. The signal from transmitter A reaches the ship 300 microseconds before the signal from transmitter B. The signals travel at a speed of 186,000 miles per second, or 0.186 mile per microsecond. Find the equation of the hyperbola with foci A and B on which the ship is located. (*Hint:* For any point on the hyperbola, the absolute value of the difference of its distances from the foci is $2a$.)

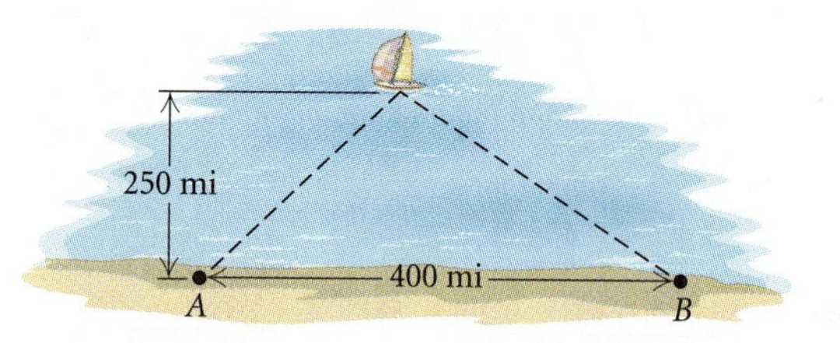

Sequences, Series, and Combinatorics 9

APPLICATION

The following sequence involves world bicycle production in years starting with 1988. We consider year 1 to be 1988. The sequence is

$$105, 95, 90, 96, 103, 108.$$

We can visualize the sequence in a table and a scatterplot, or xy-graph.

A desire to calculate odds in games of chance gave rise to the *theory of probability,* which today has many applications to business, medicine, sociology, and science.

The first part of this chapter is devoted to *sequence* and *series.* For example, when we list world bicycle production for various years, a *sequence* is being formed. When the members of a sequence are numbers, we can think of adding them. Such a sum is called a *series.* We also study a method of proof known as *mathematical induction,* which enables us to prove many important mathematical results.

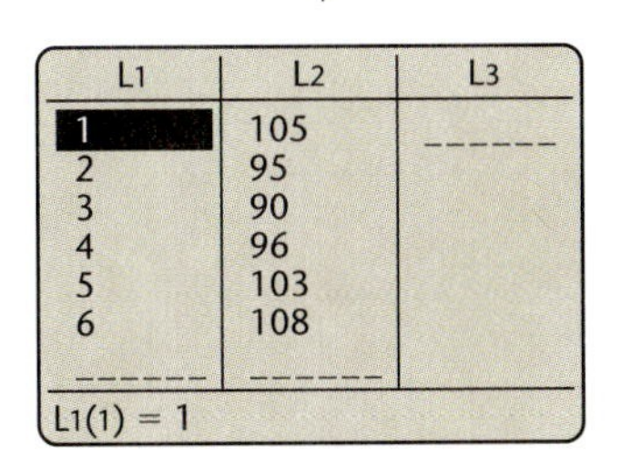

L1	L2	L3
1	105	------
2	95	
3	90	
4	96	
5	103	
6	108	
------	------	

L1(1) = 1

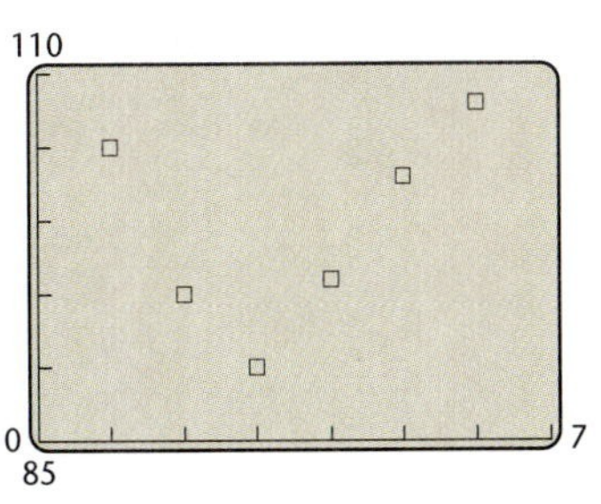

9.1 Sequences and Series
9.2 Arithmetic Sequences and Series
9.3 Geometric Sequences and Series
9.4 Mathematical Induction
9.5 Combinatorics: Permutations
9.6 Combinatorics: Combinations
9.7 The Binomial Theorem
9.8 Probability
SUMMARY AND REVIEW

9.1 Sequences and Series

- *Find terms of sequences given the nth term.*
- *Look for a pattern in a sequence and try to determine a general term.*
- *Convert between sigma notation and other notation for a series.*
- *Given a recursively defined sequence, construct terms.*

In this section, we discuss sets of numbers, considered in order, and their sums.

Sequences

Suppose that \$1000 is invested at 6%, compounded annually. The amounts to which the account will grow after 1 yr, 2 yr, 3 yr, 4 yr, and so on, are as follows:

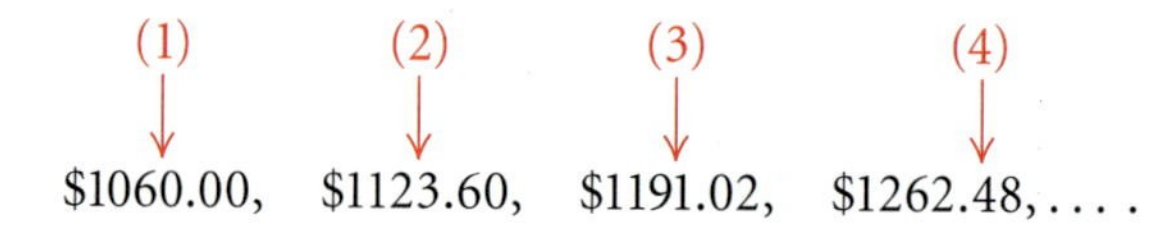

X	Y1	
1	1060	
2	1123.6	
3	1191	
4	1262.5	
5	1338.2	
6	1418.5	
7	1503.6	
X = 7		

(We found these numbers using the compound interest formula, $A = P(1 + i)^n$, expressed as $y_1 = 1000(1.06)^x$, in order to create the table on a grapher. Note the rounding error in the table at left.) Note that we can think of this ordered list of numbers as a function such that the input **1** corresponds to the number \$1060.00, **2** to the number \$1123.60, **3** to the number \$1191.02, **4** to the number \$1262.48, and so on. A **sequence** is thus a *function*, where the domain is a set of consecutive positive integers beginning with 1. (Sometimes sequences start with 0, but we will seldom consider such cases.)

If we continue to compute the amounts of money in the account forever, we obtain an **infinite sequence**:

\$1060.00, \$1123.60, \$1191.02, \$1262.48, \$1338.23, \$1418.52, \$1503.63, \$1593.85, ...

The three dots at the end ". . ." indicate that the sequence goes on without stopping. If we stop after a certain number of years, we obtain a **finite sequence**:

\$1060.00, \$1123.60, \$1191.02, \$1262.48.

Sequences

An *infinite sequence* is a function having for its domain the set of positive integers:

$$\{1, 2, 3, 4, 5, \ldots\}.$$

A *finite sequence* is a function having for its domain a set of positive integers:

$$\{1, 2, 3, 4, 5, \ldots, n\}$$

for some positive integer n.

As another example, consider the sequence given by the formula

$$a(n) = 2^n, \quad \text{or} \quad a_n = 2^n.$$

The notation a_n means the same as $a(n)$ but is more commonly used with sequences. Some of the function values (also known as the **terms** of the sequence) are as follows:

$$a_1 = 2^1 = 2,$$
$$a_2 = 2^2 = 4,$$
$$a_3 = 2^3 = 8,$$
$$a_4 = 2^4 = 16,$$
$$a_5 = 2^5 = 32.$$

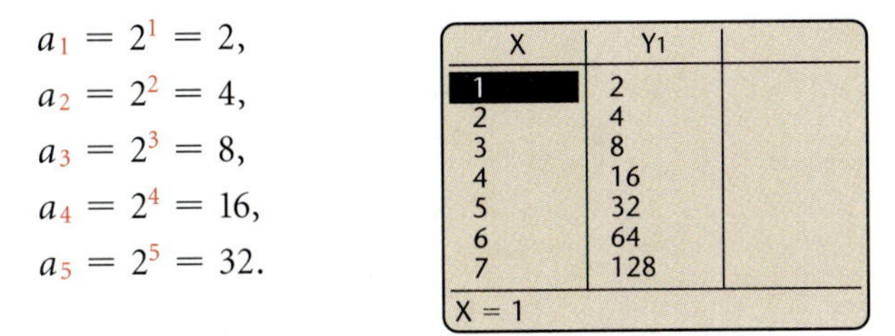

X	Y1
1	2
2	4
3	8
4	16
5	32
6	64
7	128

X = 1

The first term of the sequence is a_1, the fifth term is a_5, and the nth term, or **general term**, is a_n. This sequence can also be denoted in the following ways:

a) 2, 4, 8, . . . ;

b) $a_1, a_2, a_3, a_4, \ldots, a_n, \ldots$;

c) $2, 4, 8, \ldots, 2^n, \ldots$.

Example 1 Find the first 4 terms and the 37th term of the sequence whose general term is given by

$$a_n = \frac{(-1)^n}{(n + 1)}.$$

SOLUTION We have the following:

$$a_1 = \frac{(-1)^1}{1 + 1} = -\frac{1}{2},$$

$$a_2 = \frac{(-1)^2}{2 + 1} = \frac{1}{3},$$

$$a_3 = \frac{(-1)^3}{3 + 1} = -\frac{1}{4},$$

$$a_4 = \frac{(-1)^4}{4 + 1} = \frac{1}{5},$$

$$a_{37} = \frac{(-1)^{37}}{37 + 1} = -\frac{1}{38}.$$

Thus the first 4 terms are $-\frac{1}{2}, \frac{1}{3}, -\frac{1}{4}$, and $\frac{1}{5}$, and the 37th term is $-\frac{1}{38}$.

Note in Example 1 that the power $(-1)^n$ causes the signs of the terms to alternate between positive and negative, depending on whether n is even or odd.

Finding the General Term

When only the first few terms of a sequence are known, we do not know for sure what the general term is, but we can make a prediction by looking for a pattern.

Example 2 For each of the following sequences, predict the general term.

a) 1, 4, 9, 16, 25, ...

b) 1, $\sqrt{2}$, $\sqrt{3}$, 2, ...

c) −1, 3, −9, 27, −81, ...

d) 2, 4, 8, ...

SOLUTION

RESULT	REASON
a) n^2	These are squares of consecutive positive integers, so the general term may be n^2.
b) $\sqrt{n}$	These are square roots of consecutive integers, so the general term may be $\sqrt{n}$.
c) $(-1)^n[3^{n-1}]$	These are powers of 3 with alternating signs, so the general term may be $(-1)^n[3^{n-1}]$.
d) Uncertain	If we see the pattern of powers of 2, we will see 16 as the next term and guess 2^n for the general term. Then the sequence could be written with more terms as 2, 4, 8, 16, 32, 64, 128, ... If we see that we can get the second term by adding 2, the third term by adding 4, and the next term by adding 6, and so on, we will see 14 as the next term. A general term for the sequence is then $n^2 - n + 2$, and the sequence can be written with more terms as 2, 4, 8, 14, 22, 32, 44, 58, ...

Example 2(d) illustrates that, in fact, you can never be certain about the general term. The fewer the given terms, the greater the uncertainty.

Visualizing Sequences

There are many ways in which we can see and use sequences. The following sequence involves world bicycle production in years starting with 1988. We consider year 1 to be 1988. The sequence is

105, 95, 90, 96, 103, 108.

We can visualize the sequence in a table, a bar graph, and a scatterplot, or *xy*-graph, as follows.

TABLE

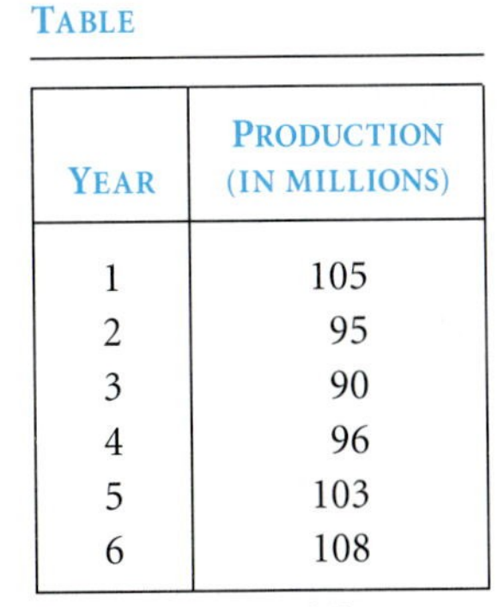

YEAR	PRODUCTION (IN MILLIONS)
1	105
2	95
3	90
4	96
5	103
6	108

Source: UN Interbike Directory.

BAR GRAPH

xy-GRAPH (SCATTERPLOT)

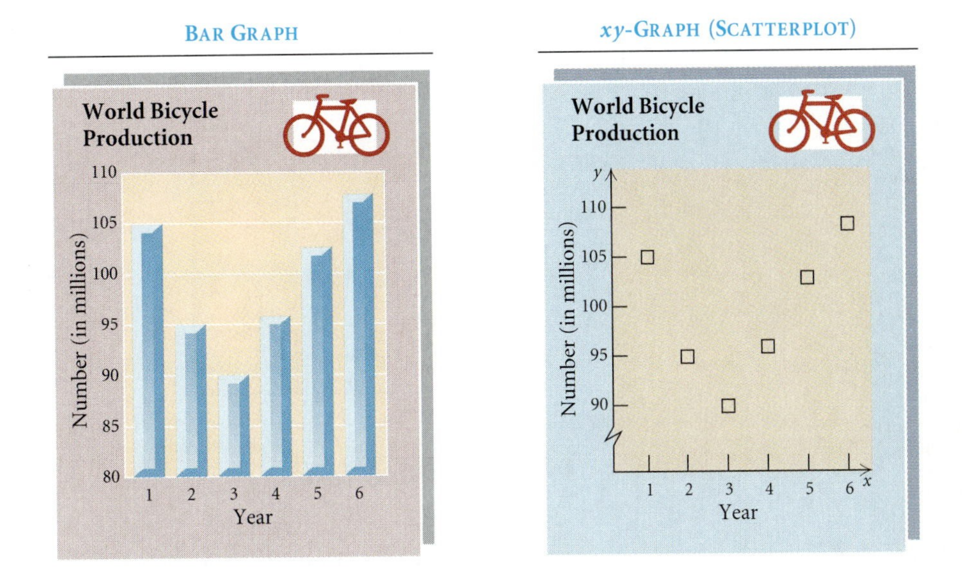

On a grapher, we can create lists and tables and draw graphs. Some graphers have special procedures in order to deal with sequences. For example, we might use a SEQUENCE mode, which allows us to give start (first) and stop (last) values of n. Generally, we start with a value of 1, but there might be situations in which we would start with 0. Then a STEP would be specified, which establishes the distance between values of n. Generally, we use Step = 1, giving 1, 2, 3, 4, ..., but there may be situations in which we might want to count by two's: 0, 2, 4, 6, ..., and so on. The window in a SEQUENCE mode will generally allow us to make these adjustments in sequences and in size of viewing window. Consult the grapher manual for more details. If available, use DOT mode rather than CONNECTED mode.

Program

LOAN: This program can be used to compute various parts of mortgage payments. Payments made to repay a loan form a sequence.

Another way to work with sequences on graphers is to use a TABLE feature or a software package that creates spreadsheets, such as Lotus 1-2-3 or Excel. A sequence is specified by a formula, a first value and a last value of n, and a step size. We can handle sequences in much the same way as we do other functions. The only difference is often that instead of entering an expression such as "$x/(x + 1)$" at an "$f(x) =$" or a "$y =$" prompt, we must enter a formula such as "$n/(n + 1)$" as a "$U_n =$" prompt. We can write the general term of a sequence by considering a function whose domain is, say, the entire set of real numbers, such as

$$f(x) = x^2 + 1,$$

and restricting its domain to the natural numbers to consider it as

$$a_n = f(n) = n^2 + 1.$$

Example 3 Construct a table of values and a graph for the first 10 terms of the sequence whose general term is given by

$$a_n = \frac{n}{n + 1}.$$

Use both a grapher and a spreadsheet program, if available.

SOLUTION We use {1, 2, 3, 4, 5, 6, 7, 8, 9, 10} as the domain.

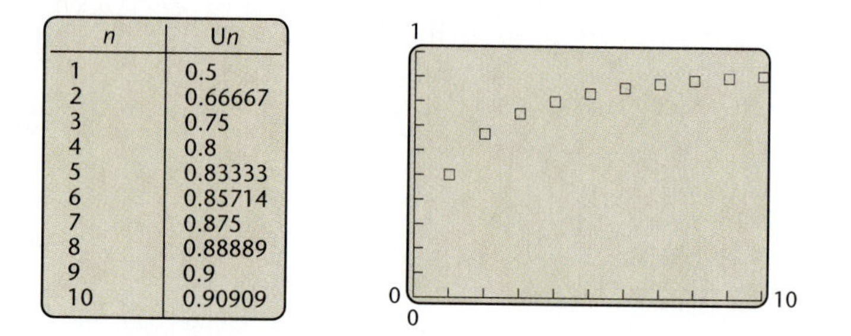

n	Un
1	0.5
2	0.66667
3	0.75
4	0.8
5	0.83333
6	0.85714
7	0.875
8	0.88889
9	0.9
10	0.90909

Sums and Series

Series

Given the infinite sequence

$$a_1, a_2, a_3, a_4, \ldots, a_n, \ldots,$$

the sum of the terms

$$a_1 + a_2 + a_3 + \cdots + a_n + \cdots$$

is called an *infinite series.* A *partial sum* is the sum of the first n terms:

$$a_1 + a_2 + a_3 + \cdots + a_n.$$

A partial sum is also called a *finite series,* or *nth partial sum,* and is denoted S_n.

Consider the sequence

$$3, 5, 7, 9, \ldots, 2n + 1, \ldots.$$

We construct some partial sums:

$S_1 = 3,$ This is the first term of the given sequence.

$S_2 = 3 + 5 = 8,$ The sum of the first two terms (second partial sum)

$S_3 = 3 + 5 + 7 = 15,$ The sum of the first three terms (third partial sum)

$S_4 = 3 + 5 + 7 + 9 = 24.$ The sum of the first four terms (fourth partial sum)

Note that if we write these partial sums in order, we create a new sequence:

$$3, 8, 15, 24, \ldots.$$

Example 4 For the sequence $-2, 4, -6, 8, -10, 12, -14, \ldots,$ find each of the following.

a) S_3

b) S_5

SOLUTION

a) $S_3 = -2 + 4 + (-6) = -4$

b) $S_5 = -2 + 4 + (-6) + 8 + (-10) = -6$

Sigma Notation: Σ

The Greek letter Σ (sigma) can be used to simplify notation when the general term of a series is a formula. For example, the sum of the first four terms of the sequence $3, 5, 7, 9, \ldots, 2k + 1, \ldots$ can be named as follows, using what is called **sigma notation**, or **summation notation**:

$$\sum_{k=1}^{4} (2k + 1).$$

This is read "the sum as k goes from 1 to 4 of $(2k + 1)$." The letter k is called the **index of summation**. Sometimes the index of summation starts at a number other than 1, and sometimes other letters are used rather than k.

Example 5 Find and evaluate each of the following sums.

a) $\sum_{k=1}^{5} k^3$

b) $\sum_{k=0}^{4} (1)^k 5^k$

c) $\sum_{i=8}^{11} \left(2 + \frac{1}{i}\right)$

SOLUTION

a) We replace k by 1, 2, 3, 4, and 5. Then we add the results.

$$\sum_{k=1}^{5} k^3 = 1^3 + 2^3 + 3^3 + 4^3 + 5^3 = 1 + 8 + 27 + 64 + 125 = 225$$

b)
$$\sum_{k=0}^{4} (-1)^k 5^k = (-1)^0 5^0 + (-1)^1 5^1 + (-1)^2 5^2 + (-1)^3 5^3 + (-1)^4 5^4$$
$$= 1 - 5 + 25 - 125 + 625 = 521$$

c)
$$\sum_{i=8}^{11} \left(2 + \frac{1}{i}\right) = \left(2 + \frac{1}{8}\right) + \left(2 + \frac{1}{9}\right) + \left(2 + \frac{1}{10}\right) + \left(2 + \frac{1}{11}\right)$$
$$= 8\frac{1691}{3960}$$

Some graphers have a SUM SEQ feature that will generate partial sums of sequences automatically.

Example 6 Write sigma notation for each of the following.

a) $-2 + 4 - 6 + 8 - 10$

b) $1 + 4 + 8 + 16 + 32 + 64 + \cdots$

c) $x + \frac{x^2}{2} + \frac{x^3}{3} + \frac{x^4}{4}$

SOLUTION

RESULT	REASON
a) $-2 + 4 - 6 + 8 - 10 = \sum_{k=1}^{5} (-1)^k(2k)$	These are even integers with alternating signs. Therefore, the general term is $(-1)^k(2k)$, beginning with $k = 1$.
b) $1 + 4 + 8 + 16 + 32 + 64 + \cdots = \sum_{k=0}^{\infty} 2^k$	This is a sum of powers of 2, and it is also an infinite series. We use the symbol ∞ to represent infinity.
c) $x + \frac{x^2}{2} + \frac{x^3}{3} + \frac{x^4}{4} = \sum_{k=1}^{4} \frac{x^k}{k}$	This is a sum of expressions x^k/k.

Recursive Definitions

A sequence may be defined by **recursive definition**. Such a definition lists the first term, or the first few terms, and then tells how to determine the remaining terms from the given terms.

Example 7 Find the first 5 terms of the sequence defined by

$$a_1 = 5, \qquad a_{k+1} = 2a_k - 3, \quad \text{for } k \geq 1.$$

SOLUTION

$$a_1 = 5,$$

$$a_2 = 2a_1 - 3 = 2 \cdot 5 - 3 = 7,$$

$$a_3 = 2a_2 - 3 = 2 \cdot 7 - 3 = 11,$$

$$a_4 = 2a_3 - 3 = 2 \cdot 11 - 3 = 19,$$

$$a_5 = 2a_4 - 3 = 2 \cdot 19 - 3 = 35.$$

The recursive definition in Example 7 can also be expressed as follows:

$$a_0 = 5, \qquad a_k = 2a_{k-1} - 3, \quad \text{for } k \geq 1,$$

where a_1 is found by substituting 1 for k.

Many graphers have the capability of working with recursively defined sequences. For Example 7, the recursive function might be entered as $U_n = 2 * U_{n-1} - 3$ with U_nStart $= 5$.

Example 8 *The Fibonacci Sequence.* One of the most famous recursively defined sequences is the *Fibonacci sequence*. In 1202, Leonardo Fibonacci (also called Leonardo da Pisa), an Italian mathematician, proposed a model for rabbit population growth as follows. We start with one pair of rabbits, one female and one male. These rabbits mature to an age of reproductivity and produce a new pair, again one female and one male. The former pair endures until each pair can reproduce, and then each produces a new pair. This pattern continues. The population of rabbits can be modeled by the following recursively defined sequence:

$$a_1 = 1, \qquad a_2 = 1, \qquad a_{k+1} = a_k + a_{k-1}, \quad \text{for } k \geq 2,$$

where $a_k =$ the total number of pairs of rabbits after $k - 2$ reproductions for $k \geq 2$. Find the first 7 terms of the Fibonacci sequence.

SOLUTION We have

$$\begin{aligned}
a_1 &= 1, \\
a_2 &= 1, \\
a_3 &= a_2 + a_1 = 1 + 1 = 2, \\
a_4 &= a_3 + a_2 = 2 + 1 = 3, \\
a_5 &= a_4 + a_3 = 3 + 2 = 5, \\
a_6 &= a_5 + a_4 = 5 + 3 = 8, \\
a_7 &= a_6 + a_5 = 8 + 5 = 13.
\end{aligned}$$

The following figure illustrates the Fibonacci sequence.

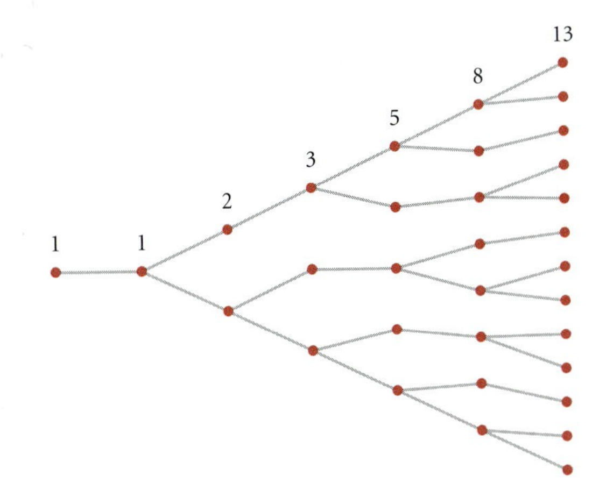

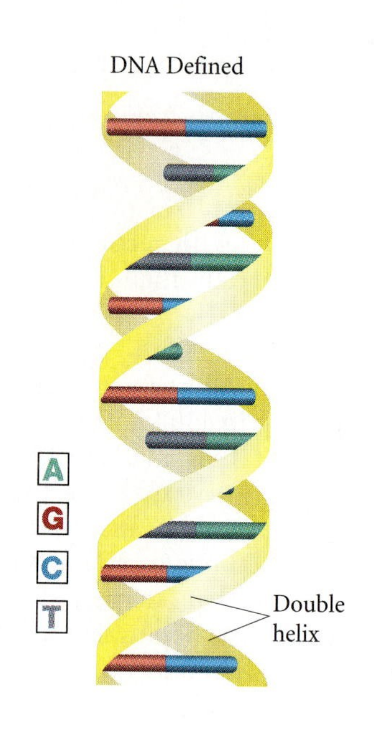

Other Applications of Sequences

DIGITAL TECHNOLOGY Digital systems use sequences of *bits*, or binary digits (0 and 1), which are used in the base-two number system. Since computer components are typically two-state mechanisms (ON–OFF), the bit is the fundamental unit of information. The compact disc (CD), fax machine, modem, and CD-ROM are other devices that use digital technology.

DNA In 1953, James Watson and Francis Crick discovered that DNA contains the "blueprint" for the construction of each individual human. The genetic code in DNA is the chemical sequence by which hereditary data are translated from genes into proteins, which serve as structural material and chemical regulators for the human body. The genetic code is composed of a sequence of four molecules: adenine (A), cytosine (C), guanine (G), and thymine (T). (These letters could just as well be represented by numbers.) The complete collection of human DNA is called the human *genome*.

9.1 Exercise Set

In each of the following, the nth term of a sequence is given. Find the first 4 terms, a_{10}, and a_{15}.

1. $a_n = 4n - 1$

2. $a_n = (n - 1)(n - 2)(n - 3)$

3. $a_n = \dfrac{n}{n - 1}, \; n \geq 2$

4. $a_n = n^2 - 1, \; n \geq 3$

5. $a_n = \dfrac{n^2 - 1}{n^2 + 1}$

6. $a_n = \left(-\dfrac{1}{2}\right)^{n-1}$

7. $a_n = (-1)^n n^2$

8. $a_n = (-1)^{n-1}(3n - 5)$

9. $a_n = 5 + \dfrac{(-2)^{n+1}}{2^n}$

10. $a_n = \dfrac{2n - 1}{n^2 + 2n}$

Find the indicated term of the given sequence.

11. $a_n = 5n - 6$; a_8

12. $a_n = (3n - 4)(2n + 5)$; a_7

13. $a_n = (2n + 3)^2$; a_6

14. $a_n = (-1)^{n-1}(4.6n - 18.3)$; a_{12}

15. $a_n = 5n^2(4n - 100)$; a_{11}

16. $a_n = \left(1 + \dfrac{1}{n}\right)^2$; a_{80}

17. $a_n = \ln e^n$; a_{67}

18. $a_n = 2 - \dfrac{1000}{n}$; a_{100}

Predict the general, or nth term, a_n, of each sequence. Answers may vary.

19. 2, 4, 6, 8, 10, ...

20. 3, 9, 27, 81, 243, ...

21. −2, 6, −18, 54, ...

22. −2, 3, 8, 13, 18, ...

23. $\frac{2}{3}, \frac{3}{4}, \frac{4}{5}, \frac{5}{6}, \frac{6}{7}, \ldots$

24. $\sqrt{2}, 2, \sqrt{6}, 2\sqrt{2}, \sqrt{10}, \ldots$

25. $1 \cdot 2, 2 \cdot 3, 3 \cdot 4, 4 \cdot 5, \ldots$

26. −1, −4, −7, −10, −13, ...

27. 0, log 10, log 100, log 1000, ...

28. $\ln e^2, \ln e^3, \ln e^4, \ln e^5, \ldots$

Find the indicated partial sums for each sequence.

29. 1, 2, 3, 4, 5, 6, 7, ...; S_3 and S_7

30. 1, −3, 5, −7, 9, −11, ...; S_2 and S_5

31. 2, 4, 6, 8, ...; S_4 and S_5

32. $1, \frac{1}{4}, \frac{1}{9}, \frac{1}{16}, \frac{1}{25}, \ldots;\ S_1$ and S_5

Find and evaluate each sum.

33. $\sum_{k=1}^{5} \frac{1}{2k}$

34. $\sum_{i=1}^{6} \frac{1}{2i+1}$

35. $\sum_{i=0}^{6} 2^i$

36. $\sum_{k=4}^{7} \sqrt{2k-1}$

37. $\sum_{k=7}^{10} \ln k$

38. $\sum_{k=1}^{4} \pi k$

39. $\sum_{k=1}^{8} \frac{k}{k+1}$

40. $\sum_{i=1}^{5} \frac{i-1}{i+3}$

41. $\sum_{i=1}^{5} (-1)^i$

42. $\sum_{k=0}^{5} (-1)^{k+1}$

43. $\sum_{k=1}^{8} (-1)^{k+1} 3k$

44. $\sum_{k=0}^{7} (-1)^k 4^{k+1}$

45. $\sum_{k=0}^{6} \frac{2}{k^2+1}$

46. $\sum_{i=1}^{10} i(i+1)$

47. $\sum_{k=0}^{5} (k^2 - 2k + 3)$

48. $\sum_{k=1}^{10} \frac{1}{k(k+1)}$

49. $\sum_{i=0}^{10} \frac{2^i}{2^i+1}$

50. $\sum_{k=0}^{3} (-2)^{2k}$

Write sigma notation.

51. $5 + 10 + 15 + 20 + 25 + \cdots$

52. $7 + 14 + 21 + 28 + 35 + \cdots$

53. $2 - 4 + 8 - 16 + 32 - 64$

54. $3 + 6 + 9 + 12 + 15$

55. $-\frac{1}{2} + \frac{2}{3} - \frac{3}{4} + \frac{4}{5} - \frac{5}{6} + \frac{6}{7}$

56. $\frac{1}{1^2} + \frac{1}{2^2} + \frac{1}{3^2} + \frac{1}{4^2} + \frac{1}{5^2}$

57. $4 - 9 + 16 - 25 + \cdots + (-1)^n n^2$

58. $9 - 16 + 25 + \cdots + (-1)^{n+1} n^2$

59. $\frac{1}{1 \cdot 2} + \frac{1}{2 \cdot 3} + \frac{1}{3 \cdot 4} + \frac{1}{4 \cdot 5} + \cdots$

60. $\frac{1}{1 \cdot 2^2} + \frac{1}{2 \cdot 3^2} + \frac{1}{3 \cdot 4^2} + \frac{1}{4 \cdot 5^2} + \cdots$

Find the first 4 terms of each recursively defined sequence.

61. $a_1 = 4,\ a_{k+1} = 1 + \frac{1}{a_k}$

62. $a_1 = 256,\ a_{k+1} = \sqrt{a_k}$

63. $a_1 = 6561,\ a_{k+1} = (-1)^k \sqrt{a_k}$

64. $a_1 = e^Q,\ a_{k+1} = \ln a_k$

65. $a_1 = 2,\ a_2 = 3,\ a_{k+1} = a_k + a_{k-1}$

66. $a_1 = -10,\ a_2 = 8,\ a_{k+1} = a_k - a_{k-1}$

Construct a table of values and a graph for the first 10 terms of each sequence. Use both a grapher and a spreadsheet program, if available.

67. $a_n = \left(1 + \frac{1}{n}\right)^n$

68. $a_n = \sqrt{n+1} - \sqrt{n}$

69. $a_1 = 2,\ a_{k+1} = \sqrt{1 + \sqrt{a_k}}$

70. $a_1 = 2,\ a_{k+1} = \frac{1}{2}\left(a_k + \frac{2}{a_k}\right)$

71. *The Gap.* The Gap, Inc., is a national clothing firm that has experienced tremendous growth (*Source:* Gap, Inc., annual report). Total sales, in millions of dollars, can be estimated by the sequence model

$$a_n = 436.27n + 1577.84,\quad n = 1, 2, 3, \ldots.$$

The year 1990 corresponds to $n = 1$.

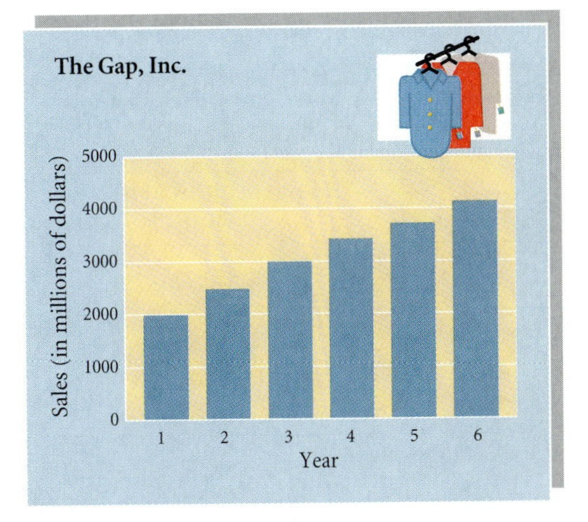

a) Find the first 10 terms of this sequence.

b) Estimate the total sales from 1990 to 1999 by evaluating the sum

$$\sum_{n=1}^{10} (436.27n + 1577.84).$$

72. *Driver Fatalities by Age.* The number of licensed drivers per 100,000 who died in motor vehicle accidents at age n can be estimated by the sequence model

$$a_n = 0.018n^2 - 1.7n + 48.7,\quad n = 15, 16, 17, \ldots, 90.$$

a) Make a table or list of the first 10 terms of the sequence.

b) Construct a graph of the sequence.

c) Find the number of fatalities per 100,000 of those drivers whose age is 16 yr, 23 yr, 50 yr, 75 yr, and 85 yr.

73. *Compound Interest.* Suppose that $1000 is invested at 6.2%, compounded annually. The value of the investment after n years is given by the sequence model

$$a_n = \$1000(1.062)^n, \quad n = 1, 2, 3, \ldots.$$

a) Find the first 10 terms of the sequence.
b) Find the amount of the investment after 20 yr.

74. *Salvage Value.* The value of an office machine is $5200. Its salvage value each year is 75% of its value the year before. Give a sequence that lists the salvage value of the machine for each year of a 10-yr period.

75. *Bacteria Growth.* A single cell of bacteria divides into two every 15 min. Suppose that the same rate of division is maintained for 4 hr. Give a sequence that lists the number of cells after successive 15-min periods.

76. *Salary Sequence.* Torrey is paid $6.30 per hour for working at Red Freight Limited. Each year he receives a $0.40 hourly raise. Give a sequence that lists Torrey's hourly salary over a 10-yr period.

77. *Household Discretionary Income by Age.* The following table contains factual data relating household discretionary income to age.

AGE, x	HOUSEHOLD DISCRETIONARY INCOME, y
25	$10,223
34	15,607
54	21,173
64	18,257
70	13,747

Source: U.S. Bureau of the Census.

a) Use the REGRESSION feature of a grapher to fit a quadratic sequence function

$$a_n = an^2 + bn + c$$

to the data.
b) Find the household discretionary income of people whose age is 16 yr, 20 yr, 40 yr, 60 yr, and 75 yr.

Skill Maintenance

78. Solve for S_n: $2S_n = n(a_1 + a_n)$.

79. Collect like terms:

$$a_1 + (a_1 + d) + (a_2 + 2d) + (a_n - 2d) + (a_n - d) + a_n.$$

Simplify.

80. $\log_a a$

81. $\log_a 1$

Synthesis

82. ◆ The Fibonacci sequence has intrigued mathematicians for centuries. In fact, a journal called the *Fibonacci Quarterly* is devoted to publishing the results pertaining to such sequences. Do some library research on the connection of the Fibonacci sequence to the idea of the "Golden Section."

83. ◆

a) Find the first few terms of the sequence $a_n = n^2 - n + 41$ and describe the pattern you observe.
b) Does the pattern you found in part (a) hold for all choices of n? Why or why not?

Find the first 5 terms of the sequence, and then find S_5.

84. $a_n = \dfrac{1}{2^n} \log 1000^n$

85. $a_n = i^n$, $i = \sqrt{-1}$

86. $a_n = \ln (1 \cdot 2 \cdot 3 \cdots n)$

For each sequence, find a formula for S_n.

87. $a_n = \ln n$

88. $a_n = \dfrac{1}{n} - \dfrac{1}{n + 1}$

9.2 Arithmetic Sequences and Series

- *For any arithmetic sequence, find the nth term when n is given and n when the nth term is given, and given two terms, find the common difference and construct the sequence.*
- *Find the sum of the first n terms of an arithmetic sequence.*
- *Insert arithmetic means between two numbers.*

If we begin with a particular first term and then add the same number successively, we obtain an **arithmetic sequence**. In this section, we study arithmetic sequences and series.

Arithmetic Sequences

Consider this sequence:

$$2, 5, 8, 11, 14, 17, \dots .$$

Note that adding 3 to any term produces the following term. In other words, the difference between any term and the preceding one is 3. This is an *arithmetic* (pronounced: ăr′ĭth-mĕt′-ĭk) *sequence.*

Arithmetic Sequence

A sequence is *arithmetic* if there exists a number d, called the *common difference,* such that

$$a_{n+1} = a_n + d, \quad \text{or} \quad a_{n+1} - a_n = d, \quad \text{for } n \geq 1.$$

Arithmetic sequences are also called **arithmetic progressions.**

Example 1 For each of the following arithmetic sequences, identify the first term, a_1, and the common difference, d.

a) $4, 9, 14, 19, 24, \dots$

b) $34, 27, 20, 13, 6, -1, -8, \dots$

c) $2, 2\frac{1}{2}, 3, 3\frac{1}{2}, 4, 4\frac{1}{2}, \dots$

SOLUTION To find the first term, a_1, we determine the first term listed. To find the common difference, d, we choose any term beyond the first and subtract the preceding term from it.

SEQUENCE	FIRST TERM, a_1	COMMON DIFFERENCE, d
a) $4, 9, 14, 19, 24, \dots$	4	$5 \quad (9 - 4 = 5)$
b) $34, 27, 20, 13, 6, -1, -8, \dots$	34	$-7 \quad (20 - 27 = -7)$
c) $2, 2\frac{1}{2}, 3, 3\frac{1}{2}, 4, 4\frac{1}{2}, \dots$	2	$\frac{1}{2} \quad \left(2\frac{1}{2} - 2 = \frac{1}{2}\right)$

We obtain the common difference by subtracting a_1 from a_2. Had we subtracted a_2 from a_3 or a_3 from a_4, we would have obtained the same values for d. Thus we can check by adding d to each term in a sequence to see if we progress correctly to the next term.

CHECK:

a) $4 + 5 = 9, \quad 9 + 5 = 14, \quad 14 + 5 = 19, \quad 19 + 5 = 24$

b) $34 + (-7) = 27, \quad 27 + (-7) = 20, \quad 20 + (-7) = 13,$
$13 + (-7) = 6, \quad 6 + (-7) = -1, \quad -1 + (-7) = -8$

c) $2 + \frac{1}{2} = 2\frac{1}{2}, \quad 2\frac{1}{2} + \frac{1}{2} = 3, \quad 3 + \frac{1}{2} = 3\frac{1}{2}$

To find a formula for the general, or nth, term of any arithmetic sequence, we denote the common difference by d, write out the first few terms, and look for a pattern:

$a_1,$

$a_2 = a_1 + d,$

$a_3 = a_2 + d = (a_1 + d) + d = a_1 + 2d$ — Substituting for a_2

$a_4 = a_3 + d = (a_1 + 2d) + d = a_1 + 3d$ — Substituting for a_3

Note that the coefficient of d in each case is 1 less than the subscript.

Generalizing, we obtain the following formula.

nth Term of an Arithmetic Sequence

The *nth term* of an arithmetic sequence is given by

$$a_n = a_1 + (n - 1)d, \quad \text{for any } n \geq 1.$$

Program

AP: This program finds the nth term of an arithmetic sequence.

Example 2 Find the 14th term of the arithmetic sequence 4, 7, 10, 13, ...

SOLUTION We first note that $a_1 = 4$, $d = 3$, and $n = 14$. Then using the formula for the nth term, we obtain

$$a_n = a_1 + (n - 1)d$$
$$a_{14} = 4 + (14 - 1) \cdot 3 = 4 + 13 \cdot 3 = 4 + 39 = 43.$$

The 14th term is 43.

Example 3 In the sequence of Example 2, which term is 301? That is, find n if $a_n = 301$.

SOLUTION We substitute $a_n = 301$, $a_1 = 4$, and $d = 3$ into the formula for the nth term and solve for n:

$$a_n = a_1 + (n - 1)d$$
$$301 = 4 + (n - 1) \cdot 3 \quad \text{Substituting}$$
$$301 = 4 + 3n - 3$$
$$301 = 3n + 1$$
$$300 = 3n$$
$$100 = n. \quad \text{Solving for } n$$

The term 301 is the 100th term of the sequence.

Given two terms and their places in an arithmetic sequence, we can construct the sequence.

Example 4 The 3rd term of an arithmetic sequence is 8, and the 16th term is 47. Find a_1 and d and construct the sequence.

SOLUTION We know that $a_3 = 8$ and $a_{16} = 47$. Thus we would have to add d 13 times to get 47 from 8. That is,

$$8 + 13d = 47. \quad a_3 \text{ and } a_{16} \text{ are 13 terms apart.}$$

Solving $8 + 13d = 47$, we obtain

$$13d = 39$$
$$d = 3.$$

Since $a_3 = 8$, we subtract d twice to get a_1. Thus,

$$a_1 = 8 - 2 \cdot 3 = 2.$$

The sequence is 2, 5, 8, 11, ... Note that we could also subtract d 15 times from a_{16} in order to find a_1.

In general, d should be subtracted $n - 1$ times from a_n in order to find a_1.

Interactive Discovery

Graph the first 10 terms of each sequence whose general term is given. Which are arithmetic? Why? Can you determine whether a sequence is arithmetic by looking at its general term? its graph?

$$a_n = 2n + 3, \quad a_n = n^2 - 100,$$
$$a_n = -0.5n + 20, \quad a_n = \ln n$$

The formula $a_n = a_1 + (n - 1)d$ and the preceding Interactive Discovery lead us to the following result:

A sequence is arithmetic $\longleftrightarrow$ the general term is a linear function.

Sum of the First n Terms of an Arithmetic Sequence

Consider the arithmetic sequence

$$3, 5, 7, 9, \ldots .$$

When we add the first 4 terms of the sequence, we get S_4, which is

$$3 + 5 + 7 + 9, \quad \text{or} \quad 24.$$

This sum is called an **arithmetic series**. To find a formula for the sum of the first n terms, S_n, of an arithmetic series, we first denote an arithmetic sequence, as follows:

This term is two terms back from the last. If you add d to this term, the result is the next-to-last term, $a_n - d$.

$$a_1, \quad (a_1 + d), \quad (a_1 + 2d), \quad \ldots, \quad (a_n - 2d), \quad (a_n - d), \quad a_n.$$

This is the next-to-last term. If you add d to this term, the result is a_n.

Then S_n is given by

$$S_n = a_1 + (a_1 + d) + (a_1 + 2d) + \cdots + (a_n - 2d) + (a_n - d) + a_n. \tag{1}$$

Reversing the order of the addition gives us

$$S_n = a_n + (a_n - d) + (a_n - 2d) + \cdots + (a_1 + 2d) + (a_1 + d) + a_1. \tag{2}$$

If we add corresponding terms of each side of equations (1) and (2), we get

$$\begin{aligned} 2S_n = {} & [a_1 + a_n] + [(a_1 + d) + (a_n - d)] + [(a_1 + 2d) + (a_n - 2d)] \\ & + \cdots + [(a_n - 2d) + (a_1 + 2d)] \\ & + [(a_n - d) + (a_1 + d)] + [a_n + a_1]. \end{aligned}$$

There are n pairs of square brackets.

This simplifies to

$$\begin{aligned} 2S_n = {} & [a_1 + a_n] + [a_1 + a_n] + [a_1 + a_n] + \cdots + [a_n + a_1] \\ & + [a_n + a_1] + [a_n + a_1]. \end{aligned}$$

Since $a_1 + a_n$ is being added n times, it follows that

$$2S_n = n(a_1 + a_n),$$

from which we get the following formula.

Sum of the First n Terms

The sum of the first n terms of an arithmetic sequence is given by

$$S_n = \frac{n}{2}(a_1 + a_n).$$

Example 5 Find the sum of the first 100 natural numbers.

SOLUTION The sum is

$$1 + 2 + 3 + \cdots + 99 + 100.$$

This is the sum of the first 100 terms of the arithmetic sequence for which

$$a_1 = 1, \qquad a_n = 100, \quad \text{and} \quad n = 100.$$

Thus substituting into the formula

$$S_n = \frac{n}{2}(a_1 + a_n),$$

we get

$$S_{100} = \frac{100}{2}(1 + 100) = 50(101) = 5050.$$

The sum of the first 100 natural numbers is 5050.

Example 6 Find the sum of the first 15 terms of the arithmetic sequence 4, 7, 10, 13, ...

SOLUTION Note that $a_1 = 4$, $d = 3$, and $n = 15$. Before using the formula

$$S_n = \frac{n}{2}(a_1 + a_n),$$

we find the last term, a_{15}:

$$\begin{aligned} a_{15} &= 4 + (15 - 1)3 \qquad \text{Substituting into the formula } a_n = a_1 + (n-1)d \\ &= 4 + 14 \cdot 3 = 46. \end{aligned}$$

Thus,

$$\begin{aligned} S_{15} &= \frac{15}{2}(4 + 46) \qquad \text{Substituting into } S_n = \frac{n}{2}(a_1 + a_n) \\ &= \frac{15}{2}(50) = 375. \end{aligned}$$

The sum of the first 15 terms is 375.

Example 7 Find the sum: $\sum_{k=1}^{130} (4k + 5)$.

SOLUTION It is helpful to first write out a few terms:

$$9 + 13 + 17 + \cdots.$$

It appears that this is an arithmetic series coming from an arithmetic sequence with $a_1 = 9$, $d = 4$, and $n = 130$. Before using the formula

$$S_n = \frac{n}{2}(a_1 + a_n),$$

we find the last term, a_{130}:

$$\begin{aligned} a_{130} &= 9 + (130 - 1)4 \qquad \text{Substituting into the formula } a_n = a_1 + (n-1)d \\ &= 9 + 129 \cdot 4 = 525. \end{aligned}$$

Thus,

$$S_{130} = \frac{130}{2}(9 + 525) \qquad \text{Substituting into } S_n = \frac{n}{2}(a_1 + a_n)$$
$$= 34{,}710.$$

Applications

The translation of some applications and problem-solving situations may involve arithmetic sequences or series. Let's look at some examples.

Example 8 *Hourly Wages.* Gloria accepts a job, starting with an hourly wage of \$14.25, and is promised a raise of 15¢ per hour every 2 months for 5 yr. At the end of 5 yr, what will Gloria's hourly wage be?

SOLUTION

1. **Familiarize.** It helps to first write down the hourly wage for several 2-month time periods:

   Beginning: \$14.25,
   After two months: 14.40,
   After four months: 14.55,
   and so on.

   What appears is a sequence of numbers:

   14.25, 14.40, 14.55, ...

   Is it arithmetic? Yes, because adding \$0.15 each time gives us the next term.

   We ask ourselves what we know about arithmetic sequences. The pertinent formulas are

   $$a_n = a_1 + (n - 1)d \quad \text{and} \quad S_n = \frac{n}{2}(a_1 + a_n).$$

   In this case, we are not looking for a sum, so it is probably the first formula that will lead to our answer. We want to know the last term of an arithmetic sequence, so we need to know a_1, n, and d. From our list above, we see that

   $$a_1 = 14.25 \quad \text{and} \quad d = 0.15.$$

   What is n? That is, how many terms are in the sequence? Each year there are 6 raises, since Gloria gets a raise every 2 months. There are 5 yr, so the total number of raises will be $5 \cdot 6$, or 30. There will be 31 terms: the original wage and 30 increased rates.

2. **Translate.** We want to find $a_n = a_1 + (n - 1)d$ for the arithmetic sequence in which $a_1 = 14.25$, $d = 0.15$, and $n = 31$.

3. **Carry out.** Substituting gives us

   $$a_{31} = 14.25 + (31 - 1) \cdot 0.15 = \$18.75.$$

4. **Check.** We can check the calculations. We can also calculate in a slightly different way for another check. For example, at the end of 1 yr, there will be 6 raises, for a total raise of \$0.90. At the end of 5 yr, the total raise will be 5 × \$0.90, or \$4.50. If we add that to the original wage of \$14.25, we obtain \$18.75. The answer checks.
5. **State.** At the end of 5 yr, Gloria's hourly wage will be \$18.75.

Example 8 is one in which the calculations or the translation could be done in a number of ways. There is often a variety of ways in which a problem can be solved. In this chapter, we will concentrate on the use of sequences and series and their related formulas.

Example 9 *Total in a Stack.* A stack of telephone poles has 30 poles in the bottom row. There are 29 poles in the second row, 28 in the next row, and so on. How many poles are in the stack if there are 5 poles in the top row?

SOLUTION

1. **Familiarize.** A picture will help in this case. The following figure shows the ends of the poles and the way in which they stack. There are 30 poles on the bottom, and we see that there are fewer in each succeeding row. How many poles are there altogether?

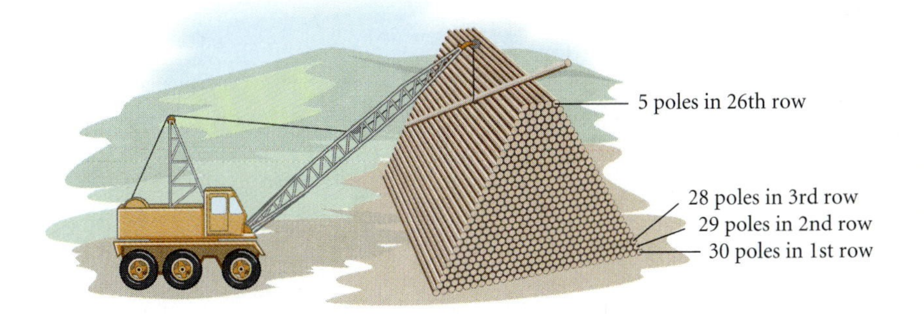

Since the number of poles goes from 30 in a row up to 5 in the top row, there must be 26 rows. We want the sum

$$30 + 29 + 28 + \cdots + 5.$$

Thus we have an arithmetic series. We recall, or look up if necessary, the formula

$$S_n = \frac{n}{2}(a_1 + a_n).$$

2. **Translate.** We want to find the sum of the first 26 terms of an arithmetic sequence in which $a_1 = 30$ and $a_{26} = 5$.
3. **Carry out.** Substituting, we get

$$S_{26} = \frac{26}{2}(30 + 5) = 455.$$

4. **Check.** In this case, we can check by repeating the calculations. A longer, harder way would be to do the entire addition: $30 + 29 + 28 + \cdots + 5$.

5. **State.** There are 455 poles in the stack.

Arithmetic Means

If p, m, and q form an arithmetic sequence, it can be shown (see Exercise 66) that $m = (p + q)/2$. The number m is the **arithmetic mean**, or **average**, of p and q. Given two numbers p and q, if we find k other numbers $m_1, m_2, \ldots, m_k$ such that

$$p, m_1, m_2, \ldots, m_k, q$$

forms an arithmetic sequence, we say that we have "inserted k arithmetic means between p and q."

Example 10 Insert three arithmetic means between 4 and 13.

SOLUTION We look for numbers m_1, m_2, and m_3 such that $4, m_1, m_2, m_3, 13$ is an arithmetic sequence. In this case, $a_1 = 4$, $n = 5$, and $a_5 = 13$. We then use the formula $a_n = a_1 + (n - 1)d$ and solve for d:

$$a_n = a_1 + (n - 1)d;$$

$$d = \frac{a_n - a_1}{n - 1}$$

$$= \frac{13 - 4}{5 - 1} = 2\frac{1}{4}.$$

Then using $d = 2\frac{1}{4}$, we compute the values of m_1, m_2, and m_3:

$$m_1 = a_1 + d = 4 + 2\tfrac{1}{4} = 6\tfrac{1}{4},$$
$$m_2 = m_1 + d = 6\tfrac{1}{4} + 2\tfrac{1}{4} = 8\tfrac{1}{2},$$
$$m_3 = m_2 + d = 8\tfrac{1}{2} + 2\tfrac{1}{4} = 10\tfrac{3}{4}.$$

9.2 Exercise Set

Find the first term and the common difference.

1. 3, 8, 13, 18, ...
2. \$1.08, \$1.16, \$1.24, \$1.32, ...
3. $9, 5, 1, -3, \ldots$
4. $-8, -5, -2, 1, 4, \ldots$
5. $\frac{3}{2}, \frac{9}{4}, 3, \frac{15}{4}, \ldots$
6. $\frac{3}{5}, \frac{1}{10}, -\frac{2}{5}, \ldots$
7. \$316, \$313, \$310, \$307, ...
8. Find the 11th term of the arithmetic sequence 0.07, 0.12, 0.17, ...
9. Find the 12th term of the arithmetic sequence 2, 6, 10, ...
10. Find the 17th term of the arithmetic sequence 7, 4, 1, ...

11. Find the 14th term of the arithmetic sequence $3, \frac{7}{3}, \frac{5}{3}, \ldots$.

12. Find the 13th term of the arithmetic sequence \$1200, \$964.32, \$728.64, ...

13. Find the 10th term of the arithmetic sequence \$2345.78, \$2967.54, \$3589.30, ...

14. In the sequence of Exercise 9, what term is the number 106?

15. In the sequence of Exercise 8, what term is the number 1.67?

16. In the sequence of Exercise 10, what term is -296?

17. In the sequence of Exercise 11, what term is -27?

18. Find a_{20} when $a_1 = 14$ and $d = -3$.

19. Find a_1 when $d = 4$ and $a_8 = 33$.

20. Find d when $a_1 = 8$ and $a_{11} = 26$.

21. Find n when $a_1 = 25$, $d = -14$, and $a_n = -507$.

22. In an arithmetic sequence, $a_{17} = -40$ and $a_{28} = -73$. Find a_1 and d. Write the first 5 terms of the sequence.

23. In an arithmetic sequence, $a_{17} = \frac{25}{3}$ and $a_{28} = \frac{95}{6}$. Find a_1 and d. Write the first 5 terms of the sequence.

24. Find the sum of the first 14 terms of the series $11 + 7 + 3 + \cdots$.

25. Find the sum of the first 20 terms of the series $5 + 8 + 11 + 14 + \cdots$.

26. Find the sum of the first 300 natural numbers.

27. Find the sum of the first 400 even natural numbers.

28. Find the sum of the odd numbers 1 to 199, inclusive.

29. Find the sum of the multiples of 7 from 7 to 98, inclusive.

30. Find the sum of all multiples of 4 that are between 14 and 523.

31. If an arithmetic series has $a_1 = 2$, $d = 5$, and $n = 20$, what is S_n?

32. If an arithmetic series has $a_1 = 7$, $d = -3$, and $n = 32$, what is S_n?

Find the sum.

33. $\sum_{k=1}^{40} (2k + 3)$

34. $\sum_{k=5}^{20} 8k$

35. $\sum_{k=0}^{19} \frac{k - 3}{4}$

36. $\sum_{k=2}^{50} (2000 - 3k)$

37. $\sum_{k=12}^{57} \frac{7 - 4k}{13}$

38. $\sum_{k=101}^{200} (1.14k - 2.8) - \sum_{k=1}^{5} \left(\frac{k + 4}{10}\right)$

39. *Pole Stacking.* How many poles will be in a stack of telephone poles if there are 50 in the first layer, 49 in the second, and so on, with 6 in the top layer?

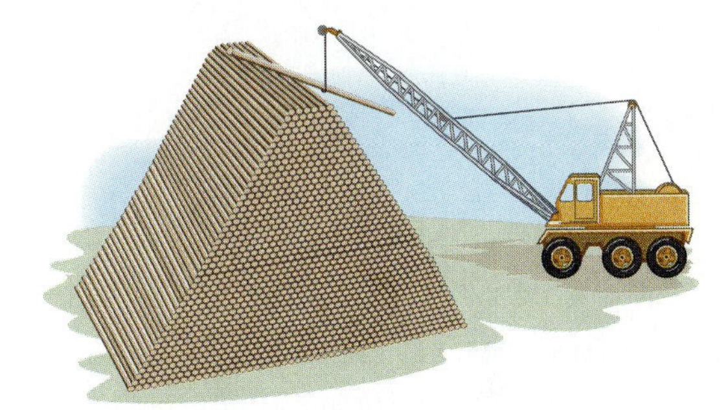

40. *Investment Return.* Max is an investment counselor. He sets up an investment situation for a client that will return \$5000 the first year, \$6125 the second year, \$7250 the third year, and so on, for 25 yr. How much is received from the investment altogether?

41. *Garden Plantings.* A gardener is making a planting in the shape of a trapezoid. It will have 35 plants in the front row, 31 in the second row, 27 in the third row, and so on. If the pattern is consistent, how many plants will there be in the last row? How many plants are there altogether?

42. *Band Formation.* A formation of a marching band has 14 marchers in the front row, 16 in the second row, 18 in the third row, and so on, for 25 rows. How many marchers are in the last row? How many marchers are there altogether?

43. *Total Savings.* If 10¢ is saved on October 1, 20¢ is saved on October 2, 30¢ on October 3, and so on, how much is saved during the 31 days of October?

44. *Parachutist Free Fall.* It has been found that when a parachutist jumps from an airplane, the distances, in feet, which the parachutist falls in each successive second before pulling the ripcord to release the parachute are as follows:

$$16, 48, 80, 112, 144, \ldots.$$

Is this sequence arithmetic? What is the common difference? What is the total distance fallen after 10 sec?

45. *Theater Seating.* Theaters are often built with more seats per row as the rows move toward the back. Suppose that the first balcony of a theater has 28 seats in the first row, 32 in the second, 36 in the third, and so on, for 20 rows. How many seats are in the first balcony altogether?

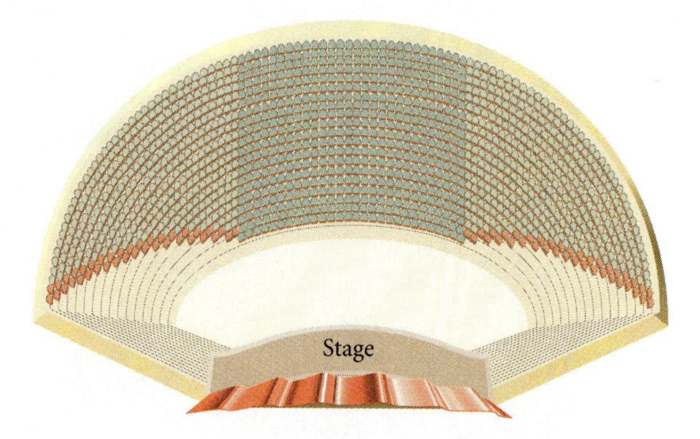

46. Insert four arithmetic means between 4 and 13.

47. Insert three arithmetic means between -3 and 5.

48. Insert ten arithmetic means between 27 and 300.

49. *Raw Material Production.* In an industrial situation, it took 3 units of raw materials to produce 1 unit of a product. The raw material needs thus formed the sequence

$$3, 6, 9, \ldots, 3n, \ldots.$$

Is this sequence arithmetic? What is the common difference?

50. *Small Group Interaction.* In a social science study, Stephan found the following data regarding an interaction measurement r_n for groups of size n.*

n	r_n
3	0.5908
4	0.6080
5	0.6252
6	0.6424
7	0.6596
8	0.6768
9	0.6940
10	0.7112

Is this sequence arithmetic? What is the common difference?

*Stephan, Frederick, "The Relative Rate of Communication Between Members of Small Groups," *American Sociological Review,* **17** (1952): 482–486.

Skill Maintenance

Simplify.

51. $64\left(-\frac{1}{2}\right)^9$

52. $4 \cdot 5^6$

53. $\frac{0.01(2^{30} - 1)}{2 - 1}$

54. Solve for S_n: $S_n - rS_n = a_1 - a_1r^n$.

Synthesis

55. ◈ The sum of the first n terms of an arithmetic sequence is given by

$$S_n = \frac{n}{2}[2a_1 + (n - 1)d].$$

Compare this formula to

$$S_n = \frac{n}{2}(a_1 + a_n).$$

Discuss the reasons for the use of one formula over the other.

56. ◈ It is said that as a young child, the mathematician Karl F. Gauss (1777–1855) was able to compute the sum $1 + 2 + 3 + \cdots + 100$ very quickly in his head to the amazement of a teacher. Explain how Gauss might have done this had he possessed some knowledge of arithmetic sequences and series. Then give a formula for the sum of the first n natural numbers.

57. Find a formula for the sum of the first n odd natural numbers:

$$1 + 3 + 5 + \cdots + (2n - 1).$$

58. Find three numbers in an arithmetic sequence such that the sum of the first and third is 10 and the product of the first and second is 15.

59. Find the first term and the common difference for the arithmetic sequence for which

$$a_2 = 40 - 3q \quad \text{and} \quad a_4 = 10p + q.$$

60. Find the first 10 terms of the arithmetic sequence for which

$$a_1 = \$8760 \quad \text{and} \quad d = -\$798.23.$$

Then find the sum of the first 10 terms.

61. The zeros of this polynomial function form an arithmetic sequence. Find them.

$$f(x) = x^4 + 4x^3 - 84x^2 - 176x + 640$$

62. Insert enough arithmetic means between 1 and 50 so that the sum of the resulting series will be 459.

63. Suppose that the lengths of the sides of a right triangle form an arithmetic sequence. Prove that the triangle is similar to a right triangle whose sides have lengths 3, 4, and 5.

64. *Straight-Line Depreciation.* A company buys an office machine for $5200 on January 1 of a given year. The machine is expected to last for 8 yr, at the end of which time its **trade-in value**, or **salvage value** will be $1100. If the company's accountant figures the decline in value to be the same each year, then its **book values**, or **salvage values**, after t years, $0 \leq t \leq 8$, form an arithmetic sequence given by

$$a_t = C - t\left(\frac{C - S}{N}\right),$$

where $C =$ the original cost of the item ($5200), $N =$ the number of years of expected life (8), and $S =$ the salvage value ($1100).

a) Find the formula for a_t for the straight-line depreciation of the office machine.

b) Find the salvage value after 0 yr, 1 yr, 2 yr, 3 yr, 4 yr, 7 yr, and 8 yr.

65. Prove that the general term of an arithmetic sequence is a linear function.

66. Prove that if p, m, and q form an arithmetic sequence, then

$$m = \frac{p + q}{2}.$$

9.3 Geometric Sequences and Series

- *Identify the common ratio of a geometric sequence, and find a given term and the sum of the first n terms.*
- *Find the sum of an infinite geometric series, if it exists.*

If we begin with a particular first term and then multiply by the same number successively, we obtain a **geometric sequence**. In this section, we study geometric sequences and series.

Geometric Sequences

Consider this sequence:

$$2, 6, 18, 54, 162, \ldots .$$

Note that multiplying each term by 3 produces the next term. Sequences in which each term can be multiplied by a certain number to get the next term are called **geometric**. We call this number the **common ratio** because it can be found by dividing any term by the preceding term.

Geometric Sequence

A sequence is *geometric* if there is a number r, called the *common ratio*, such that

$$\frac{a_{n+1}}{a_n} = r, \quad \text{or} \quad a_{n+1} = a_n r, \quad \text{for any integer } n \geq 1.$$

A geometric sequence is also called a **geometric progression**.

Example 1 For each of the following geometric sequences, identify the common ratio.

a) 3, 6, 12, 24, 48, ...

b) $1, -\frac{1}{2}, \frac{1}{4}, -\frac{1}{8}, \ldots$

c) \$5200, \$3900, \$2925, \$2193.75, ...

d) \$1000, \$1060, \$1123.60, ...

SOLUTION

SEQUENCE	COMMON RATIO
a) 3, 6, 12, 24, 48, ...	2 $\left(\frac{6}{3} = 2, \frac{12}{6} = 2, \text{ and so on}\right)$
b) $1, -\frac{1}{2}, \frac{1}{4}, -\frac{1}{8}, \ldots$	$-\frac{1}{2}$ $\left(\frac{-\frac{1}{2}}{1} = -\frac{1}{2}, \frac{\frac{1}{4}}{-\frac{1}{2}} = -\frac{1}{2}, \text{ and so on}\right)$
c) \$5200, \$3900, \$2925, \$2193.75, ...	0.75 $\left(\frac{\$3900}{\$5200} = 0.75, \frac{\$2925}{\$3900} = 0.75, \text{ and so on}\right)$
d) \$1000, \$1060, \$1123.60, ...	1.06 $\left(\frac{\$1060}{\$1000} = 1.06, \frac{\$1123.60}{\$1060} = 1.06, \text{ and so on}\right)$

We now find a formula for the general, or nth, term of a geometric sequence. Let a_1 be the first term and r the common ratio. The first few terms are as follows:

$$a_1,$$
$$a_2 = a_1 r,$$
$$a_3 = a_2 r = (a_1 r)r = a_1 r^2, \quad \text{Substituting } a_1 r \text{ for } a_2$$
$$a_4 = a_3 r = (a_1 r^2)r = a_1 r^3. \quad \text{Substituting } a_1 r^2 \text{ for } a_3$$

Note that the exponent is 1 less than the subscript.

Generalizing, we obtain the following.

***n*th Term of a Geometric Sequence**

The *nth term* of a geometric sequence is given by

$$a_n = a_1 r^{n-1}, \quad \text{for any integer } n \geq 1.$$

Program

GP: This program finds the nth term of a geometric sequence.

Example 2 Find the 7th term of the geometric sequence 4, 20, 100, ...

SOLUTION We first note that

$$a_1 = 4 \quad \text{and} \quad n = 7.$$

To find the common ratio, we can divide any term (other than the first) by the preceding term. Since the second term is 20 and the first is 4, we get

$$r = \tfrac{20}{4}, \quad \text{or} \quad 5.$$

Then using the formula $a_n = a_1 r^{n-1}$, we have

$$a_7 = 4 \cdot 5^{7-1} = 4 \cdot 5^6 = 4 \cdot 15{,}625 = 62{,}500.$$

Thus the 7th term is 62,500.

Example 3 Find the 10th term of the geometric sequence 64, −32, 16, −8, ...

SOLUTION We first note that

$$a_1 = 64, \qquad n = 10, \quad \text{and} \quad r = \frac{-32}{64}, \text{ or } -\frac{1}{2}.$$

Then using the formula $a_n = a_1 r^{n-1}$, we have

$$a_{10} = 64 \cdot \left(-\frac{1}{2}\right)^{10-1} = 64 \cdot \left(-\frac{1}{2}\right)^9 = 2^6 \cdot \left(-\frac{1}{2^9}\right) = -\frac{1}{8}.$$

Thus the 10th term is $-\frac{1}{8}$.

Sum of the First n Terms of a Geometric Sequence

Next we develop a formula for the sum S_n of the first n terms of a geometric sequence:

$$a_1, a_1r, a_1r^2, a_1r^3, \ldots, a_1r^{n-1}, \ldots.$$

The associated geometric series is given by

$$S_n = a_1 + a_1r + a_1r^2 + a_1r^3 + \cdots + a_1r^{n-1}. \tag{1}$$

We want to find a formula that allows us to find this sum without a great amount of adding. If we multiply on both sides of equation (1) by r, we have

$$rS_n = a_1r + a_1r^2 + a_1r^3 + a_1r^4 + \cdots + a_1r^n. \tag{2}$$

Subtracting corresponding terms of equation (2) from equation (1), we see that the red terms add to 0, leaving

$$S_n - rS_n = a_1 - a_1r^n,$$

or

$$S_n(1 - r) = a_1(1 - r^n). \qquad \text{Factoring}$$

Dividing on both sides by $1 - r$ gives us the following formula.

Sum of the First n Terms

The sum of the first n terms of a geometric sequence is given by

$$S_n = \frac{a_1(1 - r^n)}{1 - r}, \quad \text{for any } r \neq 1.$$

Example 4 Find the sum of the first 7 terms of the geometric sequence 3, 15, 75, 375, ...

SOLUTION We first note that

$$a_1 = 3, \qquad n = 7, \quad \text{and} \quad r = \tfrac{15}{3}, \text{ or } 5.$$

Then using the formula

$$S_n = \frac{a_1(1 - r^n)}{1 - r},$$

we have

$$S_7 = \frac{3(1 - 5^7)}{1 - 5} = \frac{3(1 - 78{,}125)}{-4}$$
$$= 58{,}593.$$

Thus the sum of the first 7 terms is 58,593.

Example 5 Find the sum: $\sum_{k=1}^{11} (0.3)^k$.

SOLUTION This is a geometric series with $a_1 = 0.3$, $r = 0.3$, and $n = 11$. Thus,

$$S_{11} = \frac{0.3(1 - 0.3^{11})}{1 - 0.3}$$
$$\approx 0.42857.$$

Infinite Geometric Series

Suppose we consider the sum of the terms of an infinite geometric sequence, such as 2, 4, 8, 16, 32, ... We get what is called an **infinite geometric series**:

$$2 + 4 + 8 + 16 + 32 + \cdots.$$

We want to examine what happens to the sums S_n as n gets larger and larger. We can use a grapher to complete a table and a graph of the sequence S_n.

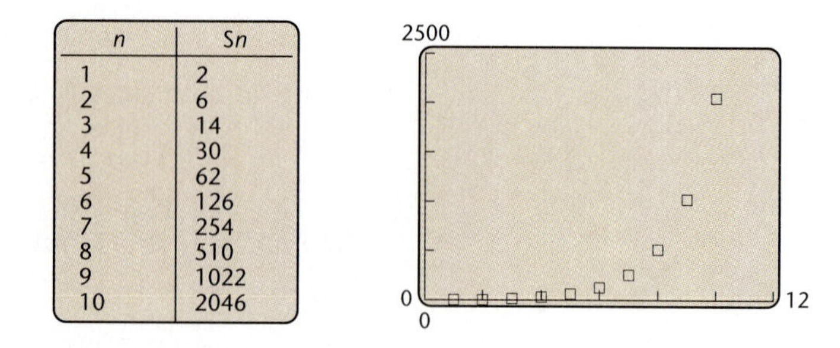

n	Sn
1	2
2	6
3	14
4	30
5	62
6	126
7	254
8	510
9	1022
10	2046

We could extend the table and the graph out for much larger values of n—for example, $S_{100} \approx 2.535 \times 10^{30}$. Note that as n gets larger and larger, the sum of the first n terms, S_n, becomes larger and larger without bound. That is, $S_n \to \infty$.

Many infinite sums get closer and closer to some specific number. Here is an example:

$$\frac{1}{2} + \frac{1}{4} + \frac{1}{8} + \frac{1}{16} + \cdots + \frac{1}{2^n} + \cdots.$$

To examine what happens to the sums S_n as n gets larger and larger, we again use a grapher and complete the following table and graph of the sequence S_n.

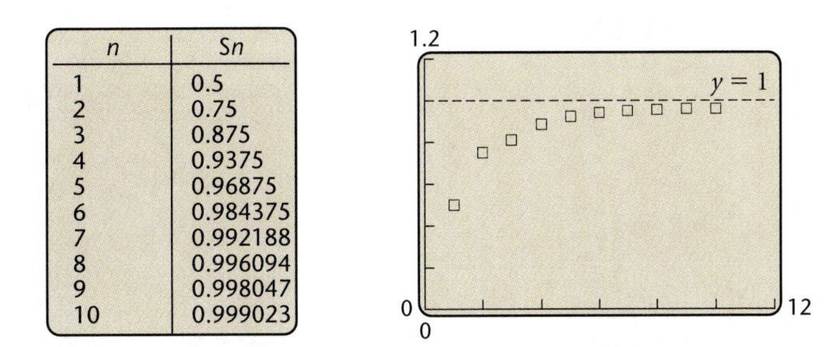

n	Sn
1	0.5
2	0.75
3	0.875
4	0.9375
5	0.96875
6	0.984375
7	0.992188
8	0.996094
9	0.998047
10	0.999023

We could extend the table and the graph out for much larger values of n—for example, $S_{25} \approx 0.9999999702$. Note that the value of S_n is less than 1 for any value of n, but as n gets larger and larger, the values of S_n get closer and closer to 1. That is, $S_n \to 1$ and 1 is the **sum of the infinite geometric series**. The sum of an infinite geometric series, if it exists, is denoted S_∞. In this case, $S_\infty = 1$.

Interactive Discovery

Use a grapher that creates tables and graphs sequences. For each of the following infinite series, make a table and a graph and try to find S_∞, if it exists. Look for a pattern connecting the value of the common ratio r with the existence of a value for S_∞.

SERIES	COMMON RATIO, r	S_∞
$1 + 7 + 49 + 343 + \cdots$		
$1 + (-1) + 1 + (-1) + \cdots$		
$\frac{1}{2} + \frac{1}{4} + \frac{1}{8} + \frac{1}{16} + \frac{1}{32} + \cdots$		
$625 + 250 + 100 + 40 + \cdots$		

It can be shown (but we will not do it here) that the sum of the terms of an infinite geometric series exists if and only if $|r| < 1$ (that is, the absolute value of the common ratio is less than 1).

To find a formula for the sum of an infinite geometric series, we first consider the sum of the first n terms:

$$S_n = \frac{a_1(1 - r^n)}{1 - r} = \frac{a_1 - a_1 r^n}{1 - r}. \qquad \text{Using the distributive law}$$

Let's look at values of r^n for a value of r for which $-1 < r < 1$ and see what happens when n gets larger and larger.

Interactive Discovery

Use a grapher that creates tables and graphs sequences. Let $r = 0.65$. Complete the following table and graph of the sequence whose general term is $(0.65)^n$.

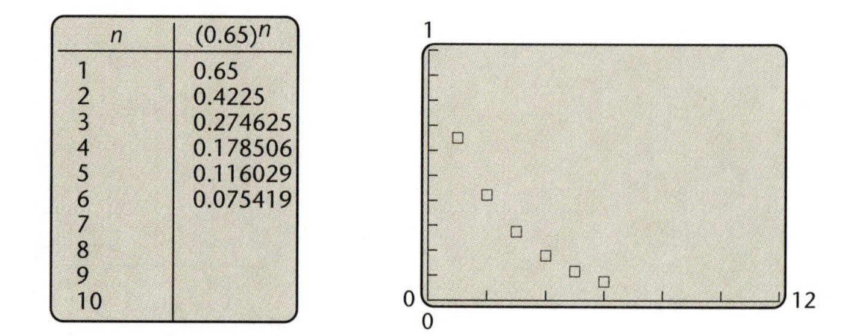

n	$(0.65)^n$
1	0.65
2	0.4225
3	0.274625
4	0.178506
5	0.116029
6	0.075419
7	
8	
9	
10	

Describe what happens to $(0.65)^n$ as values of n get larger and larger.

We see that values of r^n get closer and closer to 0 as n gets larger. As r^n gets closer and closer to 0, so does $a_1 r^n$. Thus, S_n gets closer and closer to $a_1/(1 - r)$.

Limit or Sum of an Infinite Geometric Series

When $|r| < 1$, the limit or sum of an infinite geometric series is given by

$$S_\infty = \frac{a_1}{1 - r}.$$

Example 6 Determine whether each of the following infinite geometric series has a limit. If a limit exists, find it.

a) $1 + 3 + 9 + 27 + \cdots$

b) $-2 + 1 - \frac{1}{2} + \frac{1}{4} - \frac{1}{8} + \cdots$

SOLUTION

a) Here $r = 3$, so $|r| = |3| = 3$, and since $|r| > 1$, the series *does not* have a limit.

b) Here $r = -\frac{1}{2}$, so $|r| = \left|-\frac{1}{2}\right| = \frac{1}{2}$, and since $|r| < 1$, the series *does* have a limit. We find the limit by

$$S_\infty = \frac{a_1}{1-r} = \frac{-2}{1-\left(-\frac{1}{2}\right)} = \frac{-2}{\frac{3}{2}} = -\frac{4}{3}.$$

If you wish, you can also check these results on a grapher using the TABLE and SEQUENCE GRAPH features.

Example 7 Find fractional notation for 0.78787878..., or $0.\overline{78}$.

SOLUTION We can express this as

$$0.78 + 0.0078 + 0.000078 + \cdots.$$

Then we see that this is an infinite geometric series, where $a_1 = 0.78$ and $r = 0.01$. Since $|r| < 1$, this series has a limit:

$$S_\infty = \frac{a_1}{1-r} = \frac{0.78}{1-0.01} = \frac{0.78}{0.99} = \frac{78}{99}, \quad \text{or} \quad \frac{26}{33}.$$

Thus fractional notation for 0.78787878... is $\frac{26}{33}$. You can check this on your calculator.

Applications

The translation of some application and problem-solving situations may involve geometric sequences or series. Examples 9 and 10 in particular show applications in business and economics.

Example 8 *A Daily Doubling Salary.* Suppose someone offered you a job for the month of September (30 days) under the following conditions. You will be paid \$0.01 for the first day, \$0.02 for the second, \$0.04 for the third, and so on, doubling your previous day's salary each day. How much would you earn? (Would you take the job? Make a conjecture before reading further.)

SOLUTION

1. **Familiarize.** You earn \$0.01 the first day, \$0.01(2) the second day, \$0.01(2)(2) the third day, and so on.
2. **Translate.** The amount earned is the geometric series

$$\$0.01 + \$0.01(2) + \$0.01(2^2) + \$0.01(2^3) + \cdots + \$0.01(2^{29}),$$

where

$$a_1 = \$0.01, \qquad r = 2, \quad \text{and} \quad n = 30.$$

3. **Carry out.** Using the formula

$$S_n = \frac{a_1(1-r^n)}{1-r},$$

we have

$$S_{30} = \frac{\$0.01(1-2^{30})}{1-2} \approx \$10{,}737{,}418.23.$$

4. **Check.** The calculations can be repeated.
5. **State.** The pay exceeds $10.7 million for the month. Most people would probably take the job!

Example 9 *The Amount of an Annuity.* An **annuity** is a sequence of equal payments made at equal time intervals. The payments do earn interest. For example, fixed deposits in a savings account are an example of an annuity. Suppose someone makes a sequence of yearly deposits of $1000 each in a savings account on which interest is compounded annually at 8%. The total amount in the account at the end of 5 yr is called the **amount of the annuity**. Find that amount.

SOLUTION

1. **Familiarize.** The following time diagram can help visualize the problem. Note that no deposit is made until the end of the first year.

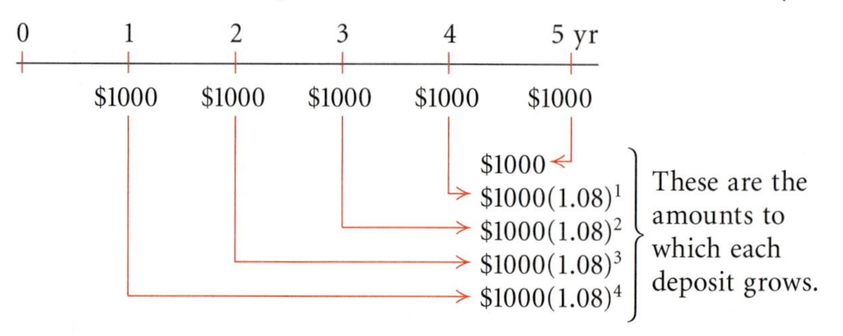

2. **Translate.** The amount of the annuity is the geometric series

$$\$1000 + \$1000(1.08)^1 + \$1000(1.08)^2 + \$1000(1.08)^3 + \$1000(1.08)^4,$$

where

$$a_1 = \$1000, \quad n = 5, \quad \text{and} \quad r = 1.08.$$

3. **Carry out.** We could simply carry this out by adding the terms, but we want to make use of the formulas developed in this section. (If we had 40 terms in the sequence, it would be more difficult to add the terms.) Using the formula

$$S_n = \frac{a_1(1 - r^n)}{1 - r},$$

we have

$$S_5 = \frac{\$1000(1 - 1.08^5)}{1 - 1.08}$$
$$\approx \$5866.60.$$

4. **Check.** The calculations can be repeated.
5. **State.** The amount of the annuity is $5866.60.

Example 10 *The Economic Multiplier.* The NCAA finals have a tremendous economic effect on the economy of the host city. Suppose 30,000 people visit the city and spend \$500 each while there. Then assume that 80% of that money is spent again in the city, and then 80% of that money is spent again, and so on. This is known as the **economic multiplier effect**. Find the total effect on the economy.

SOLUTION

1. **Familiarize.** According to certain economic theory, the money that this effectively puts into the economy can be calculated as the sum of an infinite geometric series as follows. The *initial effect* is 30,000 · \$500, or \$15,000,000.
2. **Translate.** The *total effect* is given by the series

$$\$15{,}000{,}000 + \$15{,}000{,}000(0.80) + \$15{,}000{,}000(0.80^2) + \cdots.$$

3. **Carry out.** Using the formula for the sum of an infinite geometric series, we have

$$S_\infty = \frac{a_1}{1 - r} = \frac{\$15{,}000{,}000}{1 - 0.80} = \$75{,}000{,}000.$$

4. **Check.** The calculations can be repeated.
5. **State.** The total effect on the economy of the spending from the tournament is \$75,000,000.

Do you see why cities work so hard to be chosen to host the NCAA finals?

9.3 Exercise Set

Find the common ratio.

1. 2, 4, 8, 16, ...
2. 18, −6, 2, $-\frac{2}{3}$, ...
3. −1, 1, −1, 1, ...
4. −8, −0.8, −0.08, −0.008, ...
5. $\frac{2}{3}, -\frac{4}{3}, \frac{8}{3}, -\frac{16}{3}, \ldots$
6. 75, 15, 3, $\frac{3}{5}$, ...
7. 6.275, 0.6275, 0.06275, ...
8. $\frac{1}{x}, \frac{1}{x^2}, \frac{1}{x^3}, \ldots$
9. $5, \frac{5a}{2}, \frac{5a^2}{4}, \frac{5a^3}{8}, \ldots$
10. \$780, \$858, \$943.80, \$1038.18, ...

Find the indicated term.

11. 2, 4, 8, 16, ...; the 7th term
12. 2, −10, 50, −250, ...; the 9th term
13. 2, $2\sqrt{3}$, 6, ...; the 9th term
14. 1, −1, 1, −1, ...; the 57th term
15. $\frac{7}{625}, -\frac{7}{25}, \ldots$; the 23rd term
16. \$1000, \$1060, \$1123.60, ...; the 5th term

Find the nth, or general, term.

17. 1, 3, 9, ...
18. 25, 5, 1, ...
19. 1, −1, 1, −1, ...
20. −2, 4, −8, ...
21. $\frac{1}{x}, \frac{1}{x^2}, \frac{1}{x^3}, \ldots$
22. $5, \frac{5a}{2}, \frac{5a^2}{4}, \frac{5a^3}{8}, \ldots$
23. Find the sum of the first 7 terms of the geometric series

$$6 + 12 + 24 + \cdots.$$

24. Find the sum of the first 10 terms of the geometric series

$$16 - 8 + 4 - \cdots.$$

25. Find the sum of the first 9 terms of the geometric series

$$\frac{1}{18} - \frac{1}{6} + \frac{1}{2} - \cdots.$$

26. Find the sum of the geometric series

$$-8 + 4 + (-2) + \cdots + \left(-\frac{1}{32}\right).$$

Find the sum, if it exists. If you wish, you can also check the results on a grapher using the TABLE *and* SEQUENCE GRAPH *features.*

27. $4 + 2 + 1 + \cdots$

28. $7 + 3 + \frac{9}{7} + \cdots$

29. $25 + 20 + 16 + \cdots$

30. $100 - 10 + 1 - \frac{1}{10} + \cdots$

31. $8 + 40 + 200 + \cdots$

32. $-6 + 3 - \frac{3}{2} + \frac{3}{4} - \cdots$

33. $0.6 + 0.06 + 0.006 + \cdots$

34. $\sum_{k=0}^{10} 3^k$

35. $\sum_{k=1}^{11} 15\left(\frac{2}{3}\right)^k$

36. $\sum_{k=0}^{50} 200(1.08)^k$

37. $\sum_{k=1}^{\infty} \left(\frac{1}{2}\right)^{k-1}$

38. $\sum_{k=1}^{\infty} 2^k$

39. $\sum_{k=1}^{\infty} 12.5^k$

40. $\sum_{k=1}^{\infty} 400(1.0625)^k$

41. $\sum_{k=1}^{\infty} \$500(1.11)^{-k}$

42. $\sum_{k=1}^{\infty} \$1000(1.06)^{-k}$

43. $\sum_{k=1}^{\infty} 16(0.1)^{k-1}$

44. $\sum_{k=1}^{\infty} \frac{8}{3}\left(\frac{1}{2}\right)^{k-1}$

Find fractional notation.

45. $0.131313\ldots$, or $0.\overline{13}$

46. $0.2222\ldots$, or $0.\overline{2}$

47. $8.999\overline{9}$

48. $6.1616\overline{16}$

49. $3.4125\overline{125}$

50. $12.7809\overline{809}$

51. *Bouncing Ping-Pong Ball.* A ping-pong ball is dropped from a height of 16 ft and always rebounds $\frac{1}{4}$ of the distance fallen.

a) How high does it rebound the 6th time?

b) Find the total sum of the rebound heights of the ball.

52. *Daily Doubling Salary.* Suppose someone offered you a job for the month of February (28 days) under the following conditions. You will be paid \$0.01 the 1st day, \$0.02 the 2nd, \$0.04 the 3rd, and so on, doubling your previous day's salary each day. How much would you earn altogether?

53. *Bungee Jumping.* A bungee jumper rebounds 60% of the height jumped. A bungee jump is made using a cord that stretches to 200 ft.

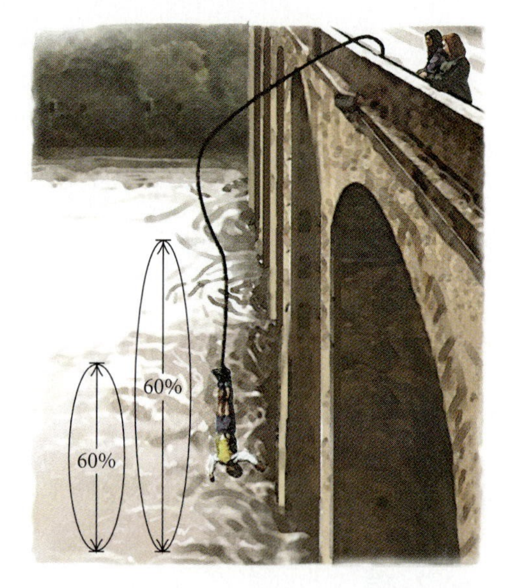

a) After jumping and then rebounding 9 times, how far has a bungee jumper traveled upward (the total rebound distance)?

b) About how far will a jumper have traveled upward (bounced) before coming to rest?

54. *Population Growth.* Gaintown has a present population of 100,000, and the population is increasing by 3% each year.

a) What will the population be in 15 yr?

b) How long will it take for the population to double?

55. *Amount of an Annuity.* To create a college fund, a parent makes a sequence of 18 yearly deposits of \$1000 each in a savings account on which interest is compounded annually at 6.2%. Find the amount of the annuity.

56. *Amount of an Annuity.* A sequence of yearly payments of P dollars is invested at the end of each of N years at interest rate i, compounded annually. The total amount in the account, or the amount of the annuity, is V.

a) Show that
$$V = \frac{P[(1+i)^N - 1]}{i}.$$

b) Suppose that interest is compounded n times per year and deposits are made every compounding period. Show that the formula for V is then given by
$$V = \frac{P\left[\left(1+\frac{i}{n}\right)^{nN} - 1\right]}{i/n}.$$

57. *Loan Repayment.* A family borrows \$120,000. The loan is to be repaid in 13 yr at 12% interest, compounded annually. How much will be repaid at the end of 13 yr?

58. *Doubling Paper Folds.* A piece of paper is 0.01 in. thick. It is cut and stacked repeatedly in such a way that its thickness is doubled each time for 20 times. How thick is the result?

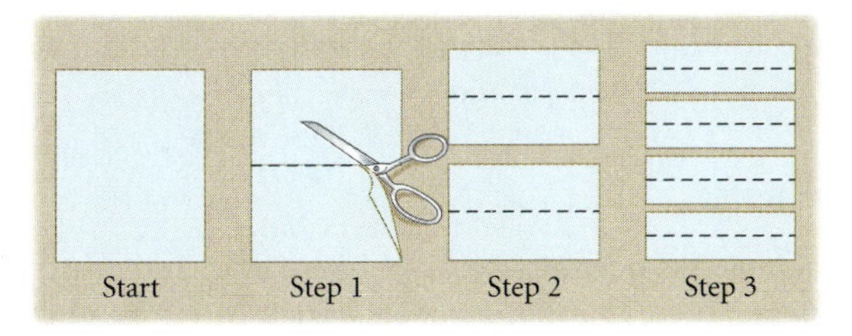

59. *The Economic Multiplier.* The government is making a \$13,000,000,000 expenditure for educational improvement. If 85% of this is spent again, and so on, what is the total effect on the economy?

60. *Advertising Effect.* The cereal company Box-o-Vim is about to market a new low-fat great-tasting cereal in a city of 5,000,000 people. They plan an advertising campaign that they think will induce 30% of the people to buy the product. They estimate that if those people like the product, they will induce 30% · 30% · 5,000,000 more to buy the product, and those will induce 30% · 30% · 30% · 5,000,000, and so on. In all, how many people will buy the product as a result of the advertising campaign? What percentage of the population is this?

Skill Maintenance

Simplify.

61. $k^2 + [2(k+1) - 1]$

62. $\dfrac{2^k - 1}{2^k} + \dfrac{1}{2^{k+1}}$

63. $2k(k+1) + 4(k+1)$

64. $\dfrac{3k(k+1)}{2} + 3(k+1)$

65. $\dfrac{k}{k+1} + \dfrac{1}{(k+1)(k+2)}$

66. $\dfrac{k^2(k+1)^2}{4} + (k+1)^3$

Synthesis

67. ◆ Write a problem for a classmate to solve. Devise the problem so that a geometric series is involved and the solution is: "The total amount in the bank is \$900(1.08)^{40}, or about \$19,552."

68. ◆ The infinite series
$$S_\infty = 2 + \frac{1}{2} + \frac{1}{2\cdot 3} + \frac{1}{2\cdot 3\cdot 4} + \frac{1}{2\cdot 3\cdot 4\cdot 5} + \cdots$$
is not geometric, but it does have a sum. Consider S_1, S_2, S_3, S_4, S_5, and S_6. Construct a table and a graph of the sequence. Expand the sequence of sums, if needed. Make a conjecture about the value of S_∞ and explain your reasoning.

69. Prove that
$$\sqrt{3} - \sqrt{2}, \quad 4 - \sqrt{6}, \quad \text{and} \quad 6\sqrt{3} - 2\sqrt{2}$$
form a geometric sequence.

70. Consider the sequence
$$4,\ 20.4,\ 104.04,\ 530.604,\ \ldots.$$
What is the error in using $a_{277} = 4(5.1)^{276}$ to find the 277th term?

71. Consider the sequence
$$x+3,\ x+7,\ 4x-2,\ \ldots.$$

a) If the sequence is arithmetic, find x and then determine each of the 3 terms and the 4th term.

b) If the sequence is geometric, find x and then determine each of the 3 terms and the 4th term.

72. Find the sum of the first n terms of

$$1 + x + x^2 + \cdots .$$

73. Find the sum of the first n terms of

$$x^2 - x^3 + x^4 - x^5 + \cdots .$$

In Exercises 74 and 75, assume that $a_1, a_2, a_3, \ldots$ is a geometric sequence.

74. Prove that $a_1^2, a_2^2, a_3^2, \ldots,$ is a geometric sequence.

75. Prove that $\ln a_1, \ln a_2, \ln a_3, \ldots,$ is an arithmetic sequence.

76. Prove that $5^{a_1}, 5^{a_2}, 5^{a_3}, \ldots,$ is a geometric sequence, if $a_1, a_2, a_3, \ldots,$ is an arithmetic sequence.

77. The sides of a square are 16 cm long. A second square is inscribed by joining the midpoints of the sides, successively. In the second square, we repeat the process, inscribing a third square. If this process is continued indefinitely, what is the sum of all the areas of all the squares? (*Hint:* Use an infinite geometric series.)

9.4 Mathematical Induction

- *List the statements of an infinite sequence that is defined by a formula.*
- *Do proofs by mathematical induction.*

In this section, we learn to prove a sequence of mathematical statements using a procedure called *mathematical induction.*

Sequences of Statements

Infinite sequences of statements occur often in mathematics. In an infinite sequence of statements, there is a statement for each natural number. For example, consider the sequence of statements represented by the following:

"For each x between 0 and 1, $0 < x^n < 1$."

Let's think of this as $S(n)$, or S_n. Substituting natural numbers for n gives a sequence of statements. We list a few of them.

Statement 1, S_1: For x between 0 and 1, $0 < x^1 < 1$.

Statement 2, S_2: For x between 0 and 1, $0 < x^2 < 1$.

Statement 3, S_3: For x between 0 and 1, $0 < x^3 < 1$.

Statement 4, S_4: For x between 0 and 1, $0 < x^4 < 1$.

In this context, the symbols S_1, S_2, S_3, and so on, do not represent sums.

Example 1 List the first four statements in the sequence obtainable from each of the following.

a) $\log n < n$

b) $1 + 3 + 5 + \cdots + (2n - 1) = n^2$

SOLUTION

a) This time, S_n is "$\log n < n$."

S_1: $\log 1 < 1$

S_2: $\log 2 < 2$

S_3: $\log 3 < 3$

S_4: $\log 4 < 4$

b) This time, S_n is "$1 + 3 + 5 + \cdots + (2n - 1) = n^2$."

S_1: $1 = 1^2$

S_2: $1 + 3 = 2^2$

S_3: $1 + 3 + 5 = 3^2$

S_4: $1 + 3 + 5 + 7 = 4^2$

Proving Infinite Sequences of Statements

We now develop a method of proof, called **mathematical induction**, which we can use to try to prove that all statements in an infinite sequence of statements are true. The statements usually have the form:

"For all natural numbers n, S_n",

where S_n is some mathematical sentence such as those of the preceding examples. Of course, we cannot prove each statement of an infinite sequence individually. Instead, we try to show that "whenever S_k holds, then S_{k+1} must hold." We abbreviate this as $S_k \rightarrow S_{k+1}$. (This is also read "*If* S_k, *then* S_{k+1}," or "S_k *implies* S_{k+1}.") Suppose that we could somehow establish that this holds for all natural numbers k. Then we would have the following:

$S_1 \rightarrow S_2$ meaning "if S_1 is true, then S_2 is true";

$S_2 \rightarrow S_3$ meaning "if S_2 is true, then S_3 is true";

$S_3 \rightarrow S_4$ meaning "if S_3 is true, then S_4 is true";

and so on, indefinitely.

Even knowing that $S_k \rightarrow S_{k+1}$, we would still not be certain whether there is *any* k for which S_k is true. All we would know is that "if S_k is true, then S_{k+1} is true." Suppose now that S_k is true for some k, say,

$k = 1$. We then must have the following.

S_1 is true.	**We have verified, or proved, this.**
$S_1 \longrightarrow S_2$	**This means that whenever S_1 is true, S_2 is true.**
Therefore, S_2 is true.	
$S_2 \longrightarrow S_3$	**This means that whenever S_2 is true, S_3 is true.**
Therefore, S_3 is true.	
and so on.	

We conclude that S_n is true for all natural numbers n.

This leads us to the principle of mathematical induction, which we use to prove the types of statements considered here.

The Principle of Mathematical Induction

We can prove an infinite sequence of statements S_n by showing the following.

(1) *Basis step.* S_1 is true.

(2) *Induction step.* For all natural numbers k, $S_k \rightarrow S_{k+1}$.

Mathematical induction is analogous to lining up a sequence of dominoes. The induction step tells us that if any one domino is knocked over, then the one next to it will be hit and knocked over. The basis step tells us that the first domino can indeed be knocked over. Note that in order for all dominoes to fall, *both* conditions must be satisfied.

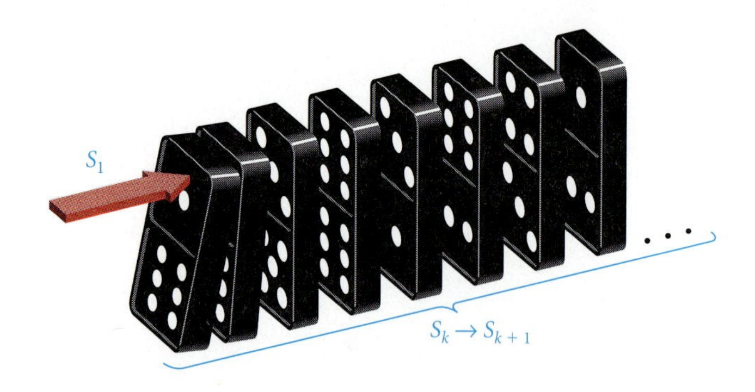

When you are learning to do proofs by mathematical induction, it is helpful to first write out S_n, S_1, S_k, and S_{k+1}. This helps to identify what is to be assumed and what is to be deduced.

Example 2 Prove: For every natural number n,

$$1 + 3 + 5 + \cdots + (2n - 1) = n^2.$$

PROOF We first list S_n, S_1, S_k, and S_{k+1}.

S_n: $1 + 3 + 5 + \cdots + (2n - 1) = n^2$

S_1: $1 = 1^2$

S_k: $1 + 3 + 5 + \cdots + (2k - 1) = k^2$

S_{k+1}: $1 + 3 + 5 + \cdots + (2k - 1) + [2(k + 1) - 1] = (k + 1)^2$

(1) *Basis step.* S_1, as listed, is true.

(2) *Induction step.* We let k be any natural number. We assume S_k to be true and try to show that it implies that S_{k+1} is true. Now S_k is

$$1 + 3 + 5 + \cdots + (2k - 1) = k^2.$$

Starting with the left side of S_{k+1} and substituting k^2 for $1 + 3 + 5 + \cdots + (2k - 1)$, we have

$$\underbrace{1 + 3 + \cdots + (2k - 1)} + [2(k + 1) - 1]$$

$$= k^2 + [2(k + 1) - 1] = k^2 + 2k + 1 = (k + 1)^2.$$

We have derived S_{k+1} from S_k. Thus we have shown that for all natural numbers k, $S_k \to S_{k+1}$. This completes the induction step. It and the basis step tell us that the proof is complete.

Example 3 Prove: For every natural number n,

$$\frac{1}{2} + \frac{1}{4} + \frac{1}{8} + \cdots + \frac{1}{2^n} = \frac{2^n - 1}{2^n}.$$

PROOF We first list S_n, S_1, S_k, and S_{k+1}.

S_n: $\frac{1}{2} + \frac{1}{4} + \frac{1}{8} + \cdots + \frac{1}{2^n} = \frac{2^n - 1}{2^n}$

S_1: $\frac{1}{2} = \frac{2^1 - 1}{2^1}$

S_k: $\frac{1}{2} + \frac{1}{4} + \frac{1}{8} + \cdots + \frac{1}{2^k} = \frac{2^k - 1}{2^k}$

S_{k+1}: $\frac{1}{2} + \frac{1}{4} + \frac{1}{8} + \cdots + \frac{1}{2^k} + \frac{1}{2^{k+1}} = \frac{2^{k+1} - 1}{2^{k+1}}$

(1) *Basis step.* We show S_1 to be true as follows:

$$\frac{2^1 - 1}{2^1} = \frac{2 - 1}{2} = \frac{1}{2}.$$

(2) *Induction step.* We let k be any natural number. We assume S_k to be true and try to show that it implies that S_{k+1} is true. Now S_k is

$$\frac{1}{2} + \frac{1}{4} + \frac{1}{8} + \cdots + \frac{1}{2^k} = \frac{2^k - 1}{2^k}.$$

Starting with the left side of S_{k+1} and substituting

$$\frac{2^k - 1}{2^k} \quad \text{for} \quad \frac{1}{2} + \frac{1}{4} + \cdots + \frac{1}{2^k},$$

we have

$$\underbrace{\frac{1}{2} + \frac{1}{4} + \frac{1}{8} + \cdots + \frac{1}{2^k}} + \frac{1}{2^{k+1}}$$

$$= \frac{2^k - 1}{2^k} + \frac{1}{2^{k+1}} = \frac{2^k - 1}{2^k} \cdot \frac{2}{2} + \frac{1}{2^{k+1}} = \frac{(2^k - 1) \cdot 2 + 1}{2^{k+1}}$$

$$= \frac{2^{k+1} - 1}{2^{k+1}}.$$

We have derived S_{k+1} from S_k. Thus we have shown that for all natural numbers k, $S_k \rightarrow S_{k+1}$. This completes the induction step. It and the basis step tell us that the proof is complete.

Example 4 Prove: For every natural number n, $n < 2^n$.

PROOF We first list S_n, S_1, S_k, and S_{k+1}.

S_n: $n < 2^n$

S_1: $1 < 2^1$

S_k: $k < 2^k$

S_{k+1}: $k + 1 < 2^{k+1}$

(1) *Basis step.* S_1, as listed, is true since $2^1 = 2$ and $1 < 2$.

(2) *Induction step.* We let k be any natural number. We assume S_k to be true and try to show that it implies that S_{k+1} is true. Now

$k < 2^k$	**This is S_k.**
$2k < 2 \cdot 2^k$	**Multiplying on both sides by 2**
$2k < 2^{k+1}$	**Adding exponents on the right**
$k + k < 2^{k+1}$.	**Rewriting $2k$ as $k + k$**

Since k is any natural number, we know that $1 \leq k$. Thus,

$k + 1 \leq k + k$. **Adding k on both sides**

Putting the results $k + 1 \leq k + k$ and $k + k < 2^{k+1}$ together gives us

$k + 1 < 2^{k+1}$. **This is S_{k+1}.**

We have derived S_{k+1} from S_k. Thus we have shown that for all natural numbers k, $S_k \rightarrow S_{k+1}$. This completes the induction step. It and the basis step tell us that the proof is complete.

Interactive Discovery

Use a TABLE feature or spreadsheet to make a table for a sequence given by $y = 2^n$ for values for n from 1 to 100. Does the inequality of Example 4 hold? Then check the inequality using a grapher.

Mathematicians often spend considerable time searching for a rule, or sequence of statements, that can be proved using mathematical induction. Although we have not discussed how such rules are found, an interesting geometric interpretation may have been used to discover the formula in Example 2:

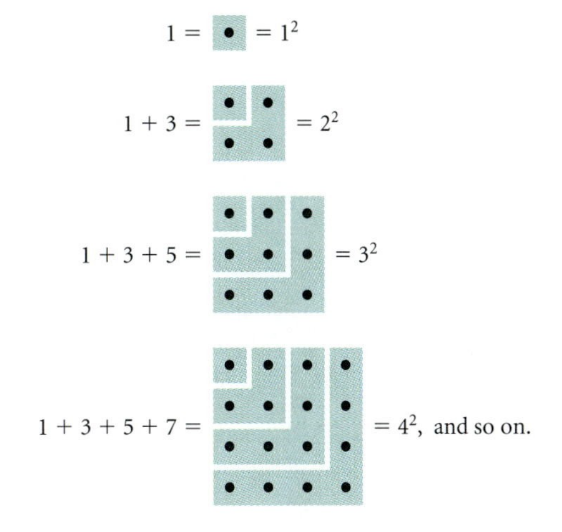

9.4 Exercise Set

List the first five statements in the sequence obtainable from each of the following. Determine whether each of the statements is true or false.

1. $n^2 < n^3$

2. $n^2 - n + 41$ is prime. Find a value for n for which the statement is false.

3. A polygon of n sides has $[n(n - 3)]/2$ diagonals.

4. The sum of the angles of a polygon of n sides is $(n - 2) \cdot 180°$.

Use mathematical induction to prove each of the following.

5. $2 + 4 + 6 + \cdots + 2n = n(n + 1)$

6. $4 + 8 + 12 + \cdots + 4n = 2n(n + 1)$

7. $1 + 5 + 9 + \cdots + (4n - 3) = n(2n - 1)$

8. $3 + 6 + 9 + \cdots + 3n = \dfrac{3n(n + 1)}{2}$

9. $2 + 4 + 8 + \cdots + 2^n = 2(2^n - 1)$

10. $2 \leq 2^n$

11. $n < n + 1$

12. $3^n < 3^{n+1}$

13. $2n \leq 2^n$

14. $\dfrac{1}{1 \cdot 2} + \dfrac{1}{2 \cdot 3} + \cdots + \dfrac{1}{n(n + 1)} = \dfrac{n}{n + 1}$

15. $\dfrac{1}{1 \cdot 2 \cdot 3} + \dfrac{1}{2 \cdot 3 \cdot 4} + \dfrac{1}{3 \cdot 4 \cdot 5} + \cdots + \dfrac{1}{n(n + 1)(n + 2)} = \dfrac{n(n + 3)}{4(n + 1)(n + 2)}$

16. If x is any real number greater than 1, then for any natural number n, $x \leq x^n$.

The following formulas can be used to find sums of powers of natural numbers. Use mathematical induction to prove each of the following.

17. $1 + 2 + 3 + \cdots + n = \dfrac{n(n + 1)}{2}$

18. $1^2 + 2^2 + 3^2 + \cdots + n^2 = \dfrac{n(n + 1)(2n + 1)}{6}$

19. $1^3 + 2^3 + 3^3 + \cdots + n^3 = \dfrac{n^2(n+1)^2}{4}$

20. $1^4 + 2^4 + 3^4 + \cdots + n^4 = \dfrac{n(n+1)(2n+1)(3n^2+3n-1)}{30}$

21. $1^5 + 2^5 + 3^5 + \cdots + n^5 = \dfrac{n^2(n+1)^2(2n^2+2n-1)}{12}$

Use mathematical induction to prove each of the following.

22. $\displaystyle\sum_{i=1}^{n} (3i - 1) = \frac{n(3n+1)}{2}$

23. $\displaystyle\sum_{i=1}^{n} i(i + 1) = \frac{n(n+1)(n+2)}{3}$

24. $\left(1 + \frac{1}{1}\right)\left(1 + \frac{1}{2}\right)\left(1 + \frac{1}{3}\right)\cdots\left(1 + \frac{1}{n}\right) = n + 1$

25. The sum of n terms of an arithmetic sequence:
$a_1 + (a_1 + d) + (a_1 + 2d) + \cdots + [a_1 + (n-1)d] = \frac{n}{2}[2a_1 + (n-1)d]$

Skill Maintenance

Simplify.

26. $\dfrac{8 \cdot 7 \cdot 6 \cdot 5 \cdot 4 \cdot 3 \cdot 2 \cdot 1}{4 \cdot 3 \cdot 2 \cdot 1}$

27. $\dfrac{8 \cdot 7 \cdot 6 \cdot 5 \cdot 4 \cdot 3 \cdot 2 \cdot 1}{5 \cdot 4 \cdot 3 \cdot 2 \cdot 1}$

28. $\dfrac{10 \cdot 9 \cdot 8 \cdot 7 \cdot 6 \cdot 5 \cdot 4 \cdot 3 \cdot 2 \cdot 1}{7 \cdot 6 \cdot 5 \cdot 4 \cdot 3 \cdot 2 \cdot 1}$

29. $\dfrac{10 \cdot 9 \cdot 8 \cdot 7 \cdot 6 \cdot 5 \cdot 4 \cdot 3 \cdot 2 \cdot 1}{8 \cdot 7 \cdot 6 \cdot 5 \cdot 4 \cdot 3 \cdot 2 \cdot 1}$

Synthesis

Use mathematical induction to prove each of the following.

30. The sum of n terms of a geometric sequence:

$$a_1 + a_1r + a_1r^2 + \cdots + a_1r^{n-1} = \frac{a_1 - a_1r^n}{1 - r}$$

31. $x + y$ is a factor of $x^{2n} - y^{2n}$

32. $x + y$ is a factor of $x^{2n-1} + y^{2n-1}$

33. $\dfrac{x^n - y^n}{x - y} = x^{n-1} + yx^{n-2} + \cdots + y^{n-2}x + y^{n-1}$

34. ◆ Write an explanation of the idea behind mathematical induction for a fellow student.

35. ◆ Find two statements not considered in this section that are not true for all natural numbers. Then try to find where a proof by mathematical induction fails.

Prove each of the following using mathematical induction. Do the basis step for $n = 2$.

36. For every natural number $n \geq 2$,
$$2n + 1 < 3^n.$$

37. For every natural number $n \geq 2$,
$$\log_a (b_1 b_2 \cdots b_n) = \log_a b_1 + \log_a b_2 + \cdots + \log_a b_n.$$

38. For every natural number $n \geq 2$,
$$\left(1 - \frac{1}{2^2}\right)\left(1 - \frac{1}{3^2}\right)\cdots\left(1 - \frac{1}{n^2}\right) = \frac{n+1}{2n}.$$

39. For every natural number $n \geq 2$,
$$\frac{1}{\sqrt{1}} + \frac{1}{\sqrt{2}} + \frac{1}{\sqrt{3}} + \cdots + \frac{1}{\sqrt{n}} > \sqrt{n}.$$

Prove each of the following for any complex numbers $z_1, z_2, \ldots, z_n$, where $i^2 = -1$ and $\bar{z}$ is the conjugate of z (see Section 2.2).

40. $\overline{z^n} = \bar{z}^n$

41. $\overline{z_1 + z_2 + \cdots + z_n} = \overline{z_1} + \overline{z_2} + \cdots + \overline{z_n}$

42. $\overline{z_1 z_2 \cdots z_n} = \overline{z_1} \cdot \overline{z_2} \cdots \overline{z_n}$

43. i^n is either 1, -1, i, or $-i$.

For any integers a and b, b is a factor of a if there exists an integer c such that $a = bc$. Prove each of the following for any natural number n.

44. 2 is a factor of $n^2 + n$.

45. 3 is a factor of $n^3 + 2n$.

46. 5 is a factor of $n^5 - n$.

47. 3 is a factor of $n(n+1)(n+2)$

48. *The Tower of Hanoi Problem.* There are three pegs on a board. On one peg are n disks, each smaller than the one on which it rests. The problem is to move this pile of disks to another peg. The final order must be the same, but you can move only one disk at a time and can never place a larger disk on a smaller one.

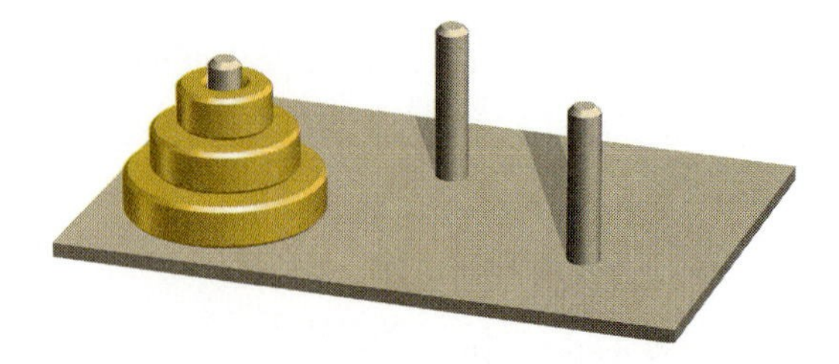

a) What is the *least* number of moves needed to move 3 disks? 4 disks? 2 disks? 1 disk?
b) Conjecture a formula for the *least* number of moves needed to move n disks. Prove it by mathematical induction.

49. Consider this statement: For every natural number n, $n = n + 1$.

a) Can you prove the basis step? If so, do it.
b) Can you prove the induction step? If so, do it.
c) Is the statement true? This illustrates the need for the basis step in an induction proof.

9.5 Combinatorics: Permutations

- *Evaluate factorial and permutation notation and solve related applications.*

In order to study probability, it is first necessary to learn about the theory of counting, called **combinatorics**. Such a study concerns itself with determining the number of ways in which a set can be arranged or combined, certain objects can be chosen, or a succession of events can occur.

Permutations

In this section, we will consider the part of combinatorics called *permutations.*

> The study of permutations involves *order* and *arrangements.*

Example 1 How many 3-letter code symbols can be formed with the letters A, B, C *without* repetition (that is, without a letter repeated)?

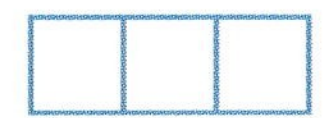

SOLUTION Examples of such symbols are ABC, CBA, ACB, and so on. Consider placing the letters in the frames shown at left. We can select any of the 3 letters for the first letter in the symbol. Once this letter has been selected, the second must be selected from the 2 remaining letters. The third letter is already determined, since only 1 possibility is left. The possibilities can be arrived at using a **tree diagram**, as shown below.

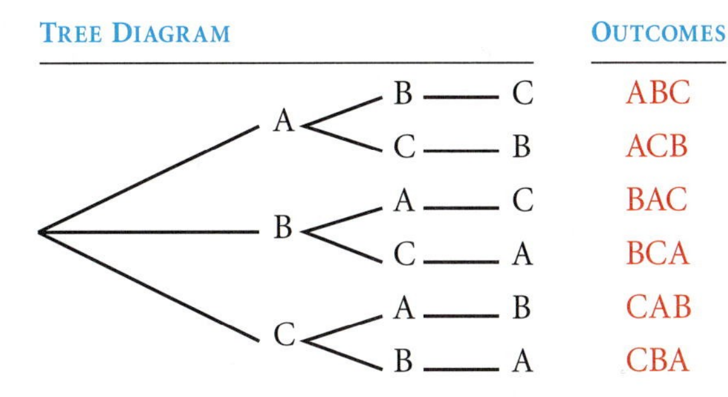

Each outcome represents one permutation of the letters A, B, C.

There are $3 \cdot 2 \cdot 1$, or 6, possibilities. The set of all of them are as follows:

{ABC, ACB, BAC, BCA, CAB, CBA}.

Suppose that we perform an experiment such as selecting letters (as in the preceding example), flipping a coin, or drawing a card. The results are called **outcomes**. An **event** is a set of outcomes. The following concerns the counting of events that occur together, or are combined.

The Fundamental Counting Principle

Given a combined action, or *event,* in which the first action can be performed in n_1 ways, the second action can be performed in n_2 ways, and so on, the total number of ways in which the combined action can be performed is the product

$$n_1 \cdot n_2 \cdot n_3 \cdot \cdots \cdot n_k.$$

Example 2 How many 3-letter code symbols can be formed with the letters A, B, C, D, and E *with* repetition (that is, allowing letters to be repeated)?

SOLUTION Since repetition is allowed, there are 5 choices for the first letter, 5 choices for the second, and 5 for the third. Thus, by the fundamental counting principle, there are $5 \cdot 5 \cdot 5$, or 125 code symbols.

Permutation

A *permutation* of a set of n objects is an ordered arrangement of all n objects.

Consider, for example, a set of 4 objects

{A, B, C, D}.

To find the number of ordered arrangements of the set, we select a first letter: There are 4 choices. Then we select a second letter: There are 3 choices. Then we select a third letter: There are 2 choices. Finally, there is 1 choice for the last selection. Thus, by the fundamental counting principle, there are $4 \cdot 3 \cdot 2 \cdot 1$, or 24, permutations of a set of 4 objects.

We can find a formula for the total number of permutations of all objects in a set of n objects. We have n choices for the first selection, $n - 1$ choices for the second, $n - 2$ for the third, and so on. For the nth selection, there is only 1 choice.

> ***A Formula for the Total Number of Permutations of n Objects***
>
> The total number of permutations of n objects, denoted ${}_nP_n$, is given by
>
> $${}_nP_n = n(n-1)(n-2)\cdots 3 \cdot 2 \cdot 1.$$

Example 3 Find each of the following.

a) ${}_4P_4$ **b)** ${}_7P_7$

SOLUTION

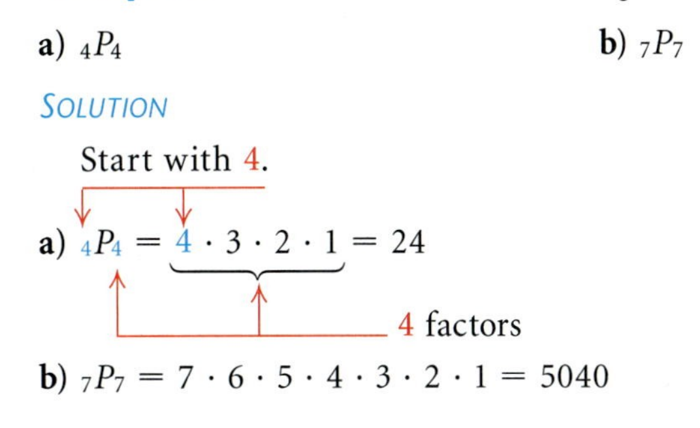

a) ${}_4P_4 = 4 \cdot 3 \cdot 2 \cdot 1 = 24$

b) ${}_7P_7 = 7 \cdot 6 \cdot 5 \cdot 4 \cdot 3 \cdot 2 \cdot 1 = 5040$

Example 4 In how many different ways can 9 packages be placed in 9 mailboxes, one package in a box?

SOLUTION We have

$${}_9P_9 = 9 \cdot 8 \cdot 7 \cdot 6 \cdot 5 \cdot 4 \cdot 3 \cdot 2 \cdot 1 = 362{,}880.$$

Factorial Notation

We will use products such as

$$7 \cdot 6 \cdot 5 \cdot 4 \cdot 3 \cdot 2 \cdot 1$$

so often that it is convenient to adopt a notation for them. For the product

$7 \cdot 6 \cdot 5 \cdot 4 \cdot 3 \cdot 2 \cdot 1$, we write $7!$, read "7 factorial."

We now define factorial notation for natural numbers and for 0.

> ***Factorial Notation***
>
> For any natural number n, we define
>
> $$n! = n(n-1)(n-2)\cdots 3 \cdot 2 \cdot 1.$$
>
> For the number 0, we define
>
> $$0! = 1.$$

We define $0!$ as 1 so that certain formulas can be stated concisely and with a consistent pattern.

Here are some examples.

$$\begin{aligned}
7! &= 7 \cdot 6 \cdot 5 \cdot 4 \cdot 3 \cdot 2 \cdot 1 = 5040\\
6! &= 6 \cdot 5 \cdot 4 \cdot 3 \cdot 2 \cdot 1 = 720\\
5! &= 5 \cdot 4 \cdot 3 \cdot 2 \cdot 1 = 120\\
4! &= 4 \cdot 3 \cdot 2 \cdot 1 = 24\\
3! &= 3 \cdot 2 \cdot 1 = 6\\
2! &= 2 \cdot 1 = 2\\
1! &= 1 = 1\\
0! &= 1 = 1
\end{aligned}$$

Interactive Discovery

Most graphers contain keys for ${}_nP_n$ and $n!$. Consult your manual. Complete the following table and look for a pattern.

n	$n!$	${}_nP_n$
1		
2		
8		
10		
11		
32		

You have probably discovered the following.

$${}_nP_n = n!$$

We will often need to manipulate factorial notation. For example, note that

$$\begin{aligned}
8! &= 8 \cdot 7 \cdot 6 \cdot 5 \cdot 4 \cdot 3 \cdot 2 \cdot 1\\
&= 8 \cdot (7 \cdot 6 \cdot 5 \cdot 4 \cdot 3 \cdot 2 \cdot 1)\\
&= 8 \cdot 7!.
\end{aligned}$$

Generalizing, we get the following.

For any natural number n, $n! = n(n - 1)!$.

By using this result repeatedly, we can further manipulate factorial notation.

Example 5 Rewrite 7! with a factor of 5!.

SOLUTION We have

$$7! = 7 \cdot 6 \cdot 5!.$$

In general, we have the following.

For any natural numbers k and n, with $k < n$,

$$n! = \underbrace{n(n-1)(n-2)\cdots[n-(k-1)]}_{k \text{ factors}} \cdot \underbrace{(n-k)!}_{n-k \text{ factors}}$$

Permutations of n Objects Taken k at a Time

Consider a set of 5 objects

$$\{A, B, C, D, E\}.$$

How many ordered arrangements are there having 3 members without repetition? Examples of such an arrangement are EBA, CAB, and BCD. There are 5 choices for the first object, 4 choices for the second, and 3 choices for the third. By the fundamental counting principle, there are

$$5 \cdot 4 \cdot 3,$$

or

60 *permutations* of a set of 5 objects taken 3 at a time.

Note that

$$5 \cdot 4 \cdot 3 = \frac{5 \cdot 4 \cdot 3 \cdot 2 \cdot 1}{2 \cdot 1}, \quad \text{or} \quad \frac{5!}{2!}.$$

Permutation of n Objects Taken k at a Time

A *permutation* of a set of n objects taken k at a time is an ordered arrangement of k objects taken from the set.

Consider a set of n objects and the selection of an ordered arrangement of k of them. There would be n choices for the first object. Then there would remain $n - 1$ choices for the second, $n - 2$ choices for the third, and so on. We make k choices in all, so there are k factors in the product. By the fundamental counting principle, the total number of permutations is

$$\underbrace{n(n-1)(n-2)\cdots[n-(k-1)].}_{k \text{ factors}}$$

We can express this in another way by multiplying by 1, as follows:

$$n(n-1)(n-2)\cdots[n-(k-1)]\cdot\frac{(n-k)!}{(n-k)!}$$
$$=\frac{n(n-1)(n-2)\cdots[n-(k-1)](n-k)!}{(n-k)!}$$
$$=\frac{n!}{(n-k)!}.$$

This gives us the following.

Formulas for the Number of Permutations of n Objects Taken k at a Time

The number of permutations of a set of n objects taken k at a time, denoted ${}_nP_k$, is given by

$$ {}_nP_k = \underbrace{n(n-1)(n-2)\cdots[n-(k-1)]}_{k \text{ factors}} \quad (1)$$
$$=\frac{n!}{(n-k)!}. \quad (2)$$

Formula (1) is most useful in applications, but formula (2) will be useful in Section 9.6.

Example 6 Compute ${}_8P_4$ using both formulas.

SOLUTION Using formula (1), we have

The 8 tells where to start.

$${}_8P_4 = \underbrace{8\cdot 7\cdot 6\cdot 5}= 1680.$$

The 4 tells how many factors.

Using formula (2), we have

$${}_8P_4 = \frac{8!}{(8-4)!}$$
$$=\frac{8!}{4!}$$
$$=\frac{8\cdot 7\cdot 6\cdot 5\cdot 4\cdot 3\cdot 2\cdot 1}{4\cdot 3\cdot 2\cdot 1}$$
$$=8\cdot 7\cdot 6\cdot 5 = 1680.$$

Example 7 *Flags of Nations.* The flags of many nations consist of three vertical stripes similar to the one shown here. For example, the flag of Ireland, shown below, has its first stripe green, second white, and third gold.

Suppose the following 9 colors are available:

{black, yellow, red, blue, white, gold, orange, pink, purple}.

How many different flags of 3 colors can be made without repetition of colors? This assumes that the order in which the stripes appear is considered.

SOLUTION We are determining the number of permutations of 9 objects taken 3 at a time. There is no repetition of colors. Using formula (1), we get

$${}_9P_3 = 9 \cdot 8 \cdot 7 = 504.$$

Example 8 *Batting Orders.* A baseball manager arranges the batting order as follows: The 4 infielders will bat first. Then the 3 outfielders, the catcher, and the pitcher will follow, not necessarily in that order. How many different batting orders are possible?

SOLUTION The infielders can bat in 4! different ways, the rest in 5! different ways. Then by the fundamental counting principle, we have

$${}_4P_4 \cdot {}_5P_5 = 4! \cdot 5!, \quad \text{or} \quad 2880 \text{ possible batting orders.}$$

You probably also have an $\boxed{{}_nP_k}$ key on your calculator or grapher. The notations $P(n, k)$ and $P_{n,k}$ are also used. Consult your manual and then check the results of Examples 6–8.

If we allow repetition, a situation like the following can occur.

Example 9 How many 5-letter code symbols can be formed with the letters A, B, C, and D if we allow a letter to occur more than once?

SOLUTION We have 5 spaces:

We can select the first letter in 4 ways, the second in 4 ways, and so on. Thus there are 4^5, or 1024 arrangements.

> The number of distinct arrangements of n objects taken k at a time, allowing repetition, is n^k.

Permutations of Sets with Nondistinguishable Objects

Consider a set of 7 marbles, 4 of which are blue and 3 of which are red. Although the marbles are all different, when they are lined up, one red marble will look just like any other red marble. In this sense, we say that the red marbles are nondistinguishable and the blue marbles are nondistinguishable.

We know that there are 7! permutations of this set. Many of them will look alike, however. We develop a formula for finding the number of distinguishable permutations.

Consider a set of n objects in which n_1 are of one kind, n_2 are of a second kind, ..., n_k are of a kth kind. The total number of permutations of the set is $n!$, but this includes many that are indistinguishable. Let N be the total number of distinguishable permutations. For each of these N permutations, there are $n_1!$ actual permutations, obtained by permuting the objects of the first kind. For each of these $N \cdot n_1!$ permutations, there are $n_2!$ nondistinguishable permutations, obtained by permuting the objects of the second kind, and so on. By the fundamental counting principle, the total number of permutations, including those that are nondistinguishable, is

$$N \cdot n_1! \cdot n_2! \cdot \cdots \cdot n_k!.$$

Then we have $N \cdot n_1! \cdot n_2! \cdot \cdots \cdot n_k! = n!$. Solving for N, we obtain

$$N = \frac{n!}{n_1! \cdot n_2! \cdot \cdots \cdot n_k!}.$$

Now, to finish our problem with the marbles, we have

$$N = \frac{7!}{4!\,3!} = \frac{7 \cdot 6 \cdot 5 \cdot 4 \cdot 3 \cdot 2 \cdot 1}{4 \cdot 3 \cdot 2 \cdot 1 \cdot 3 \cdot 2 \cdot 1} = \frac{7 \cdot 5}{1}, \quad \text{or} \quad 35$$

distinguishable permutations of the marbles.

In general:

> For a set of n objects in which n_1 are of one kind, n_2 are of another kind, ..., n_k are of a kth kind, the number of distinguishable permutations is
>
> $$\frac{n!}{n_1! \cdot n_2! \cdot \cdots \cdot n_k!}.$$

Example 10 In how many distinguishable ways can the letters of the word CINCINNATI be arranged?

SOLUTION There are 2 C's, 3 I's, 3 N's, 1 A, and 1 T for a total of 10 letters. Thus,

$$N = \frac{10!}{2! \cdot 3! \cdot 3! \cdot 1! \cdot 1!}, \quad \text{or} \quad 50{,}400.$$

The letters of the word CINCINNATI can be arranged in 50,400 distinguishable ways.

9.5 Exercise Set

Evaluate. Do your work by hand and then check it on a grapher.

1. ${}_6P_6$ **2.** ${}_4P_3$

3. ${}_{10}P_7$ **4.** ${}_{10}P_3$

5. 5! **6.** 7!

7. 0! **8.** 1!

9. $\frac{9!}{5!}$ **10.** $\frac{9!}{4!}$

11. $(8 - 3)!$ **12.** $(8 - 5)!$

13. $\frac{10!}{7!\,3!}$ **14.** $\frac{7!}{(7 - 2)!}$

15. ${}_8P_0$ **16.** ${}_{13}P_1$

Evaluate. Use a grapher.

17. ${}_{52}P_4$ **18.** ${}_{52}P_5$

Evaluate.

19. ${}_nP_3$ **20.** ${}_nP_2$

21. ${}_nP_1$ **22.** ${}_nP_0$

In each of the following exercises, give your answer using permutation notation, factorial notation, or other operations. Then evaluate.

How many permutations are there of the letters in each of the following words, if all the letters are used without repetition?

23. MARVIN **24.** JUDY

25. UNDERMOST **26.** COMBINES

27. How many permutations are there of the letters of the word UNDERMOST if the letters are taken 4 at a time?

28. How many permutations are there of the letters of the word COMBINES if the letters are taken 5 at a time?

29. How many 5-digit numbers can be formed using the digits 2, 4, 6, 8, and 9 without repetition? with repetition?

30. In how many ways can 7 athletes be arranged in a straight line?

31. How many distinguishable code symbols can be formed from the letters of the word BUSINESS? BIOLOGY? MATHEMATICS?

32. A professor is going to grade her 24 students on a curve. She will give 3 A's, 5 B's, 9 C's, 4 D's, and 3 F's. In how many ways can she do this?

33. *Phone Numbers.* How many 7-digit phone numbers can be formed with the digits 0, 1, 2, 3, 4, 5, 6, 7, 8, and 9, assuming that the first number cannot be 0 or 1? Accordingly, how many telephone numbers can there be within a given area code, before the area needs to be split with a new area code?

34. *Program Planning.* A program is planned to have 5 rock numbers and 4 speeches. In how many ways can this be done if a rock number and a speech are to alternate and a rock number is to come first?

35. Suppose the expression $a^2b^3c^4$ is rewritten without exponents. In how many ways can this be done?

36. *Coin Arrangements.* A penny, a nickel, a dime, and a quarter are arranged in a straight line.

a) Considering just the coins, in how many ways can they be lined up?
b) Considering the coins and heads and tails, in how many ways can they be lined up?

37. How many code symbols can be formed using 5 out of 6 letters of A, B, C, D, E, F if the letters:

a) are not repeated?
b) can be repeated?
c) are not repeated but must begin with D?
d) are not repeated but must begin with DE?

38. *License Plates.* A state forms its license plates by first listing a number that corresponds to the county in which the owner of the car resides (the names of the counties are alphabetized and the number is its location in that order). Then the plate lists a letter of the alphabet, and this is followed by a number from 1 to 9999. How many such plates are possible if there are 80 counties?

39. *Zip Codes.* A zip code in Dallas, Texas, is 75247. A zip code in Cambridge, Massachusetts, is 02142.

a) How many zip codes are possible if any of the digits 0 to 9 can be used?
b) If each post office has its own zip code, how many possible post offices can there be?

40. *Zip-Post Codes.* Zip-post codes use a 9-digit number like 75247-5456, in which the zip code is augmented with a post office box number.

a) How many 9-digit zip-post codes are possible?
b) There are about 266 million people in the United States. Are there enough zip-post codes?

41. *Social Security Numbers.* A social security number is a 9-digit number like 243-47-0825.

a) How many different social security numbers can there be?
b) There are about 266 million people in the United States. Can each person have a unique social security number?

Skill Maintenance

Simplify.

42. $\dfrac{8 \cdot 7 \cdot 6 \cdot 5 \cdot 4 \cdot 3 \cdot 2 \cdot 1}{3 \cdot 2 \cdot 1 \cdot 5 \cdot 4 \cdot 3 \cdot 2 \cdot 1}$

43. $\dfrac{9 \cdot 8 \cdot 7 \cdot 6 \cdot 5 \cdot 4 \cdot 3 \cdot 2 \cdot 1}{4 \cdot 3 \cdot 2 \cdot 1 \cdot 5 \cdot 4 \cdot 3 \cdot 2 \cdot 1}$

44. $\dfrac{10 \cdot 9 \cdot 8 \cdot 7 \cdot 6 \cdot 5 \cdot 4 \cdot 3 \cdot 2 \cdot 1}{4 \cdot 3 \cdot 2 \cdot 1 \cdot 6 \cdot 5 \cdot 4 \cdot 3 \cdot 2 \cdot 1}$

45. $\dfrac{44 \cdot 43 \cdot 42 \cdot 41 \cdot 40 \cdot 39}{6 \cdot 5 \cdot 4 \cdot 3 \cdot 2 \cdot 1}$

Synthesis

46. ◆ How "long" is 15!? You own 15 different books and decide to actually make up all the possible arrangements of the books on a shelf. About how long, in years, would it take you if you make one arrangement per second? Write out the reasoning you used for this problem in the form of a paragraph.

47. ◆ *Circular Arrangements.* In how many ways can the numbers on a clock face be arranged? See if you can derive a formula for the number of distinct circular arrangements of n objects. Explain your reasoning.

Solve for n.

48. ${}_nP_5 = 7 \cdot {}_nP_4$

49. ${}_nP_4 = 8 \cdot {}_{n-1}P_3$

50. ${}_nP_5 = 9 \cdot {}_{n-1}P_4$

51. ${}_nP_4 = 8 \cdot {}_nP_3$

52. Show that $n! = n(n - 1)(n - 2)(n - 3)!$.

53. *Single-Elimination Tournaments.* In a single-elimination sports tournament consisting of n teams, a team is eliminated when it loses one game. How many games are required to complete the tournament?

54. *Double-Elimination Tournaments.* In a double-elimination softball tournament consisting of n teams, a team is eliminated when it loses two games. At most, how many games are required to complete the tournament?

9.6 Combinatorics: Combinations

- *Evaluate combination notation and solve related applications.*

We now consider counting techniques in which order is not considered.

Combinations

If you play cards, you know that in most situations the *order* in which you hold cards is *not important!* That is,

The hand
is "equivalent"
to these hands.

We may sometimes make selections from a set *without regard to order.* Such selections are called **combinations**.

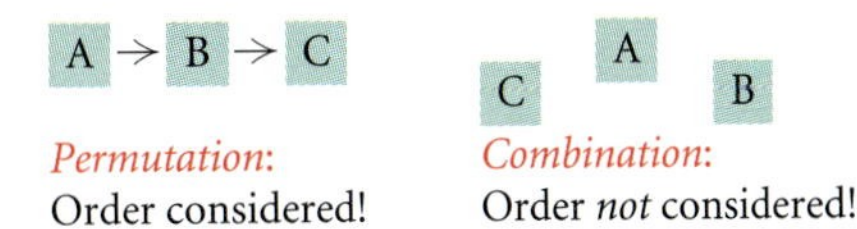
A → B → C
A
C
B
Permutation:
Order considered!
Combination:
Order *not* considered!

Example 1 Find all the combinations of 3 elements taken from the set of 5 elements {A, B, C, D, E}.

SOLUTION The combinations are

{A, B, C}, {A, B, D},
{A, B, E}, {A, C, D},
{A, C, E}, {A, D, E},
{B, C, D}, {B, C, E},
{B, D, E}, {C, D, E}.

There are 10 combinations of 5 objects taken 3 at a time.

When we find all the combinations of 5 objects taken 3 at a time, we are finding all the 3-element subsets. When a set is named, the order of the listing is *not* considered. Thus,

{A, C, B} names the same set as {A, B, C}.

Subset

The set A is a subset of B, denoted $A \subseteq B$, if every element of A is an element of B.

$A \subseteq B$

A is a subset of B.

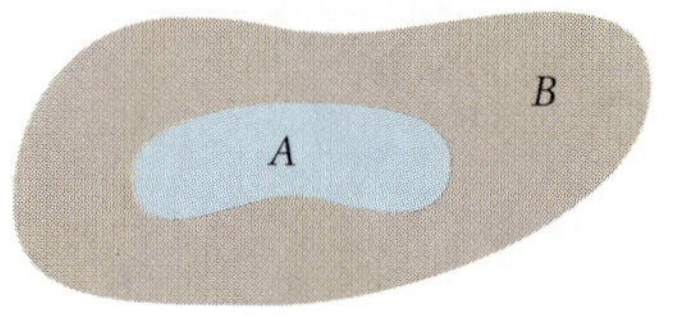

Combination

A *combination* containing k objects is a subset containing k objects.

The elements of a subset are not ordered. When thinking of *combinations,* do *not* think about order!

Example 2 Find all the subsets of {A, B, C}. Identify these as combinations. How many subsets are there in all?

SOLUTION The empty set has 0 elements in it. It is denoted ∅. The empty set is a subset of every set. In this case, it is the combination of 3 objects taken 0 at a time. There is 1 such combination.

The following are all the 1-element subsets of {A, B, C}:

{A}, {B}, {C}.

These are the combinations of 3 objects taken 1 at a time. There are 3 such combinations.

The following are all the 2-element subsets of {A, B, C}:

{A, B}, {A, C}, {B, C}.

These are the combinations of 3 objects taken 2 at a time. There are 3 such combinations.

The following are all the 3-element subsets of {A, B, C}:

{A, B, C}.

These are the combinations of 3 objects taken 3 at a time. There is only 1 such combination. A set is always a subset of itself.

The total number of subsets is $1 + 3 + 3 + 1$, or 8.

We want to develop a formula for computing the number of combinations of n objects taken k at a time without actually listing the combinations or subsets.

Combination Notation

The number of combinations of n objects taken k at a time is denoted ${}_nC_k$.

We call ${}_nC_k$ **combination notation**. We want to derive a general formula for ${}_nC_k$ for any $k \leq n$. First, it is true that ${}_nC_n = 1$, because a set with n objects has only 1 subset with n objects, the set itself. Second, ${}_nC_1 = n$, because a set with n objects has n subsets with 1 element each. Finally, ${}_nC_0 = 1$, because a set with n objects has only one subset with 0 elements, namely the empty set $\varnothing$. To consider other possibilities, let's return to Example 1 and compare the number of combinations with the number of permutations.

COMBINATIONS	PERMUTATIONS					
{A, B, C} →	ABC	BCA	CAB	CBA	BAC	ACB
{A, B, D} →	ABD	BDA	DAB	DBA	BAD	ADB
{A, B, E} →	ABE	BEA	EAB	EBA	BAE	AEB
{A, C, D} →	ACD	CDA	DAC	DCA	CAD	ADC
{A, C, E} →	ACE	CEA	EAC	ECA	CAE	AEC
{A, D, E} →	ADE	DEA	EAD	EDA	DAE	AED
{B, C, D} →	BCD	CDB	DBC	DCB	CBD	BDC
{B, C, E} →	BCE	CEB	EBC	ECB	CBE	BEC
{B, D, E} →	BDE	DEB	EBD	EDB	DBE	BED
{C, D, E} →	CDE	DEC	ECD	EDC	DCE	CED

${}_5C_3$ of these (combinations); $3! \cdot {}_5C_3$ of these (permutations)

Note that each combination of 3 objects, say {A, C, E}, yields 3!, or 6, permutations, as shown above. It follows that

$$3! \cdot {}_5C_3 = 60 = {}_5P_3 = 5 \cdot 4 \cdot 3,$$

so

$${}_5C_3 = \frac{{}_5P_3}{3!} = \frac{5 \cdot 4 \cdot 3}{3 \cdot 2 \cdot 1} = 10.$$

In general, the number of combinations of n objects taken k at a time, ${}_nC_k$, times the number of permutations of these objects, $k!$, must equal the number of permutations of n objects taken k at a time:

$$\begin{aligned} k! \cdot {}_nC_k &= {}_nP_k \\ {}_nC_k &= \frac{{}_nP_k}{k!} \\ &= \frac{1}{k!} \cdot {}_nP_k = \frac{1}{k!} \cdot \frac{n!}{(n-k)!} = \frac{n!}{k!(n-k)!}. \end{aligned}$$

Combinations of n Objects Taken k at a Time

The total number of combinations of n objects taken k at a time, denoted ${}_nC_k$, is given by

$$ {}_nC_k = \frac{n!}{k!(n-k)!}, \tag{1} $$

or

$$ {}_nC_k = \frac{{}_nP_k}{k!} = \frac{n(n-1)(n-2)\cdots[n-(k-1)]}{k!}. \tag{2} $$

Another kind of notation for ${}_nC_k$ is **binomial coefficient notation.** The reason for such terminology will be seen later.

Binomial Coefficient Notation

$$ \binom{n}{k} = {}_nC_k $$

You should be able to use either notation and either formula.

Example 3 Evaluate $\binom{7}{5}$, using formulas (1) and (2).

SOLUTION

a) By formula (1),

$$ \binom{7}{5} = \frac{7!}{5!\,2!} = \frac{7 \cdot 6 \cdot 5 \cdot 4 \cdot 3 \cdot 2 \cdot 1}{5 \cdot 4 \cdot 3 \cdot 2 \cdot 1 \cdot 2 \cdot 1} = \frac{7 \cdot 6}{2 \cdot 1} = 21. $$

b) By formula (2),

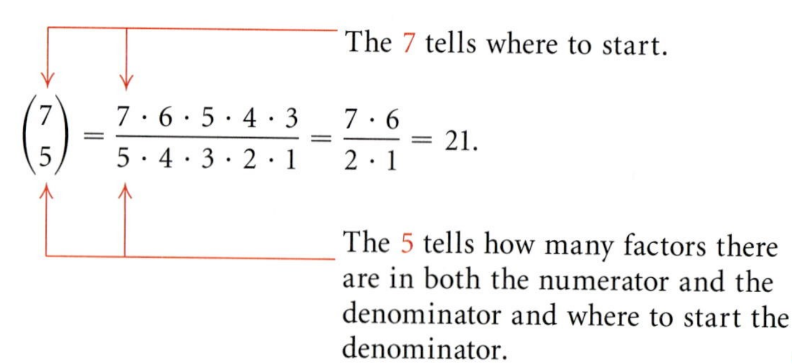

The 7 tells where to start.
The 5 tells how many factors there are in both the numerator and the denominator and where to start the denominator.

$$ \binom{7}{5} = \frac{7 \cdot 6 \cdot 5 \cdot 4 \cdot 3}{5 \cdot 4 \cdot 3 \cdot 2 \cdot 1} = \frac{7 \cdot 6}{2 \cdot 1} = 21. $$

The method in Example 3(b), using formula (2), is easier to carry out, but in some situations formula (1) becomes useful. Be sure to keep in mind that $\binom{n}{k}$ does not mean $n \div k$, or n/k.

Example 4 Evaluate $\binom{n}{0}$ and $\binom{n}{2}$.

SOLUTION We use formula (1) for the first expression and formula (2) for the second. Then

$$\binom{n}{0} = \frac{n!}{0!(n-0)!} = \frac{n!}{1 \cdot n!} = 1,$$

using formula (1), and

$$\binom{n}{2} = \frac{n(n-1)}{2!} = \frac{n(n-1)}{2}, \quad \text{or} \quad \frac{n^2-n}{2},$$

using formula (2).

Most calculators and graphers also have a $_nC_k$ key. Let's use it to make a discovery.

Interactive Discovery

Use a grapher to complete the table at right. Look for a pattern.

n	k	$_nC_k$	$_nC_{n-k}$	$\binom{n}{k}$	$\binom{n}{n-k}$
8	3				
10	7				
9	1				
7	0				
13	10				

Note that

$$\binom{7}{2} = \frac{7 \cdot 6}{2 \cdot 1} = 21,$$

so that using the result of Example 3 gives us

$$\binom{7}{5} = \binom{7}{2}.$$

This says that the number of 5-element subsets of a set of 7 objects is the same as the number of 2-element subsets of a set of 7 objects. When 5 elements are chosen from a set, one also chooses *not* to include 2 elements. To see this, consider such a set:

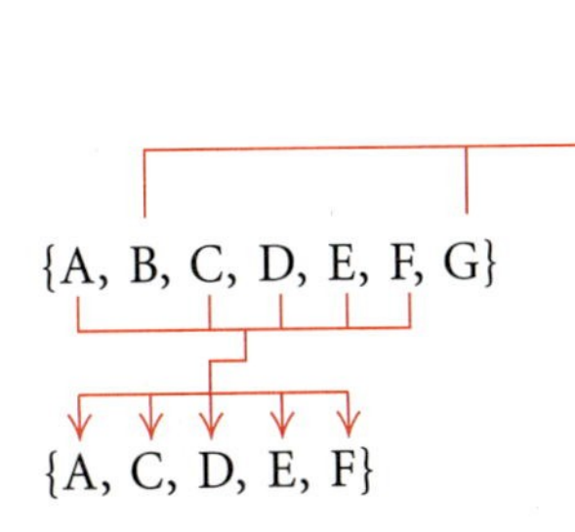

Each time we form a subset with 5 elements, we leave behind a subset with 2 elements, and vice versa.

In general, we have the following.

Subsets of Size k and of Size n − k

$$\binom{n}{k} = \binom{n}{n-k} \quad \text{and} \quad {}_nC_k = {}_nC_{n-k}$$

The number of subsets of size k of a set with n objects is the same as the number of subsets of size $n - k$. The number of combinations of n objects taken k at a time is the same as the number of combinations of n objects taken $n - k$ at a time.

This result provides an alternative way to compute. For example, it is much easier to compute ${}_{52}C_4$ than to compute ${}_{52}C_{48}$, though if you were using a grapher, the difference would not be worth considering.

We now solve problems involving combinations.

Example 5 *Michigan Lotto.* The state of Michigan runs a 6-out-of-49-number lotto twice a week that pays at least $2 million. You purchase a card for $1 and pick any 6 numbers from 1 to 49. If your numbers match those that the state draws, you win.

a) How many 6-number combinations are there for drawing?

b) Suppose it takes 10 min to pick your numbers and buy a ticket. How many tickets can you buy in 4 days?

c) How many people would you have to hire to buy all the tickets and ensure that you win?

SOLUTION

a) No order is implied here. You pick any 6 numbers from 1 to 49. Thus the number of combinations is

$$\begin{aligned} {}_{49}C_6 &= \binom{49}{6} \\ &= \frac{49 \cdot 48 \cdot 47 \cdot 46 \cdot 45 \cdot 44}{6 \cdot 5 \cdot 4 \cdot 3 \cdot 2 \cdot 1} \\ &= 13{,}983{,}816. \end{aligned}$$

b) In 4 days, there are $4 \cdot 24 \cdot 60$, or 5760 min, so you could buy 5760/10, or 576 tickets in that entire time period.

c) You would need to hire 13,983,816/576, or about 24,278 people to buy all the tickets and ensure a win. (This presumes lottery tickets can be bought 24 hours a day, which is questionable.)

Example 6 How many committees can be formed from a group of 5 governors and 7 senators if each committee consists of 3 governors and 4 senators?

SOLUTION The 3 governors can be selected in ${}_5C_3$ ways and the 4 senators can be selected in ${}_7C_4$ ways. If we use the fundamental counting principle, it follows that the number of possible committees is

$${}_5C_3 \cdot {}_7C_4 = 10 \cdot 35 = 350.$$

9.6 Exercise Set

Evaluate. Do your work by hand and then check it on a grapher.

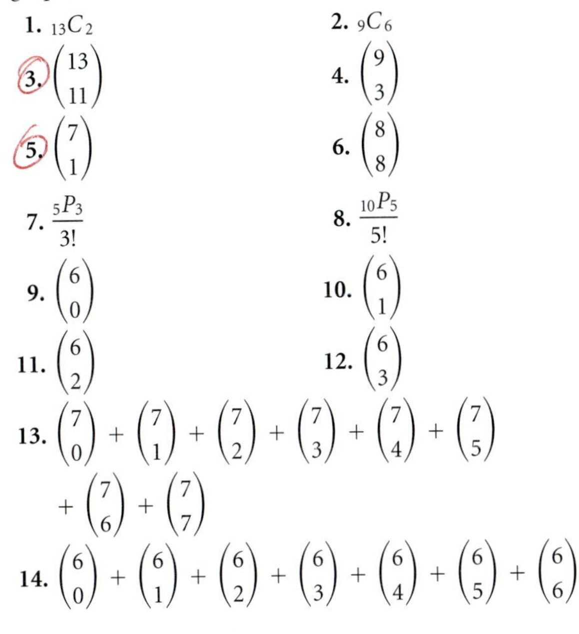

1. ${}_{13}C_2$
2. ${}_9C_6$
3. $\binom{13}{11}$
4. $\binom{9}{3}$
5. $\binom{7}{1}$
6. $\binom{8}{8}$
7. $\frac{{}_5P_3}{3!}$
8. $\frac{{}_{10}P_5}{5!}$
9. $\binom{6}{0}$
10. $\binom{6}{1}$
11. $\binom{6}{2}$
12. $\binom{6}{3}$
13. $\binom{7}{0} + \binom{7}{1} + \binom{7}{2} + \binom{7}{3} + \binom{7}{4} + \binom{7}{5} + \binom{7}{6} + \binom{7}{7}$
14. $\binom{6}{0} + \binom{6}{1} + \binom{6}{2} + \binom{6}{3} + \binom{6}{4} + \binom{6}{5} + \binom{6}{6}$

Evaluate. Use a grapher.

15. ${}_{52}C_4$
16. ${}_{52}C_5$
17. $\binom{27}{11}$
18. $\binom{37}{8}$

Evaluate.

19. $\binom{n}{1}$
20. $\binom{n}{3}$
21. $\binom{m}{m}$
22. $\binom{t}{4}$

In each of the following exercises, give an expression for the answer using permutation notation, combination notation, factorial notation, or other operations. Then evaluate.

23. *Fraternity Officers.* There are 23 students in a fraternity. How many sets of 4 officers can be selected?

24. *League Games.* How many games can be played in a 9-team sports league if each team plays all other teams once? twice?

25. *Test Options.* On a test, a student is to select 10 out of 13 questions. In how many ways can this be done?

26. *Test Options.* Of the first 10 questions on a test, a student must answer 7. Of the second 5 questions, the student must answer 3. In how many ways can this be done?

27. *Lines and Triangles from Points.* How many lines are determined by 8 points, no 3 of which are collinear? How many triangles are determined by the same points?

28. *Senate Committees.* Suppose the Senate of the United States consists of 58 Republicans and 42 Democrats. How many committees can be formed consisting of 6 Republicans and 4 Democrats?

29. *Poker Hands.* How many 5-card poker hands are possible with a 52-card deck?

30. *Bridge Hands.* How many 13-card bridge hands are possible with a 52-card deck?

31. *Baskin-Robbins Ice Cream.* Baskin-Robbins, a national firm, sells ice cream in 31 flavors.
    a) How many 2-dip cones are possible if order of flavors is to be considered and no flavor is repeated?
    b) How many 2-dip cones are possible if order is to be considered and a flavor can be repeated?
    c) How many 2-dip cones are possible if order is not considered and no flavor is repeated?

Skill Maintenance

Simplify or expand.

32. $(a + b)^0$
33. $(a + b)^1$
34. $(a + b)^2$
35. $(a + b)^3$
36. $(a - b)^3$
37. $(a + b)^4$

Synthesis

38. ◈ Explain why a "combination" lock should really be called a "permutation" lock?

39. ◈ Give an explanation that you might use with a fellow student to explain that

$$\binom{n}{k} = \binom{n}{n-k}.$$

40. *Full House.* A full house in poker consists of a pair (two of a kind) and three of a kind. How many full houses are there that consist of 3 aces and 2 queens? (See Section 9.8 for a description of a 52-card deck.)

41. *Flush.* A flush in poker consists of a 5-card hand with all cards of the same suit. How many 5-card hands (flushes) are there that consist of all diamonds?

42. There are n points on a circle. How many quadrilaterals can be inscribed with these points as vertices?

43. *League Games.* How many games are played in a league with n teams if each team plays each other team once? twice?

Solve for n.

44. $\binom{n+1}{3} = 2 \cdot \binom{n}{2}$

45. $\binom{n}{n-2} = 6$

46. $\binom{n}{3} = 2 \cdot \binom{n-1}{2}$

47. $\binom{n+2}{4} = 6 \cdot \binom{n}{2}$

48. How many line segments are determined by the n vertices of an n-agon? Of these, how many are diagonals? Use mathematical induction to prove the result for the diagonals.

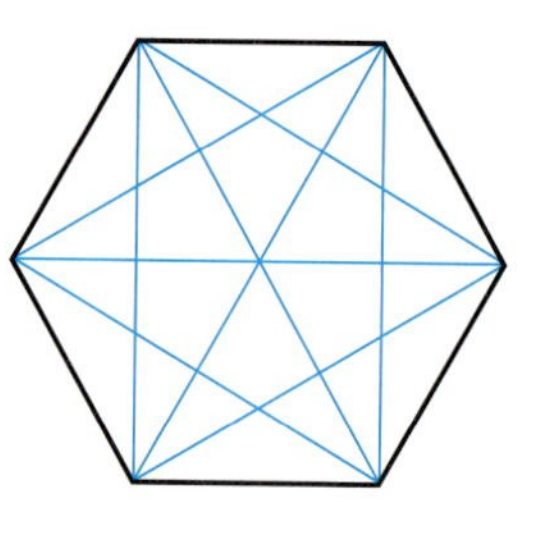

49. Prove that

$$\binom{n}{k-1} + \binom{n}{k} = \binom{n+1}{k}$$

for any natural numbers n and k, $k \leq n$.

9.7 The Binomial Theorem

- *Expand a power of a binomial using Pascal's triangle or factorial notation.*
- *Find a specific term of a binomial expansion.*
- *Find the total number of subsets of a set of n objects.*

In this section, we consider ways of expanding a binomial $(a + b)^n$.

Binomial Expansions Using Pascal's Triangle

Consider the following expanded powers of $(a + b)^n$, where $a + b$ is any binomial and n is a whole number. Look for patterns.

$$\begin{aligned}
(a + b)^0 &= 1 \\
(a + b)^1 &= a + b \\
(a + b)^2 &= a^2 + 2ab + b^2 \\
(a + b)^3 &= a^3 + 3a^2b + 3ab^2 + b^3 \\
(a + b)^4 &= a^4 + 4a^3b + 6a^2b^2 + 4ab^3 + b^4 \\
(a + b)^5 &= a^5 + 5a^4b + 10a^3b^2 + 10a^2b^3 + 5ab^4 + b^5
\end{aligned}$$

Each expansion is a polynomial. There are some patterns to be noted.

1. There is one more term than the power of the exponent, n. That is, there are $n + 1$ terms in the expansion of $(a + b)^n$.
2. In each term, the sum of the exponents is n, the power to which the binomial is raised.
3. The exponents of a start with n, the power of the binomial, and decrease to 0. The last term has no factor of a. The first term has no factor of b, so powers of b start with 0 and increase to n.
4. The coefficients start at 1 and increase through certain values about "half"-way and then decrease through these same values back to 1. Let's explore the coefficients further.

Suppose we want to find an expansion of $(a + b)^8$. The patterns we noted above indicate that there are 9 terms in the expansion:

$$a^8 + c_1a^7b + c_2a^6b^2 + c_3a^5b^3 + c_4a^4b^4 + c_5a^3b^5 + c_6a^2b^6 + c_7ab^7 + b^8.$$

How can we determine the value of each coefficient, c_i? We can answer this question in two different ways. The first method seems to be the easiest, but is not always. It involves writing down the coefficients in a triangular array, as follows. We form what is known as **Pascal's triangle**:

$$\begin{array}{lccccccccccc}
(a + b)^0: & & & & & & 1 & & & & & \\
(a + b)^1: & & & & & 1 & & 1 & & & & \\
(a + b)^2: & & & & 1 & & 2 & & 1 & & & \\
(a + b)^3: & & & 1 & & 3 & & 3 & & 1 & & \\
(a + b)^4: & & 1 & & 4 & & 6 & & 4 & & 1 & \\
(a + b)^5: & 1 & & 5 & & 10 & & 10 & & 5 & & 1
\end{array}$$

There are many patterns in the triangle. Find as many as you can.

Perhaps you discovered a way to write the next row of numbers, given the numbers in the row above it. There are always 1's on the outside. Each remaining number is the sum of the two numbers above. Let's try to find an expansion for $(a + b)^6$ by adding another row using the patterns we have discovered:

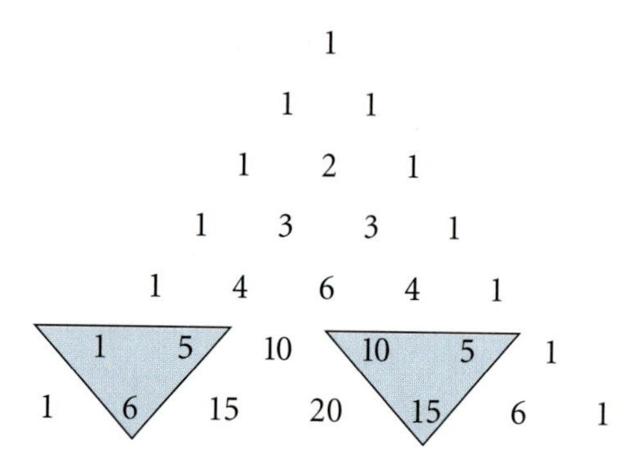

We see that in the last row

the 1st and last numbers are **1**;
the 2nd number is 1 + 5, or **6**;
the 3rd number is 5 + 10, or **15**;
the 4th number is 10 + 10, or **20**;
the 5th number is 10 + 5, or **15**; and
the 6th number is 5 + 1, or **6**.

Thus the expansion for $(a + b)^6$ is

$$(a + b)^6 = 1a^6 + 6a^5b + 15a^4b^2 + 20a^3b^3 + 15a^2b^4 + 6ab^5 + 1b^6.$$

To find an expansion for $(a + b)^8$, we complete two more rows of Pascal's triangle:

1
1 1
1 2 1
1 3 3 1
1 4 6 4 1
1 5 10 10 5 1
1 6 15 20 15 6 1
1 7 21 35 35 21 7 1
1 8 28 56 70 56 28 8 1

Thus the expansion of $(a + b)^8$ is

$$(a + b)^8 = a^8 + 8a^7b^1 + 28a^6b^2 + 56a^5b^3 + 70a^4b^4 + 56a^3b^5 + 28a^2b^6 + 8ab^7 + b^8.$$

We can generalize our results as follows.

The Binomial Theorem Using Pascal's Triangle

For any binomial $a + b$ and any natural number n,

$$(a + b)^n = c_0a^nb^0 + c_1a^{n-1}b^1 + c_2a^{n-2}b^2 + \cdots + c_{n-1}a^1b^{n-1} + c_na^0b^n,$$

where the numbers $c_0, c_1, c_2, \ldots, c_{n-1}, c_n$ are from the $(n + 1)$st row of Pascal's triangle.

Example 1 Expand: $(u - v)^5$.

SOLUTION We note that $a = u$, $b = -v$, and $n = 5$. We use the 6th row of Pascal's triangle:

$$1 \quad 5 \quad 10 \quad 10 \quad 5 \quad 1$$

Then we have

$$\begin{aligned}(u - v)^5 &= [u + (-v)]^5 \\ &= 1(u)^5 + 5(u)^4(-v)^1 + 10(u)^3(-v)^2 + 10(u)^2(-v)^3 \\ &\quad + 5(u)(-v)^4 + 1(-v)^5 \\ &= u^5 - 5u^4v + 10u^3v^2 - 10u^2v^3 + 5uv^4 - v^5.\end{aligned}$$

Note that the signs of the terms alternate between + and −. When the power of $-v$ is odd, the sign is −.

Interactive Discovery

Graph each of the following functions. See if you can discover an identity from your graphs. Confirm your results using the binomial theorem.

$$\begin{aligned}y_1 &= (x - 2)^4, \\ y_2 &= x^4 + 8x^3 + 24x^2 + 32x + 16, \\ y_3 &= x^4 - 8x^3 + 24x^2 - 32x - 16, \\ y_4 &= x^4 + 8x^3 - 24x^2 + 32x - 16, \\ y_5 &= x^4 - 8x^3 + 24x^2 - 32x + 16, \\ y_6 &= x^4 - 8x^3 + 24x^2 - 31x + 16\end{aligned}$$

Example 2 Expand: $\left(2t + \dfrac{3}{t}\right)^6$.

SOLUTION We note that $a = 2t$, $b = 3/t$, and $n = 6$. We use the 7th row of Pascal's triangle:

$$1 \quad 6 \quad 15 \quad 20 \quad 15 \quad 6 \quad 1$$

Then we have

$$\left(2t + \frac{3}{t}\right)^6 = (2t)^6 + 6(2t)^5\left(\frac{3}{t}\right)^1 + 15(2t)^4\left(\frac{3}{t}\right)^2$$

$$+ 20(2t)^3\left(\frac{3}{t}\right)^3 + 15(2t)^2\left(\frac{3}{t}\right)^4 + 6(2t)^1\left(\frac{3}{t}\right)^5 + \left(\frac{3}{t}\right)^6$$

$$= 64t^6 + 6(32t^5)\left(\frac{3}{t}\right) + 15(16t^4)\left(\frac{9}{t^2}\right) + 20(8t^3)\left(\frac{27}{t^3}\right)$$

$$+ 15(4t^2)\left(\frac{81}{t^4}\right) + 6(2t)\left(\frac{243}{t^5}\right) + \frac{729}{t^6}$$

$$= 64t^6 + 576t^4 + 2160t^2 + 4320 + 4860t^{-2}$$
$$+ 2916t^{-4} + 729t^{-6}.$$

Binomial Expansion Using Factorial Notation

Suppose we want to find the expansion of $(a + b)^{11}$. The disadvantage in using Pascal's triangle is that we must compute all the preceding rows of the table to obtain the row needed for the expansion. The following method avoids this difficulty. It will also enable us to find a specific term—say, the 8th term—without computing all the other terms of the expansion. This method is useful in such courses as finite mathematics, calculus, and statistics, and uses the *binomial coefficient notation* $\binom{n}{k}$ developed in Section 9.6.

We can restate the binomial theorem as follows.

The Binomial Theorem Using Factorial Notation

For any binomial $a + b$ and any natural number n,

$$(a + b)^n = \binom{n}{0}a^n b^0 + \binom{n}{1}a^{n-1}b^1 + \binom{n}{2}a^{n-2}b^2 + \cdots$$
$$+ \binom{n}{n-1}a^1 b^{n-1} + \binom{n}{n}a^0 b^n$$
$$= \sum_{k=0}^{n}\binom{n}{k}a^{n-k}b^k.$$

The binomial theorem can be proved by mathematical induction, but we will not do so here. This form shows why $\binom{n}{k}$ is called a *binomial coefficient.*

Example 3 Expand: $(x^2 - 2y)^5$.

SOLUTION Note that $a = x^2$, $b = -2y$, and $n = 5$. Then using the binomial theorem, we have

$$(x^2 - 2y)^5 = \binom{5}{0}(x^2)^5 + \binom{5}{1}(x^2)^4(-2y) + \binom{5}{2}(x^2)^3(-2y)^2 + \binom{5}{3}(x^2)^2(-2y)^3 + \binom{5}{4}x^2(-2y)^4 + \binom{5}{5}(-2y)^5$$

$$= \frac{5!}{0!\,5!}x^{10} + \frac{5!}{1!\,4!}x^8(-2y) + \frac{5!}{2!\,3!}x^6(-2y)^2 + \frac{5!}{3!\,2!}x^4(-2y)^3 + \frac{5!}{4!\,1!}x^2(-2y)^4 + \frac{5!}{5!\,0!}(-2y)^5$$

$$= x^{10} - 10x^8y + 40x^6y^2 - 80x^4y^3 + 80x^2y^4 - 32y^5.$$

Example 4 Expand: $\left(\frac{2}{x} + 3\sqrt{x}\right)^4$.

SOLUTION Note that $a = 2/x$, $b = 3\sqrt{x}$, and $n = 4$. Then using the binomial theorem, we have

$$\left(\frac{2}{x} + 3\sqrt{x}\right)^4 = \binom{4}{0}\left(\frac{2}{x}\right)^4 + \binom{4}{1}\left(\frac{2}{x}\right)^3(3\sqrt{x}) + \binom{4}{2}\left(\frac{2}{x}\right)^2(3\sqrt{x})^2 + \binom{4}{3}\left(\frac{2}{x}\right)(3\sqrt{x})^3 + \binom{4}{4}(3\sqrt{x})^4$$

$$= \frac{4!}{0!\,4!}\left(\frac{2}{x}\right)^4 + \frac{4!}{1!\,3!}\left(\frac{2}{x}\right)^3(3\sqrt{x}) + \frac{4!}{2!\,2!}\left(\frac{2}{x}\right)^2(3\sqrt{x})^2 + \frac{4!}{3!\,1!}\left(\frac{2}{x}\right)(3\sqrt{x})^3 + \frac{4!}{4!\,0!}(3\sqrt{x})^4$$

$$= \frac{16}{x^4} + \frac{96}{x^{5/2}} + \frac{216}{x} + 216\sqrt{x} + 81x^2.$$

Interactive Discovery

Describe how you might use a grapher to check the result of Example 4.

Finding a Specific Term

Suppose we want to determine only a particular term of an expansion. The method we have developed will allow us to find such a term without computing all the rows of Pascal's triangle or all the preceding coefficients.

Note that in the binomial theorem, $\binom{n}{0}a^nb^0$ gives us the 1st term, $\binom{n}{1}a^{n-1}b^1$ gives us the 2nd term, $\binom{n}{2}a^{n-2}b^2$ gives us the 3rd term, and so on. This can be generalized as follows.

Finding the (k + 1)st Term

The $(k + 1)$st term of $(a + b)^n$ is $\binom{n}{k}a^{n-k}b^k$.

Example 5 Find the 5th term in the expansion of $(2x - 5y)^6$.

SOLUTION First, we note that $5 = 4 + 1$. Thus, $k = 4$, $a = 2x$, $b = -5y$, and $n = 6$. Then the 5th term of the expansion is

$$\binom{6}{4}(2x)^{6-4}(-5y)^4, \quad \text{or} \quad \frac{6!}{4!\,2!}(2x)^2(-5y)^4, \quad \text{or} \quad 37{,}500x^2y^4.$$

Example 6 Find the 8th term in the expansion of $(3x - 2)^{10}$.

SOLUTION First, we note that $8 = 7 + 1$. Thus, $k = 7$, $a = 3x$, $b = -2$, and $n = 10$. Then the 8th term of the expansion is

$$\binom{10}{7}(3x)^{10-7}(-2)^7, \quad \text{or} \quad \frac{10!}{7!\,3!}(3x)^3(-128), \quad \text{or} \quad -414{,}720x^3.$$

Total Number of Subsets

Suppose a set has n objects. The number of subsets containing k elements is $\binom{n}{k}$ by a result of Section 9.6. The total number of subsets of a set is the number of subsets with 0 elements, plus the number of subsets with 1 element, plus the number of subsets with 2 elements, and so on. The total number of subsets of a set with n elements is

$$\binom{n}{0} + \binom{n}{1} + \binom{n}{2} + \cdots + \binom{n}{n}.$$

Let's find the expansion of $(1 + 1)^n$:

$$\begin{aligned}(1 + 1)^n &= \binom{n}{0} \cdot 1^n + \binom{n}{1} \cdot 1^{n-1} \cdot 1^1 + \binom{n}{2} \cdot 1^{n-2} \cdot 1^2 \\ &\quad + \cdots + \binom{n}{n} \cdot 1^n \\ &= \binom{n}{0} + \binom{n}{1} + \binom{n}{2} + \cdots + \binom{n}{n}.\end{aligned}$$

Thus the total number of subsets is $(1 + 1)^n$, or 2^n. We have proved the following.

Total Number of Subsets

The total number of subsets of a set with n elements is 2^n.

Example 7 The set {A, B, C, D, E} has how many subsets?

SOLUTION The set has 5 elements, so the number of subsets is 2^5, or 32.

Example 8 Wendy's, a national restaurant firm, offers the following condiments for its hamburgers:

{*catsup, mustard, mayonnaise, tomato, lettuce, onions, pickle, relish, cheese*}.

How many different kinds of hamburgers can Wendy's serve, excluding size of hamburger or number of patties?

SOLUTION The condiments on each hamburger is a subset of the set of all possible condiments, the empty set being a plain hamburger. The total number of possible hamburgers is

$$\binom{9}{0} + \binom{9}{1} + \binom{9}{2} + \cdots + \binom{9}{9} = 2^9 = 512.$$

Thus Wendy's serves hamburgers in 512 different ways.

We end this section with a discussion of why 0! is defined to be 1. In the binomial expansion, we want $\binom{n}{0}$ to equal 1 and we also want the definition

$$\binom{n}{k} = \frac{n!}{k!(n-k)!}$$

to hold for all whole numbers n and k, where $k \leq n$. Thus we must have

$$\binom{n}{0} = \frac{n!}{0!(n-0)!} = \frac{n!}{0!\,n!} = 1.$$

This will be satisfied if 0! is defined to be 1.

9.7 Exercise Set

Expand.

1. $(x + 5)^4$
2. $(x - 1)^4$
3. $(x - 3)^5$
4. $(x + 2)^9$
5. $(x - y)^5$
6. $(x + y)^8$
7. $(5x + 4y)^6$
8. $(2x - 3y)^5$
9. $\left(2t + \frac{1}{t}\right)^7$
10. $\left(3y - \frac{1}{y}\right)^4$
11. $(x^2 - 1)^5$
12. $(1 + 2q^3)^8$
13. $(\sqrt{5} + t)^6$
14. $(x - \sqrt{2})^6$
15. $\left(a - \frac{2}{a}\right)^9$
16. $(1 + 3)^n$
17. $(\sqrt{2} + 1)^6 - (\sqrt{2} - 1)^6$
18. $(1 - \sqrt{2})^4 + (1 + \sqrt{2})^4$
19. $(x^{-2} + x^2)^4$
20. $\left(\frac{1}{\sqrt{x}} - \sqrt{x}\right)^6$

Find the indicated term of the binomial expansion.

21. 3rd; $(a + b)^7$
22. 6th; $(x + y)^8$
23. 6th; $(x - y)^{10}$
24. 5th; $(p - 2q)^9$
25. 12th; $(a - 2)^{14}$
26. 11th; $(x - 3)^{12}$
27. 5th; $(2x^3 - \sqrt{y})^8$
28. 4th; $\left(\frac{1}{b^2} + \frac{b}{3}\right)^7$

29. Middle; $(2u - 3v^2)^{10}$

30. Middle two; $(\sqrt{x} + \sqrt{3})^5$

Determine the number of subsets of each of the following.

31. A set of 7 elements

32. A set of 6 members

33. The set of letters of the Greek alphabet, which contains 24 letters

34. The set of letters of the English alphabet, which contains 26 letters

35. What is the degree of $(x^5 + 3)^4$?

36. What is the degree of $(2 - 5x^3)^7$?

Expand each of the following, where $i^2 = -1$.

37. $(3 + i)^5$

38. $(1 + i)^6$

39. $(\sqrt{2} - i)^4$

40. $\left(\frac{\sqrt{3}}{2} - \frac{1}{2}i\right)^{11}$

41. Find a formula for $(a - b)^n$. Use sigma notation.

42. Expand and simplify:
$$\frac{(x + h)^{13} - x^{13}}{h}.$$

43. Expand and simplify:
$$\frac{(x + h)^n - x^n}{h}.$$
Use sigma notation.

Skill Maintenance

Simplify.

44. $\dfrac{\frac{13 \cdot 12}{2 \cdot 1}}{\frac{52 \cdot 51}{2 \cdot 1}}$

45. $\dfrac{\frac{6 \cdot 5}{3 \cdot 2} \cdot \frac{4 \cdot 3}{2 \cdot 1}}{\frac{10 \cdot 9 \cdot 8}{3 \cdot 2 \cdot 1}}$

Synthesis

46. ◈ Blaise Pascal (1623–1662) was a French scientist and philosopher who founded the modern theory of probability. Do some research on Pascal and see if you can find out how he discovered his famous "triangle of numbers."

47. ◈ Discuss the pros and cons of each method of finding a binomial expansion. Give examples of when you might use one method rather than the other.

Solve for x.

48. $\sum_{k=0}^{8} \binom{8}{k} x^{8-k} 3^k = 0$

49. $\sum_{k=0}^{4} \binom{4}{k} 5^{4-k} x^k = 64$

50. $\sum_{k=0}^{5} \binom{5}{k} (-1)^k x^{5-k} 3^k = 32$

51. $\sum_{k=0}^{4} \binom{4}{k} (-1)^k x^{4-k} 6^k = 81$

52. *Hitting Probability.* At one point in a recent season, Barry Bonds of the San Francisco Giants had a batting average of 0.313. Suppose he came to bat 5 times in a game. The probability (likelihood) of his getting exactly 3 hits is the 3rd term of the binomial expansion of $(0.313 + 0.687)^5$.

a) Find that term to determine the probability.

b) The probability that Bonds gets at most 3 hits in 5 at-bats is found by adding the last 4 terms of the binomial expansion of $(0.313 + 0.687)^5$. Find that probability.

53. *Probability of Being Widowed or Divorced.* The probability that a woman will be either widowed or divorced is 85%. Suppose 8 women are interviewed. The probability that exactly 5 of them will be either widowed or divorced is the 6th term of the binomial expansion of $(0.15 + 0.85)^8$.

a) Find that probability.

b) The probability that at least 6 of the women will be either widowed or divorced is found by adding the last 3 terms of the binomial expansion of $(0.15 + 0.85)^8$. Find that probability.

54. Find the middle term of $(8u + 3v^2)^{10}$.

55. Find the two middle terms of $(\sqrt{x} - \sqrt{y})^5$.

56. Find the term of
$$\left(\frac{3x^2}{2} - \frac{1}{3x}\right)^{12}$$
that does not contain x.

57. Find the middle term of $(x^2 - 6y^{3/2})^6$.

58. Find the ratio of the 4th term of
$$\left(p^2 - \frac{1}{2}p\sqrt[3]{q}\right)^5$$
to the 3rd term.

59. Find the term of
$$\left(\sqrt[3]{x} - \frac{1}{\sqrt{x}}\right)^7$$
containing $1/x^{1/6}$.

60. *Money Combinations.* A money clip contains one each of the following bills: \$1, \$2, \$5, \$10, \$20, \$50, and \$100. How many different sums of money can be formed using the bills?

Find the sum.

61. ${}_{100}C_0 + {}_{100}C_1 + \cdots + {}_{100}C_{100}$

62. ${}_nC_0 + {}_nC_1 + \cdots + {}_nC_n$

Simplify.

63. $\sum_{k=0}^{23} \binom{23}{k} (\log_a x)^{23-k} (\log_a t)^k$

64. $\sum_{k=0}^{15} \binom{15}{k} i^{30-2k}$

65. Use mathematical induction and the property

$$\binom{n}{r-1} + \binom{n}{r} = \binom{n+1}{r}$$

to prove the binomial theorem.

9.8 Probability

- *Compute the probability of a simple event.*

When a coin is tossed, we can reason that the chances, or likelihood, that it will fall heads are 1 out of 2, or the **probability** that it will fall heads is $\frac{1}{2}$. Of course, this does not mean that if a coin is tossed 10 times it will necessarily fall heads 5 times. If the coin is a "fair coin" and it is tossed a great many times, however, it will fall heads very nearly half of the time. Here we give an introduction to two kinds of probability, **experimental** and **theoretical**.

Experimental and Theoretical Probability

If we toss a coin a great number of times—say, 1000—and count the number of times it falls heads, we can determine the probability of it falling heads. If it falls heads 503 times, we would calculate the probability of it falling heads to be

$$\frac{503}{1000}, \quad \text{or} \quad 0.503.$$

This is an **experimental** determination of probability. Such a determination of probability is discovered by the observation and study of data and is quite common and very useful. Here, for example, are some probabilities that have been determined *experimentally:*

1. The probability that a woman will get breast cancer in her lifetime is $\frac{1}{11}$.
2. If you kiss someone who has a cold, the probability of your catching a cold is 0.07.
3. A person who has just been released from prison has an 80% probability of returning.

If we consider a coin and reason that it is just as likely to fall heads as tails, we would calculate the probability to be $\frac{1}{2}$. This is a **theoretical** determination of probability. Here, for example, are some probabilities that have been determined *theoretically,* using mathematics:

1. If there are 30 people in a room, the probability that two of them have the same birthday (excluding year) is 0.706.
2. While on a trip, you meet someone, and after a period of conversation, discover that you have a common acquaintance. The typical reaction, "It's a small world!", is actually not appropriate, because the probability of such an occurrence is quite high—just over 22%.

In summary, experimental and theoretical probabilities are determined by making observations and gathering data. Theoretical probabilities are determined by reasoning mathematically. Examples of experimental and theoretical probability like those above, especially those we do not expect, lead us to see the value of a study of probability. You might ask, "What is the *true* probability?" In fact, there is none. Experimentally, we can determine probabilities within certain limits. These may or may not agree with the probabilities that we obtain theoretically. There are situations in which it is much easier to determine one of these types of probabilities than the other. For example, it would be quite difficult to arrive at the probability of catching a cold using theoretical probability.

Computing Experimental Probabilities

We first consider experimental determination of probability. The basic principle we use in computing such probabilities is as follows.

Principle P (Experimental)

An experiment is performed in which n observations are made. If a situation E, or event, occurs m times out of n observations, then we say that the *experimental probability* of the event, $P(E)$, is given by

$$P(E) = \frac{m}{n}.$$

Example 1 *Sociological Survey.* The authors of this text conducted an experiment to determine the number of people who are left-handed, right-handed, or both. The results are shown in the following graph.

a) Determine the probability that a person is right-handed.
b) Determine the probability that a person is ambidextrous (uses both hands with equal ability).
c) Determine the probability that a person is left-handed.

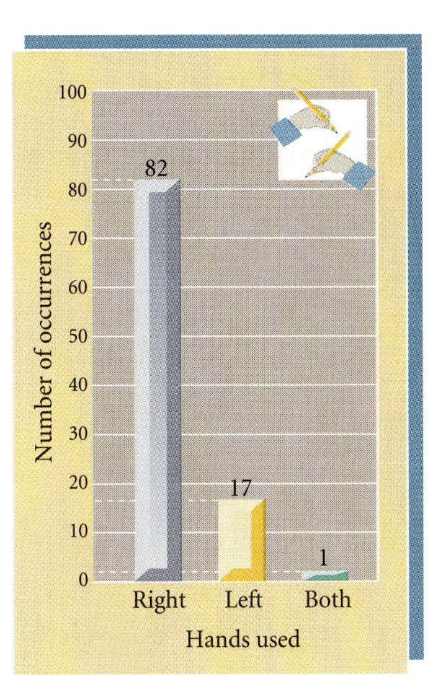

d) For most tournaments held by the Professional Bowlers Association, there are 160 bowlers. On the basis of the data in this experiment, how many of the bowlers would you expect to be left-handed?

SOLUTION

a) The number of people who are right-handed is 82, the number who are left-handed is 17, and the number who are ambidextrous is 1. The total number of observations is 82 + 17 + 1, or 100. Thus the probability that a person is right-handed is

$$P = \frac{82}{100}, \quad \text{or} \quad 0.82, \quad \text{or} \quad 82\%.$$

b) The probability that a person is ambidextrous is P, where

$$P = \frac{1}{100}, \quad \text{or} \quad 0.01, \quad \text{or} \quad 1\%.$$

c) The probability that a person is left-handed is P, where

$$P = \frac{17}{100}, \quad \text{or} \quad 0.17, \quad \text{or} \quad 17\%.$$

d) There are 160 bowlers, and from part (c) we can expect 17% to be left-handed. Since

$$17\% \text{ of } 160 = 0.17 \cdot 160 = 27.2,$$

we can expect that about 27 of the bowlers will be left-handed.

Example 2 *Quality Control.* It is very important for a manufacturer to maintain the quality of its products. In fact, companies hire quality control inspectors to ensure this process. The goal is to produce as few defective products as possible. But since a company is producing thousands of products every day, it cannot afford to check every product to see if it is defective. To find out what percentage of its products are defective, the company checks a smaller sample.

The U.S. Department of Agriculture requires that 80% of the seeds that a company produces must sprout. To find out about the quality of the seeds it produces, a company takes 500 seeds from those it has produced and plants them. It finds that 417 of the seeds sprout.

a) What is the probability that a seed will sprout?

b) Did the seeds pass government standards?

SOLUTION

a) We know that 500 seeds were planted and 417 sprouted. The probability of a seed sprouting is P, where

$$P = \frac{417}{500} = 0.834, \quad \text{or} \quad 83.4\%.$$

b) Since the percentage of seeds exceeded the 80% requirement, the company determines that it is producing quality seeds.

Example 3 *Television Ratings.* Television networks are always concerned about the percentage of homes that have TVs and are watching their programs. A sample of the homes are contacted by attaching an electronic device to the TVs of about 1400 homes across the country. Viewing information is then fed into a computer. The following are the results of a recent survey.

NETWORK	CBS	ABC	NBC	Fox	Other, or not watching
NUMBER OF HOMES WATCHING	196	224	237	126	617

What is the probability that a home was tuned to NBC during the time period? to Fox?

SOLUTION The probability that a home was tuned to NBC is P, where

$$P = \frac{237}{1400} \approx 0.169 = 16.9\%.$$

The probability that a home was tuned to Fox is P, where

$$P = \frac{126}{1400} = 0.09 = 9.0\%.$$

The percentages found in Example 3 are called *ratings.*

Theoretical Probability

Suppose we perform an experiment such as flipping a coin, throwing a dart, drawing a card from a deck, or checking an item off an assembly line for quality. The results of such an experiment are called **outcomes**. The set of all possible outcomes is called the **sample space**. An **event** is a set of outcomes, that is, a subset of the sample space.

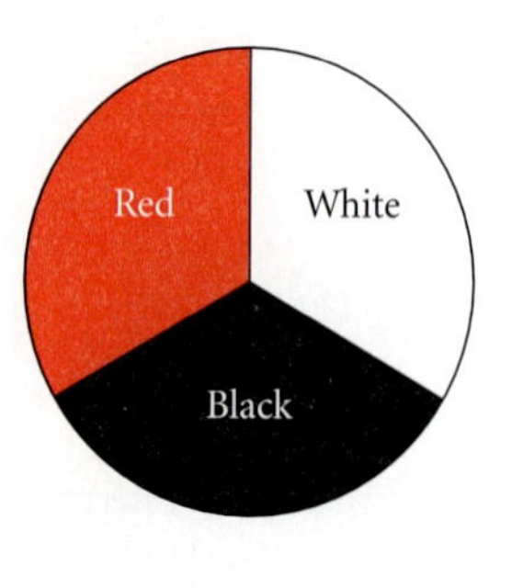

Example 4 *Dart Throwing.* Consider this dartboard. Assume that the experiment is "throwing a dart" and that the dart hits the board. Find each of the following.

a) The outcomes

b) The sample space

SOLUTION

a) The outcomes are *hitting black* (B), *hitting red* (R), and *hitting white* (W).

b) The sample space is {*hitting black, hitting red, hitting white*}, which can be simply stated as {B, R, W}.

Example 5 *Die Rolling.* A die (pl., dice) is a cube, with six faces, each containing a number of dots from 1 to 6 on each side.

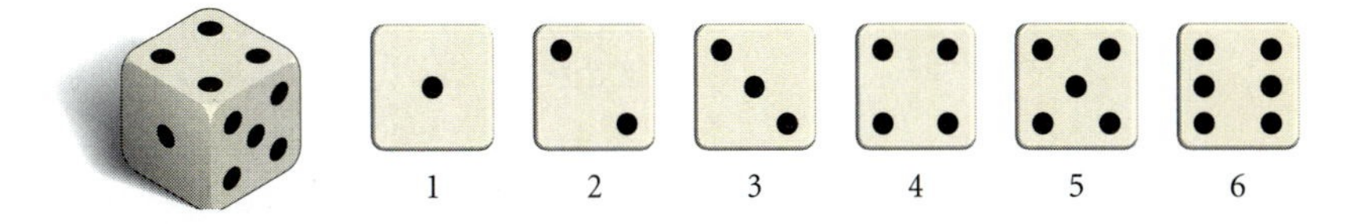

A die is rolled. Find each of the following.

a) The outcomes

b) The sample space

SOLUTION

a) The outcomes are 1, 2, 3, 4, 5, 6.

b) The sample space is {1, 2, 3, 4, 5, 6}.

We denote the probability that an event E occurs as $P(E)$. For example, "a coin falling heads" may be denoted H. Then $P(H)$ represents the probability of the coin falling heads. When all the outcomes of an experiment have the same probability of occurring, we say that they are *equally likely.* To see the distinction between events that are equally likely and those that are not, consider the dartboards shown below.

For board A, the events *hitting black, red,* and *white* are equally likely, but for board B, they are not. A sample space that can be expressed as a union of equally likely events can allow us to calculate probabilities of other events.

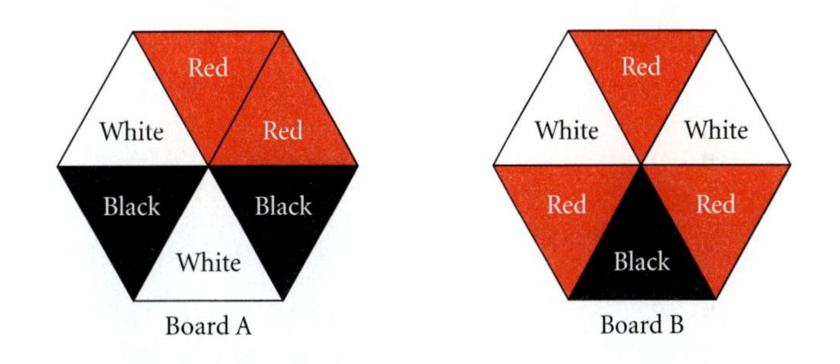

Principle P (Theoretical)

If an event E can occur m ways out of n possible equally likely outcomes of a sample space S, then the *theoretical probability* of the event, $P(E)$, is given by

$$P(E) = \frac{m}{n}.$$

Example 6 What is the probability of rolling a 3 on a die?

SOLUTION On a fair die, there are 6 equally likely outcomes and there is 1 way to roll a 3. By Principle P, $P(3) = \frac{1}{6}$.

Example 7 What is the probability of rolling an even number on a die?

SOLUTION The event is rolling an *even* number. It can occur 3 ways (getting 2, 4, or 6). The number of equally likely outcomes is 6. By Principle *P*, $P(\text{even}) = \frac{3}{6}$, or $\frac{1}{2}$.

We now use a number of examples related to a standard bridge deck of 52 cards. Such a deck is made up as shown in the following figure.

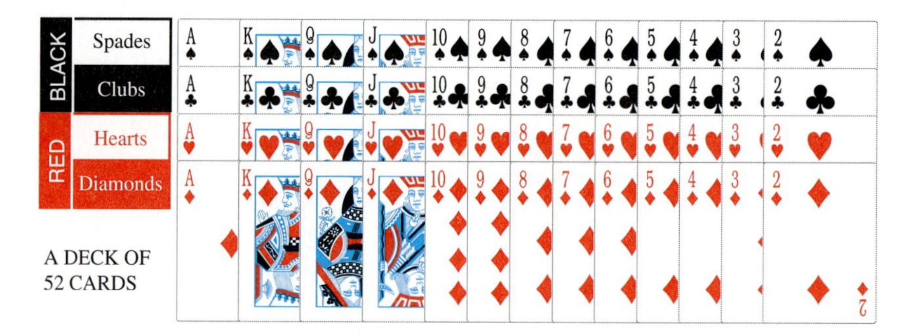

Example 8 What is the probability of drawing an ace from a well-shuffled deck of cards?

SOLUTION Since there are 52 outcomes (the number of cards in the deck), they are equally likely (from a well-shuffled deck), and there are 4 ways to obtain an ace, by Principle *P*, we have

$$P(\text{drawing an ace}) = \frac{4}{52}, \quad \text{or} \quad \frac{1}{13}.$$

Example 9 Suppose we select, without looking, one marble from a bag containing 3 red marbles and 4 green marbles. What is the probability of selecting a red marble?

SOLUTION There are 7 equally likely ways of selecting any marble, and since the number of ways of getting a red marble is 3, we have

$$P(\text{selecting a red marble}) = \frac{3}{7}.$$

The following are some results that follow from Principle *P*.

Probability Properties

a) If an event E cannot occur, then $P(E) = 0$.

b) If an event E is certain to occur (that is, every trial is a success), then $P(E) = 1$.

c) The probability that an event E will occur is a number from 0 to 1: $0 \leq P(E) \leq 1$.

For example, in coin tossing, the event that a coin will land on its edge has probability 0. The event that a coin falls either heads or tails has probability 1.

In the following examples, we use the combinatorics that we studied in Sections 9.5 and 9.6 to calculate theoretical probabilities.

Example 10 Suppose that 2 cards are drawn from a well-shuffled deck of 52 cards. What is the probability that both of them are spades?

SOLUTION The number of ways n of drawing 2 cards from a well-shuffled deck of 52 is ${}_{52}C_2$. Since 13 of the 52 cards are spades, the number of ways m of drawing 2 spades is ${}_{13}C_2$. Thus,

$$P(\text{getting 2 spades}) = \frac{m}{n} = \frac{{}_{13}C_2}{{}_{52}C_2} = \frac{78}{1326} = \frac{1}{17}.$$

Example 11 Suppose that 3 people are selected at random from a group that consists of 6 men and 4 women. What is the probability that 1 man and 2 women are selected?

SOLUTION The number of ways of selecting 3 people from a group of 10 is ${}_{10}C_3$. One man can be selected in ${}_6C_1$ ways, and 2 women can be selected in ${}_4C_2$ ways. By the fundamental counting principle, the number of ways of selecting 1 man and 2 women is ${}_6C_1 \cdot {}_4C_2$. Thus the probability that 1 man and 2 women are selected is

$$P = \frac{{}_6C_1 \cdot {}_4C_2}{{}_{10}C_3} = \frac{3}{10}.$$

Example 12 *Rolling Two Dice.* What is the probability of getting a total of 8 on a roll of a pair of dice?

SOLUTION On each die, there are 6 possible outcomes. The outcomes are paired so there are $6 \cdot 6$, or 36, possible ways in which the two can fall. (Assuming that the dice are different—say, one green and one purple—can help in visualizing this.)

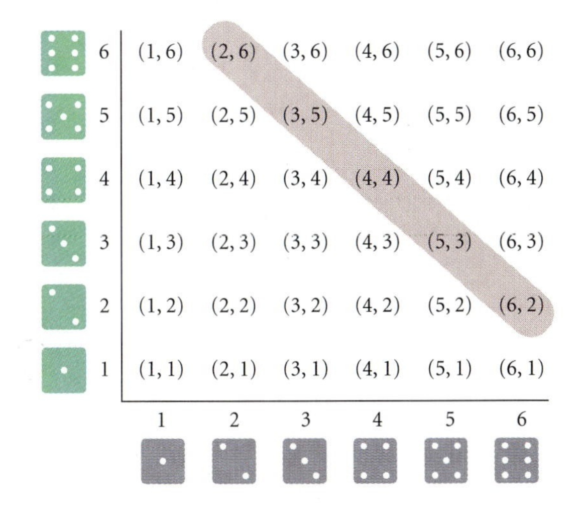

The pairs that total 8 are as shown in the figure above. Thus there are 5 possible ways of getting a total of 8, so the probability is $\frac{5}{36}$.

The Origin and Uses of Probability

A desire to calculate odds in games of chance gave rise to the theory of probability. Today the theory of probability and its closely related field, mathematical statistics, have many applications, most of them not related to games of chance. Opinion polls, with such uses as predicting elections, are a familiar example. Quality control, in which a prediction about the percentage of faulty items manufactured is made without testing them all, is an important application, among many, in business. Still other applications are in the areas of the use of DNA blood testing for genetics and crime investigation, other areas of medicine, and the kinetic theory of gases.

9.8 Exercise Set

1. *Favorite Number.* In an actual survey conducted by the authors, 100 people were polled and asked to select a number from 1 to 5. The results are shown in the following table.

NUMBER OF CHOICES	1	2	3	4	5
NUMBER WHO CHOSE THAT NUMBER	18	24	23	23	12

a) What is the probability that the number chosen is 1? 2? 3? 4? 5?

b) What general conclusion might a psychologist make from the experiment?

2. *Mason Dots®.* Made by the Tootsie Industries of Chicago, Illinois, Mason Dots® is a gumdrop candy. A box was opened by the authors and was found to contain the following number of gumdrops:

Strawberry	7
Lemon	8
Orange	9
Cherry	4
Lime	5
Grape	6

If we take one gumdrop out of the box, what is the probability of getting a lemon? lime? orange? grape? strawberry? licorice?

3. *Junk Mail.* Have you ever wondered why you receive so much junk mail? In experimental studies, the U.S. Postal Service has found that the probability that a piece of advertising is opened and read is 78%. A business sends out 15,000 pieces of advertising. How many of these can the company expect to be opened and read?

4. *Linguistics.* An experiment was conducted by the authors to determine the relative occurrence of various letters of the English alphabet. The front page of a newspaper was considered. In all, there

were 9136 letters. The number of occurrences of each letter of the alphabet is listed in the following table.

LETTER	NUMBER OF OCCURRENCES	PROBABILITY
A	853	$853/9136 \approx 9.3\%$
B	136	
C	273	
D	286	
E	1229	
F	173	
G	190	
H	399	
I	539	
J	21	
K	57	
L	417	
M	231	
N	597	
O	705	
P	238	
Q	4	
R	609	
S	745	
T	789	
U	240	
V	113	
W	127	
X	20	
Y	124	
Z	21	$21/9136 \approx 0.2\%$

a) Complete the table of probabilities with the percentage, to the nearest tenth of a percent, of the occurrence of each letter.
b) What is the probability of a vowel occurring?
c) What is the probability of a consonant occurring?

5. *Wheel of Fortune®.* The results of the experiment in Exercise 4 can be quite useful to a person playing the popular television game show *Wheel of Fortune.* Players guess letters in order to spell out a phrase, a person, or a thing.

a) What 5 consonants have the greatest probability of occurring?
b) What vowel has the greatest probability of occurring?
c) The winner of the main part of the show plays for a grand prize and at one time was allowed to guess 5 consonants and a vowel in order to discover the secret wording. The 5 consonants R, S, T, L, N, and the vowel E seemed to be chosen most often. Do the results in parts (a) and (b) support such a choice?

6. *Card Drawing.* Suppose we draw a card from a well-shuffled deck of 52 cards.

a) How many equally likely outcomes are there?

What is the probability of drawing each of the following?

b) A queen
c) A heart
d) A 7
e) A red card
f) A 9 or a king
g) A black ace

7. *Marbles.* Suppose we select, without looking, one marble from a bag containing 4 red marbles and 10 green marbles. What is the probability of selecting each of the followng?

a) A red marble
b) A green marble
c) A purple marble
d) A red or a green marble

8. *Production Unit.* The sales force of a business consists of 10 men and 10 women. A production unit of 4 people is set up at random. What is the probability that 2 men and 2 women are chosen?

9. *Coin Drawing.* A sack contains 7 dimes, 5 nickels, and 10 quarters. Eight coins are drawn at random. What is the probability of getting 4 dimes, 3 nickels, and 1 quarter?

10. *Michigan Lotto.* Twice a week, the state of Michigan runs a 6-out-of-49-number lotto that pays at least $2 million. You purchase a card for $1 and pick any 6 numbers from 1 to 49. If your numbers match those that the state draws, you win.

a) How many 6-number combinations are there for drawing?
b) You buy 1 lottery ticket. What is your probability of winning?

Five-Card Poker Hands. Suppose that 5 cards are drawn from a deck of 52 cards. What is the probability of drawing each of the following?

11. 3 sevens and 2 kings

12. 5 aces

13. 5 spades

14. 4 aces and 1 five

15. *Random-Number Generator.* Many graphers have a **random-number generator**. This feature produces a random number in the interval [0, 1]. (Consult your manual.) We can use such a feature to simulate coin flipping. A number r such that

$0 \le r \le 0.5$ would indicate heads, H. A number r such that $0.5 < r \le 1.0$ would indicate tails, T. Use a random-number generator 100 times.

a) What is the experimental probability of getting heads?

b) What is the experimental probability of getting tails?

16. *Tossing Three Coins.* Three coins are flipped. An outcome might be HTH.

a) Find the sample space.

What is the probability of getting each of the following?

b) Exactly one head

c) At most two tails

d) At least one head

e) Exactly two tails

Roulette. An American roulette wheel contains 38 slots numbered 00, 0, 1, 2, 3, . . . , 35, 36. Eighteen of the slots numbered 1–36 are colored red and 18 are colored black. The 00 and 0 slots are uncolored. The wheel is spun, and a ball is rolled around the rim until it falls into a slot. What is the probability that the ball falls in each of the following?

17. A black slot

18. A red slot

19. A red or a black slot

20. The 00 slot

21. The 0 slot

22. Either the 00 or the 0 slot (in this case, the house always wins)

23. An odd-numbered slot

24. The number 24

25. *Dartboard.* The figure below shows a dartboard. A dart is thrown and hits the board. Find the probabilities

$$P(\text{red}),\ P(\text{green}),\ P(\text{blue}),\ P(\text{yellow}).$$

Synthesis

26. ◈ Find at least one use of probability in today's newspaper. Make a report.

27. ◈ *Random Best-Selling Novels.* Sir Arthur Stanley Eddington, an astronomer, once wrote in a satirical essay that if a monkey were left alone long enough with a typewriter and typed randomly, any great novel could be replicated.

What is the probability that the following passage could have been written by a monkey? Ignore capital letters and punctuation and consider only letters and spaces.

"It was the best of times, it was the worst of times, . . ." (Charles Dickens, 1859). Explain your answer.

Five-Card Poker Hands. Suppose that 5 cards are drawn from a deck of 52 cards. For the following exercises, give both a reasoned expression and an answer.

28. *Royal Flush.* A *royal flush* consists of a 5-card hand with A-K-Q-J-10 of the same suit.

a) How many royal flushes are there?

b) What is the probability of getting a royal flush?

29. *Straight Flush.* A *straight flush* consists of 5 cards in sequence in the same suit, but excludes royal flushes. An ace can be used low, before a two, or high, following a king.

a) How many straight flushes are there?

b) What is the probability of getting a straight flush?

30. *Four of a Kind.* A *four-of-a-kind* is a 5-card hand in which 4 of the cards are of the same denomination, such as J-J-J-J-6, 7-7-7-7-A, or 2-2-2-2-5.

a) How many four-of-a-kind hands are there?
b) What is the probability of getting four of a kind?

31. *Full House.* A *full house* consists of a pair and 3 of a kind, such as Q-Q-Q-4-4.

a) How many full houses are there?
b) What is the probability of getting a full house?

32. *Three of a Kind.* A *three-of-a-kind* is a 5-card hand in which exactly 3 of the cards are of the same denomination and the other 2 are *not,* such as Q-Q-Q-10-7.

a) How many three-of-a-kind hands are there?
b) What is the probability of getting three of a kind?

33. *Flush.* An ordinary *flush* is a 5-card hand in which all the cards are of the same suit, but not all in sequence (not a straight flush or royal flush).

a) How many flushes are there?
b) What is the probability of getting a flush?

34. *Two Pairs.* A hand with *two pairs* is a hand like Q-Q-3-3-A.

a) How many are there?
b) What is the probability of getting two pairs?

35. *Straight.* An ordinary *straight* is any 5 cards in sequence, but not of the same suit—for example, 4 of spades, 5 of hearts, 6 of diamonds, 7 of hearts, and 8 of clubs.

a) How many straights are there?
b) What is the probability of getting a straight?

CHAPTER 9 Summary and Review

Important Properties and Formulas

Arithmetic Sequences and Series

General term: $a_n = a_{n-1} + d$

$a_n = a_1 + (n - 1)d$

Common difference: d

Sum of the first n terms: $S_n = \frac{n}{2}(a_1 + a_n)$

Geometric Sequences and Series

General term: $a_{n+1} = a_n r$

$a_n = a_1 r^{n-1}$

Common ratio: r

Sum of the first n terms: $S_n = \frac{a_1(1 - r^n)}{1 - r}$

Sum of an infinite geometric series:

$S_\infty = \frac{a_1}{1 - r}, \quad |r| < 1$

The Principle of Mathematical Induction

(1) *Basis step:* Prove S_1 is true.

(2) *Induction step:* Prove for all numbers k, $S_k \rightarrow S_{k+1}$.

The Fundamental Counting Principle

$n_1 \cdot n_2 \cdot n_3 \cdots n_k$

Permutations of n Objects Taken n at a Time

${}_nP_n = n! = n(n - 1)(n - 2) \cdots 3 \cdot 2 \cdot 1$

(*continued*)

Permutations of n Objects Taken k at a Time

$$_nP_k = \underbrace{n(n-1)(n-2)\cdots[n-(k-1)]}_{k \text{ factors}}$$

$$= \frac{n!}{(n-k)!}$$

Permutations of Sets With Some Nondistinguishable Objects

$$\frac{n!}{n_1! \cdot n_2! \cdot \cdots \cdot n_k!}$$

Combinations of n Objects Taken k at a Time

$$_nC_k = \binom{n}{k} = \frac{_nP_k}{k!} = \frac{n!}{k!(n-k)!}$$

$$= \frac{n(n-1)(n-2)\cdots[n-(k-1)]}{k!}$$

The Binomial Theorem

$$(a+b)^n = \sum_{k=0}^{n} \binom{n}{k} a^{n-k} b^k$$

The (k + 1)st Term of Binomial Expansion

The $(k+1)$st term of $(a+b)^n$ is $\binom{n}{k} a^{n-k} b^k$.

The total number of subsets of a set with n elements is 2^n.

Probability Principle P

$$P(E) = \frac{m}{n}$$

REVIEW EXERCISES

1. Find the first 4 terms, a_{11}, and a_{23}:

$$a_n = (-1)^n \left(\frac{n^2}{n^4 + 1} \right).$$

2. Predict the general, or nth, term. Answers may vary.

$$2, -5, 10, -17, 26, \ldots$$

3. Find and evaluate:

$$\sum_{k=1}^{4} \frac{(-1)^{k+1} 3^k}{3^k - 1}.$$

4. Construct a table of values and a graph for the first 10 terms of this sequence. Use both a grapher and a spreadsheet program, if available.

$$a_1 = 0.3, \quad a_{k+1} = 5a_k + 1$$

5. Write sigma notation:

$$0 + 3 + 8 + 15 + 24 + 35 + 48.$$

6. Find the 10th term of the arithmetic sequence

$$\tfrac{3}{4}, \tfrac{13}{12}, \tfrac{17}{12}, \ldots.$$

7. Find the 6th term of the arithmetic sequence

$$a - b, a, a + b, \ldots.$$

8. Find the sum of the first 18 terms of the arithmetic sequence

$$4, 7, 10, \ldots.$$

9. Find the sum of the first 200 natural numbers.

10. The 1st term in an arithmetic sequence is 5, and the 17th term is 53. Find the 3rd term.

11. The common difference in an arithmetic sequence is 3. The 10th term is 23. Find the first term.

12. For a geometric sequence, $a_1 = -2$, $r = 2$, and $a_n = -64$. Find n and S_n.

13. For a geometric sequence, $r = \frac{1}{2}$, $n = 5$, and $S_n = \frac{31}{2}$. Find a_1 and a_n.

Find the sum of each infinite geometric series, if it exists. If you wish, you can also check the results on a grapher using the TABLE *and* SEQUENCE GRAPH *features.*

14. $25 + 27.5 + 30.25 + 33.275 + \cdots$

15. $0.27 + 0.0027 + 0.000027 + \cdots$

16. $\frac{1}{2} - \frac{1}{6} + \frac{1}{18} - \cdots$

17. Find fractional notation for $2.\overline{43}$.

18. Insert four arithmetic means between 5 and 9.

19. *Bouncing Golfball.* A golfball is dropped from a height of 30 ft to the pavement. It always rebounds three fourths of the distance that it drops. How far (up and down) will the ball have traveled when it hits the pavement for the 6th time?

20. *The Amount of an Annuity.* To create a college fund, a parent makes a sequence of 18 yearly deposits of $2000 each in a savings account on which interest is compounded annually at 5.8%. Find the amount of the annuity.

21. *Total Gift.* You receive 10¢ on the first day of the year, 12¢ on the 2nd day, 14¢ on the 3rd day, and so on.

a) How much will you receive on the 365th day?
b) What is the sum of all these 365 gifts?

22. *The Economic Multiplier.* The government is making a $24,000,000,000 expenditure for travel to Mars. If 73% of this amount is spent again, and so on, what is the total effect on the economy?

Use mathematical induction to prove each of the following.

23. For every natural number n,

$$1 + 4 + 7 + \cdots + (3n - 2) = \frac{n(3n - 1)}{2}.$$

24. For every natural number n,

$$1 + 3 + 3^2 + \cdots + 3^{n-1} = \frac{3^n - 1}{2}.$$

25. For every natural number $n \geq 2$,

$$\left(1 - \frac{1}{2}\right)\left(1 - \frac{1}{3}\right) \cdots \left(1 - \frac{1}{n}\right) = \frac{1}{n}.$$

26. *Book Arrangements.* In how many ways can 6 books be arranged on a shelf?

27. *Flag Displays.* If 9 different signal flags are available, how many different displays are possible using 4 flags in a row?

28. *Prize Choices.* The winner of a contest can choose any 8 of 15 prizes. How many different sets of prizes can be chosen?

29. *Fraternity–Sorority Names.* The Greek alphabet contains 24 letters. How many fraternity or sorority names can be formed using 3 different letters?

30. *Letter Arrangements.* In how many distinguishable ways can the letters of the word TENNESSEE be arranged?

31. *Floor Plans.* A manufacturer of houses has 1 floor plan but achieves variety by having 3 different roofs, 4 different ways of attaching the garage, and 3 different types of entrance. Find the number of different houses that can be produced.

32. *Code Symbols.* How many code symbols can be formed using 5 out of 6 of the letters of G, H, I, J, K, L if the letters:

a) cannot be repeated?
b) can be repeated?
c) cannot be repeated but must begin with K?
d) cannot be repeated but must end with IGH?

33. Determine the number of subsets of a set containing 8 members.

Expand.

34. $(m + n)^7$

35. $(x - \sqrt{2})^5$

36. $(x^2 - 3y)^4$

37. $\left(a + \dfrac{1}{a}\right)^8$

38. $(1 + 5i)^6$, where $i^2 = -1$

39. Find the 4th term of $(a + x)^{12}$.

40. Find the 12th term of $(2a - b)^{18}$. Do not multiply out the factorials.

41. *Election Poll.* Before an election, a poll was conducted to see which candidate was favored. Three people were running for a particular office. During the polling, 86 favored candidate A, 97 favored B, and 23 favored C. Assuming that the poll is a valid indicator of the election, what is the probability that the election will be won by A? B? C?

42. *Rolling Dice.* What is the probability of getting a 10 on a roll of a pair of dice? on a roll of 1 die?

43. *Drawing a Card.* From a deck of 52 cards, 1 card is drawn at random. What is the probability that it is a club?

44. *Drawing Three Cards.* From a deck of 52 cards, 3 are drawn at random without replacement. What is the probability that 2 are aces and 1 is a king?

45. *Video Rentals.* The following table contains factual data regarding annual video rental spending per VCR household.

YEAR, x	U.S. SPENDING ON VIDEO RENTALS, y (IN BILLIONS)
1. 1982	\$141
3. 1984	139
5. 1986	127
7. 1988	124
9. 1990	120
10. 1991	113
11. 1992	114
12. 1993	119
13. 1994	122
14. 1995	129

Source: Pedonis, Suhler & Associates, Media Group Research.

a) Use the REGRESSION feature to fit a quadratic sequence function $a_n = an^2 + bn + c$ to the data.
b) Use the equation to predict the annual VCR video rental per household in 1998 and in 2002.

Synthesis

46. ◈ *Chain Business Deals.* Chain letters have been outlawed by the government. Nevertheless, "chain" business deals still exist and they can be fraudulent. Suppose a salesperson is charged with the task of hiring 4 new salespersons. Each of them gives half of their profits to the person who hires them. Each of these people hires 4 new salespersons. Each of these gives half of his or her profits to the person who hired them. Half of these profits then go back to the original hiring person. Explain the lure of this business to someone who has managed several sequences of hirings. Explain the fallacy of such a business as well. Keep in mind that there are only 266 million people in this country.

47. ◈ Write an exercise for a classmate to solve. Design it so that the solution is ${}_9C_4$.

48. Explain why the following cannot be proved by mathematical induction: For every natural number n,

a) $3 + 5 + \cdots + (2n + 1) = (n + 1)^2$.
b) $1 + 3 + \cdots + (2n - 1) = n^2 + 3$.

49. Suppose that $a_1, a_2, \ldots, a_n$ and $b_1, b_2, \ldots, b_n$ are geometric sequences. Prove that $c_1, c_2, \ldots, c_n$ is a geometric sequence, where $c_n = a_n b_n$.

50. Suppose that $a_1, a_2, \ldots, a_n$ is an arithmetic sequence. Is $b_1, b_2, \ldots, b_n$ an arithmetic sequence if:

a) $b_n = |a_n|$?
b) $b_n = a_n + 8$?
c) $b_n = 7a_n$?
d) $b_n = \dfrac{1}{a_n}$?
e) $b_n = \log a_n$?
f) $b_n = a_n^3$?

51. The zeros of this polynomial function form an arithmetic sequence. Find them.

$$f(x) = x^4 - 4x^3 - 4x^2 + 16x$$

52. Write the first 3 terms of the infinite geometric series with $r = -\frac{1}{3}$ and $S_\infty = \frac{3}{8}$.

53. Simplify:

$$\sum_{k=0}^{10} (-1)^k \binom{10}{k} (\log x)^{10-k} (\log y)^k.$$

Solve for n.

54. $\binom{n}{6} = 3 \cdot \binom{n-1}{5}$

55. $\binom{n}{n-1} = 36$

56. Solve for a:

$$\sum_{k=0}^{5} \binom{5}{k} 9^{5-k} a^k = 0.$$

Appendixes

A Determinants and Cramer's Rule

- *Evaluate determinants of square matrices.*
- *Use Cramer's rule to solve systems of equations.*

Determinants of Square Matrices

With every square matrix, we associate a number called its *determinant*.

Determinant of a 2 × 2 Matrix

The *determinant* of the matrix $\begin{bmatrix} a & c \\ b & d \end{bmatrix}$ is denoted $\begin{vmatrix} a & c \\ b & d \end{vmatrix}$ and is defined as

$$\begin{vmatrix} a & c \\ b & d \end{vmatrix} = ad - bc.$$

Example 1 Evaluate: $\begin{vmatrix} \sqrt{2} & -3 \\ -4 & -\sqrt{2} \end{vmatrix}$.

SOLUTION

$$\begin{vmatrix} \sqrt{2} & -3 \\ -4 & -\sqrt{2} \end{vmatrix}$$

The arrows indicate the products involved.

$$= \sqrt{2}(-\sqrt{2}) - (-4)(-3)$$
$$= -2 - 12 = -14$$

We now consider a way to evaluate determinants of square matrices of order 3 × 3 or higher.

Minor

For a square matrix $\mathbf{A} = [a_{ij}]$, the *minor* M_{ij} of an element a_{ij} is the determinant of the matrix formed by deleting the ith row and the jth column of $\mathbf{A}$.

Example 2 For the matrix $[a_{ij}] = \begin{bmatrix} -8 & 0 & 6 \\ 4 & -6 & 7 \\ -1 & -3 & 5 \end{bmatrix}$, find each of the following.

a) M_{11}

b) M_{23}

SOLUTION

a) Delete the first row and the first column and find the determinant of the 2×2 matrix formed by the remaining elements.

$$\begin{bmatrix} -8 & 0 & 6 \\ 4 & -6 & 7 \\ -1 & -3 & 5 \end{bmatrix}$$

$$\begin{aligned} M_{11} &= \begin{vmatrix} -6 & 7 \\ -3 & 5 \end{vmatrix} \\ &= (-6) \cdot 5 - (-3) \cdot 7 \\ &= -30 - (-21) \\ &= -30 + 21 = -9 \end{aligned}$$

b) Delete the second row and the third column and find the determinant of the 2×2 matrix formed by the remaining elements.

$$\begin{bmatrix} -8 & 0 & 6 \\ 4 & -6 & 7 \\ -1 & -3 & 5 \end{bmatrix}$$

$$\begin{aligned} M_{23} &= \begin{vmatrix} -8 & 0 \\ -1 & -3 \end{vmatrix} \\ &= -8(-3) - (-1)0 = 24 \end{aligned}$$

Cofactor

For a square matrix $\mathbf{A} = [a_{ij}]$, the *cofactor* A_{ij} of an element a_{ij} is given by

$$A_{ij} = (-1)^{i+j} M_{ij},$$

where M_{ij} is the minor of a_{ij}.

Example 3 For the matrix given in Example 2, find each of the following.

a) A_{11}

b) A_{23}

SOLUTION

a) In Example 2, we found that $M_{11} = -9$. Then

$$A_{11} = (-1)^{1+1}(-9) = (1)(-9) = -9.$$

b) In Example 2, we found that $M_{23} = 24$. Then

$$A_{23} = (-1)^{2+3}(24) = (-1)(24) = -24.$$

Evaluating Determinants Using Cofactors

Consider the matrix $\mathbf{A}$ given by

$$\mathbf{A} = \begin{bmatrix} a_{11} & a_{12} & a_{13} \\ a_{21} & a_{22} & a_{23} \\ a_{31} & a_{32} & a_{33} \end{bmatrix}.$$

The determinant of the matrix, denoted $|\mathbf{A}|$, can be found by multiplying each element of the first column by its cofactor and adding:

$$|\mathbf{A}| = a_{11}A_{11} + a_{21}A_{21} + a_{31}A_{31}.$$

Because

$$A_{11} = (-1)^{1+1}M_{11} = M_{11},$$
$$A_{21} = (-1)^{2+1}M_{21} = -M_{21},$$

and $$A_{31} = (-1)^{3+1}M_{31} = M_{31},$$

we can write

$$|\mathbf{A}| = a_{11} \cdot \begin{vmatrix} a_{22} & a_{23} \\ a_{32} & a_{33} \end{vmatrix} - a_{21} \cdot \begin{vmatrix} a_{12} & a_{13} \\ a_{32} & a_{33} \end{vmatrix} + a_{31} \cdot \begin{vmatrix} a_{12} & a_{13} \\ a_{22} & a_{23} \end{vmatrix}.$$

It can be shown that we can determine $|\mathbf{A}|$ by picking *any* row or column, multiplying each element in that row or column by its cofactor, and adding. This is called *expanding* across a row or down a column. We just expanded down the first column. We now define the determinant of a square matrix of any order.

Determinant of Any Square Matrix

For any square matrix $\mathbf{A}$ of order $n \times n$ $(n > 1)$, we define the *determinant* of $\mathbf{A}$, denoted $|\mathbf{A}|$, as follows. Choose any row or column. Multiply each element in that row or column by its cofactor and add the results. The determinant of a 1×1 matrix is simply the element of the matrix. The value of a determinant will be the same no matter how it is evaluated.

Example 4 Evaluate $|\mathbf{A}|$ by expanding across the third row.

$$\mathbf{A} = \begin{bmatrix} -8 & 0 & 6 \\ 4 & -6 & 7 \\ -1 & -3 & 5 \end{bmatrix}$$

SOLUTION We have

$$\begin{aligned}
|\mathbf{A}| &= (-1)A_{31} + (-3)A_{32} + 5A_{33} \\
&= (-1)(-1)^{3+1} \cdot \begin{vmatrix} 0 & 6 \\ -6 & 7 \end{vmatrix} + (-3)(-1)^{3+2} \cdot \begin{vmatrix} -8 & 6 \\ 4 & 7 \end{vmatrix} \\
&\quad + 5(-1)^{3+3} \cdot \begin{vmatrix} -8 & 0 \\ 4 & -6 \end{vmatrix} \\
&= (-1) \cdot 1 \cdot [0 \cdot 7 - (-6)6] + (-3)(-1)[-8 \cdot 7 - 4 \cdot 6] \\
&\quad + 5 \cdot 1 \cdot [-8(-6) - 4 \cdot 0] \\
&= -[36] + 3[-80] + 5[48] \\
&= -36 - 240 + 240 \\
&= -36.
\end{aligned}$$

The value of this determinant is -36 no matter which row or column we expand upon.

Determinants can be evaluated on most graphers using the MATRIX package. After entering a matrix on the grapher, select the determinant operation from the MATRIX MATH menu and enter the name of the matrix. The grapher will return the value of the determinant of the matrix. For example, for

$$\mathbf{A} = \begin{bmatrix} 1 & 6 & -1 \\ -3 & -5 & 3 \\ 0 & 4 & 2 \end{bmatrix},$$

we have

```
det [A]
            26
```

Cramer's Rule

Determinants can be used to solve systems of linear equations. Consider a system of two linear equations:

$$a_1x + b_1y = c_1,$$
$$a_2x + b_2y = c_2.$$

Using the methods of Chapter 7, we obtain

$$x = \frac{c_1b_2 - c_2b_1}{a_1b_2 - a_2b_1} \quad \text{and} \quad y = \frac{a_1c_2 - a_2c_1}{a_1b_2 - a_2b_1}.$$

The numerators and denominators of these expressions can be written as determinants:

$$x = \frac{\begin{vmatrix} c_1 & b_1 \\ c_2 & b_2 \end{vmatrix}}{\begin{vmatrix} a_1 & b_1 \\ a_2 & b_2 \end{vmatrix}} \quad \text{and} \quad y = \frac{\begin{vmatrix} a_1 & c_1 \\ a_2 & c_2 \end{vmatrix}}{\begin{vmatrix} a_1 & b_1 \\ a_2 & b_2 \end{vmatrix}}.$$

If we let

$$D = \begin{vmatrix} a_1 & b_1 \\ a_2 & b_2 \end{vmatrix}, \quad D_x = \begin{vmatrix} c_1 & b_1 \\ c_2 & b_2 \end{vmatrix}, \quad \text{and} \quad D_y = \begin{vmatrix} a_1 & c_1 \\ a_2 & c_2 \end{vmatrix},$$

we have

$$x = \frac{D_x}{D} \quad \text{and} \quad y = \frac{D_y}{D}.$$

This procedure for solving systems of equations is known as *Cramer's rule*.

Cramer's Rule for 2 × 2 Systems

The solution of the system of equations

$$a_1x + b_1y = c_1,$$
$$a_2x + b_2y = c_2$$

is given by

$$x = \frac{D_x}{D}, \quad y = \frac{D_y}{D},$$

where

$$D = \begin{vmatrix} a_1 & b_1 \\ a_2 & b_2 \end{vmatrix}, \quad D_x = \begin{vmatrix} c_1 & b_1 \\ c_2 & b_2 \end{vmatrix},$$

$$D_y = \begin{vmatrix} a_1 & c_1 \\ a_2 & c_2 \end{vmatrix}, \quad \text{and} \quad D \neq 0.$$

Note that the denominator D contains the coefficients of x and y, in the same position as in the original equations. For x, the numerator is obtained by replacing the x-coefficients in D (the a's) by the c's. For y, the numerator is obtained by replacing the y-coefficients in D (the b's) by the c's.

Example 5 Solve using Cramer's rule:

$$2x + 5y = 7,$$
$$5x - 2y = -3.$$

SOLUTION We have

$$x = \frac{\begin{vmatrix} 7 & 5 \\ -3 & -2 \end{vmatrix}}{\begin{vmatrix} 2 & 5 \\ 5 & -2 \end{vmatrix}} = \frac{7(-2) - (-3)5}{2(-2) - 5 \cdot 5} = -\frac{1}{29},$$

$$y = \frac{\begin{vmatrix} 2 & 7 \\ 5 & -3 \end{vmatrix}}{\begin{vmatrix} 2 & 5 \\ 5 & -2 \end{vmatrix}} = \frac{2(-3) - 5 \cdot 7}{-29} = \frac{41}{29}.$$

The solution is $\left(-\frac{1}{29}, \frac{41}{29}\right)$.

Cramer's rule works only when a system of equations has a unique solution. This occurs when $D \neq 0$. If $D = 0$, $D_x = 0$, and $D_y = 0$, then the system is dependent. If $D = 0$ and D_x and/or D_y is not 0, then the system is inconsistent.

Cramer's rule can be extended to a system of n linear equations in n variables. We consider a 3×3 system.

Cramer's Rule for 3 × 3 Systems

The solution of the system of equations

$$a_1x + b_1y + c_1z = d_1,$$
$$a_2x + b_2y + c_2z = d_2,$$
$$a_3x + b_3y + c_3z = d_3$$

is given by

$$x = \frac{D_x}{D}, \quad y = \frac{D_y}{D}, \quad z = \frac{D_z}{D},$$

where

$$D = \begin{vmatrix} a_1 & b_1 & c_1 \\ a_2 & b_2 & c_2 \\ a_3 & b_3 & c_3 \end{vmatrix}, \qquad D_x = \begin{vmatrix} d_1 & b_1 & c_1 \\ d_2 & b_2 & c_2 \\ d_3 & b_3 & c_3 \end{vmatrix},$$

$$D_y = \begin{vmatrix} a_1 & d_1 & c_1 \\ a_2 & d_2 & c_2 \\ a_3 & d_3 & c_3 \end{vmatrix}, \qquad D_z = \begin{vmatrix} a_1 & b_1 & d_1 \\ a_2 & b_2 & d_2 \\ a_3 & b_3 & d_3 \end{vmatrix}, \text{ and } D \neq 0.$$

Note that the determinant D_x is obtained from D by replacing the x-coefficients by d_1, d_2, and d_3. A similar thing happens with D_y and D_z. When $D = 0$, Cramer's rule cannot be used. If $D = 0$ and D_x, D_y, and D_z are 0, the system is dependent. If $D = 0$ and one of D_x, D_y, or D_z is not 0, then the system is inconsistent.

Example 6 Solve using Cramer's rule:

$$x - 3y + 7z = 13,$$
$$x + y + z = 1,$$
$$x - 2y + 3z = 4.$$

SOLUTION We have

$$D = \begin{vmatrix} 1 & -3 & 7 \\ 1 & 1 & 1 \\ 1 & -2 & 3 \end{vmatrix} = -10, \qquad D_x = \begin{vmatrix} 13 & -3 & 7 \\ 1 & 1 & 1 \\ 4 & -2 & 3 \end{vmatrix} = 20,$$

$$D_y = \begin{vmatrix} 1 & 13 & 7 \\ 1 & 1 & 1 \\ 1 & 4 & 3 \end{vmatrix} = -6, \qquad D_z = \begin{vmatrix} 1 & -3 & 13 \\ 1 & 1 & 1 \\ 1 & -2 & 4 \end{vmatrix} = -24.$$

Then

$$x = \frac{D_x}{D} = \frac{20}{-10} = -2,$$

$$y = \frac{D_y}{D} = \frac{-6}{-10} = \frac{3}{5},$$

$$z = \frac{D_z}{D} = \frac{-24}{-10} = \frac{12}{5}.$$

The solution is $\left(-2, \frac{3}{5}, \frac{12}{5}\right)$. In practice, it is not necessary to evaluate D_z. When we have found values for x and y, we can substitute them into one of the equations to find z.

Evaluate each determinant algebraically.

1. $\begin{vmatrix} -2 & -\sqrt{5} \\ -\sqrt{5} & 3 \end{vmatrix}$

2. $\begin{vmatrix} \sqrt{5} & -3 \\ 4 & 2 \end{vmatrix}$

3. $\begin{vmatrix} x & 4 \\ x & x^2 \end{vmatrix}$

4. $\begin{vmatrix} y^2 & -2 \\ y & 3 \end{vmatrix}$

5. $\begin{vmatrix} 3 & 1 & 2 \\ -2 & 3 & 1 \\ 3 & 4 & -6 \end{vmatrix}$

6. $\begin{vmatrix} 3 & -2 & 1 \\ 2 & 4 & 3 \\ -1 & 5 & 1 \end{vmatrix}$

7. $\begin{vmatrix} x & 0 & -1 \\ 2 & x & x^2 \\ -3 & x & 1 \end{vmatrix}$

8. $\begin{vmatrix} x & 1 & -1 \\ x^2 & x & x \\ 0 & x & 1 \end{vmatrix}$

9. Use a grapher to check your answer to Exercise 1.

10. Use a grapher to check your answer to Exercise 2.

11. Use a grapher to check your answer to Exercise 5.

12. Use a grapher to check your answer to Exercise 6.

Use the following matrix for Exercises 13–20:

$$\mathbf{A} = \begin{bmatrix} 7 & -4 & -6 \\ 2 & 0 & -3 \\ 1 & 2 & -5 \end{bmatrix}.$$

13. Find M_{11}, M_{32}, and M_{22}.

14. Find M_{13}, M_{31}, and M_{23}.

15. Find A_{11}, A_{32}, and A_{22}.

16. Find A_{13}, A_{31}, and A_{23}.

17. Evaluate $|\mathbf{A}|$ by expanding across the second row.

18. Evaluate $|\mathbf{A}|$ by expanding down the second column.

19. Evaluate $|\mathbf{A}|$ by expanding down the third column.

20. Evaluate $|\mathbf{A}|$ on a grapher.

Use the following matrix for Exercises 21–27:

$$\mathbf{A} = \begin{bmatrix} 1 & 0 & 0 & -2 \\ 4 & 1 & 0 & 0 \\ 5 & 6 & 7 & 8 \\ -2 & -3 & -1 & 0 \end{bmatrix}$$

21. Find M_{41} and M_{33}.

22. Find M_{12} and M_{44}.

23. Find A_{24} and A_{43}.

24. Find A_{22} and A_{34}.

25. Evaluate $|\mathbf{A}|$ by expanding across the first row.

26. Evaluate $|\mathbf{A}|$ by expanding down the third column.

27. Evaluate $|\mathbf{A}|$ on a grapher.

28. Evaluate on a grapher:

$$\begin{vmatrix} 5 & -4 & 2 & -2 \\ 3 & -3 & -4 & 7 \\ -2 & 3 & 2 & 4 \\ -8 & 9 & 5 & -5 \end{vmatrix}.$$

Solve using Cramer's rule.

29. $-2x + 4y = 3,$
$3x - 7y = 1$

30. $5x - 4y = -3,$
$7x + 2y = 6$

31. $2x - y = 5,$
$x - 2y = 1$

32. $3x + 4y = -2,$
$5x - 7y = 1$

33. $2x + 9y = -2,$
$4x - 3y = 3$

34. $2x + 3y = -1,$
$3x + 6y = -0.5$

35. $2x + 5y = 7,$
$3x - 2y = 1$

36. $3x + 2y = 7,$
$2x + 3y = -2$

37. $3x + 2y - z = 4,$
$3x - 2y + z = 5,$
$4x - 5y - z = -1$

38. $3x - y + 2z = 1,$
$x - y + 2z = 3,$
$-2x + 3y + z = 1$

39. $3x + 5y - z = -2,$
$x - 4y + 2z = 13,$
$2x + 4y + 3z = 1$

40. $3x + 2y + 2z = 1,$
$5x - y - 6z = 3,$
$2x + 3y + 3z = 4$

41. $x - 3y - 7z = 6,$
$2x + 3y + z = 9,$
$4x + y = 7$

42. $x - 2y - 3z = 4,$
$3x - 2z = 8,$
$2x + y + 4z = 13$

43. $6y + 6z = -1,$
$8x + 6z = -1,$
$4x + 9y = 8$

44. $3x + 5y = 2,$
$2x - 3z = 7,$
$4y + 2z = -1$

Synthesis

45. ◈ Explain why the system of equations

$$a_1x + b_1y = c_1,$$
$$a_2x + b_2y = c_2$$

is either dependent or inconsistent when

$$\begin{vmatrix} a_1 & b_1 \\ a_2 & b_2 \end{vmatrix} = 0.$$

46. ◈ If the lines $a_1x + b_1y = c_1$ and $a_2x + b_2y = c_2$ are parallel, what can you say about the values of

$$\begin{vmatrix} a_1 & b_1 \\ a_2 & b_2 \end{vmatrix}, \quad \begin{vmatrix} c_1 & b_1 \\ c_2 & b_2 \end{vmatrix}, \quad \text{and} \quad \begin{vmatrix} a_1 & c_1 \\ a_2 & c_2 \end{vmatrix}?$$

Solve.

47. $\begin{vmatrix} x & 5 \\ -4 & x \end{vmatrix} = 24$

48. $\begin{vmatrix} y & 2 \\ 3 & y \end{vmatrix} = y$

49. $\begin{vmatrix} x & -3 \\ -1 & x \end{vmatrix} \geq 0$

50. $\begin{vmatrix} y & -5 \\ -2 & y \end{vmatrix} < 0$

51. $\begin{vmatrix} x + 3 & 4 \\ x - 3 & 5 \end{vmatrix} = -7$

52. $\begin{vmatrix} m + 2 & -3 \\ m + 5 & -4 \end{vmatrix} = 3m - 5$

53. $\begin{vmatrix} 2 & x & 1 \\ 1 & 2 & -1 \\ 3 & 4 & -2 \end{vmatrix} = -6$

54. $\begin{vmatrix} x & 2 & x \\ 3 & -1 & 1 \\ 1 & -2 & 2 \end{vmatrix} = -10$

Rewrite each expression using a determinant. Answers may vary.

55. $2L + 2W$

56. $\pi r + \pi h$

57. $a^2 + b^2$

58. $\frac{1}{2}h(a + b)$

59. $2\pi r^2 + 2\pi rh$

60. $x^2y^2 - Q^2$

61. Show that if a line contains the points (x_1, y_1) and (x_2, y_2), an equation of the line can be written as

$$\begin{vmatrix} x & y & 1 \\ x_1 & y_1 & 1 \\ x_2 & y_2 & 1 \end{vmatrix} = 0.$$

B Parametric Equations

- *Graph parametric equations and determine an equivalent rectangular equation.*
- *Determine parametric equations for a rectangular equation.*
- *Determine the location of a moving object at a specific time.*

Graphing Parametric Equations

Much of our graphing of curves in this text has been with rectangular equations involving only two variables, x and y. We graphed sets of ordered pairs, (x, y), where y is a function of x. In this section, we introduce a third variable, t, such that x and y are each a function of t.

Consider a point P with coordinates (x, y) in the rectangular coordinate plane. As P moves in the plane, its location at time t is given by two functions

$$x = f(t) \quad \text{and} \quad y = g(t).$$

For example, let the equations be

$$x = \frac{1}{2}t \quad \text{and} \quad y = t^2 - 3, \quad t \geq 0.$$

We restrict t to nonnegative values since t represents time. When $t = 2$, we have

$$x = \frac{1}{2} \cdot 2 = 1 \quad \text{and} \quad y = 2^2 - 3 = 1.$$

Thus after 2 sec, the coordinates of P are (1, 1). The table below lists other ordered pairs. We plot them and draw the curve for $t \geq 0$.

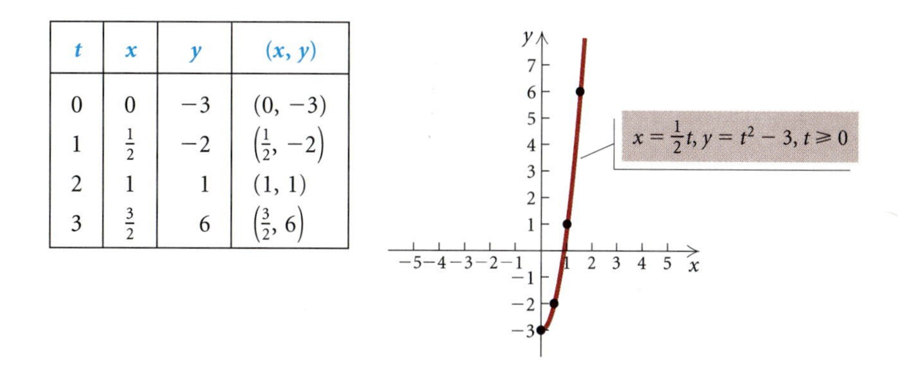

t	x	y	(x, y)
0	0	-3	$(0, -3)$
1	$\frac{1}{2}$	-2	$\left(\frac{1}{2}, -2\right)$
2	1	1	$(1, 1)$
3	$\frac{3}{2}$	6	$\left(\frac{3}{2}, 6\right)$

This curve appears to be part of a parabola. Let's verify this by finding the equivalent rectangular equation. Solving $x = \frac{1}{2}t$ for t, we get $t = 2x$. Substituting $2x$ for t in $y = t^2 - 3$, we have

$$y = (2x)^2 - 3 = 4x^2 - 3.$$

We know from Section 2.2 that this is a quadratic equation and that the graph is a parabola. Thus the curve is part of the parabola $y = 4x^2 - 3$.

The equations $x = \frac{1}{2}t$ and $y = t^2 - 3$ are called the **parametric equations** for the curve. The variable t is called the **parameter.** The equation $y = 4x^2 - 3$, with the restriction $x \geq 0$, is the equivalent rectangular equation for the curve. Since $t \geq 0$ and $x = \frac{1}{2}t$, the restriction $x \geq 0$ must be included. In this example, t represents time, but t, in general, can represent any real number in a specified interval. One advantage of parametric equations is that the restriction on the parameter t allows us to investigate a portion of the curve.

Parametric Equations

If f and g are continuous functions of t on an interval I, then the set of ordered pairs (x, y) such that $x = f(t)$ and $y = g(t)$ is a *plane curve*. The equations $x = f(t)$ and $y = f(t)$ are *parametric equations* for the curve. The variable t is the *parameter.*

Plane curves described with parametric equations can also be graphed on graphers. Consult your grapher's manual or the Graphing Calculator Manual that accompanies this text for specific instructions.

Example 1 Using a grapher, graph each of the following plane curves given their respective parametric equations and the restriction for the parameter. Then find the equivalent rectangular equation.

a) $x = t^2, y = t - 1; -1 \leq t \leq 4$

b) $x = \sqrt{t}, y = 2t + 3; 0 \leq t \leq 3$

SOLUTION

a) When using a grapher set in PARAMETRIC mode, we must set minimum and maximum values for x, y, and t.

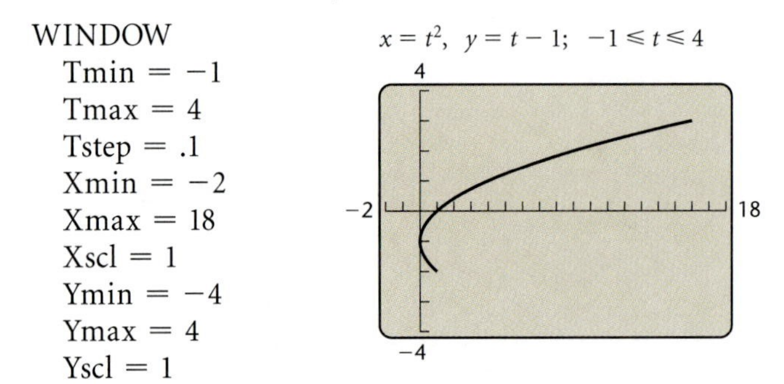

Using the TRACE feature, we can see the T, X, and Y values along the curve. To find an equivalent rectangular equation, we can solve either equation for t. Let's select the simpler equation $y = t - 1$ and solve for t:

$$y = t - 1$$
$$y + 1 = t.$$

We then substitute $y + 1$ for t in $x = t^2$:

$$x = t^2$$
$$x = (y + 1)^2. \qquad \text{Parabola}$$

This is an equation of a parabola that opens to the right. Given that $-1 \leq t \leq 4$, we have the corresponding restrictions on x and y: $0 \leq x \leq 16$ and $-2 \leq y \leq 3$. Thus the equivalent rectangular equation is

$$x = (y + 1)^2; \quad 0 \leq x \leq 16, \quad -2 \leq y \leq 3.$$

b) Using a grapher set in PARAMETRIC mode, we have

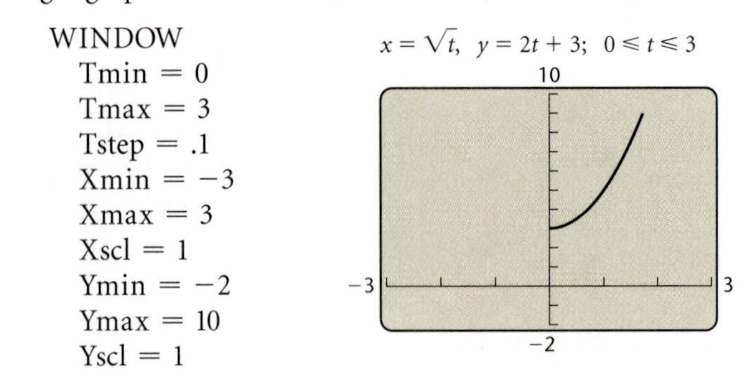

To find an equivalent rectangular equation, we first solve $x = \sqrt{t}$ for t:

$$x = \sqrt{t}$$
$$x^2 = t.$$

Then we substitute x^2 for t in $y = 2t + 3$:

$$y = 2t + 3$$
$$y = 2x^2 + 3. \quad \text{Parabola}$$

When $0 \le t \le 3$, $0 \le x \le \sqrt{3}$ and $3 \le y \le 9$. The equivalent rectangular equation is

$$y = 2x^2 + 3; \quad 0 \le x \le \sqrt{3}, \quad 3 \le y \le 9.$$

We first graphed in PARAMETRIC mode in Section 4.5 (p. 378). There we used an angle measure, usually in radians, as the parameter.

Example 2 Graph the plane curve described by $x = \cos t$ and $y = \sin t$, with t in $[0, 2\pi]$. Then determine an equivalent rectangular equation.

$x = \cos t$, $y = \sin t$; $0 \le t \le 2\pi$

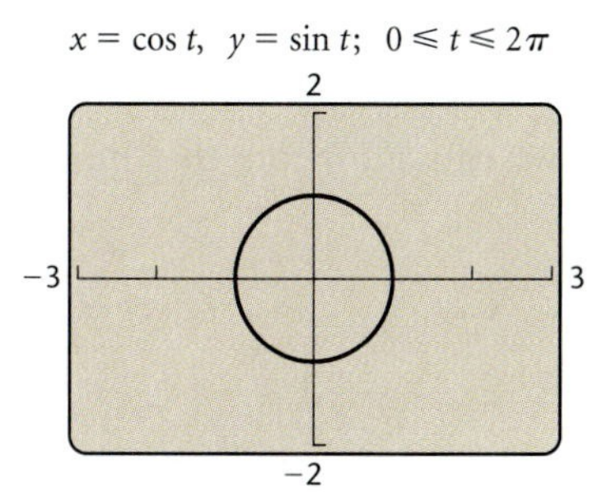

SOLUTION Using a squared window and a Tstep of $\pi/48$, we obtain the graph at left. It appears to be a circle of radius 1.

The equivalent rectangular equation can be obtained by squaring each parametric equation:

$$x^2 = \cos^2 t \quad \text{and} \quad y^2 = \sin^2 t.$$

This allows us to use the trigonometric identity $\sin^2 \theta + \cos^2 \theta = 1$. Substituting, we get

$$x^2 + y^2 = 1.$$

As expected, this is an equation of a circle with radius 1.

One advantage of graphing the unit circle parametrically is that it provides a method of finding trigonometric function values.

Example 3 Using the CALC menu and the parametric graph of the unit circle, find each of the following function values.

a) $\cos \dfrac{7\pi}{6}$ b) $\sin 4.13$

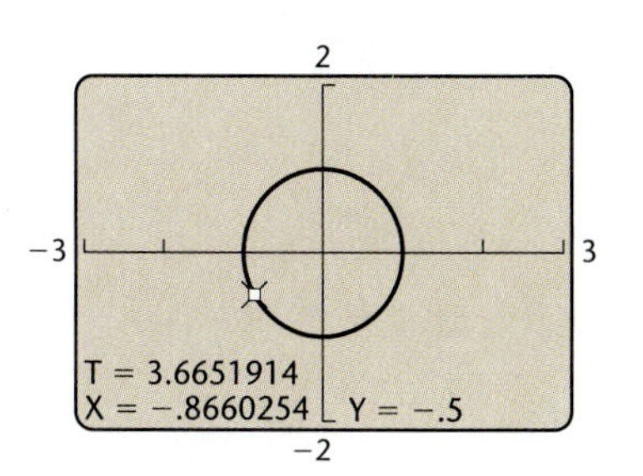

SOLUTION

a) Using the CALC menu, we enter $7\pi/6$ for t. The value of x, which is the $\cos t$, is shown on the screen. Thus,

$$\cos \frac{7\pi}{6} \approx -0.8660.$$

b) We enter 4.13 for t. The value of y, which is the $\sin t$, is shown on the screen. Thus $\sin 4.13 \approx -0.8352$.

Example 4 Graph the plane curve represented by

$$x = 5 \cos t \quad \text{and} \quad y = 3 \sin t, \quad 0 \le t \le 2\pi.$$

Then eliminate the parameter to find the rectangular equation.

SOLUTION

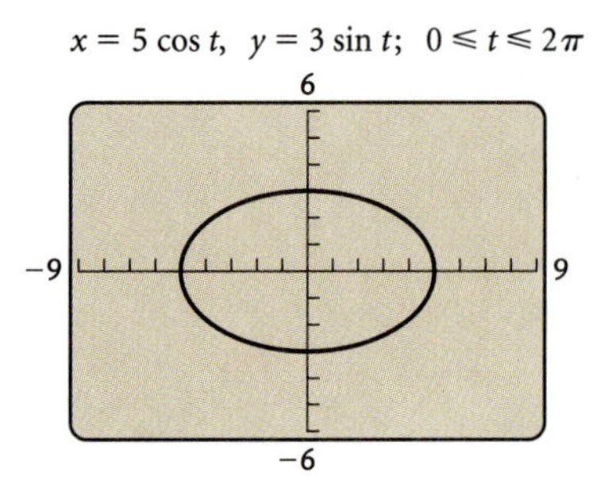

This appears to be the graph of an ellipse. To find the rectangular equation, we first solve for cos t and sin t in the parametric equations:

$$x = 5 \cos t \qquad y = 3 \sin t$$

$$\frac{x}{5} = \cos t, \qquad \frac{y}{3} = \sin t.$$

Using the identity $\sin^2 \theta + \cos^2 \theta = 1$, we can substitute to eliminate the parameter:

$$\sin^2 t + \cos^2 t = 1$$

$$\left(\frac{y}{3}\right)^2 + \left(\frac{x}{5}\right)^2 = 1 \qquad \textbf{Substituting}$$

$$\frac{x^2}{25} + \frac{y^2}{9} = 1. \qquad \textbf{Ellipse}$$

The rectangular form of the equation confirms that the graph is an ellipse centered at the origin with vertices at (5, 0) and (−5, 0).

Determining Parametric Equations for a Given Rectangular Equation

Many sets of parametric equations can represent the same plane curve. In fact, there are infinitely many such equations.

Example 5 Find four sets of parametric equations for the parabola

$$y = 4 - (x + 3)^2.$$

SOLUTION

If $x = t$, then $y = 4 - (t + 3)^2$.

If $x = t - 3$, then $y = 4 - (t - 3 + 3)^2$, or $4 - t^2$.

If $x = \dfrac{t}{3}$, then $y = 4 - \left(\dfrac{t}{3} + 3\right)^2$, or $-\dfrac{t^2}{9} - 2t - 5$.

In each of the above three sets, t is any real number. The following equations restrict t to any number greater than or equal to 0.

$$\text{If } x = \sqrt{t} - 3, \text{ then } y = 4 - (\sqrt{t} - 3 + 3)^2, \text{ or } 4 - t;\ t \geq 0.$$

Applications

The motion of an object can be described with parametric equations.

Example 6 *Motion of a Baseball.* A baseball player can throw a ball from a height of 5.9 ft with an initial velocity of 100 ft/sec at an angle of 41° with the horizontal. Neglecting air resistance, we know that the parametric equations for the path of the ball are

$$x = (100 \cos 41°)t \quad \text{and} \quad y = 5.9 + (100 \sin 41°)t - 16t^2,$$

where x and y are in feet and t is in seconds. Since t represents the time that the ball is in the air, we must have t greater than or equal to 0.

a) Graph the plane curve.

b) Determine the location of the ball after 0.5 sec, 1 sec, and 2 sec.

c) Find a rectangular equation that describes the curve.

SOLUTION

a) Set the grapher to DEGREE mode and enter the parametric equations.

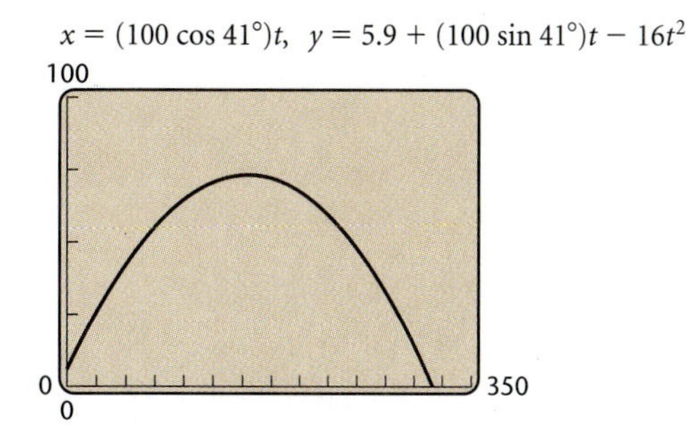

b) Using the TRACE feature, we find that

when $T = 0.5$, $(x, y) = (37.735479, 34.702951)$,
when $T = 1$, $(x, y) = (75.470958, 55.505903)$, and
when $T = 2$, $(x, y) = (150.94192, 73.111806)$.

In each ordered pair, the x-coordinate represents the horizontal distance and the y-coordinate represents the corresponding vertical height.

c) To find the equivalent rectangular equation, we solve $x = (100 \cos 41°)t$ for t to obtain

$$t = \frac{x}{100 \cos 41°}.$$

Then we substitute $x/(100 \cos 41°)$ for t in the equation for y:

$$y = 5.9 + (100 \sin 41°)\frac{x}{100 \cos 41°} - 16\left(\frac{x}{100 \cos 41°}\right)^2$$

$$y = 5.9 + x \tan 41° - \frac{16}{100^2 \cos^2 41°}x^2. \quad \textbf{Rectangular equation}$$

Note that this is a quadratic equation, as expected, since the graph is a parabola.

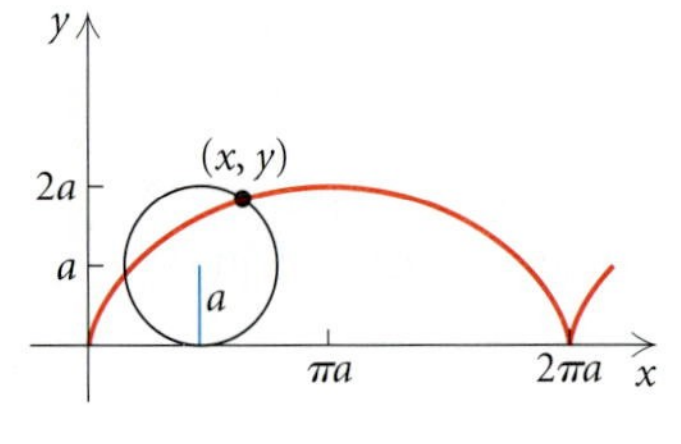

The path of a fixed point on the circumference of a circle as it rolls along a line is called a **cycloid**. For example, a point on the rim of a bicycle wheel traces out a cycloid curve.

Example 7 Graph the cycloid described by the parametric equations

$$x = 3(t - \sin t) \quad \text{and} \quad y = 3(1 - \cos t); \quad 0 \le t \le 6\pi.$$

SOLUTION

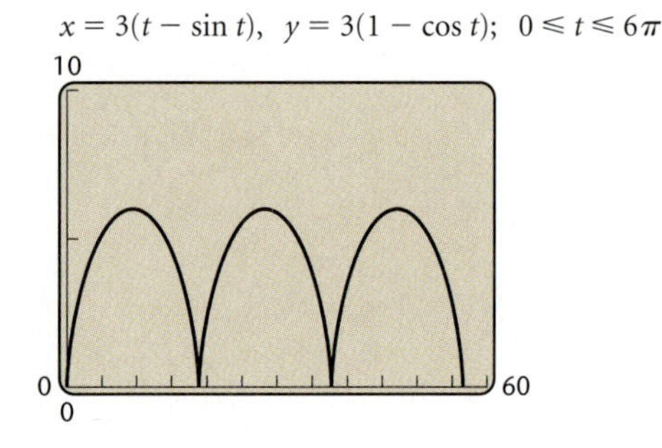

B Exercise Set

Using a grapher, graph the plane curve given by the set of parametric equations and the restriction for the parameter. Then find the equivalent rectangular equation.

1. $x = \frac{1}{2}t,\ y = 6t - 7;\ -1 \le t \le 6$

2. $x = t,\ y = 5 - t;\ -2 \le t \le 3$

3. $x = t^3,\ y = t - 4;\ -1 \le t \le 10$

4. $x = \sqrt{t},\ y = 2t + 3;\ 0 \le t \le 8$

5. $x = t^2,\ y = \sqrt{t};\ 0 \le t \le 4$

6. $x = t^3 + 1,\ y = t;\ -3 \le t \le 3$

7. $x = t + 3,\ y = \frac{1}{t+3};\ -2 \le t \le 2$

8. $x = 2t^3 + 1,\ y = 2t^3 - 1;\ -4 \le t \le 4$

9. $x = 2t - 1,\ y = t^2;\ -3 \le t \le 3$

10. $x = \frac{1}{3}t,\ y = t;\ -5 \le t \le 5$

11. $x = e^{-t},\ y = e^t;\ -\infty < t < \infty$

12. $x = 2 \ln t,\ y = t^2;\ 0 < t < \infty$

13. $x = 3 \cos t,\ y = 3 \sin t;\ 0 \le t \le 2\pi$

14. $x = 2 \cos t,\ y = 4 \sin t;\ 0 \le t \le 2\pi$

15. $x = \cos t,\ y = 2 \sin t;\ 0 \le t \le 2\pi$

16. $x = 2 \cos t,\ y = 2 \sin t;\ 0 \le t \le 2\pi$

17. $x = \sec t,\ y = \cos t;\ -\frac{\pi}{2} < t < \frac{\pi}{2}$

18. $x = \sin t,\ y = \csc t;\ 0 < t < \pi$

19. $x = 1 + 2 \cos t,\ y = 2 + 2 \sin t;\ 0 \le t \le 2\pi$

20. $x = 2 + \sec t,\ y = 1 + 3 \tan t;\ 0 < t < \frac{\pi}{2}$

Using a parametric graph of the unit circle, find the function value.

21. $\sin \frac{\pi}{4}$

22. $\cos \frac{2\pi}{3}$

23. $\cos \frac{17\pi}{12}$

24. $\sin \frac{4\pi}{5}$

25. $\tan \frac{\pi}{5}$

26. $\tan \frac{2\pi}{7}$

27. $\cos 5.29$

28. $\sin 1.83$

Find two sets of parametric equations for the rectangular equation.

29. $y = 4x - 3$

30. $y = x^2 - 1$

31. $y = (x - 2)^2 - 6x$

32. $y = x^3 + 3$

33. *Motion of a Ball.* A ball is thrown from a height of 6.25 ft with an initial velocity of 80 ft/sec at an angle of 35° with the horizontal. The parametric equations for the path of the ball are

$x = (80 \cos 35°)t$, and
$y = 6.25 + (80 \sin 35°)t - 16t^2$,

where x and y are in feet and t is in seconds.

a) Find the height of the ball after 1 sec.
b) Find the total horizontal distance traveled by the ball.
c) How long does it take the ball to hit the ground?

34. For Exercise 33, determine the rectangular equation describing the path of the ball.

35. In Example 6, use a grapher to determine the maximum height that the ball reaches.

36. The path of a projectile ejected with an initial velocity of 200 ft/sec at an angle of 55° with the horizontal is described with the equations

$x = (210 \cos 55°)t$ and $y = (210 \sin 55°)t - 16t^2$,

where $t \ge 0$.

a) Find the height after 4 sec.
b) Determine the equivalent rectangular equation for this curve.

Graph each cycloid for t in the indicated interval.

37. $x = 2(t - \sin t),\ y = 2(1 - \cos t);\ 0 \le t \le 4\pi$

38. $x = 4t - 4 \sin t,\ y = 4 - 4 \cos t;\ 0 \le t \le 6\pi$

39. $x = t - \sin t,\ y = 1 - \cos t;\ -2\pi \le t \le 2\pi$

40. $x = 5(t - \sin t),\ y = 5(1 - \cos t);\ -4\pi \le t \le 4\pi$

Synthesis

41. Graph the curve described by

$x = 3 \cos t$ and $y = 3 \sin t,\ 0 \le t \le 2\pi.$

As t increases, the path of the curve is generated counterclockwise. How can this set of equations be changed so that the curve is generated clockwise?

42. Graph the plane curve described by

$x = \cos^3 t$ and $y = \sin^3 t,\ 0 \le t \le 2\pi.$

Then find the equivalent rectangular equation.

Answers

Introduction to Graphs and Graphers

1.

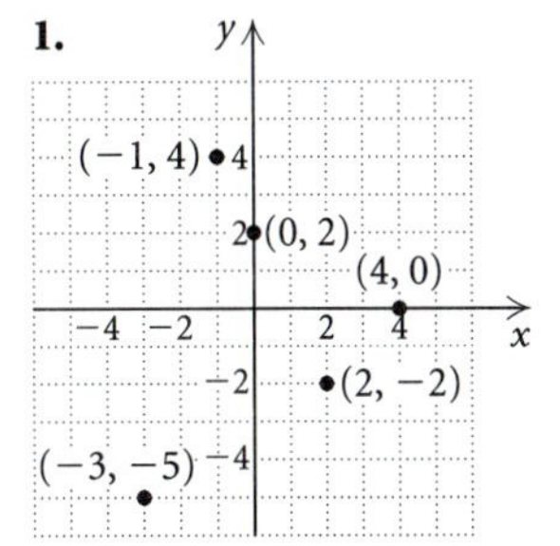

3.

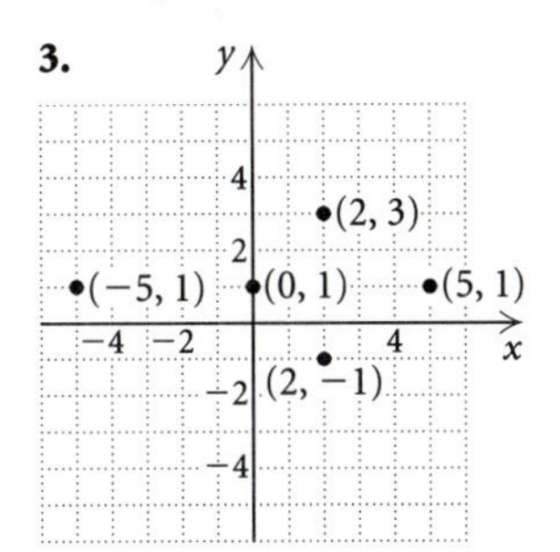

5. Yes; no **7.** No; yes **9.** No; yes

11.

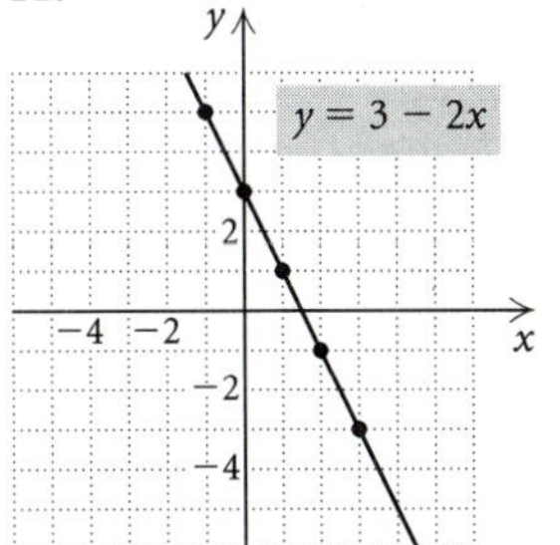

13.

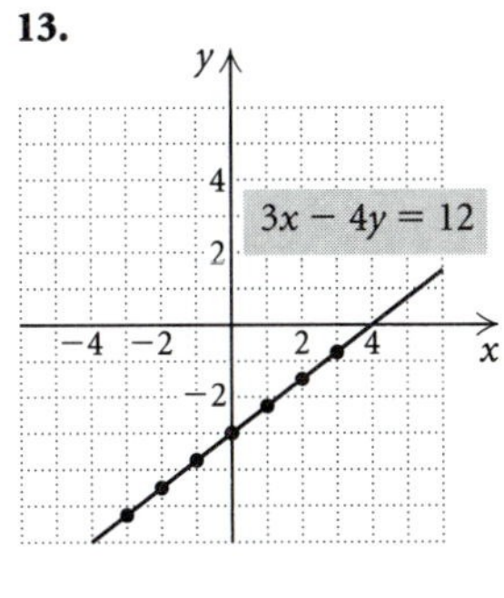

15.

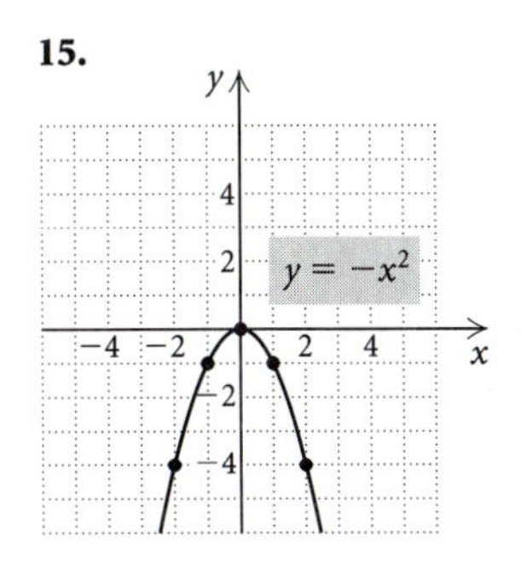

17.

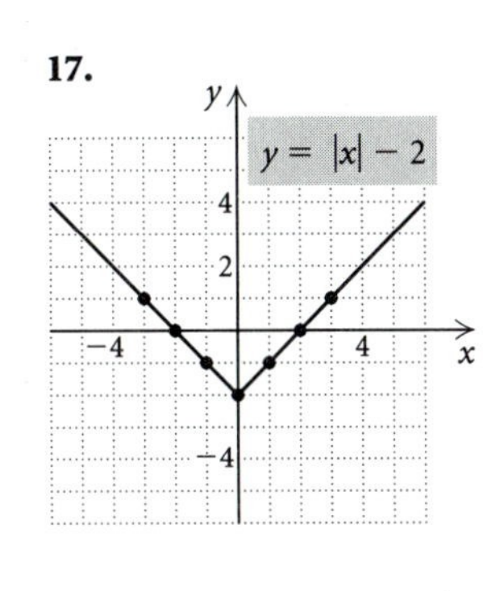

19.

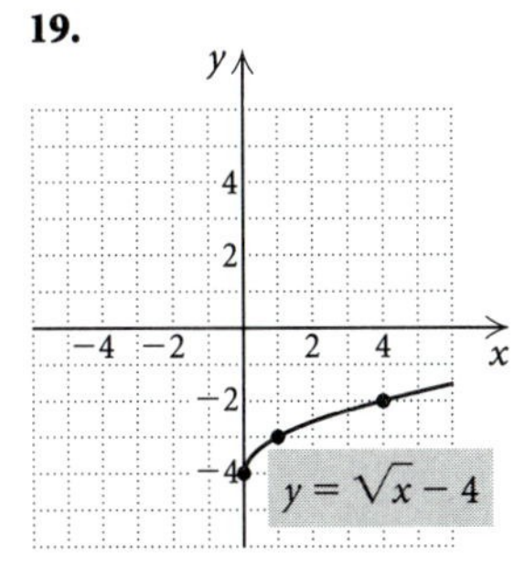

21.

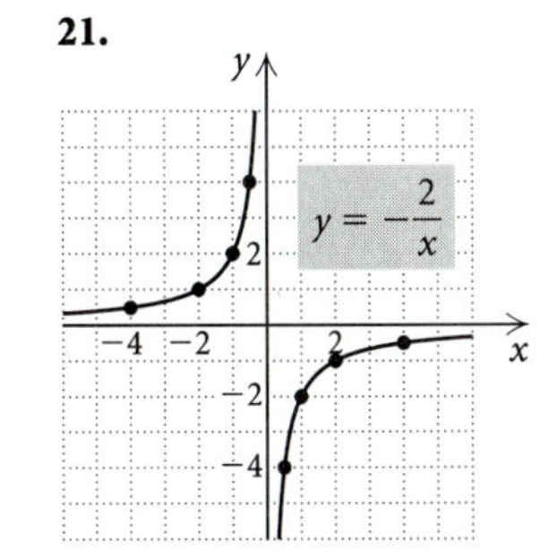

23. (f) **25.** (c) **27.** (b) **29.** (d)

31.

$y = \frac{1}{5}x + 2$

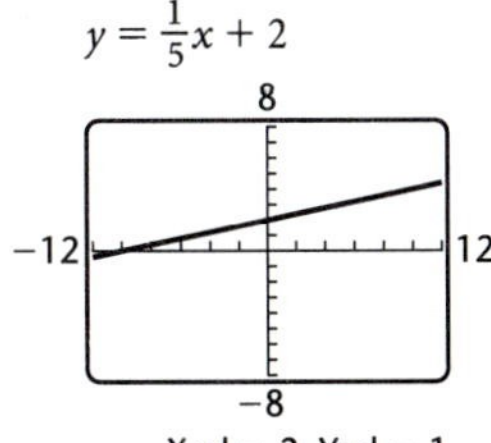
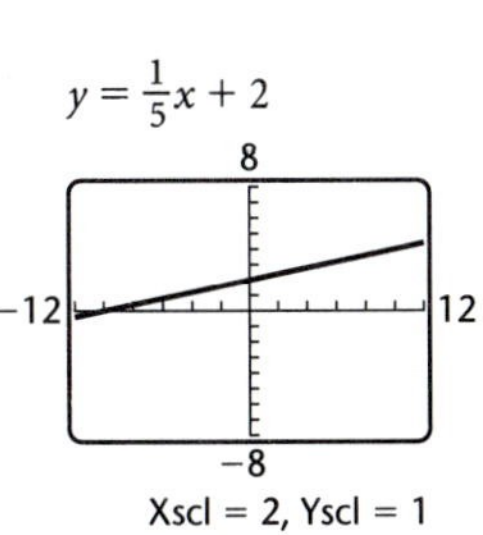
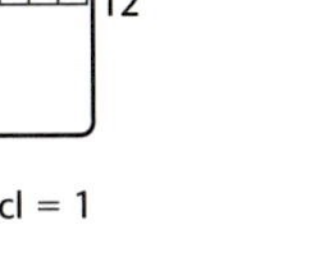

Xscl = 2, Yscl = 1

33.

$x = y^2 + 2$

$y_1 = \sqrt{x - 2}$ and $y_2 = -\sqrt{x - 2}$

Xscl = 1, Yscl = 1

35.

$y = \frac{-18}{x}$

Xscl = 5, Yscl = 5

37.

$y = x^2 - 4x + 3$

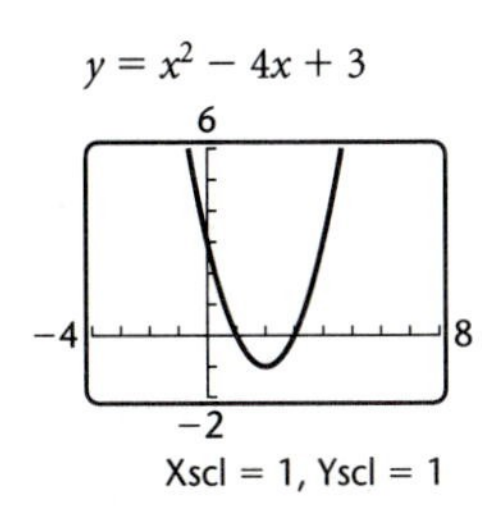

Xscl = 1, Yscl = 1

39. $y = \sqrt{x} - 2$

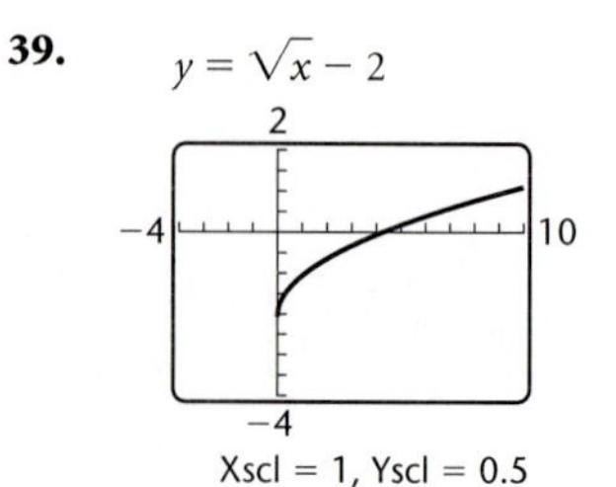

Xscl = 1, Yscl = 0.5

41. $y^2 = 4 + x$

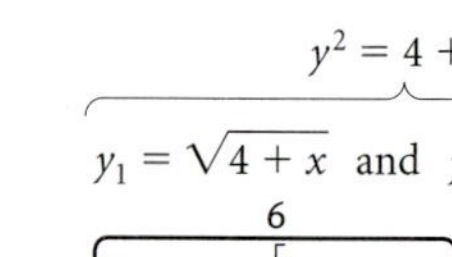

$y_1 = \sqrt{4 + x}$ and $y_2 = -\sqrt{4 + x}$

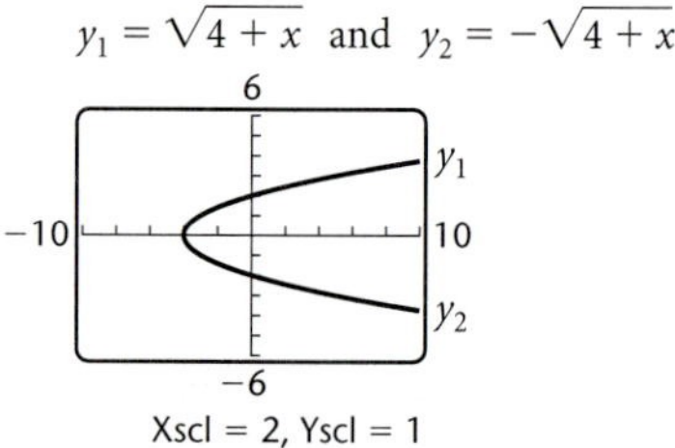

Xscl = 2, Yscl = 1

43. **(a)** Does not show curvature; **(b)** best; **(c)** tick marks too close; **(d)** does not show curvature; ticks too close
45. **(a)** Does not show curvature; **(b)** best; **(c)** does not show curvature as well as (b); **(d)** cannot see the graph
47. Error, error, .6245, 1, 1.261, 1.4697, 1.6462; −32, −40, −53.33, −80, −160, error, 160 **49.** Yes **51.** No
53. Yes **55.** No **57.** Yes **59.** −8 **61.** −0.882
63. −5.8 **65.** 20.376 **67.** −0.104 **69.** ±1.959
71. 1.489, 5.673 **73.** −2, 3
75. (−4.378, 0), (2.545, 0), (−1.167, 0)
77. (−0.929, −3.388), (1.080, −2.240), (2.848, −1.229)
79. (−2.809, −2.528), (0, 0), (2.809, 2.528)
81. Answer depends on the grapher and its features.
83. $y = x^2 + 1$; x: ±9, ±11; y: 26

Chapter R

Exercise Set R.1

1. $\sqrt[3]{8}$, 0, 9, $\sqrt{25}$
3. $\sqrt{7}$, 5.242242224..., $-\sqrt{14}$, $\sqrt[5]{5}$, $\sqrt[3]{4}$
5. −12, $5.\overline{3}$, $-\frac{7}{3}$, $\sqrt[3]{8}$, 0, −1.96, 9, $4\frac{2}{3}$, $\sqrt{25}$, $\frac{5}{7}$ **7.** $-\frac{3}{4}$
9. −5 **11.** True **13.** False **15.** True **17.** False
19. False **21.** True **23.** True **25.** True **27.** False
29. 7.1 **31.** $\frac{5}{4}$ **33.** $8|b|$ **35.** $5|xz|$, or $5|x| \cdot |z|$
37. $0.02\left|\frac{x}{y}\right|$, or $\frac{0.02|x|}{|y|}$ **39.** 11 **41.** 6 **43.** 5.4
45. Commutative property of multiplication
47. Multiplicative identity property
49. Associative property of multiplication
51. Commutative property of multiplication
53. Commutative property of addition
55. Multiplicative inverse property **57.** Identity
59. Not an identity **61.** Identity **63.** ◆
65. 0.124124412444... **67.** −0.00999
69. The hypotenuse of a right triangle with legs of lengths 1 unit and 3 units has a length of $\sqrt{10}$ units.

Exercise Set R.2

1. 5^2, or 25 **3.** 1 **5.** 7^{-1}, or $\frac{1}{7}$ **7.** $6x^5$
9. $15a^{-1}b^5$, or $\frac{15b^5}{a}$ **11.** $72x^5$ **13.** b^3 **15.** x^3y^{-3}, or $\frac{x^3}{y^3}$
17. $8xy^{-5}$, or $\frac{8x}{y^5}$ **19.** $8a^3b^6$ **21.** $16x^{12}$ **23.** $\frac{c^2d^4}{25}$
25. $8^5a^{20}b^{-25}c^{10}$, or $\frac{8^5a^{20}c^{10}}{b^{25}}$ **27.** 4.05×10^5
29. 3.9×10^{-7} **31.** 1.6×10^{-5} **33.** 0.000083
35. 20,700,000 **37.** 4,690,000,000,000 **39.** 1.395×10^3
41. 8.0×10^{-14} **43.** 1.332×10^{14} **45.** About $\$1.2 \times 10^7$
47. 2 **49.** 2048 **51.** 5 **53.** $2883.67 **55.** $8763.54
57. ◆ **59.** x^{8t} **61.** t^{8x} **63.** $9x^{2a}y^{2b}$ **65.** $660.29
67. For $x < -1$ or $x > 1$

Exercise Set R.3

1. $-5y^4, 3y^3, 7y^2, -y, -4$; 4 **3.** $3a^4b, -7a^3b^3, 5ab, -2$; 6
5. $3x^2y - 5xy^2 + 7xy + 2$ **7.** $3x + 2y - 2z - 3$
9. $-2x^2 + 6x - 2$ **11.** $x^4 - 3x^3 - 4x^2 + 9x - 3$
13. $2a^4 - 2a^3b - a^2b + 4ab^2 - 3b^3$ **15.** $x^2 + 2x - 15$
17. $2a^2 + 13a + 15$
19. The graphs of $y_1 = (2x + 3)(x + 5)$ and $y_2 = 2x^2 + 13x + 15$ appear to coincide.
21. $4x^2 + 8xy + 3y^2$ **23.** $x^2 + 10x + 25$
25. $25y^2 - 30y + 9$
27. The graphs of $y_1 = (5x - 3)^2$ and $y_2 = 25x^2 - 30x + 9$ appear to coincide. **29.** $4x^2 + 12xy + 9y^2$
31. $4x^4 - 12x^2y + 9y^2$ **33.** $a^2 - 9$
35. The graphs of $y_1 = (x + 3)(x - 3)$ and $y_2 = x^2 - 9$ appear to coincide. **37.** $9x^2 - 4y^2$
39. $4x^2 + 12xy + 9y^2 - 16$ **41.** $x^4 - 1$ **43.** ◆
45. $a^{2n} - b^{2n}$ **47.** $a^{2n} + 2a^nb^n + b^{2n}$ **49.** $x^6 - 1$
51. $x^{a^2-b^2}$ **53.** $a^2 + b^2 + c^2 + 2ab + 2ac + 2bc$
55. $a^4 + 4a^3b + 6a^2b^2 + 4ab^3 + b^4$

Exercise Set R.4

1. $2(x - 5)$ **3.** $3x^2(x^2 - 3)$ **5.** $4(a^2 - 3a + 4)$
7. $(b - 2)(a + c)$ **9.** $(x + 3)(x^2 + 6)$
11. $(y - 3)(y + 2)(y - 2)$
13. The graphs of $y_1 = x^3 + 3x^2 + 6x + 18$ and $y_2 = (x + 3)(x^2 + 6)$ appear to coincide.
15. $(p + 2)(p + 4)$ **17.** $(2n - 7)(n + 8)$
19. $(y^2 - 7)(y^2 + 3)$
21. The graphs of $y_1 = x^2 + 6x + 8$ and $y_2 = (x + 2)(x + 4)$ appear to coincide.
23. $(3x + 5)(3x - 5)$ **25.** $4x(y^2 + z)(y^2 - z)$

27. $(y - 3)^2$ **29.** $(1 - 4x)^2$
31. The graphs of $y_1 = 1 - 8x + 16x^2$ and $y_2 = (1 - 4x)^2$ appear to coincide. **33.** $(x + 2)(x^2 - 2x + 4)$
35. $(m - 1)(m^2 + m + 1)$
37. The graphs of $y_1 = x^3 - 1$ and $y_2 = (x - 1)(x^2 + x + 1)$ appear to coincide.
39. $3ab(6a - 5b)$ **41.** $(x - 4)(x^2 + 5)$
43. $8(x + 2)(x - 2)$ **45.** Prime
47. The graphs of $y_1 = x^3 - 4x^2 + 5x - 20$ and $y_2 = (x - 4)(x^2 + 5)$ appear to coincide.
49. $(x + 4)(x + 5)$ **51.** $(y - 1)(y - 5)$
53. $(2a + 1)(a + 4)$ **55.** $(3x - 1)(2x + 3)$ **57.** $(y - 9)^2$
59. $(xy - 7)^2$ **61.** $4a(x + 7)(x - 2)$
63. $3(z - 2)(z^2 + 2z + 4)$
65. $2ab(2a^2 + 3b^2)(4a^4 - 6a^2b^2 + 9b^4)$
67. The graphs of $y_1 = 6x^2 + 7x - 3$ and $y_2 = (3x - 1)(2x + 3)$ appear to coincide. **69.**
71. $(y^2 + 12)(y^2 - 7)$ **73.** $\left(y + \frac{4}{7}\right)\left(y - \frac{2}{7}\right)$
75. $(t + 0.9)(t - 0.3)$ **77.** $h(3x^2 + 3xh + h^2)$
79. $(x - 4)(x + 8)$ **81.** $(3a - 3b - 2)(a - b + 4)$
83. $(x^n + 8)(x^n - 3)$ **85.** $(x + b)(x + a)$ **87.** $\left(\frac{1}{2}t - \frac{2}{5}\right)^2$
89. $(5y^m + x^n - 1)(5y^m - x^n + 1)$
91. $3(x^n - 2y^m)(x^{2n} + 2x^ny^m + 4y^{2m})$
93. $y(y - 1)^2(y - 2)$

EXERCISE SET R.5

1. $\{x \mid x$ is a real number$\}$
3. $\{x \mid x$ is a real number and $x \neq 0$ and $x \neq 1\}$
5. $\{x \mid x$ is a real number and $x \neq 2$ and $x \neq -2$ and $x \neq -5\}$
7.

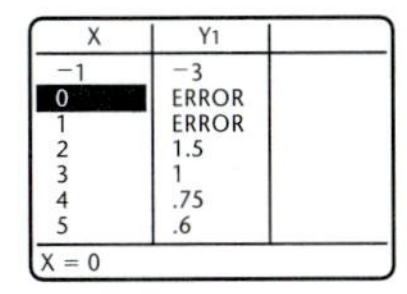

9. $\dfrac{1}{x - y}$ **11.** $\dfrac{(x + 5)(2x + 3)}{7x}$ **13.** $\dfrac{a + 2}{a - 5}$ **15.** $m + n$
17. $\dfrac{3(x - 4)}{2(x + 4)}$ **19.** $\dfrac{1}{x + y}$ **21.** $\dfrac{x - y - z}{x + y + z}$
23. The graphs of $y_1 = \dfrac{x^2 - 2x - 35}{2x^3 - 3x^2} \cdot \dfrac{4x^3 - 9x}{7x - 49}$ and $y_2 = \dfrac{(x + 5)(2x + 3)}{7x}$ appear to coincide. **25.** 1
27. $\dfrac{y - 2}{y - 1}$ **29.** $\dfrac{x + y}{2x - 3y}$ **31.** $\dfrac{3x - 4}{(x + 2)(x - 2)}$
33. $\dfrac{-y + 10}{(y - 5)(y + 4)}$ **35.** $\dfrac{4x - 8y}{(x + y)(x - y)}$
37. $\dfrac{3x - 4}{(x - 2)(x - 1)}$ **39.** $\dfrac{5a^2 + 10ab - 4b^2}{(a - b)(a + b)}$
41. $\dfrac{11x^2 - 18x + 8}{(2 + x)(2 - x)^2}$ **43.** 0
45. The graphs of $y_1 = \dfrac{3}{x + 2} + \dfrac{2}{x^2 - 4}$ and $y_2 = \dfrac{3x - 4}{(x + 2)(x - 2)}$ appear to coincide. **47.** $\dfrac{x + y}{x}$
49. $\dfrac{a^2 - 1}{a^2 + 1}$ **51.** $\dfrac{c^2 - 2c + 4}{c}$ **53.** $\dfrac{xy}{x - y}$ **55.** $x - y$
57. $\dfrac{3(x - 1)^2(x + 2)}{(x + 3)(x - 3)(10 - x)}$ **59.** $\dfrac{1 + a}{1 - a}$ **61.** $\dfrac{b + a}{b - a}$
63. **65.** $2x + h$ **67.** $3x^2 + 3xh + h^2$ **69.** x^5
71. $\dfrac{(n + 1)(n + 2)(n + 3)}{2 \cdot 3}$ **73.** $\dfrac{x^3 + 2x^2 + 11x + 20}{(x + 1)(4 + 2x)}$

EXERCISE SET R.6

1. 11 **3.** $4|x|$ **5.** $|b + 1|$ **7.** $-3x$ **9.** $|x - 2|$ **11.** 2
13. $6\sqrt{5}$ **15.** $3\sqrt[3]{2}$ **17.** $8\sqrt{2}|c|d^2$
19. The values of $y_1 = \sqrt{x^2 - 4x + 4}$ and $y_2 = |x - 2|$ appear to be the same for any value of x. **21.** $2x^2y\sqrt{6}$
23. $3x\sqrt[3]{4y}$ **25.** $2(x + 4)\sqrt[3]{(x + 4)^2}$ **27.** $\dfrac{m^2n^4}{2}$ **29.** 2
31. $\dfrac{1}{2x}$ **33.** $\dfrac{4a\sqrt[3]{a}}{3b}$ **35.** $\dfrac{x\sqrt{7x}}{6y^3}$ **37.** $51\sqrt{2}$
39. $-2x\sqrt{2} - 12\sqrt{5x}$ **41.** 1 **43.** $4 + 2\sqrt{3}$
45. The graphs of $y_1 = 8\sqrt{2x^2} - 6\sqrt{20x} - 5\sqrt{8x^2}$ and $y_2 = -2x\sqrt{2} - 12\sqrt{5x}$ appear to coincide.
47. About 13,709.5 ft **49.** **(a)** $h = \dfrac{a}{2}\sqrt{3}$; **(b)** $A = \dfrac{a^2}{4}\sqrt{3}$
51. 8 **53.** $\dfrac{\sqrt{6}}{3}$ **55.** $\dfrac{\sqrt[3]{10}}{2}$ **57.** $\dfrac{2\sqrt[3]{18}}{3}$ **59.** $\dfrac{9 - 3\sqrt{5}}{2}$
61. $\dfrac{6\sqrt{m} + 6\sqrt{n}}{m - n}$ **63.** $\dfrac{6}{5\sqrt{3}}$ **65.** $\dfrac{7}{\sqrt[3]{98}}$ **67.** $\dfrac{11}{\sqrt{33}}$
69. $\dfrac{76}{27 - 9\sqrt{3} + 3\sqrt{5} - \sqrt{15}}$ **71.** $\dfrac{a - b}{3a(\sqrt{a} - \sqrt{b})}$
73. $\sqrt[4]{x^3}$ **75.** 8 **77.** $\dfrac{1}{5}$ **79.** $a\sqrt[4]{\dfrac{a}{b^3}}$
81. $13^{5/4}$, or $13\sqrt[4]{13}$ **83.** $20^{2/3}$ **85.** $11^{1/6}$ **87.** $5^{5/6}$
89. 4 **91.** $8a^2$ **93.** $\dfrac{x^3}{3b^{-2}}$, or $\dfrac{x^3b^2}{3}$ **95.** $x\sqrt[3]{y}$
97. The graphs of $y_1 = (2x^{3/2})(4x^{1/2})$ and $y_2 = 8x^2$ appear to coincide. **99.** $\sqrt[6]{288}$ **101.** $\sqrt[12]{x^{11}y^7}$ **103.** $a\sqrt[6]{a^5}$
105. $(a + x)\sqrt[12]{(a + x)^{11}}$ **107.** $a^{-2}b^{-5}(b^{10} - a^5)$
109. $x^{-1/3}y^{-1/4}(y - x)$ **111.** $(2x - 3)^{-3}(x + 1)^{1/4}(3x - 2)$
113. **115.** $\dfrac{(2 + x^2)\sqrt{1 + x^2}}{1 + x^2}$ **117.** $a^{a/2}$
119. **(a)** $\{0, 1\}$; **(b)** $\{x \mid x > 1\}$; **(c)** $\{x \mid 0 < x < 1\}$

EXERCISE SET R.7

1. $\{4\}$ **3.** $\{8\}$ **5.** $\left\{\frac{4}{5}\right\}$ **7.** $\left\{-\frac{3}{2}\right\}$ **9.** $\left\{\frac{3}{2}, \frac{2}{3}\right\}$ **11.** $\left\{\frac{2}{3}, -1\right\}$
13. $\{0, 3\}$ **15.** $\left\{-\frac{1}{3}, 0, 2\right\}$ **17.** $\left\{-1, 1, -\frac{1}{7}\right\}$ **19.** $\left\{\frac{20}{9}\right\}$
21. $\{286\}$ **23.** $\{6\}$ **25.** $\varnothing$ **27.** $\varnothing$ **29.** $\varnothing$

31. $\{x \mid x \text{ is a real number and } x \neq 0 \text{ and } x \neq 6\}$ **33.** $\{2\}$
35. $\left\{\frac{5}{3}\right\}$ **37.** $\{\pm\sqrt{2}\}$ **39.** $\varnothing$ **41.** $\{-6\}$ **43.** $\{7\}$
45. $\varnothing$ **47.** $\{1\}$ **49.** $\{3, 7\}$ **51.** $\{-8\}$ **53.** $\{81\}$
55. $\{-7, 7\}$ **57.** $\varnothing$ **59.** $\{-3, 5\}$ **61.** $\left\{-\frac{1}{3}, \frac{1}{3}\right\}$ **63.** $\{0\}$
65. $\left\{-1, -\frac{1}{3}\right\}$ **67.** $\{-24, 44\}$ **69.** $\{-2, 4\}$ **71.** $b = \dfrac{2A}{h}$
73. $w = \dfrac{P - 2l}{2}$ **75.** $T_1 = \dfrac{P_1 V_1 T_2}{P_2 V_2}$ **77.** $R_2 = \dfrac{RR_1}{R_1 - R}$
79. $p = \dfrac{Fm}{m - F}$ **81.** ◈ **83.** ◈ **85.** $\{1\}$
87. $\{1, 5\}$ **89.** $\{-1\}$

Exercise Set R.8

1. $\{x \mid x > 3\}$, or $(3, \infty)$ **3.** $\left\{x \mid x \geq -\frac{5}{12}\right\}$, or $\left[-\frac{5}{12}, \infty\right)$
5. $\left\{y \mid y \geq \frac{22}{13}\right\}$, or $\left[\frac{22}{13}, \infty\right)$ **7.** $\left\{x \mid x \leq \frac{15}{34}\right\}$, or $\left(-\infty, \frac{15}{34}\right]$
9. $\{x \mid x < 1\}$, or $(-\infty, 1)$
11. The graph of $y_1 = 4x(x - 2)$ lies below the graph of $y_2 = 2(2x - 1)(x - 3)$ for $x < 1$. **13.** $[-3, 3]$
15. $[-14, -11)$ **17.** $(-\infty, -4]$ **19.** $(-\infty, 3.8)$
21. $(-\infty, 7) \cup (7, \infty)$ **23.** $(0, 5)$ **25.** $[-9, -4)$
27. $[x, x + h]$ **29.** (p, ∞) **31.** $[-3, 3)$ **33.** $[8, 10]$
35. $[-7, -1]$ **37.** $\left(-\frac{3}{2}, 2\right)$ **39.** $(1, 5]$ **41.** $\left(-\frac{11}{3}, \frac{13}{3}\right]$
43. $(-\infty, -2] \cup (1, \infty)$ **45.** $\left(-\infty, -\frac{7}{2}\right] \cup \left[\frac{1}{2}, \infty\right)$
47. $(-\infty, 9.6) \cup (10.4, \infty)$ **49.** $\left(-\infty, -\frac{57}{4}\right] \cup \left[-\frac{55}{4}, \infty\right)$
51. The graph of $y_1 = 6 - 2x$ lies both on or above the graph of $y_2 = -4$ and below the graph of $y_3 = 4$ for $1 < x \leq 5$.
53. $(-7, 7)$;

−7 0 7

55. $(-\infty, -4.5] \cup [4.5, \infty)$;

−4.5 0 4.5

57. $(-17, 1)$ **59.** $(-\infty, -17] \cup [1, \infty)$ **61.** $\left(-\frac{1}{4}, \frac{3}{4}\right)$
63. $\left(-\frac{1}{3}, \frac{1}{3}\right)$ **65.** $[-6, 3]$ **67.** $(-\infty, 4.9) \cup (5.1, \infty)$
69. $\left[-\frac{1}{2}, \frac{7}{2}\right]$ **71.** $\left[-\frac{7}{3}, 1\right]$ **73.** $(-\infty, -8) \cup (7, \infty)$ **75.** $\varnothing$
77. The graph of $y_1 = |x + 8|$ lies on or above the graph of $y_2 = 9$ for $x \geq 1$ or $x \leq -17$. **79.** ◈ **81.** ◈
83. $\left(-\frac{1}{4}, \frac{5}{9}\right]$ **85.** $\left(-\infty, \frac{1}{2}\right)$ **87.** $\{-\infty, \infty\}$ **89.** $\varnothing$

Exercise Set R.9

1. \$0.832 billion on employee benefits; \$3.239 billion on wages and salaries **3.** Length: 93 m; width: 68 m
5. About 11.5 sec **7.** 2 cm **9.** 26°, 130°, 24°
11. 280,526 **13.** Castulo: 8 km/h; Linette: 15 km/h
15. Amtrak: 80 mph; Central Railway: 66 mph
17. 98.3 mi **19.** \$650 **21.** 86 or higher
23. Mileages greater than $33\frac{1}{3}$ **25.** Calls longer than 2 hr
27. Gross sales greater than \$18,000 **29.** ◈
31. 32 mph **33.** $53\frac{6}{23}$ mph **35.** $51\frac{3}{7}$ mph **37.** \$16

Review Exercises, Chapter R

1. [R.1] 12, −3, −1, −19, 31, 0 **2.** [R.1] 12, 31
3. [R.1] $-43.89, 12, -3, -\frac{1}{5}, -1, -\frac{4}{3}, 7\frac{2}{3}, -19, 31, 0$
4. [R.1] $-43.89, 12, -3, -\frac{1}{5}, \sqrt{7}, \sqrt[3]{10}, -1, -\frac{4}{3}, 7\frac{2}{3}, -19, 31, 0$ **5.** [R.1] $\sqrt{7}, \sqrt[3]{10}$ **6.** [R.1] 12, 31, 0 **7.** [R.1] 3.5
8. [R.1] $\frac{5}{6}a^2|b|$ **9.** [R.1] 10 **10.** [R.1] Not an identity
11. [R.2] 117 **12.** [R.2] −10 **13.** [R.2] 3,261,000
14. [R.2] 0.00041 **15.** [R.2] 1.432×10^{-2}
16. [R.2] 4.321×10^4 **17.** [R.2] 7.8125×10^{-22}
18. [R.2] 5.46×10^{-32} **19.** [R.2] $-14a^{-2}b^7$, or $\dfrac{-14b^7}{a^2}$
20. [R.2] $6x^9y^{-6}z^6$, or $\dfrac{6x^9z^6}{y^6}$ **21.** [R.2] 3 **22.** [R.2] −2
23. [R.5] $\dfrac{b}{a}$ **24.** [R.5] $\dfrac{x + y}{xy}$ **25.** [R.6] −4
26. [R.6] $25x^4 - 10\sqrt{2}x^2 + 2$ **27.** [R.6] $13\sqrt{5}$
28. [R.3] $x^3 + t^3$
29. [R.3] $125a^3 + 300a^2b + 240ab^2 + 64b^3$
30. [R.3] $8xy^4 - 9xy^2 + 4x^2 + 2y - 7$
31. [R.4] $(x^2 - 3)(x + 2)$
32. [R.4] $3a(2a + 3b^2)(2a - 3b^2)$ **33.** [R.4] $(x + 12)^2$
34. [R.4] $x(9x - 1)(x + 4)$
35. [R.4] $(2x - 1)(4x^2 + 2x + 1)$
36. [R.4] $(3x^2 + 5y^2)(9x^4 - 15x^2y^2 + 25y^4)$
37. [R.4] $6(x + 2)(x^2 - 2x + 4)$
38. [R.4] $(x - 1)(2x + 3)(2x - 3)$
39. [R.4] $(3x - 5)^2$ **40.** [R.4] $3(6x^2 - x + 2)$
41. [R.4] $(3x + y + 4)(3x + y - 1)$ **42.** [R.5] 3
43. [R.5] $\dfrac{x - 5}{(x + 5)(x + 3)}$ **44.** [R.6] $y^3\sqrt[6]{y}$
45. [R.6] $\sqrt[3]{(a + b)^2}$ **46.** [R.6] $b\sqrt[5]{b^2}$ **47.** [R.6] $\dfrac{m^4n^2}{3}$
48. [R.6] $\dfrac{x - y}{x + 2\sqrt{xy} + y}$ **49.** [R.6] $x^{-1/2}y^{-3/4}(2x - 3y)$
50. [R.6] $(x - 2)^{-3/4}(3x + 5)^{3/2}(4x + 3)$
51. [R.6] About 18.8 ft **52.** [R.7] $\{4\}$ **53.** [R.7] $\{-1\}$
54. [R.7] $\left\{-\frac{5}{2}, \frac{1}{3}\right\}$ **55.** [R.7] $\left\{-2, \frac{4}{3}\right\}$ **56.** [R.7] $\left\{\frac{27}{7}\right\}$
57. [R.7] $\left\{-\frac{1}{2}, \frac{9}{4}\right\}$ **58.** [R.7] $\{0, 3\}$ **59.** [R.7] $\{5\}$
60. [R.7] $\{1, 7\}$ **61.** [R.7] $\{-8, 1\}$ **62.** [R.8] $\left[-\frac{4}{3}, \frac{4}{3}\right)$
63. [R.8] $\left(\frac{2}{5}, 2\right]$ **64.** [R.8] $(-\infty, \infty)$
65. [R.8] $\left(-\infty, -\frac{5}{3}\right] \cup [1, \infty)$ **66.** [R.8] $\left(-\frac{2}{3}, 1\right)$
67. [R.8] $(-\infty, -6] \cup [-2, \infty)$ **68.** [R.7] $h = \dfrac{V^2}{2g}$
69. [R.7] $t = \dfrac{ab}{a + b}$ **70.** [R.9] 30 ft, 40 ft
71. [R.9] 94% **72.** [R.9] 6 mph **73.** [R.9] 80 km/h
74. [R.9] Mileages greater than 25

75. [R.9] Calls less than 2 hr
76. [R.9] ◈ Nisha is taking a geography class in which there are 4 tests. She has scores of 76, 82, and 78 on the first three tests. To get a B, she must average at least 80 on the 4 tests. What scores on the last test will give Nisha a B?
77. [R.5], [R.7] ◈ Two expressions are equivalent if they have the same value for all numbers in both domains. Two equations are equivalent if they have the same solution set.
78. [R.3] $x^{2n} + 6x^n - 40$ **79.** [R.3] $t^{2a} + 2 + t^{-2a}$
80. [R.3] $y^{2b} - z^{2c}$ **81.** [R.3] $a^{3n} - 3a^{2n}b^n + 3a^nb^{2n} - b^{3n}$
82. [R.4] $(y^n + 8)^2$ **83.** [R.4] $(x^t - 7)(x^t + 4)$
84. [R.4] $m^{3n}(m^n - 1)(m^{2n} + m^n + 1)$ **85.** [R.7] 256

Chapter 1

EXERCISE SET 1.1

1. Yes **3.** Yes **5.** No **7.** Yes **9.** Yes **11.** Yes
13. No **15.** Function; domain: {2, 3, 4}; range: {10, 15, 20}
17. Not a function; domain: {−7, −2, 0}; range: {3, 1, 4, 7}
19. Function; domain: {−2, 0, 2, 4, −3}; range: {1}
21. $f(-1) = 2$; $f(0) = 0$; $f(1) = -2$
23. **(a)** 1; **(b)** 6; **(c)** 22; **(d)** $3x^2 + 2x + 1$; **(e)** $3t^2 - 4t + 2$; **(f)** $6a + 3h - 2$
25. **(a)** 8; **(b)** −8; **(c)** $-x^3$; **(d)** $27y^3$; **(e)** $8 + 12h + 6h^2 + h^3$; **(f)** $3a^2 + 3ah + h^2$
27. **(a)** $\frac{1}{8}$; **(b)** 0; **(c)** does not exist; **(d)** $\frac{81}{53}$; **(e)** $\frac{x + h - 4}{x + h + 3}$; **(f)** $\frac{7}{(x + h + 3)(x + 3)}$
29. 0; does not exist; does not exist as a real number; $\frac{1}{\sqrt{3}}$ or $\frac{\sqrt{3}}{3}$ **31.** All real numbers **33.** $\{x \mid x \neq 0\}$
35. $\left\{x \mid x \geq -\frac{4}{7}\right\}$ **37.** $\{x \mid x \neq 5 \text{ and } x \neq -1\}$
39. $\{x \mid -3 \leq x \leq 3\}$
41. Domain: all real numbers; range: $[0, \infty)$
43. Domain: $[-3, 3]$; range: $[0, 3]$
45. Domain: all real numbers; range: all real numbers
47. Domain: $(-\infty, 7]$; range: $[0, \infty)$
49. Domain: $(-\infty, 0) \cup (0, \infty)$; range: $(-\infty, 0) \cup (0, \infty)$
51. No **53.** Yes **55.** Yes **57.** No
59. Function; domain: [0, 5]; range: [0, 3]
61. Function; domain: $[-2\pi, 2\pi]$; range: $[-1, 1]$
63. **(a)** The domain of y_1 does not include the intervals $(-\infty, -1.2]$ and $[3.2, \infty)$. The range of y_1 appears to be [0, 2]. The domain of y_2 does not include 3.2. The range of y_2 appears to be all real numbers. **(b)** The largest output of y_1 seems to be 2, and the smallest 0. There does not appear to be a largest or smallest output of y_2.
65. 645 m; 0 m **67.** $\left\{\frac{15}{2}\right\}$ **69.** $\left\{\frac{53}{2}\right\}$ **71.** ◈
73. $\frac{1}{\sqrt{x + h} + \sqrt{x}}$ **75.** $f(x) = x$, $g(x) = x + 1$
77. 35

EXERCISE SET 1.2

1. −1.532, −0.347, 1.879 **3.** −1.414, 0, 1.414 **5.** −1, 0, 1
7. 1.5, 9.5 **9.** −2, 3
11. All numbers in the interval [−1, 2]
13. **(a)** [−5, 1]; **(b)** [3, 5]; **(c)** [1, 3]
15. **(a)** [−3, −1], [3, 5]; **(b)** [1, 3]; **(c)** [−5, −3)
17. Increasing: $[0, \infty)$; decreasing: $(-\infty, 0]$; relative minimum: 0 at $x = 0$
19. Increasing: $(-\infty, 0]$; decreasing: $[0, \infty)$; relative maximum: 5 at $x = 0$
21. Increasing: $[3, \infty)$; decreasing: $(-\infty, 3]$; relative minimum: 1 at $x = 3$
23. Increasing: [1, 3]; decreasing: $(-\infty, 1]$, $[3, \infty)$; relative maximum: −4 at $x = 3$; relative minimum: −8 at $x = 1$
25. Increasing: [−1.552, 0], $[1.552, \infty)$; decreasing: $(-\infty, -1.552]$, [0, 1.552]; relative maximum: 4.07 at $x = 0$; relative minima: −2.314 at $x = -1.552$, −2.314 at $x = 1.552$
27. **(a)**

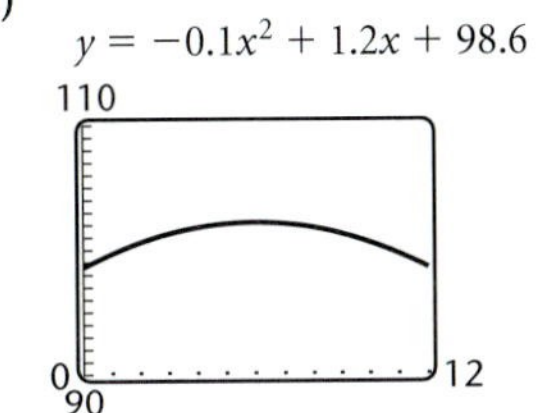

(b) 102.2; **(c)** 6 days after the illness began, the temperature is 102.2°F.
29. Increasing: [−1, 1]; decreasing: $(-\infty, -1]$, $[1, \infty)$; relative maximum: 4 at $x = 1$; relative minimum: −4 at $x = -1$
31. Increasing: [−1.414, 1.414]; decreasing: [−2, −1.414], [1.414, 2]; relative maximum: 2 at $x = 1.414$; relative minimum: −2 at $x = -1.414$

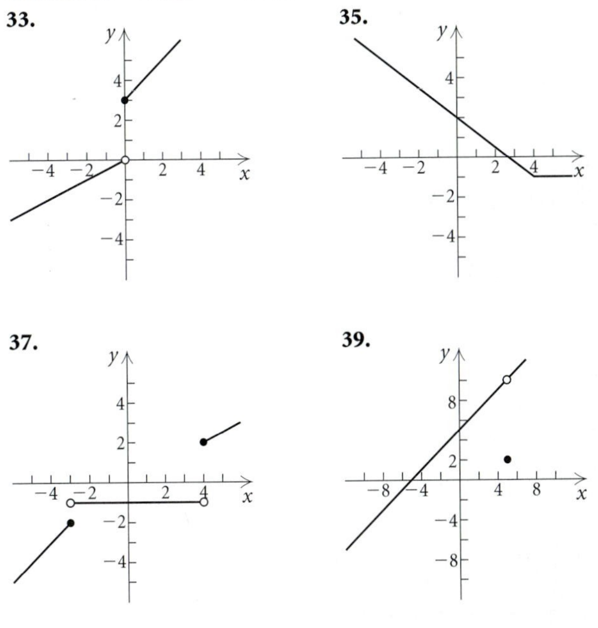

41.

43.

45.

$$y = \begin{cases} \sqrt[3]{x}, & \text{for } x \leqslant -1, \\ x^2 - 3x, & \text{for } -1 < x < 4, \\ \sqrt{x-4}, & \text{for } x \geqslant 4 \end{cases}$$

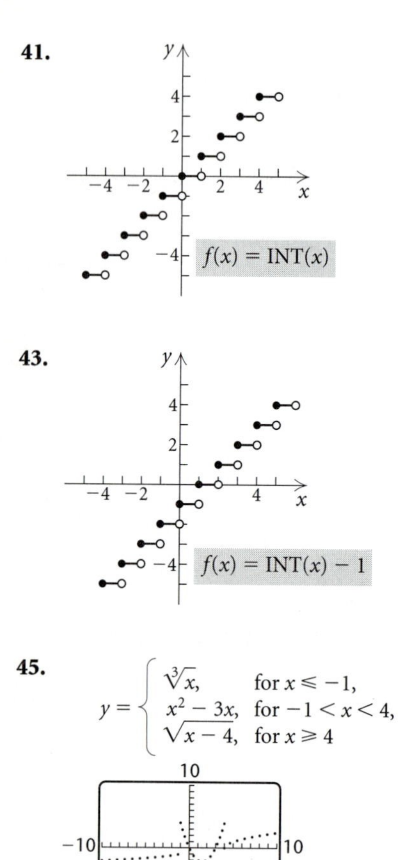

47.

$$y = \begin{cases} \sqrt[3]{x+27}, & \text{for } x < 1, \\ \left|2 - \dfrac{x}{5}\right|, & \text{for } x \geqslant 1 \end{cases}$$

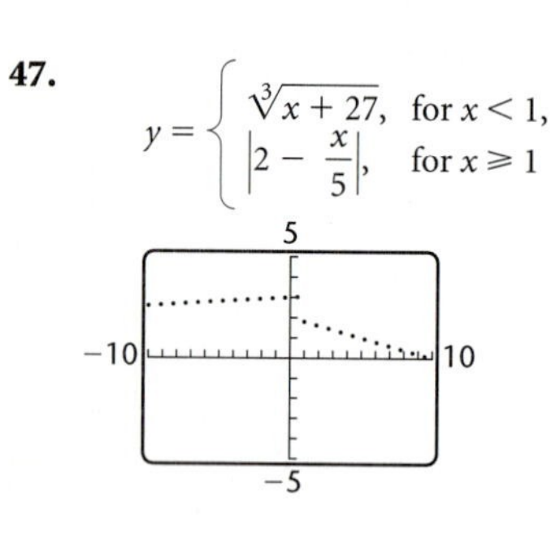

49. $d(t) = \sqrt{(120t)^2 + (400)^2}$ **51.** $A(x) = 24x - x^2$

53. $A(w) = 10w - \dfrac{w^2}{2}$ **55.** $d(s) = \dfrac{14}{s}$

57. **(a)** $V(x) = 4x^3 - 48x^2 + 144x$, or $4x(x-6)^2$;
(b) $\{x \mid 0 < x < 6\}$;
(c) $y = 4x^3 - 48x^2 + 144x$ **(d)** 8 cm by 8 cm by 2 cm

59. **(a)** $A(x) = x\sqrt{256 - x^2}$; **(b)** $\{x \mid 0 < x < 16\}$;
(c) $y = x\sqrt{256 - x^2}$

(d) 11.314 ft by 11.314 ft **61.** 7 **63.** ◈
65. Increasing: $[-1.933, 1.933]$; decreasing: $(-\infty, -1.933]$, $[1.933, \infty)$; relative maximum: 58.242 at $x = 1.933$; relative minimum: -58.242 at $x = -1.933$
67. Increasing: $[-5, -2]$, $[4, \infty)$; decreasing: $(-\infty, -5]$, $[-2, 4]$; relative maximum: 560 at $x = -2$; relative minima: 425 at $x = -5$, -304 at $x = 4$
69. **(a)**

(b) $C(t) = 2(\text{INT}(t) + 1),\ t > 0$
71. $\{x \mid -5 \le x < -4 \text{ or } 5 \le x < 6\}$
73. **(a)** $h(r) = \dfrac{30 - 5r}{3}$; **(b)** $V(r) = \pi r^2\left(\dfrac{30 - 5r}{3}\right)$;
(c) $V(h) = \pi h\left(\dfrac{30 - 3h}{5}\right)^2$

Exercise Set 1.3

1. **(a)** Yes; **(b)** yes; **(c)** yes **3.** **(a)** Yes; **(b)** no; **(c)** no
5. $-\frac{3}{5}$ **7.** $\frac{1}{5}$ **9.** $-\frac{5}{3}$ **11.** Not defined **13.** $2a + h$
15. 6.7% grade; $y = 0.067x$
17. $\frac{3}{5}$; $(0, -7)$
19. $-\frac{1}{2}$; $(0, 5)$
21. $-\frac{3}{2}$; $(0, 5)$
23. $\frac{4}{3}$; $(0, -7)$
25. $y = \frac{2}{9}x + 4$ **27.** $y = -4x - 7$
29. $y = -4.2x + \frac{3}{4}$ **31.** $y = \frac{2}{9}x + \frac{19}{3}$
33. $y = 3x - 5$ **35.** $y = -\frac{3}{5}x - \frac{17}{5}$
37. $y = -3x + 2$ **39.** $y = -\frac{1}{2}x + \frac{7}{2}$
41. Parallel **43.** Perpendicular
45. $y = \frac{2}{7}x + \frac{29}{7}$; $y = -\frac{7}{2}x + \frac{31}{2}$
47. $y = -0.3x - 2.1$; $y = \frac{10}{3}x + \frac{70}{3}$
49. $y = -\frac{3}{4}x + \frac{1}{4}$; $y = \frac{4}{3}x - 6$ **51.** $x = 3$; $y = -3$
53. **(a)** $W(h) = 3.5h - 110$; **(b)** 107 lb;

(c) $y = 3.5x - 110$

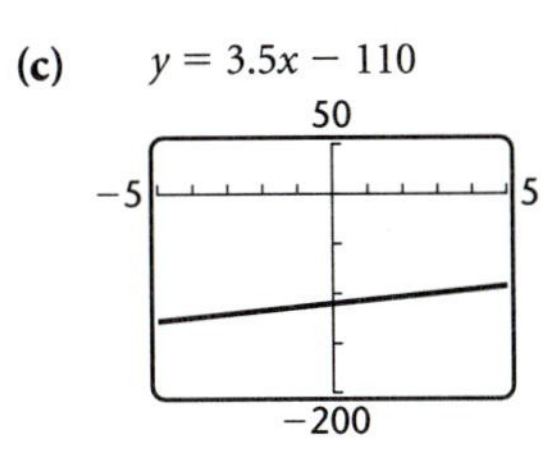

(d) $\{h \mid h > 31.43\}$, or $(31.43, \infty)$

55. (a) 115, 75, 135, 179; **(b)** $y = 2x + 115$

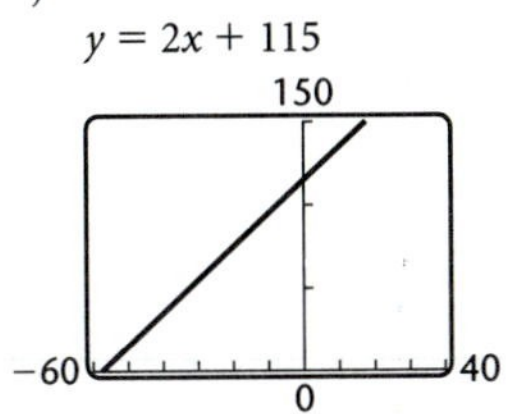

(c) Below $-57.5°$, stopping distance is negative; above $32°$, ice doesn't form.

57. (a) $\frac{11}{10}$. For each mile per hour faster that the car travels, it takes $\frac{11}{10}$ ft longer to stop; **(b)** 6, 11.5, 22.5, 55.5, 72;

(c)

$$y = \frac{11x + 5}{10}$$

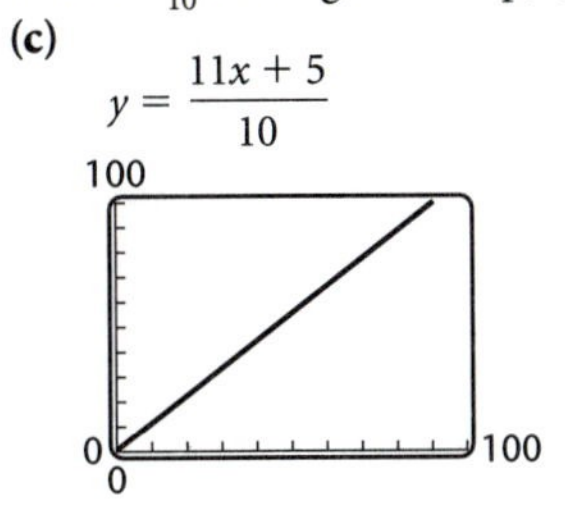

(d) $\{r \mid r > 0\}$, or $(0, \infty)$. If r is allowed to be 0, the function says that a stopped car travels $\frac{1}{2}$ ft before stopping.

59. $\dfrac{\text{Height of corn in feet}}{\text{Number of weeks after planting}}$

61. $C(t) = 60 + 40t$; $y = 60 + 40x$

63. $C(x) = 800 + 3x$; $y = 800 + 3x$

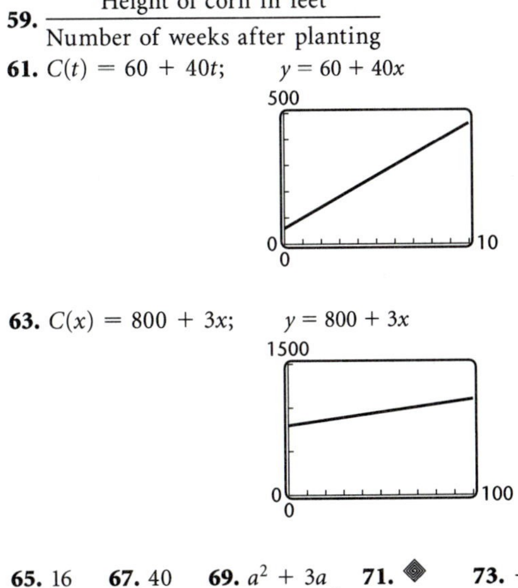

65. 16 **67.** 40 **69.** $a^2 + 3a$ **71.** ◈ **73.** -7.75

75. $C(F) = \frac{5}{9}(F - 32)$; $F(C) = \frac{9}{5}C + 32$ **77.** False

79. False **81.** $f(x) = x + b$

EXERCISE SET 1.4

1. Yes **3.** No **5.** No **7.** Yes

9. (a) $y = 2.6x + 49.7$, 75.7; **(b)** $r = 0.9856$, a good fit

11. (a) $y = 645.7x + 9799.2$; \$14,319.10; \$15,610.50; \$18,193.30; **(b)** $r = 0.9994$, a good fit; **(c)** The president's assertion is about \$1000, or 6%, higher than the estimate from the model.

13. (a) $M = 0.2H + 156$; **(b)** 164, 169, 171, 173; **(c)** $r = 1$; the regression line fits the data perfectly and should be a good predictor.

15. (a) 22; **(b)** -22; **(c)** $4a^3 - 5a$; **(d)** $-4a^3 + 5a$ **17.** ◈

19. (a) $D = 0.000008d + 11.9$, $r = 0.0282$; **(b)** The model does not fit the data and would not be a good predictor.

EXERCISE SET 1.5

1. $\sqrt{10}$, 3.162 **3.** $\sqrt{45}$, 6.708 **5.** $\sqrt{14 + 6\sqrt{2}}$, 4.742

7. 6.5 **9.** Yes

11. The distances between the points are $\sqrt{10}$, 10, and $\sqrt{90}$. Since $(\sqrt{10})^2 + (\sqrt{90})^2 = 10^2$, the three points are vertices of a right triangle. **13.** $\left(\frac{9}{2}, \frac{15}{2}\right)$ **15.** $\left(\frac{15}{2}, 2\right)$

17. $\left(\frac{\sqrt{3} - \sqrt{6}}{2}, -\frac{\sqrt{5}}{2}\right)$

19.

$\left(-\frac{1}{2}, \frac{3}{2}\right), \left(\frac{7}{2}, \frac{1}{2}\right), \left(\frac{5}{2}, \frac{9}{2}\right), \left(-\frac{3}{2}, \frac{11}{2}\right)$; no

21. $\left(\frac{\sqrt{7} + \sqrt{2}}{2}, -\frac{1}{2}\right)$

23. Square the window; for example, $[-12, 9, -4, 10]$

25. $(x - 2)^2 + (y - 3)^2 = \frac{25}{9}$

27. $(x + 1)^2 + (y - 4)^2 = 25$

29. $(x - 2)^2 + (y - 1)^2 = 169$

31. $(x + 2)^2 + (y - 3)^2 = 4$

33. $(0, 0)$; 2; $x^2 + y^2 = 4$

$y_1 = \sqrt{4 - x^2}$, $y_2 = -\sqrt{4 - x^2}$

35. (0, 3); 4;

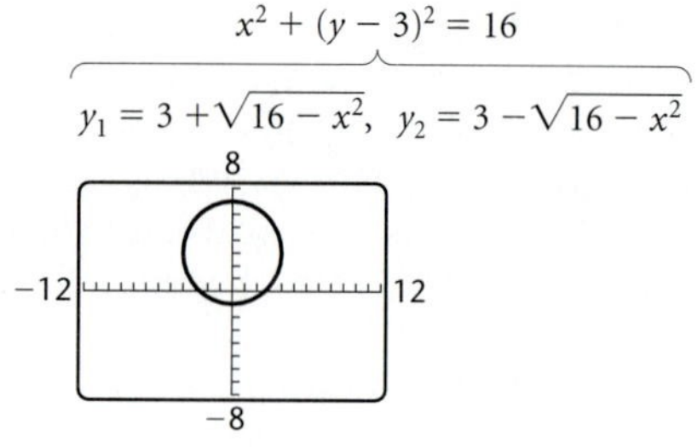

$x^2 + (y - 3)^2 = 16$
$y_1 = 3 + \sqrt{16 - x^2}$, $y_2 = 3 - \sqrt{16 - x^2}$
8
-12
12
-8

37. (1, 5); 6;

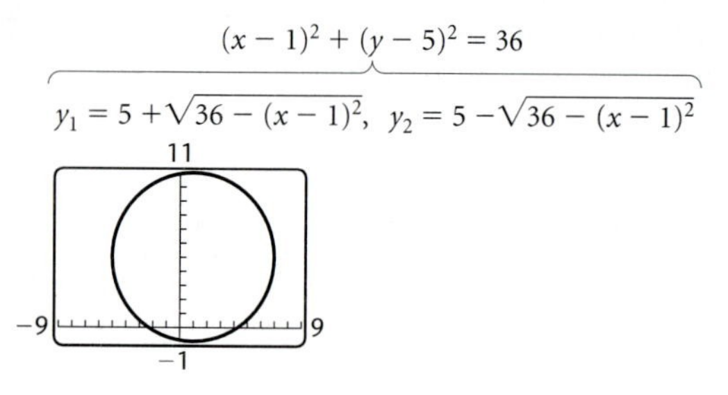

$(x - 1)^2 + (y - 5)^2 = 36$
$y_1 = 5 + \sqrt{36 - (x - 1)^2}$, $y_2 = 5 - \sqrt{36 - (x - 1)^2}$
11
-9
9
-1

39. (−4, −5); 3;

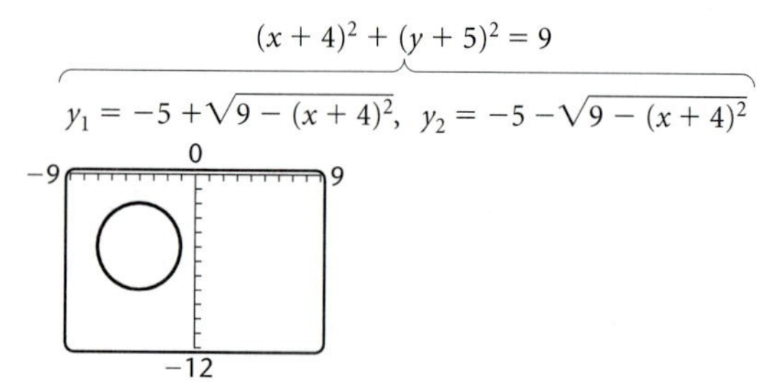

$(x + 4)^2 + (y + 5)^2 = 9$
$y_1 = -5 + \sqrt{9 - (x + 4)^2}$, $y_2 = -5 - \sqrt{9 - (x + 4)^2}$
0
-9
9
-12

41. −35 **43.** $2ah + h^2$

45. $\sqrt{h^2 + h + 2a - 2\sqrt{a^2 + ah}}$, $\left(\dfrac{2a + h}{2}, \dfrac{\sqrt{a} + \sqrt{a + h}}{2}\right)$

47. ◆ **49.** $(x - 2)^2 + (y + 7)^2 = 36$

51. $a_1 = 2.7$ ft, $a_2 = 37.3$ ft **53.** Yes **55.** Yes

57. (0, 4) **59.** Let $P_1 = (x_1, y_1)$, $P_2 = (x_2, y_2)$, and $M = \left(\dfrac{x_1 + x_2}{2}, \dfrac{y_1 + y_2}{2}\right)$. Let $d(AB)$ denote the distance from point A to point B.

(a)
$$d(P_1M) = \sqrt{\left(\frac{x_1 + x_2}{2} - x_1\right)^2 + \left(\frac{y_1 + y_2}{2} - y_1\right)^2} = \frac{1}{2}\sqrt{(x_2 - x_1)^2 + (y_2 - y_1)^2};$$

$$d(P_2M) = \sqrt{\left(\frac{x_1 + x_2}{2} - x_2\right)^2 + \left(\frac{y_1 + y_2}{2} - y_2\right)^2} = \frac{1}{2}\sqrt{(x_1 - x_2)^2 + (y_1 - y_2)^2} = \frac{1}{2}\sqrt{(x_2 - x_1)^2 + (y_2 - y_1)^2} = d(P_1M).$$

(b)
$$d(P_1M) + d(P_2M) = \frac{1}{2}\sqrt{(x_2 - x_1)^2 + (y_2 - y_1)^2} + \frac{1}{2}\sqrt{(x_2 - x_1)^2 + (y_2 - y_1)^2} = \sqrt{(x_2 - x_1)^2 + (y_2 - y_1)^2} = d(P_1P_2).$$

Exercise Set 1.6

1. x-axis, no; y-axis, yes; origin, no
3. x-axis, yes; y-axis, no; origin, no
5. x-axis, no; y-axis, no; origin, yes
7. x-axis, no; y-axis, yes; origin, no
9. x-axis, no; y-axis, no; origin, no
11. x-axis, no; y-axis, yes; origin, no
13. x-axis, no; y-axis, no; origin, yes
15. x-axis, no; y-axis, no; origin, yes
17. x-axis, yes; y-axis, yes; origin, yes
19. x-axis, no; y-axis, yes; origin, no
21. x-axis, yes; y-axis, yes; origin, yes
23. x-axis, no; y-axis, no; origin, no
25. x-axis, no; y-axis, no; origin, yes **27.** Even **29.** Odd
31. Neither **33.** Odd **35.** Even **37.** Odd **39.** Even
41. Odd **43.** Neither **45.** Odd **47.** Neither **49.** Even
51. $3a^2 - 32a + 85$ **53.** $3a^2 + 22a + 40$ **55.** ◆
57. Odd **59.** Neither **61.** Odd
63. x-axis, yes; y-axis, yes; origin, yes
65. x-axis, yes; y-axis, no; origin, no
67.

$y_1 = \frac{1}{2}x^4 - 5x^3 + 2$,
$y_2 = \frac{1}{2}x^4 + 5x^3 + 2$

200
−15 15
−600

The graph of $f(x)$ is the reflection of the graph of $g(x)$ across the y-axis.

Exercise Set 1.7

1. Start with the graph of $g(x) = x^2$. Shift it up 1 unit.

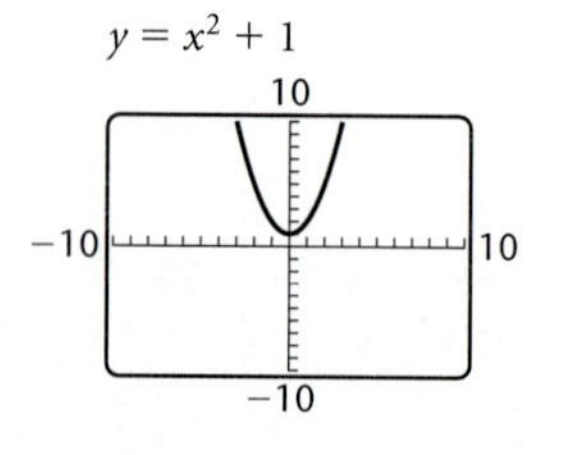

3. Start with the graph of $g(x) = 1/x$. Shift it up 3 units.

$y = \frac{1}{x} + 3$

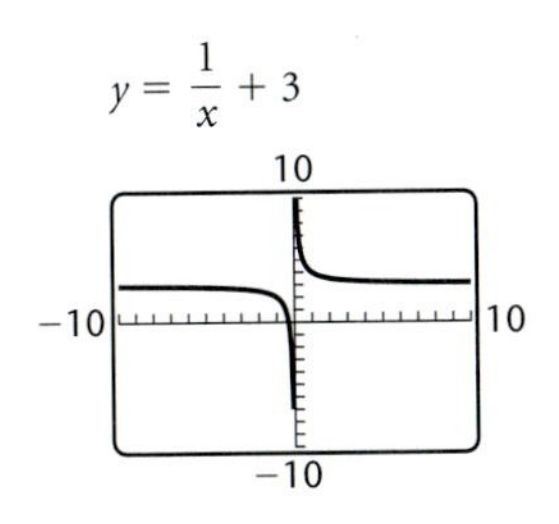

5. Start with the graph of $g(x) = \sqrt{x}$. Shift it down 5 units.

$y = \sqrt{x} - 5$

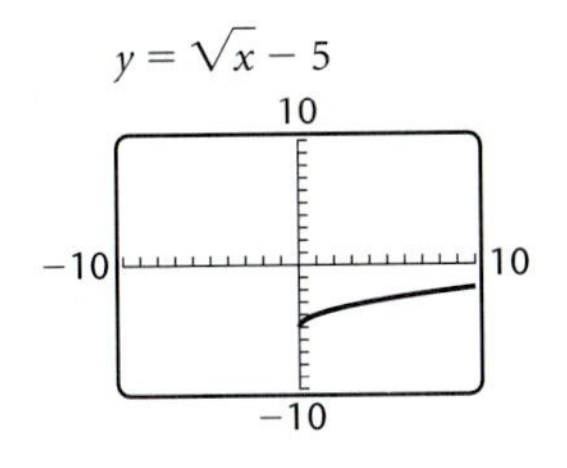

7. Start with the graph of $g(x) = x^2$. Reflect it across the x-axis.

$y = -x^2$

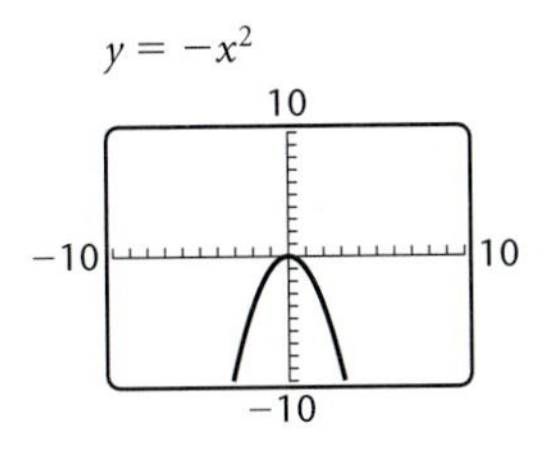

9. Start with the graph of $g(x) = |x|$. Shift it right 3 units.

$y = |x - 3|$

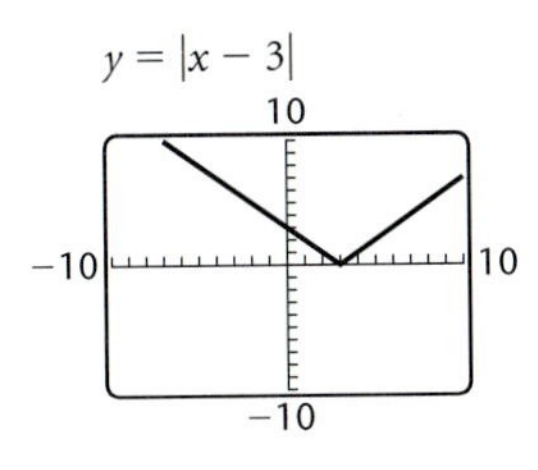

11. Start with the graph of $f(x) = x^3$. Shift it left 5 units.

$y = (x + 5)^3$

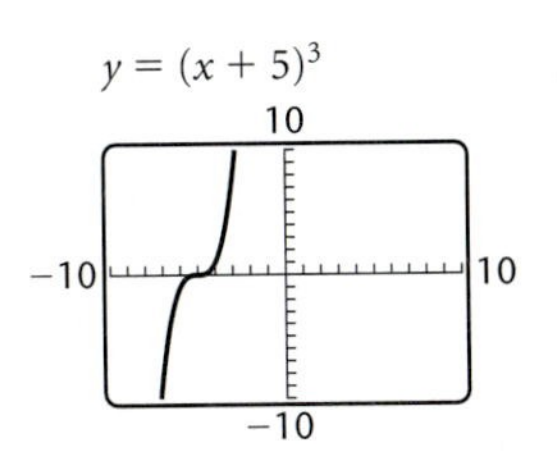

13. Start with the graph of $f(x) = x^2$. Shift it left 1 unit.

$y = (x + 1)^2$

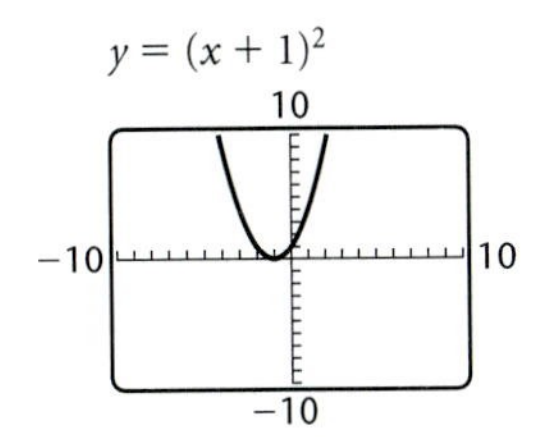

15. Start with the graph of $g(x) = \sqrt{x}$. Shrink it vertically by multiplying each y-coordinate by $\frac{1}{2}$.

$y = \frac{1}{2}\sqrt{x}$

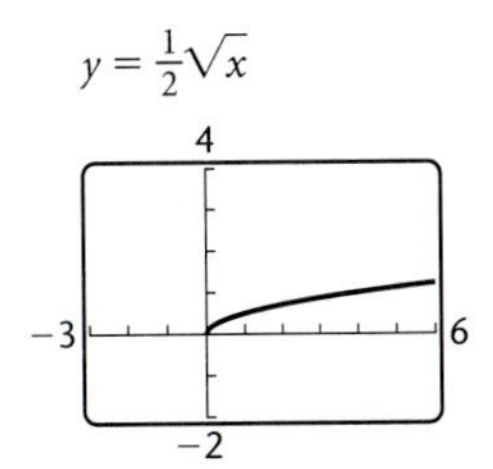

17. Start with the graph of $f(x) = 1/x$. Stretch it vertically by multiplying each y-coordinate by 2.

$y = \frac{2}{x}$

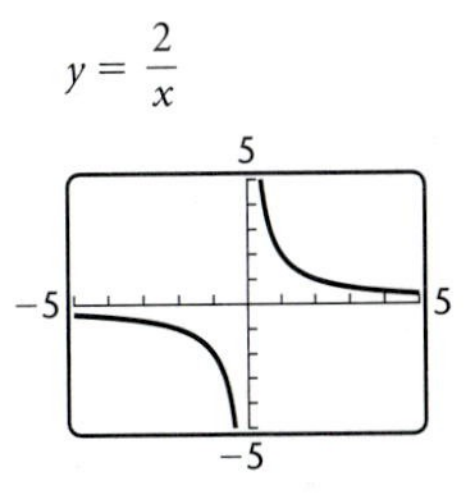

19. Start with the graph of $g(x) = \sqrt{x}$. Stretch it vertically by multiplying each y-coordinate by 3, then shift it down 5 units.

$y = 3\sqrt{x} - 5$

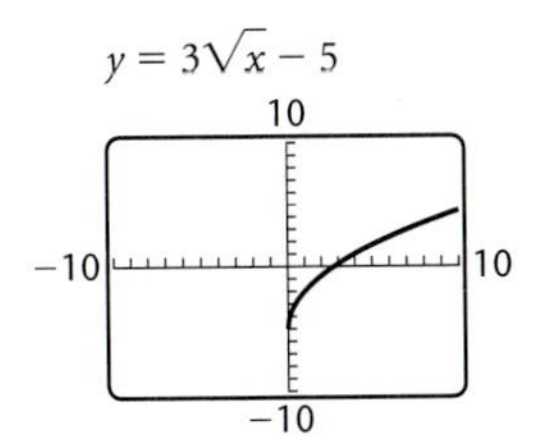

21. Start with the graph of $g(x) = x^2$. Shrink it vertically by multiplying each y-coordinate by $\frac{1}{2}$, then shift it right 3 units.

$y = \frac{1}{2}(x - 3)^2$

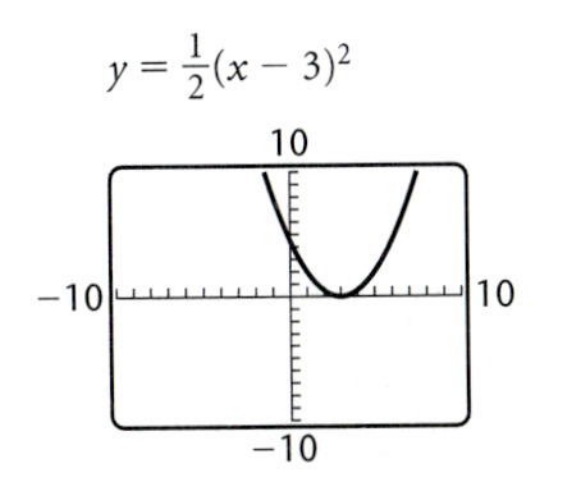

23. Start with the graph of $f(x) = |x|$. Stretch it horizontally by multiplying each x-coordinate by 3, then shift it down 4 units.

$y = \left|\frac{1}{3}x\right| - 4$

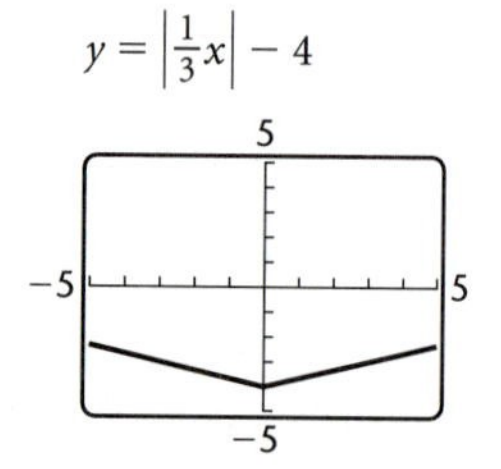

25. Start with the graph of $g(x) = x^2$. Shift it left 5 units and down 4 units.

$y = (x + 5)^2 - 4$

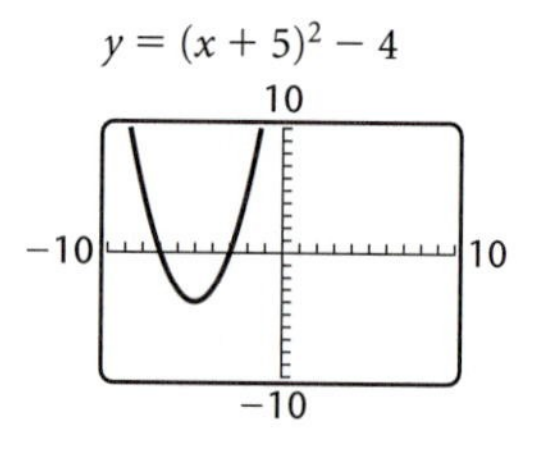

27. Start with the graph of $g(x) = x^2$. Shift it right 5 units, shrink it vertically by multiplying each y-coordinate by $\frac{1}{4}$, and reflect it across the x-axis.

$y = -\frac{1}{4}(x - 5)^2$

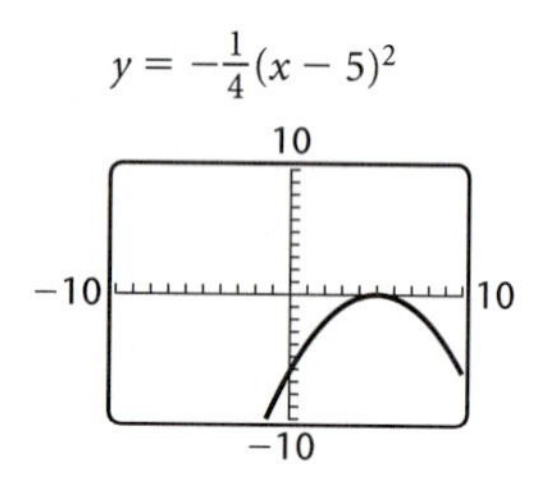

29. Start with the graph of $g(x) = 1/x$. Shift it left 3 units and up 2 units. Use DOT mode.

$y = \dfrac{1}{x + 3} + 2$

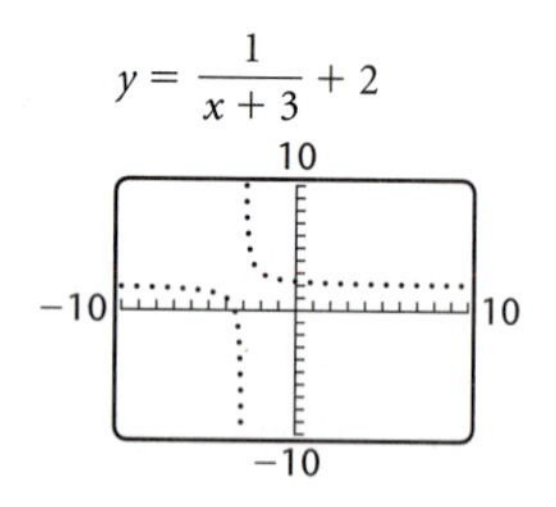

31. Start with the graph of $g(x) = x^3$. Shift it left 4 units and up 3 units.

$y = (x + 4)^3 + 3$

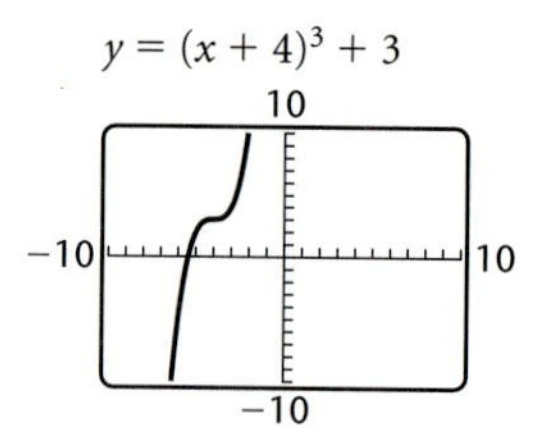

33. Start with the graph of $g(x) = \sqrt{x}$. Reflect it across the y-axis; then shift it up 5 units.

$y = \sqrt{-x} + 5$

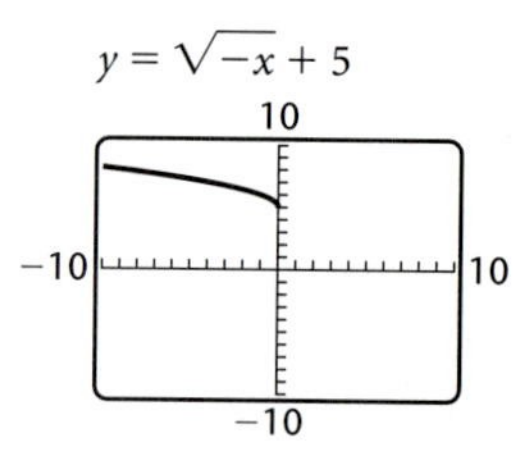

35. Start with the graph of $g(x) = x^2$. Shift it left 4 units, stretch it vertically by multiplying each y-coordinate by 3, and then shift it down 3 units.

$y = 3(x + 4)^2 - 3$

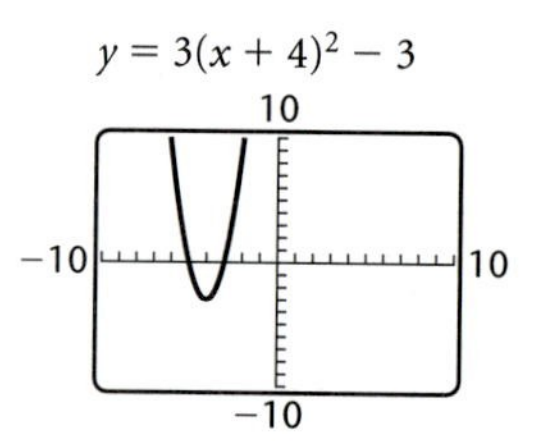

37. Start with the graph of $g(x) = x^3$. Shift it right 3 units, shrink it vertically by multiplying each y-coordinate by 0.43, and then shift it up 2.4 units.

$y = 0.43(x - 3)^3 + 2.4$

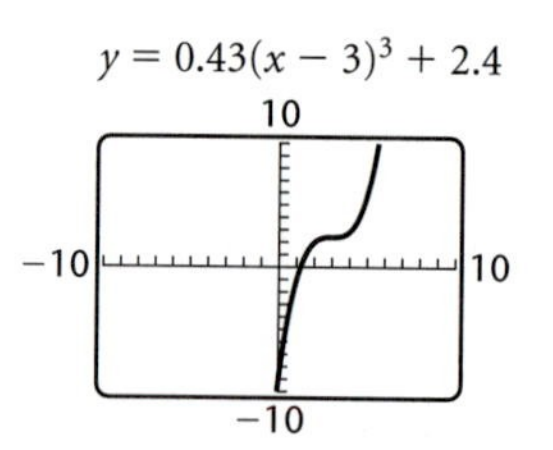

39. Start with the graph of $f(x) = x^2$. Shift it left 5.2 units, stretch it vertically by multiplying each y-coordinate by 2.8, and then shift it up 1.1 units.

$y = 2.8(x + 5.2)^2 + 1.1$

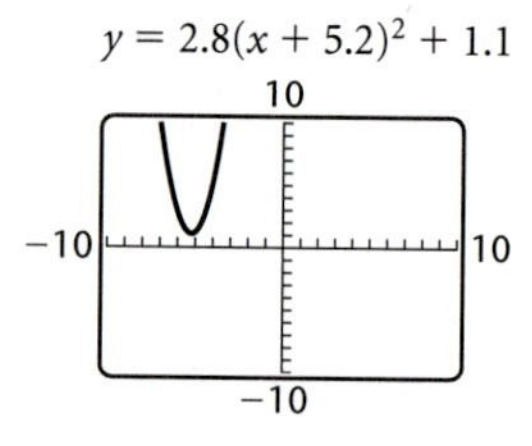

41. $f(x) = -(x - 8)^2$ **43.** $f(x) = |x + 7| + 2$

45. $f(x) = \frac{1}{2x} - 3$ **47.** $f(x) = -(x - 3)^2 + 4$

49. $f(x) = \sqrt{-(x + 2)} - 1$ **51.** $f(x) = 0.83(x + 4)^3$

53.

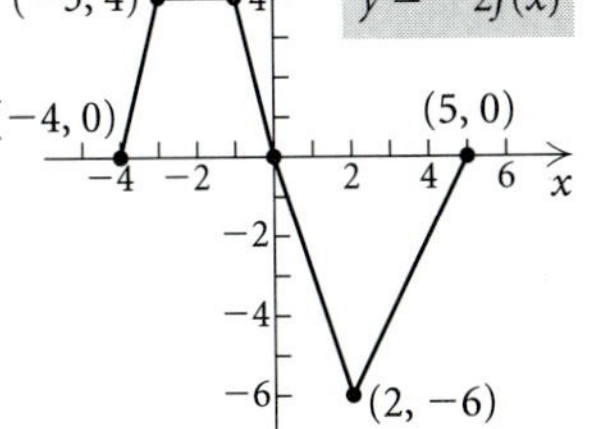

55.

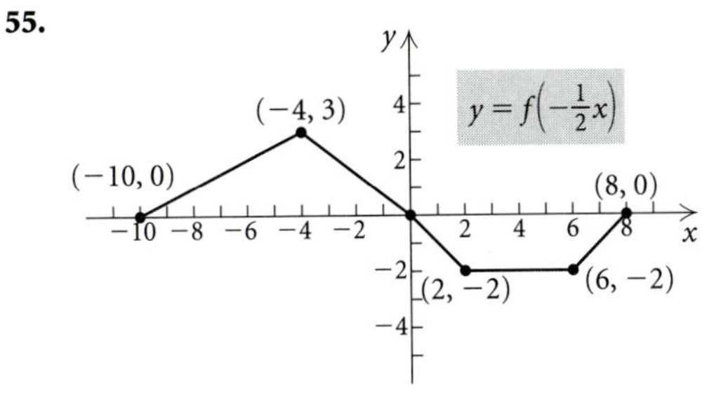

57.

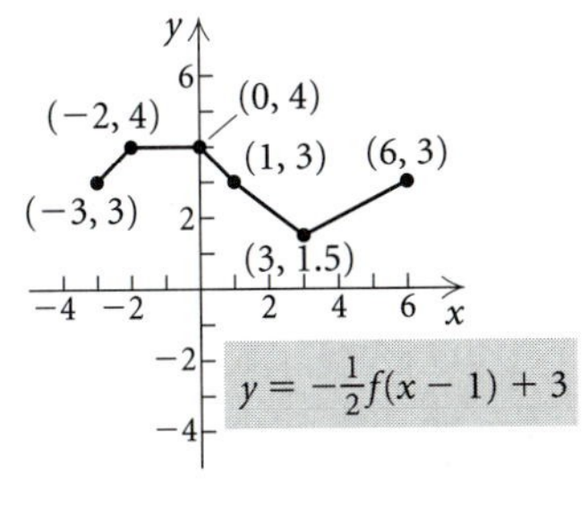

59. (f) **61.** (f) **63.** (d) **65.** (c) **67.** Use a grapher.
69. Use a grapher.
71. $f(-x) = 2(-x)^4 - 35(-x)^3 + 3(-x) - 5 = 2x^4 + 35x^3 - 3x - 5 = g(x)$ **73.** $g(x) = x^3 - 3x^2 + 2$
75. $k(x) = (x + 1)^3 - 3(x + 1)^2$ **77.** 6 **79.** 16
81.  **83.** $g(x) = 7 - \sqrt{x - 4}$
85. Start with the graph of $g(x) = \text{INT}(x)$. Shift it right $\frac{1}{2}$ unit. Domain: all real numbers; range: all integers.
87. Start with the graph of $g(x) = \text{INT}(x)$. Shift it left 2 units, stretch it vertically by multiplying each y-coordinate by 3, and shift it up 1 unit. Domain: all real numbers; range: all integers that are 1 more than a multiple of 3.
89. Start with the graph of $f(x) = 1/x$. Reflect across the x-axis the portion of the graph for which the y-coordinates are negative. Domain: all real numbers except 0; range: $(0, \infty)$
91. Start with the graph of $g(x) = |x|$. Shift it down 5 units; then reflect across the x-axis the portion of the graph for which the y-coordinates are negative. Domain: all real numbers; range: $[0, \infty)$ **93.** (3, 8); (3, 6); $\left(\frac{3}{2}, 4\right)$

EXERCISE SET 1.8

1. 33 **3.** 78 **5.** 5 **7.** 2 **9.** 0 **11.** 0
13. (a) Domain of f, all real numbers; domain of g, $[-4, \infty)$; domain of $f + g$, $f - g$, and fg, $[-4, \infty)$; domain of ff, all real numbers; domain of f/g, $(-4, \infty)$; domain of g/f, $[-4, 3) \cup (3, \infty)$; domain of $f \circ g$, $[-4, \infty)$; domain of $g \circ f$, $[-1, \infty)$; **(b)** $(f + g)(x) = x - 3 + \sqrt{x + 4}$; $(f - g)(x) = x - 3 - \sqrt{x + 4}$; $(fg)(x) = (x - 3)\sqrt{x + 4}$; $(ff)(x) = x^2 - 6x + 9$; $(f/g)(x) = (x - 3)/\sqrt{x + 4}$; $(g/f)(x) = \sqrt{x + 4}/(x - 3)$; $(f \circ g)(x) = \sqrt{x + 4} - 3$; $(g \circ f)(x) = \sqrt{x + 1}$
15. (a) Domain of f, g, $f + g$, $f - g$, fg, and ff, all real numbers; domain of f/g, all real numbers except -3 and $\frac{1}{2}$; domain of g/f, all real numbers except 0; domain of $f \circ g$ and $g \circ f$, all real numbers;
(b) $(f + g)(x) = x^3 + 2x^2 + 5x - 3$; $(f - g)(x) = x^3 - 2x^2 - 5x + 3$; $(fg)(x) = 2x^5 + 5x^4 - 3x^3$; $(ff)(x) = x^6$; $(f/g)(x) = x^3/(2x^2 + 5x - 3)$; $(g/f)(x) = (2x^2 + 5x - 3)/x^3$; $(f \circ g)(x) = (2x^2 + 5x - 3)^3$; $(g \circ f)(x) = 2x^6 + 5x^3 - 3$
17. $(f + g)(x) = 2x^2 - 2$; $(f - g)(x) = -6$; $(fg)(x) = x^4 - 2x^2 - 8$; $(f/g)(x) = (x^2 - 4)/(x^2 + 2)$; $(f \circ g)(x) = x^4 + 4x^2$
19. $(f + g)(x) = \sqrt{x - 7} + x^2 - 25$; $(f - g)(x) = \sqrt{x - 7} - x^2 + 25$; $(fg)(x) = \sqrt{x - 7}\,(x^2 - 25)$; $(f/g)(x) = \sqrt{x - 7}/(x^2 - 25)$; $(f \circ g)(x) = \sqrt{x^2 - 32}$
21. Domain of F: [0, 9]; domain of G: [3, 10]; domain of $F + G$: [3, 9]
23. $\{x \mid 3 \le x \le 9 \textit{ and } x \ne 6 \textit{ and } x \ne 8\}$, or $[3, 6) \cup (6, 8) \cup (8, 9]$
25.

G − F

27. $(f \circ g)(x) = (g \circ f)(x) = x$
29. $(f \circ g)(x) = (g \circ f)(x) = x$
31. $(f \circ g)(x) = 20$; $(g \circ f)(x) = 0.05$
33. $(f \circ g)(x) = |x|$; $(g \circ f)(x) = x$
35. $(f \circ g)(x) = (g \circ f)(x) = x$
37. $(f \circ g)(x) = x^3 - 2x^2 - 4x + 6$; $(g \circ f)(x) = x^3 - 5x^2 + 3x + 8$

39. $f(x) = x^5$; $g(x) = 4 + 3x$

41. $f(x) = \dfrac{1}{x}$; $g(x) = (x - 2)^4$

43. $f(x) = \dfrac{x - 1}{x + 1}$; $g(x) = x^3$

45. $f(x) = x^6$; $g(x) = \dfrac{2 + x^3}{2 - x^3}$

47. $f(x) = \sqrt{x}$; $g(x) = \dfrac{x - 5}{x + 2}$

49. $f(x) = x^3 - 5x^2 + 3x - 1$; $g(x) = x + 2$

51. **(a)** $P(x) = -0.4x^2 + 57x - 13$; **(b)** $R(100) = 2000$; $C(100) = 313$; $P(100) = 1687$; **(c)** Use a grapher.

53. **(a)** $r(t) = 3t$; **(b)** $A(r) = \pi r^2$; **(c)** $(A \circ r)(t) = 9\pi t^2$; the area of the ripple in terms of time t **55.** $[-1, 2]$

57. -21 **59.** ◈

61. **63.**

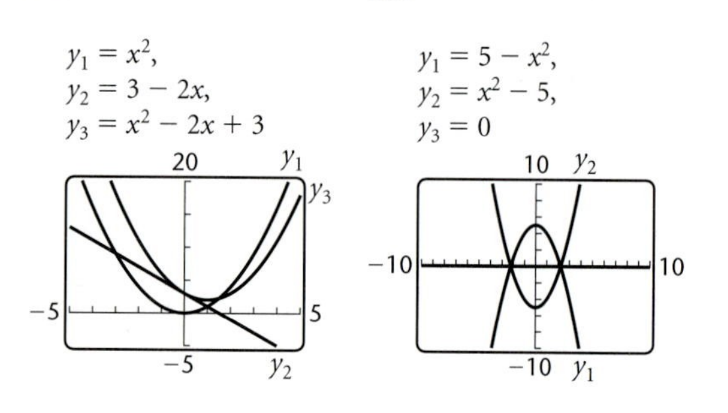

65. Domain of $F \circ G$, $[3, 10]$; domain of $G \circ G$, $[5, 8]$; domain of $F \circ F$, $[0, 6] \cup [8, 9]$ **67.** -2 **69.** $\dfrac{x - 1}{x}$; x

71. True **73.** True

75. $E(-x) = \dfrac{f(-x) + f(-(-x))}{2} = \dfrac{f(-x) + f(x)}{2} = E(x)$

77. **(a)** $E(x) + O(x) = \dfrac{f(x) + f(-x)}{2} + \dfrac{f(x) - f(-x)}{2} = \dfrac{2f(x)}{2} = f(x)$; **(b)** $f(x) = \dfrac{-22x^2 + \sqrt{x} + \sqrt{-x} - 20}{2} + \dfrac{8x^3 + \sqrt{x} - \sqrt{-x}}{2}$

Review Exercises, Chapter 1

1. [1.1] Not a function; domain: $\{3, 5, 7\}$; range: $\{1, 3, 5, 7\}$

2. [1.1] Function; domain: $\{-2, 0, 1, 2, 7\}$; range: $\{-7, -4, -2, 2, 7\}$ **3.** [1.1] No **4.** [1.1] Yes **5.** [1.1] No

6. [1.1] Yes **7.** [1.1] All real numbers **8.** [1.1] $\left(-\infty, \frac{7}{3}\right]$

9. [1.1] $\{x \mid x \neq 5 \text{ and } x \neq 1\}$

10. [1.1] $\{x \mid x \neq -4 \text{ and } x \neq 4\}$

11. [1.1] Domain: $[-5, 3]$; range: $[0, 4]$

12. [1.1] Domain: all real numbers; range: $[-5, \infty)$

13. [1.1] Domain: all real numbers; range: all real numbers

14. [1.1] Domain: all real numbers; range: $[-19, \infty)$

15. [1.1] **(a)** -3; **(b)** 9; **(c)** $a^2 - 3a - 1$; **(d)** $2x + h - 1$

16. [1.1] No **17.** [1.1] Yes **18.** [1.1] No **19.** [1.1] Yes

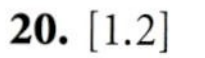

20. [1.2]

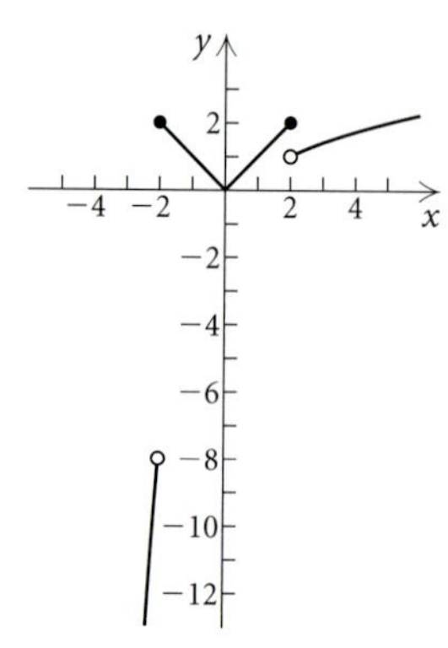

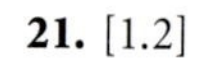

21. [1.2]

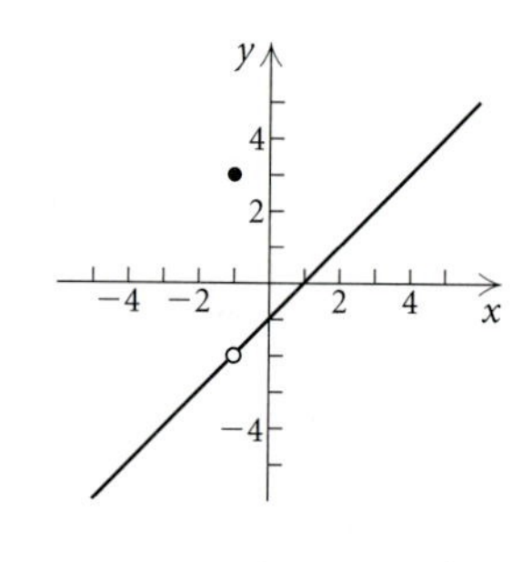

22. [1.2]

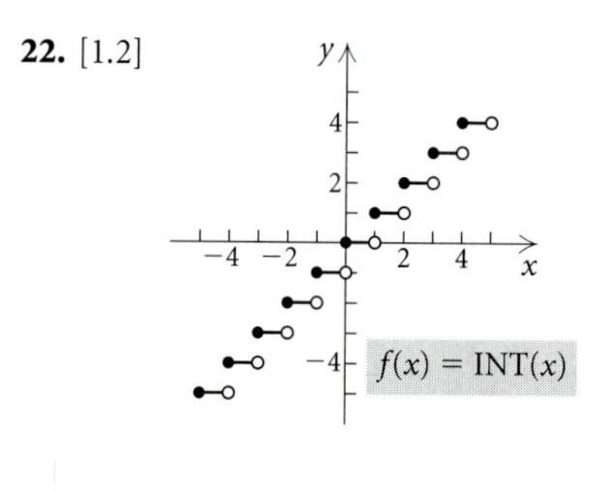

23. [1.2]

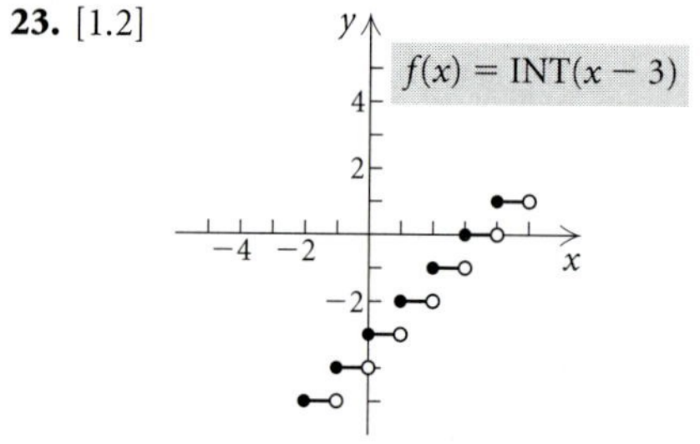

24. [1.1] **(a)** The domain of y_1 does not include the intervals $(-\infty, -5.1]$ and $[3.1, \infty)$. The domain of y_2 appears to be all real numbers. The range of y_1 appears to be $[0, 4]$. The range of y_2 appears to be $[-5, 5]$. **(b)** The largest output of y_1 appears to be 4, and the largest output of y_2 to be 5. The smallest output of y_1 seems to be 0, and the smallest output of y_2 to be -5.

25. [1.2] **(a)** -7.029; **(b)** increasing: $(-\infty, -0.860]$, $[6.339, \infty)$; decreasing: $[-0.860, 6.339]$; **(c)** relative maximum: $(-0.860, 710.094)$; relative minimum: $(6.339, 504.836)$

26. [1.2] **(a)** $-5.732, -2.268$; **(b)** increasing: $(-\infty, -4]$; decreasing: $[-4, \infty)$; **(c)** relative maximum: $(-4, 15)$

27. [1.2] **(a)** $-5, 0, 3$; **(b)** increasing: $[-3.589, 2.089]$; decreasing: $[-5, -3.589]$, $[2.089, 3]$; **(c)** relative maximum: $(2.089, 4.247)$; relative minimum: $(-3.589, -8.755)$

28. [1.2] $A(x) = 2x\sqrt{4 - x^2}$

29. [1.2] **(a)** $A(x) = x^2 + \dfrac{432}{x}$; **(b)** $(0, \infty)$;

(c)

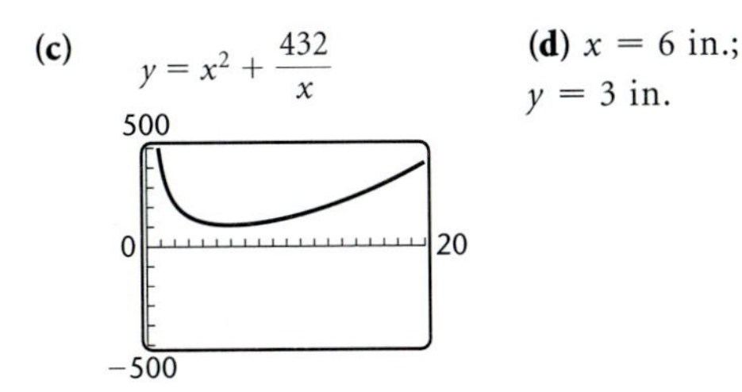

(d) $x = 6$ in.; $y = 3$ in.

30. [1.3] **(a)** Yes; **(b)** no; **(c)** no, strictly speaking, but data might be modeled by a linear regression function.
31. [1.3] **(a)** Yes; **(b)** yes; **(c)** yes **32.** [1.3] -2; $(0, -7)$
33. [1.3] $y + 4 = -\frac{2}{3}(x - 0)$ **34.** [1.3] $y + 1 = 3(x + 2)$
35. [1.3] $y - 1 = \frac{1}{3}(x - 4)$, or $y + 1 = \frac{1}{3}(x + 2)$
36. [1.3] Parallel **37.** [1.3] Neither
38. [1.3] Perpendicular **39.** [1.3] $y = -\frac{2}{3}x - \frac{1}{3}$
40. [1.3] $y = \frac{3}{2}x - \frac{5}{2}$
41. [1.3] $C(t) = 25 + 20t$; $y = 25 + 20x$ (window: 0 to 20, 0 to 500) $145

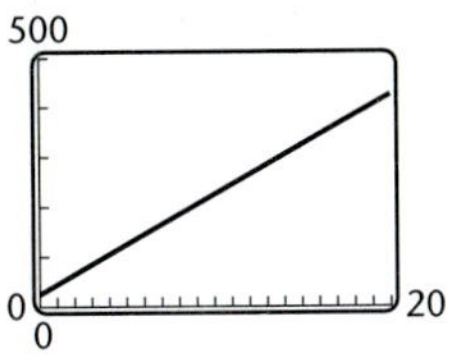

42. [1.3] **(a)** 70°C, 220°C, 10,020°C;
(b)

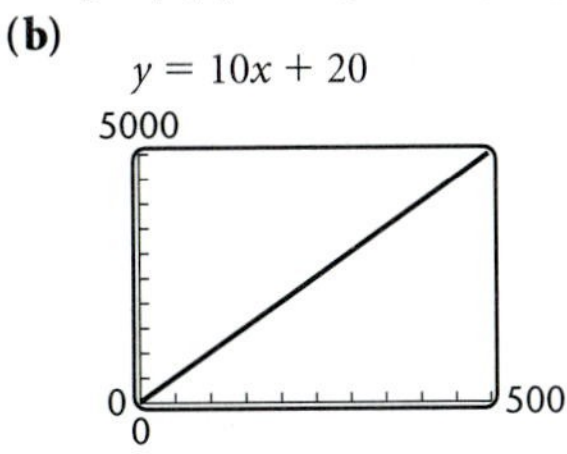

(c) [0, 5600]
43. [1.4] **(a)** $y = 2.4x + 17.5$, $36.7 billion, $41.5 billion, $65.5 billion; **(b)** $r = 0.9938$, a good fit
44. [1.5] $\sqrt{34} \approx 5.831$ **45.** [1.5] $\left(\frac{1}{2}, \frac{11}{2}\right)$
46. [1.5] $(x + 2)^2 + (y - 6)^2 = 13$ **47.** [1.5] $(-1, 3)$; 4
48. [1.5] $(x - 2)^2 + (y - 4)^2 = 26$
49. [1.6] x-axis, yes; y-axis, yes; origin, yes
50. [1.6] x-axis, yes; y-axis, yes; origin, yes
51. [1.6] x-axis, no; y-axis, no; origin, no
52. [1.6] x-axis, no; y-axis, yes; origin, no
53. [1.6] x-axis, no; y-axis, no; origin, yes
54. [1.6] x-axis, no; y-axis, yes; origin, no **55.** [1.6] Even
56. [1.6] Even **57.** [1.6] Odd **58.** [1.6] Even
59. [1.6] Even **60.** [1.6] Neither **61.** [1.6] Odd
62. [1.6] Even **63.** [1.6] Even **64.** [1.6] Odd
65. [1.7] $f(x) = (x + 3)^2$
66. [1.7] $f(x) = -\sqrt{x - 3} + 4$
67. [1.7] $f(x) = 2|x - 3|$

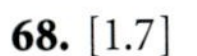
68. [1.7]
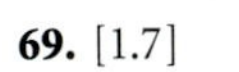
69. [1.7]

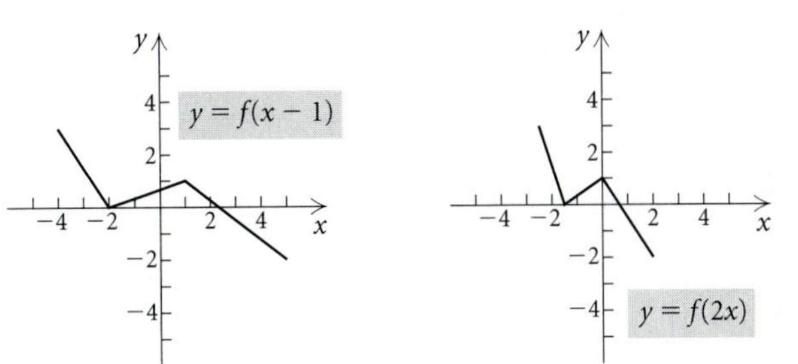

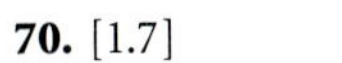
70. [1.7]
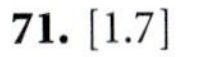
71. [1.7]

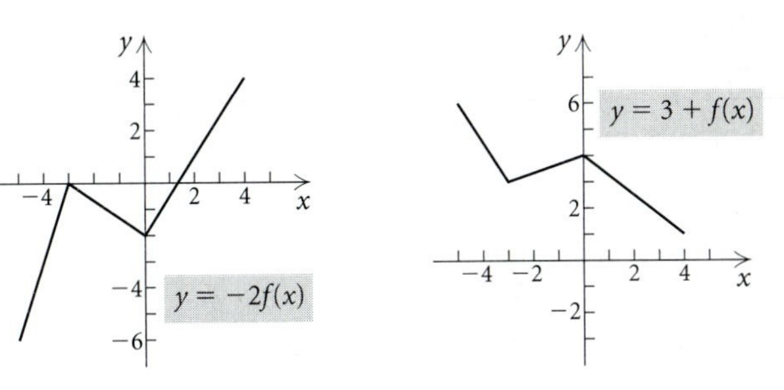

72. [1.8] **(a)** Domain of f, $\{x | x \neq 0\}$; domain of g, all real numbers; domain of $f + g$, $f - g$, and fg, $\{x | x \neq 0\}$; domain of f/g, $\left\{x | x \neq 0 \textit{ and } x \neq \frac{3}{2}\right\}$; domain of $f \circ g$, $\left\{x | x \neq \frac{3}{2}\right\}$; domain of $g \circ f$, $\{x | x \neq 0\}$;
(b) $(f + g)(x) = \frac{4}{x^2} + 3 - 2x$; $(f - g)(x) = \frac{4}{x^2} - 3 + 2x$; $fg(x) = \frac{12}{x^2} - \frac{8}{x}$; $(f/g)(x) = \frac{4}{x^2(3 - 2x)}$; $(f \circ g)(x) = \frac{4}{(3 - 2x)^2}$; $(g \circ f)(x) = 3 - \frac{8}{x^2}$
73. [1.8] **(a)** Domain of f, g, $f + g$, $f - g$, and fg, all real numbers; domain of f/g, $\left\{x | x \neq \frac{1}{2}\right\}$; domain of $f \circ g$ and $g \circ f$, all real numbers; **(b)** $(f + g)(x) = 3x^2 + 6x - 1$; $(f - g)(x) = 3x^2 + 2x + 1$; $fg(x) = 6x^3 + 5x^2 - 4x$; $(f/g)(x) = \frac{3x^2 + 4x}{2x - 1}$; $(f \circ g)(x) = 12x^2 - 4x - 1$; $(g \circ f)(x) = 6x^2 + 8x - 1$
74. [1.8] $P(x) = -0.5x^2 + 105x - 6$
75. [1.8] $f(x) = \sqrt{x}$, $g(x) = 5x + 2$. Answers may vary.
76. [1.8] $f(x) = 4x^2 + 9$, $g(x) = 5x - 1$. Answers may vary.
77. [1.1], [1.7] ◈ **(a)** $4x^3 - 2x + 9$; **(b)** $4x^3 + 24x^2 + 46x + 35$; **(c)** $4x^3 - 2x + 42$. **(a)** Adds 2 to each function value; **(b)** adds 2 to each input before finding a function value; **(c)** adds the output for 2 to the output for x

78. [1.7] ◆ In the graph of $y = f(cx)$, the constant c stretches or shrinks the graph of $y = f(x)$ horizontally. The constant c in $y = cf(x)$ stretches or shrinks the graph of $y = f(x)$ vertically. For $y = f(cx)$, the x-coordinates of $y = f(x)$ are divided by c; for $y = cf(x)$, the y-coordinates of $y = f(x)$ are multiplied by c.
79. [1.7] ◆ **(a)** To draw the graph of y_2 from the graph of y_1, reflect across the x-axis the portions of the graph for which the y-coordinates are negative. **(b)** To draw the graph of y_2 from the graph of y_1, draw the portion of the graph of y_1 to the right of the y-axis; then draw its reflection across the y-axis. **80.** [1.1] $\{x \mid x < 0\}$
81. [1.1] $\{x \mid x \neq -3 \text{ and } x \neq 0 \text{ and } x \neq 3\}$
82. [1.6] Let $f(x)$ and $g(x)$ be odd functions. Then by definition, $f(-x) = -f(x)$, or $f(x) = -f(-x)$, and $g(-x) = -g(x)$, or $g(x) = -g(-x)$. Thus, $(f + g)(x) = f(x) + g(x) = -f(-x) + [-g(-x)] = -[f(-x) + g(-x)] = -(f + g)(-x)$ and $f + g$ is odd.
83. [1.7] Reflect the graph of $y = f(x)$ across the x-axis and then across the y-axis.

Chapter 2

Exercise Set 2.1

1. (a) -2, 3; **(b)** none; **(c)** -6.25 at $x = 0.5$; **(d)** domain: all real numbers; range: $[-6.25, \infty)$
3. (a) -5, -1, 1; **(b)** 16.901 at $x = -3.431$; **(c)** -5.049 at $x = 0.097$; **(d)** domain: all real numbers; range: all real numbers
5. (a) -1.831, -0.856, 3.188; **(b)** 26.864 at $x = 1.703$; **(c)** -2.160 at $x = -1.370$; **(d)** domain: all real numbers; range: all real numbers
7. (a) -4.796, 0, 4.796; **(b)** 0.042 at $x = -2.769$; **(c)** -0.042 at $x = 2.769$; **(d)** domain: all real numbers; range: all real numbers

9.

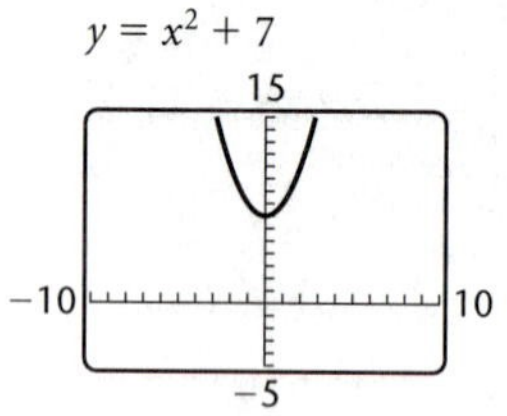

$y = x^2 + 7$
15
-10 10
-5

11.

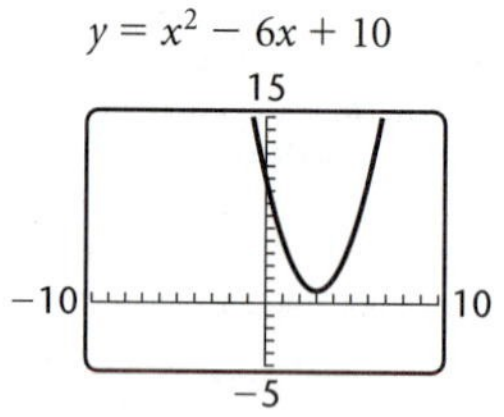

$y = x^2 - 6x + 10$
15
-10 10
-5

13.

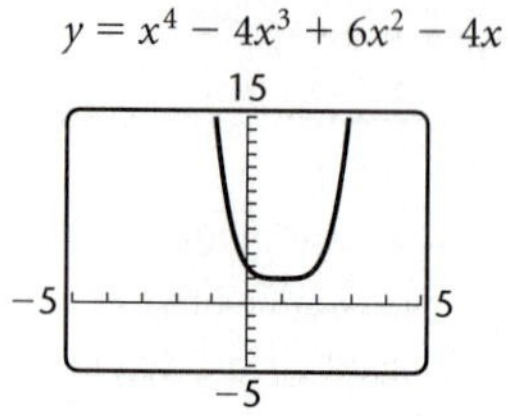

$y = x^4 - 4x^3 + 6x^2 - 4x + 3$
15
-5 5
-5

15. $\sqrt{17}i$ **17.** $7i$ **19.** $-9i$ **21.** $6 - 2\sqrt{21}i$
23. $-2\sqrt{19} + 5\sqrt{5}i$ **25.** $-\sqrt{55}$ **27.** $\frac{5}{4}i$ **29.** $2 + 11i$
31. $5 - 12i$ **33.** $5 + 9i$ **35.** $5 + 4i$ **37.** $5 + 7i$
39. $13 - i$ **41.** $-11 + 16i$ **43.** $35 + 14i$ **45.** $-14 + 23i$
47. 41 **49.** $12 + 16i$ **51.** $-45 - 28i$
53. $\dfrac{-4\sqrt{3} + 10}{41} + \dfrac{5\sqrt{3} + 8}{41}i$ **55.** $-\dfrac{14}{13} + \dfrac{5}{13}i$
57. $\dfrac{15}{146} + \dfrac{33}{146}i$ **59.** $-\dfrac{1}{2} + \dfrac{1}{2}i$ **61.** $-\dfrac{1}{2} - \dfrac{13}{2}i$
63. $-i$ **65.** $-i$ **67.** 1 **69.** i **71.** 625
73. $(x + 4)^2$ **75.** $(x - 5)^2$ **77.** ◆ **79.** $\frac{12}{5} - \frac{1}{5}i$
81. $\frac{8}{29} + \frac{9}{29}i$ **83.** True **85.** True **87.** $a^2 + b^2$

Exercise Set 2.2

1. $\{2, 6\}$ **3.** $\left\{\frac{1}{7}, 6\right\}$ **5.** $\{-9, 9\}$ **7.** $\left\{-\frac{10}{3}, \frac{10}{3}\right\}$
9. $\{-\sqrt{10}i, \sqrt{10}i\}$ **11.** $\{-12, 6\}$ **13.** $\{2, 4\}$ **15.** $\{-7, 1\}$
17. $\{4 \pm \sqrt{7}\}$ **19.** $\{-4 \pm 3i\}$ **21.** $\left\{-2, \frac{1}{3}\right\}$
23. $\left\{\dfrac{3 \pm \sqrt{13}}{4}\right\}$ **25.** $\left\{\dfrac{5 \pm \sqrt{57}}{4}\right\}$ **27.** $\{-3, 5\}$
29. $\left\{-1, \dfrac{2}{5}\right\}$ **31.** $\left\{\dfrac{5 \pm \sqrt{7}}{3}\right\}$ **33.** $\left\{-\dfrac{1}{2} \pm \dfrac{\sqrt{7}}{2}i\right\}$
35. $\left\{\dfrac{4 \pm \sqrt{31}}{5}\right\}$ **37.** $\left\{\dfrac{5}{6} \pm \dfrac{\sqrt{23}}{6}i\right\}$ **39.** No **41.** Yes
43. No **45.** $\{-0.702, 5.702\}$ **47.** $\{-1.535, 0.869\}$
49. $\left\{\frac{5}{2}, 3\right\}$ **51.** $\{-8, 64\}$ **53.** $\{-2, -1, 1, 2\}$
55. $\left\{-\frac{3}{2}, -1, \frac{1}{2}, 1\right\}$ **57.** $\{16\}$
59. (a) $(4, -4)$; **(b)** $x = 4$; **(c)** minimum value: -4
61. (a) $\left(\frac{7}{2}, -\frac{1}{4}\right)$; **(b)** $x = \frac{7}{2}$; **(c)** minimum value: $-\frac{1}{4}$
63. (a) $\left(-\frac{3}{2}, \frac{7}{2}\right)$; **(b)** $x = -\frac{3}{2}$; **(c)** minimum value: $\frac{7}{2}$
65. (a) $\left(\frac{1}{2}, \frac{3}{2}\right)$; **(b)** $x = \frac{1}{2}$; **(c)** maximum value: $\frac{3}{2}$
67. (f) **69.** (b) **71.** (h) **73.** (c)
75. (a) $(3, -4)$; **(b)** minimum value: -4; **(c)** $[-4, \infty)$; **(d)** increasing: $[3, \infty)$; decreasing: $(-\infty, 3]$
77. (a) $(-1, -18)$; **(b)** minimum value: -18; **(c)** $[-18, \infty)$; **(d)** increasing: $[-1, \infty)$; decreasing: $(-\infty, -1]$
79. (a) $\left(5, \frac{9}{2}\right)$; **(b)** maximum value: $\frac{9}{2}$; **(c)** $\left(-\infty, \frac{9}{2}\right]$; **(d)** increasing: $(-\infty, 5]$; decreasing: $[5, \infty)$
81. (a) $(-1, 2)$; **(b)** minimum value: 2; **(c)** $[2, \infty)$; **(d)** increasing: $[-1, \infty)$; decreasing: $(-\infty, -1]$
83. $\left\{-\frac{21}{4}, -1\right\}$ **85.** 6 mi **87.** ◆
89. $\left\{\dfrac{-1 \pm \sqrt{4\sqrt{2} + 1}}{2}\right\}$ **91.** $\{3 \pm \sqrt{5}\}$
93. $\left\{\dfrac{-5 + \sqrt{61}}{18}\right\}$ **95.** $\{19\}$
97. $\left\{\dfrac{1}{2} - \dfrac{\sqrt{7}}{2}i, \dfrac{1}{2} + \dfrac{\sqrt{7}}{2}i, -2 - \sqrt{2}, -2 + \sqrt{2}\right\}$

99. (a) 2; **(b)** $\frac{11}{2}$ **101. (a)** 2; **(b)** $1 - i$ **103.** -236.25
105.

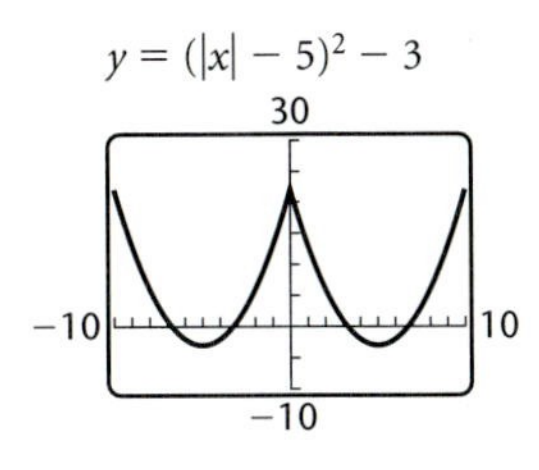

EXERCISE SET 2.3

1. 0.866 sec **3.** \$3240 **5. (a)** 18.75%; **(b)** 10%
7. 3.5 in. **9.** $b = h = 10$ cm **11.** About 350.6 ft
13. \$797 when $x = 40$ **15.** 4800 yd^2 **17.** (b) **19.** (c)
21. (a) **23.** (b) **25.** (b)
27. (a) Yes; **(b)** no; **(c)** third differences are all -384;
(d) cubic; **(e)** $f(x) = -x^3 - x^2 + x + 5$; **(f)** $f(17) = -5180$; $f(21) = -9676$
29. (a) Yes; **(b)** no; **(c)** fourth differences are all 1944;
(d) quartic; **(e)** $f(x) = x^4 - 10x^2 + x - 1$; $f(13) = 26{,}883$; $x = -2.825, -1.500, 1.707$, or 2.618 when $y = -19.94$
31. (a) $f(x) = -0.8630952381x^2 + 12.125x + 36.76785714$;
(b) 77.3 yr, 77.5 yr
33. (a) $f(x) = -2.775187775x^3 + 45.77200577x^2 - 190.2173752x + 2084.949495$;
(b) $y = -2.775187775x^3 + 45.77200577x^2 - 190.2173752x + 2084.949495$

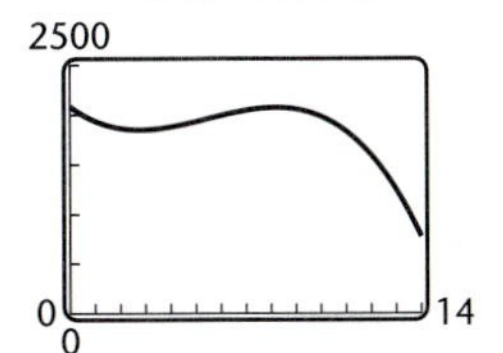

(c) $-\$2694$; **(d)** $f(x) = 1.586902681x^4 - 44.03465747x^3 + 400.5581051x^2 - 1316.011477x + 3075.176768$;
(e) $y = 1.586902681x^4 - 44.03465747x^3 + 400.5581051x^2 - 1316.011477x + 3075.176768$

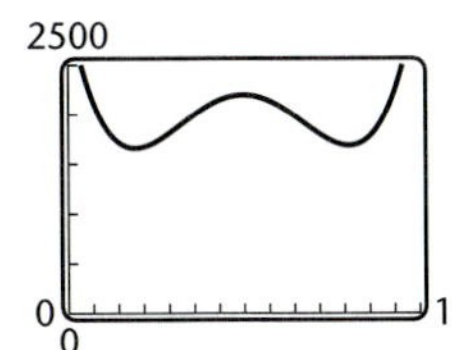

(f) \$18,944
35. $x^4 + 5x^3 + 2x - 4$ **37.** $5x^4 - 17x^3 + 11x^2 - 16x + 3$
39. ◈ **41.** 7%
43. (a) $f(x) = 0.8586956522x^3 - 0.8452380952x^2 - 0.7546583851x + 31.10766046$; $f(-4) \approx -34.354$; $f(4) \approx 69.522$; x is about -1.950 when y is 23;
(b) $g(x) = 1.558569182x^4 + 1.807389937x^3 - 15.7629717x^2 - 9.835017969x + 42.25965858$; $g(-4) \approx 112.713$; $g(4) \approx 265.379$; x is about -3.301 or -1.512 or 0.899 or 2.754 when $y = 23$
45. (a)

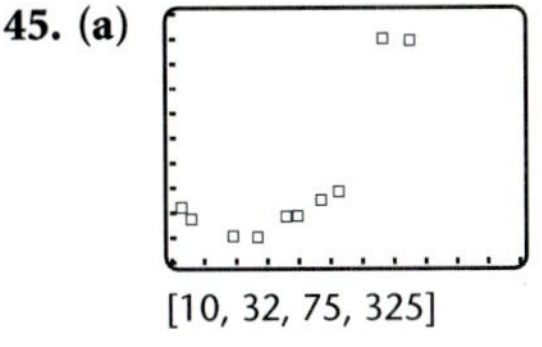

(b) Linear seems inappropriate; quadratic, cubic, quartic seem possible. **(c)** *Quadratic:* $y = 2.031904548x^2 - 59.04179598x + 527.2818092$; *cubic:* $y = -0.0273804413x^3 + 3.488800304x^2 - 83.76947384x + 660.2911579$; *quartic:* $y = -0.0403607597x^4 + 2.842908639x^3 - 70.96865626x^2 + 749.2437238x - 2721.562456$
(d) Quadratic

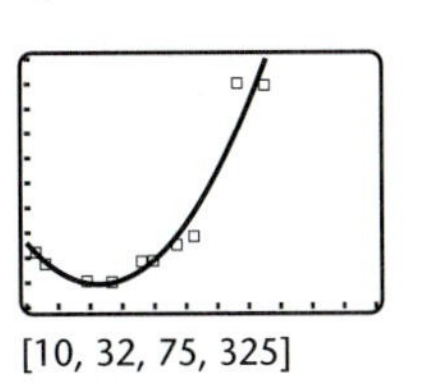

Cubic

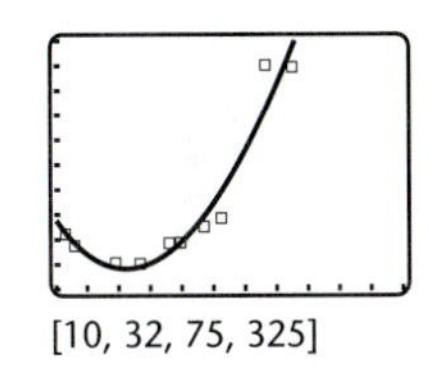

Quartic

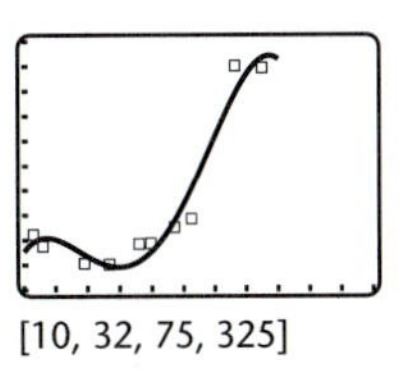

The quadratic and cubic equations seem to fit best. They indicate that, for an advertising budget of about \$16 million or greater, box office revenue increases as the budget increases. The quartic equation indicates that revenue declines sharply as the advertising budget increases beyond about \$25.5 million.
(e) \$584.74 million, \$547.86 million, $-\$49.72$ million

EXERCISE SET 2.4

1. Yes; no; no **3. (a)**
5. $P(x) = (x - 2)(x^2 + 8x + 15) + 0$
7. $P(x) = (x - 3)(x^2 + 9x + 26) + 48$
9. $P(x) = (x + 2)(x^2 - 2x + 4) - 16$
11. $P(x) = (x^2 + 4)(x^2 + 5) + 0$
13. $P(x) = (2x^2 - x + 1) \cdot \left(\frac{5}{2}x^5 + \frac{5}{4}x^4 - \frac{5}{8}x^3 - \frac{39}{16}x^2 - \frac{29}{32}x + \frac{113}{64}\right) + \frac{171x - 305}{64}$
15. $Q(x) = 2x^3 + x^2 - 3x + 10$, $R(x) = -42$
17. $Q(x) = x^2 - 4x + 8$, $R(x) = -24$
19. $Q(x) = x^3 + x^2 + x + 1$, $R(x) = 0$
21. $Q(x) = 2x^3 + x^2 + \frac{7}{2}x + \frac{7}{4}$, $R(x) = -\frac{1}{8}$

23. $Q(x) = x^3 + x^2y + xy^2 + y^3$, $R(x) = 0$ **25.** 0; -60; 0
27. 5,935,988; -772 **29.** 0; 0; 65; $1 - 12\sqrt{2}$ **31.** Yes; no
33. No; no **35.** $f(x) = (x - 1)(x + 2)(x + 3)$; 1, -2, -3
37. $f(x) = (x - 2)(x - 5)(x + 1)$; 2, 5, -1
39. $f(x) = (x - 2)(x - 3)(x + 4)$; 2, 3, -4
41. $f(x) = (x - 1)(x - 2)(x - 3)(x + 5)$; 1, 2, 3, -5
43. $y = \frac{3}{4}x + \frac{21}{4}$ **45.** $y = -\frac{2}{7}x + \frac{33}{7}$ **47.**
49. **(a)** $x + 5$, $x + 3$, $x - 4$, $x - 6$, $x - 7$;
(b) $P(x) = -(x + 5)(x + 3)(x - 4)(x - 6)(x - 7)$;
(c) yes; two examples are $f(x) = c \cdot P(x)$; for any nonzero constant c, and $g(x) = (x - a)P(x)$; **(d)** no
51. $\frac{14}{3}$ **53.** 0, 0.9636 **55.** $\{-1 \pm \sqrt{7}\}$ **57.** Yes
59. $x - 1 + 2i$, R $-9 + 4i$

Exercise Set 2.5

1. -3, multiplicity 2; 1, multiplicity 1
3. 0, multiplicity 3; 1, multiplicity 2; -4, multiplicity 1
5. $\pm\sqrt{3}$, ± 1, each has multiplicity 1
7. -3, -1, 1, each has multiplicity 1
9. $f(x) = x^3 - 6x^2 - x + 30$
11. $f(x) = x^3 + 3x^2 + 4x + 12$
13. $f(x) = x^3 - \sqrt{3}x^2 - 2x + 2\sqrt{3}$
15. $-3 - 4i$, $4 + \sqrt{5}$ **17.** $f(x) = x^3 - 4x^2 + 6x - 4$
19. $f(x) = x^3 - 5x^2 + 16x - 80$
21. $f(x) = x^4 + 4x^2 - 45$ **23.** i, 2, 3 **25.** $1 + 2i$, $1 - 2i$
27. ± 1 **29.** $\pm 1, \pm 2, \pm\frac{1}{3}, \pm\frac{1}{5}, \pm\frac{2}{3}, \pm\frac{2}{5}, \pm\frac{1}{15}, \pm\frac{2}{15}$
31. **(a)** Rational: -3, other: $\pm\sqrt{2}$; **(b)** $-3, -\sqrt{2}, \sqrt{2}$;
(c) $f(x) = (x + 3)(x + \sqrt{2})(x - \sqrt{2})$
33. **(a)** Rational: -2, 1; other: none; **(b)** -2, 1;
(c) $f(x) = (x + 2)(x - 1)^2$
35. **(a)** Rational: -1; other: $3 \pm 2\sqrt{2}i$; **(b)** $-1, 3 \pm 2\sqrt{2}i$;
(c) $f(x) = (x + 1)(x - 3 - 2\sqrt{2}i)(x - 3 + 2\sqrt{2}i)$
37. **(a)** Rational: $-\frac{1}{5}$, 1; other: $\pm 2i$ **(b)** $-\frac{1}{5}, 1, \pm 2i$;
(c) $f(x) = (5x + 1)(x + 1)(x + 2i)(x - 2i)$
39. **(a)** Rational: $-2, -1$; other: $3 \pm \sqrt{13}$;
(b) $-2, -1, 3 \pm \sqrt{13}$;
(c) $f(x) = (x + 2)(x + 1)(x - 3 - \sqrt{13})(x - 3 + \sqrt{13})$
41. **(a)** Rational: 2; other: $1 \pm \sqrt{3}$; **(b)** $2, 1 \pm \sqrt{3}$;
(c) $f(x) = (x - 2)(x - 1 - \sqrt{3})(x - 1 + \sqrt{3})$
43. **(a)** Rational: -2; other: $1 \pm \sqrt{3}i$; **(b)** $-2, 1 \pm \sqrt{3}i$;
(c) $f(x) = (x + 2)(x - 1 - \sqrt{3}i)(x - 1 + \sqrt{3}i)$
45. **(a)** Rational: $\dfrac{1}{2}$; other: $\dfrac{1 \pm \sqrt{5}}{2}$; **(b)** $\dfrac{1}{2}, \dfrac{1 \pm \sqrt{5}}{2}$;
(c) $f(x) = \dfrac{1}{3}\left(x - \dfrac{1}{2}\right)\left(x - \dfrac{1 + \sqrt{5}}{2}\right)\left(x - \dfrac{1 - \sqrt{5}}{2}\right)$
47. None **49.** None **51.** None **53.** -2, 1, 2
55. $f(-5) = -18$ and $f(-4) = 7$. By the intermediate value theorem, since $f(-5)$ and $f(-4)$ have opposite signs, then $f(x)$ has a zero between -5 and -4.
57. $\{-2\}$ **59.** $\frac{12}{13}$

61.
63. **(a)** $\left\{-1, \frac{1}{2}, 3\right\}$; **(b)** $\left\{0, \frac{3}{2}, 4\right\}$; **(c)** $\left\{-3, -\frac{3}{2}, 1\right\}$; **(d)** $\left\{-\frac{1}{2}, \frac{1}{4}, \frac{3}{2}\right\}$
65. By the rational zeros theorem, only ± 1 and ± 5 can be rational solutions of $x^2 - 5 = 0$. Since none of them is a solution, the equation has no rational solutions. But $\sqrt{5}$ is a solution, so $\sqrt{5}$ must be irrational.
67. $-8, -\frac{3}{2}, 4, 7, 15$

Exercise Set 2.6

1. (d); $x = 2$, $x = -2$, $y = 0$
3. (e); $x = 2$, $x = -2$, $y = 0$
5. (c); $x = 2$, $x = -2$, $y = 8x$
7.

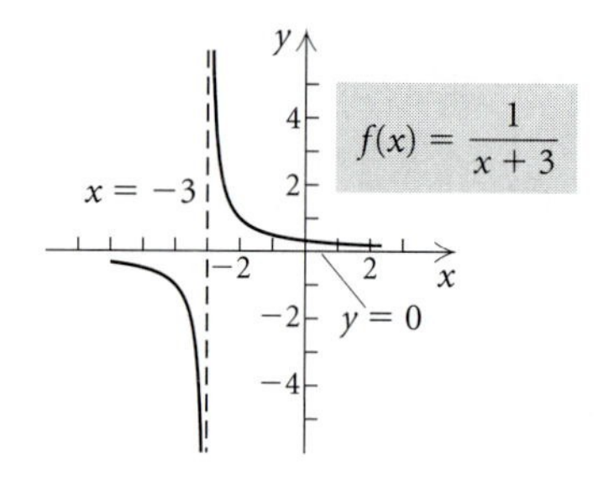

9. **11.**

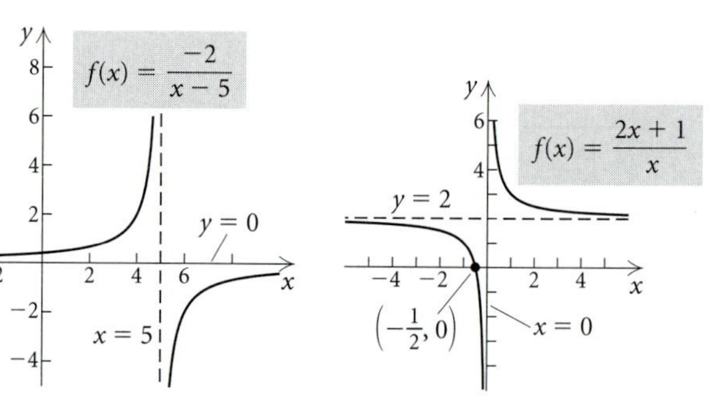

13. **15.**

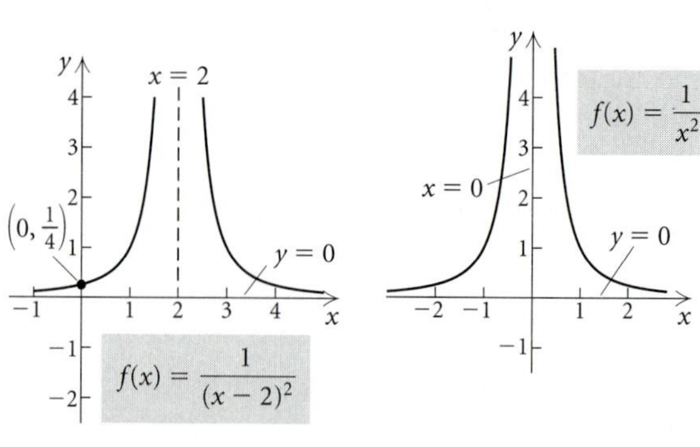

17. **19.**

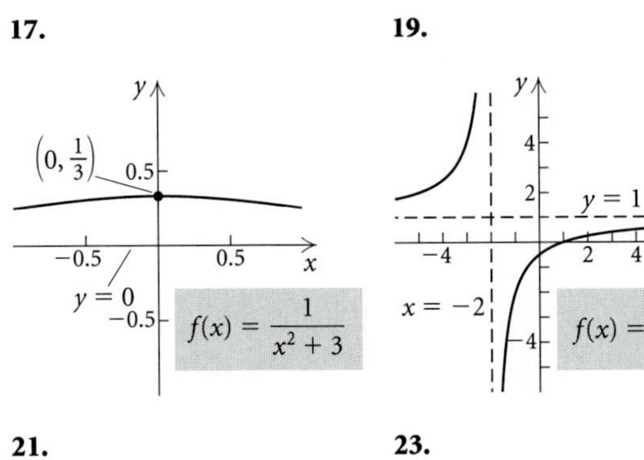

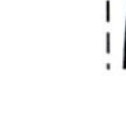

21.

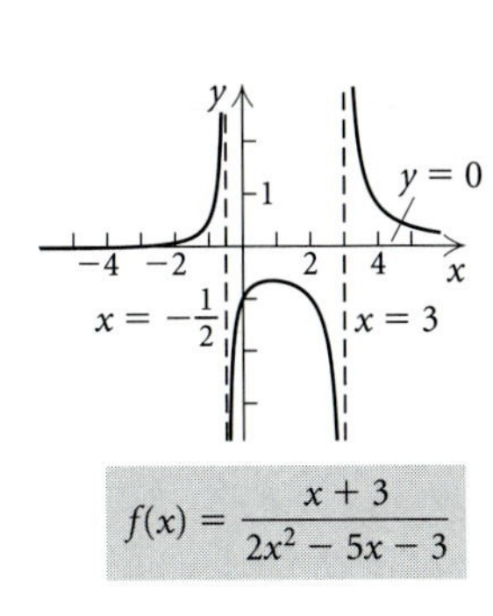

23.

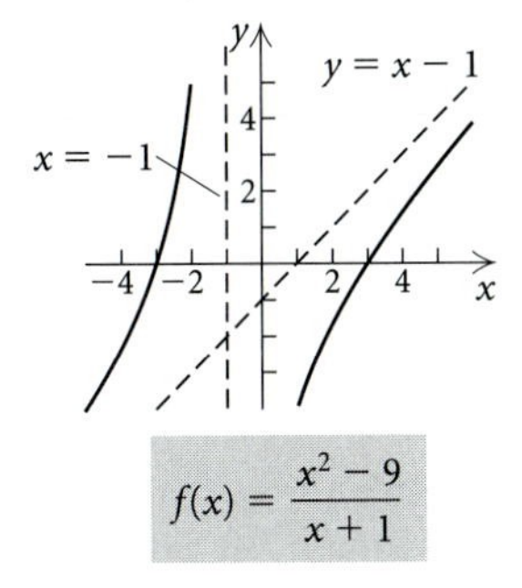

25.

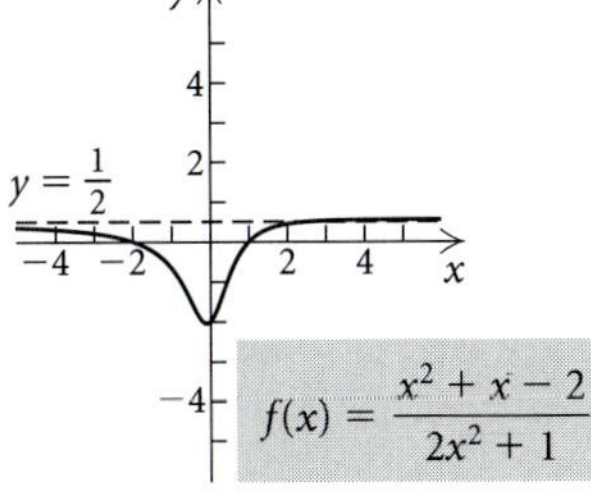

27.

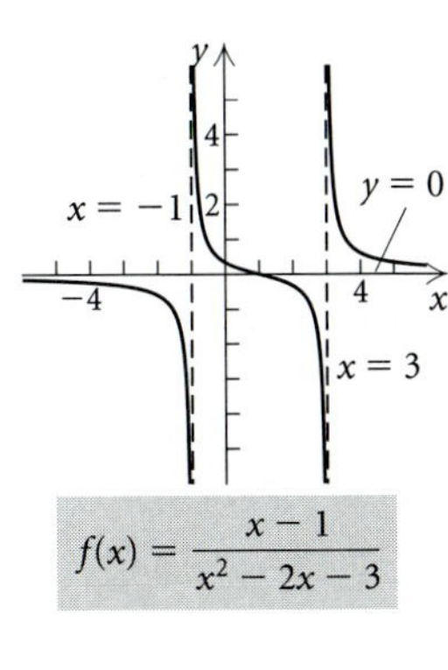

29.

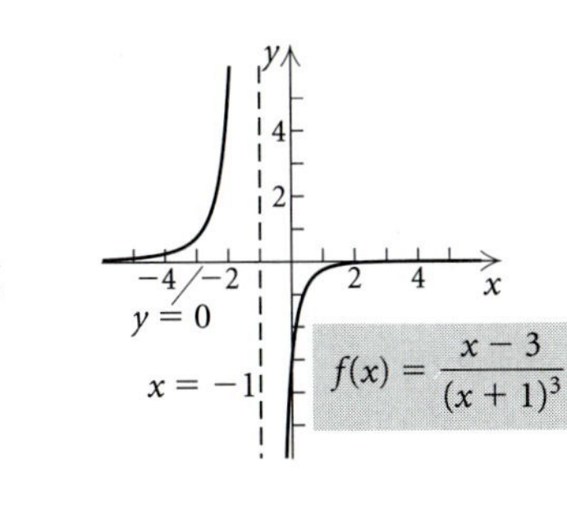

31.

$x = 0$

$f(x) = \dfrac{x^3 + 1}{x}$

33.

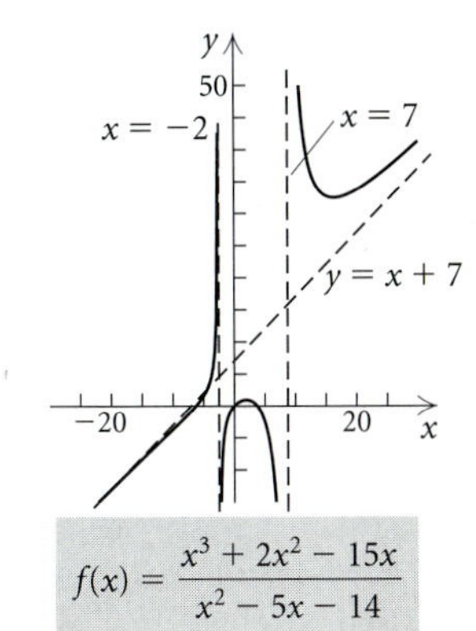

35.

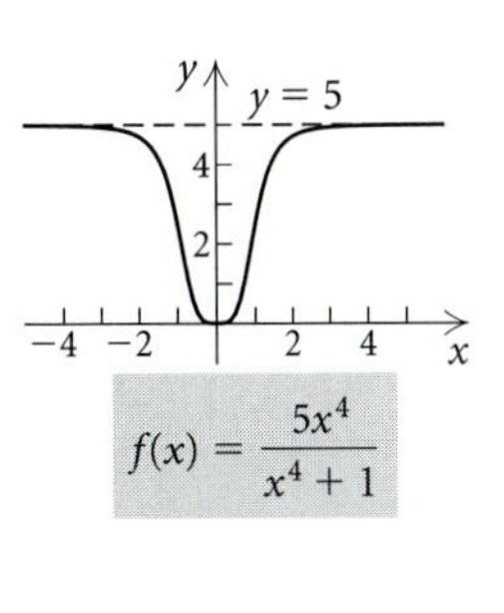

37.

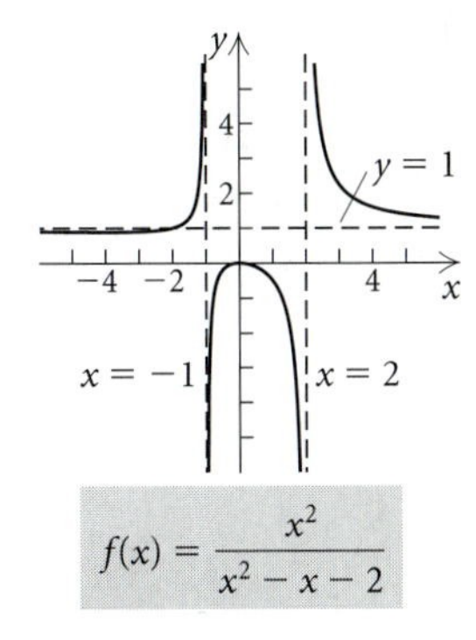

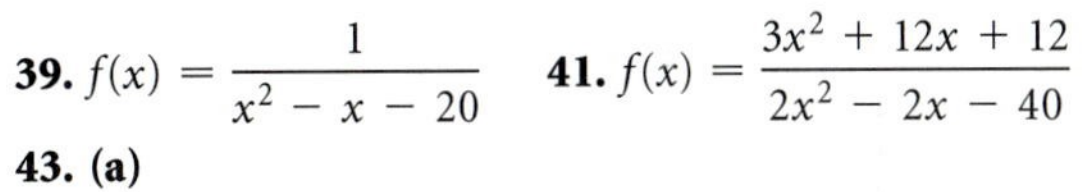

39. $f(x) = \dfrac{1}{x^2 - x - 20}$ **41.** $f(x) = \dfrac{3x^2 + 12x + 12}{2x^2 - 2x - 40}$

43. (a)

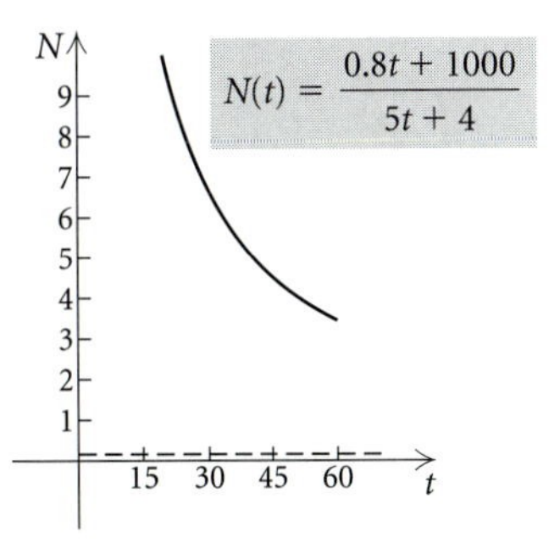

$N(t) \longrightarrow 0.16$ as $t \longrightarrow \infty$; **(b)** The medication never completely disappears from the body; a trace amount remains.

45. (a)

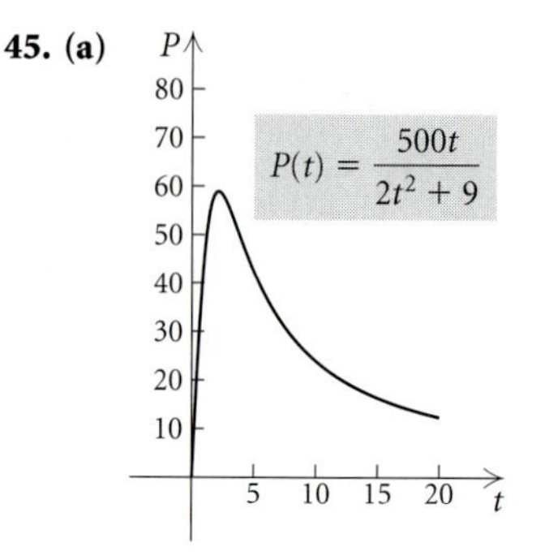

(b) $P(0) = 0$; $P(1) = 45{,}455$; $P(3) = 55{,}556$; $P(8) = 29{,}197$ **(c)** $P(t) \longrightarrow 0$ as $t \longrightarrow \infty$; **(d)** In time, no one lives in Lordsburg. **(e)** 58,926 **47.** 22.5 W/m^2 **49.** $\{x \mid x > 3\}$ **51.** $\{3\}$ **53.** ◈ **55.** ◈ **57.** $y = x^3 + 4$ **59.** **61.**

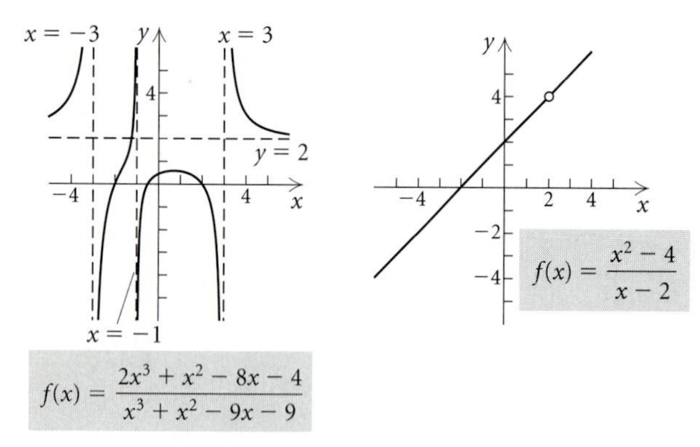

63.

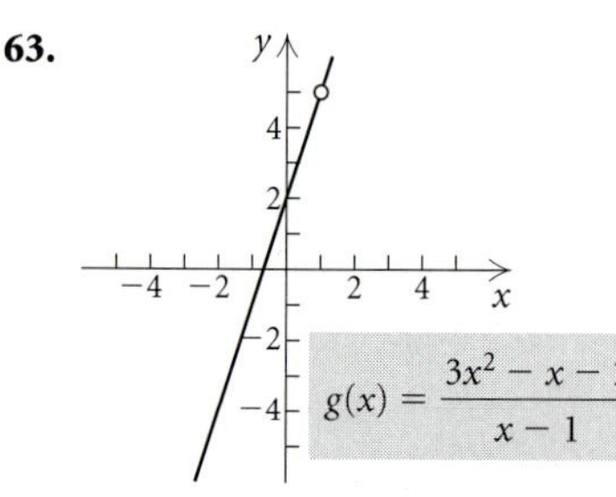

65. ◈ **67.** $(-\infty, -3) \cup (7, \infty)$

Exercise Set 2.7

1. $(-\infty, -5) \cup (-3, 2)$ **3.** $(-2, 0] \cup (2, \infty)$ **5.** $(-4, 1)$ **7.** $(-\infty, -2] \cup [4, \infty)$ **9.** $(-\infty, -2) \cup (1, \infty)$ **11.** $(-\infty, -5) \cup (5, \infty)$ **13.** $(-\infty, -2] \cup [2, \infty)$ **15.** $(-\infty, 3) \cup (3, \infty)$ **17.** $\varnothing$ **19.** $\left(-\infty, -\frac{5}{4}\right] \cup [0, 3]$ **21.** $[-3, -1] \cup [1, \infty)$ **23.** $(-\infty, -2) \cup (1, 3)$ **25.** $[-\sqrt{2}, -1] \cup [\sqrt{2}, \infty)$ **27.** $(-\infty, -1] \cup \left[\frac{3}{2}, 2\right]$ **29.** $(-\infty, 5]$ **31.** $(-\infty, -1.680) \cup (2.154, 5.526)$ **33.** $(-4, \infty)$ **35.** $\left(-\frac{5}{2}, \infty\right)$ **37.** $\left(-3, -\frac{1}{5}\right] \cup (1, \infty)$

39. $(-\infty, -3) \cup \left[\frac{5 - \sqrt{105}}{10}, -\frac{1}{3}\right) \cup \left[\frac{5 + \sqrt{105}}{10}, \infty\right)$ **41.** $\left(2, \frac{7}{2}\right]$ **43.** $(1 - \sqrt{2}, 0) \cup (1 + \sqrt{2}, \infty)$ **45.** $(-\infty, -3) \cup (1, 3) \cup \left[\frac{11}{3}, \infty\right)$ **47.** $(-\infty, \infty)$ **49.** $\left(-3, \frac{1 - \sqrt{61}}{6}\right) \cup \left(-\frac{1}{2}, 0\right) \cup \left(\frac{1 + \sqrt{61}}{6}, \infty\right)$ **51.** $(-1, 0) \cup \left(\frac{2}{7}, \frac{7}{2}\right)$ **53.** $[-6 - \sqrt{33}, -5) \cup [-6 + \sqrt{33}, 1) \cup (5, \infty)$ **55.** $(0.408, 2.449)$ **57. (a)** $\{x \mid 10 < x < 200\}$; **(b)** $\{x \mid 0 < x < 10 \text{ or } x > 200\}$ **59.** $\{n \mid 9 \le n \le 23\}$ **61.** $b = \frac{a + 7}{3}$ **63.** 14 **65.** ◈ **67.** $\{3\}$ **69.** $(-\infty, \infty)$ **71.** $[-\sqrt{5}, \sqrt{5}]$ **73.** $\left[-\frac{3}{2}, \frac{3}{2}\right]$ **75.** $\left(-\infty, -\frac{1}{4}\right) \cup \left(\frac{1}{2}, \infty\right)$ **77.** $(-4, -2) \cup (-1, 1)$ **79.** $x^2 + x - 12 < 0$; answers may vary

Review Exercises, Chapter 2

1. [2.1] **(a)** $-2.637, 1.137$; **(b)** $(-0.75, 7.125)$; **(c)** none; **(d)** domain: all real numbers; range: $(-\infty, 7.125]$ **2.** [2.1] **(a)** $-3, -1.414, 1.414$; **(b)** $(-2.291, 2.303)$; **(c)** $(0.291, -6.303)$; **(d)** domain: all real numbers; range: all real numbers **3.** [2.1] **(a)** 0, 1, 2; **(b)** $(0.610, 0.202)$; **(c)** $(0, 0)$, $(1.640, -0.620)$; **(d)** domain: all real numbers; range: $[-0.620, \infty)$ **4.** [2.1] $-2\sqrt{10}i$ **5.** [2.1] $-4\sqrt{15}$ **6.** [2.1] $-\frac{7}{8}$ **7.** [2.1] $-18 - 26i$ **8.** [2.1] $\frac{11}{10} + \frac{3}{10}i$ **9.** [2.1] $1 - 4i$ **10.** [2.1] $2 - i$ **11.** [2.1] $-i$ **12.** [2.1] 3^{28} **13.** [2.2] $x^2 - 3x + \frac{9}{4} = 18 + \frac{9}{4}$; $\left(x - \frac{3}{2}\right)^2 = \frac{81}{4}$; $x = \frac{3}{2} \pm \frac{9}{2}$; $\{-3, 6\}$ **14.** [2.2] $x^2 - 4x = 2$; $x^2 - 4x + 4 = 2 + 4$; $(x - 2)^2 = 6$; $x = 2 \pm \sqrt{6}$; $\{2 - \sqrt{6}, 2 + \sqrt{6}\}$ **15.** [2.2] $\left\{-2, \frac{4}{3}\right\}$ **16.** [2.2] $\{1 - 3i, 1 + 3i\}$ **17.** [2.2] $\{-3, 6\}$ **18.** [2.2] $\{1\}$ **19.** [2.2] $\left\{\frac{-1 - \sqrt{5}}{2}, \frac{-1 + \sqrt{5}}{2}, \frac{1 - \sqrt{5}}{2}, \frac{1 + \sqrt{5}}{2}\right\}$ **20.** [2.2] $\{-\sqrt{3}, 0, \sqrt{3}\}$ **21.** [2.5] $\left\{-2, -\frac{2}{3}, 3\right\}$ **22.** [2.5] $\{-5, -2, 2\}$ **23.** [2.2] **(a)** $\left(\frac{3}{8}, -\frac{7}{16}\right)$; **(b)** $x = \frac{3}{8}$; **(c)** maximum: $-\frac{7}{16}$; **(d)** $\left(-\infty, -\frac{7}{16}\right]$ **24.** [2.2] **(a)** $(1, -2)$; **(b)** $x = 1$; **(c)** minimum: -2; **(d)** $[-2, \infty)$ **25.** [2.2] (d) **26.** [2.2] (c) **27.** [2.2] (b) **28.** [2.2] (a) **29.** [2.3] **(a)** 4%; **(b)** 5% **30.** [2.3] $35 - 5\sqrt{33}$ ft, or about 6.3 ft **31.** [2.3] 6 ft by 6 ft **32.** [2.3] $\frac{15 - \sqrt{115}}{2}$ cm, or about 2.1 cm **33.** [2.3] 2.1 cm

34. [2.3] **(a)** *Linear:* $f(x) = 0.5408695652x - 30.30434783$; *quadratic:* $f(x) = 0.0030322581x^2 - 0.5764516129x + 57.53225806$; *cubic:* $f(x) = 0.0000247619x^3 - 0.0112857143x^2 + 2.002380952x - 82.14285714$; **(b)** the cubic function; **(c)** 298, 498
35. [2.4] $4x^2 - \frac{14}{5}x - \frac{17}{25}$, R 284/25
36. [2.4] $2x^3 + x + 1$, R $2x + 6$ **37.** [2.4] $\{-5, 1, 2\}$
38. [2.4] $\{-2 - \sqrt{11}, -5, -2 + \sqrt{11}, 4\}$
39. [2.4] $x^2 + 7x + 22$, R 120
40. [2.4] $x^3 + x^2 + x + 1$, R 0 **41.** [2.4] 120
42. [2.4] 0
43. [2.4] -3 is a root; $P(x) = (x + 5)(x + 3)(x - 1)$; other roots: -5, 1
44. [2.4] -2 is a root; $P(x) = (x + 2)(x + 1) \cdot [x - (-3 + \sqrt{10})][x - (-3 - \sqrt{10})]$
other roots: -1, $-3 \pm \sqrt{10}$
45. [2.5] $x^3 - x^2 - 10x - 8 = 0$; answers may vary
46. [2.5] $x^4 + 3x^3 - 7x^2 - 9x + 12 = 0$; answers may vary
47. [2.5] $x^6 - 13x^5 + 62x^4 - 132x^3 - 45x^2 + 745x - 858 = 0$; answers may vary
48. [2.5] $-\frac{5}{8}x^3 + \frac{5}{8}x^2 + \frac{25}{4}x + 5 = 0$
49. [2.5] $\frac{1}{2}x^4 + x^3 - \frac{11}{2}x^2 - 8x + 12 = 0$
50. [2.5] $-\frac{4}{5}x^3 - \frac{12}{5}x^2 + \frac{48}{5}x + 8 = 0$
51. [2.5] $\{-1, 2, 3\}$ **52.** [2.5] $\{-5, -3, 1\}$
53. [2.5] $\{-3, 2 - \sqrt{11}, 2 + \sqrt{11}, 3\}$
54. [2.5] **(a)** $\{-10, 1\}$; **(b)** $y = x^3 + 8x^2 - 19x + 10$

250
−15 5
−50

(c) $P(x) = (x - 1)^2(x + 10)$
55. [2.5] **(a)** $\{-4, 0, 3, 4\}$;
(b)
$y = x^6 + x^5 - 28x^4 - 16x^3 + 192x^2$

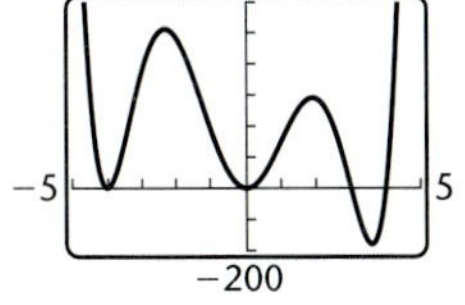

(c) $P(x) = x^2(x + 4)^2(x - 3)(x - 4)$
56. [2.6]

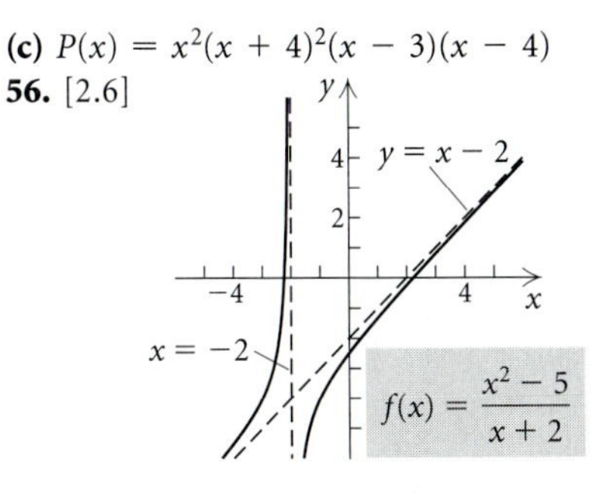

57. [2.6]

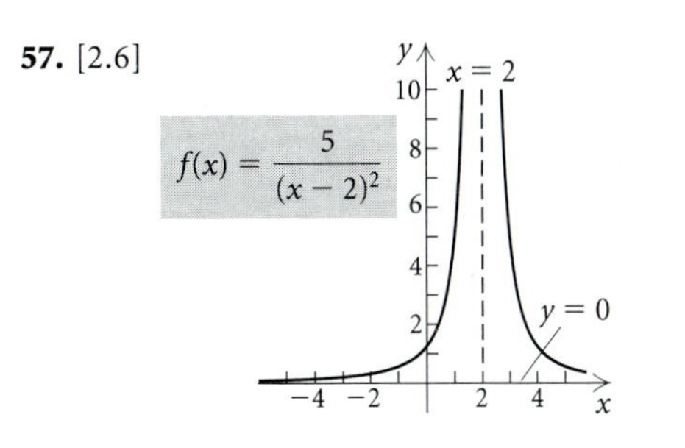

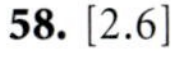
58. [2.6]

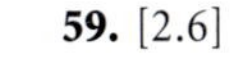
59. [2.6]

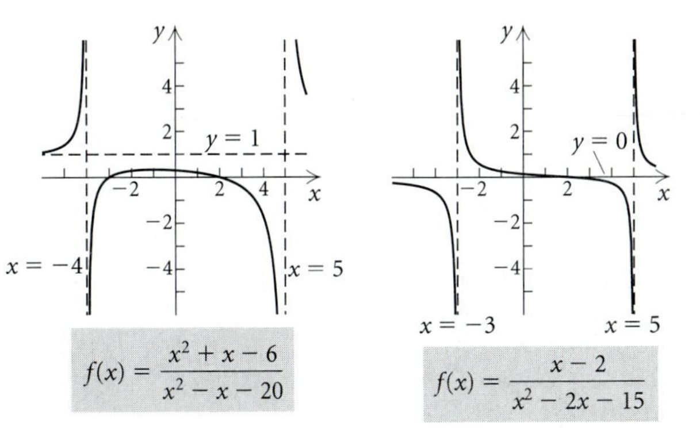

60. [2.6] $f(x) = \dfrac{1}{x^2 - x - 6}$ **61.** [2.6] $f(x) = \dfrac{4x^2 + 12x}{x^2 - x - 6}$
62. [2.6] **(a)**

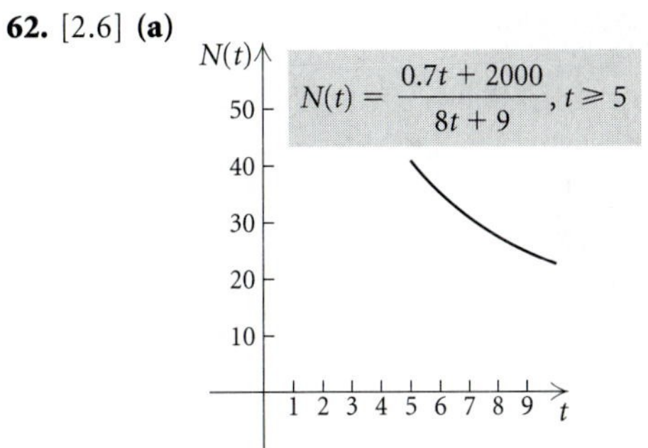

$N(t) \longrightarrow 0.0875$ as $t \longrightarrow \infty$; **(b)** The medication never completely disappears from the body; a trace amount remains. **63.** [2.7] $(-3, 3)$ **64.** [2.7] $\left(-\infty, -\frac{1}{2}\right) \cup (2, \infty)$
65. [2.7] $(-4, 1) \cup (2, \infty)$ **66.** [2.7] $\left(-\infty, -\frac{14}{3}\right) \cup (-3, \infty)$
67. [2.7] $(-5, -3.386] \cup (-2, 2) \cup (4, \infty)$
68. [2.3], [2.7] **(a)** 324 ft, 2.5 sec; **(b)** $t = 7$; **(c)** $(2, 3)$
69. [2.7] $\left[\dfrac{5 - \sqrt{15}}{2}, \dfrac{5 + \sqrt{15}}{2}\right]$
70. [2.1, [2.6] ◈ A polynomial function is a function that can be defined by a polynomial expression. A rational function is a function that can be defined as a quotient of two polynomials.
71. [2.6] ◈ Vertical asymptotes occur at any x-values that make the denominator zero. The graph of a rational

function does not cross any vertical asymptotes. Horizontal asymptotes occur when the degree of the numerator is less than or equal to the degree of the denominator. Oblique asymptotes occur when the degree of the numerator is 1 greater than the degree of the denominator. Graphs of rational functions may cross horizontal or oblique asymptotes.

72. [2.5] Yes **73.** [2.3] 9%

74. [2.7] $(-\infty, -1 - \sqrt{6}] \cup [-1 + \sqrt{6}, \infty)$

75. [2.7] $\left(-\infty, -\frac{1}{2}\right) \cup \left(\frac{1}{2}, \infty\right)$ **76.** [2.5] $\{1 + i, 1 - i, i, -i\}$

77. [2.7] $(-\infty, 2)$

78. [2.5] $(x - 1)\left(x + \frac{1}{2} - \frac{\sqrt{3}}{2}i\right)\left(x + \frac{1}{2} + \frac{\sqrt{3}}{2}i\right)$

79. [2.4] 7 **80.** [2.4] -4 **81.** [2.1] $(-\infty, -5] \cup [2, \infty)$

82. [2.1] $(-\infty, 1.1] \cup [2, \infty)$ **83.** [2.6] $\left(-1, \frac{3}{7}\right)$

Chapter 3

EXERCISE SET 3.1

1. $\{(8, 7), (8, -2), (-4, 3), (-8, 8)\}$

3. $\{(-1, -1), (4, -3)\}$ **5.** $x = 4y - 5$ **7.** $y^3x = -5$

9.

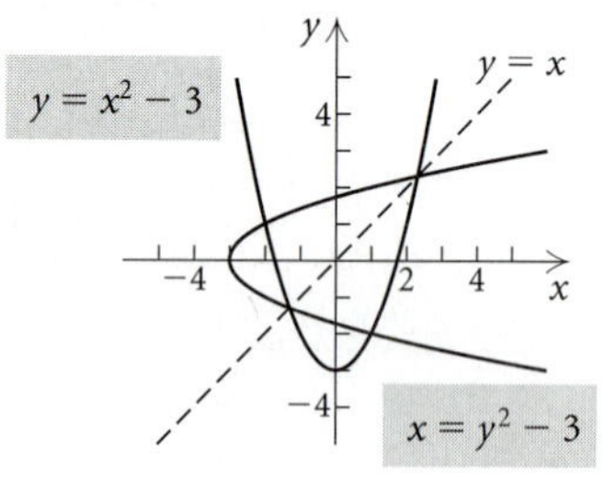

11.

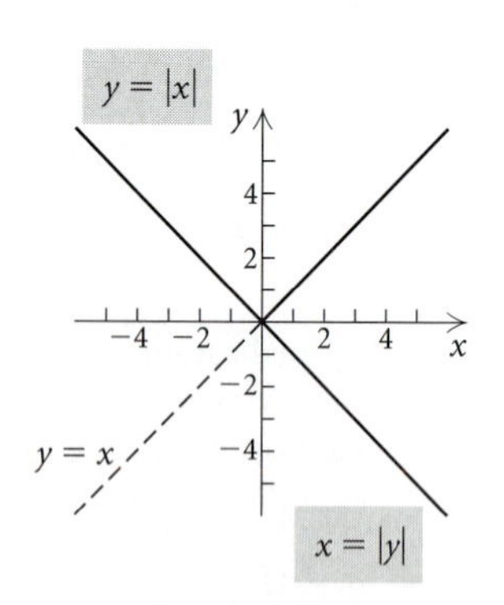

13. Yes **15.** No **17.** No **19.** Yes **21.** Yes **23.** No

25. No **27.** Yes

29. $y_1 = 0.8x + 1.7,$ $y_2 = \dfrac{x - 1.7}{0.8}$

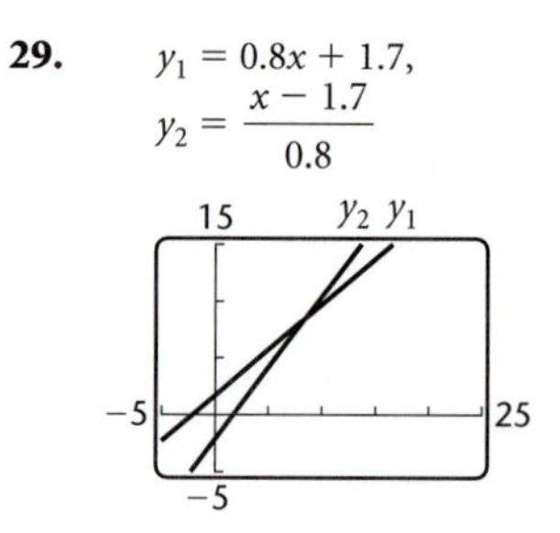

Domain and range of both f and f^{-1}: all real numbers

31. $y_1 = \frac{1}{2}x - 4,$ $y_2 = 2x + 8$

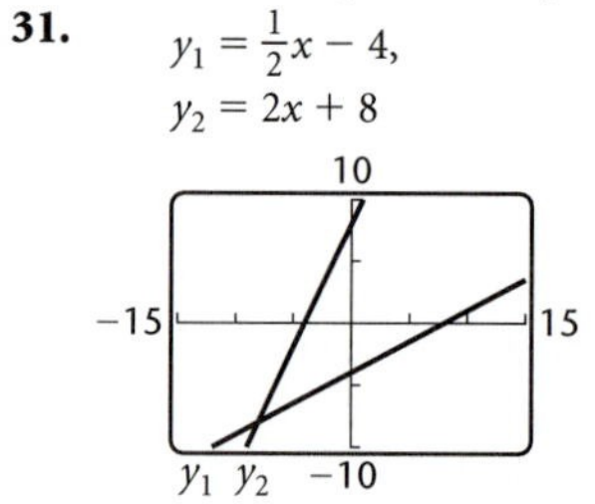

Domain and range of both f and f^{-1}: all real numbers

33. $y_1 = \sqrt{x - 3},$ $y_2 = x^2 + 3, x \geq 0$

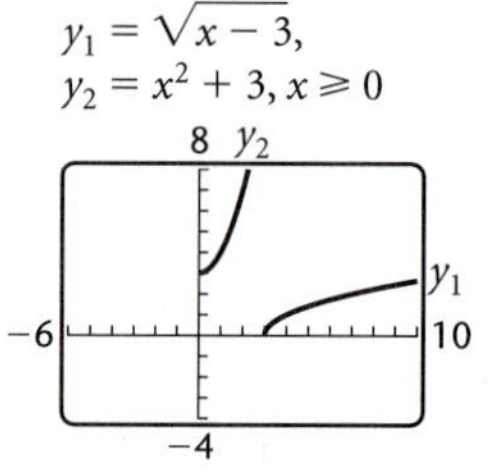

Domain of f: $[3, \infty]$; range of f: $[0, \infty)$; domain of f^{-1}: $[0, \infty)$; range of f^{-1}: $[3, \infty)$

35. $y_1 = x^2 - 4, x \geq 0;$ $y_2 = \sqrt{4 + x}$

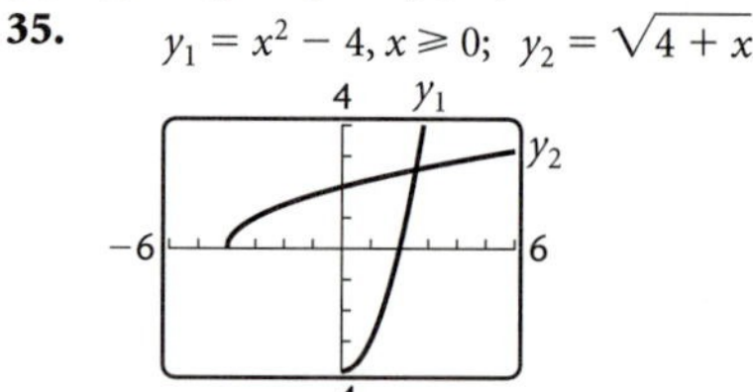

Domain of f: $[0, \infty)$; range of f: $[-4, \infty)$; domain of f^{-1}: $[-4, \infty)$; range of f^{-1}: $[0, \infty)$

37. $y_1 = (3x - 9)^3,$ $y_2 = \dfrac{\sqrt[3]{x} + 9}{3}$

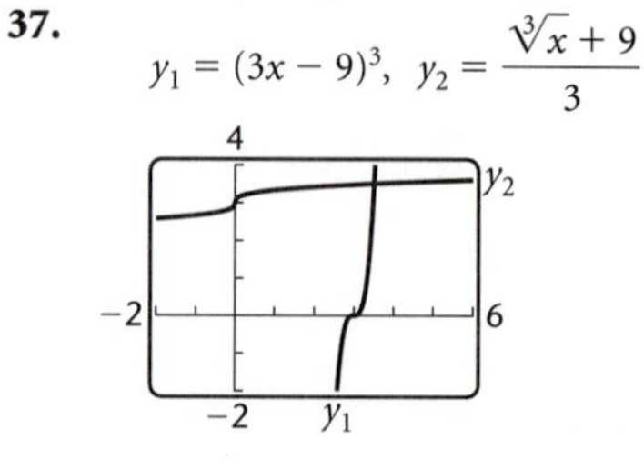

Domain and range of both f and f^{-1}: all real numbers

39. **(a)** One-to-one; **(b)** $f^{-1}(x) = x - 4$

41. **(a)** One-to-one; **(b)** $f^{-1}(x) = \dfrac{1}{2}x$

43. **(a)** One-to-one; **(b)** $f^{-1}(x) = \dfrac{4}{x} - 7$

45. **(a)** One-to-one; **(b)** $f^{-1}(x) = \dfrac{3x + 4}{x - 1}$

47. **(a)** One-to-one; **(b)** $f^{-1}(x) = \sqrt[3]{x + 1}$

49. **(a)** Not one-to-one

51. **(a)** One-to-one; **(b)** $f^{-1}(x) = \sqrt{\dfrac{x + 2}{5}}$, $x \geq 18$

53. **(a)** One-to-one; **(b)** $f^{-1}(x) = x^2 - 1$, $x \geq 0$

55. $\frac{1}{3}x$ **57.** $-x$ **59.** $x^3 + 5$

61.
$$\begin{aligned}(f^{-1} \circ f)(x) &= f^{-1}(f(x)) \\ &= f^{-1}\left(\tfrac{7}{8}x\right) \\ &= \tfrac{8}{7}\left(\tfrac{7}{8}x\right) = x; \\ (f \circ f^{-1})(x) &= f(f^{-1}(x)) \\ &= f\left(\tfrac{8}{7}x\right) \\ &= \tfrac{7}{8}\left(\tfrac{8}{7}x\right) = x\end{aligned}$$

63.
$$\begin{aligned}(f^{-1} \circ f)(x) &= f^{-1}\left(\frac{1 - x}{x}\right) \\ &= \frac{1}{\dfrac{1 - x}{x} + 1} \\ &= \frac{1}{\dfrac{1}{x}} = x; \\ (f \circ f^{-1})(x) &= f\left(\frac{1}{x + 1}\right) \\ &= \frac{1 - \dfrac{1}{x + 1}}{\dfrac{1}{x + 1}} \\ &= \frac{\dfrac{x + 1 - 1}{x + 1}}{\dfrac{1}{x + 1}} = x\end{aligned}$$

65. 5; a

67. **(a)** 36, 40, 44, 52, 60; **(b)** $g^{-1}(x) = \dfrac{x}{2} - 12$;
(c) 8, 10, 14, 18

69. **(a)** 0.5, 11.5, 22.5, 55.5, 72;
(b), **(d)** $y_1 = \dfrac{11x + 5}{10}$, $y_2 = \dfrac{10x - 5}{11}$

(c) $D^{-1}(r) = \dfrac{10r - 5}{11}$; the speed, in miles per hour, that the car is traveling when the reaction distance is r feet

71. **73.**

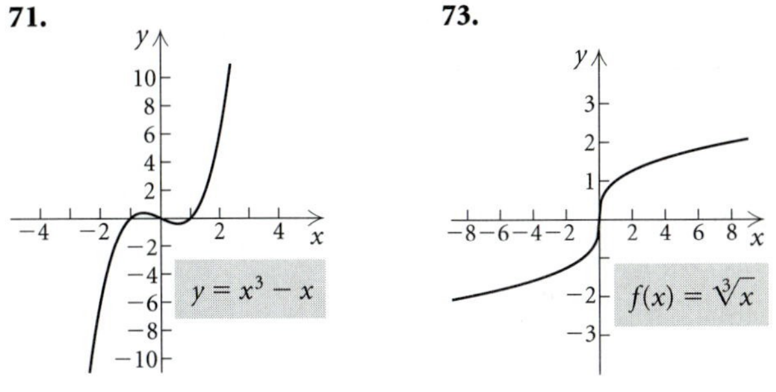

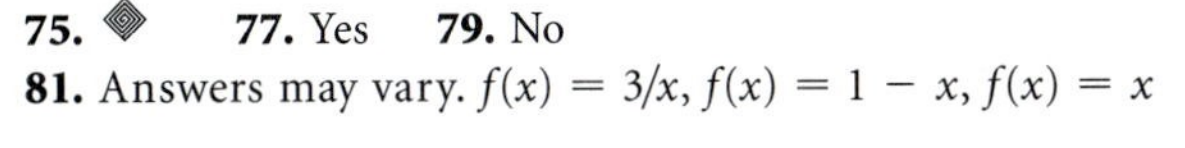

75. ◈ **77.** Yes **79.** No

81. Answers may vary. $f(x) = 3/x$, $f(x) = 1 - x$, $f(x) = x$

EXERCISE SET 3.2

1. 54.5982 **3.** 0.0856

5.

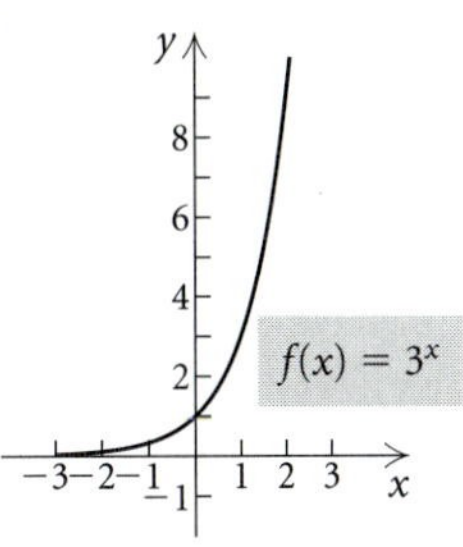

7.

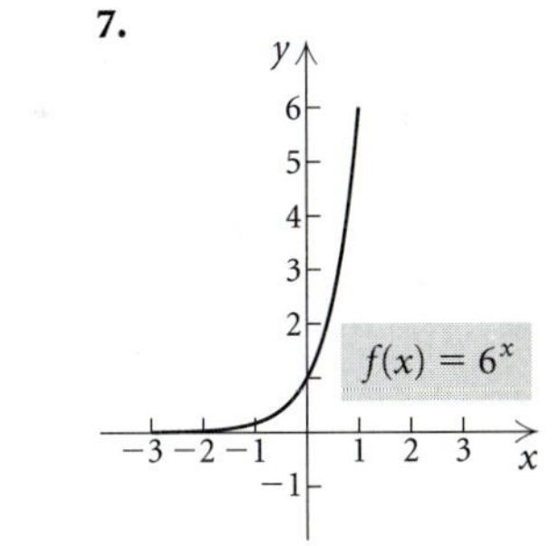

9.

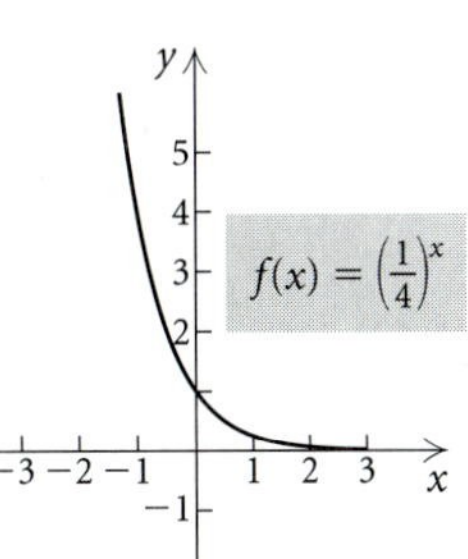

11.

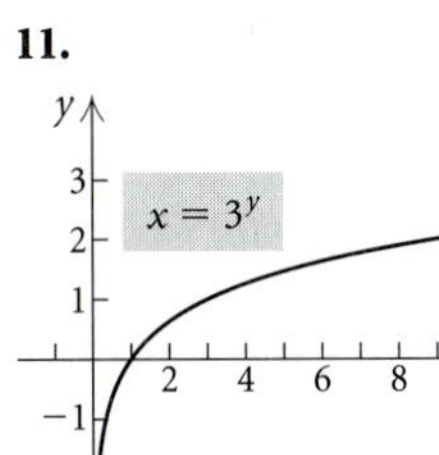

13. **15.**

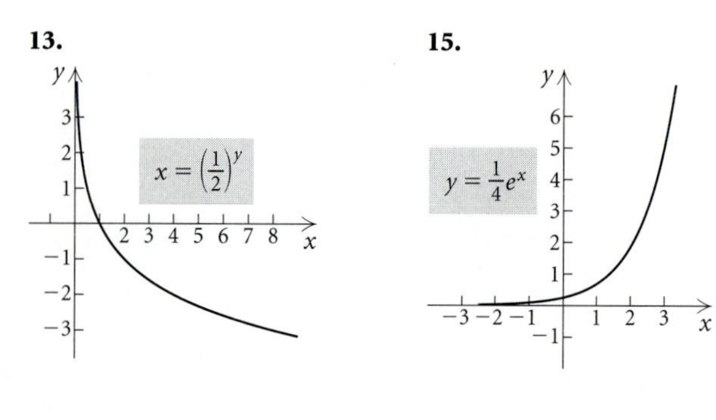

17.

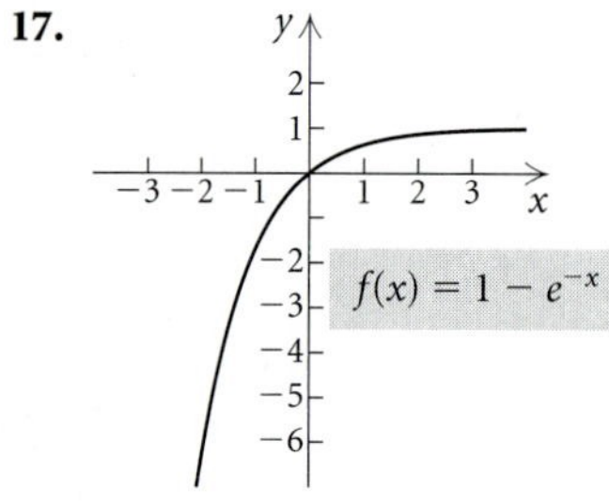

19. Shift the graph of $y = 2^x$ left 1 unit.

$y = 2^{x+1}$

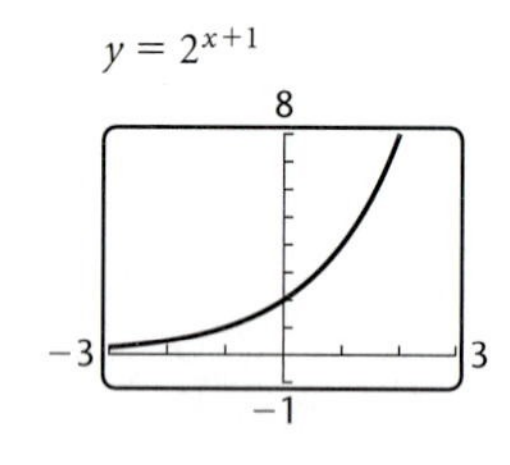

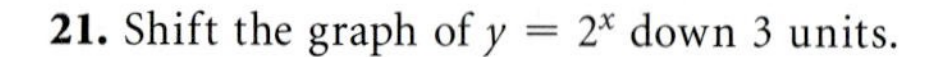

21. Shift the graph of $y = 2^x$ down 3 units.

$y = 2^x - 3$

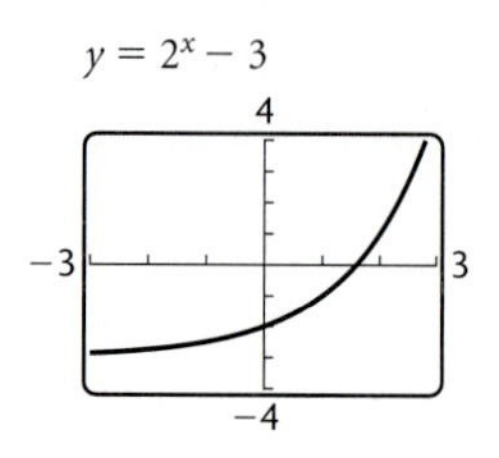

23. Reflect the graph of $y = 3^x$ across the y-axis, then across the x-axis, and then shift it up 4 units.

$y = 4 - 3^{-x}$

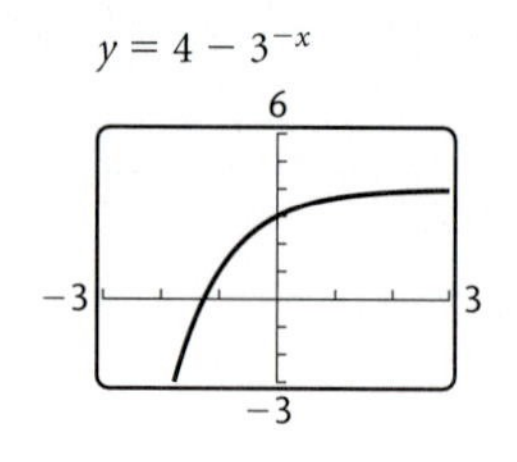

25. Shift the graph of $y = \left(\frac{3}{2}\right)^x$ right 1 unit.

$y = \left(\frac{3}{2}\right)^{x-1}$

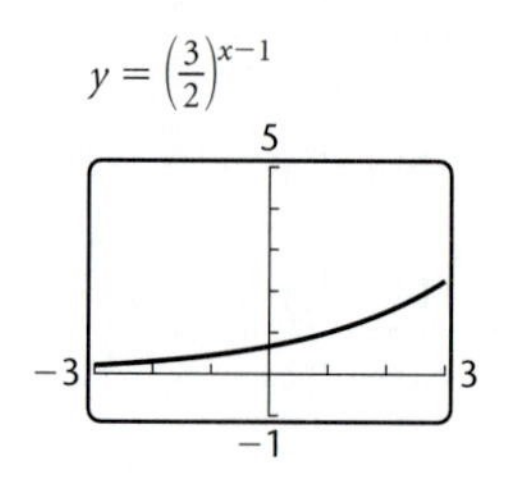

27. Shift the graph of $y = 2^x$ left 3 units, and then down 5 units.

$y = 2^{x+3} - 5$

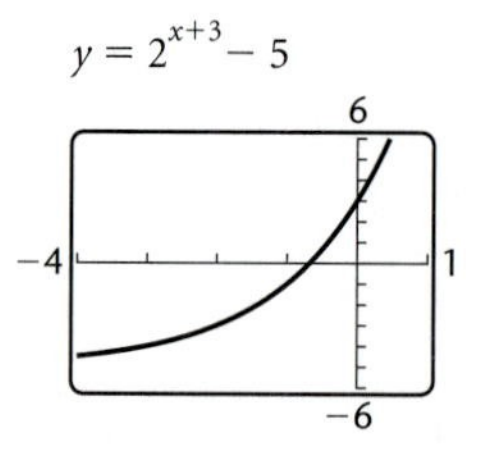

29. Shrink the graph of $y = e^x$ horizontally.

$y = e^{2x}$

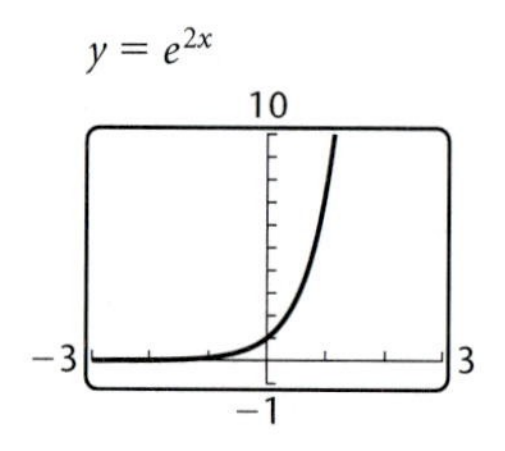

31. Shift the graph of $y = e^x$ left 1 unit, then reflect it across the y-axis.

$y = e^{-x+1}$

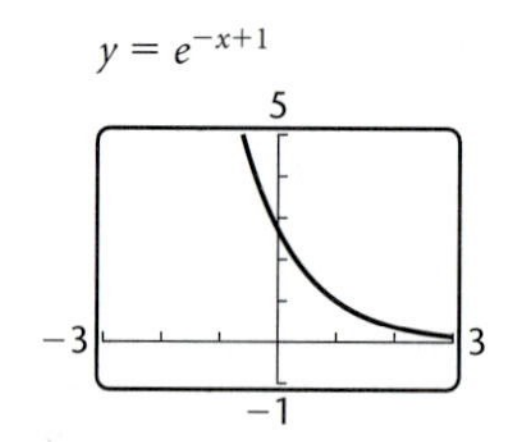

33. Reflect the graph of $y = e^x$ across the y-axis, then across the x-axis, then shift it up 1 unit, and then stretch it vertically.

$y = 2(1 - e^{-x}), x > 0$

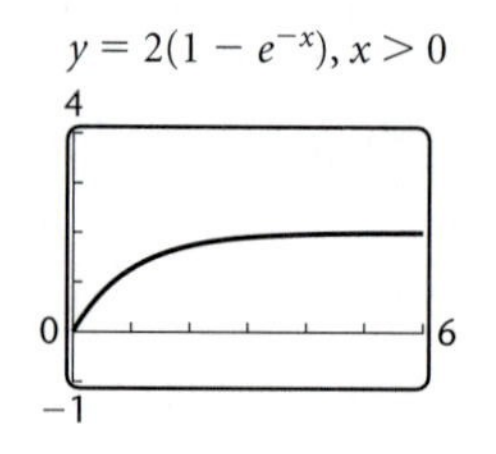

35. **(a)** 1,475,789; **(b)** 5,669,391;
(c) $y = 100{,}000(1.4)^x$ **(d)** 6.8 yr

37. **(a)** 350,000; 233,333; 69,136; 6070;
(b) $y = 350{,}000\left(\frac{2}{3}\right)^x$ **(c)** after about 13 yr

39. **(a)** \$5800, \$4640, \$3712, \$1900.54, \$622.77;
(b) $y = 5800(0.8)^x$ **(c)** after 11 yr

41. **(a)** About 50.0 billion ft^3, 62.0 billion ft^3, 78.2 billion ft^3; **(b)** $y = 46.6(1.018)^x$ **(c)** After about 39 yr

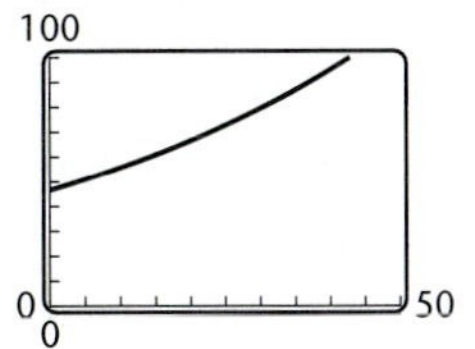

43. **(a)** About 63%; **(b)** $y = 100(1 - e^{-0.04x})$

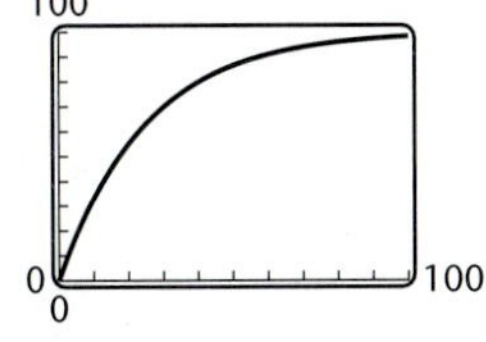

(c) after 58 days

45. (c) **47.** (a) **49.** (l) **51.** (g) **53.** (i) **55.** (k)
57. (m) **59.** (1.481, 4.090)
61. (−0.402, −1.662), (1.051, 2.722) **63.** {4.448}
65. $(0, \infty)$ **67.** {2.294, 3.228} **69.** 1 **71.** 1000 **73.** 2
75. ◈ **77.** ◈ **79.** $y = 1$

81. **(a)** $y = e^{-x^2}$

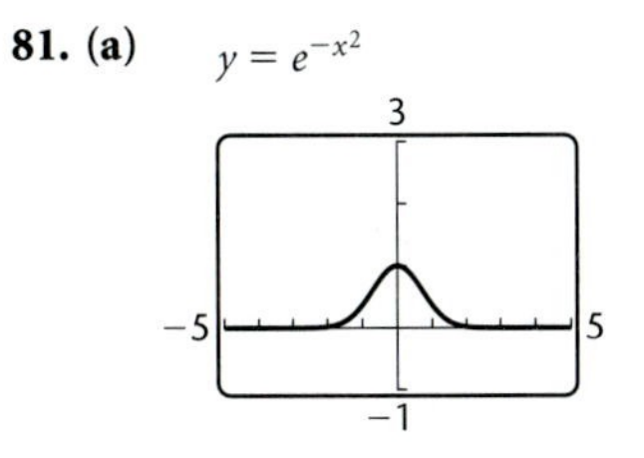

(b) none; **(c)** relative maximum: 1 at $x = 0$

EXERCISE SET 3.3

1. **3.**

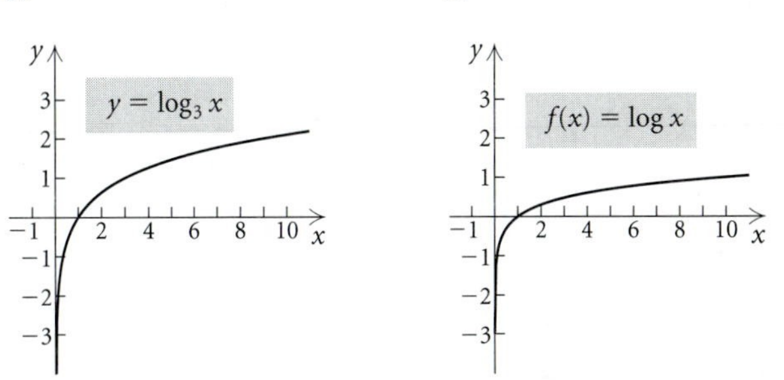

5. 4 **7.** 3 **9.** −3 **11.** −2 **13.** 0 **15.** 1
17. $\log_{10} 1000 = 3$ **19.** $\log_8 2 = \frac{1}{3}$ **21.** $\log_e t = 3$
23. $\log_e 7.3891 = 2$ **25.** $\log_p 3 = k$ **27.** $5^1 = 5$
29. $10^{-2} = 0.01$ **31.** $e^{3.4012} = 30$ **33.** $a^{-x} = M$
35. $a^x = T^3$ **37.** 0.4771 **39.** 2.7259 **41.** −0.2441
43. Does not exist **45.** 0.6931 **47.** 6.6962
49. Does not exist
51. 3.3219 **53.** −0.2614 **55.** 0.7384
57. Translate the graph of $y = \log_2 x$ left 3 units. Domain: $(-3, \infty)$; vertical asymptote: $x = -3$;

$$y = \frac{\log(x+3)}{\log 2}$$

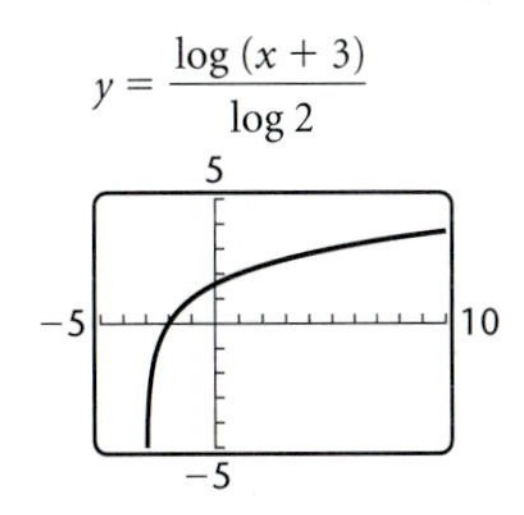

59. Translate the graph of $y = \log_3 x$ down 1 unit. Domain: $(0, \infty)$; vertical asymptote: $x = 0$;

$$y = \frac{\log x}{\log 3} - 1$$

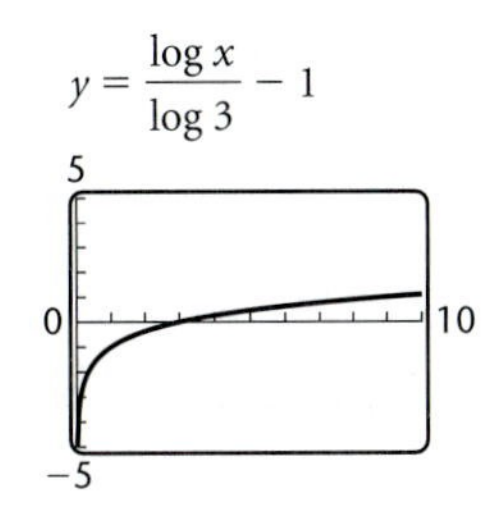

61. Stretch the graph of $y = \ln x$ vertically. Domain: $(0, \infty)$; vertical asymptote: $x = 0$;

$y = 4 \ln x$

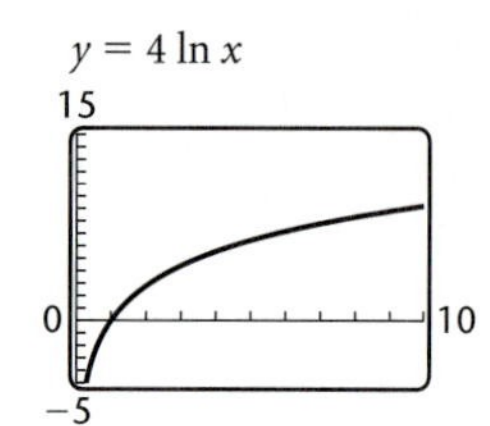

63. Reflect the graph of $y = \ln x$ across the x-axis and translate it up 2 units. Domain: $(0, \infty)$; vertical asymptote: $x = 0$;

$y = 2 - \ln x$

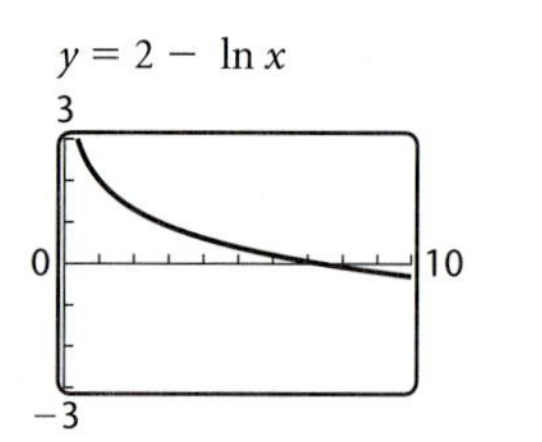

65. $y_1 = 3^x$, $y_2 = \dfrac{\log x}{\log 3}$ **67.** $y_1 = \log x$, $y_2 = 10^x$

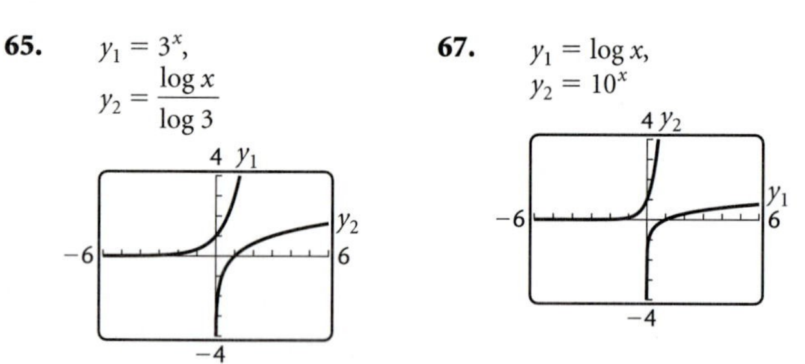

69. **(a)** 3.1 ft/sec; **(b)** 3.0 ft/sec; **(c)** 2.3 ft/sec; **(d)** 1.5 ft/sec
71. **(a)** 78%; **(b)** 67.5%, 57%;
(c) $y = 78 - 15 \log (x + 1), x \geqslant 0$ **(d)** after 73 months

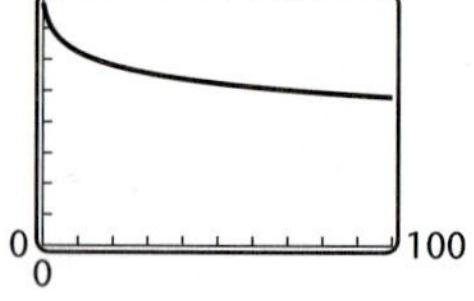

73. **(a)** 10^{-7}; **(b)** 4.0×10^{-6}; **(c)** 6.3×10^{-4}; **(d)** 1.6×10^{-5}
75. **(a)** 34 decibels; **(b)** 64 decibels; **(c)** 60 decibels; **(d)** 90 decibels
77. y^{24} **79.** 10^{4t} **81.** ◆ **83.** 3 **85.** $(0, \infty)$
87. $(-\infty, 0) \cup (0, \infty)$ **89.** $\left(-\frac{5}{2}, -2\right)$ **91.** (d) **93.** (b)

95. **(a)** $y = x \ln x$

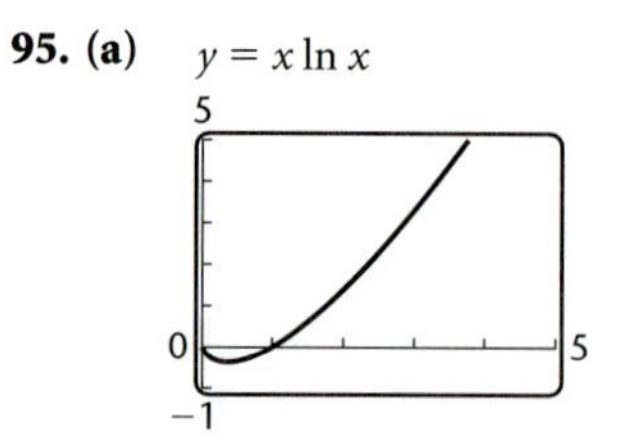

(b) 1; **(c)** relative minimum: -0.368 at $x = 0.368$
97. **(a)** $y = \dfrac{\ln x}{x^2}$

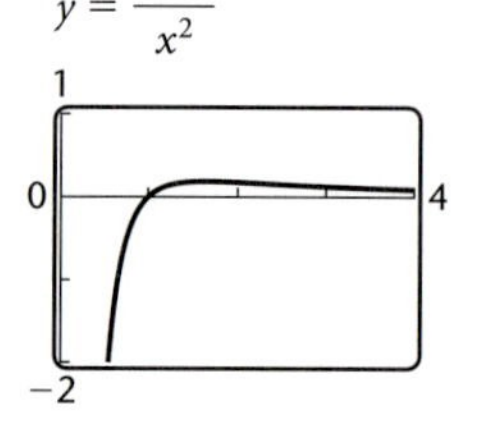

(b) 1; **(c)** relative maximum: 0.184 at $x = 1.649$

Exercise Set 3.4

1. $\log_3 81 + \log_3 27$ **3.** $\log_5 5 + \log_5 125$
5. $\log_t 8 + \log_t Y$ **7.** $3 \log_b t$ **9.** $8 \log y$ **11.** $-6 \log_c K$
13. $\log_t M - \log_t 8$ **15.** $\log_a x - \log_a y$
17. $\log_a 6 + \log_a x + 5 \log_a y + 4 \log_a z$
19. $2 \log_b p + 5 \log_b q - 4 \log_b m - 9$
21. $3 \log_a x - \frac{5}{2} \log_a p - 4 \log_a q$
23. $2 \log_a m + 3 \log_a n - \frac{3}{4} - \frac{5}{4} \log_a b$ **25.** $\log_a 150$
27. $\log 100 = 2$ **29.** $\log_a x^{-5/2}y^4$, or $\log_a \dfrac{y^4}{x^{5/2}}$ **31.** $\ln x$
33. $\ln (x - 2)$ **35.** $\ln \dfrac{x}{(x^2 - 25)^3}$ **37.** $\ln \dfrac{2^{11/5}x^9}{y^8}$
39. -0.5108 **41.** -1.6094 **43.** $\frac{1}{2}$ **45.** 2.6094
47. 4.3174 **49.** 3 **51.** $|x - 4|$ **53.** $4x$ **55.** w **57.** $8t$
59. $\{4\}$ **61.** $\{5\}$ **63.** ◆ **65.** $\{4\}$ **67.** $\log_a (x^3 - y^3)$
69. $\frac{1}{2} \log_a (x - y) - \frac{1}{2} \log_a (x + y)$ **71.** 7 **73.** True
75. True **77.** True **79.** -2 **81.** 3
83.

$$\log_a \left(\frac{x + \sqrt{x^2 - 5}}{5} \cdot \frac{x - \sqrt{x^2 - 5}}{x - \sqrt{x^2 - 5}}\right)$$
$$= \log_a \frac{5}{5(x - \sqrt{x^2 - 5})}$$
$$= -\log_a (x - \sqrt{x^2 - 5})$$

Exercise Set 3.5

1. $\{4\}$ **3.** $\left\{\frac{3}{2}\right\}$ **5.** $\{5.044\}$ **7.** $\left\{\frac{5}{2}\right\}$ **9.** $\left\{-3, \frac{1}{2}\right\}$
11. $\{0.959\}$ **13.** $\{6.908\}$ **15.** $\{84.191\}$ **17.** $\{-1.710\}$
19. $\{2.844\}$ **21.** $\{-1.567, 1.567\}$ **23.** $\{0.347\}$ **25.** $\{625\}$
27. $\{0.0001\}$ **29.** $\{e\}$ **31.** $\left\{\frac{22}{3}\right\}$ **33.** $\{10\}$ **35.** $\left\{\frac{1}{63}\right\}$
37. $\{5\}$ **39.** $\left\{\frac{21}{8}\right\}$ **41.** $\{0.367\}$ **43.** $\{0.621\}$

45. $\{-1.532\}$ **47.** $\{7.062\}$ **49.** $\{2.444\}$
51. $(4.093, 0.786)$ **53.** $(7.586, 6.684)$
55. $\{-3i, 3i, -2, 2\}$
57. **(a)** $y = 400.27x + 10{,}655.15$, \$14,257.6 million;
(b) $y = -58.325x^2 + 691.895x + 10{,}363.525$,
\$11,866.3 million
59. **61.** $\{10\}$ **63.** $\{1, e^4\}$, or $\{1, 54.598\}$ **65.** $\left\{\frac{1}{3}, 27\right\}$
67. $\{1, e^2\}$, or $\{1, 7.389\}$ **69.** $\{0, 0.431\}$ **71.** $\{-9, 9\}$
73. $\{e^{-2}, e^2\}$, or $\{0.135, 7.389\}$ **75.** $\left\{\frac{7}{4}\right\}$ **77.** $(56.598, \infty)$
79. $\{5\}$ **81.** $a = \frac{2}{3}b$ **83.** 88

EXERCISE SET 3.6

1. **(a)** $P(t) = 6.6e^{0.028t}$; **(b)** 7.2 billion, 9.5 billion;
(c) in 6.9 yr; **(d)** 24.8 yr
3. **(a)** 19.8 yr; **(b)** 1% per year; **(c)** 6.93 yr; **(d)** 17.3 yr;
(e) 2.8% per year; **(f)** 3.5% per year
5. After about 69 yr
7. **(a)** $P(t) = 10{,}000e^{0.054t}$; **(b)** \$10,555; \$11,140; \$13,100;
\$17,160; **(c)** 12.8 yr
9. About 5135 yr
11. **(a)** 23.1% per minute; **(b)** 3.15% per year; **(c)** 7.2 days;
(d) 11 yr; **(e)** 2.8% per year; **(f)** 0.015% per year; **(g)** 0.003%
per year
13. **(a)** $k \approx 0.016$; $P(t) = 80e^{-0.016t}$; **(b)** about 63 lb per
person; **(c)** 86.6 yr
15. **(a)** 96; **(b)** 286, 1005, 1719, 1971, 1997;
(c) $y = \dfrac{2000}{1 + 19.9e^{-0.6x}}$

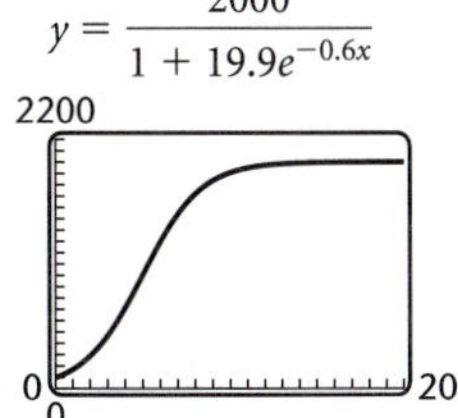

17. **(a)** 400, 520, 1214, 2059, 2396, 2478;
(b) $y = \dfrac{2500}{1 + 5.25e^{-0.32x}}$

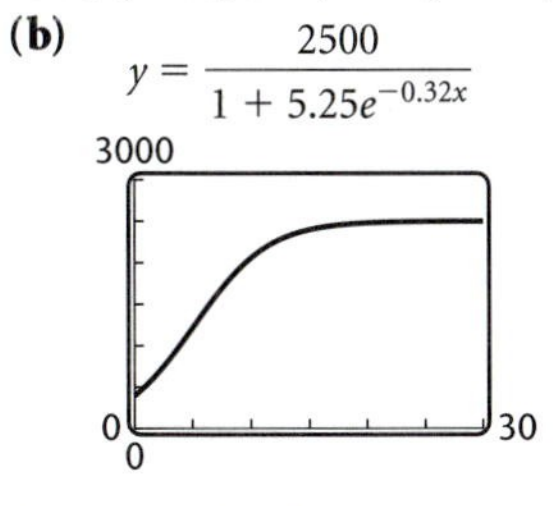

19. (f) **21.** (b) **23.** (c)
25. **(a)** Yes; **(b)** $y = 0.785(1.490)^x$, or $y = 0.785e^{0.399x}$;
(c) \$63.1 billion; **(d)** 1997

27. **(a)** Exponential function: $y = (107.9)(1.034)^x$, or
$y = (107.9)e^{0.033x}$; power function: $y = 110.8x^{0.077}$;
(b) $y_1 = (107.9)(1.034)^x$, $y_2 = 110.8x^{0.077}$

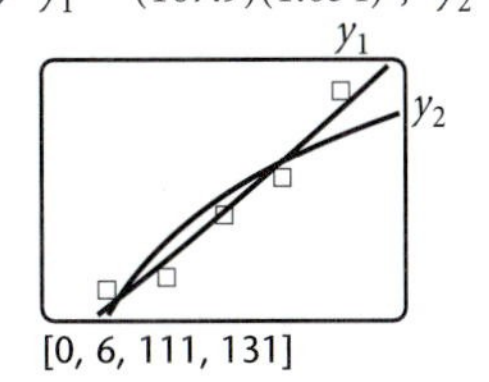

[0, 6, 111, 131]

(c) exponential function: \$178; power function: \$136;
(d) The exponential function has a higher correlation
coefficient and seems to be following data better.
29. $\sqrt{3}$ **31.** $\sqrt{3365}$ **33.** **35.** \$14,182.70
37. \$166.16 **39.** 46.7°F **41.** $t = -\dfrac{L}{R}\left[\ln\left(1 - \dfrac{iR}{V}\right)\right]$
43. Linear

REVIEW EXERCISES, CHAPTER 3

1. [3.1] $\{(-2.7, 1.3), (-3, 8), (3, -5), (-3, 6), (-5, 7)\}$
2. [3.1] **(a)** $x = 3y^2 + 2y - 1$; **(b)** $0.8y^3 - 5.4x^2 = 3y$
3. [3.1] **(a)** Yes; **(b)** $f^{-1}(x) = x^2 + 6, x \geq 0$
4. [3.1] **(a)** Yes; **(b)** $f^{-1}(x) = \sqrt[3]{x + 8}$ **5.** [3.1] **(a)** No
6. [3.1] **(a)** Yes; **(b)** $f^{-1}(x) = \ln x$ **7.** [3.1] 657
8. [3.2] (c) **9.** [3.3] (a) **10.** [3.3] (b) **11.** [3.2] (f)
12. [3.2] (e) **13.** [3.3] (d) **14.** [3.3] $4^2 = x$
15. [3.3] $\log_e 80 = x$ **16.** [3.5] $\{16\}$ **17.** [3.5] $\left\{\frac{1}{5}\right\}$
18. [3.5] $\{4.382\}$ **19.** [3.5] $\{2\}$ **20.** [3.5] $\left\{\frac{1}{2}\right\}$
21. [3.5] $\{5\}$ **22.** [3.5] $\{4\}$ **23.** [3.5] $\{9\}$ **24.** [3.5] $\{1\}$
25. [3.5] $\{3.912\}$ **26.** [3.4] $\log_b \dfrac{x^3\sqrt{z}}{y^4}$
27. [3.4] $\ln(x^2 - 4)$ **28.** [3.4] $\frac{1}{4}\ln w + \frac{1}{2}\ln r$
29. [3.4] $\frac{2}{3}\log M - \frac{1}{3}\log N$ **30.** [3.4] 0.477
31. [3.4] 1.699 **32.** [3.4] -0.699 **33.** [3.4] 0.233
34. [3.4] $-5k$ **35.** [3.4] $-6t$ **36.** [3.6] 12.8 yr
37. [3.6] 2.3% **38.** [3.6] About 2623 yr **39.** [3.3] 5.6
40. [3.3] 30 decibels
41. [3.6] **(a)** $P(t) = 52e^{0.028t}$; **(b)** 164 million, 287 million;
(c) in 62.6 yr; **(d)** 24.8 yr
42. [3.6] **(a)** $k = 0.304$; **(b)** $P(t) = 7e^{0.304t}$;
(c) $y = 7e^{0.304x}$

50,000
0
0
30

(d) 86.7 billion, 727.9 billion; **(e)** 1993

43. [3.3] **(a)** 2.3 ft/sec; **(b)** 8,553,143

44. [3.6] **(a)** Yes; **(b)** $y = (2.329)1.753^x$, or $y = 2.329e^{0.561x}$; **(c)** 207.7 million; **(d)** in 6.7 yr

45. [3.4] By the product rule, $\log_2 x + \log_2 5 = \log_2 5x$, not $\log_2 (x + 5)$. Also, substituting various numbers for x shows that both sides of the inequality are indeed unequal. You could also graph each side and show that the graphs do not coincide.

46. [3.1] The inverse of a function $f(x)$ is written $f^{-1}(x)$, whereas $[f(x)]^{-1}$ means $\dfrac{1}{f(x)}$.

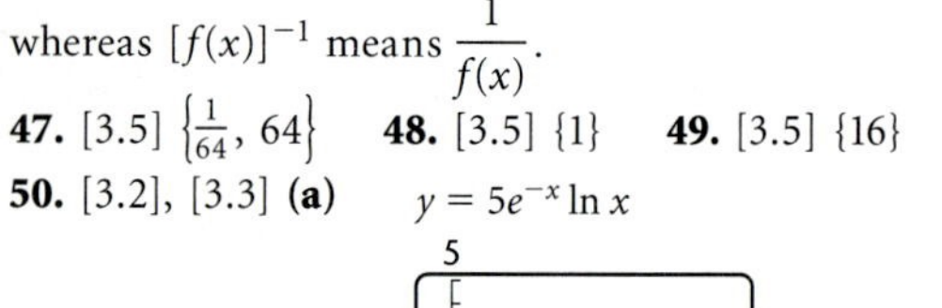

47. [3.5] $\left\{\frac{1}{64}, 64\right\}$ **48.** [3.5] {1} **49.** [3.5] {16}

50. [3.2], [3.3] **(a)** $y = 5e^{-x} \ln x$

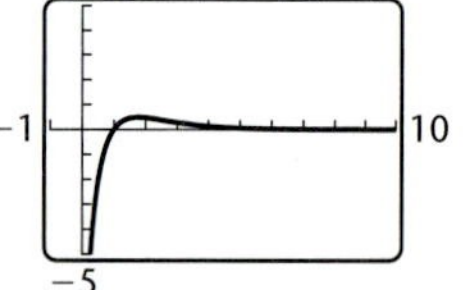

(b) Relative maximum: 0.486 at $x = 1.763$, (1.763, 0.486)

51. [3.1] No **52.** [3.3] $(1, \infty)$

53. [3.2] (−0.393, 1.143), (0.827, 1.495)

Chapter 4

Exercise Set 4.1

1. $\sin \phi = \dfrac{15}{17}$, $\cos \phi = \dfrac{8}{17}$, $\tan \phi = \dfrac{15}{8}$, $\csc \phi = \dfrac{17}{15}$, $\sec \phi = \dfrac{17}{8}$, $\cot \phi = \dfrac{8}{15}$

3. $\sin \alpha = \dfrac{\sqrt{3}}{2}$, $\cos \alpha = \dfrac{1}{2}$, $\tan \alpha = \sqrt{3}$, $\csc \alpha = \dfrac{2\sqrt{3}}{3}$, $\sec \alpha = 2$, $\cot \alpha = \dfrac{\sqrt{3}}{3}$

5. $\sin \phi = \dfrac{7\sqrt{65}}{65}$, $\cos \phi = \dfrac{4\sqrt{65}}{65}$, $\tan \phi = \dfrac{7}{4}$, $\csc \phi = \dfrac{\sqrt{65}}{7}$, $\sec \phi = \dfrac{\sqrt{65}}{4}$, $\cot \phi = \dfrac{4}{7}$

7. $\cos \theta = \dfrac{7}{25}$, $\tan \theta = \dfrac{24}{7}$, $\csc \theta = \dfrac{25}{24}$, $\sec \theta = \dfrac{25}{7}$, $\cot \theta = \dfrac{7}{24}$

9. $\sin \phi = \dfrac{2\sqrt{5}}{5}$, $\cos \phi = \dfrac{\sqrt{5}}{5}$, $\csc \phi = \dfrac{\sqrt{5}}{2}$, $\sec \phi = \sqrt{5}$, $\cot \phi = \dfrac{1}{2}$

11. $\sin \theta = \dfrac{2}{3}$, $\cos \theta = \dfrac{\sqrt{5}}{3}$, $\tan \theta = \dfrac{2\sqrt{5}}{5}$, $\sec \theta = \dfrac{3\sqrt{5}}{5}$, $\cot \theta = \dfrac{\sqrt{5}}{2}$ **13.** $\dfrac{\sqrt{2}}{2}$ **15.** 2 **17.** $\dfrac{\sqrt{3}}{3}$ **19.** $\dfrac{1}{2}$

21. 127.3 ft **23.** 9.72° **25.** 35.83° **27.** 3.03° **29.** 49.65° **31.** 17°36′ **33.** 83°1′30″ **35.** 11°45′ **37.** 47°49′36″ **39.** 0.6293 **41.** 0.0737 **43.** 1.2765 **45.** 0.7621 **47.** 0.9336 **49.** 12.4288 **51.** 1.0000 **53.** 1.7032 **55.** 30.8° **57.** 12.5° **59.** 64.4° **61.** 46.5° **63.** 25.2° **65.** 38.6° **67.** 45° **69.** 60° **71.** 45°

73. $\cos 20° = \sin 70° = \dfrac{1}{\sec 20°}$

75. $\tan 52° = \cot 38° = \dfrac{1}{\cot 52°}$

77. $\sin 25° \approx 0.4226$, $\cos 25° \approx 0.9063$, $\tan 25° \approx 0.4663$, $\csc 25° \approx 2.366$, $\sec 25° \approx 1.103$, $\cot 25° \approx 2.145$ **79.** 1.52

81. −32.03 **83.** 27.5 **85.** **87.** 0.6534

89. Area $= \frac{1}{2}ab$. But $a = c \sin A$, so Area $= \frac{1}{2}bc \sin A$.

Exercise Set 4.2

1. $F = 60°$, $d = 3$, $f \approx 5.2$ **3.** $A = 22.7°$, $a \approx 52.7$, $c \approx 136.6$ **5.** $P = 47°38'$, $n \approx 34.4$, $p \approx 25.4$

7. $B = 2°17'$, $b \approx 0.39$, $c = 9.74$

9. $A \approx 77.2°$, $B \approx 12.8°$, $a \approx 439$

11. $B = 42.42°$, $a \approx 35.7$, $b \approx 32.6$

13. $B = 55°$, $a \approx 28.0$, $c \approx 48.8$

15. $A \approx 62.4°$, $B \approx 27.6°$, $a \approx 3.56$ **17.** About 62.2 ft **19.** About 2.5 ft **21.** About 606 ft **23.** 96.7 cm **25.** 599 ft **27.** 8 km **29.** 275 ft **31.** 24 km

33. $10\sqrt{2}$, or about 14.142 **35.** $\dfrac{\sqrt{3}}{2}$ **37.** $\dfrac{\sqrt{3}}{3}$ **39.**

41. 3.3 **43.** Cut so that $\theta = 79.38°$ **45.** $\theta \approx 27°$

Exercise Set 4.3

1. III **3.** III **5.** I **7.** III **9.** II **11.** II

13. 434°, 794°, −286°, −646°

15. 475.3°, 835.3°, −244.7°, −604.7°

17. 180°, 540°, −540°, −900° **19.** 72.89°, 162.89°

21. 77°56′46″, 167°56′46″ **23.** 44.8°, 134.8°

25. $\sin \beta = \dfrac{5}{13}$, $\cos \beta = -\dfrac{12}{13}$, $\tan \beta = -\dfrac{5}{12}$, $\csc \beta = \dfrac{13}{5}$, $\sec \beta = -\dfrac{13}{12}$, $\cot \beta = -\dfrac{12}{5}$

27. $\sin \phi = -\dfrac{2\sqrt{7}}{7}$, $\cos \phi = -\dfrac{\sqrt{21}}{7}$, $\tan \phi = \dfrac{2\sqrt{3}}{3}$, $\csc \phi = -\dfrac{\sqrt{7}}{2}$, $\sec \phi = -\dfrac{\sqrt{21}}{3}$, $\cot \phi = \dfrac{\sqrt{3}}{2}$

29. $\sin \theta = -\dfrac{2\sqrt{13}}{13}$, $\cos \theta = \dfrac{3\sqrt{13}}{13}$, $\tan \theta = -\dfrac{2}{3}$

31. $\sin \theta = \dfrac{5\sqrt{41}}{41}$, $\cos \theta = \dfrac{4\sqrt{41}}{41}$, $\tan \theta = \dfrac{5}{4}$

33. $\cos \theta = -\dfrac{2\sqrt{2}}{3}$, $\tan \theta = \dfrac{\sqrt{2}}{4}$, $\csc \theta = -3$, $\sec \theta = -\dfrac{3\sqrt{2}}{4}$, $\cot \theta = 2\sqrt{2}$

35. $\sin\theta = -\frac{\sqrt{5}}{5}$, $\cos\theta = \frac{2\sqrt{5}}{5}$, $\tan\theta = -\frac{1}{2}$, $\csc\theta = -\sqrt{5}$, $\sec\theta = \frac{\sqrt{5}}{2}$

37. $\sin\phi = -\frac{4}{5}$, $\tan\phi = -\frac{4}{3}$, $\csc\phi = -\frac{5}{4}$, $\sec\phi = \frac{5}{3}$, $\cot\phi = -\frac{3}{4}$ **39.** $-\frac{\sqrt{3}}{2}$ **41.** 1 **43.** 0 **45.** $-\frac{\sqrt{2}}{2}$

47. 2 **49.** $\sqrt{3}$ **51.** $-\frac{\sqrt{3}}{3}$ **53.** Undefined **55.** −1

57. $\sqrt{3}$ **59.** $\frac{\sqrt{2}}{2}$ **61.** $-\sqrt{2}$ **63.** 1 **65.** 0 **67.** 0

69. 0 **71.** Positive: cos, sec; negative: sin, csc, tan, cot
73. Positive: tan, cot; negative: sin, csc, cos, sec
75. Positive: sin, csc; negative: cos, sec, tan, cot
77. Positive: all
79. $\sin 319° = -0.6561$, $\cos 319° = 0.7547$, $\tan 319° = -0.8693$, $\csc 319° = -1.5242$, $\sec 319° = 1.3250$, $\cot 319° = -1.1504$
81. $\sin 115° = 0.9063$, $\cos 115° = -0.4226$, $\tan 115° = -2.1445$, $\csc 115° = 1.1034$, $\sec 115° = -2.3663$, $\cot 115° = -0.4663$ **83.** East: 130 km; south: 75 km
85. 223 km **87.** −1.1585 **89.** −1.4910 **91.** 0.8771
93. 0.4352 **95.** 0.9563 **97.** 2.9238 **99.** 275.4°
101. 200.1° **103.** 288.1° **105.** 72.6° **107.** 44.18°
109. 66 ft/sec **111.** **113.** 19.625 in.

EXERCISE SET 4.4

1.

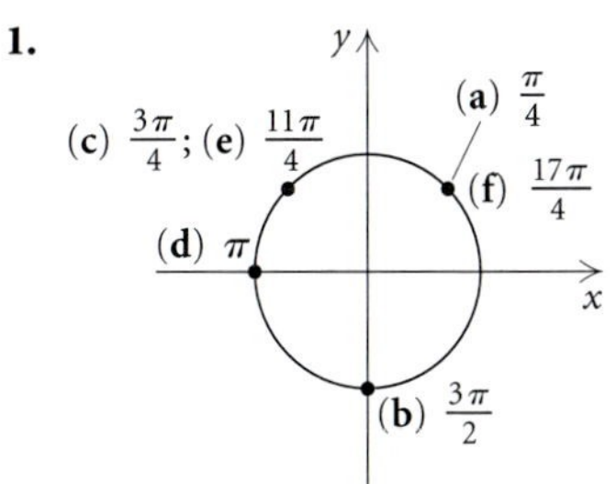

3.

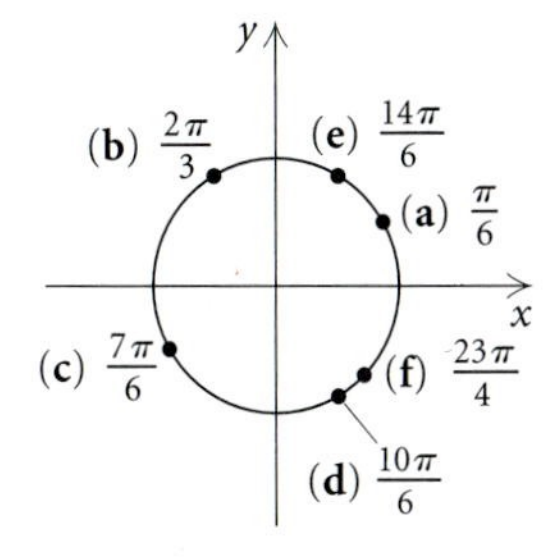

5. M: $\frac{2\pi}{3}$, $-\frac{4\pi}{3}$; N: $\frac{3\pi}{2}$, $-\frac{\pi}{2}$; P: $\frac{5\pi}{4}$, $-\frac{3\pi}{4}$; Q: $\frac{11\pi}{6}$, $-\frac{\pi}{6}$

7.

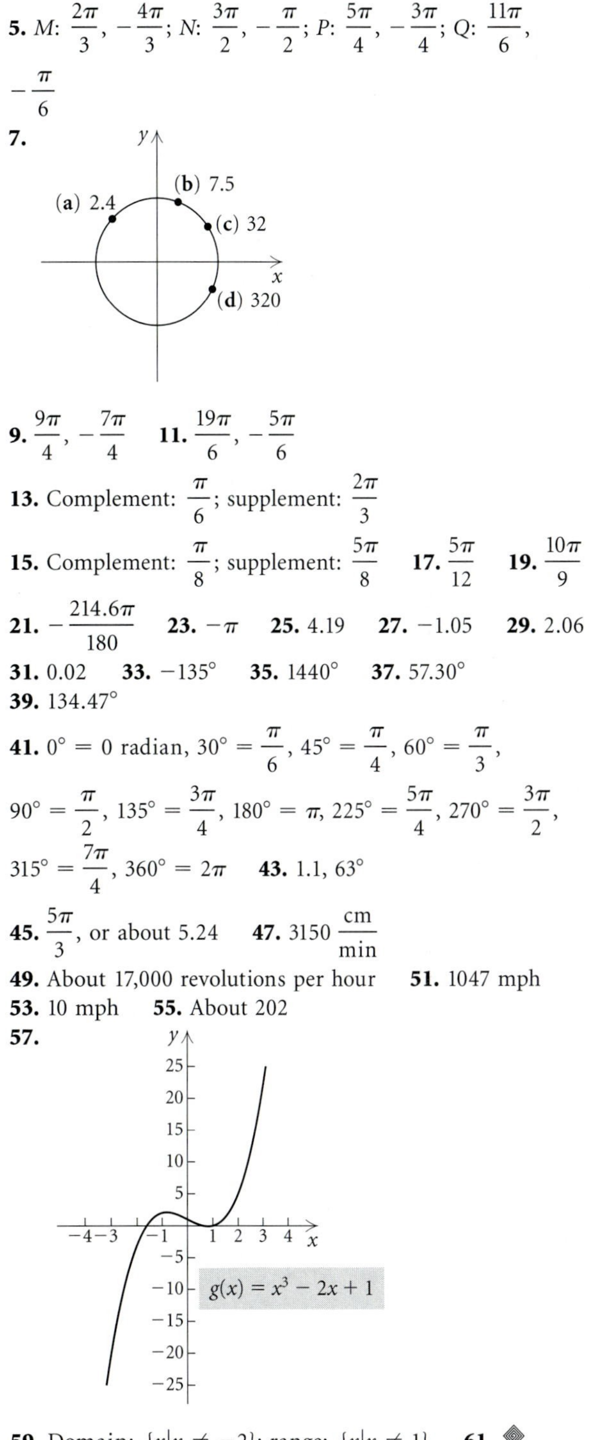

9. $\frac{9\pi}{4}$, $-\frac{7\pi}{4}$ **11.** $\frac{19\pi}{6}$, $-\frac{5\pi}{6}$

13. Complement: $\frac{\pi}{6}$; supplement: $\frac{2\pi}{3}$

15. Complement: $\frac{\pi}{8}$; supplement: $\frac{5\pi}{8}$ **17.** $\frac{5\pi}{12}$ **19.** $\frac{10\pi}{9}$

21. $-\frac{214.6\pi}{180}$ **23.** $-\pi$ **25.** 4.19 **27.** −1.05 **29.** 2.06

31. 0.02 **33.** −135° **35.** 1440° **37.** 57.30°
39. 134.47°

41. $0° = 0$ radian, $30° = \frac{\pi}{6}$, $45° = \frac{\pi}{4}$, $60° = \frac{\pi}{3}$, $90° = \frac{\pi}{2}$, $135° = \frac{3\pi}{4}$, $180° = \pi$, $225° = \frac{5\pi}{4}$, $270° = \frac{3\pi}{2}$, $315° = \frac{7\pi}{4}$, $360° = 2\pi$ **43.** 1.1, 63°

45. $\frac{5\pi}{3}$, or about 5.24 **47.** $3150\ \frac{\text{cm}}{\text{min}}$

49. About 17,000 revolutions per hour **51.** 1047 mph
53. 10 mph **55.** About 202

57.

59. Domain: $\{x|x \neq -2\}$; range: $\{x|x \neq 1\}$ **61.**
63. **65.** 111.7 km, 69.8 mi **67.** (**a**) 5°37′30″; (**b**) 19°41′15″
69. 1.676 radians/sec **71.** 1.46 nautical miles

EXERCISE SET 4.5

1. **(a)** $\left(-\frac{3}{4}, -\frac{\sqrt{7}}{4}\right)$; **(b)** $\left(\frac{3}{4}, \frac{\sqrt{7}}{4}\right)$; **(c)** $\left(\frac{3}{4}, -\frac{\sqrt{7}}{4}\right)$

3. **(a)** $\left(\frac{2}{5}, \frac{\sqrt{21}}{5}\right)$; **(b)** $\left(-\frac{2}{5}, -\frac{\sqrt{21}}{5}\right)$; **(c)** $\left(-\frac{2}{5}, \frac{\sqrt{21}}{5}\right)$

5. $\left(\frac{\sqrt{2}}{2}, -\frac{\sqrt{2}}{2}\right)$ **7.** 0 **9.** $\sqrt{3}$ **11.** 0 **13.** $-\frac{\sqrt{3}}{2}$

15. 1 **17.** $\frac{\sqrt{3}}{2}$ **19.** $-\frac{\sqrt{2}}{2}$ **21.** 0 **23.** 0 **25.** 0.4816

27. 1.3065 **29.** −2.1599 **31.** 1 **33.** −1.1747 **35.** −1

37. −0.7071 **39.** 0 **41.** 0.8391

43. **(a)**

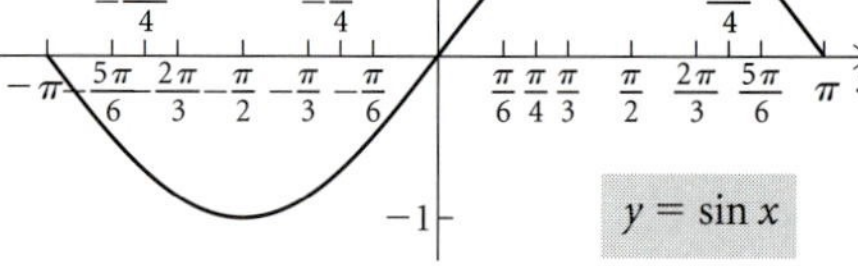

(b)

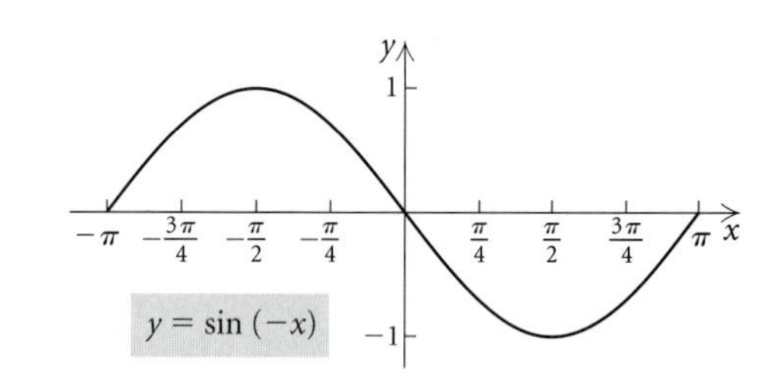

(c) same as (b); **(d)** the same **45.** **(a)** See Exercise 43(a);

(b)

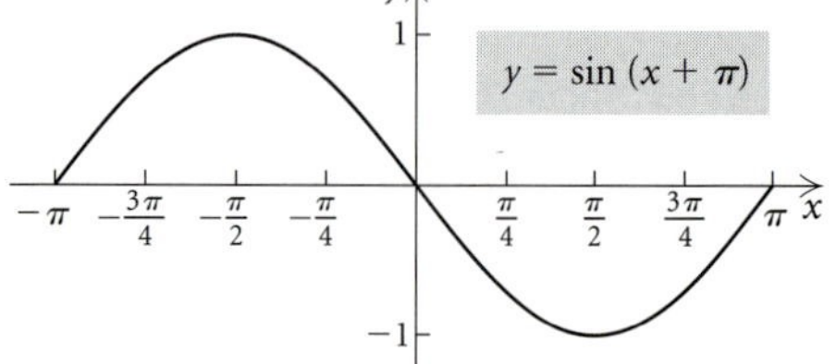

(c) same as (b); **(d)** the same

47. **(a)**

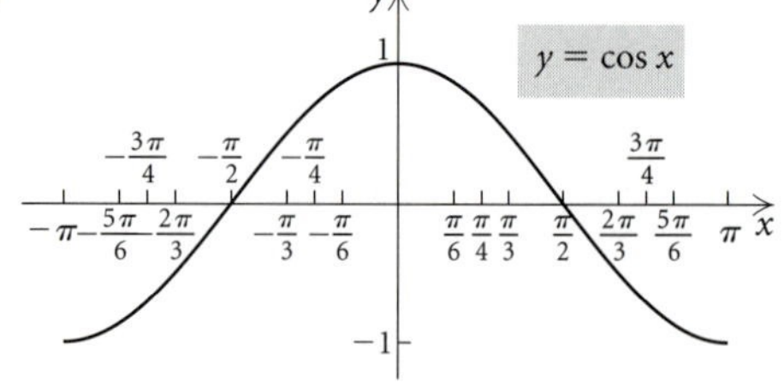

(b)

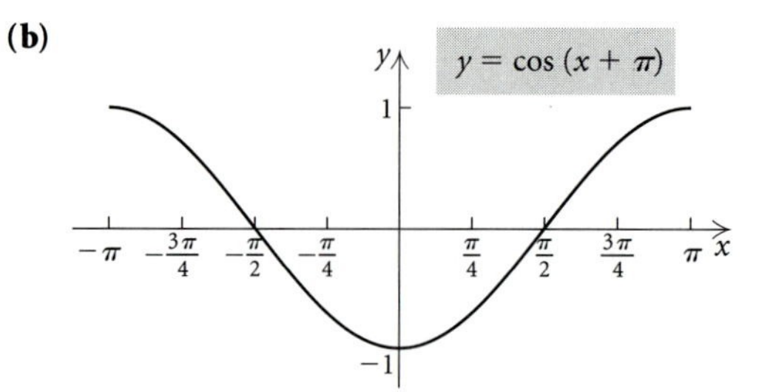

(c) same as (b); **(d)** the same **49.** Even: cosine, secant; odd: sine, tangent, cosecant, cotangent **51.** Positive: I, III; negative: II, IV **53.** Positive: I, IV; negative: II, III **55.** Stretch the graph of f vertically, then shift it down 3 units.

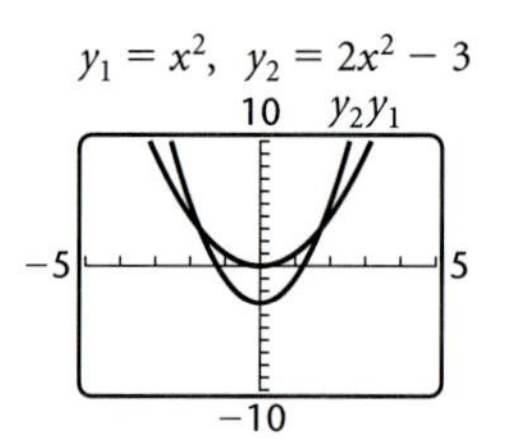

57. Shift the graph of f to the right 4 units, shrink it vertically, then shift it up 1 unit.

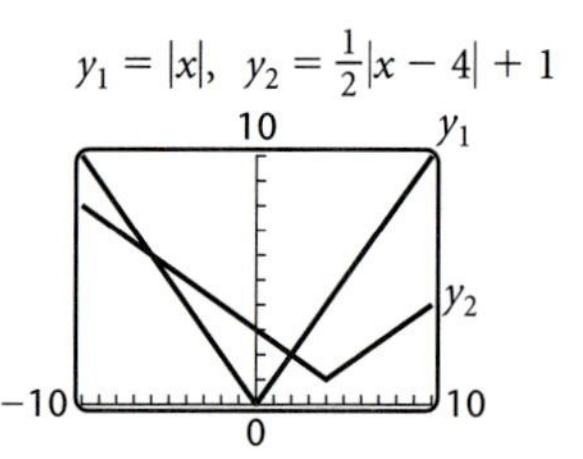

59. $y = -(x - 2)^3 - 1$ **61.** ◆ **63.** $\cos x$ **65.** $\sin x$ **67.** $\sin x$ **69.** $-\cos x$ **71.** $-\sin x$

73. **(a)** $\frac{\pi}{2} + 2k\pi,\ k \in \mathbb{Z}$; **(b)** $(2k + 1)\pi,\ k \in \mathbb{Z}$;

(c) $k\pi,\ k \in \mathbb{Z}$; **(d)** $\left(-\frac{3\pi}{4} + 2k\pi, \frac{\pi}{4} + 2k\pi\right),\ k \in \mathbb{Z}$

75. Domain: $(-\infty, \infty)$; range: $[0, 1]$; period: π;

amplitude: $\frac{1}{2}$ **77.** $\left[-\frac{\pi}{2} + 2k\pi, \frac{\pi}{2} + 2k\pi\right],\ k \in \mathbb{Z}$

79. $\left\{x \mid x \neq \frac{\pi}{2} + k\pi,\ k \in \mathbb{Z}\right\}$ **81.** 1

83.

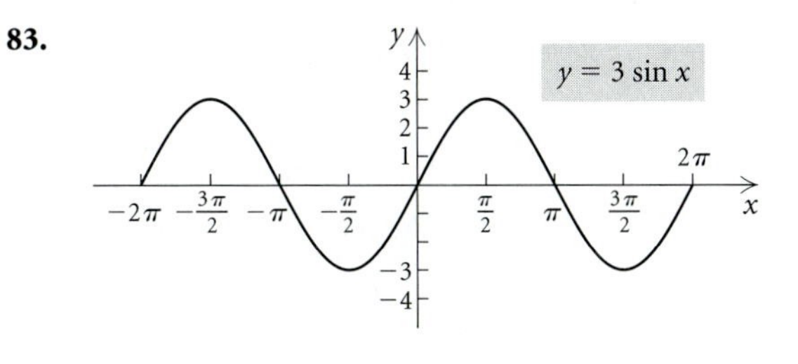

85.

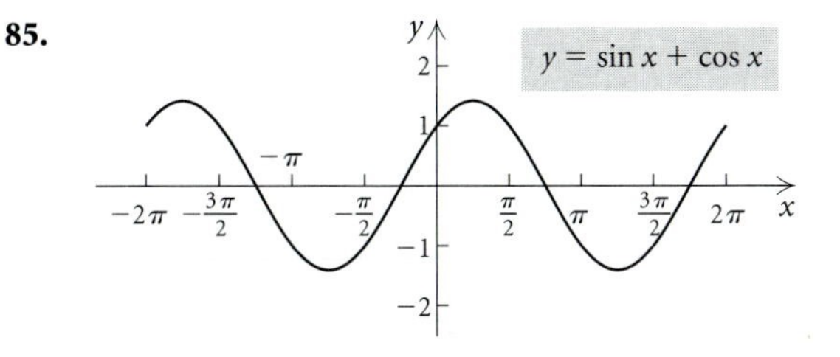

EXERCISE SET 4.6

1. Amplitude: 1; period: 2π; phase shift: 0

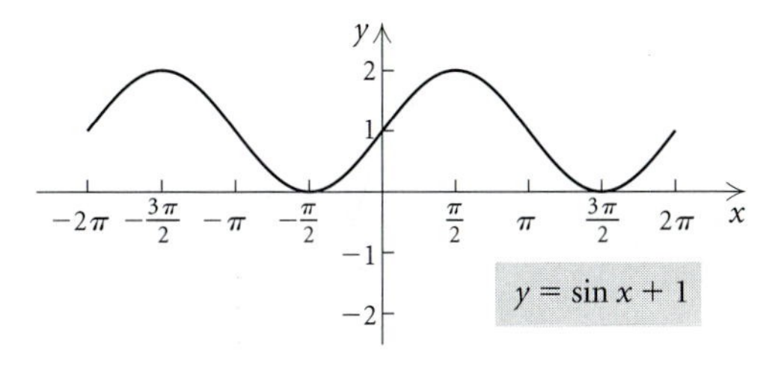

3. Amplitude: 3; period: 2π; phase shift: 0

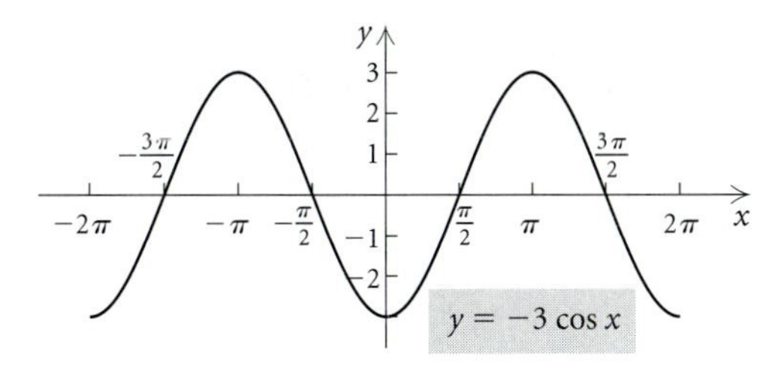

5. Amplitude: 2; period: 4π; phase shift: 0

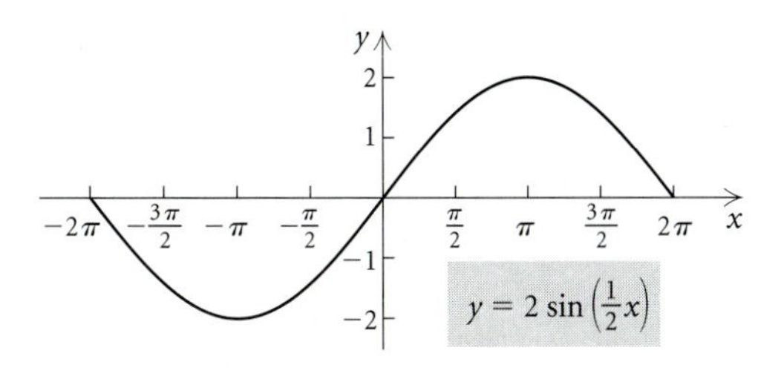

7. Amplitude: $\frac{1}{2}$; period: 2π; phase shift: $-\frac{\pi}{2}$

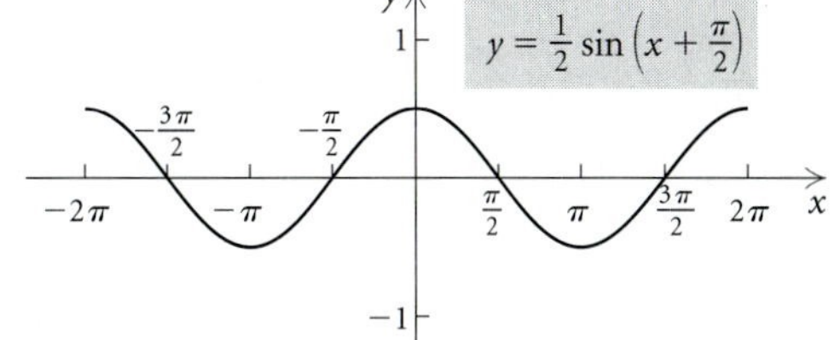

9. Amplitude: 3; period: 2π; phase shift: π

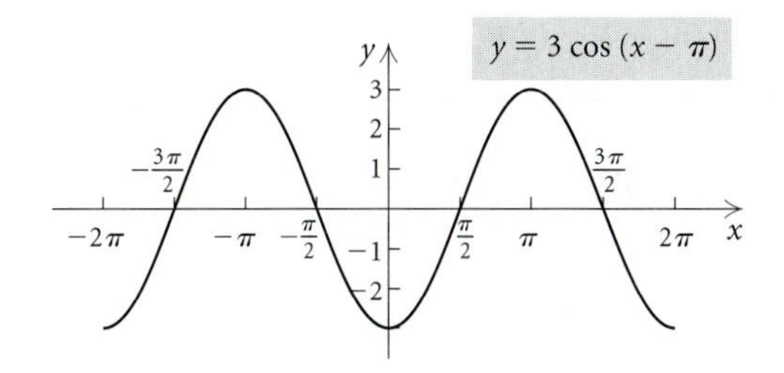

11. Amplitude: $\frac{1}{3}$; period: 2π; phase shift: 0

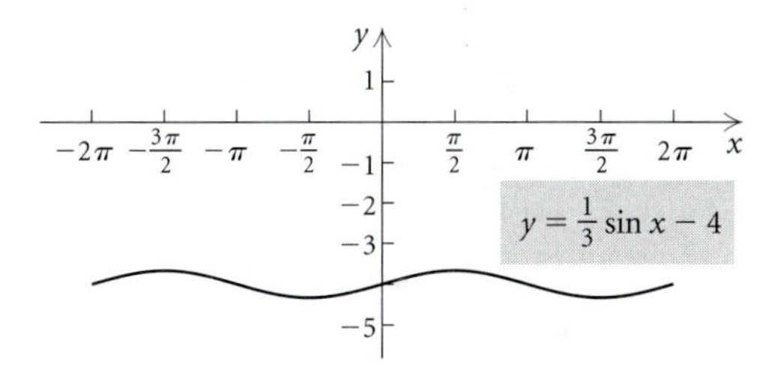

13. Amplitude: 1; period: 2π; phase shift: 0

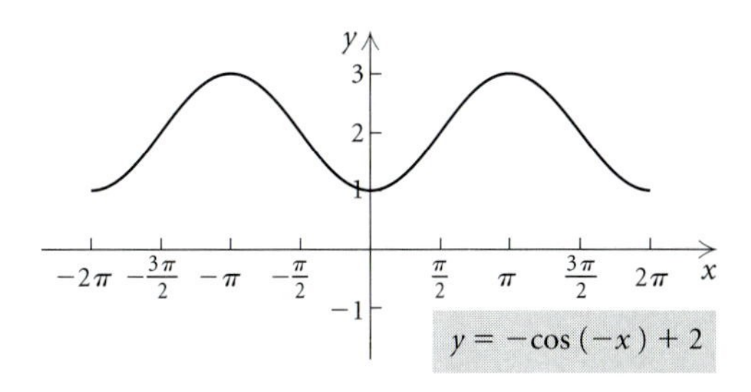

15. Amplitude: 2; period: 4π; phase shift: π

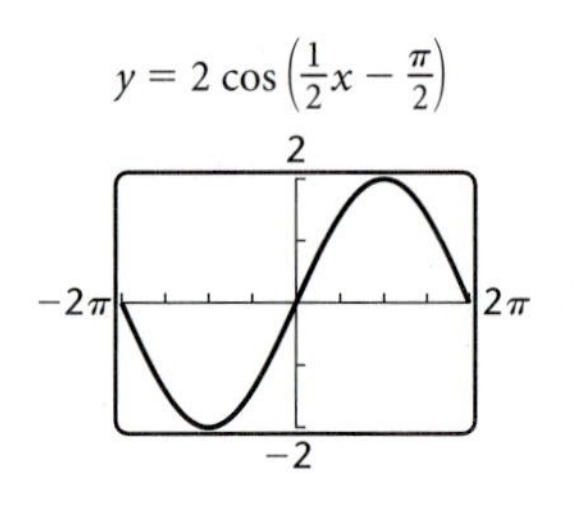

17. Amplitude: $\dfrac{1}{2}$; period: π; phase shift: $-\dfrac{\pi}{4}$

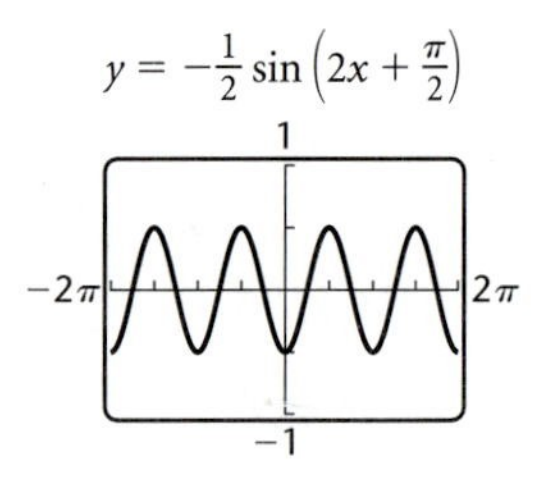

19. Amplitude: 3; period: 2; phase shift: $\dfrac{3}{\pi}$

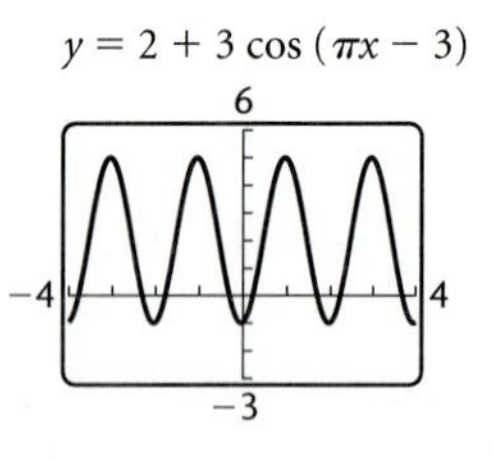

21. Amplitude: $\dfrac{1}{2}$; period: 1; phase shift: 0

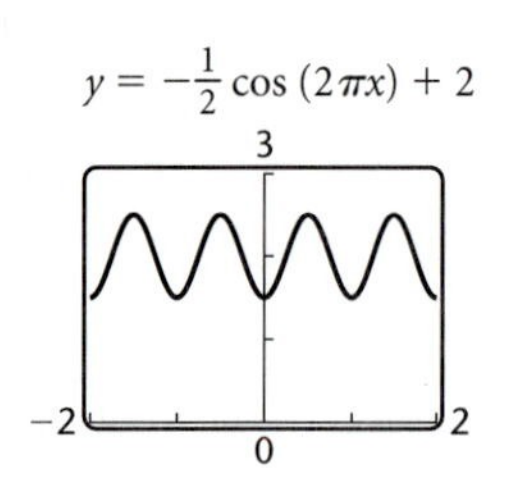

23. Amplitude: 1; period: 4π; phase shift: π

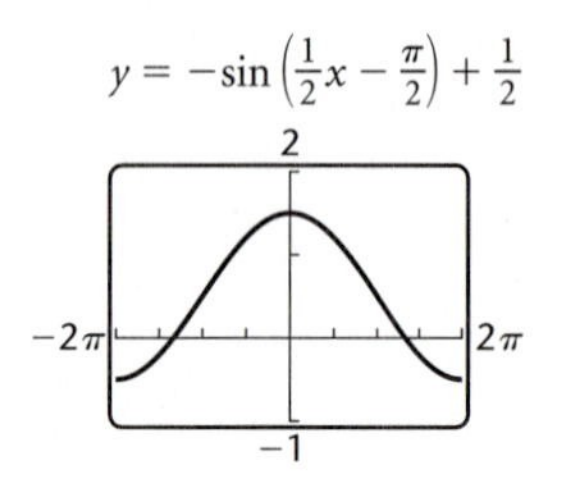

25. Amplitude: 1; period: 1; phase shift: 0

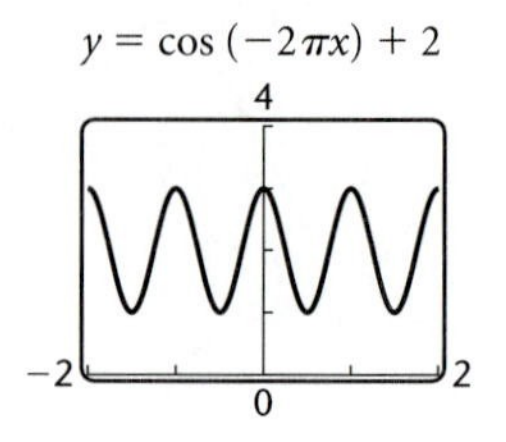

27. Amplitude: $\dfrac{1}{4}$; period: 2; phase shift: $\dfrac{4}{\pi}$

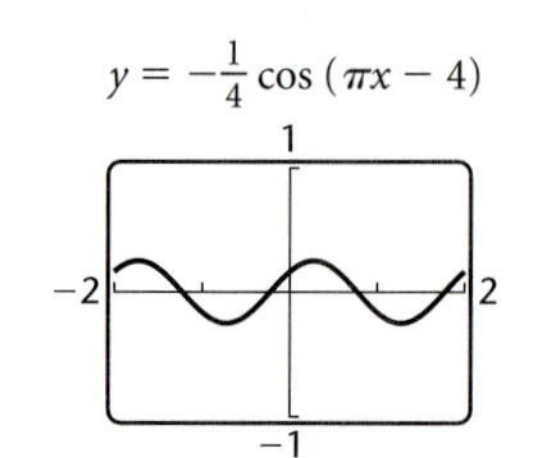

29. (b) **31.** (h) **33.** (a) **35.** (f)

37.

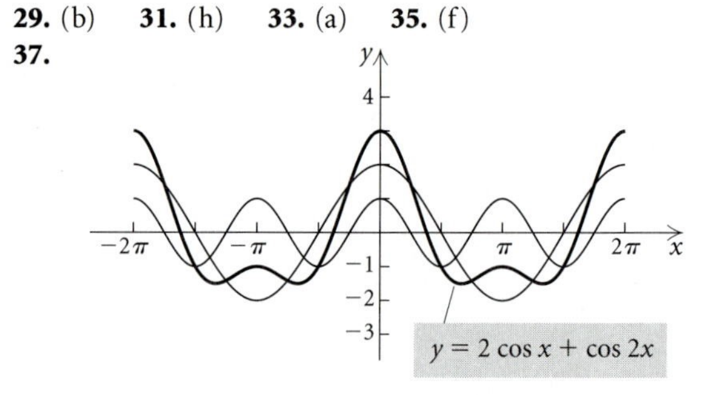

39.

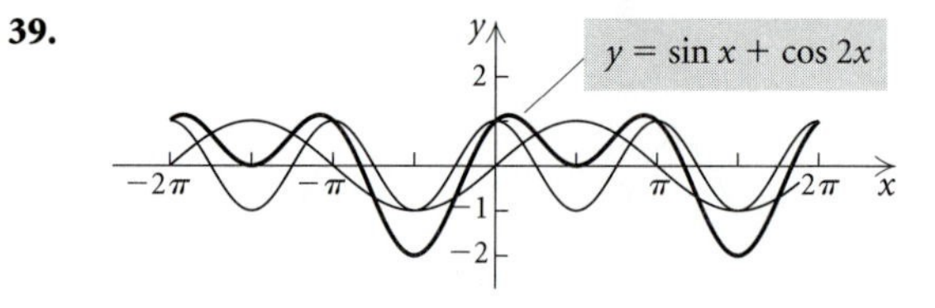

41.

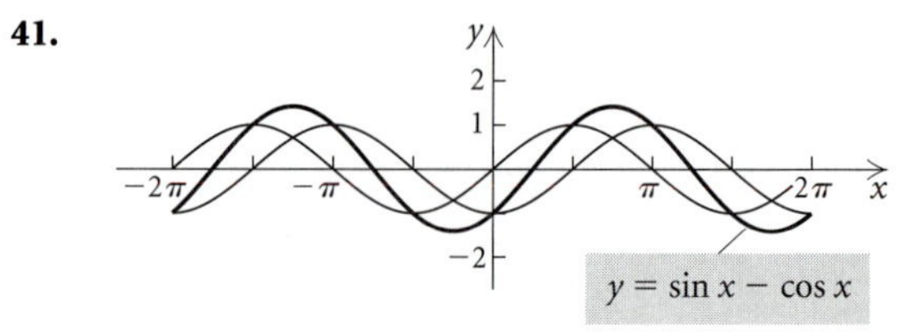

43.

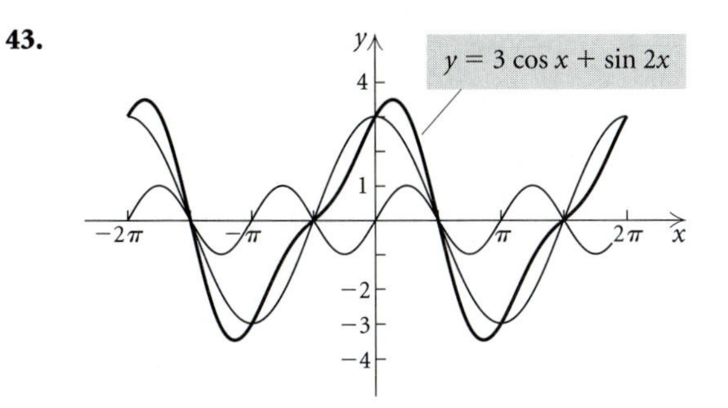

45. $y = x + \sin x$

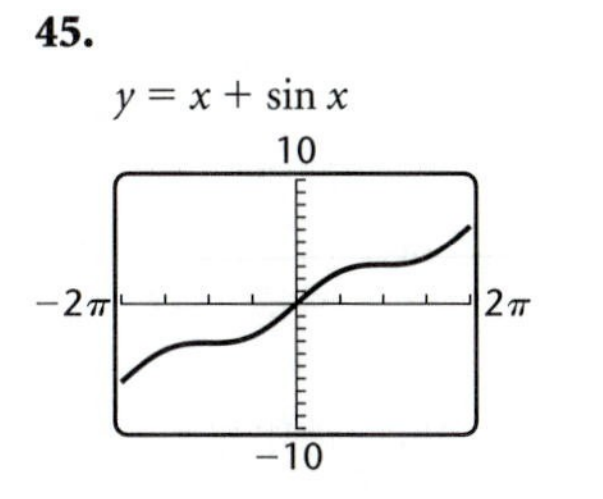

47. $y = \cos 2x + 2x$

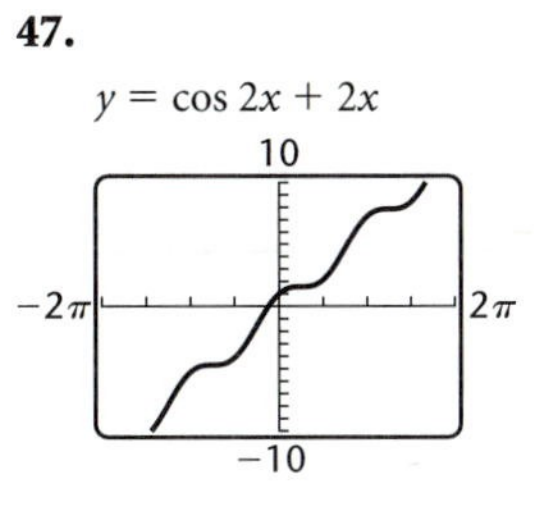

49. $y = 4 \cos 2x - 2 \sin x$

51. $\dfrac{6x + 4}{(x + 1)(x - 1)}$ **53.** $\dfrac{x - 4}{x - 1}$ **55.** ◈

57. **(a)** $y = 7 \sin(-2.6180x + 0.5236) + 7$; **(b)** \$10,500, \$13,062

59.

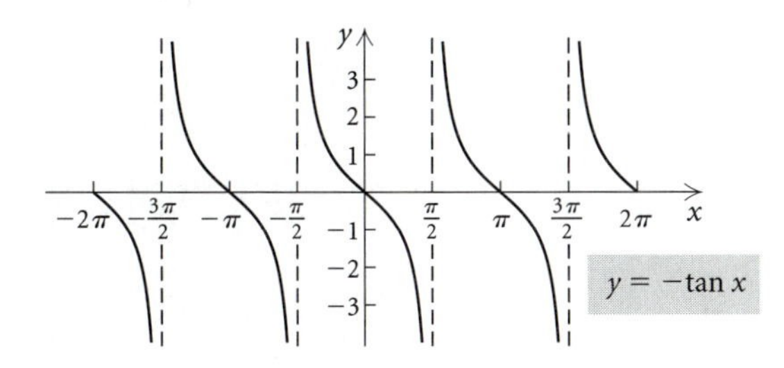

61.

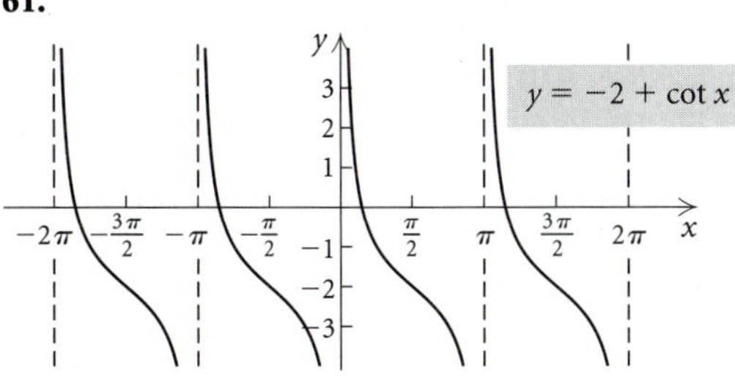

63.

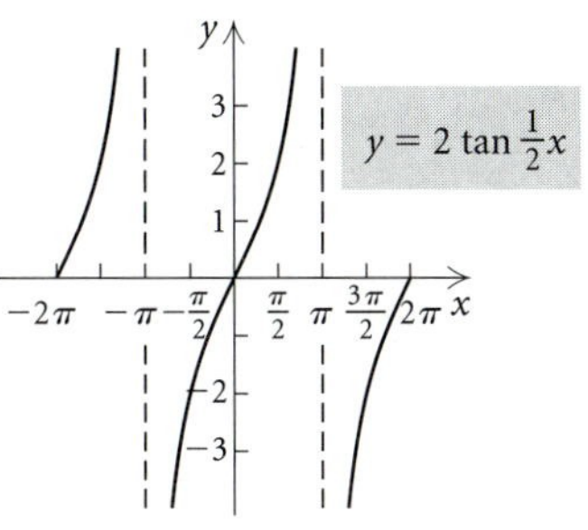

65.

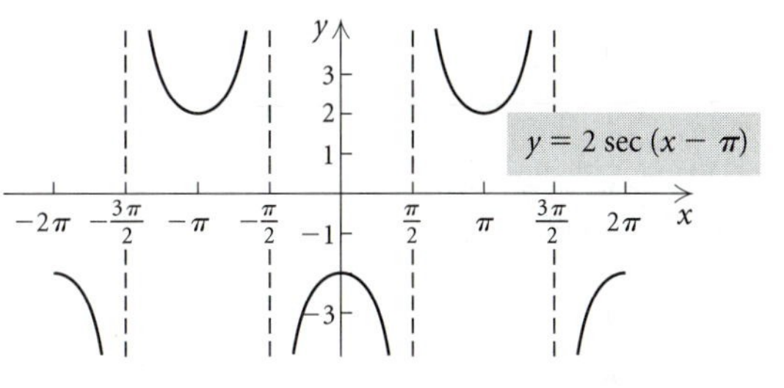

67.

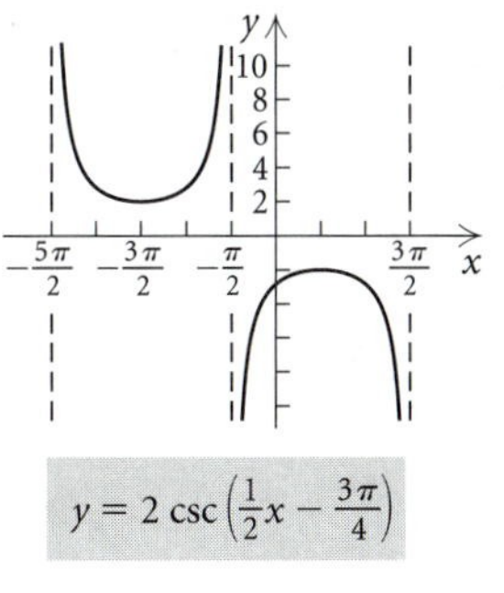

69. $y = e^{-x/2} \cos x$

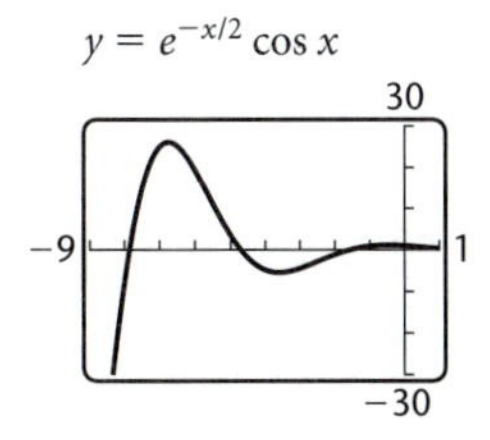

71. $y = 0.6x^2 \cos x$

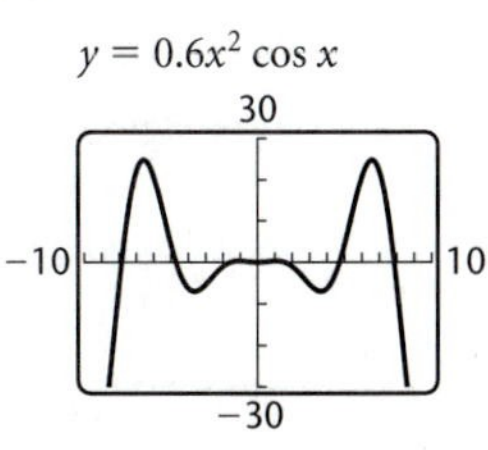

73. $y = |\tan x|$

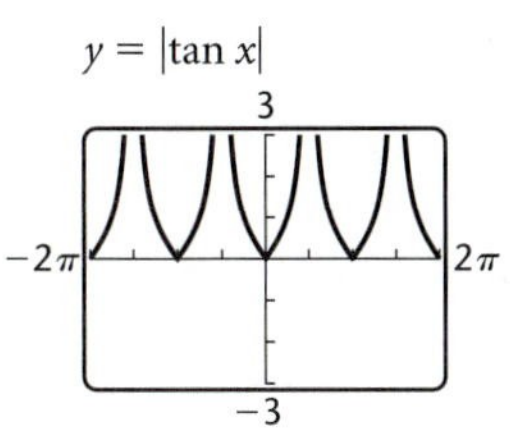

75. $-9.42, -6.28, -3.14, 3.14, 6.28, 9.42$
77. $-3.14, 0, 3.14$
79. (a) $y = 101.6 + 3\sin\left(\frac{\pi}{8}x\right)$

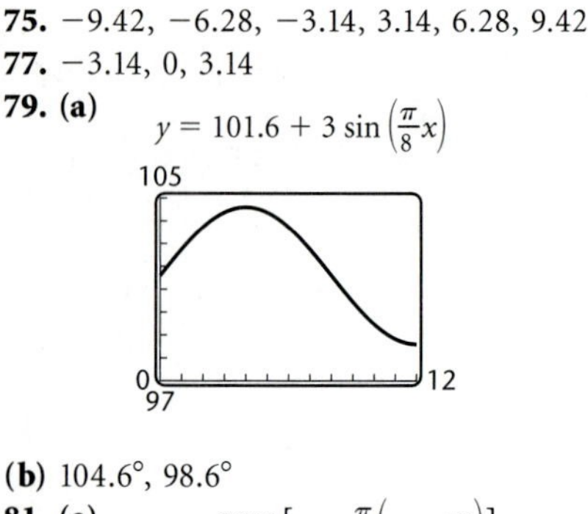

(b) 104.6°, 98.6°
81. (a) $y = 3000\left[\cos \frac{\pi}{45}(x - 10)\right]$

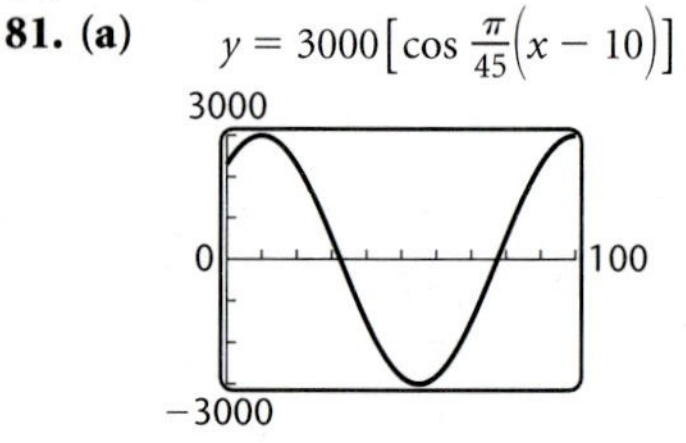

(b) amplitude: 3000; period: 90; phase shift: 10
83. (a) $y = 6e^{-0.8x}\cos(6\pi x) + 4$

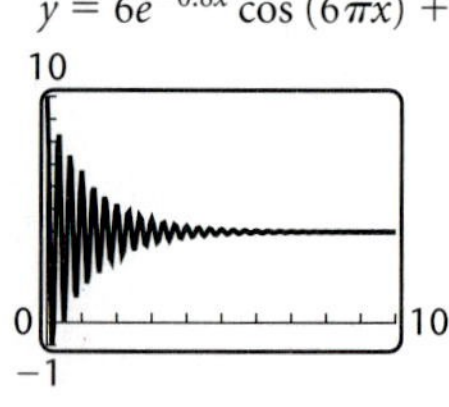

(b) 4 in.

Review Exercises, Chapter 4

1. [4.1] $\sin\theta = \frac{3\sqrt{73}}{73}$, $\cos\theta = \frac{8\sqrt{73}}{73}$, $\tan\theta = \frac{3}{8}$, $\csc\theta = \frac{\sqrt{73}}{3}$, $\sec\theta = \frac{\sqrt{73}}{8}$, $\cot\theta = \frac{8}{3}$ **2.** [4.1] $\frac{\sqrt{2}}{2}$
3. [4.1] $\frac{\sqrt{3}}{3}$ **4.** [4.3] $-\frac{\sqrt{2}}{2}$ **5.** [4.3] $\frac{1}{2}$
6. [4.3] Undefined **7.** [4.3] $-\sqrt{3}$ **8.** [4.1] 22°16′12″
9. [4.1] 47.56° **10.** [4.3] 0.4452 **11.** [4.3] 1.1315
12. [4.3] 0.9498 **13.** [4.3] −0.9092 **14.** [4.3] −1.5282
15. [4.3] −0.2778 **16.** [4.3] 205.3° **17.** [4.3] 47.2°
18. [4.1] 60° **19.** [4.1] 60°
20. [4.1] $\sin 30.9° \approx 0.5135$, $\cos 30.9° \approx 0.8581$, $\tan 30.9° \approx 0.5985$, $\csc 30.9° \approx 1.9474$, $\sec 30.9° \approx 1.1654$, $\cot 30.9° \approx 1.6709$ **21.** [4.2] $b \approx 4.5$, $A \approx 58.1°$, $B \approx 31.9°$
22. [4.2] $A = 38.83°$, $b \approx 37.9$, $c \approx 48.6$ **23.** [4.2] 1748 m
24. [4.2] 14 ft **25.** [4.3] 425°, −295° **26.** [4.4] $\frac{\pi}{3}, -\frac{5\pi}{3}$
27. [4.3] Complement: 76.6°; supplement: 166.6°
28. [4.4] Complement: $\frac{\pi}{3}$; supplement: $\frac{5\pi}{6}$

29. [4.3] $\sin\theta = \frac{3\sqrt{13}}{13}$, $\cos\theta = \frac{-2\sqrt{13}}{13}$, $\tan\theta = -\frac{3}{2}$, $\csc\theta = \frac{\sqrt{13}}{3}$, $\sec\theta = -\frac{\sqrt{13}}{2}$, $\cot\theta = -\frac{2}{3}$
30. [4.3] $\sin\theta = -\frac{2}{3}$, $\cos\theta = -\frac{\sqrt{5}}{3}$, $\cot\theta = \frac{\sqrt{5}}{2}$, $\sec\theta = -\frac{3\sqrt{5}}{5}$, $\csc\theta = -\frac{3}{2}$ **31.** [4.3] About 1743 mi
32. [4.4]

33. [4.4] II, $\frac{121}{150}\pi$, 2.534 **34.** [4.4] IV, $-\frac{\pi}{6}$, −0.524
35. [4.4] 270° **36.** [4.4] 171.89° **37.** [4.4] $\frac{7\pi}{4}$, or 5.5 cm
38. [4.4] 2.25, 129° **39.** [4.4] About 37.9 ft/min
40. [4.4] 497,829 radians/hr
41. [4.5] $\left(\frac{3}{5}, \frac{4}{5}\right), \left(-\frac{3}{5}, -\frac{4}{5}\right), \left(-\frac{3}{5}, \frac{4}{5}\right)$
42. [4.5] −1 **43.** [4.5] 1 **44.** [4.5] $-\frac{\sqrt{3}}{2}$ **45.** [4.5] $\frac{1}{2}$
46. [4.5] $\frac{\sqrt{3}}{3}$ **47.** [4.5] −1 **48.** [4.5] −0.9056
49. [4.5] 0.9218 **50.** [4.5] Undefined **51.** [4.5] 4.3813
52. [4.5] −6.1685 **53.** [4.5] 0.8090
54. [4.5]

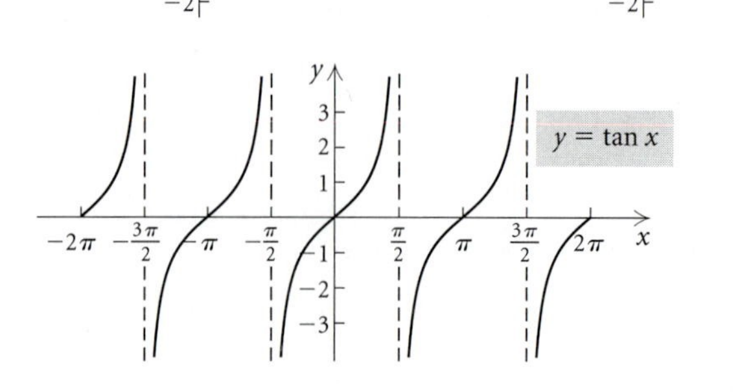

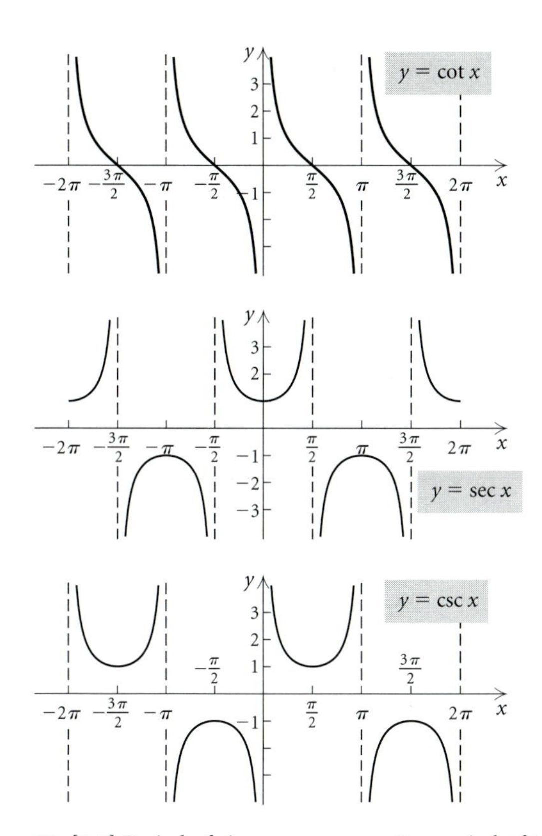

55. [4.5] Period of sin, cos, sec, csc: 2π; period of tan, cot: π

56. [4.5]

FUNCTION	DOMAIN	RANGE
Sine	$(-\infty, \infty)$	$[-1, 1]$
Cosine	$(-\infty, \infty)$	$[-1, 1]$
Tangent	$\left\{x \mid x \neq \frac{\pi}{2} + k\pi, k \in \mathbb{Z}\right\}$	$(-\infty, \infty)$

57. [4.3]

FUNCTION	I	II	III	IV
Sine	+	+	−	−
Cosine	+	−	−	+
Tangent	+	−	+	−

58. [4.6] Amplitude: 1; period: 2π; phase shift: $-\frac{\pi}{2}$

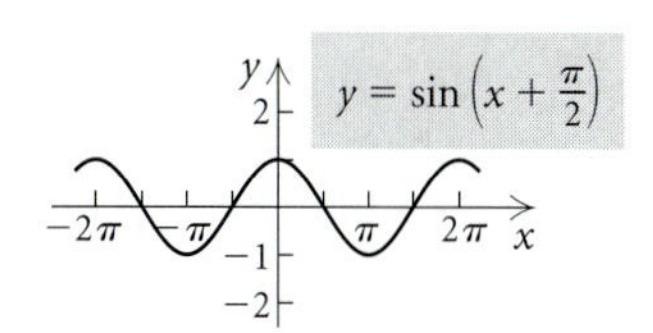

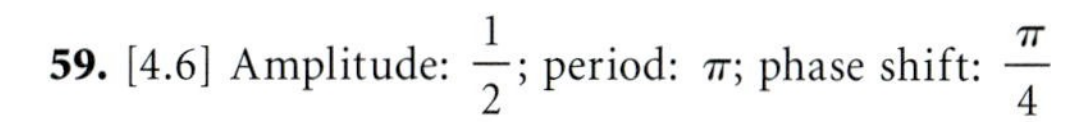

59. [4.6] Amplitude: $\frac{1}{2}$; period: π; phase shift: $\frac{\pi}{4}$

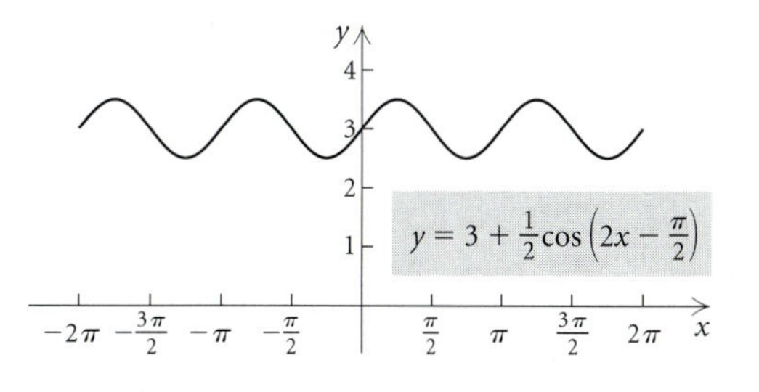

60. [4.6] (d) **61.** [4.6] (a) **62.** [4.6] (c) **63.** [4.6] (b)

64. [4.6]

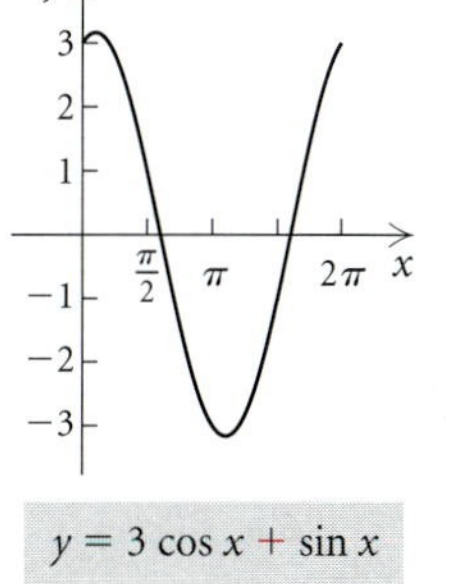

65. [4.1], [4.4] ◈ Both degrees and radians are units of angle measure. A degree is defined to be $\frac{1}{360}$ of one complete positive revolution. Degree notation has been in use since Babylonian times. Radians are defined in terms of intercepted arc length on a circle, with one radian being the measure of the angle for which the arc length equals the radius. There are 2π radians in one complete revolution.
66. [4.5] ◈ The graph of the cosine function is shaped like a continuous wave, with "high" points at $y = 1$ and "low" points at $y = -1$. The maximum value of the cosine function is 1, and it occurs at all points where $x = 2k\pi$, $k \in \mathbb{Z}$.
67. [4.6] ◈ When x is very large or very small, the amplitude of the function becomes small. The dimensions of the window must be adjusted to be able to see the shape of the graph. Also, when x is close to 0, the function is undefined, but this may not be obvious from the graph. **68.** [4.5] ◈ No; $\sin x$ is never greater than 1.

69. [4.5] All values

70. [4.6] Domain $(-\infty, \infty)$; range $[-3, 3]$; period 4π

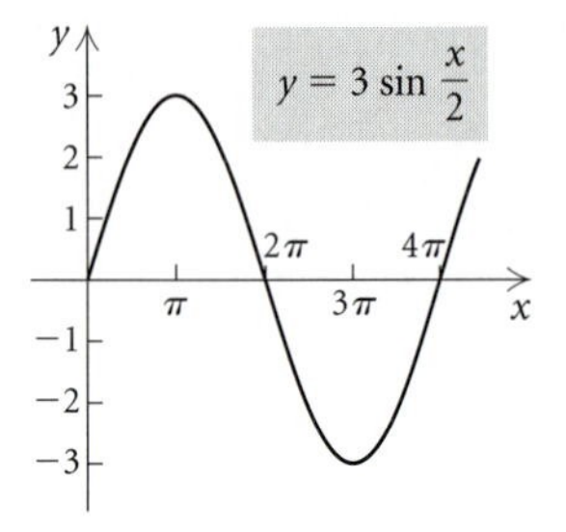

71. [4.6] $y_2 = 2 \sin\left(x + \frac{\pi}{2}\right) - 2$

72. [4.6] The domain consists of the intervals $\left(-\frac{\pi}{2} + 2k\pi, \frac{\pi}{2} + 2k\pi\right)$, $k \in \mathbb{Z}$.

73. [4.3] $\cos x = -0.7890$, $\tan x = -0.7787$, $\cot x = -1.2842$, $\sec x = -1.2674$, $\csc x = 1.6276$

Chapter 5

Exercise Set 5.1

1. $\sin^2 x - \cos^2 x$ **3.** $\sin y + \cos y$
5. $1 - 2 \sin \phi \cos \phi$ **7.** $\sin^3 x + \csc^3 x$
9. $\cos x (\sin x + \cos x)$
11. $(\sin x + \cos x)(\sin x - \cos x)$
13. $(2 \cos x + 3)(\cos x - 1)$
15. $(\sin x + 3)(\sin^2 x - 3 \sin x + 9)$ **17.** $\tan x$
19. $\sin x + 1$ **21.** $\frac{2 \tan t + 1}{3 \tan t + 1}$ **23.** 1
25. $\frac{5 \cot \phi}{\sin \phi + \cos \phi}$ **27.** $\frac{1 + 2 \sin s + 2 \cos s}{\sin^2 s - \cos^2 s}$
29. $\frac{5(\sin \theta - 3)}{3}$ **31.** $\sin x \cos x$
33. $\sqrt{\cos \alpha}\,(\sin \alpha - \cos \alpha)$ **35.** $1 - \sin y$
37. $\frac{\sqrt{\sin x \cos x}}{\cos x}$ **39.** $\frac{\sqrt{2} \cot y}{2}$ **41.** $\frac{\cos x}{\sqrt{\sin x \cos x}}$
43. $\frac{1 + \sin y}{\cos y}$ **45.** $\cos \theta = \frac{\sqrt{a^2 - x^2}}{a}$, $\tan \theta = \frac{x}{\sqrt{a^2 - x^2}}$
47. $\sin \theta = \frac{\sqrt{x^2 - 9}}{x}$, $\cos \theta = \frac{3}{x}$ **49.** $\sin \theta \tan \theta$
51. $\frac{\sqrt{6} - \sqrt{2}}{4}$ **53.** $\frac{\sqrt{3} + 1}{1 - \sqrt{3}}$, or $-2 - \sqrt{3}$
55. $\frac{\sqrt{6} + \sqrt{2}}{4}$ **57.** $\sin 59° \approx 0.8572$
59. $\tan 52° \approx 1.2799$ **61.** 0 **63.** $-\frac{7}{25}$ **65.** -1.5790
67. $2 \sin \alpha \cos \beta$ **69.** $\cos u$ **71.** All real numbers
73. 1.9417 **75.** ◈ **77.** 0°; the lines are parallel
79. $\frac{3\pi}{4}$, or 135° **81.** 22.83°

83.
$$\frac{\cos (x + h) - \cos x}{h} = \frac{\cos x \cos h - \sin x \sin h - \cos x}{h} = \frac{\cos x \cos h - \cos x}{h} - \frac{\sin x \sin h}{h} = \cos x\left(\frac{\cos h - 1}{h}\right) - \sin x\left(\frac{\sin h}{h}\right)$$

85. Let $x = \frac{\pi}{5}$. Then $\frac{\sin 5x}{x} = \frac{\sin \pi}{\pi/5} = 0 \neq \sin 5$. Answer may vary.

87. Let $\alpha = \frac{\pi}{4}$. Then $\cos (2\alpha) = \cos \frac{\pi}{2} = 0$, but $2 \cos \alpha = 2 \cos \frac{\pi}{4} = \sqrt{2}$. Answer may vary.

89. Let $x = \frac{\pi}{6}$. Then $\frac{\cos 6x}{\cos x} = \frac{\cos \pi}{\cos \frac{\pi}{6}} = \frac{-1}{\sqrt{3}/2} \neq 6$. Answer may vary.

91. $\frac{6 - 3\sqrt{3}}{9 + 2\sqrt{3}} \approx 0.0645$ **93.** 168.7°

95. $\cos 2\theta = \cos^2 \theta - \sin^2 \theta$, or $1 - 2 \sin^2 \theta$, or $2 \cos^2 \theta - 1$

97. $\tan\left(x + \frac{\pi}{4}\right) = \frac{\tan x + \tan \frac{\pi}{4}}{1 - \tan x \tan \frac{\pi}{4}} = \frac{1 + \tan x}{1 - \tan x}$

99. $\sin (\alpha + \beta) + \sin (\alpha - \beta) = \sin \alpha \cos \beta + \cos \alpha \sin \beta + \sin \alpha \cos \beta - \cos \alpha \sin \beta = 2 \sin \alpha \cos \beta$

Exercise Set 5.2

1. **(a)** $\tan \frac{3\pi}{10} \approx 1.3763$, $\csc \frac{3\pi}{10} \approx 1.2361$, $\sec \frac{3\pi}{10} \approx 1.7013$, $\cot \frac{3\pi}{10} \approx 0.7266$; **(b)** $\sin \frac{\pi}{5} \approx 0.5878$, $\cos \frac{\pi}{5} \approx 0.8090$, $\tan \frac{\pi}{5} \approx 0.7266$, $\csc \frac{\pi}{5} \approx 1.7013$, $\sec \frac{\pi}{5} \approx 1.2361$, $\cot \frac{\pi}{5} \approx 1.3763$

3. **(a)** $\cos \theta = -\frac{2\sqrt{2}}{3}$, $\tan \theta = -\frac{\sqrt{2}}{4}$, $\csc \theta = 3$, $\sec \theta = -\frac{3\sqrt{2}}{4}$, $\cot \theta = -2\sqrt{2}$; **(b)** $\sin\left(\frac{\pi}{2} - \theta\right) = -\frac{2\sqrt{2}}{3}$, $\cos\left(\frac{\pi}{2} - \theta\right) = \frac{1}{3}$, $\tan\left(\frac{\pi}{2} - \theta\right) = -2\sqrt{2}$, $\csc\left(\frac{\pi}{2} - \theta\right) = -\frac{3\sqrt{2}}{4}$, $\sec\left(\frac{\pi}{2} - \theta\right) = 3$,

$\cot\left(\frac{\pi}{2} - \theta\right) = -\frac{\sqrt{2}}{4}$; **(c)** $\sin\left(\theta - \frac{\pi}{2}\right) = \frac{2\sqrt{2}}{3}$,
$\cos\left(\theta - \frac{\pi}{2}\right) = \frac{1}{3}$, $\tan\left(\theta - \frac{\pi}{2}\right) = 2\sqrt{2}$,
$\csc\left(\theta - \frac{\pi}{2}\right) = \frac{3\sqrt{2}}{4}$, $\sec\left(\theta - \frac{\pi}{2}\right) = 3$,
$\cot\left(\theta - \frac{\pi}{2}\right) = \frac{\sqrt{2}}{4}$

5. $\sec\left(x + \frac{\pi}{2}\right) = -\csc x$ **7.** $\tan\left(x - \frac{\pi}{2}\right) = -\cot x$

9. $\sin 2\theta = \frac{24}{25}$, $\cos 2\theta = -\frac{7}{25}$, $\tan 2\theta = -\frac{24}{7}$; II

11. $\sin 2\theta = \frac{24}{25}$, $\cos 2\theta = -\frac{7}{25}$, $\tan 2\theta = -\frac{24}{7}$; II

13. $\sin 2\theta = -\frac{120}{169}$, $\cos 2\theta = \frac{119}{169}$, $\tan 2\theta = -\frac{120}{119}$; IV

15. $\cos 4x = 1 - 8\sin^2 x\cos^2 x$, or $\cos^4 x - 6\sin^2 x\cos^2 x + \sin^4 x$, or $8\cos^4 x - 8\cos^2 x + 1$

17. $\frac{\sqrt{2 + \sqrt{3}}}{2}$ **19.** $\frac{\sqrt{2 + \sqrt{2}}}{2}$ **21.** $2 + \sqrt{3}$

23. 0.6421 **25.** 0.1735

27. (d); $\frac{\cos 2x}{\cos x - \sin x} = \frac{\cos^2 x - \sin^2 x}{\cos x - \sin x}$

$= \frac{(\cos x + \sin x)(\cos x - \sin x)}{\cos x - \sin x}$

$= \cos x + \sin x$

29. (d); $\frac{\sin 2x}{2\cos x} = \frac{2\sin x\cos x}{2\cos x} = \sin x$ **31.** $\cos x$ **33.** 1

35. $\cos 2x$ **37.** 8 **39.** $\sin^2 x$ **41.** $-\cos^2 x$ **43.** ◆

45. $\sin 141° \approx 0.6293$, $\cos 141° \approx -0.7772$, $\tan 141° \approx -0.8097$, $\csc 141° \approx 1.5891$, $\sec 141° \approx -1.2867$, $\cot 141° \approx -1.2350$

47. $-\cos x(1 + \cot x)$ **49.** $\cot^2 y$

51. $\sin\theta = -\frac{15}{17}$, $\cos\theta = -\frac{8}{17}$, $\tan\theta = \frac{15}{8}$

53. **(a)** 9.80359 m/sec^2; **(b)** 9.80180 m/sec^2;
(c) $g = 9.78049(1 + 0.005264\sin^2\phi + 0.000024\sin^4\phi)$;
(d) g is greatest at 90°N and 90°S (at the poles); g is least at 0° (at the equator).

Exercise Set 5.3

1.

$\sec x - \sin x\tan x$	$\cos x$
$\frac{1}{\cos x} - \sin x \cdot \frac{\sin x}{\cos x}$	
$\frac{1 - \sin^2 x}{\cos x}$	
$\frac{\cos^2 x}{\cos x}$	
$\cos x$	

3.

$\frac{1 - \cos x}{\sin x}$	$\frac{\sin x}{1 + \cos x}$
	$\frac{\sin x}{1 + \cos x} \cdot \frac{1 - \cos x}{1 - \cos x}$
	$\frac{\sin x\,(1 - \cos x)}{1 - \cos^2 x}$
	$\frac{\sin x\,(1 - \cos x)}{\sin^2 x}$
	$\frac{1 - \cos x}{\sin x}$

5.

$\frac{1 + \tan\theta}{1 - \tan\theta} + \frac{1 + \cot\theta}{1 - \cot\theta}$	0
$\frac{1 + \frac{\sin\theta}{\cos\theta}}{1 - \frac{\sin\theta}{\cos\theta}} + \frac{1 + \frac{\cos\theta}{\sin\theta}}{1 - \frac{\cos\theta}{\sin\theta}}$	
$\frac{\frac{\cos\theta + \sin\theta}{\cos\theta}}{\frac{\cos\theta - \sin\theta}{\cos\theta}} + \frac{\frac{\sin\theta + \cos\theta}{\sin\theta}}{\frac{\sin\theta - \cos\theta}{\sin\theta}}$	
$\frac{\cos\theta + \sin\theta}{\cos\theta} \cdot \frac{\cos\theta}{\cos\theta - \sin\theta} + \frac{\sin\theta + \cos\theta}{\sin\theta} \cdot \frac{\sin\theta}{\sin\theta - \cos\theta}$	
$\frac{\cos\theta + \sin\theta}{\cos\theta - \sin\theta} + \frac{\sin\theta + \cos\theta}{\sin\theta - \cos\theta}$	
$\frac{\cos\theta + \sin\theta}{\cos\theta - \sin\theta} - \frac{\cos\theta + \sin\theta}{\cos\theta - \sin\theta}$	
0	

7.

$\dfrac{\cos^2\alpha + \cot\alpha}{\cos^2\alpha - \cot\alpha}$	$\dfrac{\cos^2\alpha\tan\alpha + 1}{\cos^2\alpha\tan\alpha - 1}$
$\dfrac{\cos^2\alpha + \dfrac{\cos\alpha}{\sin\alpha}}{\cos^2\alpha - \dfrac{\cos\alpha}{\sin\alpha}}$	$\dfrac{\cos^2\alpha\dfrac{\sin\alpha}{\cos\alpha} + 1}{\cos^2\alpha\dfrac{\sin\alpha}{\cos\alpha} - 1}$
$\dfrac{\cos\alpha\left(\cos\alpha + \dfrac{1}{\sin\alpha}\right)}{\cos\alpha\left(\cos\alpha - \dfrac{1}{\sin\alpha}\right)}$	$\dfrac{\sin\alpha\cos\alpha + 1}{\sin\alpha\cos\alpha - 1}$
$\dfrac{\cos\alpha + \dfrac{1}{\sin\alpha}}{\cos\alpha - \dfrac{1}{\sin\alpha}}$	
$\dfrac{\dfrac{\sin\alpha\cos\alpha + 1}{\sin\alpha}}{\dfrac{\sin\alpha\cos\alpha - 1}{\sin\alpha}}$	
$\dfrac{\sin\alpha\cos\alpha + 1}{\sin\alpha\cos\alpha - 1}$	

9.

$\dfrac{2\tan\theta}{1 + \tan^2\theta}$	$\sin 2\theta$
$\dfrac{2\tan\theta}{\sec^2\theta}$	$2\sin\theta\cos\theta$
$\dfrac{2\sin\theta}{\cos\theta}\cdot\dfrac{\cos^2\theta}{1}$	
$2\sin\theta\cos\theta$	

11.

$1 - \cos 5\theta\cos 3\theta - \sin 5\theta\sin 3\theta$	$2\sin^2\theta$
$1 - [\cos 5\theta\cos 3\theta + \sin 5\theta\sin 3\theta]$	$1 - \cos 2\theta$
$1 - \cos(5\theta - 3\theta)$	
$1 - \cos 2\theta$	

13.

$2\sin\theta\cos^3\theta + 2\sin^3\theta\cos\theta$	$\sin 2\theta$
$2\sin\theta\cos\theta(\cos^2\theta + \sin^2\theta)$	$2\sin\theta\cos\theta$
$2\sin\theta\cos\theta$	

15.

$\dfrac{\tan x - \sin x}{2\tan x}$	$\sin^2\dfrac{x}{2}$
$\dfrac{1}{2}\left[\dfrac{\dfrac{\sin x}{\cos x} - \sin x}{\dfrac{\sin x}{\cos x}}\right]$	$\dfrac{1 - \cos x}{2}$
$\dfrac{1}{2}\dfrac{\sin x - \sin x\cos x}{\cos x}\cdot\dfrac{\cos x}{\sin x}$	
$\dfrac{1 - \cos x}{2}$	

17.

$\sin(\alpha + \beta)\sin(\alpha - \beta)$	$\sin^2\alpha - \sin^2\beta$
$(\sin\alpha\cos\beta + \cos\alpha\sin\beta)(\sin\alpha\cos\beta - \cos\alpha\sin\beta)$	$1 - \cos^2\alpha - (1 - \cos^2\beta)$
$\sin^2\alpha\cos^2\beta - \cos^2\alpha\sin^2\beta$	$\cos^2\beta - \cos^2\alpha$
$\cos^2\beta(1 - \cos^2\alpha) - \cos^2\alpha(1 - \cos^2\beta)$	
$\cos^2\beta - \cos^2\alpha\cos^2\beta - \cos^2\alpha + \cos^2\alpha\cos^2\beta$	
$\cos^2\beta - \cos^2\alpha$	

19.

$\tan\theta(\tan\theta + \cot\theta)$	$\sec^2\theta$
$\tan^2\theta + \tan\theta\cot\theta$	
$\tan^2\theta + 1$	
$\sec^2\theta$	

21.

$\dfrac{1 + \cos^2 x}{\sin^2 x}$	$2\csc^2 x - 1$
$\dfrac{1}{\sin^2 x} + \dfrac{\cos^2 x}{\sin^2 x}$	
$\csc^2 x + \cot^2 x$	
$\csc^2 x + \csc^2 x - 1$	
$2\csc^2 x - 1$	

23.

$\dfrac{1 + \sin x}{1 - \sin x} + \dfrac{\sin x - 1}{1 + \sin x}$	$4\sec x\tan x$
$\dfrac{(1 + \sin x)^2 - (1 - \sin x)^2}{1 - \sin^2 x}$	$4\cdot\dfrac{1}{\cos x}\cdot\dfrac{\sin x}{\cos x}$
$\dfrac{(1 + 2\sin x + \sin^2 x) - (1 - 2\sin x + \sin^2 x)}{\cos^2 x}$	$\dfrac{4\sin x}{\cos^2 x}$
$\dfrac{4\sin x}{\cos^2 x}$	

25.

$\cos^2\alpha\cot^2\alpha$	$\cot^2\alpha - \cos^2\alpha$
$(1 - \sin^2\alpha)\cot^2\alpha$	
$\cot^2\alpha - \sin^2\alpha\cdot\dfrac{\cos^2\alpha}{\sin^2\alpha}$	
$\cot^2\alpha - \cos^2\alpha$	

27.

$2\sin^2\theta\cos^2\theta + \cos^4\theta$	$1 - \sin^4\theta$
$\cos^2\theta(2\sin^2\theta + \cos^2\theta)$	$(1 + \sin^2\theta)(1 - \sin^2\theta)$
$\cos^2\theta(\sin^2\theta + \sin^2\theta + \cos^2\theta)$	$(1 + \sin^2\theta)(\cos^2\theta)$
$\cos^2\theta(\sin^2\theta + 1)$	

29.

$\dfrac{1 + \sin x}{1 - \sin x}$	$(\sec x + \tan x)^2$
$\dfrac{1 + \sin x}{1 - \sin x}\cdot\dfrac{1 + \sin x}{1 + \sin x}$	$\left(\dfrac{1}{\cos x} + \dfrac{\sin x}{\cos x}\right)^2$
$\dfrac{(1 + \sin x)^2}{1 - \sin^2 x}$	$\dfrac{(1 + \sin x)^2}{\cos^2 x}$
$\dfrac{(1 + \sin x)^2}{\cos^2 x}$	

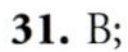

31. B;

$\dfrac{\cos x + \cot x}{1 + \csc x}$	$\cos x$
$\dfrac{\dfrac{\cos x}{1} + \dfrac{\cos x}{\sin x}}{1 + \dfrac{1}{\sin x}}$	
$\dfrac{\sin x \cos x + \cos x}{\sin x} \cdot \dfrac{\sin x}{\sin x + 1}$	
$\dfrac{\cos x\,(\sin x + 1)}{\sin x + 1}$	
$\cos x$	

33. A;

$\sin x \cos x + 1$	$\dfrac{\sin^3 x - \cos^3 x}{\sin x - \cos x}$
	$\dfrac{(\sin x - \cos x)(\sin^2 x + \sin x \cos x + \cos^2 x)}{\sin x - \cos x}$
	$\sin^2 x + \sin x \cos x + \cos^2 x$
	$\sin x \cos x + 1$

35. C;

$\dfrac{1}{\cot x \sin^2 x}$	$\tan x + \cot x$
$\dfrac{1}{\dfrac{\cos x}{\sin x} \cdot \sin^2 x}$	$\dfrac{\sin x}{\cos x} + \dfrac{\cos x}{\sin x}$
$\dfrac{1}{\cos x \sin x}$	$\dfrac{\sin^2 x + \cos^2 x}{\cos x \sin x}$
	$\dfrac{1}{\cos x \sin x}$

37. **(a)**, **(d)**

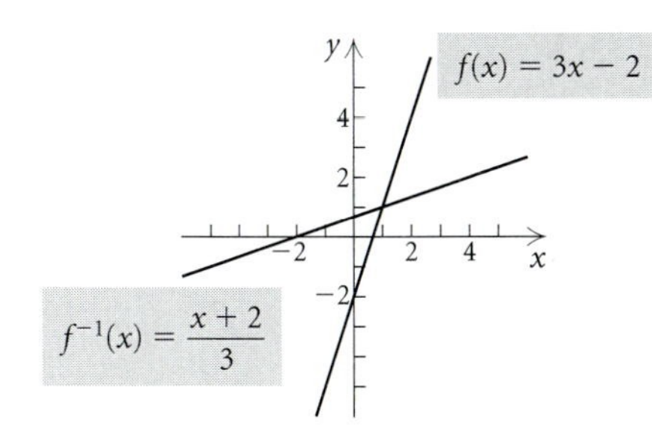

(b) yes; **(c)** $f^{-1}(x) = \dfrac{x + 2}{3}$

39. **(a)**, **(d)**

$f(x) = x^2 - 4,\ x \geq 0$

$f^{-1}(x) = \sqrt{x + 4}$

(b) yes; **(c)** $f^{-1}(x) = \sqrt{x + 4}$ **41.** ◈

43.

$\ln \lvert \tan x \rvert$	$-\ln \lvert \cot x \rvert$
$\ln \left\lvert \dfrac{1}{\cot x} \right\rvert$	
$\ln \lvert 1 \rvert - \ln \lvert \cot x \rvert$	
$0 - \ln \lvert \cot x \rvert$	
$-\ln \lvert \cot x \rvert$	

45. $\log(\cos x - \sin x) + \log(\cos x + \sin x)$
$= \log[(\cos x - \sin x)(\cos x + \sin x)]$
$= \log(\cos^2 x - \sin^2 x) = \log \cos 2x$

47.
$$\frac{1}{\omega C(\tan\theta + \tan\phi)} = \frac{1}{\omega C\left(\dfrac{\sin\theta}{\cos\theta} + \dfrac{\sin\phi}{\cos\phi}\right)} = \frac{1}{\omega C\left(\dfrac{\sin\theta\cos\phi + \sin\phi\cos\theta}{\cos\theta\cos\phi}\right)} = \frac{\cos\theta\cos\phi}{\omega C \sin(\theta + \phi)}$$

EXERCISE SET 5.4

1. $-\dfrac{\pi}{3}$, $-60°$ **3.** $\dfrac{\pi}{4}$, $45°$ **5.** $\dfrac{\pi}{4}$, $45°$ **7.** 0, 0°

9. $\dfrac{\pi}{6}$, $30°$ **11.** $\dfrac{\pi}{6}$, $30°$ **13.** $-\dfrac{\pi}{6}$, $-30°$

15. $-\dfrac{\pi}{6}$, $-30°$ **17.** $\dfrac{\pi}{2}$, $90°$ **19.** $\dfrac{\pi}{3}$, $60°$

21. 0.352, 20.2° **23.** 1.292, 74.0° **25.** 2.946, 168.8°

27. −0.160, −9.2° **29.** 0.829, 47.5° **31.** −0.960, −55.0°

33. $\sin^{-1}$: $[-1, 1]$; $\cos^{-1}$: $[-1, 1]$; $\tan^{-1}$: $(-\infty, \infty)$

35. $\theta = \sin^{-1}\left(\dfrac{2000}{d}\right)$ **37.** 0.3 **39.** $\dfrac{\pi}{4}$ **41.** $\dfrac{\pi}{5}$

43. $-\dfrac{\pi}{3}$ **45.** $\dfrac{1}{2}$ **47.** 1 **49.** $\dfrac{\pi}{3}$ **51.** $\dfrac{\sqrt{11}}{33}$

53. $\dfrac{a}{\sqrt{a^2+9}}$ **55.** $\dfrac{\sqrt{q^2-p^2}}{p}$ **57.** $\dfrac{p}{3}$ **59.** $\dfrac{\sqrt{3}}{2}$

61. $-\dfrac{\sqrt{2}}{10}$ **63.** $xy + \sqrt{(1-x^2)(1-y^2)}$ **65.** 0.9861

67. $\left\{0, \dfrac{5}{2}\right\}$ **69.** $\{\pm 2, \pm 3i\}$ **71.** $\{27\}$ **73.** ◆ **75.** ◆

77.

$\sin^{-1} x + \cos^{-1} x$	$\dfrac{\pi}{2}$
$\sin(\sin^{-1} x + \cos^{-1} x)$	$\sin \dfrac{\pi}{2}$
$[\sin(\sin^{-1} x)][\cos(\cos^{-1} x)] +$ $[\cos(\sin^{-1} x)][\sin(\cos^{-1} x)]$	1
$x \cdot x + \sqrt{1-x^2} \cdot \sqrt{1-x^2}$	
$x^2 + 1 - x^2$	
1	

79.

$\sin^{-1} x$	$\tan^{-1} \dfrac{x}{\sqrt{1-x^2}}$
$\sin(\sin^{-1} x)$	$\sin\left(\tan^{-1} \dfrac{x}{\sqrt{1-x^2}}\right)$
x	x

81.

$\arcsin x$	$\arccos \sqrt{1-x^2}$
$\sin(\arcsin x)$	$\sin(\arccos \sqrt{1-x^2})$
x	x

83. $\theta = \arctan \dfrac{y+h}{x} - \arctan \dfrac{y}{x}$; 38.7°

Exercise Set 5.5

1. $\dfrac{\pi}{6} + 2k\pi, \dfrac{11\pi}{6} + 2k\pi$, or $30° + k \cdot 360°, 330° + k \cdot 360°$

3. $\dfrac{2\pi}{3} + k\pi$, or $120° + k \cdot 180°$ **5.** 98.09°, 261.91°

7. $\dfrac{4\pi}{3}, \dfrac{5\pi}{3}$ **9.** $\dfrac{\pi}{4}, \dfrac{3\pi}{4}, \dfrac{5\pi}{4}, \dfrac{7\pi}{4}$ **11.** $\dfrac{\pi}{6}, \dfrac{5\pi}{6}, \dfrac{3\pi}{2}$

13. $\dfrac{\pi}{6}, \dfrac{\pi}{2}, \dfrac{3\pi}{2}, \dfrac{11\pi}{6}$ **15.** 109.47°, 120°, 240°, 250.53°

17. $0, \dfrac{\pi}{4}, \dfrac{3\pi}{4}, \pi, \dfrac{5\pi}{4}, \dfrac{7\pi}{4}$ **19.** 139.81°, 220.19°

21. 37.22°, 169.35°, 217.22°, 349.35° **23.** $0, \pi, \dfrac{7\pi}{6}, \dfrac{11\pi}{6}$

25. $0, \dfrac{\pi}{2}, \pi, \dfrac{3\pi}{2}$ **27.** $0, \pi$ **29.** $\dfrac{3\pi}{4}, \dfrac{7\pi}{4}$

31. $\dfrac{2\pi}{3}, \dfrac{4\pi}{3}, \dfrac{3\pi}{2}$ **33.** $\dfrac{\pi}{4}, \dfrac{3\pi}{4}, \dfrac{5\pi}{4}, \dfrac{7\pi}{4}$ **35.** $\dfrac{\pi}{12}, \dfrac{5\pi}{12}$

37. 0.967, 1.853, 4.109, 4.994 **39.** $\dfrac{2\pi}{3}, \dfrac{4\pi}{3}$

41. 1.114, 2.773 **43.** 0.515 **45.** 0.422, 1.756

47. $B = 35°$, $b \approx 140.7$, $c \approx 245.4$ **49.** 36 **51.** ◆

53. $\dfrac{\pi}{3}, \dfrac{2\pi}{3}, \dfrac{4\pi}{3}, \dfrac{5\pi}{3}$ **55.** $\dfrac{\pi}{3}, \dfrac{4\pi}{3}$ **57.** 0

59. $e^{3\pi/2 + 2k\pi}$, where k (an integer) ≤ -1

61. 1.24 days, 6.76 days **63.** 16.5°N **65.** 1 **67.** 0.1923

Review Exercises, Chapter 5

1. [5.1] $\csc^2 x$ **2.** [5.1] 1 **3.** [5.1] $\tan^2 y - \cot^2 y$

4. [5.1] $\dfrac{(\cos^2 x + 1)^2}{\cos^2 x}$ **5.** [5.1] $\csc x (\sec x - \csc x)$

6. [5.1] $(3 \sin y + 5)(\sin y - 4)$

7. [5.1] $(10 - \cos u)(100 + 10 \cos u + \cos^2 u)$ **8.** [5.1] 1

9. [5.1] $\dfrac{1}{2} \sec x$ **10.** [5.1] $\dfrac{3 \tan x}{\sin x - \cos x}$

11. [5.1] $\dfrac{3 \cos y + 3 \sin y + 2}{\cos^2 y - \sin^2 y}$ **12.** [5.1] 1

13. [5.1] $\dfrac{1}{4} \cot x$ **14.** [5.1] $\sin x + \cos x$

15. [5.1] $\dfrac{\cos x}{1 - \sin x}$ **16.** [5.1] $\dfrac{\cos x}{\sqrt{\sin x}}$ **17.** [5.1] $3 \sec \theta$

18. [5.1] $\cos x \cos \dfrac{3\pi}{2} - \sin x \sin \dfrac{3\pi}{2}$

19. [5.1] $\dfrac{\tan 45° - \tan 30°}{1 + \tan 45° \tan 30°}$

20. [5.1] $\cos(27° - 16°)$, or $\cos 11°$ **21.** [5.1] $\dfrac{-\sqrt{6} - \sqrt{2}}{4}$

22. [5.1] $2 - \sqrt{3}$ **23.** [5.1] -0.3745 **24.** [5.2] $-\sin x$

25. [5.2] $\sin x$ **26.** [5.2] $-\cos x$

27. [5.2] **(a)** $\sin \alpha = -\dfrac{4}{5}$, $\tan \alpha = \dfrac{4}{3}$, $\cot \alpha = \dfrac{3}{4}$, $\sec \alpha = -\dfrac{5}{3}$, $\csc \alpha = -\dfrac{5}{4}$; **(b)** $\sin\left(\dfrac{\pi}{2} - \alpha\right) = -\dfrac{3}{5}$, $\cos\left(\dfrac{\pi}{2} - \alpha\right) = -\dfrac{4}{5}$, $\tan\left(\dfrac{\pi}{2} - \alpha\right) = \dfrac{3}{4}$, $\cot\left(\dfrac{\pi}{2} - \alpha\right) = \dfrac{4}{3}$, $\sec\left(\dfrac{\pi}{2} - \alpha\right) = -\dfrac{5}{4}$, $\csc\left(\dfrac{\pi}{2} - \alpha\right) = -\dfrac{5}{3}$; **(c)** $\sin\left(\alpha + \dfrac{\pi}{2}\right) = -\dfrac{3}{5}$, $\cos\left(\alpha + \dfrac{\pi}{2}\right) = \dfrac{4}{5}$, $\tan\left(\alpha + \dfrac{\pi}{2}\right) = -\dfrac{3}{4}$, $\cot\left(\alpha + \dfrac{\pi}{2}\right) = -\dfrac{4}{3}$, $\sec\left(\alpha + \dfrac{\pi}{2}\right) = \dfrac{5}{4}$, $\csc\left(\alpha + \dfrac{\pi}{2}\right) = -\dfrac{5}{3}$ **28.** [5.2] $-\sec x$

29. [5.2] $\tan 2\theta = \frac{24}{7}$, $\cos 2\theta = \frac{7}{25}$, $\sin 2\theta = \frac{24}{25}$; I

30. [5.2] $\frac{\sqrt{2-\sqrt{2}}}{2}$

31. [5.2] $\sin 2\beta = 0.4261$, $\cos \frac{\beta}{2} = 0.9940$, $\cos 4\beta = 0.6369$

32. [5.2] $\cos x$ **33.** [5.2] 1 **34.** [5.2] $\sin 2x$

35. [5.2] $\tan 2x$

36. [5.3]

$\frac{1-\sin x}{\cos x}$	$\frac{\cos x}{1+\sin x}$
$\frac{1-\sin x}{\cos x}\cdot\frac{\cos x}{\cos x}$	$\frac{\cos x}{1+\sin x}\cdot\frac{1-\sin x}{1-\sin x}$
$\frac{\cos x-\sin x\cos x}{\cos^2 x}$	$\frac{\cos x-\sin x\cos x}{1-\sin^2 x}$
	$\frac{\cos x-\sin x\cos x}{\cos^2 x}$

37. [5.3]

$\frac{1+\cos 2\theta}{\sin 2\theta}$	$\cot\theta$
$\frac{1+2\cos^2\theta-1}{2\sin\theta\cos\theta}$	$\frac{\cos\theta}{\sin\theta}$
$\frac{\cos\theta}{\sin\theta}$	

38. [5.3]

$\frac{\tan y+\sin y}{2\tan y}$	$\cos^2\frac{y}{2}$
$\frac{1}{2}\left[\frac{\frac{\sin y+\sin y\cos y}{\cos y}}{\frac{\sin y}{\cos y}}\right]$	$\frac{1+\cos y}{2}$
$\frac{1}{2}\left[\frac{\sin y\,(1+\cos y)}{\cos y}\cdot\frac{\cos y}{\sin y}\right]$	
$\frac{1+\cos y}{2}$	

39. [5.3]

$\frac{\sin x-\cos x}{\cos^2 x}$	$\frac{\tan^2 x-1}{\sin x+\cos x}$
	$\frac{\frac{\sin^2 x}{\cos^2 x}-1}{\sin x+\cos x}$
	$\frac{\sin^2 x-\cos^2 x}{\cos^2 x}\cdot\frac{1}{\sin x+\cos x}$
	$\frac{\sin x-\cos x}{\cos^2 x}$

40. [5.3] B;

$\csc x-\cos x\cot x$	$\sin x$
$\frac{1}{\sin x}-\cos x\,\frac{\cos x}{\sin x}$	
$\frac{1-\cos^2 x}{\sin x}$	
$\frac{\sin^2 x}{\sin x}$	
$\sin x$	

41. [5.3] D;

$\frac{1}{\sin x\cos x}-\frac{\cos x}{\sin x}$	$\frac{\sin x\cos x}{1-\sin^2 x}$
$\frac{1}{\sin x\cos x}-\frac{\cos^2 x}{\sin x\cos x}$	$\frac{\sin x\cos x}{\cos^2 x}$
$\frac{1-\cos^2 x}{\sin x\cos x}$	$\frac{\sin x}{\cos x}$
$\frac{\sin^2 x}{\sin x\cos x}$	
$\frac{\sin x}{\cos x}$	

42. [5.3] A;

$\frac{\cot x-1}{1-\tan x}$	$\frac{\csc x}{\sec x}$
$\frac{\frac{\cos x}{\sin x}-\frac{\sin x}{\sin x}}{\frac{\cos x}{\cos x}-\frac{\sin x}{\cos x}}$	$\frac{\frac{1}{\sin x}}{\frac{1}{\cos x}}$
$\frac{\cos x-\sin x}{\sin x}\cdot\frac{\cos x}{\cos x-\sin x}$	$\frac{1}{\sin x}\cdot\frac{\cos x}{1}$
$\frac{\cos x}{\sin x}$	$\frac{\cos x}{\sin x}$

43. [5.3] C;

$\frac{\cos x+1}{\sin x}+\frac{\sin x}{\cos x+1}$	$\frac{2}{\sin x}$
$\frac{(\cos x+1)^2+\sin^2 x}{\sin x\,(\cos x+1)}$	
$\frac{\cos^2 x+2\cos x+1+\sin^2 x}{\sin x\,(\cos x+1)}$	
$\frac{2\cos x+2}{\sin x\,(\cos x+1)}$	
$\frac{2(\cos x+1)}{\sin x\,(\cos x+1)}$	
$\frac{2}{\sin x}$	

44. [5.4] $-\frac{\pi}{6}$, $-30°$ **45.** [5.4] $\frac{\pi}{6}$, $30°$ **46.** [5.4] $\frac{\pi}{4}$, $45°$

47. [5.4] 0, $0°$ **48.** [5.4] 1.792, $102.7°$

49. [5.4] 0.398, 22.8° **50.** [5.4] $\frac{1}{2}$ **51.** [5.4] $\frac{\sqrt{3}}{3}$

52. [5.4] $\frac{\pi}{7}$ **53.** [5.4] $\frac{\sqrt{2}}{2}$ **54.** [5.4] $\frac{3}{\sqrt{b^2+9}}$

55. [5.4] $-\frac{7}{25}$

56. [5.5] $\frac{3\pi}{4} + 2k\pi, \frac{5\pi}{4} + 2k\pi$, or $135° + k \cdot 360°$, $225° + k \cdot 360°$ **57.** [5.5] $\frac{\pi}{3} + k\pi$, or $60° + k \cdot 180°$

58. [5.5] $\frac{\pi}{6}, \frac{5\pi}{6}, \frac{7\pi}{6}, \frac{11\pi}{6}$

59. [5.5] $\frac{\pi}{4}, \frac{\pi}{2}, \frac{3\pi}{4}, \frac{5\pi}{4}, \frac{3\pi}{2}, \frac{7\pi}{4}$ **60.** [5.5] $\frac{2\pi}{3}, \pi, \frac{4\pi}{3}$

61. [5.5] $0, \pi$ **62.** [5.5] $\frac{\pi}{4}, \frac{3\pi}{4}, \frac{5\pi}{4}, \frac{7\pi}{4}$

63. [5.5] $0, \frac{\pi}{2}, \pi, \frac{3\pi}{2}$ **64.** [5.5] $\frac{7\pi}{12}, \frac{23\pi}{12}$

65. [5.5] 0.864, 2.972, 4.006, 6.114

66. [5.5] −4.488, −2.074, 4.917 **67.** [5.5] −0.515

68. [5.3] ◆

(a)
$$\begin{aligned} 2\cos^2 x - 1 &= \cos 2x = \cos^2 x - \sin^2 x \\ &= 1 \cdot (\cos^2 x - \sin^2 x) \\ &= (\cos^2 x + \sin^2 x)(\cos^2 x - \sin^2 x) \\ &= \cos^4 x - \sin^4 x; \end{aligned}$$

(b)
$$\begin{aligned} \cos^4 x - \sin^4 x &= (\cos^2 x + \sin^2 x)(\cos^2 x - \sin^2 x) \\ &= 1 \cdot (\cos^2 x - \sin^2 x) \\ &= \cos^2 x - \sin^2 x = \cos 2x \\ &= 2\cos^2 x - 1; \end{aligned}$$

(c)

$2\cos^2 x - 1$	$\cos^4 x - \sin^4 x$
$\cos 2x$	$(\cos^2 x + \sin^2 x)(\cos^2 x - \sin^2 x)$
	$1 \cdot (\cos^2 x - \sin^2 x)$
	$\cos^2 x - \sin^2 x$
	$\cos 2x$

Answer may vary. Method 2 may be the more efficient because it involves straightforward factorization and simplification. Method 1(a) requires a "trick" such as multiplying by a particular expression equivalent to 1.

69. [5.4] ◆ The ranges of the inverse trigonometric functions are restricted in order that they might be functions.

70. [5.1] ◆ The expression $\tan(x + 450°)$ can be simplified using the sine and cosine sum formulas, but cannot be simplified using the tangent sum formula because $\tan 450°$ is undefined.

71. [5.1] 108.4°

72. [5.1]
$$\begin{aligned} \cos(u + v) &= \cos u \cos v - \sin u \sin v \\ &= \cos u \cos v - \cos\left(\frac{\pi}{2} - u\right)\cos\left(\frac{\pi}{2} - v\right) \end{aligned}$$

73. [5.2] $\cos^2 x$

74. [5.2] $\sin\theta = \sqrt{\frac{1}{2} + \frac{\sqrt{6}}{5}}$; $\cos\theta = \sqrt{\frac{1}{2} - \frac{\sqrt{6}}{5}}$; $\tan\theta = \sqrt{\frac{5 + 2\sqrt{6}}{5 - 2\sqrt{6}}}$

75. [5.3] $\ln e^{\sin t} = \log_e e^{\sin t} = \sin t$

76. [5.4]

y
π
$y = \sec^{-1} x$
$\frac{\pi}{2}$
−3 −2 −1 1 2 3 x

77. [5.4] Let $x = \frac{\sqrt{2}}{2}$. Then $\tan^{-1}\frac{\sqrt{2}}{2} \approx 0.6155$ and $\frac{\sin^{-1}\frac{\sqrt{2}}{2}}{\cos^{-1}\frac{\sqrt{2}}{2}} = \frac{\frac{\pi}{4}}{\frac{\pi}{4}} = 1$. **78.** [5.5] $\frac{\pi}{2}, \frac{3\pi}{2}$

Chapter 6

Exercise Set 6.1

1. $A = 121°$, $a \approx 33$, $c \approx 14$ **3.** $B \approx 57.4°$, $C \approx 86.1°$, $c \approx 40$, or $B \approx 122.6°$, $C \approx 20.9°$, $c \approx 14$
5. $B \approx 44°24'$, $A \approx 74°26'$, $a \approx 33.3$
7. $A = 110.36°$, $a \approx 5$ mi, $b \approx 3$ mi
9. $B \approx 83.78°$, $A \approx 12.44°$, $a \approx 12.30$ yd
11. $B \approx 14.7°$, $C \approx 135.0°$, $c \approx 28.04$ cm
13. No solution **15.** $B = 125.27°$, $b \approx 302$ m, $c \approx 138$ m
17. 8.2 ft² **19.** 12 yd² **21.** 596.98 ft² **23.** 787 ft²
25. About 12.86 ft, or 12 ft, 10 in. **27.** About 51 ft
29. From A: about 35 mi; from B: about 66 mi
31. About 60 mi
33. 1.348, 77.2° **35.** 18.24° **37.** ◆
39. Use the formula for the area of a triangle and the law of sines.

$$K = \frac{1}{2}bc\sin A \quad \text{and} \quad b = \frac{c\sin B}{\sin C},$$
$$\text{so} \quad K = \frac{c^2 \sin A \sin B}{2\sin C}.$$
$$K = \frac{1}{2}ab\sin C \quad \text{and} \quad b = \frac{a\sin B}{\sin A},$$
$$\text{so} \quad K = \frac{a^2 \sin B \sin C}{2\sin A}.$$
$$K = \frac{1}{2}bc\sin A \quad \text{and} \quad c = \frac{b\sin C}{\sin B},$$
$$\text{so} \quad K = \frac{b^2 \sin A \sin C}{2\sin B}.$$

41.

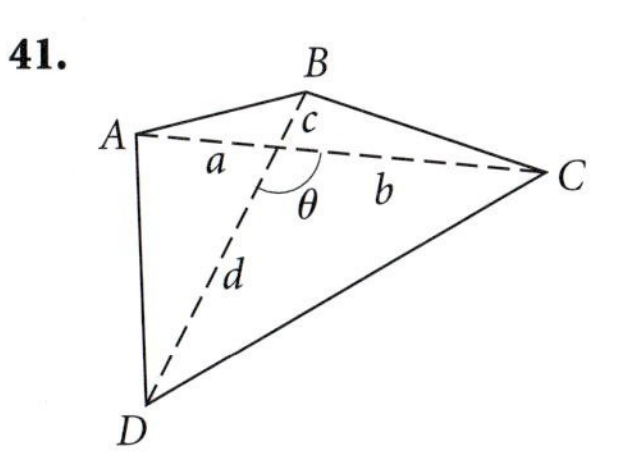

For the quadrilateral $ABCD$, we have:

$$\begin{aligned}\text{Area} &= \frac{1}{2}bd \sin \theta + \frac{1}{2}ac \sin \theta \\ &\quad + \frac{1}{2}ad(\sin 180^\circ - \theta) + \frac{1}{2}bc \sin (180^\circ - \theta) \\ &= \frac{1}{2}(bd + ac + ad + bc) \sin \theta \\ &= \frac{1}{2}(a + b)(c + d) \sin \theta \\ &= \frac{1}{2}d_1 d_2 \sin \theta,\end{aligned}$$

where $d_1 = a + b$ and $d_2 = c + d$.

Exercise Set 6.2

1. $a \approx 15, B \approx 24°, C \approx 126°$
3. $A \approx 36.18°, B \approx 43.53°, C \approx 100.29°$
5. $b \approx 75$ m, $A \approx 94°51', C \approx 12°29'$
7. $A \approx 24.15°, B \approx 30.75°, C \approx 125.10°$
9. No solution
11. $A \approx 79.93°, B \approx 53.55°, C \approx 46.52°$
13. $c \approx 45.17$ mi, $A \approx 89.3°, B \approx 42.0°$
15. $a \approx 13.9$ in., $B \approx 36.127°, C \approx 90.417°$
17. Law of sines; $C = 98°, a \approx 96.7, c \approx 101.9$
19. Law of cosines; $A \approx 73.71°, B \approx 51.75°, C \approx 54.54°$
21. Cannot be solved
23. Law of cosines; $A \approx 33.71°, B \approx 107.08°, C \approx 39.21°$
25. About 367 ft **27.** About 1.5 mi
29. About 37 nautical mi **31.** About 912 km
33. **(a)** About 16 ft; **(b)** about 122 ft^2
35. About 4.7 cm **37.** 5 **39.** $\frac{\sqrt{2}}{2}$ **41.** $-\frac{1}{2}$ **43.** ◈
45. About 9386 ft
47. $A = \frac{1}{2}a^2 \sin \theta$; when $\theta = 90°$

Exercise Set 6.3

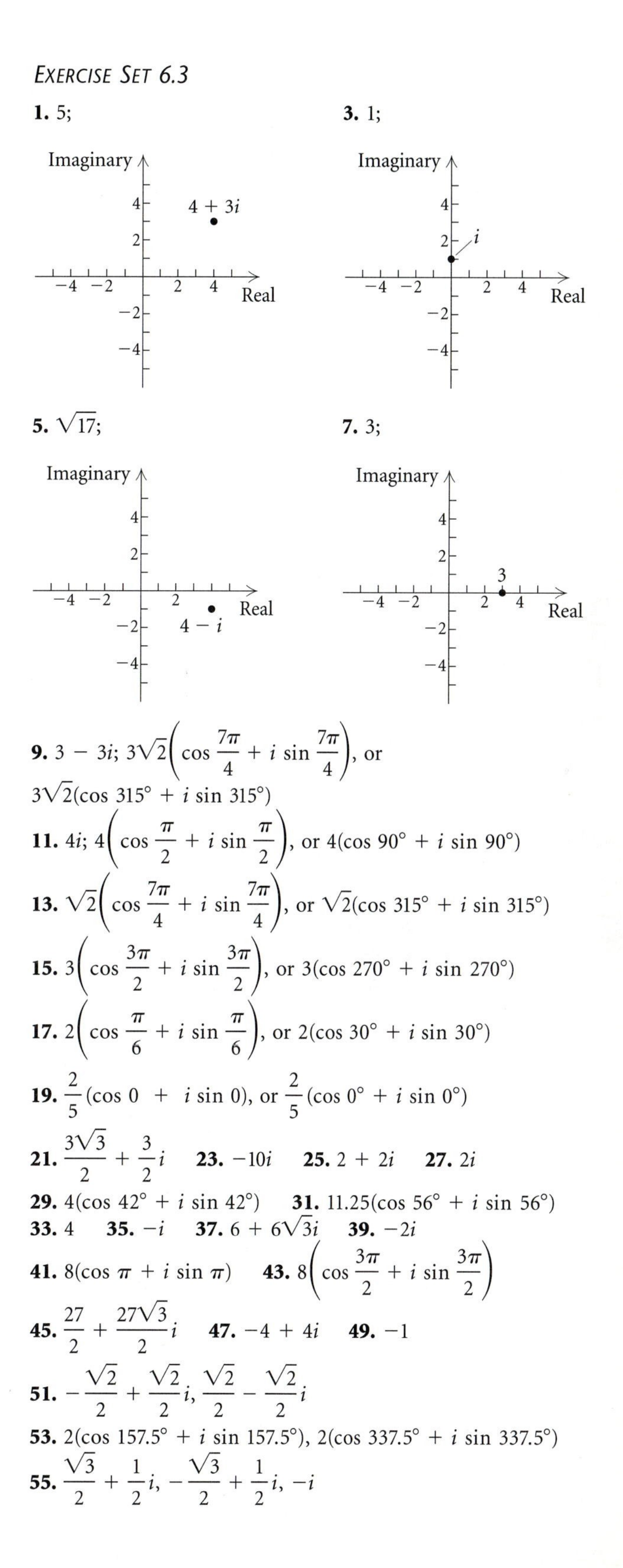

1. 5; **3.** 1;

5. $\sqrt{17}$; **7.** 3;

9. $3 - 3i$; $3\sqrt{2}\left(\cos \frac{7\pi}{4} + i \sin \frac{7\pi}{4}\right)$, or $3\sqrt{2}(\cos 315° + i \sin 315°)$

11. $4i$; $4\left(\cos \frac{\pi}{2} + i \sin \frac{\pi}{2}\right)$, or $4(\cos 90° + i \sin 90°)$

13. $\sqrt{2}\left(\cos \frac{7\pi}{4} + i \sin \frac{7\pi}{4}\right)$, or $\sqrt{2}(\cos 315° + i \sin 315°)$

15. $3\left(\cos \frac{3\pi}{2} + i \sin \frac{3\pi}{2}\right)$, or $3(\cos 270° + i \sin 270°)$

17. $2\left(\cos \frac{\pi}{6} + i \sin \frac{\pi}{6}\right)$, or $2(\cos 30° + i \sin 30°)$

19. $\frac{2}{5}(\cos 0 + i \sin 0)$, or $\frac{2}{5}(\cos 0° + i \sin 0°)$

21. $\frac{3\sqrt{3}}{2} + \frac{3}{2}i$ **23.** $-10i$ **25.** $2 + 2i$ **27.** $2i$

29. $4(\cos 42° + i \sin 42°)$ **31.** $11.25(\cos 56° + i \sin 56°)$

33. 4 **35.** $-i$ **37.** $6 + 6\sqrt{3}i$ **39.** $-2i$

41. $8(\cos \pi + i \sin \pi)$ **43.** $8\left(\cos \frac{3\pi}{2} + i \sin \frac{3\pi}{2}\right)$

45. $\frac{27}{2} + \frac{27\sqrt{3}}{2}i$ **47.** $-4 + 4i$ **49.** -1

51. $-\frac{\sqrt{2}}{2} + \frac{\sqrt{2}}{2}i, \frac{\sqrt{2}}{2} - \frac{\sqrt{2}}{2}i$

53. $2(\cos 157.5° + i \sin 157.5°)$, $2(\cos 337.5° + i \sin 337.5°)$

55. $\frac{\sqrt{3}}{2} + \frac{1}{2}i, -\frac{\sqrt{3}}{2} + \frac{1}{2}i, -i$

57. $\sqrt[3]{4}(\cos 110° + i \sin 110°)$, $\sqrt[3]{4}(\cos 230° + i \sin 230°)$, $\sqrt[3]{4}(\cos 350° + i \sin 350°)$

59. $2, 2i, -2, -2i$;

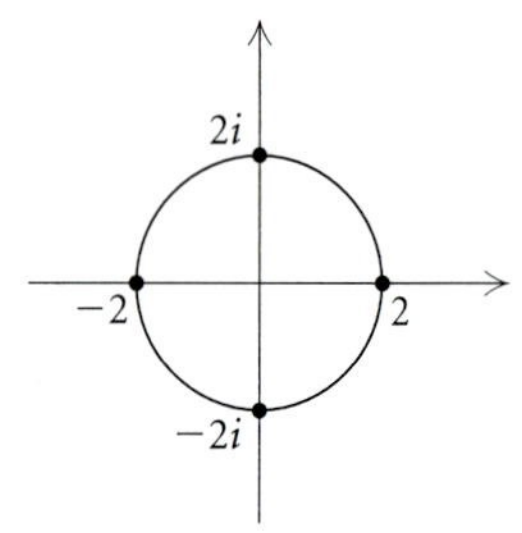

61. $\cos 36° + i \sin 36°$, $\cos 108° + i \sin 108°$, -1, $\cos 252° + i \sin 252°$, $\cos 324° + i \sin 324°$;

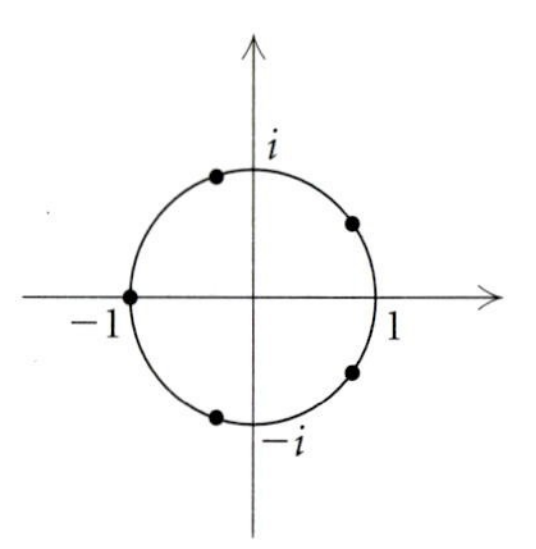

63. $\sqrt[10]{8}$, $\sqrt[10]{8}(\cos 36° + i \sin 36°)$, $\sqrt[10]{8}(\cos 72° + i \sin 72°)$, $\sqrt[10]{8}(\cos 108° + i \sin 108°)$, $\sqrt[10]{8}(\cos 144° + i \sin 144°)$, $-\sqrt[10]{8}$, $\sqrt[10]{8}(\cos 216° + i \sin 216°)$, $\sqrt[10]{8}(\cos 252° + i \sin 252°)$, $\sqrt[10]{8}(\cos 288° + i \sin 288°)$, $\sqrt[10]{8}(\cos 324° + i \sin 324°)$

65. $\dfrac{\sqrt{3}}{2} + \dfrac{1}{2}i, i, -\dfrac{\sqrt{3}}{2} + \dfrac{1}{2}i, -\dfrac{\sqrt{3}}{2} - \dfrac{1}{2}i, -i, \dfrac{\sqrt{3}}{2} - \dfrac{1}{2}i$

67. $1, -\dfrac{1}{2} + \dfrac{\sqrt{3}}{2}i, -\dfrac{1}{2} - \dfrac{\sqrt{3}}{2}i$

69. $\cos 67.5° + i \sin 67.5°$, $\cos 157.5° + i \sin 157.5°$, $\cos 247.5° + i \sin 247.5°$, $\cos 337.5° + i \sin 337.5°$

71. $\dfrac{\sqrt{3}}{2} + \dfrac{1}{2}i, i, -\dfrac{\sqrt{3}}{2} + \dfrac{1}{2}i, -\dfrac{\sqrt{3}}{2} - \dfrac{1}{2}i, -i, \dfrac{\sqrt{3}}{2} - \dfrac{1}{2}i$

73. $15°$ **75.** $\dfrac{11\pi}{6}$ **77.** $3\sqrt{5}$ **79.** ◈

81. $-\dfrac{1 + \sqrt{3}}{2} + \dfrac{1 + \sqrt{3}}{2}i, -\dfrac{1 - \sqrt{3}}{2} + \dfrac{1 - \sqrt{3}}{2}i$

83. $\cos \theta - i \sin \theta$

85. $z = a + bi$, $|z| = \sqrt{a^2 + b^2}$; $\bar{z} = a - bi$, $|\bar{z}| = \sqrt{a^2 + (-b)^2} = \sqrt{a^2 + b^2}$, $\therefore |z| = |\bar{z}|$

87.
$$\begin{aligned} |(a + bi)^2| &= |a^2 - b^2 + 2abi| \\ &= \sqrt{(a^2 - b^2)^2 + 4a^2b^2} \\ &= \sqrt{a^4 + 2a^2b^2 + b^4} = a^2 + b^2, \end{aligned}$$
$$|a + bi|^2 = (\sqrt{a^2 + b^2})^2 = a^2 + b^2$$

89.
$$\begin{aligned} \frac{z}{w} &= \frac{r_1(\cos \theta_1 + i \sin \theta_1)}{r_2(\cos \theta_2 + i \sin \theta_2)} \\ &= \frac{r_1}{r_2}(\cos (\theta_1 - \theta_2) + i \sin (\theta_1 - \theta_2)), \end{aligned}$$
$$\begin{aligned} \left|\frac{z}{w}\right| &= \sqrt{\left[\frac{r_1}{r_2}\cos (\theta_1 - \theta_2)\right]^2 + \left[\frac{r_1}{r_2}\sin (\theta_1 - \theta_2)\right]^2} \\ &= \sqrt{\frac{r_1^2}{r_2^2}} = \frac{|r_1|}{|r_2|}; \end{aligned}$$
$$\begin{aligned} |z| &= \sqrt{(r_1 \cos \theta_1)^2 + (r_1 \sin \theta_1)^2} \\ &= \sqrt{r_1^2} = |r_1|; \end{aligned}$$
$$\begin{aligned} |w| &= \sqrt{(r_2 \cos \theta_2)^2 + (r_2 \sin \theta_2)^2} \\ &= \sqrt{r_2^2} = |r_2|; \end{aligned}$$
Then $\left|\dfrac{z}{w}\right| = \dfrac{|r_1|}{|r_2|} = \dfrac{|z|}{|w|}$.

91.

$z + \bar{z} = 3$

$\frac{3}{2}$

EXERCISE SET 6.4

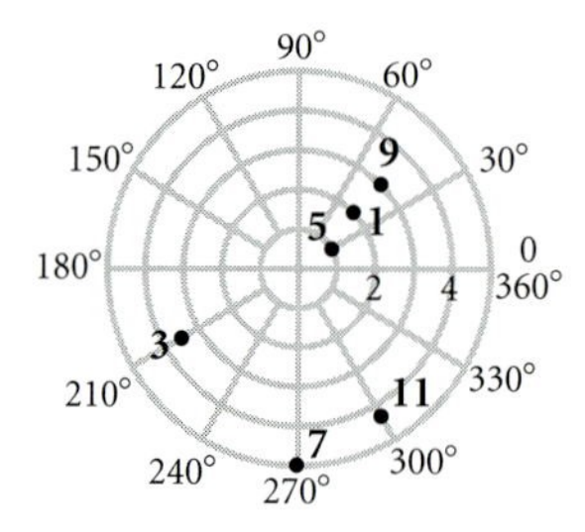

13. *A*: $(4, 30°)$, $(4, 390°)$, $(-4, 210°)$; *B*: $(5, 300°)$, $(5, -60°)$, $(-5, 120°)$; *C*: $(2, 150°)$, $(2, 510°)$, $(-2, 330°)$; *D*: $(3, 225°)$, $(3, -135°)$, $(-3, 45°)$; answers may vary

15. $(3, 270°)$, $\left(3, \dfrac{3\pi}{2}\right)$ **17.** $(6, 300°)$, $\left(6, \dfrac{5\pi}{3}\right)$

19. $(8, 330°)$, $\left(8, \dfrac{11\pi}{6}\right)$ **21.** $(2, 225°)$, $\left(2, \dfrac{5\pi}{4}\right)$

23. $(7.616, 66.8°)$, $(7.616, 1.166)$

25. $(4.643, 132.9°)$, $(4.643, 2.320)$

27. $\left(\dfrac{5}{2}, \dfrac{5\sqrt{3}}{2}\right)$ **29.** $\left(-\dfrac{3\sqrt{2}}{2}, -\dfrac{3\sqrt{2}}{2}\right)$

31. $\left(-\dfrac{3}{2}, -\dfrac{3\sqrt{3}}{2}\right)$ **33.** $(-1, \sqrt{3})$ **35.** $(2.19, -2.05)$

37. $(1.30, -3.99)$ **39.** $r(3 \cos \theta + 4 \sin \theta) = 5$

41. $r \cos \theta = 5$ **43.** $r = 6$ **45.** $r^2 \cos^2 \theta = 25r \sin \theta$
47. $r^2 \sin^2 \theta - 5r \cos \theta - 25 = 0$ **49.** $x^2 + y^2 = 25$
51. $y = 2$ **53.** $y^2 = -6x + 9$
55. $x^2 - 9x + y^2 - 7y = 0$ **57.** $x = 5$
59. **61.**

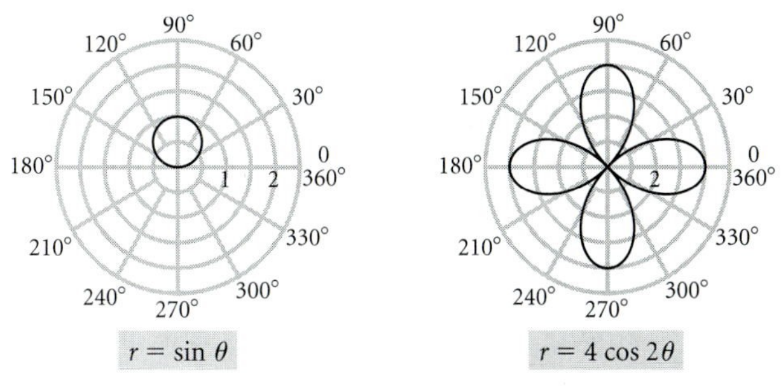

63. (d) **65.** (g) **67.** (j) **69.** (b) **71.** (e) **73.** (k)
75. {12}
77.

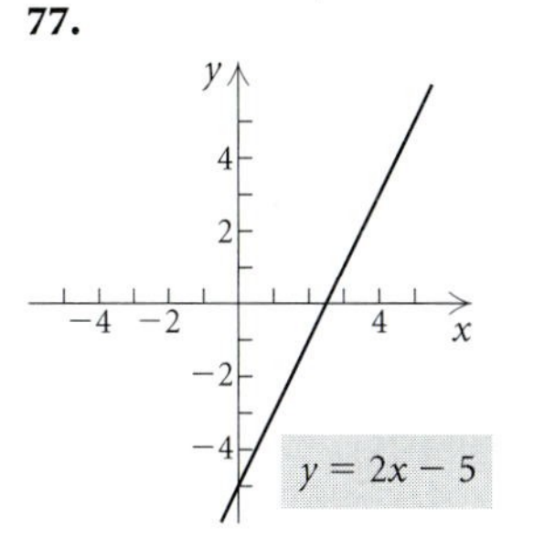

79.

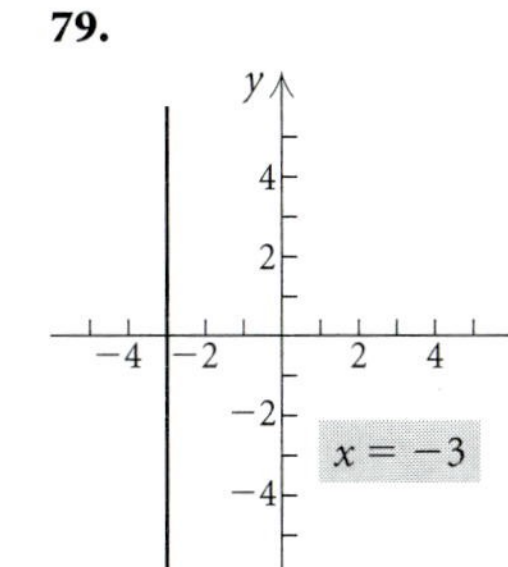

81. ◈ **83.** $y^2 = -4x + 4$
85.

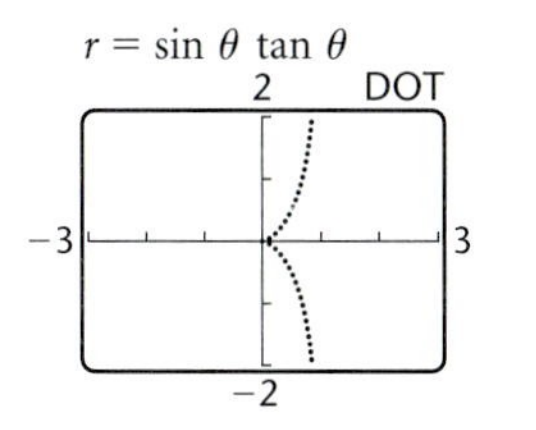

87.

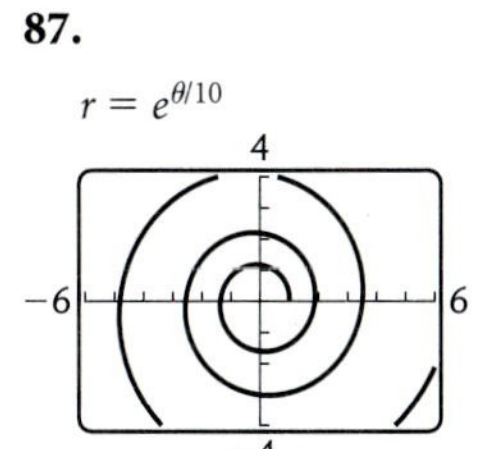

89.

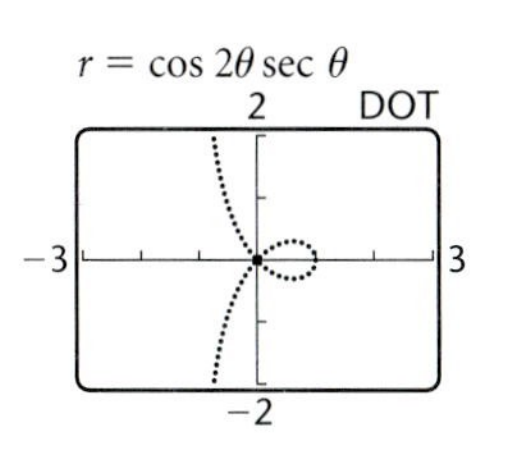

91.

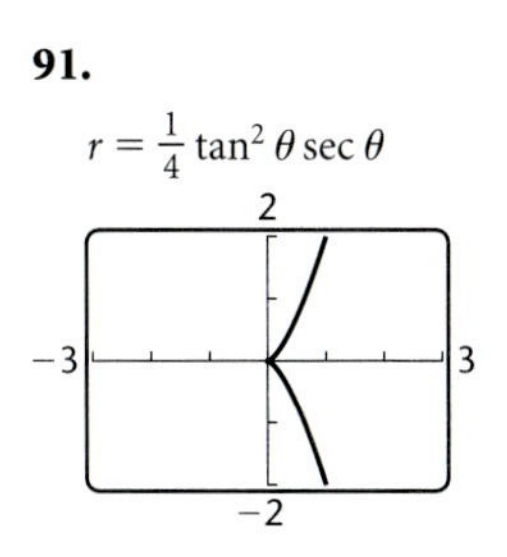

EXERCISE SET 6.5

1. 57.0, 38° **3.** 18.4, 37° **5.** 20.9, 58° **7.** 68.3, 18°

9.

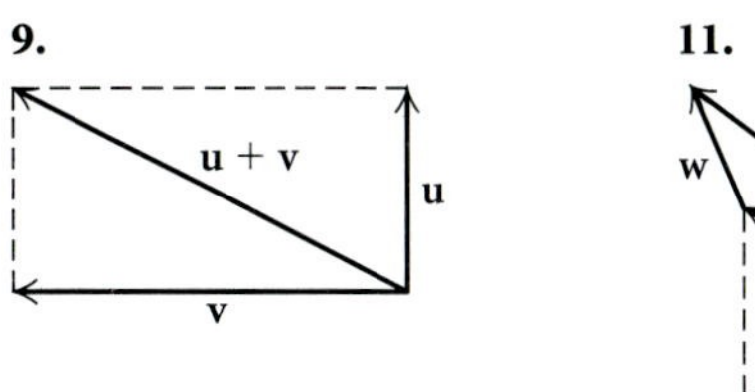

11.

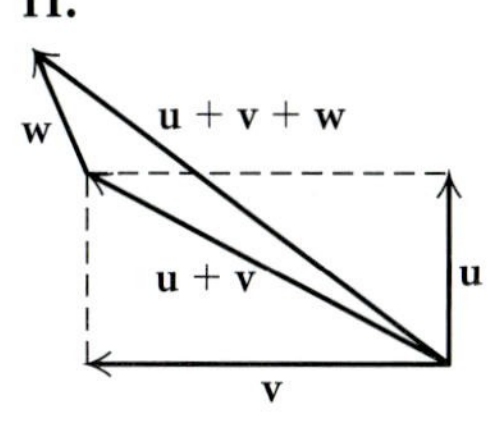

13. **(a)** $\mathbf{w} = \mathbf{u} + \mathbf{v}$; **(b)** $\mathbf{v} = \mathbf{w} - \mathbf{u}$
15. $\langle -5, -7 \rangle = -5\mathbf{i} - 7\mathbf{j}$ **17.** $-7\mathbf{i} + 5\mathbf{j}$ **19.** $8\mathbf{i} - \mathbf{j}$
21. $-5\mathbf{i} + 5\mathbf{j}$ **23.** $\mathbf{0}$ **25.** $(5, 8)$ **27.** $215.17\mathbf{i} + 65.78\mathbf{j}$
29. Horizontal: 390 lb forward; vertical: 675.5 lb up
31. 55 N, 55° **33.** 70.7 east, 70.7 south **35.** 11 ft/sec, 63°
37. 174 nautical mi, S15°E **39.** 60° **41.** $19\mathbf{i} + 36\mathbf{j}$
43. $14\mathbf{i} + 8\mathbf{j}$ **45.** -8 **47.** 17.9 **49.** 36
51.

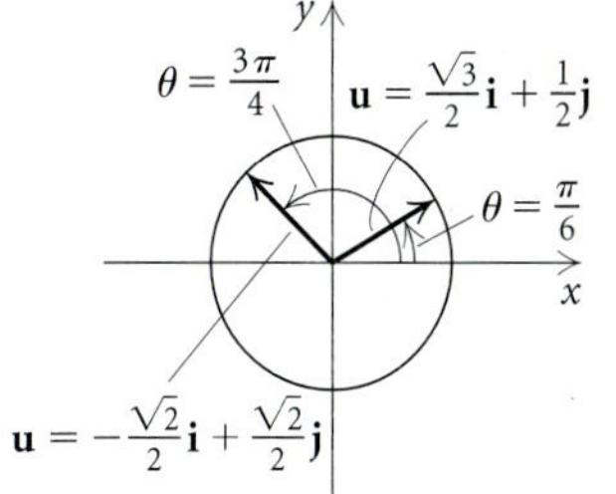

53. $\mathbf{u} = -\frac{\sqrt{2}}{2}\mathbf{i} - \frac{\sqrt{2}}{2}\mathbf{j}$ **55.** $\mathbf{u} = -\frac{\sqrt{10}}{10}\mathbf{i} + \frac{3\sqrt{10}}{10}\mathbf{j}$

57. $\sqrt{13}\left(\frac{2\sqrt{13}}{13}\mathbf{i} - \frac{3\sqrt{13}}{13}\mathbf{j}\right)$

59.

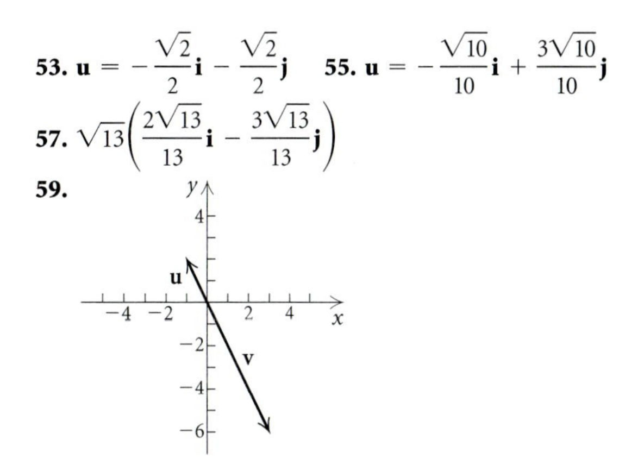

61. $(-0.286, 1.143)$ **63.** $(2, 5), (-3, -10)$ **65.** ◈
67. $\frac{3}{5}\mathbf{i} - \frac{4}{5}\mathbf{j}, -\frac{3}{5}\mathbf{i} + \frac{4}{5}\mathbf{j}$
69. **(a)** (4.950, 4.950); **(b)** (0.950, −1.978) **71.** **0**

REVIEW EXERCISES, CHAPTER 6

1. [6.2] $A \approx 153°$, $B \approx 18°$, $C \approx 9°$
2. [6.1] $A = 118°$, $a \approx 37$ in., $c \approx 24$ in.
3. [6.1] $B = 14°50'$, $a \approx 2523$ m, $c \approx 1827$ m
4. [6.1] No solution **5.** [6.1] 33 m^2 **6.** [6.1] 13.72 ft^2
7. [6.1] 67 ft^2 **8.** [6.2] 92.1°, 33.0°, 54.8°
9. [6.1] 419 ft **10.** [6.2] About 650 km

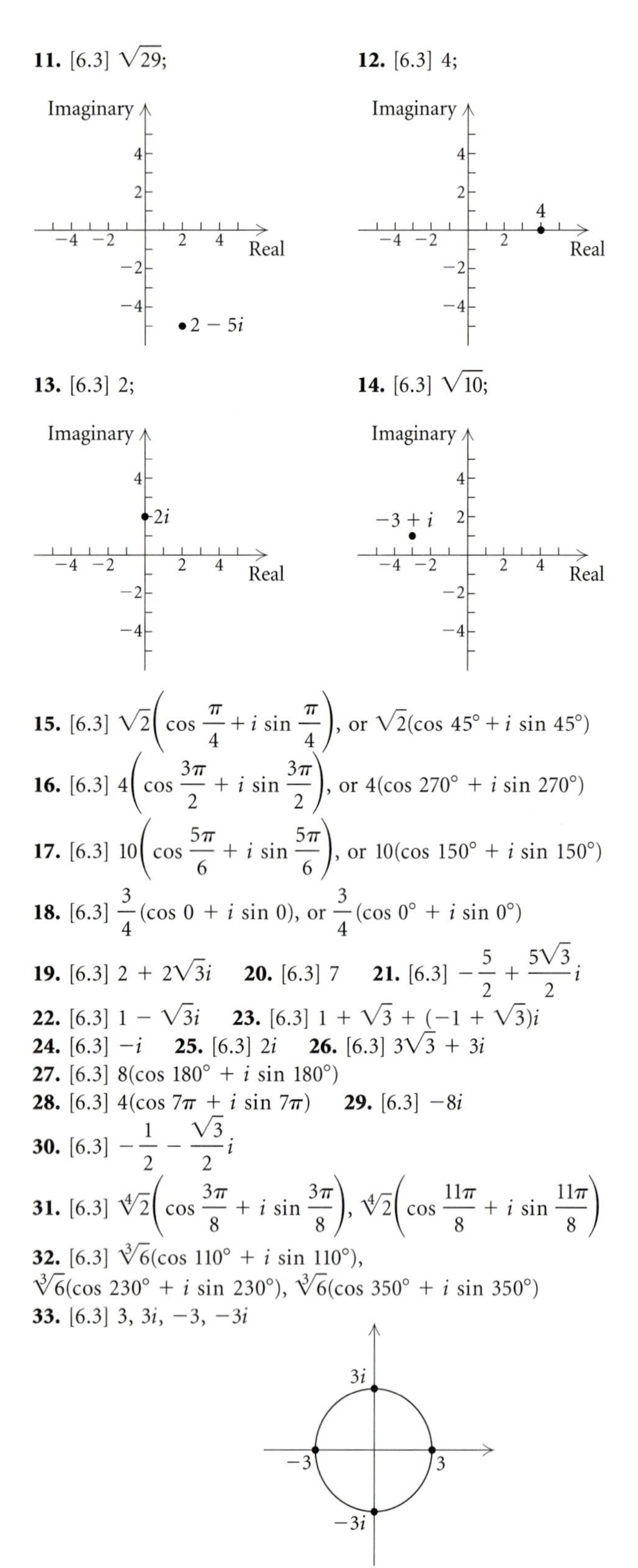

11. [6.3] $\sqrt{29}$;

12. [6.3] 4;

13. [6.3] 2;

14. [6.3] $\sqrt{10}$;

15. [6.3] $\sqrt{2}\left(\cos\frac{\pi}{4} + i\sin\frac{\pi}{4}\right)$, or $\sqrt{2}(\cos 45^\circ + i\sin 45^\circ)$

16. [6.3] $4\left(\cos\frac{3\pi}{2} + i\sin\frac{3\pi}{2}\right)$, or $4(\cos 270^\circ + i\sin 270^\circ)$

17. [6.3] $10\left(\cos\frac{5\pi}{6} + i\sin\frac{5\pi}{6}\right)$, or $10(\cos 150^\circ + i\sin 150^\circ)$

18. [6.3] $\frac{3}{4}(\cos 0 + i\sin 0)$, or $\frac{3}{4}(\cos 0^\circ + i\sin 0^\circ)$

19. [6.3] $2 + 2\sqrt{3}i$ **20.** [6.3] 7 **21.** [6.3] $-\frac{5}{2} + \frac{5\sqrt{3}}{2}i$

22. [6.3] $1 - \sqrt{3}i$ **23.** [6.3] $1 + \sqrt{3} + (-1 + \sqrt{3})i$

24. [6.3] $-i$ **25.** [6.3] $2i$ **26.** [6.3] $3\sqrt{3} + 3i$

27. [6.3] $8(\cos 180^\circ + i\sin 180^\circ)$

28. [6.3] $4(\cos 7\pi + i\sin 7\pi)$ **29.** [6.3] $-8i$

30. [6.3] $-\frac{1}{2} - \frac{\sqrt{3}}{2}i$

31. [6.3] $\sqrt[4]{2}\left(\cos\frac{3\pi}{8} + i\sin\frac{3\pi}{8}\right)$, $\sqrt[4]{2}\left(\cos\frac{11\pi}{8} + i\sin\frac{11\pi}{8}\right)$

32. [6.3] $\sqrt[3]{6}(\cos 110^\circ + i\sin 110^\circ)$, $\sqrt[3]{6}(\cos 230^\circ + i\sin 230^\circ)$, $\sqrt[3]{6}(\cos 350^\circ + i\sin 350^\circ)$

33. [6.3] $3, 3i, -3, -3i$

34. [6.3] $1, \cos 72^\circ + i\sin 72^\circ, \cos 144^\circ + i\sin 144^\circ, \cos 216^\circ + i\sin 216^\circ, \cos 288^\circ + i\sin 288^\circ$

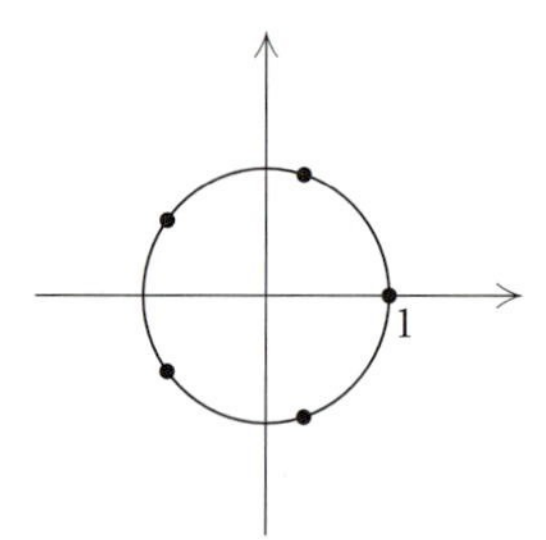

35. [6.3] $\cos 22.5^\circ + i\sin 22.5^\circ, \cos 112.5^\circ + i\sin 112.5^\circ, \cos 202.5^\circ + i\sin 202.5^\circ, \cos 292.5^\circ + i\sin 292.5^\circ$

36. [6.3] $\frac{1}{2} + \frac{\sqrt{3}}{2}i, -1, \frac{1}{2} - \frac{\sqrt{3}}{2}i$

37. [6.4] A: $(5, 120^\circ), (5, 480^\circ), (-5, 300^\circ)$; B: $(3, 210^\circ), (-3, 30^\circ), (-3, 390^\circ)$; C: $(4, 60^\circ), (4, 420^\circ), (-4, 240^\circ)$; D: $(1, 300^\circ), (1, -60^\circ), (-1, 120^\circ)$; answers may vary

38. [6.4] $(8, 135^\circ), \left(8, \frac{3\pi}{4}\right)$ **39.** [6.4] $(5, 270^\circ), \left(5, \frac{3\pi}{2}\right)$

40. [6.4] $(5.385, 111.8^\circ), (5.385, 1.951)$

41. [6.4] $(4.964, 147.8^\circ), (4.964, 2.579)$

42. [6.4] $\left(\frac{3\sqrt{2}}{2}, \frac{3\sqrt{2}}{2}\right)$ **43.** [6.4] $(3, 3\sqrt{3})$

44. [6.4] $(1.93, -0.52)$ **45.** [6.4] $(-1.86, -1.35)$

46. [6.4] $r(5\cos\theta - 2\sin\theta) = 6$ **47.** [6.4] $r\sin\theta = 3$

48. [6.4] $r = 3$ **49.** [6.4] $r^2\sin^2\theta - 4r\cos\theta - 16 = 0$

50. [6.4] $x^2 + y^2 = 36$ **51.** [6.4] $x^2 + 2y = 1$

52. [6.4] $y^2 - 6x = 9$ **53.** [6.4] $x^2 - 2x + y^2 - 3y = 0$

54. [6.4] (b) **55.** [6.4] (d) **56.** [6.4] (a) **57.** [6.4] (c)

58. [6.5] 13.7, 71° **59.** [6.5] 98.7, 15°

60. [6.5]

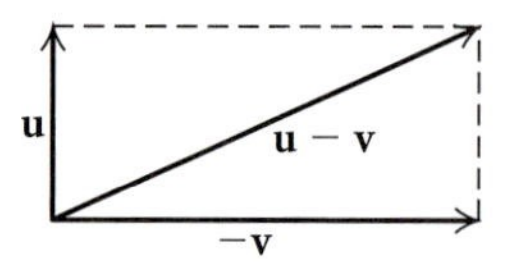

61. [6.5]

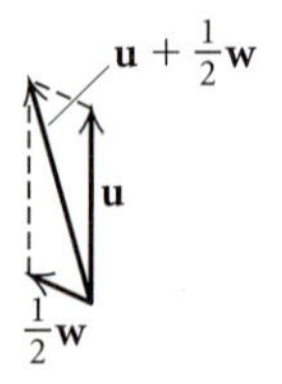

62. [6.5] 666.7 N, 36° **63.** [6.5] 29 km/h, 329°

64. [6.5] 102.4 nautical mi, S43°E **65.** [6.5] $34\mathbf{i} - 55\mathbf{j}$

66. [6.5] $\mathbf{i} - 12\mathbf{j}$ **67.** [6.5] $5\sqrt{2}$

68. [6.5] $3\sqrt{65} + \sqrt{109}$ **69.** [6.5] $-5\mathbf{i} + 5\mathbf{j}$

70. [6.5] **71.** [6.5]

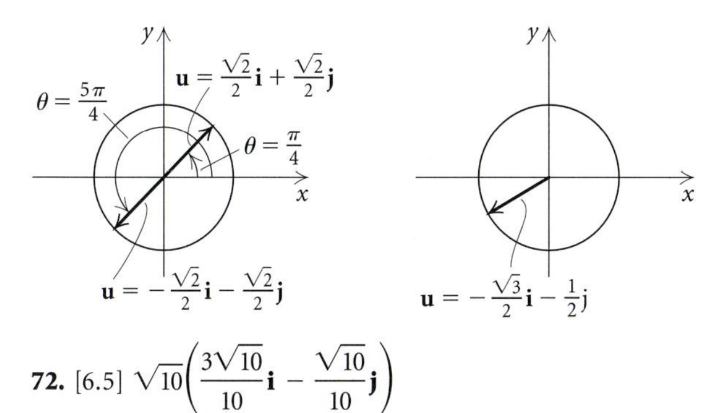

72. [6.5] $\sqrt{10}\left(\frac{3\sqrt{10}}{10}\mathbf{i} - \frac{\sqrt{10}}{10}\mathbf{j}\right)$

73. [6.1], [6.2] ◈ A triangle has no solution when a sine or cosine value found is less than -1 or greater than 1. A triangle also has no solution if the sum of the angle measures calculated is greater than 180°. A triangle has only one solution if only one possible answer is found, or if one of the possible answers has an angle sum greater than 180°. A triangle has two solutions when two possible answers are found and neither results in an angle sum greater than 180°.
74. [6.1] ◈ When only the lengths of the sides of a triangle are known, the law of sines cannot be used to find the first angle because there will be two unknowns in the equation. To use the law of sines, you must know the measure of at least one angle and the length of the side opposite it.

75. [6.1] 50.52°, 129.48° **76.** [6.5] $\frac{36}{13}\mathbf{i} + \frac{15}{13}\mathbf{j}$
77. [6.4] $y^2 = 4x + 4$

Chapter 7

EXERCISE SET 7.1

1. (c) **3.** (f) **5.** (b) **7.** $(-1, 3)$ **9.** $(-0.5, 1.75)$
11. No solution **13.** $(5, 4)$ **15.** $(1, -3)$ **17.** $(2, -2)$
19. $\left(\frac{39}{11}, -\frac{1}{11}\right)$ **21.** $(1, -1)$
23. $(1, 3)$; consistent, independent
25. $(-4, -2)$; consistent, independent
27. $(-3, 0)$; consistent, independent
29. $(4y + 2, y)$ or $\left(x, \frac{1}{4}x - \frac{1}{2}\right)$; consistent, dependent
31. $(10, 8)$; consistent, independent
33. $(1, 1)$; consistent, independent
35. 10 free rentals, 38 boxes of popcorn
37. Adult: \$6, child: \$3 **39.** (15, \$100) **41.** 140
43. 6000
45. $1\frac{1}{2}$ servings of spaghetti, 2 servings of lettuce
47. Boat: 20 km/h, stream: 3 km/h
49. \$6000 at 7%, \$9000 at 9%
51. 6 lb of French roast, 4 lb of Kenyan **53.** \$21,428.57
55. -19

57.

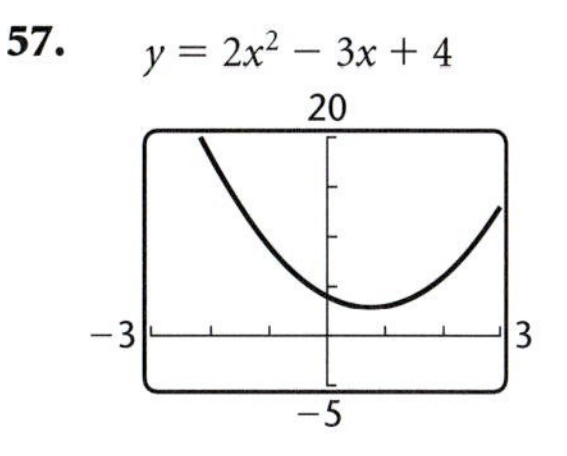

59. ◈ **61.** 4 km
63. First train: 36 km/h, second train: 54 km/h **65.** \$96
67. $m = -\frac{4}{3}, b = \frac{1}{3}$ **69.** City: 261 mi, highway: 204 mi

EXERCISE SET 7.2

1. $(3, -2, 1)$ **3.** $(-3, 2, 1)$ **5.** $\left(2, \frac{1}{2}, -2\right)$
7. No solution **9.** $\left(\frac{11y + 19}{5}, y, \frac{9y + 11}{5}\right)$
11. $\left(\frac{1}{2}, \frac{2}{3}, -\frac{5}{6}\right)$ **13.** $(1, -2, 4, -1)$
15. Under 10 lb: 60; 10 lb up to 15 lb: 70; 15 lb or more: 20
17. $1\frac{1}{4}$ servings of beef, 1 baked potato, $\frac{3}{4}$ serving strawberries
19. 4%: \$1300; 6%: \$900; 7%: \$2800
21. Orange juice: \$1; bagel: \$1.25; coffee: \$0.75
23. Airways: 351 billion; bus: 32 billion; railroad: 22 billion
25. Par-3: 4; par-4: 10; par-5: 4
27. **(a)** $f(x) = \frac{12}{247}x^2 - \frac{384}{247}x + 43$; **(b)** 58.5%
29. **(a)** $f(x) = -\frac{135}{238}x^2 + \frac{2659}{238}x + 539$; **(b)** 364 lb
31. $4x - 6y$ **33.** $x + 2y$ **35.** ◈ **37.** $\left(-1, \frac{1}{5}, -\frac{1}{2}\right)$
39. 180° **41.** $3x + 4y + 2z = 12$
43. $y = -4x^3 + 5x^2 - 3x + 1$
45. Adults: 5; students: 1; children: 94

EXERCISE SET 7.3

1. 3×2 **3.** 1×4 **5.** 3×3 **7.** $\left[\begin{array}{rr|r} 2 & -1 & 7 \\ 1 & 4 & -5 \end{array}\right]$

9. $\left[\begin{array}{rrr|r} 1 & -2 & 3 & 12 \\ 2 & 0 & -4 & 8 \\ 0 & 3 & 1 & 7 \end{array}\right]$

11. $3x - 5y = 1,$
$x + 4y = -2$

13. $2x + y - 4z = 12,$
$3x \quad + 5z = -1,$
$x - y + z = 2$

15. $\left(\frac{3}{2}, \frac{5}{2}\right)$ **17.** $\left(-\frac{63}{29}, -\frac{114}{29}\right)$ **19.** $\left(-1, \frac{5}{2}\right)$ **21.** $(0, 3)$
23. No solution **25.** $(3y - 2, y)$ **27.** $(-1, 2, -2)$
29. $\left(\frac{3}{2}, -4, 3\right)$ **31.** $(-1, 6, 3)$
33. $\left(\frac{1}{2}z + \frac{1}{2}, -\frac{1}{2}z - \frac{1}{2}, z\right)$ **35.** $(r - 2, -2r + 3, r)$
37. No solution **39.** $(1, -3, -2, -1)$ **41.** 1:00 A.M.
43. \$8000 at 8%; \$12,000 at 10%; \$10,000 at 12% **45.** 8

47. -9 **49.** ◈ **51.** $y = 3x^2 + \frac{5}{2}x - \frac{15}{2}$

53. $\begin{bmatrix} 1 & 5 \\ 0 & 1 \end{bmatrix}, \begin{bmatrix} 1 & 0 \\ 0 & 1 \end{bmatrix}$ **55.** $\left(-\frac{4}{3}, -\frac{1}{3}, 1\right)$

57. $\left(-\frac{14}{13}z - 1, \frac{3}{13}z - 2, z\right)$ **59.** $(-3, 3)$

EXERCISE SET 7.4

1. $x = -3, y = 5$ **3.** $x = -1, y = 1$ **5.** $\begin{bmatrix} -2 & 7 \\ 6 & 2 \end{bmatrix}$

7. $\begin{bmatrix} 1 & 3 \\ 2 & 6 \end{bmatrix}$ **9.** $\begin{bmatrix} 9 & 9 \\ -3 & -3 \end{bmatrix}$ **11.** $\begin{bmatrix} 11 & 13 \\ 5 & 3 \end{bmatrix}$

13. $\begin{bmatrix} -4 & 3 \\ -2 & -4 \end{bmatrix}$ **15.** $\begin{bmatrix} 17 & 9 \\ -2 & 1 \end{bmatrix}$ **17.** $\begin{bmatrix} 0 & 0 \\ 0 & 0 \end{bmatrix}$

19. $\begin{bmatrix} 1 & 2 \\ 4 & 3 \end{bmatrix}$

21. (a) [150 80 40]; **(b)** [157.5 84 42]; **(c)** [307.5 164 82], the total budget for each area in June and July

23. (a) $C = [140 \quad 27 \quad 3 \quad 13 \quad 64]$, $P = [180 \quad 4 \quad 11 \quad 24 \quad 662]$, $B = [50 \quad 5 \quad 1 \quad 82 \quad 20]$; **(b)** $C + 2P + 3B = [650 \quad 50 \quad 28 \quad 307 \quad 1448]$, the total nutritional values of a meal of 1 serving of chicken, 1 cup of potato salad, and 3 broccoli spears

25. $\begin{bmatrix} 1 \\ 40 \end{bmatrix}$ **27.** $\begin{bmatrix} -10 & 28 \\ 14 & -26 \\ 0 & -6 \end{bmatrix}$ **29.** Not defined

31. $\begin{bmatrix} 3 & 16 & 3 \\ 0 & -32 & 0 \\ -6 & 4 & 5 \end{bmatrix}$

33. (a) $\begin{bmatrix} 45.29 & 6.63 & 10.94 & 7.42 & 8.01 \\ 53.78 & 4.95 & 9.83 & 6.16 & 12.56 \\ 47.13 & 8.47 & 12.66 & 8.29 & 9.43 \\ 51.64 & 7.12 & 11.57 & 9.35 & 10.72 \end{bmatrix}$; **(b)** [65 48 93 57]; **(c)** [12,851.86 1862.1 3019.81 2081.9 2611.56]; **(d)** the total cost, in cents, for each item for the day's meals

35. (a) $\begin{bmatrix} 8 & 15 \\ 6 & 10 \\ 4 & 3 \end{bmatrix}$; **(b)** [3 1.50 2]; **(c)** [41 66]; **(d)** the total cost, in dollars, of ingredients for each coffee shop

37. (a) [6 4.50 5.20]; **(b)** **PS** = [95.80 150.60]

39. $\begin{bmatrix} 2 & -3 \\ 1 & 5 \end{bmatrix}\begin{bmatrix} x \\ y \end{bmatrix} = \begin{bmatrix} 7 \\ -6 \end{bmatrix}$ **41.** $\begin{bmatrix} 1 & 1 & -2 \\ 3 & -1 & 1 \\ 2 & 5 & -3 \end{bmatrix}\begin{bmatrix} x \\ y \\ z \end{bmatrix} = \begin{bmatrix} 6 \\ 7 \\ 8 \end{bmatrix}$

43. $\begin{bmatrix} 3 & -2 & 4 \\ 2 & 1 & -5 \end{bmatrix}\begin{bmatrix} x \\ y \\ z \end{bmatrix} = \begin{bmatrix} 17 \\ 13 \end{bmatrix}$

45. $\begin{bmatrix} -4 & 1 & -1 & 2 \\ 1 & 2 & -1 & -1 \\ -1 & 1 & 4 & -3 \\ 2 & 3 & 5 & -7 \end{bmatrix}\begin{bmatrix} w \\ x \\ y \\ z \end{bmatrix} = \begin{bmatrix} 12 \\ 0 \\ 1 \\ 9 \end{bmatrix}$ **47.** $\{9\}$ **49.** $\{-7\}$

51. ◈

53. $(\mathbf{A} + \mathbf{B})(\mathbf{A} - \mathbf{B}) = \begin{bmatrix} -2 & 1 \\ 2 & -1 \end{bmatrix}$; $\mathbf{A}^2 - \mathbf{B}^2 = \begin{bmatrix} 0 & 3 \\ 0 & -3 \end{bmatrix}$

55. $(\mathbf{A} + \mathbf{B})(\mathbf{A} - \mathbf{B}) = \begin{bmatrix} -2 & 1 \\ 2 & -1 \end{bmatrix}$

$= \mathbf{A}^2 + \mathbf{BA} - \mathbf{AB} - \mathbf{B}^2$

57. $\mathbf{A} + \mathbf{B} =$

$$\begin{bmatrix} a_{11} + b_{11} & a_{12} + b_{12} & a_{13} + b_{13} & \cdots & a_{1n} + b_{1n} \\ a_{21} + b_{21} & a_{22} + b_{22} & a_{23} + b_{23} & \cdots & a_{2n} + b_{2n} \\ a_{31} + b_{31} & a_{32} + b_{32} & a_{33} + b_{33} & \cdots & a_{3n} + b_{3n} \\ \vdots & \vdots & \vdots & & \vdots \\ a_{m1} + b_{m1} & a_{m2} + b_{m2} & a_{m3} + b_{m3} & \cdots & a_{mn} + b_{mn} \end{bmatrix}$$

$$= \begin{bmatrix} b_{11} + a_{11} & b_{12} + a_{12} & b_{13} + a_{13} & \cdots & b_{1n} + a_{1n} \\ b_{21} + a_{21} & b_{22} + a_{22} & b_{23} + a_{23} & \cdots & b_{2n} + a_{2n} \\ b_{31} + a_{31} & b_{32} + a_{32} & b_{33} + a_{33} & \cdots & b_{3n} + a_{3n} \\ \vdots & \vdots & \vdots & & \vdots \\ b_{m1} + a_{m1} & b_{m2} + a_{m2} & b_{m3} + a_{m3} & \cdots & b_{mn} + a_{mn} \end{bmatrix}$$

$= \mathbf{B} + \mathbf{A}$

59. $(kl)\mathbf{A} = \begin{bmatrix} (kl)a_{11} & (kl)a_{12} & (kl)a_{13} & \cdots & (kl)a_{1n} \\ (kl)a_{21} & (kl)a_{22} & (kl)a_{23} & \cdots & (kl)a_{2n} \\ (kl)a_{31} & (kl)a_{32} & (kl)a_{33} & \cdots & (kl)a_{3n} \\ \vdots & \vdots & \vdots & & \vdots \\ (kl)a_{m1} & (kl)a_{m2} & (kl)a_{m3} & \cdots & (kl)a_{mn} \end{bmatrix}$

$$= \begin{bmatrix} k(la_{11}) & k(la_{12}) & k(la_{13}) & \cdots & k(la_{1n}) \\ k(la_{21}) & k(la_{22}) & k(la_{23}) & \cdots & k(la_{2n}) \\ k(la_{31}) & k(la_{32}) & k(la_{33}) & \cdots & k(la_{3n}) \\ \vdots & \vdots & \vdots & & \vdots \\ k(la_{m1}) & k(la_{m2}) & k(la_{m3}) & \cdots & k(la_{mn}) \end{bmatrix}$$

$$= k\begin{bmatrix} la_{11} & la_{12} & la_{13} & \cdots & la_{1n} \\ la_{21} & la_{22} & la_{23} & \cdots & la_{2n} \\ la_{31} & la_{32} & la_{33} & \cdots & la_{3n} \\ \vdots & \vdots & \vdots & & \vdots \\ la_{m1} & la_{m2} & la_{m3} & \cdots & la_{mn} \end{bmatrix}$$

$= k(l\mathbf{A})$

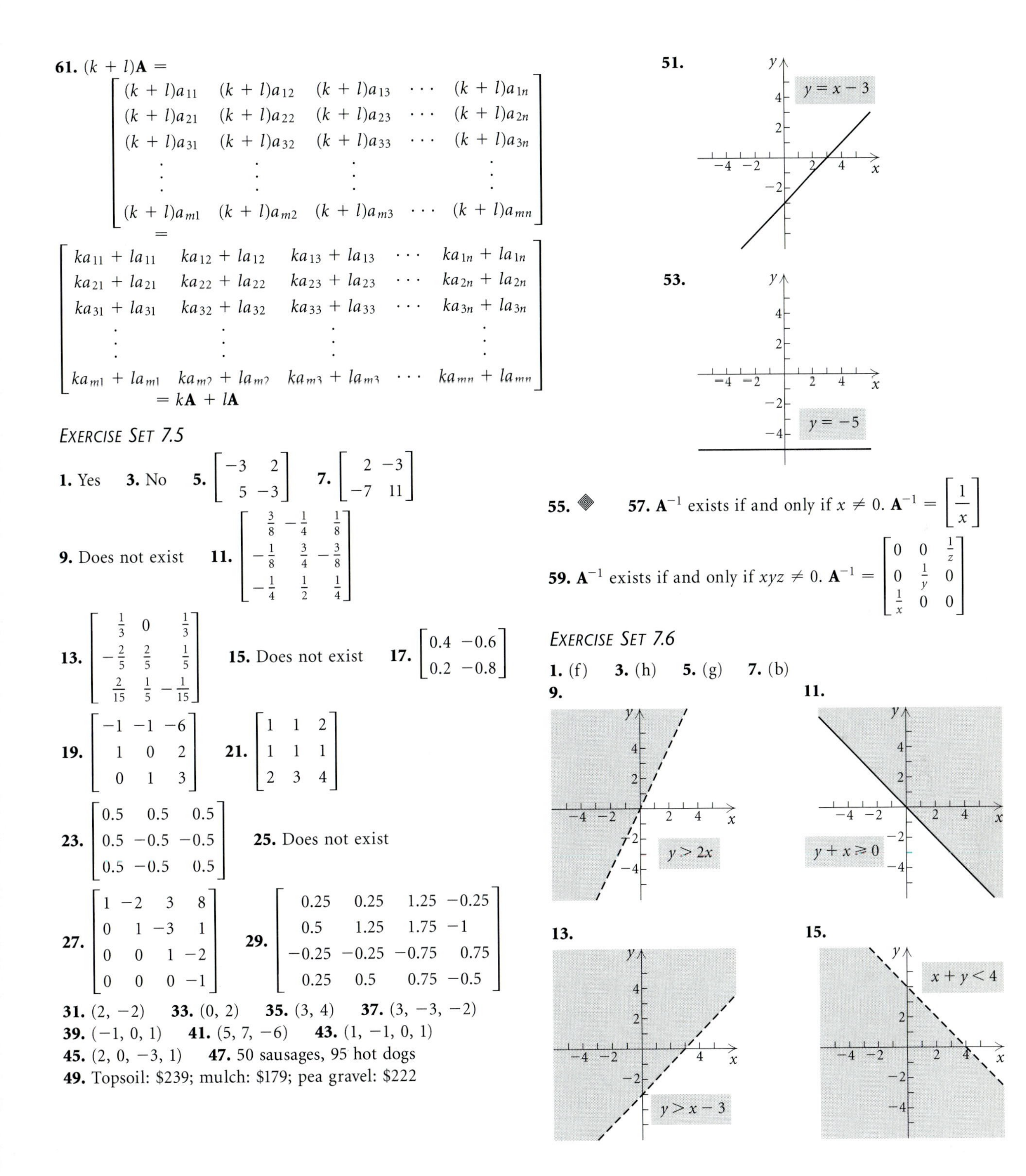

61. $(k + l)\mathbf{A} =$

$$\begin{bmatrix} (k+l)a_{11} & (k+l)a_{12} & (k+l)a_{13} & \cdots & (k+l)a_{1n} \\ (k+l)a_{21} & (k+l)a_{22} & (k+l)a_{23} & \cdots & (k+l)a_{2n} \\ (k+l)a_{31} & (k+l)a_{32} & (k+l)a_{33} & \cdots & (k+l)a_{3n} \\ \vdots & \vdots & \vdots & & \vdots \\ (k+l)a_{m1} & (k+l)a_{m2} & (k+l)a_{m3} & \cdots & (k+l)a_{mn} \end{bmatrix}$$

$$= \begin{bmatrix} ka_{11}+la_{11} & ka_{12}+la_{12} & ka_{13}+la_{13} & \cdots & ka_{1n}+la_{1n} \\ ka_{21}+la_{21} & ka_{22}+la_{22} & ka_{23}+la_{23} & \cdots & ka_{2n}+la_{2n} \\ ka_{31}+la_{31} & ka_{32}+la_{32} & ka_{33}+la_{33} & \cdots & ka_{3n}+la_{3n} \\ \vdots & \vdots & \vdots & & \vdots \\ ka_{m1}+la_{m1} & ka_{m2}+la_{m2} & ka_{m3}+la_{m3} & \cdots & ka_{mn}+la_{mn} \end{bmatrix}$$

$= k\mathbf{A} + l\mathbf{A}$

EXERCISE SET 7.5

1. Yes **3.** No **5.** $\begin{bmatrix} -3 & 2 \\ 5 & -3 \end{bmatrix}$ **7.** $\begin{bmatrix} 2 & -3 \\ -7 & 11 \end{bmatrix}$

9. Does not exist **11.** $\begin{bmatrix} \frac{3}{8} & -\frac{1}{4} & \frac{1}{8} \\ -\frac{1}{8} & \frac{3}{4} & -\frac{3}{8} \\ -\frac{1}{4} & \frac{1}{2} & \frac{1}{4} \end{bmatrix}$

13. $\begin{bmatrix} \frac{1}{3} & 0 & \frac{1}{3} \\ -\frac{2}{5} & \frac{2}{5} & \frac{1}{5} \\ \frac{2}{15} & \frac{1}{5} & -\frac{1}{15} \end{bmatrix}$ **15.** Does not exist **17.** $\begin{bmatrix} 0.4 & -0.6 \\ 0.2 & -0.8 \end{bmatrix}$

19. $\begin{bmatrix} -1 & -1 & -6 \\ 1 & 0 & 2 \\ 0 & 1 & 3 \end{bmatrix}$ **21.** $\begin{bmatrix} 1 & 1 & 2 \\ 1 & 1 & 1 \\ 2 & 3 & 4 \end{bmatrix}$

23. $\begin{bmatrix} 0.5 & 0.5 & 0.5 \\ 0.5 & -0.5 & -0.5 \\ 0.5 & -0.5 & 0.5 \end{bmatrix}$ **25.** Does not exist

27. $\begin{bmatrix} 1 & -2 & 3 & 8 \\ 0 & 1 & -3 & 1 \\ 0 & 0 & 1 & -2 \\ 0 & 0 & 0 & -1 \end{bmatrix}$ **29.** $\begin{bmatrix} 0.25 & 0.25 & 1.25 & -0.25 \\ 0.5 & 1.25 & 1.75 & -1 \\ -0.25 & -0.25 & -0.75 & 0.75 \\ 0.25 & 0.5 & 0.75 & -0.5 \end{bmatrix}$

31. (2, −2) **33.** (0, 2) **35.** (3, 4) **37.** (3, −3, −2)
39. (−1, 0, 1) **41.** (5, 7, −6) **43.** (1, −1, 0, 1)
45. (2, 0, −3, 1) **47.** 50 sausages, 95 hot dogs
49. Topsoil: \$239; mulch: \$179; pea gravel: \$222

51.

53.

55. ◈ **57.** $\mathbf{A}^{-1}$ exists if and only if $x \neq 0$. $\mathbf{A}^{-1} = \begin{bmatrix} \frac{1}{x} \end{bmatrix}$

59. $\mathbf{A}^{-1}$ exists if and only if $xyz \neq 0$. $\mathbf{A}^{-1} = \begin{bmatrix} 0 & 0 & \frac{1}{z} \\ 0 & \frac{1}{y} & 0 \\ \frac{1}{x} & 0 & 0 \end{bmatrix}$

EXERCISE SET 7.6

1. (f) **3.** (h) **5.** (g) **7.** (b)

9.

11.

13.

15.

17.

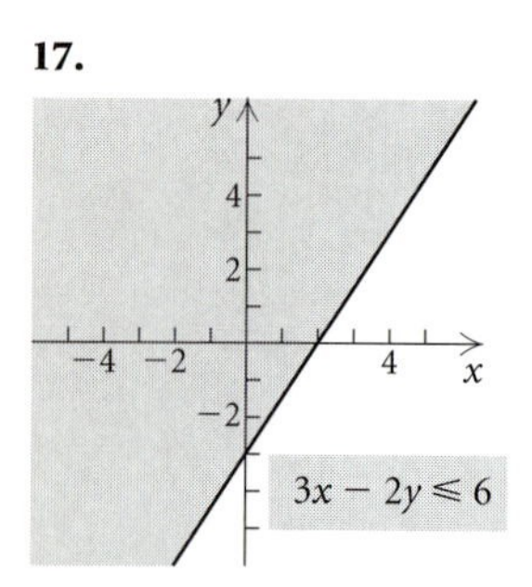

19.

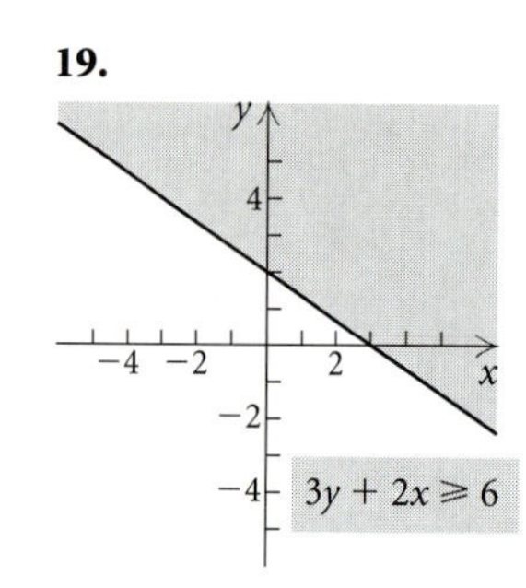

21.

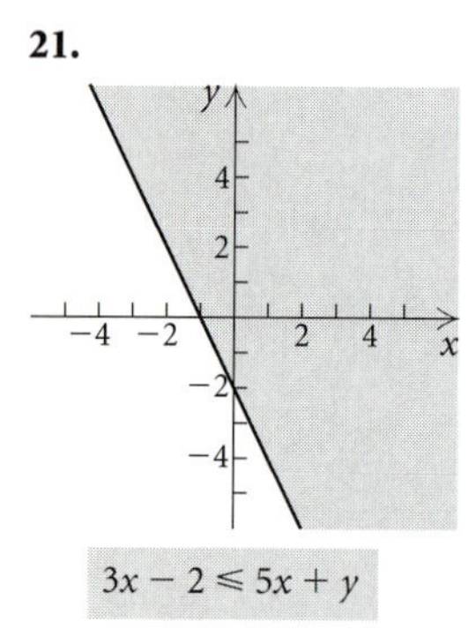

23.

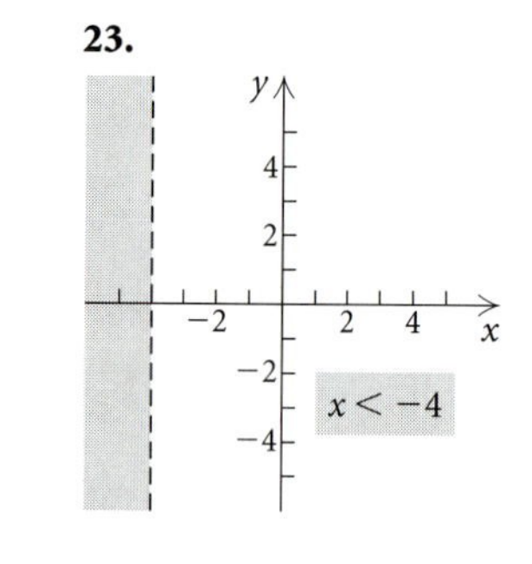

25.

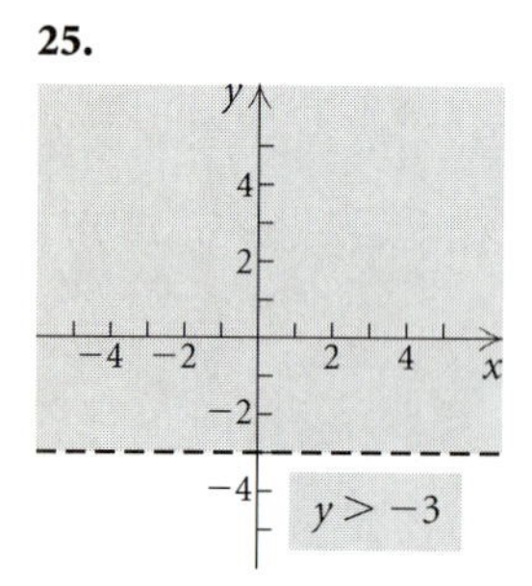

27.

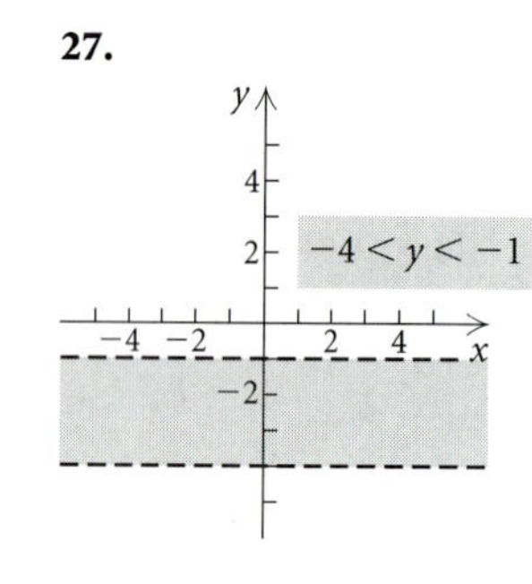

29.

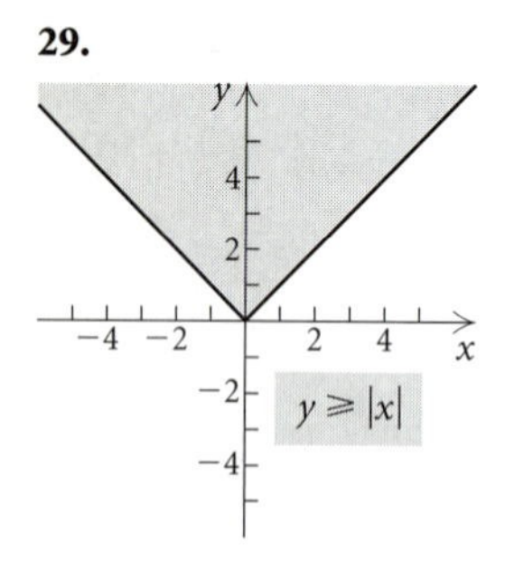

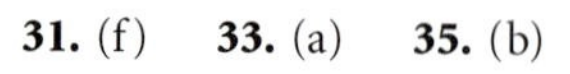

31. (f) **33.** (a) **35.** (b)

37.

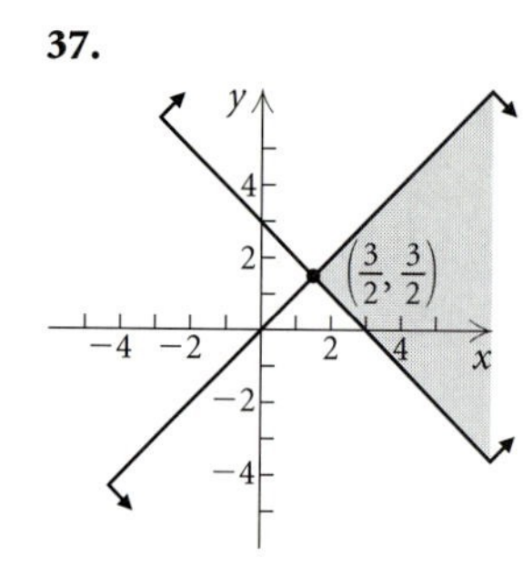

39.

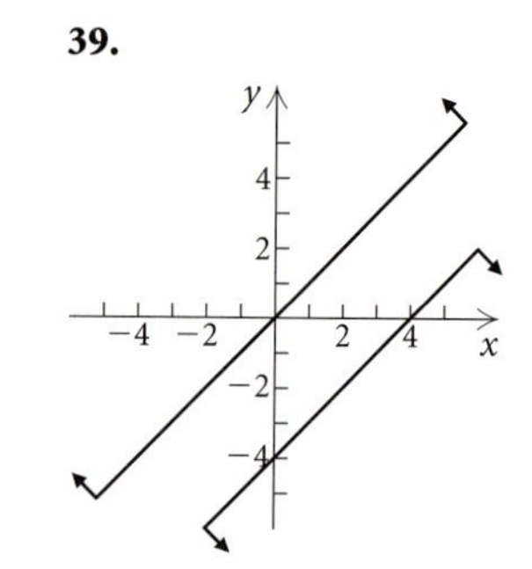

41.

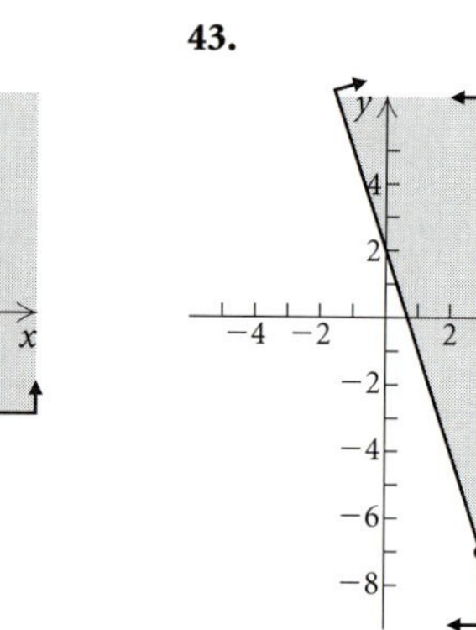

43.

45.

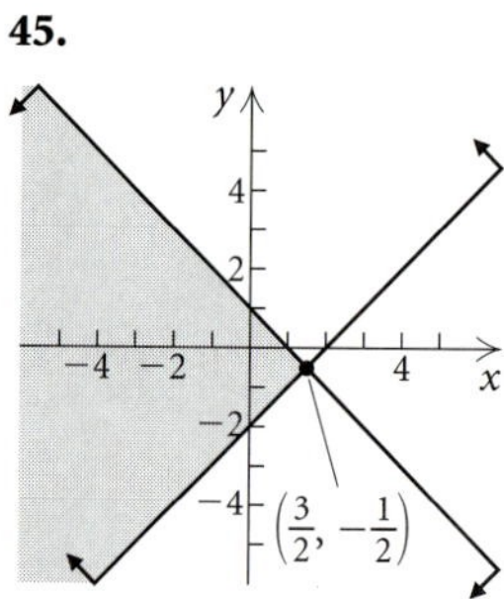
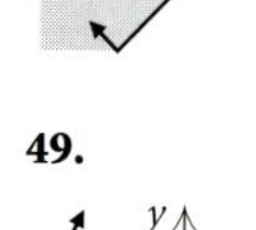

47.

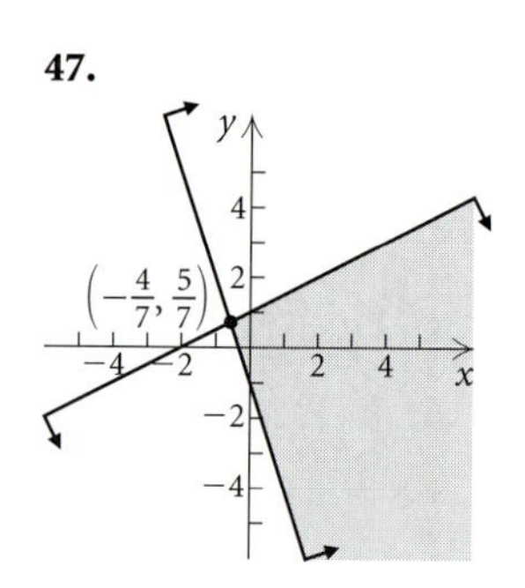

49.

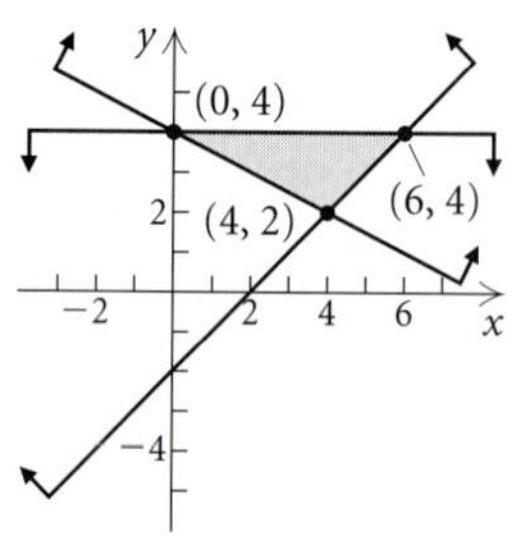

51.

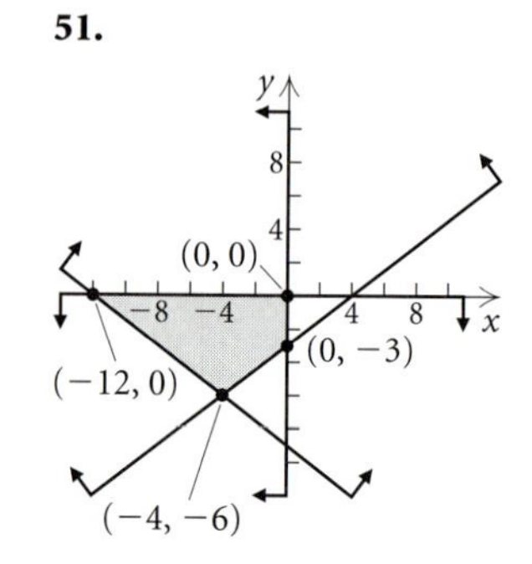

53.

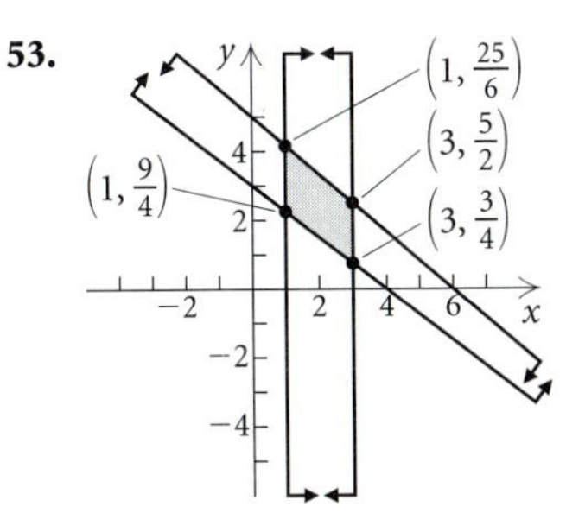

55. Maximum: 179 when $x = 7$ and $y = 0$; minimum: 48 when $x = 0$ and $y = 4$

57. Maximum: 216 when $x = 0$ and $y = 6$; minimum: 0 when $x = 0$ and $y = 0$

59. Maximum income of \$18 when 100 of each type of biscuit is made

61. Maximum profit of \$11,000 is achieved by producing 100 units of lumber and 300 units of plywood.

63. Minimum cost of $\$36\frac{12}{13}$ is achieved by using $1\frac{11}{13}$ sacks of soybean meal and $1\frac{11}{13}$ sacks of oats.

65. Maximum income of \$3110 is achieved when \$22,000 is invested in corporate bonds and \$18,000 is invested in municipal bonds.

67. Minimum cost of \$460 thousand is achieved using 30 P_1's and 10 P_2's.

69. Maximum profit per day of \$192 is achieved when 2 knit suits and 4 worsted suits are made.

71. Minimum weekly cost of \$19.05 when 1.5 lb of meat and 3 lb of cheese are used

73. Maximum total number is 800, or 550 of A and 250 of B.

75. $(x - 5)(x + 3)$ 77. $\dfrac{5x - 10}{(x + 1)(x - 4)}$ 79. ◈

81. 83.

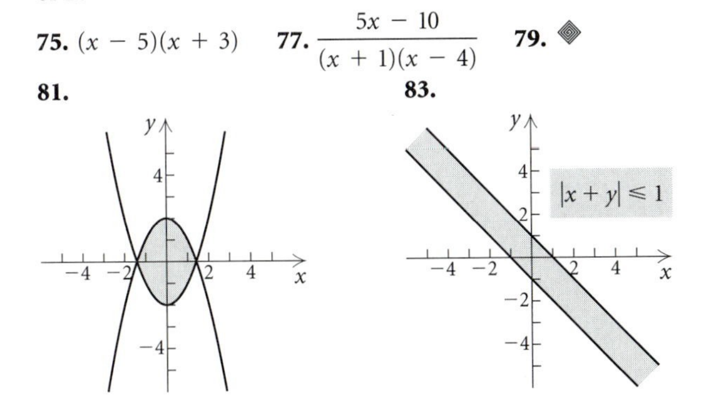

85.

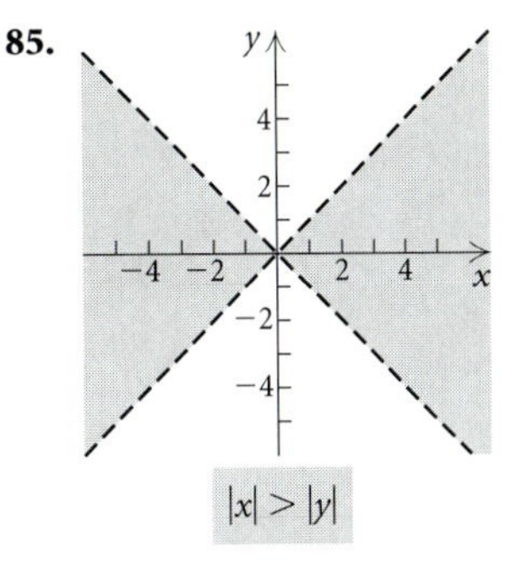

87. Maximum income of \$28,500 is achieved by making 30 less expensive assemblies and 30 more expensive assemblies.

EXERCISE SET 7.7

1. $\dfrac{2}{x - 3} - \dfrac{1}{x + 2}$ 3. $-\dfrac{4}{3x - 1} + \dfrac{5}{2x - 1}$

5. $-\dfrac{3}{x - 2} + \dfrac{2}{x + 2} + \dfrac{4}{x + 1}$

7. $-\dfrac{3}{(x + 2)^2} - \dfrac{1}{x + 2} + \dfrac{1}{x - 1}$ 9. $\dfrac{3}{x - 1} - \dfrac{4}{2x - 1}$

11. $x - 2 - \dfrac{\frac{11}{4}}{(x + 1)^2} + \dfrac{\frac{17}{16}}{x + 1} - \dfrac{\frac{17}{16}}{x - 3}$

13. $\dfrac{3x + 5}{x^2 + 2} - \dfrac{4}{x - 1}$ 15. $-\dfrac{2}{x + 2} + \dfrac{10}{(x + 2)^2} + \dfrac{3}{2x - 1}$

17. $3x + 1 + \dfrac{2}{2x - 1} + \dfrac{3}{x + 1}$ 19. $-\dfrac{1}{x - 3} + \dfrac{3x}{x^2 + 2x - 5}$

21. $\dfrac{5}{3x + 5} - \dfrac{3}{x + 1} + \dfrac{4}{(x + 1)^2}$ 23. $\dfrac{8}{4x - 5} + \dfrac{3}{3x + 2}$

25. $\dfrac{2x - 5}{3x^2 + 1} - \dfrac{2}{x - 2}$ 27. $(2x + 5y)(2x - 5y)$

29. $(x + 4)^2$ 31. ◈ 33. $-\dfrac{\frac{1}{2a^2}x}{x^2 + a^2} + \dfrac{\frac{1}{4a^2}}{x - a} + \dfrac{\frac{1}{4a^2}}{x + a}$

35. $-\dfrac{3}{25(\ln x + 2)} + \dfrac{3}{25(\ln x - 3)} + \dfrac{7}{5(\ln x - 3)^2}$

37. ◈

REVIEW EXERCISES, CHAPTER 7

1. [7.1] (a) 2. [7.1] (e) 3. [7.1] (h) 4. [7.1] (d)
5. [7.6] (b) 6. [7.6] (g) 7. [7.6] (c) 8. [7.6] (f)
9. [7.1] $(-2, -2)$ 10. [7.1] $(-5, 4)$ 11. [7.1] No solution
12. [7.1] $(y - 2, y)$, or $(x, x + 2)$ 13. [7.2] No solution
14. [7.2] $(0, 0, 0)$ 15. [7.2] $(-5, 13, 8, 2)$
16. [7.1], [7.2] Consistent: 9, 10, 12, 14, 15; the others are inconsistent.
17. [7.1], [7.2] Dependent: 12; the others are independent.
18. [7.3] $(1, 2)$ 19. [7.3] $(-3, 4, -2)$
20. [7.3] $\left(\dfrac{z}{2}, -\dfrac{z}{2}, z\right)$ 21. [7.3] $(-4, 1, -2, 3)$

22. [7.1] 31 nickels, 44 dimes
23. [7.1] \$1600 at 10%, \$3400 at 10.5%
24. [7.2] 1 bagel, $\frac{1}{2}$ serving cream cheese, 2 bananas
25. [7.2] 74.5, 68.5, 82
26. [7.2] **(a)** $f(x) = -\frac{23}{350}x^2 + \frac{171}{350}x + \frac{69}{10}$; **(b)** 2.1 lb

27. [7.4] $\begin{bmatrix} 0 & -1 & 6 \\ 3 & 1 & -2 \\ -2 & 1 & -2 \end{bmatrix}$ **28.** [7.4] $\begin{bmatrix} -3 & 3 & 0 \\ -6 & -9 & 6 \\ 6 & 0 & -3 \end{bmatrix}$

29. [7.4] $\begin{bmatrix} -1 & 1 & 0 \\ -2 & -3 & 2 \\ 2 & 0 & -1 \end{bmatrix}$ **30.** [7.4] $\begin{bmatrix} -2 & 2 & 6 \\ 1 & -8 & 18 \\ 2 & 1 & -15 \end{bmatrix}$

31. [7.4] Not possible **32.** [7.4] $\begin{bmatrix} 2 & -1 & -6 \\ 1 & 5 & -2 \\ -2 & -1 & 4 \end{bmatrix}$

33. [7.4] $\begin{bmatrix} 3 & -2 & -6 \\ 3 & 8 & -4 \\ -4 & -1 & 5 \end{bmatrix}$ **34.** [7.4] $\begin{bmatrix} -2 & -1 & 18 \\ 5 & -3 & -2 \\ -2 & 3 & -8 \end{bmatrix}$

35. [7.4] **(a)** $\begin{bmatrix} 46.1 & 5.9 & 10.1 & 8.5 & 11.4 \\ 54.6 & 4.6 & 9.6 & 7.6 & 10.6 \\ 48.9 & 5.5 & 12.7 & 9.4 & 9.3 \\ 51.3 & 4.8 & 11.3 & 6.9 & 12.7 \end{bmatrix}$
(b) [32 19 43 38];
(c) [6564.7 695.1 1481.1 1082.8 1448.7];
(d) the total cost, in cents, for each item for the day's meal

36. [7.5] $\begin{bmatrix} -\frac{1}{2} & 0 \\ \frac{1}{6} & \frac{1}{3} \end{bmatrix}$ **37.** [7.5] $\begin{bmatrix} 0 & 0 & \frac{1}{4} \\ 0 & -\frac{1}{2} & 0 \\ \frac{1}{3} & 0 & 0 \end{bmatrix}$

38. [7.5] $\begin{bmatrix} 1 & 0 & 0 & 0 \\ 0 & \frac{1}{9} & \frac{5}{18} & 0 \\ 0 & -\frac{1}{9} & \frac{2}{9} & 0 \\ 0 & 0 & 0 & 1 \end{bmatrix}$

39. [7.4] $\begin{bmatrix} 3 & -2 & 4 \\ 1 & 5 & -3 \\ 2 & -3 & 7 \end{bmatrix}\begin{bmatrix} x \\ y \\ z \end{bmatrix} = \begin{bmatrix} 13 \\ 7 \\ -8 \end{bmatrix}$ **40.** [7.5] $(-8, 7)$

41. [7.5] $(1, -2, 5)$ **42.** [7.5] $(2, -1, 1, -3)$
43. [7.6]

$y \le 3x + 6$

44. [7.6] **45.** [7.6]

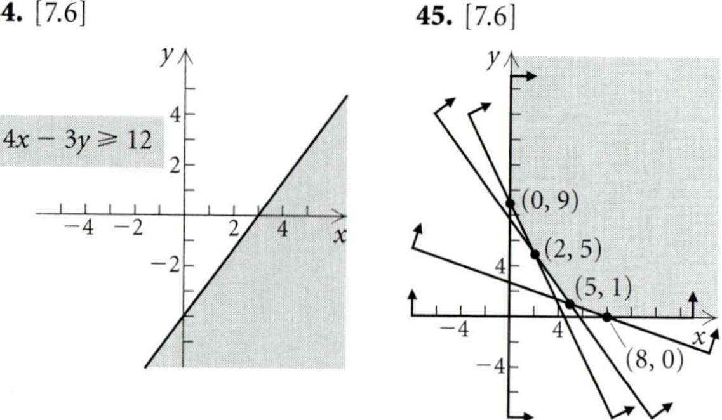

46. [7.6] Minimum = 52 at (2, 4); maximum = 92 at (2, 8)
47. [7.6] Maximum score of 96 when 0 group A questions and 8 group B questions are answered
48. [7.7] $\frac{5}{x+1} - \frac{5}{x+2} - \frac{5}{(x+2)^2}$
49. [7.7] $\frac{2}{2x-3} - \frac{5}{x+4}$
50. [7.1] During a holiday season, a caterer sold a total of 55 food trays. She sold 15 more seafood trays than cheese trays. How many of each were sold?
51. [7.4] In general, $(\mathbf{AB})^2 \neq \mathbf{A}^2\mathbf{B}^2$. $(\mathbf{AB})^2 = \mathbf{ABAB}$ and $\mathbf{A}^2\mathbf{B}^2 = \mathbf{AABB}$. Since matrix multiplication is not commutative, $\mathbf{BA} \neq \mathbf{AB}$, so $(\mathbf{AB})^2 \neq \mathbf{A}^2\mathbf{B}^2$.
52. [7.2] 12%: \$10,000; 13%: \$12,000; $14\frac{1}{2}$%: \$18,000
53. [7.1] $\left(\frac{5}{18}, \frac{1}{7}\right)$ **54.** [7.2] $\left(1, \frac{1}{2}, \frac{1}{3}\right)$
55. [7.6] **56.** [7.6]

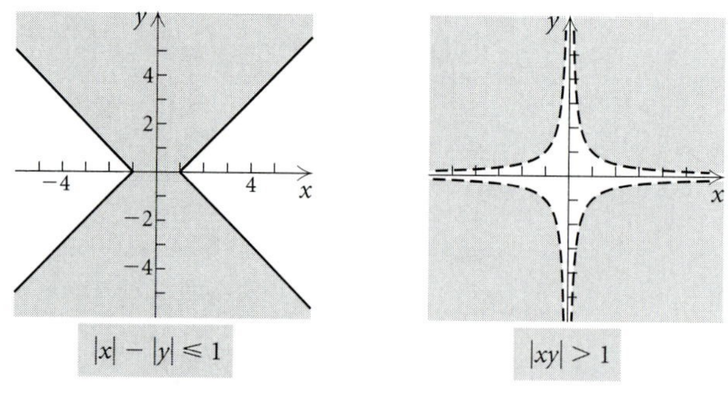

Chapter 8

Exercise Set 8.1

1. (f) **3.** (b) **5.** (d)

7. V: $(0, 0)$; F: $(0, 5)$; D: $y = -5$

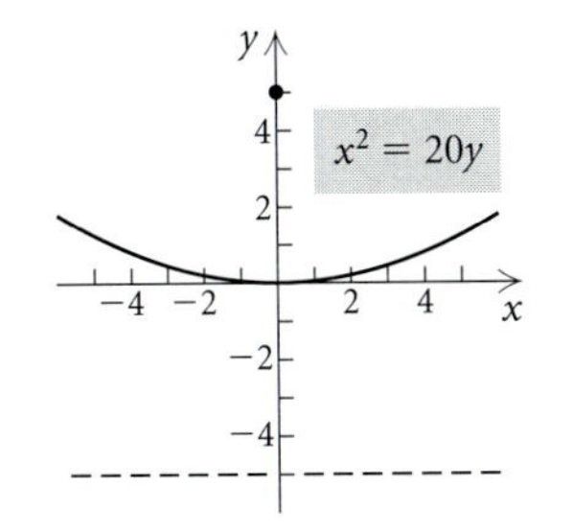

9. V: $(0, 0)$; F: $\left(-\frac{3}{2}, 0\right)$; D: $x = \frac{3}{2}$

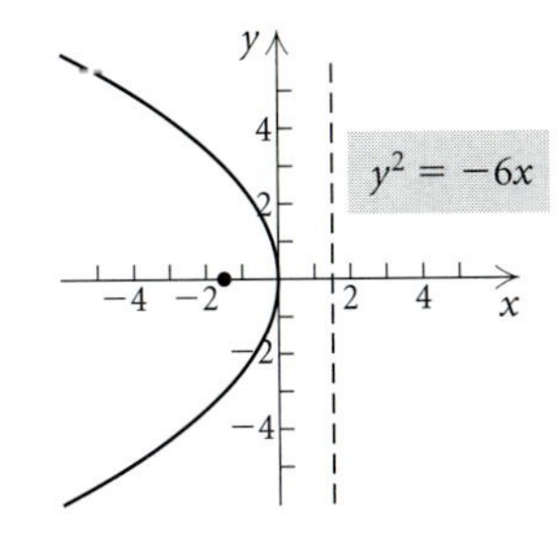

11. V: $(0, 0)$; F: $(0, 1)$; D: $y = -1$

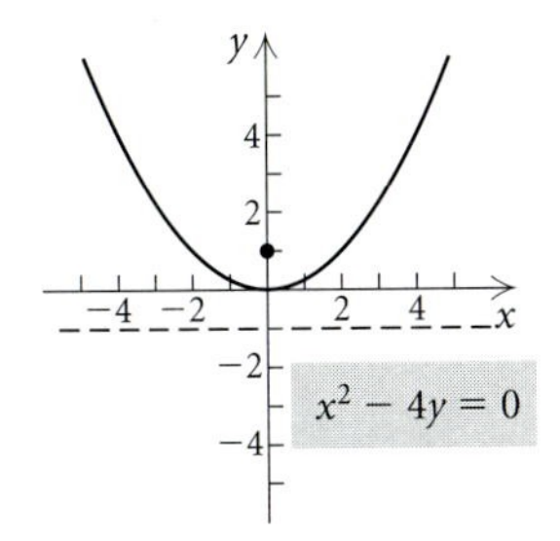

13. V: $(0, 0)$; F: $\left(\frac{1}{8}, 0\right)$; D: $x = -\frac{1}{8}$

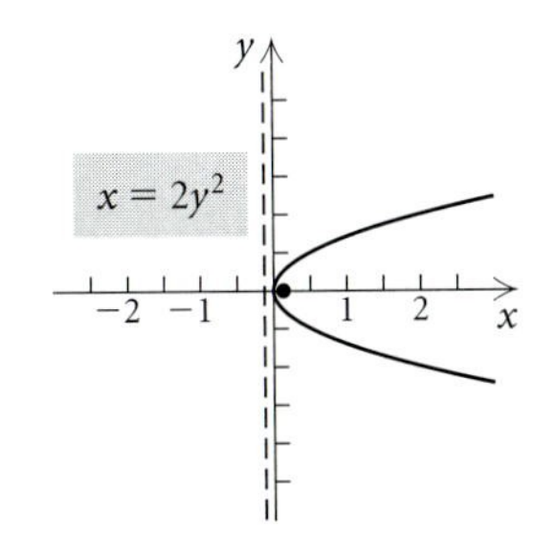

15. $y^2 = 16x$ **17.** $x^2 = -4\pi y$

19. $(y - 2)^2 = 14\left(x + \frac{1}{2}\right)$

21. V: $(-2, 1)$; F: $\left(-2, -\frac{1}{2}\right)$; D: $y = \frac{5}{2}$

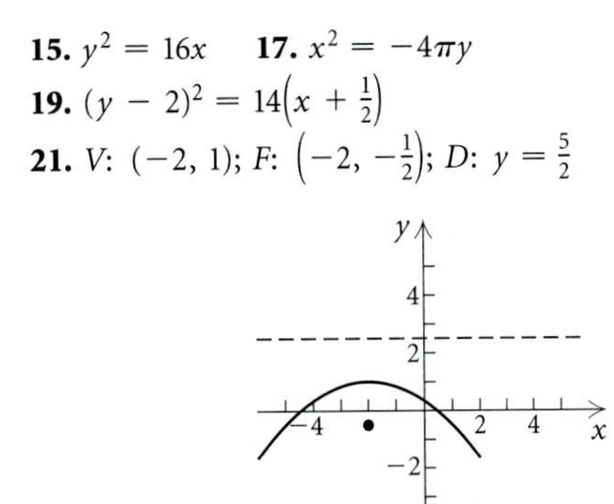

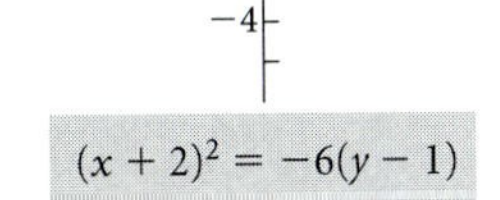

23. V: $(-1, -3)$; F: $\left(-1, -\frac{7}{2}\right)$; D: $y = -\frac{5}{2}$

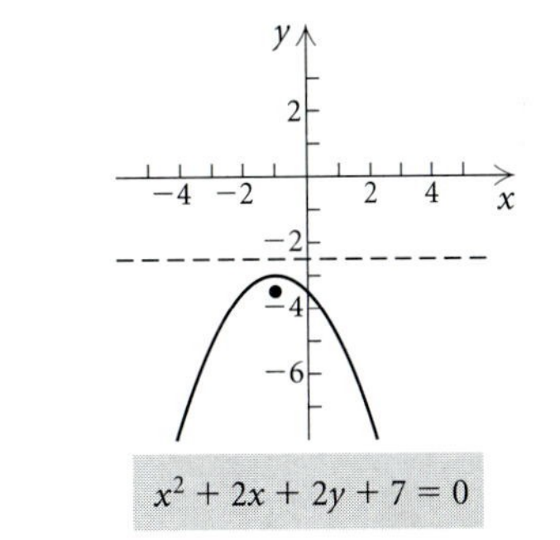

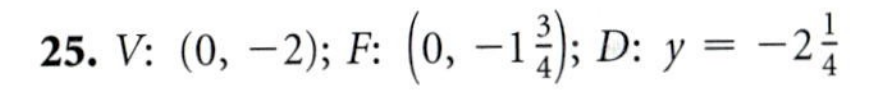

25. V: $(0, -2)$; F: $\left(0, -1\frac{3}{4}\right)$; D: $y = -2\frac{1}{4}$

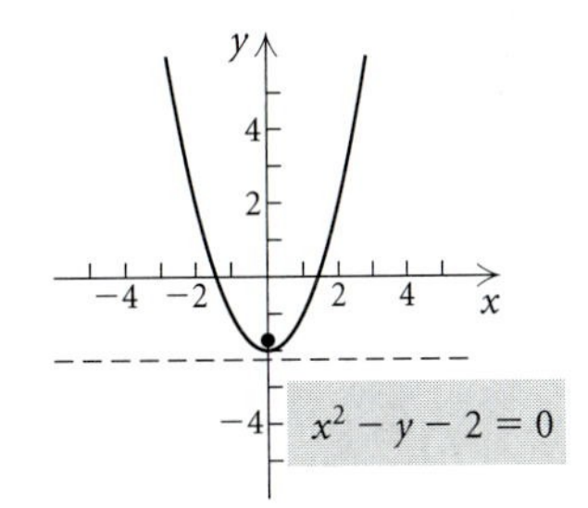

27. V: $(-2, -1)$; F: $\left(-2, -\frac{3}{4}\right)$; D: $y = -1\frac{1}{4}$

$y = x^2 + 4x + 3$

29. V: $\left(5\frac{3}{4}, \frac{1}{2}\right)$; F: $\left(6, \frac{1}{2}\right)$; D: $x = 5\frac{1}{2}$

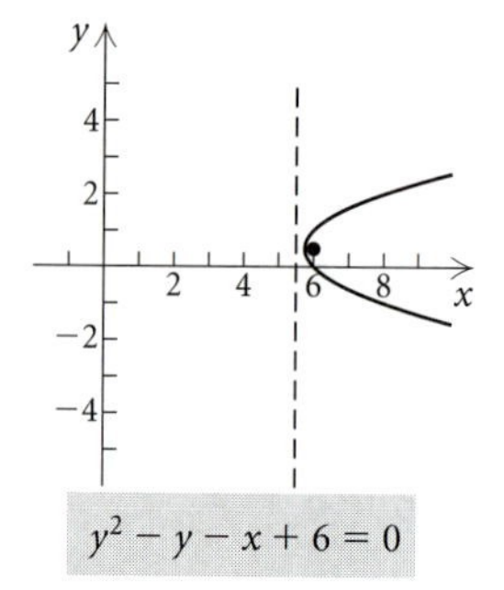

$y^2 - y - x + 6 = 0$

31. **(a)** $y^2 = 16x$; **(b)** $3\frac{33}{64}$ ft **33.** $\frac{2}{3}$ ft, or 8 in.
35. $(x + 5)^2$ **37.** $(1, -2)$; 3 **39.** ◈
41. $(x + 1)^2 = -4(y - 2)$
43. V: $(0.867, 0.348)$; F: $(0.867, -0.191)$; D: $y = 0.887$
45. 10 ft, 11.6 ft, 16.4 ft, 24.4 ft, 35.6 ft, 50 ft

Exercise Set 8.2

1. (b) **3.** (d) **5.** (a)
7. $(7, -2)$; 8

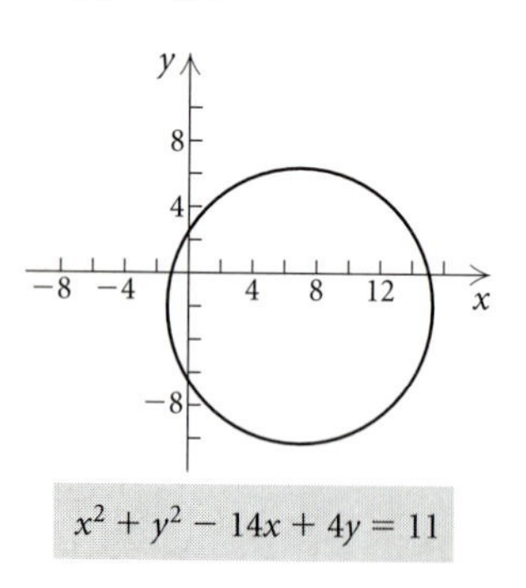

$x^2 + y^2 - 14x + 4y = 11$

9. $(-2, 3)$; 5

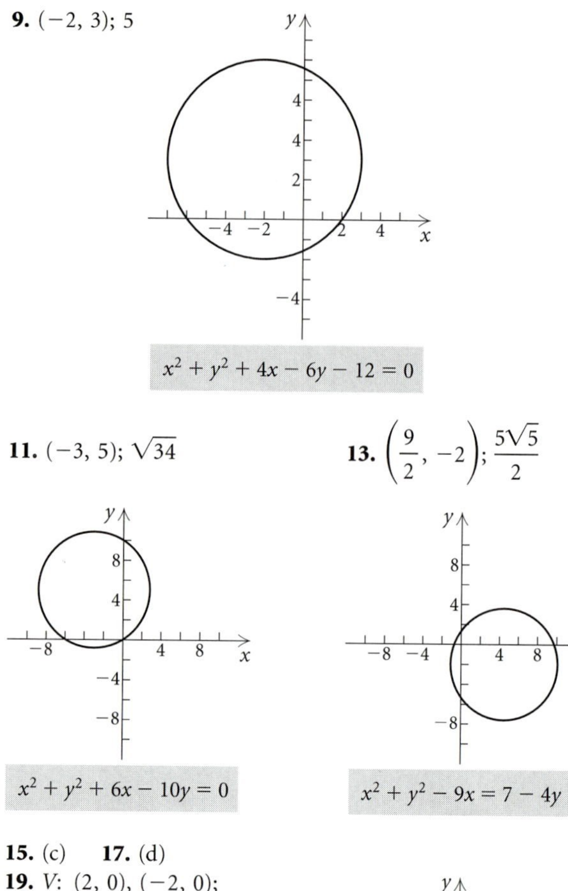

$x^2 + y^2 + 4x - 6y - 12 = 0$

11. $(-3, 5)$; $\sqrt{34}$

$x^2 + y^2 + 6x - 10y = 0$

13. $\left(\frac{9}{2}, -2\right)$; $\frac{5\sqrt{5}}{2}$

$x^2 + y^2 - 9x = 7 - 4y$

15. (c) **17.** (d)
19. V: $(2, 0)$, $(-2, 0)$;
F: $(\sqrt{3}, 0)$, $(-\sqrt{3}, 0)$

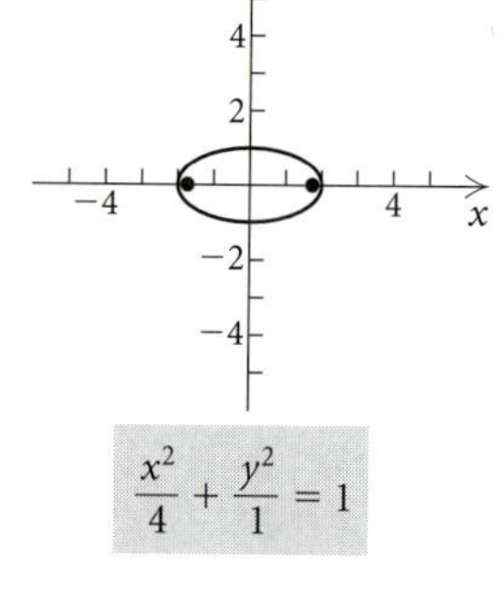

$\frac{x^2}{4} + \frac{y^2}{1} = 1$

21. V: $(0, 4)$, $(0, -4)$;
F: $(0, \sqrt{7})$, $(0, -\sqrt{7})$

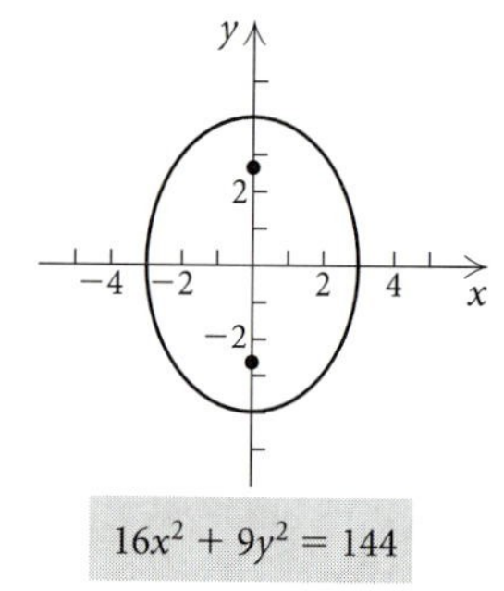

$16x^2 + 9y^2 = 144$

23. V: $(-\sqrt{3}, 0)$, $(\sqrt{3}, 0)$; F: $(-1, 0)$, $(1, 0)$

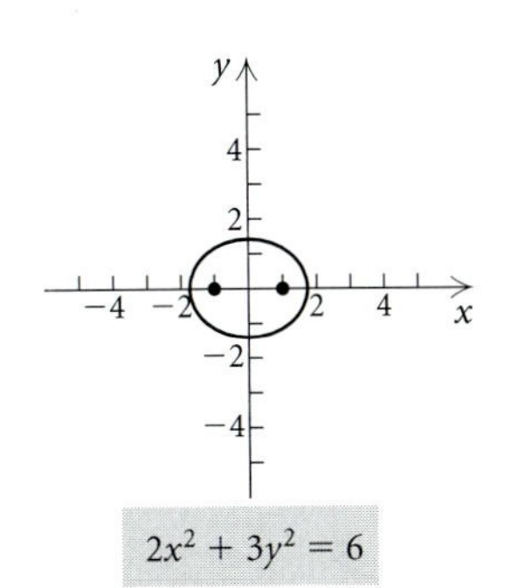

$2x^2 + 3y^2 = 6$

25. V: $\left(-\frac{1}{2}, 0\right), \left(\frac{1}{2}, 0\right)$; F. $\left(-\frac{\sqrt{5}}{6}, 0\right), \left(\frac{\sqrt{5}}{6}, 0\right)$

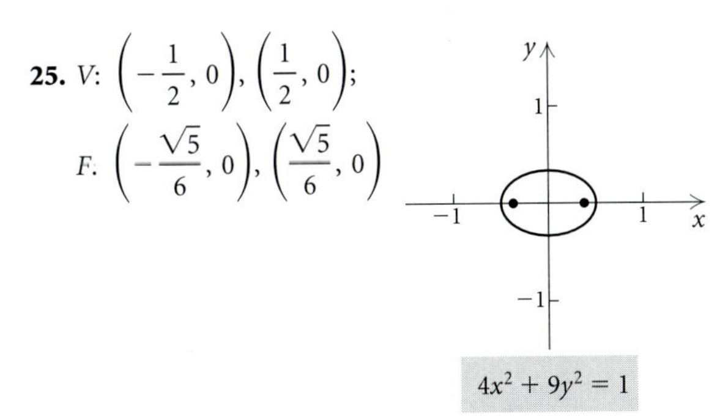

$4x^2 + 9y^2 = 1$

27. $\frac{x^2}{49} + \frac{y^2}{40} = 1$ **29.** $\frac{x^2}{25} + \frac{y^2}{64} = 1$ **31.** $\frac{x^2}{9} + \frac{y^2}{5} = 1$

33. C: $(1, 2)$; V: $(4, 2)$, $(-2, 2)$; F: $(1 + \sqrt{5}, 2)$, $(1 - \sqrt{5}, 2)$

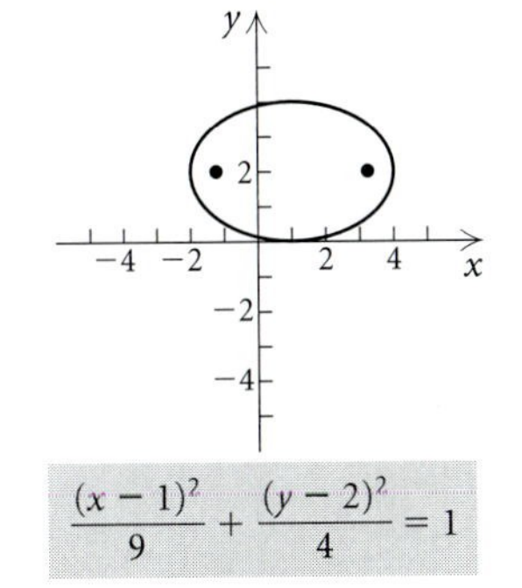

$\frac{(x-1)^2}{9} + \frac{(y-2)^2}{4} = 1$

35. C: $(-3, 5)$; V: $(-3, 11)$, $(-3, -1)$; F: $(-3, 5 + \sqrt{11})$, $(-3, 5 - \sqrt{11})$

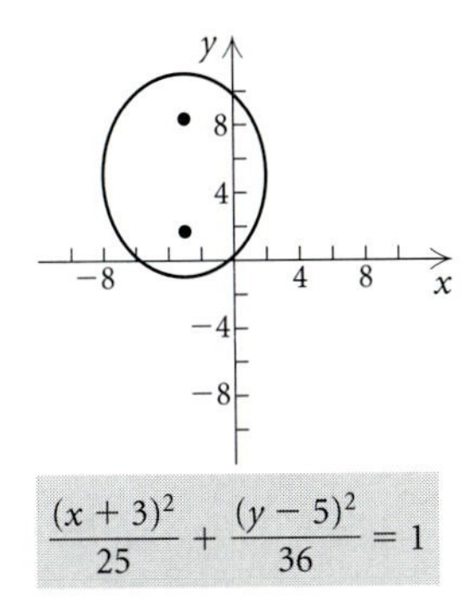

$\frac{(x+3)^2}{25} + \frac{(y-5)^2}{36} = 1$

37. C: $(-2, 1)$; V: $(-10, 1)$, $(6, 1)$; F: $(-6, 1)$, $(2, 1)$

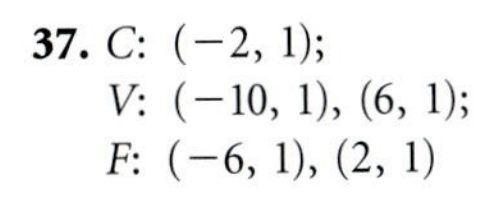

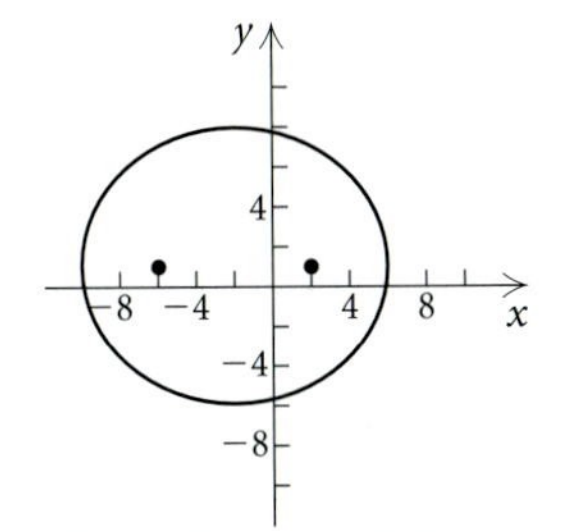

$3(x+2)^2 + 4(y-1)^2 = 192$

39. C: $(2, -1)$; V: $(-1, -1)$, $(5, -1)$; F: $(2 + \sqrt{5}, -1)$, $(2 - \sqrt{5}, -1)$

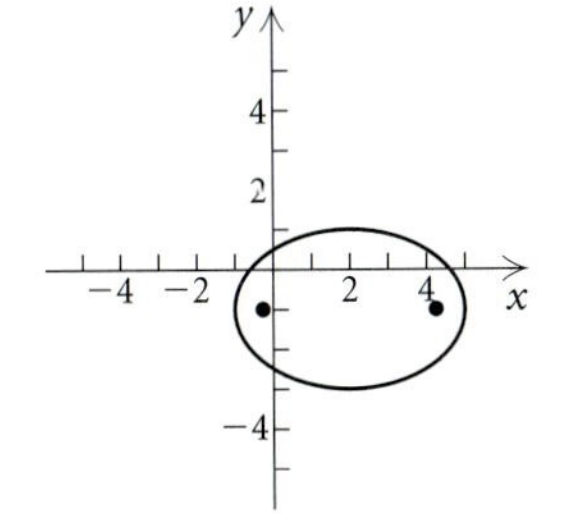

$4x^2 + 9y^2 - 16x + 18y - 11 = 0$

41. C: $(1, 1)$; V: $(1, 3)$, $(1, -1)$; F: $(1, 1 + \sqrt{3})$, $(1, 1 - \sqrt{3})$

$4x^2 + y^2 - 8x - 2y + 1 = 0$

43. Example 2; $\frac{3}{5} < \frac{\sqrt{12}}{4}$ **45.** $\frac{x^2}{15} + \frac{y^2}{16} = 1$

47. $\frac{x^2}{2500} + \frac{y^2}{144} = 1$ **49.** 2×10^6 mi

51.

$y = \frac{5}{2}x$

53.

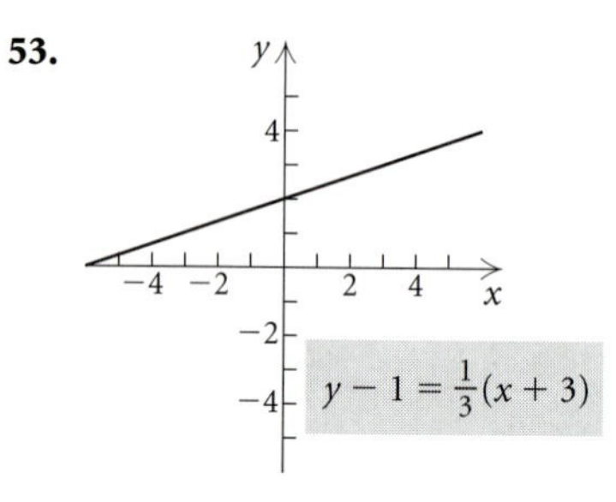

$y - 1 = \frac{1}{3}(x + 3)$

55. ◈ **57.** $\frac{(x-3)^2}{4} + \frac{(y-1)^2}{25} = 1$

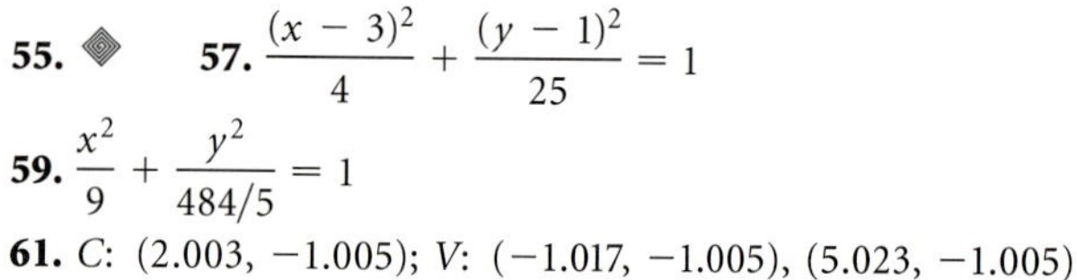

59. $\frac{x^2}{9} + \frac{y^2}{484/5} = 1$

61. C: $(2.003, -1.005)$; V: $(-1.017, -1.005)$, $(5.023, -1.005)$

63. About 9.1 ft

Exercise Set 8.3

1. (b) **3.** (c) **5.** (a) **7.** $\frac{y^2}{9} - \frac{x^2}{16} = 1$ **9.** $\frac{x^2}{4} - \frac{y^2}{9} = 1$

11. C: $(0, 0)$; V: $(2, 0)$, $(-2, 0)$; F: $(2\sqrt{2}, 0)$, $(-2\sqrt{2}, 0)$; A: $y = x$, $y = -x$

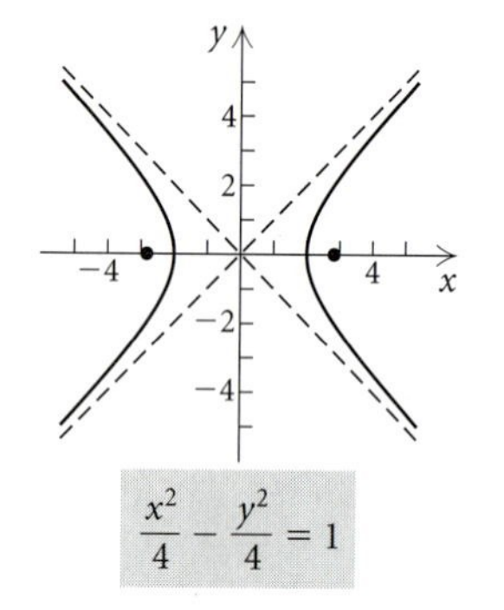

$\frac{x^2}{4} - \frac{y^2}{4} = 1$

13. C: $(2, -5)$; V: $(-1, -5)$, $(5, -5)$; F: $(2 - \sqrt{10}, -5)$, $(2 + \sqrt{10}, -5)$; A: $y = -\frac{x}{3} - \frac{13}{3}$, $y = \frac{x}{3} - \frac{17}{3}$

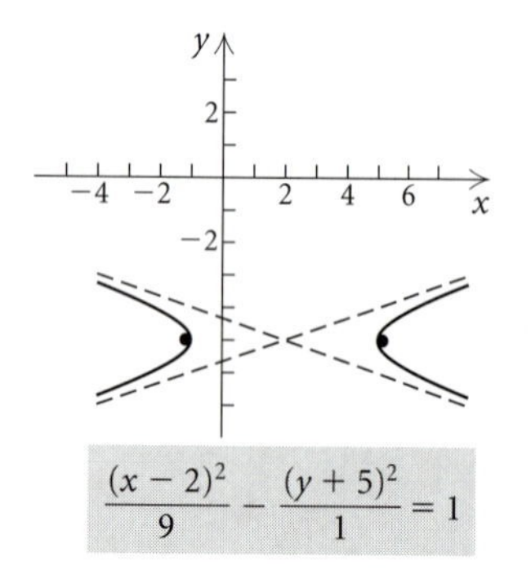

$\frac{(x-2)^2}{9} - \frac{(y+5)^2}{1} = 1$

15. C: $(-1, -3)$; V: $(-1, -1)$, $(-1, -5)$; F: $(-1, -3 + 2\sqrt{5})$, $(-1, -3 - 2\sqrt{5})$; A: $y = \frac{1}{2}x - \frac{5}{2}$, $y = -\frac{1}{2}x - \frac{7}{2}$

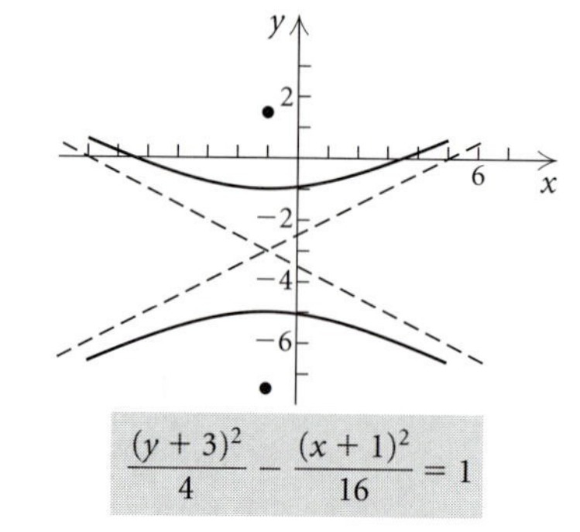

$\frac{(y+3)^2}{4} - \frac{(x+1)^2}{16} = 1$

17. C: $(0, 0)$; V: $(-2, 0)$, $(2, 0)$; F: $(-\sqrt{5}, 0)$, $(\sqrt{5}, 0)$; A: $y = -\frac{1}{2}x$, $y = \frac{1}{2}x$

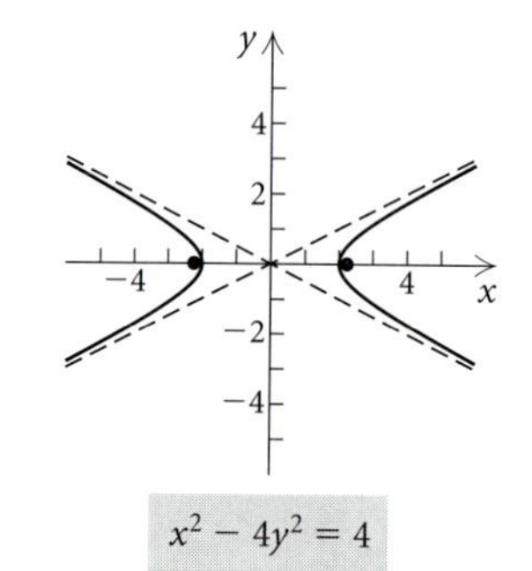

$x^2 - 4y^2 = 4$

19. C: $(0, 0)$; V: $(0, -3)$, $(0, 3)$; F: $(0, -3\sqrt{10})$, $(0, 3\sqrt{10})$; A: $y = \frac{1}{3}x$, $y = -\frac{1}{3}x$

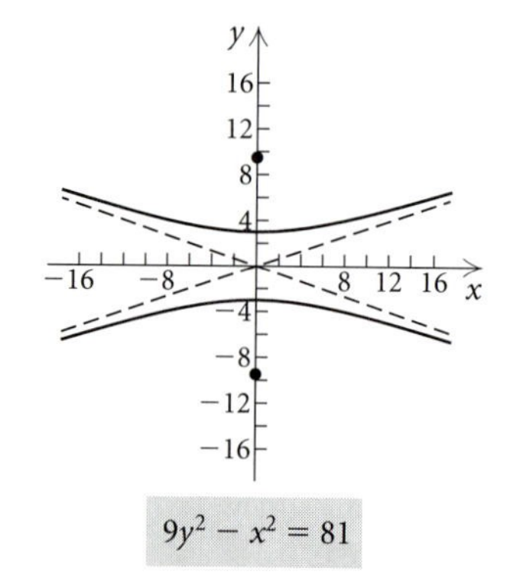

$9y^2 - x^2 = 81$

21. C: $(0, 0)$; V: $(-\sqrt{2}, 0)$, $(\sqrt{2}, 0)$; F: $(-2, 0)$, $(2, 0)$; A: $y = x$, $y = -x$

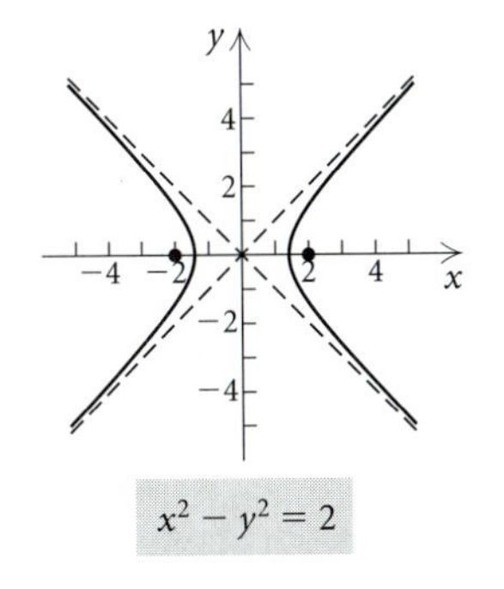

$x^2 - y^2 = 2$

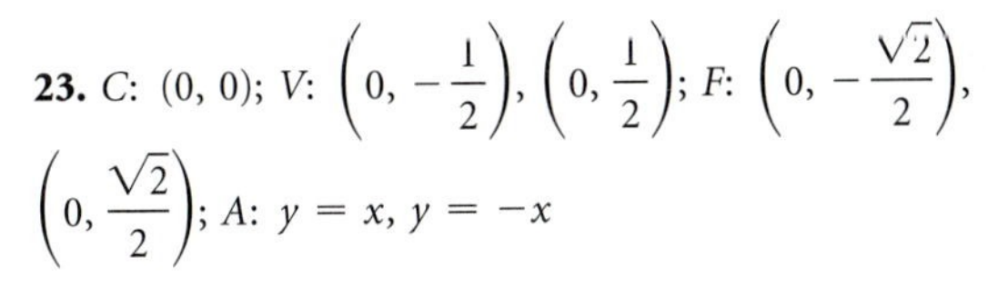

23. C: $(0, 0)$; V: $\left(0, -\dfrac{1}{2}\right)$, $\left(0, \dfrac{1}{2}\right)$; F: $\left(0, -\dfrac{\sqrt{2}}{2}\right)$, $\left(0, \dfrac{\sqrt{2}}{2}\right)$; A: $y = x$, $y = -x$

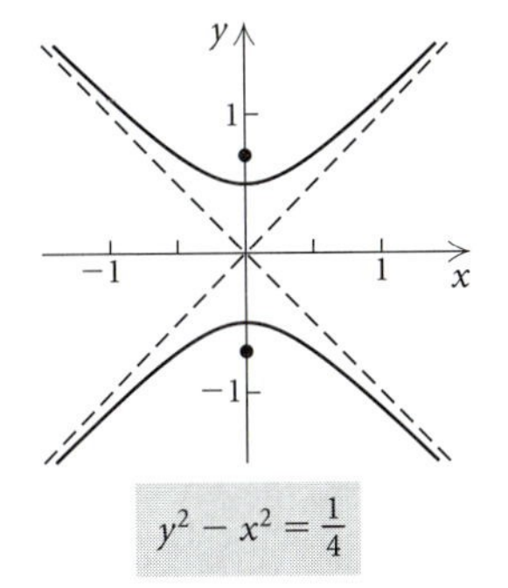

$y^2 - x^2 = \frac{1}{4}$

25. C: $(1, -2)$; V: $(0, -2)$, $(2, -2)$; F: $(1 - \sqrt{2}, -2)$, $(1 + \sqrt{2}, -2)$; A: $y = -x - 1$, $y = x - 3$

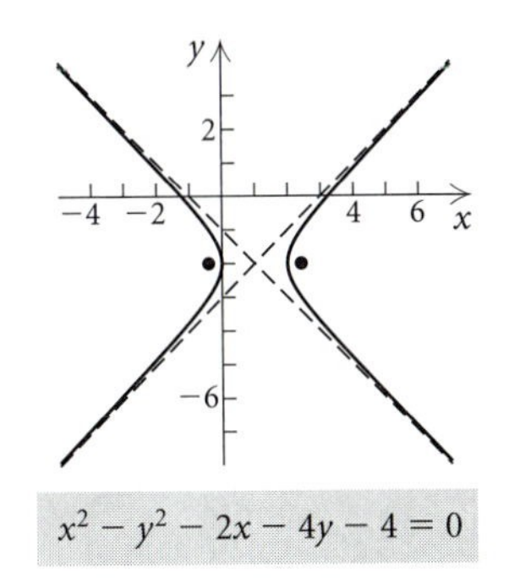

$x^2 - y^2 - 2x - 4y - 4 = 0$

27. C: $\left(\frac{1}{3}, 3\right)$; V: $\left(-\frac{2}{3}, 3\right)$, $\left(\frac{4}{3}, 3\right)$; F: $\left(\frac{1}{3} - \sqrt{37}, 3\right)$, $\left(\frac{1}{3} + \sqrt{37}, 3\right)$; A: $y = 6x + 1$, $y = -6x + 5$

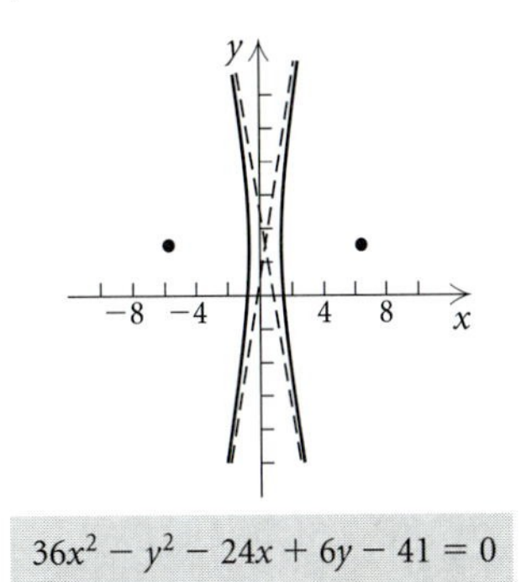

$36x^2 - y^2 - 24x + 6y - 41 = 0$

29. C: $(3, 1)$; V: $(3, 3)$, $(3, -1)$; F: $(3, 1 + \sqrt{13})$, $(3, 1 - \sqrt{13})$; A: $y = \frac{2}{3}x - 1$, $y = -\frac{2}{3}x + 3$

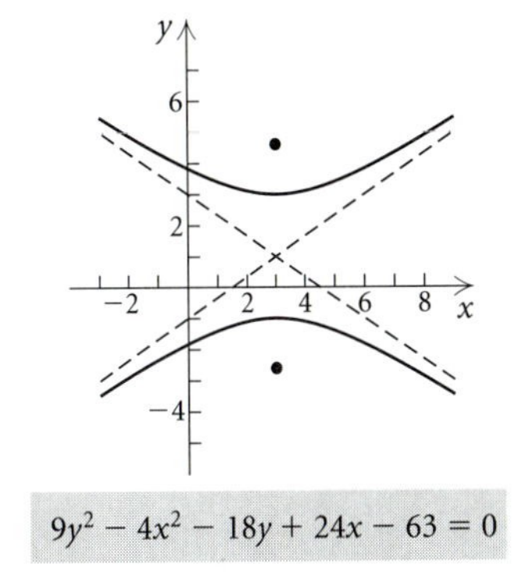

$9y^2 - 4x^2 - 18y + 24x - 63 = 0$

31. C: $(1, -2)$; V: $(2, -2)$, $(0, -2)$; F: $(1 + \sqrt{2}, -2)$, $(1 - \sqrt{2}, -2)$; A: $y = x - 3$, $y = -x - 1$

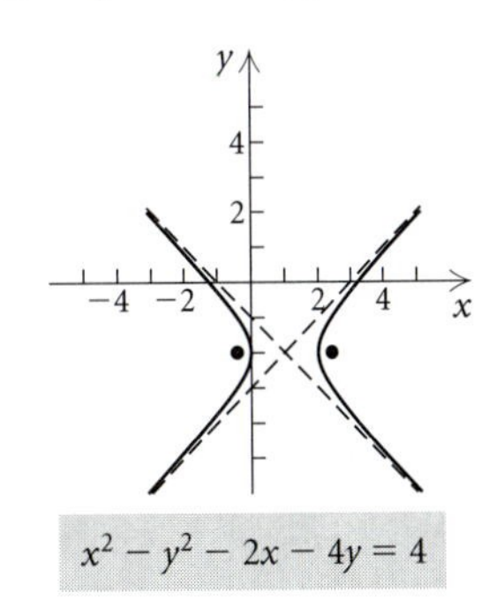

$x^2 - y^2 - 2x - 4y = 4$

33. C: $(-3, 4)$; V: $(-3, 10)$, $(-3, -2)$; F: $(-3, 4 + 6\sqrt{2})$, $(-3, 4 - 6\sqrt{2})$; A: $y = x + 7$, $y = -x + 1$

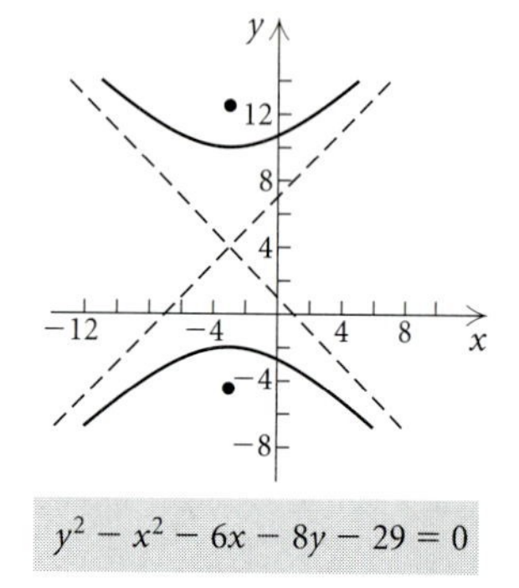

35. Example 3; $\frac{\sqrt{5}}{1} > \frac{5}{4}$ 37. $\frac{x^2}{9} - \frac{(y-7)^2}{16} = 1$

39. $\frac{y^2}{25} - \frac{x^2}{11} = 1$ 41. $(6, -1)$ 43. $(2, -1)$ 45. ◈

47. $\frac{(y+5)^2}{9} - (x-3)^2 = 1$

49. C: $(-1.460, -0.957)$; V: $(-2.360, -0.957)$, $(-0.560, -0.957)$; A: $y = -1.429x - 3.043$, $y = 1.429x + 1.129$

51. $\frac{x^2}{345.96} - \frac{y^2}{22{,}154.04} = 1$

Exercise Set 8.4

1. (e) 3. (c) 5. (b) 7. $(-4, -3)$, $(3, 4)$
9. $(0, 2)$, $(3, 0)$ 11. $(-5, 0)$, $(4, 3)$, $(4, -3)$
13. $(3, 0)$, $(-3, 0)$ 15. $(0, -3)$, $(4, 5)$ 17. $(-2, 1)$
19. $(3, 4)$, $(-3, -4)$, $(4, 3)$, $(-4, -3)$

21. $\left(\frac{6\sqrt{21}}{7}, \frac{4i\sqrt{35}}{7}\right)$, $\left(\frac{6\sqrt{21}}{7}, -\frac{4i\sqrt{35}}{7}\right)$, $\left(-\frac{6\sqrt{21}}{7}, \frac{4i\sqrt{35}}{7}\right)$, $\left(-\frac{6\sqrt{21}}{7}, -\frac{4i\sqrt{35}}{7}\right)$

23. $(3, 2)$, $\left(4, \frac{3}{2}\right)$

25. $\left(\frac{5+\sqrt{70}}{3}, \frac{-1+\sqrt{70}}{3}\right)$, $\left(\frac{5-\sqrt{70}}{3}, \frac{-1-\sqrt{70}}{3}\right)$

27. $(\sqrt{2}, \sqrt{14})$, $(-\sqrt{2}, \sqrt{14})$, $(\sqrt{2}, -\sqrt{14})$, $(-\sqrt{2}, -\sqrt{14})$

29. $(1, 2)$, $(-1, -2)$, $(2, 1)$, $(-2, -1)$

31. $\left(\frac{15+\sqrt{561}}{8}, \frac{11-3\sqrt{561}}{8}\right)$, $\left(\frac{15-\sqrt{561}}{8}, \frac{11+3\sqrt{561}}{8}\right)$

33. $\left(\frac{7-\sqrt{33}}{2}, \frac{7+\sqrt{33}}{2}\right)$, $\left(\frac{7+\sqrt{33}}{2}, \frac{7-\sqrt{33}}{2}\right)$

35. $(3, 2)$, $(-3, -2)$, $(2, 3)$, $(-2, -3)$

37. $\left(\frac{5-9\sqrt{15}}{20}, \frac{-45+3\sqrt{15}}{20}\right)$, $\left(\frac{5+9\sqrt{15}}{20}, \frac{-45-3\sqrt{15}}{20}\right)$ 39. $(3, -5)$, $(-1, 3)$

41. $(8, 5)$, $(-5, -8)$ 43. $(3, 2)$, $(-3, -2)$
45. $(2, 1)$, $(-2, -1)$, $(1, 2)$, $(-1, -2)$

47. $\left(4 + \frac{3i\sqrt{6}}{2}, -4 + \frac{3i\sqrt{6}}{2}\right)$, $\left(4 - \frac{3i\sqrt{6}}{2}, -4 - \frac{3i\sqrt{6}}{2}\right)$

49. $(3, \sqrt{5})$, $(-3, -\sqrt{5})$, $(\sqrt{5}, 3)$, $(-\sqrt{5}, -3)$

51. $\left(\frac{8\sqrt{5}}{5}i, \frac{3\sqrt{105}}{5}\right)$, $\left(\frac{8\sqrt{5}}{5}i, -\frac{3\sqrt{105}}{5}\right)$, $\left(-\frac{8\sqrt{5}}{5}i, \frac{3\sqrt{105}}{5}\right)$, $\left(-\frac{8\sqrt{5}}{5}i, -\frac{3\sqrt{105}}{5}\right)$

53. $(2, 1)$, $(-2, -1)$, $\left(-i\sqrt{5}, \frac{2i\sqrt{5}}{5}\right)$, $\left(i\sqrt{5}, -\frac{2i\sqrt{5}}{5}\right)$

55. 6 cm by 8 cm 57. 4 in. by 5 in. 59. 1 m by $\sqrt{3}$ m
61. 30 yd by 75 yd 63. 16 ft, 24 ft 65. $\frac{24}{5}$ 67. $\frac{23}{25}$

69. ◈ 71. $(x-2)^2 + (y-3)^2 = 1$ 73. $\frac{x^2}{4} + y^2 = 1$

75. $\left(x + \frac{5}{13}\right)^2 + \left(y - \frac{32}{13}\right)^2 = \frac{5365}{169}$

77. There is no number x such that $\frac{x^2}{a^2} - \frac{\left(\frac{b}{a}x\right)^2}{b^2} = 1$, because the left side simplifies to $\frac{x^2}{a^2} - \frac{x^2}{a^2}$, which is 0.

79. $\left(\frac{1}{2}, \frac{1}{4}\right)$, $\left(\frac{1}{2}, -\frac{1}{4}\right)$, $\left(-\frac{1}{2}, \frac{1}{4}\right)$, $\left(-\frac{1}{2}, -\frac{1}{4}\right)$

81. Factor: $x^3 + y^3 = (x+y)(x^2 - xy + y^2)$. We know that $x + y = 1$, so $(x+y)^2 = x^2 + 2xy + y^2 = 1$, or $x^2 + y^2 = 1 - 2xy$. We also know that $xy = 1$, so $x^2 + y^2 = 1 - 2 \cdot 1 = -1$. Then $x^3 + y^3 = 1 \cdot (-1 - 1) = -2$.

83. $(2, 4)$, $(4, 2)$ 85. $(3, -2)$, $(-3, 2)$, $(2, -3)$, $(-2, 3)$

87. $\left(\frac{2\log 3 + 3\log 5}{3(\log 3 \cdot \log 5)}, \frac{4\log 3 - 3\log 5}{3(\log 3 \cdot \log 5)}\right)$

89. $(1.564, 2.448)$, $(0.138, 0.019)$ 91. $(0.871, 1.388)$
93. $(1.146, 3.146)$, $(-1.841, 0.159)$
95. $(0.965, 4402.33)$, $(-0.965, -4402.33)$
97. $(2.112, -0.109)$, $(-13.041, -13.337)$
99. $(400, 1.431)$, $(-400, 1.431)$, $(400, -1.431)$, $(-400, -1.431)$

Review Exercises, Chapter 8

1. [8.1] (d) 2. [8.2] (a) 3. [8.2] (e) 4. [8.3] (g)
5. [8.2] (b) 6. [8.2] (f) 7. [8.1] (h) 8. [8.3] (c)
9. [8.1] $x^2 = -6y$ 10. [8.1] F: $(-3, 0)$; V: $(0, 0)$; D: $x = 3$
11. [8.1] V: $(-5, 8)$; F: $\left(-5, \frac{15}{2}\right)$; D: $y = \frac{17}{2}$

12. [8.2] C: $(2, -1)$; V: $(-3, -1)$, $(7, -1)$; F: $(-1, -1)$, $(5, -1)$;

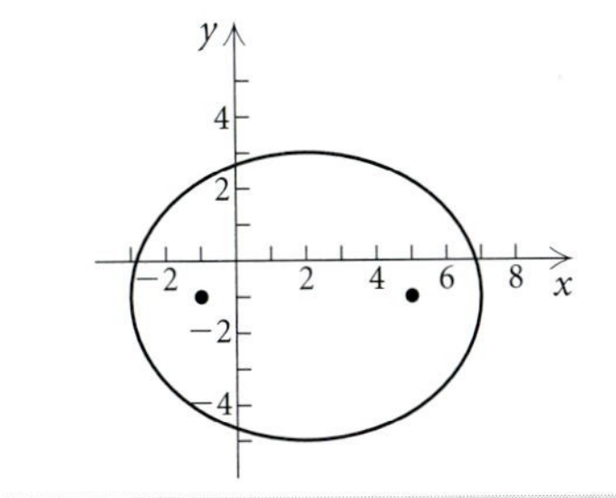

13. [8.2] $\dfrac{x^2}{9} + \dfrac{y^2}{16} = 1$

14. [8.3] C: $\left(-2, \dfrac{1}{4}\right)$; V: $\left(0, \dfrac{1}{4}\right)$, $\left(-4, \dfrac{1}{4}\right)$;
F: $\left(-2 + \sqrt{6}, \dfrac{1}{4}\right)$, $\left(-2 - \sqrt{6}, \dfrac{1}{4}\right)$;
A: $y - \dfrac{1}{4} = \dfrac{\sqrt{2}}{2}(x + 2)$, $y - \dfrac{1}{4} = -\dfrac{\sqrt{2}}{2}(x + 2)$

15. [8.1] 0.167 ft **16.** [8.4] $(-8\sqrt{2}, 8)$, $(8\sqrt{2}, 8)$

17. [8.4] $\left(3, \dfrac{\sqrt{29}}{2}\right)$, $\left(-3, \dfrac{\sqrt{29}}{2}\right)$, $\left(3, -\dfrac{\sqrt{29}}{2}\right)$, $\left(-3, -\dfrac{\sqrt{29}}{2}\right)$ **18.** [8.4] $(7, 4)$ **19.** [8.4] $(2, 2)$, $\left(\frac{32}{9}, -\frac{10}{9}\right)$

20. [8.4] $(0, -3)$, $(2, 1)$

21. [8.4] $(4, 3)$, $(4, -3)$, $(-4, 3)$, $(-4, -3)$

22. [8.4] $(-\sqrt{3}, 0)$, $(\sqrt{3}, 0)$, $(-2, 1)$, $(2, 1)$

23. [8.4] $\left(-\frac{3}{5}, \frac{21}{5}\right)$, $(3, -3)$

24. [8.4] $(6, 8)$, $(6, -8)$, $(-6, 8)$, $(-6, -8)$

25. [8.4] $(2, 2)$, $(-2, -2)$, $(2\sqrt{2}, \sqrt{2})$, $(-2\sqrt{2}, -\sqrt{2})$

26. [8.4] 7, 4 **27.** [8.4] 7 m by 12 m **28.** [8.4] 4, 8

29. [8.4] 32 cm, 20 cm **30.** [8.4] 11 ft, 3 ft

31. [8.4] ◈ An algebraic solution of a nonlinear system of equations may be preferable to a graphical solution if there are complex-number solutions or if any of the equations is difficult to enter on a grapher. However, there are nonlinear systems that are difficult or impossible to solve using algebraic methods. Approximations of the real-number solutions of many of those systems may be found using graphical methods.

32. [8.2] ◈ The equation of a circle can be written as

$$\frac{(x - h)^2}{a^2} + \frac{(y - k)^2}{b^2} = 1,$$

where $a = b = r$, the radius of the circle. In an ellipse, $a > b$, so a circle is not a special type of ellipse.

33. [8.2] $x^2 + \dfrac{y^2}{9} = 1$ **34.** [8.4] $\dfrac{8}{7}, \dfrac{7}{2}$

35. [8.2], [8.4] $(x - 2)^2 + (y - 1)^2 = 100$

36. [8.3] $\dfrac{x^2}{778.41} - \dfrac{y^2}{39{,}221.59} = 1$

Chapter 9

EXERCISE SET 9.1

1. 3, 7, 11, 15; 39; 59 **3.** $2, \frac{3}{2}, \frac{4}{3}, \frac{5}{4}; \frac{10}{9}; \frac{15}{14}$

5. $0, \frac{3}{5}, \frac{4}{5}, \frac{15}{17}; \frac{99}{101}; \frac{112}{113}$ **7.** $-1, 4, -9, 16; 100; -225$

9. 7, 3, 7, 3; 3; 7 **11.** 34 **13.** 225 **15.** $-33{,}880$

17. 67 **19.** $2n$ **21.** $(-1)^n \cdot 2 \cdot 3^{n-1}$ **23.** $\dfrac{n + 1}{n + 2}$

25. $n(n + 1)$ **27.** $\log 10^{n-1}$, or $n - 1$ **29.** 6; 28

31. 20; 30 **33.** $\frac{1}{2} + \frac{1}{4} + \frac{1}{6} + \frac{1}{8} + \frac{1}{10} = \frac{137}{120}$

35. $1 + 2 + 4 + 8 + 16 + 32 + 64 = 127$

37. $\ln 7 + \ln 8 + \ln 9 + \ln 10 = \ln (7 \cdot 8 \cdot 9 \cdot 10) = \ln 5040 \approx 8.5252$

39. $\frac{1}{2} + \frac{2}{3} + \frac{3}{4} + \frac{4}{5} + \frac{5}{6} + \frac{6}{7} + \frac{7}{8} + \frac{8}{9} = \frac{15{,}551}{2520}$

41. $-1 + 1 - 1 + 1 - 1 = -1$

43. $3 - 6 + 9 - 12 + 15 - 18 + 21 - 24 = -12$

45. $2 + 1 + \frac{2}{5} + \frac{1}{5} + \frac{2}{17} + \frac{1}{13} + \frac{2}{37} = \frac{157{,}351}{40{,}885}$

47. $3 + 2 + 3 + 6 + 11 + 18 = 43$

49. $\frac{1}{2} + \frac{2}{3} + \frac{4}{5} + \frac{8}{9} + \frac{16}{17} + \frac{32}{33} + \frac{64}{65} + \frac{128}{129} + \frac{256}{257} + \frac{512}{513} + \frac{1024}{1025} \approx 9.736$

51. $\sum_{k=1}^{\infty} 5k$ **53.** $\sum_{k=1}^{6} (-1)^{k+1}2^k$ **55.** $\sum_{k=1}^{6} (-1)^k \dfrac{k}{k + 1}$

57. $\sum_{k=2}^{n} (-1)^k k^2$ **59.** $\sum_{k=1}^{\infty} \dfrac{1}{k(k + 1)}$ **61.** $4, 1\frac{1}{4}, 1\frac{4}{5}, 1\frac{5}{9}$

63. $6561, -81, 9i, -3\sqrt{i}$ **65.** 2, 3, 5, 8

67.

n	U_n
1	2
2	2.25
3	2.3704
4	2.4414
5	2.4883
6	2.5216
7	2.5465
8	2.5658
9	2.5812
10	2.5937

69.

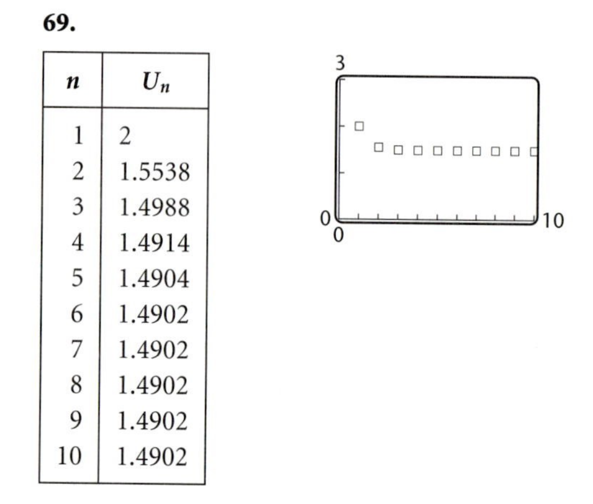

n	U_n
1	2
2	1.5538
3	1.4988
4	1.4914
5	1.4904
6	1.4902
7	1.4902
8	1.4902
9	1.4902
10	1.4902

71. (a) 2014.11, 2450.38, 2886.65, 3322.92, 3759.19, 4195.46, 4631.73, 5068, 5504.27, 5940.54; **(b)** \$39,773.25 million
73. (a) 1062, 1127.84, 1197.77, 1272.03, 1350.90, 1434.65, 1523.60, 1618.07, 1718.39, 1824.93; **(b)** \$3330.35
75. 1, 2, 4, 8, 16, 32, 64, 128, 256, 512, 1024, 2048, 4096, 8192, 16,384, 32,768, 65,536
77. (a) $a_n = -17.748n^2 + 1784.95n - 23{,}768.85$;
(b) \$246.86, \$4830.95, \$19,232.35, \$19,435.35, \$10,269.90
79. $2a_1 + a_2 + 3a_n$ **81.** 0 **83.** ◆ **85.** $i, -1, -i, 1, i; i$
87. $\ln (1 \cdot 2 \cdot 3 \cdot \cdots \cdot n)$

EXERCISE SET 9.2

1. $a_1 = 3, d = 5$ **3.** $a_1 = 9, d = -4$ **5.** $a_1 = \frac{3}{2}, d = \frac{3}{4}$
7. $a_1 = \$316, d = -\$3$ **9.** $a_{12} = 46$ **11.** $a_{14} = -\frac{17}{3}$
13. $a_{10} = \$7941.62$ **15.** 33rd **17.** 46th **19.** $a_1 = 5$
21. $n = 39$ **23.** $a_1 = \frac{1}{3}; d = \frac{1}{2}; \frac{1}{3}, \frac{5}{6}, \frac{4}{3}, \frac{11}{6}, \frac{7}{3}$ **25.** 670
27. 160,400 **29.** 735 **31.** 990 **33.** 1760 **35.** $\frac{65}{2}$
37. $-\frac{6026}{13}$ **39.** 1260 **41.** 3; 171 **43.** 4960¢, or \$49.60
45. 1320 **47.** $-1, 1, 3$ **49.** Yes; 3 **51.** $-\frac{1}{8}$
53. 10,737,418.23 **55.** ◆ **57.** n^2
59. $a_1 = -5p - 5q + 60, d = 5p + 2q - 20$
61. $-10, -4, 2, 8$
63. Sides are a, $a + d$, $a + 2d$; and $a^2 + (a + d)^2 = (a + 2d)^2$. Solving, we get $d = \dfrac{a}{3}$. Thus the sides are a, $\dfrac{4a}{3}$, and $\dfrac{5a}{3}$ in the ratio 3:4:5.
65. $a_n = a_1 + (n - 1)d$; $a_n = f(n) = d \cdot n + (a_1 - d)$, where $f(n)$ is a linear function of n with slope d and y-intercept $(a_1 - d)$.

EXERCISE SET 9.3

1. 2 **3.** -1 **5.** -2 **7.** 0.1 **9.** $\dfrac{a}{2}$ **11.** 128 **13.** 162

15. $7(25)^{20}$ **17.** 3^{n-1} **19.** $(-1)^{n-1}$ **21.** $\dfrac{1}{x^n}$ **23.** 762

25. $\frac{4921}{18}$ **27.** 8 **29.** 125 **31.** Does not exist **33.** $\frac{2}{3}$
35. $29\frac{38{,}569}{59{,}049}$ **37.** 2 **39.** Does not exist **41.** $\$4545.\overline{45}$
43. $\frac{160}{9}$ **45.** $\frac{13}{99}$ **47.** 9 **49.** $\frac{34{,}091}{9990}$
51. (a) $\frac{1}{256}$ ft; **(b)** $5\frac{1}{3}$ ft **53. (a)** About 297 ft; **(b)** 300 ft
55. \$31,497.57 **57.** \$523,619.17 **59.** \$86,666,666,667
61. $k^2 + 2k + 1$ **63.** $2k^2 + 6k + 4$ **65.** $\dfrac{k + 1}{k + 2}$ **67.** ◆
69. $(4 - \sqrt{6})/(\sqrt{3} - \sqrt{2}) = 2\sqrt{3} + \sqrt{2}$,
$(6\sqrt{3} - 2\sqrt{2})/(4 - \sqrt{6}) = 2\sqrt{3} + \sqrt{2}$;
there exists a common ratio, $2\sqrt{3} + \sqrt{2}$; thus the sequence is geometric.
71. (a) $\frac{13}{3}; \frac{22}{3}, \frac{34}{3}, \frac{46}{3}, \frac{58}{3}$;
(b) $-\frac{11}{3}; -\frac{2}{3}, \frac{10}{3}, -\frac{50}{3}, \frac{250}{3}$ or 5; 8, 12, 18, 27
73. $S_n = \dfrac{x^2(1 - (-x)^n)}{x + 1}$
75. $\dfrac{a_{n+1}}{a_n} = r$, so $\ln \dfrac{a_{n+1}}{a_n} = \ln r$. But $\ln \dfrac{a_{n+1}}{a_n} = \ln a_{n+1} - \ln a_n = \ln r$. Thus, $\ln a_1, \ln a_2, \ldots$, is an arithmetic sequence with common difference $\ln r$.
77. 512 cm^2

EXERCISE SET 9.4

1. $1^2 < 1^3$, false; $2^2 < 2^3$, true; $3^2 < 3^3$, true; $4^2 < 4^3$, true; $5^2 < 5^3$, true **3.** A polygon of 3 sides has $\dfrac{3(3 - 3)}{2}$ diagonals. True; A polygon of 4 sides has $\dfrac{4(4 - 3)}{2}$ diagonals. True; A polygon of 5 sides has $\dfrac{5(5 - 3)}{2}$ diagonals. True; A polygon of 6 sides has $\dfrac{6(6 - 3)}{2}$ diagonals. True; A polygon of 7 sides has $\dfrac{7(7 - 3)}{2}$ diagonals. True.

5.
S_n: $2 + 4 + 6 + \cdots + 2n = n(n + 1)$
S_1: $2 = 1(1 + 1)$
S_k: $2 + 4 + 6 + \cdots + 2k = k(k + 1)$
S_{k+1}: $2 + 4 + 6 + \cdots + 2k + 2(k + 1) = (k + 1)(k + 2)$
1. *Basis step:* S_1 true by substitution.
2. *Induction step:* Assume S_k. Deduce S_{k+1}.
Starting with the left side of S_{k+1}, we have
$$\begin{aligned} &2 + 4 + 6 + \cdots + 2k + 2(k + 1) \\ &= k(k + 1) + 2(k + 1) \quad \textbf{By } S_k \\ &= (k + 1)(k + 2). \quad \textbf{Distributive law} \end{aligned}$$

7. S_n: $1 + 5 + 9 + \cdots + (4n - 3) = n(2n - 1)$
S_1: $1 = 1(2 \cdot 1 - 1)$
S_k: $1 + 5 + 9 + \cdots + (4k - 3) = k(2k - 1)$
S_{k+1}: $1 + 5 + 9 + \cdots + (4k - 3) + [4(k + 1) - 3] = (k + 1)[2(k + 1) - 1] = (k + 1)(2k + 1)$
1. *Basis step*: S_1 true by substitution.
2. *Induction step*: Assume S_k. Deduce S_{k+1}.

Starting with the left side of S_{k+1}, we have

$$\underbrace{1 + 5 + 9 + \cdots + (4k - 3)} + [4(k + 1) - 3]$$
$$= k(2k - 1) + [4(k + 1) - 3] \quad \textbf{By } S_k$$
$$= 2k^2 - k + 4k + 4 - 3$$
$$= (k + 1)(2k + 1).$$

9. S_n: $2 + 4 + 8 + \cdots + 2^n = 2(2^n - 1)$

S_1: $2 = 2(2 - 1)$

S_k: $2 + 4 + 8 + \cdots + 2^k = 2(2^k - 1)$

S_{k+1}: $2 + 4 + 8 + \cdots + 2^k + 2^{k+1} = 2(2^{k+1} - 1)$

1. *Basis step*: S_1 is true by substitution.
2. *Induction step*: Assume S_k. Deduce S_{k+1}.
   Starting with the left side of S_{k+1}, we have

$$\underbrace{2 + 4 + 8 + \cdots + 2^k} + 2^{k+1}$$
$$= 2(2^k - 1) + 2^{k+1} \quad \textbf{By } S_k$$
$$= 2^{k+1} - 2 + 2^{k+1}$$
$$= 2 \cdot 2^{k+1} - 2$$
$$= 2(2^{k+1} - 1).$$

11. 1. *Basis step:* Since $1 < 1 + 1$, S_1 is true.
2. *Induction step:* Assume S_k. Deduce S_{k+1}. Now

$$k < k + 1 \quad \textbf{By } S_k$$
$$k + 1 < k + 1 + 1; \quad \textbf{Adding 1}$$
$$\therefore k + 1 < k + 2.$$

13. 1. *Basis step:* Since $2 = 2$, S_1 is true.
2. *Induction step:* Let k be any natural number. Assume S_k. Deduce S_{k+1}.

$$2k \leq 2^k \quad \textbf{By } S_k$$
$$2 \cdot 2k \leq 2 \cdot 2^k \quad \textbf{Multiplying by 2}$$
$$4k \leq 2^{k+1}$$

Since $1 \leq k$, $k + 1 \leq k + k$, or $k + 1 \leq 2k$.
Then $2(k + 1) \leq 4k$.
Thus, $2(k + 1) \leq 4k \leq 2^{k+1}$, so $2(k + 1) \leq 2^{k+1}$.

15.

S_n: $$\frac{1}{1 \cdot 2 \cdot 3} + \frac{1}{2 \cdot 3 \cdot 4} + \frac{1}{3 \cdot 4 \cdot 5} + \cdots + \frac{1}{n(n + 1)(n + 2)} = \frac{n(n + 3)}{4(n + 1)(n + 2)}$$

S_1: $$\frac{1}{1 \cdot 2 \cdot 3} = \frac{1(1 + 3)}{4 \cdot 2 \cdot 3}$$

S_k: $$\frac{1}{1 \cdot 2 \cdot 3} + \frac{1}{2 \cdot 3 \cdot 4} + \cdots + \frac{1}{k(k + 1)(k + 2)} = \frac{k(k + 3)}{4(k + 1)(k + 2)}$$

S_{k+1}: $$\frac{1}{1 \cdot 2 \cdot 3} + \frac{1}{2 \cdot 3 \cdot 4} + \cdots + \frac{1}{k(k + 1)(k + 2)} + \frac{1}{(k + 1)(k + 2)(k + 3)} = \frac{(k + 1)(k + 1 + 3)}{4(k + 1 + 1)(k + 1 + 2)} = \frac{(k + 1)(k + 4)}{4(k + 2)(k + 3)}$$

1. *Basis step*: Since $\frac{1}{1 \cdot 2 \cdot 3} = \frac{1}{6}$ and $\frac{1(1 + 3)}{4 \cdot 2 \cdot 3} = \frac{1 \cdot 4}{4 \cdot 2 \cdot 3} = \frac{1}{6}$, S_1 is true.
2. *Induction step*: Assume S_k. Deduce S_{k+1}.
   Add $\frac{1}{(k + 1)(k + 2)(k + 3)}$ on both sides of S_k and simplify the right side.
   Only the right side is shown here.

$$\frac{k(k + 3)}{4(k + 1)(k + 2)} + \frac{1}{(k + 1)(k + 2)(k + 3)}$$
$$= \frac{k(k + 3)(k + 3) + 4}{4(k + 1)(k + 2)(k + 3)}$$
$$= \frac{k^3 + 6k^2 + 9k + 4}{4(k + 1)(k + 2)(k + 3)}$$
$$= \frac{(k + 1)^2(k + 4)}{4(k + 1)(k + 2)(k + 3)}$$
$$= \frac{(k + 1)(k + 4)}{4(k + 2)(k + 3)}$$

17.

S_n: $1 + 2 + 3 + \cdots + n = \frac{n(n + 1)}{2}$

S_1: $1 = \frac{1(1 + 1)}{2}$

S_k: $1 + 2 + 3 + \cdots + k = \frac{k(k + 1)}{2}$

S_{k+1}: $1 + 2 + 3 + \cdots + k + (k + 1) = \frac{(k + 1)(k + 2)}{2}$

1. *Basis step*: S_1 true by substitution.
2. *Induction step*: Assume S_k. Deduce S_{k+1}.
   Starting with the left side of S_{k+1}, we have

$$\underbrace{1 + 2 + 3 + \cdots + k} + (k + 1)$$
$$= \frac{k(k + 1)}{2} + (k + 1) \quad \textbf{By } S_k$$
$$= \frac{k(k + 1) + 2(k + 1)}{2} \quad \textbf{Adding}$$
$$= \frac{(k + 1)(k + 2)}{2}. \quad \textbf{Distributive law}$$

19. 1. *Basis step*. S_1: $1^3 = \frac{1^2(1 + 1)^2}{4} = 1$. True.
2. *Induction step*: Assume S_k. Deduce S_{k+1}.

$$S_k:\quad 1^3 + 2^3 + \cdots + k^3 = \frac{k^2(k+1)^2}{4}$$

$$1^3 + 2^3 + \cdots + (k+1)^3 = \frac{k^2(k+1)^2}{4} + (k+1)^3 \qquad \textbf{By } S_k$$
$$= \frac{(k+1)^2}{4}[k^2 + 4(k+1)]$$
$$= \frac{(k+1)^2(k+2)^2}{4}$$

21.

1. *Basis step.* S_1: $1^5 = \dfrac{1^2(1+1)^2(2\cdot 1^2 + 2\cdot 1 - 1)}{12}$. True.

2. *Induction step.* Assume S_k: $1^5 + 2^5 + \cdots + k^5 = \dfrac{k^2(k+1)^2(2k^2+2k-1)}{12}$.

Then $1^5 + 2^5 + \cdots + k^5 + (k+1)^5$
$$= \frac{k^2(k+1)^2(2k^2+2k-1)}{12} + (k+1)^5$$
$$= \frac{k^2(k+1)^2(2k^2+2k-1) + 12(k+1)^5}{12}$$
$$= \frac{(k+1)^2(2k^4 + 14k^3 + 35k^2 + 36k + 12)}{12}$$
$$= \frac{(k+1)^2(k+2)^2(2k^2+6k+3)}{12}$$
$$= \frac{(k+1)^2(k+1+1)^2(2(k+1)^2 + 2(k+1) - 1)}{12}.$$

23.

1. *Basis step.* S_1: $1(1+1) = \dfrac{1(1+1)(1+2)}{3}$. True.

2. *Induction step.* Assume
S_k: $1(1+1) + 2(2+1) + \cdots + k(k+1) = \dfrac{k(k+1)(k+2)}{3}$.

Then $1(1+1) + 2(2+1) + \cdots + k(k+1) + (k+1)(k+1+1)$
$$= \frac{k(k+1)(k+2)}{3} + (k+1)(k+2)$$
$$= \frac{(k+1)(k+2)(k+3)}{3}$$
$$= \frac{(k+1)(k+1+1)(k+1+2)}{3}.$$

25. 1. *Basis step:* Since $\frac{1}{2}[2a_1 + (1-1)d] = \frac{1}{2}\cdot 2a_1 = a_1$, S_1 is true.
2. *Induction step:* Assume S_k. Deduce S_{k+1}. Starting with the left side of S_{k+1}, we have

$$\underbrace{a_1 + (a_1 + d) + \cdots + [a_1 + (k-1)d]}_{\frac{k}{2}[2a_1 + (k-1)d]} + [a_1 + kd]$$
$$= \frac{k}{2}[2a_1 + (k-1)d] + [a_1 + kd] \qquad \textbf{By } S_k$$
$$= \frac{k[2a_1 + (k-1)d]}{2} + \frac{2[a_1 + kd]}{2}$$
$$= \frac{2ka_1 + k(k-1)d + 2a_1 + 2kd}{2}$$
$$= \frac{2a_1(k+1) + k(k-1)d + 2kd}{2}$$
$$= \frac{2a_1(k+1) + (k-1+2)kd}{2}$$
$$= \frac{2a_1(k+1) + (k+1)kd}{2}$$
$$= \frac{k+1}{2}[2a_1 + kd].$$

27. 336 **29.** 90

31.

1. *Basis step.* S_1: $x + y$ is a factor of $x^2 - y^2$. True.
S_2: $x + y$ is a factor of $x^4 - y^4$. True.

2. *Induction step.* Assume S_{k-1}: $x + y$ is a factor of $x^{2(k-1)} - y^{2(k-1)}$.
Then $x^{2(k-1)} - y^{2(k-1)} = (x+y)Q(x)$ for some polynomial Q.
Assume S_k: $x + y$ is a factor of $x^{2k} - y^{2k}$.
Then $x^{2k} - y^{2k} = (x+y)P(x)$ for some polynomial P.

$$x^{2(k+1)} - y^{2(k+1)}$$
$$= (x^{2k} - y^{2k})(x^2 + y^2) - (x^{2(k-1)} - y^{2(k-1)})(x^2y^2)$$
$$= (x+y)P(x)(x^2+y^2) - (x+y)Q(x)(x^2y^2)$$
$$= (x+y)[P(x)(x^2+y^2) - Q(x)(x^2y^2)]$$

so $x + y$ is a factor of $x^{2(k+1)} - y^{2(k+1)}$.

33.

1. *Basis step.* S_2: $\dfrac{x^2 - y^2}{x - y} = x + y$. True.
S_3: $\dfrac{x^3 - y^3}{x - y} = x^2 + yx + y^2$. True.

2. *Induction step.* Assume S_{k-1}: $\dfrac{x^{k-1} - y^{k-1}}{x - y} = x^{k-2} + yx^{k-3} + \cdots + y^{k-3}x + y^{k-2}$

Assume S_k: $\dfrac{x^k - y^k}{x - y} = x^{k-1} + yx^{k-2} + \cdots + y^{k-2}x + y^{k-1}$.

$$\frac{x^{k+1} - y^{k+1}}{x - y} = \frac{(x^k - y^k)(x+y)}{x - y} - \frac{xy(x^{k-1} - y^{k-1})}{x - y}$$
$$= (x^{k-1} + yx^{k-2} + \cdots + y^{k-2}x + y^{k-1})(x+y)$$
$$- xy(x^{k-2} + yx^{k-3} + \cdots + y^{k-3}x + y^{k-2})$$
$$= x^k + yx^{k-1} + \cdots + y^{k-2}x^2 + y^{k-1}x$$
$$+ x^{k-1}y + y^2x^{k-2} + \cdots + y^{k-1}x + y^k$$
$$- (x^{k-1}y + y^2x^{k-2} + \cdots + y^{k-2}x^2 + y^{k-1}x)$$
$$= x^k + yx^{k-1} + \cdots + y^{k-1}x + y^k$$

35. ◆

37.

S_2: $\log_a(b_1b_2) = \log_a b_1 + \log_a b_2$

S_k: $\log_a(b_1b_2 \ldots b_k) = \log_a b_1 + \log_a b_2 + \cdots + \log_a b_k$

S_{k+1}: $\log_a(b_1b_2 \ldots b_{k+1}) = \log_a b_1 + \log_a b_2 + \cdots + \log_a b_{k+1}$

1. *Basis step*: S_2 is true by the properties of logarithms.
2. *Induction step*: Let k be a natural number $k \geq 2$. Assume S_k. Deduce S_{k+1}.

$$\begin{aligned}\log_a(b_1b_2\ldots b_{k+1}) & \quad \textbf{Left side of } S_{k+1}\\ &= \log_a(b_1b_2\ldots b_k) + \log_a b_{k+1} \quad \textbf{By } S_2\\ &= \log_a b_1 + \log_a b_2 + \cdots + \log_a b_k + \log_a b_{k+1}\end{aligned}$$

39. 1. *Basis step:* $\frac{1}{\sqrt{1}} + \frac{1}{\sqrt{2}}$

$$= 1 + \frac{1}{\sqrt{2}}$$

$$= \frac{\sqrt{2}+1}{\sqrt{2}} > \frac{2}{\sqrt{2}} = \sqrt{2} \quad \textbf{Since } \sqrt{2} > 1$$

2. *Induction step*: Assume S_k. Deduce S_{k+1}.

$$\frac{1}{\sqrt{1}} + \frac{1}{\sqrt{2}} + \frac{1}{\sqrt{3}} + \cdots + \frac{1}{\sqrt{k}} > \sqrt{k} \quad \textbf{By } S_k$$

Therefore, adding $\frac{1}{\sqrt{k+1}}$, we get

$$\frac{1}{\sqrt{1}} + \frac{1}{\sqrt{2}} + \frac{1}{\sqrt{3}} + \cdots + \frac{1}{\sqrt{k}} + \frac{1}{\sqrt{k+1}} > \sqrt{k} + \frac{1}{\sqrt{k+1}}.$$

Then

$$\sqrt{k} + \frac{1}{\sqrt{k+1}} = \frac{\sqrt{k}\cdot\sqrt{k+1}+1}{\sqrt{k+1}} > \frac{k+1}{\sqrt{k+1}} = \sqrt{k+1}.$$

41. S_2: $\overline{z_1 + z_2} = \bar{z}_1 + \bar{z}_2$:

$$\begin{aligned}\overline{(a+bi)+(c+di)} &= \overline{(a+c)+(b+d)i}\\ &= (a+c)-(b+d)i\\ \overline{(a+bi)}+\overline{(c+di)} &= a - bi + c - di\\ &= (a+c)-(b+d)i.\end{aligned}$$

S_k: $\overline{z_1 + z_2 + \cdots + z_k} = \bar{z}_1 + \bar{z}_2 + \cdots + \bar{z}_k$.

$$\begin{aligned}&\overline{(z_1+z_2+\cdots+z_k)+z_{k+1}}\\ &= \overline{(z_1+z_2+\cdots+z_k)} + \overline{z_{k+1}} \quad \textbf{By } S_2\\ &= \bar{z}_1+\bar{z}_2+\cdots+\bar{z}_k+\bar{z}_{k+1} \quad \textbf{By } S_k\end{aligned}$$

43. S_1: i is either i or -1 or $-i$ or 1.
S_k: i^k is either i or -1 or $-i$ or 1.
$i^{k+1} = i^k \cdot i$ is then $i \cdot i = -1$ or $-1 \cdot i = -i$ or $-i \cdot i = 1$ or $1 \cdot i = i$.

45. S_1: 3 is a factor of $1^3 + 2 \cdot 1$.
S_k: 3 is a factor of $k^3 + 2k$, i.e., $k^3 + 2k = 3 \cdot m$.
S_{k+1}: 3 is a factor of $(k+1)^3 + 2(k+1)$.
Consider

$$\begin{aligned}(k+1)^3 + 2(k+1) &= k^3 + 3k^2 + 5k + 3\\ &= (k^3 + 2k) + 3k^2 + 3k + 3\\ &= 3m + 3(k^2 + k + 1). \quad \textbf{A multiple of 3}\end{aligned}$$

47. S_1: 3 is a factor of $1(1+1)(1+2)$.
S_k: 3 is a factor of $k(k+1)(k+2)$, or $k(k^2+3k+2)$.

$$\begin{aligned}&(k+1)(k+1+1)(k+1+2)\\ &= (k+1)(k+2)(k+3)\\ &= (k^2+3k+2)(k+3)\\ &= k(k^2+3k+2) + 3(k^2+3k+2)\end{aligned}$$

By S_k, 3 is a factor of $k(k^2+3k+2)$; hence 3 is a factor of the righthand side, so 3 is a factor of $(k+1)(k+2)(k+3)$.

49. (a) *Basis step:* $1 = 1 + 1$. Cannot be proved.
(b) *Induction step:* Assume S_k: $k = k + 1$. Deduce S_{k+1}.

$$\begin{aligned}k &= k + 1 \quad \textbf{By } S_k\\ k + 1 &= k + 2 \quad \textbf{Adding 1}\\ &= (k+1) + 1\end{aligned}$$

(c) No

Exercise Set 9.5

1. 720 **3.** 604,800 **5.** 120 **7.** 1 **9.** 3024 **11.** 120 **13.** 120 **15.** 1 **17.** 6,497,400 **19.** $n(n-1)(n-2)$ **21.** n **23.** $6! = 720$ **25.** $9! = 362,880$ **27.** ${}_9P_4 = 3024$ **29.** ${}_5P_5 = 120$; $5^5 = 3125$ **31.** $\frac{8!}{3!} = 6720$; $\frac{7!}{2!} = 2520$; $\frac{11!}{2!\,2!\,2!} = 4,989,600$ **33.** $8 \cdot 10^6 = 8,000,000$; 8 million **35.** $\frac{9!}{2!\,3!\,4!} = 1260$ **37. (a)** ${}_6P_5 = 720$; **(b)** $6^5 = 7776$; **(c)** $1 \cdot {}_5P_4 = 120$; **(d)** $1 \cdot 1 \cdot {}_4P_3 = 24$ **39. (a)** 10^5, or 100,000; **(b)** 100,000 **41. (a)** $10^9 = 1,000,000,000$; **(b)** yes **43.** 126 **45.** 7,059,052 **47.** ◆ **49.** 8 **51.** 11 **53.** $n - 1$

Exercise Set 9.6

1. 78 **3.** 78 **5.** 7 **7.** 10 **9.** 1 **11.** 15 **13.** 128 **15.** 270,725 **17.** 13,037,895 **19.** n **21.** 1 **23.** ${}_{23}C_4 = 8855$ **25.** ${}_{13}C_{10} = 286$ **27.** $\binom{8}{2} = 28$; $\binom{8}{3} = 56$ **29.** $\binom{52}{5} = 2,598,960$ **31. (a)** ${}_{31}P_2 = 930$; **(b)** $31^2 = 961$; **(c)** ${}_{31}C_2 = 465$ **33.** $a + b$ **35.** $a^2 + 3a^2b + 3ab^2 + b^3$ **37.** $a^4 + 4a^3b + 6a^2b^2 + 4ab^3 + b^4$ **39.** ◆ **41.** $\binom{13}{5} = 1287$ **43.** $\binom{n}{2}$; $2\binom{n}{2}$ **45.** 4 **47.** 7

49. $\binom{n}{k-1} + \binom{n}{k}$

$$\begin{aligned}&= \frac{n!}{(k-1)!(n-k+1)!}\cdot\frac{k}{k} + \frac{n!}{k!(n-k)!}\cdot\frac{(n-k+1)}{(n-k+1)}\\ &= \frac{n!(k+(n-k+1))}{k!(n-k+1)!}\\ &= \frac{(n+1)!}{k!(n-k+1)!} = \binom{n+1}{k}\end{aligned}$$

EXERCISE SET 9.7

1. $x^4 + 20x^3 + 150x^2 + 500x + 625$
3. $x^5 - 15x^4 + 90x^3 - 270x^2 + 405x - 243$
5. $x^5 - 5x^4y + 10x^3y^2 - 10x^2y^3 + 5xy^4 - y^5$
7. $15{,}625x^6 + 75{,}000x^5y + 150{,}000x^4y^2 + 160{,}000x^3y^3 + 96{,}000x^2y^4 + 30{,}720xy^5 + 4096y^6$
9. $128t^7 + 448t^5 + 672t^3 + 560t + 280t^{-1} + 84t^{-3} + 14t^{-5} + t^{-7}$
11. $x^{10} - 5x^8 + 10x^6 - 10x^4 + 5x^2 - 1$
13. $125 + 150\sqrt{5}t + 375t^2 + 100\sqrt{5}t^3 + 75t^4 + 6\sqrt{5}t^5 + t^6$
15. $a^9 - 18a^7 + 144a^5 - 672a^3 + 2016a - 4032a^{-1} + 5376a^{-3} - 4608a^{-5} + 2304a^{-7} - 512a^{-9}$
17. $140\sqrt{2}$ **19.** $x^{-8} + 4x^{-4} + 6 + 4x^4 + x^8$ **21.** $21a^5b^2$
23. $-252x^5y^5$ **25.** $-745{,}472a^3$ **27.** $1120x^{12}y^2$
29. $-1{,}959{,}552u^5v^{10}$ **31.** 128 **33.** 2^{24}, or 16,777,216
35. 20 **37.** $-12 + 316i$ **39.** $-7 - 4\sqrt{2}i$
41. $\sum_{k=0}^{n}\binom{n}{k}(-1)^k a^{n-k}b^k$ **43.** $\sum_{k=1}^{n}\binom{n}{k}x^{n-k}h^{k-1}$ **45.** $\frac{1}{4}$
47. ◆ **49.** $-5 \pm 2\sqrt{2}$ **51.** 9, 3
53. **(a)** 0.08386; **(b)** 0.89479 **55.** $10x^{3/2}y$, $-10xy^{3/2}$
57. $-4320x^6y^{9/2}$ **59.** $-\dfrac{35}{x^{1/6}}$ **61.** 2^{100} **63.** $[\log_a (xt)]^{23}$
65. 1. *Basis step*: Since $a + b = (a + b)^1$, S_1 is true.
2. *Induction step*: Let S_k be the statement of the binomial theorem with n replaced by k. Multiply both sides of S_k by $(a + b)$ to obtain

$$\begin{aligned}(a+b)^{k+1} &= \left[a^k + \cdots + \binom{k}{r-1}a^{k-(r-1)}b^{r-1} + \binom{k}{r}a^{k-r}b^r + \cdots + b^k\right](a+b)\\ &= a^{k+1} + \cdots + \left[\binom{k}{r-1} + \binom{k}{r}\right]a^{(k+1)-r}b^r + \cdots + b^{k+1}\\ &= a^{k+1} + \cdots + \binom{k+1}{r}a^{(k+1)-r}b^r + \cdots + b^{k+1}.\end{aligned}$$

This proves S_{k+1}, assuming S_k. Hence S_n is true for $n = 1, 2, 3, \ldots$.

EXERCISE SET 9.8

1. **(a)** 0.18, 0.24, 0.23, 0.23, 0.12; **(b)** Opinions may vary, but it seems that people tend not to pick the first or last numbers.
3. 11,700 **5.** **(a)** T, S, R, N, L; **(b)** E; **(c)** yes
7. **(a)** $\frac{2}{7}$; **(b)** $\frac{5}{7}$; **(c)** 0; **(d)** 1 **9.** $\dfrac{350}{31{,}977}$ **11.** $\dfrac{1}{108{,}290}$
13. $\dfrac{33}{66{,}640}$ **15.** Answers will vary. **17.** $\frac{9}{19}$ **19.** $\frac{18}{19}$
21. $\frac{1}{38}$ **23.** $\frac{9}{19}$
25. $P(\text{red}) = \frac{1}{3}$, $P(\text{green}) = \frac{2}{9}$, $P(\text{blue}) = \frac{1}{6}$, $P(\text{yellow}) = \frac{5}{18}$
27. ◆ **29.** **(a)** 36; **(b)** 1.39×10^{-5}
31. **(a)** $(13 \cdot {}_4C_3) \cdot (12 \cdot {}_4C_2) = 3744$; **(b)** 0.00144
33. **(a)** $4 \cdot \binom{13}{5} - 4 - 36 = 5108$; **(b)** 0.00197
35. **(a)** $\binom{10}{1}\binom{4}{1}\binom{4}{1}\binom{4}{1}\binom{4}{1}\binom{4}{1} - 4 - 36 = 10{,}200$; **(b)** 0.00392

REVIEW EXERCISES, CHAPTER 9

1. [9.1] $-\frac{1}{2}, \frac{4}{17}, -\frac{9}{82}, \frac{16}{257}$; $-\frac{121}{14{,}642}$; $-\frac{529}{279{,}842}$
2. [9.1] $(-1)^{n+1}(n^2 + 1)$ **3.** [9.1] $\frac{3}{2} - \frac{9}{8} + \frac{27}{26} - \frac{81}{80} = \frac{417}{1040}$
4. [9.1]

n	U_n
1	0.3
2	2.5
3	13.5
4	68.5
5	343.5
6	1718.5
7	8593.5
8	42968.5
9	214843.5
10	1074218.5

1,100,000
0
0
10

5. [9.1] $\sum_{k=1}^{7} k^2 - 1$ **6.** [9.2] $3\frac{3}{4}$ **7.** [9.2] $a + 4b$
8. [9.2] 531 **9.** [9.2] 20,100 **10.** [9.2] 11 **11.** [9.2] -4
12. [9.3] $n = 6$, $S_n = -126$ **13.** [9.3] $a_1 = 8$, $a_5 = \frac{1}{2}$
14. [9.3] Does not exist **15.** [9.3] $\frac{3}{11}$ **16.** [9.3] $\frac{3}{8}$
17. [9.3] $\frac{241}{99}$ **18.** [9.2] $5\frac{4}{5}, 6\frac{3}{5}, 7\frac{2}{5}, 8\frac{1}{5}$ **19.** [9.3] 167.3 ft
20. [9.3] \$60,653.05 **21.** [9.2] **(a)** \$7.38; **(b)** \$1365.10
22. [9.3] \$88,888,888,889
23. [9.4]

$$\begin{aligned}S_n&: \quad 1 + 4 + 7 + \cdots + (3n-2) = \frac{n(3n-1)}{2}\\ S_1&: \quad 1 = \frac{1(3-1)}{2}\\ S_k&: \quad 1 + 4 + 7 + \cdots + (3k-2) = \frac{k(3k-1)}{2}\\ S_{k+1}&: \quad 1 + 4 + 7 + \cdots + [3(k+1) - 2]\\ &\qquad = 1 + 4 + 7 + \cdots + (3k-2) + (3k+1)\\ &\qquad = \frac{(k+1)(3k+2)}{2}\end{aligned}$$

1. *Basis step*: $1 = \dfrac{2}{2} = \dfrac{1(3-1)}{2}$ is true.

2. *Induction step*: Assume S_k. Add $(3k + 1)$ to both sides.

$$1 + 4 + 7 + \cdots + (3k - 2) + (3k + 1)$$
$$= \frac{k(3k-1)}{2} + (3k+1)$$
$$= \frac{k(3k-1)}{2} + \frac{2(3k+1)}{2}$$
$$= \frac{3k^2 - k + 6k + 2}{2}$$
$$= \frac{3k^2 + 5k + 2}{2}$$
$$= \frac{(k+1)(3k+2)}{2}$$

24. [9.4] S_1: $1 = \frac{3^1 - 1}{2}$; S_2: $1 + 3 = \frac{3^2 - 1}{2}$

S_k: $1 + 3 + 3^2 + \cdots + 3^{k-1} = \frac{3^k - 1}{2}$

2. *Induction step*: Assume S_k. Add 3^k on both sides.

$$1 + 3 + \cdots + 3^{k-1} + 3^k$$
$$= \frac{3^k - 1}{2} + 3^k = \frac{3^k - 1}{2} + 3^k \cdot \frac{2}{2}$$
$$= \frac{3 \cdot 3^k - 1}{2} = \frac{3^{k+1} - 1}{2}$$

25. [9.4]

S_n: $\left(1 - \frac{1}{2}\right)\left(1 - \frac{1}{3}\right)\cdots\left(1 - \frac{1}{n}\right) = \frac{1}{n}$

S_2: $\left(1 - \frac{1}{2}\right) = \frac{1}{2}$

S_k: $\left(1 - \frac{1}{2}\right)\left(1 - \frac{1}{3}\right)\cdots\left(1 - \frac{1}{k}\right) = \frac{1}{k}$

S_{k+1}: $\left(1 - \frac{1}{2}\right)\left(1 - \frac{1}{3}\right)\cdots\left(1 - \frac{1}{k}\right)\left(1 - \frac{1}{k+1}\right) = \frac{1}{k+1}$

1. *Basis step*: S_2 is true by substitution.
2. *Induction step*: Assume S_k. Deduce S_{k+1}. Starting with the left side of S_{k+1}, we have

$$\underbrace{\left(1 - \frac{1}{2}\right)\left(1 - \frac{1}{3}\right)\cdots\left(1 - \frac{1}{k}\right)}\left(1 - \frac{1}{k+1}\right)$$
$$= \frac{1}{k} \cdot \left(1 - \frac{1}{k+1}\right). \quad \textbf{By } S_k$$
$$= \frac{1}{k} \cdot \left(\frac{k+1-1}{k+1}\right)$$
$$= \frac{1}{k} \cdot \frac{k}{k+1}$$
$$= \frac{1}{k+1}. \quad \textbf{Simplifying}$$

26. [9.5] $6! = 720$ **27.** [9.5] $9 \cdot 8 \cdot 7 \cdot 6 = 3024$

28. [9.6] $\binom{15}{8} = 6435$ **29.** [9.5] $24 \cdot 23 \cdot 22 = 12{,}144$

30. [9.5] $\frac{9!}{1!\,4!\,2!\,2!} = 3780$ **31.** [9.5] 36

32. [9.5] **(a)** ${}_6P_5 = 720$; **(b)** $6^5 = 7776$; **(c)** ${}_5P_4 = 120$; **(d)** ${}_3P_2 = 6$

33. [9.7] 256

34. [9.7] $m^7 + 7m^6n + 21m^5n^2 + 35m^4n^3 + 35m^3n^4 + 21m^2n^5 + 7mn^6 + n^7$

35. [9.7] $x^5 - 5\sqrt{2}x^4 + 20x^3 - 20\sqrt{2}x^2 + 20x - 4\sqrt{2}$

36. [9.7] $x^8 - 12x^6y + 54x^4y^2 - 108x^2y^3 + 81y^4$

37. [9.7] $a^8 + 8a^6 + 28a^4 + 56a^2 + 70 + 56a^{-2} + 28a^{-4} + 8a^{-6} + a^{-8}$

38. [9.7] $-6624 + 16{,}280i$

39. [9.7] $220a^3x^9$ **40.** [9.7] $-\binom{18}{11}128a^7b^{11}$

41. [9.8] $\frac{86}{206} \approx 0.42$, $\frac{97}{206} \approx 0.47$, $\frac{23}{206} \approx 0.11$ **42.** [9.8] $\frac{1}{12}$, 0

43. [9.8] $\frac{1}{4}$ **44.** [9.8] $\frac{6}{5525}$

45. [9.1] **(a)** $a_n = 0.359n^2 - 6.965n + 151.913$; **(b)** \$137 billion, \$164 billion

46. ◈ [9.3] Someone who has managed several sequences of hiring has a considerable income from the sales of the people in the lower levels. However, with a finite population, it will not be long before the salespersons in the lowest level have no one to hire and no one to sell to.

47. ◈ [9.6] A list of 9 candidates for an office is to be narrowed down to 4 candidates. In how many ways can this be done?

48. [9.4] S_1 fails for both (a) and (b).

49. [9.3] $\frac{a_{k+1}}{a_k} = r_1$, $\frac{b_{k+1}}{b_k} = r_2$, so $\frac{a_{k+1}b_{k+1}}{a_kb_k} = r_1r_2$, a constant

50. [9.2] **(a)** No (unless a_n is all positive or all negative); **(b)** yes; **(c)** yes; **(d)** no (unless a_n is constant); **(e)** no (unless a_n is constant); **(f)** no (unless a_n is constant)

51. [9.2] -2, 0, 2, 4 **52.** [9.3] $\frac{1}{2}$, $-\frac{1}{6}$, $\frac{1}{18}$

53. [9.6] $\left(\log \frac{x}{y}\right)^{10}$ **54.** [9.6] 18 **55.** [9.6] 36

56. [9.7] -9

Appendixes

EXERCISE SET A

1. -11 **3.** $x^3 - 4x$ **5.** -109 **7.** $-x^4 + x^2 - 5x$

9. The answer checks. **11.** The answer checks.

13. $M_{11} = 6$, $M_{32} = -9$, $M_{22} = -29$

15. $A_{11} = 6$, $A_{32} = 9$, $A_{22} = -29$ **17.** -10 **19.** -10

21. $M_{41} = -14$, $M_{33} = 20$ **23.** $A_{24} = 15$, $A_{43} = 30$

25. 110 **27.** 110 **29.** $\left(-\frac{25}{2}, -\frac{11}{2}\right)$ **31.** $(3, 1)$

33. $\left(\frac{1}{2}, -\frac{1}{3}\right)$ **35.** $(1, 1)$ **37.** $\left(\frac{3}{2}, \frac{13}{14}, \frac{33}{14}\right)$ **39.** $(3, -2, 1)$

41. $(1, 3, -2)$ **43.** $\left(\frac{1}{2}, \frac{2}{3}, -\frac{5}{6}\right)$ **45.** **47.** ± 2

49. $(-\infty, -\sqrt{3}] \cup [\sqrt{3}, \infty)$ **51.** -34 **53.** 4

55. Answers may vary.
$$\begin{vmatrix} L & -W \\ 2 & 2 \end{vmatrix}$$

57. Answers may vary.
$$\begin{vmatrix} a & b \\ -b & a \end{vmatrix}$$

59. Answers may vary.
$$\begin{vmatrix} 2\pi r & 2\pi r \\ -h & r \end{vmatrix}$$

61. Evaluate the determinant and compare with the two-point equation of a line.

APPENDIX B

1. $y = 12x - 7,\ -\frac{1}{2} \le x \le 3$

3. $y = \sqrt[3]{x} - 4,\ -1 \le x \le 1000$ **5.** $x = y^4,\ 0 \le x \le 16$

7. $y = \frac{1}{x},\ 1 \le x \le 5$ **9.** $y = \frac{1}{4}(x + 1)^2,\ -7 \le x \le 5$

11. $y = \frac{1}{x},\ x > 0$ **13.** $x^2 + y^2 = 9,\ -3 \le x \le 3$

15. $x^2 + \frac{y^2}{4} = 1,\ -1 \le x \le 1$ **17.** $y = \frac{1}{x},\ x \ge 1$

19. $(x - 1)^2 + (y - 2)^2 = 4,\ -1 \le x \le 3$ **21.** 0.7071

23. -0.2588 **25.** 0.7265 **27.** 0.5460

29. Answers may vary. $x = t,\ y = 4t - 3;\ x = \frac{t}{4} + 3,\ y = t + 9$

31. Answers may vary. $x = t,\ y = (t - 2)^2 - 6t;\ x = t + 2,\ y = t^2 - 6t - 12$

33. **(a)** About 36.1 ft; **(b)** about 196.6 ft; **(c)** about 3 sec

35. About 73.1 ft

37. $x = 2(t - \sin t),\ y = 2(1 - \cos t);\ 0 \le t \le 4\pi$

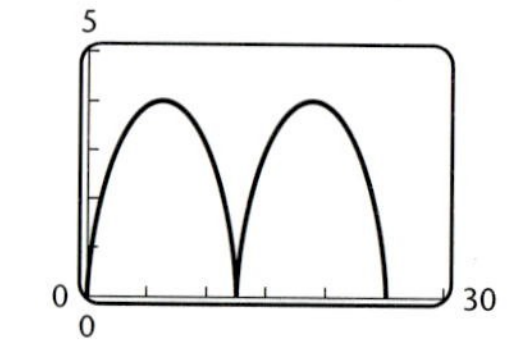

39. $x = t - \sin t,\ y = 1 - \cos t;\ -2\pi \le t \le 2\pi$

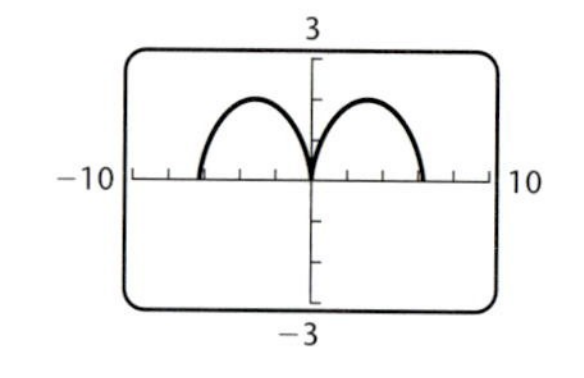

41. $x = 3\cos t,\ y = -3\sin t$

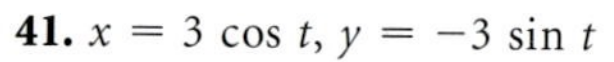

Index of Applications

Astronomy

Distance between points on the earth, 372
Distance to a star, 28
Earth's orbit, 31, 607
Hyperbolic mirror, 616
Linear speed of the earth, 371
Linear speed of an earth satellite, 367
Linear speed at the equator, 371
Lunar crater, 474
Measuring the radius of the earth, 343
Planet diameters vs. distance from sun, 133
Satellite location, 403, 409, 458

Biology/Health/Life Sciences

AIDS in the U.S., 261
Bacteria growth, 640
Cases of skin cancer in the U.S., 132
Cholesterol level—risk of heart attack, 247, 312
Growth of AIDS, 272
Growth of bacteria *Escherichia coli*, 272
Heart arrhythmia, 124
Hours of sleep vs. death rate, 201
Ibuprofen in the bloodstream, 171, 196
Ideal weight, 125
Life expectancy of men, 132
Life expectancy of women, 128, 129
Limited population growth in a lake, 315
Maximizing animal support in a forest, 576
Maximum heart rate, 133
Medical dosage, 236, 248
Melanoma deaths, 2
Minimizing nutrition cost, 576
Number of births, 2
Nutrition, 530, 538, 557, 587
Population growth of rabbits, 313
Spread of an epidemic, 315
Stranglehold of nicotine, 131
Temperature during an illness, 107, 235, 244, 403, 458
Territorial area of an animal, 97
Threshold weight, 196

Business

Advertising, 273, 286
Advertising effect, 107, 316, 661
Advertising effect on movie revenue, 207
Advertising expense, 547
Airline bumping rates, 205
Angular speed on a printing press, 371
Cellular phone boom, 261, 310
Chain business deals, 708
Coffee mixtures, 531
Commissions, 531
Corporate fax expense, 77
Crude steel production, 539
Dow Chemical, 77
Easel display, 340
FaxMax, 112
Food service management, 557, 558, 587
Height of a tree, 340
Hourly wages, 646
Junk mail, 702
Mail-order business, 530, 531, 538
Number of cellular phones, 128, 249
Periodic sales, 403
Pie sales, 79
Produce, 557
Production, 552
Quality control, 697
Retail hardware sales, 31
Revenues of Packard Bell, 272
Sales, 402, 565
Sales promotions, 530
Salvage value, 272, 640
Tee-shirt sales, 531
Television ratings, 698
The Gap, Inc., 639
Total revenues of Microsoft, 261, 316
Total sales of Goodyear, 302
Video rentals, 316, 708
World bike production, 633

Chemistry

Antifreeze mixtures, 531
Boiling point and elevation, 97
Fahrenheit and Celsius temperatures, 126
Mass of a neutron, 28
Newton's law of cooling, 317
Nuclear disintegration, 31
pH of substances in chemistry, 286
When was the murder committed?, 317

Construction

Arch of a circle—carpentry, 139
Beam deflection, 214
Box construction, 77
Carpentry, 607, 624
Chesapeake Bay Bridge-Tunnel, 31
Construction of picnic pavilions, 342
Cost of material, 109
Fencing, 624
Floor plans, 707
Ladder safety, 329
Landscaping, 624
Norman window, 204
Play space, 109
Pole stacking, 649

Construction (continued)

Ramps for handicapped, 115
Road grade, 124
Sandbox, 519
Sidewalk width, 246
Swimming pool, 139
Theatre seating, 650
Total in a stack, 647

Consumer Applications

Acceptance of seat belt laws, 315
Baskin-Robbins ice cream, 685
Book buying growth in the U.S., 132
Budget, 557
Concert ticket prices, 530
Cost of a service call, 78, 82
Cost of snack food, 538
Cost of tuition—state universities, 132
Dress sizes, 165
Insurance plans, 78
Lodging boom, 131, 205
Mason Dots, 702
Maximizing mileage, 575
Mortgage payments, 31
Moving costs, 78
Multimedia personal computers, 319
Price of personal computers, 206
Price of school supplies, 565
Salary plans, 78
Snack mixtures, 531
Stamp purchase, 547
Time of return, 547
Value of a horse, 531

Economics

Allocation of resources, 577
Amount of an annuity, 658, 661, 707
Average cost, 236
Borrowing, 547
Compound interest, 30, 31, 78, 266, 640
Cost, 565
Daily doubling salary, 657, 660
Dairy profit, 555
Distribution of money, 79
Economic multiplier, 659, 661, 707
Equilibrium point, 521
Fixed costs, 126
Growth of a stock, 273
Interest in a college trust fund, 272
Interest compounded annually, 203, 246
Interest compounded continuously, 305, 313, 314
Investment, 531, 535, 538, 565, 587, 624
Investment return, 649
Loan repayment, 661
Maximizing income, 575
Maximizing profit, 204, 572, 575, 576
Minimizing cost, 204
Minimizing power line costs, 110
Minimizing salary cost, 576
Minimizing transportation cost, 576
National debt, 28
Present value, 317
Production cost, 558, 559
Profit, 558
Raw material production, 650
Salary sequence, 640
Stocks and gold, 237
Straight-line depreciation, 125, 651
Supply and demand, 317, 528, 530
Total cost, 126
Total cost, revenue, and profit, 165
Total profit, 244
Total savings, 649
Value of Manhattan Island, 313

Engineering

Bridge arch, 608
Bridge expansion, 56
Bridge supports, 607
Canyon depth, 483
Headlight mirror, 597
Satellite dish, 597
Spotlight, 597, 627
Suspension bridge, 598
Windmill power, 203

Environment

Atmospheric pressure, 317
Cloud height, 336
Damage to the ozone layer, 207
Daylight hours, 402
Determining the speed of a river, 371
Distance to a forest fire, 337
Eagle's flight, 517
Earthquake magnitude, 284, 285
Fire tower, 475
Height of a weather balloon, 341
Lightning, 84
Lightning detection, 83, 341
Oil demand, 131
Poachers, 482
Recycling aluminum cans, 272
Sand Dunes National Park, 340
Sources of water, 2
Timber demand, 272
Water wave, 403

General Interest

Band formation, 649
Book arrangements, 707
Circular arrangements, 678
Code symbols, 707
Coin arrangements, 677
Coins, 587
Doubling paper folds, 661
Flag displays, 707
Fraternity officers, 685
Fraternity–sorority names, 707
Garden plantings, 649
Hands of a clock, 372
Letter arrangements, 707
License plates, 677
Money combinations, 695
Number of handshakes, 244
Numerical relationship, 624, 625, 627
Phone numbers, 677
Prize choices, 707
Program planning, 677
Random best-selling novels, 704
Rotating beacon, 403
Seats of a ferris wheel, 358
Social security numbers, 678
Total gift, 707
Tower of Hanoi Problem, 668
Zip codes, 677
Zip-post codes, 678

Geometry

Aerial photography, 336
Angle of elevation, 341
Angle of inclination, 446
Angle measure, 77
Angles between lines, 422
Area, 474
Area of an inscribed rectangle, 109, 110
Area of a parallelogram, 475
Area of a quadrilateral, 476
Area of the Peace Monument, 473
Banner design, 624
Boarding stable, 474
Box dimensions, 624
Diameter of a pipe, 342
Dimensions of a box, 247
Dimensions of a piece of land, 621
Dimensions of a rectangle, 627
Distance across a pond, 52
Distance across a river, 332

Distance between towns, 341
Ellipse, The, 607
Enclosing an area, 340
Field microphone, 597
Finding the depth of a well, 198
Finding the height of a cliff, 204
Finding the height of an elevator shaft, 204
Garden, 519
Garden area, 108
Garden dimensions, 77
Gas tank volume, 109
Gears, 475
Height of a building, 341
Height of a kite, 341
Height of a mural, 447
Heron's formula, 483
Inscribed octagon, 340
Inscribed rhombus, 108
Knife bevel, 480
Legs of a right triangle, 82
Length of an antenna, 341
Lines and triangles from points, 685
Maximizing area, 197, 204
Maximizing box dimensions, 247
Maximizing volume, 204, 247
Minimizing area, 207
Minimizing surface area, 237
Nuclear cooling tower, 616
Number of diagonals, 244
Office dimensions, 624
Pamphlet design, 624
Perimeter, 627
Picture frame dimensions, 77, 623
Radius of a circle, 628
Seed test plots, 624
Sign dimensions, 624
Storage area, 105
Student recreation building, 589
Swimming pool, 483
Tablecloth area, 108
Tee-shirt design, 340
Test plot dimensions, 77
Triangular flag, 108
Volume of a box, 109
Volume of an inscribed cylinder, 110

Physics

Acceleration due to gravity, 431, 458
Alternating current, 437
Angular speed of a gear wheel, 372
Angular speed of a pulley, 372
Beer–Lambert Law, 317
Bouncing golf ball, 707
Bouncing ping-pong ball, 660
Damped oscillations, 403
Electrical theory, 437
Electricity, 317
Height of a thrown object, 244, 248
Hooke's law, 126
Hot-air balloon, 516
Intensity of light, 237
Linear speeds on a carousel, 371
Loudness of sound, 286
Luggage wagon, 516
Parachutist free fall, 649
Pressure at sea depth, 125
Projectile motion, 203
Ripple spread, 165
Rock concert, 474
Sound of an airplane, 342
Speed of sound in air, 94
Time of a free fall, 77
Walking speed, 283, 285, 319
Wheelbarrow, 516

Social Sciences

Anthropology estimates, 123, 125
Carbon dating, 309, 314, 317
Election poll, 707
Flags of nations, 674
House of Representatives, 126
Linguistics, 702
Radioactive decay—archaeology, 314
Senate committees, 685
Small group interaction, 650
Sociological survey, 696

Sports/Entertainment

Average price of a movie ticket, 97
Baseball bunt, 482
Batting orders, 675
Bridge hands, 685
Bungee jumping, 660
Card drawing, 703
Circus guy wire, 422
Circus highwire act, 482
Comiskey Park, 338
Dart throwing, 698
Dartboard, 704
Die rolling, 698
Distance between bases, 332
Drawing a card, 707
Drawing three cards, 707
Double-elimination tournaments, 678
Female Olympic athletes, 313
Five-card poker hands, 703, 704
Flush, 686, 705
Four of a kind, 705
Full house, 686, 705
Games in a sports league, 203
Golf, 539
Golf distance finder, 109
Hiking at the Grand Canyon, 335
Hitting probability, 694
In-line skater, 482
League games, 685, 686
Marbles, 703
Museum admission prices, 530
Musical instruments, 321
Poker hands, 685
Rolling dice, 707
Rolling two dice, 701
Roulette, 704
Royal flush, 704
Safety line to raft, 340
Single-elimination tournaments, 678
Slow-pitch softball, 483
Straight, 705
Straight flush, 704
Survival trip, 483
Ted Williams—war years, 134
Three of a kind, 705
Tour de France, 371
Two pairs, 705
Value of Eddie Murray's baseball card, 314
Vertical leap, 203
Wheel of Fortune, 703

Statistics/Demographics

Average vacation days—U.S. worker, 19
Bread consumption, 260
Cheese consumption, 538
Coin drawing, 703
Decline in beef consumption, 314
Decline of long-playing records, 314
Dress sizes, and Italy, 259
Driver fatalities by age, 205, 639
Favorite number, 702
Fibonacci sequence, 637
Forgetting, 286, 316
HMO enrollment, 77
Household discretionary income by age, 640
Ice milk consumption, 587
Limited population growth, 307
Maximizing a test score, 588
Michigan lotto, 684, 703
Milk consumption, 539
Number of marriages, 539

Statistics/Demographics (continued)

Opinion on capital punishment, 315
Population of Brazil, 319
Population growth, 237, 244, 248, 304, 313, 660
Population growth, Virgin Islands, 313
Probability of being widowed or divorced, 694
Production unit, 703
Random-number generator, 703
Senior profile, 131
Shoe size and life expectancy, 206
Study time vs. grades, 133
Test options, 685
Test scores, 75, 78, 82, 587
Tossing three coins, 704
Typing speed, 273
U.S. vs. Australian vacation times, 73
World population growth, 306, 313

Transportation

Aerial navigation, 357
Airline passenger bumping, 206
Airplane distance, 108, 165
Angle of depression, 446
Angle of revolution, 368, 370
Angular speed of a capstan, 368
Average speed, 79
Bicycling speed, 78
Boat speed, 78
Bus chartering, 259
Car distance, 104
Car rental, 78, 82
Cost of a new car, 539
Cost of operating an automobile at various speeds, 536
Cost of a taxi ride, 78
Distance from an airport, 56
Distance from a lighthouse, 341
Distance traveled, 78, 79
Height of a hot-air balloon, 327
John Deere tractor, 371
Linear speed, 370
Lobster boat, 341
Mackinac Island, 475
Motion, 82, 527, 531, 532
Nautical mile, 431, 458
Navigation, 617, 628
Passenger transportation, 538
Reaction distance, 260
Reaction time, 125
Reconnaissance airplane, 475, 484
Rescue mission, 463, 468
Rising balloon, 108
Speed, 74, 79
Stopping distance on glare ice, 125
Train speed, 78
Transcontinental railroad, 540
Truck speed, 78
Valve cap on a bicycle, 358

Index

A

Abscissa, 2
Absolute value, 21, 22
  of a complex number, 485, 486, 518
  equations with, 63, 80
  inequalities with, 70, 80
  properties, 22, 79
  and roots, 50
Absolute value function, 147
Absorption, coefficient of, 317
Acceleration due to gravity, 431, 458
Acute angle, 322, 345
Addition
  of complex numbers, 176
  of exponents, 26, 80
  of functions, 158, 167
  of matrices, 549, 553, 585
  of ordinates, 398
  of polynomials, 33
  of radical expressions, 52
  of rational expressions, 44
  of vectors, 507, 511, 518
Addition principle
  for equations, 58, 80
  for inequalities, 67, 80
Addition properties, 23
Additive identity
  for matrices, 551, 553, 585
  for real numbers, 23
Additive inverse
  for matrices, 550, 553, 585
  for real numbers, 23
Algebra, fundamental theorem of, 215, 245
Algebra of functions, 158–164, 167
Ambiguous case, 469
Amount of an annuity, 658
Amplitude, 382, 391, 396
Angle of depression, 335, 336
Angle of elevation, 335, 336
Angles, 343
  acute, 322, 345
  between lines, 422
  complementary, 330, 345
  coterminal, 344
  of inclination, 446
  initial side, 343
  obtuse, 345
  reference, 352, 353
  right, 345
  as a rotation, 343
  in standard position, 344
  straight, 345
  supplementary, 345
  terminal side, 344
  vertex, 343
Angular speed, 366
Annuity, 658
Anthropology estimates, 123, 125
Applications, *see Index of Applications*
Area
  of a parallelogram, 475
  of a quadrilateral, 476
  territorial, 97
  of a triangle, 473, 518
Argument, 486
Arithmetic means, 648
Arithmetic progressions, *see* Arithmetic sequences
Arithmetic sequences, 641, 705
  common difference of, 641
  *n*th term, 642
  sums, 644
Arithmetic series, 643, 705
ASK mode, 10
Associative properties, 23
  for matrices, 553, 556, 585
Astronomical unit, 133
Asymptotes
  of exponential functions, 262
  horizontal, 226, 230, 232, 245
  of a hyperbola, 610
  oblique, 231, 232, 245
  slant, 231, 232, 245
  vertical, 227, 228, 230, 232, 245
Atmospheric pressure, 317
Augmented matrix, 541, 562
Average, 648
Average speed, 79
Axes, 2. *See also* Axis.
  of an ellipse, 600
  of a hyperbola, 608, 610
Axis
  conjugate, 610
  imaginary, 484
  major, 600
  minor, 600
  polar, 497
  real, 484
  of symmetry, 590
  transverse, 608

B

Back substitution, 524
Base
  change of, 280, 294, 318
  of an exponential function, 261
  in exponential notation, 26
  of a logarithmic function, 276
Base *e* logarithm, 279
Base–exponent property, 296, 318
Base 10 logarithm, 278
Basis step, 664
Bearing, 337, 357
Beer–Lambert law, 317
Bel, 286
Bevel, 480

Binomial, *see* Binomials
Binomial coefficient notation, 682
Binomial expansion
  using factorial notation, 690, 706
  $(k + 1)$st term, 692, 706
  using Pascal's triangle, 687–690
Binomial theorem, 689, 690, 706
Binomials, 33
  product of, 34, 35, 80
Bit (binary digit), 638
Boiling point and elevation, 97
Book value, 651
Break-even point, 530

C

Canceling, 42, 43
Carbon dating, 308, 309
Cardioid, 501
Carry out, 73
Cartesian coordinate system, 2
Catenary, 596
Center
  of circle, 137, 598
  of ellipse, 600
  of hyperbola, 608
Change-of-base formula, 280, 294, 318
Checking the solution, 59, 62, 69, 73, 187, 454
Circles, 137, 167, 590, 598, 626
  unit, 139, 358, 719
Circular functions, 373, 405
Cissoid, 505
Closed interval, 68
Coefficient, 32
  binomial, 682
  of correlation, 130, 311, 312
  leading, 32, 172
Coefficient of absorption, 317
Coefficient matrix, 541
Cofactor, 710
Cofunction, 330
  identities, 331, 404, 423, 424, 459
Collecting like (or similar) terms, 33
Column matrix, 555
Columns of a matrix, 541
Combinations, 680–685, 706
Combinatorics, 669. *See also* Combinations; Permutations.
Combining like (or similar) terms, 33
Common difference, arithmetic sequence, 641
Common factors, 36
Common logarithms, 278
Common ratio, geometric sequence, 651
Commutative properties, 23
  for matrices, 553, 585
Complementary angles, 330, 345
Completing the square, 181, 190, 593, 599, 604, 613
Complex numbers, 173, 175, 176, 245
  absolute value of, 485, 486, 518
  addition, 176
  conjugates of, 178, 218, 245
  division of, 178, 179, 489, 518
  graphs of, 484
  imaginary part, 175
  multiplication of, 176, 177, 488, 518
  polar notation, 486, 518
  powers of, 491, 518
  pure imaginary, 175
  real part, 175
  roots of, 492, 518
  subtraction, 176
  trigonometric notation, 486, 518
Complex rational expressions, 45
Components of vectors, 509, 510
Composition of functions, 160–164, 167
  and inverse functions, 257, 443, 444, 460
Compound inequality, 69, 70
Compound interest, 30, 31, 78, 80, 203, 246, 266, 305
Conic sections, 590
  circles, 590, 598
  ellipses, 590, 600
  hyperbolas, 590, 608
  parabolas, 590
Conjugate axis of hyperbola, 610
Conjugates
  of complex numbers, 178, 245
    as zeros, 218
  of radical expressions, 53
Conjunction, 69
CONNECTED mode, 16
Consistent system of equations, 523
Constant, variation, 126, 237
Constant function, 99, 100, 111, 127, 147
Constant term, 32
Constraints, 572
Continuous functions, 221
Coordinate system
  Cartesian, 2
  polar, 497
Coordinates of a point, 2
  polar, 497
Correlation, coefficient of, 130, 311, 312
Correspondence, 84, 85, 86
Cosecant function, 323, 347, 373, 404
  graph, 385
  properties, 386
Cosine function, 322, 323, 347, 373, 404, 405
  graph, 379, 389–398
  properties, 383, 405
Cosines, law of, 477, 518
Costs, 126
Cotangent function, 323, 347, 373, 404
  graph, 385
  properties, 386
Coterminal angles, 344
Counting principle, fundamental, 670, 705
Cramer's rule, 713, 714
Critical value, 241, 242
Cube root, 49
Cube root function, 147
Cubes, sums and differences, factoring, 39, 80
Cubic function, 201
Cubic regression, 206
Cubing function, 147
Curve, plane, 717
Curve fitting, 128–130, 200–203, 310–312
Cycloid, 722

D

Decay, exponential, 308, 318
Decay rate, 308
Decibel, 286
Decimals, 20
Decomposing functions, 163, 164
Decomposition, partial fractions, 577, 578, 585
Decreasing function, 99, 100
Degree
  of a polynomial, 32, 33
  of a polynomial function, 172
  of a term, 33
Degree measure, 344
  converting to radians, 362, 363, 405
  decimal form, 327
  DMS form, 327
DeMoivre, Abraham, 490
DeMoivre's theorem, 491, 518
Denominator
  least common, 44, 60
  rationalizing, 53

Dependent system of equations, 523, 526, 534, 545
Dependent variable, 5
Depreciation, straight-line, 125, 651
Depression, angle of, 335, 336
Descartes, René, 2
Descending order, 32
Determinants, 709
  evaluating, 709, 710, 711
Diagonals of a polygon, 244
Difference. *See also* Subtraction.
  of cubes, factoring, 39, 80
  of functions, 158, 167
  of logarithms, 290, 294, 318
  of squares, factoring, 37, 38, 80
Difference identities, 417, 419, 459
Digital technology, 638
Dimensions of a matrix, 541
Direct variation, 126
Direction, 337, 357
  of a vector, 506, 514
Directrix of a parabola, 590
Discriminant, 185
Disjunction, 70
Displacement, 505
Distance. *See also* Distance formula.
  of a fall, 77
  on the number line, 22
  projectile motion, 203
  on the unit circle, 359
Distance formula. *See also* Distance.
  for motion, 74
  between points, 136, 167
Distributive property, 23
  for matrices, 553, 556, 585
Dividend, 210
Division
  of complex numbers, 178, 179, 489, 518
  of exponential expressions, 26, 80
  of functions, 158, 167
  of polynomials, 209, 245
  of radical expressions, 50, 80
  of rational expressions, 44
  synthetic, 211
Divisor, 209, 210
DNA, 638
Domain
  of a function, 84, 85, 88, 92, 94, 224
  of a rational expression, 41
  of a relation, 86
  restricting, 258
DOT mode, 16
Double-angle identities, 425, 426, 459
Doubling time, 304, 306

E

e, 269
Earthquake magnitude, 284
Eccentricity
  of an ellipse, 607
  of a hyperbola, 615
Economic multiplier effect, 659
Element
  in a function's domain, 84
  of a set, 21
Elevation, angle of, 335, 336
Elevation and boiling point, 97
Elimination
  Gaussian, 532, 543
  Gauss–Jordan, 544
Ellipse, 590, 600
  applications, 605
  center, 600
  eccentricity, 607
  equations, 601, 603, 604, 626
  foci, 600
  graphs, 601, 603, 604
  major axis, 600
  minor axis, 600
  parametric equation, 720
  vertices, 600
  x-, y-intercepts, 600
Empty set, 64
Endpoints, 68
Entries of a matrix, 541
Equality
  of matrices, 548
  of vectors, 506
Equally likely outcomes, 699
Equations, 58
  with absolute value, 63, 80
  of circles, 137, 167, 598, 626
  of ellipses, 601, 603, 604, 626
  equivalent, 58
  exponential, 277, 296
  graphs of, 3
  of hyperbolas, 609, 612, 613, 626
  identity, 11
  of inverse relations, 250
  linear, 58, 117, 532
  logarithmic, 277, 299
  logistic, 307
  matrix, 556
  of parabolas, 591, 593, 625
  parametric, 717
  point–slope, 118, 167
  polar to rectangular, 499
  quadratic, 58, 180
  in quadratic form, 184
  radical, 61
  rational, 60
  rectangular to polar, 499
  reducible to quadratic, 186
  related, 238, 567
  slope–intercept, 116, 167
  solutions of, 3, 58
  solving, 13, 58–64, 181–188, 297–301, 448–456
  solving systems of, 522, 523, 532, 541, 544, 563, 617
  systems of, 522, 532, 617
  trigonometric, 448
  two-point, 119, 167
Equilateral triangle, 56
Equilibrium point, 528
Equilibrium price, 317, 528
Equivalent equations, 58
  systems, 525
Equivalent expressions, 43
Equivalent inequalities, 67
Euler, Leonhard, 269
Evaluating a determinant, 709, 710, 711
Even functions, 143, 144, 167
Even roots, 50, 80
Event, 670, 698
Expanding a determinant, 711
Experimental probability, 695, 696, 706
Exponential decay model, 308
Exponential equations, 277, 296
Exponential function, 261, 318
  base, 261
  converting from base b to base e, 311
  graphs, 262, 270
  inverses, 265. *See also* Logarithmic functions.
  properties, 264
  y-intercept, 264
Exponential growth model, 303, 318
  and doubling time, 306
  growth rate, 303
Exponential regression, 311
Exponents
  integer, 25, 26
  properties of, 26, 80
  rational, 54
Expressions
  complex rational, 45
  equivalent, 43
  radical, 49
  rational, 40
  with rational exponents, 54, 55

F

Factor, 209
Factor, common, 36
Factor theorem, 212, 245
Factorial notation, 671
Factoring
completely, 38
differences of cubes, 39, 80
differences of squares, 37, 38, 80
by grouping, 36
rational exponents, expressions with, 55
polynomials, 36–39, 80, 212
sums of cubes, 39, 80
terms with common factors, 36
of trigonometric expressions, 413
trinomials, 37, 38, 80
Factors of polynomials, 209, 212
Familiarize, 72
Feasible solutions, 572
Fibonacci sequence, 637
Finite sequence, 630
Finite series, 634
First coordinate, 2
Fitting a curve to data, 128–130, 202, 206, 310, 311, 402
Fixed costs, 126
Focus
of an ellipse, 600
of a hyperbola, 608
of a parabola, 590
FOIL, 34
Force, 506
Formulas, 64
change-of-base, 280, 294, 318
diagonals of a polygon, 244
distance (between points), 136, 167
distance (for motion), 74
for functions, 94
Heron's, 483
for inverses of functions, 255
midpoint, 136, 167
quadratic, 184, 245
solving for a given variable, 64
Fractional equations, *see* Rational equations.
Fractional exponents, 54
Fractional expressions, 40. *See also* Rational expressions.
Fractions, partial, 577, 585
Functions, 84, 85
absolute value, 147
algebra of, 158–164, 167
circular, 373, 405
composite, 160–164, 167
constant, 99, 100, 111, 127, 147
continuous, 221
as correspondences, 84, 85
cosecant, 323
cosine, 323
cotangent, 323
cube root, 147
cubic, 201
cubing, 147
decomposing, 163, 164
decreasing, 99, 100
defined piecewise, 101
difference of, 158, 167
domain, 84, 85, 88, 92, 94
element, 84
even, 143, 144, 167
exponential, 261, 318
formulas for, 94
graphs, 90–92
greatest integer, 103, 147
horizontal-line test, 253
identity, 111, 112
increasing, 99, 100
input, 87, 94
inverses of, 252
linear, 111, 112, 127, 201
logarithmic, 275, 276
notation for, 86
objective, 572
odd, 143, 144, 167
one-to-one, 252, 318
output, 87, 94
periodic, 381
piecewise, 101
polynomial, 172, 245
product of, 158, 167
quadratic, 127, 147, 181, 201, 245
quartic, 201
quotient of, 158, 167
range, 84, 85, 92, 94
rational, 224, 245
reciprocal, 147
relative maxima and minima, 100
sawtooth, 400
secant, 323
sequence, 630
sine, 323
square root, 147
squaring, 127, 147
sum of, 158, 167
tangent, 323
transformations, 147–155, 167
translations, 147–149, 167
trigonometric, *see* Trigonometric functions.
inverse, 440, 442, 459
values, 87, 88, 89, 211, 212
vertical-line test, 91
zeros of, 98, 209, 215–222, 245
Fundamental counting principle, 670, 705
Fundamental theorem of algebra, 215, 245

G

Games in a sports league, 203
Gauss, Karl Friedrich, 532, 544
Gaussian elimination, 532, 543
Gauss–Jordan elimination, 544
General term of a sequence, 631, 632
Genome, 638
Geometric progressions, *see* Geometric sequences
Geometric sequences, 651, 705
common ratio, 651
infinite, 654
*n*th term, 652
sums, 653
Geometric series, 654
sum, 655, 656
Grad, 372
Grade, 115
Graphers, 1
CONNECTED mode, 16
and domain, 92
DOT mode, 16
DRAW feature, 138
and function values, 89
and identities, 11
INTERSECT feature, 14
notation, 8
and piecewise functions, 102
pixels, 104
programs, 10, 183, 211, 280, 334, 471, 493, 535, 633, 642, 652
and range, 92
and relative maxima and minima, 101
SIMULTANEOUS mode, 156
SOLVE feature, 14
and solving equations, 13
TABLE feature, 10
TRACE feature, 12
viewing window, 6
squaring, 9
standard, 7
and zeros of functions, 99
ZOOM feature, 13

Graphs, 2, 3
  addition of ordinates, 398
  asymptotes, *see* Asymptotes
  circles, 137, 138
  complex numbers, 484
  cosecant function, 385
  cosine function, 379, 389–398
  cotangent function, 385
  ellipses, 601, 603, 604
  equations, 3
  exponential functions, 262, 270
  functions, 90–92
  hand-drawn, 4
  hole in, 18, 237
  horizontal lines, 112
  hyperbolas, 609, 612, 613
  inequalities, 68, 568
  intercepts, 13, 98
  intersection, points of, 14
  inverse relations, 250
  inverse trigonometric functions, 440
  inverses of exponential functions, 265
  linear functions and equations, 111–114
  linear inequalities, 567, 568, 585
  logarithmic functions, 275, 276, 280–283
  of ordered pairs, 2
  parabolas, 188–193, 590, 593
  polar coordinates, 497
  polar equations, 501
  quadratic functions, 188–192
  rational functions, 230–234
  reflection, 141, 150, 155, 167
  secant function, 386
  sine function, 379, 389–398
  solving equations using, 522
  sums, 398
  systems of equations, 522
  systems of inequalities, 570, 571
  tangent function, 384
  transformations, 147–155, 167, 188, 189, 389, 405
  vertical lines, 112
Greater than (>), 21
Greatest integer function, 103, 147
Grouping, factoring by, 36
Growth model, exponential, 303, 318
Growth rate, exponential, 303, 306

H

Half-angle identities, 428, 459
Half-life, 308
Half-open interval, 68
Half-plane, 567
Handshakes, number of, 244
Hang time, 203
Heron's formula, 483
Hole in a graph, 18, 237
Hooke's law, 126
Horizontal asymptotes, 226, 230, 232, 245
Horizontal component, 509
Horizontal line, 112, 123, 167
  slope, 114
Horizontal stretching and shrinking, 152, 155, 167, 189
Horizontal translation, 148, 154, 167, 189
Horizontal-line test, 253
Hyperbola, 590, 608
  applications, 614
  asymptotes, 610
  center, 608
  conjugate axis, 610
  eccentricity, 615
  equations, 609, 612, 613, 626
  foci, 608
  graphs, 609, 612, 613
  transverse axis, 608
  vertices, 608
Hypotenuse, 52, 322

I

i, 174
  powers of, 177
Identities, 11
  additive
    matrix, 551
    real numbers, 23
  checking on a grapher, 11
  cofunction, 331, 404
  multiplicative
    matrix, 560
    real numbers, 23
  proving, 431, 432
  Pythagorean, 411, 412, 459
  trigonometric, 410–412, 417, 419, 423–426, 428, 459
    proving, 431, 432
Identity function, 111, 112
Identity matrix, 560
Imaginary axis, 484
Imaginary numbers, 174, 175, 245, 484, 485
Imaginary part, 175
Inclination, angle of, 446
Inconsistent system of equations, 523, 535, 546
Increasing functions, 99, 100
Independent system of equations, 523
Independent variable, 5
Index
  of a radical, 49
  of summation, 635
Induction, mathematical, 663, 664, 705
Induction step, 664
Inequalities, 66
  with absolute value, 70, 80
  compound, 69, 70
  graphing, 68, 568
  linear, 67, 566
  polynomial, 238
  quadratic, 238
  rational, 241
  solutions, 66, 67
  solving, 66, 67, 69, 70, 238–242
  systems of, 570
Infinite sequence, 630
Infinite series, 634, 654
Initial side of an angle, 343
Inner product, 517
Input, 87, 94
Integers, 20
  as exponents, 25, 26
Intercepts, 13, 98, 116
Interest, 30, 31, 78, 80, 203, 246, 248, 266
  compounded continuously, 305
Intermediate value theorem, 222, 245
INTERSECT feature, 14
Intersection, points of, 14
Interval notation, 68
INT(x), 103
Inverse relation, 250
Inverse variation, 237
Inverses
  additive
    of matrices, 550
    of real numbers, 23
  function, 252
    and composition, 257, 443, 444, 460
    exponential, 265
    formula for, 255
    and reflection across $y = x$, 256
    trigonometric, 440, 442, 459
  multiplicative
    of matrices, 560, 561
    of real numbers, 23
  relation, 250

Invertible matrix, 563
Irrational numbers, 20
Irreducible quadratic factor, 578, 586
Isosceles triangle, 325

J

Jordan, Wilhelm, 544

L

Latitude, 430
Law of cosines, 477, 518
Law of sines, 466, 518
Leading coefficient, 32, 172
Leading 1, 543
Least common denominator (LCD), 44, 60
Legs of a right triangle, 52
Less than (<), 21
Less than or equal to (≤), 21
Libby, Willard E., 308
Light-year, 28
Like terms, 33
Limit, infinite geometric series, 656
Limited population growth, 307, 308, 318
Limiting value, 307, 308
Line of symmetry, 189. *See also* Lines.
Linear equations, 58, 117, 532
Linear functions, 111, 112, 127, 147, 201
  applications, 123
Linear inequalities, 67, 566
  systems of, 570
Linear programming, 572, 585
Linear regression, 128–130
Linear speed, 366, 367, 405
Lines
  angle between, 422
  graphs of, 111–114
  horizontal, 112, 123, 167
  parallel, 120, 123, 167
  perpendicular, 121, 123, 167
  point–slope equation, 118, 123, 167
  slope of, 112–114, 116, 167
  slope–intercept equation, 116, 123, 167
  two-point equation, 119, 123, 167
  vertical, 112, 123, 167
Lithotripter, 605
ln, 279
Log, 278
Logarithm, 275. *See also* Logarithmic functions.
Logarithm function, *see* Logarithmic functions.
Logarithmic equations, 277, 299
Logarithmic functions, 275, 276
  base *e*, 279
  base 10, 278
  change of base, 280, 294, 318
  common, 278
  domain, 276
  on a grapher, 278
  graphs, 275, 276, 280–283
  natural, 279
  properties of, 278, 288, 289, 290, 293, 294, 318
  range, 276
  x-intercept, 276
Logarithmic spiral, 505
Logistic equation, 307, 318
LORAN, 614
Loudness of sound, 286

M

Magnitude of a vector, 506, 514
Main diagonal, 541
Major axis of an ellipse, 600
Mathematical induction, 663, 664, 705
Mathematical models, 111, 127
Matrices, 541
  addition of, 549, 553, 585
  additive identity, 551
  additive inverses of, 550
  augmented, 541, 562
  coefficient, 541
  cofactor, 710
  column, 555
  columns of, 541
  determinant of, 709, 711
  dimensions, 541
  entries, 541
  equal, 548
  identity, multiplicative, 560
  inverses of, 560, 561
  invertible, 563
  main diagonal of, 541
  minor, 709
  multiplying, 553, 556, 585
    by a scalar, 551, 553, 585
  nonsingular, 563
  opposite of, 550
  order, 541
  row, 555
  row-echelon form, 543, 585
  row-equivalent, 541, 585
  rows of, 541
  and scalar multiplication, 551, 553, 585
  singular, 563
  square, 541
  subtraction, 549
  and systems of equations, 556, 563
  zero, 551
Matrix, *see* Matrices.
Matrix equations, 556
Maximum
  linear programming, 572
  quadratic function, 190
  relative, 100
Mean, arithmetic, 648
Midpoint formula, 136, 167
Mil, 372
Minimum
  linear programming, 572
  quadratic function, 190
  relative, 100
Minor, 709
Minor axis of an ellipse, 600
Minute, 327
Models, mathematical, 111, 127, 196, 303, 307, 308, 311, 318, 536, 621. *See also* Regression.
Monomial, 33
Mortgage payment formula, 31
Motion, projectile, 203, 721
Multiplication
  complex numbers, 176, 177, 488, 518
  of exponential expressions, 26, 80
  of exponents, 26, 80
  of functions, 158, 167
  of matrices, 553, 556, 585
    and scalars, 551, 553, 585
  of polynomials, 34, 35, 80
  properties, 23
  of radical expressions, 50, 80
  of rational expressions, 44
  scalar, 507, 512, 518, 553, 585
Multiplication principle
  for equations, 58, 80
  for inequalities, 67, 80
Multiplicative identity
  matrices, 560
  real numbers, 23
Multiplicative inverse
  matrices, 560, 561
  real numbers, 23
Multiplicity of zeros, 216

N

*n*th root, 49
*n*th roots of unity, 494
*n*th term of a sequence, 642
Nappes of a cone, 590
Natural logarithms, 279

Natural numbers, 20
Nautical mile, 431, 458
Negative exponents, 26
Negative rotation, 343
Newton's Law of Cooling, 317
Nondistinguishable objects, permutations of, 676, 706
Nonlinear systems of equations, 617
Nonnegative root, 49
Nonsingular matrix, 563
Notation
  absolute value, 22
  binomial coefficient, 682
  combination, 681
  exponential, 26
  factorial, 671
  function, 86
  grapher, 8
  interval, 68
  inverse function, 252
  for matrices, 548
  polar, 486, 518
  radical, 49
  scientific, 27
  set, 41
  sigma, 635
  summation, 635
  trigonometric, 486, 518
Number line, 21, 22
Numbers
  complex, 173, 175, 176, 245
  imaginary, 174, 175, 245, 484, 485
  integers, 20
  irrational, 20
  natural, 20
  rational, 20
  real, 20, 21
  whole, 20
Numerator, rationalizing, 53

O

Objective function, 572
Oblique asymptotes, 231, 232, 245
Oblique triangles, 464
Obtuse angle, 345
Odd functions, 143, 144, 167
Odd roots, 50, 80
One-to-one function, 252, 318
Open interval, 68
Opposite of a matrix, 550
Order
  of a matrix, 541
  of operations, 29
  of real numbers, 21
Ordered pair, 2
Ordered triple, 532
Ordinate(s), 2
  addition of, 398
Origin, 2
  symmetry with respect to, 141, 142, 143, 167
Oscilloscope, 398
Outcomes, 670, 698
  equally likely, 699
Output, 87, 94

P

Parabola, 188, 590
  applications, 596
  directrix, 590
  equations, 591, 593, 625
    parametric, 717–721
  focus, 590
  graphs, 188–193, 590, 593
  line of symmetry, 189
  vertex, 189, 193
Parallel lines, 120, 123, 167
Parallelogram, area of, 475
Parallelogram law, 508
Parameter, 717
Parametric equations, 717
  and motion, 721
Partial fractions, 577, 585
Partial sum, 634
Pascal's triangle, 687
Peanut, 505
Period, 381, 393, 396
Periodic functions, 381
Perimeter, 64
Permutations, 670
  of n objects, n at a time, 671, 672, 705
  of n objects, k at a time, 673, 674, 706
    allowing repetition, 675
  of nondistinguishable objects, 676, 706
Perpendicular lines, 121, 123, 167
pH, 286
Phase shift, 394, 395, 396
Pi (π), 20
Piecewise-defined function, 101
Pixels, 104
Plane curve, 717
Plotting a point, 3, 484, 497
Point, coordinates of, 2
Point–slope equation, 118, 123, 167
Points of intersection, 14
Polar axis, 497
Polar coordinate system, 497
Polar coordinates, 497
  converting to rectangular coordinates, 498, 499
Polar equations, 499
  graphing, 501
Polar notation, 486, 518
Pole, 497
Polygon, number of diagonals, 244
Polynomial difference theorem, 201
Polynomial function, 172, 245. *See also* Polynomials.
  coefficients, 172
  degree, 172
  domain, 172
  leading coefficient, 172
  as models, 196–203
  values, 210–212
  zeros of, 209, 215–222
Polynomial inequalities, 238
  solving, 238–240
Polynomials. *See also* Polynomial function.
  addition of, 33
  binomial, 33
  coefficients, 32
  constant term, 32
  degree of, 32, 33
  division of, 209, 245
  factoring, 36–39, 80
  factors of, 209, 212
  monomial, 33
  multiplication of, 34, 35, 80
  in one variable, 32
  prime, 39
  in several variables, 33
  subtraction of, 33
  term of, 32, 33
  trinomial, 33
Population growth, 303, 318
  limited, 307
Positive rotation, 343
Power models, 311
Power rule
  for exponents, 26, 80
  for logarithms, 289, 294, 318
Powers
  of complex numbers, 491, 518
  of i, 177
  principle of, 61, 80
Present value, 317
Pressure at sea depth, 125
Prime polynomial, 39
Principal root, 49
Principle(s)
  addition, 58, 67, 80
  fundamental counting, 670, 705

Principle(s) (*continued*)
  of mathematical induction, 664, 705
  multiplication, 58, 67, 80
  P (experimental probability), 696, 706
  P (theoretical probability), 699, 706
  of powers, 61, 80
  of square roots, 58, 80
  of zero products, 58, 80
Probability, 695
  experimental, 695, 696, 706
  properties, 700
  theoretical, 695, 699, 706
Problem-solving process, 72, 73, 80
Product. *See also* Multiplication.
  of functions, 158, 167
  inner, 517
  raised to a power, 26, 80
  scalar, 512, 551
Product rule
  for exponents, 26, 80
  for logarithms, 288, 294, 318
Profit, total, 204
Programming, linear, 572, 585
Programs for graphers, 10, 183, 211, 280, 334, 471, 493, 535, 633, 642, 652
Progressions, *see* Sequences
Projectile motion, 203
Properties
  of absolute value, 22, 79
  of addition, 23
  base–exponent, 296, 318
  of cosecant function, 386
  of cosine function, 383, 405
  of cotangent function, 386
  of exponents, 26, 80
  of logarithms, 288, 289, 290, 293, 294, 318
  probability, 700
  of radicals, 50, 80
  of real numbers, 23, 79
  of secant function, 386
  of sine function, 383, 405
  of tangent function, 386
Proportion, *see* Variation
Proving identities, 431, 432
Pure imaginary numbers, 175
Pythagorean identities, 411, 412, 459
Pythagorean theorem, 25, 52, 79

Q
Quadrants, 2
Quadratic equations, 58, 180. *See also* Quadratic function.
  discriminant, 185
  solving, 58, 60, 181–188
Quadratic form, equation in, 186
Quadratic formula, 184, 245
Quadratic function, 127, 147, 181, 201, 245. *See also* Quadratic equations.
  discriminant of, 185
  graph, 188–192. *See also* Parabola.
  maximum, 190
  minimum, 190
Quadratic inequalities, 238
Quadratic regression, 202
Quadrilateral, area of, 476
Quartic function, 201
Quartic regression, 206
Quotient
  of functions, 158, 167
  of polynomials, 209, 210
  raised to a power, 26, 80
Quotient rule
  for exponents, 26, 80
  for logarithms, 290, 294, 318

R
Radian measure, 361, 365
  converting to degrees, 362, 363, 405
Radical equations, 61
Radical expressions. *See also* Roots.
  addition, 52
  conjugates, 53
  converting to exponential notation, 54
  division, 50, 80
  multiplication, 50, 80
  properties of, 50, 80
  rationalizing denominators or numerators, 53
  simplifying, 50
  subtraction, 52
Radical, 49
Radicand, 49
Radius, 137, 598
Raising a power to a power, 26, 80
Raising a product or quotient to a power, 26, 80
Ramps for the handicapped, 115
Random-number generator, 703
Range
  of a function, 84, 85, 92, 94
  of a relation, 86
Rational equations, 60
Rational exponents, 54
  factoring expressions with, 55
Rational expressions, 40
  addition, 44
  complex, 45
  decomposing, 577, 585
  division, 44
  domain, 41
  multiplication, 44
  simplifying, 42, 45
  subtraction, 44, 45
Rational function, 224, 245
  asymptotes, 226–228, 230–232, 245
  domain, 224
  graphs of, 230–234
Rational inequalities, 241
Rational numbers, 20
Rational zeros theorem, 219, 245
Rationalizing denominators or numerators, 53
Reaction time, 125
Real axis, 484
Real numbers, 20, 21
  properties of, 23, 79
Real part, 175
Reciprocal
  in division, 43
  of trigonometric functions, 323, 328, 404
Reciprocal function, 147
Rectangular coordinates, converting to polar coordinates, 498
Rectangular equations, 499
Recursive definition, 636
Reduced row-echelon form, 543
Reducible to quadratic, equation, 186
Reference angle, 352, 353
Reference triangle, 346, 352
Reflection, 141, 150, 155, 167
  and inverse relations and functions, 250, 251, 256
  on the unit circle, 374
Regression
  cubic, 206
  exponential, 310
  linear, 128–130
  logarithmic, 310
  logistic, 310
  power, 311
  quadratic, 202
  quartic, 206
  sine, 402
Regression line, 129
Related equation, 238, 567
Relation, 86
  domain, 86
  inverse, 250
  range, 86
Relative maxima and minima, 100
Remainder, 209, 210

Remainder theorem, 210, 245
Repeating decimals, 20
  changing to fractional notation, 657
Representing a vector, 509
Resolving a vector, 509
Resultant, 507
Richter scale, 284
Right angle, 345
Right triangles, 52, 322
  solving, 333
  and trigonometric functions, 322, 323, 404
Roots. *See also* Radical expressions.
  of complex numbers, 492, 518
  cube, 49
  even, 50, 80
  *n*th, 49
  nonnegative, 49
  odd, 50, 80
  principal, 49
  square, 49
Rotations, 343. *See also* Angles.
Row matrix, 555
Row-echelon form, 543, 585
Row-equivalent matrices, 541
Row-equivalent operations, 541, 585
Rows of a matrix, 541

S

Salvage value, 651
Sample space, 698
Sawtooth function, 400
Scalar, 507, 551
Scalar components, 510
Scalar multiplication
  and matrices, 553, 585
  and vectors, 507, 512, 518
Scalar product, 551
Scatterplot, 3
Scientific notation, 27
Secant function, 323, 347, 373, 404
  graph, 386
  properties, 386
Second, 327
Second coordinate, 2
Semicubical parabola, 505
Semiperimeter, 483
Sequences, 630
  applications, 637, 638
  arithmetic, 641, 705
  Fibonacci, 637
  finite, 630
  general term, 631, 632
  geometric, 651, 705
  infinite, 630
  partial sum, 634
  recursive definition, 636
  terms, 631
Series
  arithmetic, 643, 705
  finite, 634
  geometric, 654
  infinite, 634
Set-builder notation, 41
Sets
  element of, 21
  empty, 64
  notation, 41
  solution, 58, 67, 567
  subsets, 21, 680, 684, 692, 706
  union of, 70
Shift, phase, 394, 395, 396
Shift of a function, *see* Translation of a function
Shrinkings, 151, 152, 155, 167, 189
Sigma notation, 635
Signs of trigonometric function values, 347, 348, 405
Similar terms, 33
Similar triangles, 324
Simplifying radical expressions, 50
Simplifying rational expressions, 42
  complex, 45
Simplifying trigonometric expressions, 412
Sine function, 322, 323, 347, 404, 405
  graph, 379, 389–398
  inverse, 440, 442, 459
  properties, 383, 405
Sine regression, 402
Sines, law of, 466, 518
Singular matrix, 563
Slant asymptote, 231, 232, 245
Slope, 112–114, 116, 123, 167
  applications, 115
Slope–intercept equation, 116, 123, 167
Solution set, 58, 67, 567
Solutions
  of equations, 3, 58
  feasible, 572
  of inequalities, 66, 67
  of systems of equations, 522, 532
SOLVE feature, 14
Solving equations, 13, 58–64, 181–188, 297–301, 448–456. *See also* Solving systems of equations.
Solving formulas, 64
Solving inequalities, 66, 67, 69, 70, 238–242
Solving systems of equations, 522, 523, 532, 541, 544, 563, 617, 713, 714
Solving triangles
  oblique, 464, 477, 480
  right, 333
Sound, speed in air, 94
Speed
  angular, 366
  average, 79
  linear, 366, 367, 405
  of sound in air, 94
Spiral
  of Archimedes, 505
  logarithmic, 505
Sports league, number of games, 203
Square matrix, 541
  determinant, 709, 711
Square root, 49
Square root function, 147
Squares of binomials, 35, 80
Squaring function, 127, 147
Squaring the window, 9
Standard form
  for equation of a circle, 137, 167, 598, 626
  for equation of an ellipse, 601, 603, 604, 626
  for equation of a hyperbola, 609, 612, 613, 626
  for equation of a parabola, 591, 593, 625
Standard position
  of angles, 344
  of vectors, 514
Standard window, 7
State the answer, 73
Stopping distance, 125
Straight angle, 345
Straight-line depreciation, 125, 651
Stretchings, 151, 152, 155, 167, 189, 390, 405
Strophoid, 505
Subsets, 21, 680, 684
  total number, 692, 706
Substitution method, 523
Subtraction
  of complex numbers, 176
  of exponents, 26, 80
  of functions, 158, 167
  of matrices, 549
  of polynomials, 33
  of radical expressions, 52
  of rational expressions, 44, 45
  of vectors, 512, 518
Sum and difference, product of, 35, 80
Sum identities, 417, 419, 459
Summation notation, 635

Sums
of arithmetic sequences, 644
of cubes, factoring, 39, 80
of functions, 158, 167
of geometric series, 655, 656
graphs of, 398
of logarithms, 288, 318
partial, 634
of vectors, 507, 511, 518
Supplementary angles, 345
Supply and demand, 317, 528
Symmetry
axis of, 590
line of, 189
with respect to the origin, 141, 142, 143, 167
with respect to the x-axis, 141, 142, 167
with respect to the y-axis, 141, 142, 143, 167
Synthetic division, 211
Systems of equations, 522
consistent, 523
in decomposing rational expressions, 582
dependent, 523, 526, 534, 545
equivalent, 525
inconsistent, 523, 535, 546
independent, 523
nonlinear, 617
and partial fractions, 582
solutions, 522, 532
solving, *see* Solving systems of equations
in three or more variables, 532
in two variables, 522
Systems of inequalities, 570

T

TABLE feature, 10
Tangent function, 323, 347, 373, 404, 405
graph, 384
properties, 386
Terminal side of an angle, 344
Terminating decimal, 20
Terms of a polynomial, 32
constant, 32
degree of, 33
like (similar), 33
Terms of a sequence, 631
Territorial area, 97
Theorem
binomial, 689, 690, 706
DeMoivre's, 491, 518
factor, 212, 245
fundamental, of algebra, 215, 245
intermediate value, 222, 245
Pythagorean, 25, 52, 79
rational zeros, 219, 245
remainder, 209, 210
Theoretical probability, 695, 696, 699, 706
Threshold weight, 196
Total cost, profit, revenue, 204
Tower of Hanoi problem, 668
TRACE feature, 12
Trade-in value, 651
Transformations of functions, 147–155, 167, 188, 189
trigonometric, 389–398, 405
Translate to an equation, 73
Translation of a function, 147–149, 167, 189, 389, 405
Transverse axis of a hyperbola, 608
Tree diagram, 669
Triangles
area, 473, 518
equilateral, 56
isosceles, 325
oblique, 464
reference, 346, 352
right, 52, 322
similar, 324
solving, 333, 464, 477, 480
and trigonometric functions, 323
Trigonometric equations, 448
Trigonometric expressions, 412
Trigonometric functions
of acute angles, 323, 404
of any angle, 347, 354, 404
cofunctions, 330
composition of, 443, 444, 460
cosecant, 323, 373, 404, 405
cosine, 322, 323, 373, 404, 405
cotangent, 323, 373, 404, 405
graphs, 378–386
inverse, 440, 442, 459
reciprocal, 323, 328, 404
and right triangles, 322, 323, 404
secant, 323, 373, 404, 405
signs of, 347, 348, 405
sine, 322, 323, 373, 404, 405
tangent, 323, 373, 404, 405
for terminal side on an axis, 351
on the unit circle, 375
values for 45°, 30°, 60°, 325–327, 404
Trigonometric identities
basic, 410, 459
cofunction, 423, 424, 459
difference, 417, 419, 459
double-angle, 425, 426, 459
half-angle, 428, 459
proving, 431, 432
Pythagorean, 411, 412, 459
sum, 417, 419, 459
Trigonometric notation, 486, 518
Trinomial, 33
factoring, 37, 38, 80
Trinomial square, 38, 80
Twisted sister, 505
Two-point equation, 119, 123, 167

U

Union of sets, 70
Unit circle, 139, 358
and function values, 375
parametric equation, 719
reflections on, 374
Unit vector, 513
Unity, nth roots of, 494

V

Values, function, 87, 88, 89, 211, 212
Variable, 15
Variable costs, 126
Variation
direct, 126
inverse, 237
Variation constant, 126, 237
Vectors, 505, 506
addition of, 507, 511, 518
components of, 509, 510
direction, 506, 514
equal, 506
inner product, 517
magnitude, 506, 514
parallelogram law, 508
representing, 509
resolving, 509
resultant of, 507
scalar components, 510
scalar multiplication, 507, 512
standard position, 514
subtraction, 512, 518

sum of, 507, 511, 518
unit, 513
zero, 513
Velocity, 506
Vertex
of an angle, 343
of an ellipse, 600
of a hyperbola, 608
of a parabola, 189, 193, 590
Vertical asymptotes, 227, 228, 230, 232, 245
Vertical component, 509
Vertical leap, 203
Vertical line, 112, 123, 167
slope, 114
Vertical stretching and shrinking, 151, 155, 167, 189
Vertical translation, 147, 148, 154, 167, 189, 389, 405
Vertical-line test, 91
Vertices, *see* Vertex.
Viewing window, 6
squaring, 9
standard, 7

W

Walking speed, 283
Weight
ideal, 125
threshold, 196
Whispering gallery, 605
Whole numbers, 20
Windmill power, 203
Window, 6
squaring, 9
standard, 7

X

x-axis, 2
symmetry with respect to, 141, 142, 167
x-coordinate, 2
x-intercept, 13, 98, 600
logarithmic function, 276

Y

y-axis, 2
symmetry with respect to, 141, 142, 143, 167
y-coordinate, 2
y-intercept, 116, 600
exponential function, 264

Z

Zero, exponent of, 26
Zero matrix, 551
Zero products, principle of, 58, 80
Zero vector, 513
Zeros of functions, 98, 209, 215–222, 245
multiplicity, 216
ZOOM feature, 13

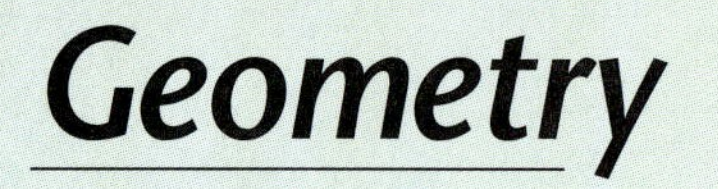

Plane Geometry

Rectangle
Area: $A = lw$
Perimeter: $P = 2l + 2w$

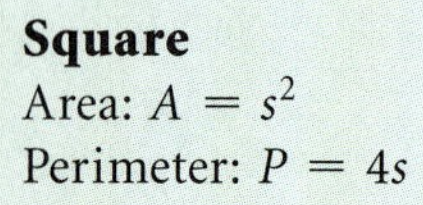
Square
Area: $A = s^2$
Perimeter: $P = 4s$

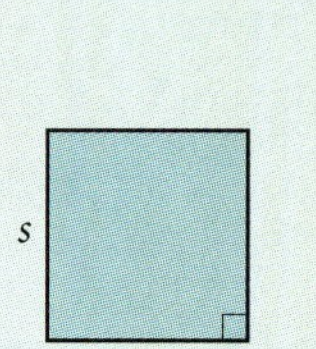

Triangle
Area: $A = \frac{1}{2}bh$

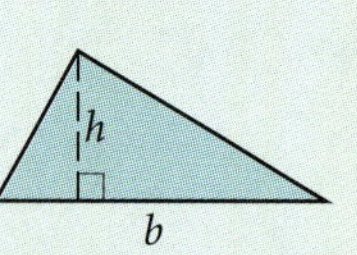

Sum of Angle Measures
$A + B + C = 180°$

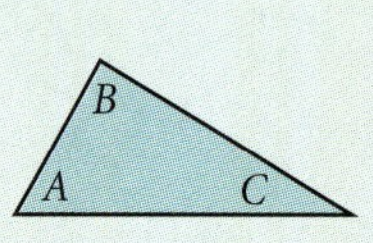

Right Triangle
Pythagorean theorem (equation):
$a^2 + b^2 = c^2$

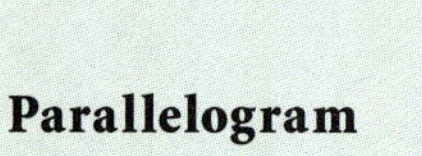
Parallelogram
Area: $A = bh$

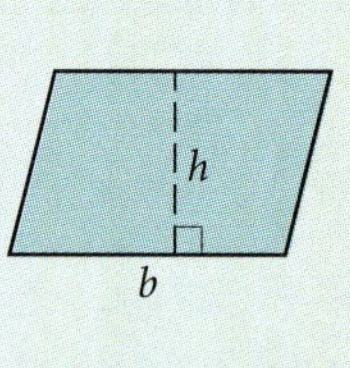

Trapezoid
Area: $A = \frac{1}{2}h(a + b)$

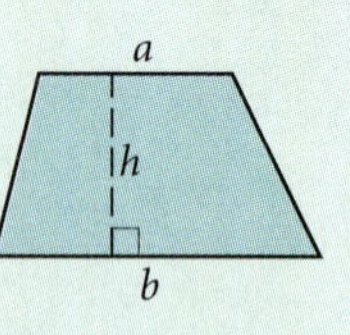

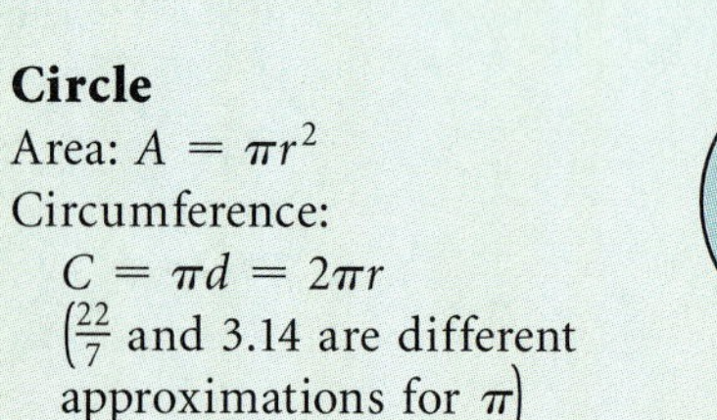
Circle
Area: $A = \pi r^2$
Circumference:
$C = \pi d = 2\pi r$
($\frac{22}{7}$ and 3.14 are different approximations for π)

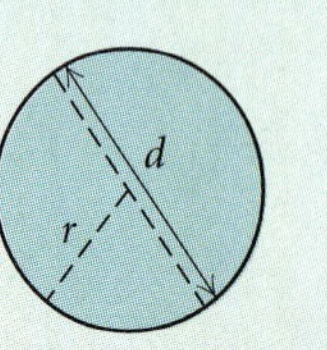

Solid Geometry

Rectangular Solid
Volume: $V = lwh$

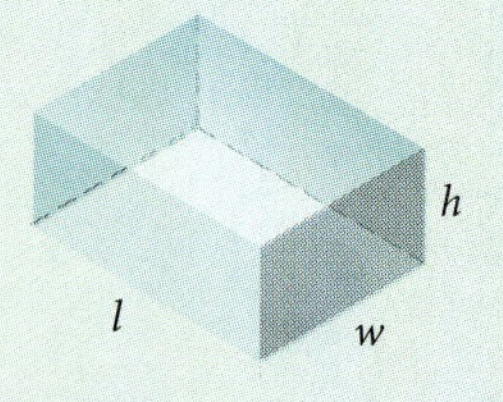

Cube
Volume: $V = s^3$

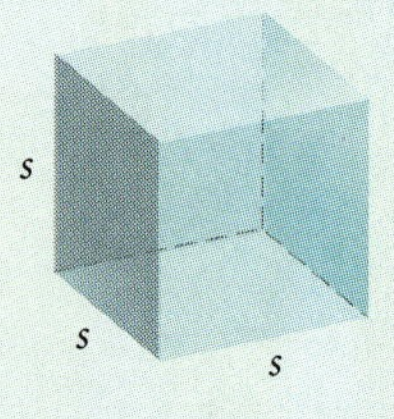

Right Circular Cylinder
Volume: $V = \pi r^2 h$
Lateral surface area:
$L = 2\pi rh$
Total surface area:
$S = 2\pi rh + 2\pi r^2$

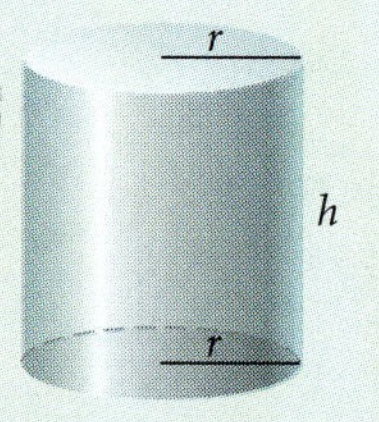

Right Circular Cone
Volume: $V = \frac{1}{3}\pi r^2 h$
Lateral surface area:
$L = \pi rs$
Total surface area:
$S = \pi r^2 + \pi rs$
Slant height:
$s = \sqrt{r^2 + h^2}$

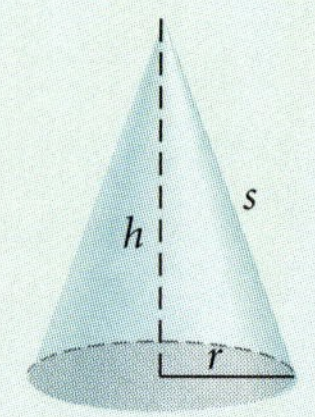

Sphere
Volume: $V = \frac{4}{3}\pi r^3$
Surface area: $S = 4\pi r^2$

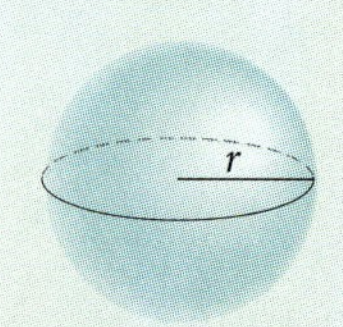

Algebra (continued)

The Distance Formula

The distance from (x_1, y_1) to (x_2, y_2) is given by

$$d = \sqrt{(x_2 - x_1)^2 + (y_2 - y_1)^2}.$$

Midpoint Formula

The midpoint of the line segment from (x_1, y_1) to (x_2, y_2) is given by

$$\left(\frac{x_1 + x_2}{2}, \frac{y_1 + y_2}{2}\right).$$

Formulas Involving Lines

The slope of the line containing points (x_1, y_1) to (x_2, y_2) is given by

$$m = \frac{y_2 - y_1}{x_2 - x_1}.$$

Slope–intercept equation: $y = f(x) = mx + b$

Horizontal line: $y = b$ or $f(x) = b$

Vertical line: $x = a$

Point–slope equation: $y - y_1 = m(x - x_1)$

Two-point equation: $y - y_1 = \frac{y_2 - y_1}{x_2 - x_1}(x - x_1)$

The Quadratic Formula

The solutions of $ax^2 + bx + c = 0$, $a \neq 0$, are given by

$$x = \frac{-b \pm \sqrt{b^2 - 4ac}}{2a}.$$

Compound Interest Formulas

Compounded n times per year: $A = P\left(1 + \frac{i}{n}\right)^{nt}$

Compounded continuously: $P(t) = P_0e^{kt}$

Properties of Exponential and Logarithmic Functions

$\log_a x = y \leftrightarrow x = a^y$ $\quad$ $a^x = a^y \leftrightarrow x = y$

$\log_a MN = \log_a M + \log_a N$ $\quad$ $\log_a M^p = p \log_a M$

$\log_a \frac{M}{N} = \log_a M - \log_a N$

$\log_b M = \frac{\log_a M}{\log_a b}$

$\log_a a = 1$ $\quad$ $\log_a 1 = 0$

$\log_a a^x = x$ $\quad$ $a^{\log_a x} = x$

Arithmetic Sequences and Series

$a_1, a_1 + d, a_1 + 2d, a_1 + 3d, \ldots$

$a_{n+1} = a_n + d$ $\quad$ $a_n = a_1 + (n - 1)d$

$S_n = \frac{n}{2}(a_1 + a_n)$

Geometric Sequences and Series

$a_1, a_1r, a_1r^2, a_1r^3, \ldots$

$a_{n+1} = a_nr$ $\quad$ $a_n = a_1r^{n-1}$

$S_n = \frac{a_1(1 - r^n)}{1 - r}$ $\quad$ $S_\infty = \frac{a_1}{1 - r}, |r| < 1$

Conic Sections

Circle: $(x - h)^2 + (y - k)^2 = r^2$

Ellipse: $\frac{(x - h)^2}{a^2} + \frac{(y - k)^2}{b^2} = 1,$

$\frac{(x - h)^2}{b^2} + \frac{(y - k)^2}{a^2} = 1$

Parabola: $(x - h)^2 = 4p(y - k),$

$(y - k)^2 = 4p(x - h)$

Hyperbola: $\frac{(x - h)^2}{a^2} - \frac{(y - k)^2}{b^2} = 1,$

$\frac{(y - k)^2}{a^2} - \frac{(x - h)^2}{b^2} = 1$

Trigonometry

Trigonometric Functions

Acute Angles

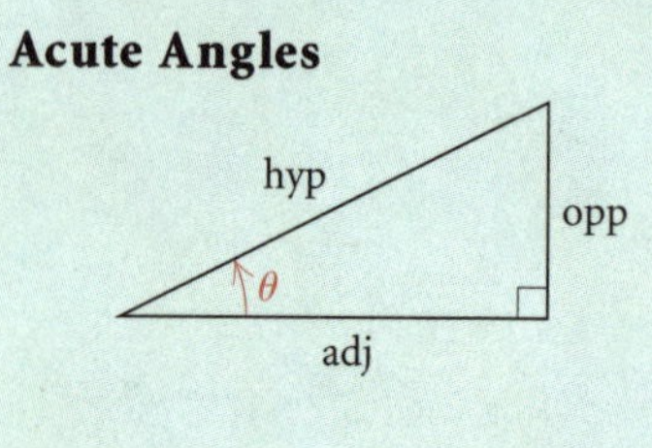

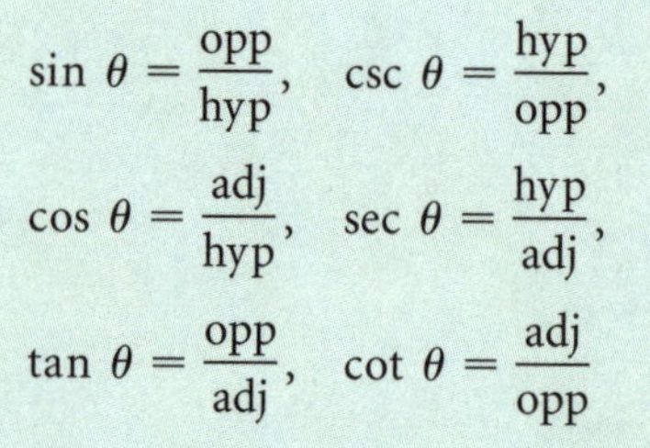

$$\sin \theta = \frac{\text{opp}}{\text{hyp}}, \quad \csc \theta = \frac{\text{hyp}}{\text{opp}},$$

$$\cos \theta = \frac{\text{adj}}{\text{hyp}}, \quad \sec \theta = \frac{\text{hyp}}{\text{adj}},$$

$$\tan \theta = \frac{\text{opp}}{\text{adj}}, \quad \cot \theta = \frac{\text{adj}}{\text{opp}}$$

Any Angle

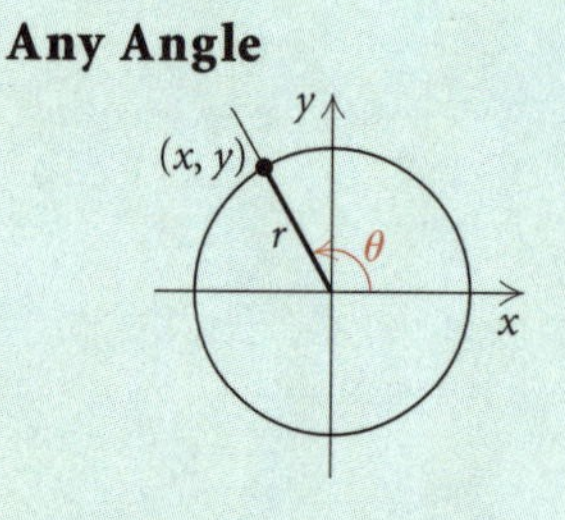

$$\sin \theta = \frac{y}{r}, \quad \csc \theta = \frac{r}{y},$$

$$\cos \theta = \frac{x}{r}, \quad \sec \theta = \frac{r}{x},$$

$$\tan \theta = \frac{y}{x}, \quad \cot \theta = \frac{x}{y}$$

Real Numbers

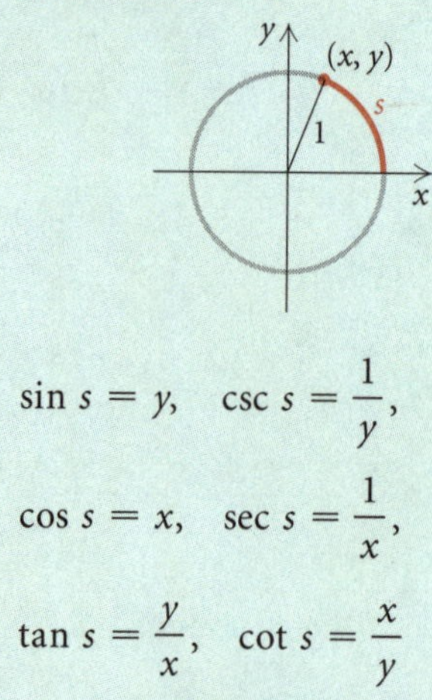

$$\sin s = y, \quad \csc s = \frac{1}{y},$$

$$\cos s = x, \quad \sec s = \frac{1}{x},$$

$$\tan s = \frac{y}{x}, \quad \cot s = \frac{x}{y}$$

Basic Trigonometric Identities

$$\sin(-x) = -\sin x,$$
$$\cos(-x) = \cos x,$$
$$\tan(-x) = -\tan x,$$

$$\tan x = \frac{\sin x}{\cos x},$$

$$\cot x = \frac{\cos x}{\sin x},$$

$$\csc x = \frac{1}{\sin x},$$

$$\sec x = \frac{1}{\cos x},$$

$$\cot x = \frac{1}{\tan x}$$

Pythagorean Identities

$$\sin^2 x + \cos^2 x = 1,$$
$$1 + \cot^2 x = \csc^2 x,$$
$$1 + \tan^2 x = \sec^2 x$$

Identities Involving $\pi/2$

$$\sin(\pi/2 - x) = \cos x,$$
$$\cos(\pi/2 - x) = \sin x, \qquad \sin(x \pm \pi/2) = \pm\cos x,$$
$$\tan(\pi/2 - x) = \cot x, \qquad \cos(x \pm \pi/2) = \mp\sin x$$

Sum and Difference Identities

$$\sin(u \pm v) = \sin u \cos v \pm \cos u \sin v,$$
$$\cos(u \pm v) = \cos u \cos v \mp \sin u \sin v,$$
$$\tan(u \pm v) = \frac{\tan u \pm \tan v}{1 \mp \tan u \tan v}$$

Double-Angle Identities

$$\sin 2x = 2 \sin x \cos x,$$
$$\cos 2x = \cos^2 x - \sin^2 x$$
$$= 1 - 2\sin^2 x$$
$$= 2\cos^2 x - 1,$$
$$\tan 2x = \frac{2 \tan x}{1 - \tan^2 x}$$

Half-Angle Identities

$$\sin \frac{x}{2} = \pm\sqrt{\frac{1 - \cos x}{2}},$$

$$\cos \frac{x}{2} = \pm\sqrt{\frac{1 + \cos x}{2}},$$

$$\tan \frac{x}{2} = \pm\sqrt{\frac{1 - \cos x}{1 + \cos x}} = \frac{\sin x}{1 + \cos x} = \frac{1 - \cos x}{\sin x}$$

(continued)